Commercial Drafting and Detailing

Commercial Drafting and Detailing

Third Edition

Alan Jefferis

Clackamas Community College, Oregon City, OR

Kenneth D. Smith, A.I.A.

Principal—Kenneth D. Smith Architect & Associates, Inc.

Australia • Brazil • Japan • Korea • Mexico • Singapore • Spain • United Kingdom • United States

DELMAR
CENGAGE Learning™

Commercial Drafting and Detailing, Third Edition
Alan Jefferis & Kenneth D. Smith

Vice President, Editorial: **Dave Garza**
Director of Learning Solutions: **Sandy Clark**
Senior Acquisitions Editor: **James Devoe**
Managing Editor: **Larry Main**
Senior Product Manager: **Sharon Chambliss**
Editorial Assistant: **Christopher Savino**
Vice President, Marketing: **Jennifer McAvey**
Executive Marketing Manager: **Deborah S. Yarnell**
Marketing Manager: **Jimmy Stephens**
Marketing Specialist: **Mark Pierro**
Production Director: **Wendy Troeger**
Production Manager: **Mark Bernard**
Content Project Manager: **David Plagenza**
Art Director: **Bethany Casey**
Technology Project Manager: **Christopher Catalina**

10% total recycled fiber

Library of Congress Control Number: 2008930578

ISBN-13: 978-1-4354-2597-2

ISBN-10: 1-4354-2597-9

Delmar
5 Maxwell Drive
Clifton Park, NY 12065-2919
USA

Cengage Learning products are represented in Canada by Nelson Education, Ltd.

For your lifelong learning solutions, visit **www.delmar.cengage.com**
Visit our corporate website at **www.cengage.com**

Printed in the United States of America
2 3 4 X X 11 10

Contents

Preface, vi

SECTION 1 CAD Drafting and Design Considerations of Commercial Structures

Chapter 1 Professional Careers and Commercial CAD Drafting, 3

Chapter 2 The CAD Drafter's Role in Office Practice and Procedure, 18

Chapter 3 Applying AutoCAD Tools to Commercial Drawings, 38

Chapter 4 Introduction to the International Building Code, 73

Chapter 5 Access Requirements for People with Disabilities, 94

SECTION 2 Building Methods and Materials of Commercial Construction

Chapter 6 Making Connections, 125

Chapter 7 Wood and Timber Framing Methods, 146

Chapter 8 Engineered Lumber Products, 166

Chapter 9 Steel Framing Methods and Materials, 186

Chapter 10 Unit Masonry Methods and Materials, 207

Chapter 11 Concrete Methods and Materials, 232

Chapter 12 Fire-Resistive Construction, 263

SECTION 3 Preparing Architectural and Civil Drawings

Chapter 13 Structural Considerations Affecting Design, 281

Chapter 14 Project Manuals and Written Specifications, 306

Chapter 15 Land Descriptions and Drawings, 317

Chapter 16 Floor Plan Components, Symbols, and Development, 344

Chapter 17 Orthographic Projection and Elevations, 384

Chapter 18 Roof Plan Components and Drawings, 403

Chapter 19 Drawing Sections, 426

Chapter 20 Interior Elevations, 450

Chapter 21 Ramp, Stair, and Elevator Drawings, 464

SECTION 4 Preparing Structural Drawings

Chapter 22 Drawing Framing Plans, 485

Chapter 23 Drawing Structural Elevations and Sections, 518

Chapter 24 Foundation Systems and Components, 547

Index, 600

Preface

COMMERCIAL DRAFTING AND DETAILING is a practical, comprehensive textbook intended to introduce students to the development of architectural and structural drawings required to develop a commercial structure. Students are expected to have previously completed either a basic drawing class or a class dealing with residential drafting as well as have a working knowledge of AutoCAD. Students will build on their knowledge gained through residential drafting and design as common materials and construction methods for commercial structures are explored. Throughout the text, students will be exposed to the work that the engineer or architect will do to develop the project. This book is in no way an attempt to transform CAD technicians into junior engineers. The work of the engineer and the architect is presented to help develop an understanding of the technician's role in the overall process of developing a set of plans.

LAYOUT

This text has two major divisions. Chapters 1 through 12 are designed to introduce the student to the materials and construction methods of commercial construction. This portion of the book is itself divided into two major sections. Each chapter provides an end-of-chapter test to assess understanding, as well as practical drawing problems to apply knowledge. Many drawing problems are intended to force the student to make use of *Sweets Catalogs*, vendor Web sites, or local suppliers to prepare for the research that is expected in most professional offices.

Section 1 introduces students to the major codes that they will encounter in an engineer or architect's office. The content of the entire book is based on the 2009 edition of the *International Building Code Handbook*. Students are also introduced to Americans with Disabilities Act (ADA) standards that affect commercial structures. Although this is information that the project designer will be responsible for, we feel it is critical for the CAD technician to understand the codes that shape design.

Section 2 deals with the materials that shape the construction process, including wood, engineered lumber, timber, steel, concrete block, poured concrete construction, and the fasteners used to connect these materials. Each chapter could serve as the core for a class dealing with each material or as an overview of the entire process.

The second half of this text is intended to be a practical approach to the development of the architectural and structural drawings. The key elements that make up the architectural and structural drawings are introduced in Chapters 15 through 24. In Chapter 15, the site drawings for four major projects are introduced, with each project reflecting one of the major building materials. These projects include the following:

- Wood: multifamily complex
- Concrete block/truss roof: retail sales
- Concrete block: warehouse with mezzanine, panelized roof
- Precast concrete: warehouse with mezzanine, truss roof

Various aspects of these projects are presented throughout the remaining chapters so that students will develop the key drawings that make up a set of architectural and structural drawings related to one structure. Portions of each of the projects contained in Chapters 15 through 24 can be accessed from the student CD in the DRAWING PROJECTS folder and used as a base to complete the assignment. The drawings are a rough draft only and need to be updated to reflect common standards; appropriate symbols, linetypes, and dimensioning methods; and notation methods.

> *Note:* The drawings contained at the end of the chapters are to be used as a guide only. You should use these drawings with the mindset that the previous CAD drafter who started the drawings was fired. Information may not be reliable. Because your name will go in the title block, you must be very careful before blindly accepting another person's layout. As you progress through the drawings

you'll find that some portions of the drawings do not match things that have been drawn on previous drawings.

Each project has errors that will need to be solved. The errors are placed in the project to force you to think, in addition to drawing. Most of the errors are so obvious that you will have no trouble finding them. If you think you have found an error, do not make changes until you have discussed the problem and possible solutions with your engineer (your instructor). It's not enough to find the mistakes. Come up with a solution that incorporates materials that have been completed in previous chapters, and coordinate your ideas with materials that will be drawn in future chapters.

ACKNOWLEDGMENTS

In addition to the staff of Delmar Cengage Learning, we're especially grateful to the following people:

Janice Ann Jefferis, Chief Encourager/Photographer

Dean K. Smith, Ginger M. Smith, Gisela Smith, and David Ambler of Kenneth D. Smith Architect & Associates, Inc.

Julie Searls for her excellent mastery of the English language and skill as a copy editor

Terrel Broiles of Architectural Drafting, Design, and Rendering of Albany, New York, for his excellent artwork in this text

Deepti Narwat and her staff at International Typesetting and Composition for her excellent work on the production of this text

The International Building Code related topics in Chapters 4, 12, and 13 were reviewed by the ICC Staff under the supervision of Hamid Naderi. The International Code Council, a member association dedicated to building safety and fire prevention, develops the codes used to construct residential and commercial buildings. Most U.S. cities, counties, and states that adopt codes choose the International Codes developed by the International Code Council, 500 New Jersey Avenue, NW, 6th Floor, Washington, DC 20001-2070 [P]1-888-ICC-SAFE (422-72330; www.iccsafe.org).

The following architects and engineers have reviewed portions of the manuscript and provided the illustrations that are used throughout the text:

G. Williamson Archer, A.I.A., Archer & Archer P.A.

Scott Beck, Scott R. Beck, Architect

Bill Berry, Berry-Nordling Engineers, Inc.

Laura Bourland, Halliday Associates

Charles J. Conlee, P.E., Conlee Engineers, Inc.

LeRoy Cook, Architectural Instructor, Clackamas Community College

Chris DiLoreto, DiLoreto Architects, LLC

Russ Hanson, H.D.N. Architects, A.I.A.

Havlin G. Kemp P.E., Van Domelen/Looijenga/McGarrigle/Knauf Consulting Engineers

Tom Kuhns, Michael and Kuhns Architects, P.C.

Ron Lee, A.I.A., Architects Barrentine, Bates & Lee, A.I.A.

Ned Peck, Peck, Smiley, Ettlin Architects

David Rogencamp, KPFF Consulting Engineers

The following associations and companies have also made major contributions to this text and would be excellent references for students who desire further information:

Robert G. Wiedyke
Director of Publications Services
American Concrete Institute
PO Box 19150
Detroit, MI 48219
Web site: www.aci-int.net

American Institute of Architects Press
1735 New York Ave, NW
Washington, DC 20006-5292
Web site: www.aia.org

American Institute of Steel Construction
1 East Wacker Dr.
Suite 3100
Chicago, IL 60601
Web site: www.aiscweb.com

Bruce D. Pooley, P.E.
Director of Technical Services
American Institute of Timber Construction
7012 S. Revere Pkwy.
Suite 140
Englewood, CO 80112
Web site: www.aitc-glulam.org

American Society for Concrete Construction
1902 Techny Ct.
Northbrook, IL 60062

Americans with Disabilities Act
Federal Register Vol 56, No 144, July 26, 1991

Concrete Masonry Association of California and Nevada
6060 Sunrise Vista Dr.
Suite 1875
Citrus Heights, CA 95610
Web site: www.amacn.org

Steven E. Ellingson, P.E.
Director of Marketing
Concrete Reinforcing Steel Institute
933 N. Plum Grove Rd.
Schaumburg, IL 60173-4758
Web site: www.crsi.org

Construction Metrication Council
National Institute of Building Sciences
1201 L St., NW
Suite 400
Washington, DC 20005
Web site: www.nibs.org

Construction Specifications Institute
601 Madison St.
Alexandria, VA 22314-1791
Web site: www.csinet.org

Mike Shultz
Engineered Wood Association (formerly the American Plywood Association)
PO Box 11700
Tacoma, WA 98411-0700
Web site: www.apawood.org

Hamid Naderi, P.E,
CBO Vice President
Product Development International Code Council
8650 Spicewood Springs
#145, Box 610 Austin, Texas 78759
Web site: www.hnaderi@iccsafe.org

Patsy Harms
Portland Cement Association
54200 Old Orchard Rd.
Skokie, IL 60077-1083
Web site: www.cement.org

Post-Tensioning Institute
1717 W. Northern Ave.
Suite 218
Phoenix, AZ 85021
Web site: www.post-tensioning.org

Sarah Carlson
Simpson Strong-Tie
4637 Chabot Dr.
Suite 200
Pleasanton, CA 94588
Web site: www.strongtie.com

Richard Wallace
Media Director
Southern Forest Products Association
PO Box 641700
Kenner, LA 70064
Web site: www.sfpa.org

R. Donald Murphy
Managing Director
Steel Joist Institute
1205 48th Ave.
North Suite A
Myrtle Beach, SC 29577-5424
Web site: www.steeljoist.org

Western Wood Products Association
1500 Yeon Building
Portland, OR 97204
Web site: www.wwpa.org

Weyerhaeuser Company
PO Box 9777
Federal Way, WA 98063-9777
Web site: www.weyerhaeuser.com

Section 1

CAD Drafting and Design Considerations of Commercial Structures

1 Professional Careers and Commercial CAD Drafting

You are about to start the exciting exploration of commercial drafting. As authors who have spent our lives in commercial design and education, we wrote this book with the expectation that you will enter the drawing field as a CAD technician, or use the skills you will acquire in this text to progress through an accredited architectural or engineering program. The exploration of commercial drawing will occur in four distinct stages, including:

- CAD drafting and design consideration of commercial structures.
- Common materials and building methods of commercial construction.
- Preparation of architectural drawings.
- Preparation of structural drawings.

This chapter opens our exploration of commercial construction by discussing:

- Common employment opportunities in the design and drafting of commercial structures.
- The effect of green construction on the design and construction of commercial structures.

EMPLOYMENT OPPORTUNITIES IN COMMERCIAL DESIGN AND DRAFTING

To many in the construction field, commercial drafting is the development of the construction drawings for any nonresidential structure. For most professionals involved in the design process, the general area of commercial drafting and construction comprises public, industrial, institutional, and commercial projects. Examples of each are listed below:

- *Public structures* are buildings or portions of structures used for assembly, education, or civic administration, such as schools, churches, stadiums, post offices, and libraries.
- *Industrial structures* are buildings or portions of structures used for manufacturing, assembly, or storage, such as factories, warehouses, and businesses involved in hazardous, toxic, or unstable materials.
- *Institutional structures* are buildings or portions of structures used by occupants with limitations caused by health, age, correctional purposes, or in which the mobility is restricted such as residential care facilities, medical facilities where 24-hour care is provided, day care facilities, jails, and prisons.
- *Commercial structures* are buildings or portions of structures used for sales, business and professional, or service transactions, including office buildings and eating and drinking establishments.

This text will look at commercial drafting as it relates to all these fields. Your role in the development of commercial drawings will vary depending on the size and structure of the firm where you work. Some design firms are very specialized and work only in one field, such as educational or institutional facilities. Some firms have several different divisions within the company for designing various areas of occupancy. Chapter 4 will introduce specific areas of construction based on the occupancy of the structure.

No matter how you define commercial construction, commercial drawing offers far more career options and challenges than residential drawing. Commercial projects are typically much larger and are completed by a team rather than an individual. In addition to the design team, engineers, architects, and CAD technicians from several different firms in a variety of construction fields complete the project.

The field of architecture has always offered many career opportunities. The use of computers and software such as AutoCAD®, Revit® Architecture, and AutoCAD® Architecture has revolutionized architectural and engineering offices and opened an almost unlimited potential for design. This chapter explores the use of AutoCAD and some of the main areas of employment in the field of commercial architecture, and explains how CAD has affected each area. Although Revit Architecture and AutoCAD Architecture play important roles in the world of construction drawings, their use lies beyond the scope of this text. Major areas of employment in the architectural field are that of CAD drafter, designer, interior designer, architect, and engineer.

CAD TECHNICIAN

A **CAD technician** is a person who draws the designs that originate with another person. Traditionally referred to as a *drafter* in the days of manual drafting, most firms and job postings now use the term *CAD technician* or use both terms interchangeably. The drafter's main responsibility is to take a drawing similar to Figure 1-1 and fill in the missing material, using acceptable office standards so that it resembles the drawing in Figure 1-2. The experience and education of the drafter affect the actual job assignments.

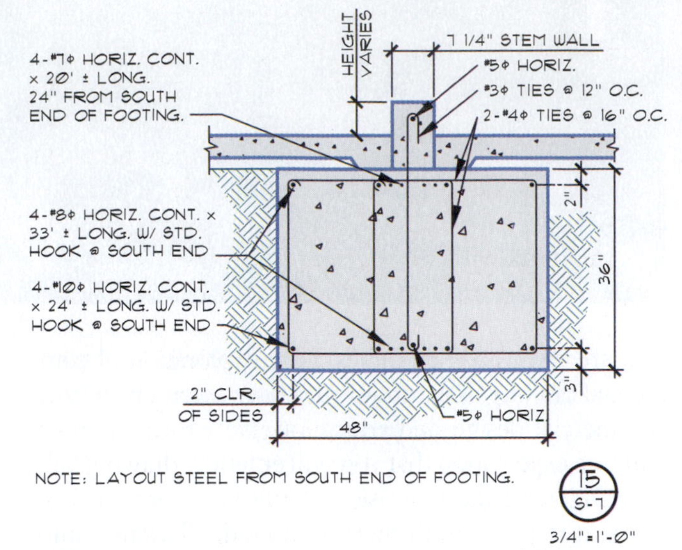

FIGURE 1-2 The CAD technician's main responsibility is to accurately convey information from the architect or engineer to those involved in the construction project using acceptable office standards. This drawing was completed by the technician using the sketch in Figure 1-1 as a guide, along with vendor catalogs.

Entry-Level Technician

An entry-level CAD technician is typically preoccupied with making corrections, running prints, and completing simple drawings while confidence is gained in the standard office procedure. In some areas of the country, and in a thriving economy, it is possible to obtain an entry-level technician intern position as a high school student. Because of the complexities involved in creating commercial drawings, most offices require college training. A new CAD technician with some college or trade school experience will typically start a new job by making corrections to existing drawings. To advance in an office you'll need to become proficient using the firm's computer standards and any special menus and list-processing language (LISP) routines needed to work efficiently. An understanding of basic construction techniques is also essential for advancement. One of the best ways to gain an understanding of typical construction practice is to spend time at construction sites. Being able to follow a project through the various stages of construction will greatly aid a new CAD drafter in gaining an understanding of the information that is entered at the keyboard.

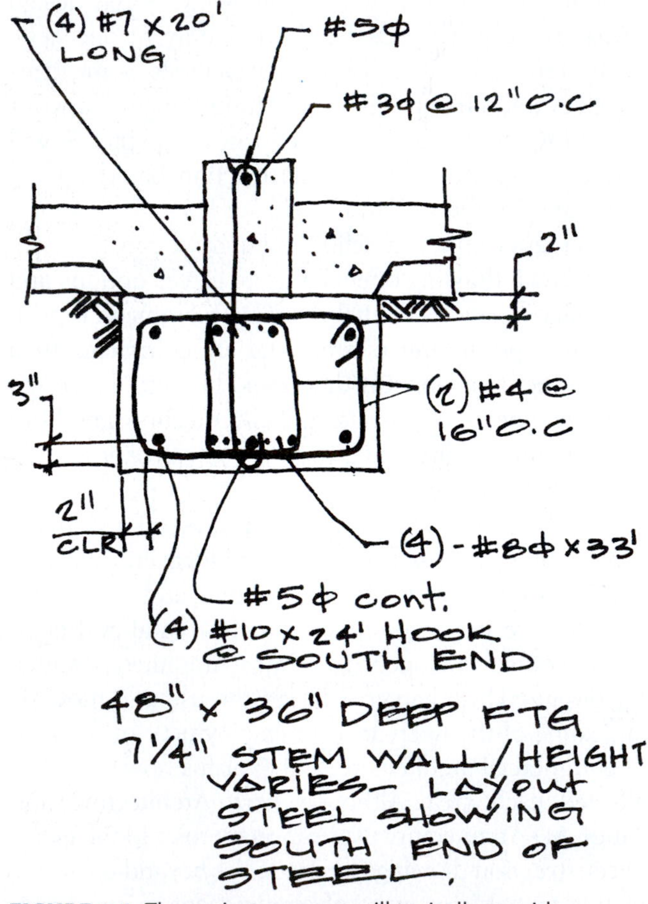

FIGURE 1-1 The project manager will typically provide a sketch for an entry-level CAD technician to use as a guide. The job of the drafter is to accurately represent all materials using the proper office formatting for lines and layers.

In addition to improving CAD drafting skills, most employers expect academic skills to increase before more challenging projects are assigned. Although most office procedures for an entry-level technician involve only basic math skills, an understanding of algebra, geometry, and trigonometry will help in career advancement. Other necessities for advancement are confidence, and the ability to work in a team. As confidence is gained, and understanding of the types of drawings being completed increases, the supervisor will be able to provide sketches with less detail and rely on the technician to research a solution based on the governing code and similar projects that have been completed in the office. The decisions involved in making drawings without sketches also require the technician to have a good understanding of what is being drawn. This understanding does not come just from a textbook. A good method to gain an understanding of what you are drafting is to spend time working at a construction site so that you understand what a craftsperson must do as a result of what you have drawn.

An additional skill required for a new technician is to be reliable. Reliability within an office is measured by the maintenance of good attendance patterns and the production of drawings as scheduled. To advance and become a leader on the drawing team will require you to become an effective manager of your time. This includes the ability to determine what drawings will need to be created, to select them from a stock library and edit them, and to estimate the time needed to complete these assignments and meet deadlines established by the team captain, the client, the lending institution, or the building department. It is also important that the drafter be able to accurately estimate the time required to complete each project. You can develop this skill while working on school projects by estimating the amount of time that will be required to complete the project prior to starting the drawing. In the planning stage, break the project into components and estimate how long each component will take. When the drawing is complete, review your estimates and use the AutoCAD TIME command to determine the actual time required to complete the drawing. Although most firms would prefer a drafter who can quickly complete a project, speed is no substitute for accuracy. Push yourself as a student to meet a self-imposed time deadline while maintaining quality and accuracy. Reliability is also important because a team completes the drawings for a structure. The ability to get along with others, to complete assigned projects in a timely manner, and to coordinate different parts of a project with others will greatly affect how fast a drafter will advance.

Don't be discouraged as you consider the typical entry-level CAD drafting position. This is the type of position you might consider during your first year in school, not as a career. Although you might aspire to design the eighth, ninth, and tenth wonders of the world, an entry-level position can offer valuable insight into the world of architecture and supplement your academic course work. Depending on the size of the office where you work, you may also spend a lot of your time as a new employee editing stock details, running prints, making deliveries, obtaining permits, and doing other office chores. Don't get the idea that a technician only does the menial chores around an office. But do be prepared, as you go to your first drafting job, to do things other than drafting.

Senior CAD Technician

A **senior CAD technician**, often referred to as a *project manager* or a *job captain,* is part of a design team in most midsized offices. This person is responsible for supervising several CAD technicians. The senior technician is expected to be familiar with the building codes that govern the project and to maintain legal building standards. In small firms, the senior technician is typically in charge of producing the working drawings and is often expected to perform simple beam calculations and preliminary wind and seismic studies. Other job duties might include assigning projects for the drawing team, site visitations, and conferences with municipal building officials. A senior CAD drafter is often instrumental in assigning file and layer names, linetype, and other CAD drawing standards and requirements to be used throughout the project.

In many offices that draw multifamily and light commercial projects, senior CAD drafters are part of a larger design team made up of other drafters and supervised by an architect or engineer. The drafter is expected to select material from vendor catalogs such as *Sweets Catalogs* and to apply basic information from *Architectural Graphics* or *Time-Saver Standards* to specific items within each project. To advance as a senior drafter and be a good team leader, you will also need to develop skills that promote a sense of success among your teammates. Although it is against the law to discriminate on the basis of race, color, religion, gender, sexual orientation, age, marital status, or disability, moving beyond the law and creating a friendly and productive work environment is a critical skill for a team leader.

Educational Requirements

The education required to be a CAD technician ranges from a degree from an accredited junior college or technical school to a graduate degree from an accredited school of architecture. Basic writing, math, art, and drafting classes will enhance job opportunities. Computer classes such as keyboarding and LISP will also be helpful. Classes offering a complete mastery of AutoCAD's 2D commands are a must and update classes in the most recent release of AutoCAD are often required. Once you are established within the firm, continuing education that will improve job performance is often a company benefit.

Employment Opportunities

Firms of all sizes hire drafters to help complete drawing projects. Many opportunities exist for converting hand-drawn details and drawings to AutoCAD. In addition

EASY BEING GREEN

A green or sustainable building is a structure that is designed, built, renovated, operated, or reused in an ecological and resource-efficient manner. Green buildings are designed to meet certain objectives such as protecting the health of the occupants; using energy, water, and other resources more efficiently; and reducing the overall impact to the environment. As a senior CAD technician, or with each of the following design positions to be described, you will be expected to provide input regarding materials and construction methods. Much of your time will also be spent investigating the use of environmentally friendly products suitable for each specific project. This will require good use of the Internet, good research skills, and a thorough understanding of the project to determine a product's compatibility with the project. As a new technician, one of your main responsibilities will be to become familiar with the Leadership in Energy and Environmental Design (LEED) Green Building Rating System and certification process.

to jobs in architectural and engineering offices, many architectural equipment suppliers such as heating, air-conditioning, electrical, and plumbing companies employ CAD drafters. Construction subcontractors also hire CAD technicians to create their shop drawings. Firms that manufacture the actual construction components such as trusses, steel, and concrete beams also employ drafters to produce the needed drawings to manufacture their components. Due to their computer skills and construction knowledge, CAD technicians are often hired as sales representatives by many firms. Utility companies such as cable television, telephone, water, and gas companies all employ large numbers of CAD technicians to draw maps of individual routes of service. City, county, state, and federal agencies also hire large numbers of CAD drafters to update municipal records and projects such as parcel and subdivision maps, sewer layouts, and roadway projects. Figure 1-3

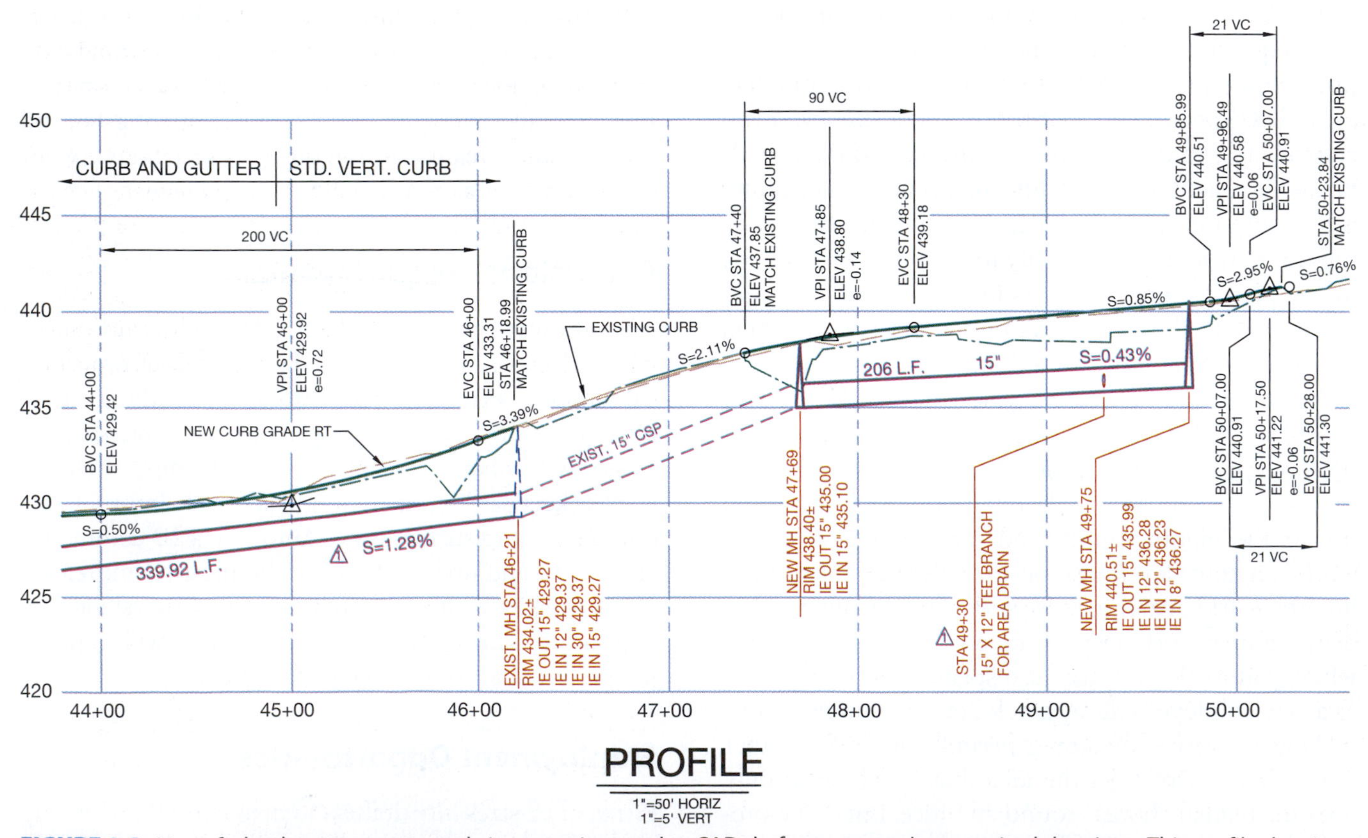

FIGURE 1-3 Many federal, state, county, and city agencies employ CAD drafters to complete required drawings. This profile drawing was completed by a CAD technician employed by a city sanitation department. *Courtesy of Oregon City.*

shows an example of a sewer drawing prepared by a CAD technician working for a city sanitation department. Information about CAD drafting opportunities in related construction trades can be obtained from the American Design Drafting Association. This organization includes members who work in areas such as heating and air-conditioning, plumbing, and electrical design. Contact the American Design Drafting Association for further information:

American Design Drafting Association
P.O. Box 11937
Columbia, SC 29211
803-771-0008
Web site: www.adda.org

DESIGNER

In some states a designer is a drafter who, by experience or state certification, is allowed to design multifamily and small commercial projects. The size of the project is dictated by the state. State laws vary on the types of buildings a designer may assume responsibility for without the supervision of an architect or engineer. In some areas of the country, anyone can decide that he or she is a building designer. In most states a designer working with commercial projects is required to work under the direct supervision of a licensed architect or engineer.

Because of increasing liability problems, many states now require any person using the title *designer* to be licensed. The American Institute of Building Design (AIBD) now oversees designers in 40 states throughout the United States. Contact the AIBD for further information:

American Institute of Building Design
P.O. Box 1148
Pacific Palisades, CA 64057
Web site: www.aibd.org

INTERIOR DESIGNER

Interior designers work with the architect to optimize and harmonize the interior design of structures. In addition to health and safety concerns, interior designers help plan how a space will be used, the amount of light that will be required, acoustics, seating, storage, and work areas. The elements of interior design include the consideration of how the visual, tactile, and auditory senses of the occupants will be impacted. An interior designer must have an aesthetic, practical, and technical appreciation for how people use and respond to these elements, and how the elements interact with one another.

Designers must also be knowledgeable about the many types and characteristics of furnishings, accessories, and ornaments used in creating interiors. Furniture, lighting, carpeting and floor covering, paint and wall covering, glass, wrought metal, fixtures, art, and artifacts are just some of the many items and materials designers select from. In addition, they must be familiar with the various styles of design, art, and architecture and their history.

Educational Requirements

Although a college degree is currently not a requirement, the trend among employers and in states that have licensing requirements is to require a degree from an accredited institution. This can range from training in a two-year program to earn an associate's degree or certificate, to a four- or five-year program leading to a bachelor's (BA, BS, BFA) or master's (MA, MS, MFA) degree. The option chosen may depend on the licensing requirements in your state and whether you have completed a degree in another field.

In the United States, interior designers are registered by title. People cannot represent themselves using the title *interior designer* or *registered interior designer* unless they have met the requirements for education, experience, and examination as set forth in the statutes established by the National Council for Interior Design Qualification (NCIDQ). Candidates who apply to take the NCIDQ examination must demonstrate an acceptable level of professional work experience and completion of related course work. The minimum examination requirements include two years of formal interior design experience and four years of full-time work experience in the practice of interior design. Passage of the examination is required in 20 jurisdictions in the United States and 8 provinces in Canada that regulate the profession of interior design. For further information about careers in the field of interior design, contact:

American Society of Interior Designers (ASID)
202-546-3480
Web site: www.asid.org

International Interior Design Association
Web site: www.iida.org

ARCHITECT

An **architect** is a person who is licensed by individual states to design and supervise the construction of structures. Although a few work full-time in the design or supervision of residential projects, most work in multifamily and commercial design because of the design challenges, job security, pay, and fringe benefits. Figure 1-4 shows a project designed by an architect. An architect is responsible for the design of a structure, from the preliminary drawings through the construction process, and must be able to coordinate the desires and limitations of the client, the financial restraints of the project, the physical setting of the project, and the structural elements of the material to be used.

Use of the title *architect* is legally restricted by each state. In order to be a licensed architect, a person must obtain a master's degree from an accredited school of architecture, complete three years of work experience for a licensed architect, and then pass the state test to demonstrate competency in several areas related to architectural practice. In some states, an architect's license can be obtained by gaining practical experience under the direct supervision of an architect and then passing the written exams. The length of time varies with each state, but typically five to eight years of work experience under the direct supervision of an architect are required before the state examinations may be taken.

FIGURE 1-4 The entryway to the Interfirst II project by S.O.M. Architects of Houston, Texas, is representative of the wide range of projects architects design and draw. *Courtesy MERO Structures, Inc.*

Educational Requirements

Because of the wide range of areas in which an architect must be competent, the educational requirements are very diverse. High school and two-year college students can prepare for a degree program by taking classes in fine arts, math, science, and social science. Many two-year drafting programs offer drafting classes that can be used for credit in four- or five-year architectural programs. A student planning to transfer to a four-year program should verify with the new college which classes can be transferred. Preparation to enter a degree program in architecture should include the following:

- Fine arts classes such as drawing, sketching, design, and art, along with architectural history, will help the future architect develop an understanding of the cultural significance of structures and help transform ideas into reality.
- Math and science, including algebra, geometry, trigonometry, and physics, will provide a stable base for the advanced structural classes that will be required.
- Sociology, psychology, cultural anthropology, and classes dealing with human environments will help develop an understanding of the people who will use the structure.
- Literature and philosophy courses will help prepare students to read, write, and think clearly about abstract concepts.

In addition to formal study, students should discuss with local architects the opportunities and possible disadvantages that may await them in pursuing the study and practice of architecture.

Areas of Study

The study of architecture is not limited to the design of buildings. Although the architectural curriculum typically is highly structured for the first two years of study, students begin to specialize in an area of interest during the third year of the program. Students may branch from architecture into related fields such as urban planning, interior architecture, and landscape architecture.

- *Urban planners* study the relationship among the components within a city.
- *Interior architects* work specifically with the interior of a structure to ensure that all aspects of the building will be functional.
- *Landscape architects* specialize in relating the exterior of a structure to the environment.

Employment Opportunities

An architect performs the tasks of many professionals, including designer, artist, project manager, and construction supervisor.

As described by the American Institute of Architects (AIA), common positions that architects fill include:

Technical staff—CAD operators; drafters; consulting engineers such as mechanical, electrical, and structural engineers; landscape architects; interior designers.

Intern—Unlicensed architectural graduate with less than three years of experience. Common responsibilities include developing design and technical solutions under the supervision of an architect.

Architect I—Licensed architect with three to five years of experience. The job description typically includes responsibility for a specific portion of a project within the parameters set by a supervisor.

Architect II—Licensed architect with six to eight years of experience. The job description typically includes responsibility for the daily design and technical development of a project.

Architect III—Licensed architect with eight to ten years of experience. The job description typically includes responsibility for the management of major projects.

Manager—Licensed architect with more than ten years of experience. Job duties include management of several projects, project teams, client contacts, project scheduling, and budgeting.

Associate—Senior management architect, but not an owner in the firm. This person is responsible for major departments and their functions.

Principal—Owner or partner in an architectural firm.

The work of an architect is not limited to the design of structures. Architects are often employed by local and state governments to work as urban planners, planning whole areas rather than a single structure. Some architects, such as interior planners, lighting specialists, or audio experts, specialize in specific areas of design. Contact the AIA for more information about the role of an architect:

American Institute of Architects
1735 New York Avenue, NW
Washington, DC 20006
Web site: www.aia.org

Contact the following organizations for related information about specialties in the field of architecture:

American Society of Landscape Architects, Inc.
636 Eye Street, NW
Washington, DC 20001-3736
202-898-2444
Web site: www.asla.org

American Society of Interior Design
Web site: www.asid.org

ENGINEER

The title ***engineer*** encompasses a wide variety of professions. Structural, electrical, mechanical, and civil engineers are typically involved in the design of a structure. *Structural engineers* are licensed professionals who specialize in the design of the structural skeleton of a structure. Figure 1-5 shows the framework of a structure that is the result of an engineer's calculations and drawings.

FIGURE 1-5 The steel skeleton for a structure can be seen before the nonstructural materials are applied. Engineers are responsible for the design of the structural skeleton. *Courtesy Janice Jefferis.*

Electrical engineers work with architects and structural engineers and are responsible for the design of lighting, power supply, and communications systems. This includes supervision of the design and installation of specific lighting fixtures, telephone services, and requirements for computer networking.

Mechanical engineers are responsible for the sizing and layout of the heating, ventilation, and air-conditioning systems (HVAC) and details similar to those shown in Figure 1-6. This involves working with the project architect, who determines the number of occupants and the heating and cooling loads, and designing the route that conditioned air will take through the building. The mechanical engineer also plans the size and locations of plumbing lines for fresh and wastewater as well as lines for other liquids or gases that might be required for specific occupancies.

Civil engineers are responsible for the design and supervision of a wide variety of construction projects such as subdivision plans, as well as grading and utility plans for construction projects such as buildings, highways, bridges, sanitation facilities, and water treatment plants. Civil engineers are often directly employed by construction companies to oversee the construction of large projects and to verify that the specifications of the design architects and engineers have been carried out.

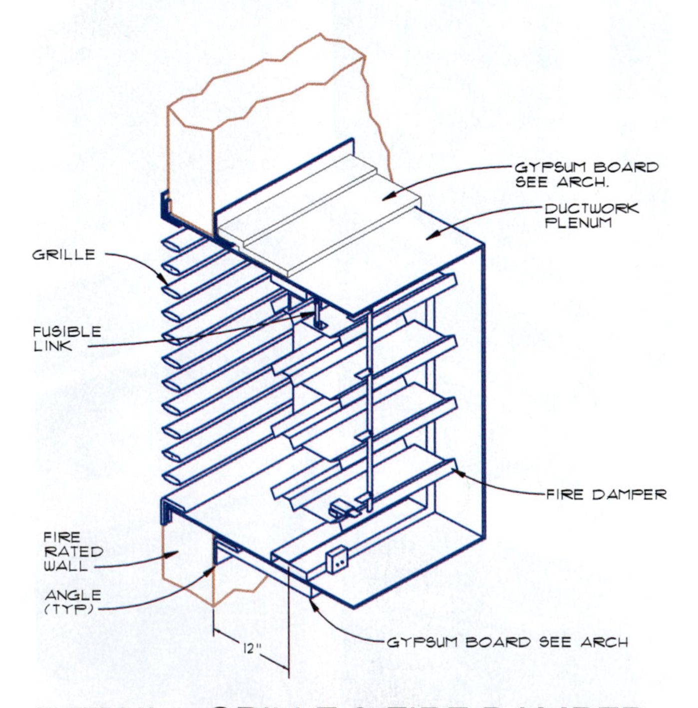

FIGURE 1-6 Mechanical engineers are responsible for the design of the HVAC system and all required drawings and details. *Courtesy Tereasa Jefferis, Systems Designer; Manfull Curtis, Consulting Engineer.*

Educational Requirements

A license is required to function as an engineer. The license can be applied for after several years of practical experience, or after obtaining a bachelor's degree and three years of practical experience. Success in any of the engineering fields requires a high degree of proficiency in math and science, including courses in physics, mechanics, print reading and architecture, mathematics, and material sciences. Similar to the requirements for becoming an architect, engineers are required to complete five years of education at an approved college or university, followed by successful completion of a state-administered examination. Certification can also be accomplished by training under a licensed engineer and then successfully completing the examination. Contact these organizations for additional information about the field of engineering:

American Society of Civil Engineers (ASCE)
1801 Alexander Bell Drive
Reston, VA 20191-4400
800-548-2723
Web site: www.asce.org

American Consulting Engineers Council (ACEC)
1015 15th Street, NW, Suite 802
Washington, DC 20005
202-347-7474
Web site: www.acec.org

American Society of Heating, Refrigerating, and Air-Conditioning Engineers, Inc. (ASHRAE)
1791 Tullie Circle NE
Atlanta, GA 30329-2305
800-527-4723
Web site: www.ashrae.org

Illuminating Engineering Society of North America (IESNA)
120 Wall Street, 17th Floor
New York, NY 10005
212-248-5000
Web site: www.iesna.org

RELATED FIELDS

In addition to the drawing and design aspects of construction drawings, a drafting background can lead to several other careers such as illustrator, model maker, specifications writer, and inspector. A drafting background can also be useful in careers such as sales representative, contractor, and estimator. Although these jobs do not require drafting as a job requirement, knowledge of drafting fundamentals will help in the performance of job requirements.

Illustrator

Commercial drawing requires the use of a rendering such as that seen in Figure 1-7. By combining artistic talent with a basic understanding of architectural principles, an illustrator is able to produce drawings that realistically show a proposed structure. Traditionally, graphite, ink, and watercolor have been the main tools of an illustrator. Many firms are now adopting the use of AutoCAD's solid modeling programs and third-party software to create computer-generated renderings. Drawings created in 3D allow the viewpoint to be altered, providing a realistic look at the structure from any vantage point. Illustrators also use *virtual reality* to represent how a structure will blend with its environment and to allow clients to walk through proposed buildings and see how components can be arranged.

Model Maker

In addition to presentation drawings, many architectural firms use models of building projects to help convey design concepts. Models similar to the one in Figure 1-8 are often used for public presentations to groups such as zoning commissions, citizens' advisory committees, and other regulatory commissions. Model makers need basic drafting skills to interpret the working drawings so that the model can be an accurate representation of the proposed structure. Model makers can be employed within a large architectural firm or by a company that specializes in modeling.

Specifications Writer

Most architectural and engineering firms have an internal staff that specializes in written documentation. The written specifications that accompany the working drawings, environmental impact reports, and addendums to the permit forms are all written by a specifications writer. An understanding of basic drafting principles, a mastery of grammar and basic English, keyboarding, word processing, and spreadsheet skills are all essential for a specifications writer. A thorough understanding of the Construction Specifications Institute's (CSI) chapter format, which will be introduced in Chapter 14, is also required.

Inspector

Municipalities require that the plans and the construction process be inspected to ensure that the required codes for public safety have been met. A plans examiner must be licensed by the state to certify a thorough understanding of the construction process and common engineering practices. Most states require a degree in architecture or engineering for a state or city inspector

FIGURE 1-7 Drawings are often prepared by an architectural illustrator to help convey design concepts. *Courtesy Dean K. Smith, Kenneth D. Smith Architect & Associates, Inc.*

FIGURE 1-8 Models are used for public displays to convey design ideas. *Courtesy KOIN Center, Olympia & York Properties, Inc (Oregon).*

or plans examiner. Most architectural and engineering firms also require job inspection as part of their design process to ensure that contractors have rigidly followed the plans and specifications. This also ensures that conflicts between the numerous firms involved in the construction process can be resolved to the benefit of the client. An inspector or job supervisor is not required to be a licensed architect or engineer, but a thorough understanding of the drawings and the construction process is essential.

GREEN CONSTRUCTION

Reduce, reuse, recycle! No matter what area of the design or construction field you enter, an important aspect of your career will be a building mind-set that revolves around these three words. Whether it is called *earth-friendly, green, ecological,* or *sustainable* construction, the concept is to build in a manner that will produce a structure that uses energy efficiently, that uses materials that have a low impact on the environment, and that contributes to a healthier workplace.

One of the leaders in the development of sustainable construction is the Leadership in Energy and Environmental Design Green Building Rating System, referred to as LEED. **LEED** is a rating system used to evaluate key areas of building projects, such as:

- Sustainable sites
- Water efficiency
- Indoor environmental quality
- Energy and atmosphere
- Material and resources
- Innovation and the design process

Other key organizations that are leaders in green construction will be introduced in Chapter 4, and information related to specific materials will be discussed throughout the text. As a new CAD technician, you may not be making the decisions on how to increase the sustainability of a structure, but in order to advance in an office, you must understand key green issues.

Section 2 of this text will examine common framing methods found in commercial construction. Environmentally friendly framing is not just a matter of selecting green materials; once selected, these green products must be used in a manner that will reduce the environmental impact of the structure. Products that are not considered green can be used in a structure, but they must benefit the building owner and the occupants in a manner that contributes to its sustainability. Creating an environmentally friendly structure requires the matching of materials to a specific design and site that minimizes the effect on that site. Five questions should be considered that will affect the selection of materials to make a sustainable structure:

1. Can products be selected that are made from environmentally friendly materials?
2. Can products be selected because of what they do not contain?
3. Will the products to be used reduce the environmental impact during construction?
4. Will the products to be used reduce the environmental impact of operating the building?
5. Will the products to be used contribute to a safe, healthy indoor environment?

Environmentally Friendly Materials

One of the major considerations that affect framing materials is the selection of products and construction materials that are made from environmentally friendly components. The materials used to produce a building product, and where those materials came from, are key determinants in labeling a product a green building material. Points to consider in selecting building materials include products that can be salvaged, products with recycled content, certified green quick-growth products, agricultural waste materials, and products that require minimal processing.

Salvaged Products

A major goal of sustainable construction is to reuse a product whenever possible instead of producing a new one. Common salvaged materials used in buildings include bricks, millwork, framing lumber, plumbing fixtures, and period hardware. Each of these materials is sold on a local or regional basis by salvage yards. Depending on the occupancy rating of a structure, this may not be possible on larger commercial projects.

Products with Recycled Content

Recycled content is an important feature of many green products because materials are more likely to be diverted from landfills. Industrial by-products such as iron-ore slag can be used to make mineral wool insulation, fly ash can be used to make concrete, and PVC pipe scraps can be used to make shingles.

Certified Wood Products

Third-party forest certification, based on standards developed by the Forest Stewardship Council (FSC), will help to ensure that wood products come from well-managed forests. Wood products must go through a certification process to carry an FSC stamp. Manufactured wood products can meet the FSC certification requirements with less than 100% certified wood content.

Quick Growth and Waste By-Products

The use of quick-growth products for framing materials allows old- and second-growth trees to remain in the forest. Rapidly renewable materials are made with wood from tree farms with a harvest rotation of approximately 10 years to make products such as LVL, OSB, and laminated beams. Interior finish products are also produced from agricultural crops or their waste products. Examples of green products made from agricultural crops include linoleum, form-release agents made from plant oils, natural paints, textile fabrics made from coir and jute, cork, organic cotton, wool, and sisal. Building products can also be produced from agricultural waste products made from straw (the stems left after harvesting cereal grain), rice hulls, and citrus oil. These products can be used to improve the interior environment and create an ecologically friendly structure.

Engineered lumber products are an example of a green building product. Engineered lumber products are made by turning small pieces of wood into framing members. Structural engineered members are made from fast-growing tree species grown on tree farms specifically for the purpose of being used to make structural materials. Depending on the product to be made, sawdust, wood scraps, small pieces of lumber, or whole pieces of sawn lumber can be joined by adhesives applied under heat and pressure to produce engineered building products. Common engineered products found throughout a structure include laminated veneered lumber, oriented strand board, engineered studs, I-joists, and laminated beams. Production of engineered framing products offers three benefits: (1) efficient use of each log that enters the mill, (2) predictable, superior structural quality, and (3) reduced construction waste at the job site. See Chapter 8 for additional information.

Minimally Processed Products

Products that are minimally processed can be green because of low energy use and low risk of chemical releases during manufacture. These can include wood products, agricultural or nonagricultural plant products, and mineral products such as natural stone and slate shingles.

Removing Materials to Become Earth Friendly

Some building products are considered green, not because of their content, but because they allow material savings elsewhere, or they are better alternatives to conventional products containing harmful chemicals. Chemicals that deplete the ozone, CCA wood preservative, polyvinyl chloride (PVC), and polycarbonate are products that should be avoided in a structure, but products with these chemicals may be considered earth friendly because the products have significant environmental benefits. Some examples include drywall clips that allow the elimination of corner studs, engineered lumber that reduces lumber waste, the piers for a joist floor system that minimizes concrete use compared to a post-and-beam system, and concrete pigments that eliminate the need for conventional finish flooring by using the concrete slabs as the finished floor.

Reducing the Impact of Construction

Some building products produce their environmental benefits by avoiding pollution or other environmental impacts during construction. Products that reduce the impacts of new construction include various erosion-control products, foundation products that eliminate the need for excavation, and exterior stains that result in lower volatile organic compound (VOC) emissions into the atmosphere. The greatest impact from construction can come from careful design that creates a home that suits the site, and a foundation system that requires minimal excavation. See Chapters 15 for methods of reducing site impact during construction.

Reducing the Impact after Construction

The ongoing environmental impact that results from operating a structure will far outweigh the impact associated with the construction phase. It is important during the design phase to select components that will reduce heating and cooling loads and conserve energy, select fixtures and equipment that conserve water, choose materials with exceptional durability or low-maintenance requirements, select products that prevent pollution or reduce waste, and specify products that reduce or eliminate pesticide treatments.

Reducing Energy Demands

Structurally insulated panels (SIPs), insulated concrete forms (ICFs), autoclaved aerated concrete (AAC) blocks, and high-performance windows are examples of materials that can be used during construction to reduce HVAC loads over the life of the structure. Other energy-consuming equipment such as water heaters, furnaces, and basic office equipment should be carefully selected for their ability to conserve energy after construction. Most office equipment is now rated by Energy Star™ standards that have been adopted nationally. Energy Star is a government-backed program designed to help consumers obtain energy efficiency. Using compact fluorescent lamps and occupancy or day lighting control equipment can save additional energy. Installing efficient equipment as the structure is constructed will reduce energy needs over the life of the structure.

Renewable Energy

Equipment that uses renewable energy rather than fossil fuels and conventional electricity is highly beneficial from an environmental standpoint. Examples for smaller projects include solar water heaters and photovoltaic systems. Natural gas fuel cells or cells that use other fossil fuels such as a hydrogen source are considered green because emissions are lower than that of the combustion-based equipment they replace.

Conserving Water

All toilets are required to meet the federal water efficiency standards. Other products, such as rainwater storage systems will also contribute to making a structure earth friendly.

Reducing Maintenance

Products that reduce maintenance make a structure environmentally attractive because those products need to be replaced less frequently, or their maintenance has very low impact. Sometimes, durability is a contributing factor to the green designation but not enough to distinguish the product as green on its own. Included in this category are such products as fiber-cement siding, fiberglass windows, slate shingles, and vitrified clay waste pipes.

Preventing Pollution

Methods of controlling substances from entering the environment contribute to making a structure earth friendly. Alternative wastewater disposal systems reduce groundwater pollution by decomposing organic wastes more effectively. Porous paving products and vegetated roofing systems result in less storm water runoff and thereby reduce surface water pollution. Providing convenient recycling centers within the structure allows the occupants to safely store recyclables for collection.

Eliminating Pesticide Treatments

Although they may be needed to increase livability in smaller wood-framed structures, periodic pesticide treatment around buildings can be a significant health and environmental hazard. The use of products such as termite barriers, borate-treated building products, and bait systems that eliminate the need for pesticide application all contribute to a sustainable structure.

Contributing to the Environment

Product selection has a significant effect on the quality of the interior environment. Green building products that help ensure a healthy interior living space can be separated into several categories including products that don't release pollutants, products that block the spread of indoor contaminants, and products that warn occupants of health hazards.

Nonpolluting Products

One of the dangers of modern construction is the ability to make a structure practically airtight. This in itself is not a problem, but the adhesives in most products can be harmful when constant air changes are not provided.

Products that don't release significant pollutants into a structure contribute to an earth-friendly home. Interior products that contribute to improving the interior environment include zero- and low-VOC paints, caulks, and adhesives, as well as products with very low emissions, such as nonformaldehyde manufactured wood products.

Blocking, Removing, and Warning of Contaminants

Certain materials and products are green because they prevent contaminants from entering the interior environment. Linoleum is available that helps control microbial growth. Coated duct board is available that helps control mold growth, and products are available for blocking the entry of mold-laden air into a duct system. Other products can help remove pollutants from the shoes of people entering the structure. Each of these types of products can help provide an earth-friendly interior environment.

Once contaminants have entered the structure, several products are available to warn the occupants. Examples of warning systems include carbon monoxide (CO) detectors and lead paint test kits. Once occupants are aware of environmental dangers, several products are available to remove the contaminants. Ventilation products, filters, radon mitigation equipment, and other equipment can remove pollutants or introduce fresh air.

ADDITIONAL READING

The following Web sites can be used as a resource to help you keep current with changes in professional design careers.

ADDRESS	COMPANY/ ORGANIZATION
www.acec.org	American Consulting Engineers Council
www.aceee.org	American Council for Energy Efficient Economy
www.aiaonline.com	American Institute of Architects
www.ashrae.org	American Society of Heating, Refrigerating, and Air-Conditioning Engineers
www.asla.org	American Society of Landscape Architects
www.aspenational.com	American Society of Professional Engineers
www.aeinstitute.org	Architectural Engineering Institute
www.usace.army.mil	Army Corp of Engineers
www.awa-la.org	Association for Women in Architecture
www.athenasmi.ca	Athena Sustainable Materials Institute
www.buildinggreen.com	Building Green Inc. (This site contains a listing of site-related links based on LEED standards and CSI formats.)
www.crbt.org	Center for Resourceful Building Technology
www.corrim.org	Consortium for Research of Renewable Industrial Materials
www.thebluebook.com	Construction Information Network
www.csinet.org	Construction Specifications Institute
www.eere.energy.gov	Department of Energy Integrated Building for Energy Efficiency
www.dbia.org	Design-Build Institute of America
www.enercept.com	Enercept, Inc. (SIPs)
www.energydesignresources.com	Energy Design Resources
www.eeba.org	Energy and Environmental Building Association
www.energystar.gov	Energy Star (appliance energy standards)
www.ei.org	Engineering Information
www.iesna.org	Illuminating Engineering Society of North America
www.ieee.org	Institute of Electrical and Electronics Engineers
www.thegbi.org	Green Building Initiative

www.greenguard.org	Green Guard Environmental Institute
www.greenbuilder.com	Greenbuilder
www.buildinggreen.com	GreenSpec
www.noma.net	National Organization of Minority Architects
www.nsbe.org	National Society of Black Engineers
www.nspe.org	National Society of Professional Engineers
www.naima.org	North American Insulation Manufacturers Association
www.pge.com	Pacific Energy Center
www.raic.org	Royal Architectural Institute of Canada
www.scscertified.com	Scientific Certification Systems
www.sara-national.org	Society of American Registered Architects
www.oneshpe.org	Society of Hispanic Professional Engineers
www.same.org	Society of Military Engineers
www.housingzone.com	Sustainable Buildings Industry Council
www.usgbc.org	U.S. Green Building Council
www.wbdg.org	Whole Building Design Guide
www.beconstructive.com	Wood Promotion Network

REVIEW ACTIVITIES

Activity 1-1 Contact your state's branch of the AIA and list the requirements for becoming an architect in your state.

Activity 1-2 Visit the "Information for Consumers" section of the AIA Web site and list the three areas of information found under "Commercial."

Activity 1-3 List the benefits of using an architect, as described in the "Institutional" area of the AIA Web site.

Activity 1-4 Visit the National Society of Professional Engineers Web site and briefly describe three current news items that are listed.

Activity 1-5 Use a search engine to list and describe a minimum of five Web sites for structural engineers.

KEY TERMS

Architect
CAD technician
CAD technician, senior
Certified Wood Products
Commercial structures
Engineer
Engineer, civil
Engineer, electrical
Engineer, mechanical
Environmentally friendly materials
Green construction
Interior designer
LEED

CHAPTER 1 TEST Professional Careers and Commercial CAD Drafting

QUESTIONS

Answer the following questions with short complete statements. Using a word processor, type the chapter title, question number, and a short complete statement for each question.

Question 1-1 List the main job responsibilities of a senior CAD drafter.

Question 1-2 List several skills or assets needed to become a senior CAD drafter.

Question 1-3 List and describe nine careers in which CAD drafting could be helpful.

Question 1-4 List three duties a junior CAD technician might perform.

Question 1-5 Why would a student in an architecture program need to take classes in social science?

Question 1-6 List the Web sites of five architectural or engineering associations in your area.

Question 1-7 Why are engineering students required to have such a strong understanding of math?

Question 1-8 Explain the term *VOC* and explain how it relates to construction.

Question 1-9 Why are engineered lumber products considered earth friendly?

Question 1-10 Contact one member of the AIBD, AIA, or ASCE and determine specific requirements for education, employment, and advancement in your area.

The CAD Drafter's Role in Office Practice and Procedure

2

The office that you work in, the size of the staff, and the type of project being drawn will all affect the role of the drafter. This chapter explores common roles of a drafter and common types of drawings included in a set of plans. Because most graduates of two-year technical and junior college programs work for architectural and engineering firms, these will be the only types of drawings explored in this text. This chapter will explore:

- Common office procedures you're likely to encounter as a new technician.
- The types of drawings you're likely to encounter in an architect's or engineer's office.
- Common drawing set organization.
- Using the engineer's calculations, vendor catalogs, and codes.
- Common drawing page layout and project coordination.

COMMON OFFICE PRACTICE

The design of a structure starts with the selection of the design team. The client typically selects the architectural firm based on several common reasons. The reputation of the design firm is typically of utmost importance. The relationship with previous clients and the ability to meet the design requirements in a timely and economical manner are key elements in any designer's reputation. Previous experience with projects similar to the structure to be designed is also an important consideration. Although many large firms work with a variety of structures and materials, some smaller architectural firms are very specialized in the type of structures they design. For instance, some firms work primarily with educational structures. Some firms work only with specific materials such as concrete tilt-up structures or with rigid steel frames.

Recommendations from a realtor, builder, or previous client, as well as marketing also play an important part in the selection of an architectural firm.

Once selected, the architectural team can provide valuable assistance to the client in the selection of the construction site and in the refinement of the design criteria. With an understanding of the basic design requirements, analysis of the potential site based on access, climate, and zoning laws can begin. Chapters 4 and 5 will provide further information on how building codes and accessibility standards affect the design process. Other areas such as the shape of the potential job site, availability of utilities, topography, soil conditions, drainage patterns, and proposed building materials also must be considered in the initial stages of design. With a specific site purchased, design studies can be done to determine the best possible merging of the client's desires with the limitations of the site and the financial constraints of the project.

The Role of the CAD Drafter

The experience of the CAD drafter, the number of CAD drafters assigned to a project, and the complexity of the project will dictate when the CAD technician will become involved in the project. The CAD technician may be brought into the project at a very early stage to set up sheet templates to meet specific needs of the projects. As the template is established, drafters need to be sure to use established scaling values. See Appendix A for common construction drawing scales. Typically, a CAD drafter will become involved in the completion of the **preliminary drawings** that make up the design presentation. This includes work on the site plan, floor plan, elevations, and sections so that the key elements of the design can be presented to the client. One of the first needs of the designer is a layout of the site. Working with drawings provided by a civil engineer, the architectural team can start the preliminary site plan. On simpler projects the technician may need to draw the lot boundary and other items such as the street and curb (see Chapter 15).

After the designer has developed a concept site plan, floor plan, and exterior elevations, the technician may be directed to clean up the drawings. This work might include developing a site or building section, calculating areas, numbering parking spaces, or adding text such as room names, reference grid lines, and dimensions. The technician may also be asked to assist in preparing presentation drawings by adding color, texture, and foliage to the various drawings. These drawings will be used to present and explain the project to the client, control committees, planning commissions, lenders, and other similar groups.

Design Development

The next step for the technician will probably be to further refine the drawings to work out drawing conflicts, and to get preliminary input from consultants to make sure that the necessary space has been provided for their needs. This is a time to carefully review the project again to make sure that the project still complies with code requirements. One major item that should be checked is the exit path (see Chapter 4). Another item to review is the requirements for firewalls for occupancy or area separation. Once these reviews have been done, background drawings required by consultants need to be prepared. Consultants for the project may include a civil engineer, landscape architect, electrical engineer, mechanical engineer, and structural engineer, among others. These consultants will add to the drawings created by the architectural technician or incorporate the technician's drawings into their projects. This basic design information from the consultants will then be returned to the architect so that any necessary adjustments can be made.

Construction Documents

Once all the adjustments have been incorporated into the preliminary drawings, the technician begins the biggest job, preparing the working drawings. Most likely the first thing to be done is issuing new base sheets to the consultants. Getting the revised base sheets out to the consultants first will allow them to be developing their drawings while the technician is completing the **architectural drawings.** The aim here is to have the consultants' drawings completed simultaneously while staff members at the architect's office complete the architectural drawings.

After the consultants' drawings have been distributed, the technician can continue developing all of the architectural working drawing sheets. During this process, changes may develop that impact the information that has been sent to the consultants. It is very important to keep communication lines open, informing the consultants of these changes and distributing new base sheets if necessary. Information may also flow the other way. The consultants may need updates to be made to the design to make their system work. The mechanical engineer may need a shaft added to adequately distribute the heating, ventilation, air-conditioning (HVAC); the structural engineer may need to add a sheer wall or column. This information will need to be incorporated into the drawings and distributed to the other consultants. The use of external referenced drawings can greatly aid in this communication.

Depending on the size and complexity of the project, a structural engineering firm will work in conjunction with the architectural firm to analyze the forces that will be imposed on the structure. Chapter 13 will introduce common structural considerations affecting the design of a structure. The structural engineer might provide calculations, framing plans, structural elevations and sections, and details for the architectural firm to use to complete the project or might provide a complete set of **structural drawings** to accompany the architectural drawings. In either case, CAD drafters will be required to ensure that materials required to meet all of the mathematical calculations of the engineer have been incorporated into the drawings. CAD drafters will also be required to incorporate information supplied by several other firms including mechanical, electrical, plumbing, lighting, and interior design firms.

Permit and Bidding Stages

Once the working drawings have been completed and the consultants' drawings have been received, the drawings are assembled, printed, and submitted to the building department. The building department will check the drawings and frequently will require corrections to be done. The technician will make these corrections so the project can be resubmitted to the building department. Once the corrections have been completed and the drawings rechecked by the building department, a building permit can be issued.

At some time during this process, the plans will be issued to contractors so that they can prepare a bid for the project. It is normal for the plans to be issued prior to the building department completing their plan check. When changes and corrections are made to the plans it will be necessary to inform the bidding contractors of those changes so that they can make appropriate price adjustments to their bid.

Construction Phase

After the contractors have turned in their bids, one of them will be selected as the general contractor to build the project. The building permit will not normally be issued until the general contractor has been named. During the construction process, the general contractor will normally submit shop drawings to the architect's office. *Shop drawings* are prepared by subcontractors and show a level of detail beyond that which is in the working drawings. The technician may be asked to check the shop drawings to see if they are in general compliance with the working drawings. During the construction process, unanticipated conditions may arise that necessitate changes to the working drawings. The technician may be asked to prepare the drawings for these changes.

TYPES OF DRAWINGS

Commercial projects often contain five major types of drawings:

- *Procurement drawings* are issued for bidding or negotiating before an agreement is signed.
- *Contract drawings* describe the work required to complete the project.
- *Resource drawings* show existing conditions or new work not included in the project.
- *Addenda drawings* and *modification drawings*—Both types of drawings are known as *supplemental drawings*. Addenda drawings are used to amend the work during the bidding process but prior to the awarding of the contract. Modification drawings are used to alter or revise an element of the working drawings after the drawings have been released for construction.

The majority of the information that is provided in Sections 3 and 4 is devoted to preparing contract drawings.

Drawing Organization

Commercial drawing sets often include more than 100 sheets of 24" × 36" or 30" × 42" drawings. Because of the size and scope of most commercial projects, drawings are arranged by subsets of related information. Based on the **National CAD Standards** (NCS), these common drawing groups, and their order within the drawing set include:

G—General

H—Hazardous Materials

V—Survey Mapping

B—Geotechnical

C—Civil

L—Landscape

S—Structural

A—Architectural

I—Interiors

Q—Equipment

F—Fire Protection

P—Plumbing

D—Process

M—Mechanical

E—Electrical

T—Telecommunications

R—Resource

X—Other Disciplines

Z—Contractor/Shop Drawings

O—Operations

The letter that precedes each subset name will be used in the page numbering system to reference the subset. The NCS refer to this letter as the *sheet type designator*. Page numbering will be introduced later in this chapter. Not all of these subsets will be required for each project, and some projects may require additional subsets depending on the scope, size, and complexity of the project. The most common drawing subsets and their order in the completed drawing set include:

C—Civil

S—Structural

A—Architectural

I—Interiors

P—Plumbing

M—Mechanical

E—Electrical

Separate consulting firms that are coordinated by the architectural firm usually prepare these drawings. On smaller projects, the architectural team may prepare these drawings based on recommendations from consulting firms.

Note: Even though the NCS recommends that structural drawings be placed before the architectural drawings, in order to help better understand the process, the architectural drawings will precede the structural drawings in Sections 3 and 4 of this text.

Drawing Set Organization

Regardless of the source of origination, CAD drafters will be employed to develop each type of drawing. The first page of most commercial plans is a title page, which contains a table of contents similar to the one seen in Figure 2-1. Other items that should be placed on the title page include:

- The name and contact information of the project owner.
- The name, logo, and contact information for the project architect and each consulting firm responsible for some aspect of the plans.

INDEX OF DRAWINGS

T1 — TITLE SHEET, SITE DEMOLITION & SURVEY

CIVIL

C1 — SITE UTILITIES
C2 — EROSION CONTROL PLAN & DETAILS
C3 — PUBLIC UTILITY PLAN
C4 — UTILITY DETAILS

LANDSCAPE

L1 — IRRIGATION SYSTEM PLAN & LEGEND
L2 — IRRIGATION SYSTEM DETAILS & NOTES
L3 — PLANTING PLAN, DETAILS & LEGEND

STRUCTURAL

S1 — FOUNDATION PLAN
S2 — FOUNDATION DETAILS
S3 — SECOND FLOOR FRAMING PLAN
S4 — ROOF FRAMING PLAN
S5 — DETAILS
S6 — DETAILS

ARCHITECTURAL

A1 — SITE PLAN & DETAILS
A2 — GRADING PLAN & DETAILS
A3 — FIRST FLOOR PLAN, SCHEDULES & DETAIL
A4 — SECOND FLOOR PLAN, SCHEDULES & DET.
A5 — ENLARGED PLANS, INTERIOR ELEV. & DET.
A6 — EXTERIOR ELEVATIONS & DETAILS
A7 — BUILDING & WALL SECTIONS & DETAILS
A8 — ROOF PLAN & DETAILS
A9 — DETAILS
A10 — REFLECTED CEILING PLAN & DETAILS

MECHANICAL

M1 — FIRST FLOOR PLUMBING PLAN, LEGEND
M2 — SECOND FLOOR PLUMBING PLAN, NOTE
M3 — DETAILS, SCHEDULES
M4 — FIRST FLOOR H.V.A.C PLAN, LEGENDS
M5 — SECOND FLOOR H.V.A.C. PLAN
M6 — ROOF MOUNTED H.V.A.C. EQUIP. PLAN
M7 — H.V.A.C. DETAILS, SCHEDULES

ELECTRICAL

E1 — NOTES, LEGEND, RISER
E2 — FIRST FLOOR LIGHTING PLAN
E3 — SECOND FLOOR LIGHTING PLAN
E4 — FIRST FLOOR POWER PLAN
E5 — SECOND FLOOR POWER PLAN
E6 — ROOF MOUNTED EQUIP. POWER PLAN
E7 — DETAILS, SCHEDULES
E8 — FIRST FLR. COMMUNICATIONS PLAN, LEGEN
E9 — SECOND FLOOR COMMUNICATIONS PLAN
E10 — H.V.A.C. DETAILS, SCHEDULES

FOOD SERVICE

FS1 — EQUIPMENT PLAN, SCHEDULES
FS2 — PLUMBING PLAN, MECHANICAL PLAN
FS3 — EQUIPMENT, DETAILS,

FIGURE 2-1 The title page, the first sheet of each project, typically contains a table of contents to help print readers use the drawings effectively. *Courtesy Architects Barrentine, Bates & Lee, A.I.A.*

- A rendering or a photograph (for franchise structures) of the project.
- A list of abbreviations, project data, a location map, and general notes relevant to the entire project.

Drawing Set Page Designators

For a very simple project that will be completely drawn by the architectural team, drawings within the project can be numbered in successive order using a designator such as A-1, A-2, A-3. Most projects require a more thorough system of sheet designators. Common sheet designators for a single-level structure include:

0—General information such as symbols, legends, and notes specific to the subset

1—Plan views

2—Elevations

3—Sections

4—Large-scale views such as plans, elevations, stair sections, or sections that are not details

5—Details

6—Schedules and diagrams

7—User-defined sheets that do not fall into other categories such as typical detail sheets

8—User-defined sheets that do not fall into other categories

9—3D drawings including isometric perspectives and photographs

This numbering system would be used for each subset throughout the project.

For multiple-level structures, a modified version of this system is used. The categories remain the same, but instead of using a single digit, a number based on 100 is used. Floor plans for a three-level structure would be numbered:

101—First floor plan

102—Second floor plan

103—Third floor plan

With this expanded system, 1 represents a floor plan, and the 01 represents the first floor. A number starting with 00 is not permitted by the NCS. These numbers are also used throughout each of the subsets to aid in coordination. Sheet A-102 would contain the architectural floor plan for the second level; sheet S-102 would contain the second structural framing plan; and E-102 would contain the electrical plan for the second level.

Civil Drawings

Drawings that are related to the construction site typically follow the title page. Each drawing is numbered in successive order starting with **C-1** (**Civil**) or **L-1** (**Landscape**). Figure 2-2 shows an example of a commercial site plan.

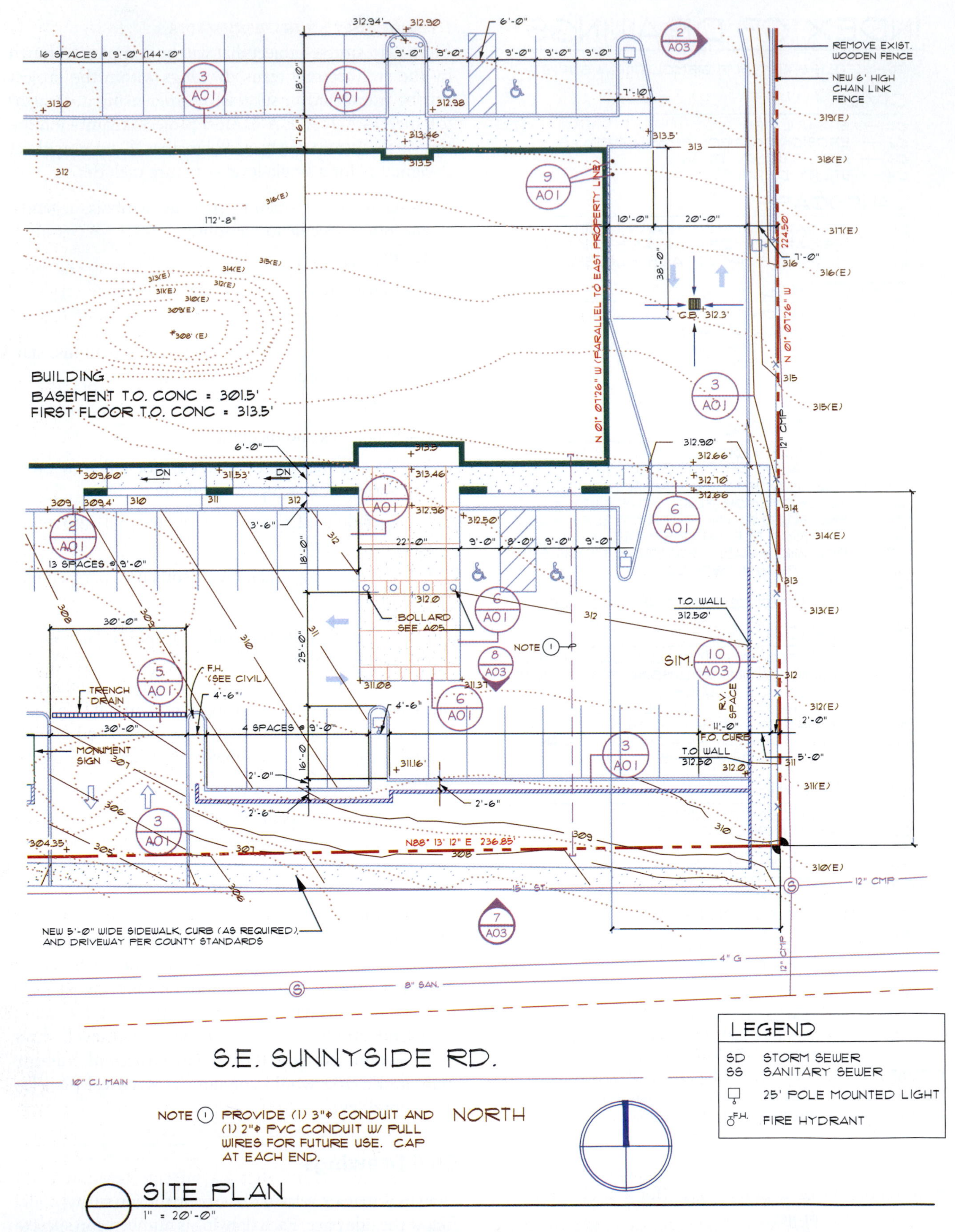

FIGURE 2-2 A commercial site plan can be completed by a CAD technician working for an architect or a civil engineer. *Courtesy Peck, Smiley, Ettlin Architects.*

Depending on the complexity of the project and the site, the site drawings might include an existing site plan, a proposed site plan, a grading plan, a utility plan, an irrigation or sprinkler plan, and a landscape plan.

Each of these drawings will be introduced in Chapter 15.

Architectural Drawings

The architectural drawings are placed in the working drawings in the subset represented by the letter A. The architectural drawings are the drawings that describe the size and shape of a structure. They are prepared by or under the direct supervision of an architect. Common architectural drawings for a commercial project include floor plans, enlarged floor plans, elevations, wall sections, roof plan, reflected ceiling plan, interior elevations, finish schedules, and interior details. Each type of architectural drawing will be introduced in Section 3.

Structural Drawings

As their name implies, these are the drawings used to construct the skeleton of the structure. Engineers or CAD drafters working directly under the supervision of the engineer prepare these drawings. Structural drawings will be discussed in Section 4 and are placed in the working drawings in the subset represented by the letter S. Structural drawings include the framing plans, foundation plan, and related sections and details. Figure 2-3 shows an example of a framing plan drawn by CAD drafters working in an engineering firm.

Cabinet and Fixture Drawings

On structures with few cabinets, the interior details and elevations similar to Figure 2-4 are part of the architectural drawings. On structures such as a restaurant or a medical facility that will contain a magnitude of specialty **cabinet (or trim) drawings**, these drawings will be in their own section and be in successive order starting with **I-1** (for **Interiors**). This section will also include details covering interior trim and specialized equipment, as well as cabinets. Figure 2-5 shows an example of interior drawings for a commercial kitchen.

Electrical Drawings

The **electrical drawings** include the electrical plans, lighting plans, equipment plans, and related schedules and details needed to completely specify the electrical requirements of the structure. For simple projects, CAD technicians working for the architectural firm might complete the electrical drawings. Electrical drawings for larger projects are prepared by CAD drafters working under the direct supervision of a licensed electrical engineer. These drawings will be successively arranged starting with drawing **E-1** or **E-101**. The preparation of electrical drawings and details for commercial plans will not be discussed in this text.

Mechanical Drawings

The drawings that are used to show the movement of air throughout the structure make up the **mechanical drawing** portion of a project. These drawings will be successively arranged starting with **M-1** or **M-101**. Several schedules and details are also typically very instrumental to these drawings. The preparation of mechanical drawings and details will not be discussed in this text.

Plumbing Drawings

The **plumbing drawings** are used to show how fresh water and wastewater will be routed throughout the structure. Plumbing drawings, schedules, and details are successively arranged starting with page **P-1** or **P-101**. The preparation of plumbing drawings and details will not be discussed in this text.

WORKING WITH CALCULATIONS

With so many different types of drawings and so many different firms contributing to the structure, an inexperienced drafter could easily get lost in the developmental process of a structure. To ensure that all required information is incorporated in the drawings, architects and engineers typically provide a set of **calculations** and sketches for the drawing team to use in order to complete the plans. The calculations for the structural drawings are usually required to be signed by an architect or engineer and must be submitted to the building department with the completed drawings in order for a building permit to be obtained.

Figure 2-6 shows a portion of one page of an engineer's calculations. Typically, calculations are divided into three sections that include a problem statement, mathematical solution, and the material to be used to meet the imposed stress. In Figure 2-6, the engineer is determining the size of two beams and how they will be connected. Although it is important that the drafter understand basic structural principles to advance in the field, the technician will not be expected to do structural calculations. Sections 3 and 4 will introduce concepts the drafter should be familiar with to complete

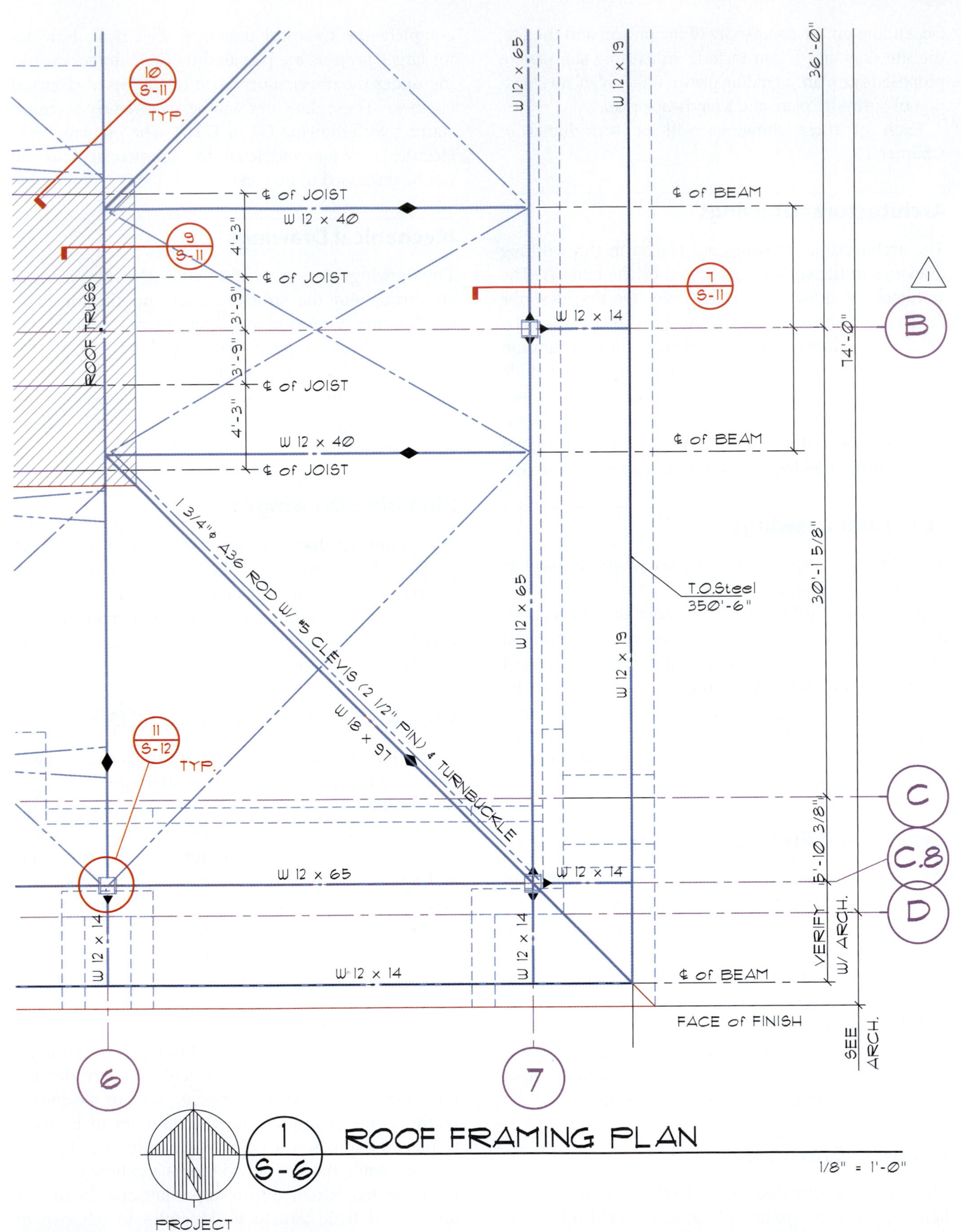

FIGURE 2-3 The structural drawings show how to construct the skeleton of a structure. *Courtesy Van Domelen/Looijenga/McGarrigle/Knauf Consulting Engineers.*

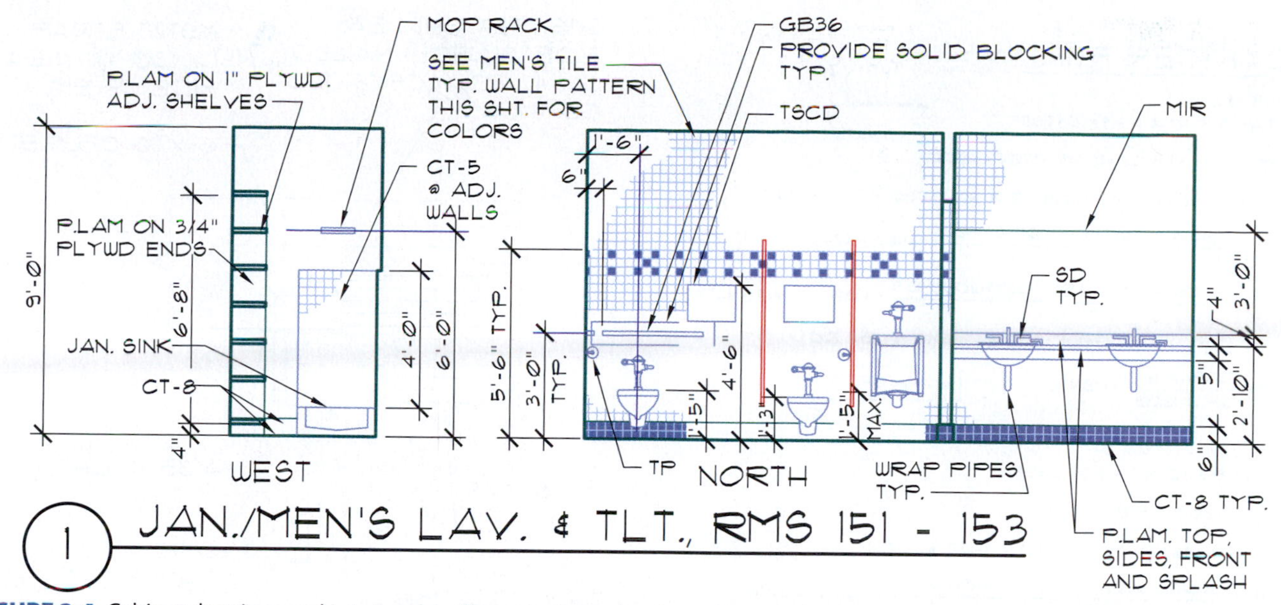

FIGURE 2-4 Cabinet drawings and interior elevations are drawn by the architectural team. *Courtesy Michael & Kuhns Architects, P. C.*

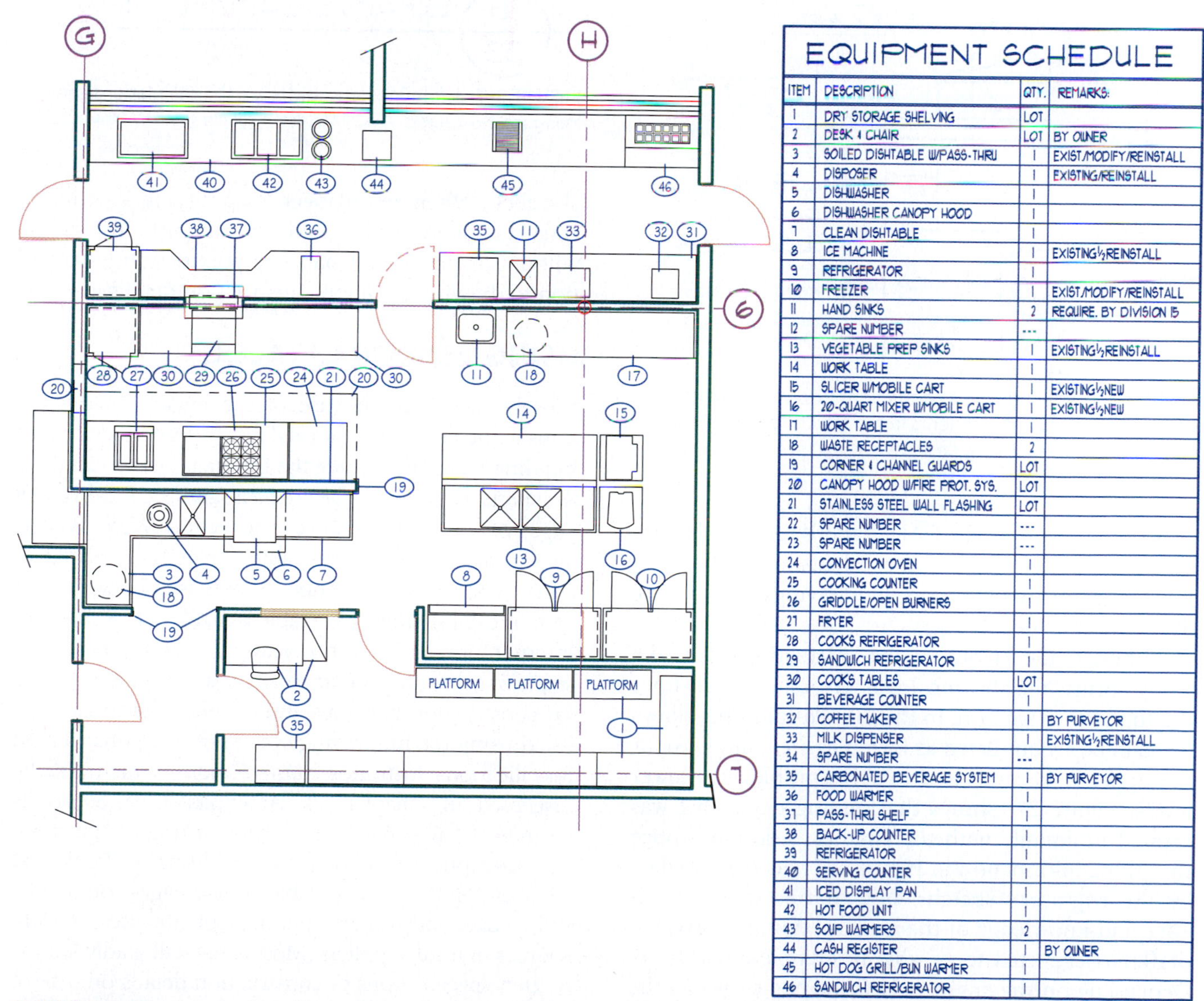

EQUIPMENT SCHEDULE

ITEM	DESCRIPTION	QTY.	REMARKS:
1	DRY STORAGE SHELVING	LOT	
2	DESK & CHAIR	LOT	BY OWNER
3	SOILED DISHTABLE W/PASS-THRU	1	EXIST/MODIFY/REINSTALL
4	DISPOSER	1	EXISTING/REINSTALL
5	DISHWASHER	1	
6	DISHWASHER CANOPY HOOD	1	
7	CLEAN DISHTABLE	1	
8	ICE MACHINE	1	EXISTING½REINSTALL
9	REFRIGERATOR	1	
10	FREEZER	1	EXIST/MODIFY/REINSTALL
11	HAND SINKS	2	REQUIRE. BY DIVISION 15
12	SPARE NUMBER	---	
13	VEGETABLE PREP SINKS	1	EXISTING½REINSTALL
14	WORK TABLE	1	
15	SLICER W/MOBILE CART	1	EXISTING½NEW
16	20-QUART MIXER W/MOBILE CART	1	EXISTING½NEW
17	WORK TABLE	1	
18	WASTE RECEPTACLES	2	
19	CORNER & CHANNEL GUARDS	LOT	
20	CANOPY HOOD W/FIRE PROT. SYS.	LOT	
21	STAINLESS STEEL WALL FLASHING	LOT	
22	SPARE NUMBER	---	
23	SPARE NUMBER	---	
24	CONVECTION OVEN	1	
25	COOKING COUNTER	1	
26	GRIDDLE/OPEN BURNERS	1	
27	FRYER	1	
28	COOKS REFRIGERATOR	1	
29	SANDWICH REFRIGERATOR	1	
30	COOKS TABLES	LOT	
31	BEVERAGE COUNTER	1	
32	COFFEE MAKER	1	BY PURVEYOR
33	MILK DISPENSER	1	EXISTING½REINSTALL
34	SPARE NUMBER	---	
35	CARBONATED BEVERAGE SYSTEM	1	BY PURVEYOR
36	FOOD WARMER	1	
37	PASS-THRU SHELF	1	
38	BACK-UP COUNTER	1	
39	REFRIGERATOR	1	
40	SERVING COUNTER	1	
41	ICED DISPLAY PAN	1	
42	HOT FOOD UNIT	1	
43	SOUP WARMERS	2	
44	CASH REGISTER	1	BY OWNER
45	HOT DOG GRILL/BUN WARMER	1	
46	SANDWICH REFRIGERATOR	1	

FIGURE 2-5 Drawings to describe special equipment in a structure are often drawn by drafters employed by a consulting firm. *Courtesy Halliday Associates.*

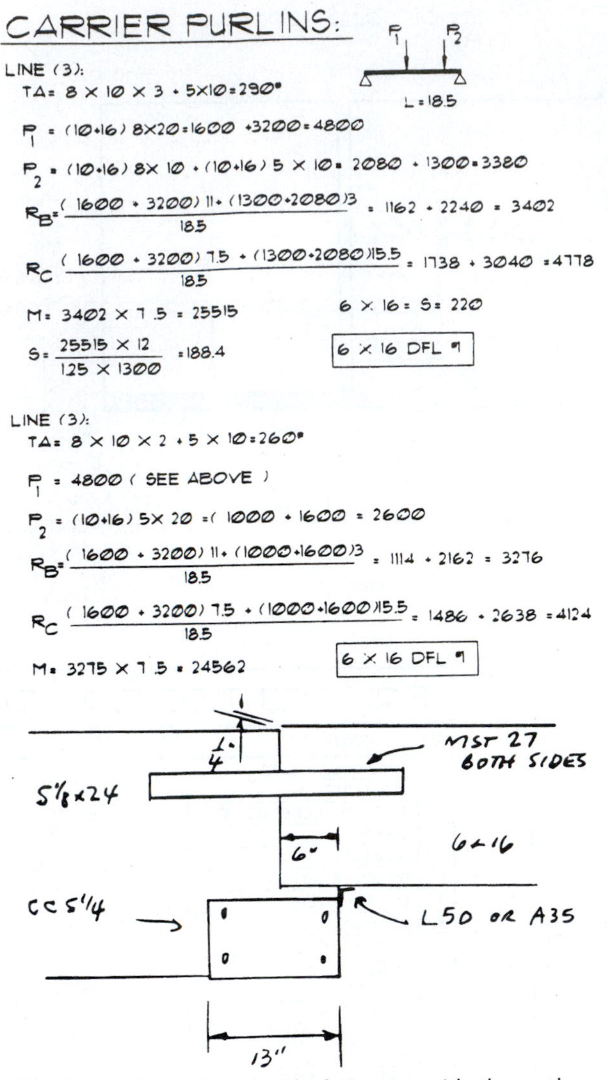

FIGURE 2-6 The engineer's calculations provide the mathematical proof that components in the structure will resist all loads and stresses. *Courtesy Dean Smith, Kenneth D. Smith Architects & Associates, Inc.*

structural drawings. The CAD drafter's role is to be sure that the third portion of the calculations—the solution—is correctly placed on the drawings and to draw the connecting detail.

Although the layout of the calculations can take many forms, usually the loads are determined from the top of the structure to the foundation. This allows loads to be accumulated as formulas are completed for the upper levels and used to solve problems at lower levels. Figure 2-7 shows the framing detail that was created to comply with the engineer's design. Notice that some information in Figure 2-7 was not specified by the engineer's sketch. As a drafter gains experience and knowledge of the construction industry, the drafter is expected to be able to complete the detail. Depending on the engineer and the experience of the drafting team, the calculations may or may not contain

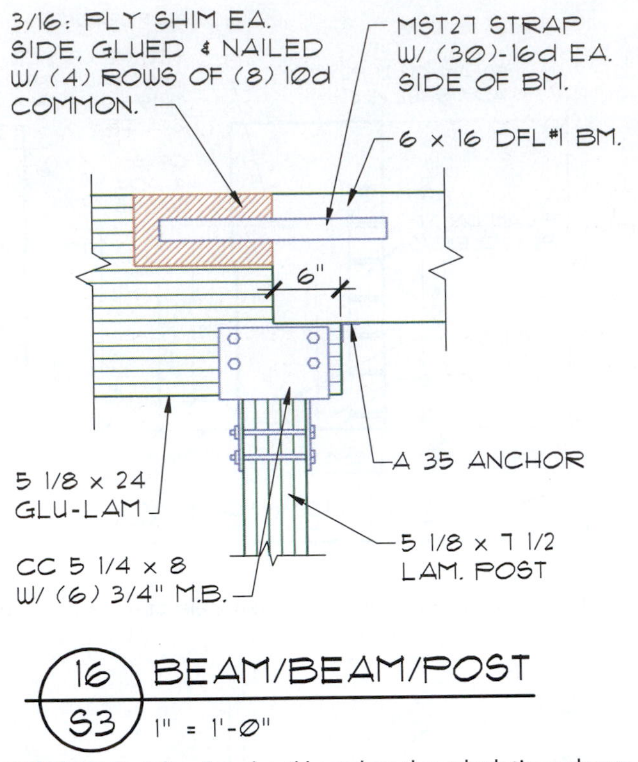

FIGURE 2-7 A framing detail based on the calculations shown in Figure 2-6. *Courtesy Ginger M. Smith, Kenneth D. Smith Architects & Associates, Inc.*

sketches. When an engineer is specifying a common construction method to a skilled team, usually a sketch will not be provided. Chapter 6 provides an introduction to drawing common construction details.

USING VENDOR CATALOGS

Closely related to the use of the engineer's calculations is the use of vendor catalogs. Common sources of building materials include the Internet, *Sweets Catalogs*, *Architects' First Source for Products*, and the local yellow pages of your phone book. The World Wide Web is an excellent source for obtaining materials related to every phase of construction. *Sweets* is a collection of vendor catalogs that is available in book, CD-ROM, and Internet formats. Published yearly, *Sweets* has become a staple of every architectural firm as a convenient source for construction materials and supplies. Information is listed using the numbering system of the Construction Specifications Institute's *MasterFormat*, which will be discussed in Chapter 14. *Architects' First Source* is an index of materials with a short listing of the product descriptions and performance features. Book and Web site options are available. Yellowpages.com or the yellow pages of your area phone book are also excellent sources of local suppliers. Most firms will gladly supply written specifications or construction details on disk or CD-ROM to help ensure the use of their products.

As seen in Figure 2-6, the engineer has specified a particular column cap to be used. Notice that no information is supplied about the cap other than the manufacturer and the model number. The drafter is expected to be able to research a specific product and completely specify the product on the working drawings. Figure 2-8 shows a sample listing from the Simpson Strong-Tie brochure for standard column caps. The sizes and bolting for the required connector can be obtained from the company's Web site, or from their display in *Sweets Catalogs*.

WORKING WITH CODES

In addition to using calculations and vendor catalogs, a CAD technician will also be required to have a working knowledge of the building code that governs the municipality where the structure will be built. Chapter 4 introduces basic code considerations for design and construction.

Building codes address two general categories in the design of a structure. To ensure public safety, most aspects of a code relate directly to either fire protection

These products are available with additional corrosion protection. Additional products on this page may also be available with this option, check with Simpson for details.

Model No. (CC shown ECC/ECCU similar)	Beam Width	Dimensions		L				Fasteners	Beam				Allowable Loads	Down		Uplift			Code Ref.	CCO Model No. (No Legs)	ECCO Model No. (No Legs)
		W_1	W_2	CC	ECC	ECCU	H_1	Size	CC	ECC	ECCU	Post	CC	ECC/ ECCU	CC 133	CC 160	ECCU 133/160				
CC3¼-4	3⅛	3¼	3⅝	11	7½	9½	6½	⅝	4	2	4	2	16980	6125	3035	3640	1010		20, 142	CCO3¼	ECCO3¼
CC3¼-6	3⅛	3¼	5½	11	7½	9½	6½	⅝	4	2	4	2	19250	9625	3035	3640	1010				
CC44	4x	3⅝	3⅝	7	5½	6½	4	⅝	2	1	2	2	15310	7655	1220	1465	205			CCO4	ECCO4
CC46	4x	3⅝	5½	11	8½	9½	6½	⅝	4	2	4	2	24060	12030	2330	2800	740			CCO4/6	ECCO4/6
CC48	4x	3⅝	7½	11	8½	9½	6½	⅝	4	2	4	2	24060	16405	2330	2800	740		170		
CC5¼-4	5⅛	5¼	3⅝	13	9½	10½	8	¾	4	2	4	2	26635	10045	6305	7530	2735		20, 80, 142	CCO5¼	ECCO5¼
CC5¼-6	5⅛	5¼	5½	13	9½	10½	8	¾	4	2	4	2	28190	15785	6275	7530	2735				
CC5¼-8	5⅛	5¼	7½	13	9½	10½	8	¾	4	2	4	2	37310	21525	6275	7530	2735				
CC64	6x	5½	3⅝	11	7½	9½	6½	⅝	4	2	4	2	28586	12030	3365	4040	1165			CCO6	ECCO6
CC66	6x	5½	5½	11	7½	9½	6½	⅝	4	2	4	2	30250	18905	3365	4040	1165				
CC68	6x	5½	7½	11	9½	9½	6½	⅝	4	2	4	2	37810	25780	3365	4040	1165				ECCO68
CC6-7⅛	6x	5½	7⅛	11	9½	9½	6½	⅝	4	2	4	2	37810	24060	3365	4040	1165		170		
CC7⅛-4	7	7⅛	3⅝	13	10½	10½	8	¾	4	2	4	2	34736	18375	6260	7510	4855			CCO7⅛	ECCO7⅛
CC7⅛-6	7	7⅛	5½	13	10½	10½	8	¾	4	2	4	2	58500	28875	6320	7585	4855				
CC7⅛-7⅛	7	7⅛	7⅛	13	10½	10½	8	¾	4	2	4	2	57750	36750	6320	7585	4855				
CC7⅛-8	7	7⅛	7½	13	10½	10½	8	¾	4	2	4	2	52500	36750	6320	7585	4855				
CC74	6¾	6⅞	3⅝	13	10½	10½	8	¾	4	2	4	2	33490	13230	6270	7525	3605			CCO7	ECCO7
CC76	6¾	6⅞	5½	13	10½	10½	8	¾	4	2	4	2	37125	20790	6270	7525	3605		20, 80, 142		
CC77	6¾	6⅞	6⅞	13	10½	10½	8	¾	4	2	4	2	49140	25515	6270	7525	3605				
CC78	6¾	6⅞	7½	13	10½	10½	8	¾	4	2	4	2	49140	28350	6270	7525	3605				
CC86	8x	7½	5½	13	10½	10½	8	¾	4	2	4	2	41250	23100	6200	7440	2625			CCO8	ECCO8
CC88	8x	7½	7½	13	10½	10½	8	¾	4	2	4	2	54600	31500	6200	7440	2625				
CC96	8¾	8⅞	5½	13	10½	10½	8	¾	4	4	4	2	48125	26950	6260	7515	4670			CCO9	ECCO9
CC98	8¾	8⅞	7½	13	10½	10½	8	¾	4	4	4	2	63700	36750	6260	7515	4670				
CC106	10x	9½	5½	13	10½	10½	8	¾	4	4	4	2	52250	29260	6260	7515	3325			CCO10	ECCO10

1. Post sides are assumed to lie in the same vertical plane as the beam sides.
2. Loads may not be increased for short-term loading.
3. Downloads are determined using Fc⊥ equal to: 560 psi for glulam sizes and CC86, CC88 and CC106; 750 psi for 7⅛" size; 625 psi for all others; reduce where end grain bearing or buckling capacity of the column, or other criteria are limiting.
4. Uplift loads have been increased for earthquake or wind load durations with no further increase allowed; reduce where other load durations govern. Uplift loads are limited by the beam shear capacity per 2005 NDS except CC76, CC78, and CC96 through CC106.
5. Beam splices with CC's must be detailed by the Designer to transfer tension loads between spliced members by means other than the column cap.
6. CC uplift loads do not apply to splice conditions.
7. Beam depth must be greater than H_1.
8. Structural composite lumber columns have sides that show either the wide face or the edges of the lumber strands/veneers. Values in the tables reflect installation into the wide face.
9. For 5¼" engineered lumber, use CC 6X or ECC 6X models.

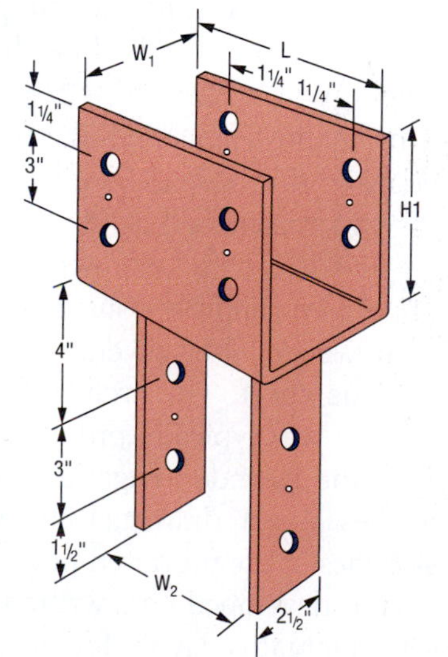

FIGURE 2-8 CAD drafters are often required to consult vendor catalogs to find needed information to complete drawings. *Courtesy Simpson Strong-Tie Co., Inc.*

or the ability of a material or portion of a structure to resist collapsing under the loads that will be resisted.

The first concern of building codes, fire resistance, is broken down into three general areas: (1) the tendency of a material to fail by combustion, (2) the tendency of a material to lose its strength once a fire has broken out, and (3) methods of fire containment. You will be introduced to basic code requirements as you progress throughout this text.

The second major concern of the building code is the ability of the materials and the structure as a whole to resist loads. Building codes typically use either the working stress method or the factored load method to assure the safety of a structure. Each method deals with the resistance of loads from stress on materials and the predetermined ability of a material to resist stress. Chapter 13 will provide an introduction to forces that affect the design of a structure. Subsequent chapters will explain how these stresses are resisted throughout a structure.

Although the architect or engineer will complete the majority of work with codes, it is important for the drafter to have a basic understanding of the code. The architect will determine the size, height, and type of materials to be used in construction, based on a thorough knowledge of the code. The drafter will need an understanding of the code to comprehend many of the components that will be added to a drawing. This will require knowledge of basic code layout and, specifically, an understanding of the chapters dealing with basic building requirements based on a specific type of construction.

DRAWING PLACEMENT

Just as you should develop an outline before writing a research paper, thought should be given to the drawings needed to describe a construction project. The drawing layout must be one of the first things considered in planning a project. General office practice is to plan the drawing requirements and placement prior to starting the working drawings. The general order of a set of drawings was introduced earlier in this chapter. As the drawing layout is being planned, the placement on the page should be carefully considered. Great care must be taken, however, when several drawings will be placed on the same sheet. The architect, engineer, or project coordinator will typically provide new drafters with a sketch of the intended page layout along with individual sketches of each drawing to be placed on the page. These sketches of the proposed project are similar to an outline that is developed for a written project. The sketches provide a tentative layout for the placement of each plan, elevation, section, schedule, and detail that is anticipated.

When all drawings are complete and printed, they will be assembled into a set and bound along the left margin. The larger the sheets used, the more likely that drawings, notes, or parts of drawings placed on the left side of the sheet will not be seen. Whenever only one drawing is to be placed on a sheet, it should be placed near the right margin, leaving any blank space on the left (or binding) side. This same guideline should be used if smaller drawings, such as details, are to be placed with a large drawing. Placing the details on the left could cost the print reader needless time in skimming the drawings because the details might not be seen. Details placed on the right side can be clearly seen while the reader thumbs through the drawing. It is true that the print reader should be careful when examining a set of drawings; however, anything the drafter can do to improve communication should be done.

Another consideration in placing details is to place them as close as possible to where they occur, still being careful that small drawings will not be lost. The layout in Figure 2-9 places the detail close to its source and in a location where it can't be hidden by other pages. Wherever possible on symmetrical drawings, place the detail symbol where the right side can be seen.

When positioning drawings together, it is important that the limits of each drawing not interfere with its neighboring drawing. Lettering from one drawing should not be intertwined with a neighboring drawing. Some offices share text between two drawings, as seen in Figure 2-10. Text that relates only to the drawing on the right should be separated to avoid confusion. Other offices use a layout similar to Figure 2-11, which places details in neat rows with no shared text. Another option is to use a grid system similar to Figure 2-12. All boxes do not need to be the same size, but the box edges must align. Care must be taken to number the detail grids in an orderly manner. Order can vary, with the numbering of details starting in any one of the corners, but the same order of progression should be used throughout the project. Text assigned to details' symbols will be further discussed as the drawings are being coordinated.

Keep in mind that there is no one correct way to arrange details. When entire sheets will be filled with details, it is important to have a logical order to their arrangement. Figure 2-13 shows a section of a brick wall. Notice that three details are required to completely explain this component. These details have been placed in an order that matches their arrangement within the section.

Another consideration in placing details throughout the drawing set is to place them in groups according to the labor force that will do the work. Roof drawings should not be mixed with concrete drawings because

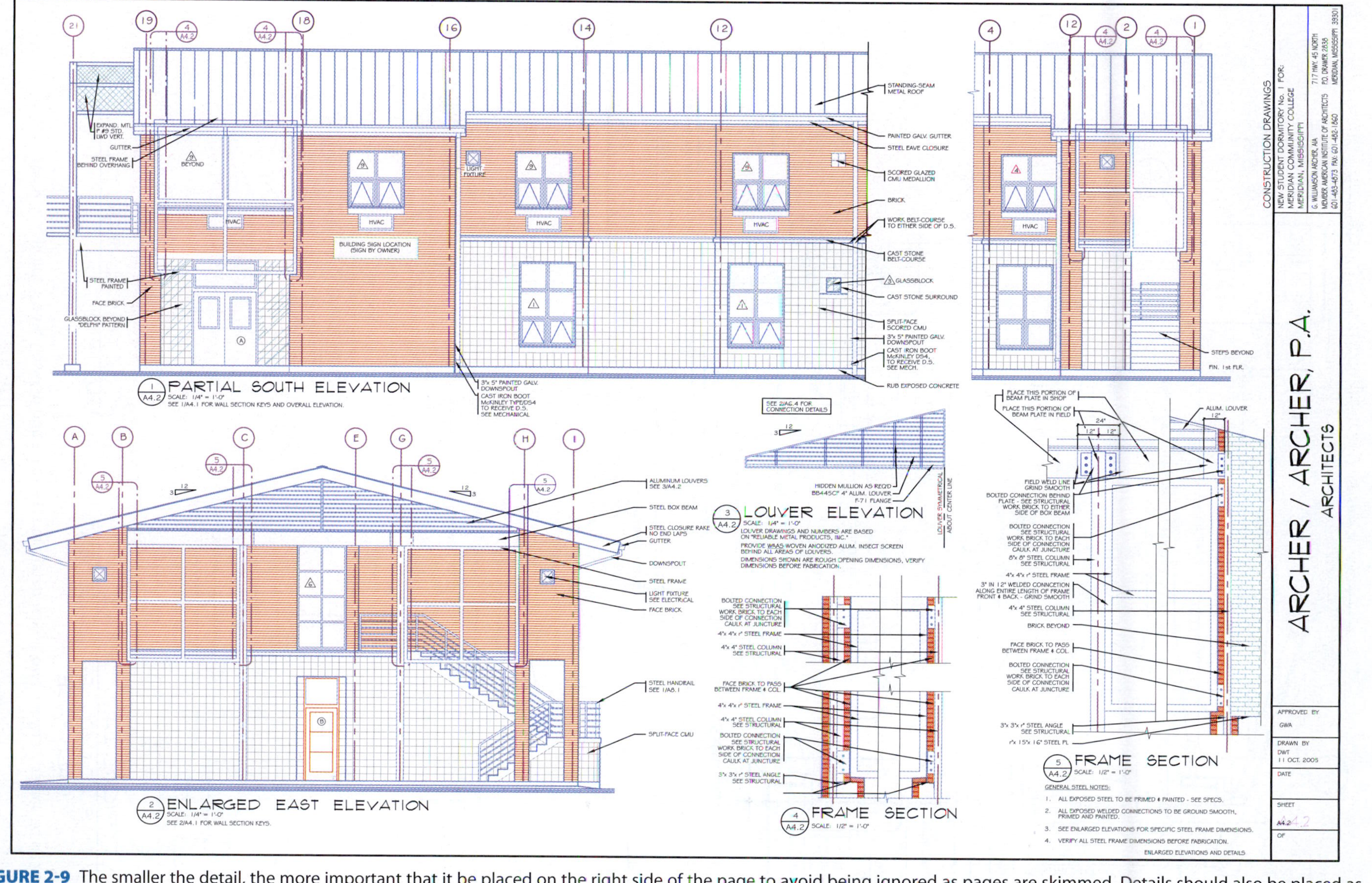

FIGURE 2-9 The smaller the detail, the more important that it be placed on the right side of the page to avoid being ignored as pages are skimmed. Details should also be placed as close as possible to the drawing that they reference. *Courtesy G. Williamson Archer A.I.A., Archer & Archer P.A.*

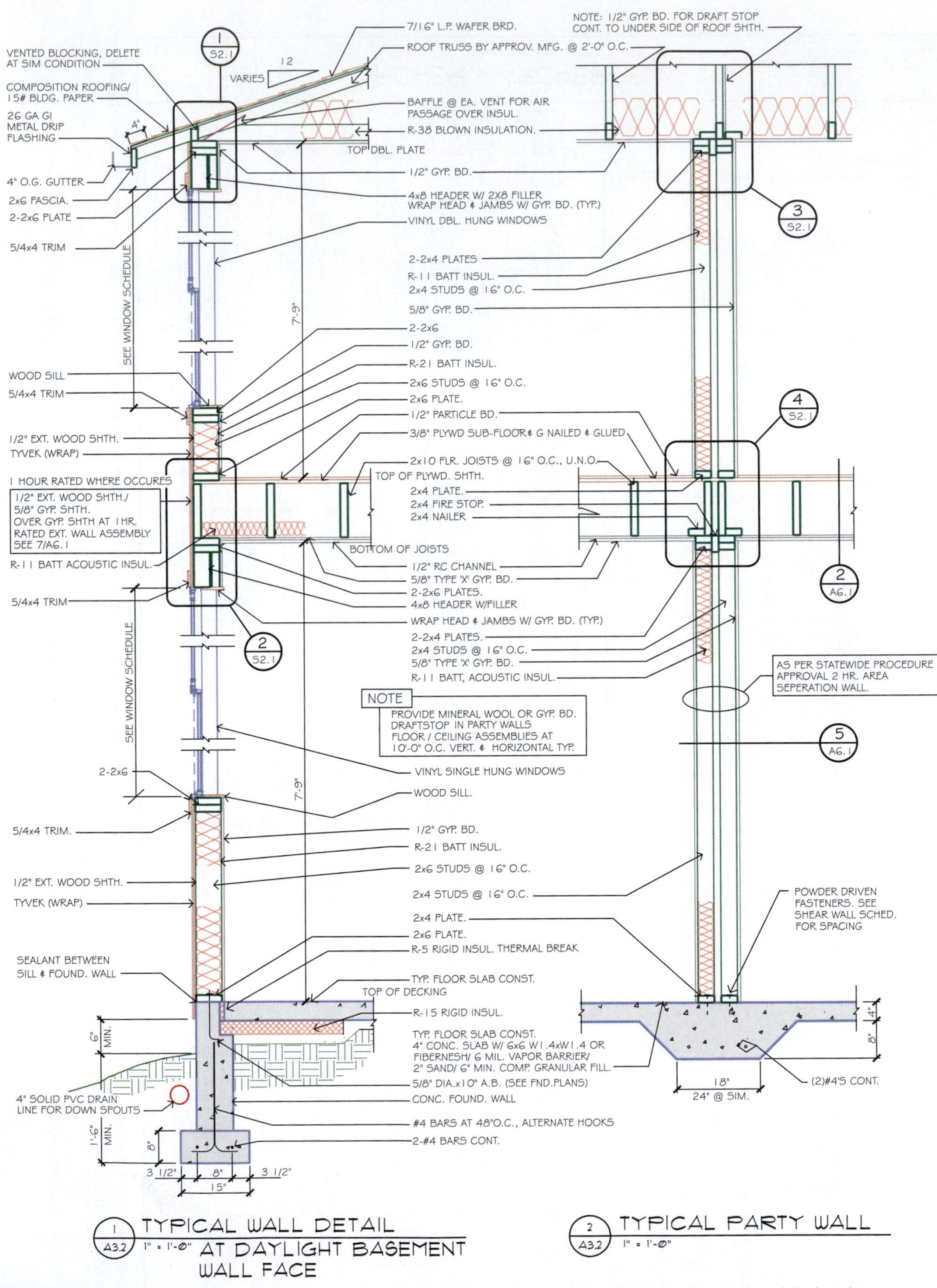

FIGURE 2-10 Text can be shared between two similar details to save space and drawing time. *Courtesy Scott R. Beck, Architect.*

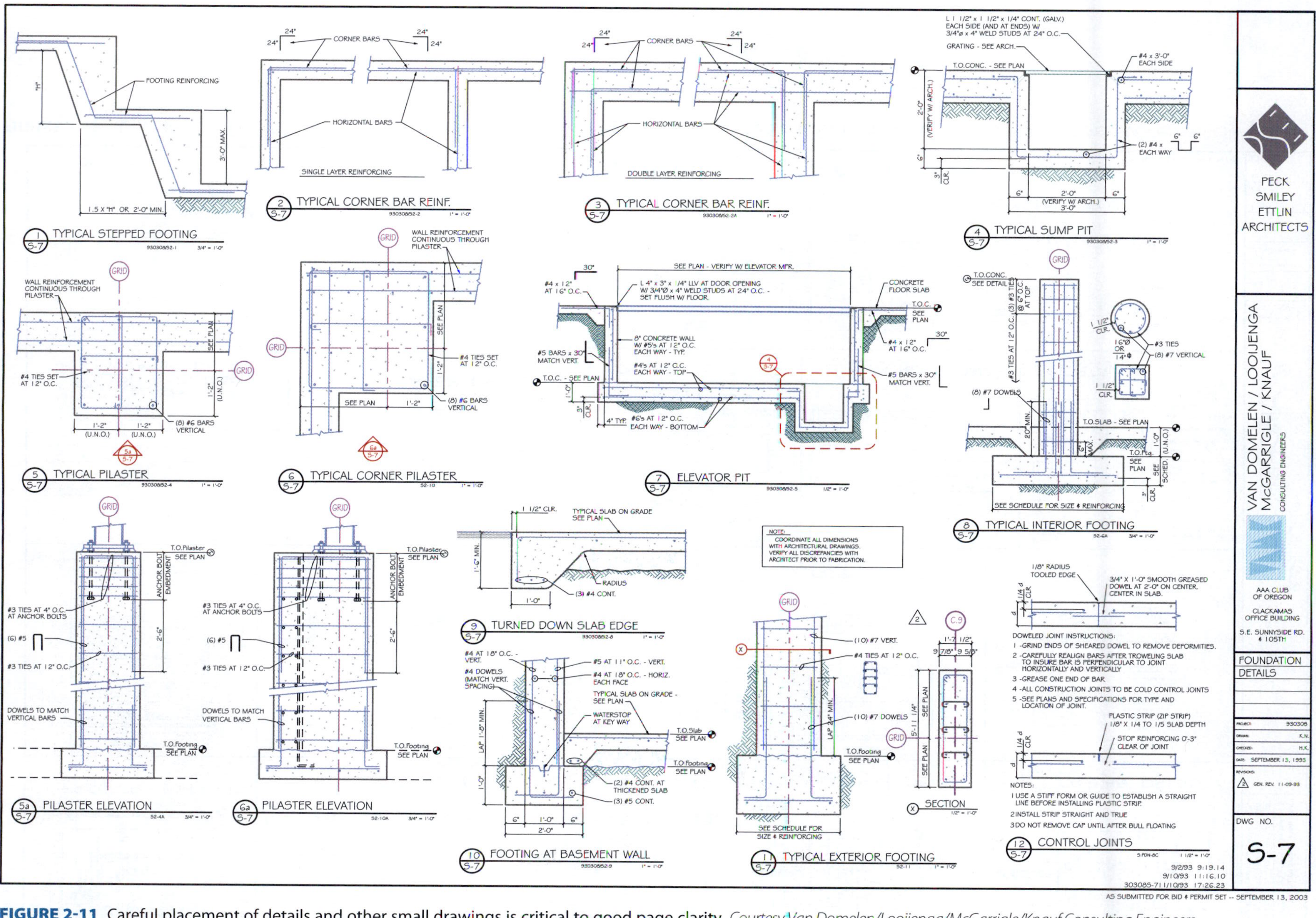

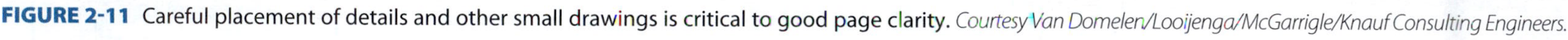

FIGURE 2-11 Careful placement of details and other small drawings is critical to good page clarity. *Courtesy Van Domelen/Looijenga/McGarrigle/Knauf Consulting Engineers.*

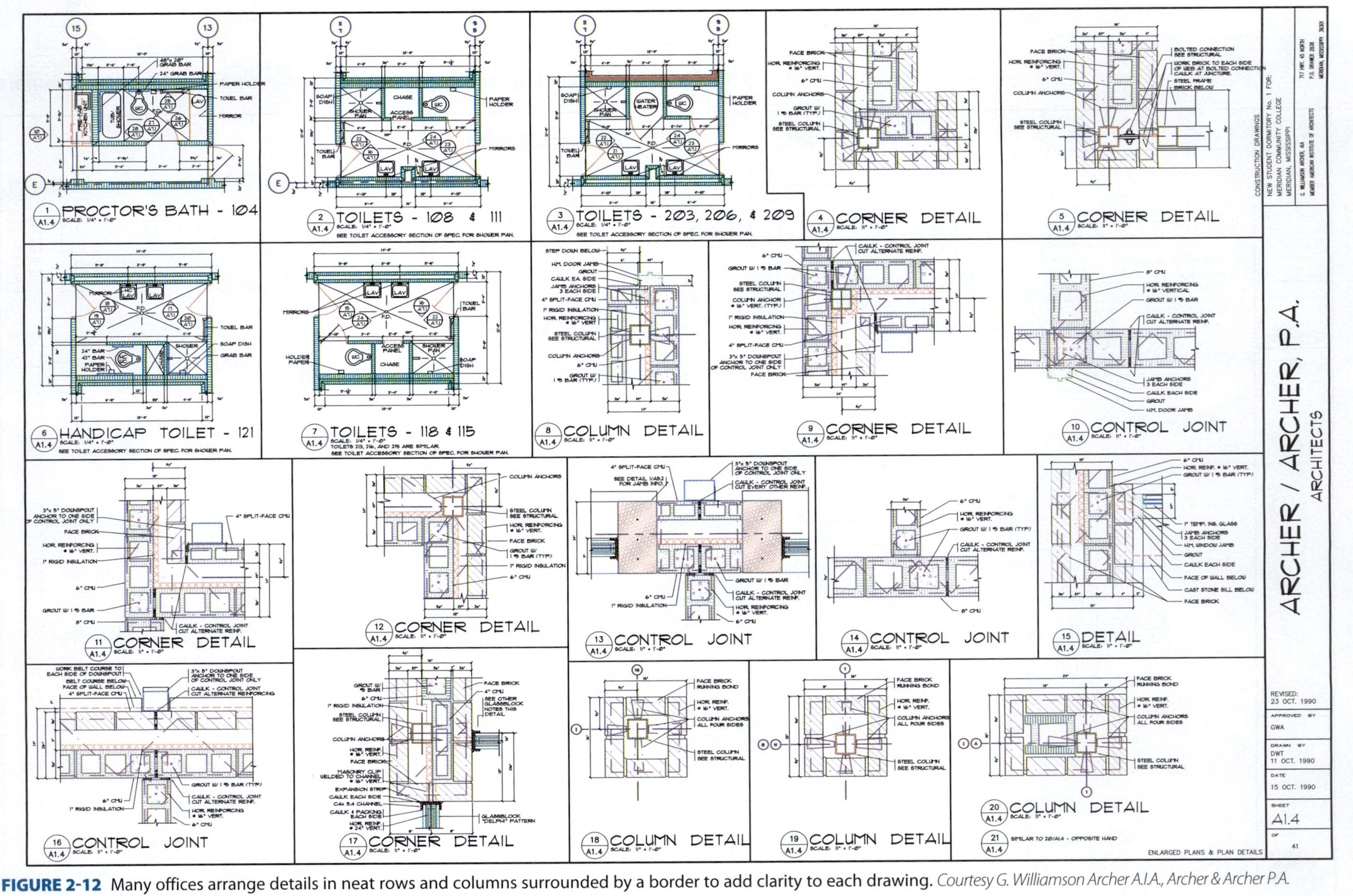

FIGURE 2-12 Many offices arrange details in neat rows and columns surrounded by a border to add clarity to each drawing. *Courtesy G. Williamson Archer A.I.A., Archer & Archer P.A.*

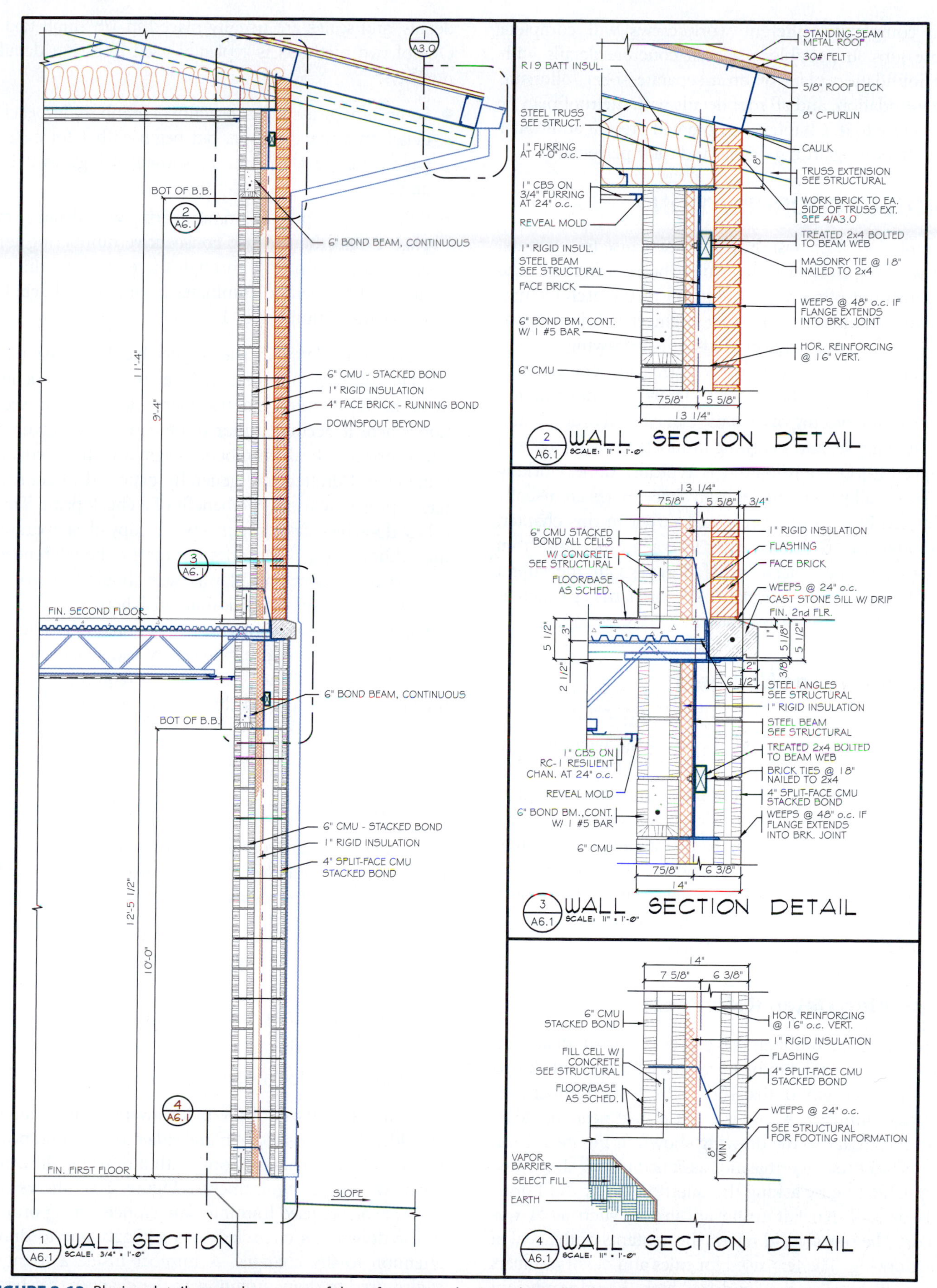

FIGURE 2-13 Placing details near the source of the reference and using a logical order of presentation can aid the print reader.
Courtesy G. Williamson Archer A.I.A., Archer & Archer P.A.

two completely different work crews will complete those jobs. If possible, place all concrete details with the foundation plan, or on a separate sheet following the foundation, and all roof details with the roof plan or very close to it. Chapters 3 and 6 will present additional information regarding details and their placement.

PROJECT COORDINATION

Completion of the last drawing is no cause to celebrate. All working drawings must now be coordinated. The architectural office will arrange all of the architectural drawings and each consulting engineer will be responsible for the coordination of his or her drawings.

Consulting firms will provide electronic copies ready to be added to the project. **Drawing coordination** involves the placing of each page in its final order within the drawing set and assigning numbers to each page and detail. Project coordination is typically done working with a small paper copy of each sheet. Pages are usually arranged in the order presented earlier in this chapter. Although this order will vary for each office and from job to job, the placement of drawings in the subset should match the National CAD Standards as much as possible to aid the print readers.

Assigning Page Numbers

A space in the lower-right corner of the title block is generally reserved for the display of the page number. Two methods of page numbering have been introduced. As long as the drawings communicate effectively, they can be arranged in either order, based on office procedure. Generally, offices try to maintain a similar order for each of their jobs so that bidders and construction crews will have a sense of familiarity with their plans. Sheets that contain abbreviations or symbols should be placed very early in the drawing set.

Assigning Detail References

Once the page numbers have been assigned, details on each specific page can be numbered. Each detail should have a detail symbol that shows the detail number over the page number. A title is often placed near the detail symbol similar to the drawing shown in Figure 2-13. If the project manager has not assigned a title, determine a suitable title by asking the question "Why did I draw this detail?" No fair using an answer such as "I was paid." The title should reflect the contents or the goal of the drawing. The text used for titles and detail numbers is usually between 1/8" and 1/4" high. A uniform height should be used throughout the entire project. Pages, details, and scales are usually placed in 1/8"-high text. One of two methods is typically used to assign detail numbers:

- Some offices assign consecutive numbers for each detail on a specific page, but begin with 1 for every page containing a detail, as shown in Figures 2-11 and 2-12.
- Other offices opt to begin numbering with detail 1 on page 1, and then to use consecutive numbering for all subsequent details through the end of the drawing set. This method eliminates having several details with a detail number of 1.

Once every detail in the project has been assigned a detail and page number, each reference to the detail must be represented on the plan, elevation, or section where it occurs. Junior drafters are often given a check print with all numbers assigned to be filled in. Experienced drafters are generally expected to coordinate a project without the benefit of a check print. This is best done by working with a paper copy of each drawing. Although the detail does not have to be referenced to every occurrence on a plan view, it should be marked so that there is no confusion about where a particular detail occurs. Referencing every occurrence of the detail would often decrease drawing clarity and take unnecessary drawing time. The print reader is required to make intelligent decisions regarding where every occurrence of the detail will take place.

REVISIONS

No matter how carefully a set of drawings is prepared, portions of the drawing set might need to be revised. Common causes for revisions are:

- Changes in building codes after the drawings were completed but before the building permits were issued.
- Changing owner or tenant requirements.
- Changes occurring at the job site.
- Errors by the design team.

To make all users of the plans aware of the change, an addendum is issued. An *addendum* is a written notification of the changes, along with a drawing showing the new design requirements. Figure 2-14 shows an example of a revised framing plan. Notice that a portion of the drawing is encircled by a squiggly line to draw attention to the changes. A number inside a triangle is also used to draw attention to the changes that have been made from earlier prints.

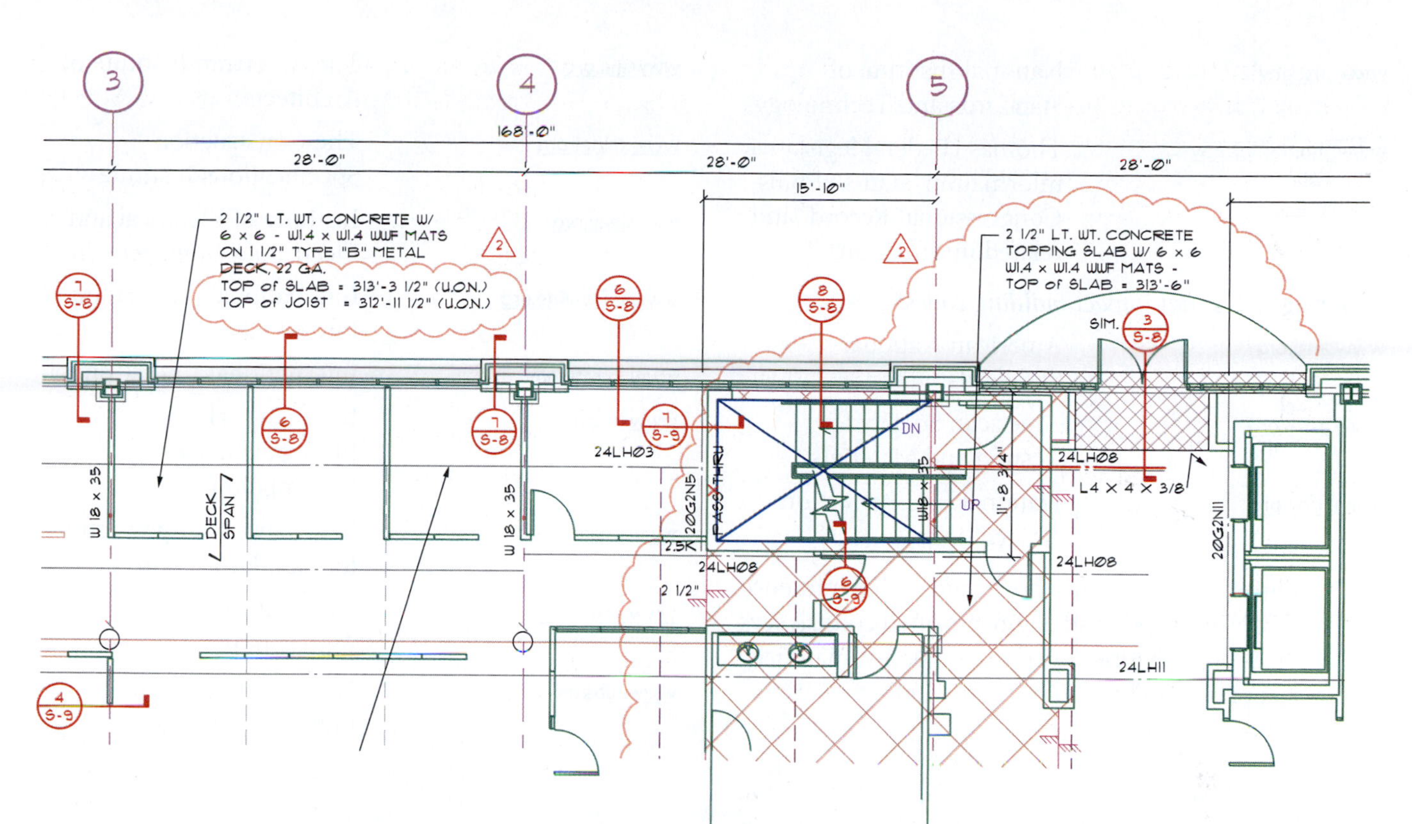

FIGURE 2-14 Revisions are often enclosed by a squiggly line to draw attention to the change. The number 2 inside the triangle refers to a notation in the title block. In this case, floor elevations have been added in two places on the drawing since the last printing. *Courtesy Van Domelen/Looijenga/McGarrigle/Knauf Consulting Engineers.*

The number relates to a note that is then placed somewhere on the page to explain the required changes. Many offices have a specific portion of the title block or the drawing area reserved for revision references or explanations.

ADDITIONAL READING

The following Web sites can be used as a resource to help you keep current with changes in professional design careers.

ADDRESS	COMPANY/ ORGANIZATION
Major research tools:	
www.afsonl.com	Architects' First Source for Products
www.sweets.com	Sweets Building Product Information
Major testing labs, quality assurance, and inspection agencies:	
www.apawood.org	APA—The Engineered Wood Association
www.etl.go.jp	ETL Testing Services
www.fmglobal.com/researchstandard testing/research.html	Factory Mutual Research Corporation
www.itsglobal.com	Intertek Testing Services
www.ul.com	Underwriters Laboratories Inc.
www.wwpa.org	Western Wood Products Association
Other U.S. and federal agencies related to construction:	
www.access-board.gov	U.S. Architectural and Transportation Barriers Compliance Board
www.eren.doe.gov	U.S. Department of Energy's Energy Efficiency and Renewable Energy Network
www.epa.gov	U.S. Environmental Protection Agency
www.nara.gov/fedreg	Federal Register—The Government Printing Office

www.nist.gov	National Institute of Standards and Technology
www.thomas.loc.gov	Thomas (Federal legislative information, status of bills, Congressional Record, and related information)

Other agencies that impact building codes:

www.ansi.org	American National Standards Institute
www.astm.org	American Society for Testing and Materials
www.nfpa.org	National Fire Protection Association

In addition to these regulatory agencies that influence construction, architects and engineers often consult several related organizations to ensure quality materials. These agencies and their Web sites include:

www.aia.org	The American Institute of Architects
www.csinet.org	The Construction Specifications Institute
www.ieee.org	Institute of Electrical and Electronics Engineers, Inc.
www.iafconline.org	International Association of Fire Chiefs
www.iccsafe.org	International Code Council
www.iec.ch	International Electrotechnical Commission
www.iso.ch	International Organization for Standardization
www.ncma.org	National Concrete Masonry Association
www.nibs.org	National Institute of Building Sciences

KEY TERMS

Architectural drawings
Cabinet drawings
Calculations
Civil drawings
Construction documents
Drawing coordination
Electrical drawings
Mechanical drawings
National CAD Standards
Plumbing drawings
Preliminary drawings
Structural drawings

CHAPTER 2 TEST

The CAD Drafter's Role in Office Practice and Procedure

QUESTIONS

Answer the following questions with short complete statements. Using a word processor, type the chapter title, question number, and a short complete statement for each question. Use material in the Appendix found on the CD as necessary.

Question 2-1 List the three major portions of a set of calculations.

Question 2-2 How is information arranged in the *Sweets Catalogs*?

Question 2-3 List the four main national building codes.

Question 2-4 Where should small details be located in the drawing area? Explain your answer.

Question 2-5 Explain the guidelines for labeling details.

Question 2-6 Describe methods of arranging details on a page.

Question 2-7 What does drawing coordination entail?

Question 2-8 List and describe two common methods of assigning page numbers.

Question 2-9 List two methods of describing revisions.

Question 2-10 What is an addendum?

Use Figures 2-7 and 2-8 to answer the following questions:

Question 2-11 List the quantity and size of bolts to be used to connect a CC68 to a 6 × 8 column.

Question 2-12 How many bolts will be needed to connect a CCO68 to a 6 ×12 beam?

Question 2-13 How much uplift can a CC88 resist?

Question 2-14 When would a CCO cap be useful and how is it attached?

Question 2-15 How many pounds can an ECC106 support?

Use Figure 2-6 to answer the following questions:

Question 2-16 The letter L represents the length of the beam. What length of beam will be used?

Question 2-17 The letter P is used to represent the loads on the beam. What loads will this beam support?

Question 2-18 List the size of the beams to be supported.

Question 2-19 List three connectors that will be used at this beam intersection.

Question 2-20 List the size, quantity, and location of the bolts to be used with the column cap.

This chapter will provide an overview and practical application of the computer skills necessary to produce a set of drawings. It is assumed that you have completed an entry-level CAD drafting class prior to working through this text and that you are working with AutoCAD 2004 or newer. This chapter will help you utilize your CAD skills to prepare a set of commercial drawings. Consideration will be given to:

- The use of computers in architectural and engineering offices.
- Managing the drawing environment.
- Managing drawing properties with a template.
- Managing efficient use of drawing templates, assembling drawings for plotting at multiple scales, using the XREF command, working with multiple documents, and using DesignCenter.
- Management of drawing information.
- Management of drawing files and folders.

CAD INFLUENCES IN ARCHITECTURAL AND ENGINEERING OFFICES

The use of computers in the construction industry has had a substantial impact on the design and documentation of a structure. Given the magnitude of the project size and the number of professionals to be coordinated, the use of computers has greatly aided the management and production of architectural projects. This text assumes you have mastered basic CAD skills and will attempt to help you understand how the skills taught in those classes relate to construction drawings.

Working Drawings

Computers are a valuable tool for drawing and coordinating the working drawings of a structure. Figure 3-1 shows a detail of open-web steel floor trusses intersecting a steel column that was completed by a CAD drafter.

Mathematical Calculations

Third-party software programs that work within AutoCAD are available to provide many of the mathematical calculations required for structural analysis. Programs have been available for several years to design wood, steel, and concrete beams and supports. Computer models of a structure can also be created to analyze how specific features of a structure, and the entire structure as a whole, will react under various loading conditions. These studies can be used to determine how a structure will respond to wind pressure, floodwater, or an earthquake of a specific magnitude. Software programs are also available for designing static and fatigue calculations to predict repair and maintenance cost. Third-party software is available to provide computer analysis of heat loss. Many programs will also show the average outdoor temperature and the average amount of energy needed to reach specific temperatures within the structure. Cost comparisons can be determined based on predicted weather patterns, cost of a specific fuel, and predicted efficiency of a heating unit.

Three-Dimensional Drawings

Software innovations by AutoCAD and third-party developers provide the ability to draw 3D drawings similar to Figure 3-2. These drawings can be used to help develop the working drawings as well as the artistic drawings for preliminary planning. In addition to 3D and wire-frame drawings, one of the biggest advances in the architectural field is the use of solid models. Once drawn, the structure can be rotated so that the client can view the structure at any angle; also, the drawing can depict the shade that would be generated at specific times of the day, at any specified time of the year.

Three-dimensional models can also be used to provide the client with a walk-through of the structure

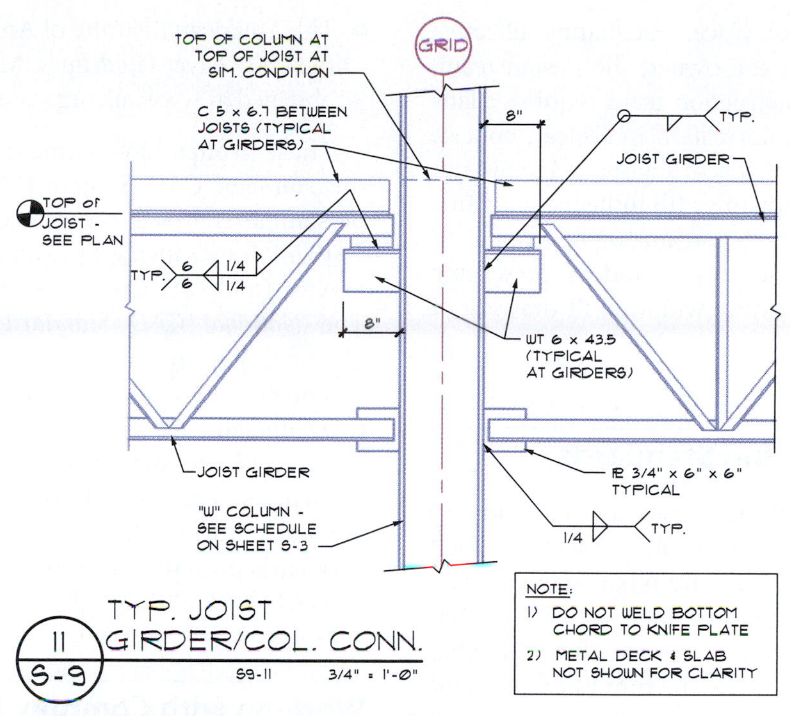

FIGURE 3-1 Computers are used to aid CAD drafters, architects, and engineers in nearly every phase of drawing development. *Courtesy Peck, Smiley, Ettlin Architects.*

FIGURE 3-2 Three-dimensional drawings can be drawn and rendered using AutoCAD and third-party software. *Courtesy Ginger M. Smith, Kenneth D. Smith Architects & Associates, Inc.*

before the working drawings are even started. These models can save huge amounts of time by providing insight to placement of heating ducts, electrical conduits, and various piping runs. Placement problems, which may not be easily identified in 2D drawings, often can be identified in 3D drawings.

MANAGING THE DRAWING ENVIRONMENT

A well-drawn set of plans doesn't just happen because you've worked really hard. Long hours and hard work may be required, but they will not assure that effective communication can take place. Facilitating effective communication between the owner, the design team, consultants, and the construction team requires plans that accurately present information in a clear, concise format similar to the drawing in Figure 3-3. This can be accomplished by complying with industry standards for presenting information, by organizing material in an orderly fashion within the project, and by presenting information in a uniform way within each drawing of the project.

National CAD Drawing Standards

Unlike other areas of drafting, there is no one standard way of doing things in the field of architecture. Each office has its own standard. The office standard can even vary based on time constraints, or the fees that have been set for the project. The need for a drawing standard has been hindered by the mindset that architects were part artist. As computers, the Internet, and the World Wide Web dominate the office setting, most professionals have found that a standard method for creating drawings is a must. A uniform drawing standard offers the following advantages as offices share information:

- Consistent display of information for all projects, regardless of the project type or client.
- Seamless transfer of information among team members and consulting architects, engineers, and design professionals.
- Reduced preparation time for translation of electronic data files between different proprietary software file formats; predictable file translation results.
- Reduced data file formatting and setup time.
- Reduced staff training time to teach "office standards."

Several organizations have worked to provide the industry with a uniform drawing standard. Among the leaders in developing a CAD standard are:

- The National Institute of Building Sciences publishes the *National CAD Standard*. More information can be obtained at www.nibs.org.
- The Construction Specifications Institute (CSI) publishes the *Uniform Drawing Standard (UDS)*—a standard consisting of eight modules that cover drawing set organization, sheet organization, drafting conventions, terms and abbreviations, symbols, notations, and general regulatory information. More information can be obtained at www.csinet.org.
- The American Institute of Architects (AIA) publishes the *CAD Layer Guidelines*. More information can be obtained at www.aia.org.

These groups have come together and developed the *National CAD Standard*. This standard is a set of standards based on AIA *CAD Layer Guidelines*; *CSI Uniform Drawing System*, modules 1–8; and Tri-Service (and U.S. Coast Guard) Plotting Guidelines. The *National CAD Standard* is published by the National Institute of Building Sciences. The most recent edition of this standard is the *U.S. National CAD Standard, Version 4.0*. This text will present key features of the standard. Verify with your instructor what standards will govern your projects. If you are allowed to modify one of the base models or the standards presented in this text, just remember to be consistent.

Working with Complex Drawings

One of the biggest challenges a new CAD drafter will face when completing commercial projects is the difference in size and complexity when compared to residential projects. Combined with the fact that the drafter is now working on a project as part of a team, good computer skills will be essential. Three skills that will be essential in working with the drawings include: (1) using the XREF command to create drawings, (2) creating views to move about a drawing, and (3) using layouts to display drawings.

External Referenced Drawings

External referencing is executed by using the AutoCAD XREF command. The command provides an effective method for working with related drawings when multiple firms will be collaborating on a project. Floor-related drawings provide an excellent example of drawings that can be referenced. A CAD technician working for the architect can be working on the floor plan while other consultants are working on the structural, electrical, or mechanical plans. A technician for the architectural team can draw information in a base drawing that will be reflected on all other plan drawings. Copies of this drawing file can then be given electronically to other firms that will develop related drawings.

An externally referenced drawing is similar to a wblock, in that it is displayed each time the master drawing is accessed. This drawing is not stored as part of the master drawings file. It is brought into other

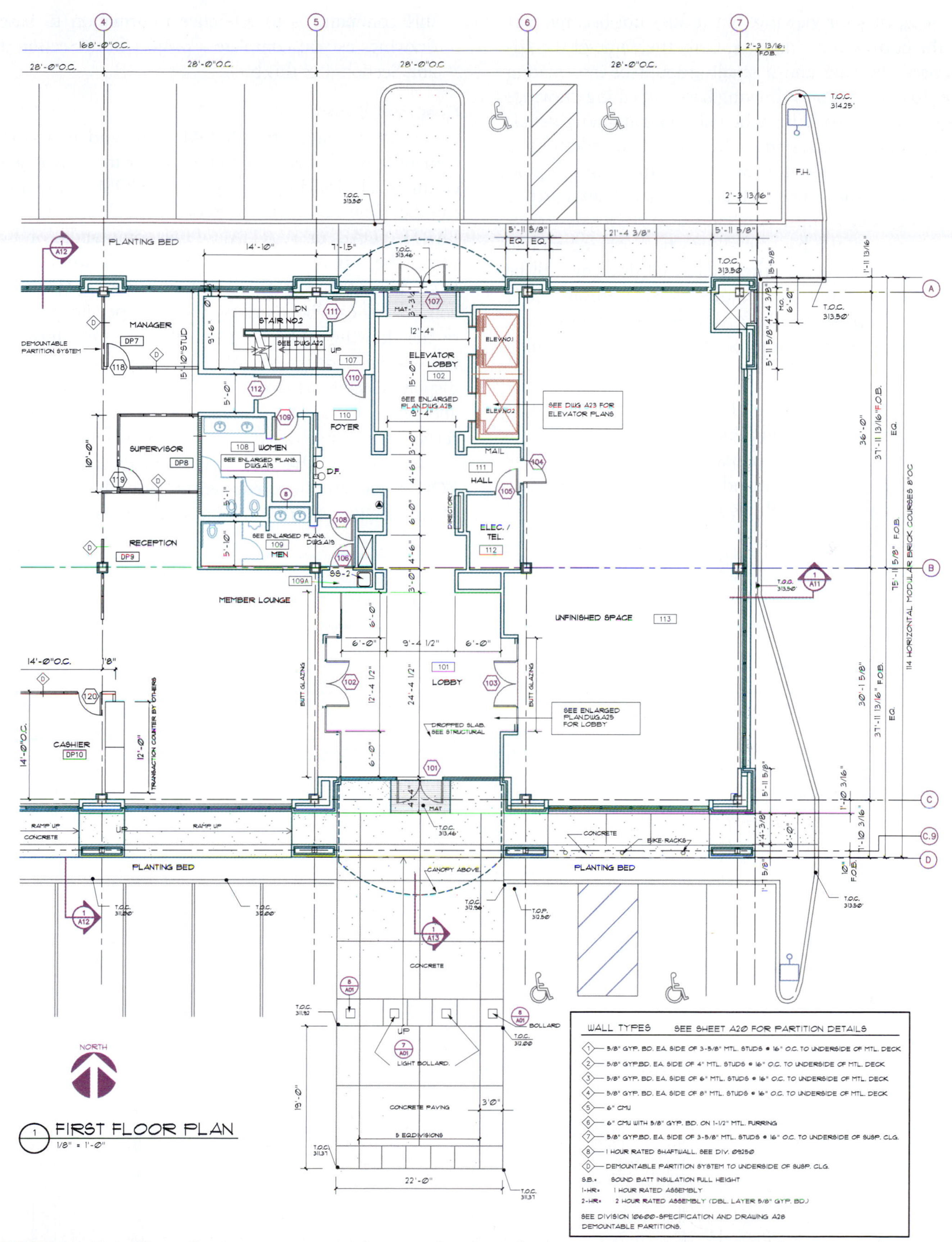

FIGURE 3-3 To facilitate effective communication between the owner, the design team, consultants, and the construction team requires plans that accurately present information in a clear, concise format. *Courtesy Peck, Smiley, Ettlin Architects.*

drawing files for viewing, but it does not become part of the current drawing base. Only the name of the referenced drawing and a small amount of information are stored in the new drawing base. Attaching drawings using XREF also helps when sheets containing details with multiple scales must be plotted. In addition to simplifying the assembly of multiple drawings for plotting, this method ensures that stock details are current. If a stock detail must be changed to reflect new code requirements, the base detail can be corrected and all drawings that contain the referenced drawing will be automatically updated. Figures 3-4a, 3-4b, and 3-4c show examples of a base drawing and information that has been attached to create the floor and framing drawings.

Note: Because the use of referenced drawings is so prevalent when compiling commercial drawings, students should be prepared to demonstrate skill with the XREF command when applying for a job. One of the best methods for improving your use of this command is to reference information to base drawings as you complete a project from Section 3 and Section 4 of this book.

Creating Views

On small drawings, the ZOOM command is a convenient way to navigate around a drawing. On larger commercial drawings using the ZOOM command can be slow. You can overcome this by using the PARTIALOAD or PARTIALOPEN commands, or by using the VIEW command. When you work with large files, the PARTIALOAD and PARTIALOPEN commands can be used to allow a specific portion of the drawing to be accessed. If the entire drawing is to be opened, the VIEW command can be used to display specific areas of the drawing. The command allows specific areas to be named, saved, and quickly retrieved and magnified. Figure 3-5a shows a partial sheet of details that are unreadable because of the ZOOM magnification required to show the full width of the details. Figure 3-5b shows the results of using

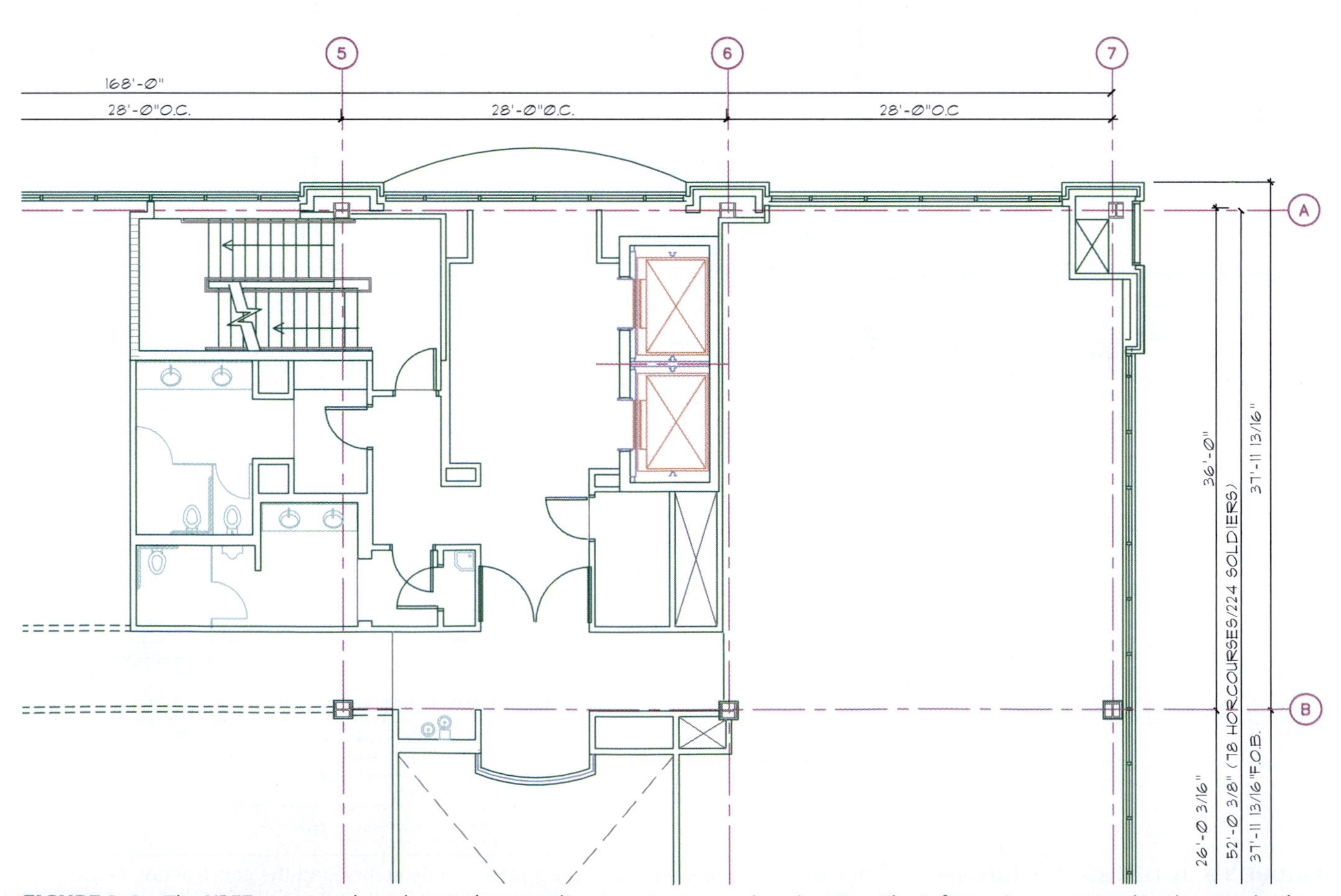

FIGURE 3-4a The XREF command can be used to attach a drawing to another drawing. The information contained in the attached drawing is not stored in the base drawings. Each time the base drawing is updated, all drawings that are attached to it will also be updated. *Courtesy Peck, Smiley, Ettlin Architects.*

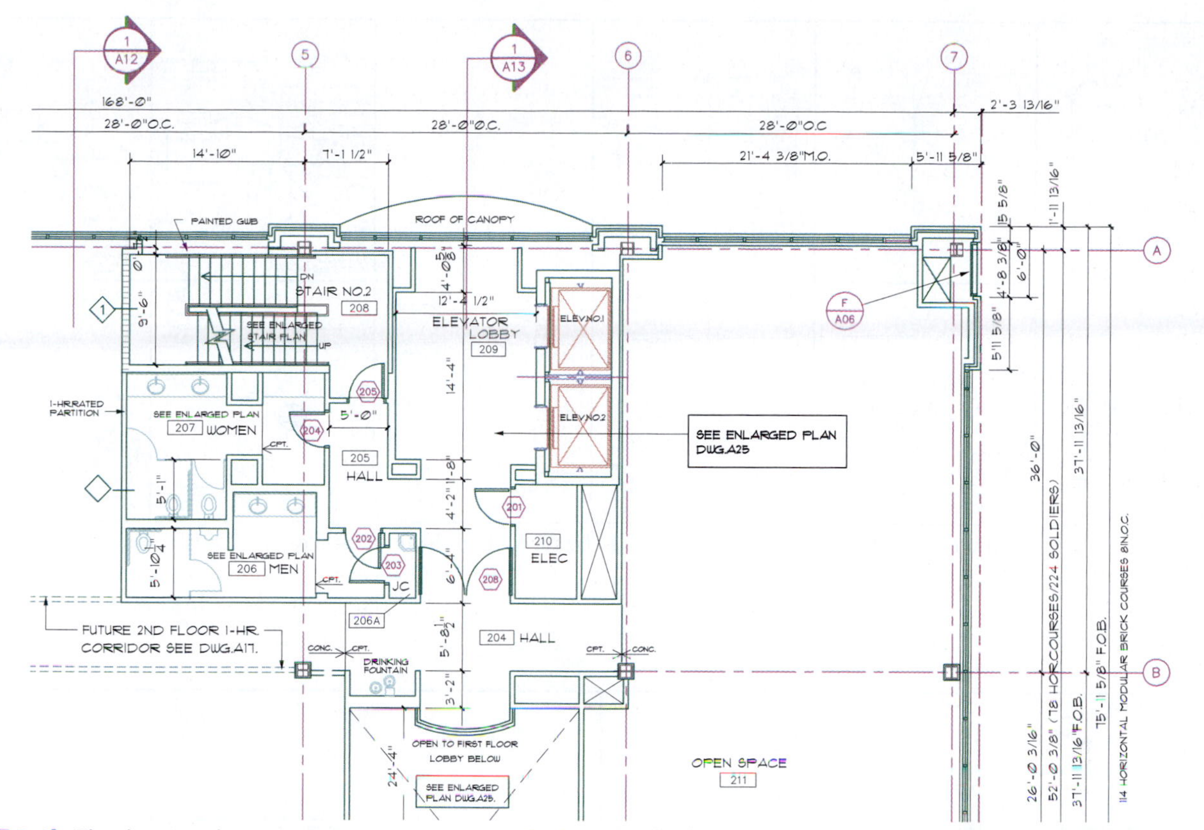

FIGURE 3-4b The drawing shown in Figure 3-4a is used as the base for the floor plan shown here as well as the drawings that will be prepared by the mechanical, electrical, and plumbing contractors. *Courtesy Peck, Smiley, Ettlin Architects.*

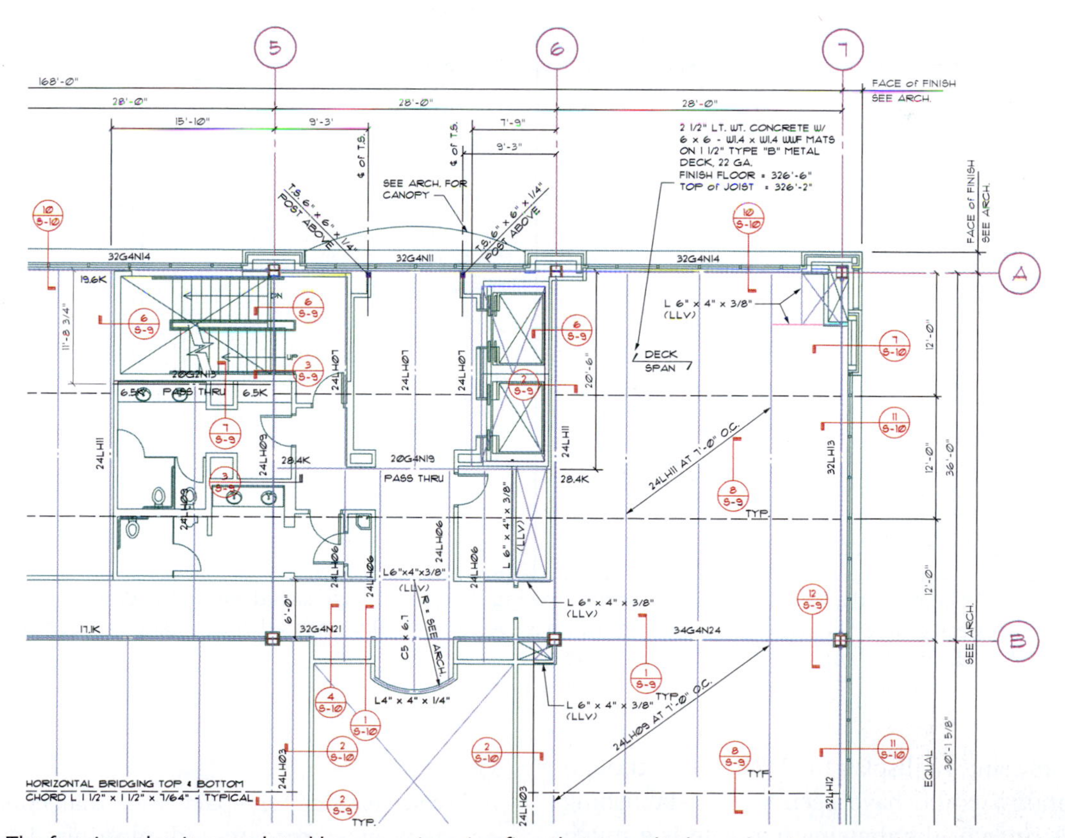

FIGURE 3-4c The framing plan is completed by an engineering firm that uses the base drawing created by the architectural team. The drawing is attached using XREF, and then the needed material is added to the drawing by the engineering team. *Courtesy Domelen/Looijeng/McGarrigle/Knauf Consulting Engineers..*

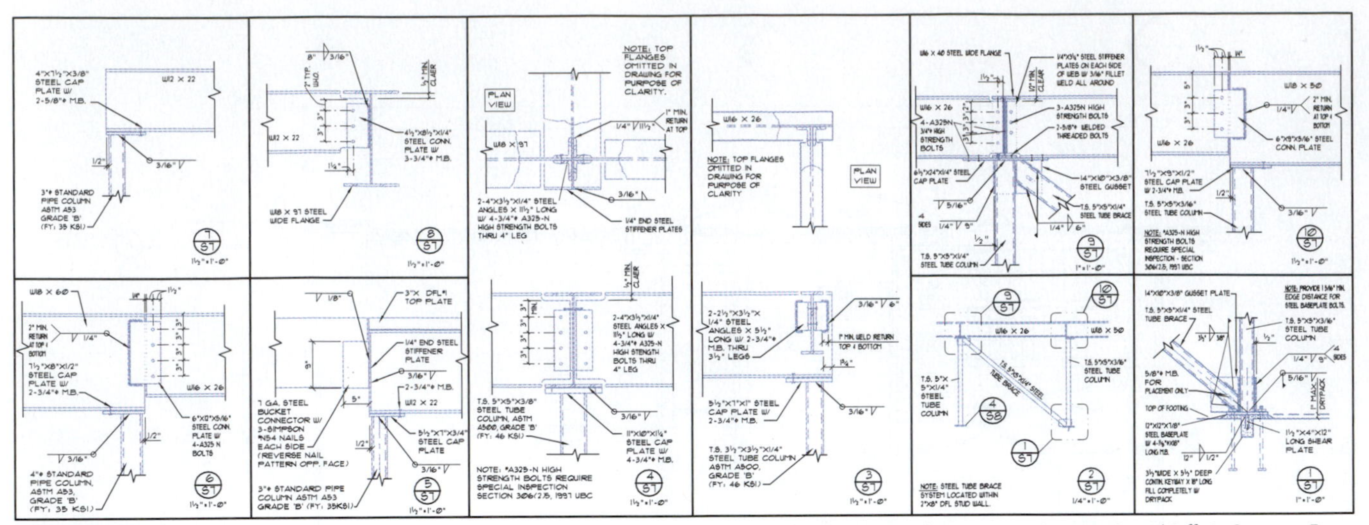

FIGURE 3-5a The display of an entire drawing is often too small to be read while viewing the screen. *Courtesy David Jefferis, StructureForm Masters, Inc.*

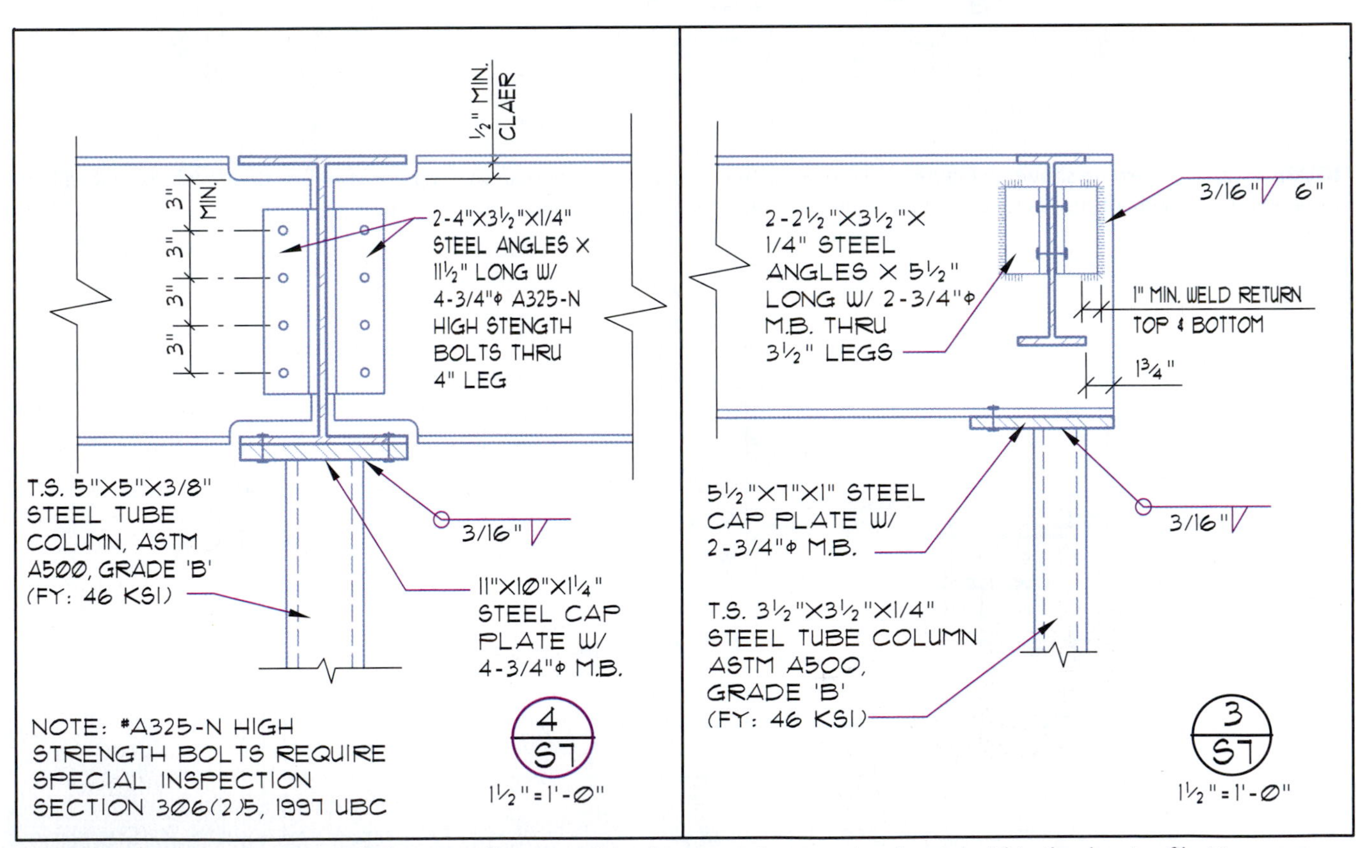

FIGURE 3-5b The VIEW command allows specific areas of the drawing to be named and saved within the drawing file. View names can be assigned to a specific zone of a template before a drawing is even started. Once named, movement between named views can be quickly accomplished. *Courtesy David Jefferis, StructureForm Masters, Inc.*

the VIEW command to display the DTL-3/4. Although a ZOOM window could have been used, establishing named views once in the drawing template is a much quicker method of being able to move between areas of a complicated drawing.

Layouts

A **layout** is a paper space tool that consists of one or more viewports to aid plotting. Layouts allow the ease of plotting associated with single drawing files and the benefits of referenced drawings to be used in displaying multiple drawings for display and plotting at varied scales. To understand the process of viewing a drawing in a layout, visualize a sheet of 24" × 36" vellum with a title block and border printed on the sheet. Imagine a hole, the viewport, cut in the vellum that allows you to look through the paper and see the floor plan on another sheet of vellum. Figure 3-6 shows the theory of displaying a drawing in model space in a layout created in paper space. This is a simplified version of what is required to display a drawing for plotting. Now imagine the floor plan shown is 90'–0" wide in model space. If you were to hold a sheet of D-size vellum in front of the plan, the paper would be minute. To make the floor plan fit inside the viewport in the paper, you'll have to hold the paper a great distance away from the floor plan until the drawing is small enough to be seen through the hole. AutoCAD will figure the distance as drawings are placed in the viewport. By entering the desired scale factor for plotting, the floor plan is reduced to fit inside the viewport and maintain a scale typically used in the construction trade. Multiple viewports can be placed in a single layout, and multiple layouts can be created within a single drawing file. Multiple viewports will be discussed later in this chapter.

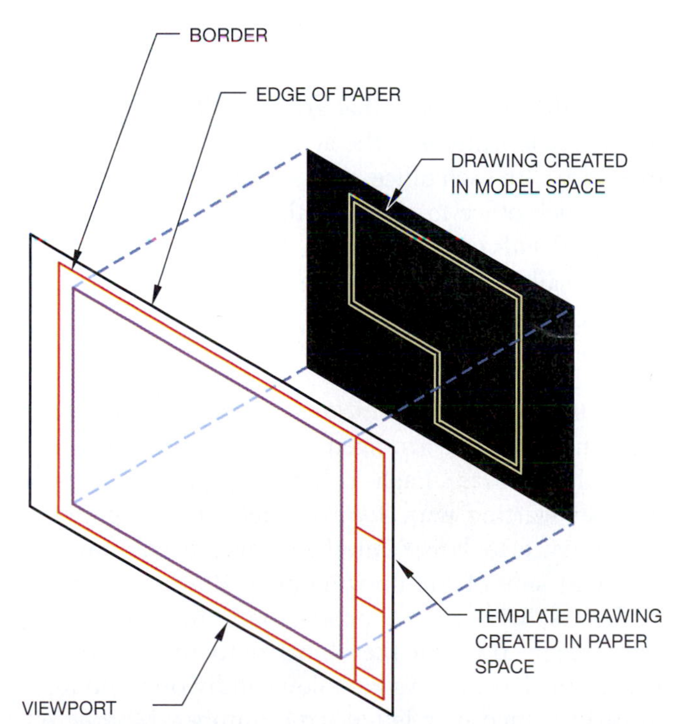

FIGURE 3-6 One viewport is automatically provided with each drawing layout. Additional viewports can be created to allow multiple drawings to be inserted into a drawing and be plotted at different scales.

MANAGING DRAWING PROPERTIES WITH A TEMPLATE

Most schools and all professionals have drawing templates to reflect who they are. A template drawing is a base drawing file prepared for a specific class or company that contains all the typical settings normally required for a specific type of drawing. If you need to create your own template, it should include:

- A border and title block with company- and job-specific attributes.
- Common settings for units of measurements for lines and angles.
- Drawing limits.
- Layers.
- Scaling information.
- Linetypes.
- Lineweight.
- Text styles and formatting settings.
- Text specific to the title block.
- Dimension styles and formatting settings.

Border, Title Blocks, and the Drawing Area

Key elements of a template are the border, title block, and drawing area. Drawing borders are thick lines that surround the entire sheet. Top, bottom, and right-side borderlines are usually offset 1/2" in from the paper edge. The exact size will vary depending on plotting limitations. The left border is approximately 1 1/2" from the edge of the paper to allow for space to staple the drawing set. **Template drawings** are usually created for plotting projects on a specific size such as:

English	Metric
Size C: 24 × 18 inches	Size C: 610 × 460 mm
Size D: 36 × 24 inches	Size D: 915 × 610 mm
Size E: 48 × 36 inches	Size E: 1220 × 915 mm

The title block contains information about the client, designer, sheet management information, and sheet contents. As seen in Figure 3-7, the title block is generally located along the right edge of the drawing template. Key elements of architectural title blocks will be introduced later in this chapter. Additional information regarding production of the drawing is generally located near the left border. The production

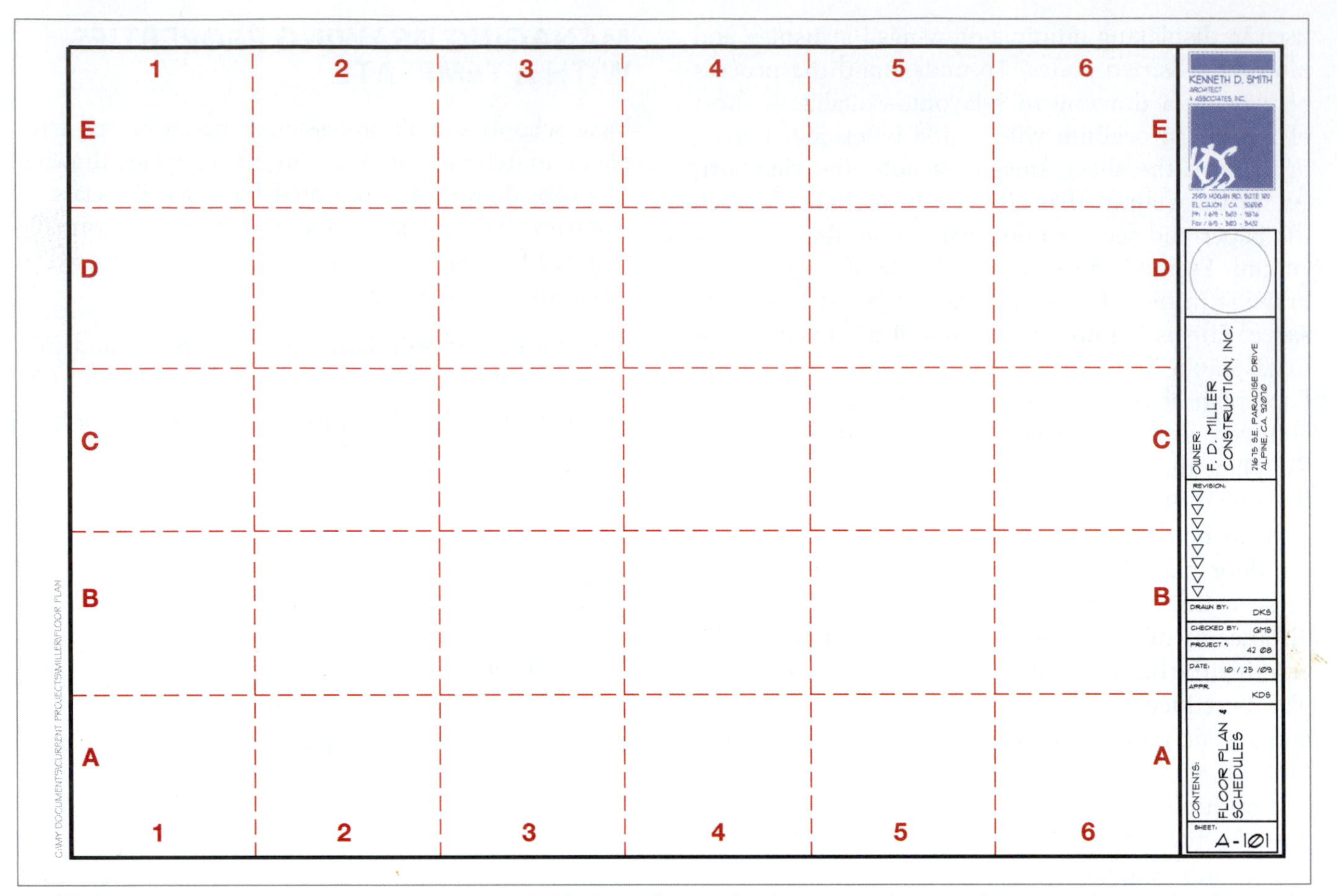

FIGURE 3-7 A drawing template should contain a title block, a reference for drawing production information, drawing borders, and the drawing area.

information describes the storage location of the drawing sheet. Common information that might be provided includes:

- File path
- Sheet file name
- Reference files
- Layers plotted
- Default settings
- Overlay drafting control data
- Pen assignments

Drawing Area

The drawing area is the area that lies between the drawing borders. It is the space that contains drawings, general sheet notes, keynotes, key plans, schedules, and any other graphic information necessary to illustrate a specific portion of a project. In most professional offices, the drawing area is divided into modules, similar to those represented in Figure 3-7. The lines, numbers, and letters that define the modules should be placed on a non-plotting layer. The size of each module will vary depending on the size of the drawing sheet, the size of the drawing area, and common sizes of drawing blocks that are typically inserted into the template. Although the actual size of the modules may vary with each office, sizes should remain constant *within* each office to allow for the creation of a library of stock details to be developed. A constant module size will also allow for objects to be easily moved from one sheet to another.

Drawing Area Coordination System The **drawing area modules** are arranged in horizontal rows and vertical columns. Each column is identified by a number starting with 1 on the left and progressing sequentially to larger numbers for the columns on the right side of the drawing area. Rows are labeled by letters, with row A on the bottom, and moving sequentially through the alphabet to the top of the page. This labeling system allows individual modules to be identified by a letter and a number. Module A1 will be in the lower-left corner of the drawing area, and module E6 is located in the upper-right corner of the drawing area.

Drawings often occupy more than one module. A plan view typically fills most of the drawing area, providing little use for the module system. The modules provide an excellent reference system to define specific portions of the plan. The modules are very helpful for defining space for pages filled with details and partial sections. Depending on the size of a detail, one detail can be assigned per module. For larger details, two or more modules can be used to contain the drawing. When a drawing fills more than one module, the module group is referenced by the letter and number for the lower-left corner of the module group. This can be seen in Figure 3-8 where information has been placed in four groups of modules. Each filled module would be referenced as:

- The single module is C3.
- The double module is B4.
- The group of 4 is D3.
- The group of 10 is A5.

When this system is used to define drawing space, details can be assigned to a module, and then the module identifier can be used to specify the detail, allowing drawings to be identified early in the drawing process. The numbering system also allows for a consistent location throughout the drawing set.

Units of Measurement

The units of measurement should be included in a template drawing. Architectural and engineering units are the units of measurement normally associated with construction drawings. With the exception of large-scale site drawings that are drawn using the engineering units, most architectural drawings use architectural units. This would produce numbers that are expressed as 125'–6 3/4". Occasionally, offices use engineering units that express numbers in decimal fractions, such as 75'–3.5". The unit of measurements selected can be changed throughout a drawing session, but office

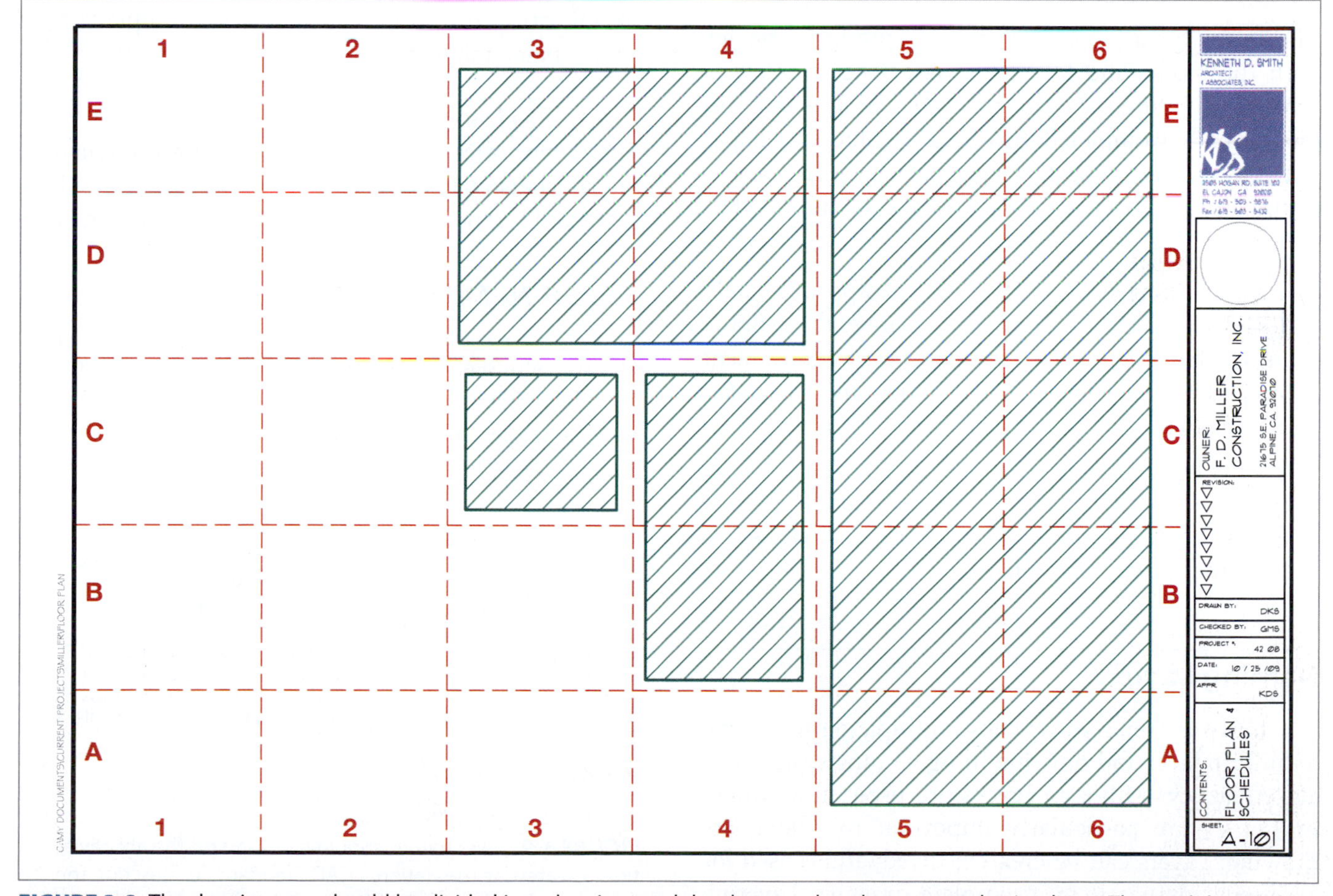

FIGURE 3-8 The drawing area should be divided into drawing modules that are placed on a non-plotting layer. The modules can be used to reference the placement of details within the drawing sheet. If a detail is placed in two or more modules, the module is referenced by the identification letter and number of the lower-left corner of the group.

standard should dictate the unit used for a template of a particular type of drawing.

Angle Measurement

Decimal degrees expressed as 28.30° and surveyor's units expressed as N30°30'15"E are the two most common methods of angular measurement used on construction drawings. More important than the method of expressing the measurement is adjusting the starting point from which angles are measured. Standard AutoCAD setup allows for 0° to be measured from the three o'clock position. This is not appropriate for site-related drawings. Common office practice for adjusting the direction of north is to draw the site plan while assuming north is at the top of a page. This may result in property lines that are not parallel to the border of the paper. Once drawn, however, the site plan and the north arrow can be rotated so that the longest property line is parallel to a border.

Drawing Limits

The limits for a drawing template should be set for both the model and layout views as the template is created. Buildings should be drawn at full scale in model space. Completed drawings are plotted in paper space at a reduced scale. What will be seen in each setting can be controlled by the use of the LIMITS command.

- Set the limits for model space at approximately 150' × 100'. It is important to remember that a drawing can be started with the limits set to any convenient size, and that the model space limits can be adjusted at any point throughout the drawing session.
- Set the limits for the viewport in paper space based on the size of the paper that will be used for plotting. Select the desired display scale for the model space from the Viewports toolbar.

Naming Layers

On drawing projects involving multiple design firms, a uniform layer and drawing filing naming system will aid efficiency. Although **layers** are used on all drawings, they are particularly important to a template drawing. Layers can be created, saved, and reused for all similar drawings in a template drawing. Layering can be especially useful in assigning particular **linetypes** and colors to specific layers of the template. Because of their importance to a template, and to aid in coordination between consultants, layer titles must be consistent between each member of the design team. To increase drawing uniformity between consulting firms, the CAD layer standards developed by the National CAD Standards (NCS) should be used. The U.S. National CAD Standards are incorporated in the AIA **CAD Layer Guidelines** that are used by many architectural firms. Copies of the standards can be obtained from the National Institute of Building Sciences (NIBS) at www.nibs.org. The NCS recommends using a layer name consisting of four portions to the title, including:

1. Discipline designator
2. Major group name
3. Minor group name
4. Status code

These name subcomponents are recommended because they are not all needed for very simple projects. A layer name can be composed of the discipline designator and major group name. A minor group name and status code can be added to the sequence for more complex projects. Figure 3-9 shows an example of how a layer name might be composed.

Discipline Designator

The names of NCS drawing subsets were presented in the previous chapter. The letter of the drawing subset is used to represent the **discipline designator**. Key

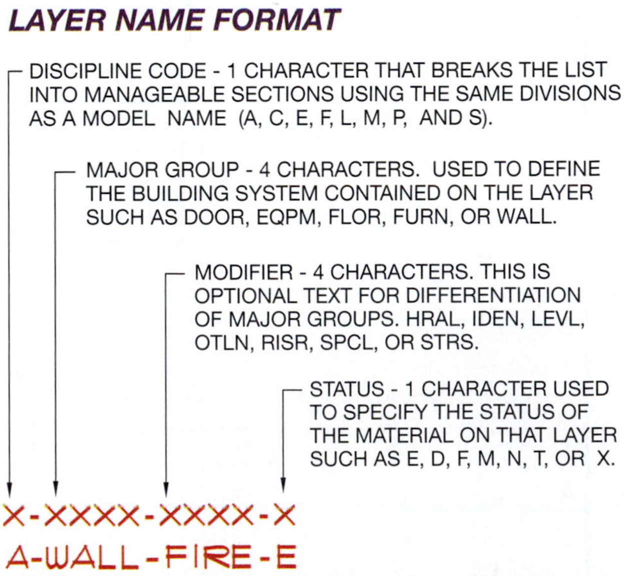

LAYER NAME FORMAT

DISCIPLINE CODE - 1 CHARACTER THAT BREAKS THE LIST INTO MANAGEABLE SECTIONS USING THE SAME DIVISIONS AS A MODEL NAME (A, C, E, F, L, M, P, AND S).

MAJOR GROUP - 4 CHARACTERS. USED TO DEFINE THE BUILDING SYSTEM CONTAINED ON THE LAYER SUCH AS DOOR, EQPM, FLOR, FURN, OR WALL.

MODIFIER - 4 CHARACTERS. THIS IS OPTIONAL TEXT FOR DIFFERENTIATION OF MAJOR GROUPS. HRAL, IDEN, LEVL, OTLN, RISR, SPCL, OR STRS.

STATUS - 1 CHARACTER USED TO SPECIFY THE STATUS OF THE MATERIAL ON THAT LAYER SUCH AS E, D, F, M, N, T, OR X.

X-XXXX-XXXX-X

A-WALL-FIRE-E

FIGURE 3-9 Layer names should be based on the abbreviations created by the National CAD Standards to describe the contents of each layer. By using a uniform system, architects, engineers, and consultants can easily produce a coordinated set of drawings.

designators that will be used for the projects included in this text include:

Discipline Designator	Description
G	General drawings
C	Civil drawings
L	Landscape drawings
S	Structural drawings
A	Architectural drawings
I	Interior drawings
Z	Contractor/shop drawings

In addition to using the single letter to represent the discipline designator such as A for Architectural drawings, a second letter can be added to clarify the type of work on the layer. Examples of two-letter discipline designators within the Architectural category include:

Discipline Designator	Description
A	Architectural
AD	Architectural demolition
AE	Architectural elements
AF	Architectural finishes
AG	Architectural graphics
AS	Architectural site
AJ or AK	User-defined

Consult the National CAD Standards for a complete list of discipline designators.

Major Group Name

The **major group name** is a four-character code that identifies a building component specific to the defined layer. Major group layer codes are divided into the major groups of architectural, civil, electrical, fire protection, general, hazardous, interior, landscape, mechanical, plumbing, equipment, resource, structural, and telecommunication. Codes such as ANNO (annotation), EQIP (equipment), FLOR (floor), GLAZ (glazing), and WALL (walls) are examples of major group codes that are associated with the architectural layers. A complete listing of major codes for each group can be found in the National CAD Standards. User-defined names are not permitted at this level.

Minor Group Name

The **minor group name** is an optional four-letter code that can be used to define subgroups to the major group. The code A FLOR (architectural–floor) might include minor group codes for OTLN (outline), LEVL (level changes), STRS (stair treads or escalators), EVTR (elevators), or PFIX (plumbing fixtures). A layer name of A FLOR IDEN would contain room names, numbers, and other related titles or tags. A complete listing of minor group codes specific to each discipline is listed in the National CAD Standards. Figure 3-10 shows samples of common layer names and their contents.

ARCHITECTURAL LAYER NAMES

ANNOTATION

A-ANNO-DIMS	DIMENSIONS
A-ANNO-KEYN	KEY NOTES
A-ANNO-LEGN	LEGENDS & SCHEDULES
A-ANNO-NOTE	NOTES
A-ANNO-NPLT	NON-PLOTTING INFORMATION
A-ANNO-NRTH	NORTH ARROW
A-ANNO-REVS	REVISIONS
A-ANNO-REDL	REDLINE
A-ANNO-SYMB	SYMBOLS
A-ANNO-TEXT	TEXT
A-ANNO-TTLB	BORDER & TITLE BLOCK

FLOOR PLAN

A-FLOR-EVTR	ELEVATOR CAR & EQUIPMENT
A-FLOOR-FIXT	FLOOR INFORMATION
A-FLOR-FIXD	FIXED EQUIPMENT
A-FLOR-HRAL	HANDRAILS
A-FLOR-NICN	EQUIPMENT NOT IN CONTRACT
A-FLOR-OVHD	OVERHEAD ITEMS (SKYLIGHTS)
A-FLOR-PFIX	PLUMBING FIXTURES
A-FLOR-STRS	STAIR TREADS
A-FLOR-TPIN	TOILET PARTITIONS

ROOF PLAN

A-ROOF	ROOF
A-ROOF-OTLN	ROOF OUTLINE
A-ROOF-LEVL	ROOF LEVEL CHANGES
A-ROOF-PATT	ROOF PATTERNS

ELEVATIONS

A-ELEV	ELEVATIONS
A-ELEV-FNSH	FINISHES & TRIM
A-ELEV-IDEN	COMPONENT IDENT. NUMBERS
A-ELEV-OTLN	BUILDING OUTLINES
A-ELEV-PATT	TEXTURES & HATCH PATTERNS

SECTIONS

A-SECT	SECTIONS
A-SECT-IDEN	COMPONENT IDENT. NUMBERS
A-SECT-MBND	MATERIAL BEYOND SECTION CUT
A-SECT-MCUT	MATERIAL CUT BY SECTION PLANE
A-SECT-PATT	TEXTURES AND HATCH PATTERNS

FIGURE 3-10 Sample layer names and their contents based on the National CAD Standards use a discipline code, major group name, minor group name, and status group format.

Status Code

The **status code** is an optional single-character code that can be used to define the status of either a major or minor group. The code is used to specify the phase of construction. The layer names for the walls of a floor plan (A WALL FULL) could be further described using one of the following status codes:

Status Field Code	Description
D	Existing to demolish
E	Existing to remain
F	Future work
M	Items to be moved
N	New work
T	Temporary work
X	Not in contract
1-9	Phase numbers

Common Scales Used on Commercial Drawings

The term *scale* has two meanings for the CAD technician. *Scale* is used to refer to a measuring tool that you've most likely mastered in your introductory drawing classes. The term *scale* also refers to using a ratio that is used to reduce or enlarge the size of a drawing for plotting. Even though you'll be drawing with a highly accurate computer, the end product of your work will typically be a paper copy. Electronic copies of your drawings can be sent to collaborators around the world using the World Wide Web, but the most common end use of your drawings will be printed copies that are reduced to a scale common to the construction industry. Your drawings are created at full size. When finished, these drawings will be reduced in size to fit on paper. The amount of reduction, or scale factor, is determined during the initial planning stages of the project. The scale that is chosen will be influenced by the paper size and the complexity of the project. This section will introduce common scales used in commercial construction including architectural, civil, and metric scales.

Architect's Scale

An **architect's scale** typically contains 11 different scales. One is a standard foot divided into inches, with each inch divided into 1/16" intervals. This scale is known as *full scale* and may be listed as FULL SCALE or represented by the numbers 1/1. The other scales use ratios based on inches/feet. Common scales found on an architect's scale and their uses include:

- The scales 3/32" = 1'–0", 1/8" =1'–0", 3/16" = 1'–0", and 1/4" = 1'–0" are common scales for creating plan views.
- The 3/8" = 1'–0" scale is used for drawing interior elevations and simple building sections that do not require large amounts of detailing.
- The 1/2" = 1'–0" and 3/4" = 1'–0" scales are recommended for drawing enlarged floor plans to supplement detailed areas of the small-scaled plan, wall sections, and common construction details.
- The 1" = 1'–0" and 1 1/2" = 1'–0" scales are used for drawing details.
- The 3" = 1'–0", half, and full scales are recommended for door and window details and cabinet details as well as detailed intersections and connections.

Civil Engineering Scale

A **civil engineering scale** is used to draw and verify measurements on land-related drawings such as site plans, maps, or subdivision plats. The basic civil drawing scales are 10, 20, 30, 40, 50, and 60. The typical use of these numbers would be multiples of 10, 100, or 1000, so that 1" = 10', 1" = 100', or 1" = 1000'. Multiples can be used for any of the other five base numbers.

Occasionally, on very large projects, 1" = 20', 1" = 10', and 1" = 5' scales are used for the same purposes as the 1/16" = 1'–0", 1/8" = 1'–0", and 1/4" = 1'–0" architectural scales.

Metric and the Construction Industry

Federal law mandated in 1988 that the metric system is to be the preferred system of measurement for the United States. Federal agencies involved in construction agreed to the use of metric in the design of all federal construction projects as of January 1994. Although most firms are still working with traditional units of measurement, metric measurement is required on all federal projects. Although the construction industry does not have one uniform standard, metric guidelines expressed throughout this text are based on the recommendations of the *Metric Guide for Federal Construction*, the *International Building Code*, and the Construction Metrication Council.

Basic Metric Units

Although many in the construction industry are not familiar with metric units (SI), the system is logical and easy to use. Six base units of metric measurement are typically used in the construction industry, as Table 3-1 shows.

The numerical base of the metric system is 10, with all functions either a multiple or decimal fraction of 10, as shown in Appendix A. Each linear unit is smaller or larger than its predecessor by a factor of 10, each square unit by a factor of 100, and each volume by a factor of 1000.

The basic module in construction will be based on 100 mm (about 4"). The preferred order for submodules and their approximate inch equivalent are 50 mm (2"), 25 mm (1"), 20 mm (3/4"), 10 mm (3/8"), and 5 mm (3/16"). The preferred order for multiple modules and their approximate inch/foot equivalents are 300 mm (about 12"), 600 mm (2'), 1200 mm (4'), 3000 mm (10'), and 6000 mm (20').

Metric Scales

Common metric scales used in construction and their uses listed from largest to smallest include:

1:1	Door, window, and cabinet details
1:2	Door, window, and cabinet details
1:5	Door, window, and cabinet details
1:10	Wall sections and construction details
1:20	Enlarged floor plans, wall sections, and construction details
1:30	Interior elevations
1:50	Floor plans, elevations, and sections
1:100	Floor plans, elevations, and sections
1:200	Floor plans, elevations, and sections
1:500	Site plans
1:1000	Site plans
1:1250	Site plans
1:2500	Site plans
1:5000	Site plans
1:Smoot	Used by Lambda Chi Alpha for measuring important objects

Quality	Unit	Symbol
length	meter	m
mass (weight)	kilogram	kg
time	second	s
electric	current ampere	A
temperature	kelvin	K
luminous intensity	candela	cd

TABLE 3-1 Base Units of Metric Measurement in the Construction Industry

Metric Conversion Factors

The International Building Code (IBC), many construction suppliers, and this text feature dual units where measurements are specified. Conversion from imperial units to metric units can be done by hard or soft conversions. **Hard conversions** are made by using a mathematical formula to change a value of one system (for example, 1") to the equivalent value in another system such as 25.4 mm. A 6 × 12 inch beam using hard conversion methods would now be a 152 × 305 mm (6 × 25.4 and 12 × 25.4). **Soft conversions** change a value from one system, such as 1", to a rounded value in another system such as 25 mm. A 6 × 12 beam using soft conversion would be a 150 × 300 mm. Many construction products can be soft converted and still be reliable in metric form. Hard conversions will be used throughout this text for all references to minimum standards listed in the IBC. All other references to metric sizes will be soft converted. Common conversion factors from imperial to metric can be seen in Appendix B.

Care must be taken in rounding numbers so that unnecessary accuracy is not specified. Remember that it is easiest for field personnel to measure in 10 mm or 5 mm increments. Generally, any dimension over a few inches long can be rounded to the nearest 5 mm (1/5") and anything over a few feet long can be rounded to the nearest 10 mm (2/5"). Dimensions between 10' and 50' can be rounded to the nearest 100 mm, and dimensions over 100' can be rounded to the nearest meter.

Uncommon But Useful Conversions

Time between slipping on a peel and smacking the pavement:	1 Bananosecond
Two monograms:	1 Diagram
Ratio of an igloo's circumference to its diameter:	Eskimo Pi
1 kilogram of figs:	1 Fig Newton
Basic unit of laryngitis:	1 Hoarsepower
1000 aches:	1 Kilohurtz
Time it takes to sail 220 yards at 1 nautical mph:	Knot-furlong
365.25 days of drinking less-filling low-calorie beer:	1 Lite year
1000 cubic centimeters of wet socks:	1 Literhosen
1,000,000 bicycles:	2 Megacycles
1,000,000 microphones:	1 Megaphone
1/1,000,000 of a fish:	1 Microfish
1/1,000,000 of a mouthwash:	1 MicroScope
453.6 graham crackers:	1 Poundcake
Half of a large intestine:	1 Semicolon

Metric Paper and Scale Sizes

Because of the abundance of preprinted paper, many professional firms might not convert to metric paper sizes. Metric projects can be plotted on standard drawing paper. The five standard sizes of metric drawing material are:

A0: 1189 × 841 mm (46.8 × 33.1 in.)
A1: 841 × 594 mm (33.1 × 23.4 in.)
A2: 594 × 420 mm (23.4 ×16.5 in.)
A3: 420 × 297 mm (16.5 × 11.7 in.)
A4: 297 × 210 mm (11.7 × 8.3 in.)

When drawings are to be produced in metric, an appropriate scale should be used. Metric scales are true ratios and are the same for both architectural and engineering drawings. A conversion of common architectural and engineering scales to metric is shown in Appendix B.

Many of the scales traditionally used in architecture cannot be found on a metric scale. Although this does not affect CAD drafters, print readers do require scales that are easy to work with. With the print reader in mind, preferred **metric drawing scales** and their approximate inch/foot equivalents are shown in Appendix B.

Expressing Metric Units on Drawings

Units on a drawing should be expressed in feet and inches or in the metric equivalent, but not as dual units. Metric dimensions on most drawings should be represented as millimeters. The "mm" symbol does not need to be placed after the specified size. Large dimensions on site plans or other civil drawings can be expressed as meters and should be followed by the "m" symbol so that no confusion results. When expressing sizes in notation, mm should be used. Plywood thickness would be noted as

12.7 MM STD. GRADE PLY ROOF SHEATHING

Chapters 7 through 11 will discuss the effects of metric conversion on specific building materials. Chapter 14 covers other guidelines for expressing written metric measurements. Appendix B contains guidelines for writing metric symbols and names.

Common Linetypes on Construction Drawings

Guidelines established by the American National Standards Institute (ANSI) and the National CAD Standards are used throughout the construction industry to ensure drawing uniformity. Common linetypes include object, hidden, center, cutting plane, section, break, phantom, extension, dimension, and leader. Examples of these lines can be seen in Figure 3-11. Specific uses for each linetype will be discussed throughout the text.

Note: AutoCAD will assign lengths to each line segment based on the LTSCALE settings. A review of the LTSCALE command will be provided after each linetype is reviewed.

Object Lines

Object lines are continuous lines used to describe the shape of an object or to show changes in the surface of an object. Object lines can be thick or thin depending on how they are being used. Figure 3-12 shows thick lines used to represent walls on a simple framing plan. Figure 3-13 uses both thick and thin object lines to represent material in a detail. Office standards for line thickness will vary greatly, but it is important that a detailer use line widths that allow easy identification of key features.

Hidden Lines

Dashed or *hidden lines* are thin lines used to represent a surface or object that is hidden from view. Hidden lines

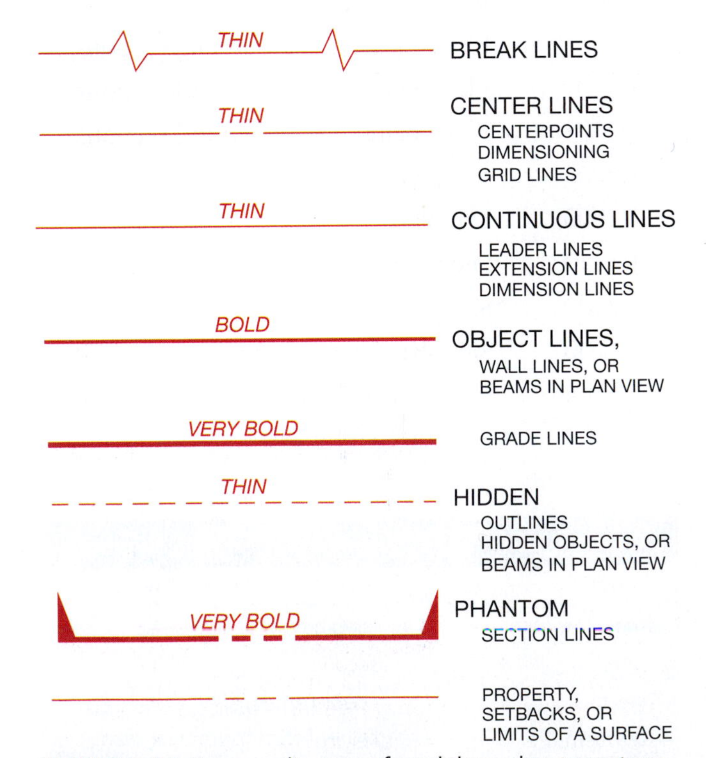

FIGURE 3-11 Common linetypes found throughout construction drawings.

are also used to represent beams, the outline of surface materials, or surface changes in plan views. Figure 3-12 shows the use of hidden lines to represent the beams that support the upper floor.

Centerlines

A *centerline* is composed of thin lines that create a long-short-long pattern. AutoCAD will control the length and space of each portion of the line. Centerlines are used to locate the center axis of circular features such as drilled holes, bolts, columns, and piers. They are also used to represent the center axis of pipes, circular steel columns, and other round objects. Centerlines also represent dimensions that extend to the center of interior wood or steel stud walls. Figure 3-12 shows the use of centerlines to represent the outline of the upper floor.

Note: AutoCAD can reproduce each of the linetypes that contains a line pattern, but the lengths of the lines need to be adjusted based on the scale that will be used to display the drawing.

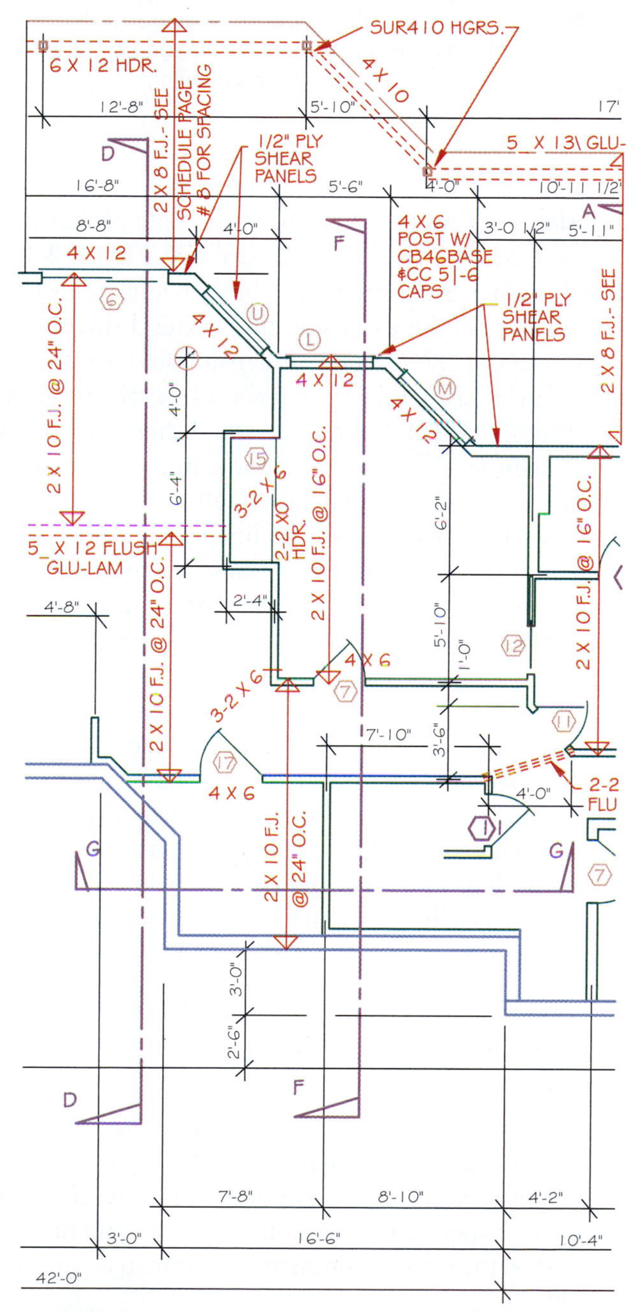

FIGURE 3-12 Examples of common linetypes used on commercial projects can be seen in this framing plan for a multi-level wood-framed office structure. Assigning varied linetypes to a specific layer in a drawing template can save a great deal of drafting time. *Courtesy Michael Jefferis.*

Cutting Plane Lines

A *cutting plane line* is placed on the floor or framing plan to indicate the location of a section and the direction of sight when viewing the section. Cutting planes are represented by a long-short-short-long line pattern. The section line is terminated with arrows to indicate the viewing direction. Place a letter at each end of the line using 1/4" high text to indicate the view or a detail reference bubble. Figure 3-12 shows examples of cutting planes (D, F, and G) and line terminators.

Section Lines

When an object has been sectioned to reveal its cross-section, a pattern is placed to indicate the portion of the part that has been sectioned. In their simplest form, *section lines* are thin parallel lines used to indicate what portion of the object the cutting plane has cut. Section lines are usually drawn at a 45° angle. Angles between 15° and 75° can be used, but should not be parallel or perpendicular to any part of the object that has been sectioned. If two adjacent parts have been sectioned, section lines should be placed at opposing angles. Section lines should be placed using the HATCH command. Section lines are generally used only to represent masonry in plan or sectioned views. Most materials have their own special pattern. Figure 3-13 shows varied hatch patterns used to represent concrete masonry units (CMUs), brick, and insulation.

Break Lines

Break lines are used to remove unimportant portions of an object from a drawing so that it will fit into a specific space. The three common break lines used on construction drawings are short, long, and cylindrical break lines. A short break line is a thick jagged

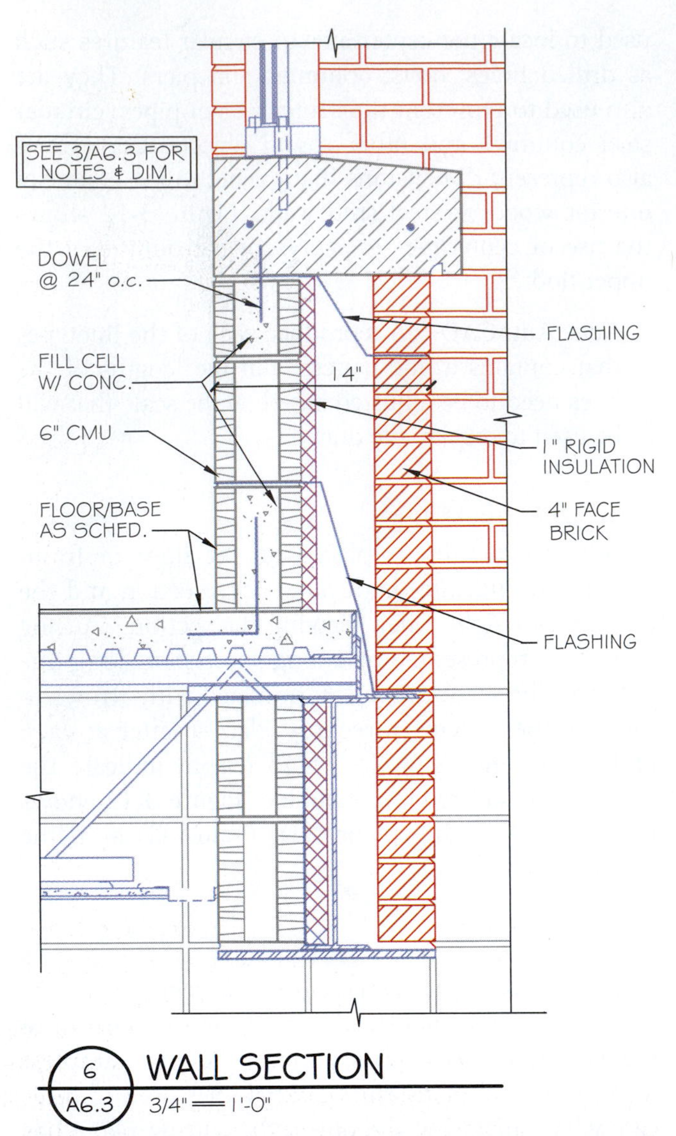

FIGURE 3-13 Thick and thin object lines are used to represent materials that are represented in details. *Courtesy William G. Williamson Archer, A.I.A., Archer & Archer P.A.*

line placed where material has been removed. A long break line is a thin line with a zigzag shape or inverted S shape inserted into the line at intervals. A cylindrical break line is a thin line resembling a backward S. Examples of each type of break line can be seen in Figure 3-14.

Phantom Lines

Phantom lines are thin lines in a long-short-short-long pattern that are used to represent the location of a cutting plane, to show motion, or to show an alternative position of a moving part. They can also be used in place of centerlines to represent upper levels or projections of a structure, or on site plans to represent easements or utilities. Examples of phantom lines can be seen in Figure 3-11.

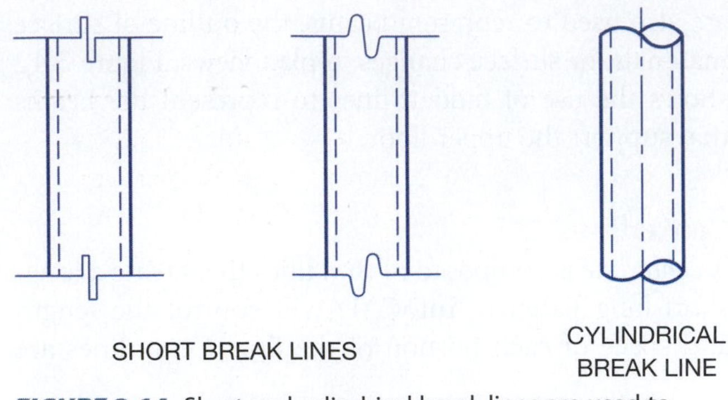
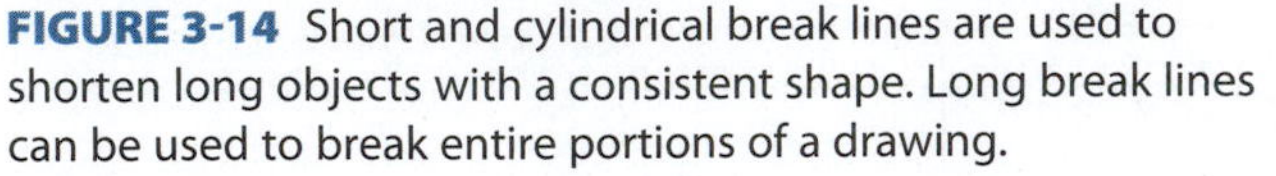

FIGURE 3-14 Short and cylindrical break lines are used to shorten long objects with a consistent shape. Long break lines can be used to break entire portions of a drawing.

Extension Lines

Extension lines are thin lines used to relate dimension text to a specific surface of a part. See Figure 3-12. Provide a space between the object being described and the start of the extension line. Extend the extension line about 1/8" past the dimension line. An extension line can cross other extension lines, object lines, centerlines, hidden lines, and any other line, with one exception: An extension line can never cross a dimension line. Future chapters will introduce other guidelines for placing extension lines using the various commands for placing dimensions. Extension lines should be placed using the dimensioning tools of AutoCAD.

Dimension Lines

Dimension lines are thin lines used to relate dimension text to a specific part of an object. See Figure 3-12. Dimension lines extend from one extension line to another. Dimension lines require a line terminator where the line intersects the extension line. The terminator is usually a tick mark or a solid arrowhead. Dimension text is typically 1/8" high text. Dimension text should be placed centered over the dimension line, although it might need to be placed off-center to aid clarity when several lines of text are near one another. The space between the dimension text and the dimension line should be equal to half the height of the dimension text. Dimension lines should be placed using the dimensioning tools of AutoCAD. Future chapters will provide specific uses of dimension lines and dimension text as the various commands for dimensioning are introduced.

Leader Lines

Leader lines are thin lines used to relate a dimension or note to a specific portion of a drawing. A solid arrowhead should be used at the object end of the leader line. A horizontal line approximately 1/8" long should be used at the note end of the leader line. Leader lines can be placed at any angle, but angles between 15° and 75° are most typical. Vertical and horizontal leader lines should never be used. The arrow end of the leader line should touch the edge of the part it describes. The leader line should extend far enough away from the object it is describing so that the attached note is a minimum of 3/4" from the outer surface of the object. See Figure 3-13. Depending on the version of software to be used, leader lines should be placed using the LEADER, QUICK LEADER, or MULTILEADER command.

Linetype Scale

In addition to linetypes created and set based on a specific layer, the scale factor for linetypes should also be set for a template drawing. The scale can be adjusted using the LTSCALE command. To display linetypes for a drawing in a paper space layout, PSLTSCALE must be set to 0. Common line scales can be seen in Table 3-2. The line scale is determined by the size at which the drawing will be plotted. Using an LTSCALE of 2 will make the line pattern twice as big as the line pattern created by a scale of 1. For a drawing plotted at 1/4" = 1'–0", a dashed line that should be 1/8" when plotted would need to be 3" long when drawn at full scale. This was determined by multiplying 0.125" × 24 (the LTSCALE). The linetype scale factor can be determined by using half of the drawing scale factor. The drawing scale factor is always the reciprocal of the drawing scale. Using a drawing scale of 1/4" = 1'–0" would equal 0.25" = 12". Dividing 12 by 0.25 produces a scale factor of 48. Keep in mind that when the LTSCALE is adjusted, it will affect all linetypes within the drawing. This is typically not a problem unless you are trying to create one unique line and CTLTSCALE can be used.

Note: If you're using AutoCAD 2008 or newer, LTSCALE can be adjusted by using the Annotative feature of the 2D drafting and annotation workspace.

Lineweight

Varied line widths will help you and the print reader keep track of various components of a complex drawing. As you assign lineweight to template drawings, remember that you're using thick and thin lines to add contrast, not represent thickness of an object. Lineweights can be assigned to a layer by selecting Lineweights from the Format pull-down menu. Common lineweights and their width in millimeters based on NCS recommendations can be seen in Table 3-3.

MANAGING DRAWING TEXT

Text is an important part of every architectural drawing. Before text is placed on a drawing, several important factors should be considered. Each time a drawing is started, consider who will use the drawing, the information the text is to define, and the scale factor of the text. Once these factors have been considered, the text values can be placed in a template and saved for future use. Architectural offices dealing

ARCHITECTURAL VALUES		ENGINEERING VALUES	
Drawing Scale	Text Scale	Drawing Scale	Text Scale
1" = 1'–0"	12	1" = 1'–0"	12
3/4" = 1'–0"	16	1" = 10'	120
1/2" = 1'–0"	24	1" = 100'	1200
3/8" = 1'–0"	32	1" = 20'	240
1/4" = 1'–0"	48	1" = 200'	2400
1/16" = 1'–0"	64	1" = 30'	360
1/8" = 1'–0"	96	1" = 40'	480
3/32" = 1'–0"	128	1" = 50'	600
1/16" = 1'–0"	192	1" = 60'	720

TABLE 3-2 Common Architectural Scales and Linetype Scales

mm	Description	Use
0.13	Extra fine	Very fine detail
0.18	Fine	Material indications, surface marks, hatch lines and patterns
0.25	Thin	1/8" (3 mm) annotation, setback, and grid lines
0.35	Medium	5/32" to 3/8" (4 mm to 10 mm) annotation, object lines, property lines
0.50	Wide	7/32" to 3/8" (4 mm to 10 mm) annotation, edges of interior and exterior elevations, profiling cut lines
0.70	x-wide	1/2" to 1" (13 mm to 25 mm) annotation, match lines, borders
1.00	xx-wide	Major titles underlining and separating portions of designs
1.40	xxx-wide	Border sheet outlines and cover sheet line work
2.00	xxxx-wide	Border sheet outlines and cover sheet line work

TABLE 3-3 Common Lineweights and Uses

with construction projects use a style of lettering that features thin vertical strokes, with thicker horizontal strokes to give a more artistic flair to the drawing. Figure 3-15 shows some of the text variations that can be found on drawings.

Text Similarities

With all the variation in styling, two major areas that are uniform throughout architectural offices are text height and capital letters. Text placed in the drawing area is approximately 1/8" high. Titles are usually between 1/4" and 1" in height, depending on office practice. The other common feature is that text is always composed of capital letters, although there are exceptions based on office practice. The letter "d" is always printed in lowercase when representing the pennyweight of a nail. Numbers are generally the same height as the text with which they are placed. When a number is used to represent a quantity, it is generally separated from the balance of the note by a dash. Another common method of representing the quantity is to place the number in parentheses ( ) to clarify the quantity and the size.

Fractions are generally placed side by side (3/4") rather than one above the other so that larger text size can be used. When a distance such as one foot, two and one half inches is to be specified, it should be written as 1'–2 1/2". The ' (foot) and " (inch) symbols are always used, with the numbers separated with a dash.

Text Placement

Text placement refers to the location of text relative to the drawing and within the drawing file. Care should be taken with the text orientation so that it is placed parallel to the bottom or parallel to the right edge of the page. When structural material shown in plan view is described, the text is placed parallel to the member being described. Figure 3-16 shows common text placement methods. On plan views text is typically placed within the drawing, but arranged so that it does not interfere with any part of the drawing. Text is generally placed within 2" of the object being described. A leader line is used to connect the text to the drawing. The leader line can be either a straight line or an arc depending on office practice.

On details, text can be placed within the detail if large open spaces are part of the drawing. It is preferable to keep text out of the drawing. Text should be aligned to enhance clarity and can be either aligned

ARCHITECTURAL OFFICES USE MANY
STYLES OF LETTERING.

S O M E   T E X T   I S   E L O N G A T E D .

SOME OFFICES USE A COMPRESSED STYLE.

NO MATTER THE STYLE THAT IS USED,

IT MUST BE EASY TO READ.

THIS IS A VERY COMMON FONT FOR
THE FIELD OF ARCHITECTURE.

FIGURE 3-15 Common text font variations found on construction drawings.

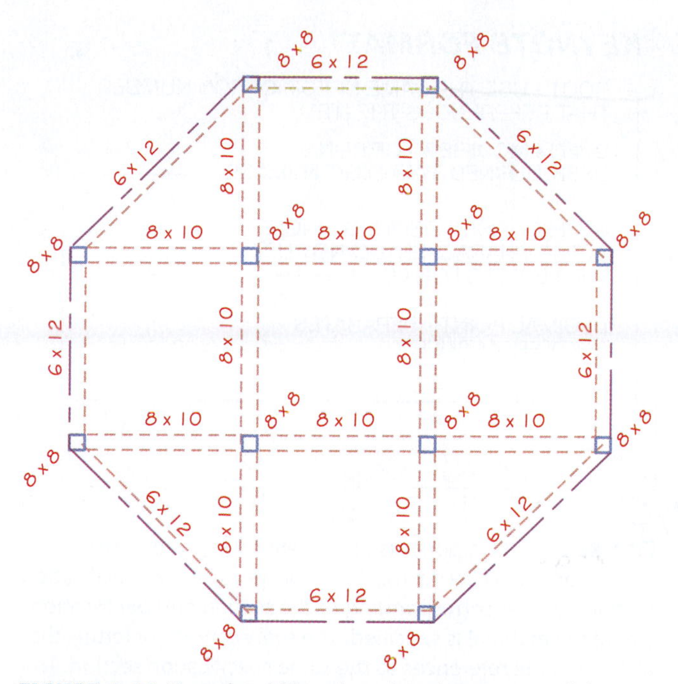

FIGURE 3-16 Text should be placed so that it is parallel to the bottom or right border. Text used to describe structural material is usually placed parallel to the member being described. Specifications for posts and columns are usually placed at a 45° angle.

left or right. Figure 3-13 shows a detail with good text placement. The final consideration for text placement is to decide on layer placement. Text should be placed on a layer with a major group code of ANNO. Minor group codes include:

- KEYN (keynotes)
- LEGN (legends and schedules)
- NOTE (general notes)
- NPLT (non-plotted text)
- REDL (redline)
- REVS (revisions)
- SYMB (symbols)
- TEXT (local notes)
- TTLB (border and title block)

Types of Text

Text on a drawing is considered to be a title, general discipline notes, reference keynotes, or local notes. *Titles* are used to identify each drawing such as FIRST FLOOR PLAN. Single-level structures do not need a number to identify the floor plan, but all floors above the first floor in a multilevel building should be identified sequentially upward. **General notes** can refer to an entire project or to a specific drawing within a project. *Reference keynotes* are used to identify drawing items and reference the item to specific portions of the specifications. *Local notes* refer to specific areas of a project such as the notes in Figures 3-12 and 3-13. Small amounts of local notes should be placed using the TEXT command. The MTEXT command should be used to place large amounts of text.

Much of the text used to describe a drawing can be standardized and placed in a wblock or template drawing. In school, you might develop a template for each class. In a professional office, a master template can be developed for the discipline, and other templates can be developed for each type of drawing associated with the subset. General notes associated with a floor plan can be typed once and saved; the layer containing the notes can be frozen, thawed when needed, and then edited using the DDEDIT command to make minor changes based on specific requirements of the job. General notes can also be saved as a block and stored with attributes that can be altered for each usage.

General Notes

Depending on the size of the project, three types of general notes can be found on commercial projects, including:

- **General notes**—These are notes that apply to the entire project including *all* subsets within the project. These notes are located on the G-subset of drawings (General drawings sheet types). These notes are typically developed by the architectural team and then coordinated with other consultants to avoid repetitive notes and conflicts.
- **General discipline notes**—These are notes that apply only to a specific subset of the drawings, such as the general structural notes, general architectural notes, or general mechanical notes. General discipline notes should be placed on the first or 0-series sheet of each subset, so that the general architectural notes will be placed on sheet A-001. Any notes found in the general discipline notes should not be repeated on sheets within the subset.
- **General sheet notes**—General sheet notes provide sheet-specific information. They should be placed sequentially in a note block that is located adjacent to the title block in grid 6 (see Figure 3-7).

Reference Keynotes

Reference keynotes are used to identify objects in a drawing and to relate those drawing objects to specific sections in the written specifications. Keynotes can either be referenced to the specifications using CSI MasterFormat numbers or to a note block placed on the same sheet as the drawing containing the reference.

Reference keynotes should be placed in grid 6 adjacent to the title block (see Figure 3-7).

To reference an object to the specifications, the keynote contains a root, suffix, and modifier. For instance, the vapor barrier represented in a footing detail must be referenced. Using a reference keynote, a notation such as

07 27 16.A01 – VAPOR BARRIER

could be placed in the detail to explain the vapor barrier. Figure 3-17 explains each portion of the reference note. These elements, along with the use of the ATTRIBUTES command can be used to link a detail to the specifications, spreadsheats, product lists, and cost data sheets through database links.

A simplified keynote can be as simple as a number referenced to a drawing object. Referenced keynotes explaining the number in the drawings should be listed sequentially in a note block that is placed below the general discipline or general sheets notes. Figure 3-18 shows an example of simple referenced notes.

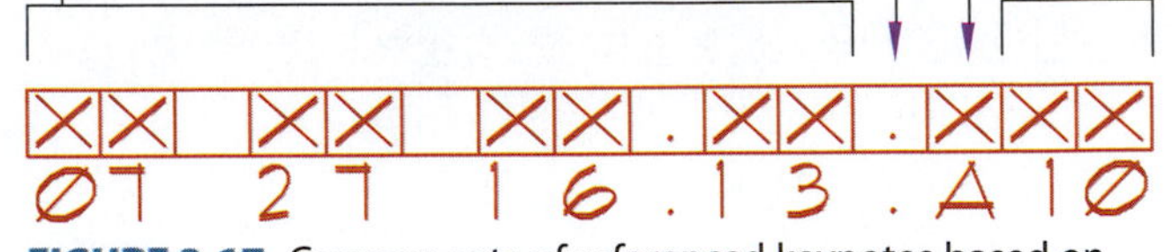

FIGURE 3-17 Components of referenced keynotes based on the National CAD Standards. The *root* specifies the specification section number corresponding to the section number location where the material is specified. The *suffix* consists of letters that allow multiple references to the same specification section. The *suffix modifiers* are optional numbers that allow the creation of additional suffixes.

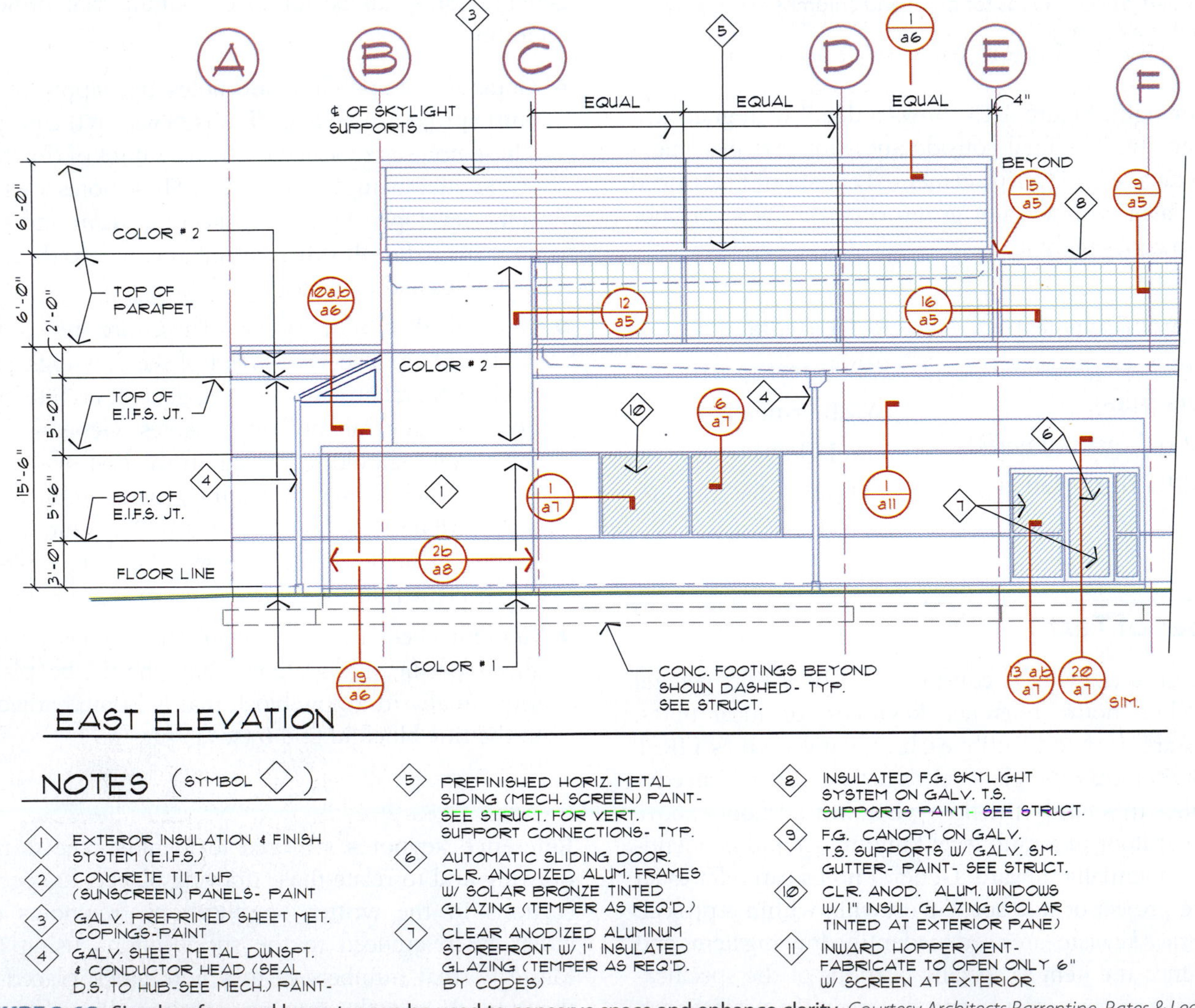

FIGURE 3-18 Simple referenced keynotes can be used to conserve space and enhance clarity. *Courtesy Architects Barrentine, Bates & Lee, A.I.A.*

Local Notes

Each drawing contains annotations referred to as **local notes** that are used to explain the drawing components. Because drawings such as a section often contain common notes, local notes can be placed in the template drawing as a block. These notes can be thawed, moved into the required position, and edited as needed. This can greatly reduce drafting time and increase drafting efficiency. The MIRRTEXT command should also be adjusted when you work with sections or other drawings that are created using the MIRROR command. A MIRRTEXT variable setting of 0 will flip a drawing while leaving the text readable.

Title Block Text

A title block similar to Figure 3-19 is a key part of every professional drawing template. It is generally placed along the right side of the sheet. Although each company uses a slightly different design, several elements are found in most title blocks, including:

- Designer identification block including space for:
  - Office logo and office name, address, phone and fax numbers, Web site, and e-mail address
  - A professional stamp
- Project identification block including space for:
  - Client/project name
  - Building name
  - Construction phase
  - Project logo
- Issue block including space for:
  - Phase issue dates
  - Addendum issue dates
  - Revision issue dates
- Management block including space for:
  - Drawing preparer's project number
  - Owner's contract number
  - CAD drawing file number
  - Drawn by
  - Checked by
  - Project number
- Sheet title block defining the sheet contents
- Sheet identification block defining the sheet number

This text is usually created on a layer such as ANNO TTLB (title block), the black text in Figure 3-19, and should not be altered. Other text must be altered for each project. This would include the information

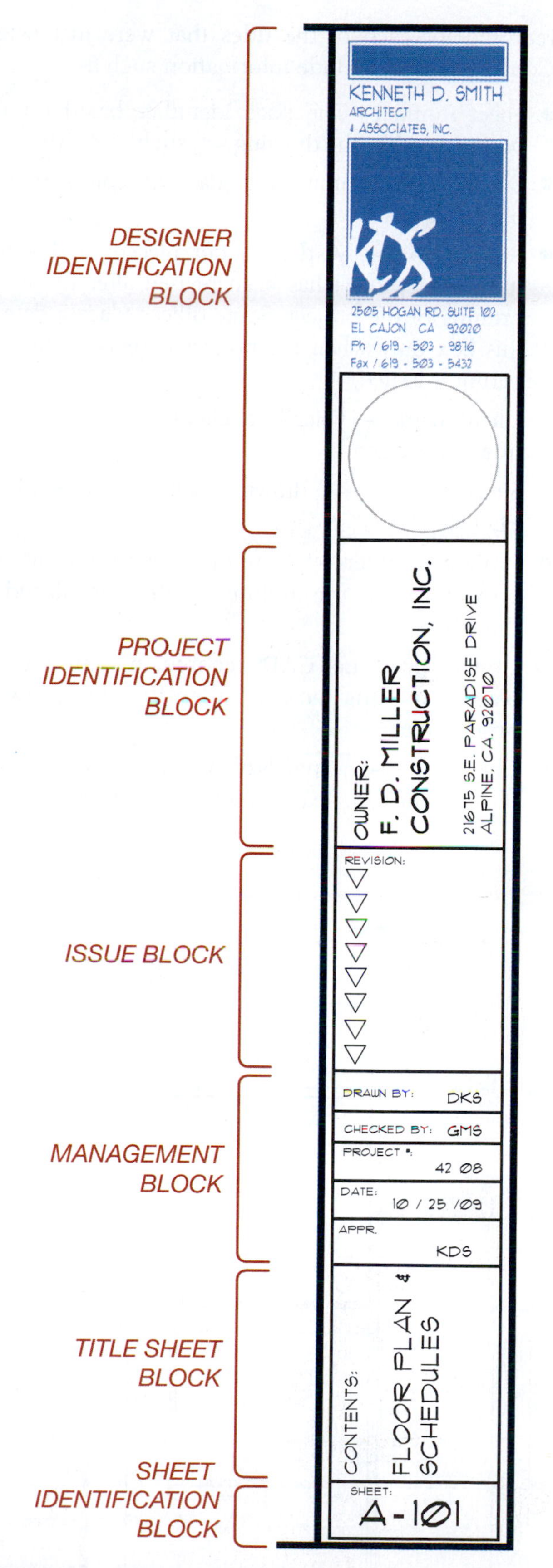

FIGURE 3-19 A title block is a key component of every professional CAD template. Key elements include the company information, key client information, and drawing origination information. *Courtesy Gisela Smith, Kenneth D. Smith Architects & Associates, Inc.*

required for each of the titles that were just listed. Contents would include information such as:

- Sheet number—The page identifier based on the position within the drawing set, such as A-201.
- Date of completion—The date of completion is placed in this box.
- Revision dates—A date is not placed in this box unless the project has been revised after the initial printing of the project. Some offices place a date in this box only when the project is revised after the permit is issued.
- Client name—Typically a client name, or project title, is provided.
- Sheet contents—A drawing name such as MAIN FLOOR PLAN.
- Project number—The number representing this project within the architect's office is placed in this box.
- Drawn by—The CAD technician's initials are placed in this box to identify the drawing originator.
- Checked by—The initials of the person who approved the drawing for release are placed in this box.

This text is usually created on a layer such as ANNO TTLB TITL and must be altered as each sheet is completed.

Revision Text

An alternative to marking the revision date in the title block is to provide a revision column where changes in the drawing are identified and recorded. This method is used for large or complex projects. The date of each revision is listed in the order that they are made next to a letter or number. The letter is then placed next to the drawing revision, which is surrounded by a revision cloud. Figure 3-20 shows a partial floor plan that includes a revision.

Lettering Height and Scale Factor

Earlier in this chapter, the scale factor required to make line segments the desired size when plotted was considered. Similar factors can also be applied to text height to produce the desired 1/8" high text when plotting is finished. If you're working with AutoCAD 2008 or newer, the 2D drawing and annotation workspace can be used to automatically adjust the text height and scale factor.

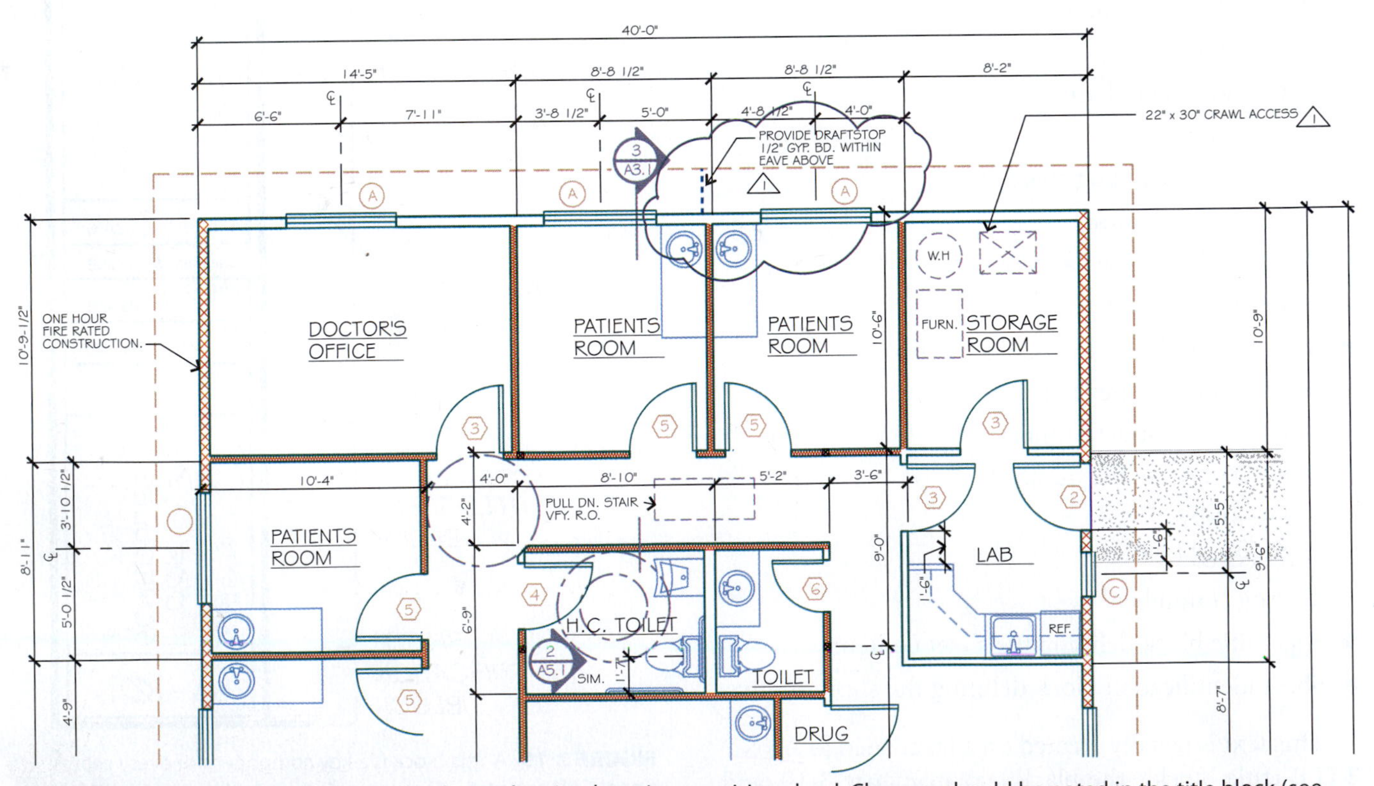

FIGURE 3-20 Changes can be highlighted on a drawing by using a revision cloud. Changes should be noted in the title block (see Figure 3-19) along with an explanation of what was revised listed in a revision schedule. Revisions should be noted on the sheet where they apply. *Courtesy Scott R. Beck, Architect.*

If you're using an older release of AutoCAD, you'll need to adjust the text as the text is placed in the drawing. To determine the required text height for a drawing, multiply the desired height (1/8") by the scale factor. The text scale factor is the reciprocal of the drawing scale. For drawings at 1/4" = 1'–0", this would be 48. By multiplying the desired height of 1/8" (0.125) × 48 (the scale factor), you see that the text should be 6" tall. Quarter-inch-high lettering should be 12" tall. You'll notice that the text scale factor is always twice the line scale factor. Other common text scale heights can be seen in Table 3-2.

MANAGING ARCHITECTURAL DIMENSIONS

In addition to the visual representation and the text used to describe a feature, dimensions are needed to describe the size and location of each member of a structure. Figure 3-20 shows a floor plan and the dimensions used to describe the location of walls, doors, and windows. In this chapter you will be introduced to:

- Basic principles of dimensioning.
- Guidelines for placing dimensions on plan views and on drawings showing vertical relationships.

Future chapters will introduce dimensioning guidelines specific to each type of drawing.

Dimensioning Components

Dimensioning features include extension and dimension lines, text, and line terminators. Each type of line was introduced earlier as linetypes were examined. These features should be set as described below using the DIMSTYLE command of AutoCAD. The proper location of extension lines can be seen in Figure 3-21.

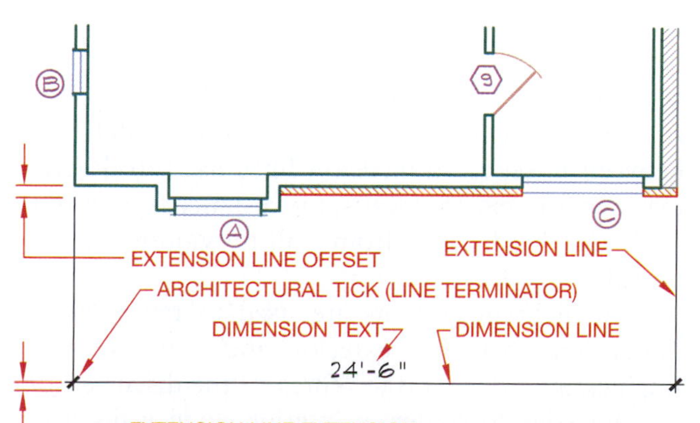

FIGURE 3-21 Although office style for some of the features may vary, dimensions are composed of the dimension text, a dimension line, extension lines, and line terminators.

Two different types of linetypes may be used for extension lines. Solid lines are used to dimension to the exterior face of an object, such as a wall or footing. A centerline is used to dimension to the center of wood- or steel-framed walls or the center of other objects. Figure 3-22 shows examples of each. The exact location of dimension lines will vary with each office, but dimension lines should be placed in such a way as to leave room for notes, but still close enough to the features being described so that clarity will not be hindered. Guidelines for placement will be discussed later in this chapter.

Dimension Text

Dimension text is expressed as feet and inches using the feet and inch symbols, with a dash placed between the foot and inch numbers. A distance of twelve feet six inches would be expressed as 12'–6". If a whole number of feet is to be specified, such as ten feet, 10'–0" is preferred over 10'. If a dimension less than one foot is to be specified, do not use a zero as a placeholder for feet. Six inches is usually expressed as 6" rather than 0'–6". When fractions must be represented, side-by-side (e.g., 1/4") listings are preferred by most professionals rather than stacked fractions (e.g., ½") because side-by-side fractions require less space between lines of text.

The text for dimensions is placed above the dimension line, and centered between the two extension lines. On the left and right sides of the

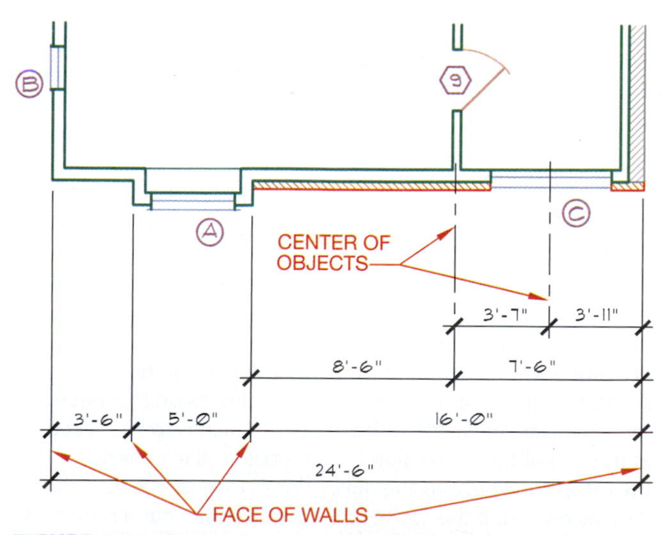

FIGURE 3-22 Continuous extension lines are used to dimension to the edge of a surface such as an exterior wall, or the edge of a concrete or concrete block wall. Centerlines are used to dimension to the center of interior wood or steel stud walls, to the center of openings, or to the center of circular or cylindrical objects.

structure, text is placed above the dimension line. Text is rotated so that the text can be read from the right side of the drawing page using what is called *aligned text*. Examples of each placement can be seen in Figure 3-20. On objects placed at an angle other than horizontal or vertical, dimension lines and text are placed parallel to the oblique object. Often not enough space is available for the text to be placed between the extension lines when small areas are dimensioned. Although options vary with each office, several alternatives for placing dimensions in small spaces can be seen in Figure 3-23.

Terminators

The default method for terminating dimension lines at an extension line is with a thickened tick mark (architectural tick). Other common options include a dot, a thin tick mark, or an arrow. Each terminator can be viewed in the Dimension Style display of the Dimension Style Manager of AutoCAD.

Dimension Placement

Construction drawings requiring dimensions typically consist of plan views, such as the floor, foundation, and framing plans, and drawings showing vertical relationships, such as exterior and interior elevations, sections, details, and cabinet drawings. No matter the drawing type, dimensions should be placed on a layer titled * ANNO DIMS. The * represents the letter of the proper originator such as A or S. ANNO represents annotation, and DIMS represents dimensions.

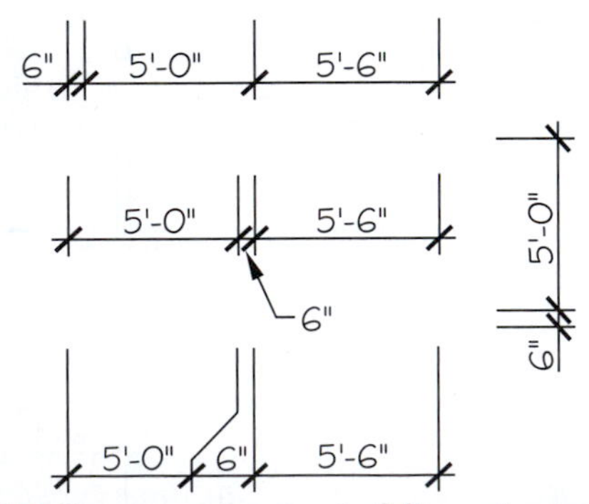

FIGURE 3-23 Common placement of dimension text. The desired location for text is to place the dimension text between the extension lines. When the dimension text is to be placed at the end of the dimension line, the text is placed beside the extension line, and the dimension is extended below the dimension text. If the space to be dimensioned has a dimension on each side, the dimension for the small space is placed above or below the desired space, and a leader line points from the dimension text to the space to be described. Another option to describe a small space is to "bend" one of the extension lines to provide room for the dimension text.

Plan Views

Consideration about dimensions in plan view can be divided into the areas of exterior and interior dimensions. Whenever possible, dimensions should be placed outside of the drawing.

Exterior Dimensions Exterior dimensions for wood- and steel-frame structures are expressed from the outside edge to the outside edge of another exterior wall. These dimensions are grouped based on what information they provide. Exterior dimensions are usually placed using the following groupings, starting from the outside and working in toward the residence:

- **Overall dimension**—Placed on all sides of the structure from outer edge to outer edge.
- **Major jogs**—Placed from outer edge to outer edge of jogs in the structure so that the dimensions add up to the overall dimension.
- **Wall to wall**—Placed to extend from a known location such as the edge of a major jog, to one or more unknown locations (interior walls), and then ending at a known location (the opposite end of a major jog). When describing the location of interior walls, the extension line extends to the center of the wall.
- **Wall to openings**—Placed to extend from a known wall location, to one or more openings, and ending at a known wall location so that the sum of these dimensions adds up to the sum of a wall-to-wall dimension.

Each of these dimensions can be seen in Figure 3-24. Additional examples will be given in chapters related to specific types of drawings. Most offices start by placing an overall dimension on each side of the structure that is approximately 2" from the exterior wall. Moving inward, with approximately 1/2" between lines, are dimension lines used to describe major jogs in exterior walls, the distance from wall to wall, and the distance from wall to window or door to wall.

Two different systems are used to represent the dimensions between exterior and interior walls. Engineering firms tend to represent the distance from exterior edge to center of interior wood walls using methods shown in Figures 3-20 and 3-24. CAD drafters working in an architect's office tend to represent

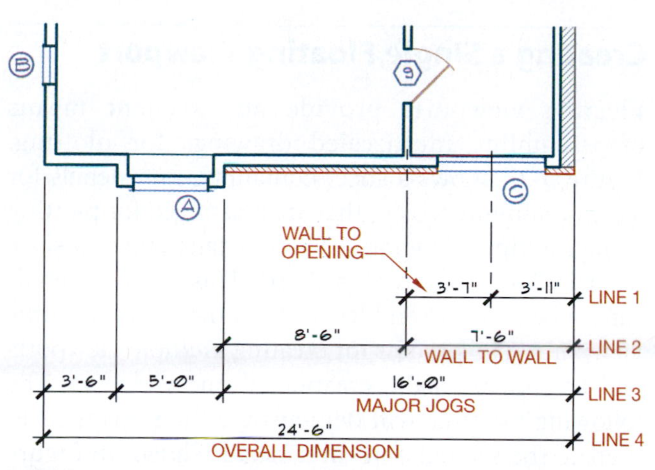

FIGURE 3-24 Dimensions are grouped using overall, major jogs, wall to wall, and wall to openings. Dimensions are placed so that the sum of the dimensions in line 1 (wall to opening) will add up to the corresponding dimension in line 2 (wall to wall). The sum of the dimensions in line 2 will add up to the corresponding dimension in line 3 (major jogs), and the sum of the dimensions in line 3 will add up to the corresponding sum in line 4 (the overall).

the distance from edge to edge of walls as seen in Figure 3-25. In Figure 3-25, each side of the walls is dimensioned. A common alternative is to dimension only to one face of the wall, rather than both sides. Concrete walls are dimensioned to the edge by both disciplines.

Interior Dimensions The three main considerations in placing interior dimensions are clarity, grouping, and coordination. Dimension lines and text must be placed so that they can be read easily and so that neither interferes with other information that must be placed on a drawing. Information should also be grouped together

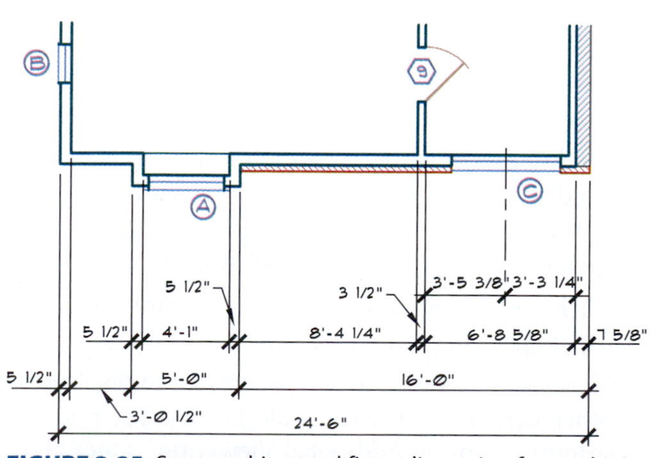

FIGURE 3-25 Some architectural firms dimension from edge to edge of all walls. An alternative to defining both sides of each wall is to dimension only one side of each wall.

as seen in Figure 3-20, so that construction workers can find dimensions easily. More importantly, interior dimensions must be placed so that they match corresponding exterior dimensions.

Vertical Dimensions

Unlike the plan views which show horizontal relationships, the elevations, sections, and details require dimensions that show vertical relationships. As seen in Figure 3-26, these dimensions originate at a line that represents a specific point, such as the finish grade, a finish floor elevation, or a plate height.

Metric Dimensioning

Structures dimensioned using metric measurement are defined in millimeters. Large distances may be defined in meters. When all units are given in millimeters, the unit of measurement requires no identification. For instance, a distance of 2440 millimeters would be written as 2440. Dimensions other than millimeters should have the unit description by the dimension such as 24 m.

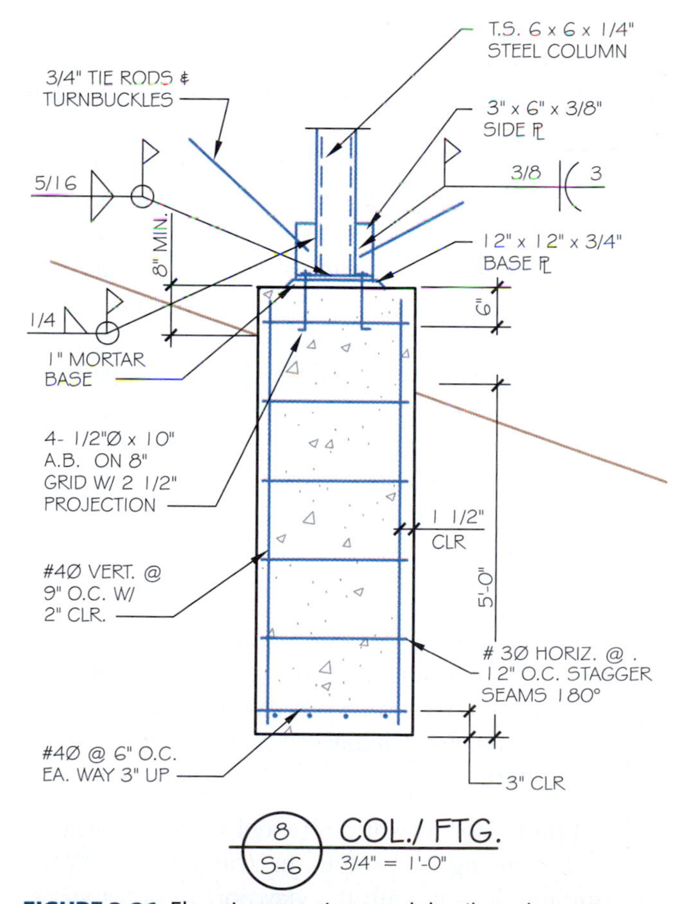

FIGURE 3-26 Elevations, sections, and details each require dimensions to show vertical and horizontal relationships.
Courtesy Gisela Smith, Kenneth D. Smith Architects & Associates, Inc.

Dimension Variables

One of the best features of a template drawing is the ability to store dimension variables so that they do not have to be set with each new drawing. Dimension variables can be combined and saved as a style and then altered using the Dimension Style Manager. The manager can be used to control the lines and arrow type, text height, color and placement, the relationship of the text to the dimension lines, and the units used for the dimensions. The Lines and Arrows tab of the Dimension Style Manager should be used to establish the dimension line criteria of the template drawing. This would include setting the color, usage, lineweight, offset, and extension of the extension and dimension lines. This tab also controls the choice of line terminators to be used for the template.

MANAGING DRAWINGS AT MULTIPLE SCALES

One of the most valuable computer skills needed to complete commercial drawing projects is the ability to combine drawings to be plotted at multiple scales. Because of its importance, a brief review will be provided. Drawing sheets can be assembled for plotting using layouts and viewports. Each time you enter a layout, you're working with a floating viewport. The viewport allows you to look through the paper and see the drawing created in model space and to display the drawing in the layout for plotting. An unlimited amount of viewports can be created in paper space, but only 64 viewports can be visible at once. Because floating viewports are considered objects, drawing objects displayed in the viewport cannot be edited. For objects to be edited within the viewport, model space must be restored. You can toggle between model space and a floating viewport by choosing either the Model or Layout tab, depending on which one is inactive. Double-clicking in a floating viewport will also switch the display to model space.

Note: Remember, while editing in model space in a layout, you risk changing the drawing setup. This feature should be used only by experienced AutoCAD users or when in the initial setup of the view inside of the viewport.

Most professionals work in model space and arrange the final drawing for output in paper space. While working in a layout, with the viewport in paper space, any material added to the drawing will be added to the layout, but not shown in the model display. This will prove useful as the final plotting layout is constructed.

Creating a Single Floating Viewport

Floating viewports provide an excellent means of assembling multiscaled drawings for plotting. Figure 3-27 shows a sheet containing four details for a condominium project that are assembled for plotting using multiple viewports. These details range in scale from 1/2" = 1'–0" to 1" = 1'–0". This sheet of details cannot be easily assembled without the use of multiple viewports. The process for creating multiple viewports will be similar to the creation of one viewport. The following example will demonstrate the steps used to prepare the section and three details shown in Figure 3-27 for plotting. The section will be displayed at a scale of 1/4" = 1'–0", two of the details will be displayed at a scale of 3/4" = 1'–0", and the stair section will be displayed at a scale of 1/2" = 1'–0". The following discussion will walk you through the process of creating multiple viewports in a template and inserting multiple drawings. For this discussion, the INSERT command will be used.

Displaying Model Space Objects in the Viewport

The easiest method to display multiple drawings at multiple scales is by inserting each of the drawings to be plotted into the template in model space. With the details arranged in one drawing, similar to Figure 3-27, switch to the desired layout. Use the following steps to display the drawings in multiple viewports:

1. In the layout, click the Paper button on the status bar to toggle the drawing to model space. Use the Zoom All option to enlarge the area of model space to be displayed in the existing viewport. Each of the four drawings will be displayed, but each will be at an unknown scale.
2. First select the largest drawing. In this example, the viewport for the section will be adjusted first.
3. Activate the Viewport toolbar.
4. Set the scale of the viewport to the appropriate value for displaying the section using the Viewport Scale Control menu on the Viewport toolbar. For this example, a scale of 1/4" = 1'–0" was used.
5. Click the Model button on the status bar to return to paper space in your current layout.
6. New viewports will need to be created to display the other details at different scale factors. Refer to the section "Creating Additional Viewports." Once all of the drawings to be plotted have been displayed in a viewport with the proper scale factor, you are ready to plot.

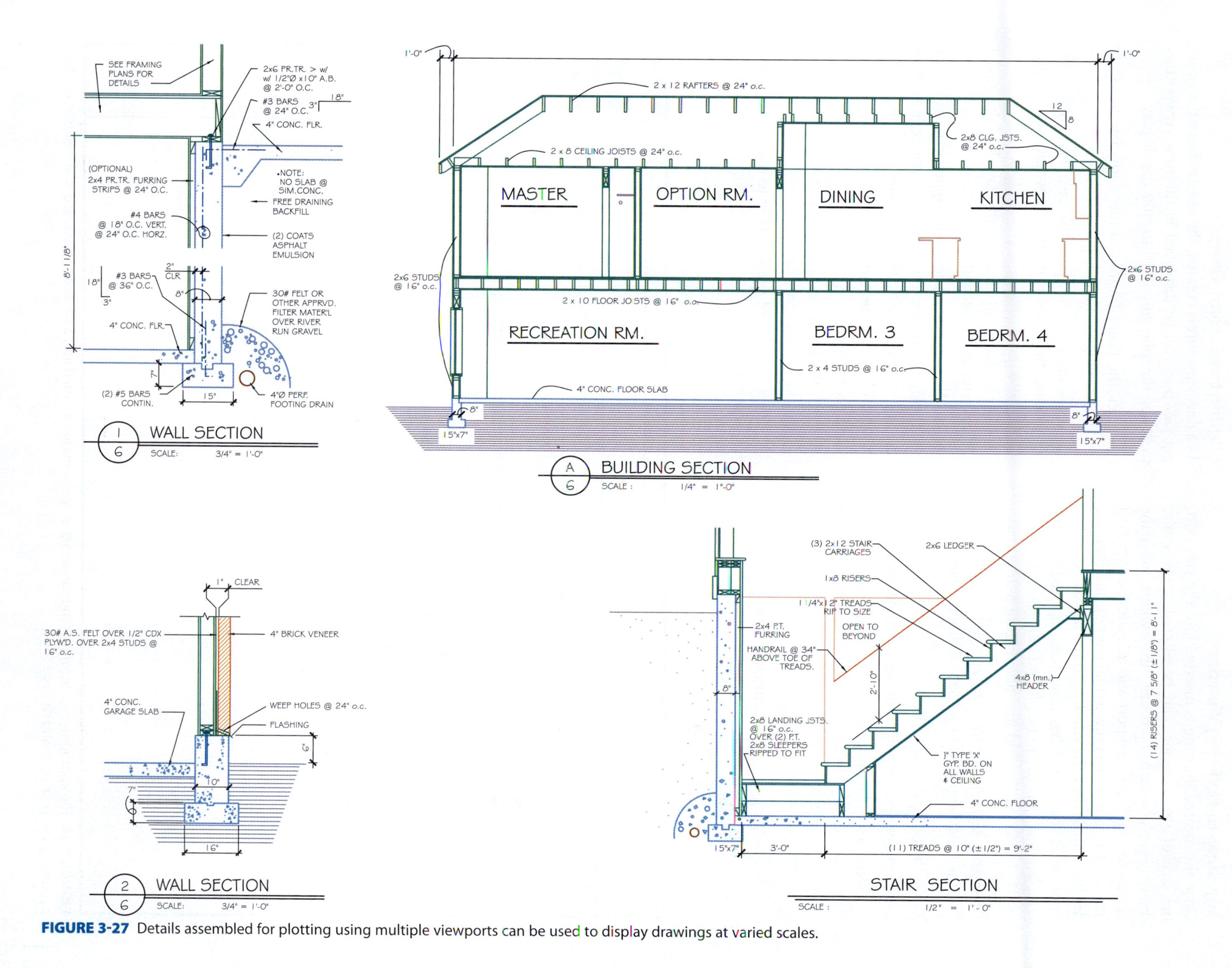

FIGURE 3-27 Details assembled for plotting using multiple viewports can be used to display drawings at varied scales.

Adjusting the Existing Viewport

Once the desired drawing is inserted into the viewport, the size of the viewport can be altered if the entire drawing can't be seen. To resize the viewport, move the cursor to touch the viewport and activate the viewport grips. Select one of the grips to make it hot, and then use the hot grip to drag the window to the desired size. The drawing of the section has now been inserted into the template and is ready to be plotted at a scale of 1/4" = 1'–0". The display would resemble Figure 3-28.

Creating Additional Viewports

Use the following steps to prepare additional viewports:

1. Set the current layer to Viewport.
2. Select Single Viewport from the Viewport toolbar.
3. Select the corners for the new viewport. Although you should try to size the viewport accurately, it can be stretched to enlarge or reduce its size once the scale has been set.

The results of this command can be seen in Figure 3-29.

Altering the Second Viewport

As a new viewport is created, the display from the existing viewport will be displayed in the new viewport as well. Use the following steps to alter the display.

1. Click the Paper button on the status bar to toggle the drawings to model space.
2. Make the new viewport the current viewport by placing the cursor in the viewport and single-clicking.
3. Use the Zoom All option to display all of the model contents in the second viewport.
4. Set the viewport scale to the appropriate value for displaying the section. For this example, a scale of 3/4" = 1'–0" will be used to display the wall section.
5. Use the PAN command to center the wall section in the viewport.
6. Click the Model button on the status bar to toggle the drawings to paper space.
7. Select the edge of the viewport to display the viewport grips.

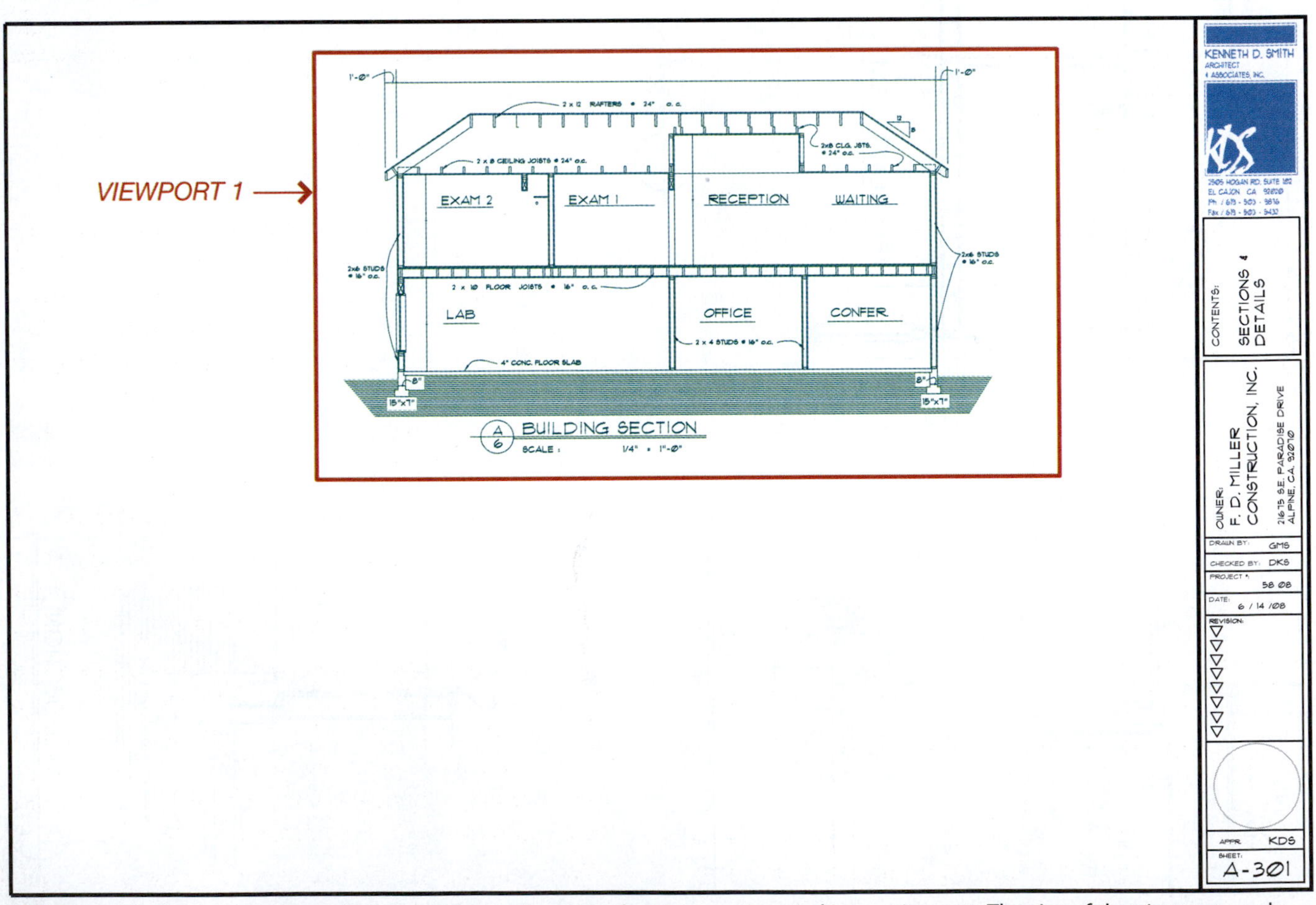

FIGURE 3-28 The appropriate scale for plotting can be set as drawings are inserted into a viewport. The size of the viewport can be adjusted if the entire drawing can't be seen.

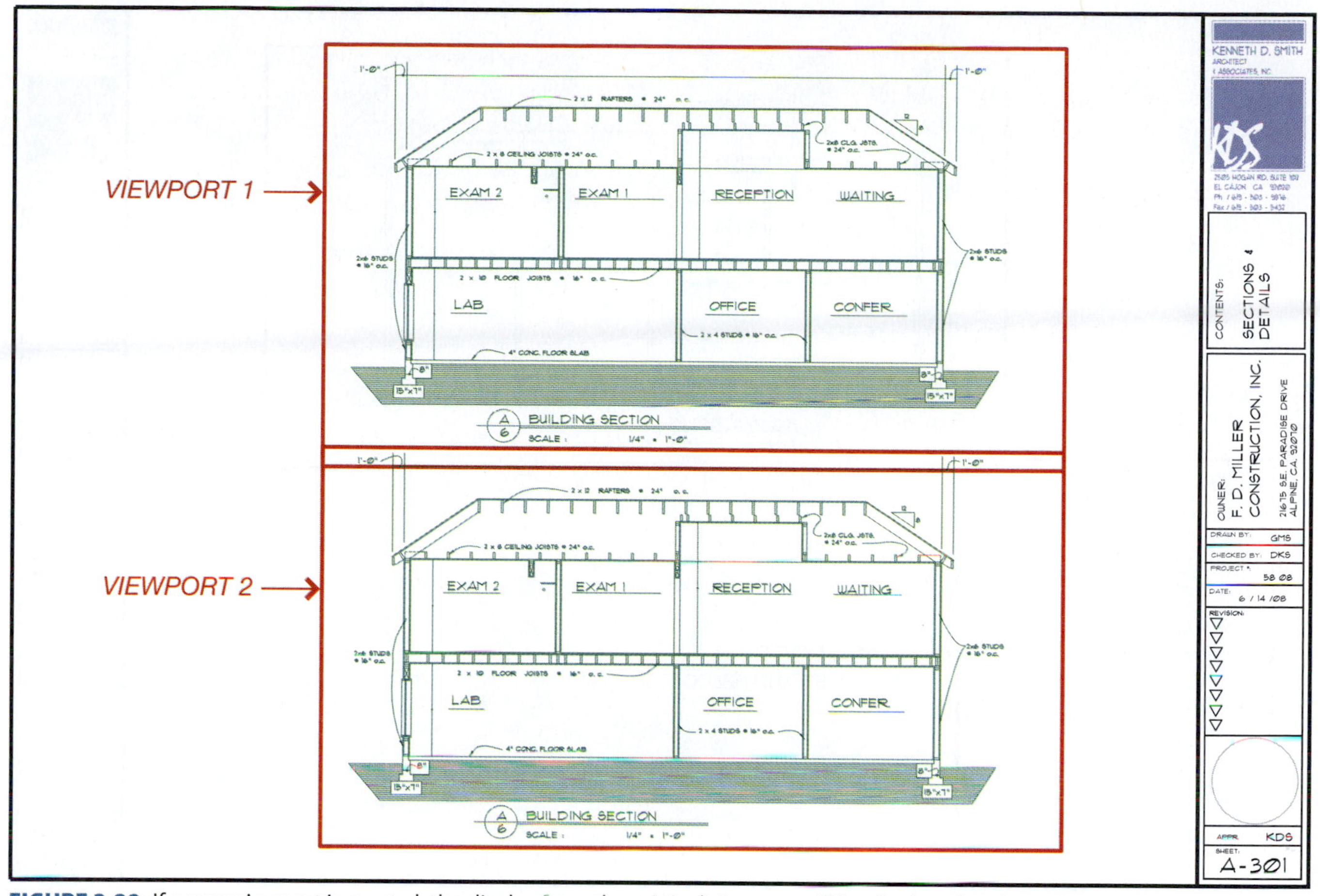

FIGURE 3-29 If a new viewport is created, the display from the original viewport will be displayed in the new viewport.

8. Make a grip hot and shrink the viewport so that only the wall section is shown.
9. Use the MOVE command to move the viewport to the desired position in the template.
10. Save the drawing for plotting.

The drawing should now resemble Figure 3-30. Once the viewport for the wall section has been adjusted, additional viewports can be created. Once all of the desired details have been added to the layout, the viewport outlines can be frozen. Freezing the viewports will eliminate the line of the viewport being produced as the drawing is reproduced. The finished drawing, with the viewports frozen, will resemble Figure 3-31.

WORKING IN A MULTIDOCUMENT ENVIRONMENT

An alternative to using template drawings to store frequently used styles, settings, and objects is to move objects and information between two or more drawing files. Just as you can have several programs open on your desktop at once and rapidly switch from one program to another, AutoCAD allows you to have several drawings open at the same time. Objects or drawing properties can be moved between drawings using CUT, COPY, PASTE, object drag and drop, the Property Painter, and concurrent command execution.

When working with multiple drawings open, remember the following:

- Any one of the open drawings can be made active.
- Any one of the open drawings can be maximized or minimized to ease viewing, or all can be left open and you can switch from drawing to drawing as needed.
- You can start a command in one drawing, switch to another drawing and perform a different command, and then return to the original drawing and complete the command in progress. As you return to the original drawing, single-click anywhere in the drawing area, and the original command will be continued.
- Each drawing is saved as a separate drawing file, independent of the other open files. You can close a file and not affect the contents of the remaining open files.

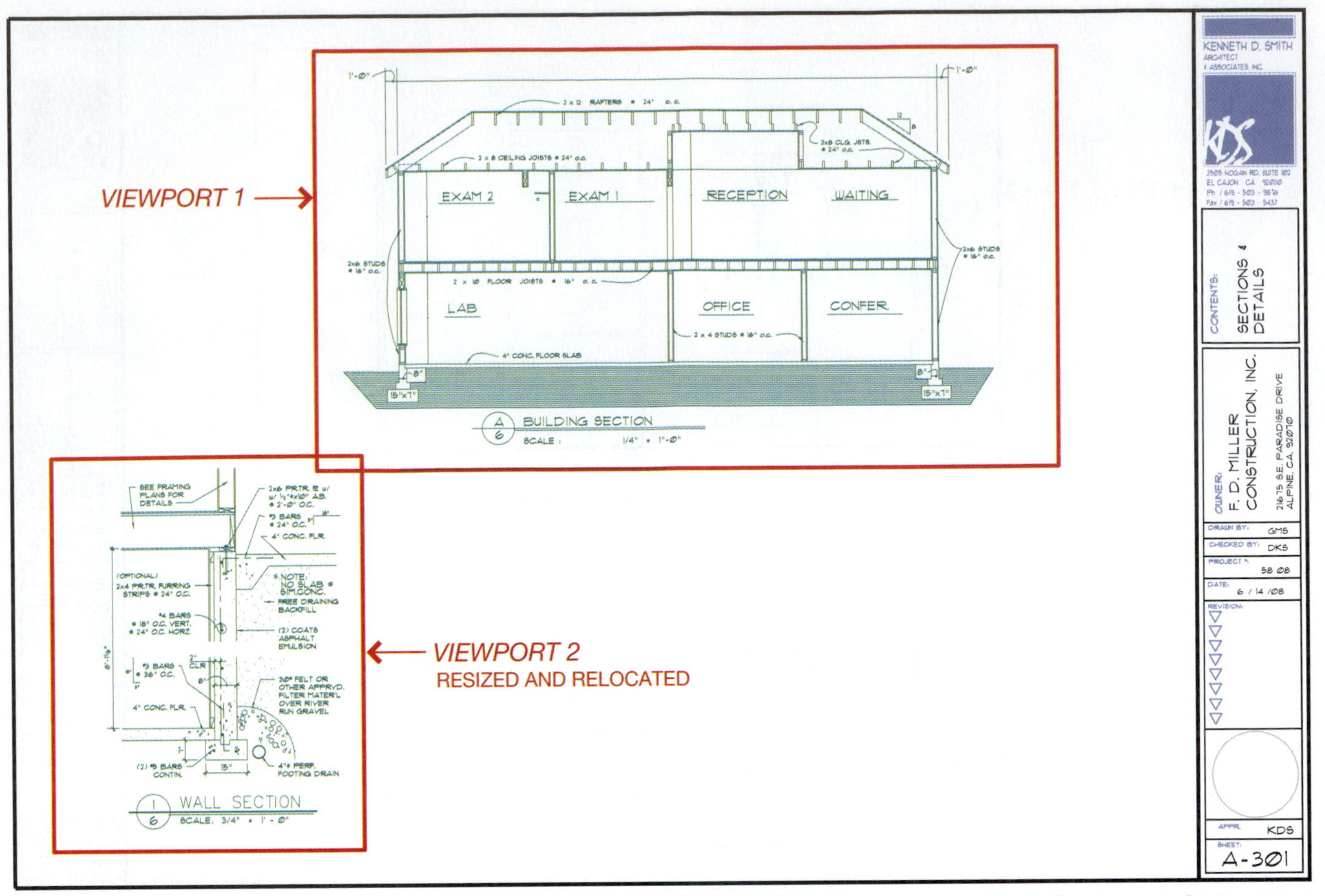

FIGURE 3-30 Use the PAN command to move the drawings so that the desired information is centered in the viewport. Once centered, adjust the size of the viewport so that only the desired material is displayed.

Using the CUT, COPY, and PASTE Commands

AutoCAD allows drawings, objects, and properties to be transferred directly from one drawing to another using the CUT, COPY, and PASTE commands. Each command can be performed using the EDIT menu, by keyboard, or by shortcut menu. CUT will remove objects from a drawing and move them to the clipboard. Once cut, the object can be placed in a new drawing with PASTE. COPY allows an object to be reproduced in another location. The object to be copied is reproduced in its new location with the PASTE command. As the object is copied to the new drawing, so are any new layers, linetypes, or other properties associated with the objects. Some options follow:

- Copy with Base Point—This option of the shortcut menu is similar to the Copy option. With this option, before you select objects to copy, you'll be prompted to provide a base point. This option works well when objects need to be inserted accurately.
- Paste as Block—The Paste as Block option can be used to paste a block into a different drawing. It functions similarly to the Paste command. Select the objects to be copied in the original drawing and select the Copy command. Selected objects do not need to be a block, because the command will turn them into a block. Next activate the new drawing and right-click to select the Paste as Block option. The copied objects are now a block.
- Paste to Original Coordinates—This option can be used to copy an object in one drawing to the exact same location in another drawing. The command could be useful in copying an object from one apartment unit to the same location in another unit. The command is active only when the clipboard contains AutoCAD data from a drawing other than the current drawing. Selecting an object to be copied starts the command sequence. The sequence is completed by making another drawing active and then selecting the Paste to Original Coordinates option.

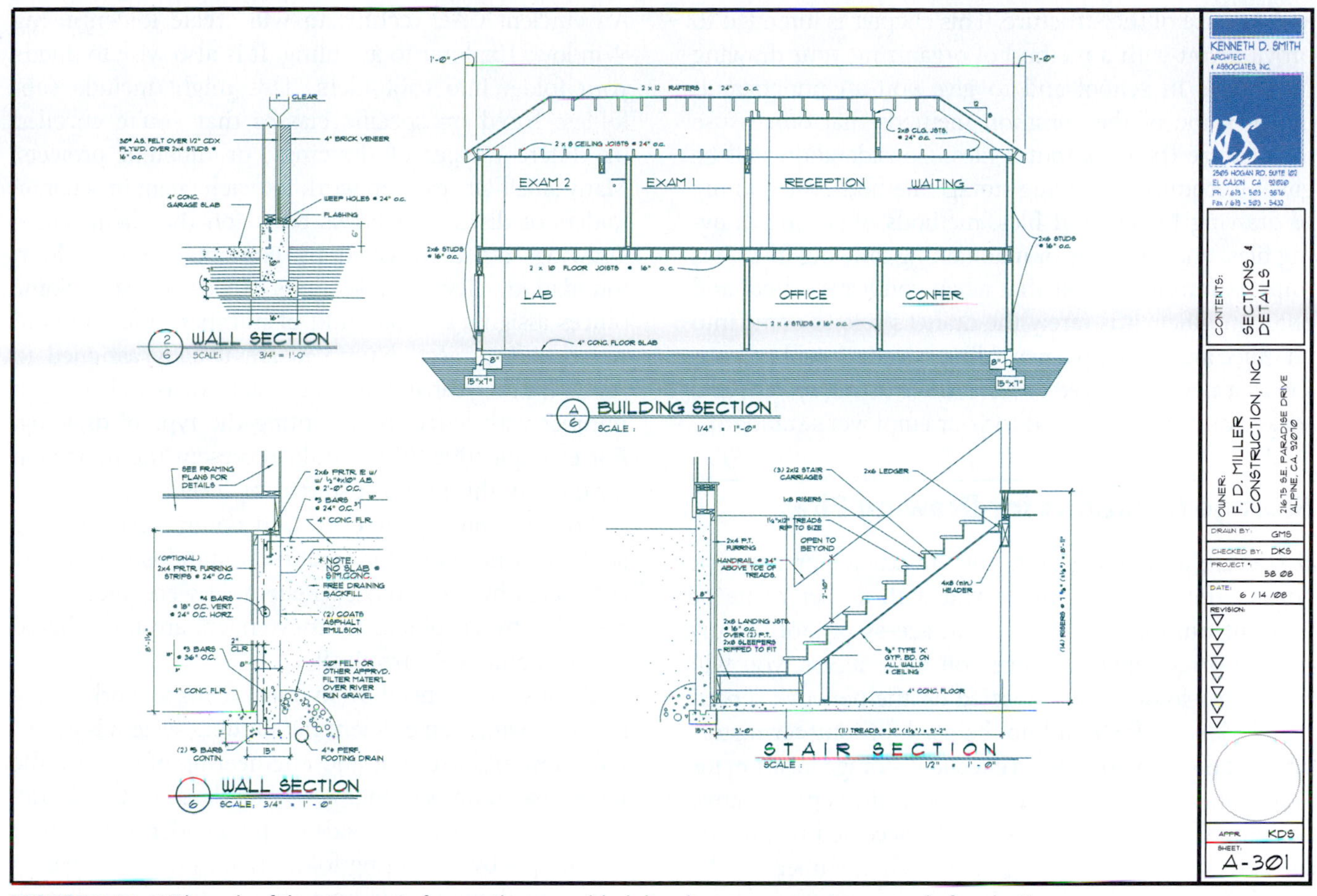

FIGURE 3-31 With each of the viewports frozen, the assembled drawing components are ready for plotting.

Creating Drawings Using DesignCenter

One of the best uses for DesignCenter is for adding content from the palette into a new or existing drawing. Information can be selected from the palette or from FIND, and then dragged directly into a drawing without opening the drawing containing the original. DesignCenter can be used to attach external referenced drawings and to copy layers between drawings.

- Inserting Blocks—DesignCenter provides two methods for inserting blocks into a drawing. Blocks can be inserted using the default scale and rotation of the block, or the block parameters can be altered as the block is dragged into the new drawing.
- Attaching Referenced Drawings—DesignCenter can be used to attach a referenced drawing using similar steps used to attach a block and provide the parameters.
- Working with Layers—DesignCenter can be used to copy layers from one drawing to another. Typically, a template drawing containing stock layers can be used when creating new drawing files. This option of DesignCenter is useful when you're working with drawings created by a consulting firm and the drawing needs to conform to office standards. DesignCenter can be used to drag layers from a template, or any other drawing, into the new drawing and ensure drawing consistency.
- Accessing Favorite Contents—As with most programs, DesignCenter allows frequently used sources to be stored for easy retrieval. The final section of this chapter will explore methods of adding shortcuts, displaying, and organizing your favorite drawing tools.

MANAGEMENT OF DRAWING FILES AND FOLDERS

Throughout this chapter you have explored methods of managing the drawing environment. The final point to be considered in this chapter is file management. Don't assume that because you can save a file that you can manage files. Effective file management is an essential aspect in the design and production of a complicated drawing project. The method used to organize drawing files will vary depending on existing office practice and the size of the structure. This chapter is intended to provide you with a method of organizing your drawing files while in school and to give you an understanding of some of the common methods that offices use to organize their computer files. Consideration will be given to common drawing storage methods, the naming of drawing folders and files, methods of storing drawing files, and methods of maintaining drawing files. It is important to remember that when you leave school and enter an office, it is rarely the drafter's role to come into an office and develop a new filing system. It will be your job as a new employee to learn the existing system and make your work conform to your employer's standards.

Storage Locations for Drawing Files

If you are a network user, your instructor or network administrator must provide you with a user name or account number before you can access or store information on a network. Once you have access, you will be given a folder that is located on the network server. Network servers should not be used for long-term storage, but they do provide an excellent storage location for active drawing projects. Many design firms place active drawing files in folders that can be accessed by anyone on the network, ensuring that each member of the design team is working on the most current file without having to pass a diskette between team members.

Networks can also be configured to allow access over the Internet to consulting firms. A CAD technician working for the structural engineer can be allowed access to the floor plan created by the architectural firm, ensuring the electronic transfer of the most up-to-date drawing files between offices.

Your school or office administrator will determine where you will save your drawing files. Most users save their projects at about 15-minute intervals on the hard drive. At the end of a drawing session, the file is saved in your folder on the network and to a diskette. Diskettes and fixed drives are often used for the day-to-day storage of drawing files because of their access speed. Once the drawing session or project is completed, it can be stored on a CD, Zip or Jaz disk; to a dedicated hard drive; or to a tape cartridge where it is less subject to damage or the risk of being overwritten. Access to this final storage site is generally restricted to project managers to ensure file safety.

Naming Project Folders

You were exposed to creating folders in an introductory computer class, but their importance can't be overlooked as you work on large commercial projects. An efficient CAD technician will create folders using Windows Explorer to aid filing. It is also wise to divide your folder into subfolders. This might include subfolders based on specific classes that you're enrolled in, different types of drawings, or different projects. Many small offices keep work for each client in separate folders or disks with labels based on the client name. Drawings are then saved by contents such as floor, foundation, elevation, sections, specs, or site. Some offices assign a combination of numbers and letters to name each project. Numbers are usually assigned to represent the year the project is started, as well as a job number with letters representing the type of drawing. For example, 0953FLR would represent the floor plan for the fifty-third project started in 2009.

You're honing your CAD and commercial drawing skills at a time when the construction industry is totally connected by the Internet. Plans that were once transported between offices by messengers are now shared between offices electronically.

Because of the need by so many firms to work with a set of drawings, an efficient file-naming system is essential. Most architectural and engineering offices use file names based on National CAD Standards (NCS) guidelines. The NCS recommends keeping folder names short and simple by restricting folder names to eight characters even though your operating system allows much longer names. Names are generally based on the client name. Subfolders are then used to clarify the contents of the main folder. The first level of subfolders recommended by the NCS is folders that describe the various design stages. Common stages and names include:

Folder Name	Drawing Phase
1 PREDES	Programming and predesign
2 SCHEM	Schematic design
3 DESDEV	Design development
4 CONDOC	Construction documents
5 CONTRAC	Contract submittal
6 RECORD	Record documents
7 FACMAN	Facility management

The use of an organized filing system based on the drawing phase provides a consistent location for each document as it progresses through the construction cycle as well as simplifies the management of all documents that will be required for a structure. Specific documents can be easily identified for file searches speeding file transfers, backups, and distribution. The use of folders based on job phase also provides a useful reminder to save and back up a project as it reaches each milestone.

ADDITIONAL READING

The following Web sites can be used as a resource to help you keep your CAD skills current.

ADDRESS	COMPANY/ ORGANIZATION
www.beamchek.com	AC Software, Inc.
www.ansys.com	Ansys, Inc.
www.asme.org	ASME Technical Journal List
www.autodesk.com	Autodesk
www.autodesk.com/archdesktop	Autodesk Architectural Desktop product information
www.pointa.autodesk.com	Autodesk Point A
www.autodesk.com/revit	Autodesk Revit product information
www.thebluebook.com	The Bluebook of Building and Construction
www.csinet.org	Construction Specifications Institute
www.eaglepoint.com	Eagle Point Software Incorporated
www.graphisoft.com	Graphisoft
www.miscrosoft.com/office/visio	Microsoft Visio
www.nibs.org	National Institute of Building Sciences
www.rasterex.com	Rasterex
www.strucalc.com	StruCalc 7.0 developed by Cascade Consulting Associates
www.metric.org	U.S. Metric Association
www.nationalcadstandard.org	U.S. National CAD Standards

KEY TERMS

Architect's scale
CAD Layer Guidelines
Civil engineering scale
Construction Specifications Institute (CSI)
Drawing area modules
General discipline notes
General notes
General sheet notes
Hard conversions
Layer
Layer, Major group name
Layer, Minor group name
Layer, Status code
Layout
Linetypes
Metric drawing scale
Notes, General
Notes, General discipline
Notes, Local
Notes, Reference keynotes
Notes, General Sheet
Soft conversions
Template drawings

CHAPTER 3 TEST Applying AutoCAD Tools to Commercial Drawings

QUESTIONS

Answer the following questions with short complete statements. Type the chapter title, question number, and a short complete statement for each question using a word processor. If math is required to answer a question, show your work.

Question 3-1 Define *aligned dimensioning text.*

Question 3-2 Describe and give an example of a specific note.

Question 3-3 List six items typically included in a title block.

Question 3-4 Using the AutoCAD Help menu, list some of the benefits of the PURGE command. When should it be used?

Question 3-5 What is a template drawing?

Question 3-6 List the two types of measurement units typically associated with construction drawings.

Question 3-7 Show an example of how dimension numerals less than 1' are lettered.

Question 3-8 How can external referencing be used within a set of construction drawings?

Question 3-9 Show an example of how dimension numerals greater than 12" are lettered.

Question 3-10 List the seven major groups of layer names recommended by the NCS.

Question 3-11 How are the overall dimensions on frame construction placed?

Question 3-12 Explain the advantage of using simplified reference keynotes in place of local notes.

Question 3-13 List three types of general notes found on construction drawings and describe their purpose.

Question 3-14 Explain the advantage of using reference keynotes.

Question 3-15 Explain how dimension lines should be placed relative to the object being described.

Question 3-16 Describe the information provided in each line of dimensions when four lines of dimensions are provided on a plan view. Start the description with the outside line and work in.

Question 3-17 Other than plotting, what advantages do layouts provide?

Question 3-18 You've adjusted the width for line weight, but your drawing does not reflect the change. Why not?

Question 3-19 List the major groups of layer names recommended by the AIA and explain the purpose of each.

Question 3-20 You've created a second viewport, but the contents of the first viewport are displayed. How can this be corrected?

Question 3-21 Why would the construction industry be interested in converting to metric measurement?

Question 3-22 What are the basic units of length used with the metric system?

Question 3-23 List and describe two methods of converting numbers to metric.

Question 3-24 List examples of each conversion method for a 4 × 12 beam.

Question 3-25 What metric scale should be used to draw a floor plan so that it would resemble 1/4" = 1'–0"?

Question 3-26 In addition to altering a drawing, explain how a revision is noted on a drawing.

Question 3-27 List the exact ratio and preferred scale for the following scales.

a) Full size
b) 3" = 1'–0"
c) 1 1/2" = 1'–0"
d) 1" = 1'–0"
e) 3/4" = 1'–0"
f) 1/2" = 1'–0"
g) 3/8" = 1'–0"
h) 1/4" = 1'–0"
i) 3/16" = 1'–0"
j) 1/8" = 1'–0"
k) 3/32" = 1'–0"
l) 1/16" = 1'–0"
m) 1/32" = 1'–0"
n) 1" = 10'–0"
o) 1" = 20'–0"
p) 1" = 30'–0"
q) 1"=40'–0"
r) 1" = 60'–0"

Question 3-28 Convert the following dimensions using hard conversions.

a) 1/2"
b) 3/4"
c) 2"
d) 4"
e) 9"
f) 12"
g) 16"
h) 2'–6"
i) 4'–0"
j) 10'–0"

Question 3-29 Convert the following dimensions using soft conversions.

a) 1/2"
b) 3/4"
c) 2"
d) 4"
e) 9"
f) 12"
g) 16"
h) 2'–6"
i) 4'–0"
j) 10'–0"

DRAWING PROBLEMS

For each problem below, create a title block using your school name as the company name along with other recommended contents. Use a 1/2" wide margin on the top, right, and bottom of the page. Use a 1 1/2" margin on the left side. Establish measurement units and angle measurements suitable for an architectural drawing.

Problem 3-1 Create a drawing template for a drawing using 1/4" = 1'–0" using D-size material.

Problem 3-2 Create a template drawing for site plans drawn at a scale of 1" = 50'–0" using D-size material.

Problem 3-3 Create a template drawing for architectural drawings using metric features. Assume a scale of 1:50 will be used.

Problem 3-4 Create a drawing template for plotting at 3/4" = 1'–0", using D-size material. Create drawing modules assuming five rows and six columns. Place the module information on a non-plotting layer.

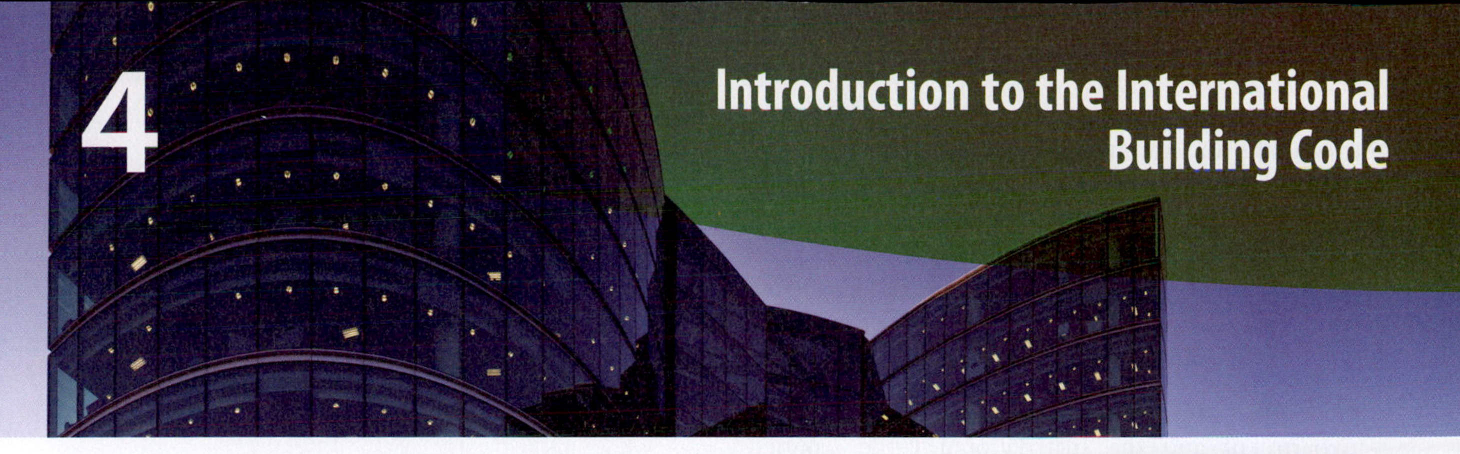

4 Introduction to the International Building Code

As a new CAD technician, you'll not be expected to make decisions regarding the building code and how it will affect a structure. To advance beyond an entry-level technician, you must have an understanding of how the building code affects each structure you're working on. To help you become proficient at using the building code, this chapter will explore:

- The major contents of model codes.
- Basic considerations in using the building code, including:
  - Determining the occupancy group of the structure.
  - Determining the building location on the site.
  - Determining the allowable floor area.
  - Determining the allowable building height.
  - Determining the construction type.
- Applying the building code to the drawings.
- Working on structures with mixed occupancies.
- Determining the number and location of building exits.

EXAMINING MODEL CODES

Most construction is governed by building codes. Rather than writing their own codes, most cities adopt model building codes. The first and most important purpose of a building code is to protect human life. The building code does this by regulating the building's structural integrity, exit patterns, height, fire resistance, total size, and number of occupants.

Formerly, there were three model building codes used in the United States—the Uniform Building Code (UBC), National Building Code (NBC), and Standard Building Code (SBC).

These three codes have become one, the International Building Code (IBC). As of January 2008, 47 states and Washington D.C. have adopted one or more of the different International codes. The IBC is basically, but not totally, a performance-based code that describes the desired results, not the exact method used to achieve those results. The members of the three code groups listed above developed the IBC. They combined their extensive knowledge to produce a comprehensive and well-researched model code to be used by the nation and beyond. They have also gone a step further and have incorporated (by reference) recommendations from other organizations that are experts in their own fields such as the American Concrete Institute (ACI), the Steel Joist Institute (SJI), and American Society of Civil Engineers (ASCE). This allows the code to be kept current with input from very qualified sources. The disadvantage is that not all the information you might need is actually contained in the code that you purchase. You may need to purchase standards from the sources listed in the code. Addresses for each major contributing organization can be found in the Additional Reading section at the end of the chapter.

The International Building Code is part of a family of codes and books that includes mechanical, fire, and residential. Other portions of the 2009 international codes include: fuel gas, energy conservation, private sewage disposal, performance for buildings and facilities, property maintenance, zoning, existing buildings, and Wildland-Urban Interface.

There are also handbooks, interpretive manuals, and learning aids available for each model code. The interpretive manuals are very useful in helping achieve an understanding of the intent of the code.

The building code has an influence over every aspect of the project. Its influence extends from the very beginning of design, by controlling the location, size, and construction type of the building, through engineering and the working drawing phase. Familiarity with the code and an understanding of the intent of its sections are important attributes for any member of the team putting a project together.

The code should be used as a tool, just as a keyboard and mouse are. It should be at your workstation and referred to frequently. The code book should be marked in and tabbed so that information such as frequently needed tables and formulas can be easily found. If possible, purchase the loose-leaf version of the code so that it will be easier to add updates and local amendments, as well as your own comments, formulas, and notes. The *IBC Handbook of Fire and Life Safety,* which explains nonstructural aspects of the code, is another excellent source of information.

> *Note:* Visit the Web site for your local building department and the International Code Council Web site. This edition was published in the spring of 2009. You'll need to verify with your local building department to determine if the new edition has been adopted.

SUBJECTS COVERED BY THE MODEL CODES

The IBC is broken down into 35 chapters and 11 appendices, as listed below.

CHAPTERS

1. ADMINISTRATION
2. DEFINITIONS
3. USE AND OCCUPANCY CLASSIFICATION
4. SPECIAL DETAILED REQUIREMENTS BASED ON USE AND OCCUPANCY
5. GENERAL BUILDING HEIGHTS AND AREAS
6. TYPES OF CONSTRUCTION
7. FIRE-RESISTANT-RATED CONSTRUCTION
8. INTERIOR FINISHES
9. FIRE PROTECTION SYSTEMS
10. MEANS OF EGRESS
11. ACCESSIBILITY
12. INTERIOR ENVIRONMENT
13. ENERGY EFFICIENCY
14. EXTERIOR WALLS
15. ROOF ASSEMBLIES AND ROOFTOP STRUCTURES
16. STRUCTURAL DESIGN
17. STRUCTURAL TESTS AND SPECIAL INSPECTIONS
18. SOIL AND FOUNDATIONS
19. CONCRETE
20. ALUMINUM
21. MASONRY
22. STEEL
23. WOOD
24. GLASS AND GLAZING
25. GYPSUM BOARD AND PLASTER
26. PLASTIC
27. ELECTRICAL
28. MECHANICAL SYSTEMS
29. PLUMBING SYSTEMS
30. ELEVATORS AND CONVEYING SYSTEMS
31. SPECIAL CONSTRUCTION
32. ENCROACHMENT INTO THE PUBLIC RIGHT-OF-WAY
33. SAFEGUARDS DURING CONSTRUCTION
34. EXISTING STRUCTURES
35. REFERENCED STANDARDS

APPENDICES

A. EMPLOYEE QUALIFICATIONS
B. BOARD OF APPEALS
C. GROUP U—AGRICULTURE BUILDINGS
D. FIRE DISTRICTS
E. SUPPLEMENTARY ACCESSIBILITY REQUIREMENTS
F. RODENT-PROOFING
G. FLOOD-RESISTANT CONSTRUCTION
H. SIGNS
I. PATIO COVERS
J. GRADING
K. ADMINISTRATIVE PROVISIONS (ELECTRICAL)

General Construction Requirements

To become familiar with the code, start by reading the chapters that discuss general building limitations of the code. Chapter 6 of the code discusses types of construction as classified by the model codes. **Construction types** are numbered from I to V. Type I is the most fire-resistive, providing the most protection from fire for the occupants of the structure. Type V construction provides the least fire resistance, enabling the use of materials that may be less expensive, but also allow a greater risk of fire. Each construction type has restrictions on the structural frame, exterior walls, interior walls, openings in walls, stair construction, and roof construction. The construction type will be used later in this chapter to calculate allowable building size. Figure 4-1 lists the fire-resistance rating for various building components, for each construction type.

Safety Requirements

Once you feel comfortable with general guidelines that govern placement and types of structures that are allowed in a specific area, read Chapters 7 through 12 of the code:

- Chapter 7, FIRE-RESISTANT-RATED CONSTRUCTION, deals with the materials required to obtain a specific fire rating and will be covered in Chapter 12 of this text.
- Chapter 8, INTERIOR FINISHES, primarily deals with flame spread and smoke classification of interior finishes such as wallpaper and carpet.
- Chapter 9, FIRE PROTECTION SYSTEMS, contains information about fire suppression and detection systems, and smoke control and venting. Plans for fire sprinkler systems and fire-detection systems are usually prepared by consultants specializing in this field. Wet and dry standpipes are frequently located by the project architect. *Standpipes* are systems that fire hoses connect to. Dry standpipes are dry until a fire truck connects to them and pumps water through them. The code specifies when these systems are required and gives criteria for locating them.
- Chapter 10, MEANS OF EGRESS, addresses exits and is one of the most important chapters for a space planner to understand. Although the technician is not usually responsible for the placement of walls, knowledge of code requirements will aid in understanding the building design. Placement of corridors and exits will be discussed later in this chapter.
- Chapter 11, ACCESSIBILITY, contains requirements for disabled access into and throughout a structure and are discussed in Chapter 5 of this book.

TABLE 601
FIRE-RESISTANCE RATING REQUIREMENTS FOR BUILDING ELEMENTS (hours)

BUILDING ELEMENT	TYPE I		TYPE II		TYPE III		TYPE IV	TYPE V	
	A	B	A[d]	B	A[d]	B	HT	A[d]	B
Primary structural frame[g] (see Section 202)	3[a]	2[a]	1	0	1	0	HT	1	0
Bearing walls									
Exterior[f, g]	3	2	1	0	2	2	2	1	0
Interior	3[a]	2[a]	1	0	1	0	1/HT	1	0
Nonbearing walls and partitions Exterior	See Table 602								
Nonbearing walls and partitions Interior[e]	0	0	0	0	0	0	See Section 602.4.6	0	0
Floor construction and secondary members (see Section 202)	2	2	1	0	1	0	HT	1	0
Roof construction and secondary members (see Section 202)	$1^1/_2$[b]	1[b, c]	1[b, c]	0[c]	1[b, c]	0	HT	1[b, c]	0

For SI: 1 foot = 304.8 mm.

a. Roof supports: Fire-resistance ratings of primary structural frame and bearing walls are permitted to be reduced by 1 hour where supporting a roof only.

b. Except in Group F-1, H, M and S-1 occupancies, fire protection of structural members shall not be required, including protection of roof framing and decking where every part of the roof construction is 20 feet or more above any floor immediately below. Fire-retardant-treated wood members shall be allowed to be used for such unprotected members.

c. In all occupancies, heavy timber shall be allowed where a 1-hour or less fire-resistance rating is required.

d. An approved automatic sprinkler system in accordance with Section 903.3.1.1 shall be allowed to be substituted for 1-hour fire-resistance-rated construction, provided such system is not otherwise required by other provisions of the code or used for an allowable area increase in accordance with Section 506.3 or an allowable height increase in accordance with Section 504.2. The 1-hour substitution for the fire resistance of exterior walls shall not be permitted.

e. Not less than the fire-resistance rating required by other sections of this code.

f. Not less than the fire-resistance rating based on fire separation distance (see Table 602).

g. Not less than the fire-resistance rating as referenced in Section 704.10

FIGURE 4-1 The fire resistance of structural elements must be considered as the project is designed. The architect will determine the required rating for each element using Table 601 from the IBC. The designer must then devise a way to achieve that rating.

TABLE 1507.9.8
WOOD SHAKE WEATHER EXPOSURE AND ROOF SLOPE

ROOFING MATERIAL	LENGTH (inches)	GRADE	EXPOSURE (inches) 4:12 PITCH OR STEEPER
Shakes of naturally durable wood	18 24	No. 1 No. 1	7.5 10[a]
Preservative-treated taper sawn shakes of Southern yellow pine	18 24	No. 1 No. 1	7.5 10
	18 24	No. 2 No. 2	5.5 7.5
Taper sawn shakes of naturally durable wood	18 24	No. 1 No. 1	7.5 10
	18 24	No. 2 No. 2	5.5 7.5

For SI: 1 inch = 25.4 mm.
a. For 24-inch by 0.375-inch handsplit shakes, the maximum exposure is 7.5 inches.

FIGURE 4-2 The primary focus of Chapter 15 of the IBC is roof coverings. Table 1507.9.8 of the IBC shows the allowable exposure for different lengths and types of wood shakes on a 4 in 12 or greater roof pitch. The table will guide the project designer as specifications are developed for roof materials. *Courtesy 2009* International Building Code, *copyright © 2009. Washington, DC. International Code Council. Reproduced with permission. All rights reserved. www.iccsafe.org.*

- Chapter 12, INTERIOR ENVIRONMENT, contains minimum requirements for light and ventilation, including natural, artificial, and mechanical types. It also discusses minimum room sizes and ceiling heights.

Building Systems

Several chapters in the code describe roof, wall, and floor systems. The primary focus of Chapter 15, ROOF ASSEMBLIES AND ROOFTOP STRUCTURES, describes roof coverings. Minimum roof slope and maximum exposure to the weather are outlined with tables similar to Figure 4-2. The fire classification of the roof, as shown in Figure 4-3, is also discussed as well as roof flashing and roof drainage.

Chapter 16, STRUCTURAL DESIGN, lists loads and loading conditions that a building must be designed to resist. The chapter is primarily intended for the engineer, but knowledge of the loads acting

TABLE 1505.1[a, b]
MINIMUM ROOF COVERING CLASSIFICATION FOR TYPES OF CONSTRUCTION

IA	IB	IIA	IIB	IIIA	IIIB	IV	VA	VB
B	B	B	C[c]	B	C[c]	B	B	C[c]

For SI: 1 foot = 304.8 mm, 1 square foot = 0.0929 m^2.
a. Unless otherwise required in accordance with the *International Wildland-Urban Interface Code* or due to the location of the building within a fire district in accordance with Appendix D.
b. Nonclassified roof coverings shall be permitted on buildings of Group R-3 and Group U occupancies, where there is a minimum fire-separation distance of 6 feet measured from the leading edge of the roof.
c. Buildings that are not more than two stories above grade plane and having not more than 6,000 square feet of projected roof area and where there is a minimum 10-foot fire-separation distance from the leading edge of the roof to a lot line on all sides of the building, except for street fronts or public ways, shall be permitted to have roofs of No. 1 cedar or redwood shakes and No. 1 shingles.

FIGURE 4-3 Roofs are classified for their ability to resist fire. Different building types are required to have different classifications of roofs for fire safety reasons. Table 1505.1 shows this requirement. *Courtesy 2009* International Building Code, *copyright © 2009. Washington, DC. International Code Council. Reproduced with permission. All rights reserved. www.iccsafe.org.*

on each portion of the structure will add to the technician's understanding of the project. Figure 4-4 shows required live loads for which different types of occupancies must be designed. Chapter 13 of this text will explore how the engineer makes use of this information in designing the structure. Snow loads, wind loads, and earthquake loads will be discussed later in this chapter as you are introduced to how the architect begins the preliminary design of a structure.

TABLE 1607.1
MINIMUM UNIFORMLY DISTRIBUTED LIVE LOADS, L_o, AND MINIMUM CONCENTRATED LIVE LOADS[g]

OCCUPANCY OR USE	UNIFORM (psf)	CONCENTRATED (lbs.)
1. Apartments (see residential)	—	—
2. Access floor systems Office use Computer use	 50 100	 2,000 2,000
3. Armories and drill rooms	150	—
4. Assembly areas and theaters Fixed seats (fastened to floor) Follow spot, projections and control rooms Lobbies Movable seats Stages and platforms Other assembly areas	 60 50 100 100 125 100	—
5. Balconies (exterior) and decks[h]	Same as occupancy served	—
6. Bowling alleys	75	—
7. Catwalks	40	300
8. Cornices	60	—
9. Corridors, except as otherwise indicated	100	—
10. Dance halls and ballrooms	100	—
11. Dining rooms and restaurants	100	—
12. Dwellings (see residential)	—	—
13. Elevator machine room grating (on area of 4 in^2)	—	300
14. Finish light floor plate construction (on area of 1 in^2)	—	200
15. Fire escapes On single-family dwellings only	100 40	—
16. Garages (passenger vehicles only) Trucks and buses	40 See Section 1607.6	Note a
17. Grandstands (see stadium and arena bleachers)	—	—
18. Gymnasiums, main floors and balconies	100	—
19. Handrails, guards and grab bars	See Section 1607.7	
20. Hospitals Corridors above first floor Operating rooms, laboratories Patient rooms	 80 60 40	 1,000 1,000 1,000
21. Hotels (see residential)	—	—
22. Libraries Corridors above first floor Reading rooms Stack rooms	 80 60 150[b]	 1,000 1,000 1,000

TABLE 1607.1—continued
MINIMUM UNIFORMLY DISTRIBUTED LIVE LOADS, L_o, AND MINIMUM CONCENTRATED LIVE LOADS[g]

OCCUPANCY OR USE	UNIFORM (psf)	CONCENTRATED (lbs.)
23. Manufacturing Heavy Light	 250 125	 3,000 2,000
24. Marquees	75	—
25. Office buildings Corridors above first floor File and computer rooms shall be designed for heavier loads based on anticipated occupancy Lobbies and first-floor corridors Offices	 80 — 100 50	 2,000 — 2,000 2,000
26. Penal institutions Cell blocks Corridors	 40 100	—
27. Residential One- and two-family dwellings Uninhabitable attics without storage[i] Uninhabitable attics with limited storage[i, j, k] Habitable attics and sleeping areas All other areas Hotels and multifamily dwellings Private rooms and corridors serving them Public rooms and corridors serving them	 10 20 30 40 40 100	—
28. Reviewing stands, grandstands and bleachers	Note c	
29. Roofs All roof surfaces subject to maintenance workers Awnings and canopies Fabric construction supported by a lightweight rigid skeleton structure All other construction Ordinary flat, pitched, and curved roofs Primary roof members, exposed to a work floor Single panel point of lower chord of roof trusses or any point along primary structural members supporting roofs: Over manufacturing, storage warehouses, and repair garages All other occupancies Roofs used for other special purposes Roofs used for promenade purposes Roofs used for roof gardens or assembly purposes	 5 nonreducible 20 20 Note 1 60 100	 300 2,000 300 Note 1
30. Schools Classrooms Corridors above first floor First-floor corridors	 40 80 100	 1,000 1,000 1,000
31. Scuttles, skylight ribs and accessible ceilings	—	200
32. Sidewalks, vehicular driveways and yards, subject to trucking	250[d]	8,000[e]
33. Skating rinks	100	—

FIGURE 4-4 How the structure will be used will determine the live loads that must be designed for by the structural engineer. Live loads are an estimate of the load that will be imposed by such things as the occupants, furniture, and equipment. If architects think that these loads might be exceeded, they should inform the engineers. *Courtesy* 2009 International Building Code, *copyright © 2009. Washington, DC. International Code Council. Reproduced with permission. All rights reserved. www.iccsafe.org.*

(continued)

TABLE 1607.1—continued
MINIMUM UNIFORMLY DISTRIBUTED LIVE LOADS, L_o, AND MINIMUM CONCENTRATED LIVE LOADS[g]

OCCUPANCY OR USE	UNIFORM (psf)	CONCENTRATED (lbs.)
34. Stadiums and arenas Bleachers Fixed seats (fastened to floor)	 100[c] 60[c]	 —
35. Stairs and exits One- and two-family dwellings All other	 40 100	Note f
36. Storage warehouses (shall be designed for heavier loads if required for anticipated storage) Heavy Light	 250 125	
37. Stores Retail First floor Upper floors Wholesale, all floors	 100 75 125	 1,000 1,000 1,000
38. Vehicle barrier systems	See Section 1607.7.3	
39. Walkways and elevated platforms (other than exitways)	60	—
40. Yards and terraces, pedestrians	100	—

For SI: 1 inch = 25.4 mm, 1 square inch = 645.16 mm², 1 square foot = 0.0929 m², 1 pound per square foot = 0.0479 kN/m², 1 pound = 0.004448 kN, 1 pound per cubic foot = 16 kg/m³

a. Floors in garages or portions of buildings used for the storage of motor vehicles shall be designed for the uniformly distributed live loads of Table 1607.1 or the following concentrated loads: (1) for garages restricted to passenger vehicles accommodating not more than nine passengers, 3,000 pounds acting on an area of 4.5 inches by 4.5 inches; (2) for mechanical parking structures without slab or deck which are used for storing passenger vehicles only, 2,250 pounds per wheel.

b. The loading applies to stack room floors that support nonmobile, double-faced library bookstacks, subject to the following limitations:
   1. The nominal bookstack unit height shall not exceed 90 inches;
   2. The nominal shelf depth shall not exceed 12 inches for each face; and
   3. Parallel rows of double-faced bookstacks shall be separated by aisles not less than 36 inches wide.

c. Design in accordance with the ICC 300.

d. Other uniform loads in accordance with an approved method which contains provisions for truck loadings shall also be considered where appropriate.

e. The concentrated wheel load shall be applied on an area of 4.5 inches by 4.5 inches.

f. Minimum concentrated load on stair treads (on area of 4 square inches) is 300 pounds.

g. Where snow loads occur that are in excess of the design conditions, the structure shall be designed to support the loads due to the increased loads caused by drift buildup or a greater snow design determined by the building official (see Section 1608). For special-purpose roofs, see Section 1607.11.2.2.

h. See Section 1604.8.3 for decks attached to exterior walls.

i. Attics without storage are those where the maximum clear height between the joist and rafter is less than 42 inches, or where there are not two or more adjacent trusses with the same web configuration capable of containing a rectangle 42 inches high by 2 feet wide, or greater, located within the plane of the truss. For attics without storage, this live load need not be assumed to act concurrently with any other live load requirements.

j. For attics with limited storage and constructed with trusses, this live load need only be applied to those portions of the bottom chord where there are two or more adjacent trusses with the same web configuration capable of containing a rectangle 42 inches high by 2 feet wide or greater, located within the plane of the truss. The rectangle shall fit between the top of the bottom chord and the bottom of any other truss member, provided that each of the following criteria is met:
   i. The attic area is accessible by a pull-down stairway or framed opening in accordance with Section 1209.2, and
   ii. The truss shall have a bottom chord pitch less than 2:12.
   iii. Bottom chords of trusses shall be designed for the greater of actual imposed dead load or 10 psf, uniformly distributed over the entire span.

k. Attic spaces served by a fixed stair shall be designed to support the minimum live load specified for habitable attics and sleeping rooms.

l. Roofs used for other special purposes shall be designed for appropriate loads as approved by the building official.

FIGURE 4-4 *(continued)* How the structure will be used will determine the live loads that must be designed for by the structural engineer. Live loads are an estimate of the load that will be imposed by such things as the occupants, furniture, and equipment. If architects think that these loads might be exceeded, they should inform the engineers. *Courtesy 2009* International Building Code, *copyright © 2009. Washington, DC. International Code Council. Reproduced with permission. All rights reserved. www.iccsafe.org.*

Chapter 18, SOIL AND FOUNDATIONS, describes foundations and retaining walls. Foundation investigations, which are usually done by a soils testing lab, are discussed. Requirements for standard foundations, such as minimum thickness of foundation walls (see Figure 4-5) and allowable foundation loads are also addressed. A substantial portion of this chapter covers the design of pile foundations. Chapters 11 and 24 of this text will explain how each of these systems affects a CAD drafter.

Building Materials

Once you feel at ease with the guidelines of building systems, work through the chapters that deal with specific types of materials. Wood is one of the most common materials of light construction and is covered in Chapter 23 of the IBC. Although you may be familiar with parts of these chapters from experience with residential construction, commercial construction will require a greater depth of understanding. This chapter has a lot of technical design information for the engineer, but it also has many tables and standard construction provisions that can be used by the technician. Chapters 7 and 8 of this text will explore wood, engineered lumber, and timber construction methods.

Chapter 22 of the IBC, STEEL AND ITS REFERENCE STANDARDS, is a very technical chapter intended for the engineer involved with the structural steel. Chapter 19 of the IBC, CONCRETE AND ITS REFERENCE STANDARDS, is also very complex material. That chapter sets the rules for the use of concrete and discusses highly technical subjects such as concrete mixes and seismic performance of concrete. Minimum required slab thickness, formwork, and durability are also addressed. An example of an important piece of information for the CAD drafter is shown in Figure 4-6. This is a portion of a table showing the minimum requirements for covering steel reinforcing with concrete. Unfortunately, this table is no longer contained directly in the code. The code has made reference to a document published by ACI, in this case ACI 318 Section 7.5, so you will need access to that document to obtain all of the appropriate information for dealing with concrete. This information is critical to developing proper details for reinforced concrete construction. Chapters 9 and 11 of this text will introduce materials and construction methods generally associated with steel and concrete construction.

Note: Chapter 35, REFERENCED STANDARDS, provides a list of the standards referenced throughout the building code. See the Additional Reading section at the end of this chapter for a partial list of these resources.

TABLE 1805.5(2)
8-INCH MASONRY FOUNDATION WALLS WITH REINFORCEMENT WHERE d ≥ 5 INCHES[a, b, c]

MAXIMUM WALL HEIGHT (feet-inches)	MAXIMUM UNBALANCED BACKFILL HEIGHT[d] (feet-inches)	VERTICAL REINFORCEMENT — Soil classes and lateral soil load[a] (psf per foot below natural grade)		
		GW, GP, SW and SP soils 30	GM, GC, SM, SM-SC and ML soils 45	SC, ML-CL and Inorganic CL soils 60
7-4	4-0 (or less)	#4 at 48″ o.c.	#4 at 48″ o.c.	#4 at 48″ o.c.
	5-0	#4 at 48″ o.c.	#4 at 48″ o.c.	#4 at 48″ o.c.
	6-0	#4 at 48″ o.c.	#5 at 48″ o.c.	#5 at 48″ o.c.
	7-4	#5 at 48″ o.c.	#6 at 48″ o.c.	#7 at 48″ o.c.
8-0	4-0 (or less)	#4 at 48″ o.c.	#4 at 48″ o.c.	#4 at 48″ o.c.
	5-0	#4 at 48″ o.c.	#4 at 48″ o.c.	#4 at 48″ o.c.
	6-0	#4 at 48″ o.c.	#5 at 48″ o.c.	#5 at 48″ o.c.
	7-0	#5 at 48″ o.c.	#6 at 48″ o.c.	#7 at 48″ o.c.
	8-0	#5 at 48″ o.c.	#6 at 48″ o.c.	#7 at 48″ o.c.
8-8	4-0 (or less)	#4 at 48″ o.c.	#4 at 48″ o.c.	#4 at 48″ o.c.
	5-0	#4 at 48″ o.c.	#4 at 48″ o.c.	#5 at 48″ o.c.
	6-0	#4 at 48″ o.c.	#5 at 48″ o.c.	#6 at 48″ o.c.
	7-0	#5 at 48″ o.c.	#6 at 48″ o.c.	#7 at 48″ o.c.
	8-8	#6 at 48″ o.c.	#7 at 48″ o.c.	#8 at 48″ o.c.
9-4	4-0 (or less)	#4 at 48″ o.c.	#4 at 48″ o.c.	#4 at 48″ o.c.
	5-0	#4 at 48″ o.c.	#4 at 48″ o.c.	#5 at 48″ o.c.
	6-0	#4 at 48″ o.c.	#5 at 48″ o.c.	#6 at 48″ o.c.
	7-0	#5 at 48″ o.c.	#6 at 48″ o.c.	#7 at 48″ o.c.
	8-0	#6 at 48″ o.c.	#7 at 48″ o.c.	#8 at 48″ o.c.
	9-4	#7 at 48″ o.c.	#8 at 48″ o.c.	#9 at 48″ o.c.
10-0	4-0 (or less)	#4 at 48″ o.c.	#4 at 48″ o.c.	#4 at 48″ o.c.
	5-0	#4 at 48″ o.c.	#4 at 48″ o.c.	#5 at 48″ o.c.
	6-0	#4 at 48″ o.c.	#5 at 48″ o.c.	#6 at 48″ o.c.
	7-0	#5 at 48″ o.c.	#6 at 48″ o.c.	#7 at 48″ o.c.
	8-0	#6 at 48″ o.c.	#7 at 48″ o.c.	#8 at 48″ o.c.
	9-0	#7 at 48″ o.c.	#8 at 48″ o.c.	#9 at 48″ o.c.
	10-0	#7 at 48″ o.c.	#9 at 48″ o.c.	#9 at 48″ o.c.

For SI: 1 inch = 25.4 mm, 1 foot = 304.8 mm, 1 pound per square foot per foot = 0.157kPa/m.

a. For design lateral soil loads, see Section 1610. Soil classes are in accordance with the Unified Soil Classification System and design lateral soil loads are for moist soil conditions without hydrostatic pressure.

b. Provisions for this table are based on construction requirements specified in Section 1805.5.2.2.

c. For alternative reinforcement, see Section 1805.5.3.

d. For height of unbalanced backfill, see Section 1805.5.1.2.

FIGURE 4-5 Table 1805.5(2) from the IBC shows the required reinforcing for masonry foundation walls of different heights with different unbalanced backfill heights. *Courtesy 2009* International Building Code, *copyright © 2009. Washington, DC. International Code Council. Reproduced with permission. All rights reserved. www.iccsafe.org.*

Structural design of masonry construction is addressed in IBC Chapter 21, MASONRY. Required reinforcing, grout and mortar strength, and construction with glass block and screen block are examined. Each of these subjects will be discussed in Chapter 10 of this text.

IBC Chapter 24, GLASS AND GLAZING, deals with structural requirements for safety glass and glass. It covers vertical and sloped glazing as well as unusual uses for glass such as glass handrails and racquetball courts.

RECOMMEND MINIMUM CONCRETE COVER

CONCRETE EXPOSURE	MINIMUM COVER Inches
1. Concrete cast against and permanently exposed to earth	3
2. Concrete exposed to earth or weather No. 6 through No. 18 bar No. 5 bar, W31 or D31 wire, and smaller	 2 1 1/2
3. Concrete not exposed to weather or in contact with ground Slabs, walls, joists: No. 14 and No. 18 bars No. 11 bar and smaller Beams, columns: Primary reinforcement, ties, stirrups, spirals Shells, folded plate members: No. 6 bar and larger No. 5 bar, W31 or D31 wire, and smaller	 1 1/2 3/4 1 1/2 3/4 1/2

FIGURE 4-6 Common recommended reinforcing steel coverage. Drafters must complete details that must meet or exceed minimum concrete cover for steel reinforcing. The code requirements for coverage of steel reinforcement are contained in section 7.7 of ACI 318 (referenced by, but not included in the current edition of the IBC).

USING THE CODE

To effectively use the code during the initial design phase, five basic classifications of the building must be determined:

- Occupancy group
- Location on the site
- Floor area
- Height or number of stories
- Construction type

There is no set order for determining these classifications when designing a building. The starting point depends on what is already known about the project.

Occupancy Classification

A good way to begin checking a project for code compliance is to determine what occupancy classification the building will fall into. Basic occupancy classifications are listed in IBC Chapter 3, USE AND OCCUPANCY CLASSIFICATION. The IBC defines 10 major occupancy classifications, and then breaks these classifications into subclassifications. All uses within a building must be categorized as one of these subclassifications as shown in Table 4-1. Remember that the following list is a summary of an entire chapter of the IBC, it is not all-inclusive; there are exceptions listed in the code. Two other things to consider when determining occupancy:

- If a use is not listed then it must be classified in the group that the occupancy most nearly resembles.
- Buildings may have more than one occupancy classification.

Location on the Property

Zoning ordinances will have a significant impact on where the building is located on the property. Where the building is located on the lot will also have an impact on the construction of the building. The location of the building in relationship to other buildings or the lot lines is very important in regard to fire safety and will affect the requirements for the fire resistance of exterior walls and the openings in those walls. The closer two buildings are located together, the more likely it is that a fire in one will damage the other. Larger space between buildings will also provide greater access for fire fighting equipment. The code has provisions for increasing the allowable size of a building when large yards are provided. For the purpose of area increases, a yard includes adjacent public streets, alleys, and other similar areas. Yards on adjacent lots do not count.

Figure 4-7 (Section 506 of IBC) contains the formulas for calculating the positive impact of large yards around the building on the allowable size of the building.

Assembly occupancies (A) represent buildings used for public assembly purposes such as theaters and churches, Including but not limited to:	
A-1	Structures used for the production and viewing such as motion picture theaters, symphony and concert halls, and television studios that have live audiences.
A-2	Structures used for consuming food and drink such as banquet halls, nightclubs, restaurants, and bars.
A-3	Structures used or intended for worship, recreation, or amusement; and assembly uses not listed elsewhere such as amusement arcades, art galleries, bowling alleys, places of religious worship, community halls, courtrooms, dance halls (without food and drink), funeral parlors, gymnasiums, indoor swimming pools without spectator seating, libraries, museums, and pool halls.
A-4	Structures used or intended for viewing indoor sporting events and activities such as arenas, skating rinks, swimming pools, and tennis courts.
A-5	Structures used for viewing outdoor sporting events and activities such as amusement park structures, bleachers, grandstands, and stadiums.
Business occupancies (B) represent buildings used for nonhazardous business, such as an office building or buildings that provide the following services:	
	Animal hospitals, banks, barber or beauty shops, car washes, civic administration, outpatient clinics, dry cleaners and laundry, educational occupancies above the twelfth grade, laboratories, car showrooms, post offices, offices offering professional services such as an architect, engineer, physician, or attorneys.
Education occupancies (E) represent buildings used for educational purposes by six or more occupants through the twelfth grade including day care.	
Factory occupancies (F) represent buildings used for industrial factories, including but limited to the following uses:	
F-1	Factory uses that involve moderately hazardous uses that are not classified as H, such as aircraft, appliances, bakeries, bicycles, cameras, canvas, clothing, electronics, food processing, furniture, leather products, millwork, plastic products, shoes, textiles, tobacco, and cabinet making.
F-2	Factory uses that involve low hazardous uses that involve the manufacturing of noncombustible materials such as beverages containing up to 12% alcohol, brick, ceramics, foundries, glass products, gypsum, ice, and metal products.
Hazardous occupancies (H) represent buildings involving highly hazardous occupancies, such as areas where large volumes of flammable liquids or highly toxic materials are stored or used including:	
H-1	Buildings containing materials that pose a detonation hazard.
H-2	Buildings that contain a deflagration or accelerated burning hazard.
H-3	Buildings that contain materials that readily support combustion or a physical hazard.
H-4	Buildings that contain health hazards such as corrosives and toxic or highly toxic materials.
H-5	Semiconductor fabrication facilities.
Institutional occupancies (I) represent buildings for institutional uses where people have restricted movement due to physical health or age or because they are being detained.	
I-1	Uses such as residential board and care, assisted living facilities, halfway houses, group homes, social rehabilitee facilities, alcohol or drug centers, and convalescent facilities.
I-2	24-hour medical care facility uses such as hospitals, nursing homes, mental hospitals, and detoxification facilities.
I-3	Building used by people who are under constraint such as prisons, jails reformatories, correction centers, and prerelease centers.
Mercantile occupancies (M) represent buildings used for sales of merchandise such as department stores and markets, gas stations, retail or wholesale stores.	
Residential occupancies (R) represent buildings used for single or multifamily residential usage.	
R-1	Residential occupancies containing sleeping units primarily for short-term stays such as hotels and motels.
R-2	Multiple residential uses for long-term stays such as apartment houses, boarding houses, convents, dormitories, hotels, and vacation time-share properties.
R-3	Single- or two-unit residences or care facilities with no more than 5 people for less than 24 hours and congregate living facilities with less than 16 people.
R-4	Residential care or assisted living facilities with more than 5 but not more than 16 occupants plus staff.
Storage occupancies (S) classify buildings used for storage of moderate- and low-hazard materials	
S-1	Storage of moderate-hazard materials, such as those products produced in an F-1 manufacturing plant.
S-2	Storage of low-hazard materials, such as aircraft hangars and those products produced in an F-2 manufacturing plant.
Utility occupancies (U) are miscellaneous structures such as agriculture buildings, aircraft hangars accessory to a one- or two-family residence, barns, carports, fences more than 6′ high, grain silos accessory to a residential occupancy, greenhouses, livestock shelters, private garages, retaining walls, sheds, stables, tanks, and towers.	

TABLE 4-1 Occupancy Classifications of the IBC

SECTION 506
BUILDING AREA MODIFICATIONS

506.1 General. The *building areas* limited by Table 503 shall be permitted to be increased due to frontage (I_f) and *automatic sprinkler system* protection (I_s) in accordance with the following:

$$A_a = \{A_t + [A_t \times I_f] + [A_t \times I_s]\} \quad \textbf{(Equation 5-1)}$$

where:

A_a = Allowable *building area* per *story* (square feet).

A_t = Tabular *building area* per *story* in accordance with Table 503 (square feet).

I_f = Area increase factor due to frontage as calculated in accordance with Section 506.2.

I_s = Area increase factor due to sprinkler protection as calculated in accordance with Section 506.3.

506.2 Frontage increase. Every building shall adjoin or have access to a *public way* to receive a *building area* increase for frontage. Where a building has more than 25 percent of its perimeter on a *public way* or open space having a minimum width of 20 feet (6096 mm), the frontage increase shall be determined in accordance with the following:

$$I_f = [F/P - 0.25]W/30 \quad \textbf{(Equation 5-2)}$$

where:

I_f = Area increase due to frontage.

F = Building perimeter that fronts on a *public way* or open space having 20 feet (6096 mm) open minimum width (feet).

P = Perimeter of entire building (feet).

W = Width of *public way* or open space (feet) in accordance with Section 506.2.1.

506.2.1 Width limits. The value of W shall be at least 20 feet (6096 mm). Where the value of W varies along the perimeter of the building, the calculation performed in accordance with Equation 5-2 shall be based on the weighted average of each portion of *exterior wall* and open space where the value of W is greater than or equal to 20 feet (6096 mm). Where the value of W exceeds 30 feet (9144 mm), a value of 30 feet (9144 mm) shall be used in calculating the weighted average, regardless of the actual width of the open space. Where two or more buildings are on the same lot, W shall be measured from the exterior face of a building to the exterior face of an opposing building, as applicable.

Exception: The value of W divided by 30 shall be permitted to be a maximum of 2 when the building meets all requirements of Section 507 except for compliance with the 60-foot (18 288 mm) *public way* or *yard* requirement, as applicable.

506.2.2 Open space limits. Such open space shall be either on the same lot or dedicated for public use and shall be accessed from a street or *approved fire lane*.

506.3 Automatic sprinkler system increase. Where a building is equipped throughout with an *approved automatic sprinkler system* in accordance with Section 903.3.1.1, the *building area* limitation in Table 503 is permitted to be increased by an additional 200 percent ($I_s = 2$) for buildings with more than one *story above grade plane* and an additional 300 percent ($I_s = 3$) for buildings with no more than one *story above grade plane*. These increases are permitted in addition to the height and *story* increases in accordance with Section 504.2.

Exception: The *building area* limitation increases shall not be permitted for the following conditions:

1. The *automatic sprinkler system* increase shall not apply to *buildings* with an occupancy in Group H-1.
2. The *automatic sprinkler system* increase shall not apply to the *building area* of an occupancy in Group H-2 or H-3. For *buildings* containing such occupancies, the allowable *building area* shall be determined in accordance with Section 508.4.2, with the sprinkler system increase applicable only to the portions of the building not classified as Group H-2 or H-3.
3. *Fire-resistance rating* substitution in accordance with Table 601, Note d.

506.4 Single occupancy buildings with more than one story. The total allowable *building area* of a single occupancy building with more than one *story above grade plane* shall be determined in accordance with this section. The actual aggregate *building area* at all *stories* in the building shall not exceed the total allowable *building area*.

Exception: A single basement need not be included in the total allowable *building area*, provided such basement does not exceed the area permitted for a building with no more than one *story above grade plane*.

506.4.1 Area determination. The total allowable *building area* of a single occupancy building with more than one *story above grade plane* shall be determined by multiplying the allowable *building area* per *story* (A_a), as determined in Section 506.1, by the number of *stories above grade plane* as listed below:

1. For buildings with two *stories above grade plane*, multiply by 2;
2. For buildings with three or more *stories above grade plane*, multiply by 3; and
3. No *story* shall exceed the allowable *building area* per *story* (A_a), as determined in Section 506.1, for the occupancies on that *story*.

Exceptions:

1. Unlimited area buildings in accordance with Section 507.
2. The maximum area of a building equipped throughout with an *automatic sprinkler system* in accordance with Section 903.3.1.2 shall be determined by multiplying the allowable area per *story* (A_a), as determined in Section 506.1, by the number of *stories above grade plane*.

FIGURE 4-7 The size of a structure can often be increased because of the number and size of yards, or the use of fire sprinklers. The formulas provided by the IBC in Section 506 can be used for calculating the allowable building size, taking yards and sprinklers into account. *Courtesy 2009* International Building Code, *copyright © 2009. Washington, DC. International Code Council. Reproduced with permission. All rights reserved. www.iccsafe.org.*

With the increase in the size of yards, fire safety is increased and therefore the code allows larger buildings to be built.

It might be beneficial to increase the size of a yard to avoid required fire-resistive construction. Columns 3 and 4 of Figure 4-8 (Table 602 of IBC) call for fire-resistive walls when yards become less than a given size. Openings in walls, such as doors and windows, may need to be fire rated (protected) when they open onto small yards. This is discussed in IBC Chapter 7, FIRE-RESISTANT-RATED CONSTRUCTION.

Allowable Floor Area, Height and Number of Stories, Type of Construction

Allowable floor area, height and number of stories, and construction type are three factors that need to be worked out together. After the building has been laid out on the site, you will usually have a building footprint determined. The number of stories needed to achieve the required size of the building for the project can then be determined.

The next challenge is to find an allowable way to construct the building. The code will classify a building according to the materials it is built with. Buildings made of wood frame construction in general are not allowed to be as large or as tall as buildings built of concrete and steel, because wood is not as stable and permanent as steel and concrete. Buildings that have their structural members protected from fire are allowed to be larger and taller than buildings without such fire protection. Referring to Figure 4-1, it can be seen that construction types with an "A" following are required to be more fire-resistant than those with a "B." Type III-A is required to be more fire-resistant than Type III-B. The code has tables that specify allowable areas for a variety of construction types. These values will be different for each occupancy classification. Figure 4-9 shows the square footage allowed by the IBC for each occupancy group and each construction type. According to that table, a "B" occupancy of Type III-A construction has a **basic allowable area** of 28,500 square feet. If the "B" occupancy is built of Type III-B construction its basic allowable area is only 19,000 square feet. Why would anyone build with Type III-B construction instead of Type III-A construction? Because Type III-B is less expensive to build than Type III-A.

Notice that the allowable building sizes listed above were referred to as "basic allowable area." These values are a starting point, a basic building area. The basic allowable building area can be increased based on the size of the yard around the building and if fire sprinklers are added, as described in Figure 4-7.

APPLYING THE CODE

The information that has just been described can be applied to the site plan in Figure 4-10. The building is to be a professional office building, two stories high, with 30,000 square feet on each floor. Table 4-1, which is a summary of IBC Section 304, lists office buildings as a Type B occupancy. Now it must be determined how

TABLE 602
FIRE-RESISTANCE RATING REQUIREMENTS FOR EXTERIOR WALLS BASED ON FIRE SEPARATION DISTANCE[a, e]

FIRE SEPARATION DISTANCE = X (feet)	TYPE OF CONSTRUCTION	OCCUPANCY GROUP H[f]	OCCUPANCY GROUP F-1, M, S-1[g]	OCCUPANCY GROUP A, B, E, F-2, I, R, S-2[g], U[b]
X < 5[c]	All	3	2	1
5 ≤ X <10	IA Others	3 2	2 1	1 1
10 ≤ X < 30	IA, IB IIB, VB Others	2 1 1	1 0 1	1[d] 0 1[d]
X ≥ 30	All	0	0	0

For SI: 1 foot = 304.8 mm.
a. Load-bearing exterior walls shall also comply with the fire-resistance rating requirements of Table 601.
b. For special requirements for Group U occupancies, see Section 406.1.2.
c. See Section 706.1.1 for party walls.
d. Open parking garages complying with Section 406 shall not be required to have a fire-resistance rating.
e. The fire-resistance rating of an exterior wall is determined based upon the fire separation distance of the exterior wall and the story in which the wall is located.
f. For special requirements for Group H occupancies, see Section 415.3.
g. For special requirements for Group S aircraft hangars, see Section 412.4.1.

FIGURE 4-8 In addition to the fire-resistance ratings shown in Figure 4-1, exterior walls are required to have fire-resistance ratings based on the building occupancy classification, construction type, and the fire separation distance per Table 602. The fire-separation distance is the distance to a real or imaginary property line. If the building faces public open space such as a street, the distance can be measured to the centerline. *Courtesy 2009 International Building Code, copyright © 2009. Washington, DC. International Code Council. Reproduced with permission. All rights reserved. www.iccsafe.org.*

TABLE 503
ALLOWABLE BUILDING HEIGHTS AND AREAS[a]
Building height limitations shown in feet above grade plane. Story limitations shown as stories above grade plane.
Building area limitations shown in square feet, as determined by the definition of "Area, building," per story

		TYPE OF CONSTRUCTION								
		TYPE I		TYPE II		TYPE III		TYPE IV	TYPE V	
		A	B	A	B	A	B	HT	A	B
	HEIGHT(feet)	UL	160	65	55	65	55	65	50	40
GROUP		STORIES(S) AREA (A)								
A-1	S	UL	5	3	2	3	2	3	2	1
	A	UL	UL	15,500	8,500	14,000	8,500	15,000	11,500	5,500
A-2	S	UL	11	3	2	3	2	3	2	1
	A	UL	UL	15,500	9,500	14,000	9,500	15,000	11,500	6,000
A-3	S	UL	11	3	2	3	2	3	2	1
	A	UL	UL	15,500	9,500	14,000	9,500	15,000	11,500	6,000
A-4	S	UL	11	3	2	3	2	3	2	1
	A	UL	UL	15,500	9,500	14,000	9,500	15,000	11,500	6,000
A-5	S	UL	UL	UL	UL	UL	UL	UL	UL	UL
	A	UL	UL	UL	UL	UL	UL	UL	UL	UL
B	S	UL	11	5	3	5	3	5	3	2
	A	UL	UL	37,500	23,000	28,500	19,000	36,000	18,000	9,000
E	S	UL	5	3	2	3	2	3	1	1
	A	UL	UL	26,500	14,500	23,500	14,500	25,500	18,500	9,500
F-1	S	UL	11	4	2	3	2	4	2	1
	A	UL	UL	25,000	15,500	19,000	12,000	33,500	14,000	8,500
F-2	S	UL	11	5	3	4	3	5	3	2
	A	UL	UL	37,500	23,000	28,500	18,000	50,500	21,000	13,000
H-1	S	1	1	1	1	1	1	1	1	NP
	A	21,000	16,500	11,000	7,000	9,500	7,000	10,500	7,500	NP
H-2[d]	S	UL	3	2	1	2	1	2	1	1
	A	21,000	16,500	11,000	7,000	9,500	7,000	10,500	7,500	3,000
H-3[d]	S	UL	6	4	2	4	2	4	2	1
	A	UL	60,000	26,500	14,000	17,500	13,000	25,500	10,000	5,000
H-4	S	UL	7	5	3	5	3	5	3	2
	A	UL	UL	37,500	17,500	28,500	17,500	36,000	18,000	6,500
H-5	S	4	4	3	3	3	3	3	3	2
	A	UL	UL	37,500	23,000	28,500	19,000	36,000	18,000	9,000
I-1	S	UL	9	4	3	4	3	4	3	2
	A	UL	55,000	19,000	10,000	16,500	10,000	18,000	10,500	4,500
I-2	S	UL	4	2	1	1	NP	1	1	NP
	A	UL	UL	15,000	11,000	12,000	NP	12,000	9,500	NP
I-3	S	UL	4	2	1	2	1	2	2	1
	A	UL	UL	15,000	10,000	10,500	7,500	12,000	7,500	5,000
I-4	S	UL	5	3	2	3	2	3	1	1
	A	UL	60,500	26,500	13,000	23,500	13,000	25,500	18,500	9,000
M	S	UL	11	4	2	4	2	4	3	1
	A	UL	UL	21,500	12,500	18,500	12,500	20,500	14,000	9,000
R-1	S	UL	11	4	4	4	4	4	3	2
	A	UL	UL	24,000	16,000	24,000	16,000	20,500	12,000	7,000
R-2	S	UL	11	4	4	4	4	4	3	2
	A	UL	UL	24,000	16,000	24,000	16,000	20,500	12,000	7,000
R-3	S	UL	11	4	4	4	4	4	3	3
	A	UL	UL	UL	UL	UL	UL	UL	UL	UL
R-4	S	UL	11	4	4	4	4	4	3	2
	A	UL	UL	24,000	16,000	24,000	16,000	20,500	12,000	7,000
S-1	S	UL	11	4	2	3	2	4	3	1
	A	UL	48,000	26,000	17,500	26,000	17,500	25,500	14,000	9,000
S-2[b, c]	S	UL	11	5	3	4	3	5	4	2
	A	UL	79,000	39,000	26,000	39,000	26,000	38,500	21,000	13,500
U[c]	S	UL	5	4	2	3	2	4	2	1
	A	UL	35,500	19,000	8,500	14,000	8,500	18,000	9,000	5,500

For SI: 1 foot = 304.8 mm, 1 square foot = $0.0929\ m^2$.

A = building area per story, S = stories above grade plane, UL = Unlimited, NP = Not permitted.

a. See the following sections for general exceptions to Table 503:
1. Section 504.2, Allowable building height and story increase due to automatic sprinkler system installation.
2. Section 506.2, Allowable building area increase due to street frontage.
3. Section 506.3, Allowable building area increase due to automatic sprinkler system installation.
4. Section 507, Unlimited area buildings.

b. For open parking structures, see Section 406.3.

c. For private garages, see Section 406.1.

d. See Section 415.5 for limitations.

FIGURE 4-9 The basic size and height of a structure is affected by the fire resistance of the materials used. The architect must consider the materials of construction very early in the design process. Type I-A construction is the most restrictive construction, meaning that materials used are the least likely to burn. Type V-B is the least restrictive construction, allowing materials that have a high risk of fire damage to be used. Table 503 shows the basic or tabular value for the allowable area of a building based on the construction type and occupancy. *Courtesy 2009 International Building Code, copyright © 2009. Washington, DC. International Code Council. Reproduced with permission. All rights reserved. www.iccsafe.org.*

to achieve 30,000 square feet per floor. Referring to Figure 4-9 (Table 503) under B, it can be seen that only Type I-A, Type I-B, Type II-A, and Type-IV construction have large enough basic allowable areas for this building. These types of construction are relatively expensive; therefore, less costly answers should be explored. As mentioned earlier, the basic allowable area can be increased by installing fire sprinklers and by having large yards around the building.

Based on Figure 4-7 (Section 506 of the IBC), the building size can be increased if it has enough of its walls fronting on yards larger than 20' (6100 mm). The building in Figure 4-10 has 452' of wall facing a yard larger than 20' (212' [192+20] of south walls +150' of east wall +90' of west wall). In the second formula shown in Figure 4-7, 452 will be *F*. The building has a total of 724' of exterior wall (perimeter). In the formula, 724 will be *P*.

Note: When considering the size of yards for increasing the allowable building area, there is often confusion about how much of the yard may be used when a yard faces a street or when the yard is between two buildings on the same lot. Often designers and building officials try to use the assumed property line that is established for measuring the distance to openings. This assumed property line is not used for yard as applied to area increases. Yards between two buildings on the same lot are measured from building to building. You get to use the entire yard width. For yards facing streets, you may use the entire yard on your site plus the entire width of the street. Not just to the centerline, but the entire street.

Once the values for *F* and *P* have been determined, the sizes of the three yards that are being used for the area increase need to be examined. If the building meets the following criteria per IBC Sections 506.1 and 507, then the maximum yard that can be used is 60'; otherwise the maximum yard that can be used is 30'.

1. The building is one-story, is not sprinklered, and the occupancies are limited to F-2 or S-2 occupancies.
2. The building is one-story, is sprinklered, is limited to B, F, M, or S occupancy; or is one-story, sprinklered, A-4 occupancy, not of Type V construction.
3. The building is two-story and is limited to B, F, M, or S occupancies.

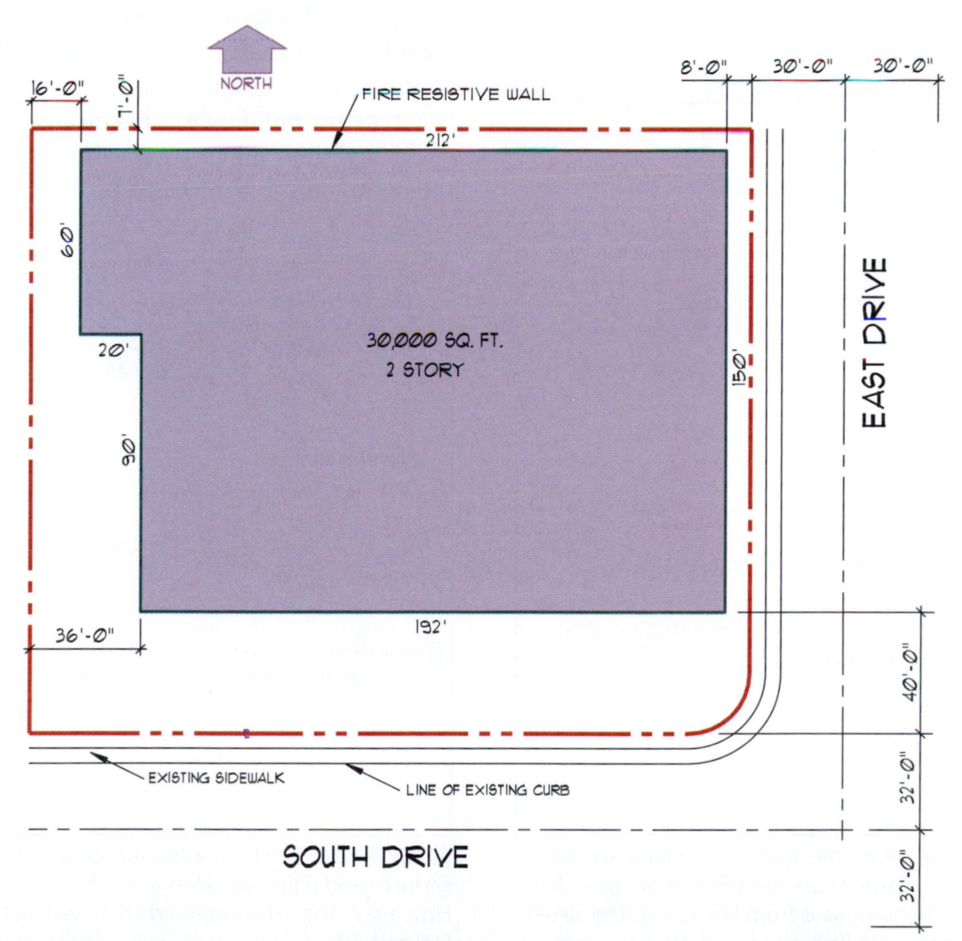

FIGURE 4-10 The site plan for a proposed structure can be useful in visualizing construction requirements.

4. An A-3 occupancy building of Type II construction that is sprinklered, does not have a stage other than a platform, is within 21" of grade, and has all exits accessible.
5. Some buildings containing a limited amount of H occupancies (see IBC Sections 507.8 and 507.9).
6. An E occupancy one-story building meeting the requirements of IBC 507.10.
7. For motion picture theaters and covered malls, see IBC 507.11 and 507.12.

A two-story non-sprinklered office building is limited to the 30' maximum yard. All of the yards that are greater than 20' are also greater than 30' so in the second formula of Figure 4-7, 30 will be W. The calculations in Figure 4-11 show that the basic allowable area can be increased by 37%.

Now go to Table 503 (Figure 4-9) and look at the available options. By multiplying the allowable area of a Type V-B building by 1.37, a 12,330 square foot area is allowed (9000 × 1.37); with Type V-A, 24,660 square feet are allowed. Neither of the numbers is large enough. Using the same process with Types III-B and II-B results in 26,030 and 31,510, respectively. Type II-B is large enough. Type III-A is also large enough at 39,045 sq ft. There are now six construction types from which to choose.

Another option for increasing the allowable area of the building is to install a fire sprinkler system. This option has several other benefits, such as allowing longer dead-end corridors and unrated corridors. The first formula in Figure 4-7 gives the maximum allowable area after considering yard increases and fire sprinkler increases. This formula looks rather complicated at first, but it really is not that difficult to use. To determine I_f, W must first be determined. In this example the maximum value of 60' may be used because Section 506.2.1 has an exception. This exception references Section 507. This building meets the conditions described in Section 507.4 (the requirement for 60' yards on all sides is waved by section 506.2.1). Because one of the yards is less than 60', a weighted average calculation must be used. Figure 4-12 illustrates the calculation for the least restrictive construction method, Type V-B construction. Because this example results in an area greater than what is needed, it will not be necessary to work out the allowable area for any other construction type. Using a spreadsheet to solve this formula allows you to explore several solutions in a very short period of time. A spreadsheet has been provided on the Student CD in the Resource Material/Chapter 4 folder. Try using it with Figure 4-10. If you do not use the west yard, you will actually get a larger allowable area.

Reading Figure 4-8 (Table 602 of the IBC) for a B occupancy classification, it can be seen that walls facing a yard smaller than 10' (6100 mm) will have to be built of fire-resistive construction no matter what construction type the building is. This means that the north wall of the building will have to be fire-resistive. A portion of the west wall will have to be fire-resistive for many

$I_f = [F/P - 0.25]W/30$

$F = 452$

$P = 724$

$W = 30$

Divide 724 into 452

$I_f = [452/724 - 0.25]\ 30/30$

Divide 30 into 30

The answer rounded to 2 decimal places is close enough

$I_f = [.62 - 0.25]1$

Work inside the brackets

$I_f = [.37]1$

Multiply .37 x 1

$I_f = .37$

FIGURE 4-11 The basic allowable area of a building may be increased if unoccupied yards of greater than 20' are provided around the building. The formula is from Figure 4-7. The values entered are based on the building in Figure 4-10.

Wall yard smaller than allowed maximum for calculations x width of yard

Weighted Average for $W = (90 \times 36 + 362 \times 60) / 452$

Perimeter facing yards greater than 20' – length shown above

$W = (3240 + 21720) / 452$

$W = 24960 / 452$

$W = 55.22$

Allowable Area

$A_a = A_t + [A_t \times I_f] + [A_t \times I_s]$

$A_t = 9000$ from Figure 4.9 for a B occupancy of Type V-B construction

$I_f = [452 / 724 - 0.25]\ 55.22 / 30 = .6890$

$I_s = 2$ From Figure 4.7 506.3 for a multi story building

Work inside the brackets first

$A_a = 9000 + [9000 \times .6890] + [9000 \times 2]$

$A_a = 9000 + [6201] + [18000]$

$A_a = 33201$

FIGURE 4-12 The basic allowable area of a building can also be increased if fire sprinklers are installed. The formula is from Figure 4-7. The values entered are based on the building in Figure 4-10.

construction types but not for Type V-B. Chapter 12 of this text will introduce methods to achieve specific fire ratings. The east wall of the building will not have to be fire-resistive (unless a fire rated construction type is chosen for the entire building) even though it is only 8' (3050 mm) from the property line, because one half of the street can be counted as part of the yard (the entire width of the street is counted for yard increases). Openings on the north side of the buildings will be allowed, but they must be fire-resistive (protected). This is controlled by IBC Section 704.8. Portions of the west wall will have restrictions on openings. The restrictions on both the north and west walls will vary with the construction type chosen. Again, unrestricted unprotected openings are allowed on the east side of the building because one half of the street may be used as a part of the required yard.

MIXED OCCUPANCY

Buildings often contain more than one occupancy classification. This complicates their design in several ways. If an 8000 square foot community hall is to be provided on the lower floor of the office building just examined, you will have a building with both a B and an A3 occupancy. Referring to Figure 4-13 (Table 508.4 of the IBC), it can be determined that a two-hour fire separation between the two occupancies is required if the building is not sprinklered. A one-hour wall is required if the building is sprinklered. This means that the walls surrounding the community hall and the floor ceiling assembly over the community hall will have to be constructed to resist a fire for at least two hours in a non-sprinklered building and one hour in a sprinklered building. Note that there are provisions in the code to build without occupancy separation walls. There are descriptions of various fire-resistive constructions in the codes and in other sources. Chapter 12 of this text further explores options for creating fire assemblies. If the entire building is built to the requirements of the more restrictive M occupancy, the area separation wall is not required.

Determining Mixed Occupancy Size

One of the other aspects affected by mixed occupancy is the allowable area of the building. The building's allowable size is not as apparent as with a single occupancy type. In the previous example, using Figure 4-10, the community hall is an A3 occupancy and the office

TABLE 508.4
REQUIRED SEPARATION OF OCCUPANCIES (HOURS)

	A[d], E		I-1, I-3, I-4		I-2		R		F-2, S-2[b], U		B, F-1, M, S-1		H-1		H-2		H-3, H-4, H-5	
OCCUPANCY	**S**	**NS**	**S**	**NS**	**S**	**NS**	**S**	**NS**	**S**	**NS**	**S**	**NS**	**S**	**NS**	**S**	**NS**	**S**	**NS**
A[d], E	N	N	1	2	2	NP	1	2	N	1	1	2	NP	NP	3	4	2	3[a]
I-1, I-3, I-4	—	—	N	N	2	NP	1	NP	1	2	1	2	NP	NP	3	NP	2	NP
I-2	—	—	—	—	N	N	2	NP	2	NP	2	NP	NP	NP	3	NP	2	NP
R	—	—	—	—	—	—	N	N	1[c]	2[c]	1	2	NP	NP	3	NP	2	NP
F-2, S-2[b], U	—	—	—	—	—	—	—	—	N	N	1	2	NP	NP	3	4	2	3[a]
B, F-1, M, S-1	—	—	—	—	—	—	—	—	—	—	N	N	NP	NP	2	3	1	2[a]
H-1	—	—	—	—	—	—	—	—	—	—	—	—	N	NP	NP	NP	NP	NP
H-2	—	—	—	—	—	—	—	—	—	—	—	—	—	—	N	NP	1	NP
H-3, H-4, H-5	—	—	—	—	—	—	—	—	—	—	—	—	—	—	—	—	1[e, f]	NP

For SI: 1 square foot = 0.0929 m^2.

S = Buildings equipped throughout with an automatic sprinkler system installed in accordance with Section 903.3.1.1.

NS = Buildings not equipped throughout with an automatic sprinkler system installed in accordance with Section 903.3.1.1.

N = No separation requirement.

NP = Not permitted.

a. For Group H-5 occupancies, see Section 903.2.5.2.
b. The required separation from areas used only for private or pleasure vehicles shall be reduced by 1 hour but to not less than 1 hour.
c. See Section 406.1.4.
d. Commercial kitchens need not be separated from the restaurant seating areas that they serve.
e. Separation is not required between occupancies of the same classification.
f. For H-5 occupancies, see Section 415.8.2.2.

FIGURE 4-13 When there will be more than one type of occupancy classification in a building, the required separation of common walls and floor systems must be determined using IBC Table 508.4. Please note that there are exceptions in the code that eliminate the requirement for occupancy separation. *Courtesy 2009* International Building Code, *copyright © 2009. Washington, DC. International Code Council. Reproduced with permission. All rights reserved. www.iccsafe.org.*

portion of the building is a B occupancy. Each of these occupancies has a different basic allowable area for every construction type. The allowable area for each of the occupancy classifications needs to be calculated. The sum of the ratios of the actual areas to the allowable areas may not exceed 1.

Looking over the formula in Figure 4-12, the only variable that changes is *At*, the tabular area or basic allowable area. To check the building to see if Type V-B construction still works, rework the formula using an *At* of 8000 from Figure 4-9 for an A-3 occupancy. The result will be an allowable area of 22,134 for Type V-B construction.

The formula for calculating the sum of the ratios of the actual areas to the allowable areas is shown in Figure 4-14. The actual area for the B occupancy with 22,000 square feet (30,000 – 8000) is divided by the allowable area of 33,201 equals 0.663. The actual area of the A-3 occupancy of 8000 divided by the allowable area of 22,134 equals 0.361. The sum of 0.663 plus 0.361 is greater than 1 (see Figure 4-14), so Type V-B construction is not acceptable. A higher classification of construction will need to be used. Going through the same steps for Type III-B construction proves that it will work, as shown in Figure 4-14.

EXITS

After the basic exterior shape of the building has been determined, you will begin to fill it with rooms. One of the most important steps in this process is to determine the exit path. The number and size of the exits will be determined in most cases by the number of occupants in the building and the occupancy classification.

The number of occupants in the building is not necessarily determined by the number of people an architect says will be in the various rooms. Room occupancy is determined by a ratio of the room area to the code specified area per person, as shown in Figure 4-15 (IBC Table 1004.1.1), or by the number of occupants that are to use the room, whichever is greater. In some cases, the number of fixed seats in the room determines the number of occupants. Theaters and churches are examples of spaces where the number of occupants is determined by the number of seats. In most other spaces, the area of the room is divided by the factor shown in column 2 of IBC Table 1004.1.1 (Figure 4-15) to determine the number of occupants.

After determining the number of occupants in the building, the number and width of the required exits can be determined. Figure 4-16 (IBC Table 1015.1) lists the IBC requirements for two exits under most conditions. Three exits are required if the building has 501 to 1000 occupants; four exits are required for 1001 and more occupants.

The location of the exits is also an important factor. If more than one exit is required, the code requires at least two of the exits to be separated by a distance equal to or greater than one-half of the diagonal of the area served (one-third of the diagonal in a sprinklered building), as shown in Figure 4-17. Dead-end corridors over 20' (6100 mm) in length in a non-sprinklered building for most occupancies (see Figure 4-17) are not allowed. The length of a dead-end corridor can be

Mixed occupancy

$$\frac{\text{actual area 1}}{\text{allowable area 1}} + \frac{\text{actual area 2}}{\text{allowable area 2}} + \frac{\text{actual area 3}}{\text{allowable area 1}} \leq 1$$

check for Figure 4-10 with a 8000 sq ft art gallery

Type V-B construction

$$\frac{22{,}000}{33{,}201} + \frac{8{,}000}{22{,}134} = .663 + .361 = 1.024$$ this is greater than 1 and is not OK

Type III-B construction

$$\frac{25{,}000}{65{,}534} + \frac{8{,}000}{32{,}767} = .381 + .153 = .534 \leq \text{OK}$$

FIGURE 4-14 The formula for computing the allowable area of a building containing more than one occupancy classification is shown along with a sample calculation. This calculation uses information from Figures 4-10 and 4-12.

TABLE 1004.1.1
MAXIMUM FLOOR AREA ALLOWANCES PER OCCUPANT

FUNCTION OF SPACE	FLOOR AREA IN SQ. FT. PER OCCUPANT
Accessory storage areas, mechanical equipment room	300 gross
Agricultural building	300 gross
Aircraft hangars	500 gross
Airport terminal	
Baggage claim	20 gross
Baggage handling	300 gross
Concourse	100 gross
Waiting areas	15 gross
Assembly	
Gaming floors (keno, slots, etc.)	11 gross
Assembly with fixed seats	See Section 1004.7
Assembly without fixed seats	
Concentrated (chairs only—not fixed)	7 net
Standing space	5 net
Unconcentrated (tables and chairs)	15 net
Bowling centers, allow 5 persons for each lane including 15 feet of runway, and for additional areas	7 net
Business areas	100 gross
Courtrooms—other than fixed seating areas	40 net
Day care	35 net
Dormitories	50 gross
Educational	
Classroom area	20 net
Shops and other vocational room areas	50 net
Exercise rooms	50 gross
H-5 Fabrication and manufacturing areas	200 gross
Industrial areas	100 gross
Institutional areas	
Inpatient treatment areas	240 gross
Outpatient areas	100 gross
Sleeping areas	120 gross
Kitchens, commercial	200 gross
Library	
Reading rooms	50 net
Stack area	100 gross
Locker rooms	50 gross
Mercantile	
Areas on other floors	60 gross
Basement and grade floor areas	30 gross
Storage, stock, shipping areas	300 gross
Parking garages	200 gross
Residential	200 gross
Skating rinks, swimming pools	
Rink and pool	50 gross
Decks	15 gross
Stages and platforms	15 net
Warehouses	500 gross

For SI: 1 square foot = 0.0929 m^2.

FIGURE 4-15 The number of occupants in a space is usually calculated on a square footage basis when fixed seats are not installed. This table lists the maximum number of square feet that it can be assumed one person occupies for various uses. This is important for determining exit requirements. Sometimes this is important for determining the occupancy classification of a space. *Courtesy 2009* International Building Code, *copyright © 2009. Washington, DC. International Code Council. Reproduced with permission. All rights reserved. www.iccsafe.org.*

TABLE 1015.1
SPACES WITH ONE EXIT OR EXIT ACCESS DOORWAY

OCCUPANCY	MAXIMUM OCCUPANT LOAD
A, B, E[a], F, M, U	49
H-1, H-2, H-3	3
H-4, H-5, I-1, I-3, I-4, R	10
S	29

a. Day care maximum occupant load is 10.

FIGURE 4-16 If a large number of people occupy a space, more than one exit will be required. For example, a Type B occupancy with an occupancy load of 50 or less would require only one exit. If the floor area were increased to allow more than 50 occupants, two exits would be required. The number of occupants is determined by dividing the square footage of the room by the values shown in Figure 4-15. When there are more that 500 occupants, additional exits are required. *Courtesy 2009* International Building Code, *copyright © 2009. Washington, DC. International Code Council. Reproduced with permission. All rights reserved. www.iccsafe.org.*

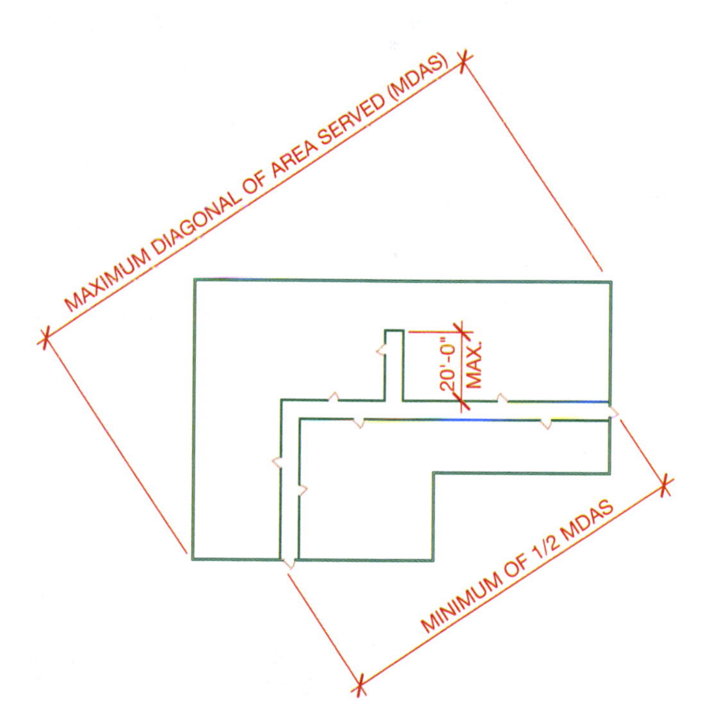

FIGURE 4-17 If more than one exit from a building is required, the exits will need to have proper separation. For non-sprinklered buildings, exits will need to be separated by a distance equal to a minimum of one-half of the maximum diagonal of the area served. In a sprinklered building this distance may be reduced to one-third of the maximum diagonal. The code limits the length of dead-end corridors when two or more exits are required. For all occupancies except for some I-3 occupancies, the dead end cannot exceed 20′ (6100 mm) in length for a building without fire sprinklers. The dead end may extend to 50′ for B and F occupancies in a building with fire sprinklers.

extended to 50' (15,240 mm) for some occupancies if the building is fully sprinklered. If the exits are not reasonably separated, there is a chance that a fire might block both exits, denying people an escape route. If the occupants of a building run down a long dead-end corridor, they will lose valuable time retracing their steps to get out of the building. If the exits are wide enough, people will be able to flow through them rapidly in an emergency.

Determining Exits

A single-story non-sprinklered building with 35,000 square feet of office space and a 900 square foot conference room will be considered. Figure 4-15 states that for the purpose of calculating exits, one occupant for every 100 square feet for business uses and one occupant for every 15 square feet of an assembly area with tables and chairs can be used. Dividing 35,000 by 100 results in 350 occupants for the office area, and dividing 900 by 15 results in 60 occupants for the conference room, for a total of 410 occupants. Per Figure 4-18 (Section 1005.1 of the IBC), the code requires 0.2" of exit width for every occupant. Multiplying 0.2" by 410 equals 82" (6'–10") of required exit. This building requires only two exits but two 3'–0" doors will not provide ample exit width. This means that if only two exits are provided, a 3'–0" door at one exit and a 4'–0" door or a pair of 3'–0" doors at the other exit would seem to be acceptable. Making one of the doors wider than the other is not always an option. The code states that when more than one exit is required, if one of the exits is blocked, it can not reduce the available exit width by more than half. With a 3' and a 4' door, if the 4' door is blocked, more than 50% of the exit would be blocked. In this case, both exits would need to be at least 3'-6" wide.

There are minimum requirements in the code for hall and door widths. Three-foot-wide exit doors are required for most conditions. In some cases, corridors as narrow as 36" are acceptable but most conditions require 44" as a minimum. Stairways have required minimum widths and a required minimum width per occupant.

THE CODE AND COMPUTERS

As you can see, using the code can be a lot of work. There is a lot of flipping back and forth between chapters. A thorough code check is important to the safety of the building's occupants. As your familiarity of the code increases, its use will become easier. You will remember what areas of the code to check. It is important that you check the code thoroughly, rather than try to commit it completely to memory, because the code is revised every three years. Your memory may dredge up an outdated version of the code section.

There are computer-based code-checking systems available for each of the model building codes. These range from a simple computerized checklist to a comprehensive analyst program. These programs can greatly speed up code checking and increase your confidence in the adequacy of your design.

SECTION 1005
EGRESS WIDTH

1005.1 Minimum required egress width. The *means of egress* width shall not be less than required by this section. The total width of *means of egress* in inches (mm) shall not be less than the total *occupant load* served by the *means of egress* multiplied by 0.3 inches (7.62 mm) per occupant for stairways and by 0.2 inches (5.08 mm) per occupant for other egress components. The width shall not be less than specified elsewhere in this code. Multiple *means of egress* shall be sized such that the loss of any one *means of egress* shall not reduce the available capacity to less than 50 percent of the required capacity. The maximum capacity required from any *story* of a building shall be maintained to the termination of the *means of egress*.

➡ **Exception:** *Means of egress* complying with Section 1028.

FIGURE 4-18 Once the number of exits has been determined, Section 1005 of the IBC can be used to determine the exit width for every occupant to be served. *Courtesy 2009* International Building Code, *copyright © 2009. Washington, DC. International Code Council. Reproduced with permission. All rights reserved. www.iccsafe.org.*

EASY BEING GREEN

Traditionally, building codes address how the environment affects a structure. Stresses resulting from gravity, wind, snow, water, and seismic activity have always been a major concern of the model codes. The Green Building Standard now addresses how the building affects the environment. The International Code Council (ICC) and the National Association of Home Builders (NAHB) have jointly developed the National Green Building Standard that promotes the flexibility of green building practices and provides a common national benchmark for builders and designers. The standards are the first to be accredited by the American National Standards Institute (ANSI). Although primarily aimed at residential construction, the standard does apply to multifamily construction. The standard requires designers and builders to address features in six categories including:

- Design considerations
- Land conservation
- Water conservation
- Material resource conservation
- Energy conservation
- Indoor and outdoor air quality

In addition to setting construction standards,ratings for bronze, silver, gold, and emerald certification levels are provided in an effort to reduce negative environmental impacts and to promote positive environmental impacts.

In addition to the Green Building Standard, the ICC and the U.S. Green Building Council (USGBC) have signed a Memorandum of Understanding between the two organizations in order to further green building practices. Because green code development is such a fluid area, it is imperative that you constantly monitor the Web sites of the ICC, USGBC, ANSI, and NAHB for the most current information on green building and developing codes. The address for each site is listed at the end of this chapter.

ADDITIONAL READING

The International Building Code and Commentary Volume I and II have excellent illustrations and interpretations of the code. The following Web sites can be used as a resource to help you keep current with changes in building codes. All four sites are sources for the 2009 IBC, as well as the other code listed with their names. They are sources for interpretive manuals and training.

ADDRESS	COMPANY/ ORGANIZATION
www.iccsafe.org	International Code Council

The following is a partial listing of the sites referred to in IBC Chapter 35, REFERENCED STANDARDS. See the IBC for additional references.

ADDRESS	COMPANY/ ORGANIZATION
www.aamanet.org	American Architectural Manufacturers Association
www.aci-int.org	American Concrete Institute
www.afanda.org	American Forest and Paper Association
www.aisc.org	American Institute of Steel Construction
www.aitc-glulam,org	American Institute of Timber Construction
www.steel.org	American Iron and Steel Institute
www.ansi.org	American National Standards Institute
www.asce.org	American Society of Civil Engineers
www.asme.org	American Society of Mechanical Engineers
www.astm.org	ASTM International
www.aws.org	American Welding Society
www.awpa.org	American Wood-Preservers Association
www.apawood.org	APA—The Engineered Wood Association
www.pwgsc.gc.ca.cgsb	Canadian General Standards Board
www.cedarbureau.org	Cedar Shake and Shingle Bureau
www.gypsum.org	Gypsum Association

www.iso.org International Standards Organization
www.masonrysociety.org Masonry Society
www.naamm.org National Association of Architectural Metal Manufacturers
www.nahb.org National Association of Home Builders
www.ncma.org National Concrete Masonry Association
www.nfpa.org National Fire Protection Association
www.post-tensioning.org Post-Tensioning Institute
www.pci.org Precast Prestressed Concrete Institute
www.mhia,org Rack Manufacturing Institute (Material Handling Industry)
www.spri.org Single-Ply Roofing Institute
www.steeljoist.org Steel Joist Institute
www.seinstitute.org Structural Engineers Institute
www.tpinst.org Truss Plate Institute
www.ul.com Underwriters Laboratories
www.usgbc.org United States Green Building Council
www.wdma.com Window and Door Manufacturers Association
www.wirereinforcementinstitute.org Wire Reinforcement, Inc.

KEY TERMS

Allowable floor area
Basic allowable area
Construction type
Hazardous materials
Occupancy classification
Occupancy, Assembly
Occupancy, Business
Occupancy, Education
Occupancy, Factory
Occupancy, Hazardous
Occupancy, Institutional
Occupancy, Mercantile
Occupancy, Mixed
Occupancy, Residential
Occupancy, Storage
Occupancy, Utility

CHAPTER 4 TEST Introduction to the International Building Code

Answer the following questions with short complete statements. Type the chapter title, question number, and a short complete statement for each question using a word processor. Answer the questions according to the code that governs your area. If math is required to answer a question, show your work.

Question 4-1 What occupancy classification would a 10-unit apartment building be?

Question 4-2 What occupancy classification would a clothing store be?

Question 4-3 In a building that requires occupancy separation, what fire separation is required between a B and an H-3 occupancy using the IBC? Between an A-2 and an E?

Question 4-4 What is the basic allowable area of an H-4 occupancy in a Type V building? In a Type III building?

Question 4-5 What is the most important purpose of the model code?

Question 4-6 Use the drawing below and the IBC to determine the maximum size allowed for a one-story building without a fire sprinkler system. Assume an occupancy of E, and a building type of II-A.

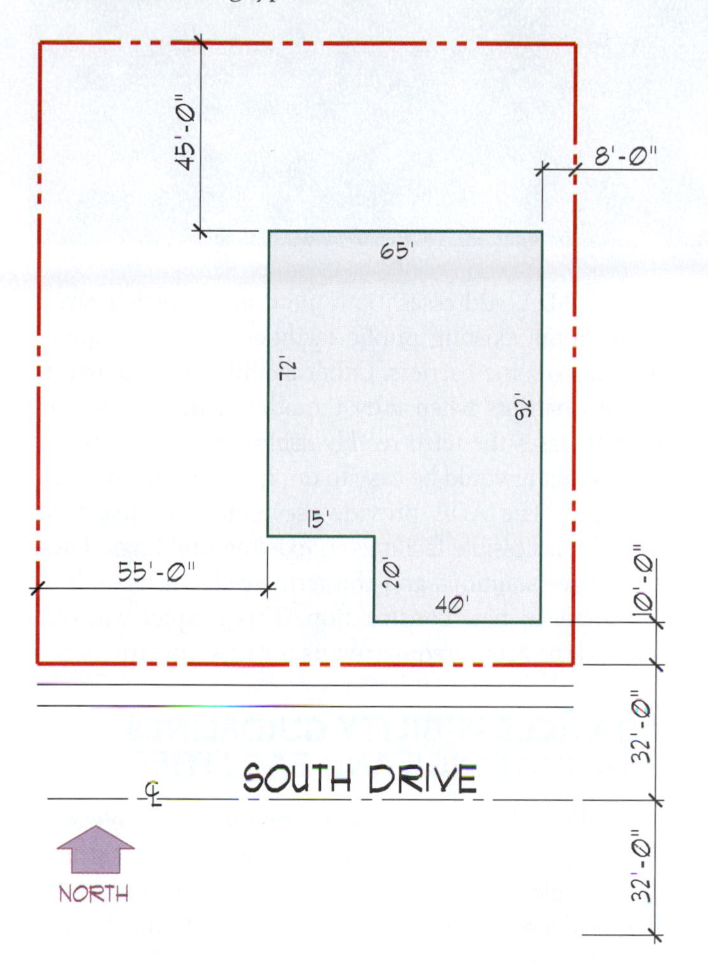

Question 4-7 What is the maximum allowable size for the building in question 4-6 if the building is provided with a fire sprinkler system and is of Type II-B construction?

Question 4-8 How far do the exits need to be separated if the maximum diagonal of the building is 85' and the building is sprinklered?

Question 4-9 Use Figure 4-10 and assume that there is no street on the east side of the property so that the distance from the building to the property line is 8'. What is the value of I_f if the building is non-sprinklered?

Question 4-10 Use the information in question 4-9. How large can the building be if it is used for a Type M occupancy and made of Type III-A construction?

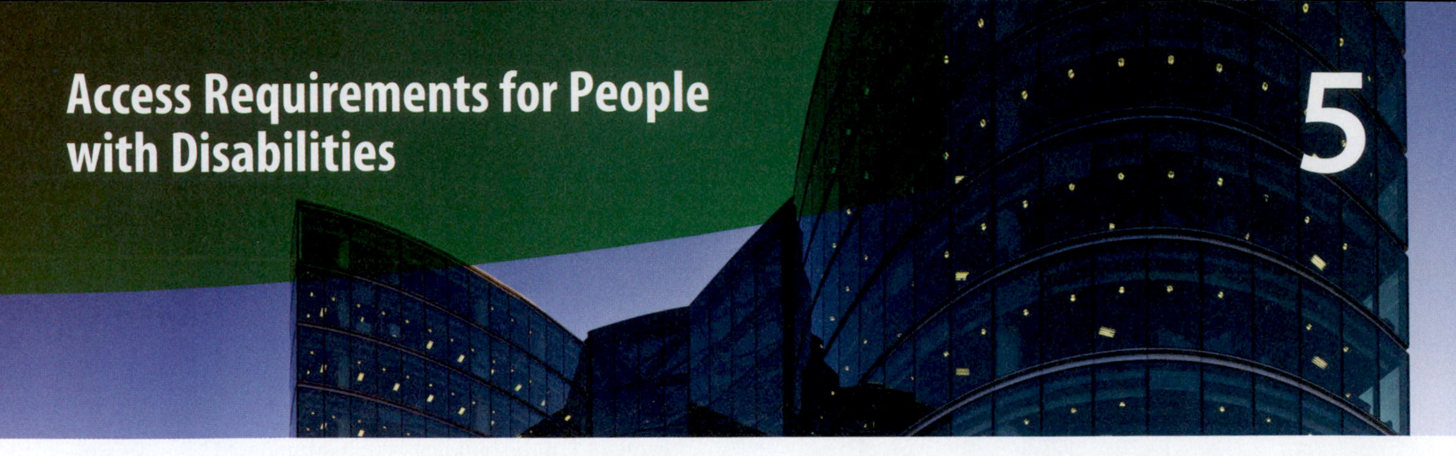

Access Requirements for People with Disabilities

Just as with the building code, as a new CAD technician you will not be expected to apply the Americans with Disabilities Act (ADA) guidelines to the design of a structure. However, to advance beyond an entry-level position, you must understand the basic concepts presented in the ADA. To help you become proficient at using the standards, this chapter will explore the ADA as it is intended to address the requirements of new construction regarding:

- Parking requirements.
- Methods of egress.
- Restrooms and bathing facilities.
- Requirements based on occupancy.

Note: The United States Access Board published changes to the 1994 edition of the Americans with Disabilities Act in 2009 at a date that was too late to be addressed in this text. Verify current ADA requirements at the ADA home page at the following address: http://www.ADA.gov.

The Americans with Disabilities Act, otherwise known as the ADA, is a federal law requiring all new public accommodations and commercial facilities to be accessible and usable by people with disabilities. This landmark law was driven by the fact that medical and scientific technology has allowed people to survive previously fatal conditions and live on with significant disabilities. The ADA gives these citizens a better chance to live productive and fulfilling lives.

This law has had a significant impact on building and site design. It does not guarantee people with disabilities access to every square foot of every new building, but it does greatly increase the building space they can utilize in a comfortable and convenient manner. Access is required to the ground floor of almost every nonresidential building, and it is required to all floors of most public and medical buildings and buildings over three stories in height. Make note that public buildings include retail stores, restaurants, theaters, and other privately owned buildings that the general public utilizes. Public buildings also include public-owned facilities such as libraries and courthouses.

The ADA addresses both new and existing structures. Some existing public facilities are now required to remove some barriers. Other facilities are required to remove barriers when they remodel or do an addition. In both cases the term *readily achievable* is used, which means that it would be easy to do, not creating an undue hardship. The ADA provides several alternative solutions for accessible facilities in existing buildings. These alternative solutions and the term *readily achievable* do not apply to new construction. This chapter will only address the access requirements for new construction.

ADA ACCESSIBILITY GUIDELINES FOR BUILDINGS AND FACILITIES

The ADA addresses the requirements of people with disabilities. Many of the requirements are to accommodate people in wheelchairs, but the ADA also addresses the problems of the elderly, the blind, the hearing impaired, and those with other forms of physical disabilities. It is important to realize that the ADA may not be the only set of regulations governing access. Several states had access laws prior to the ADA that might still be in effect with requirements more stringent than those of the ADA. The requirements of the ADA do not necessarily provide the ideal or optimum conditions for people with disabilities, only the minimum acceptable requirements. Meeting ADA requirements does not excuse the designer from meeting the requirements of any other governing building code.

The primary concerns addressed by the ADA are access to the building, access to the building spaces it contains, and the use of facilities such as restrooms, drinking fountains, and telephones. A specific outline of ADA requirements, *Accessibility Guidelines for Buildings and Facilities,* is available at www.access-board.gov.

PARKING REQUIREMENTS

Special parking spaces must be provided for people with disabilities. This parking must be relatively flat (2% maximum slope in every direction including diagonally) and must be located as close as possible to the entrance of the building. There must be an accessible walkway from the parking space to the building. The parking spaces must be marked with the international symbol of accessibility shown in Figure 5-1.

Parking spaces must be of a size adequate to accommodate the special needs of people with disabilities. At least one space needs to be 8' (2438 mm) wide with an 8' (2438 mm) wide loading space or 11' (3350 mm) wide with a 5' (1525 mm) wide loading space to accommodate a van with a wheelchair lift, as shown in Figure 5-2. Other parking spaces for disabled drivers are required to provide only a 5' (1525 mm) wide loading space to allow a wheelchair to be brought alongside a car. The minimum vertical clearance for parking spaces for the disabled is 8' (2438 mm). If there is a curb at the parking space, there will be a need for a curb ramp, as shown in Figure 5-3.

The CAD technician will have to show the proper size and number of parking spaces on the site plan based on the criteria listed in Table 5-1.

Parking details will normally be stock details stored as wblocks or contained on a standard sheet with other ADA details. Site grades can influence the location of accessible parking spaces. The space must be on an **accessible route** of travel and meet rigid requirements for slope in both directions. The required slope of the parking space and the accessible path of travel will need to be coordinated with the civil engineer for the project.

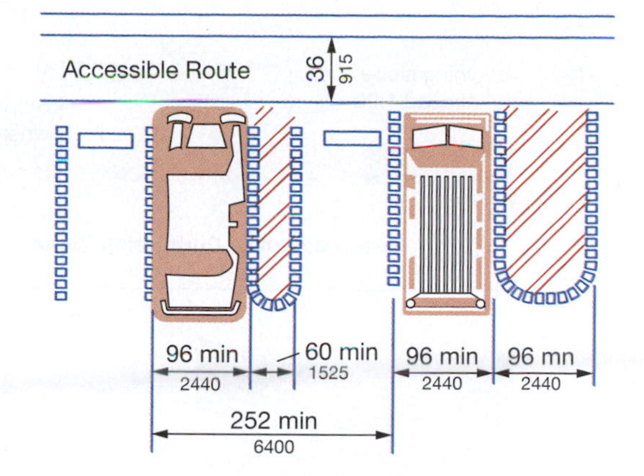

Van Accessible Space at End Row

FIGURE 5-2 Normal parking spaces for the disabled must have a 5' (1525 mm) wide loading space. Van spaces are required to have an 8' (2438 mm) wide loading space. The parking space itself is required to be 8' (2438 mm) wide, although some jurisdictions may require the parking space to be wider. The loading space must be on the passenger side of the car unless it is shared by two parking spaces. *Courtesy Uniform Federal Accessibility Standards.*

METHODS OF EGRESS

Many of the ADA requirements deal with access to and throughout a structure. Major areas to be considered by the design team include walks and hallways, ramps, stairs, elevators, lifts, areas of rescue assistance, and doors.

Walks and Hallways

There must be an accessible route from public transportation, parking spaces, loading zones, and the public sidewalk to the building. The route shall lead to a primary building entrance, not a service entrance, and must be at least 36" wide (915 mm). A wider route of 48" (1220 mm) is recommended so that someone walking can pass a wheelchair. Turns in the walk or hall must be able to accommodate a wheelchair, as shown in Figures 5-4a and 5-4b. If the route is less than 60" wide (1525 mm), there must

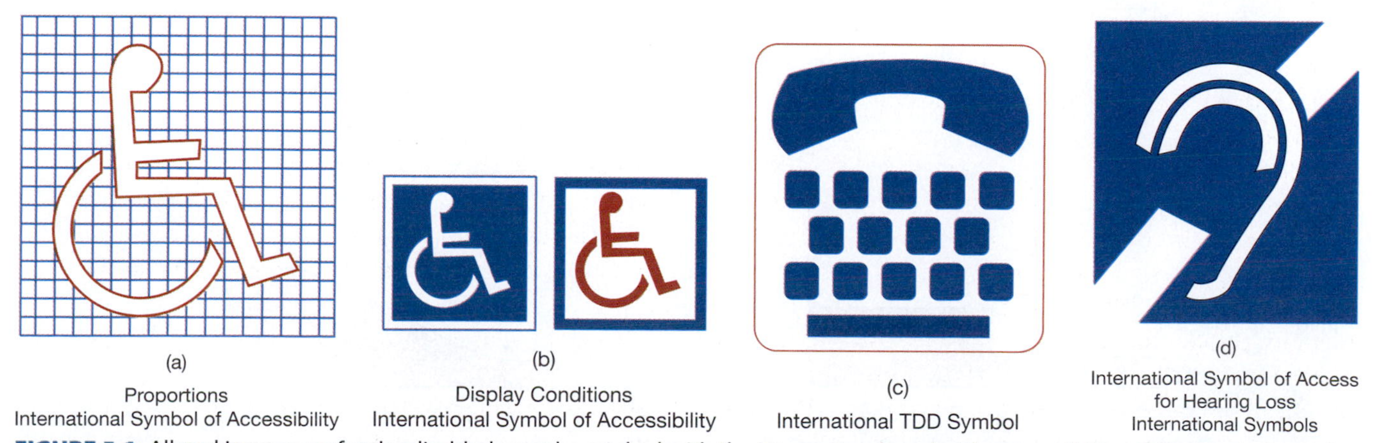

(a) Proportions International Symbol of Accessibility

(b) Display Conditions International Symbol of Accessibility

(c) International TDD Symbol

(d) International Symbol of Access for Hearing Loss International Symbols

FIGURE 5-1 All parking spaces for the disabled must be marked with the international symbol for accessibility. Other symbols used to designate special provisions include the telecommunication device for the deaf (TDD) and hearing loss symbols. *Courtesy Federal Register/Vol. 56 No.144/July 26, 1991.*

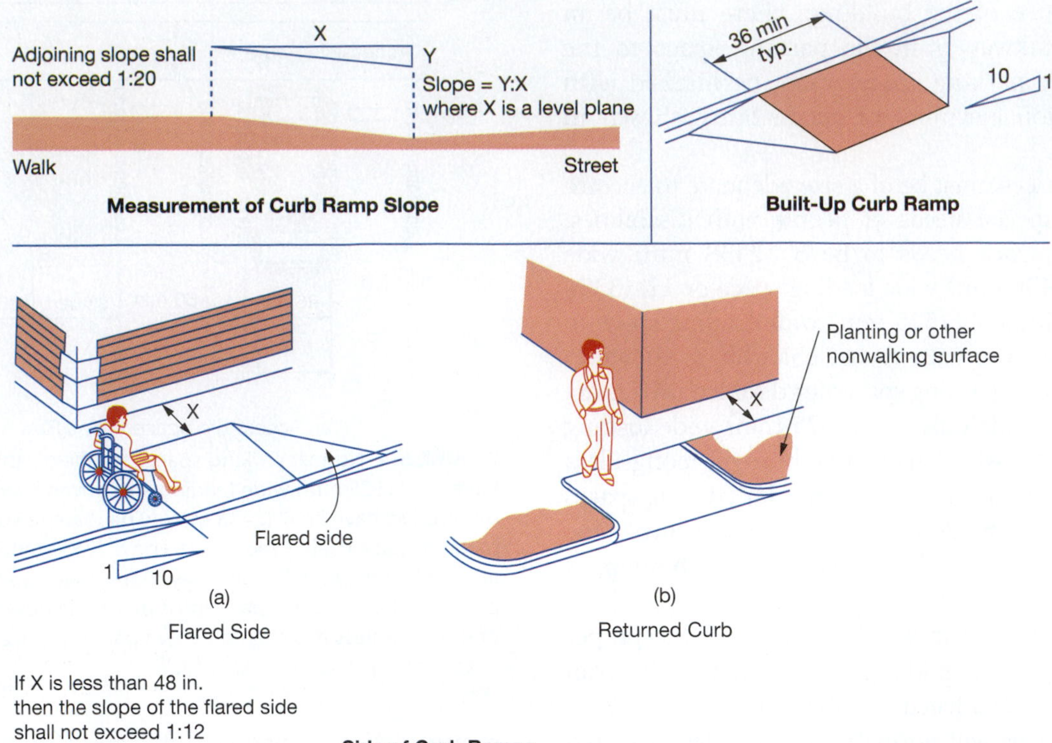

FIGURE 5-3 Curb ramps must be provided to get from the parking space to the adjacent walk (path of travel). These ramps may not project into the required parking space. The path of travel must not go sideways across the ramp or the 2% allowable cross fall will be exceeded. *Courtesy Uniform Federal Accessibility Standards.*

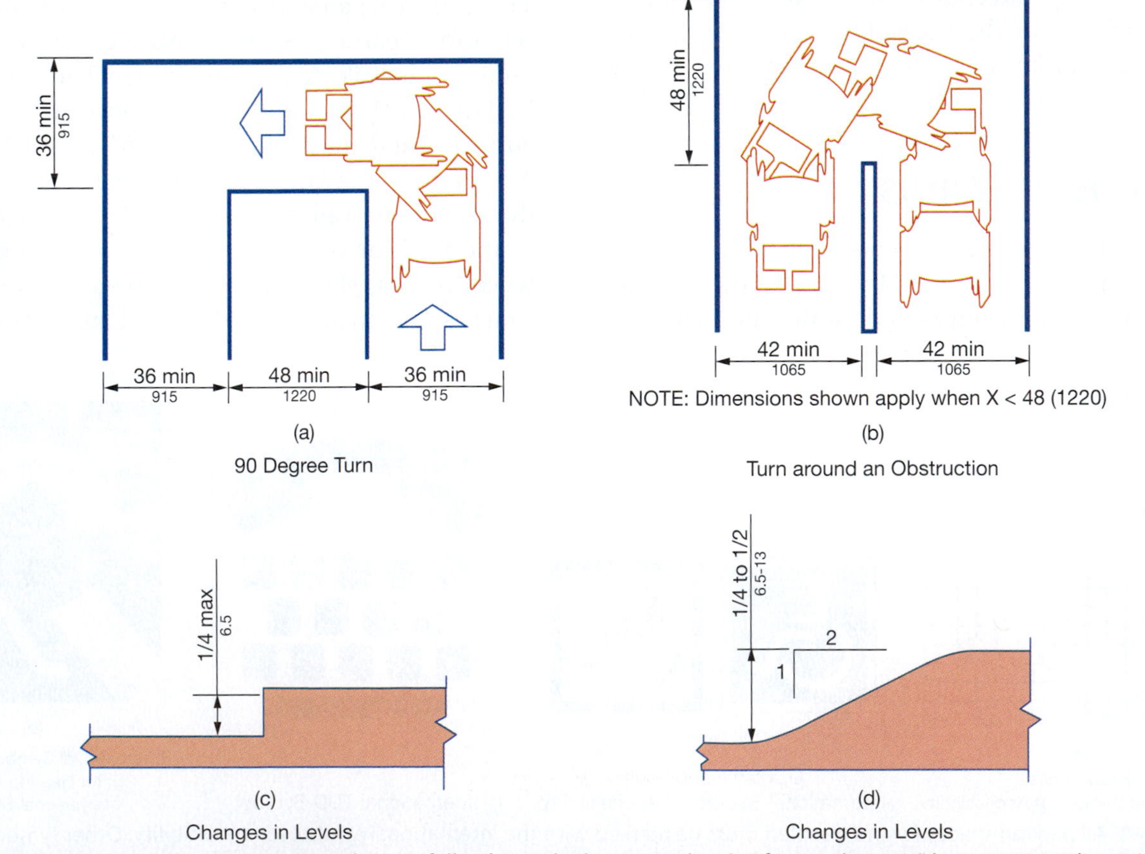

FIGURE 5-4 Turns in a walk or hallway must be carefully planned. Changes in levels of more than 1/4" but not exceeding 1/2" must be sloped. Abrupt changes of more than 1/2" are not allowed. *Courtesy Uniform Federal Accessibility Standards.*

Based on the Uniform Federal Accessibility Standards

Total Parking in Lot	Required Minimum Number of Accessible Spaces
1 to 25	1
26 to 50	2
51 to 75	3
76 to 100	4
101 to 150	5
151 to 200	6
201 to 300	7
301 to 400	8
401 to 500	9
501 to 1000	2% of total
1001 or more	20 plus 1 for each 100 sq. ft. or fraction over 1001

TABLE 5-1 Required Minimum Number of Handicapped Spaces Based on Total Parking

be a 60" × 60" (1525 × 1525 mm) minimum wide passing space located no further than 200' (60, 960 mm) apart. The route must have a minimum vertical clearance of 80" (2030 mm). The surface of the route must be slip-resistant and must also be a reasonably firm surface (no plush carpet or pad, etc.). Projections along the access route could be hazardous to the visually impaired. Therefore projections are limitations, as shown in Figure 5-5.

Changes in level along the accessible route of up to 1/2" (13 mm) shall be per Figures 5-4c and 5-4d. Changes in elevation of more than 1/2" (13 mm) must be via curb ramp, elevator, or lift. If the walk slopes more than 5% (1 in 20) in the direction of travel, it is considered a ramp and shall comply with the requirements of a ramp. The cross slope, the sideways slope of the path of travel route, may not slope more than 1 in 50, or 2%.

CAD technicians must draw the walks to their proper width and destination and assure proper maneuvering

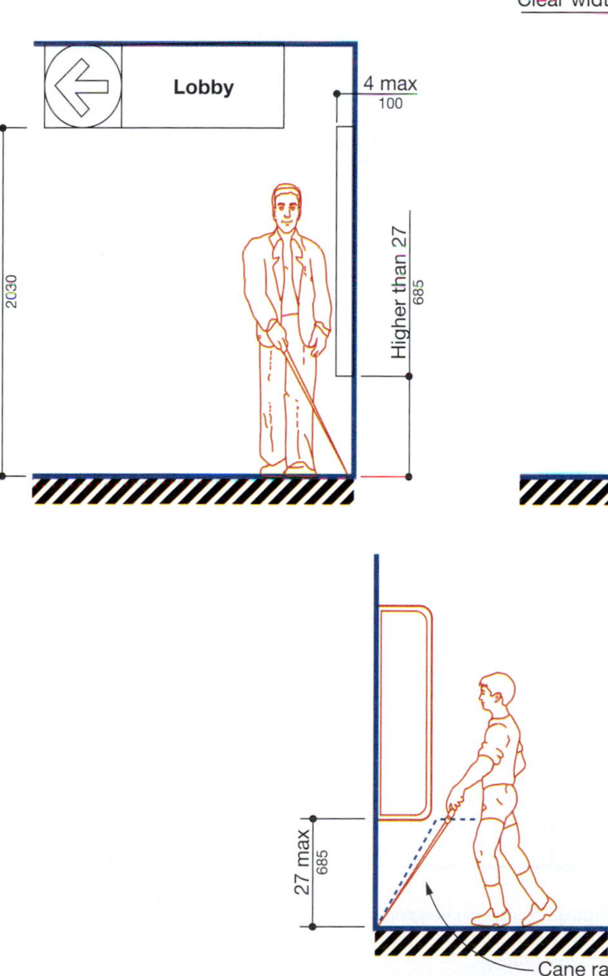

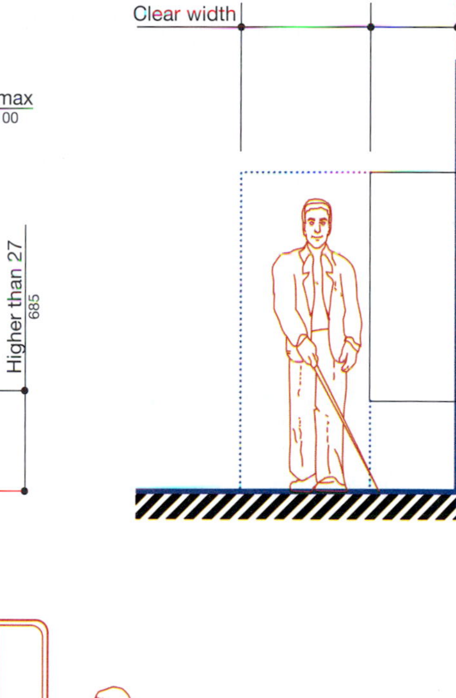

FIGURE 5-5 There are limitations placed on projections along the path of travel to protect the visually impaired. *Courtesy Uniform Federal Accessibility Standards.*

(continued)

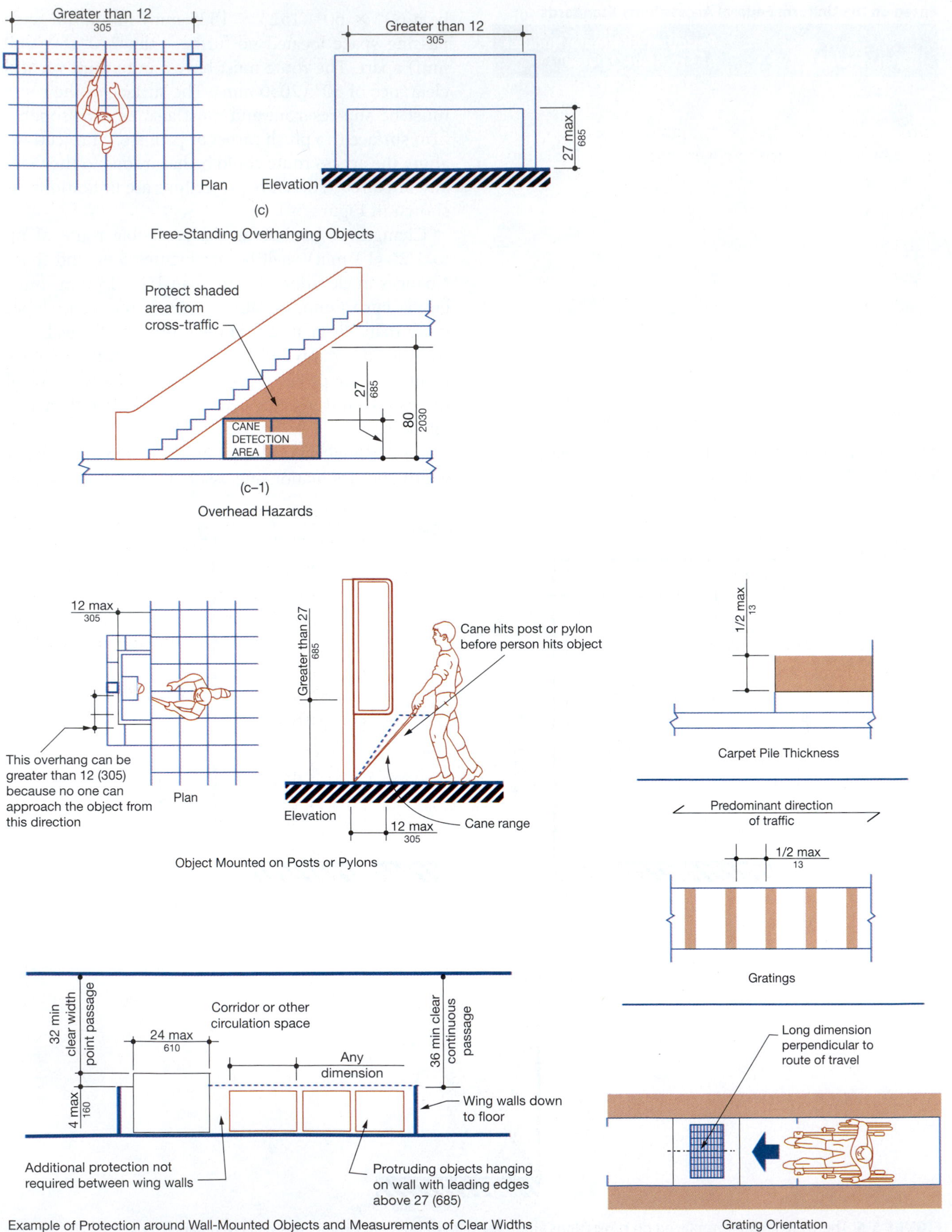

FIGURE 5-5 *(continued)* There are limitations placed on projections along the path of travel to protect the visually impaired. *Courtesy Uniform Federal Accessibility Standards.*

space for turns. Slope of walks, both in the direction of travel and cross fall, must be coordinated with the civil engineer. If ramps or stairs are required, they need to be properly detailed to meet the requirements outlined in the following sections. Most offices will have standard details, reflecting the requirements of the ADA that can be referenced to cover many of the conditions that occur on a site or within the accessible route in the building.

Ramps

Curb ramps, as the name implies, are intended to make the transition through the change in level caused by a normal curb. These ramps shall be a minimum of 36" (915 mm) wide and shall be sloped as shown in Figure 5-3. Curb ramps are required to have a detectable warning at their top edge, as shown in Figure 5-6. Curb ramps do not require handrails or guardrails.

Other ramps are usually longer than curb ramps. Like a walk, ramps are required to be a minimum of 36" (915 mm) wide. It is desirable to make the ramps wider, especially if this is also the primary access for people who are not disabled. Any access route having a slope of 1 in 20 (5%) or greater is considered to be a ramp. A ramp should be as shallow a slope as possible and may not exceed a slope of 1 in 12 (8.33%). The maximum change in height between landings is 30" (760 mm). Level landings are required at each end of a ramp. They must be at least as wide as the ramp and be 60" (1525 mm) long in the direction of travel (see Figure 5-7). If the ramp changes direction, the landing must be 60" × 60" (1525 × 1525 mm) minimum. If there is a doorway on the landing, all requirements for maneuvering at a door must be met. The requirements for maneuvering at doors are discussed later in this chapter.

Guardrails must be provided on the open side of a ramp, as shown in Figure 5-8. Unless adjacent to a seating area, ramps rising more than 6" (150 mm) require a handrail on both sides. The handrail shall be constructed per Figure 5-9, projecting 1 1/2" (38 mm) clear of the wall and extending 12" (300 mm) beyond the ends of the ramp, except for a continuous inside handrail on a switchback or dogleg ramp as shown in Figure 5-10. The handrail shall be continuous and be mounted 34" to 38" (860 to 960 mm) above the ramp surface.

Required ramps frequently fail to receive proper attention during the design stage. Total rise must be considered and, if the ramp must rise over 30" (760 mm), a landing must be inserted. Landings at changes in direction must be properly sized. Ramps and landings take up a surprising amount of room. It is important to

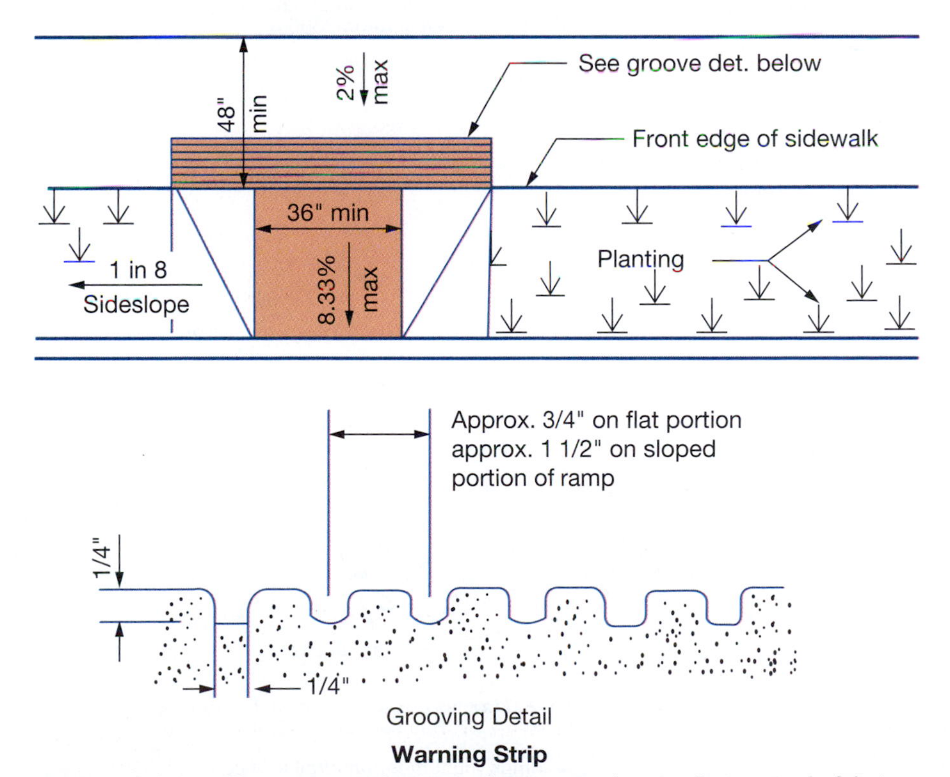

FIGURE 5-6 Ramps must have a detectable warning at their top edge to warn the visually impaired of their presence. *Courtesy Uniform Federal Accessibility Standards.*

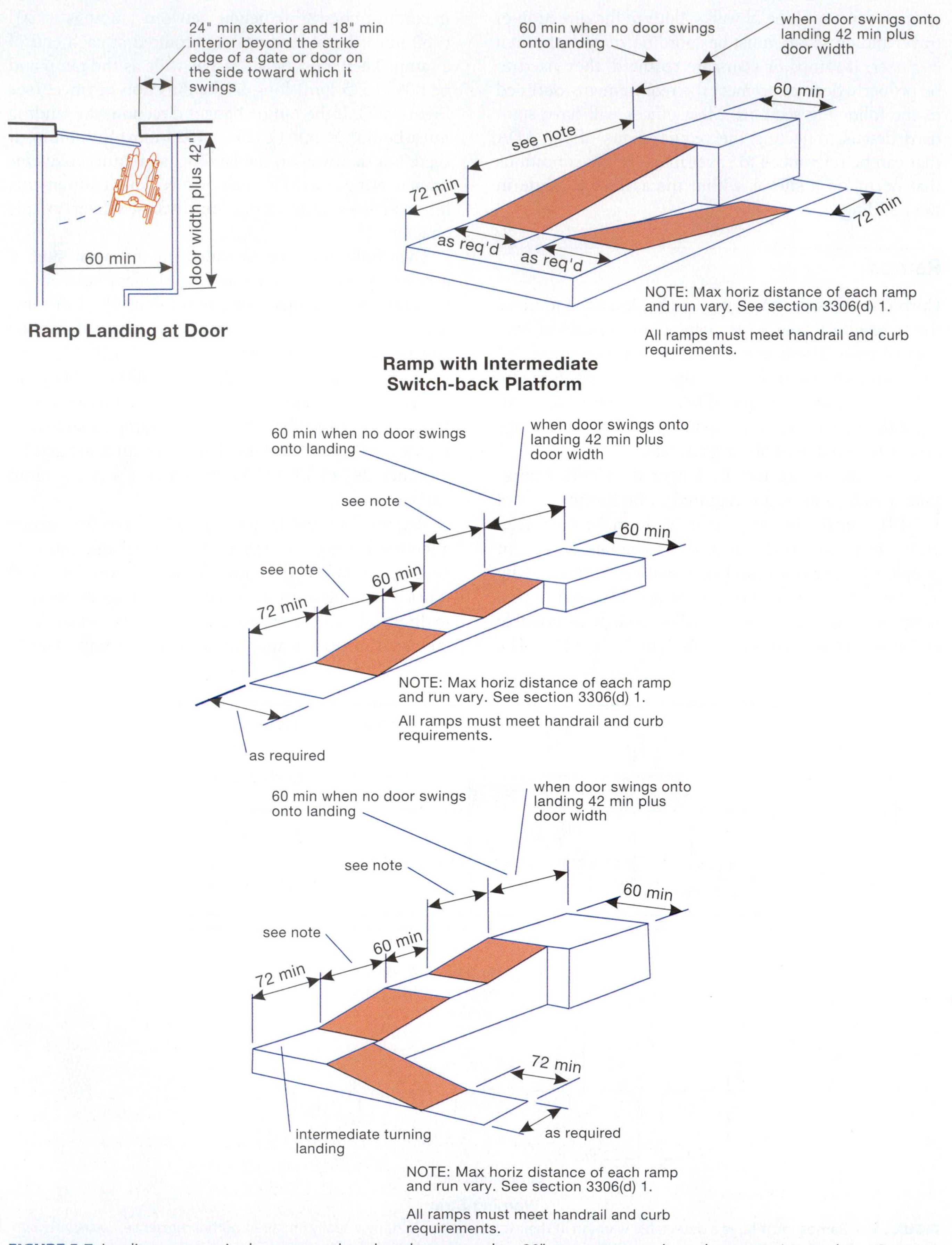

FIGURE 5-7 Landings are required on ramps when they slope more than 30", turn corners, or have doors entering on them. *Courtesy Uniform Federal Accessibility Standards.*

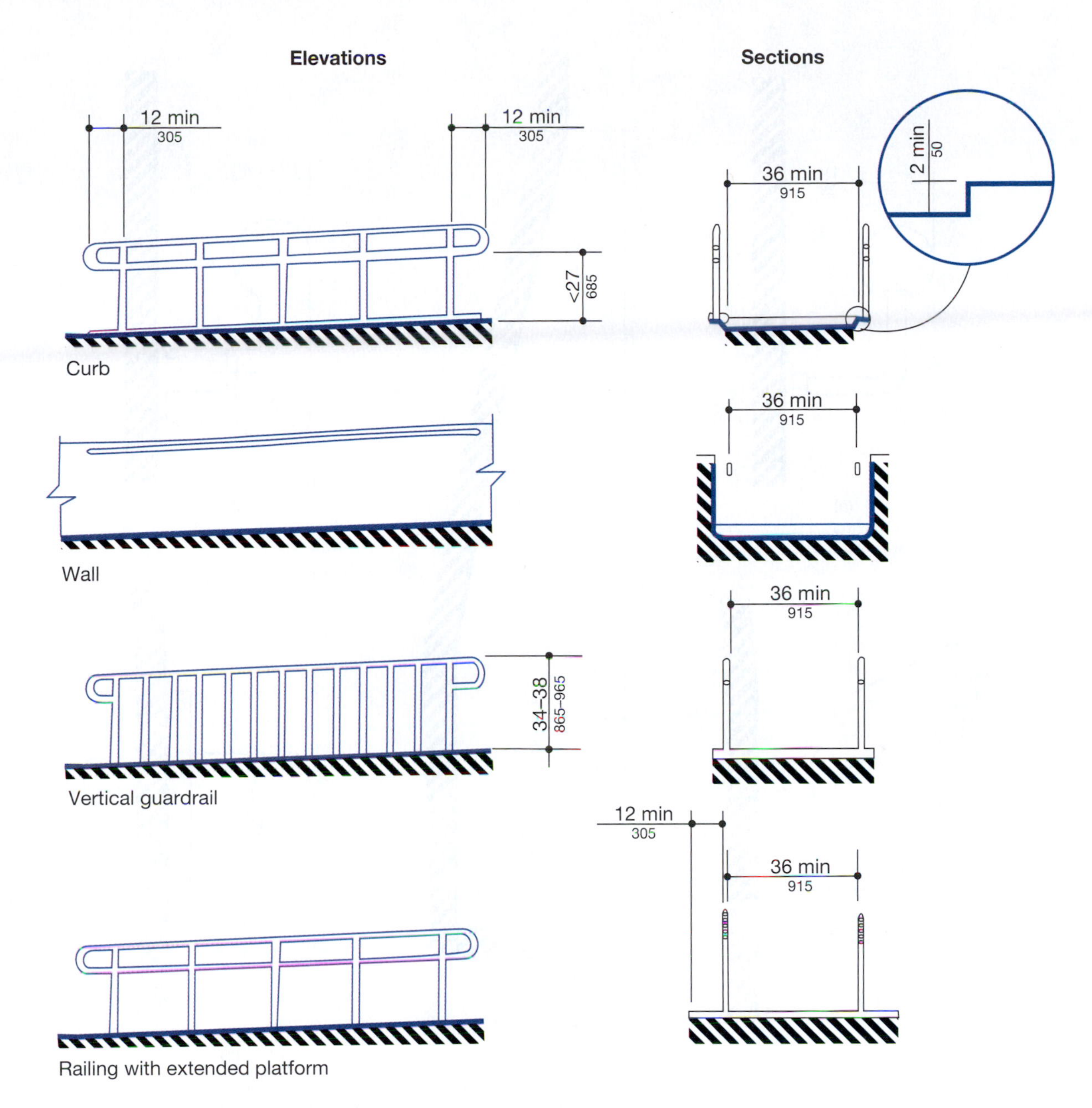

FIGURE 5-8 Guardrails are required on the open side of a ramp. Handrails are required on both sides of a ramp. *Courtesy Uniform Federal Accessibility Standards.*

analyze the total amount of room needed early in the drawing process so that any deficiencies can be corrected with minimum effort.

Stairs

Stairways serving a floor that is not otherwise accessible to people with disabilities shall meet the following conditions in addition to the requirements of the building code. Treads shall be a uniform width not less than 11" (280 mm) and risers shall be a uniform height not to exceed 7". When nosings are provided, they shall be constructed as shown in Figure 5-11. Stairways shall not have open risers. Open risers can be confusing to people with poor sight and can be a problem for cane use by the blind. Handrails shall be provided, measuring 34" to 38" (865 to 965 mm) high on both sides of a stair, and shall be constructed as shown in Figures 5-10 and 5-11. Handrails shall be continuous and uninterrupted by newel posts or other obstructions. For additional information on stairs, see Chapter 21.

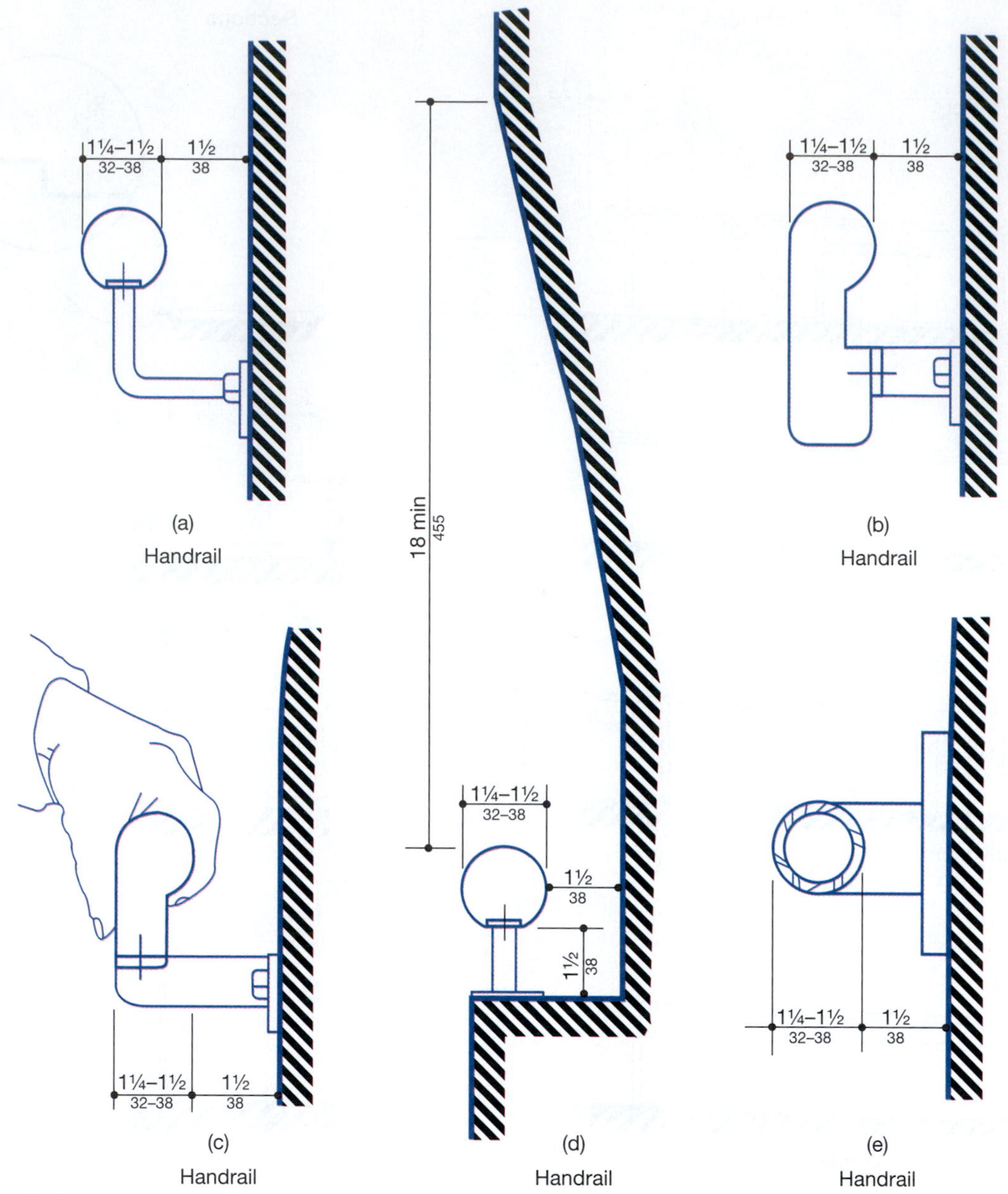

FIGURE 5-9 Handrails at ramps and stairs are required to have a shape that is easy to grip and have a sufficient clearance from adjacent surfaces to allow for easy access. *Courtesy Uniform Federal Accessibility Standards.*

Elevators

An elevator is usually the easiest way to provide access to upper floors when access is required above the ground floor. If an elevator is installed, it must meet the ADA requirements whether the elevator was required or not. Every floor that the elevator serves is accessible and will need to meet access requirements whether they were otherwise required or not. The elevator must be automatic and have proper call buttons, car control, illumination, door timing, and a reopening device as shown and called for in Figure 5-12. The elevator floor is a part of the access route so it must be similar to the floor described in walks and hallways. The shape and size of the elevator must be such that it is usable by a wheelchair, as shown in Figure 5-13. Much of the drawings required for elevators are done by the elevator manufacturer or supplier. The CAD technician's responsibility is normally limited to details concerning interior finishes of the elevator.

Lifts

Platform or wheelchair lifts may be used in place of a ramp or elevator in some instances. If they are used, they must allow for unassisted entry, operation, and exit. They must also meet elevator and escalator safety requirements.

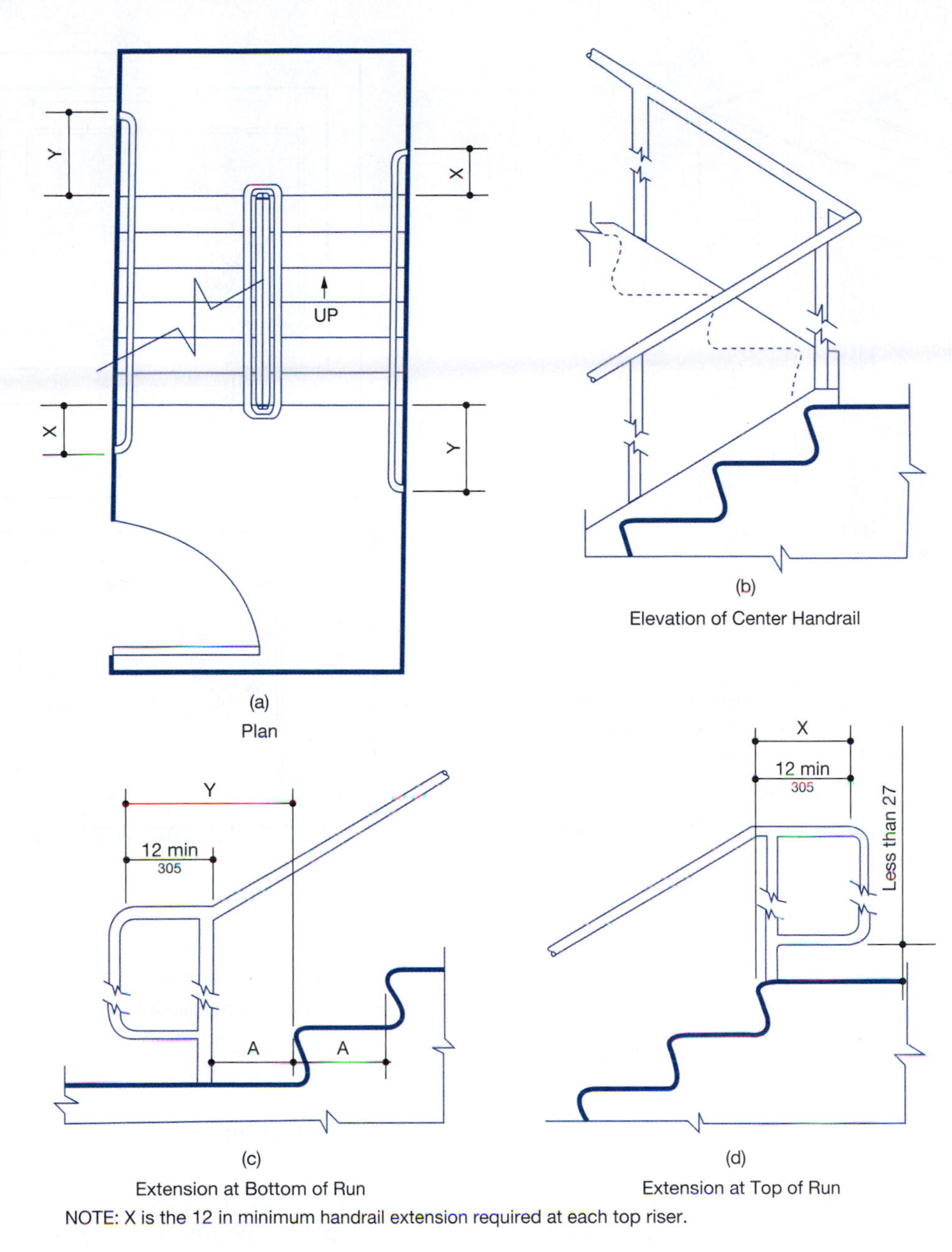

FIGURE 5-10 Stair handrails at ramps and stairs are required to have extensions at their top and bottom. *Courtesy Uniform Federal Accessibility Standards.*

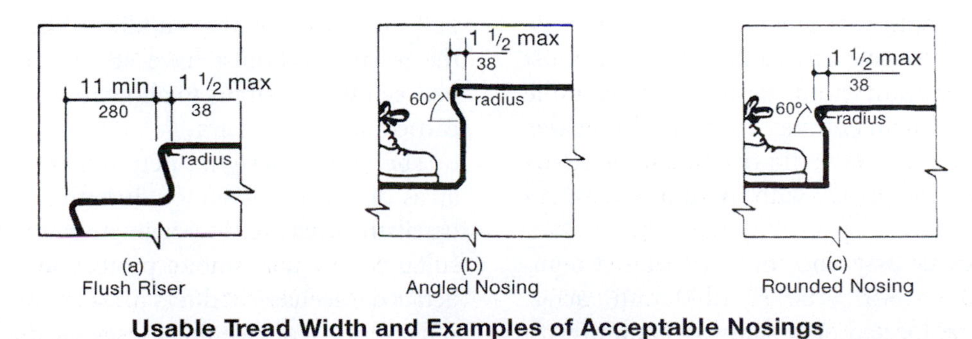

FIGURE 5-11 The size of nosing is limited. *Courtesy Uniform Federal Accessibility Standards.*

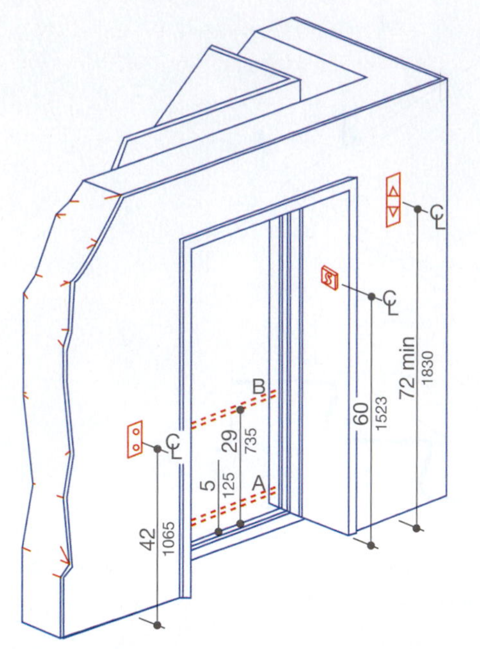

NOTE: The automatic door reopening device is activated if an object passes through either line A or line B. Line A and line B represent the vertical locations of the door reopening device not requiring contact.

Hoistway and Elevator Entrances

FIGURE 5-12 An elevator must be automatic and have proper call buttons, car control, illumination, door timing, and a reopening device to meet the demands of ADA requirements. *Courtesy Uniform Federal Accessibility Standards.*

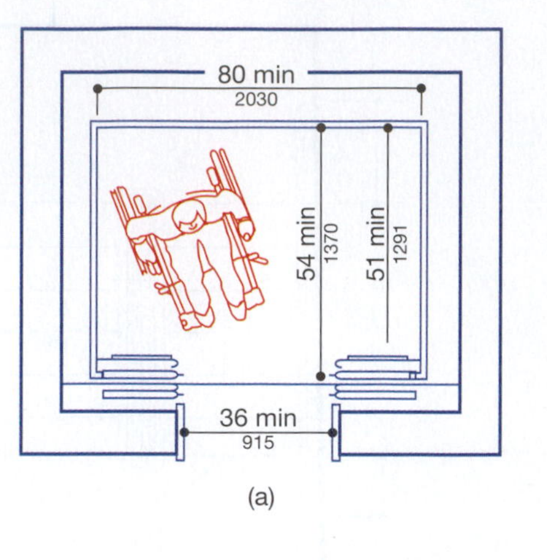

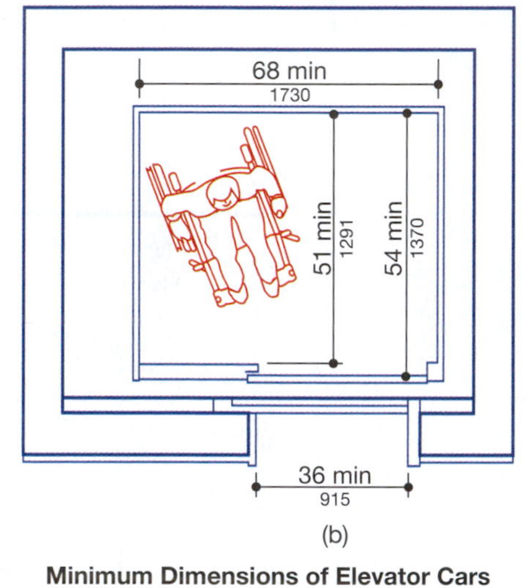

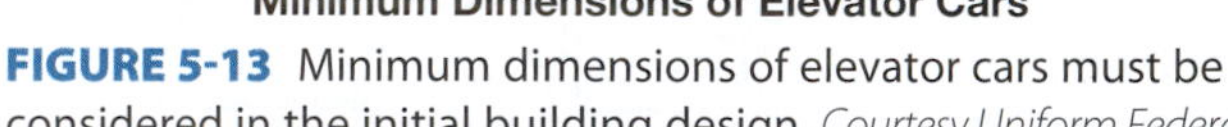

Minimum Dimensions of Elevator Cars

FIGURE 5-13 Minimum dimensions of elevator cars must be considered in the initial building design. *Courtesy Uniform Federal Accessibility Standards.*

Areas of Rescue Assistance

Elevators do not provide a good means of exit from a building in an emergency situation such as a fire. This presents a substantial exit problem when an elevator is the means of access to the building. People with disabilities have been given a way of getting into the building but no way of getting out. The solution to this problem is an area of rescue assistance where people with disabilities can wait safely until rescue crews come to their aid. This area must be on or near the exit route so that the rescue crew has quick and easy access to its occupants. The space needs to be out of the normal exit flow so that its occupants do not interfere with the exiting of others and must provide a smoke-free and fire-safe environment. There must be a two-way communication system between the rescue area and the main entrance to the building area so that rescue workers can verify that someone needs their assistance or the people with disabilities can cry for help.

The area of rescue assistance must provide a minimum of two 30" × 48" (760 × 1200 mm) areas. These areas can be located on a stairway landing with a smoke-proof enclosure, on an exterior exit balcony located adjacent to an exit stairway, or in a one-hour fire-rated corridor immediately adjacent to an exit enclosure. They can be located in an elevator lobby that is separated from the balance of the building by two-hour fire-resistive construction when the elevator shaft is properly pressurized. They can also be located on a stairway landing within an exit enclosure that is vented to the exterior and separated from the rest of the building by a one-hour door. There must be at least one 30" × 48" (760 × 1200 mm) rescue location for each 200 occupants of the building. The stairway serving the rescue area must have 48" (1220 mm) minimum between the handrails to allow room for someone to be carried down the stairs.

The project designer determines the location of the areas of rescue assistance. It is the technician's job to detail the areas and coordinate with the proper consultants. Fire and smoke protection must be properly detailed. Mechanical ducts must be properly dampered. A damper is a device that closes the duct so that smoke is not blown into the space through the mechanical system. Electrical devices such as communication systems that are provided in the rescue area must also be detailed throughout the drawings.

Doors

People with disabilities must have access to all areas on floors that are required to be accessible. At least one door to each space must meet the following requirements:

- Doors must provide a minimum of 32" (810 mm) clear in the 90° position as shown in Figure 5-14. This normally requires a 36" (915 mm) door.
- Doors must open with one hand and without requiring a tight grip, tight pinching, or twisting of the wrist. This normally requires lever-operated door handles, push-type mechanisms, or U-shaped handles. Note that this requirement also applies to sliding and pocket doors.
- The hardware may not be mounted more than 48" (1220 mm) above the floor. Interior hinged doors, sliding doors, and folding doors must open with a maximum force of 5 pounds (2.3 kg). This is important with regard to automatic closing devices. It is recommended that doors have a 10" (255 mm) high kickplate on the bottom.

Maneuvering space at doors shall be as shown in Figures 5-15a and 5-15b. The required width in front of doors can have an influence on hall or corridor widths. Doors also have a required clear width on the swing side. This can affect widths of halls and alcoves when doors are located at their ends. Assuring that these clearances are provided is often the responsibility of the technician. If a door is placed in a manner that does not allow for the required clearances, the technician should point this out to the designer so that the proper adjustments to the design may be made.

RESTROOMS AND BATHING FACILITIES

The ADA requires at least one restroom to be accessible to people with disabilities for each sex, per accessible floor. In some cases a unisex restroom can satisfy this requirement. The door to the restrooms must meet all door requirements shown in Figure 5-15, including the distance between doors at the vestibule. All fixtures intended for people with disabilities shall be on an accessible route. This means that there must be adequate access to the fixtures, and the restroom must have room for a wheelchair to maneuver. Figure 5-16

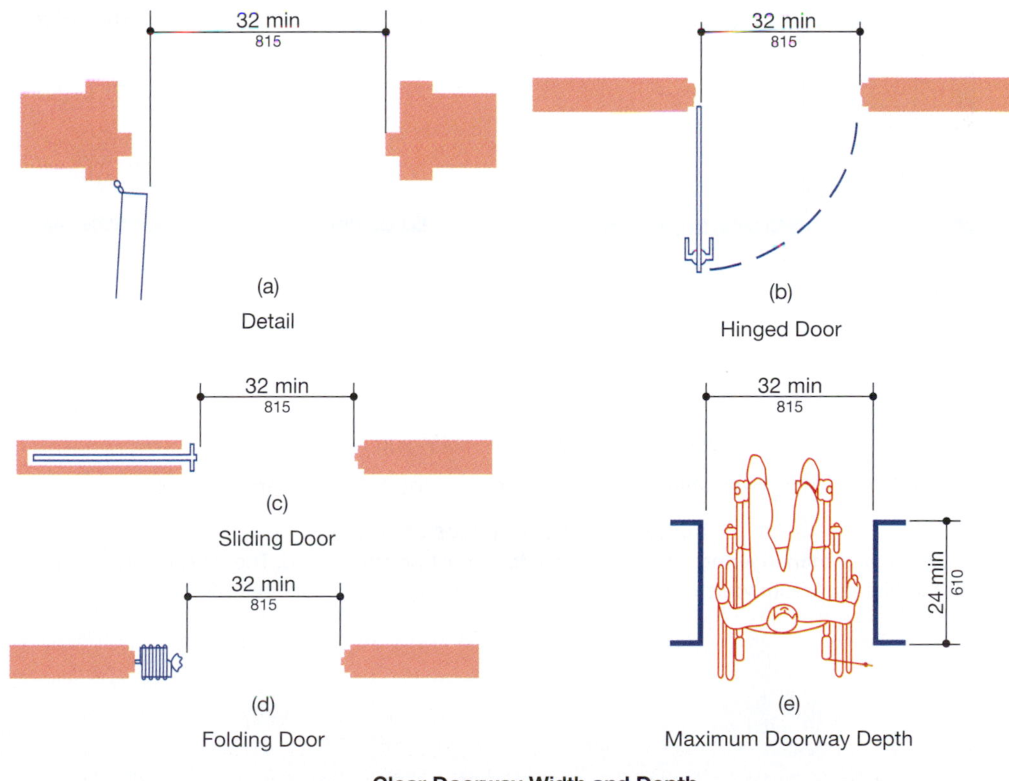

FIGURE 5-14 Minimum clear doorway width and depth. *Courtesy Uniform Federal Accessibility Standards.*

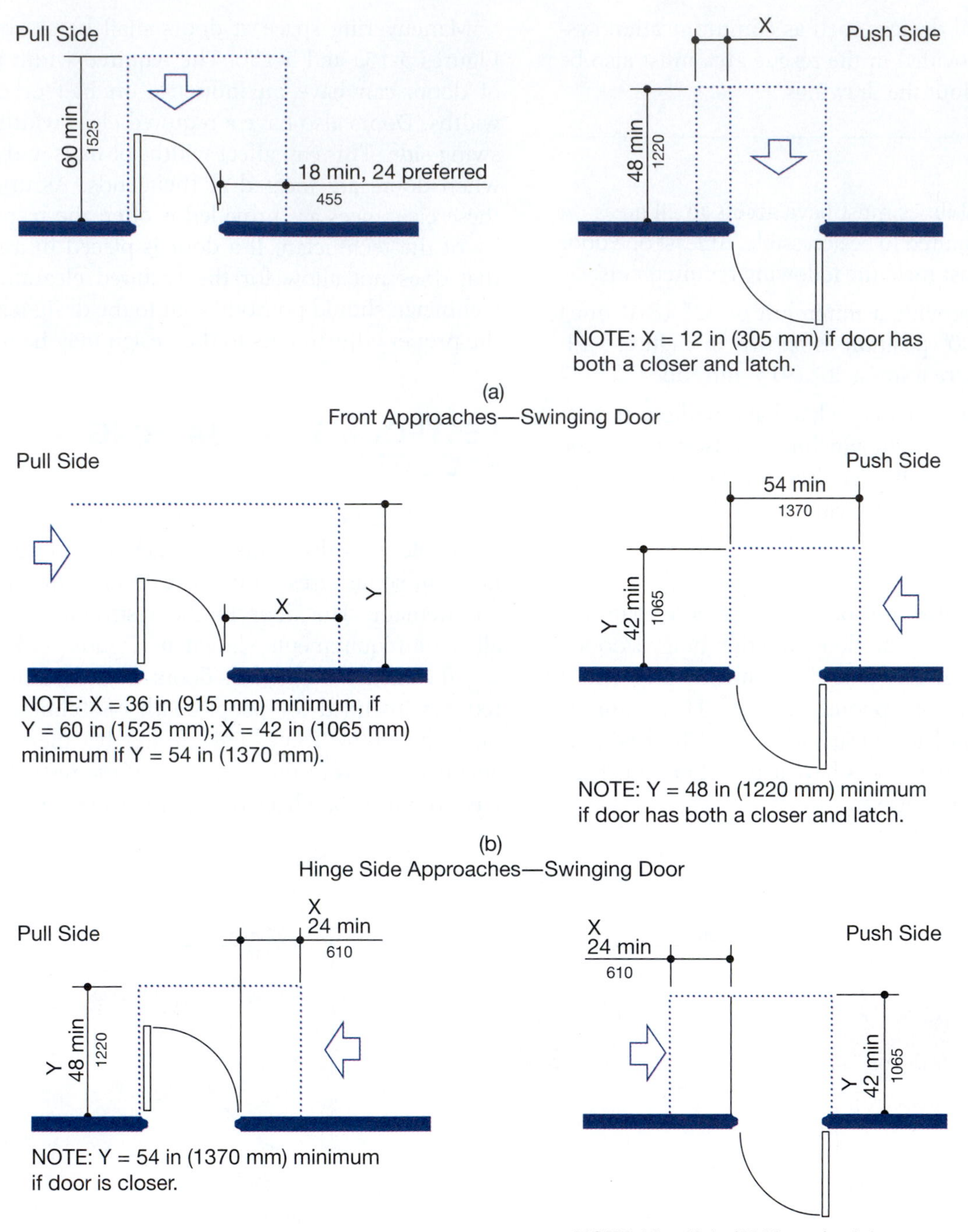

FIGURE 5-15 To allow for easy maneuvering, the area near a door must be considered. The size and shape of the maneuvering space varies depending on the direction from which the door is approached. *Courtesy Uniform Federal Accessibility Standards.*

(continued)

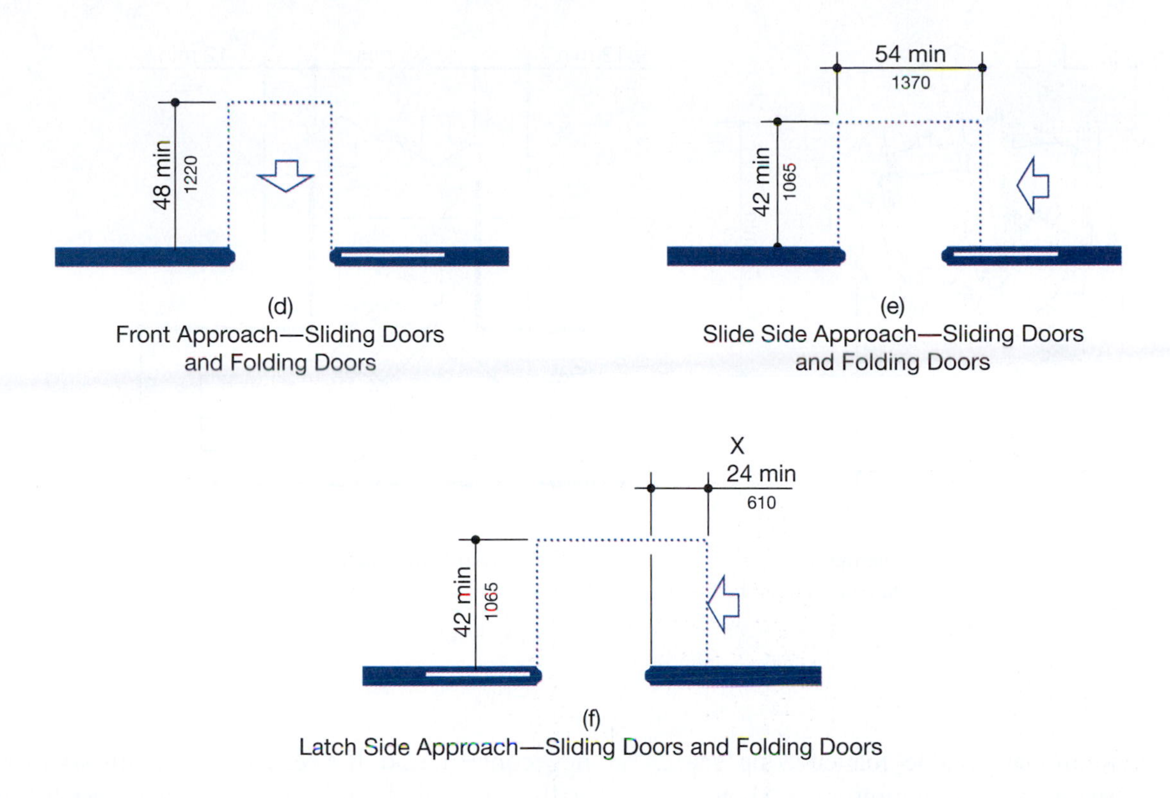

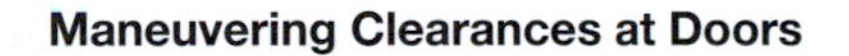

NOTE: All doors in alcoves shall comply with the clearances for front approaches.

Maneuvering Clearances at Doors

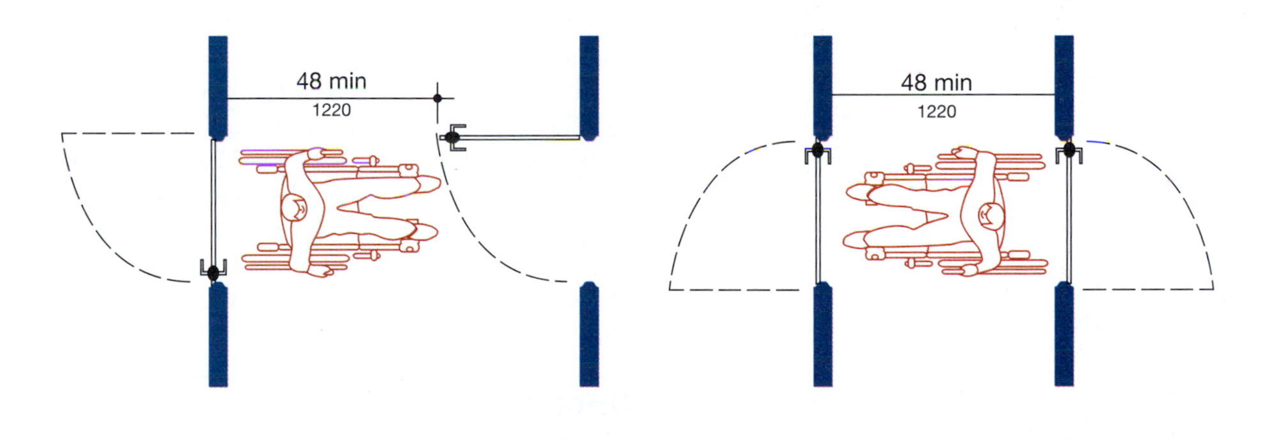

Two Hinged Doors in Series

FIGURE 5-15 *(continued)* To allow for easy maneuvering, the area near a door must be considered. The size and shape of the maneuvering space varies depending on the direction from which the door is approached. *Courtesy Uniform Federal Accessibility Standards.*

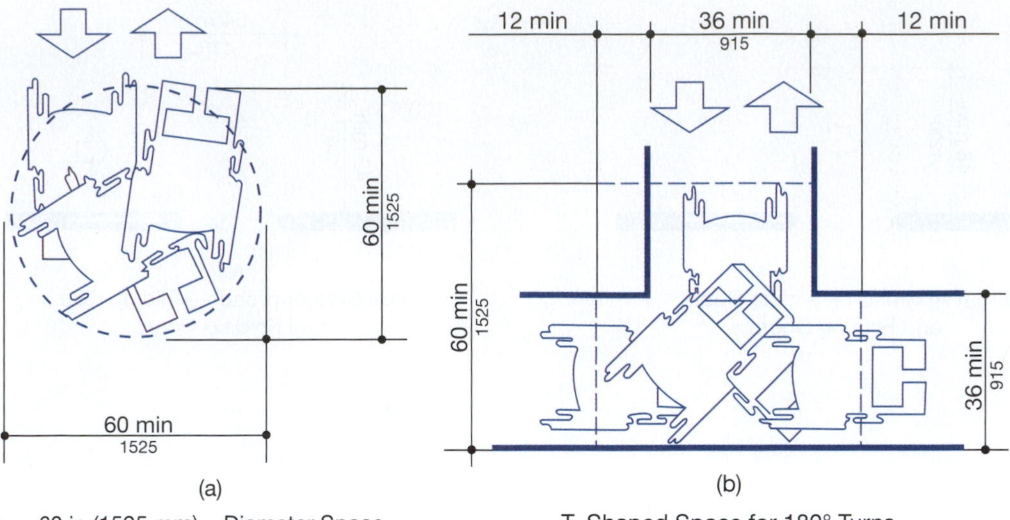

FIGURE 5-16 Restrooms and other tight spaces must have adequate turning space for a wheelchair. *Courtesy Uniform Federal Accessibility Standards.*

shows the minimum acceptable maneuvering space. A preferred maneuvering configuration is shown in Figure 5-17.

Toilet Stalls

At least one toilet stall in each restroom shall meet the requirements shown in Figure 5-18a. These stalls may be reversed to allow for either a left- or right-handed approach. Alternate stall placement such as the layout shown in Figure 5-18b is not allowed in new construction. If a restroom contains six or more stalls, in addition to the standard accessible stall, one of these stalls must be at least 36" (915 mm) wide with a door that swings outward. The front and one side partition of the stall must be at least 9" (230 mm) above the floor unless the stall is longer than 60" (1525 mm). A single-person restroom with a toilet and a sink without a stall shall meet the requirements shown in Figure 5-19. Details that show the door swing in relationship to the stalls should be provided to insure that required clearances are maintained.

The toilet or water closet shall be 17" to 19" (430 to 485 mm) to the top of the toilet seat. Stalls for wall-mounted toilets can be 3" (75 mm) shorter than those for floor-mounted toilets. Grab bars, shown in Figure 5-18, shall meet the requirements illustrated in Figure 5-9e. The toilet-paper dispenser shall be 19" to 36" (485 to 915 mm) above the floor.

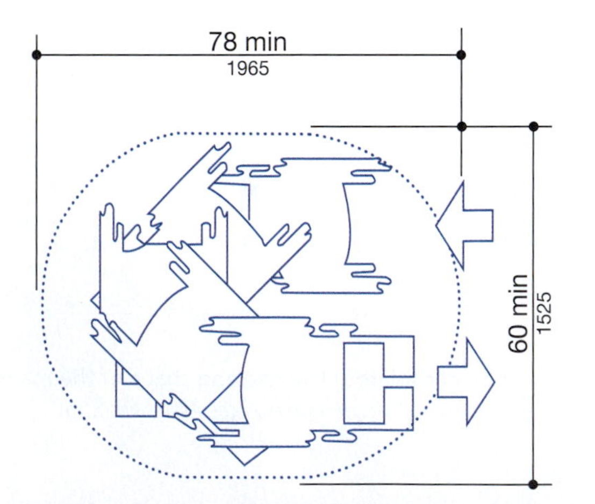

FIGURE 5-17 The preferred wheelchair turning space exceeds the minimum required turning radius. *Courtesy Uniform Federal Accessibility Standards.*

Urinals

If urinals are provided, at least one shall meet the following requirements: It shall be stall-type or wall-hung with an elongated rim at a maximum of 17" (430 mm) above the floor. A clear floor space of 30" × 48" (815 × 1220 mm) shall be provided in front of the urinals. This clear space can overlap the accessible route. Partitions that do not project beyond the rim can be installed with 29" (740 mm) minimum clearance.

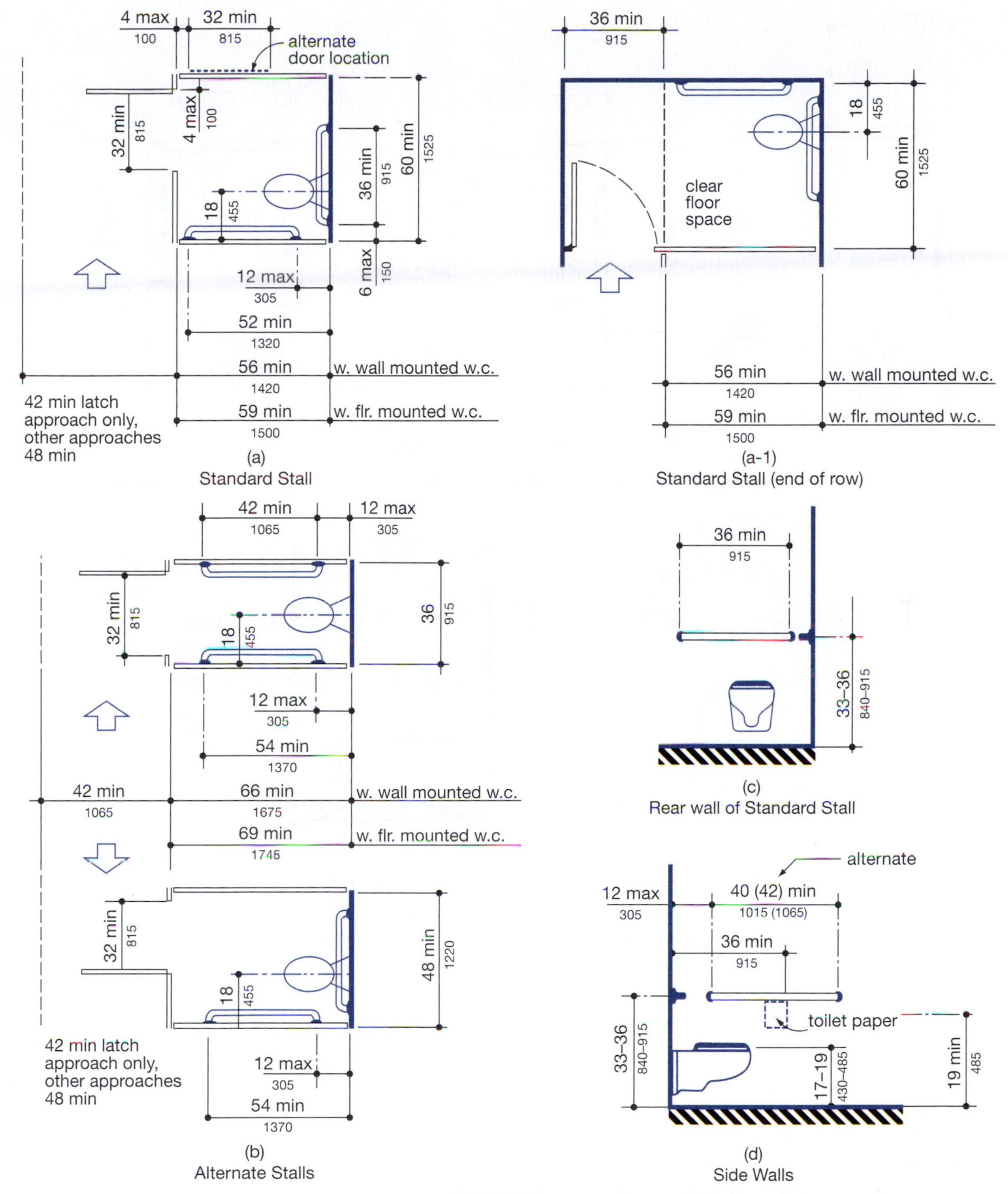

FIGURE 5-18 The requirements for toilet stalls. The alternate stall sizes shown in (b) are not allowed in new construction. *Courtesy Uniform Federal Accessibility Standards.*

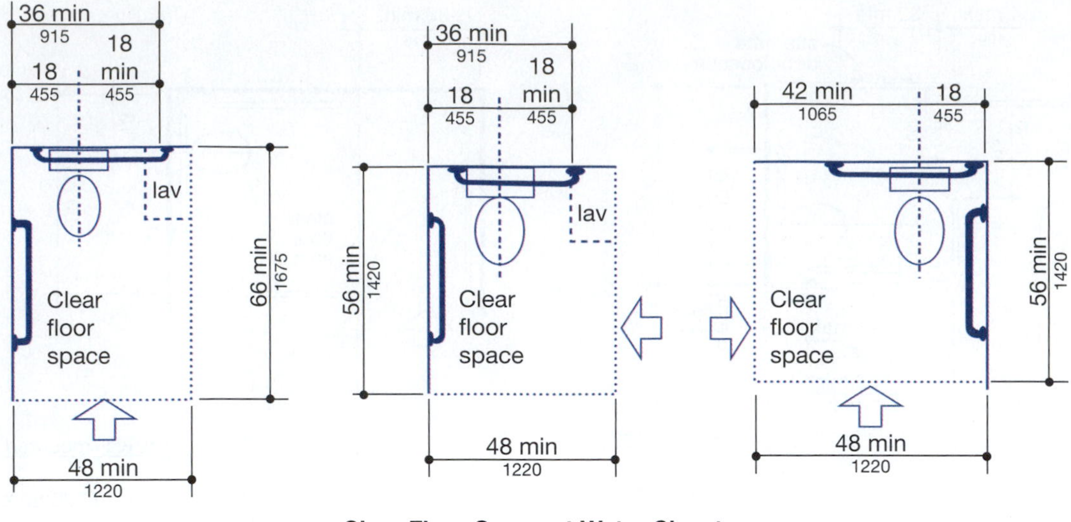

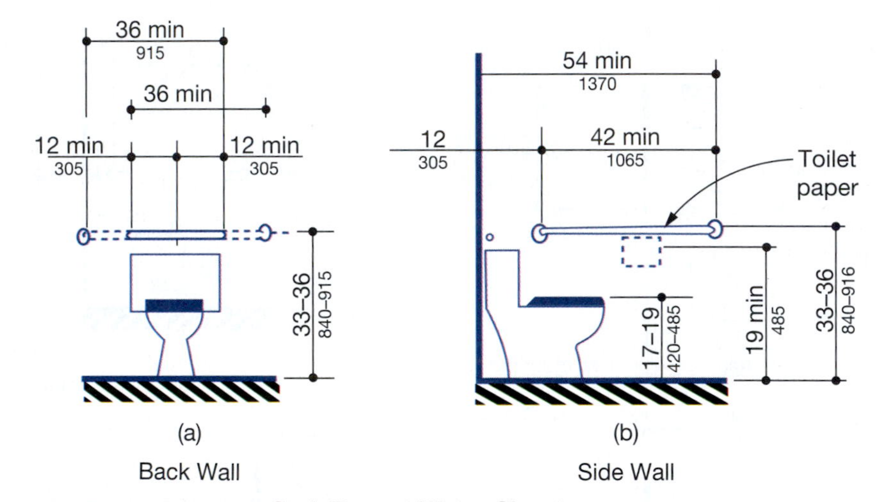

FIGURE 5-19 Design standards for a single toilet that is not in a stall. *Courtesy Uniform Federal Accessibility Standards.*

Lavatories and Mirrors

If lavatories and mirrors are provided, at least one shall meet the following requirements: Lavatories and mirrors shall conform to the requirements shown in Figure 5-20. The clear space can overlap the accessible route. Hot water and drainpipes under lavatories shall be insulated or otherwise configured to protect against contact. There shall be no sharp or abrasive surfaces under the lavatories. Faucets shall be operable with one hand and without requiring a tight grip, tight pinching, or twisting of the wrist. They shall be operable with a maximum force of 5 pounds (2.3 kg). Lever-operated, push-type, and electrically operated faucets are examples of acceptable designs. If self-closing valves are used, the faucet shall remain open for at least 10 seconds.

Bathtubs

If bathtubs are provided, at least one shall meet the requirements:

- Provide seats, grab bars, and clear space in front of the tub, as shown in Figure 5-21.
- Doors shall not swing into the required clear space. Grab bars shall be placed per Figure 5-9e.
- Faucets shall have controls as described above for lavatories and shall be located as shown in Figure 5-21.

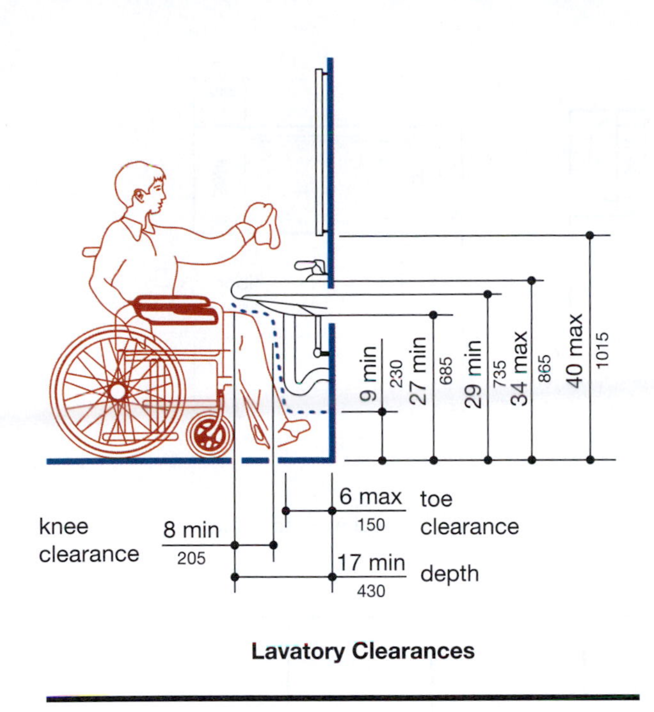

Lavatory Clearances

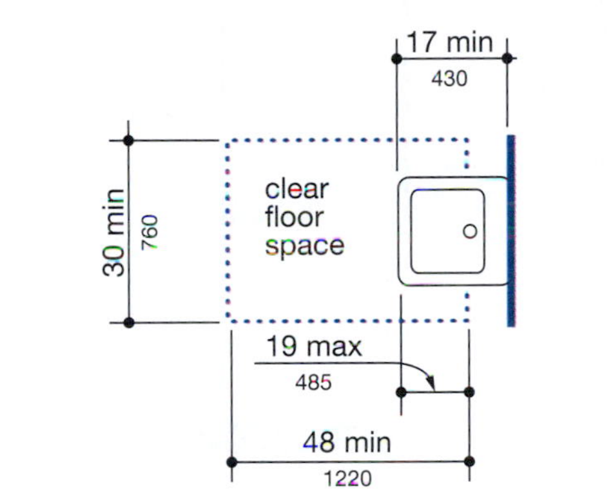

Clear Floor Space at Lavatories

FIGURE 5-20 Placement and clearances at lavatories and mirrors. *Courtesy Uniform Federal Accessibility Standards.*

- The seat must be securely mounted. See "Handrails, Grab Bars, and Tub and Shower Seats" later in this chapter for the structural requirements of the seat and grab bars.
- Tub enclosures may not obstruct the controls or transfer onto the seat.
- Enclosures shall not have tracks mounted on the rim of the tub.

Shower Stalls

If shower stalls are provided, at least one shall meet the following requirements:

- Shower stalls for hotels, motels, and other transient lodging facilities shall be provided as shown in Figure 5-22.
- All other accessible showers shall be the size and shape shown in Figure 5-23. Doors shall not swing into the required clear space.
- Seats shall be located between 17" and 19" (430 and 485 mm) above the floor and shall be per Figure 5-24.

When a seat is provided for in the 30" × 60" (760 × 1525 mm) shower configuration, it shall fold down and be located on the wall adjacent to the controls. This will allow a wheelchair to roll into the shower or a person to use the built-in seat and be able to reach the controls. See Figure 5-21 and Figure 5-9e for grab bar locations. See "Handrails, Grab Bars, and Tub and Shower Seats" later in this chapter for the structural requirements of the seat and grab bars.

Faucets shall have controls as described previously for lavatories and the controls shall be located as shown in Figure 5-25. Tub enclosures may not obstruct the controls or restrict transfer onto the seat. Shower spray units are required to be attached to a minimum hose length of 60" (1525 mm). They are required to be mounted in such a manner as to be usable as a handheld unit or a wall-mounted unit. If an unmonitored shower site is subject to vandalism, a fixed shower head mounted 48" (1220 mm) above the floor can be used in lieu of a hand-held unit.

Curbs are not allowed at the entry to a 30" × 60" (760 × 1525 mm) shower. If a curb is installed at a 36" × 36" (915 × 915 mm) shower, it may not be higher than 1/2" (13 mm). The requirement not to have a curb or to have a very small curb means that the structural floor must be recessed at the shower to accommodate the slope of the shower floor.

Toilet Room Accessories

At least one paper towel dispenser, air hand dryers, toilet seat cover dispensers, and other similar fixtures in each restroom shall be located from 48" to 54" (1220 to 1370 mm) from the floor, as shown in Figure 5-26. Wall-mounted accessories need to be detailed for height and reach limit. Check the approach to accessories to verify whether front or side reach limits should be used.

Medicine Cabinets

If medicine cabinets are provided, they shall be located with a usable shelf no higher than 44" (1120 mm) above the floor. They shall have an accessible floor space in front of them.

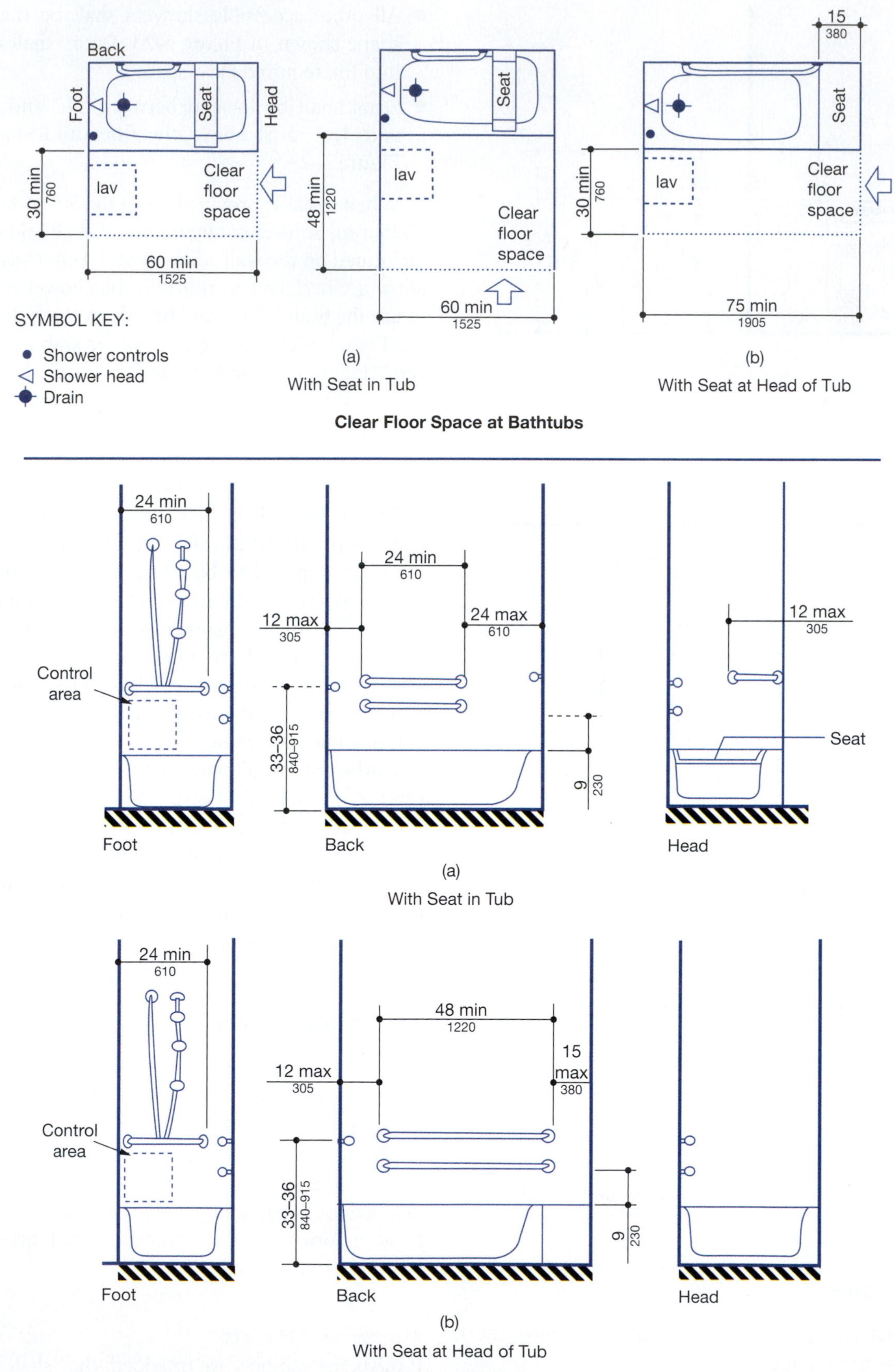

FIGURE 5-21 Required sizes and locations of bathtubs and grab bars. Note that lavatories may protrude into the required clear space for a tub. *Courtesy Uniform Federal Accessibility Standards.*

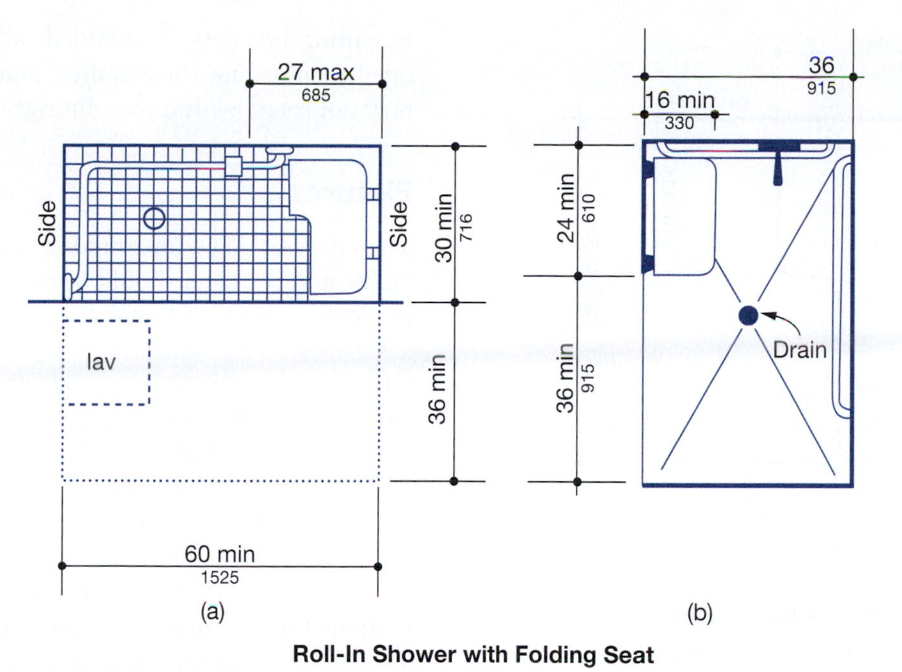

FIGURE 5-22 A roll-in shower with folding seat. The configurations in this figure are required for shower stalls in hotels, motels, and other transient lodging. *Courtesy Uniform Federal Accessibility Standards.*

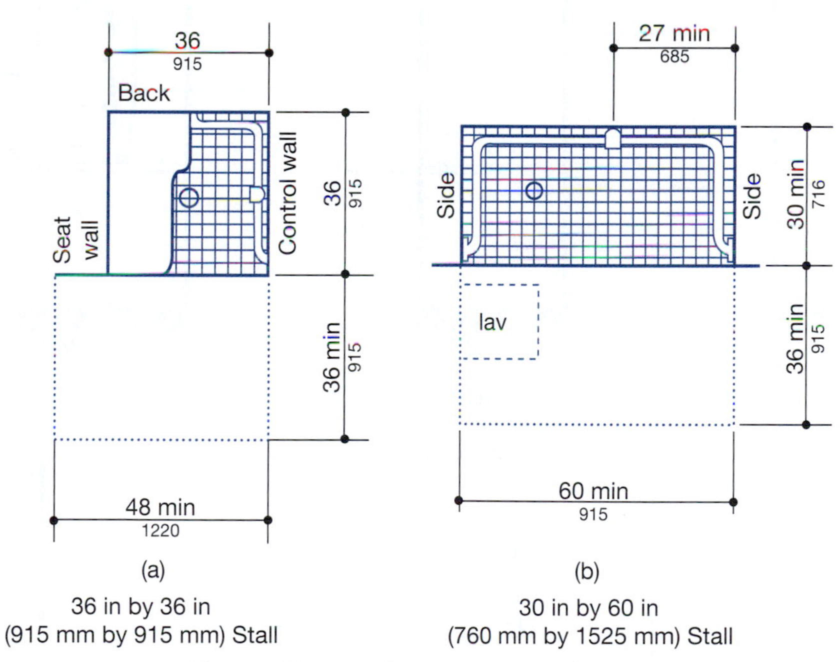

FIGURE 5-23 Minimum shower size and clearances for showers in facilities other than those listed in Figure 5-22. *Courtesy Uniform Federal Accessibility Standards.*

Handrails, Grab Bars, and Tub and Shower Seats

Handrails and grab bars shall be 1 1/4" to 1 1/2" (32 to 38 mm) in diameter or a shape that provides an equivalent gripping surface and shall have a 1 1/2" (38 mm) clear space between them and adjacent surfaces, as shown in Figure 5-9. Handrails, grab bars, and the surface behind them shall be free of sharp edges. Edges shall have a minimum radius of 1/8" (3 mm).

Grab bars and shower seats shall be capable of safely supporting a 250 lb (113 kg) load at any point. This will require a secure surface for mounting on and secure

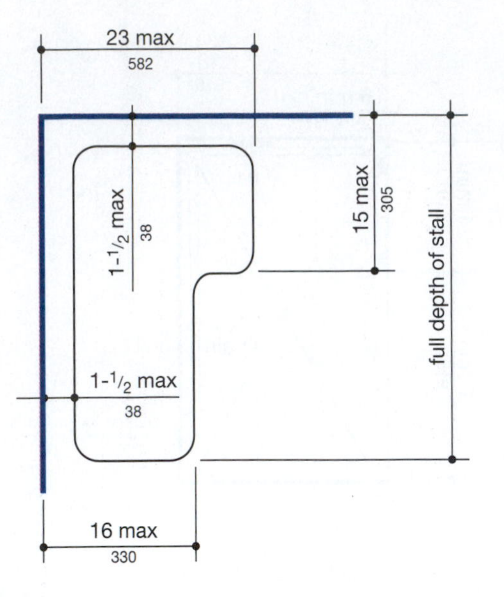

FIGURE 5-24 Required shower seat dimensions. *Courtesy Uniform Federal Accessibility Standards.*

mounting brackets. Standard details should be developed to illustrate the required attachment. Grab bars may not rotate within their fittings.

Fixtures

The effect of ADA on drinking fountains, telephones, tables, and counters should be considered in the design process.

Drinking Fountains

At least half the drinking fountains provided on an accessible floor shall meet the requirements illustrated in Figure 5-27. These drinking fountains must also have a proper water flow direction and height. Because of the water flow requirements, it is important to choose a drinking fountain or water cooler that is certified by the manufacturer to meet ADA requirements. There must also be drinking fountains at a height appropriate for those people who cannot stoop over.

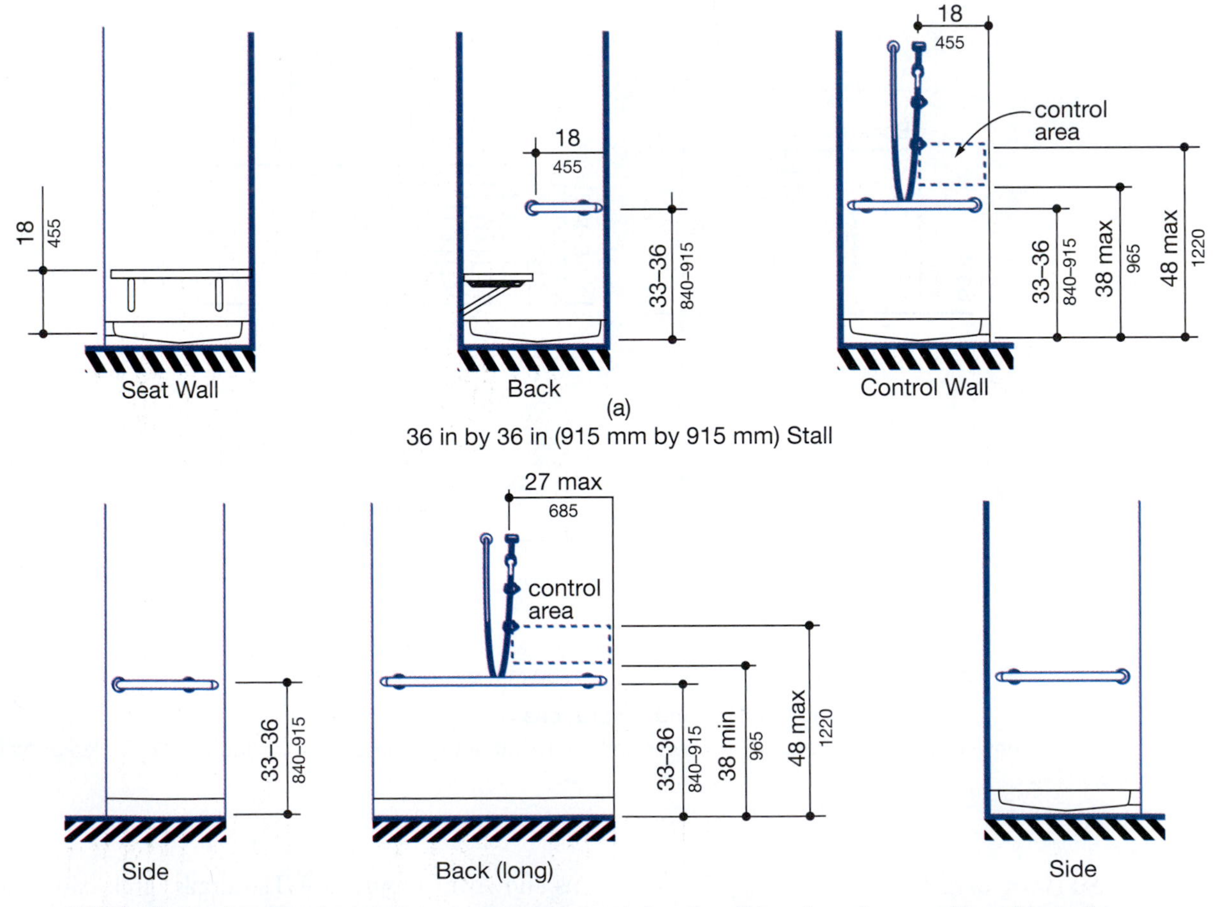

FIGURE 5-25 Minimum requirements for grab bars in showers and required control location. *Courtesy Uniform Federal Accessibility Standards.*

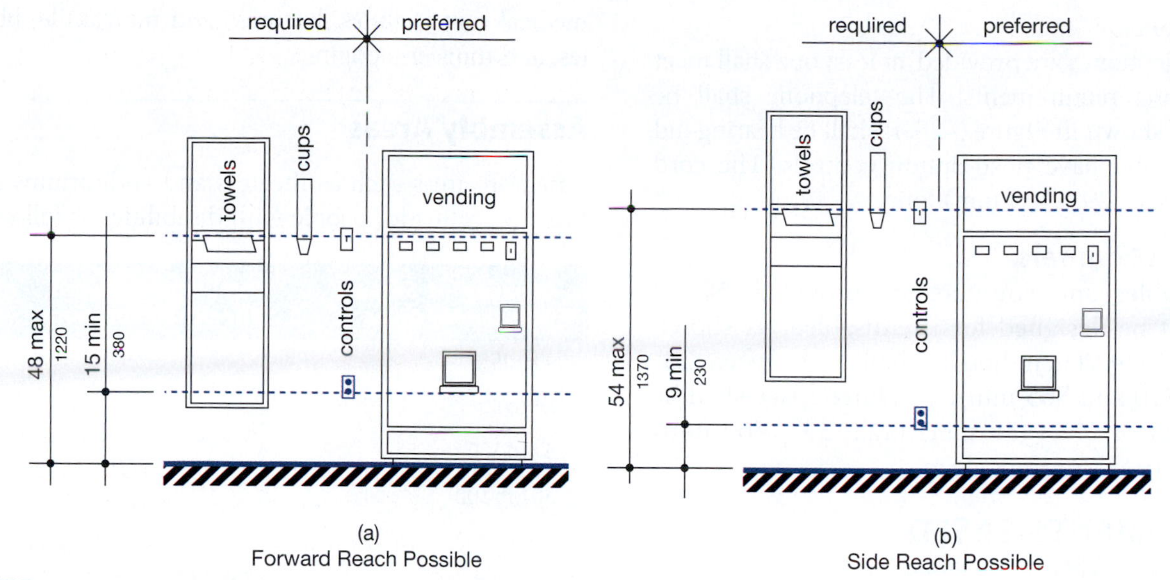

FIGURE 5-26 Mounting guidelines for bathroom accessories. *Courtesy Uniform Federal Accessibility Standards.*

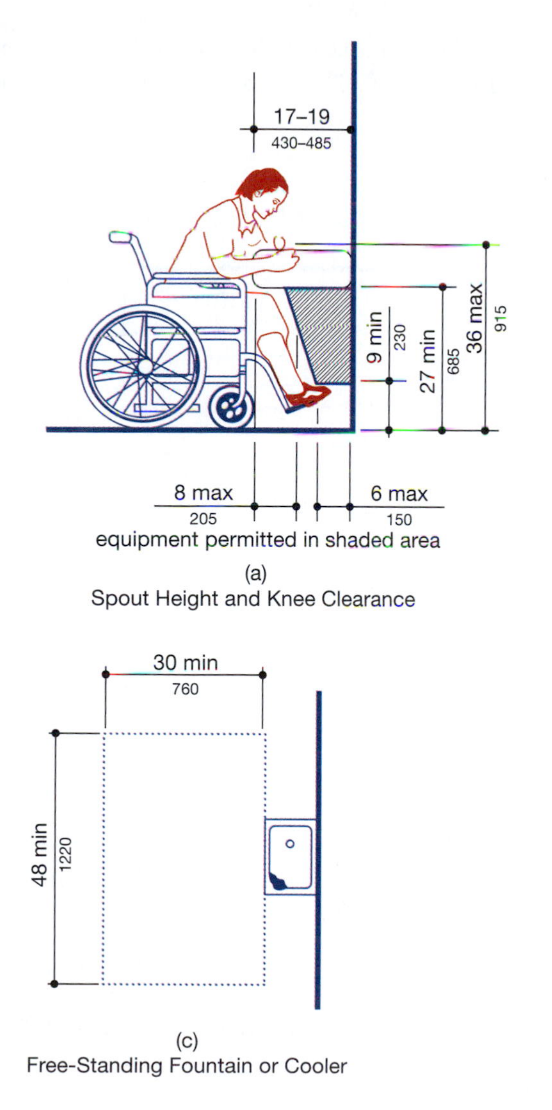

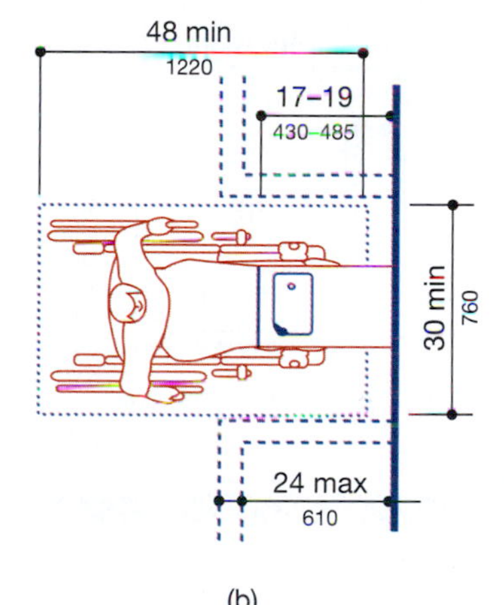

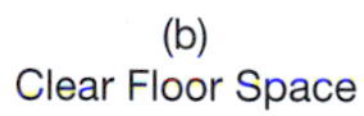

(b)
Clear Floor Space

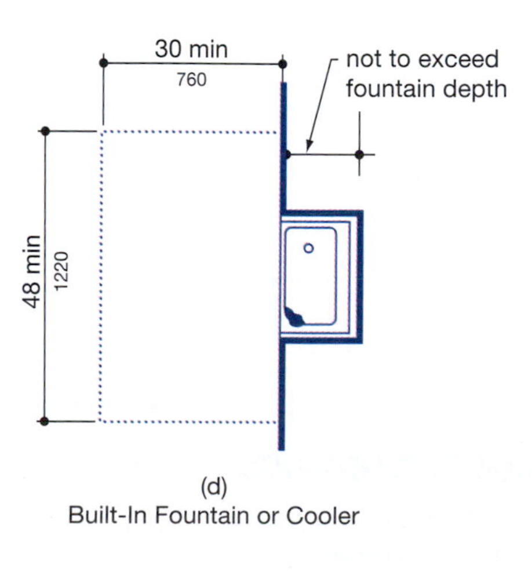

(d)
Built-In Fountain or Cooler

Drinking Fountains and Water Coolers

FIGURE 5-27 Requirements for placement of a drinking fountain. *Courtesy Uniform Federal Accessibility Standards.*

Telephones

If public telephones are provided, at least one shall meet the following requirements: The telephone shall be installed as shown in Figure 5-28. It shall be hearing-aid compatible and have push-button controls. The cord shall be at least 29" (735 mm) long.

Tables and Counters

If fixed tables and counters are provided, 5% of them must be designed for the disabled, as shown in Figure 5-29. Their height shall be between 28" and 34" (710 and 865 mm). The knee space shall be a minimum of 27" (430 mm) high, 30" (760 mm) wide, and 19" (485 mm) deep.

REQUIREMENTS BASED ON OCCUPANCY

Design parameters are also established based on the usage of a structure. Occupancies addressed by the ADA include assembly areas, restaurants and cafeterias, medical care facilities, business and mercantile, libraries, and transient lodging.

Assembly Areas

Assembly areas such as theaters and auditoriums shall provide seating for people with disabilities as follows:

Capacity of Seating in Assembly Areas	Number of Required Wheelchair Locations
4 to 25	1
26 to 50	2
51 to 300	4
301 to 500	6
501 or more	6 + 1 for each 100 spaces (or portion thereof)

These spaces shall conform to the shape shown in Figure 5-30. These seating locations shall be distributed

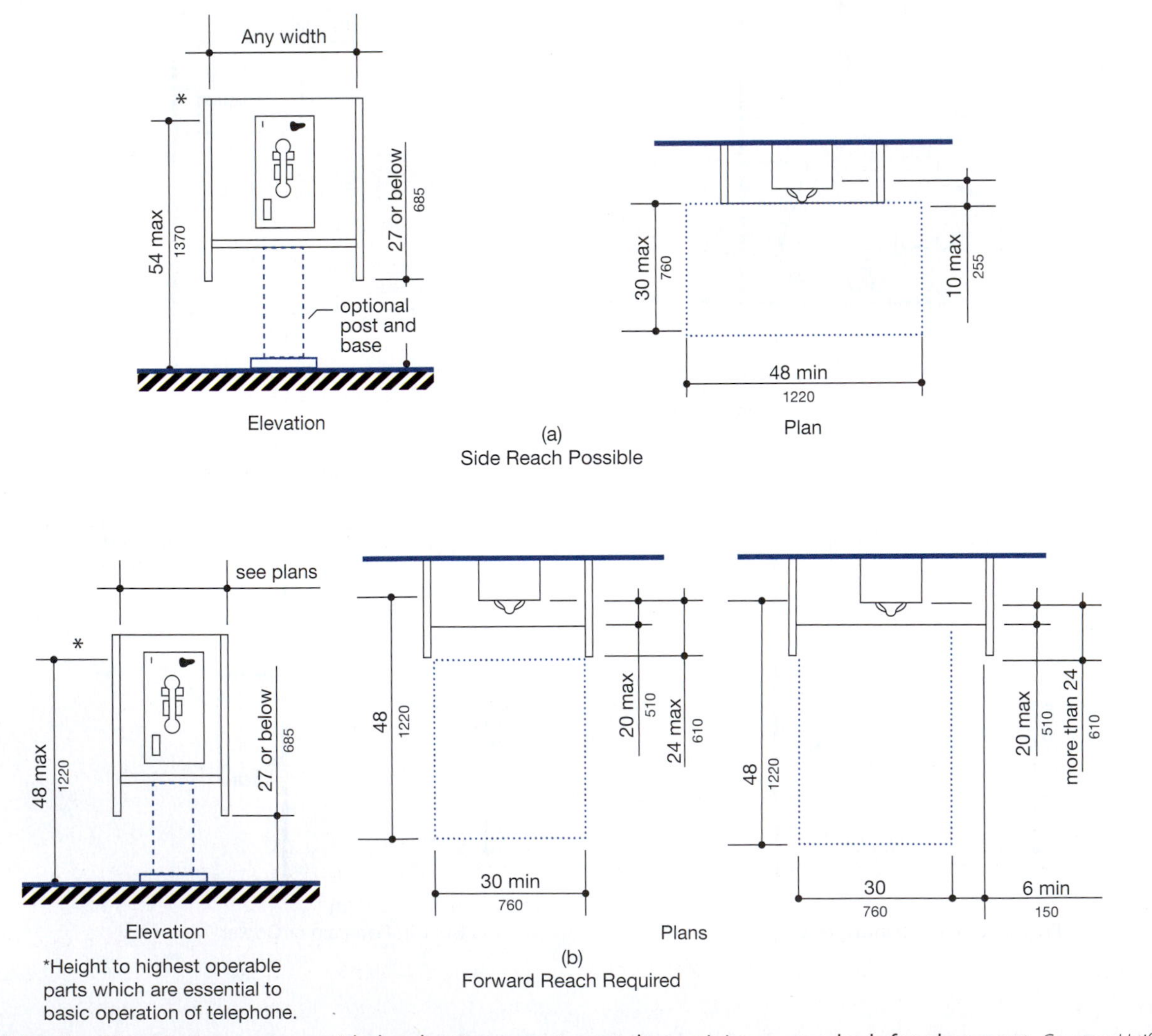

FIGURE 5-28 If public telephones are provided, as least one must meet these minimum standards for placement. *Courtesy Uniform Federal Accessibility Standards.*

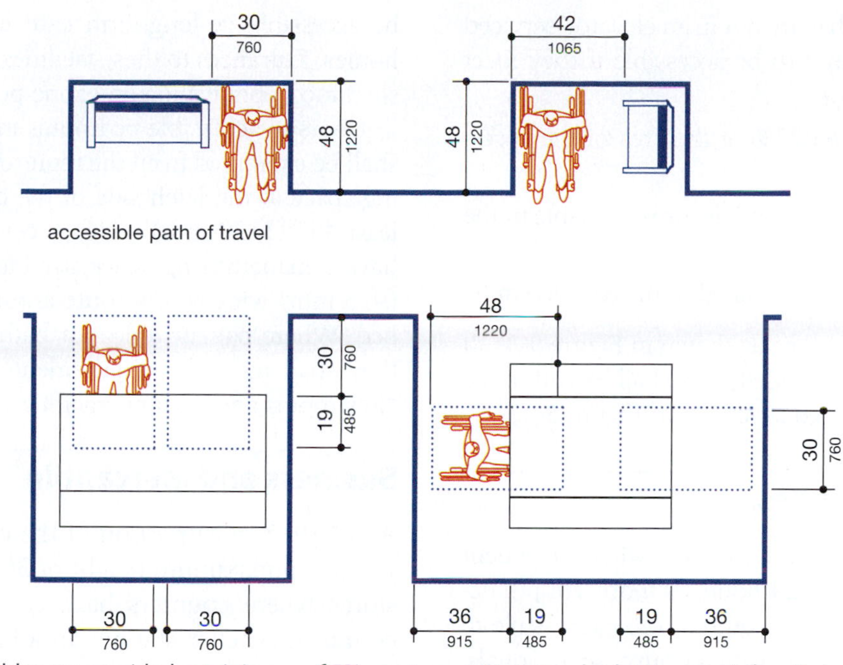

FIGURE 5-29 If fixed tables are provided, a minimum of 5% must meet ADA standards. *Courtesy Uniform Federal Accessibility Standards.*

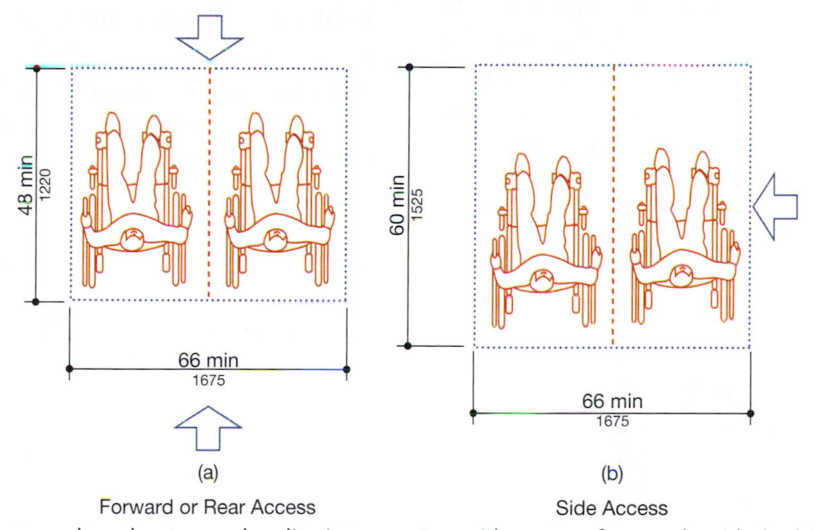

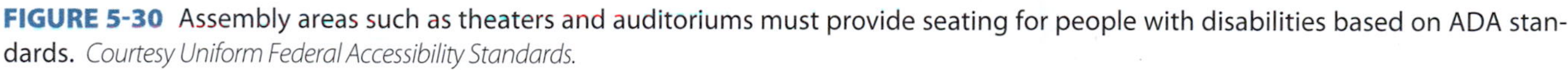

FIGURE 5-30 Assembly areas such as theaters and auditoriums must provide seating for people with disabilities based on ADA standards. *Courtesy Uniform Federal Accessibility Standards.*

in such a manner as to provide the disabled with a choice of admission prices and line of sight. Seating must be accessible and adjacent to at least one companion fixed seat. When the seating capacity exceeds 300, wheelchair spaces shall be provided in more than one place. Removable seats can be temporarily placed in wheelchair spaces. In addition to the wheelchair spaces, 1% of all fixed seats must be aisle seats with no armrest on the aisle side.

Accessible viewing positions can be clustered for bleachers, balconies, and other areas having sight lines that require slopes of greater than 5%. Equivalent accessible viewing positions can be located on levels having accessible egress.

Restaurants and Cafeterias

Where tables or counters are provided for dining or drinking, 5% shall be accessible as described in the previous section "Tables and Counters." Where food or drink is served at counters higher than 34" (865 mm), a 5' (1525 mm) section shall be provided matching the requirements of Figure 5-29 or at a table in the same area. Seating for people with disabilities should be distributed in smoking and nonsmoking areas. All dining and drinking areas, except some mezzanines, shall be accessible. This includes raised, sunken, and outside areas. Raised platforms such as those provided for the head table at a banquet shall be accessible to the disabled. Mezzanines that are not in an elevator-serviced building are not required to be accessible if they meet the following conditions:

- They are smaller than 33% of the area of the accessible area.
- The same level of service and decor is available in the accessible area.
- The accessible area is not limited to the disabled only.

Food-service lines shall be as shown in Figure 5-31. If self-service shelves are provided, they shall be within the reach range of the disabled as shown in Figure 5-32.

Medical Care Facilities

This section covers medical facilities where a patient might stay for a period of 24 hours or more. All public- and common-use areas are required to be accessible to people with disabilities. In general-purpose hospitals, psychiatric facilities, and detoxification facilities, 10% of the patient bedrooms and toilets are required to be accessible. In hospitals and rehabilitation facilities that specialize in treating conditions that affect mobility, all patient bedrooms and toilets are required to be accessible. At least 50% of patient bedrooms and toilets shall be accessible in long-term care facilities and nursing homes. Entrances to these facilities shall be covered and shall incorporate a loading zone per Figure 5-33.

Doors to accessible bedrooms in acute-care hospitals shall be exempted from the requirements for maneuvering space at the latch side of the door if the door is at least 44" (1120 mm) wide. Accessible bedrooms shall have a maneuvering space per Figure 5-17, and a 36" (915 mm) wide access route around three sides of the bed. Where patient toilets or bathrooms are provided, they shall meet the requirements discussed earlier in "Restrooms and Bathing Facilities."

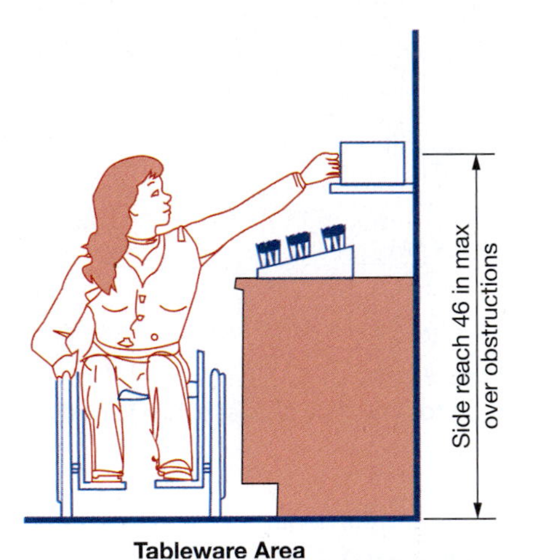

FIGURE 5-31 Guidelines for tableware areas in food-service facilities.

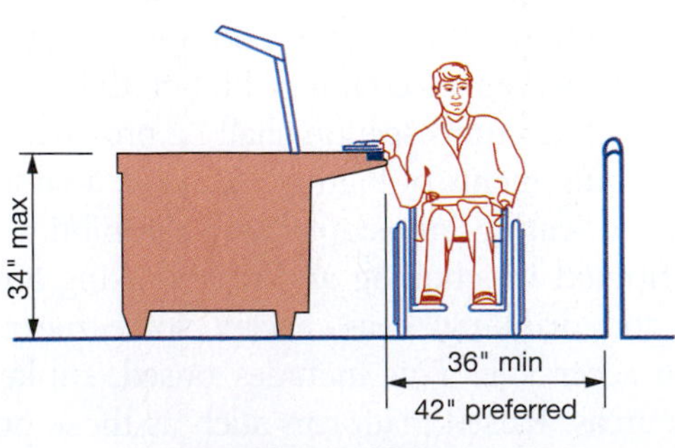

FIGURE 5-32 Minimum sizes for food-service areas.

Business and Mercantile

A 36" (915 mm) portion of the counter shall be provided at a maximum height of 36" (915 mm) in retail stores where counters have cash registers, at ticket counters, at teller stations in a bank, and at registration counters in hotels and motels. In lieu of this requirement, an auxiliary counter with a maximum height of 36" (915 mm) can be provided in close proximity to the main counter. Accessible checkout aisles shall be provided per Table 5-2 unless the selling space is less than 5000 square feet. When there is less than 5000 square feet of selling space, one accessible checkout aisle shall be provided.

Security bollards or similar devices used to prevent the removal of shopping carts from the store shall not prevent access or egress to people in wheelchairs. An alternate entry that is equally convenient to that provided for the ambulatory population is acceptable.

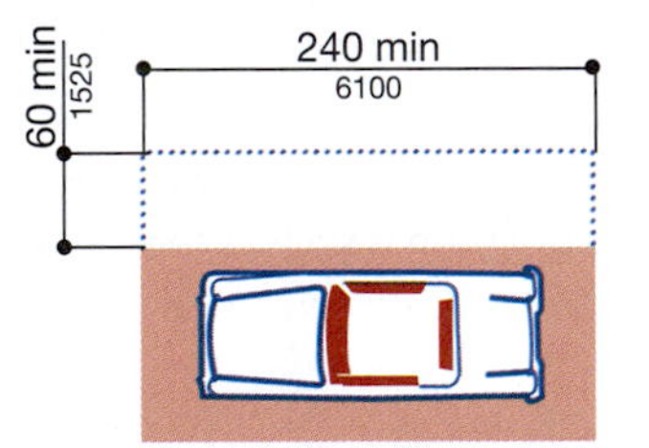

FIGURE 5-33 Loading zones for patients at a medical facility. This space is required to be covered. *Courtesy Uniform Federal Accessibility Standards.*

Total Checkout Aisles of Each Design	Minimum Number of Accessible Checkout Aisles of Each Design
1–4	1
5–6	2
8–15	3
More than 15	3 + 20% of additional aisles

TABLE 5-2 Checkout Aisles.

Libraries

In reading and study areas, at least 5% of fixed seating, tables, and study carrels shall meet the requirements for tables and counters. At least one lane at each checkout area shall meet the requirements for counters with cash registers, as described previously in "Business and Mercantile." Card catalogs and magazine displays shall comply with Figure 5-34 and stacks shall comply with Figure 5-35.

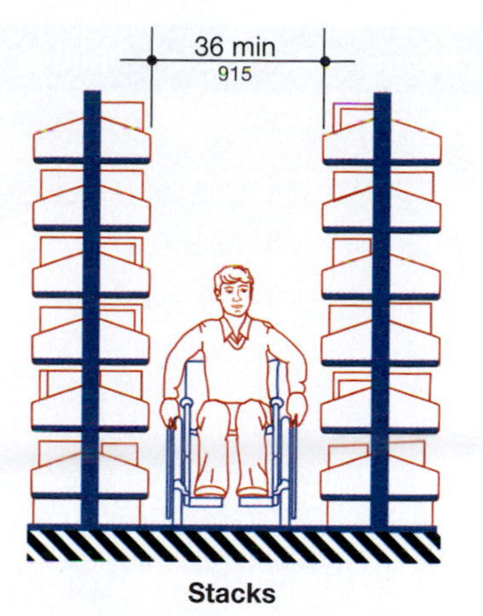

FIGURE 5-35 Placement requirements for book stacks in a library. *Courtesy Uniform Federal Accessibility Standards.*

Transient Lodging

All public- and common-use areas in hotels, motels, inns, boarding houses, dormitories, resorts, and other similar lodging places are required to be accessible. A portion of the sleeping rooms or suites is required to be accessible per Table 5-3. The roll-in shower required by this table shall conform to Figure 5-23. Accessible rooms or suites shall be dispersed among the various classes of accommodations. Selection can be limited to rooms for multiple occupancies if they are offered at single-occupancy prices for people with disabilities needing single occupancy.

Accessible rooms shall be on an accessible route. They shall have maneuvering space as shown in Figure 5-16. There shall be a 36" (915 mm) space on both sides of a single bed or a 36" (915 mm) space between two beds. In accessible suites, the living rooms, dining areas, covered parking areas, one bathroom, and one sleeping area shall be accessible. Balconies and terraces shall be accessible unless higher thresholds or a change in level is necessary to protect the unit from wind or water damage (equivalent facilities must be provided). If storage cabinets are provided, at least one type of each shall be constructed per Figure 5-36.

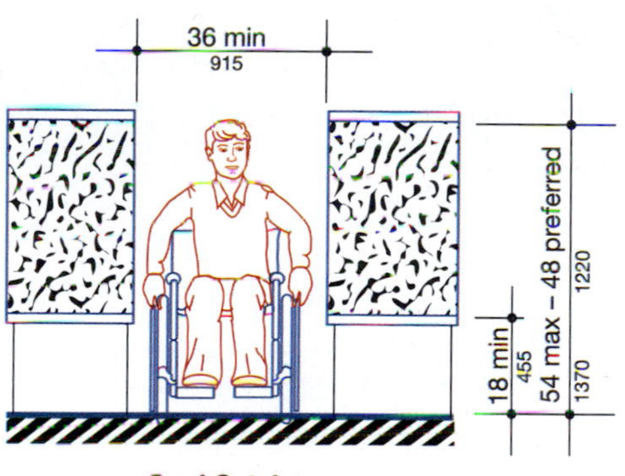

FIGURE 5-34 Standards for placement of card catalogs or other freestanding displays. *Courtesy Uniform Federal Accessibility Standards.*

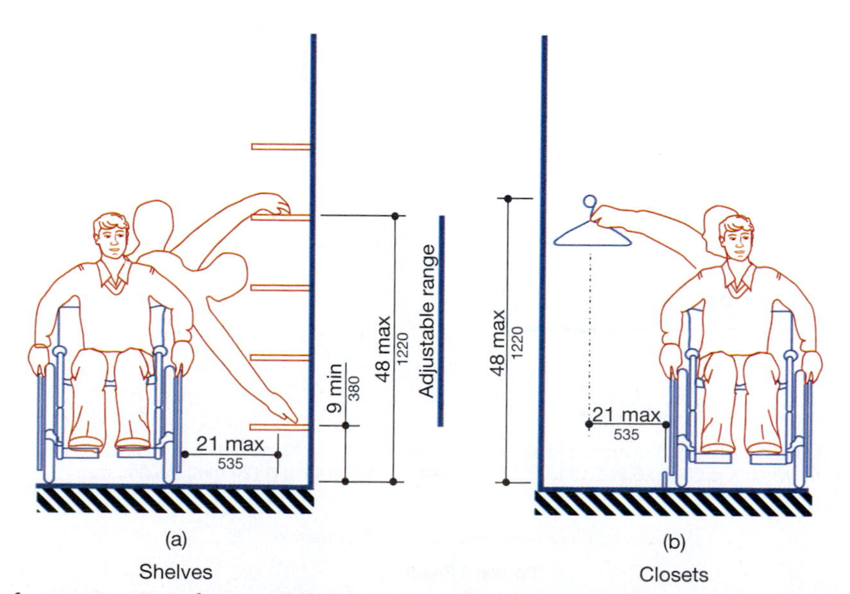

FIGURE 5-36 Guidelines for construction of storage cabinets in accessible sleeping rooms or suites at hotels, motels, inns, boarding houses, dormitories, resorts, and other similar lodging places. *Courtesy Uniform Federal Accessibility Standards.*

Number of Rooms	Accessible	Roll-In Showers
1 to 25	1	
26 to 50	2	
51 to 75	3	1
76 to 100	4	1
101 to 150	5	2
151 to 200	6	2
201 to 300	7	3
301 to 400	8	4
401 to 500	9	4 + 1 for each additional 100 over 400

TABLE 5-3 Required Accessible Rooms Based on the Total Number of Rooms Based on Uniform Federal Accessibility Standards.

Kitchens and wet bars, when provided, shall have counters no higher than 34" (865 mm) and at least 50% of the cabinets and refrigerator/freezer space shall be within the reach range shown in Figures 5-37 and 5-38. The space shall be designed to allow for the operation of cabinet and appliance doors so that all cabinets and appliances are accessible and usable.

Rooms shall be provided per Table 5-4 for the hearing impaired, in addition to the rooms required for the disabled by Table 5-3. Rooms or suites required to accommodate persons with hearing impairments shall have auxiliary visual devices to alert occupants of danger and phone calls. Phones shall have volume controls and an electrical outlet within 4' (1220 mm) of the telephone connection.

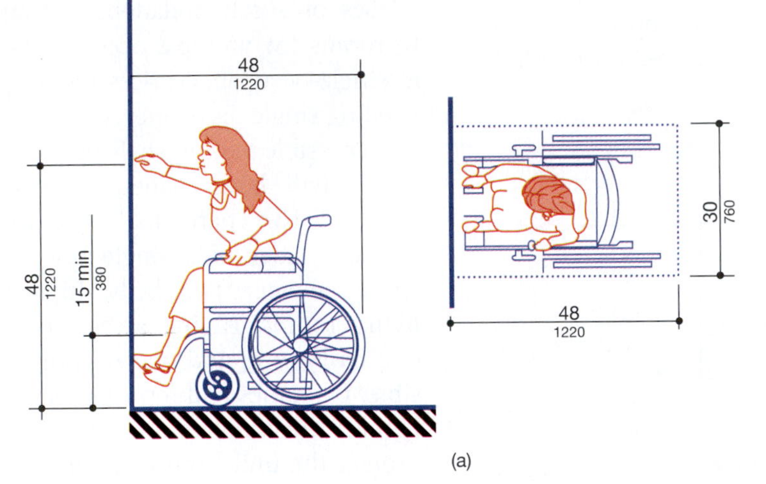

(a)

High Forward Reach Limit

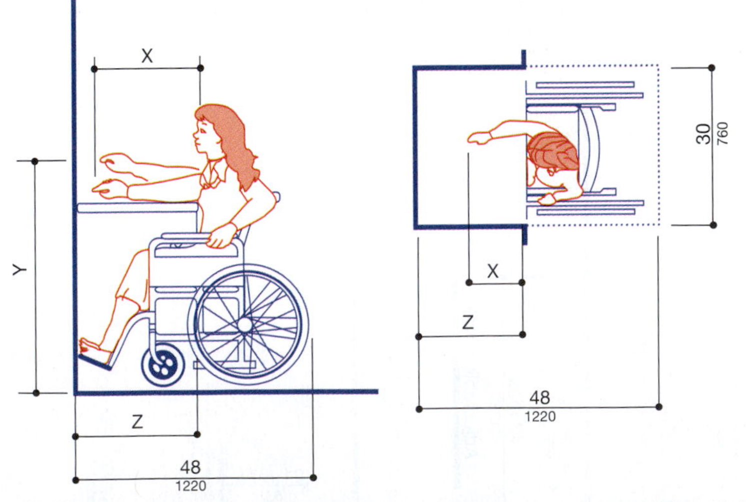

NOTE: X shall be ≤ 25 in (635 mm); Z shall be ≥ X. When X < 20 in (510 mm), then Y shall be 48 in (1220 mm) maximum. When X is 20 to 25 in (510 to 635 mm), then Y shall be 44 in (1120 mm) maximum.

(b)

Maximum Forward Reach over an Obstruction

Forward Reach

FIGURE 5-37 Kitchens and wet bars, when provided, shall have counters no higher than 34" (865 mm) and at least 50% of the cabinets and refrigerator/freezer space shall be within the specified reach range. *Courtesy Uniform Federal Accessibility Standards.*

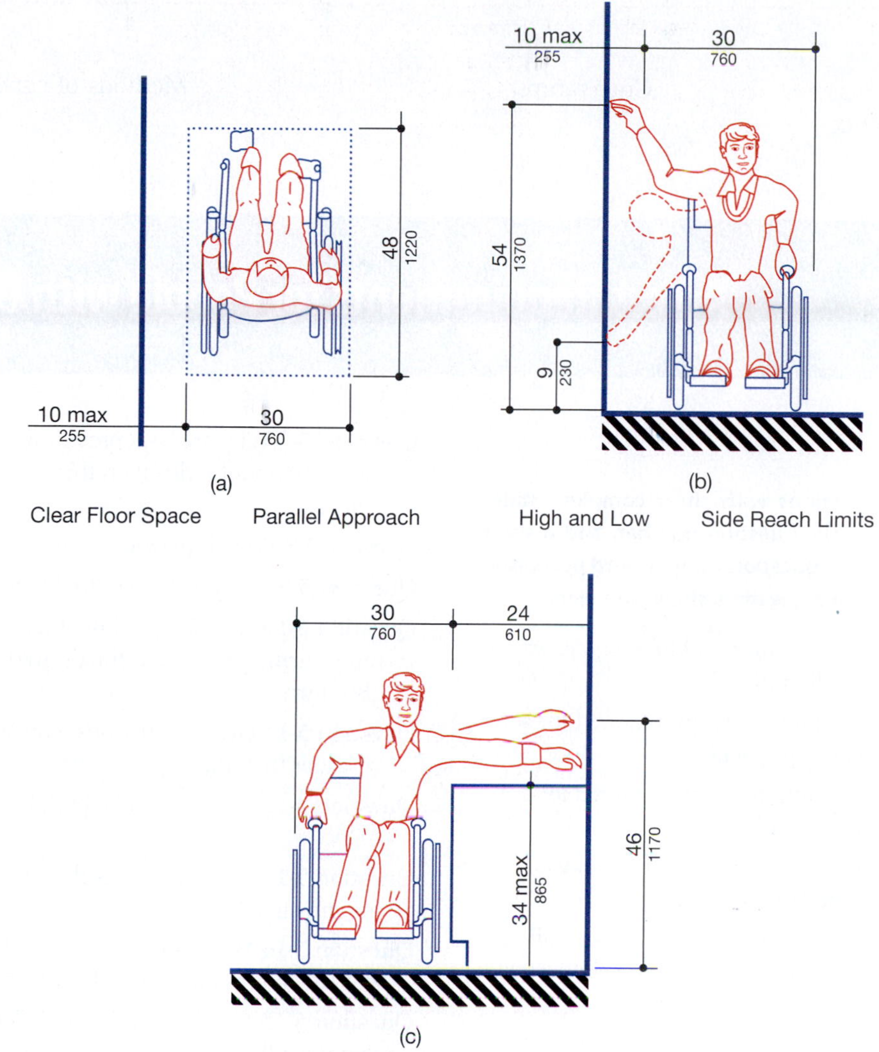

FIGURE 5-38 Design standards for side reach. *Courtesy Uniform Federal Accessibility Standards.*

Number of Rooms	Accessible Rooms
1 to 25	1
26 to 50	2
51 to 75	3
76 to 100	4
101 to 150	5
151 to 200	6
201 to 300	7
301 to 400	8
401 to 500	9
501 to 1000	2% of total
1001 and over	20 + 1 for each 100 over 1000

TABLE 5-4 Number of Rooms to Accommodate the Hearing Impaired Based on Uniform Federal Accessibility Standards.

ADDITIONAL READING

The following Web sites can be used as a resource to help get additional information about the ADA and to keep you current with changes.

ADDRESS	COMPANY/ ORGANIZATION
www.access-board.gov/adaag/html/adaag.htm	Federal Government ADA Accessibility Guidelines for Buildings and Facilities
www.usdoj.gov/crt/ada/adahom1.htm	Federal Government ADA

KEY TERMS

Accessible route
Areas of rescue assistance
Curb ramps
Methods of egress

CHAPTER 5 TEST Access Requirements for People with Disabilities

QUESTIONS

Answer the following questions with short complete statements. Type the chapter title, question number, and a short complete statement for each question using a word processor. If math is required to answer a question, show your work.

Question 5-1 Does a two-story office building require an elevator for people with disabilities?

Question 5-2 What is the minimum width required for a single van-accessible parking space?

Question 5-3 What is the minimum depth required for a disabled-accessible toilet space?

Question 5-4 What is the maximum slope of a walkway (in the direction of travel) before it becomes a ramp?

Question 5-5 What is the maximum allowable cross fall (the slope perpendicular to the direction of travel) in decimals of a foot for a 4'–0" wide accessible route?

Question 5-6 What is the maximum height allowed for a single run of a ramp?

Question 5-7 Why are laws providing access for the disabled more important today than they might have been 60 years ago?

Question 5-8 What is the minimum length of a 26" high ramp?

Question 5-9 What is the required height of a toilet seat?

Question 5-10 Explain the difference between the minimum turning room and the desired turning room for a 180° turn.

Question 5-11 Give two reasons why an access route should be wider than the required 36".

Question 5-12 Define a public building within the scope of ADA.

Question 5-13 What aids does the ADA require for the hearing impaired?

Question 5-14 What is the maximum height of a stair riser and the minimum length of a stair tread?

Question 5-15 List three types of physical disabilities that must be considered as the plans for a structure are developed.

Section 2

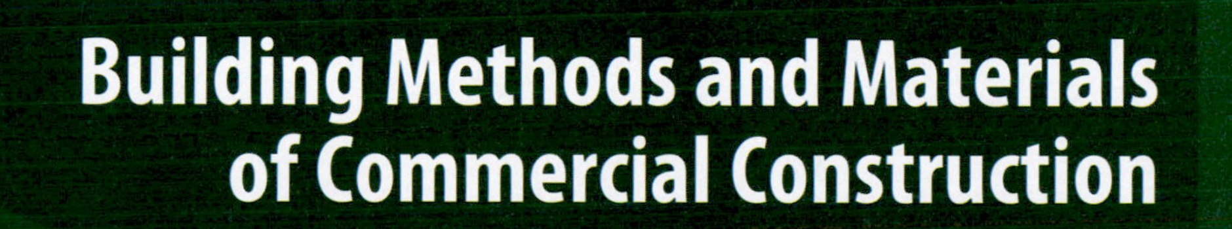

Building Methods and Materials of Commercial Construction

6 Making Connections

Section 2 will examine common commercial construction materials including wood, timber, engineered products, steel, concrete, and concrete blocks. Regardless of the materials being connected, the type of connection is of the utmost importance to the structural integrity of a building. Connections must be fast, economical, and safe. The connection of materials is critical if each element is being used to transfer gravity or lateral loads, or the forces of uplift from one member to another. Structural connectors are specified on the framing plans, wall sections, and specific details. This chapter will introduce common connectors as well as standard methods of representing each type of fastener. Topics to be explored include:

- Structural connectors used with wood and engineered products including adhesives, nails, staples, power-driven anchors, screws, and prefabricated metal connectors.
- Structural connectors used with large wood members including bolts, timber connectors, and prefabricated metal connectors.
- Welded connections.
- Representing structural connectors and other materials in details.

STRUCTURAL CONNECTORS USED WITH WOOD AND ENGINEERED PRODUCTS

Wood, timber, and engineered wood products are usually held together by the use of adhesives, nails, staples, and screws.

Adhesives

Adhesives are a key ingredient in the building industry. Materials such as plywood, OSB, I-joists, and laminated beams all depend on adhesives. Adhesives are also used within the structure to strengthen wood framing connections. Subflooring is often glued to floor joists to stiffen the floor system and help eliminate squeaking. Gusset plates for light trusses are often glued at chord and web connections to strengthen joints. Adhesives are also used to attach plastic piping and finishing materials such as paneling, moldings, and countertops. Structural adhesives should be specified in the written specifications, the framing plan, and specific details.

Nails

Nails are often used for the fabrication of wood-to-wood members when their thickness is less than 1 1/2" (40 mm). Thicker wood assemblies are normally bolted. The type and size of a nail to be used, as well as the placement and nailing pattern, will all affect the strength of the joint. Consideration should be given to how and where nailing specifications will be placed.

Types of Nails

Nails specified by engineers in the calculations for framing include common, deformed, box, and spikes. Most nails are made from stainless steel, but copper and aluminum are also used. Figure 6-1 shows typical types of nails used for construction. Only nails used for structural purposes will be considered at this time. A common nail is used for most rough framing applications. **Common nails** have a single head, smooth shank, and diamond tip. **Box nails** are slightly thinner and have less holding power than common nails. Box nails are used because they generate less resistance in penetration and are less likely to split the lumber. Box nails come in sizes up to 16d. **Spikes** are nails larger than 20d. For further explanation on the measurements of nails, see the upcoming section "Nail Sizes." Nailing specifications are found throughout the written specifications, on framing plans, and in details.

Nail Components

Nails consist of a head, shank, and tip. The head and the shank are often specified by the engineer in the calculations to meet a specific need.

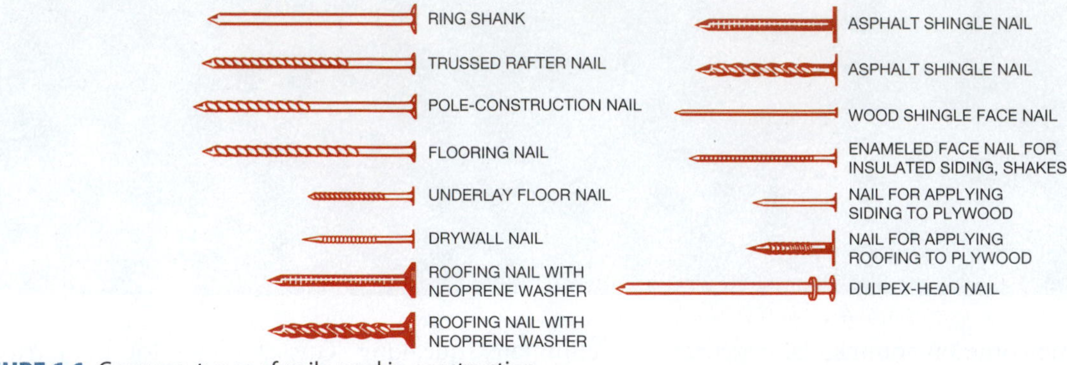

FIGURE 6-1 Common types of nails used in construction.

Head The *head* of most nails used for framing is usually a flat round head. Siding and wood trim are typically installed with nails that have a small cylindrically shaped head. Nails intended for easy removal typically have two heads—a lower head approximately 3/8" (10 mm) below the upper head. The nail is driven in until the first head makes contact, and the second head protrudes for future removal. Boards used for concrete forms are often nailed with double-headed nails.

Shank Nails are specified by the length of the shank. The *shank* of a nail is the body of a nail. It is usually smooth, but ribs or deformations can be added to the shank to increase holding power. Deformations are typically placed perpendicular to the length of the shaft. If material is removed from the shank, a valley is created. Fluted nails have valleys that spiral around the length of the shank.

Tip The *tip* of most nails has a diamond-shaped point, but a chiseled point or a flat tip can be used. The sharper the tip, the less damage done to the surrounding wood fiber, resulting in greater resistance to withdrawal. A **pilot hole** can be predrilled to further reduce wood damage and further increase holding strength. Notes for predrilling are often found on architectural drawings where appearance is critical.

Finish The type of finish of the nail depends on how the nail is manufactured and how it will be used. Most nails have a smooth finish and are referred to as *bright finished* nails. Nails that are tempered by a heating process have a slightly blue finish. A resin coating called a *cement finish* can be applied to a nail to increase the holding power of the nail. A zinc coating is added to nails to increase the holding power and to increase resistance to corrosion. Zinc-coated or galvanized nails reduce the stains that result on siding when uncoated nails are used. If a special finish is required, it is typically added as a footnote to the nailing schedule or placed in the written specifications.

Nail Sizes

Nail sizes are described by the term *penny*, which is represented by the symbol *d* and used to describe standard nails from 2d through 60d. **Penny** is a weight classification of nails, which compares the number of pounds of nails per 1000 nails. For example, 1000 8d nails weigh 8 pounds. Figure 6-2 shows common sizes of nails. Nails are required to penetrate into the supporting member by half the depth of the supporting member. If two 2 × (50×) members are being attached, a nail would be required to penetrate 3/4" (20 mm) into the lower member. The engineer will determine the nail size to be specified, and then the size, spacing, and quantity of nails to be used are specified in structural details.

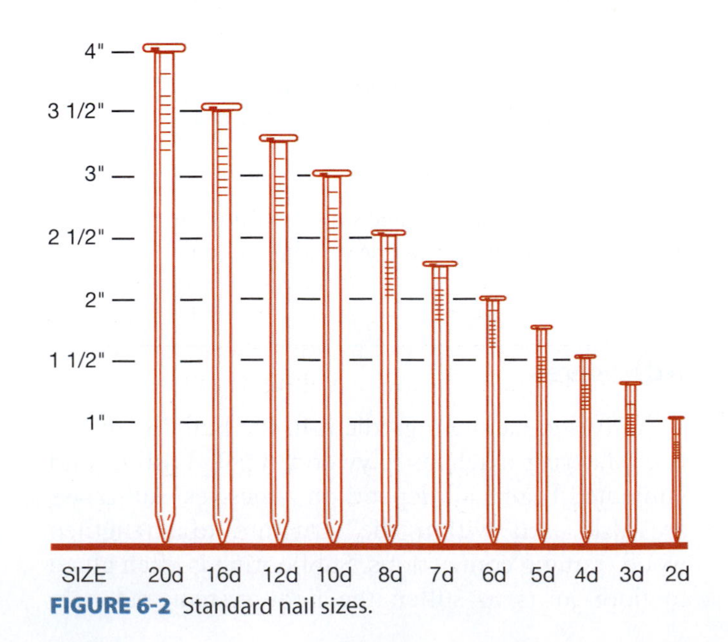

FIGURE 6-2 Standard nail sizes.

Nailing Specifications

Nails smaller than 20d are typically specified by the penny size and spacing. They are used for continuous joints such as a plate attached to a floor system. An example of a nailing note found on a framing plan or detail would be as follows:

3 × 6 DFPT SILL W/ 20d's @ 4" o.c.

meaning that a 3 × 6 Douglas Fir pressure-treated sill is to be attached to another member with 20 penny (4" long) nails placed at 4-inch spacing.

Nails are specified by a quantity and penny size for repetitive joints, such as a joist-to-sill connection. The specification on a detail would simply read (3)-8d's and point to the area in the detail where the nails will be placed. Depending on the scale of the drawing, the nails may or may not be shown. Specifications for spikes are given in a similar manner to nails, but the penny size is replaced by the spike diameter.

Most wood-frame projects include a schedule based on the International Building Code (IBC) drawing requirements. Figure 6-3 shows the IBC schedule for specifying the method, quantity, and size of nailing to be used. Nails are generally not specified in details if the connection is listed in the nailing schedule. Specifications are often given for the location of nails to the end or edge of a piece of lumber. If nails are too close to the edge or end of a piece of lumber, splitting will occur. Minimum edge distances are often included in the nailing specifications to reduce wood failure.

Nailing Placement

Nailing placements are described by the manner in which they are driven into the members being connected. Common methods include face, end, toe, and blind nailing. Figure 6-4 shows each method of driving nails. The type of nailing to be used is determined by how accessible the head of the nail is during construction and the type of stress to be resisted. The engineer will specify if nailing other than that recommended by the nailing schedule of the prevailing code is to be used. Nails that are required to resist shear are strongest when perpendicular to the grain. Nails that are placed parallel to the end grain, such as end nailing, are weakest, and tend to pull out as stress is applied. Common nailing options are listed below.

- **Face nailing** is the process of driving a nail through the face or surface of one board into the face of another. Face nailing is used in connecting sheathing to rafters or studs, in nailing a plate to the floor sheathing, or in nailing a let-in brace to a stud.
- **End nailing** is the process of driving a nail through the face of one member into the end of another member. A plate is end-nailed into the studs as a wall is assembled.
- **Toe nailing** drives a nail through the face of one board into the face of another. With face and end nailing, nails are typically driven in approximately 90° to the face. Toe nailing drives the nail at approximately a 30° angle. Rafter-to-top-plate or header-to-trimmer connections are examples of toe-nailed joints.
- **Blind nailing** is used where it is not desirable to see the nail head. Attaching wood flooring to the subfloor is done with blind nailing. Nails are driven at approximately a 45° angle through the tongue of the flooring and hidden by the next piece of flooring.

Nailing Patterns

Nails for sheathing and other large areas of nailing often refer to nail placement along an edge, boundary, or field nailing. Figure 6-5 shows an example of each location. *Edge nailing* refers to nails placed at the edge of a sheet of plywood. **Field nailing** refers to nails placed into the supports for a sheet of plywood excluding the edges. **Boundary nailing** is the nailing at the edge of a specific area of plywood.

Staples

Power-driven nails have greatly increased the speed and ease at which nails can be inserted. As technology advanced, staples started to replace nailing for some applications. Staples are most often used for connecting asphalt roofing and attaching sheathing to roof, wall, and floor supports. Required size and spacing of staples is listed in the IBC Nailing Schedule.

Power-Driven Anchors

Metal anchors can be used to attach wood or metal to masonry. They range from 1/4" to 1/2" (6 to 13 mm) diameter, in lengths from 3/4" to 6" (20 to 150 mm). **Power-driven anchors** are made from heat-treated steel and inserted by a powder charge from a gun-like device. They are typically used where it would be difficult to insert the **anchor bolts** at the time of the concrete pour. Anchors can also be used to join wood-to-steel construction.

Screws

Screws are often specified for use in wood connections that must be resistant to withdrawal. Three common

NAILING SCHEDULE

CONNECTION	NAILING[1]
1 JOIST TO SILL OR GIRDER, TOE NAIL	3- 8d 3- 3" 14 ga. STAPLE
2. BRIDGING TO JOIST, TOE NAIL EA. END	2- 8d 2- 3" 14 ga. STAPLE
3 1 x 6 (25 X 150) SUBFLOOR OR LESS TO EACH JOIST, FACE NAIL	2-8d
4. WIDER THAN 1 x 6 (25 x 150) SUBFLOOR TO EACH JOIST, FACE NAIL	3-8d
5. 2" (50) SUBFLOOR TO JOIST OR GIRDER BLIND AND FACE NAIL	2-16d
6. SOLE PLATE TO JOIST OR BLOCKING FACE NAIL SOLE PLATE TO JOIST OR BLOCKING AT BRACED WALL PANELS	16d @ 16" (400 mm) O.C. 3" 14 ga. STAPLE @ 12" 3- 16d PER 16" (400 mm) 3" 14 ga. STAPLE PER 16"
7. TOP OR SOLE PLATE TO STUD, END NAIL	2-16d 3- 3" 14 ga. STAPLE
8. STUD TO SOLE PLATE, TOE NAIL	4-8d, TOE NAIL 3- 3" 14 ga. STAPLE or 2-16d END NAIL 3- 3" 14 ga. STAPLE
9. DOUBLE STUDS, FACE NAIL	16d @ 24" (600 mm) O.C. 3" 14 ga. STAPLE @ 8" O.C.
10. DOUBLE TOP PLATE, FACE NAIL DOUBLE TOP PLATE, LAP SPLICE	16d @ 16" (400mm) O.C. 3" 14 ga. STAPLE @ 12" O.C. or 8-16d 12- 3" 14 ga. STAPLE
11. BLOCKING BTWN. JOIST OR RAFTERS TO TOP PLATE, TOE NAIL	3- 8d 3- 3" 14 ga. STAPLE
12. RIM JOIST TO TOP PLATE, TOE NAIL	8d @ 6" (150 mm) O.C. 3" 14 ga. STAPLE @ 6" O.C.
13. TOP PLATES, TAPS & INTERSECTIONS, FACE NAIL	2-16d 3- 3" 14 ga. STAPLE
14. CONTINUED HEADER, TWO PIECES	16d @ 16" (400 mm) O.C. ALONG EDGE
15 CEILING JOIST TO PLATE, TOE NAIL	3-8d 5- 3" 14 ga. STAPLE
16. CONTINUOUS HEADER TO STUD, TOE NAIL	4-8d
17. CEILING JOIST, LAPS OVER PARTITIONS, FACE NAIL	3-16d (SEE TABLE 2308.10.4.1) 4- 3" 14 ga. STAPLE
18. CEILING JOIST TO PARALLEL RAFTERS, FACE NAIL	3-16d 4- 3" 14 ga. STAPLE
19. RAFTERS TO PLATE, TOE NAIL	3-8d 3- 3" 14 ga. STAPLE
20. 1" (25mm) BRACE TO EA. STUD & PLATE FACE NAIL	2-8d 2- 3" 14 ga. STAPLE
21. 1 x 8 (25 x 203mm) SHEATHING OR LESS TO EACH BEARING, FACE NAIL	2-8d
22. WIDER THAN 1 x 8 (25 X 203mm) SHEATHING TO EACH BEARING, FACE NAIL	3-8d
23. BUILT-UP CORNER STUDS	16d @ 24" (600 mm) O.C. 3" 14 ga. STAPLE @ 16" O.C.
24. BUILT-UP GIRDER AND BEAMS	20d @ 32" (800 mm) O.C. FACE NAIL @ TOP/BOTTOM 3" 14 ga. STAPLE @ 24" O.C. & STAGGER 2-20d @ ENDS & @ EACH SPLICE. 3- 3" 14 ga. STAPLE
25. 2" PLANKS	2-16d AT EACH BEARING
26. COLLAR TIE TO RAFTER	3-10d / 4-3" 14 ga. STAPLE
27. JACK RAFTER TO HIP	3-10d / 4-3" 14 ga. STAPLE 2-16d / 3-3" 14 ga. STAPLE
28. RAFT. TO 2x RIDGE BEAM, TOE OR FACE NAIL.	2-16d / 3-3" 14 ga. STAPLE
29. JOIST TO RIM JOIST	3-16d / 5-3" 14 ga. STAPLE
30. LEDGER STRIP	3-16d / 4-3" 14 ga. STAPLE

CONNECTION	NAILING
31. WOOD STRUCTURAL PANELS AND PARTICLEBOARD:[2] SUBFLOOR, ROOF AND WALL SHEATHING (TO FRAMING).	
1/2" OR LESS	6d[3] / 1 3/4" 16 ga.[15]
19/32 - 3/4"	8d[4] OR 6d[5] / 2" 16 ga.[16]
7/8 - 1"	8d[3]
1 1/8 - 1 1/4"	10d[4] OR 8d[5]
COMBINATION SUBFLOOR-UNDERLAYMENT (TO FRAMING) 1" = 25.4 mm)	
3/4 AND LESS	6d[5]
7/8" - 1"	8d[5]
1 1/8 - 1 1/4"	10d[4] OR 8d[5]
32. PANEL SIDING (TO FRAMING): 1/2" (13 mm)	6d[6]
5/8" (16 mm)	8d[6]
33. FIBERBOARD SHEATHING:[7] 1/2" (13 mm)	No. 11 ga.[8] 6d[4] No. 16 ga.[9]
25/32" (20 mm)	No. 11 ga.[8] 8d[4] No. 16 ga.[9]
34. INTERIOR PANELING	
1/4" (6.4 mm)	4d[10]
3/8" (9.5 mm)	6d[11]

1. COMMON OR BOX NAILS MAY BE USED EXCEPT WHERE OTHERWISE STATED.
2. NAILS SPACED @ 6" (150 mm) ON CENTER @ EDGES, 12" (300 mm) AT INTERMEDIATE SUPPORTS EXCEPT 6" (150 mm) AT ALL SUPPORTS WHERE SPANS ARE 48" 1200 mm OR MORE. FOR NAILING OF WOOD STRUCTURAL PANEL AND PARTICLEBOARD DIAPHRAGMS AND SHEAR WALLS, REFER TO SECTION 2305. NAIL FOR WALL SHEATHING MAY BE COMMON, BOX, OR CASING.
3. COMMON OR DEFORMED SHANK.
4. COMMON.
5. DEFORMED SHANK..
6. CORROSION-RESISTANT SIDING OR CASING NAILS.
7. FASTENERS SPACED 3" (75 mm) O.C. AT EXTERIOR EDGES AND 6" (150 mm) O.C. AT INTERMEDIATE SUPPORTS.
8. CORROSION-RESISTANT ROOFING NAILS W/ 7/16" ⌀ (11 mm) HEAD & 1 1/2" (38 mm) LENGTH FOR 1/2" (13 mm) SHEATHING AND 1 3/4" (44 mm) LENGTH (FOR 25/32" (20 mm) SHEATHING.
9. CORROSION-RESISTANT STAPLES WITH NOMINAL 7/16" (11 mm) CROWN AND 1 1/8" (28 mm) LENGTH FOR 1/2" (13 mm) SHEATHING AND 1 1/2" (38 mm) LENGTH (FOR 25/32" (20 mm) SHEATHING. PANEL SUPPORTS @ 16" O.C.(400 MM) (20" (500 mm) IF STRENGTH AXIS IN THE LONG DIRECTION OF THE PANEL, UNLESS OTHERWISE MARKED.
10. CASING OR FINISHED NAILS SPACED 6" (150 mm) ON PANEL EDGES, 12" (300 mm) @ INTER. SUPPORTS.
11. - PANEL SUPPORTS @ 24" (600 mm). CASING OR FINISH NAILS SPACED 6" (150 mm) ON PANEL EDGES, 12" (300 mm) AT INTERMEDIATE SUPPORTS.
12. FOR ROOF SHEATHING APPLICATIONS, 8d NAILS ARE THE MINIMUM REQUIRED FOR WOOD STRUCTURAL PANELS.
13. STAPLES SHALL HAVE A MINIMUM CROWN OF 7/16 (11 mm).
14. FOR ROOF SHEATHING APPLICATIONS, FASTENERS SPACED 4" (100 mm) O.C. AT EDGES, 8" (200 mm) O.C. AT INTERMEDIATE SUPPORTS.
15. FASTENERS SPACED 4" (100 mm) O.C. AT EDGES, 8"(200 mm) AT INTERMEDIATE SUPPORTS FOR SUBFLOOR AND WALL SHEATHING AND 3" (75 mm) O.C. AT EDGES, 6" (150 mm) AT INTERMEDIATE SUPPORTS FOR ROOF SHEATHING.
16. FASTENERS SPACED 4" (100 mm) O.C. AT EDGE, 8" (200 mm) AT INTERMEDIATE.

FIGURE 6-3 A nailing schedule can be used to describe common wood connections. Any connection listed in the schedule does not need to have the required nailing specified in a detail. The schedule is usually placed on a sheet containing other general discipline notes.

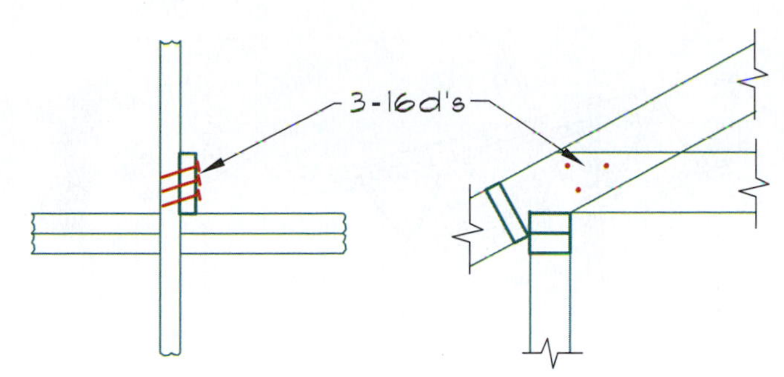

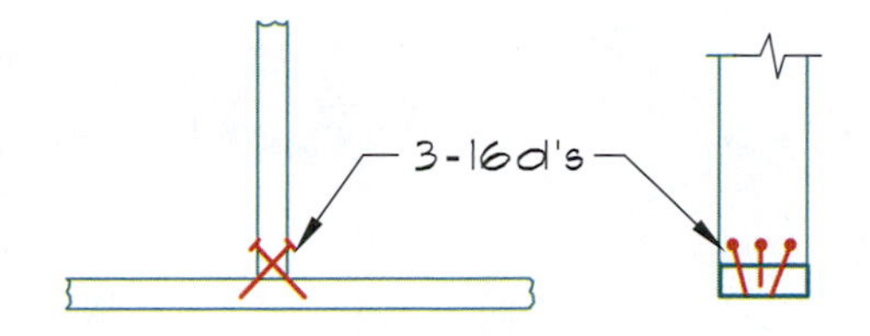

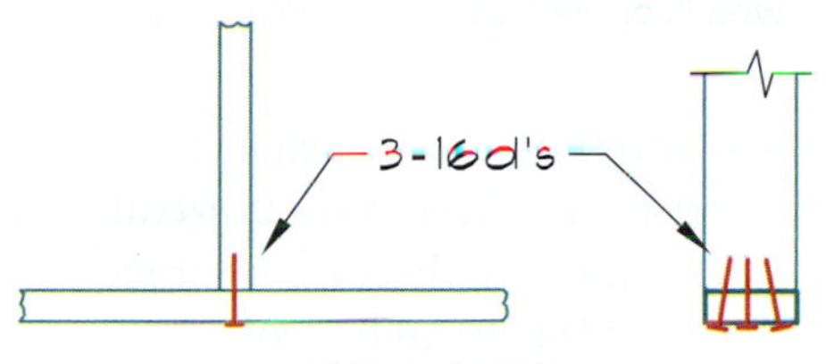

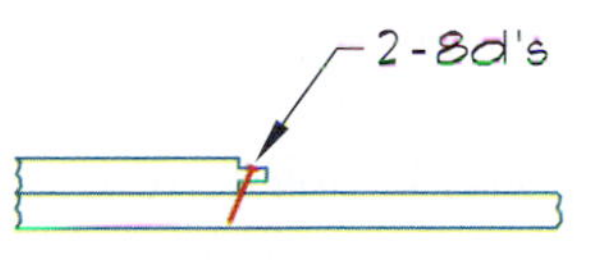

FIGURE 6-4 Four methods of placing nails.

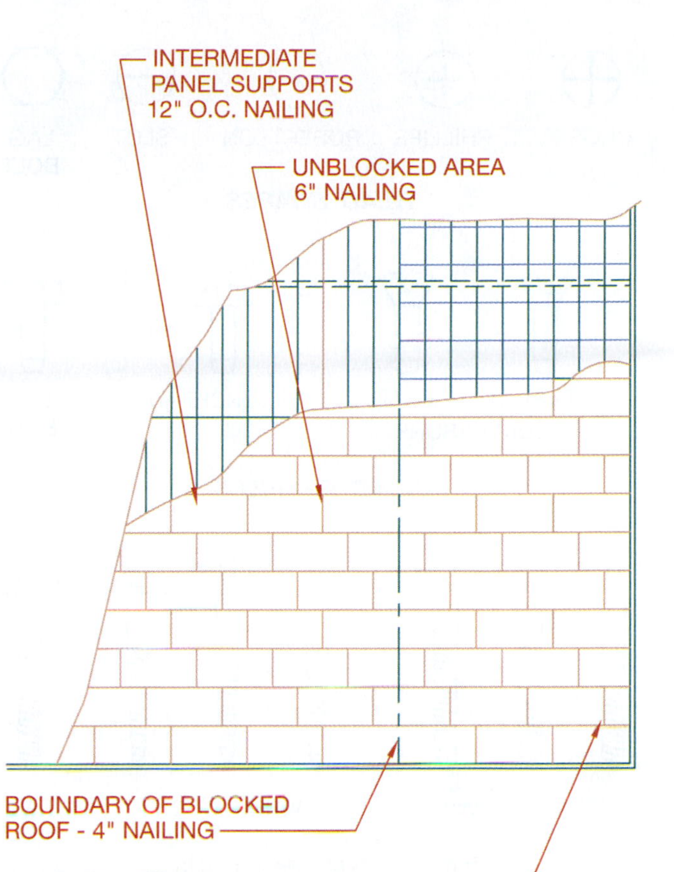

FIGURE 6-5 Nail specifications often refer to placement as edge, boundary, and field or intermediate nailing.

screws are used throughout the construction industry. Each is identified by its head shape (see Figure 6-6).

- **Flathead (or countersunk) screws** are typically specified in the architectural drawings for finish work where a nail head is not desirable.
- **Roundhead screws** are used at lumber connections where a nail head is tolerable. Roundhead screws are also used to connect lightweight metal to wood. Both flathead and roundhead screws are designated by a gauge that specifies a diameter in inches, the length in inches, and a head shape. A typical specification for a 3" long flathead wood screw with a diameter of #10 might read as follows:

#10 × 3" F.H.W.S.

- Lengths of screws range from 1/4" to 6" (6 to 150 mm).
- **Lag screws** have a hexagon- or square-shaped head designed to be tightened by a wrench rather than a screwdriver. Lag screws are used for lumber connections 1 1/2" (40 mm) and thicker. Lag bolts are available in lengths from 1" to 12" (25 to 300 mm) with diameters ranging from 1/4" to 1 1/4" (6 to 30 mm).

A pilot hole that is smaller than the screw size is often provided to aid in the placement of screws and to reduce wood splitting. A washer is used with a lag bolt to guard against crushing wood fibers near the bolt. A pilot hole that is approximately three-fourths of the shaft diameter is often specified to reduce wood damage and increase the resistance to withdrawal.

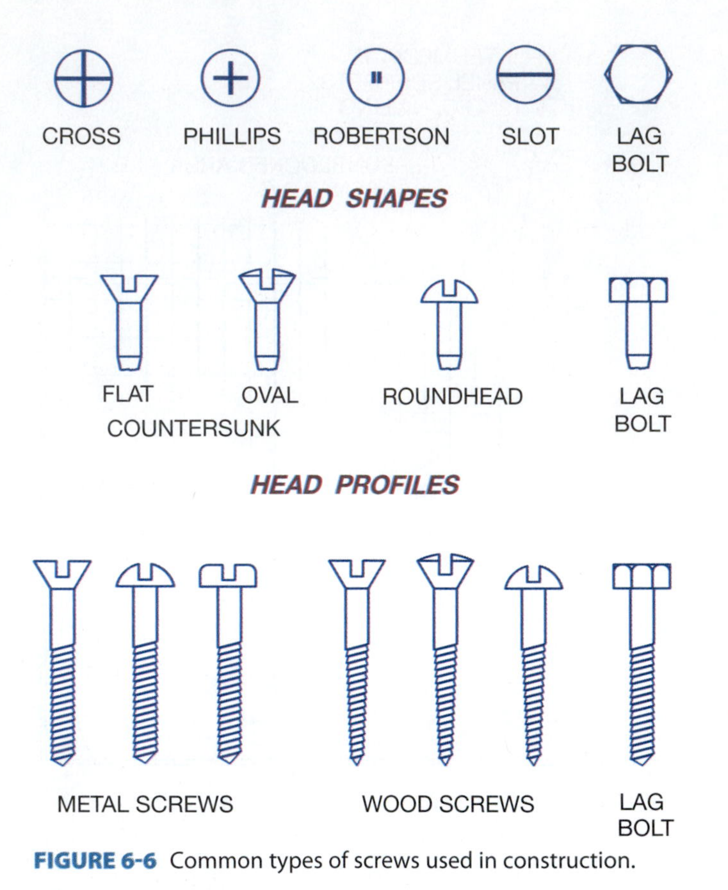

FIGURE 6-6 Common types of screws used in construction.

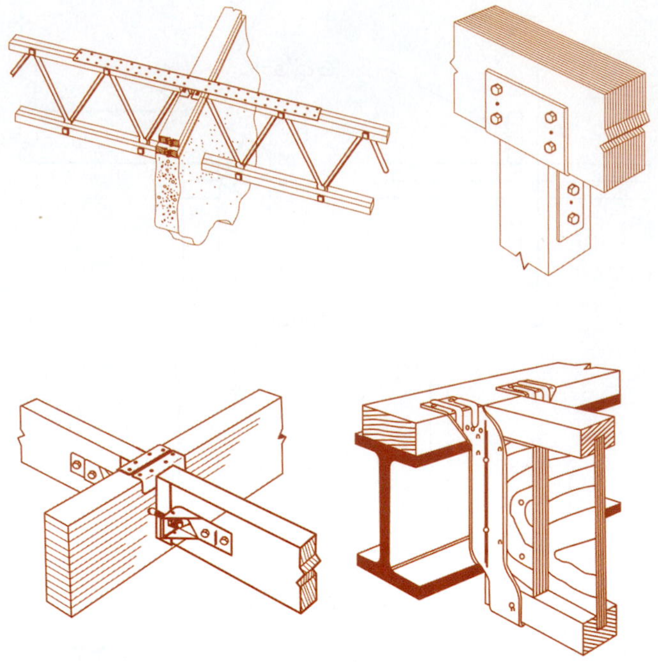

FIGURE 6-7 A wide variety of metal connectors are used to strengthen wood connections. *Courtesy Simpson Strong-Tie Co., Inc.*

Metal Framing Connectors

Premanufactured metal connectors by companies such as Simpson Strong-Tie are used at many wood connections to strengthen nailed connections. Joist hangers, post caps, post bases, and straps are some of the most common lightweight metal hangers used with wood construction. Each is shown in Figure 6-7. Each connector comes in a variety of gauges of metal, ranging from 12 to 16 gauge, with sizes to fit a wide variety of lumber. Metal connectors are typically specified on the framing plans, sections, and details by listing the model number and type of connector. A metal connector specification for connecting (2)- 2 × 12 joists to a beam would be:

HHUS212-2TF JST. HGR

The supplier is typically specified in the written specifications or in a general note such as the following:

ALL METAL HANGERS AND CONNECTORS TO BE
SIMPSON STRONG-TIE COMPANY, INC. OR EQUAL.

Depending on the connection, the nails or bolts used with the metal fastener may or may not be specified. If no specification is given, it is assumed all nail holes in the connector will be filled. If bolts are to be used, they will normally be specified with the connector, based on the manufacturer's recommendations. Nails associated with metal connectors are typically labeled with the letter *n* instead of *d*. These nails are equal to their *d* counterpart, but the length has been modified by the manufacturer to fit the metal hanger.

STRUCTURAL CONNECTORS USED WITH LARGE WOOD MEMBERS

Just as with lumber connections, large wood members such as beams, headers and girders, and timber connections must also be able to resist the stress placed on them. Bolts, metal connectors, and timber connectors are the most common methods of connecting timbers.

Bolts and Washers

Bolts used in the construction industry include anchor bolts, carriage bolts, and machine bolts. Each can be seen in Figure 6-8. Washers are used under the head and nut for most bolting applications. Washers keep the head and nut from pulling through the lumber and reduce lumber damage by spreading the stress from the bolt across more wood fibers. Typically a circular washer, specified by its diameter, is used.

Anchor Bolts

An anchor bolt is an L-shaped bolt that is inserted into concrete and used to bolt lumber to the concrete (Figure 6-8a). The short leg of the L is inserted into the concrete and provides resistance from withdrawal. The upper end of the long leg is threaded to receive a

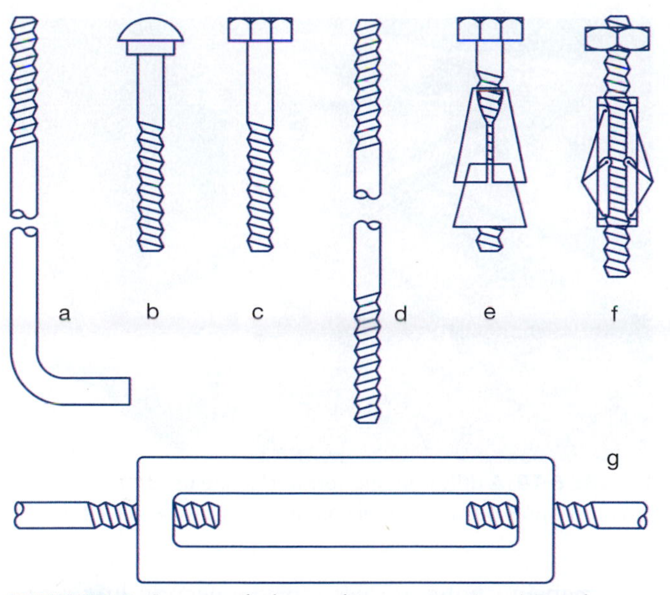

FIGURE 6-8 Common bolts used in construction: (a) anchor bolt; (b) carriage bolt; (c) machine bolt; (d) drift bolt (threaded rod); (e) expansion bolt; (f) toggle bolt; (g) turnbuckle and threaded rods.

nut. A 2" (50 mm) washer is typically used with anchor bolts. Anchor bolts can be abbreviated on the structural drawings with the letters *A.B.*, along with a specification including the size of the member to be connected, the bolt diameter, length, embedment into concrete, spacing, and washer size. A typical note might resemble the following:

2 × 6 DFPT SILL W/ 5/8"Ø × 12" A.B. @48" O.C. W/ 2" Ø WASHERS. PROVIDE 9" MINIMUM EMBEDMENT.

When anchor bolts are used to attach lumber to the side of a concrete wall, a specification is usually given to stagger the bolts in relation to the edge of the lumber to be attached as well as the bolt spacing. A typical specification might resemble:

3 ×12 DFPT LEDGER W/ 3/4" Ø × 8" A.B. @ 32" O.C. STAGGERED 3" UP/DN WITH 2" Ø WASHERS.

Carriage Bolts

A carriage bolt is used for connecting steel and other metal members as well as timber connections. **Carriage bolts** (Figure 6-8b) have a rounded head with the lower portion of the shaft threaded. Directly below the head, at the upper end of the shaft is a square shank. As the shank is pulled into the lumber, it will keep the bolt from spinning as the nut is tightened. Diameters range from 1/4" to 1" (6 to 25 mm) with lengths typically available to 12" (300 mm).

The specification for a carriage bolt will be similar to an anchor bolt except for the designation of the bolt type.

Machine Bolts

Bolts with a hexagonal head and a threaded shaft are described as **machine bolts** (M.B.). Machine bolts (Figure 6-8c) are divided into the classifications of unfinished or high-strength bolts. Common bolts are made from A-307 low-carbon steel and are used for attaching steel-to-steel, steel-to-wood, or wood-to-wood connections. Common bolts are often used in steel joints to provide a temporary connection while field welds are completed. Bolts are assumed to be common unless high-strength bolts are specified and are referred to with a note such as:

USE (3)-3/4" Ø × 8" M.B. @ 3" O.C. W/ 1 1/2" Ø WASHERS.

The strength of bolts will be called out in a general note on the framing plans or on pages of details containing bolted connections. Although the engineer will determine the bolt locations based on the stress to be resisted, bolts for wood members are usually placed:

- 1 1/2" (40 mm) from an edge parallel to the grain
- 3" (75 mm) minimum from the edge when perpendicular to the grain
- 1 1/2" (40 mm) from the edge of steel members
- 2" (50 mm) from the edge of concrete

The pilot hole for the bolts in wood is usually not specified but is assumed to be 1/16" (2 mm) bigger than the bolt shaft. If an elongated hole is used to allow slight movement for field adjustment, the hole and the bolt will be specified with a note such as:

3/4" Ø × 8" M.B. THRU 7/8" × 1 1/2" SLOTTED HOLE W/ 2" Ø WASHER EA. SIDE.

High-Strength Bolts

Bolts used for connecting timber and structural steel are referred to as **high-strength bolts** and come in several common compositions. These bolts are manufactured with an ASTM (American Society for Testing Materials) or Institute of Steel Construction Specifications grading number on the head. Common specifications used in commercial construction include the following:

- A-325 high-strength steel bolts, which come in three classifications:
  - Type 1 bolts are made from medium-carbon steel in sizes from 1/2" to 1 1/2" (13 to 40 mm) diameter.
  - Type 2 bolts are made from low-carbon steel in sizes of 1/2" to 1" (13 to 25 mm) diameter.
  - Type 3 bolts are made of weathering steel.

- A-490 high-strength, medium-carbon steel bolts
- A-441 high-strength, low-alloy steel bolts
- A-242 corrosion-resistant high-strength low-alloy steel bolts

High-strength bolts are usually specified to be tightened with a pneumatic impact wrench. Chapter 9 will discuss steel requirements and specifications.

Miscellaneous Bolt Types

Several other types of bolts are used for special construction circumstances. These include studs, drift bolts, expansion bolts, and toggle bolts.

- **Studs** are bolts that have no head. A stud is welded to a steel beam so that a wood plate can be bolted to the beam.
- **Drift bolts** (Figure 6-8d) are steel rods that have been threaded.
- **Threaded rods** can be driven into one wood member with another member bolted to the threaded protrusion. Threaded rods can also be used to span between metal connectors on two separate beams (as seen in Figure 6-9) or to connect concrete panels. Threaded rods are also used with steel construction to provide lateral support (see Figure 6-10).

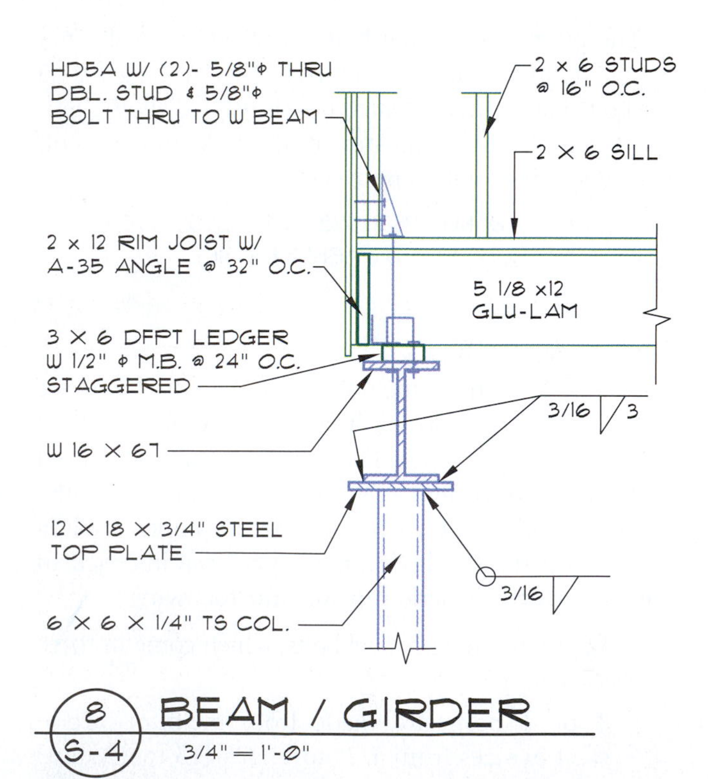

FIGURE 6-9 Framing anchors connected by drift bolts or metal straps are often used to tie structural members of one level to those of another level. In this detail, an HD5A anchor with a 5/8" diameter bolt will be used to tie the wood wall to the steel girder.

FIGURE 6-10 A drift bolt and turnbuckle are used to provide lateral support between steel columns. *Courtesy Megan Jefferis.*

- **Expansion bolts** are used for connecting lumber to masonry. Expansion bolts have a special expanding sleeve (Figure 6-8e) that will expand once inserted into a hole to increase holding power.
- **Toggle bolts** have a nut that is designed to expand once inserted through a hole, so that it cannot be removed. Toggle bolts (Figure 6-8f) are used where one end of the bolt may not be accessible due to construction parameters.

Timber Connectors

Timber connections are often reinforced by metal rings, plates, and disks that are embedded into each piece of lumber at the joint to resist sliding. The engineer will determine the size and number of connectors per joint based on the stress that must be resisted. A groove is typically precut into adjoining pieces of timber to hide the metal connector. The connector is placed between the members to be joined and forced into the groove as bolts are tightened. A toothed ring is forced into each member by pressure, usually from hammering. Shear plates are set in pairs to join lumber. The face of a shear plate is set flush with each timber to be joined and then bolted together.

Metal Framing Connectors

Premanufactured metal connectors made by companies such as Simpson or Tyco are used at many timber-to-timber connections to strengthen nailed connections. Beam hangers, post caps and bases, and straps are some of the most common metal connectors used with timber construction. Examples of premanufactured connectors can be seen in Figure 6-11. These connectors are usually made from metal, ranging from 3 to 7

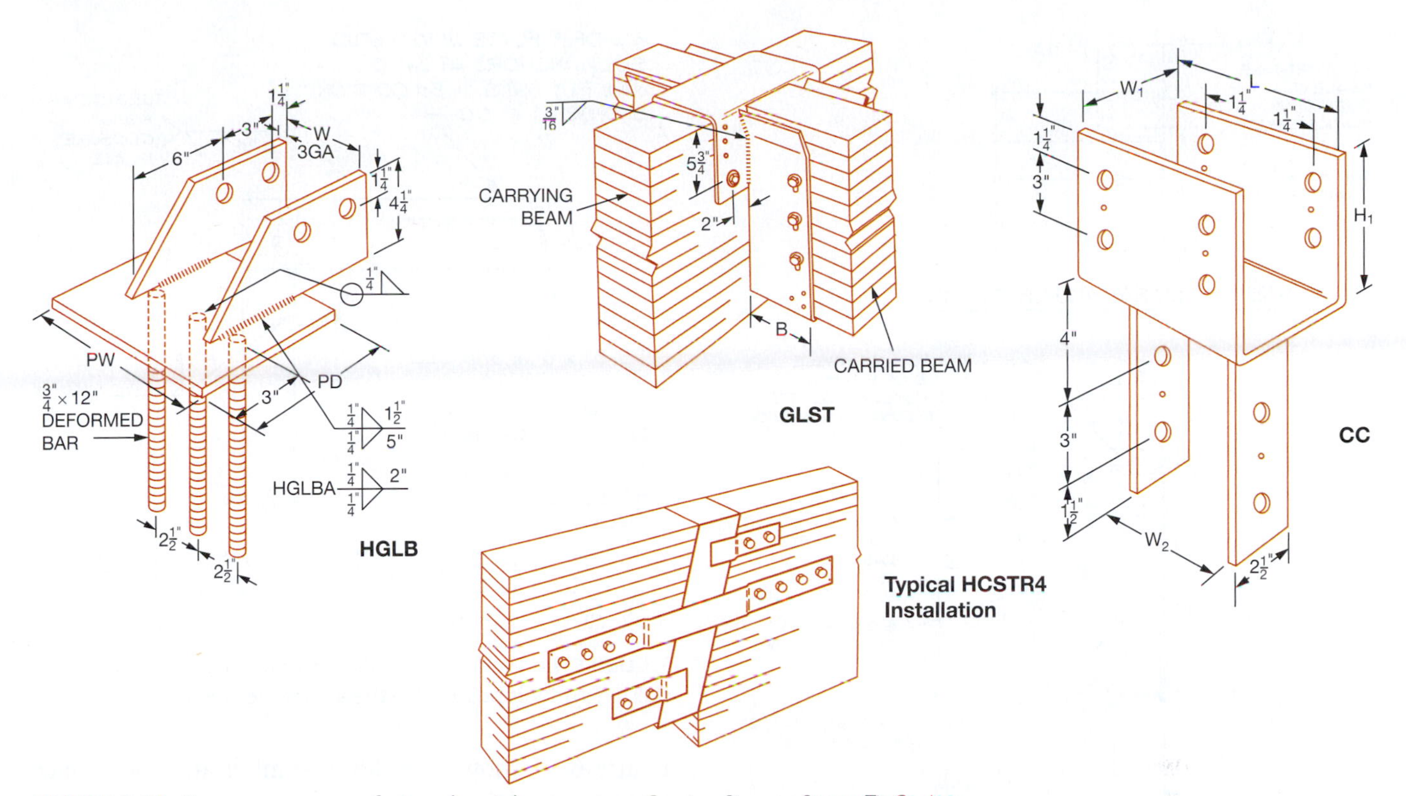

FIGURE 6-11 Common premanufactured metal connectors. *Courtesy Simpson Strong-Tie Co., Inc.*

gauge. Timber connectors are often held in place by bolting rather than nailing because of the heavy loads being carried. Figure 6-12 shows a detail specifying a premanufactured metal connector and the required nailing.

Because the hanger is premanufactured, the drafter must only specify the size of the hanger and the required nailing or bolting size and quantity. The location of

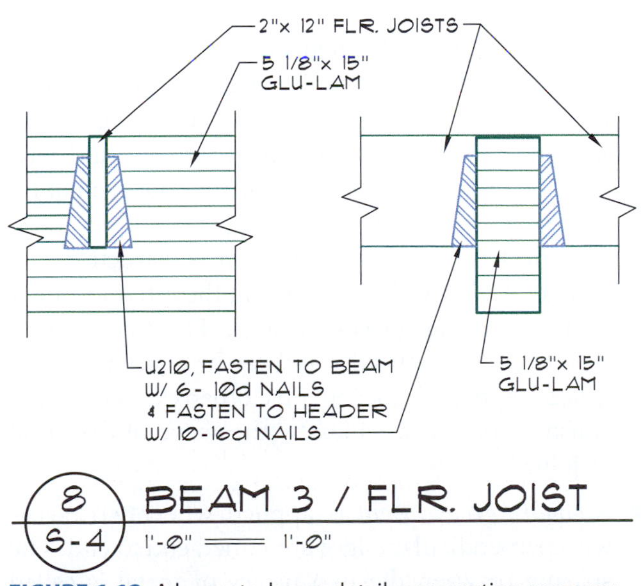

FIGURE 6-12 A beam-to-beam detail representing premanufactured metal hangers does not require dimensions to locate hangers' nails or bolts.

bolts is not required because the holes for them are predrilled. Metal connectors are typically specified on the framing plans, sections, and details by listing the model number and type of connector, just as with lightweight metal connectors. When seen in end view, the metal connector is represented by pairs of parallel diagonal lines.

If premanufactured connectors are not available that can support the required loads, the engineer must design a steel connector to be fabricated for a specific joint. These connectors are assembled in a shop and shipped to the job site. The CAD technician will need to specify the material, welds, and exact bolt locations. Figure 6-13 shows an example of a fabricated metal connector. Chapter 9 will explore the use of steel.

WELDED STEEL CONNECTIONS

Steel connections are typically bolted and welded. Carriage and machine bolts described earlier in this chapter are used to make the initial connection as steel members are positioned. Once all members of a specific portion of the structure are positioned and squared, the connections are welded to provide the required strength.

Welding is the method of providing a rigid connection between two or more pieces of steel. Through the process of welding, metal is heated to a temperature

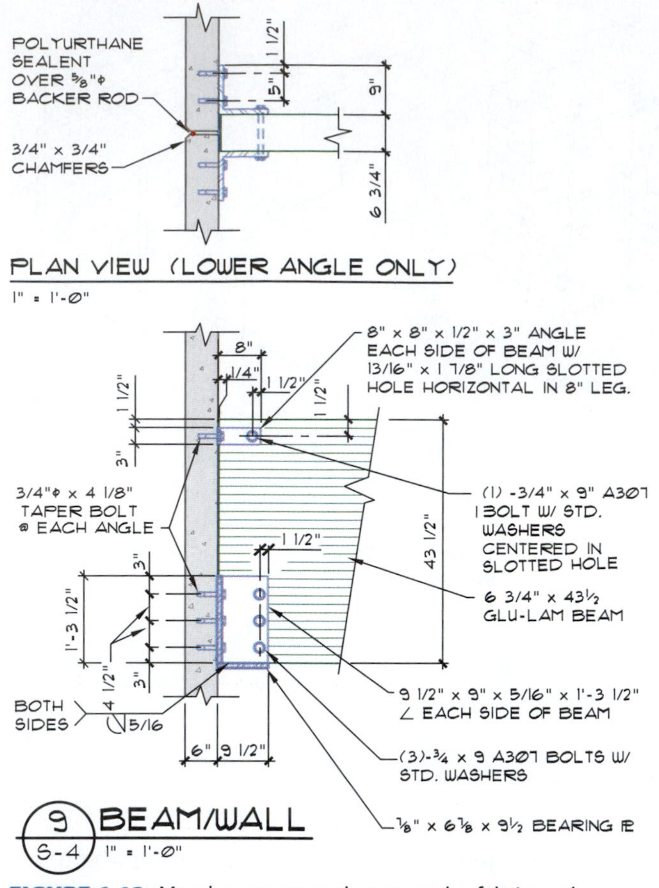

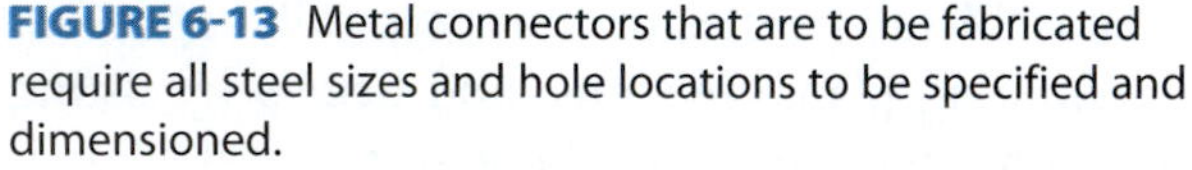
FIGURE 6-13 Metal connectors that are to be fabricated require all steel sizes and hole locations to be specified and dimensioned.

high enough to cause melting of metal. The parts that are welded become one, with the welded joint actually stronger than the original material. Welding offers better strength, better weight distribution of supported loads, and a greater resistance to shear or rotational forces than a bolted connection. There are many other welding processes available to the industry, but the most common welds in the construction field are shielded metal arc welding, gas tungsten arc welding, and gas metal arc welding. In each case, the components to be welded are placed in contact with each other, and the edges are melted to form a bond. Additional metal is also added to form a sufficient bond.

Welds are specified in details similar to Figure 6-14. A horizontal reference line is connected to the parts to be welded by an inclined line with an arrow. The arrow touches the area to be welded. It is not uncommon to see the welding line bend to point to places that are difficult to reach. The welding symbol may also have more than one leader line extending from the reference line. Information about the type of weld, the location of the weld, the welding process, and the size and length of the weld is specified on or near the reference line.

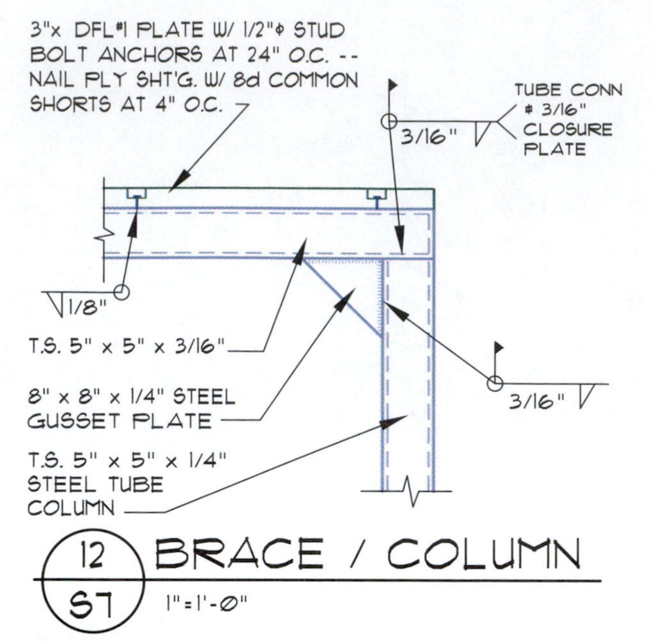

FIGURE 6-14 Welds are specified in details by the use of specialized symbols pointing to the area to be welded.

Figure 6-15 shows a welding symbol and the proper location of information.

Welded Joints

The way that steel components intersect greatly influences the method used to weld the materials together. The welding method and the joint to be used are often included in the specification. Common joints include butt, lap, tee, outside corner, and edge joints. Examples of each joint are illustrated in Figure 6-16.

Types of Welds

The type of weld is distinguished by the weld shape and/or the type of groove in the metal components that receive the weld. Welds specified on structural and architectural drawings include fillet, groove, and plug welds. The methods of joining steel and the symbol for each type can be seen in Figure 6-17.

The most common weld used in construction is a fillet weld. A **fillet weld** is formed at the internal corner of two intersecting pieces of steel. The fillet can be applied to one or both sides and can be continuous or a specified length and spacing. Five other welds common to construction are specified by the shape of the metal to be joined.

- A *square groove weld* is applied when two pieces with perpendicular edges are joined end to end. The spacing between the two pieces of metal is called the *root opening*. The root opening is shown to the left of the symbol.

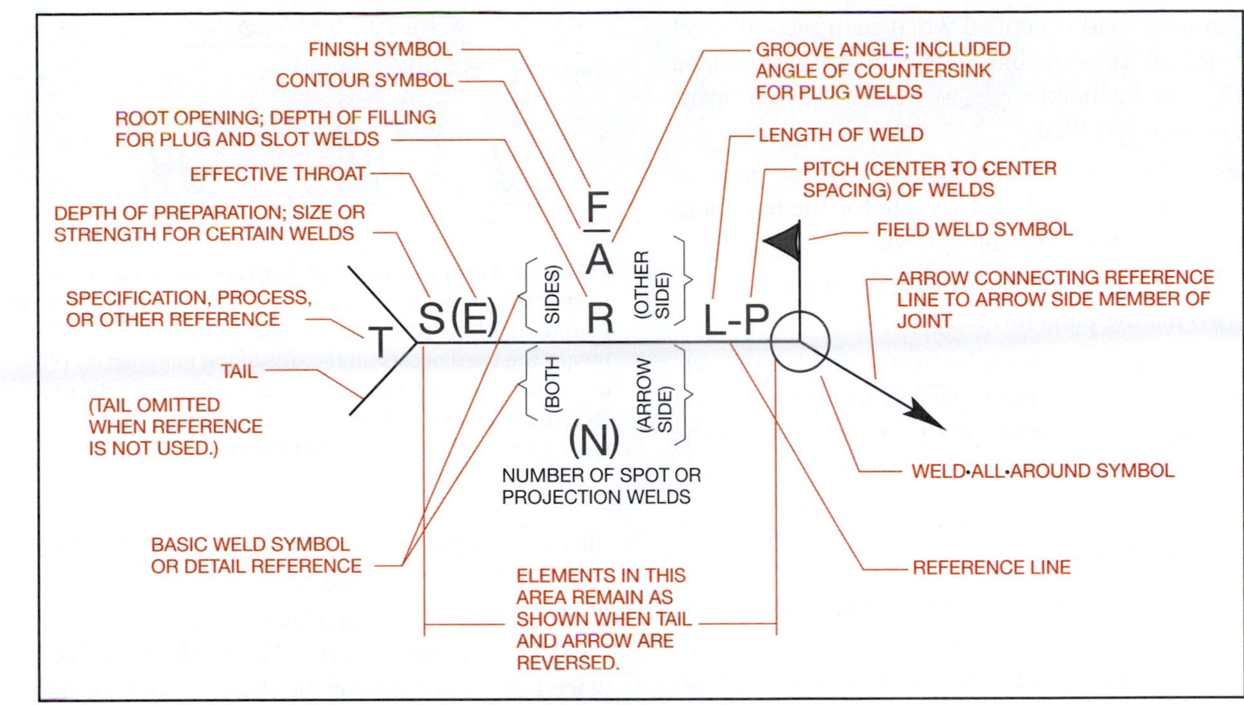

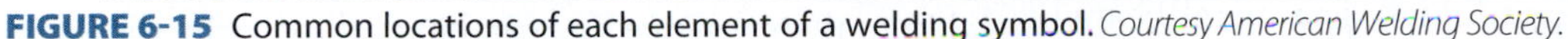

FIGURE 6-15 Common locations of each element of a welding symbol. *Courtesy American Welding Society.*

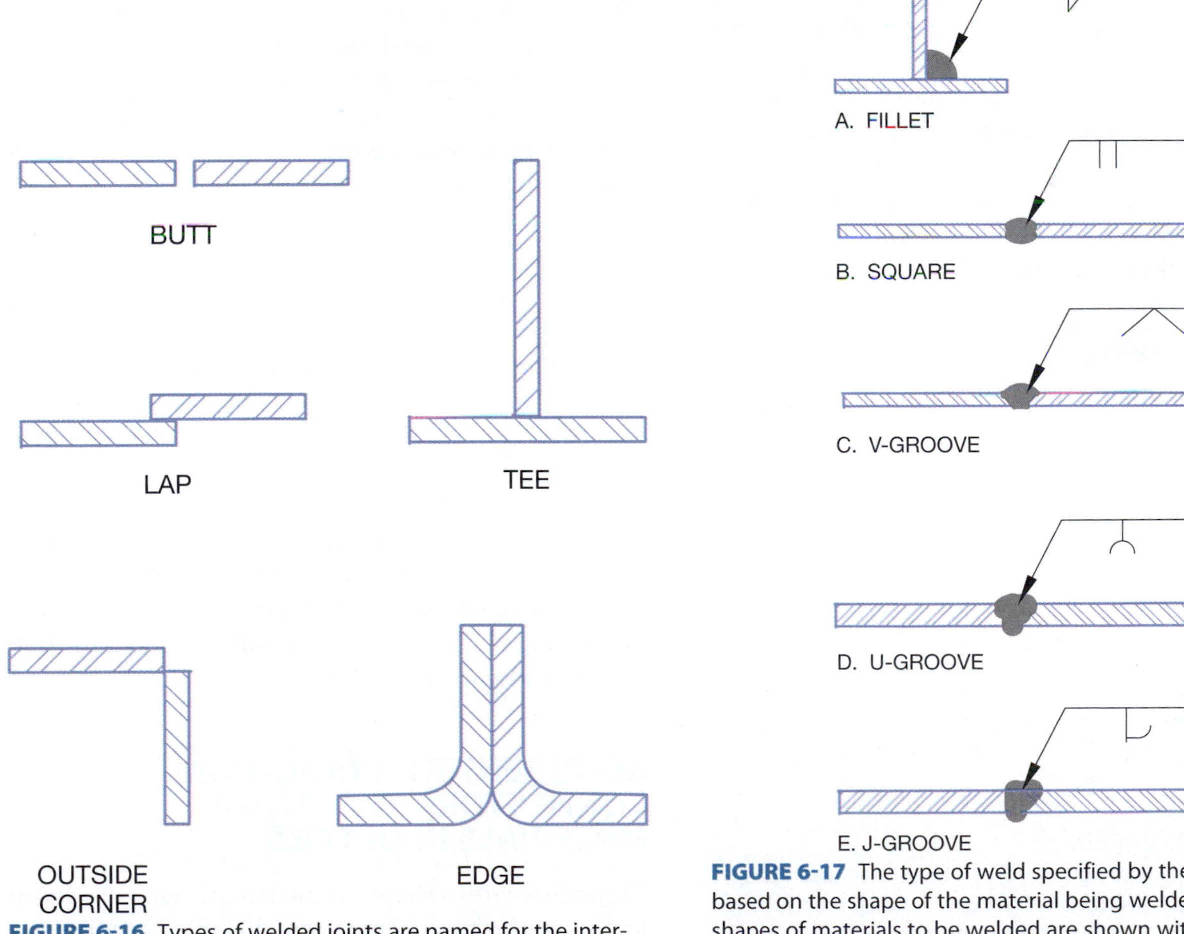

FIGURE 6-16 Types of welded joints are named for the intersections of the two mating pieces of steel.

FIGURE 6-17 The type of weld specified by the engineer is based on the shape of the material being welded. Common shapes of materials to be welded are shown with the symbol used to represent the weld.

- A **V-groove weld** is applied when each piece of steel to be joined has an inclined edge that forms the shape of a V. The included angle, as well as the root opening, is often specified.
- A **beveled weld** is created when only one piece of steel has a beveled edge. An angle for the bevel and the root opening is typically given.
- A *U-groove weld* is created when the groove between the two mating parts forms a U.
- A *J-groove weld* results when one piece has a perpendicular edge and the other has a curved grooved edge. The included angle, the root opening, and the weld size are given for both types of welds.

Welding Location

Welds specified on the structural drawings may be done in a shop away from the job site, and the components are then shipped, ready to be installed. For an item that can be easily shipped such as a metal beam connector, shop assembly is less expensive and more efficient.

For large components that must be assembled at the job site, the welding is said to be **field welded.** Two symbols used to refer to a field weld are shown in Figure 6-18. When field welds are specified, the engineer may require that these welds are to be inspected by a representative of the engineer in addition to the inspection provided by the building department. The drafter will need to provide a specification beside each welding detail indicating the need for special inspection.

Weld Placements

The placement of the welding symbol in relation to the reference line is critical in indicating where the actual weld will take place. Options include *this side*, *other side*, and *both sides*. Figure 6-19 shows the effects of placing a

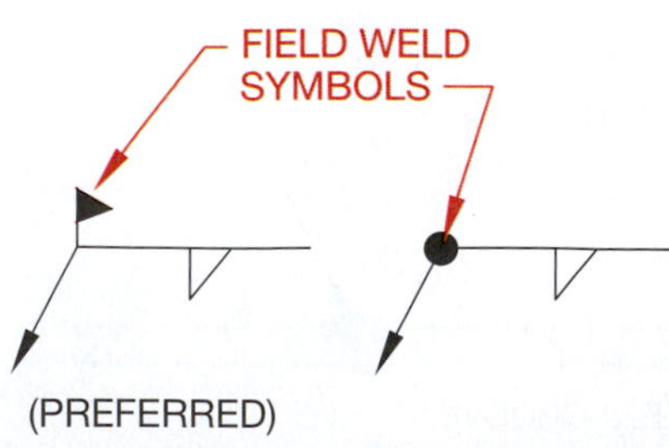

FIGURE 6-18 The welding symbol can be used to distinguish if the weld will be performed on-site or off-site by the use of a small flag or a solid circle placed at the intersection of the leader line.

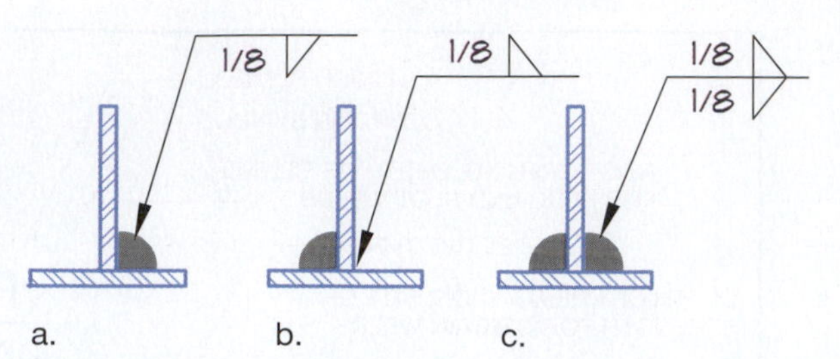

FIGURE 6-19 The welding symbol can be used to describe the relationship of the weld to the materials being welded by the placement of the symbol to the line: (a) symbol below the line means the weld occurs on *this side* of the material; (b) symbol above the line means the weld is to be placed on the *other side;* (c) symbol placed on each side of the reference line means the weld is to occur on *both sides* of the material.

fillet weld symbol on the reference line. The distinction of symbol placement can be quite helpful to the drafter if adequate space is not available on the proper side of the detail to place a symbol. The welding symbol can be placed on either side of the drawing and the relationship to the reference line can be used to clarify the exact location.

All Around

The all around symbol is used to indicate that a weld is to be placed around the entire intersection. A circle placed at the intersection of the leader and reference line indicates that a feature is to be welded *all around.* Figure 6-14 shows a column-beam intersection joined with welds requiring an all-around-weld.

Weld Length and Increment

If a weld does not surround the entire part, the length and spacing of the weld should be indicated beside the weld symbol. The number preceding the weld symbol indicates the size of the weld. When weld sizes are expressed in millimeters, they should be designated in 1-millimeter increments up to 8 mm, 2-millimeter increments from 8 to 20 mm, 5-millimeter increments from 20 to 40 mm, and 10-millimeter increments beyond 40 mm. The number following the weld symbol indicates the length each weld is to be, and the final size indicates the spacing of the weld along a continuous intersection of two mating parts.

REPRESENTING STRUCTURAL CONNECTORS AND OTHER MATERIALS IN DETAILS

Throughout this chapter you have been introduced to the methods of joining wood, timber, engineered wood products, and steel members to other structural members. The final portion of this chapter will introduce you to the skills needed to assemble the details where these connection methods will be specified. Chapter 3 introduced basic considerations for general drawing layouts. This chapter will introduce skills specific to creating details.

Drawing and Editing Details

Details are enlargements of specific areas of a structure and are typically drawn where several components intersect. Details can provide an exterior view of a specific portion of a project that lies parallel to an imaginary viewing plane or show the interior of materials as if an imaginary cutting plane had cut them. Figure 6-20 shows an example of a partial section of a hillside structure built on a piling foundation. Figure 6-21 shows two details that relate to the larger drawing in Figure 6-20. The larger drawing shows major intersections of material but does not allow specific materials to be specified clearly. This chapter will provide an introduction to the considerations for editing and drawing construction details.

Stock Details

Most offices have a library of **stock details**. These are details of repetitive items such as footings that remain the same for most of the buildings drawn by the office. Common stock foundations might include:

- 1-level concrete slab with stud walls—bearing
- 1-level concrete slab with stud walls—nonbearing
- 1-level concrete slab with steel stud walls—bearing
- 1-level concrete slab with steel stud walls—nonbearing
- 1-level concrete slab with concrete block walls—bearing

The same details would typically exist for two-level construction, and for wood-framed floor systems if appropriate for the area of construction or construction type. Using a computer, the typical detail can be drawn, with all the required dimensions, notations, and tables to describe various sizes based on loads or depth. Once complete, the detail can be stored in the office library for future use. For future projects, the detail can be inserted or referenced into the project and edited as needed. The first detail may take a few hours

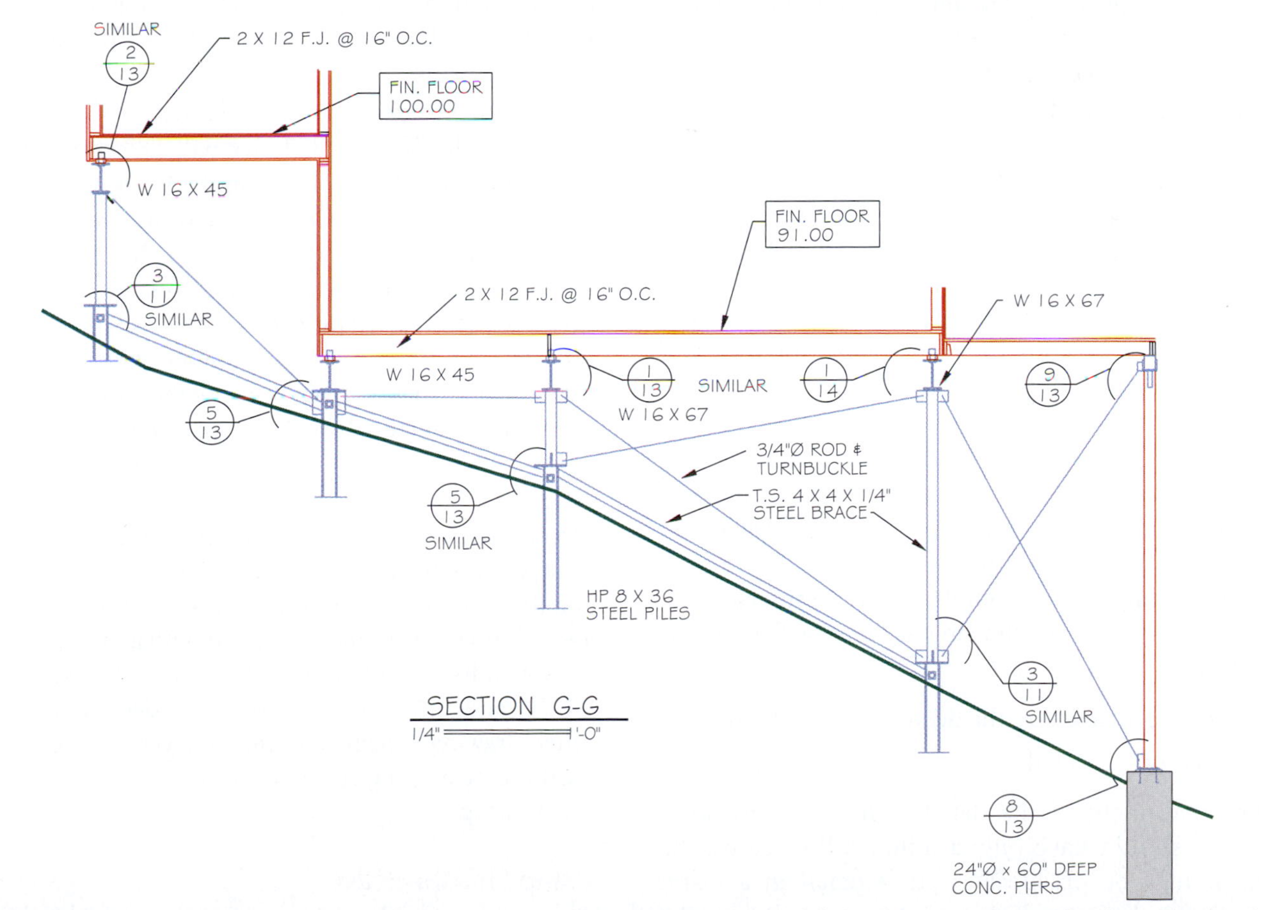

FIGURE 6-20 This partial section for a hillside structure is used to show the shape of the structure in one specific place. The details shown in Figure 6-21 provide information about specific material intersections.

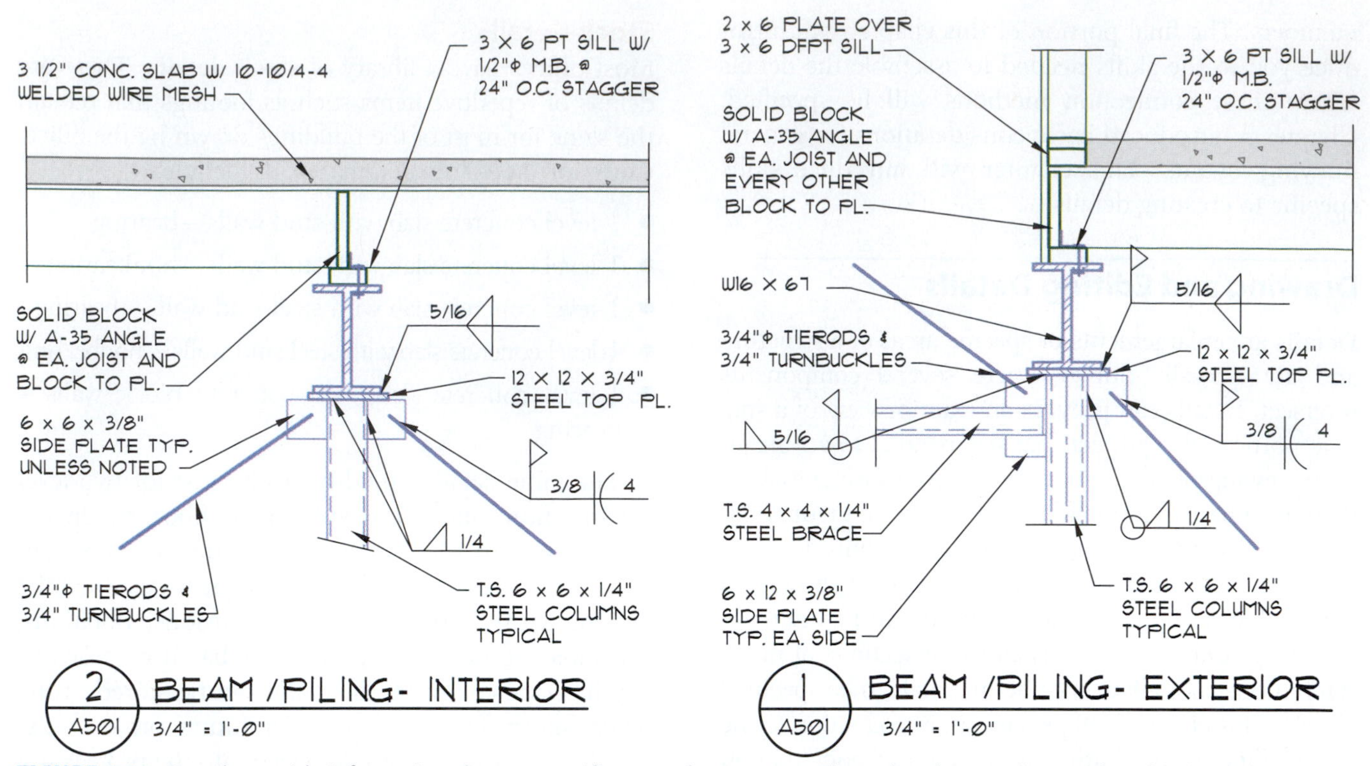

FIGURE 6-21 Details provide information about a specific area of a structure. These details are referenced to the drawing in Figure 6-20.

to complete, but only a few minutes of editing time is required for use in future projects.

Inserting and editing details is a common job for a new CAD technician. Figures 3-5a and 3-5b show details that were created by editing the master details. To successfully complete a drawing, consideration must be given to the plotting scale, material representation, layering, lineweights, symbols, annotation, and dimensions. Many of these topics were introduced in Chapter 3, but additional information must be considered as details are edited or constructed for use on a sheet containing multiple details.

Plotting Scales To make the details easier to read, they have become somewhat standardized in several areas, including scales and alignment. Details are drawn at a scale of 1/2" = 1'–0" through 3" = 1'–0" (1:16 through 1:4), depending on the complexity of the intersection. The choice of scale in drawing sections will be influenced by:

1. Size of the area where the detail is to be placed.
2. Purpose of the detail.

The placement of the detail as it relates to other drawings should have only a minor influence on the scale. It may be practical to put a detail in a blank corner of a drawing, but don't let space dictate the scale. The most important factor should be the purpose of the detail. A scale must be chosen for plotting that will allow each material to be clearly represented.

Representing and Locating Materials

The size of the detail to be drawn will dictate the amount of information to be displayed, and how the material will be represented. The smaller the plotting scale, the less information will be presented, and the fewer number of linetypes will be used. In addition to using larger plotting scales, details require more attention to line contrast and the use of more varied lineweights than other types of drawings. Careful consideration must also be given to how materials will be represented.

Adding Layers

Before details can be started, layers need to be created to separate information by material and by lineweight and linetypes. Layers should start with a prefix of DETL and information should be named with a modifier listed in Appendix C. Subnames of ANNO, DIMN, FOOT, OUTL, and SYMB will always be needed. As with any other drawing, create additional layers as needed to ensure that only layers that will be used are added to the drawing.

Using Line Contrast

Although standards will vary in each office, details will often require a minimum of four different lineweights to provide contrast between materials. Unfortunately, there is no standard of "always use this lineweight." The lineweights that are used will vary depending on the plotting scale of the detail, and the materials that must be represented. The use of 0.0 (default) for thin lines and 0.60 lines will serve as a starting point for all details to provide contrast for thin and thick lines. When drawing foundation detail, a thickness of 0.90 can be used to represent the outline of concrete, and a weight of 1.00 can be used to represent the finished grade.

Note: The goal of any detail is to clearly represent material. Because there is no set standard for lineweight, draw a few lines using varied lineweights, and then make a test plot. Line thickness should be thick enough to provide contrast, but not so thick that lines bleed into other objects.

Once the lineweights have been selected, a method for assigning lineweights will need to be established. Two common methods for assigning lineweights include:

1 Assign and name layers such as THIN, THICK, VERY THICK, and MEGA THICK.

2 Assign layer names based on materials such as WOOD, STEEL, CONCRETE, or SOIL. Lineweights can then be assigned to objects on those layers using the PROPERTIES command. See the guidelines for naming layers on the student CD.

The office where you work will determine the correct method of assigning lineweights. When plotting, grayscale can be used to provide contrast to existing and new materials by assigning gray to existing materials and black to new materials.

Representing Materials in Details

The method used to represent each material will vary depending on the scale that is used. Different methods are also used to represent continuous materials that are cut by the cutting plane or intermittent materials that lie beyond the cutting plane. Although the method of representing materials may vary with each office, it is critical that each material be distinguished from other materials to provide drawing clarity. Common materials shown in sections are shown in Figure 6-22. It is also important not to spend more time than necessary detailing materials. If a product is delivered to the site ready to be installed, minimal attention representing the product is required. If a component must be constructed at the job site, the drawings must provide enough information for all of the different tradespeople who are depending on the drawings.

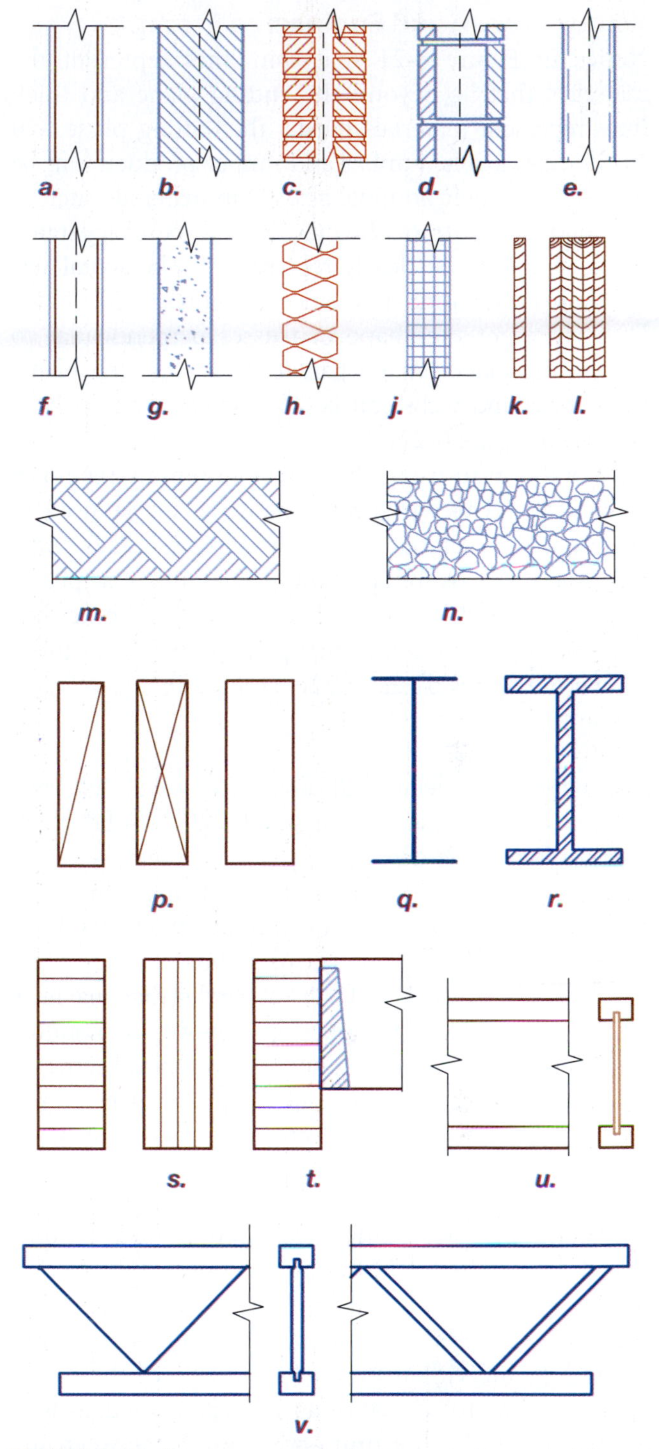

FIGURE 6-22 Common symbols for representing materials in sections and details. Materials include: (a) wood-framed wall; (b) small-scale masonry; (c) double wythe brick wall; (d) concrete masonry units; (e) steel tubes; (f) steel I or W shapes; (g) poured concrete walls; (h) batt insulation; (j) rigid insulation; (k) small-scale plywood; (l) large-scale plywood; (m) soil; (n) gravel; (p) wood and timber in end view (blocking and two methods of showing continuous members); (q) small-scale steel shapes in end view; (r) large-scale steel shapes in end view; (s) laminated timbers in end view; (t) wood member supported by metal hanger in side view on a laminated member in end view; (u) solid-web trusses in side and end views; and (v) open-web trusses in side and end views.

Wood, Timber, and Engineered Products

Notice in Figure 6-21 that thin lines represent the materials that lie beyond the cutting plane and thick lines represent materials cut by the cutting plane. On small-scale details, lumber and timber products can be drawn using their nominal size. Thin materials such as plywood in Figure 6-22k may have to be exaggerated so that they can be clearly represented. The actual size of lumber and timber should be represented. Thick lines represent the shape of trusses perpendicular to the cutting plane. When parallel to the cutting plane, the chords and webs can be represented by thin lines similar to Figure 6-22v.

In addition to using thick lines to outline members shown in end views, several methods can be used to represent the material. Figures 6-22p and 6-22s show common methods to represent materials such as plates, ledgers, and beams. Plywood, sheet rock, and other finishes are represented by hatch patterns similar to those shown in Figures 6-22a, 6-22k, and 6-22l.

Steel

The size of the detail will affect how sectioned steel members are represented. At small scale, a solid, thick line represents the desired shape of steel members (see Figure 6-22q). As the scale increases, pairs of lines can represent the desired shape of sectioned members. Pairs of thin lines (ANSI32) represent sectioned steel with a hatch pattern consisting of pairs of parallel diagonal lines (see Figure 6-22r). Thin lines representing the nominal thickness are used to represent steel columns, beams, or trusses that are beyond the cutting plane. Steel trusses are represented in a similar method to wood trusses.

Unit Masonry

Methods of representing brick and masonry products vary as the scale of the drawing increases. In small-scale sections, units are typically hatched with diagonal lines and no attempt is made to represent cavities or individual units. As the size of the drawing increases, individual units are represented, as well as cavities within the unit and grouting between the units. Individual hatch patterns are used to differentiate between the masonry unit and the grout. Steel reinforcing can be represented by either a hidden or continuous polyline. Figures 6-22b, 6-22c, and 6-22d show an example of a wall section, representing unit masonry and brick veneer. Chapter 10 will provide additional guidelines for representing masonry.

Concrete

The edges of poured members are represented by thick lines and a hatch pattern consisting of dots and small triangles (see Figure 6-22g). Chapter 11 will provide additional guidelines for representing poured concrete products.

Glazing

Glass is represented in details by a single line or pairs of lines depending on the drawing scale. In small-scale details, glass is generally represented by thin lines with little attention given to intersections between the glazing and window frames. As the drawing scale is increased, the detail shown when representing the glass and the frame also increases. Figure 3-11 shows glazing in its frame.

Insulation

The type of insulation will dictate how it is drawn. Batt insulation is generally represented as shown in Figure 6-22h. Depending on the complexity of the detail, the insulation may be shown across its entire drawing, or may be shown in only one portion of the drawing. When only a portion of the insulation is drawn, notes must be placed that clearly define the limits of the insulation. Rigid insulation can be represented as shown in Figure 6-22j, and can be shown using the same considerations as batt insulation. As the scale increases, insulation should be shown throughout the entire detail.

Locating Materials with Dimensions

Both vertical and horizontal dimensions may be placed on details. On small-scale drawings, the use of dimensions depends on the area being represented. The job captain will generally provide the exact dimensions for inexperienced technicians on a check print. As technicians gain experience, they are expected to determine required sizes based on similar circumstances of previous projects, or from other areas of the project.

Vertical Dimensions

Vertical heights can be represented using standard dimension methods presented in Chapter 3 or they can be represented by elevation symbols from a known point. Each type of dimension is usually placed on the outside of the drawing. Dimensions are generally given from the bottom of the sole plate, from the top of the top plate, or from the edge of a beam to the material being referenced.

Horizontal Dimensions

The use of horizontal dimensions in details varies greatly based on the material being represented. When provided, horizontal dimensions are generally located from grid lines to the desired member. Exterior wood and concrete members should be referenced to their edge. Interior wood members are referenced to a centerline. Interior concrete members are referenced to an edge. Steel members are referenced to their centers. The distance for roof overhangs and balcony projections also may be placed in details.

Drawing Symbols

Details use symbols that match those of the floor, roof, and elevation drawings to reference material. Symbols that might be found on a detail included:

- Grid markers
- Elevation markers
- Section markers
- Detail markers

Examples of each are shown in Figure 6-23. Grid markers should match, in both size and style, those used on other drawings so the details can be easily matched to other drawings. Elevations that are specified on the floor plan and elevation drawings should also be referenced to related details by use of a datum line or an elevation placed over a leader line.

Each detail is referenced to other drawings by a detail marker, which defines the page the detail is drawn on, and what is being viewed. A reference such as 3 over A–700 would indicate that the detail is drawing number 3 on page A–700.

Drawing Annotation

Annotation on details is used to specify materials and explain special installation procedures. As with other drawings, notes may be either placed as local or keyed notes. Most offices use local notes with a leader line that connects the note to the material. Local notes should be aligned to be parallel to the drawing to aid the print reader. The smaller the scale, the more generic the notes should be. For instance, on a small-scale detail, reinforcement might be specified as:

HORIZONTAL REBAR - SEE REINFORCING SCHEDULE.

This information would be referenced by complete notes in a detail drawn at a larger scale. A second method of specifying materials is to use the MasterFormat numbering system to reference materials. Using this system, the reinforcement specified above might be specified as:

03 20 00 A

Drawing Process

Because no two details are alike, one step-by-step process cannot be used for their layout. Every CAD technician should have a logical method of attacking drawing layout, however. A common method of layout would be as follows:

1. Draw the materials to be represented.
2. Detail each material.
3. Dimension required components.
4. Place all required notes.

A detail should generally be started by drawing the major component to be represented. Figure 6-24 shows the layout for a beam-column-foundation connection with the three major components represented. The detail was started by representing the foundation because it is used to support the column. Once the major

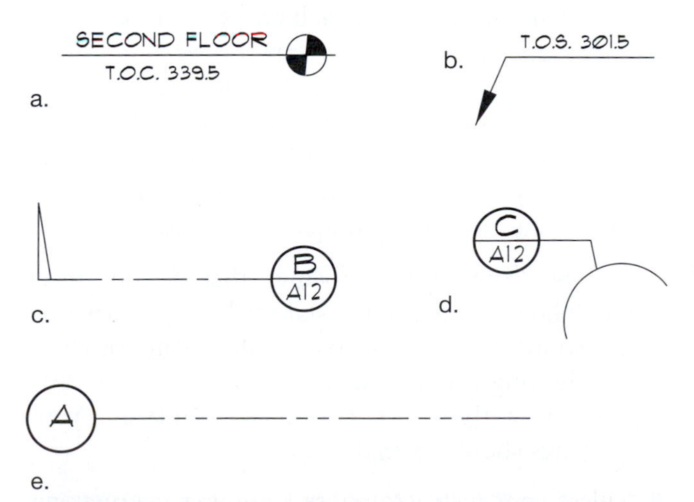

FIGURE 6-23 Symbols used on details are common to other drawings and include (a and b) elevation markers; (c) cutting plane; (d) detail reference marker; and (e) grid marker.

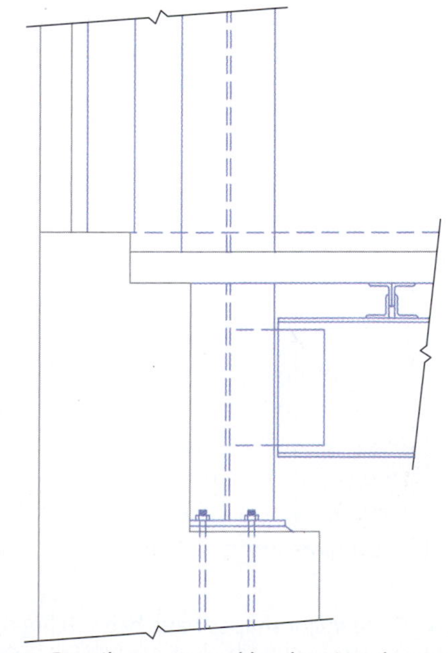

FIGURE 6-24 Details are started by drawing the outlines of major construction materials.

shapes were defined, specific details such as straps and bolt holes were added as shown in Figure 6-25. The drawing process was completed by adding varied lineweights and hatch patterns to represent each material. See Figure 6-26.

With all material drawn, dimensions should be added. Any feature that is to be fabricated must be dimensioned. Any feature that is premanufactured does not need to be dimensioned. Place dimensions so that smaller dimensions do not cross over larger dimensions. On large-scale drawings, dimensions can be placed within clear spaces of the detail. Never place dimensions in the component if clarity will be compromised.

Notes should be placed to reference all material represented in the detail. Care should be taken not to repeat information specified in the architectural drawings or within the specifications. Generally, sizes should be specified in dimension form rather than as a note. When a print reader scans a drawing looking for information, numerals on dimension lines are easier to visualize than numbers placed within fields of text. Text should be placed using the guidelines given in Chapter 2. Information presented in a note should be placed using the following format:

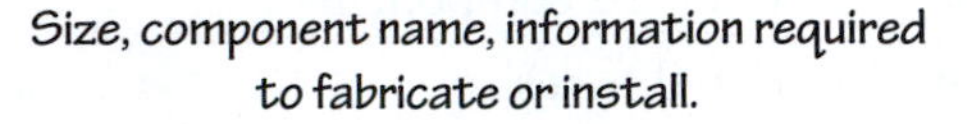

Size, component name, information required
to fabricate or install.

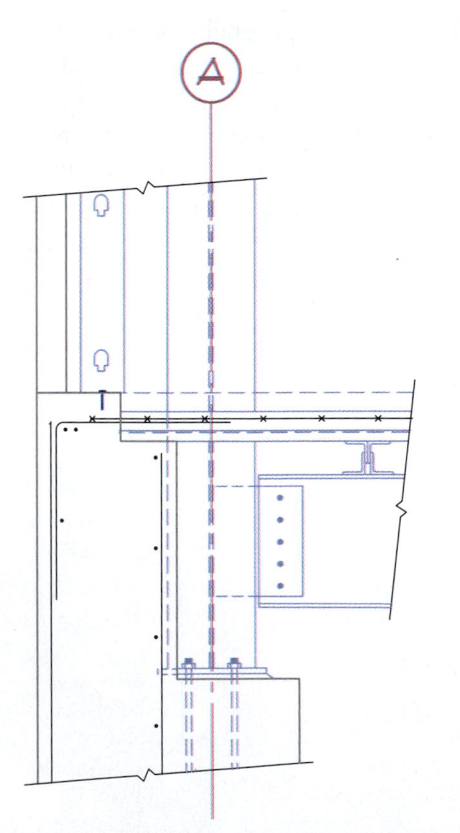

FIGURE 6-25 Once major shapes have been defined, specific details should be added.

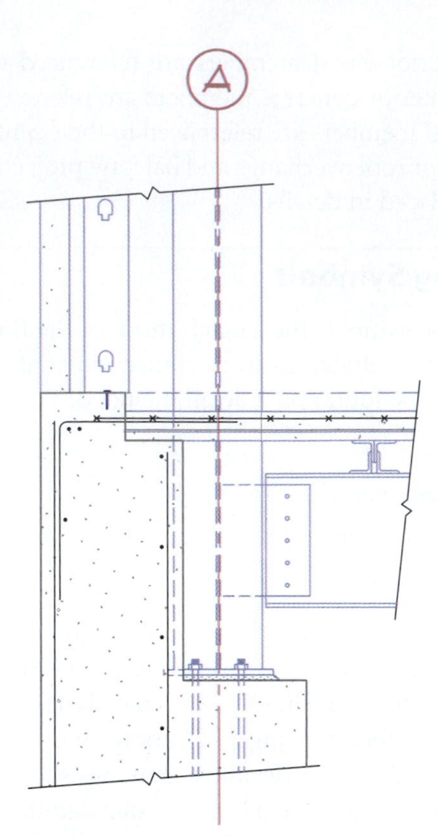

FIGURE 6-26 Line quality and hatch patterns are added to improve drawing quality.

Although the information cannot be completed at this time, a detail title, scale, and reference bubble should also be placed as the text is being provided. Figure 6-27 shows the completed detail.

DRAWING CRITERIA FOR COMPLETING DETAILS

As you progress through Section 2, you will find drawing problems at the end of each chapter. These details will be related to the projects drawn in Sections 3 and 4, and may require referencing sketches in other chapters to gain additional information. Use the following assumptions to complete the details found in each chapter in Sections 2, 3, and 4. These "assumptions" should be considered as company standards.

Warning: Submitting drawings that do not meet the following minimum standards may earn you the wrath of your supervisor, slow your progress in achieving a pay raise, or lead to your dismissal. Remember these are minimum standards. Your drawings should exceed them.

- Unless your instructor gives you other instructions, complete the following details and save them as

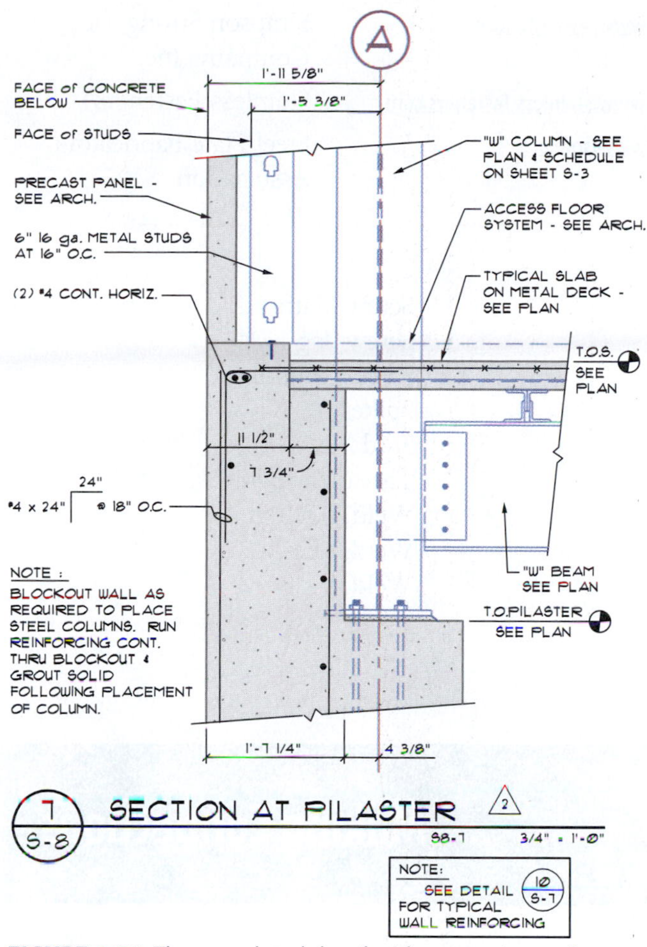

FIGURE 6-27 The completed detail with annotation and dimensions provided.

wblocks. Skeletons of most details can be found on the student CD. Use these drawings as a base to complete the assignment.

- Show all required views to describe each connection. Provide the fewest number of views to completely describe the specified connection.
- Draw each detail at a minimum scale of 3/4" = 1'–0" unless noted. Adjust all scale, lineweight, text, and dimension factors as required.
- Show and specify all connecting materials based on local standards. Base nailing on IBC standards unless your instructor tells you otherwise.
- Specify all material based on common local practice.
- Represent structural wood members that the cutting plane has passed through with bold lines.
- Use separate layers for wood, concrete, text, and dimensions.
- Use dimensions for locations where possible, instead of notes.
- Provide annotation to specify that all metal hangers are to be provided by Simpson Strong-Tie Company or equal. Obtain the needed catalog from a local supplier or from the company's Web site.
- Assume all trusses are to be provided by Weyerhaeuser. Verify all truss sizes from tables found on their Web site.
- Hatch each material with the appropriate hatch pattern.
- Use an appropriate architectural text font to label and dimension each drawing as needed. Refer to *Sweets Catalogs*, vendor catalogs, or the Internet to research needed sizes and specifications.
- Keep all text 3/4" minimum from the drawing, and use an appropriate architectural style leader line to reference the text to the drawing.
- Provide a detail marker, with a drawing title, scale, and problem number below each detail.
- Assemble the details required by your instructor into an appropriate sheet for plotting. Arrange the drawings using the guidelines presented in Chapters 2 and 3.

ADDITIONAL READING

The following Web sites can be used as a resource to help you keep current with changes in building materials.

ADDRESS	COMPANY/ ORGANIZATION
www.americanfasteners.com	American Fastener Technology Corporation
www.astm.org	ASTM International
www.americanwelding.com	American Welding Services, Inc.
www.aws.org	American Welding Society
www.atlasfasteners.com	Atlas Fasteners for Construction
www.boltdepot.com	Bolt Depot
www.grabberman.com	Grabber Construction Products
www.boltproducts.com	Metric Bolts
www.nailpower.com	Nail Power (construction adhesives)

www.nutsandbolts.com Nuts and Bolts Fasteners
www.portlandbolt.com Portland Bolt and Manufacturing Company
www.screw-products.com Screw Products
www.strongtie.com Simpson Strong-Tie Company, Inc.
www.stainless-fasteners.com Stainless Fasteners
www.spfa.org Steel Plate Fabricators Association

KEY TERMS

Bolt, anchor
Bolt, carriage
Bolt, drift
Bolt, expansion
Bolt, high-strength
Bolt, machine
Bolt, toggle
Detail, stock
Details
Field welded
Nail, box
Nail, common
Nailing, blind
Nailing, boundary
Nailing, end
Nailing, face
Nailing, field
Nailing, toe
Penny
Pilot hole
Power-driven anchors
Screw
Screw, flathead
Screw, lag
Screw, roundhead
Spikes
Stud
Threaded rods
Weld, beveled
Weld, fillet
Weld, V-groove

CHAPTER 6 TEST Making Connections

QUESTIONS

Answer the following questions with short complete statements. Type the chapter title, question number, and a short complete statement for each question using a word processor. Some answers may require the use of additional chapters in this text, vendor catalogs, or seeking out local suppliers.

Question 6-1 List two common uses for adhesives in the construction industry.

Question 6-2 What type of nail is most typically used for wood framing?

Question 6-3 What would cause a nail to have a bluish finish?

Question 6-4 How long is an 8-penny nail?

Question 6-5 Based on the IBC, how will a rafter be connected to a plate?

Question 6-6 Describe the difference between the terms *boundary, edge,* and *field.*

Question 6-7 List the nailing the IBC recommends for securing 3/4" plywood to the floor joists.

Question 6-8 List two common uses for roundhead screws.

Question 6-9 What purpose does a washer serve when used with a bolt or a screw?

Question 6-10 A 3/4" diameter anchor bolt is going to be used to join a 3 × 8 ledger to a concrete wall. What size hole should be provided in the ledger?

Question 6-11 List the names of the following bolts and provide information common to the bolt specification for using A.B., C.B., and M.B. regardless of the material being attached.

Question 6-12 List five pieces of information that might be included in an anchor bolt specification.

Question 6-13 What allows a carriage bolt to be tightened?

Question 6-14 Why do steel members need to be bolted if a weld is going to be used to attach the members?

Question 6-15 List the types of steel from which high-strength bolts are typically made.

Question 6-16 How is a split-ring connector used to hold timber together?

Question 6-17 List two major suppliers of timber connectors.

Question 6-18 Give the gauge of a Simpson HD5A hold-down anchor.

Question 6-19 What are the most common methods of connecting large steel shapes to each other?

Question 6-20 What is the CAD technician's role regarding the connection of webs to the top and bottom chords of a truss?

Question 6-21 List the three possible locations for placing a weld and give the symbol that represents each location.

Question 6-22 List five types of joints where welds can be applied.

Question 6-23 Sketch the symbol for a 3/16" fillet weld, applied to the other side, all around a column, and applied at the job site.

Question 6-24 Sketch the symbol used for a 1/4" × 3" long V-groove weld applied to this side.

Question 6-25 Sketch the symbol for a 1/8" field, fillet weld on both sides.

Question 6-26 What is a stock detail, and when would it be used?

Question 6-27 What factors influence the scale of a detail?

Question 6-28 What is a cutting plane, and how does it relate to a section?

Question 6-29 How far should nails penetrate into the supporting member?

Question 6-30 What are computer-related items that must be considered prior to starting a detail?

DRAWING PROBLEMS

Unless your instructor provides other instructions, use the "Drawing Criteria for Completing Details" section presented earlier in this chapter to complete the following details. Save each drawing as a wblock.

- **Open the appropriate detail on the student CD and save the file to an appropriate storage device.**
- **Materials are currently located on the 0 layers. Create layers based on the National CAD Standards and move each drawing component to the appropriate layer.**
- **Assume plotting in monochrome, but assign a different color to each material and drawing component to help you track each item.**
- **Assign lineweights as required to materials that will add clarity to the drawing.**
- **Provide hatch patterns as necessary to define each material.**
- **Edit the detail as required to meet the minimum standards above.**

Problem 6-1 Slab footing—1-level

Problem 6-2 CMU wall—1-level footing

Problem 6-3 Truss/stud wall—soffited eave

Problem 6-4 Truss/CMU wall—open eave

Problem 6-5 Raft/CMU wall—open eave

Edit the specified problem to create the following details:

Problem 6-6 Edit problem 6-1 to show a 2 × 6 stud wall supported on a two-story footing. Show the 15" wide footing extending 18" into the natural grade.

Problem 6-7 Edit problem 6-2 to reflect an 8" deep footing and a 5" thick concrete slab. Provide vertical wall steel with a 48" spacing, and horizontal steel with a 16" spacing.

Problem 6-8 Edit problem 6-3 to reflect a 3/12 pitch and a 36" overhang.

Problem 6-9 Edit problem 6-4 to reflect an enclosed soffit matching the soffit in problem 6-3.

Problem 6-10 Edit problem 6-5 to reflect an 8' high stud wall with double top plates and a truss roof framed to match the roof in problem 6-3.

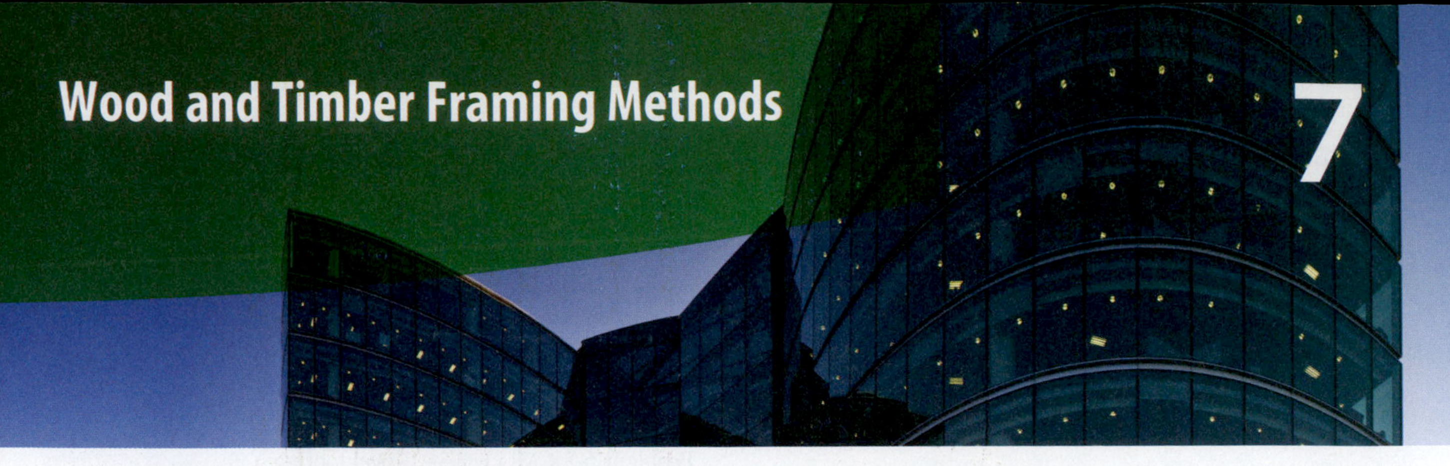

Wood and Timber Framing Methods 7

Wood products are a major material of construction for light commercial projects. Wood construction materials include sawn lumber and timber that is directly cut from trees. To effectively work with drawings that reflect either type of lumber construction, a technician must have a thorough understanding of the materials and terminology of lumber construction. This chapter will explore:

- Uses of wood in light construction.
- The terms that describe wood used in light construction.
- Wood framing.
- Timber.
- Heavy timber.

USES OF WOOD IN LIGHT CONSTRUCTION

Wood products are a valuable building resource because of their many green qualities. Products made from fast-growth trees, recycled materials, and especially engineered products reduce the demand for old-growth lumber products. Engineered wood products will be explored in Chapter 8. The use of wood in commercial projects is limited in some types of commercial projects by the loads that must be supported and the greater need to protect the occupants of the structure from fire. The use of wood is restricted by the International Building Code (IBC) based on the size of the structure and the occupancy of the structure. Multifamily housing projects, individual retail sales establishments, and office structures are common uses of **wood construction**.

Building Code Restriction of Wood

Chapter 4 introduced the effect of codes on the selection of building materials. Wood construction with no special covering is considered Type V-B. Notice in Figure 4-9 that a wood office structure (B occupancy) has a height restriction of two stories and an allowable basic area of 9000 square feet per floor. Chapter 12 will introduce materials that can be used to upgrade wood construction to a more fire-resistant type of construction. As materials are substituted to increase the fire resistance to Type III-B, the height can be increased to four levels and the size is increased to 19,000 square feet per floor. Because more products must be used to enclose the wood frame and protect it from fire, the use of wood is typically not an economical choice for framing materials in structures that demand greater fire safety. Not only is the size of each floor restricted by the type of construction, but the number of floors allowed is limited.

Loading Limitations

Notice from Figure 4-9, a Type V-B office structure has a basic allowable height of two levels. The use of Type III-B construction increases the basic allowable height to four floors, or a maximum of 55' high (16, 784 mm). The size and height of a structure will also dictate the selection of building products because of the loads that will be generated. The IBC restricts non-engineered wood bearing walls to a maximum of 10' (3050 mm) in height. Walls designed by an architect or engineer can exceed this limit based on specific load studies. Codes also require vertical wood wall members to be a minimum of 2 × 6 (50 × 150) if more than one floor level and roof level is being supported. In addition to the loads generated from gravity, the structure must resist lateral loads. Loads imposed on a mid-rise or high-rise structure from wind, seismic activity, and overturning are typically too severe to be resisted by wood structures at an economical price. Because of its flexibility, availability, and economic features, wood is a popular building material. Natural beauty and warmth make wood the material of choice for many low-level structures similar to Figure 7-1.

LUMBER TERMINOLOGY

An understanding of common wood terminology is essential to work efficiently on details and other drawings representing wood construction. The terms *wood*, *lumber*, and *timber* are often used interchangeably, but

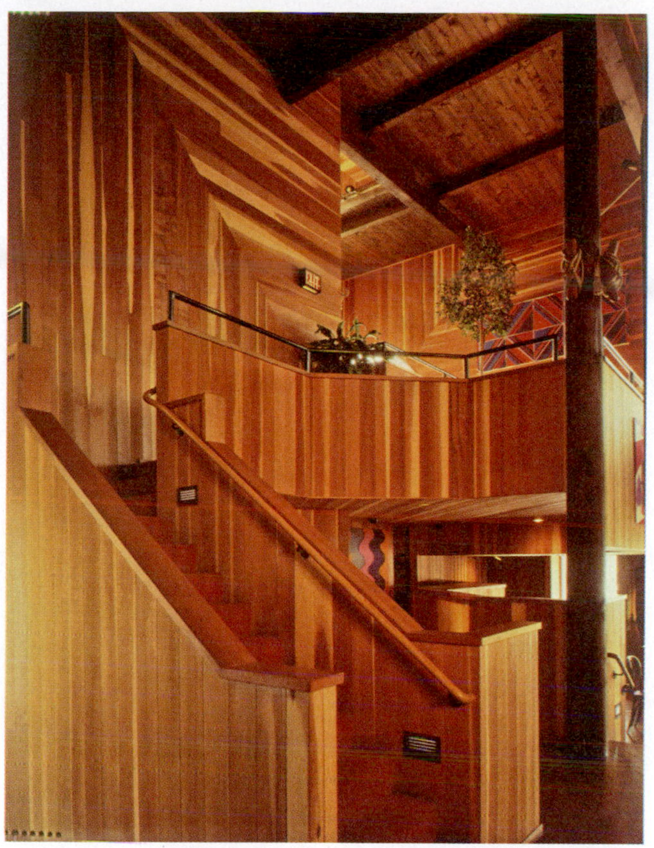

FIGURE 7-1 Both hardwoods and softwoods can be found in this design by Sandy and Babcock, AIA. *Photo by Jeremiah O. Bragstad. Courtesy California Redwood Association.*

each has its own distinct meaning. *Wood* is the material that forms the trunk of trees. *Lumber* is produced as a result of milling wood by sawing, resawing, and planing. This would include materials of 2" and 4" (50 and 100 mm) thickness and 4", 6", 8", 10", 12", and 14" (100, 150, 200, 250, 300, 350 mm) depths. Several different terms based on the **nominal size** of the lumber are used to describe lumber. These terms include the following:

- **Rough lumber** is lumber that has only been sawn, edged, and trimmed on each of its four longitudinal surfaces.
- **Dressed lumber** is lumber that has been planed on one side (S1S), two sides (S2S), or a combination of sides and edges (S1S1E), (S1S2E), or (S4S).
- **Boards** are wood products that are less than 2" (50 mm) thick but are wider than 2" (50 mm).
- **Structural lumber** is lumber that is 2" (50 mm) thick or wider nominal but less than 5" (130 mm) wide for use where working stresses are required.
- **Dimension lumber** is lumber that is 2" (50 mm) thick but less than 5" (130 mm) thick and is at least 2" (50 mm) wide.
- **Timber** is lumber that is 5" (130 mm) or thicker and 5" (130 mm) or wider.
- **Millwork** is a term used to describe products made of wood such as door and window frame components, trim, mantels, and moldings.

Lumber and timber are milled from **softwoods**, which are needle-bearing trees such as firs, pines, and spruces. Millwork is typically made from **hardwoods**, which come from leaf-bearing trees such as oak and maple. It is important to remember that the terms *hardwood* and *softwood* do not refer to the hardness or softness of the wood but to the leaf or needle. Some softwoods are harder than some hardwoods and vice versa.

Wood Structure

Each year of growth is marked by a new layer of wood being added to the outer layer of the tree. This new growth is reflected by the addition of a growth ring each year. The size of the ring is affected by the growth conditions for that year. Short, dry growing seasons result in less growth and smaller rings, while larger rings result from warmer, moist weather. Each growth ring is also divided into two portions. The inner portion of the growth ring represents the growth in springtime. This wood is usually lighter in color and the wood is composed of large thin-walled cavities. The outer portion of each ring represents summer growth. This portion of the ring is a darker color and is denser wood. Growth rings can be seen in the ends of lumber and timber.

The rate of growth affects the strength of wood. Wood with narrow growth rings typically has higher strength ratings than wood with wider rings. Wood with a high proportion of summer growth is stronger than spring-growth wood. The grading of wood for structural purposes takes into account the rate of growth and the proportion of spring to summer wood.

Wood Grain

Grain is used to describe the pattern of wood fibers in wood products. Wood that has slow growth will have narrow rings and is said to be *close grain. Coarse grain* is used to describe wood with wide rings. Lumber grain is also referred to as straight grain or cross grain. These terms relate to the direction of the wood fibers in relation to the sides of a piece of lumber and are not related to the growth rings. *Straight grain* lumber has a grain pattern that runs parallel to the piece of lumber. The grain of *cross grain* lumber runs at an angle to the edge of the lumber.

Moisture Content

Building codes and the engineering specifications set limits for the moisture content of lumber to be used throughout the project. Before wood can be used as a lumber product, it must be dried. Freshly cut wood is described as **green wood.** Green wood contains water within each cell, known as free water. Green wood also contains moisture within the walls of each cell, which is referred to as absorbed water. As wood dries, the moisture within the cell evaporates first. As the evaporation of free water is completed, the absorbed water begins to evaporate. The completion of free-water evaporation is called the *fiber saturation point*, and results when approximately 25% to 30% of total moisture content has been removed. Once the fiber saturation point is reached, wood will begin to shrink and the strength will be increased. Wood dried to a 5% moisture content can range from two to three times its original strength in bending and crushing strength. A trade-off to the increase in strength is an increase of splitting along the grain.

Removal of moisture from wood also increases its resistance to fungus. Moisture content of wood is expressed as a percentage and is determined by dividing the original weight by the oven-dried weight. Dry lumber is defined as lumber that has a moisture content of 19% or less and is represented by the symbol S-Dry. The International Building Code considers wood with moisture content higher than 19% to be green lumber.

Units of Measurement

Lumber is measured in nominal units expressed as board feet. One **board foot** is equal to 1" thick, 12" wide, and 12" long (25 × 300 × 300 mm). Lumber less than 1" (25 mm) thick is considered to be 1" (25 mm) thick. The number of board feet in a piece of lumber is determined by multiplying the nominal thickness in inches by the nominal width in feet by the length in feet. The nominal size of a piece of wood is larger than the **actual size.** Comparable sizes of nominal, seasoned, and dry wood are shown in Table 7-1.

Standard lengths of lumber come in 2' (600 mm) increments starting in 4' (1200 mm) long material. Most lumberyards stock lumber in sizes from 8' through 20' (2400 through 6100 mm). Lumber in lengths of 22' (6700 mm) is available in some sizes.

Grading

Because different species of wood have such a wide variety of qualities, lumber must be graded to be certain that the structural requirements of the building code are met. *Grading* is the process of specifying the strength of a certain species of wood after it has been sawn, planed, and seasoned. Wood can be inspected by either visual or mechanical methods. *Visual* grading occurs when lumber is sawn at a mill. With visual inspection methods, wood is evaluated by ASTM standards in cooperation with the U.S. Forest Products laboratory. Structural lumber that is tested nondestructively by machine and graded is referred to as *mechanically* evaluated lumber (MEL). Graded lumber will be stamped with a grading symbol that represents the quality, species, use, strength, and grading authority. Grading usually takes into account the size of knots or holes and their location, the size and location of splits, and the amount of warp in a piece of lumber. Most lumber is visually graded and given a designation of select structural, no. 1, no. 2, no. 3, construction, standard, utility, or stud. Wood is also divided into three classifications depending on how the lumber is to be used and its size. These classifications are as follows:

Dimensional—Dimensional lumber is either 2" or 4" (50 or 100 mm) thick and 2" (50 mm) wide and is used for joists or planks.

Beams and stringers—Lumber that has a nominal size of 5" × 8" (130 × 200 mm) or larger and is graded for strength in bending when loaded on the narrow face.

Board Thickness

Nominal	Seasoned	Dry
1" (25 mm)	25/32" (20 mm)	3/4" (19 mm)
2" (50 mm)	19/16" (40 mm)	1 1/2" (38 mm)
3" (75 mm)	29/16" (65 mm)	2 1/2" (64 mm)
4" (100 mm)	39/16" (90 mm)	3 1/2" (65 mm)

Face Width

Nominal	Seasoned	Dry
2" (50 mm)	19/16" (40 mm)	1 1/2" (38 mm)
4" (100 mm)	39/16" (90 mm)	3 1/2" (89 mm)
6" (150 mm)	5 5/8" (143 mm)	5 1/2" (140 mm)
8" (200 mm)	7 1/2" (191 mm)	7 1/4" (184 mm)
10" (250 mm)	9 1/2" (241 mm)	9 1/4" (235 mm)
12" (300 mm)	11 1/2" (292 mm)	11 1/4" (286 mm)
14" (350 mm)	13 1/2" (343 mm)	13 1/4" (337 mm)
16" (400 mm)	15 1/2" (394 mm)	15 1/4" (387 mm)

TABLE 7-1 Comparable Sizes of Nominal, Seasoned, and Dry Wood (Metric Sizes Given as Hard Conversions)

Posts and timbers—Lumber that is square or nearly square in cross section with a nominal size of 5" × 5" (130 × 130 mm) or larger and graded for use as a post or column or other uses in which bending strength is not important.

WOOD FRAMING

Wood-frame construction is often referred to as *stick construction* because the frame is made by adding one stick at a time. Most accurately known as *western platform framing*, the framing system is widely used throughout the United States. Framing members are typically 2" (50 mm) thick and spaced at 12", 16", 19.2", and 24" (300, 400, 490, and 600 mm) o.c. Use of 24" (600 mm) spacing is limited to one-level bearing walls with roof framing members directly aligned with the studs. The length of the vertical members is determined by the distance from floor to floor. Figure 7-2 shows an example of western platform construction for a two-story structure. Engineered lumber products are used throughout western platform construction to replace naturally grown lumber products. These products will be discussed in the next chapter. In order to understand the drawings involved with commercial construction, knowledge of common framing terms is essential. These terms include those used for floor, wall, and roof construction.

Floor Construction

It is important to remember that most lower floor systems used with commercial construction are concrete slabs. Wood floors are used in cooler, damper areas of the country to frame the lower floor system for multifamily dwellings. Concrete floors and foundations will be covered in Chapters 10 and 24. Basic terms used to describe ground-level wood floor systems include mudsill, floor joist, girder, and rim joist. Each can be seen in Figure 7-3. Joist, rim joist, and blocking are also used to describe upper floors framed with wood members.

FIGURE 7-2 Western platform construction for a multifamily structure. *Courtesy Southern Forest Products Association.*

Mudsills

The first piece of wood that comes in contact with the concrete foundation is the **mudsill**. It is also referred to as a *base plate* or *sill* in different areas of the country. In order to be protected from moisture in the concrete, the mudsill is required by code to be pressure-treated or made from foundation-grade cedar or redwood. A 2 × 6 (50 × 150) is typically used for a mudsill. This is specified on the foundation plan with a note and shown on foundation details.

Anchor Bolts

The mudsill is bolted to the foundation with anchor bolts to resist uplifting and sliding. *Anchor bolts* are L-shaped bolts ranging in size from 1/2" to 3/4" (13 to 19 mm) diameter for most applications. Maximum spacing of anchor bolts allowed by code is 6'–0" (1800 mm) o.c.

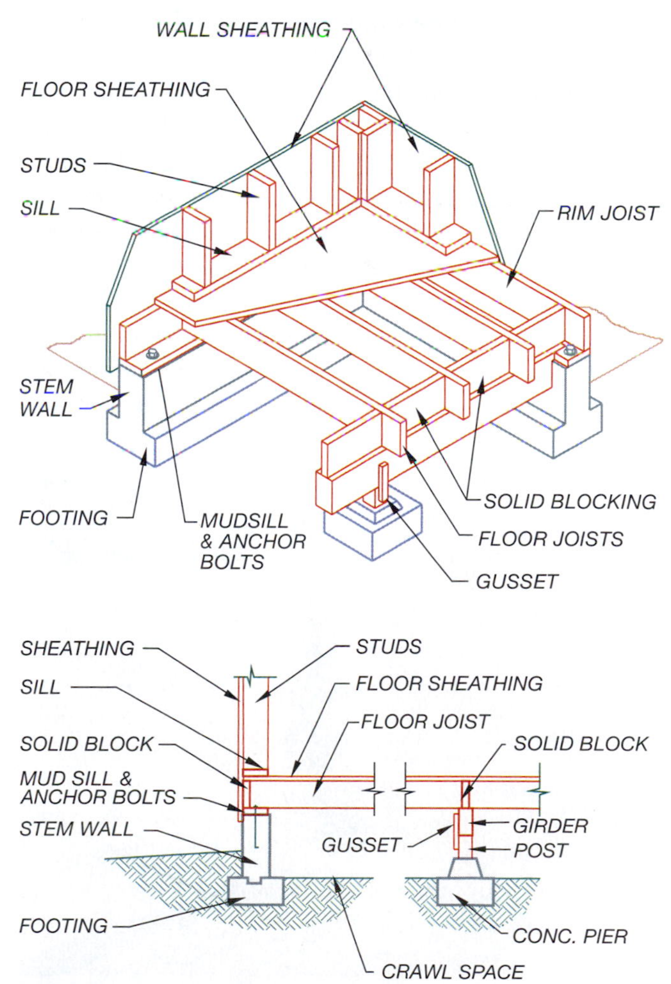

FIGURE 7-3 Conventional floor framing methods consist of floor joists supported by a mudsill resting on the concrete stem wall. Girders are used to support the joists if the distance between the stem walls exceeds the span rating of the joists.

with a minimum length of 10" (250 mm). The loads to be resisted will affect the size and spacing of the bolts. A 2" (50 mm) diameter washer is placed beneath the nut to keep the mudsill from lifting off the bolts when forces of uplifting are applied. In seismic zones, metal anchors or straps similar to those shown in Figure 7-4 will also be specified on the foundation and framing details.

Girders

Once the mudsill is in place, the girders can be set. **Girders** are beams used to support floor joists as they span across the foundation. Girders are supported by the foundation stem wall, and by wood posts at the interior of the floor system. A minimum bearing surface of 1 1/2" (38 mm) is required by code to support a girder resting on a wood support, and a 3" (75 mm) bearing surface is required if the girder is resting on concrete. The floor joists typically rest on top of the girder. A girder can also be set so that the top of the girder aligns with the top of the floor joists. When a girder is set level with the joists it is called a *flush* girder. When set level with the floor joists, the weight of the joists must be supported by metal hangers.

Girders are typically made from 4× or 6× (100× or 150×) members but they can also be constructed by 2× (50×) members joined together. If only two members are required to form the girder, they can be nailed together. If three or more 2× (50×) members are to be used, they must be bolted together with bolts that pass through the girder. Where supports need to be kept to a minimum, laminated wood beams called *glu-lams* can be used for girders. Chapter 8 will explore the use of laminated beams in construction. Steel girders are often used where foundation supports must be kept to a minimum. When steel girders are used, steel floor joists are also typically used. Steel framing methods will be introduced in Chapter 9. Regardless of the type of girder used, the width and depth are determined by the load to be supported and the span to be covered.

FIGURE 7-4 Metal anchors such as this HD5A hold-down anchor by Simpson Strong-Tie are often used to bond wood framing members to the foundation. *Courtesy Matthew Jefferis.*

Posts

Wood posts are a common method used to support girders. As a general rule of thumb, a 4 × 4 (100 × 100) post is typically used below a 4" (100 mm) wide girder, and a 6 × 6 (150 × 150) wide post is used below a 6" (150 mm) wide girder. Post sizes vary based on the load and the height of the post. Solid wood posts are limited by code to a maximum height-to-depth ratio (*l/d*). That is the quotient resulting from the length or height of the column expressed in inches, divided by the width of the column. It cannot exceed 50. This is referred to as the *slenderness ratio.* Wood posts and columns fall into three different categories based on their slenderness ratio. Short columns fail in crushing and long columns fail in bending. For intermediate columns, the failure mode is indeterminate. The engineer will determine the required size and connectors to be used.

Posts that provide a bearing surface of 1 1/2" (38 mm) minimum are required by code to support a wood member. Because a wooden post will draw moisture out of the foundation, it must rest on 55# felt, although sometimes an asphalt shingle is used. If the post is subject to uplift or lateral forces, a metal post base or strap may be specified by the engineer to firmly attach the post to the concrete. Chapter 8 will explore posts made from engineered products. Steel columns can also be used to support girders. Columns are typically rectangular, round, or in the shape of an *I.* Chapter 9 will discuss the use of steel in construction.

Floor Joists

With the mudsills and girders in place, the framing crew can set the floor joists in place. **Floor joists** are the structural members used to support the subfloor or rough floor with a maximum span of approximately 20' (6100 mm). The exact span will depend on the size, spacing, and load to be supported. Floor joists range in size from 2 × 6 to 2 × 14 (50 × 150 to 50 × 360) and are typically spaced at 12", 16", 19.2", or 24" (300, 400, 490, or 600 mm) o.c. The spacing depends on the load to be supported and distance the joist will span. Floor joists made of engineered lumber or trusses are also common. These materials will be discussed in Chapter 8. Regardless of the material, floor joists are typically shown and specified on the appropriate framing plan and in framing details. Figure 7-5 shows an example of a detail that a CAD

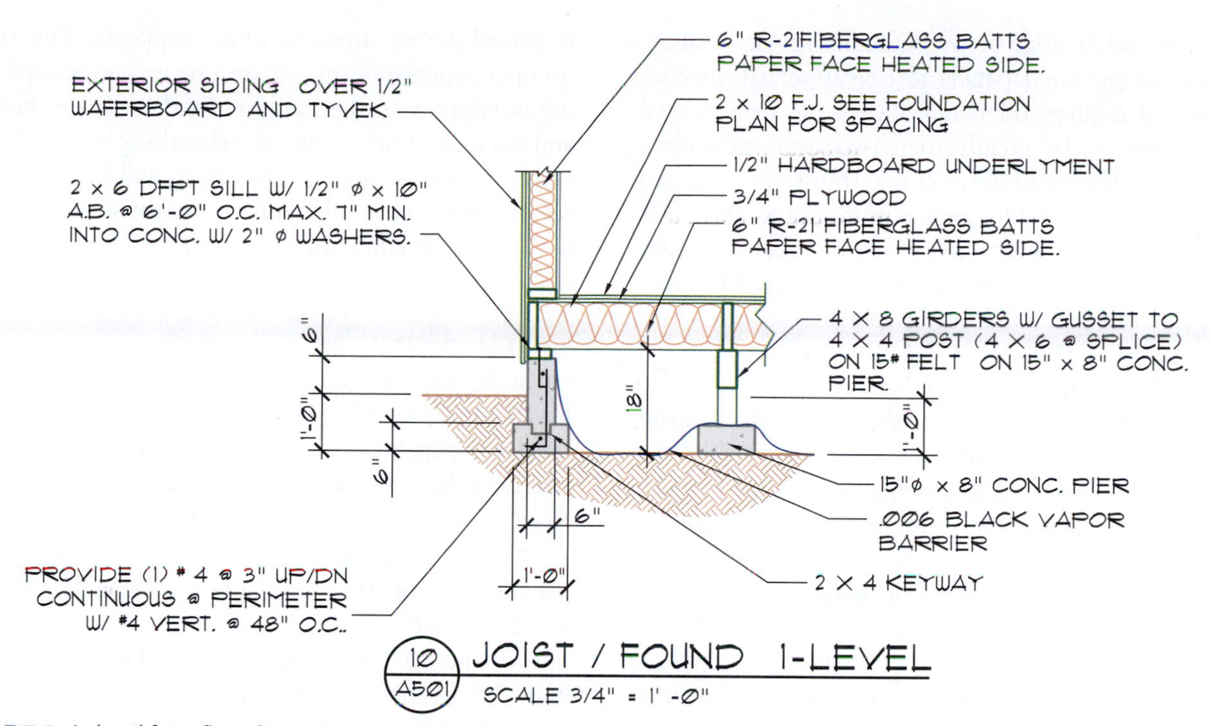

FIGURE 7-5 A detail for a floor framed with standard floor joist construction. Note that the distance between the girder and the stem wall has been greatly reduced.

drafter would be expected to complete to represent western platform floor construction.

When a bearing wall is to be supported by floor joists that are parallel to the wall, building codes require the floor joists directly below the wall to be doubled as a minimum standard, and they often require the loads to be analyzed. Because of the decreasing supply and escalating price of sawn lumber, each of the national wood manufacturers has developed several alternatives to sawn lumber for floor joists. These alternatives will be introduced in Chapter 8.

Floor Bracing

Because of the height-to-depth proportions of a joist, it will tend to roll over onto its side as loads are applied. To resist this tendency, a rim joist or blocking is used to support the edge of each joist at its exterior edge. A **rim joist** (sometimes referred to as a *band* or *header*) is aligned with the outer face of the foundation and mudsill. Sometimes a rim joist will be set around the entire perimeter and then end-nailed to the perpendicular floor joist. *Solid blocking*, an alternative to a rim joist, is a block of wood used to span between two floor joists to transfer lateral loads. The International Building Code requires blocking between floor joists at a maximum spacing of 10' (3400 mm) and beneath any bearing walls resting on the floor. Solid blocking is also placed to help provide support for plumbing lines and HVAC ducts, as seen in Figure 7-6. Lightweight metal cross bracing can be substituted for blocking to resist lateral loads in floor joists if the blocking is not required for a fire-stop. Floor blocking is specified in the written specifications and will be shown in the framing details.

Blocking is also used to provide added support to the floor sheathing. Often, walls will be reinforced to resist lateral loads. These loads are, in turn, transferred to the foundation through the floor system and are resisted by the diaphragm. A **diaphragm** is a rigid plate that acts similar to a beam and can be found in the roof, wall, or floor system. In a floor diaphragm, 2 × (50×) blocking

FIGURE 7-6 Blocking is used to provide stiffness to floor joists and to resist lateral loads. *Courtesy Benny Molina-Manriquez.*

is typically laid on edge between the floor joists to allow the edges of plywood panels to be supported. Nailing all edges of a plywood panel allows the design load to be resisted to be greatly increased over unblocked diaphragms. The engineer will determine the size and spacing of nails and blocking required, and the CAD technician will make the notations on details similar to Figure 7-7. Refer to Chapter 8, "Engineered Lumber Products," for other subflooring options.

Floor Sheathing

Floor sheathing is installed over the floor joists to form the subfloor. The subfloor provides a surface for the walls to set on. Plywood laid perpendicular to the floor joist is usually used for the subfloor. Plywood ranging from 1/2" to 1 1/8" (13 to 30 mm) thick with an APA (Engineered Wood Association formerly the American Plywood Association) grade of EXP 1 or 2, EXT, STRUCT 1 EXP 1, or STRUCT 1 EXT is typically used for floor sheathing. EXT represents exterior grade, STRUCT represents structural, and EXP represents exposure. Plywood is also printed with a number to represent the span rating, which represents the maximum spacing from center to center of supports. The span rating is listed as two numbers separated by a slash (for example, 32/16). The first number represents the maximum recommended spacing of supports if the panel is used for roof sheathing and the long dimension if the sheathing is placed across three or more supports. The second number represents the maximum recommended spacing of supports if the panel is used for floor sheathing and the long dimension of the sheathing is placed across three or more supports. Sheathing will be specified by the engineer in the calculations and must be specified in the written specifications, the wall sections, and specific details.

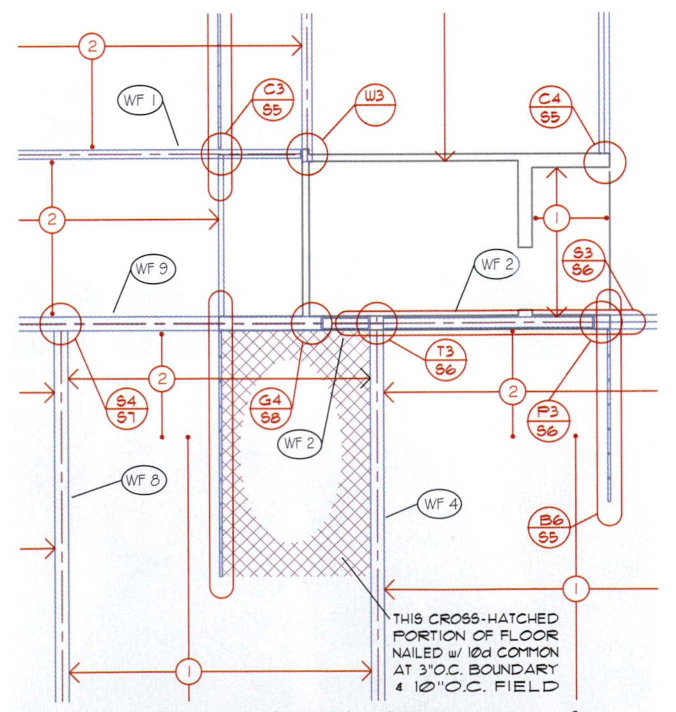

FIGURE 7-7 Blocking and special nailing patterns are often shown throughout framing drawings by a CAD technician.
Courtesy Dean Smith, Kenneth D. Smith Architect and Associates, Inc.

Floor Underlayment

Once the subfloor has been installed, an underlayment for the finish flooring is laid. The underlayment to the finish flooring is not installed until the roof, doors, and windows have been installed, making the structure weather tight. The underlayment provides a smooth surface on which to install the finished floor and is usually 3/8" or 1/2" (10 or 13 mm) APA underlayment GROUP 1, EXPOSURE 1 plywood, hardboard, or wafer board. Hardboard is referred to as medium- or high-density fiberboard (MDF or HDF) and is made from wood particles of various sizes that are bonded together with a synthetic resin under heat and pressure. APA STURD- I-FLOOR rated plywood 19/32" to 13/32" (15 to 30 mm) thick can be used to eliminate the underlayment.

Framed Wall Construction

Walls are framed using a sole plate, studs, and a top plate. Walls are assembled in a horizontal position on the floor and then tilted into place. Exterior and load-bearing walls are usually assembled and located first. Interior nonbearing walls are usually assembled after the shell of the structure is completed.

Studs

Studs are the vertical members of the wall and are typically spaced at 16" or 24" (400 or 600 mm) o.c. Occasionally, a spacing of 12" (300 mm) or smaller may be used, depending on the loads to be resisted. Typically, 2 × 4's or 2 × 6's (50 × 100's or 50 × 150's) are used to frame walls, but 2 × 8's (50 × 200's) or larger can be used to support larger loads or hide other structural members such as a steel column within the wall. Stud-length lumber can be cut to either 88 5/8", 92 5/8", or 96" (2250, 2350, or 2440 mm) long material. If a suspended ceiling is to be used, walls are often framed at 10' (3400 mm) or taller to allow for HVAC ducts and lighting fixtures.

Walls over 10' (3400 mm) high must be blocked at mid-height. Individual studs are not shown in plan view, but their location and spacing is indicated on the framing plan by the use of annotations.

Wall Sheathing

In addition to transferring loads downward, studs are also used to support the interior and exterior wall materials. Interior material is typically 1/2" or 5/8" (13 or 16 mm) gypsum board for most light construction projects. Exterior coverings are supported by 3/8" or 1/2" (9 or 13 mm) thick plywood in colder regions. Plywood sheathing should be APA rated EXP 1 or 2, EXT, STRUCT 1, EXP 1, or STRUCT 1 EXT. Fiberboard or 1/2" (50 mm) OSB can also be used for sheathing in place of plywood (see Chapter 8). Composite products such as fiber-reinforced gypsum panels are also used for sheathing interior and exterior walls. In temperate regions, unless sheathing is required for structural reasons, the exterior siding can be applied to the studs. In either case, a vapor barrier such as Tyvek is applied to the exterior side of the studs prior to installing the exterior siding. Gypsum board can be added to the exterior side of an exterior wall to increase the fire rating of the wall, as described in Chapter 12. Sheathing size and nailing is usually specified on the framing plans but is not shown. Sheathing is shown and specified in sections, framing details, and the written specifications.

Bracing

In temperate regions, and where sheathing is not required to resist lateral loads, the studs can be kept in a vertical position by the use of a let-in brace. A *let-in brace* is a board that is placed in a diagonal position across the studs. Typically, a 1 × 4 or 1 × 6 (25 × 100 or 25 × 150) is placed in a notch that is placed in the studs so that the brace is flush with the exterior face of the studs. A metal strap, which lies diagonally across the studs, can usually be substituted for the let-in brace. The IBC typically requires the brace or strap to cross a minimum of three studs and tie into the top and bottom plates of the wall. If plywood siding rated APA STURD-I-WALL is used, no underlayment or let-in braces are required. Bracing is shown on a framing plan. Let-in braces are shown on the framing plan and occasionally shown on the exterior elevations.

Sole Plate

The **sole plate** or *bottom plate* supports the studs and is used to help disperse the weight of the wall across the floor. The sole plate is also used to help hold the wall together as it is moved into its vertical position. A 2× or 3× (50× or 75×) is usually used for a sole plate.

The size is dictated by the loads to be supported and the material used for the finished floor. If a lightweight concrete is to be used over the plywood subfloor, the plate may be thicker and made from pressure-treated material. Figure 7-8 shows a detail of a wall constructed

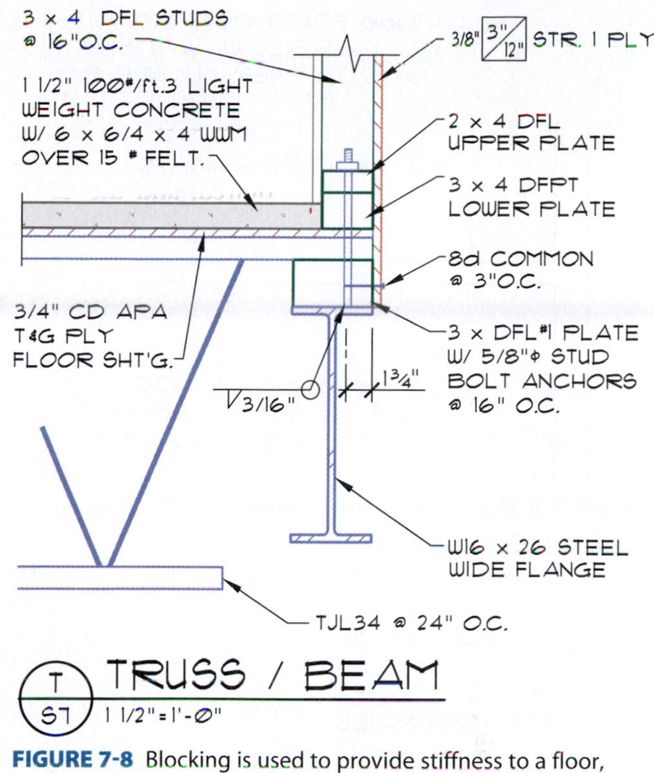

FIGURE 7-8 Blocking is used to provide stiffness to a floor, resist lateral loads, and provide support to plumbing lines.
Courtesy Ginger Smith, Kenneth D. Smith Architect and Associates, Inc.

with a double sole plate. The upper sole plate holds the studs together, and the lower sole plate is pressure-treated and protects the wall from the moisture in the concrete floor. The sole plate is end-nailed into the studs while the wall is horizontal, and then nailed to the floor system. The sole plate is nailed into the floor sheathing and the rim joist below the exterior walls and into solid blocking or double joists below interior partitions. Nailing is specified in details or by a nailing schedule. Special nailing that is required to transfer lateral loads through the sole plate to the floor system is specified by the engineer in the calculations and indicated on the framing plan in a method similar to Figure 7-9, or shown in framing details.

Top Plate

The **top plates** are the horizontal members used at the top of a wall to tie the studs together and to provide a bearing surface for the roof or upper floors. Two top plates are required for bearing walls. The lower plate is used to tie walls to each other. The upper top plate must lap the lower plate at splices by 48" (1200 mm). The upper plate can be omitted if the single plate is tied together by a steel strap at each splice. As seen in Figure 7-10, both plates can be eliminated and substituted with a flush header, if the header is connected to the top plate at each end by a steel strap. Although

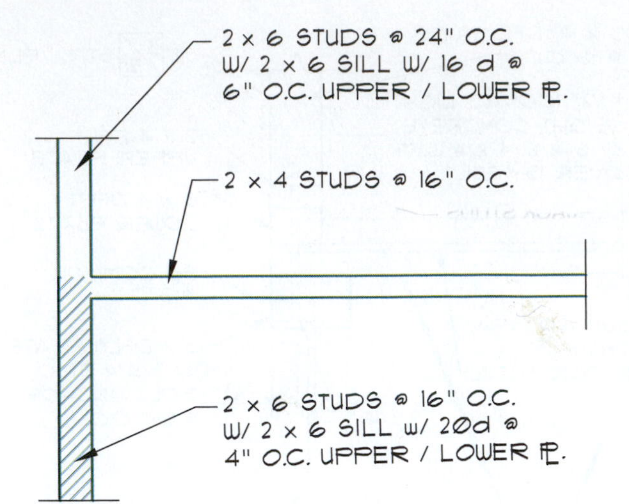

FIGURE 7-9 Blocking and special nailing patterns are often shown throughout drawings by a drafter.

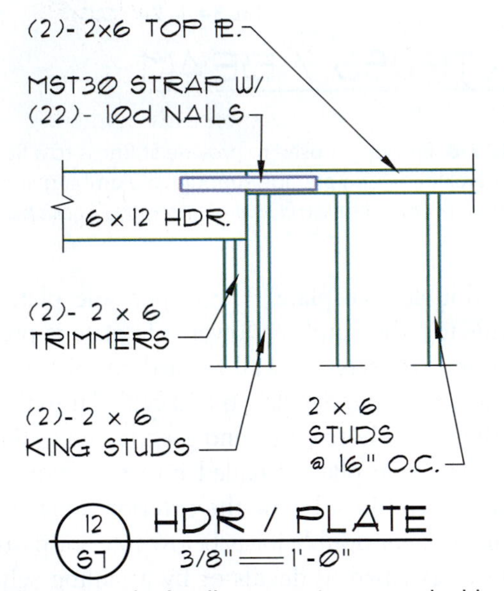

FIGURE 7-10 Standard wall construction uses a double top plate on the top of a wall. A header can be set level with the top of the top plate if a metal strap is used to tie the header to the top plate. The strap can be placed beside or on top of the header depending on the design of the engineer.

this situation is not typically used, it allows an opening to be set at its highest position in a wall. Top plates are usually shown on wall sections. A flush header is usually specified on the framing plan and in the framing details.

Framing Components for Wall Openings

Several additional terms need to be understood to frame an opening in a wall. These terms include *header, trimmer, king stud, subsill,* and *jack stud.* Each component can be seen in Figure 7-11. These components are represented on framing details and sections.

Header

A **header** is the horizontal piece of lumber used to support the loads over an opening in a wall. When a hole is framed for a door or window, one or more studs must be omitted. A header carries the weight that the missing studs would have carried. Headers are usually the same width as the studs they replace. Headers can be made from sawn lumber, engineered lumber, or steel. The load to be supported and the span will usually dictate which material is used. The size of the header is specified on the framing plan as well as in the sections and specific details.

Header Supports The header is supported at each end by a **trimmer.** It is like a regular stud except that it has been cut shorter to fit beneath the header. A trimmer can be made from one or more 2×'s (50×'s), or one 4× (100×), depending on the loads to be supported. In addition to supporting the weight of the header, a trimmer provides a nailing edge for the door or window frame, as well as the interior and exterior finish materials. A **king stud** is normally the same length and size as other full studs and is placed beside the trimmer to help resist the tendency of the trimmer to bend under its load. The king stud also provides stability from lateral pressure applied to the side of the header, which causes a hinge point at the header-trimmer connection (see Figure 7-12). Single trimmers and king studs are not specified on the framing plan. Larger or double trimmers and king studs that are required to support the header are specified on the framing plan. King studs and trimmers may also be shown on the framing details if connections are required from one level to another level, as shown in Figure 7-13.

Subsill

The **subsill** is a horizontal support placed between the trimmers to support the lower edge of a window. The subsill will be the same width as the studs it is replacing. The subsill is supported on **jack studs,** which are studs that are less than full height. Neither component is typically shown on the framing plan, but they are often shown in the architectural and structural details.

Roof Construction

Wood roof framing includes both conventional and truss framing methods. Truss construction will be considered in Chapter 8 as engineered products are examined. Each framing method has its own special terminology, but many terms also apply to both types of

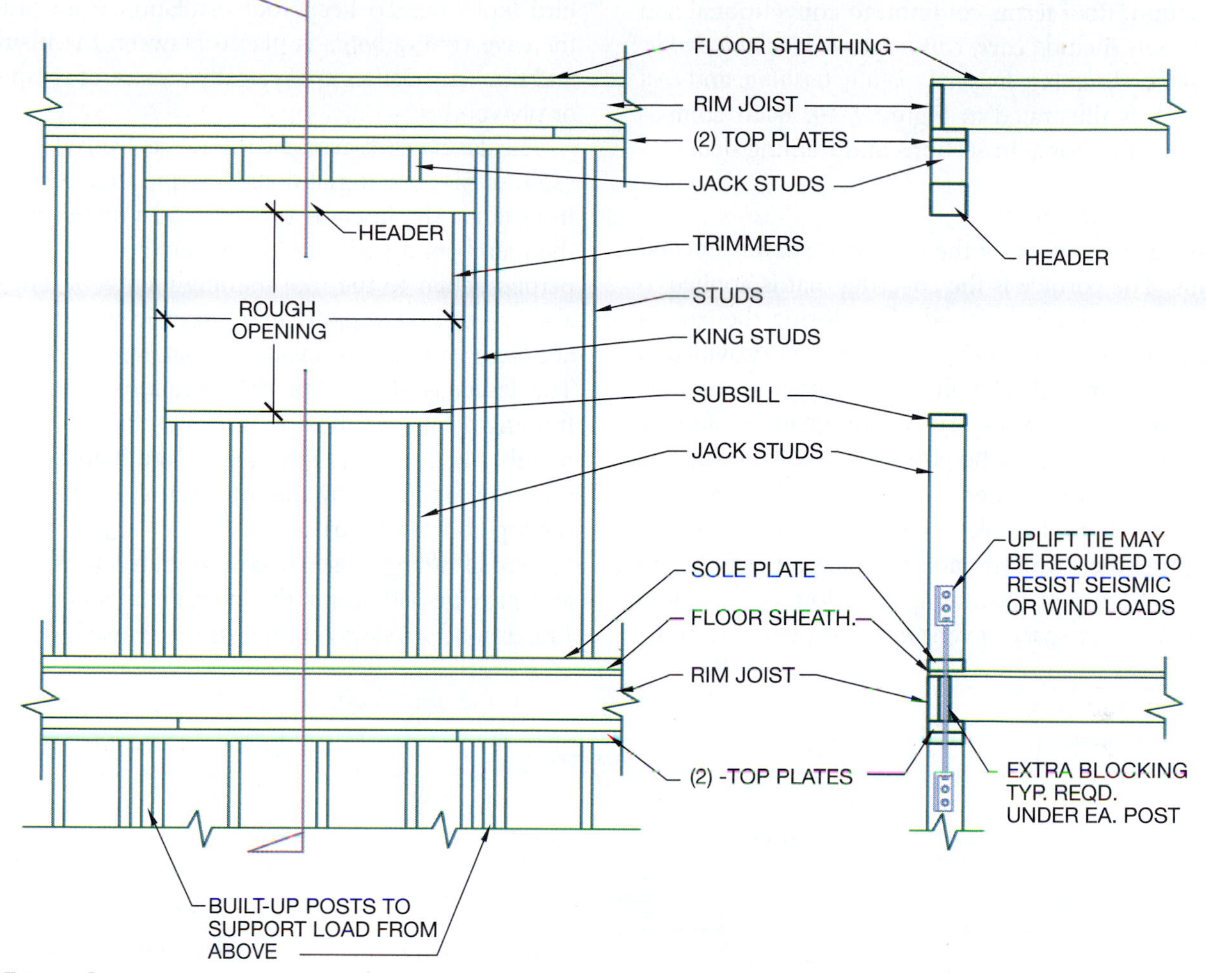

FIGURE 7-11 Construction components of a wall opening include a header, trimmer, king stud, subsill, and jack stud.

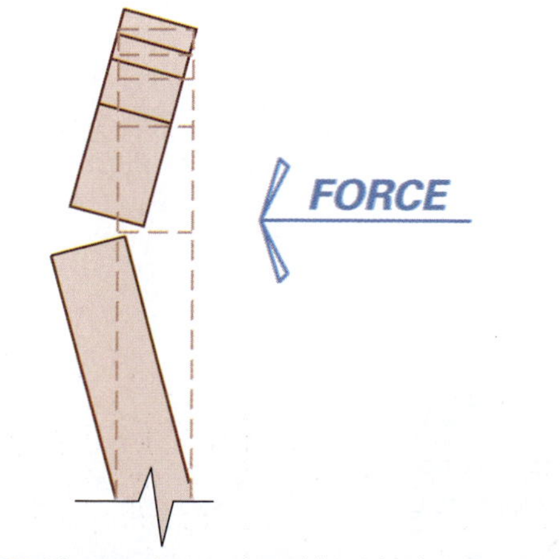

FIGURE 7-12 King studs are placed beside the trimmers to support the header to resist lateral pressure.

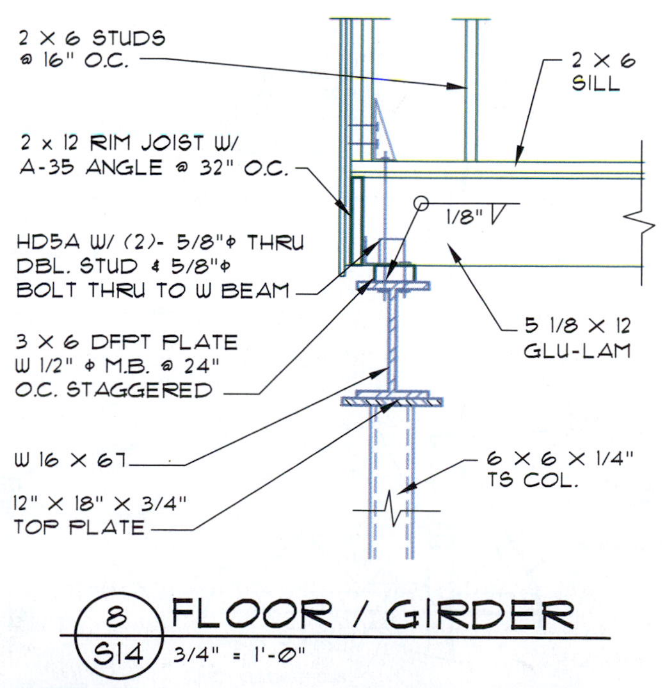

FIGURE 7-13 To resist seismic activity, king studs and trimmers are often required to be anchored to members of lower levels.

construction. Roof terms common to conventional and trussed roofs include *eave, cornice, eave blocking, baffle, fascia, ridge, sheathing, finished roofing, flashing,* and *roof pitch.* Each is illustrated in Figure 7-14. Each component is usually shown in sections and framing details.

Common Roof Terms

The **eave** is the portion of the roof that extends beyond the walls. The **cornice** is the covering that is applied to the eaves. A common method of enclosing the eave is shown in Figure 7-14. When a cornice is provided, a vent must be provided to allow cool air into the attic or rafter space. When the eave is not enclosed, an *eave block* or *bird block* must be provided to keep birds and small animals from entering the space between the framing members. The block also keeps the spacing of framing members uniform and provides an even termination to the top of the siding. To allow cool air into the attic or rafter space, a vent must be provided in the bird blocking. To keep roof insulation from plugging the eave vent, a *baffle* is placed between the insulation and the vent. Baffles are typically a piece of scrap wood or plywood.

The **fascia** is a trim board made from 1× or 2× (25× or 50×) material that is used to hide the rafter or truss tails. The fascia is often 2" (50 mm) deeper than the members it hides and runs parallel to the wall and perpendicular to the roof framing members. The fascia also serves as a support to a gutter in wetter climates, although gutters are available that replace the fascia. The fascia is shown on the elevations, sections, and framing details. At the opposite end of the roof framing member is the ridge. The ridge is the highest part of a roof and is formed by the intersection of the rafters or the top chords of a truss.

Roof sheathing is used to cover the structural members and provide a base for the exterior finishing material. Either solid or skip roof sheathing is used, depending

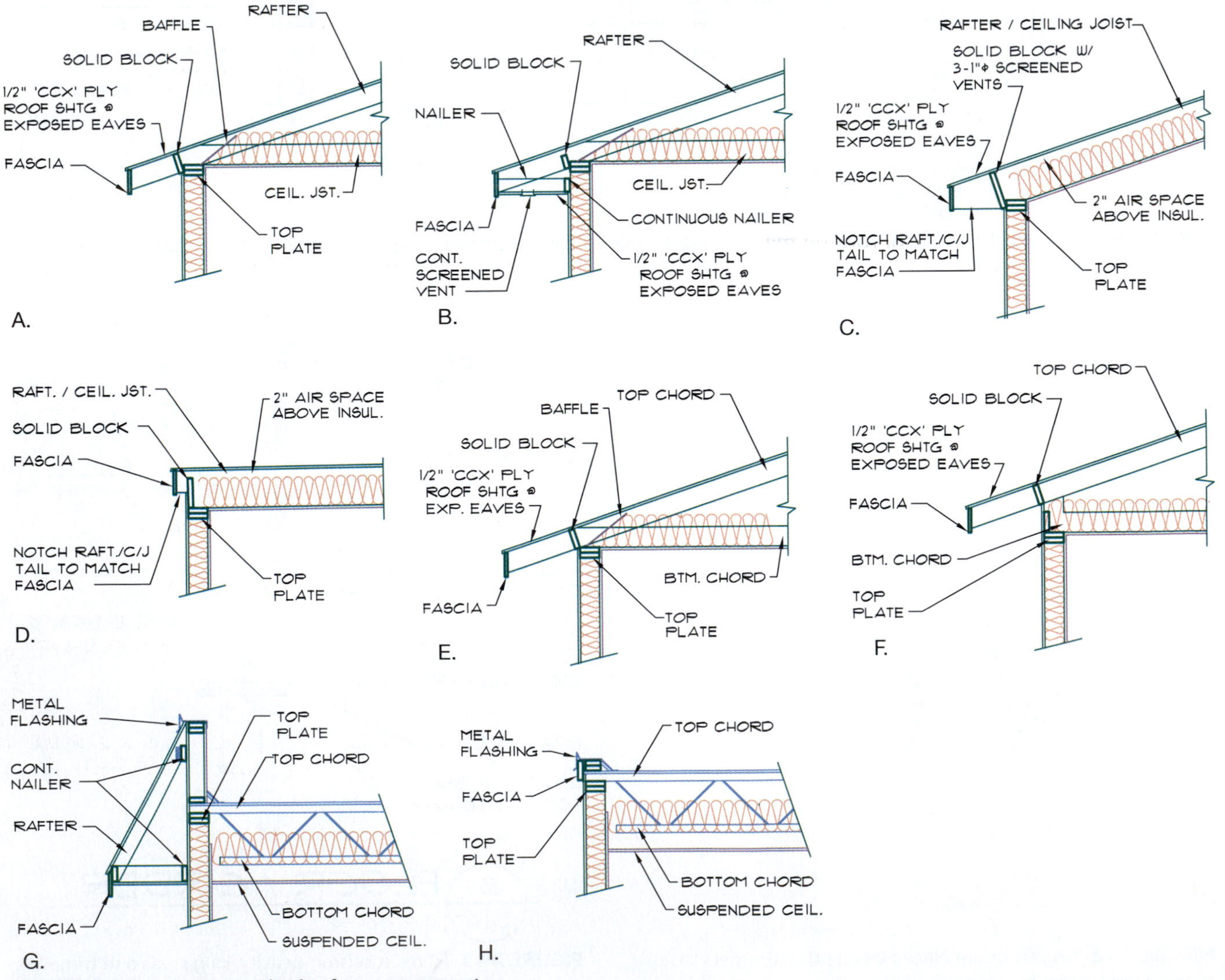

FIGURE 7-14 Common methods of eave construction.

on the type of roofing material to be used and the area of the country where the structure will be built.

Solid sheathing is typically 1/2" (13 mm) thick CDX plywood with span ratings of 24/16, 32/16, 40/20, and 48/24. OSB is also used in many areas of the country for roof sheathing that is not over an exposed eave. Because the material at the eave will be affected by moisture, CCX plywood or 1" (25 mm) T & G lumber must be used.

Skip sheathing is used on sloped roofs to support tile or cedar shake roofing. In warm, temperate climates 1 × 4's (25 × 100's) at 7" (180 mm) o.c. are placed directly over the roof framing members, as seen in Figure 7-15. In cool, damp climates the skip sheathing is applied over continuous plywood sheathing. Once the skip sheathing is in place, 15# building paper is applied as a base for the finished roofing. The *finished roofing* is the weather protection system and includes materials such as built-up hot asphalt, asphalt and fiberglass shingles, cedar shakes, concrete and clay tile, sprayed and poured concrete, foam, and metal panels. Chapter 18 will review each roofing material and the information that must be specified on the architectural and structural drawings.

Two other considerations of roof design are pitch and span. Each is illustrated in Figure 7-16. **Pitch** is used to describe the slope of the roof and is listed in terms of *rise over run*. A 3/12 roof will rise 3" for each 12" horizontal run. The type of roofing material and the local building code will affect the minimum and maximum pitch to be used. A minimum pitch of 1/4" per 12" is required for flat roofs with a built-up hot asphalt roof. A 3/12 pitch is the minimum slope recommended by most manufacturers for other roofing materials without requiring special waterproofing materials under the finished roofing. Pitch is specified on the elevations, sections, details, and roof plan.

FIGURE 7-15 Skip or spaced sheathing is used under shakes and tile roofs.

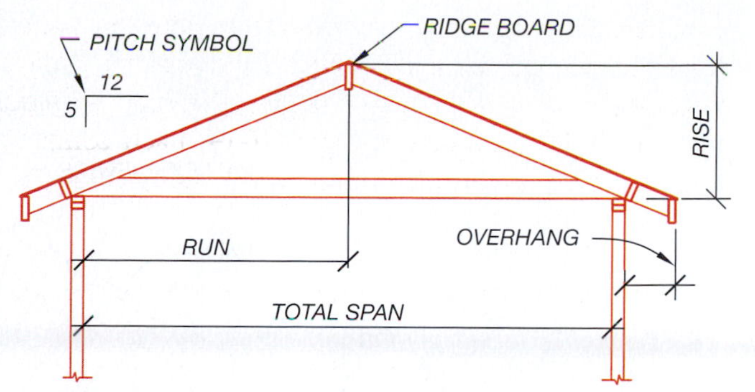

FIGURE 7-16 Roof dimensions needed for construction.

Span is used to determine the horizontal run of a piece of lumber or truss. Span is the horizontal distance from the interior edge of the required bearing points.

Conventionally Framed Roof Terms

Conventional or stick framing involves the use of wood members placed in repetitive fashion one board at a time. Key terms to be understood and represented on the structural drawings include *rafters, ridge board,* and *ceiling joist*. Each component is shown in sections and framing details. Ceiling joists, rafters, and rafter/ceiling joists are also shown on the appropriate framing plan.

Rafters

Rafters are the inclined members used to support the weight of the finished roofing. Rafters are typically spaced at 24" (600 mm) o.c.; however, 12", 16", and 19.2" (300, 400, and 490 mm) spacings are also used. The span, rafter size, and material being supported will determine the spacing. Rafters are typically 2× (50×) material with depths ranging from 6" to 12" (150 to 300 mm). Just as with floor joists, engineered products can also be used for rafters. The engineer will determine the type and size of the rafter, based on the span and the material to be supported. The location of the rafter in the roof will affect the name of the rafter. Figure 7-17 shows several different terms given to rafters based on their use in the roof frame.

A notch called a *bird's mouth* is cut into a sawn rafter where the rafter is to be supported on a wall or beam. The bird's mouth increases the bearing surface of the rafter and spreads the weight of the roof evenly over the wall. A bird's mouth is not specified within the drawings but is placed by normal construction practice. A bird's mouth is not placed in engineered materials. The materials are often required to sit on a top plate that has been cut to provide an inclined surface.

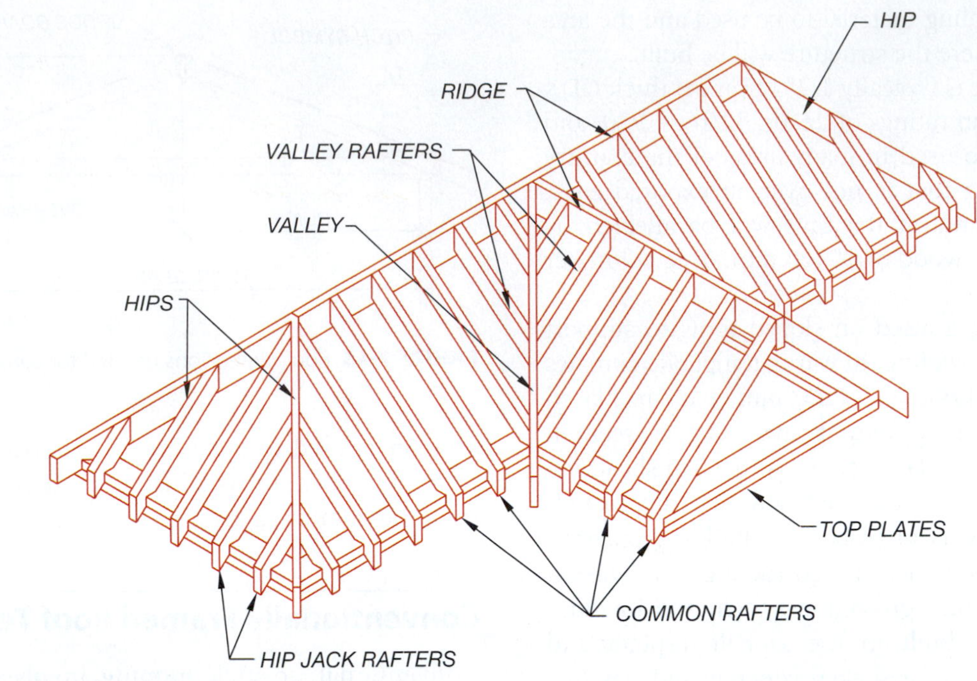

FIGURE 7-17 Roof members in conventional construction.

Roof Support

The **ridge board** is a horizontal board placed at the ridge to support the rafters. The ridge board is typically 2" (50 mm) deeper than the rafters. The ridge board does not support the weight carried by the rafters but provides a bearing surface to transfer force from a rafter to an opposing rafter. The ridge board is supported by a 2× (50×) brace that rests on a bearing wall. The brace resists any downward force and is usually a 2× (50×) member placed at 48" (1200 mm) o.c. The ridge brace must be within 45° of vertical to effectively transfer weight downward. A **purlin** can be used to reduce the span of a rafter. A purlin is placed below the rafter and supported by a purlin brace. The purlin is usually the same size as the rafters being supported. A *purlin brace* transfers the weight from the rafter and purlin into a bearing wall. A purlin brace must be set within 45° of vertical. If no bearing wall is near, a strongback can be used to support the weight from the rafters. A **strongback** is a beam placed in or above the ceiling to support roof weight. Each of these roof components is illustrated in Figure 7-18. Each term is specified in the sections.

The lower end of a rafter is nailed to the top plate of a wall while the upper end rests against the ridge. The natural tendency of the rafter is to rotate downward around the fixed lower end of the rafter. The downward rotation is resisted by the rafter on the opposite side of the ridge. If the rafter cannot rotate downward, the force of the rafter will attempt to force the bearing walls outward. This outward thrust of the rafters is resisted by a **ceiling joist**. Ceiling joists (C.J.) rest on the top plate of the wall and span between the bearing walls. The connection of the ceiling joist to the top plate and the rafters is necessary to keep the roof from sagging and the walls from separating. In addition to resisting the outward force of the rafters, ceiling joists also provide support for the finished ceiling.

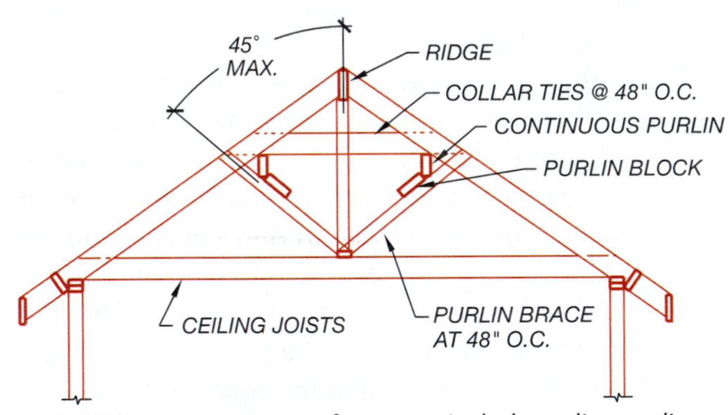

FIGURE 7-18 Common roof supports include purlins, purlin blocks, and purlin braces.

If the forces of outward thrust are great enough, a collar tie may be added to the upper third of the roof. A *collar tie* attaches to two opposing rafters and is used to resist the outward force of the rafters. The collar tie is usually the same size as the rafters, but the engineer will determine exact sizes based on the spans and loads supported by the rafters. Ceiling joists and collar ties are illustrated in Figure 7-18.

Vaulted Ceilings

The size of a rafter must be increased to support both the weight of the roof and the finished ceiling if a roof is to be vaulted. The rafters used with a vaulted ceiling are called *rafter/ceiling joists*. Rafter/ceiling joists (raft./C.J.) must be of sufficient size to contain the required insulation and still provide a minimum of 2" (50 mm) of air space above the insulation for ventilation. Rigid insulation or foam sandwich panels can be installed over the rafter/ceiling joists as an alternative method of placing the insulation. Because of the weight to be supported and the lack of interior supports, the rafter/ceiling joists will be supported by a *ridge beam* rather than just a ridge board. Because there are no ceiling joists to resist the outward force from the rafter/ceiling joists on the bearing walls, the rafters must be securely anchored to the ridge beam. Figure 7-19 shows two common methods of attaching the rafter/ceiling joists to the ridge beam. In areas subject to high winds or seismic activity, the lower end of the rafter/ceiling joist will be connected to the top plate with a metal angle. This angle helps to form a secure bond between the roof and wall.

Roof Openings

If an opening is to be framed in the roof, a header and trimmers will need to be specified. Each can be seen in Figure 7-20. A *header* at the roof level consists of two or more members nailed together to support the rafters at the upper and lower edges of the opening. The headers are usually the same size as the rafters. The *trimmers* consist of two or more rafters nailed together to support the sides of the opening.

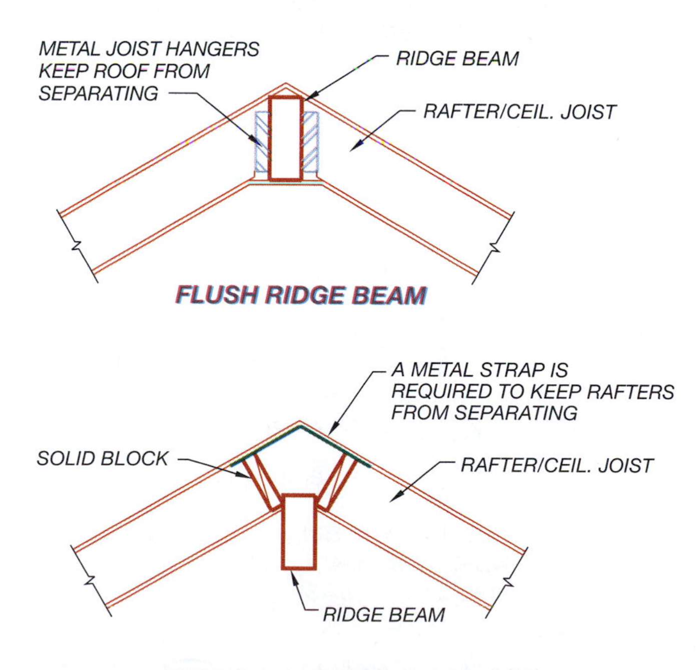

FIGURE 7-19 Common connections between the ridge beam and rafters include resting the rafters on the ridge, or hanging the rafters from the beam with metal hangers.

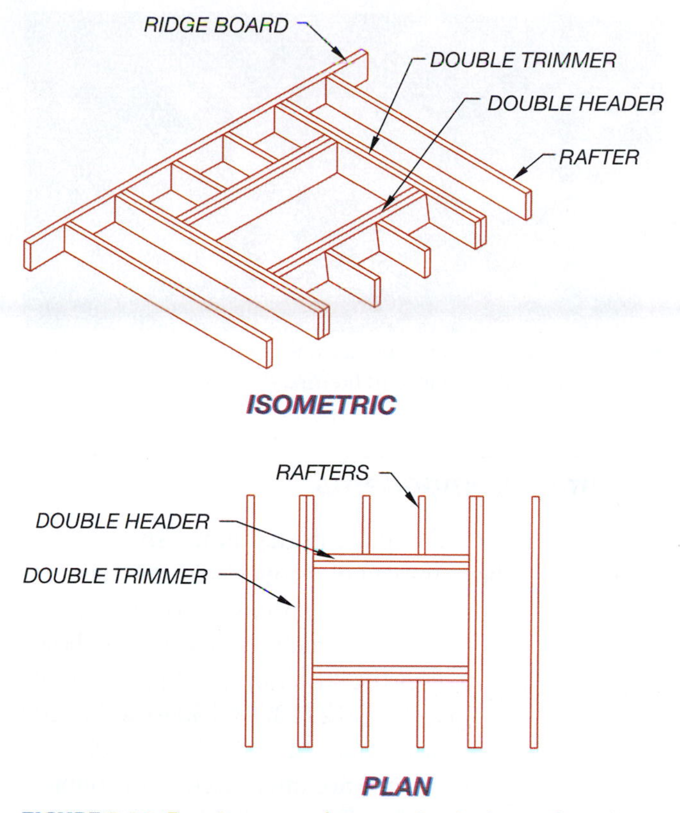

FIGURE 7-20 Framing around an opening in the roof requires the use of headers and trimmers.

TIMBER FRAMING

In addition to standard uses of wood in western platform construction, timber is often used for the structural framework of a building. Timber construction is used for both appearance and structural reasons. The term *timber* is used to describe wood members that are 5" (130 mm) and thicker, although 6" (150 mm) is typically the smallest size used. Timber construction is used to form many types of commercial structures because it affords large expanses of glass in the exterior shell. Timber has excellent structural and fire-retardant qualities. In a fire, timber will char on the exterior surfaces but will maintain its structural integrity long after an equally sized steel beam would have failed. Components of timber framing include posts, beams, and planks. In addition to framing plans showing the size of the components, details of connections are of utmost importance with timber construction. Figure 7-21 shows an example of a framed structure using timber construction.

FIGURE 7-21 Timber is often used to frame a structure for its appearance and resistance to fire. *Courtesy Timberpeg.*

Common Components

Timber can be used in place of traditional studs or roof framing members. Instead of using 2" (50 mm) studs at a 16" or 24" (400 or 600 mm) spacing, vertical supports are provided by posts for which the location is based on the design needs of the structure. Post size typically ranges from 5 × 5 to 12 × 12 (130 × 130 to 300 × 300 mm). Wood posts larger than 12 × 12 (300 × 300 mm) are available, but they are not often used, for economic reasons. Posts larger than 8 × 8 (200 × 200 mm) are typically laminated from solid material or from a combination of lumber and plywood. Horizontal beams are used to span between the post and other beams. Beam size is determined by the engineer based on the span and loads to be supported. Usually, 1 1/8" (30 mm) APA STURD-I-FLOOR 48 T & G plywood is used to span between beams. Two-inch (50 mm) T & G planks can also be used as a floor material but the exact size must be based on the live loads to be supported.

Representing Members on Drawings

The size and location of posts and beams are specified on the framing plan. Posts are represented using linetypes similar to those used to represent walls. Post size is often specified using diagonal text, so that the specification will be clear. Beams are represented with dashed or centerlines. Text to describe the beam is normally placed parallel to the beam or in a schedule. Figure 7-22 shows methods of specifying beams and posts on a framing plan.

Decking is typically specified in details and sections. Details showing the connection of post to foundation, post to beams at floor and roof levels, beam to beam, and decking to beams, are typically required. Figure 7-23 shows a beam-to-beam connection detail. Beams that are drawn in end view can be drawn using different methods, as shown in Figure 7-24. At a scale of 3/4" = 1'–0" or larger, the x is usually placed to help define the beam. The x is

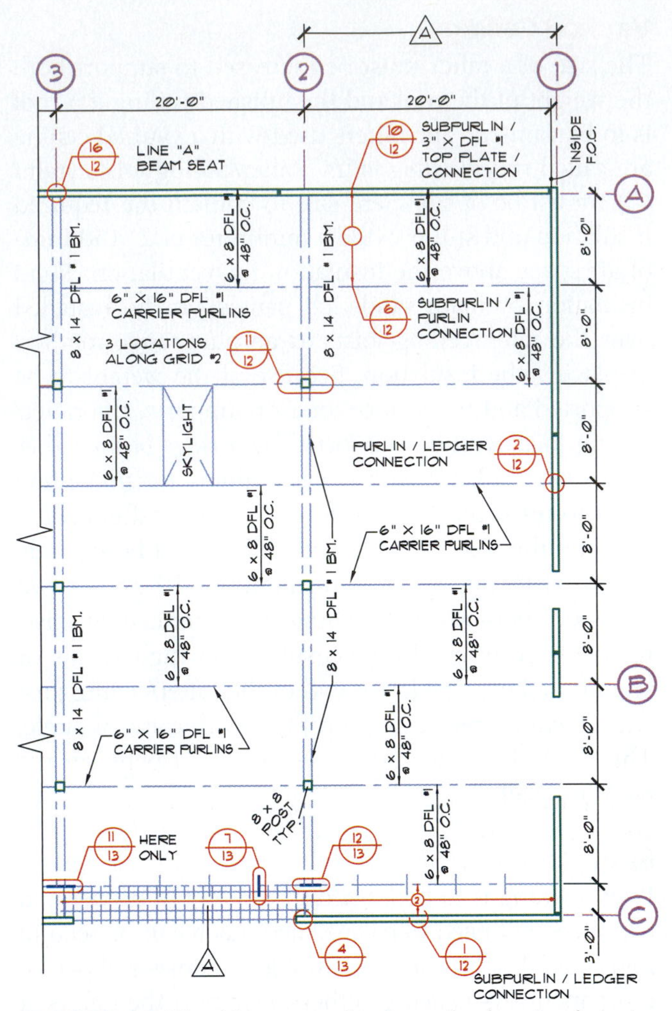

FIGURE 7-22 Wood beams can be shown in plan view by pairs of dashed lines that represent the beam width, or by a polyline. Polylines of continuous, center, or dashed linetypes will be used according to office practice.

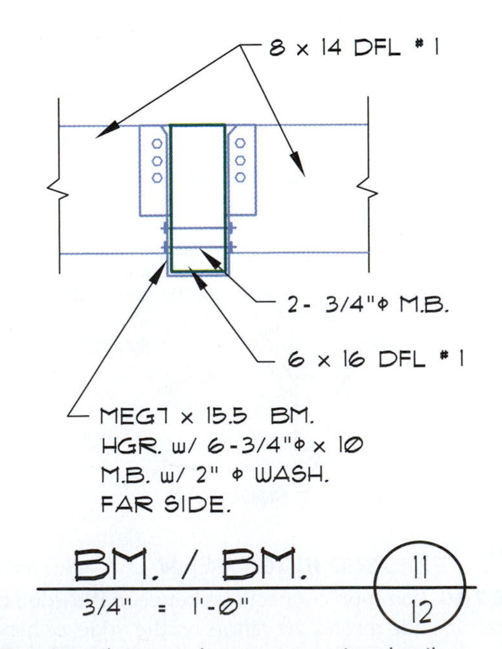

FIGURE 7-23 A beam-to-beam connection detail.

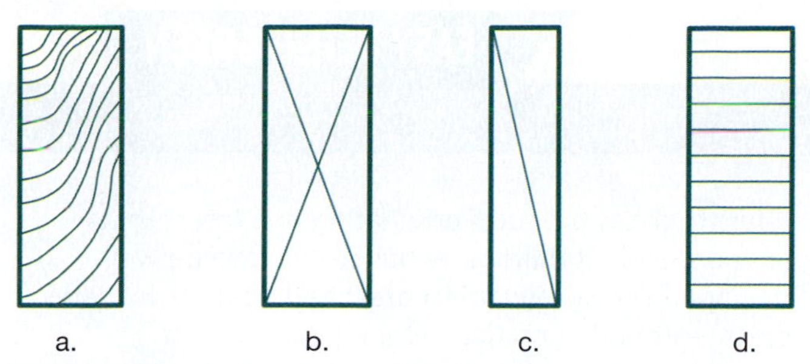

FIGURE 7-24 Beam representation in end view, for details drawn at a scale of 3/4"= 1'–0" or larger: (a) end grain; (b) continuous beam; (c) noncontinuous beam, blocking; (d) laminated beam (see Chapter 8).

also used to represent the end view of a continuous member such as a beam. One diagonal line is used to represent the end view of noncontinuous members such as blocking placed between beams. Many offices use a thick line around the beam perimeter to help distinguish beams.

HEAVY TIMBER

Heavy timber construction is similar to timber construction, except for the size of the members used. The IBC restricts the size of the members that can be used to achieve a heavy timber fire rating. The rating is referred to as a Type IV-HT in the occupancy tables of the IBC. In addition to limiting the size of members used, heavy timber construction eliminates the concealed spaces associated with western platform construction. Space between rafters, joists, or studs that is concealed by the interior finish is eliminated, leaving the framing members and the exterior shell exposed to the interior. Post size for an HT rating requires the use of 8 × 8 (200 × 200) minimum to meet the fire-resistance requirements of the building code. If no floor loads are supported, and the beam does not extend below the floor line, a post with a minimum of 6" wide × 8" deep (150 × 200 mm) may be used. If a laminated arch (see Chapter 8) is used to support floor loads, it may not be smaller than 8" (200 mm) in any direction. Laminated arches that do not support floor loads may be as small as 4" in width and 6" (100 × 150 mm) in depth. The IBC requires beams to be a minimum of 6" (150 mm) wide and 10" (250 mm) deep to provide sufficient fire protection for a floor with a Type IV-HT rating. Beams 6" (150 mm) wide and 8" (200 mm) deep are the minimum allowed to support roof loads. Floor decking must be a minimum of 3" (75 mm) deep planks with an overlay of 1" (25 mm) T & G plywood laid crosswise. Floor decking must maintain a 1/2" (13 mm) space between wall members to allow for swelling and shrinkage. Roof decking must be a minimum of 2" (50 mm) thick T & G planks. A double layer of 1" (25 mm) boards may also be used if joints are staggered.

Heavy Timber Representation

The size and location of posts and beams are specified on the framing plan using methods similar to other timber callouts. Details showing the connection of post to foundation, post to beams at floor and roof levels, beam to beam, and decking to beams, are required just as with other types of timber construction. Figure 7-25 shows a drawing created by a CAD technician to represent a beam-to-beam connection.

Timber Trusses

Similar in shape to conventional trusses (see Chapter 8), trusses can be formed of timber. Heavy timber trusses used to support floor loads must have members that are a minimum of 8" (200 mm) nominal in any direction. Connections of truss components are typically achieved by using bolted or ringed connectors (see Chapter 6). Depending on the type and amount of load to be resisted, plates similar to a gusset can be used at each web-to-chord connection. Figure 7-26 shows an example of a timber truss used to frame the roof of a condominium.

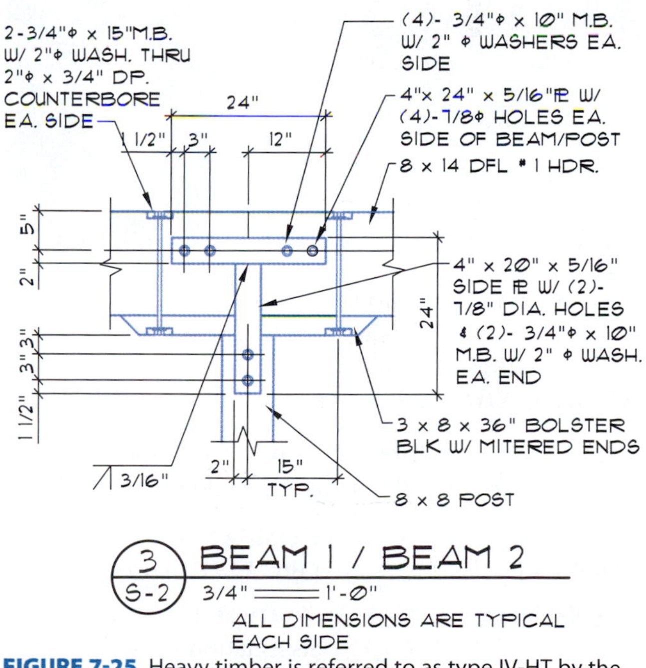

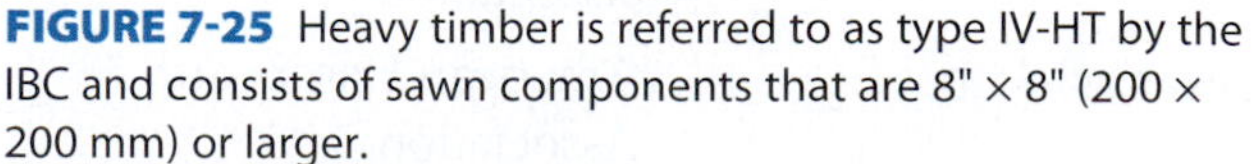

FIGURE 7-25 Heavy timber is referred to as type IV-HT by the IBC and consists of sawn components that are 8" × 8" (200 × 200 mm) or larger.

EASY BEING GREEN

When you think of wood, you might be tempted to think of the depletion of old-growth timber by large money-grabbing corporations. That would be a mistake. Each major lumber supplier is very involved in developing lumber and timber products from sustainable forests. Two nonprofit organizations provide green building ratings systems for commercial wood and timber construction in the U.S. market. These national organizations are the:

- U.S. Green Building Council that publishes the LEEDS (Leadership in Energy and Environment Designs for New Standards®) green guidelines.
- Green Building Initiative™ that features the Green Globes®.

Architects and builders use these standards to gain green certification for their projects. In contrast, the Forest Stewardship Council (FSC) provides certification for entire forests. The term ***certified forest products*** refers to those products originating in a forest that an independent third party has certified to be well-managed and sustainable. Forest certification validates on-the-ground operations employing the best management practices at a specific forest to ensure the long-term health of the total forest ecosystem. The only ratings available that meet the criteria established by the Certified Forest Products Council are those of the Forest Stewardship Council. FSC standards were developed by representatives of conservation groups, the timber industry, economic development organizations, and the general public. A forestry operation that meets FSC standards protects forest ecosystems, water quality, wildlife habitats, and local communities. To ensure the integrity of the certification, the wood and fiber from certified forests are tracked through the commercial chain from logging sites to retailers and to the end user.

FIGURE 7-26 In spite of its expense, timber is often used to frame trusses for its appearance. *Courtesy Aaron Jefferis.*

ADDITIONAL READING

The following Web sites can be used as a resource to help you keep current with changes in building materials.

ADDRESS	COMPANY/ ORGANIZATION
www.afandpa.org	American Forest and Paper Association
www.ahardbd.org	American Hardboard Association
www.alsc.org	American Lumber Standards Committee, Inc.
www.awc.org	American Wood Council
www.apawood.org	APA—The Engineered Wood Association
www.bc.com	Boise Cascade Corporation
www.canadianforestry.com	Canadian Forestry Association
www.csa.ca	Canadian Standards Association
www.cwc.ca	Canadian Wood Council
www.fscus.org	Forest Stewardship Council
www.gp.com	Georgia-Pacific Corporation
www.thegbi.org	Green Building Initiative
www.hpva.org	Hardwood Plywood and Veneer Association
www.internationalpaper.com	International Paper (high-performance building products)
www.nafb.org	National Frame Building Association
www.nlga.org	National Lumber Grades Authority

www.sfpa.org	Southern Forest Products Association
www.southernpine.com	Southern Pine Council
www.usgbc.org	U.S. Green Building Initiative
www.westernwoodstructures.com	Western Wood Structures, Inc.
www.weyerhaeuser.com	Weyerhaeuser

KEY TERMS

Board feet
Boards
Ceiling joists
Certified forest products
Cornice
Diaphragm
Eave
Fascia
Floor joists
Girders
Hardwoods
Header
Heavy timber construction
King stud
Lumber, beams and stringers
Lumber, dimension
Lumber, dressed
Lumber, post and timbers
Lumber, rough
Lumber, structural
Millwork
Mudsill
Pitch
Purlin
Rafters
Ridge board
Rim joists
Size, actual
Size, nominal
Skip sheathing
Softwoods
Sole plate
Span
Strongback
Studs
Studs, jack
Subsill
Timber
Top plates
Trimmer
Wood construction
Wood, green

CHAPTER 7 TEST Wood and Timber Framing Methods

QUESTIONS

Answer the following questions with short complete statements. Type the chapter title, question number, and a short complete statement for each question using a word processor. Some answers may require the use of vendor catalogs or seeking out local suppliers. If math is required to answer a question, show your work.

Question 7-1 How wide are girders with a conventional floor system?

Question 7-2 List four common materials suitable for beams and girders.

Question 7-3 What is the advantage of providing blocking at the edge of a floor diaphragm?

Question 7-4 What are the common span ratings for plywood suitable for roof sheathing?

Question 7-5 What grades of plywood are typically used for floor sheathing?

Question 7-6 List eight qualities typically given in an APA wood rating.

Question 7-7 List six methods of planing lumber.

Question 7-8 What is the difference between timber and lumber?

Question 7-9 Describe the two portions of a growth ring and explain their significance to strength.

Question 7-10 How would lumber with a moisture content of 20 be rated?

Question 7-11 What is the dry face width of a 2 × 12?

Question 7-12 What is the difference between a king stud and a trimmer?

Question 7-13 List four types of rafters.

Question 7-14 What is the common spacing of skip sheathing?

Question 7-15 Explain the meaning of the numbers 6/12 in relation to a roof.

Question 7-16 Explain the difference between a ridge, ridge board, and a ridge beam.

Question 7-17 List two common methods of resisting the outward thrust of a rafter.

Question 7-18 List three advantages of using timber construction.

Question 7-19 At what size do sawn posts become uneconomical and difficult to guarantee quality?

Question 7-20 What properties make heavy timber construction more fire-resistant than western platform construction?

Question 7-21 What is the minimum size of a post used to support floor loads using heavy timber?

Question 7-22 What are the minimum sizes allowed for timber trusses?

Question 7-23 What is the maximum allowable square footage and height for an office building constructed of heavy timber?

Question 7-24 Use the Internet to research and list the names of five national lumber suppliers.

Question 7-25 Use the Internet and visit five Web sites that deal with timber. Order information related to timber framing.

DRAWING PROBLEMS

Unless your instructor gives other instructions, use the minimum standards presented in the Chapter 6 section, "Drawing Criteria for Completing Details," to complete the following details. Skeletons of most details can be found on the student CD. Use these drawings as a base to complete the assignment.

Problem 7-1 Draw a detail with a front view showing double 2 × 6 trimmers and king studs resting on a 2 × 6 plate, 3/4" CD APA 42/20 T & G floor sheathing, 2 × 12 F.J. at 12" o.c. with solid blocking, resting on another 2 × 6 stud wall. Use double solid blocking under each post. Support the trimmers and king stud with a 4 × 6 post below. Use a Simpson Strong-Tie Company HD5A hold-down anchor and specify all bolt sizes.

Problem 7-2 Draw a detail showing a 6 × 14 DFL #1 beam resting on a 4 × 6 post. Select a metal cap that can resist 3500 pounds in uplift.

Problem 7-3 Show a 2 × 8 DFL rafter at 24" o.c. with a 12" overhang supporting 1/2" plywood sheathing, and composition shingles at a 6/12 pitch resting on (2) - 2 × 6 top plates resting on 2 × 6 studs at 16" o.c. Use 2 × 6 ceiling joists at 16" o.c. and specify a 1 × 6 fascia.

Problem 7-4 A 3" × 12" DFPT ledger will be connected to a concrete wall with a 5/8" diameter × 8" A.B. bolts at 24" o.c. staggered 3" u/d. Simpson Strong-Tie Company U210 hangers will be used to support 2 × 12 DFL #1 joist at 16" o.c. The joist will support 3/4" standard-grade plywood sheathing. Provide a 4" cant strip (see Chapter 18) at the rafter/wall intersection.

Problem 7-5 Draw a detail showing an 8 × 8 post connecting to concrete with an appropriate column base to resist 6000# in uplift.

Problem 7-6 Draw a detail showing an 8 × 8 post supporting an 8 ×16 header. Use a suitable CC cap to connect the beams to the post and suitable hangers to provide support for 6 × 12 purlins 48" o.c. supporting 5000# max. (1 on each side and perpendicular to the 8 × 16.)

Problem 7-7 Draw a side and front view showing a 6 × 6 post hidden in a 2 × 6 stud wall on the upper floor and resting on a 2 × 6 / 3 × 6 base plates. The lower plate is to be DFPT. The plates will rest on 1 1/8" plywood, and 2 × 12 DFL #2 floor joists at 12" o.c. The lower wall will be framed out of 2 × 6 studs with a double top plate. A 6 × 6 post will be placed directly below the upper post. A hold-down anchor that can resist at least 7000 pounds in uplift will be required.

Problem 7-8 Using Figure 7-22, draw a detail to show the beam-to-beam connections. Assume the use of wood walls and that the roof weighs 50 psf, and determine the load that will be on each beam to determine the required load that will be on each required hanger. This should include details for:

7-8-1 – 8 × 14 bm / 6 × 8 post at wall

7-8-2 – 8 × 14 bm / 8 × 8 post

7-8-3 – 6 × purlin / 8 × 14 beam

7-8-4 – 6 × purlin to 4 × 6 post in wall

7-8-5 – 6 × 8 rafters to purlins

Problem 7-9 Draw a plan view for a 32' × 60' structure framed with 2 × 6 studs at 16" o.c. at a scale of 1/8" = 1'–0". Place an 8 × 16 ridge beam in the center of the structure 16' from the side walls. Specify an 8× wood post at 20' intervals along this beam with 6 × 8 posts at the ends. Show splices in the ridge beam 24" from the post so that the center beam is 16' long. Use an 8 × 12 deep beam for the center span of the ridge beam.

Hang 6 ×12 purlins at 15'–0" o.c. on each side of the ridge beam and 6 × 12 subpurlins at 12' o.c. Hang 2 × 6 rafters at 24" o.c. between each purlin with appropriate hangers. Show enough of the roof framing to represent the framing patterns. Provide grids to represent each wall and post so that four vertical and three horizontal grids are provided. Specify the location of all members. Use a schedule wherever possible. Assume the sheet to be S-501.

Problem 7-10 Draw details to represent all of the post-to-beam and beam-to-beam connections represented in problem 7-9. Assume the sheet to be S-101. Place all details in a logical order based on Chapter 2 and place detail reference numbers by each detail and on the framing plan. Possible details would include, but are not limited to:

7-10-1—Ridge to wall

7-10-2—Ridge to post

7-10-3—Ridge to ridge

7-10-4—Beam to beam

7-10-5—Beam to purlin

7-10-6—Purlin to rafter

If all of the details fit, they may be placed on the same page as the framing plan. If all will not fit, place the details on sheet S-501.

Engineered Lumber Products 8

Engineered wood products are a major material of construction for light commercial projects. Engineered components range from products made from the sawdust generated as sawn lumber products are milled to products created using sawn lumber. These products may serve the same function as their sawn counterparts, but their benefits to the environment are unsurpassed. This chapter will explore the use of:

- Engineered materials such as oriented strand board, laminated veneer lumber, and parallel strand lumber.
- Engineered building components such as studs, joists, and I-joists.
- Laminated beams.
- Open-web trusses.

ENGINEERED MATERIALS

Engineered wood products similar to those in Figure 8-1 are divided into two major categories including structural composite lumber and laminated beams. **Structural composite lumber (SCL)** includes oriented strand lumber (OSL), laminated veneer lumber (LVL), and laminated strand lumber (LSL). These materials are created by layering dried and graded wood veneers or flakes with waterproof adhesive into blocks of material known as billets and then curing these layers in a controlled process. SCL is available in various thicknesses and widths and is easily worked in the field using conventional construction tools. **Laminated beams** are beams made from conventional 2× (50× mm) materials that are glued together to construct beams that are larger in size and length than natural sawn beams.

Oriented Strand Board

Oriented strand board (OSB) or wafer board is a common alternative to traditional plywood. Produced in huge, continuous mats, OSB is a solid-panel product of consistent quality with no laps, gaps, or voids. OSB is made from three or more layers of small strips of wood that that are arranged in cross-oriented layers, similar to plywood. The exterior layers of strips are parallel to the length of the panel, and the core layers are laid in a random pattern. The wood is then saturated with a waterproof binder and compressed under heat. This results in a structural engineered wood panel that shares many of the strength and performance characteristics of plywood. Common uses of OSB include roof, wall, and floor sheathing; as web material for solid wood I-joists; and as the structural covering for structural insulated panels (SIPs). Figure 8-2 shows the use of OSB for floor sheathing and in I-joists.

Laminated Veneer Lumber

Laminated veneer lumber (LVL) is made from ultrasonically graded Douglas fir veneers that are laminated with exterior-grade adhesives under heat and pressure, with all grains parallel to one another. LVL is primarily used for joists and rafters, hip and valley rafters, headers, beams, posts, and the flange material for I-joists. LVL is produced in widths ranging from 1 3/4" to 3 1/2" (45 to 90 mm) and depths ranging from 5 1/2" to 18" (140 to 450 mm). Lengths are generally available up to 60' (18,000 mm). Posts made of laminated veneer range in sizes from 3 1/2" to 7" (90 to 180 mm), and are being used increasingly for their ability to support loads. LVL joists are designed for single- or multispan uses that must support heavy loads. Materials made of LVL offer the superior performance and durability of other engineered products and are superior to their sawn wood counterparts. Figures 8-3 and 8-4 show the use of an LVL beam to support engineered joists.

Parallel Strand Lumber

Parallel strand lumber (PSL) is laminated from veneer strips peeled from the outermost sections of fir and southern pine. Veneers are dried to control the moisture content before being chopped into strands. The strands are then aligned to be parallel to each other, coated with resin, and compressed and heated. Typically used for beams, headers, studs, posts, and

EASY BEING GREEN

Engineered lumber products provide many environmentally friendly materials for the construction industry, providing efficient use of the available resources. Engineered wood products can be manufactured from fast-growing, underutilized, and less expensive wood species and from by-products from other production processes. This includes the use of sawdust, small chips, and unusable bits of wood created from cutting logs. Lumber can also be recycled and reused to make some engineered wood products. By using many of the production scraps from making sawn lumber products to make engineered wood, many of the natural defects found in wood are eliminated, making the engineered material stronger than natural lumber.

Another important factor that makes engineered material an earth-friendly product is the energy efficient methods by which they are manufactured. Based on statistics provided by the APA, when compared to the amount of energy required to produce 1 ton of cement, glass, steel, or aluminum to 1 ton of wood product, the manufacture of engineered wood products takes:

- 5 times less energy than the energy required to produce 1 ton of cement;
- 14 times less energy than the energy required to produce 1 ton of glass;
- 24 times less energy than the energy required to produce 1 ton of steel;
- 126 times less energy than the energy required to produce 1 ton of aluminum.

FIGURE 8-1 Because of the strength, cost, and availability, environmentally friendly engineered lumber products are a key component of most Type IV and V commercial projects. This intersection shows the use of a glued-laminated beam (right), a laminated veneer lumber (LVL) beam (center), and I-joists made from oriented strand board (OSB). *Courtesy Katja Poschwatta.*

millwork components, widths include 3 1/2", 5 1/4", and 7" (90, 130, 180 mm), with depths ranging from 7" to 18" (180 to 460 mm).

ENGINEERED COMPONENTS

Engineered products are those products made from structural composite lumber. Major construction products include studs, posts, and I-joists. Other common engineered products made from lumber products

FIGURE 8-2 Oriented strand board (OSB) is used for the floor decking and the solid-web floor joists. *Courtesy APA—The Engineered Wood Association.*

or a combination of engineered products include laminated beams, as well as open-web floor and roof trusses.

Studs and Posts

Engineered studs are made from short sections of stud-grade lumber that have had the knots and splits removed. Quick-growing, small-diameter aspen and yellow poplar are often used instead of the more traditional materials used with sawn lumber. Sections of wood are joined together with 5/8" (16 mm) finger joints. Engineered studs in 2 × 4 or 2 × 6 (50 × 100

FIGURE 8-3 A laminated veneer lumber (LVL) beam. *Courtesy APA—The Engineered Wood Association.*

FIGURE 8-4 The top and bottom chords of the solid-web floor joists and the beam that supports the floor joists are made of laminated veneer lumber (LVL). *Courtesy APA—The Engineered Wood Association.*

or 50 × 150) are available in standard lengths of 8', 9', and 10' (2440, 2740, and 3400 mm) lengths, but lengths up to 48' (14 400 mm) can be ordered. Laminated veneer and PSL studs are also available. Depending on the manufacturer, notches are generally not allowed. Sheathing must be provided in place of a let-in brace if engineered studs are used. Engineered posts similar to those in Figure 8-5 are being used increasingly for their ability to support loads. Engineered posts are available in 3 1/2", 5 1/4", and 7" (90, 135, and 180 mm) widths and depths, in lengths up to 24' (7200 mm).

I-Joists

I-joists are a high-strength, lightweight, cost-efficient alternative to sawn lumber. I-joists form a very quiet floor system because they are exceptionally stiff and uniform in size, have no crown, and do not shrink. I-joists come in depths ranging from 9 1/2" to 37" (240 to 940 mm) and are suitable for spans up to approximately 40' (12,000 mm). I-joists are able to span greater distances than comparably sized sawn joists. **Webs** can be made from plywood or (OSB). Figure 8-6a shows examples of solid-web I-joists. Figure 8-6b shows a truss made with I-joists connected to a glu-lam beam. Holes can be placed in the web to allow for HVAC ducts and electrical requirements based on the manufacturer's specifications. Figure 8-7 shows an example of a detail drawn by a CAD technician to represent I-joists.

FIGURE 8-5 Engineered posts made of SCL and laminated products are used for many projects because of the appearance, strength, and fire-resistance. *Courtesy APA—The Engineered Wood Association.*

FIGURE 8-6a Solid-web I-joists are available in a variety of depths and lengths. *Courtesy APA—The Engineered Wood Association.*

FIGURE 8-6b Solid-web floor joists supported by metal hangers are attached to a laminated beam. *Courtesy Pavel Adi Sandu.*

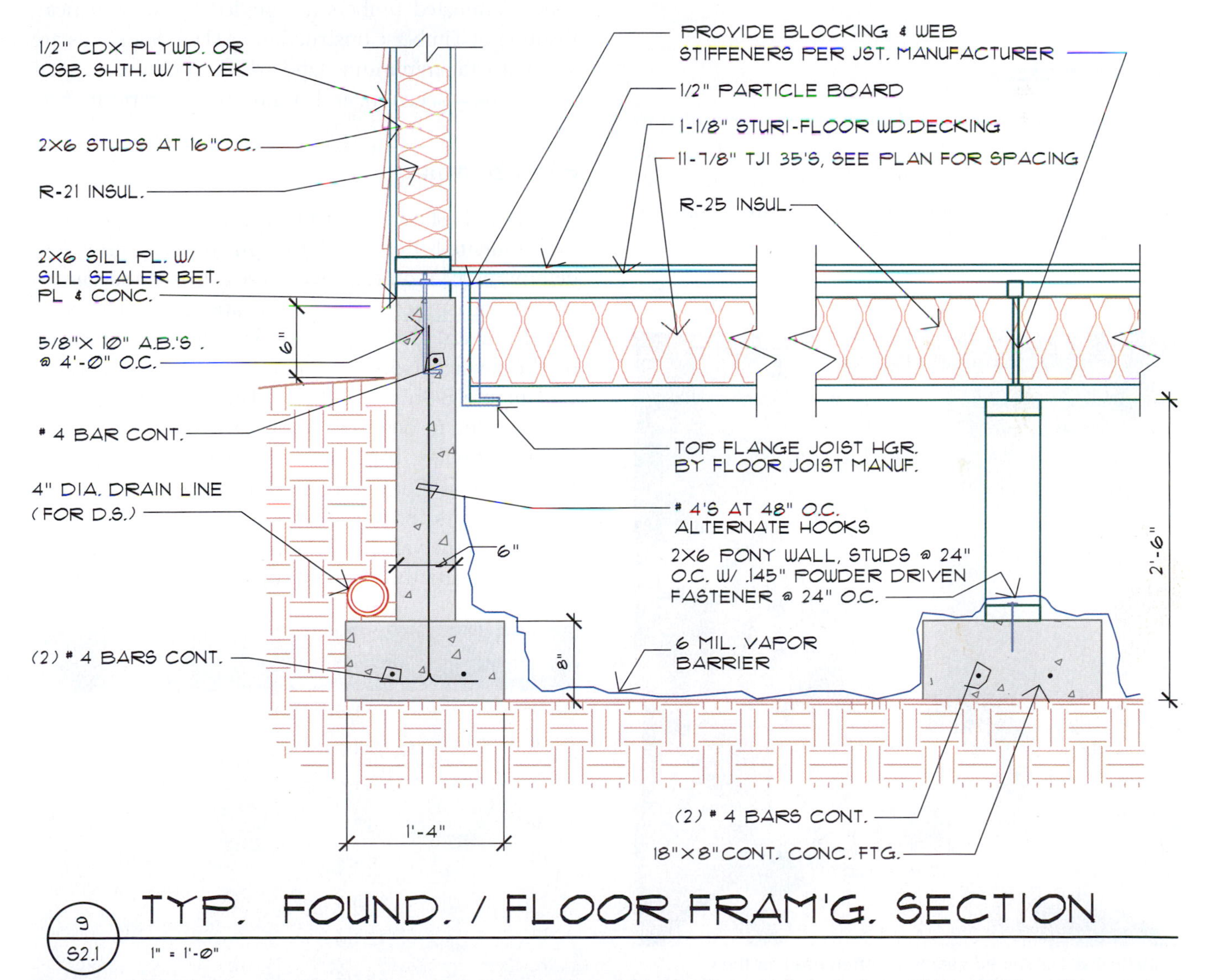

FIGURE 8-7 Drawings representing trusses and I-joists are often provided by the manufacturer, but a CAD technician will be required to edit and label the detail to meet specific aspects of each project. *Courtesy Scott R. Beck, Architect.*

LAMINATED BEAMS

Because of the strength limitations of solid wood and limited availability, glued-laminated beams, often referred to as glu-lams, are made from sawn lumber that is glued together under pressure to form a beam that is stronger than its lumber counterpart. Increased length is gained by laminating lumber that is spliced together with scarf or finger joints. Laminated timber is often used for structural framing materials because of its increased strength over sawn lumber, and because of its beauty (see Figure 8-8). Laminated beams are made from dry lumber, which offers a much higher level of dimensional stability than sawn lumber. The use of dry lumber virtually eliminates cracking, twisting, warping, and shrinking. Because of their increased size, laminated beams are installed with mobile construction equipment but can be set using conventional hand tools.

Common Sizes

Laminated beams are made from 2× (50×) material in 4", 6", 8", 10", 12", 14", and 16" (100, 150, 200, 250, 300, 350, and 400 mm) widths. Wider beams are available by special order. Once the members are laminated, the beams are planed to their finish size.

FIGURE 8-8 Laminated beams are often used for their strength, beauty, and the wide variety of shapes available. *Courtesy American Institute of Timber Construction.*

Laminated timber comes in finished widths as shown in Table 8-1 below.

Depths for simple beams range from 3" to 84" (75 to 2300 mm) in 1 1/2" (38 mm) increments. One of the features of laminated timbers is that beams do not have to conform to these standard sizes, but can be laminated to specific sizes to fit the design criteria. Figure 8-9a shows the properties of standard laminated beam sizes. These beams all have their wide face parallel to the X-X axis. Laminated beams are available with their wide face parallel to the Y-Y axis but these are uncommon. Straight beam sizes are available through most lumberyards, and custom shapes are shipped directly from the manufacturer.

Grading

As well as being graded by each of the major building codes, laminated timbers are graded by the American Institute of Timber Construction (AITC). Standards are set to provide minimum standards for production, quality control, inspection, and certification of performance.

Fiber Bending

Figure 8-9b shows a partial listing of design values for laminated timber. Although the engineer will determine all required beam sizes, it is important to understand some of the specifications, so that laminated timbers can be properly specified on the structural drawings. Notice the first beam listed in column 1 of Figure 8-9b is a 24F-V4. The 24F relates to the design value in bending, represented in column 3. In design formulas, this value is usually represented by the symbol f_b. Common values listed by most lumber manufacturers include 1600, 2000, 2200, and 2400. The higher the value, the more units of stress that can be resisted, and the stronger the beam. The value to be

Nominal Width	Western Species	Southern Pine
3	2 1/8	2 1/8
4	3 1/8	3 or 3 1/8
6	5 1/8	5 or 5 1/8
8	6 3/4	6 3/4
10	8 3/4	8 1/2
12	10 3/4	10 1/2
14	12 1/4	12
16	14 1/4	14

TABLE 8-1 Finished Widths for Laminated Timber (in Inches)

Table 5: Section Properties
Based on 1-1/2 in. thick laminations

No. of lams	d Depth (in.)	A Area (in.^{2})	I Moment of Inertia (in.^{4})	S Section Modulus (in.^{3})
2 1/2 in. Widths				
6	9	22.50	151.9	33.75
7	10-1/2	26.25	241.2	45.94
8	12	30.00	360.0	60.00
9	13-1/2	33.75	512.6	75.64
10	15	37.50	703.1	93.75
11	16-1/2	41.25	935.9	113.4
12	18	45.00	1215	135.0
13	19-1/2	48.75	1545	158.4
14	21	52.50	1929	183.8
15	22-1/2	56.25	2373	210.9
3 1/8 in. Widths				
4	6-	18.75	56.25	18.75
5	7-1/2	23.44	109.9	29.30
6	9	28.13	189.9	42.19
7	10-1/2	32.81	301.5	57.42
8	12	37.50	450.0	75.00
9	13-1/2	42.19	640.7	94.92
10	15	46.88	878.9	117.2
11	16-1/2	51.56	1170	141.8
12	18	56.25	1519	168.8
13	19-1/2	60.94	1931	198.0
14	21	65.63	2412	229.7
15	22-1/2	70.31	2966	263.7
16	24	75.00	3600	300.0
17	25-1/2	79.70	4318	338.7
18	27	84.40	5126	379.7
19	28-1/2	89.10	6028	423.0
5 1/8 in. Widths				
4	6	30.75	92.25	30.75
5	7-1/2	38.44	180.2	48.05
6	9	46.13	311.3	69.19
7	10-1/2	53.81	494.4	94.17
8	12	61.50	738.0	123.0
9	13-1/2	69.19	1051	155.7
10	15	76.88	1441	192.2
11	16-1/2	84.56	1919	232.5
12	18	92.95	2491	276.8
13	19-1/2	99.94	3167	324.8
14	21	107.6	3955	376.7
15	22-1/2	115.3	4865	432.4
16	24	123.0	5904	492.0
17	25-1/2	130.7	7082	555.4
18	27	138.4	8406	622.7
19	28-1/2	146.1	9887	693.8
20	30	153.8	11530	768.8
21	31-1/2	161.4	13350	847.5
22	33	169.1	15350	930.2
23	34-1/2	176.8	17540	1017
24	36	184.5	19930	1107
6 3/4 in. Widths				
5	7-1/2	50.63	273.3	63.28
6	9	60.75	410.1	91.13
7	10-1/2	70.88	651.2	124.0
8	12	81.00	972.0	162.0
9	13-1/2	91.33	1384	205.0

No. of lams	d Depth (in.)	A Area (in.^{2})	I Moment of Inertia (in.^{4})	S Section Modulus (in.^{3})
6 3/4 in. Widths (cont.)				
10	15	101.3	1898	253.1
11	16-1/2	111.4	2527	306.3
12	18	121.5	3281	364.5
13	19-1/2	131.6	4171	427.8
14	21	141.8	5209	496.1
15	22-1/2	151.9	6407	569.5
16	24	162.0	7776	648.0
17	25-1/2	172.1	9327	731.5
18	27	182.3	11070	820.1
19	28-1/2	192.4	13020	913.8
20	30	202.5	15190	1013
21	31-1/2	212.6	17581	1116
22	33	222.8	20210	1225
23	34-1/2	232.9	23100	1339
24	36	243.0	26240	1458
25	37-1/2	253.1	29660	1582
26	39	263.3	33370	1711
27	40-1/2	273.4	37370	1845
28	42	283.5	41670	1985
29	43-1/2	293.6	46300	2129
30	45	303.8	51260	2278
31	46-1/2	313.9	56560	2433
32	48	324.0	62210	2592
8 3/4 in. Widths				
6	9	78.75	531.6	118.1
7	10-1/2	91.88	844.1	160.8
8	12	105.0	1260	210.0
9	13-1/2	118.1	1794	265.8
10	15	131.3	2461	328.1
11	16-1/2	144.4	3276	397.0
12	18	157.5	4253	472.5
13	19-1/2	170.6	5407	554.5
14	21	183.8	6753	643.1
15	22-1/2	196.9	8306	738.3
16	24	210.0	10080	840.0
17	25-1/2	223.1	12090	948.3
18	27	236.3	14350	1063
19	28-1/2	249.4	16880	1185
20	30	262.5	19690	1313
21	31-1/2	275.6	22790	1447
22	33	288.8	26200	1588
23	34-1/2	301.9	29940	1736
24	36	315.0	34020	1890
25	37-1/2	328.1	38450	2051
26	39	341.3	43250	2218
27	40-1/2	354.4	48440	2392
28	42	367.5	54020	2573
29	43-1/2	380.6	60020	2760
30	45	393.8	66440	2953
31	46-1/2	406.9	73310	3153
32	48	420.0	80640	3360
33	49-1/2	433.1	88440	3573
34	51	446.3	96720	3793
35	52-1/2	459.4	105500	4020
36	54	472.5	114800	4253

No. of lams	d Depth (in.)	A Area (in.^{2})	I Moment of Inertia (in.^{4})	S Section Modulus (in.^{3})
8 3/4 in. Widths (cont.)				
37	55-1/2	485.6	124654	4492
38	57	498.8	135037	4738
39	58-1/2	511.9	145980	4991
40	60	525.0	157500	5250
41	61-1/2	538.1	169610	5516
42	63	551.3	182326	5788
10 3/4 in. Widths				
7	10-1/2	112.9	1037	197.5
8	12	129.0	1548	258.0
9	13-1/2	145.1	2204	326.5
10	15	161.3	3023	403.1
11	16-1/2	177.4	4024	487.8
12	18	193.5	5225	580.5
13	19-1/2	209.6	6642	681.3
14	21	225.8	8296	790.1
15	22-1/2	241.9	10200	907.0
16	24	258.0	12380	1032
17	25-1/2	274.1	14850	1165
18	27	290.3	17630	1306
19	28-1/2	306.4	20740	1455
20	30	322.5	24190	1613
21	31-1/2	338.6	28000	1778
22	33	354.8	32190	1951
23	34-1/2	370.9	36790	2133
24	36	387.0	41800	2322
25	37-1/2	403.1	47240	2520
26	39	419.3	53140	2725
27	40-1/2	435.4	59510	2939
28	42	451.5	66370	3161
29	43-1/2	467.6	73740	3390
30	45	483.8	81630	3628
31	46-1/2	499.9	90070	3874
32	48	516.0	99070	4128
33	49-1/2	532.1	108600	4390
34	51	548.3	108800	4660
35	52-1/2	564.4	129600	4938
36	54	580.5	141000	5225
37	55-1/2	596.6	153100	5519
38	57	612.8	165900	5821
39	58-1/2	628.9	179300	6132
40	60	645.0	193500	6450
41	61-1/2	661.1	208400	6777
42	63	677.3	224000	7111
43	64-1/2	693.4	240400	7454
44	66	709.5	257500	7805
45	67-1/2	725.6	275500	8163
46	69	741.8	294300	8530
47	70-1/2	757.9	313900	8905
48	72	774.0	334400	9288
49	73-1/2	790.1	355700	9679
50	75	806.3	377900	10000
51	76-1/2	822.4	401000	10500
52	78	838.5	425100	10900
53	79-1/2	854.6	450100	11300
54	81	870.8	476000	11800

FIGURE 8-9a Common sizes and properties of laminated beams. *Courtesy American Institute of Timber Construction.*

used will depend on the type of lumber to be used and will be selected by the engineer.

Grading Method

The second portion of the specification represents the method used to grade the beam. The letters V or E may be used to describe the grading procedure. *V* represents wood that has been visually inspected and *E* specifies that electronic methods of nondestructive testing have been used. The number that follows the ratings specification represents a specific combination of grades and species of material to be used to form the beam.

Design Values for Structural Glued Laminated Timber from *AITC 117-93--Design* Table 1
For normal duration of load and dry conditions of use [a]

		Bending About X-X Axis						Bending About Y-Y Axis					Axially Loaded		
		Loaded Perpendicular to Wide Faces of Laminations						Loaded Parallel to Wide Faces of Laminations							
		Extreme Fiber in Bending, F_{bx}		Compression Perpendicular to Grain, $F_{c\perp x}$											
Combination Symbol[c]	Species-Outer Laminations /Core Laminations[d]	Tension Zone Stressed in Tension psi	Compression Zone Stressed in Tension psi	Tension Face psi	Compression Face psi	Shear Parallel to Grain (Horizontal) F_{vx} psi	Modulus of Elasticity, E_x Million psi	Extreme Fiber in Bending, F_{by} psi	Compression Perpendicular to Grain, $F_{c\perp y}$ psi	Shear Parallel to Grain (Horizontal) F_{vy} psi	Shear Parallel to Grain (Horizontal) (For members with multiple piece laminations which are not edge glued), F_{vy} psi	Modulus of Elasticity, E_y Million psi	Tension Parallel to Grain, F_t psi	Compression Parallel to Grain, F_c psi	Modulus of Elasticity, E Million psi
1	2	3	4	5	6	7	8	9	10	11	12	13	14	15	16
Visually Graded Western Species															
The following two combinations are not balanced and are for either dry or wet use.															
24F-V4	DF/DF	2400	1200	650	650	165	1.8	1500	560	145	75	1.6	1100	1600	1.6
24F-V5	DF/HF			650	650	155	1.7	1350	375	140	70	1.5	1100	1450	1.5
The following combination is balanced and is intended for members continuous or cantilevered over supports and provide equal capacity in both positive and negative bending.															
24F-V8	DF/DF	2400	2400	650	650	165	1.8	1450	560	145	75	1.6	1100	1650	1.6
Visually Graded Southern Pine															
The following two combinations are not balanced and are for either dry or wet use.															
24F-V1	SP/SP	2400	1200	650	560	200	1.7	1500	560	175	90	1.5	1100	1350	1.5
24F-V3	SP/SP			650	650	200	1.8	1600	560	175	90	1.6	1150	1700	1.6
The following combination is balanced and is intended for members continuous or cantilevered over supports and provides equal capacity in both positive and negative bending.															
24F-V5	SP/SP	2400	2400	650	650	200	1.7	1600	560	175	90	1.5	1150	1700	1.5
Wet-use factors[b]		0.8	0.8	0.53	0.53	0.875	0.833	0.8	0.53	0.875	0.875	0.833	0.8	0.73	0.833

Footnotes for Example (See *AITC 117-93--Design*)

[a] The combinations in this table are applicable to members consisting of 4 or more laminations and are intended primarily for members stressed in bending due to loads applied perpendicular to the wide faces of the laminations. Design values are tabulated, however, for loading both perpendicular and parallel to the wide faces of the laminations. For combinations and design values applicable to members loaded primarily axially or parallel to the wide faces of the laminations, and for members of 2 or 3 laminations, see Table 2, *AITC 117-93--Design*.

[b] Wet-use factors apply to all species.

[c] The combinations symbols relate to a specific combination of grades and species in *AITC 117—Manufacturing* that will provide the design values shown for the combination. The first two numbers in the combination symbol correspond to the design value in bending shown in Column 3. The letter in the combination symbol (either a "V" or an "E")indicates whether the combination is made from visually graded (V) or E-rated (E) lumber in the outer zones.

[d] The symbols used for species are DF = Douglas Fir-Larch; HF = Hem-Fir; and SP = Southern Pine.

FIGURE 8-9b A partial listing of design values for laminated timbers. *Courtesy American Institute of Timber Construction.*

Material

Column 2 of Figure 8-9b represents the type or types of material to be laminated together using abbreviations such as DF/HF. Common abbreviations include:

DF Douglas Fir
DFS Douglas Fir South
HF Hem Fir
WW Soft woods
ES Eastern Spruce
AC Alaska Cedar
CSP Canadian Spruce Pine
SP Southern Pine

In the specifications, the first group of letters represents the species of wood used for the outer laminations, and the second letters represent the species used for the beam core or inner layers.

Appearance

In addition to the number/letter grading, beams are also described by their appearance. Common grades include industrial, architectural, and premium. These terms apply to the exposed surfaces of the laminated member and regulate items such as growth characteristics, inserts, wood fillers, and surfacing operations. These ratings in no way affect the structural quality of the beam. For beams that will not be exposed, industrial grades are typically specified. Architectural-grade beams can be used in exposed situations, but minor flaws in the beams will be seen. Premium-grade beams are intended for high-visibility uses where no flaws are desired.

Finish

Laminated beams can be finished similar to their wood counterparts. Sealers, stains, and paint products are common finishing products. Surface sealers are used to resist soiling, to help control grain separation, and to reduce moisture absorption. In addition to the finish that can be applied at the job site, the laminator can also apply preservative treatments. Water-borne salt chemicals or oil-borne chemicals are used to treat individual pieces of lumber prior to lamination. Oil-based products such as creosote can be used to treat beams after gluing.

Specifications

The specifications required for a laminated beam on a drawing differ from those of a sawn beam. Laminated beams are referenced on the framing plans, sections, and details, with complete information given in the project manual that accompanies the drawings. Common information included for a laminated beam specification on a framing plan includes the beam size, beam type, grade, and the type of material used to form each layer. A typical specification would resemble:

5 1/8 × 13 1/2 PURLIN DF/HF 2400 f_b – V5.

In addition to the information specified on the drawing beside the beam, the appearance, grade, and information about the finish are specified in the written specifications. A typical notation for laminated beams would resemble:

ALL LAMINATED BEAMS TO BE INDUSTRIAL
APPEARANCE UNLESS NOTED.

Drawing Representation

Laminated beams are represented in plan view just as sawn beams are with the required text parallel to the beam. Figure 8-10 shows an example of referencing laminated beams on a framing plan. Depending on the complexity of the plan, a schedule can be used to list the specifications for laminated timbers. If a schedule is used, a beam symbol is placed near the beam so that appropriate material can be easily referenced. Notice

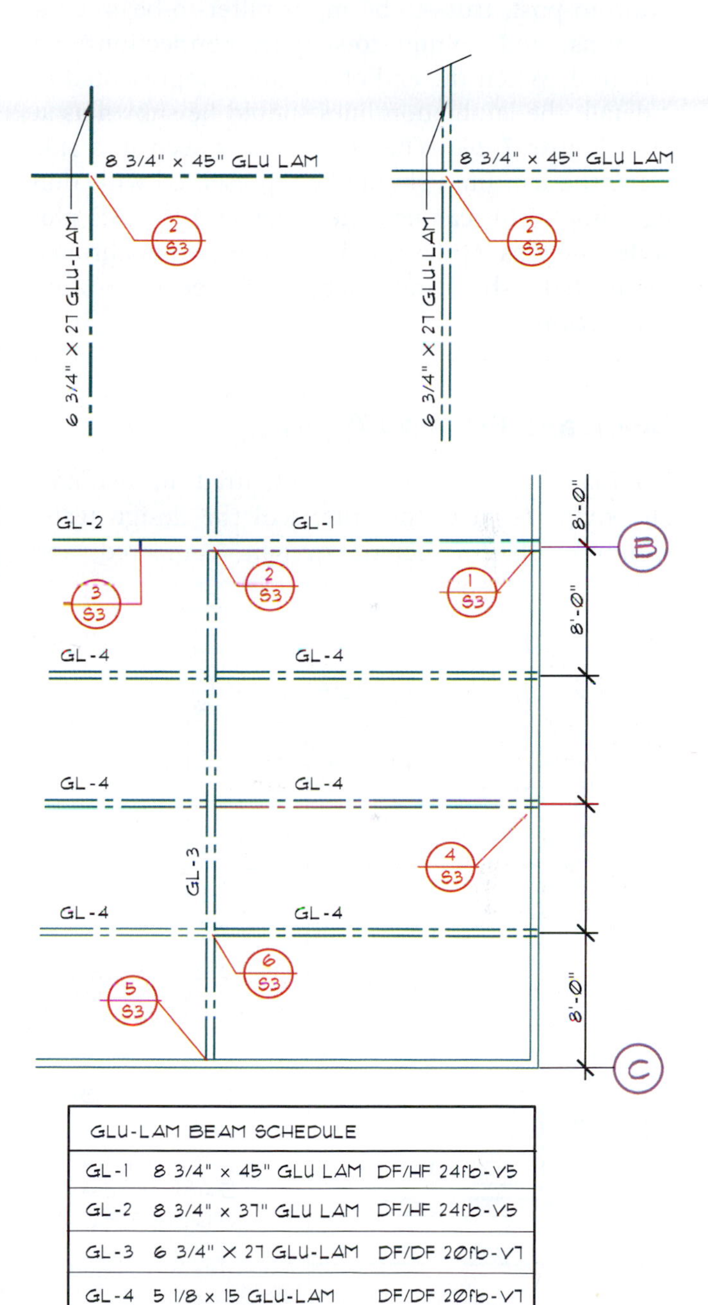

GLU-LAM BEAM SCHEDULE		
GL-1	8 3/4" x 45" GLU LAM	DF/HF 24fb-V5
GL-2	8 3/4" x 37" GLU LAM	DF/HF 24fb-V5
GL-3	6 3/4" X 27 GLU-LAM	DF/DF 20fb-V7
GL-4	5 1/8 x 15 GLU-LAM	DF/DF 20fb-V7

MANUFACTURER TO CERTIFY IN WRITING PRIOR TO INSTALLATION OF GLU-LAMINATED BEAMS THAT MINIMUM DESIGN VALUES HAVE BEEN PROVIDED. ALL BEAMS UNLESS NOTED SHALL BE Fv=165, Fc=450, fb AS NOTED, INDUSTRIAL APPEARANCE WITH EXTERIOR GLUE. PROVIDE "A.I.T.C. CERTIFICATE OF CONFORMANCE TO BUILDING DEPT. PRIOR TO ERECTION.

FIGURE 8-10 Laminated beams are represented similar to the method used to represent wood beams. As plans become more complicated, schedules are often used to maintain drawing clarity.

that many of the specifications are given in a general note. Sections and details may only refer to the beam size, grade, and material.

Similar to the use of sawn lumber, laminated beams will require the use of many details to explain connection. Typically, details showing beam-to-beam, beam-to-post, truss-to-beam, or rafter-to-beam connections, and column-to-support connections are required. When the end of a beam is represented in a detail, the lamination lines should be shown as at D in Figure 7-24. When the beam is seen in a side view, the lam lines should be represented with thin light lines. This can be done using an AutoCAD plot style such as Grayscale instead of Monochrome. Figure 8-11 shows an example of a beam-to-beam connection.

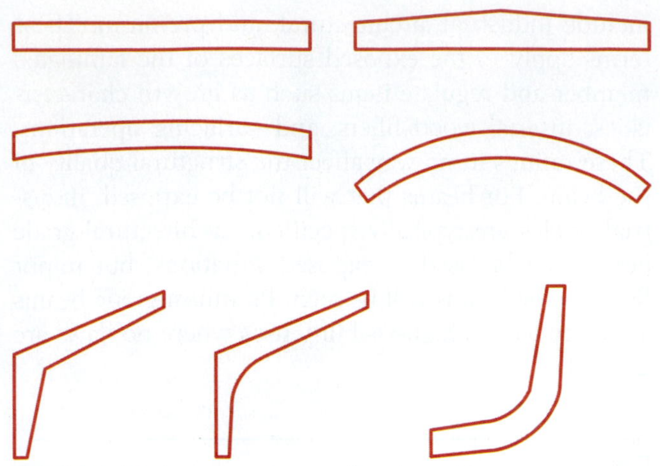

FIGURE 8-12 Common shapes of laminated beams.

Beam and Framing Types

Laminated beams can be configured into almost any shape to meet the criteria of the design team. Common shapes include straight, cambered, and arched. Figure 8-12 shows the common shapes into which laminated beams are fabricated.

Single Span

Single-span laminated beams similar to those in Figure 8-13 are used to span between two or more

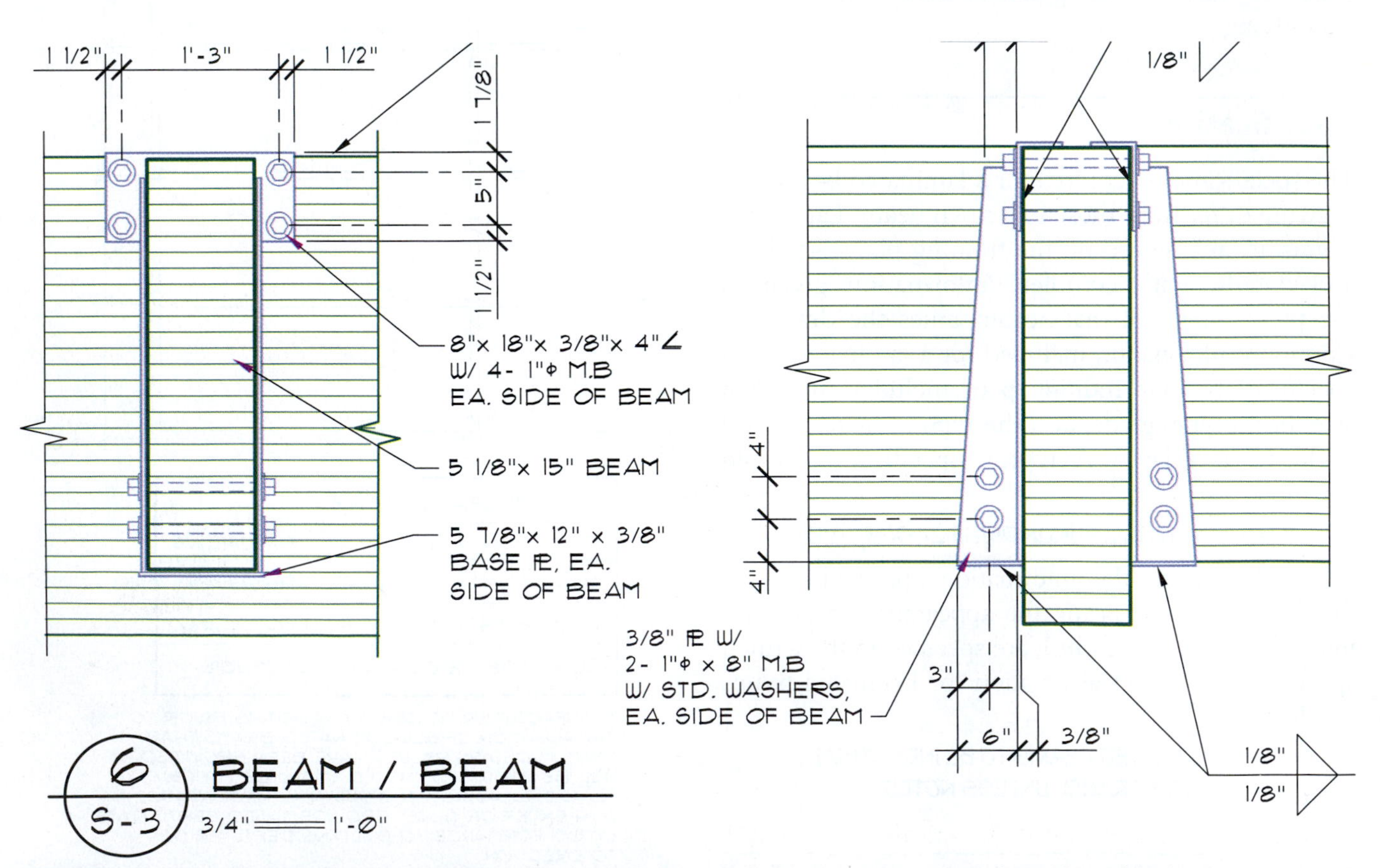

FIGURE 8-11 Laminated beam representation in detail. When the end of a beam is represented, the outline of the beam is represented by a thick line surrounding the lamination lines. When the beam is seen in a side view, the lam lines should be represented with thin light lines.

FIGURE 8-13 A single-span beam is a beam that spans between two supports and supports a uniform load at all points of the beam. The laminated beam supports OSB solid-web trusses. *Courtesy Georgia-Pacific Corporation. All rights reserved.*

supports for large openings. Laminated beams are referenced as shown in Figure 8-10. Because of their structural capabilities, laminated beams can span much larger distances than similarly sized sawn beams. Figure 8-14 shows a comparison of sawn, laminated, steel, and LVL beam sizes. Just as with sawn beams, the spacing of laminated beams is based on the strength of the beam being considered, as well as the strength of the material being supported. The architectural team will dictate the use of timber or steel, based on the design criteria.

Cantilevers

A **cantilever** is a beam that extends past its supports (see Figure 8-15). Although sawn lumber can be cantilevered, laminated members are much more likely to be cantilevered because of the greater distances they typically span. The cantilever is used to decrease the loads on the center of the beam, by increasing the downward action on the ends of the beam. As the ends of the beam are forced downward, the center is forced up, which reduces its natural tendency to sag. Figure 8-16 compares the reactions of a single-span beam and a cantilevered beam. This loading pattern is often used on timber roofs forming a panelized roof framing system.

Panelized Framing

A panelized floor or roof system typically is composed of beams placed in parallel patterns with a spacing of approximately 20' to 30' (6100 to 9100 mm). Smaller beams called *purlins* are used to span between the beams. A spacing of 8' (2400 mm) is typically used for purlins. Joists of 2" or 3" (50 or 75 mm) width lumber are used to span between the purlins and support the floor or roof sheathing. Figure 7-22 shows an example of how a panelized roof constructed of timber is referenced on a framing plan. Figure 8-17 shows a roof panel being lifted into place. Figure 8-18 shows common details for beam connections for a panelized roof.

Curved Beams

As loads and spans are increased, a **camber** or small curve is built into the beam. The camber, which is also known as a crown, is built into beams to allow for the natural deflection that occurs in a beam as loads are applied. The size of the curve will depend on the length of the beam and the size of the material used to fabricate the beam. Laminated beams can be curved so that the beam can be used in domes. Figure 8-19 shows the curved beams used to form a roof. As seen in Figure 8-20, beams can be curved and formed into almost any shape to meet the design criteria of the architectural team.

Arches, Vaults, and Domes

Arches, vaults, and domes have been used for centuries to span large areas. Arches come in many shapes and can be thought of as structural ribs with a skin placed between them. Vaults and domes can be either a ribbed structure with a skin between ribs, or they can be created as a shell. When they are created in a shell form, concrete is most often the material of choice.

Arches Timber, steel, or concrete can be used to form an arch. Common arch shapes include circular or elliptical. The most common arches in construction include the fixed arch, a two-hinged arch, and a three-hinged arch. Each can be seen in Figure 8-21. Fixed arches are typically associated with smaller spans, and two- and three-hinged arches are used for long spans. There are no real hinges used with arches. The number of hinges refers to the point where forces causing rotation are resisted by the connection method. A two-hinge arch uses one beam to form the required arch. A three-hinged arch uses two beams to complete the arch to be formed. Triple-hinged arches are often used to provide a roof structure for arenas and convention centers. As with other types of wood construction, intersections of

DESIGN CONVERSION TABLES[1] F_b = 2400 psi E = 1,800,000 psi

Sawn[4] Section Nominal Size	EQUIVALENT GLULAM SECTIONS FOR SOLID SAWN BEAM							
	Roof Beams[1, 2]				Floor Beams[1, 3]			
	Select Structural		No. 1		Select Structural		No. 1	
	Douglas Fir	Southern Pine[7]	Douglas Fir	Southern Pine[7]	Douglas Fir	Southern Pine[7]	Douglas Fir	Southern Pine[7]
3 x 8	3-1/8 x 6	3 x 6-7/8	3-1/8 x 6	3 x 5-1/2	3-1/8 x 7-1/2	3 x 6-7/8	3-1/8 x 7-1/2	3 x 6-7/8
3 x 10	3-1/8 x 7-1/2	3 x 8-1/4	3-1/8 x 6	3 x 6-7/8	3-1/8 x 9	3 x 9-5/8	3-1/8 x 9	3 x 9-5/8
3 x 12	3-1/8 x 9	3 x 9-5/8	3-1/8 x 7-1/2	3 x 8-1/4	3-1/8 x 10-1/2	3 x 11	3-1/8 x 10-1/2	3 x 11
3 x 14	3-1/8 x 9	3 x 11	3-1/8 x 7-1/2	3 x 9-5/8 *	3-1/8 x 13-1/2	3 x 13-3/4 *	3-1/8 x 13-1/2	3 x 12-3/8
4 x 6	3-1/8 x 6	3 x 6-7/8	3-1/8 x 6	3 x 5-1/2	3-1/8 x 6	3 x 6-7/8	3-1/8 x 6	3 x 6-7/8
4 x 8	3-1/8 x 7-1/2	3 x 8-1/4	3-1/8 x 6	3 x 6-7/8	3-1/8 x 9	3 x 8-1/4	3-1/8 x 7-1/2	3 x 8-1/4
4 x 10	3-1/8 x 9	3 x 11 *	3-1/8 x 7-1/2	3 x 8-1/4	3-1/8 x 10-1/2	3 x 11 *	3-1/8 x 10-1/2	3 x 9-5/8
4 x 12	3-1/8 x 10-1/2	3 x 12-3/8	3-1/8 x 9	3 x 9-5/8	3-1/8 x 12	3 x 12-3/8	3-1/8 x 12	3 x 12-3/8
4 x 14	3-1/8 x 12	3 x 13-3/4	3-1/8 x 10-1/2	3 x 11	3-1/8 x 15[9]	3 x 15-1/8	3-1/8 x 15[9]	3 x 13-3/4
4 x 16	3-1/8 x 13-1/2	3 x 15-1/8	3-1/8 x 10-1/2	3 x 12-3/8	3-1/8 x 16-1/2	3 x 16-1/2	3-1/8 x 16-1/2	3 x 16-1/2
6 x 8	5-1/8 x 7-1/2	5 x 6-7/8	5-1/8 x 7-1/2	5 x 6-7/8	5-1/8 x 7-1/2	5 x 8-1/4	5-1/8 x 7-1/2	5 x 8-1/4
6 x 10	5-1/8 x 9	5 x 8-1/4	5-1/8 x 7-1/2	5 x 8-1/4	5-1/8 x 10-1/2	5 x 9-5/8	5-1/8 x 10-1/2	5 x 9-5/8
6 x 12	5-1/8 x 10-1/2	5 x 9-5/8	5-1/8 x 9	5 x 9-5/8	5-1/8 x 12	5 x 12-3/8	5-1/8 x 12	5 x 12-3/8
6 x 14[9]	5-1/8 x 12	5 x 12-3/8 *	5-1/8 x 10-1/2	5 x 11	5-1/8 x 13-1/2	5 x 13-3/4	5-1/8 x 13-1/2	5 x 13-3/4
6 x 16	5-1/8 x 13-1/2	5 x 13-3/4	5-1/8 x 12	5 x 12-3/8	5-1/8 x 15-1/8	5 x 15-1/8	5-1/8 x 15-1/8	5 x 15-1/8
6 x 18	5-1/8 x 15	5 x 15-1/8	5-1/8 x 13-1/2	5 x 13-3/4	5-1/8 x 18	5 x 17-7/8	5-1/8 x 18	5 x 17-7/8
6 x 20	5-1/8 x 16-1/2	5 x 16-1/2	5-1/8 x 16-1/2	5 x 15-1/8	5-1/8 x 19-1/2	5 x 19-1/4	5-1/8 x 19-1/2	5 x 19-1/4
8 x 10	6-3/4 x 9	6-3/4 x 8-1/4	6-3/4 x 9	6-3/4 x 8-1/4	6-3/4 x 10-1/2	6-3/4 x 9-5/8	6-3/4 x 10-1/2	6-3/4 x 9-5/8
8 x 12	6-3/4 x 10-1/2	6-3/4 x 9-5/8	6-3/4 x 10-1/2	6-3/4 x 9-5/8	6-3/4 x 12	6-3/4 x 12-3/8	6-3/4 x 12	6-3/4 x 12-3/8
8 x 14	6-3/4 x 12	6-3/4 x 12-3/8	6-3/4 x 12	6-3/4 x 11	6-3/4 x 13-1/2	6-3/4 x 13-3/4	6-3/4 x 13-1/2	6-3/4 x 13-3/4
8 x 16	6-3/4 x 13-1/2	6-3/4 x 13-3/4	6-3/4 x 13-1/2	6-3/4 x 12-3/8	6-3/4 x 16-1/2	6-3/4 x 15-1/8	6-3/4 x 16-1/2	6-3/4 x 15-1/8
8 x 18	6-3/4 x 15	6-3/4 x 15-1/8	6-3/4 x 15	6-3/4 x 13-3/4	6-3/4 x 18	6-3/4 x 17-7/8	6-3/4 x 18	6-3/4 x 17-7/8
8 x 20	6-3/4 x 18	6-3/4 x 16-1/2	6-3/4 x 16-1/2	6-3/4 x 16-1/2	6-3/4 x 19-1/2	6-3/4 x 19-1/2	6-3/4 x 19-1/2	6-3/4 x 19-1/4
8 x 22	6-3/4 x 19-1/2	6-3/4 x 17-7/8	6-3/4 x 18	6-3/4 x 17-7/8	6-3/4 x 21	6-3/4 x 22	6-3/4 x 21	6-3/4 x 22

Steel[5] Section	EQUIVALENT GLULAM SECTIONS FOR STEEL BEAMS							
	Roof Beams[1, 2]				Floor Beams[1, 3]			
	Douglas Fir		Southern Pine[7]		Douglas Fir		Southern Pine[7]	
W 6 x 9	3-1/8 x 10-1/2 or	5-1/8 x 9	3 x 11 or	5 x 8-1/4	3-1/8 x 10-1/2 or	5-1/8 x 9	3 x 11 or	5 x 9-5/8
W 8 x 10	3-1/8 x 12	5-1/8 x 9	3 x 12-3/8	5 x 9-5/8	3-1/8 x 13-1/2	5-1/8 x 12	3 x 13-3/4	5 x 11
W 12 x 14	3-1/8 x 16-1/2	5-1/8 x 13-1/2	3 x 16-1/2	5 x 13-3/4 *	3-1/8 x 18	5-1/8 x 15	3 x 17-7/8	5 x 15-1/8
W 12 x 16	3-1/8 x 18	5-1/8 x 13-1/2	3 x 17-7/8	5 x 13-3/4	3-1/8 x 19-1/2	5-1/8 x 16-1/2	3 x 19-1/4	5 x 16-1/2
W 12 x 19	3-1/8 x 19-1/2	5-1/8 x 16-1/2	3 x 20-5/8 *	5 x 15-1/8	3-1/8 x 21	5-1/8 x 18	3 x 20-5/8	5 x 17-7/8
W 10 x 22	3-1/8 x 21	5-1/8 x 16-1/2	3 x 20-5/8	5 x 16-1/2	3-1/8 x 19-1/2	5-1/8 x 16-1/2	3 x 20-5/8	5 x 17-7/8 *
W 12 x 22	5-1/8 x 18	6-3/4 x 15	3 x 22	5 x 17-7/8 *	5-1/8 x 19-1/2	6-3/4 x 16-1/2	5 x 19-1/4	5 x 16-1/2
W 14 x 22	5-1/8 x 18	6-3/4 x 16-1/2	3 x 23-3/8	5 x 17-7/8	5-1/8 x 21	6-3/4 x 18	5 x 20-5/8	6-3/4 x 17-7/8
W 12 x 26	5-1/8 x 19-1/2	6-3/4 x 18	5 x 19-1/4	6-3/4 x 16-1/2	5-1/8 x 21	6-3/4 x 19-1/2	5 x 20-5/8	6-3/4 x 19-1/4
W 14 x 26	5-1/8 x 21	6-3/4 x 18	5 x 20-5/8	6-3/4 x 17-7/8	5-1/8 x 21	6-3/4 x 19-1/2	5 x 22	6-3/4 x 19-1/4
W 16 x 26	5-1/8 x 21	6-3/4 x 19-1/2	5 x 20-5/8	6-3/4 x 17-7/8	5-1/8 x 22-1/2	6-3/4 x 21	5 x 23-3/8	6-3/4 x 20-5/8
W 12 x 30	5-1/8 x 21	6-3/4 x 19-1/2	5 x 20-5/8	6-3/4 x 17-7/8	5-1/8 x 21	6-3/4 x 19-1/2	5 x 22	6-3/4 x 19-1/4
W 14 x 30	5-1/8 x 22-1/2	6-3/4 x 19-1/2	5 x 22	6-3/4 x 19-1/4	5-1/8 x 22-1/2	6-3/4 x 21	5 x 23-3/8	6-3/4 x 20-5/8
W 16 x 31	5-1/8 x 24	6-3/4 x 21	5 x 23-3/8	6-3/4 x 20-5/8	5-1/8 x 25-1/2	6-3/4 x 22-1/2	5 x 24-3/4	6-3/4 x 23-3/8
W 14 x 34	5-1/8 x 24	6-3/4 x 21	5 x 23-3/8	6-3/4 x 20-5/8	5-1/8 x 24	6-3/4 x 22-1/2	5 x 24-3/4	6-3/4 x 22
W 18 x 35	5-1/8 x 27	6-3/4 x 24	5 x 26-1/8	6-3/4 x 22	5-1/8 x 27	6-3/4 x 25-1/2	5 x 27-1/2	6-3/4 x 24-3/4
W 16 x 40	5-1/8 x 28-1/2	6-3/4 x 25-1/2	5 x 27-1/2	6-3/4 x 23-3/8	5-1/8 x 27	6-3/4 x 25-1/2	5 x 27-1/2	6-3/4 x 24-3/4
W 21 x 44	5-1/8 x 33	6-3/4 x 28-1/2	5 x 31-5/8 *	6-3/4 x 27-1/2	5-1/8 x 33	6-3/4 x 30	5 x 33	6-3/4 x 30-1/4
W 18 x 50	5-1/8 x 34-1/2	6-3/4 x 30	5 x 33 *	6-3/4 x 28-7/8	5-1/8 x 31-1/2	6-3/4 x 28-1/2	5 x 31-5/8	6-3/4 x 28-7/8
W 21 x 50	5-1/8 x 34-1/2	6-3/4 x 31-1/2	5 x 34-3/8 *	6-3/4 x 28-7/8	5-1/8 x 34-1/2	6-3/4 x 31-1/2	5 x 34-3/8	6-3/4 x 31-5/8
W 18 x 55	5-1/8 x 36	6-3/4 x 31-1/2	5 x 34-3/8	6-3/4 x 30-1/4	5-1/8 x 33	6-3/4 x 30	5 x 33	6-3/4 x 30-1/4
W 21 x 62		6-3/4 x 36		6-3/4 x 34-3/8		6-3/4 x 34-1/2		6-3/4 x 34-3/8

LVL[6] Section	EQUIVALENT GLULAM SECTIONS FOR LVL BEAMS							
	Roof Beams[1, 2]				Floor Beams[1, 3]			
	Douglas Fir		Southern Pine[7]		Douglas Fir		Southern Pine[7]	
2 pcs. 1-3/4 x 9-1/2	3-1/8 x 10-1/2 or	5-1/8 x 9	3 x 11 or	5 x 8-1/4	3-1/8 x 10-1/2 or	5-1/8 x 9	3 x 11 or	5 x 9-5/8
2 pcs. 1-3/4 x 11-7/8	3-1/8 x 13-1/2	5-1/8 x 10-1/2	3 x 13-3/4	5 x 11	3-1/8 x 13-1/2	5-1/8 x 12	3 x 13-3/4	5 x 11
2 pcs. 1-3/4 x 14	3-1/8 x 15	5-1/8 x 12	3 x 15-1/8	5 x 12-3/8	3-1/8 x 16-1/2	5-1/8 x 13-1/2	3 x 16-1/2 *	5 x 13-3/4
2 pcs. 1-3/4 x 16	3-1/8 x 18	5-1/8 x 13-1/2	3 x 17-7/8	5 x 13-3/4	3-1/8 x 18	5-1/8 x 15	3 x 17-7/8	5 x 15-1/8
2 pcs. 1-3/4 x 18	3-1/8 x 19-1/2	5-1/8 x 15	3 x 19-1/4	5 x 15-1/8	3-1/8 x 19-1/2	5-1/8 x 16-1/2	3 x 20-5/8	5 x 16-1/2 *
3 pcs. 1-3/4 x 9-1/2	3-1/8 x 13-1/2	5-1/8 x 10-1/2	3 x 13-3/4	5 x 11	3-1/8 x 12	5-1/8 x 10-1/2	3 x 12-3/8	5 x 11
3 pcs. 1-3/4 x 11-7/8	3-1/8 x 16-1/2	5-1/8 x 13-1/2	3 x 16-1/2	5 x 12-3/8	3-1/8 x 15	5-1/8 x 13-1/2	3 x 15-1/8	5 x 13-3/4
3 pcs. 1-3/4 x 14	3-1/8 x 19-1/2	5-1/8 x 15	3 x 19-1/4	5 x 15-1/8	3-1/8 x 18	5-1/8 x 15	3 x 17-7/8	5 x 15-1/8
3 pcs. 1-3/4 x 16	5-1/8 x 19-1/2	6-3/4 x 16-1/2	3 x 22	5 x 16-1/2	3-1/8 x 21	5-1/8 x 18	3 x 20-5/8	5 x 17-7/8
3 pcs. 1-3/4 x 18	5-1/8 x 24	6-3/4 x 19-1/2	3 x 24-3/8 *	5 x 19-1/4	3-1/8 x 22-1/2	5-1/8 x 19-1/2	3 x 23-3/8	5 x 19-1/4

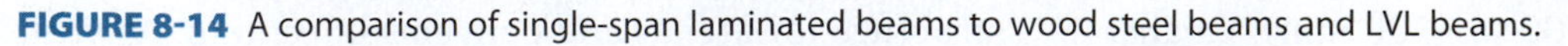

FIGURE 8-14 A comparison of single-span laminated beams to wood steel beams and LVL beams.

FIGURE 8-15 A cantilevered beam is a beam that extends past its supports. An HCA hinge connector by Simpson Strong-Tie Company is used at the cantilevered end of the large beam to support a smaller glu-lam beam. Each beam is supporting open-web floor trusses. *Courtesy LeRoy Cook.*

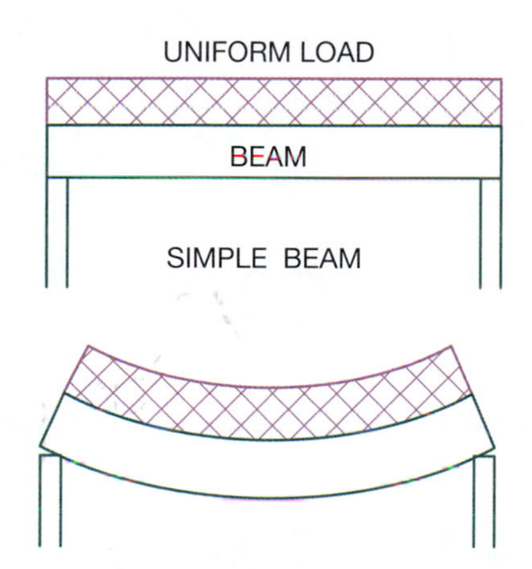

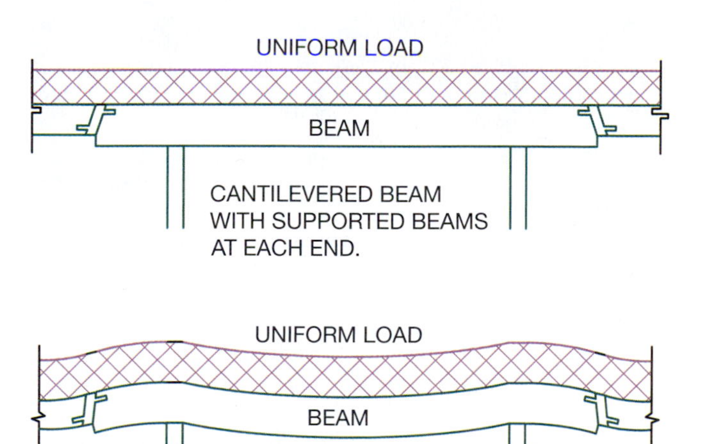

FIGURE 8-16 Beam reactions of simple and cantilevered beams. A simple beam will sag in the center. If each end of the beam is extended past its supports, the loads on the end of the beam force the center of the beam upward, helping to resist the tendency to sag in the center.

FIGURE 8-17 A panelized floor or roof system is typically composed of beams placed in parallel patterns with a spacing of approximately 20' to 30' (6100 to 9100 mm). Smaller beams called *purlins* are used to span between the beams. This roof uses open-web trusses to span between larger open-web beams, with wood purlins spanning between the larger purlins. *Courtesy APA—The Engineered Wood Association.*

structural members must be specified on the framing plans and details. Figure 8-22 shows an arch foundation connection.

Vaults Vaults similar to Figure 8-23 are produced by placing arches next to each other to form elongated shapes. Figure 8-24 shows other common methods of constructing arches. When adjacent arches are placed perpendicular to their planes, a ribbed ceiling as found in most European cathedrals is constructed. Vaults can be constructed between rectangular shapes in plan view.

Domes When an arch is rotated around itself at its crown, a dome is formed. A dome structure will form a circular shape in plan view. Figure 8-25 shows an example of a dome framed with timber.

OPEN-WEB TRUSSES

Although truss construction is considered nonconventional framing, most upper-level floors and roofs of commercial structures are framed with trusses. A *truss* is a premanufactured component made up of triangular shapes arranged in a single plane used to span large distances. Because of the triangular shapes, loads will result in stress at the intersection of members in either compression or tension. Loads are spread throughout the truss so that the ends of the chords are usually the only portions of the truss that require support. If a truss is used to frame a roof, then rafters, ceiling joists, purlins, and braces are eliminated for that portion of

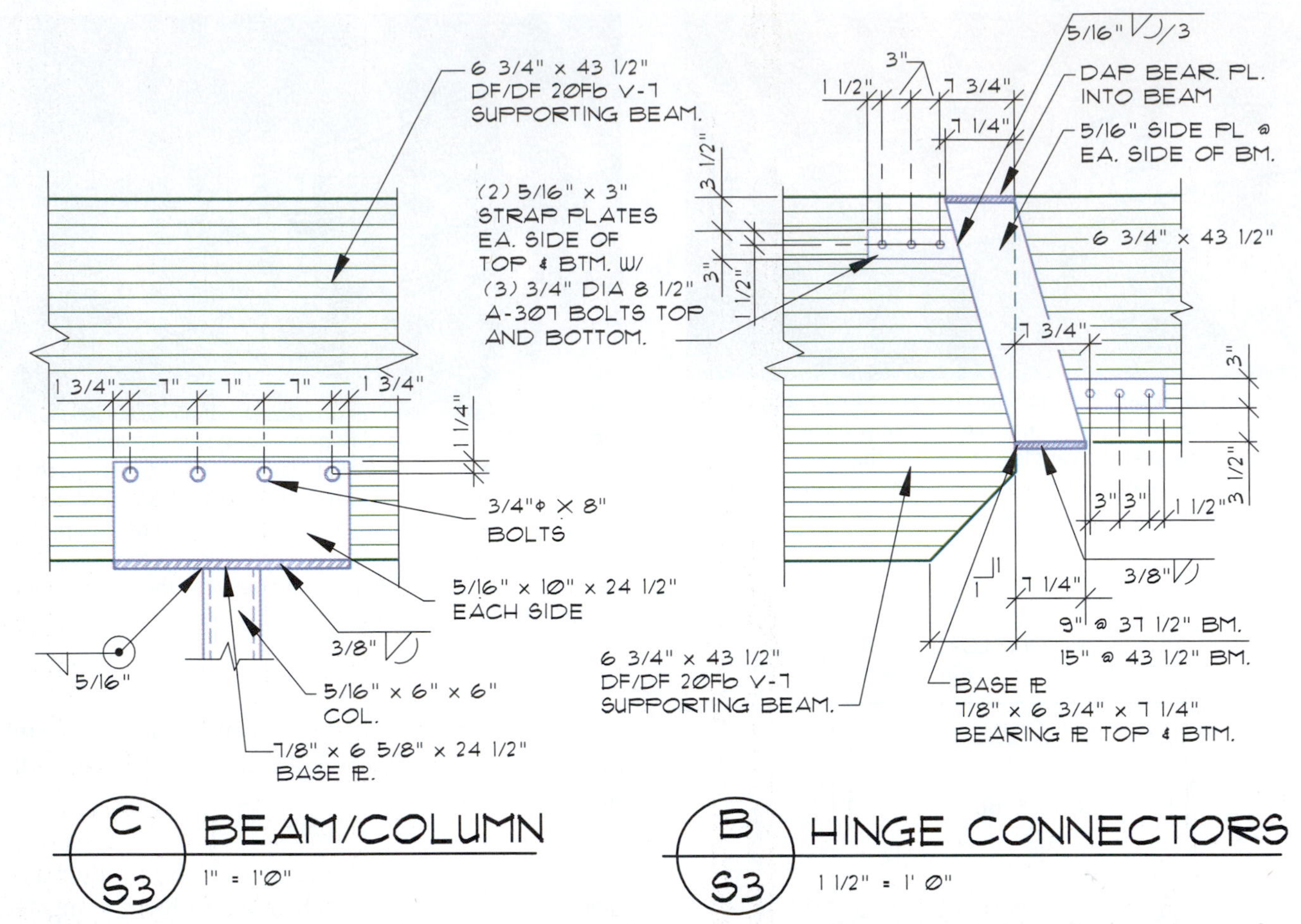

FIGURE 8-18 A beam-to-column and a beam-to-beam connection. The cantilevered beam supports the smaller beam by use of a metal support called a *saddle*.

FIGURE 8-19 Curved beams used to form a school roof. *Courtesy APA—The Engineered Wood Association.*

FIGURE 8-20 Laminated beams with an irregular curve used to form the roof of a hockey arena. *Courtesy APA—The Engineered Wood Association.*

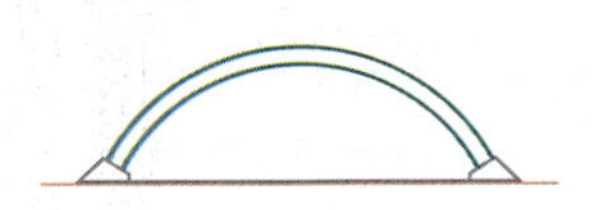

FIXED ARCH

ONE-HINGE ARCH

TWO-HINGE ARCH

THREE-HINGE ARCH

FIGURE 8-21 Common arches.

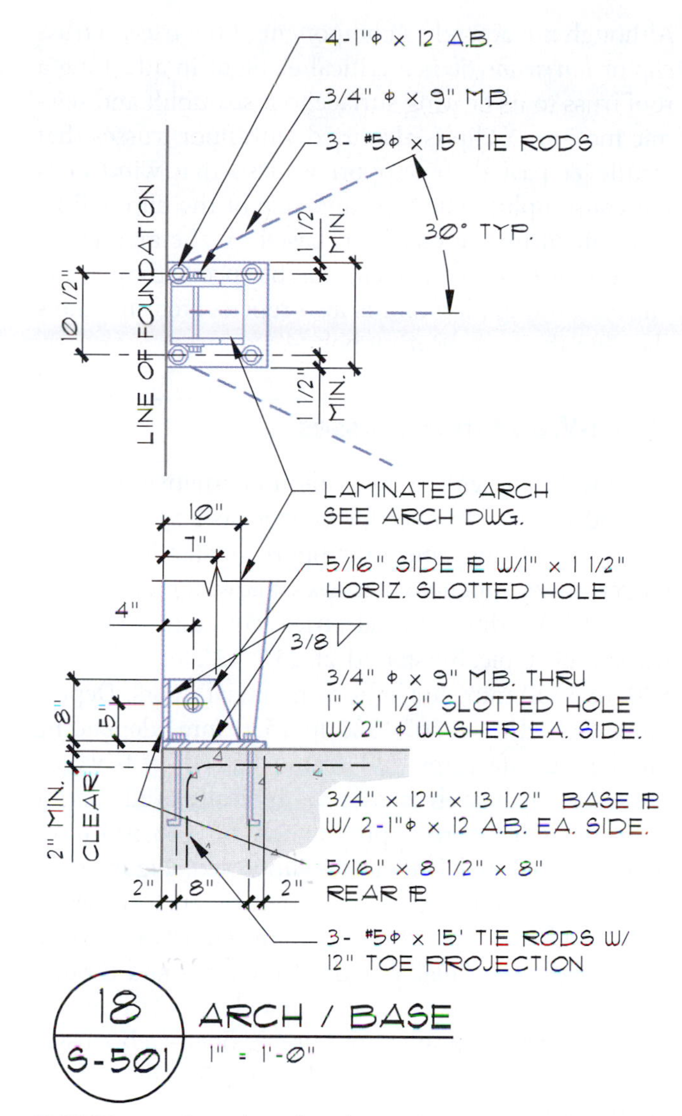

FIGURE 8-22 An arch-to-foundation detail.

FIGURE 8-23 Vaults are produced by placing arches next to each other to form elongated shapes. *Courtesy APA—The Engineered Wood Association.*

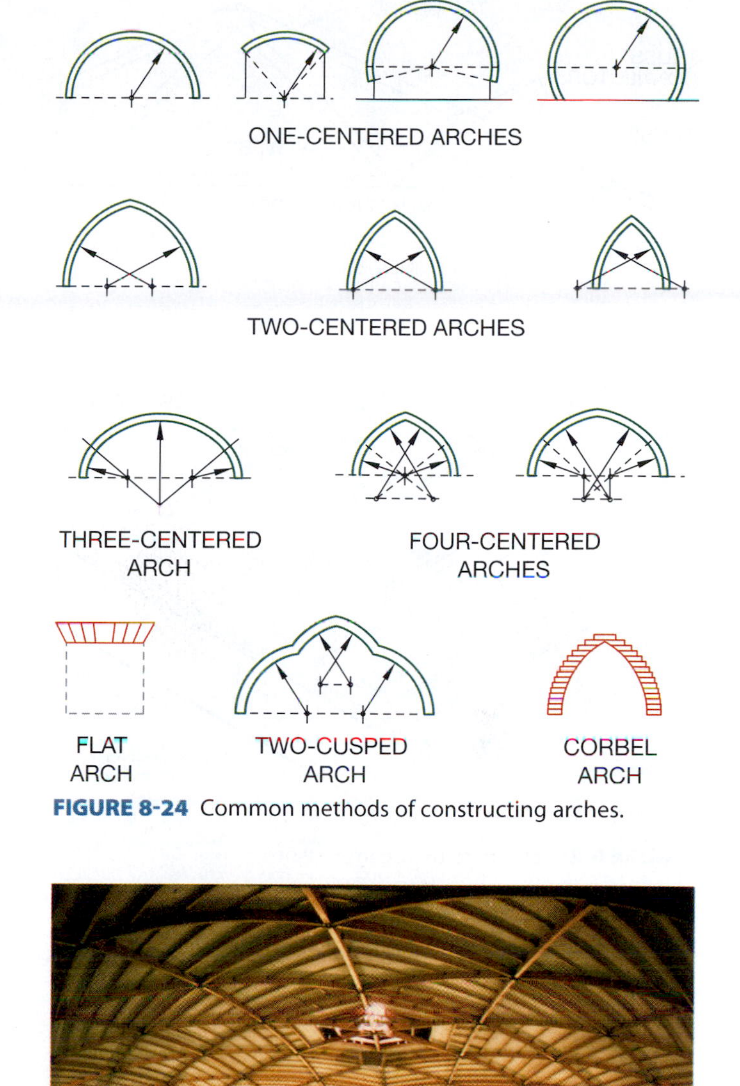

FIGURE 8-24 Common methods of constructing arches.

FIGURE 8-25 A lamella frame is a dome framed from a series of perpendicular ribs. *Courtesy American Institute of Timber Construction.*

the structure. Trusses and conventional framing systems can be combined in the same structure to form irregular shapes.

Truss Terms

Major components of a truss include the top and bottom **chords**, the webs, and gussets. Figure 8-26 shows common truss components. The *top chords* of a truss are the members that support the loads from the finished roofing material. The *bottom chord* is the lower

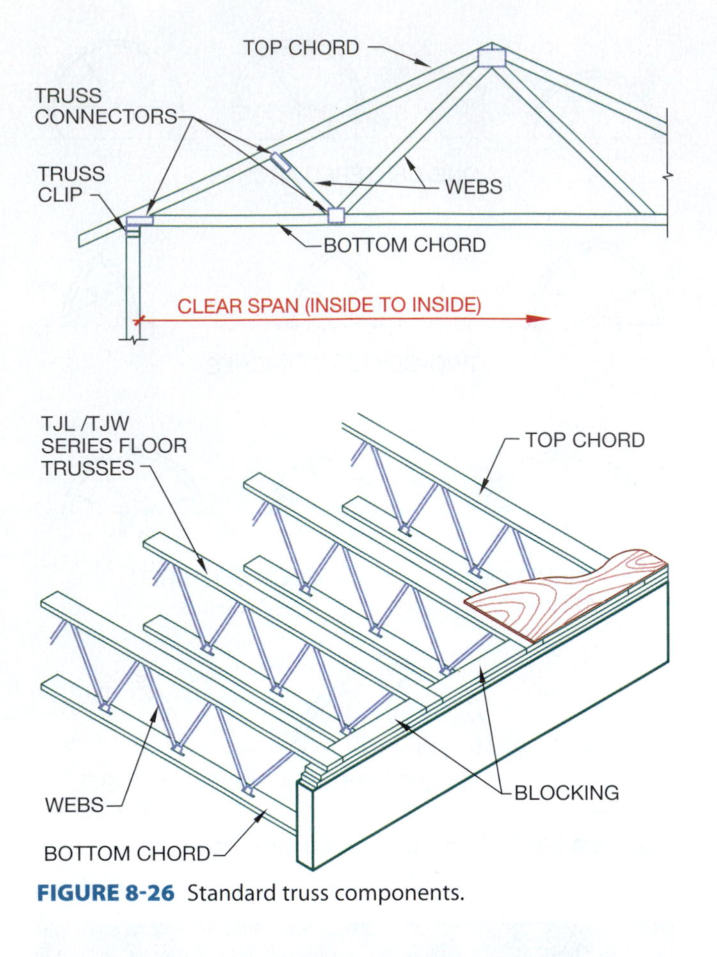

FIGURE 8-26 Standard truss components.

member of a truss and is typically a horizontal member. The *webs* are the interior members that transfer the loads between the top and bottom chords. Because a roof truss is formed as a single component, no ridge board or beam is required for support at the ridge. A *block* is placed between individual trusses to maintain a uniform spacing, to resist the tendency of the trusses to fall sideways, and to provide a nailing surface for the roof sheathing. Components of trusses are not specified on the structural drawings but are shown on the drawings supplied by the manufacturer.

The spacing of the trusses will determine the size of the chords and webs. Common spacing for trusses includes 24" and 32" (600 and 800 mm) o.c. and allows the chords and webs to be made of 2× (50×) material. As the spacing of the trusses increases beyond 32" (800 mm) o.c., timber or steel is used to frame the trusses. A *gusset* is a side plate that is often used at the intersection of a chord to a web to form a secure connection. On trusses framed from 2× (50×) material, the gusset is typically a lightweight metal plate. The truss manufacturer determines the size of the gusset and other truss components. A structural engineer designs the components of a timber-framed truss. Review Chapter 7 for examples of timber trusses.

Although not actually a component of the truss, a *truss clip* or *hurricane tie* is a critical element in attaching a roof truss to its bearing surface to resist uplift and seismic motion. A clip is also used with floor trusses that cantilever past their support wall so that wind does not cause uplift. The type and size of the clip will be determined by the engineer, based on the amount of uplift to be resisted. The clip can be specified by a note on the framing plan and is also represented in details showing the truss bearing points.

Open-Web Floor Trusses

Open-web trusses are a common alternative to using 2× (50×) sawn lumber for floor joists and rafters. In addition to being able to support greater loads over longer spans, open-web trusses allow for easy routing of HVAC ducts and electrical conduit. Open-web trusses are typically spaced at 24", 32", or 48" (600, 800, or 1220 mm) for light commercial floors. Depths range from 14" to 60" (360 to 1520 mm) depending on the manufacturer and material used to fabricate the truss. Open-web trusses are typically available for floor spans of up to approximately 60' (18,300 mm) depending on the load to be supported. Figure 8-27 shows open-web trusses supported by an LVL beam. Sawn lumber is typically used for the chords and webs, but trusses are available with LVL chords and tubular steel with a diameter of approximately 1" (25 mm) for the webs. Depending on the usage, open-web trusses may be top- or bottom-chord bearing. Rarely are both chords bearing.

When drawn in section or details, wood webs are drawn as any other wood member. Webs made of

FIGURE 8-27 Open-web floor trusses supported by an LVL header.

tubular steel are typically represented with a bold centerline drawn at a 45° angle. The exact angle is determined by the manufacturer and is usually unimportant to the detail. As the span or load on the joist increases, the size and type of material used is increased. When drawing sections showing open-web trusses, it is important to work with the manufacturer's details to find exact sizes and truss depth. Figure 8-28 shows common methods of placing the top chord. The CAD technician will need to carefully study the details provided by the manufacturer to ensure proper usage. Most truss manufacturers supply electronic copies of common truss connections that can be used for a base to complete the truss-to-wall detail.

Common Types of Roof Trusses

Figure 8-29 shows common truss shapes used in wood-frame commercial construction. Each type of truss shown with parallel chords can be built with an inclined top chord to facilitate roof design and water drainage without altering the design of the truss. It is important to remember that the exact number of webs will vary for each application and will be determined by the span of the truss. Whatever the configuration of the truss, it is shown on the framing plan as seen in Figure 8-30.

A *fink* or standard truss in Figure 8-29a is used to frame a gable roof for projects such as multifamily dwellings and retail sales outlets. The top chord is always inclined, but the interior webs can be arranged in different configurations to allow for storage space in the center of the truss (see Figure 8-29b and 8-29c). The openness of the modified fink can be useful for attic storage as well as a location for HVAC equipment. A *camber*, which is a curve, can also be built into the bottom chord (see Figure 8-29d) to assist in supporting loads.

Similar to a fink truss are *howe* and *pratt trusses* (Figure 8-29e and 8-29g). Each of the trusses can be framed with an inclined top chord or a partially inclined top chord with a horizontal chord (Figure 8-29f and 8-29h). The inclined chords are used for framing roofs, and the horizontal top chords are typically used for floor framing. The difference between pratt and howe trusses lies in the arrangement of the webs and the method of transferring loads throughout the truss. The partial flat top chord can be used to form hip roofs (see Chapter 18). Pratt trusses can also combine inclined and flat bottom chords, as seen in Figure 8-29j, to form a *camelback pratt truss*. Other flat top chord trusses include *warren trusses* (see Figure 8-29k). Warren trusses are typically used for floor trusses. The top chord can be slightly inclined for use as roof trusses that are especially useful in dry climates where roof drainage is not heavy. A *transverse bent* style truss (Figure 8-29l) has top chords with two slightly inclined top chords to facilitate drainage.

A *scissor truss* has top and bottom chords that are inclined, allowing for a vaulted ceiling to be framed. The bottom chord usually must be at a minimum of two pitches less than the top chord. If the top chord is to be set at a 6/12 pitch, the bottom chord cannot be steeper than a 4/12 pitch. This guideline can vary as the chords and webs are strengthened. Figure 8-29m shows a scissor truss with equal interior pitches. A scissor truss can be combined with a standard truss to create a flat ceiling with a vaulted area as shown in Figure 8-29n. As the configuration of the scissor truss is altered, the intersections between flat and inclined bottom chord members may require support from interior walls or beams. Additional bearing points of the truss are also based on the span.

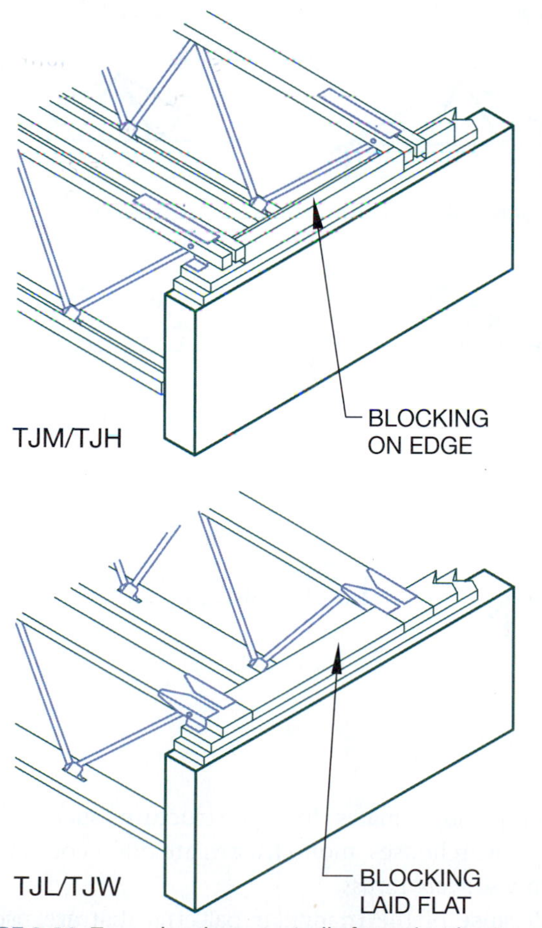

FIGURE 8-28 Truss chords are typically framed with a 2× (50×) or by engineered material supported by a metal bracket. Engineered chords are typically set to be 3 1/2" (90 mm) tall.

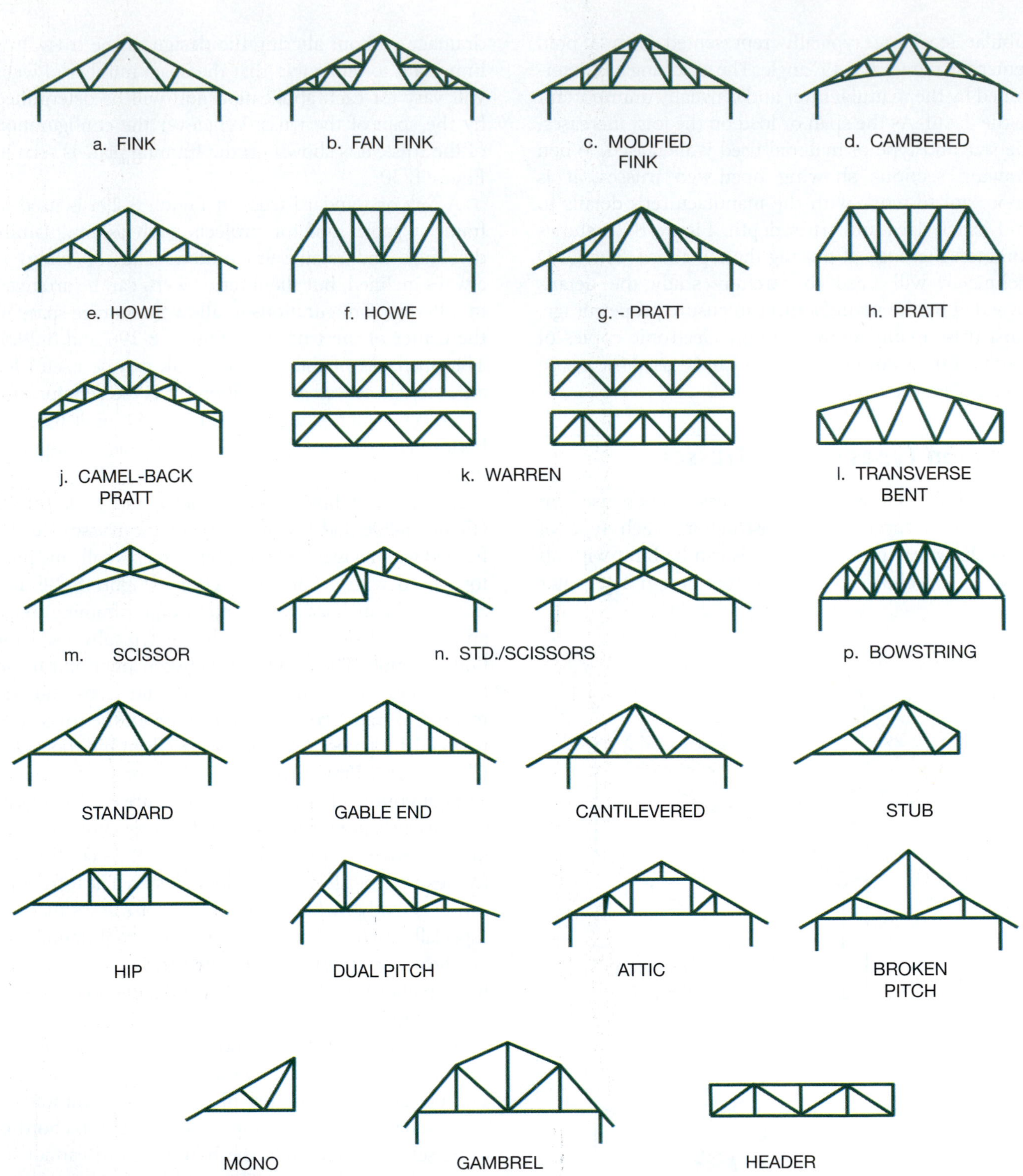

FIGURE 8-29 Common shapes and types of trusses.

A *mono truss* (Figures 8-29 and 8-30) resembles half of a standard truss. The mono truss has an inclined and horizontal chord as well as one vertical exterior web. Mono trusses can be used to blend a one-level structure into a two-level structure, or they can be altered to have a dual pitch as illustrated in Figure 8-29. The short, steep top chord works well to support solar collectors in temperate climates. In large structures such as factories or warehouses, mono trusses are often combined to form a *sawtooth truss*.

Because of the triangular patterns that are used to form a truss, the overall configuration of trusses is almost unlimited. One additional truss that may be specified on a framing plan is a *girder truss*. A girder

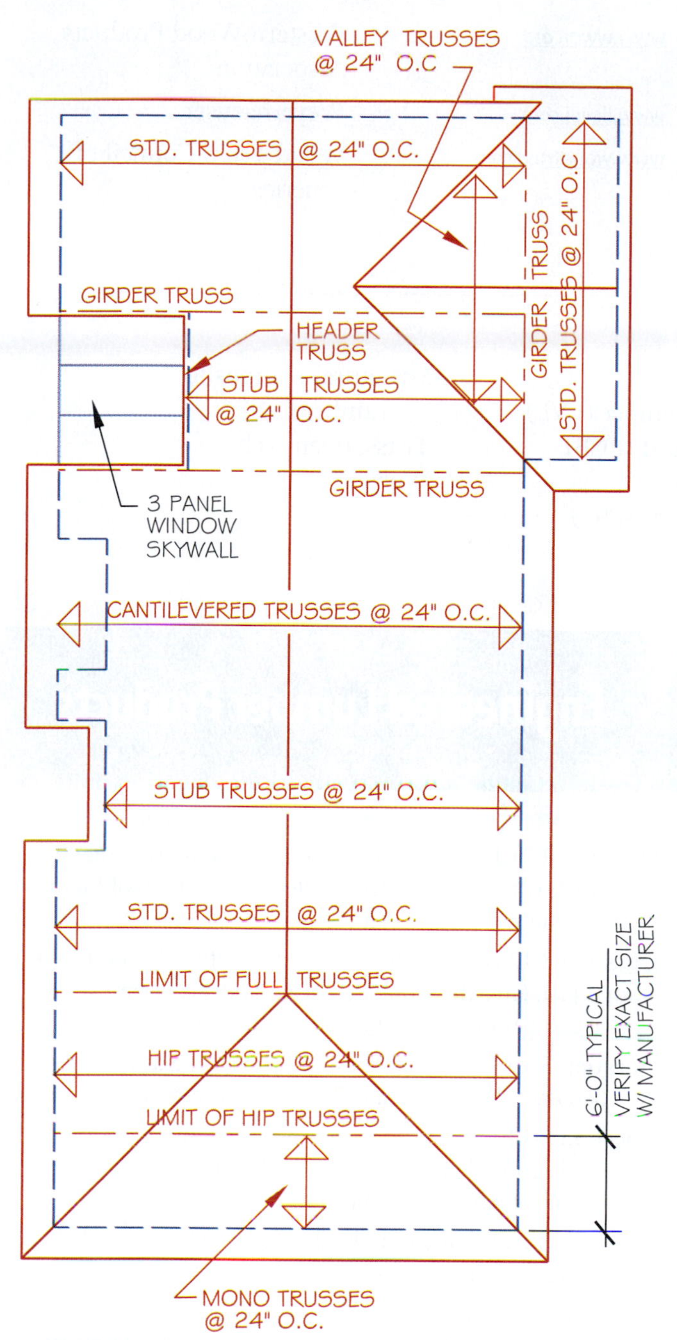

FIGURE 8-30 Standard, valley, girder, header, stub, cantilever, hip, and mono trusses can be used to form different roof shapes. Each type of truss needs to be represented on the roof framing plan.

truss is a stronger version of any of the shapes previously mentioned, and it is used to support the weight of other trusses. A girder truss allows interior support walls to be removed by providing the support for other trusses. Figure 8-30 shows a framing plan using trusses set perpendicular to each other. The girder truss lies between the two areas of trusses. The trusses that are perpendicular to the girder truss are hung from the girder truss by metal hangers. Figure 8-31 shows an example of a detail of the intersection of a girder truss

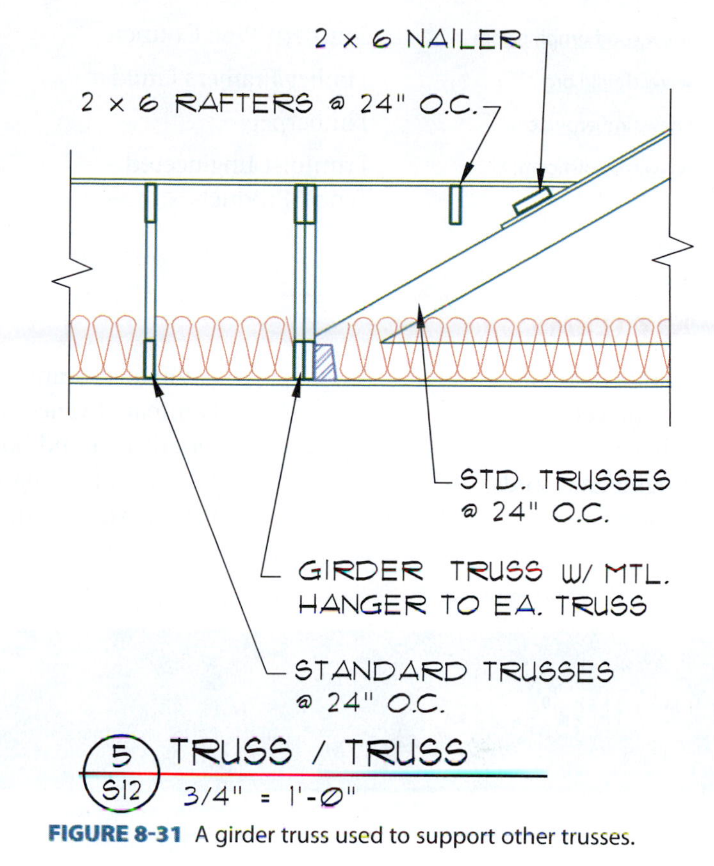

FIGURE 8-31 A girder truss used to support other trusses.

with other trusses. Notice that the void area that results in the roof between the trusses is filled in with conventional framing material.

ADDITIONAL READING

The following Web sites can be used as a resource to help you keep current with changes in building materials.

ADDRESS	COMPANY/ ORGANIZATION
www.aitc-glulam.org	American Institute of Timber Construction
www.bcewp.com	Boise Cascade Engineered Wood Products
www.forestdirectory.com	Directory of Products, Wood Science and Marketing
www.iwpawood.org	International Wood Products Association
www.lpcorp.com	Louisiana-Pacific Corporation
www.strongtie.com	Simpson Strong-Tie Company, Inc.
www.sfpa.org	Southern Forest Products Association

www.southernpine.com Southern Pine Council
www.tfguild.org Timber Framers Guild
www.timberpeg.com Timberpeg
www.trimjoist.com TrimJoist Engineered Wood Products
www.wwpa.org Western Wood Products Association
www.ilevel.com Weyerhaeuser
www.woodtruss.com Wood Truss Council of America

KEY TERMS

Camber
Cantilever
Chord
Engineered studs
I-joists
Laminated beams
Laminated veneer lumber (LVL)
Oriented strand board (OSB)
Panelized framing
Parallel strand lumber (PSL)
Structural composite lumber (SCL)
Truss, open-web
Webs

CHAPTER 8 TEST Engineered Lumber Products

QUESTIONS

Answer the following questions with short complete statements. Type the chapter title, question number, and a short complete statement for each question using a word processor. Use the appropriate charts from this chapter when possible. Some answers may require the reference material from the Simpson Strong-Tie or Weyerhaeuser Web sites. If math is required to answer a question, show your work.

Question 8-1 Define the following abbreviations: PSL, OSB, LVL, MDF, EXP, HDF, APA, EXT, and STRUCT.

Question 8-2 List common types of web materials that can be used in solid and open-web trusses.

Question 8-3 What two elements are applied to wood to make engineered wood products?

Question 8-4 List the common sizes of engineered wood studs.

Question 8-5 What resists uplift when a truss roof system is used?

Question 8-6 If an 8/12 pitch is used for a scissor truss, what is the maximum pitch for the bottom chord?

Question 8-7 List five types of laminated beams and show sketches of how they could be used in a structure.

Question 8-8 List six common finished widths of laminated beams.

Question 8-9 A laminated beam is to be 24-V5. What are the options for materials?

Question 8-10 For question 8-9, if the fir option is used, what will be the value for E?

Question 8-11 For question 8-9, if the option other than fir is used, what is the category listed in column 6 of Figure 8-9b, and what is the value?

Question 8-12 A plan calls for a 6 × 14 DFL #1 floor beam or equal. Give the laminated beam and steel beam sizes that can be substituted for the sawn beam.

Question 8-13 List two alternatives for placing annotation when describing laminated beams on a framing plan.

Question 8-14 List seven types of information typically specified for a laminated beam and then provide an example of a beam specification.

Question 8-15 What would be the depth of beams with 9, 15, and 24 laminations?

Question 8-16 List and explain the eight symbols typically used to describe the materials used to fabricate laminated beams.

Question 8-17 What grade of laminated beam should be used for a beam that will be in a highly visible location?

Question 8-18 Why is it possible, given an equal length and load, to use a smaller cantilevered beam rather than a single-span beam?

Question 8-19 What is a *hinge* as the term relates to arches?

Question 8-20 Describe how a panelized roof is made.

DRAWING PROBLEMS

Unless your instructor provides other instructions, use the minimum standards presented in the Chapter 6 section, "Drawing Criteria for Completing Details," to complete the following

details. Skeletons of most details can be found on the student CD. Use these drawings as a base to complete the assignment.

Problem 8-1 Draw a detail showing a truss with a 4/12 roof pitch (assume 2 × 6 chord material) resting on 2 × 4 studs at 16" o.c. Cover the interior with 1/2" gypsum board and the exterior side of the wall with 3/4" exterior stucco or CD-EIFS. Specify a concrete tile roof.

Problem 8-2 Draw a detail showing a scissor truss with a 6/12 roof pitch (assume 2 × 4 chord material and a 3/12 pitch for the bottom chord) resting on 2 × 6 studs at 16" o.c. Cover the interior with 1/2" gypsum board and the exterior side of the wall with horizontal beveled siding over 1/2" gypsum board and 1/2" OSB. Specify a 26-gauge metal roof over 3/4" plywood sheathing.

Problem 8-3 Draw the required details to show 9 1/2" deep TJI-110 @ 24" o.c. floor trusses resting on a 9' high 2 × 6 stud wall. Provide an LVL rim joist at the exterior end of the trusses at the top of the support wall. Show a 6 × 10 DFL # 1 window header in the wall with the bottom of the header set at 7'–0" high. Show 7/8" stucco over 1/2" plywood on the exterior side. Show 1 1/2" concrete over 55# felt over 1 1/8" plywood subfloor. Frame an upper wall with 2 × 4 studs at 16" o.c. with (2) sills (2 × 4 over 3 × 4). Specify that the lower pressure-treated plate is nailed w/ 20d's at 4" o.c. Use break lines so that the full height of each wall is not drawn.

Problem 8-4 Draw a detail showing 18" deep TJL floor trusses resting on each side of a 10' high, interior wall framed with 2 × 6 studs. Provide solid blocking between the trusses at the top of the support wall. Show a 6 × 12 door header in the wall with the bottom of the header set at 7'–0" high. Show 1/2" gypsum board on each side of the wall. Show 1 1/2" concrete over 55# felt over 1 1/8" T & G plywood subfloor.

Problem 8-5 Complete the following detail showing a 6 3/4" × 15" beam supporting 16" deep TJM trusses at 48" o.c. along each side of the beam. Use the manufacturer's specifications to show top chord bearing with butted trusses.

Problem 8-6 A 3 × 12 DFPT ledger will be connected to a 2 × 6 stud wall per nailing schedule. The wall is to extend 30" minimum above the finish roofing and cap with standard plates, and with 26-gauge flashing. Use the ledger to support TJW 18 trusses at 24" o.c. The trusses will support 5/8" standard-grade plywood sheathing and a hot mopped roof. Provide a 4" cant strip at the truss/wall intersection. Cover the exterior sides of the wall with 1/2" plywood covered with 7/8" exterior stucco. Cover the interior side of the wall with 1/2" gypsum board that extends to the ledger and provide a suspended ceiling that provides 12" clearance below the trusses.

Problem 8-7 Draw a detail showing (2) -8 3/4 × 22 1/2 glu-lam resting on a 6 × 8 post. Use a column cap sufficient to hold 4000# in uplift.

Problem 8-8a Draw details showing an 8 3/4 × 22 1/2 glu-lam that cantilevers 6' past the 8 × 8 post.

Problem 8-8b Support a 8 3/4 × 16 1/2 beam. Draw the beam-to-beam connection based on the following assumptions:

a) The center of all bolts are to be 1 1/2" from the edge of steel, with 3" minimum spacing.

b) All welded connections to be 5/16" fillet connections.

c) Side plates to be made from 5/16" × 8" wide steel.

d) Base and top plates to be 3/4" × 9 5/8" × 8 1/4" steel.

e) 3 1/2" wide × 5/16" side strap with (2)-3/4" diameter M.B. Bolts to be a minimum of 10" from beam splice. Length based on bolt placement. Place the centerline of straps 4" up from bottom, 3" down from top.

Problem 8-9 Draw a plan view for a 48' × 60' structure framed with 2 × 6 studs at 16" o.c. at a scale of 1/8" = 1'–0". Place a 6 3/4" × 16 1/2" ridge beam in the center of the structure 24' from the sidewalls. Specify 6 × 6 wood posts at 20' intervals along this beam with 4 × 6 posts at the ends hidden in the walls. Specify 4 × 6 or 4 × 4 post at each beam/wall intersection (hide in wall). Show splices in the ridge beam 48" from the post so that the center beam is only 12' long. Use a 13 1/2" deep beam for the center span.

Hang 5 1/8 × 15 glu-lam beams at 15'–0" o.c. on each side of the ridge beam. Hang 3 1/8 × 13 1/2 glu-lams at 8'–0" o.c. between each 15" glu-lam. Support these beams with (2) -2 × 6 at each wall. Hang 2 × 6 rafters at 24" o.c. between each 13 1/2" glu-lam with appropriate hangers. Show enough of the roof framing to represent the framing patterns. Provide grids to represent each wall and post so that four vertical and three horizontal grids are provided. Specify the location of all members. Use a schedule wherever possible. Assume the sheet to be S-502.

Problem 8-10 Draw details to represent the post-to-beam connections and all beam-to-beam connections represented in problem 8-9. Assume the sheet to be S-503. Place all details in a logical order based on Chapter 2 and place detail reference numbers by each detail and on the framing plan. Possible details would include, but are not limited to:

8-10-1 BF—Ridge to wall

8-10-2 BF—Ridge to post

8-10-3 BF—Ridge to ridge

8-10-4 BF—Purlins to ridge

8-10-5 BF—Rafter to purlins

If all the details fit, they may be placed on the same page as the framing plan. If they will not all fit, place the details on sheet S-503.

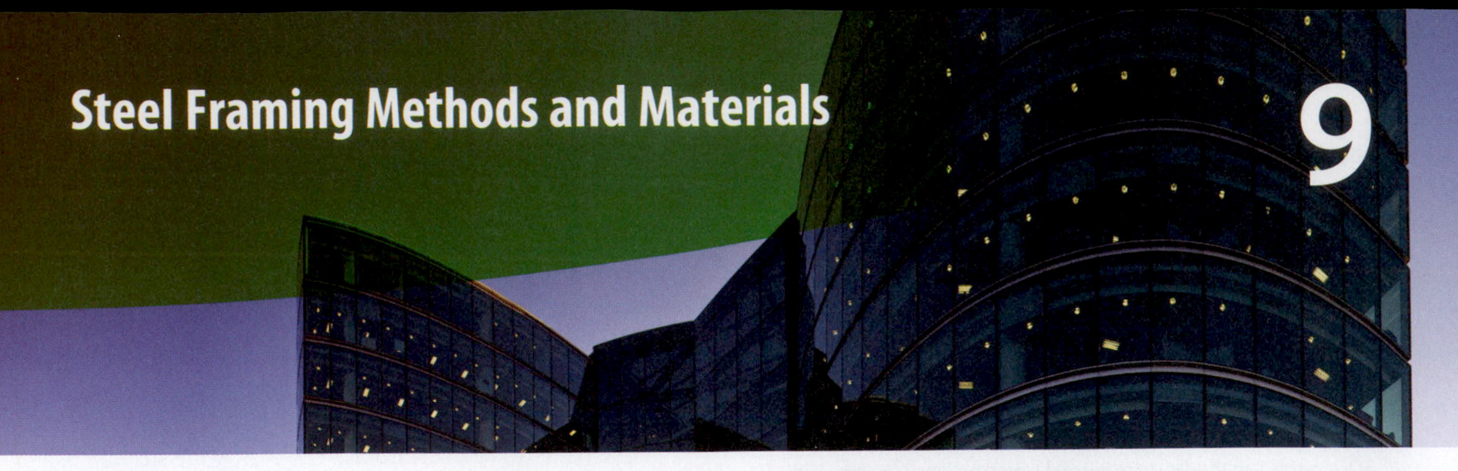

Steel is used throughout the construction industry in many sizes, shapes, and types of construction. This chapter will explore steel and the many metals and metal alloys used throughout the construction of a structure and describe common methods of representing these materials in the drawing set. Subjects to be explored include:

- Steel in construction.
- Types of metal and alloys.
- Lightweight steel framing members such as steel studs, joists, open-web joists, and decking.
- Metal buildings.
- Multilevel steel frames.

The type of metal or steel that is used will affect the type of framing used, the design of the structure, and what the CAD technician is required to place on the architectural and structural drawings. Four of the most common uses of metal and steel in construction include the use of steel framing members, metal buildings, specialty structures such as space frames, and steel rigid frames. Figure 9-1 shows examples of four uses of steel in a multilevel structure.

STEEL IN CONSTRUCTION

Although a CAD technician will not specify the contents of the steel on the structural drawings, knowledge of the contents helps a technician understand the steel used throughout a structure. Steel is a metal made from various elements, but it consists of approximately 98% iron. The other major element in steel is carbon. Although steel is composed of a maximum of only 1.5% carbon, it is carbon that controls the stiffness and hardness of steel. Steel is classified by its carbon content. The carbon contents of various steels referenced on structural drawings based on the American Society of Testing Materials International (ASTM) are as follows:

Category of Steel	Carbon Content (%)
Low-carbon steel	0.06 – 0.30
Medium-carbon steel	0.30 – 0.50
High-carbon steel	0.50 – 0.80

In addition to carbon and iron, steel contains phosphorus and sulfur. Both are limited because they affect the brittleness of the steel. Brittle steel tends to crack as loads are applied. Common elements such as silicon, manganese, copper, nickel, chromium, tungsten, molybdenum, and vanadium are also added to steel. These elements comprise a very small percentage of the chemical composition of steel but affect qualities such as strength, hardness, and corrosion resistance. The elements used for making steel are strictly controlled by the ASTM.

Structural Steel

Varied amounts of elements will produce different types of steel. Strength, hardness, corrosion resistance, brittleness, and ductility are specified on the drawings. They are all controlled by varying the elements introduced into steel. Before examining various types of steel, it is important to realize how steel reacts to stress induced from a load. Although the engineer will determine the strength of steel to be used, the CAD technician must understand the values that are contained in the calculations. Figure 9-2 shows the stages that a low-carbon steel member will go through as it resists a load.

- In the **elastic range**, a member resisting a load will deflect, but maintain its structural integrity. This area corresponds to the modulus of elasticity of a wood member.
- The **yield point** is the measurement of stress in a material, which corresponds to the limiting value of the usefulness of the material. Once steel is stressed past its yield point, it will not return to its original shape or size. The yield point is represented by the letters *Fy* in beam formulas and is placed in the

FIGURE 9-1 Many multilevel structures include steel for the frame, trusses, decking, and wall studs. *Courtesy Jordan Jefferis.*

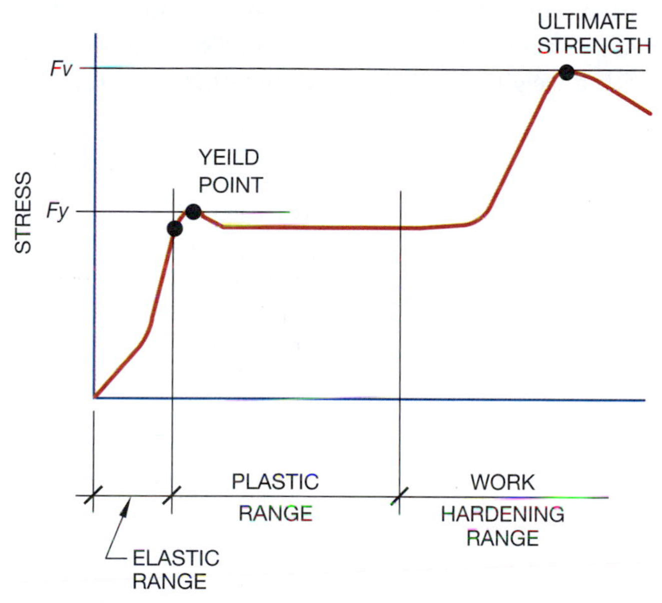

FIGURE 9-2 Typical reactions to stress in mild-carbon steel.

written specifications. The higher the yield point, the less ductile the steel.

If a material is **ductile**, it has a great ability to bend before breaking. Materials that are supported by a beam that has gone past its yield point may crack, but the beam itself will not actually fail. Once the stress has entered the *work hardening range*, steel actually increases its stress-carrying capability. As the steel deforms, the internal structure of the steel is dislocated, which makes it harder for future dislocations to occur.

The final stage in resisting stress is when the member reaches its *ultimate tensile strength* or breaking point. Once the ultimate tensile strength has been exceeded, the beam will cause structural failure.

Structural steels are identified on structural drawings by their ASTM designation number. Common numbers and their metric equivalent used for construction include A-36 (A36M), A-242 (A242M), A-441 (A441M), A-529 (A529M), A-572 (A572M), and A-588 (A588M).

ASTM A-36—Steel with the designation ASTM A-36 is the most widely used structural steel. It is a mild-carbon steel with a yield point of 36 ksi (36,000 pounds per square inch) (250 MPa).

ASTM A-242—Steel with a designation of ASTM A-242 represents a high-strength, low-alloy steel with a yield point of 50 ksi (345 MPa). It has limited amounts of carbon and manganese, which increase its welding capabilities. It is also used where self-weathering is required because of its high resistance to corrosion. A-242 steel has four times the atmospheric corrosion resistance of carbon steel without copper.

ASTM A-441—ASTM A-441 is high-strength, weldable, low-alloy structural manganese vanadium steel. Its yield point is 42 ksi (290 Mpa).

ASTM A-529—A-529 steel is high-strength carbon steel with a minimum yield point of 42 ksi (290 Mpa). With the addition of 0.02% copper to the content, A-529 steel achieves twice the atmospheric corrosion resistance of A-36 steel.

ASTM A-572—A-572 and A572M is high-strength, low-allow Columbium-Vanadium structural steel.

ASTM A-588—A-588 is high-strength, low-alloy steel with a minimum yield point of 50 ksi (345 Mpa). It is often referred to as weathering steel because it is left unpainted. It has greater durability, lighter weight and is four times greater in weather resistance than A-36 steel. Self-weathering steel is used for exposed steel structures such as access bridges or other exposed areas where maintenance would be costly.

Common Steel Shapes

Steel on the structural drawings can be divided into structural shapes, plates, bars, and cables. Each has its own distinctive use, shape, and specifications that must be understood by the drafter to complete the drawings. The standard designations of structural shapes of steel adopted by the American Institute of Steel Construction include *W, S, M, C, MC, L, WT,* and *MT*. These common shapes are illustrated in Figure 9-3. Figure 9-4 shows several of these common shapes as well as common methods of representing bolting and welding information.

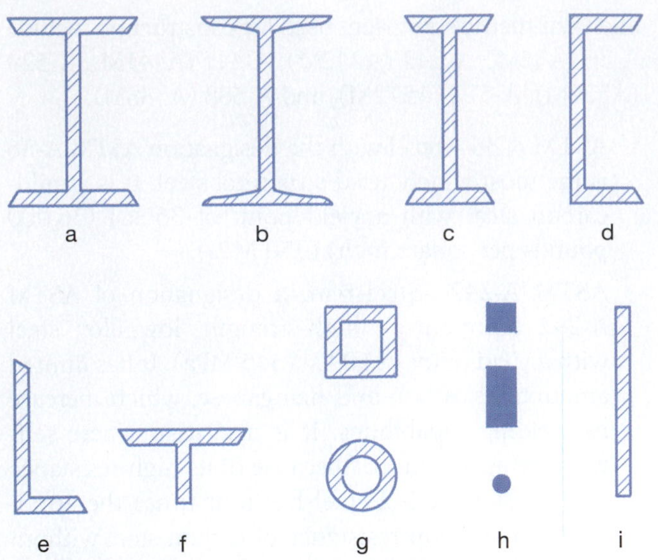

FIGURE 9-3 Common steel shapes in construction: (a) wide flange; (b) I-beam or American standard beam; (c) M shape; (d) channel; (e) angle; (f) tee; (g) tubes; (h) bars; (i) plates.

Wide Flange

The *W* or **wide-flange** steel seen in Figure 9-5 is used for columns and beams. It has the cross-sectional shape of the letter **I** and is composed of two horizontal surfaces called **flanges**, and a vertical surface called the **web.** W-shaped members are relatively narrow when compared to their depth and are used for beams and girders. Shapes that are closer to square are generally used for columns. *W-shaped* steel comes in a minimum depth of 4" with a weight of 7.5 pounds per linear foot (plf) (100 mm–0.01 kg/mm), and a maximum depth of 48" and a weight of 848 plf (1200 mm–1.26 kg/mm). An *S-shaped* beam, sometimes called an *I-beam* or **American standard beam**, has a narrower flange than a W-shape and has a slope of 1/6 for the inner face of the flange. Sizes range from a minimum of 3" deep with a weight of 5.7 plf (75 mm–0.008 kg/mm) to 24" with a weight of 120 plf (610 mm–0.18 kg/mm). The

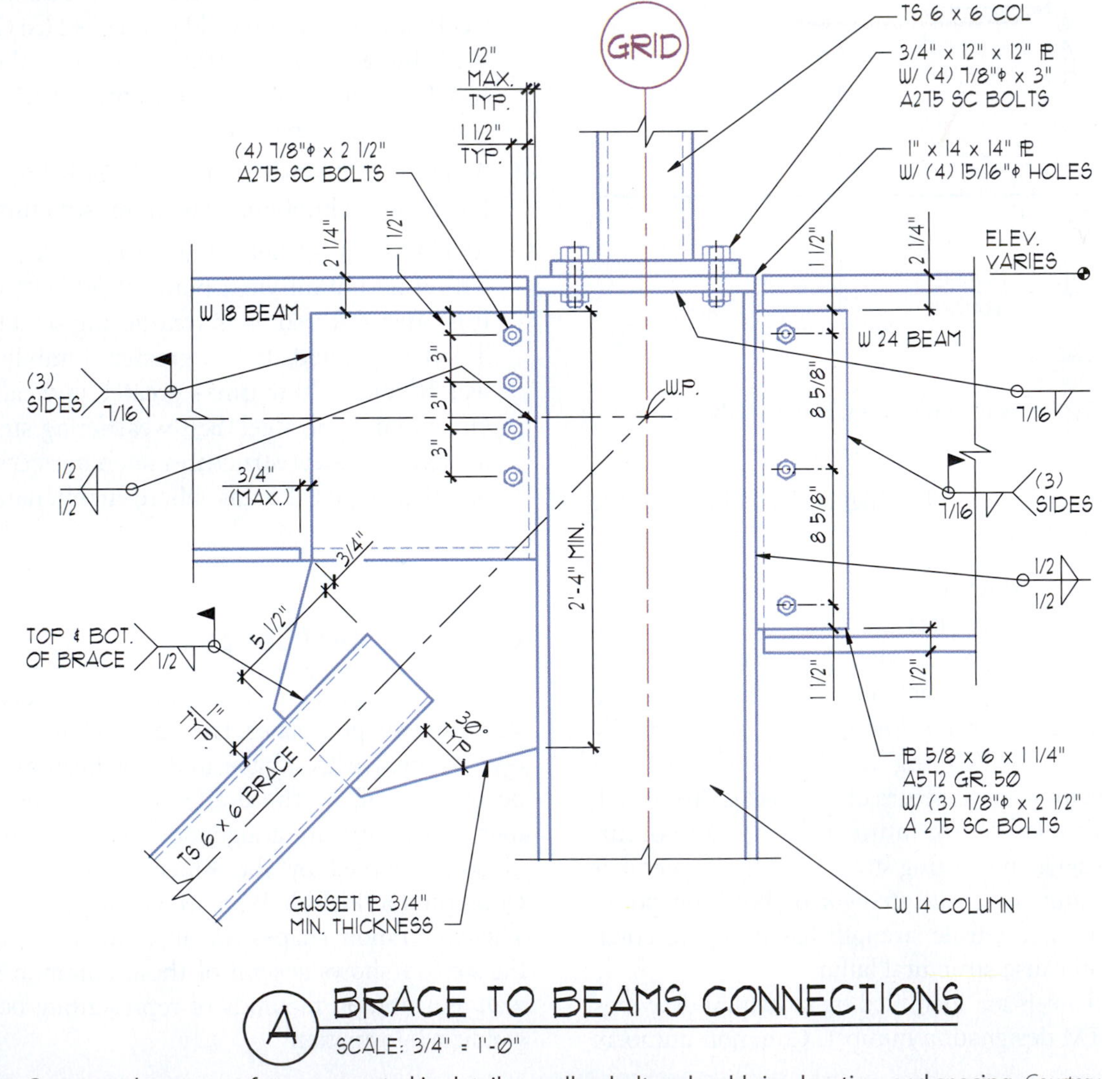

FIGURE 9-4 Common shapes are often represented in detail as well as bolt and weld size, location, and spacing. *Courtesy Jim Ellsworth.*

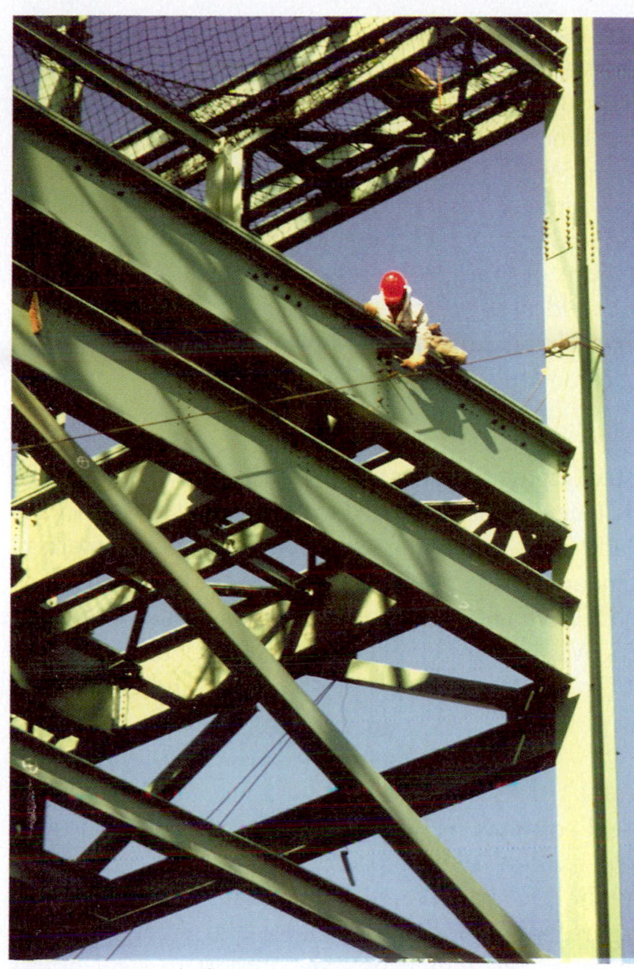

FIGURE 9-5 Wide-flange steel used for beams and columns.

M designation refers to miscellaneous shapes that can't be classified as a **W** or **S** shape.

A W-beam or S-beam can be represented on a framing plan as illustrated in Figure 9-6. The specification for the beam can be placed parallel to the beam or represented with a reference symbol and referred to a beam schedule using the same procedures used for sawn and laminated beams. If a schedule is used, separate schedules should be used for timber and steel beams because they will come from different suppliers. When steel products thicker than 1/8" (3 mm) are represented in end view in a section or detail, they are crosshatched with pairs of parallel lines at a 45° angle, as shown in Figure 9-7.

Wide-flange members are specified by numbers representing the approximate depth and the weight of the beam. For example, the W 18 × 46 beam in Figure 9-7 has an actual depth of 18.06" (460 mm). Other characteristics can be seen in Figure 9-8, a sample of a steel table published by the American Institute of Steel Construction. CAD drafters will have to refer to tables like these to gain the information needed to draw accurate details. The actual flange and web thickness are

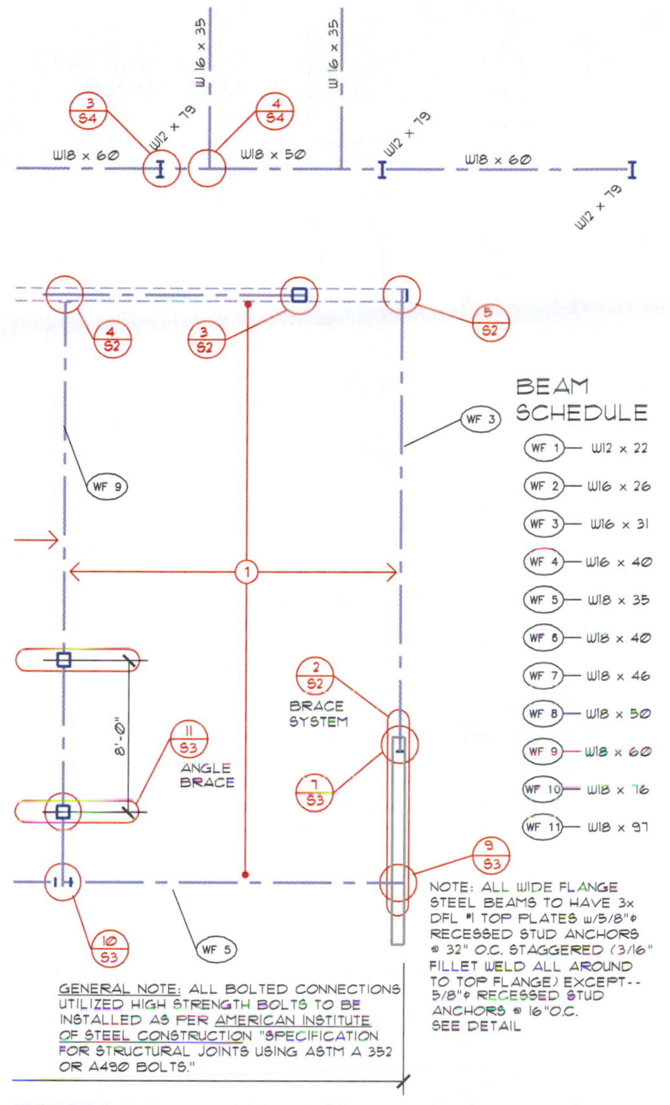

FIGURE 9-6 Representing steel beam and columns in plan view. Steel members should be represented in a schedule to keep the framing plan uncluttered.

usually slightly exaggerated so that the beam can be easily seen.

Channels

A **channel** is shaped like half of an American standard beam (Figure 9-3d). Channels are represented by the letters **C** for a standard channel and **MC** for miscellaneous shapes. Standard shapes range in size from 3" and 4.1 plf (76 mm–0.006 kg/mm) to 18" deep and 58 plf (460 mm–0.09 kg/mm). The inner face of each flange has a slope matching that of an S-shaped beam. Channels are represented on the plan views similar to a W-beam. The text is placed parallel to the channel and numbers are used to represent the approximate depth and pounds per linear foot of beam. A typical specification would resemble:

C 12 × 20.7

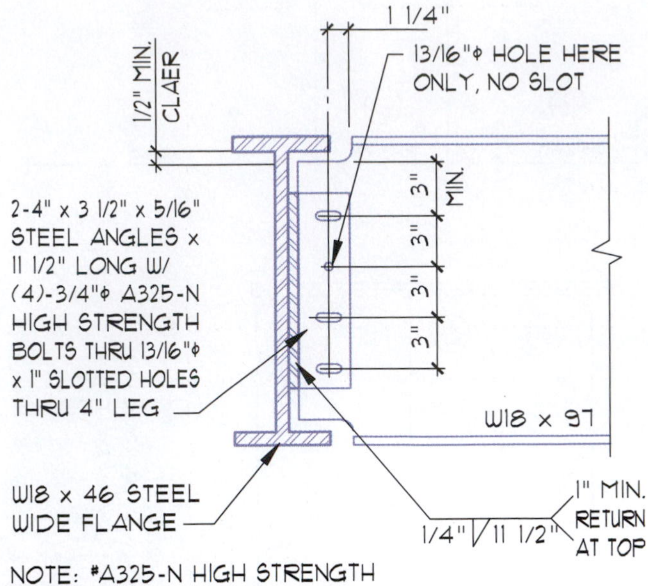

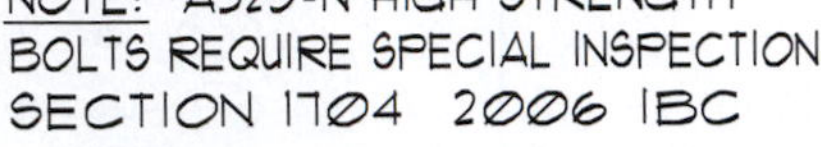

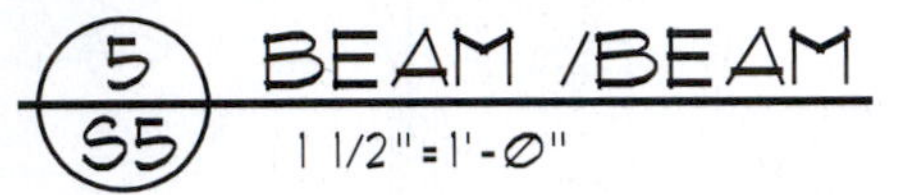

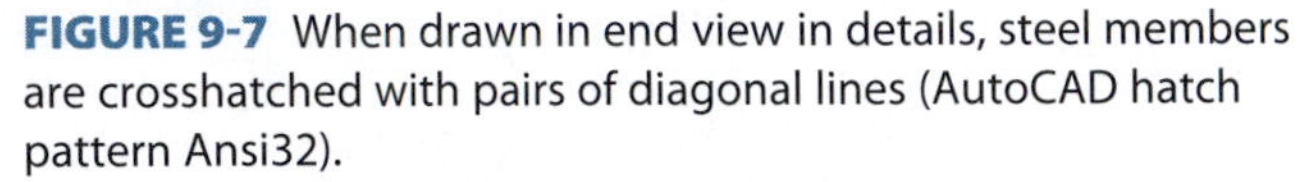
FIGURE 9-7 When drawn in end view in details, steel members are crosshatched with pairs of diagonal lines (AutoCAD hatch pattern Ansi32).

Angle

An **angle** is a piece of steel that has a 90° bend in it (Figure 9-3e). Angles may have legs of equal or unequal length. Equal length angles range from 1 × 1 × 1/8"/0.80 plf to a maximum of 8 × 8 × 1 1/8"/56.9 plf (25 × 25 × 3/0.009 kg/mm to 200 × 200 × 29/0.08 kg/ mm). Angles are represented by the letter **L** followed by numbers representing the height of each leg followed by the thickness. The weight is usually not specified, but a length may be specified after the thickness is listed. A typical specification would resemble:

L6 × 6 × 3/8 × 10'-0"

Angles with unequal leg lengths range in size from 1 3/4 × 1 1/4/1.23 to 8 × 6 × 1/44.2 plf (45 × 45 × 32/0.002 kg/ mm to 200 × 160 × 25/0.07 kg/mm).

When a wide-flange beam is used to support another wide-flange beam, an **L** is typically welded to the supporting beam and bolted to the beam being supported. Figure 9-9 shows angles used to support wide-flange beams. Figure 9-7 shows an angle used in a beam-to-beam connection. A CAD drafter would be expected to draw similar details using bolting and welding procedures introduced in Chapter 6. If a wood member is to be joined to a steel beam, it is typically done by welding bolts with no heads—called *studs*—to the flange or web of the steel member. The studs are then used to bolt a **wood plate** or ledger to the beam. *Plates* are wood members bolted to the top of a web to support wood joists and facilitate nailing. **Ledgers** are wood members that are bolted to the flange of a steel beam. Wood joists can then be hung from the ledger with a metal hanger. The engineer will determine the size and spacing of the studs based on the type and amount of loads to be resisted, as well as the size of the weld used to connect the stud to the steel beam. Figure 9-10 shows an example of a top plate bolted to a beam with a stud.

Tees

A **tee** (Figure 9-3f) has a cross-sectional area shaped like the letter **T**. Often referred to as structural tees, they are made from cutting the web of wide-flange or American standard beams in half. The size of the tee is based on the beam that it is cut from and is represented by the letters **WT**, **ST**, or **MT** followed by the height and weight of the tee. The specification for a tee would resemble:

WT 12 × 38

Steel Tubes and Columns

Steel columns or pipes are hollow circular shapes of steel used to support loads (Figure 9-3g). **Steel tubes** are hollow square or rectangular shapes of steel. Figure 9-11 shows both tubes and steel base and gusset plates, which provide support to steel beams and bracing rods. Square tubes are available in sizes ranging from 2 × 2 to 10 × 10 (50 × 50 to 250 × 250 mm) with the wall thickness ranging from 3/16" to 5/8" (5 to 16 mm) thick. Rectangular tubes range from 2 × 3 to 8 × 12 (50 × 75 to 200 × 300 mm) with a wall thickness of 3/16" to 1/2" (5 to 13 mm). Circular steel columns range from 1/2" to 12" (13 to 300 mm) with wall thickness ranging from 0.216" to 0.50" (5 to 13 mm) thick.

Tubes and columns are shown in plan view just as a wood post would be represented. When seen in a section or detail, the interior surface of the tube is represented by a hidden line as shown in Figure 9-12. The actual wall thickness is usually slightly exaggerated so it can be easily seen. Square and rectangular tubes are specified on drawings by the letters **TS** followed by the overall size and thickness. A tube specification would resemble:

TS 4" × 4" × 3/8" or TS 4" × 1/4"

Round columns are specified by the nominal diameter, and a specification of standard, extra strong, or double extra strong. A pipe would be specified as:

TS STD. 4" DIA. PIPE

Chapter 22 will include a further discussion of steel columns and tubes.

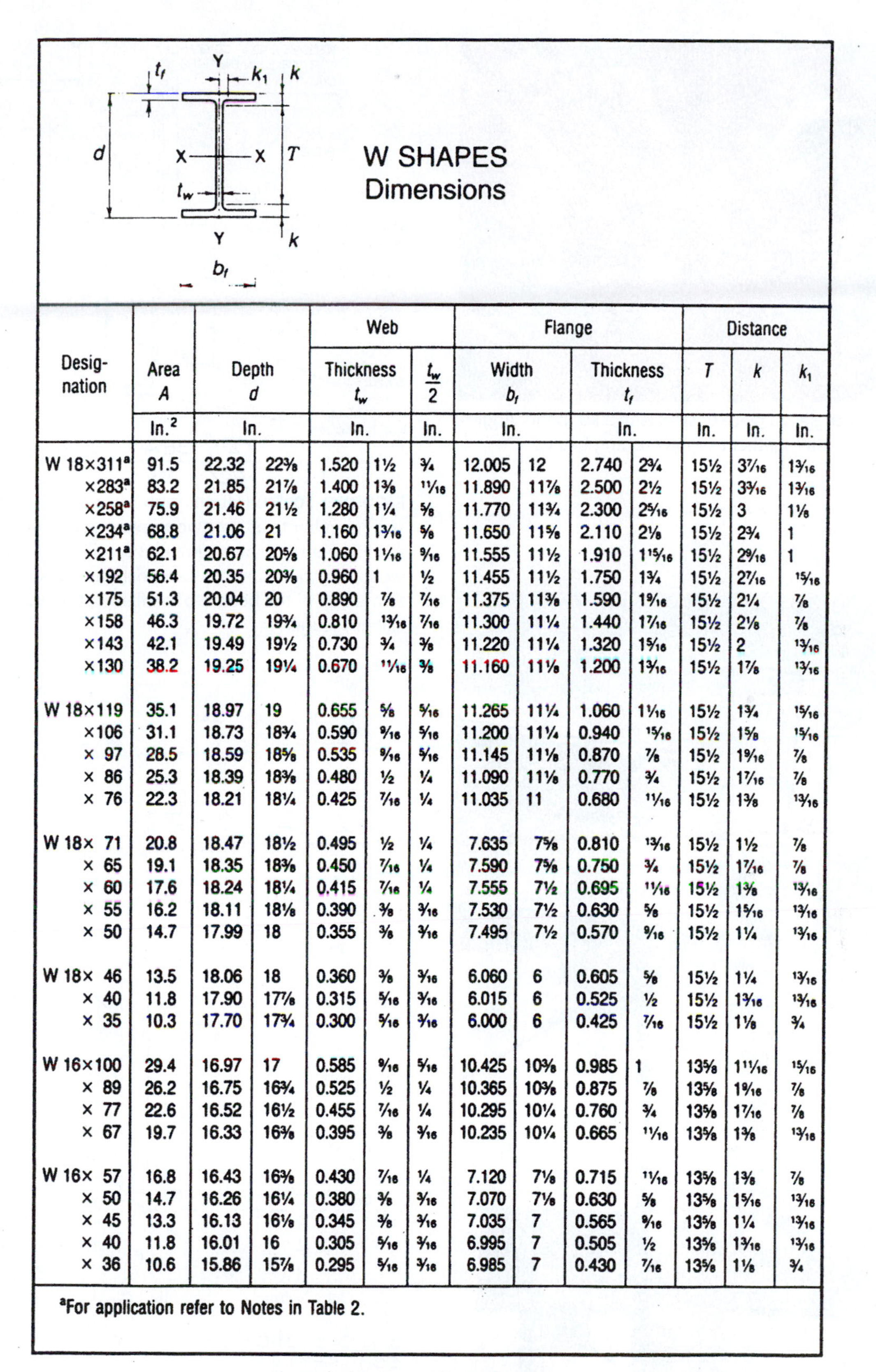

W SHAPES
Dimensions

Designation	Area A	Depth d		Web Thickness t_w		Web $\frac{t_w}{2}$	Flange Width b_f		Flange Thickness t_f		Distance T	Distance k	Distance k_1
	In.^{2}	In.		In.		In.	In.		In.		In.	In.	In.
W 18×311[a]	91.5	22.32	22 3/8	1.520	1 1/2	3/4	12.005	12	2.740	2 3/4	15 1/2	3 7/16	1 3/16
×283[a]	83.2	21.85	21 7/8	1.400	1 3/8	11/16	11.890	11 7/8	2.500	2 1/2	15 1/2	3 3/16	1 3/16
×258[a]	75.9	21.46	21 1/2	1.280	1 1/4	5/8	11.770	11 3/4	2.300	2 5/16	15 1/2	3	1 1/8
×234[a]	68.8	21.06	21	1.160	1 3/16	5/8	11.650	11 5/8	2.110	2 1/8	15 1/2	2 3/4	1
×211[a]	62.1	20.67	20 5/8	1.060	1 1/16	9/16	11.555	11 1/2	1.910	1 15/16	15 1/2	2 9/16	1
×192	56.4	20.35	20 3/8	0.960	1	1/2	11.455	11 1/2	1.750	1 3/4	15 1/2	2 7/16	15/16
×175	51.3	20.04	20	0.890	7/8	7/16	11.375	11 3/8	1.590	1 9/16	15 1/2	2 1/4	7/8
×158	46.3	19.72	19 3/4	0.810	13/16	7/16	11.300	11 1/4	1.440	1 7/16	15 1/2	2 1/8	7/8
×143	42.1	19.49	19 1/2	0.730	3/4	3/8	11.220	11 1/4	1.320	1 5/16	15 1/2	2	13/16
×130	38.2	19.25	19 1/4	0.670	11/16	3/8	11.160	11 1/8	1.200	1 3/16	15 1/2	1 7/8	13/16
W 18×119	35.1	18.97	19	0.655	5/8	5/16	11.265	11 1/4	1.060	1 1/16	15 1/2	1 3/4	15/16
×106	31.1	18.73	18 3/4	0.590	9/16	5/16	11.200	11 1/4	0.940	15/16	15 1/2	1 5/8	15/16
× 97	28.5	18.59	18 5/8	0.535	9/16	5/16	11.145	11 1/8	0.870	7/8	15 1/2	1 9/16	7/8
× 86	25.3	18.39	18 3/8	0.480	1/2	1/4	11.090	11 1/8	0.770	3/4	15 1/2	1 7/16	7/8
× 76	22.3	18.21	18 1/4	0.425	7/16	1/4	11.035	11	0.680	11/16	15 1/2	1 3/8	13/16
W 18× 71	20.8	18.47	18 1/2	0.495	1/2	1/4	7.635	7 5/8	0.810	13/16	15 1/2	1 1/2	7/8
× 65	19.1	18.35	18 3/8	0.450	7/16	1/4	7.590	7 5/8	0.750	3/4	15 1/2	1 7/16	7/8
× 60	17.6	18.24	18 1/4	0.415	7/16	1/4	7.555	7 1/2	0.695	11/16	15 1/2	1 3/8	13/16
× 55	16.2	18.11	18 1/8	0.390	3/8	3/16	7.530	7 1/2	0.630	5/8	15 1/2	1 5/16	13/16
× 50	14.7	17.99	18	0.355	3/8	3/16	7.495	7 1/2	0.570	9/16	15 1/2	1 1/4	13/16
W 18× 46	13.5	18.06	18	0.360	3/8	3/16	6.060	6	0.605	5/8	15 1/2	1 1/4	13/16
× 40	11.8	17.90	17 7/8	0.315	5/16	3/16	6.015	6	0.525	1/2	15 1/2	1 3/16	13/16
× 35	10.3	17.70	17 3/4	0.300	5/16	3/16	6.000	6	0.425	7/16	15 1/2	1 1/8	3/4
W 16×100	29.4	16.97	17	0.585	9/16	5/16	10.425	10 3/8	0.985	1	13 5/8	1 11/16	15/16
× 89	26.2	16.75	16 3/4	0.525	1/2	1/4	10.365	10 3/8	0.875	7/8	13 5/8	1 9/16	7/8
× 77	22.6	16.52	16 1/2	0.455	7/16	1/4	10.295	10 1/4	0.760	3/4	13 5/8	1 7/16	7/8
× 67	19.7	16.33	16 3/8	0.395	3/8	3/16	10.235	10 1/4	0.665	11/16	13 5/8	1 3/8	13/16
W 16× 57	16.8	16.43	16 3/8	0.430	7/16	1/4	7.120	7 1/8	0.715	11/16	13 5/8	1 3/8	7/8
× 50	14.7	16.26	16 1/4	0.380	3/8	3/16	7.070	7 1/8	0.630	5/8	13 5/8	1 5/16	13/16
× 45	13.3	16.13	16 1/8	0.345	3/8	3/16	7.035	7	0.565	9/16	13 5/8	1 1/4	13/16
× 40	11.8	16.01	16	0.305	5/16	3/16	6.995	7	0.505	1/2	13 5/8	1 3/16	13/16
× 36	10.6	15.86	15 7/8	0.295	5/16	3/16	6.985	7	0.430	7/16	13 5/8	1 1/8	3/4

[a] For application refer to Notes in Table 2.

AMERICAN INSTITUTE OF STEEL CONSTRUCTION

FIGURE 9-8 CAD technicians often need to research the sizes of steel shapes to complete details. *Courtesy American Institute of Steel Construction.*

FIGURE 9-9 An angle is a piece of steel with a 90° bend. Angles are often used to bolt together two intersecting beams.

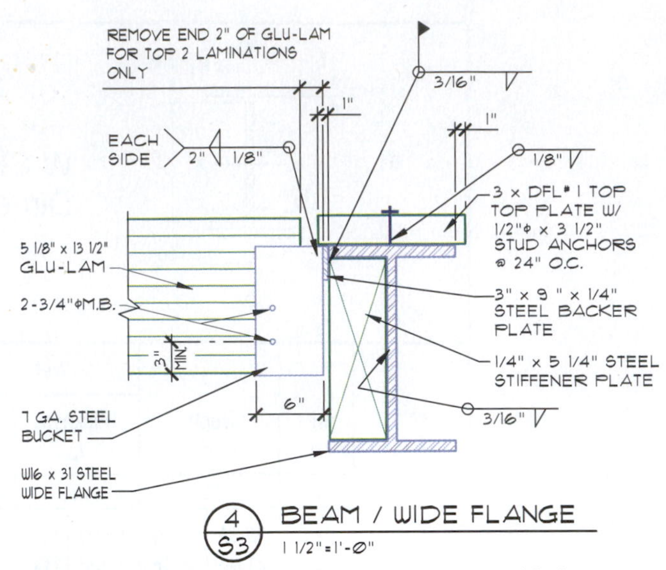

FIGURE 9-10 Studs (bolts with no heads) are welded to a steel member to allow a wood plate to be bolted to the top of a steel beam. The plate provides a nailing surface so that other wood members can be attached. If the wood member is bolted to the side of the beam (to the flange) the wood member is called a ledger.

FIGURE 9-11 Steel tubes, columns, rods, and plates are used to transfer loads into the foundation. *Courtesy Zachary Jefferis.*

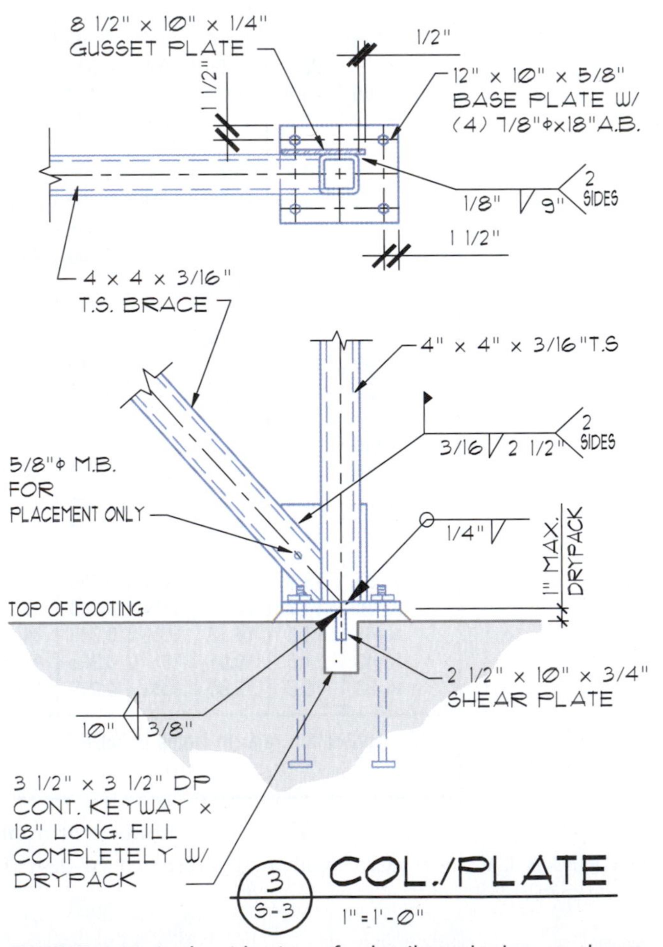

FIGURE 9-12 In the side view of a detail, steel tubes are drawn similar to wood and timber posts. The interior surface of a steel column is represented by a hidden line in order to distinguish a steel tube from a wood column.

Bars

Steel bars (Figure 9-3h) are solid members available in circular, square, and flat shapes that are used to provide reinforcing to other members. Bars are often used for lateral bracing, or as hangers to support horizontal members. Figure 9-11 shows round bars used to stabilize a column. Steel is classed as a bar if it is less than 8" (200 mm) wide. Bars are specified in 1/8" (3 mm) increments for thickness and 1/4" (6 mm) increments for widths (Figure 9-3h). Depending on the shape, bars will be specified on drawings in one of three possible methods:

Square bar	1" SQ. BAR
Flat	2" × 3/8" BAR
Round	1" DIA. BAR

See Chapters 10 and 11 for information about bars used for masonry and concrete reinforcing. Threaded barsused for fastening devices were discussed in Chapter 6.

Plates

A plate (Figure 9-3i) is a flat piece of steel 8" (200 mm) or wider ranging in thickness from 1/2" to 2" (13 to 50 mm) thick. Common uses of plates include top, base, gusset, end, stiffener plates, or as part of a fabricated hanger.

- A **top plate** provides a bearing surface for a steel beam resting on a column (see Figure 9-11). The plate is welded to the top of the column, and the beam to be supported is bolted and welded to the plate.
- A **base plate** provides support for a column resting on another steel member or for a column resting on concrete. Figure 9-12 shows a detail of a column welded to a base plate, with the plate attached to the concrete with anchor bolts.
- A **gusset plate** is a plate added to an intersection of structural members to provide support and stiffness. Figures 9-11 and 9-12 show a *gusset plate* used to provide a welding surface for the diagonal steel cable or tube.
- An **end plate** is used to provide an attachment surface at the end of a W or similar shaped beam.
- The flanges of a W-beam are often reinforced by a **stiffener plate** to provide support for members that are to be hung from the beam.
- Fabricated angles for beam hangers similar to Figure 9-13 can be constructed from flat steel plates.

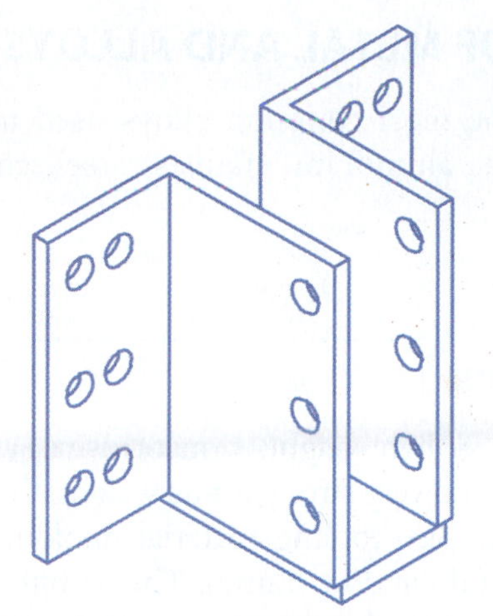

FIGURE 9-13 Steel plates used to fabricate a beam hanger.

A plate is referenced by its size followed by its thickness. A typical callout would resemble:

12 × 12 × 3/8 PL

Cable

Cables can be used in place of solid bars for lateral bracing because of their high resistance to forces in tension. A cable can be used to form an *X* between two parallel members. The cable is tightened by a turnbuckle placed at a convenient place in the cable span. Cable diameters range from 1/2" to 3 5/8" (13 to 90 mm). When specified on plans, cable smaller than 1" (25 mm) diameter is specified by giving the diameter and safe working stress limits of the cable. Larger cable sizes are specified by providing the diameter, the number of strands, and the number of wires per strand. Figure 9-14 shows an example of specifying cables on a framing plan.

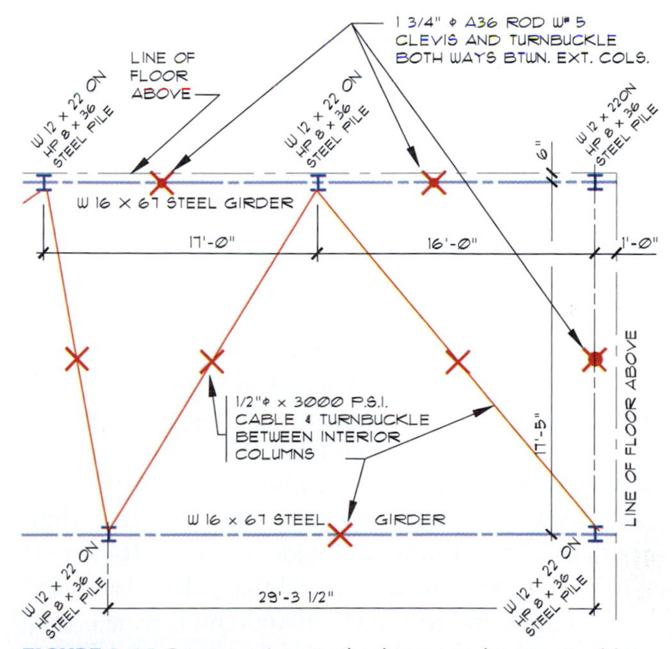

FIGURE 9-14 Representing steel columns and support cables in plan view.

TYPES OF METAL AND ALLOYS

Some of the most common alloys used in construction include aluminum, stainless steel, copper, lead, and tin.

Aluminum

Aluminum is lightweight, noncorrosive, a good heat conductor, and very strong for its weight. Common uses of aluminum are roofing material, moldings, window mullions, and window frames. The architectural drawings often include details similar to Figure 9-15 to show how the aluminum window frame will intersect with the structural shell. Adding one or more elements to aluminum improves both the hardness and strength of aluminum. The Aluminum Association classifies aluminum alloys using a series code number. If a specific class of aluminum is to be used it would be referenced in the project manual. Classes include:

2000 series	adds copper
3000 series	adds manganese
4000 series	adds silicone
5000 series	adds magnesium

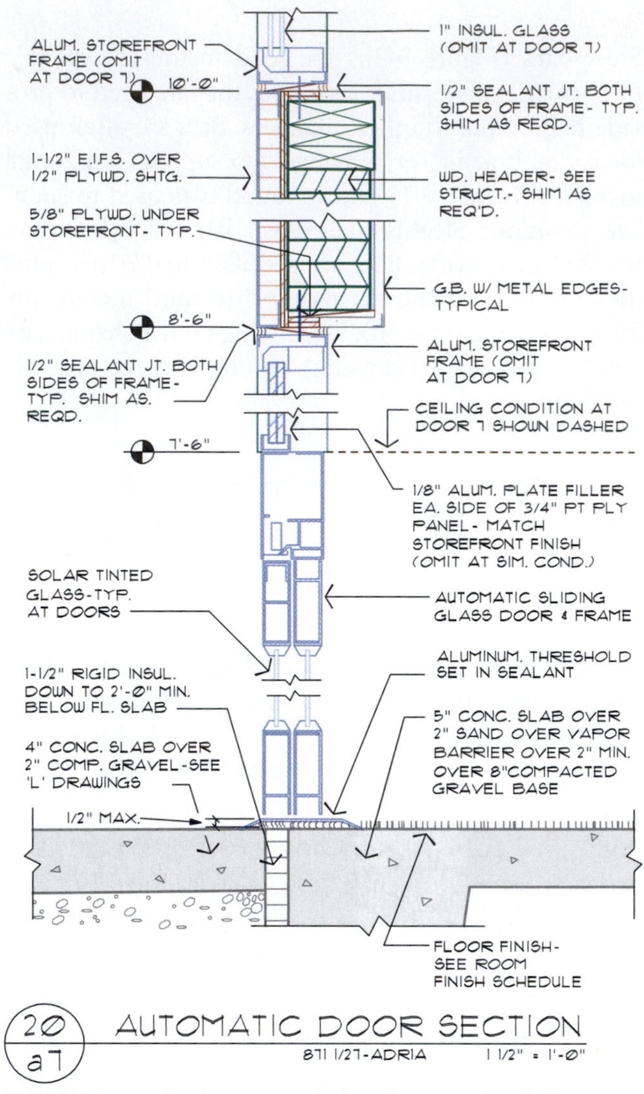

FIGURE 9-15 An aluminum window frame in detail. *Courtesy Architects Barrentine, Bates & Lee, AIA.*

Aluminum alloys are used for such products as roof and siding panels, trim, nails, structural members such as truss webs or space frames, railings, electrical wiring, and door and window frames.

Stainless Steel

Stainless steel is used where appearance and maintenance are a priority for both interior and exterior uses. It has a nonstaining finish and is used for kitchen equipment for food establishments, elevator and doorway trim, or other metal products subject to abuse or in need of a high polish. Stainless steel is usually specified on drawings by a gauge number, grade, type, and finish. Material to be made from stainless steel that is 3/16" (5 mm) or thicker or widths of greater than 10" (250 mm), is specified as a plate. Material that is thinner than 3/16" (5 mm) and wider than 24" (600 mm) is referred to as a sheet. Material that is less than 3/16" (5 mm) thick and less than 24" (600 mm) wide is classified as a strip.

Grades

Two grades of stainless steel can be included in the written specifications of a drawing. These include austenitic and ferritic grades.

- Austenitic grades of stainless steel have the highest corrosion resistance and contain both chromium and nickel. This grade is nonmagnetic.
- Ferritic grades are always magnetic. Both ferritic and austenitic grades can be hardened by heat treatment.

Types

Although there are more than forty types of stainless steel, only five types are used in the construction industry. Each type is established by the American Iron and Steel Institute and given a designation number. The five common types of stainless steel used in construction and their uses include:

Type of Stainless Steel	Uses
Type 301	Used for structural applications such as door frames.
Type 302	Used for architectural applications of stainless steel such as exterior paneling, false columns, soffits, fascias, and gutters.
Type 304	Has a lower carbon content than type 302 stainless steel. Used for many of the same applications.
Type 316	Molybdenum is added to provide a high resistance to atmospheric corrosion.
Type 430	Used for interior applications. Resistance to corrosion is less than type 302.

Finish

In addition to the grade and type of stainless steel, a finish is often specified in the project manual. Common finishes include:

Finish of Stainless Steel	Characteristics
No. 2D Special	A frosty matte finish
No. 4	A dense, bright, highly reflective finish
No. 6	A soft satin finish, moderately reflective
No. 7	Highly reflective finish

Copper

Copper is used throughout the construction industry for its ability to conduct electricity and its resistance to corrosion. Copper is used for electrical wiring, pipes for fresh water supply, sheet metal panels for roofing, shingles, gutters, and flashing.

Lead

Although not widely used throughout the construction of a structure, lead is used for flashing as well as specialty items such as shower pans or other areas that are required to be watertight. Lead is also used in hospitals and labs for protection from X rays and for plating to provide a resistance to acid.

Tin

Tin is very resistant to corrosion. The major use of tin is as a coating for sheet metal used for roofing or siding panels.

LIGHTWEIGHT STEEL FRAMING

Lightweight steel products are used throughout the framing of many structures that require increased fire protection over Type V construction. Common uses of steel include studs, joists, trusses, decking, and lath.

Steel Studs

Steel studs are often used to help meet the requirements of Types I, II, and III construction. Depending on the covering material, steel studs can achieve fire ratings of between one and three hours (see Chapter 12). Steel studs offer lightweight, noncombustible, corrosion-resistant framing for interior partitions and interior and exterior load-bearing walls up to five stories in height. Steel products also offer greater dimensional stability and a level surface without problems such as attack by termites, rotting, shrinkage, splitting, or warping associated with wood construction.

Studs are designed for rapid assembly and are predrilled for electrical and plumbing conduits (see Figure 9-16). The standard 24" (600 mm) spacing reduces the number of studs required by about one-third when compared with traditional wood framing. Stud width ranges from 3 5/8" to 10" (90 to 250 mm) but can be manufactured in any width. Stock lengths include 8', 9', 10', 12', and 16' (2400, 2700, 3100, 3700, and 4900 mm), with custom lengths typically available up to 40' (12,200 mm). Studs are made of steel, ranging from 12 to 25 gauge, with a yield point of 40 ksi. The engineer selects the gauge to be used based on the loads to be supported and the usage. The shape of the studs varies with the load to be supported. Figure 9-17 shows common shape and size variations of studs used for bearing and nonbearing walls.

Steel studs are mounted in a channel at the top and bottom of the wall. This channel serves a similar function to the plates and sill of wood-frame construction.

FIGURE 9-16 Steel studs are designed for rapid assembly and are predrilled for electrical and plumbing materials. *Courtesy Matthew Jefferis.*

The material used for channels typically has a yield point of 33 ksi. Horizontal bridging may be placed through the predrilled holes in the studs and then welded to the stud to serve the same function as blocking in a stud wall.

Stud Specifications

Section properties and steel specifications vary among manufacturers; therefore, the material to be specified on the plans will vary. Specific catalogs developed by the Metal Stud Manufacturers Association should be consulted to determine the structural properties of the desired stud. The architect or engineer determines the size of stud to be used, but the drafter often needs to consult vendor materials to completely specify the studs. As a minimum, the gauge and usage are specified on the framing plan, details, and sections. If a manufacturer is to be specified, the callout will include the size, style, gauge, and manufacturer of the stud. An example of a steel stud specification would read:

362SJ20 STEEL STUDS BY UNIMAST

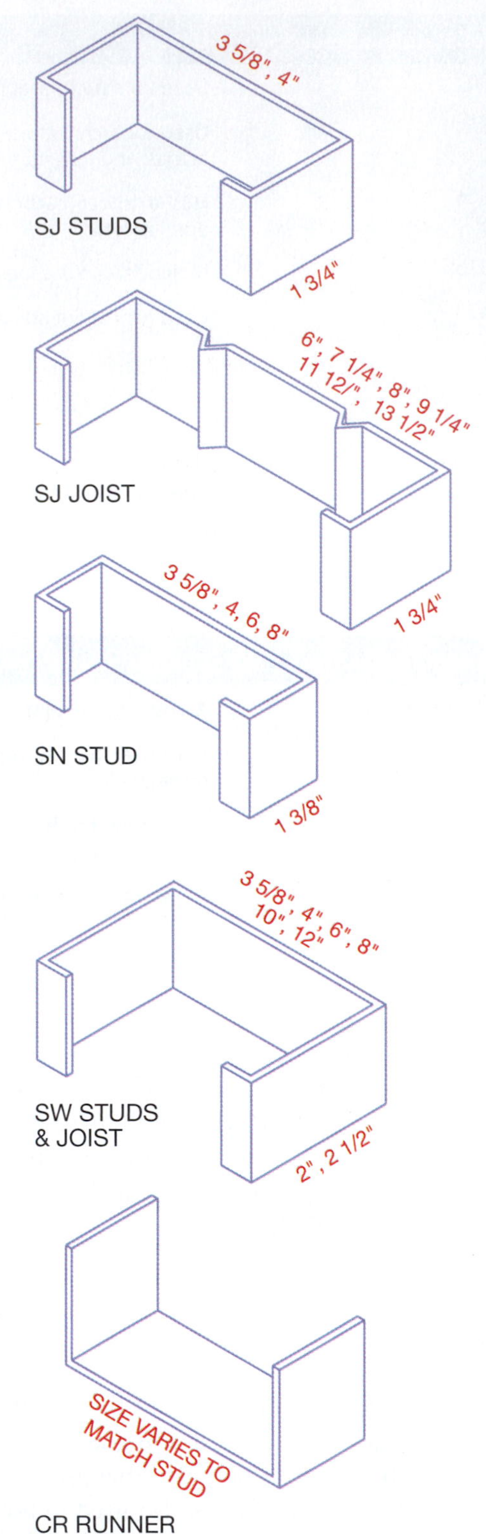

FIGURE 9-17 Common sizes and shapes of steel studs and channels.

Steel Joists

Steel joists offer the same benefits over their wood counterparts, just as steel studs do. Joists are available in 3 5/8", 4", 6", 7 1/4", 8", 9 1/4", 10", 11 1/2", 12", and 13 1/2" (90, 100, 150, 185, 200, 235, 250, 290, 305, and 340 mm) depths. The gauge and yield point for joists are similar to these values for studs. Many joists are manufactured so that a joist may be placed around another joist, or *nested*. Nested joists allow the strength of a joist to be greatly increased without increasing the size of the framing members. A nested joist is often used to support a bearing wall above the floor. If a post is to be supported by a metal joist, a steel stiffener plate can be placed between the flanges of the joist at the bearing point to prevent bending of the web. Joists are available in lengths up to approximately 40' (12 200 mm) depending on the manufacturer.

Open-Web Steel Joists

Open-web steel joists are manufactured in the same configurations as open-web wood trusses. In comparison to their wood counterparts, open-web steel joists offer the advantage of being able to support greater loads over longer spans with less dead load than solid members. Figure 9-18 shows an example of light steel trusses used to support floor loads. These joists are usually made of small angles, round bars, or tees, in depths ranging from 8" to 72" (200 to 1800 mm).

Open-web joists are referred to by their series number, which defines their use. The Steel Joist Institute (SJI) designations include K, LH, and DLH. All sizes listed below are based on SJI specifications for steel joists. K series joists are parallel-web members that range in depth from 8" to 30" (200 to 760 mm) in 2" (50 mm) increments, for spans up to approximately 60' (18 300 mm). LH (long span) and DLH (deep long span) series of joists are made from steel with a yield point of 50 ksi. Depths range from 18" to 48" (460 to 1200 mm) for LH series and from 52" to 72" (1320 to 1830 mm) for DLH series joists. Spans range up to 96' (29 300 mm) for LH series joists to up to 144' (43 900 mm) for DLH joists.

Steel joists are specified on the framing plans, structural details, and sections by providing the approximate depth, the series designation, the chord size, and the spacing. An open-web steel joist specification would resemble:

14K4 OPEN-WEB STEEL JOIST AT 32" O.C.

FIGURE 9-18 Steel open-web beams and trusses are often used to support floor and ceiling loads. Steel decking is typically placed over the trusses to provide support for concrete floors. *Courtesy Megan Jefferis.*

Just as with a wood truss, span tables are available to determine safe working loads of steel joists. The engineer determines the truss to be used, but the CAD drafter is often required to find information to complete the truss specification.

Steel Decking

Corrugated steel decking is supported by steel channels or open-web joists used for most lightweight floor systems (see Figure 9-18). Decking is produced in a variety of shapes, depths, and gauges. Common shapes are shown in Figure 9-19. Common rib depths include 1 1/2", 3", and 4 1/2" (40, 75, and 115 mm) deep with lengths available up to 30' (9000 mm). The engineer determines the required rib depth based on the load to be supported and the lateral load to be resisted. Steel decking is attached to the framing system with screws or welds. Decking is covered with a minimum of 1 1/2" (40 mm) of concrete above the ribs to provide the finished floor. Decking is specified in the project manual and in sections and details by giving the manufacturer, type of decking, depth, and gauge. A typical specification would resemble:

METAL DECKING TO BE VERCO, TYPE 'N',
3"-20GA. GALV. DECKING

Steel Lath

Galvanized metal lath in 27" × 96" (690 × 2440 mm) sheets are typically used for backing for plaster, gypsum lath, fiberboard, or similar plaster bases used to cover steel products. Other metal products include drywall clips, drywall corner bead, wire, and screws.

METAL BUILDINGS

Prefabricated or rigid-frame buildings are common in many areas of the country because they provide fast erection time compared with other types of construction.

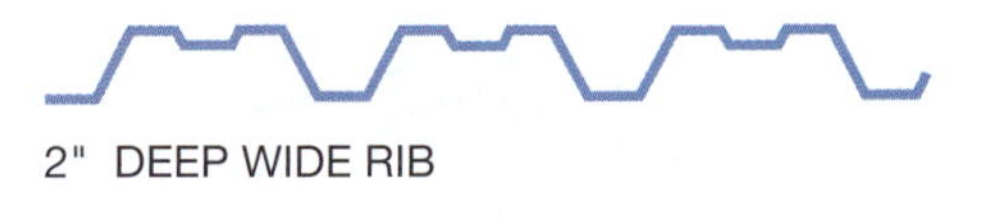

3" DEEP WIDE RIB

FIGURE 9-19 Common corrugated steel decking shapes.

The frame, roof, and wall components are standardized and sold as modular units with given spans, wall heights, and lengths, to reduce cost and eliminate wasted materials.

Figure 9-20 shows a steel-framed structure being erected.

Typical Components

The structural system is made up of a frame that supports the walls, roof, and all externally applied loads. There are several different types of structural systems used. The two most common shapes are shown in Figure 9-21. Frame members are normally made from plate and bar stock with a yield point of 50 ksi (345 Mpa). The size of members is determined by engineers working for the fabricator. Tapering of members allows the minimum amount of material to be used, while still maintaining the required area to resist the loads to be supported. End plates are welded or bolted to each end of the vertical members. The plate at the base is predrilled to allow for bolting the steel to the concrete support (see Figure 9-22). The plate at the top of the frame is used to bolt the vertical to the inclined member. The vertical wall member is bolted to what will become the inclined roof member to form one rigid member. This frame member is similar to the three-hinged laminated arch discussed in Chapter 8.

The wall system is made of horizontal girts attached to the vertical frame. Usually, the girts are a channel or a Z that is welded to the frame. Girts are typically welded to the outer flange of the frame, but they can also be welded between the frames if the wall thickness must be kept to a minimum. Girts are shown in Figure 9-23.

FIGURE 9-20 A rigid-frame structure consists of the frame, girts, and metal siding. *Courtesy David Jefferis.*

FIGURE 9-22 Predrilled end plates are welded to the top and bottom of each portion of the frame. *Courtesy Michael Jefferis.*

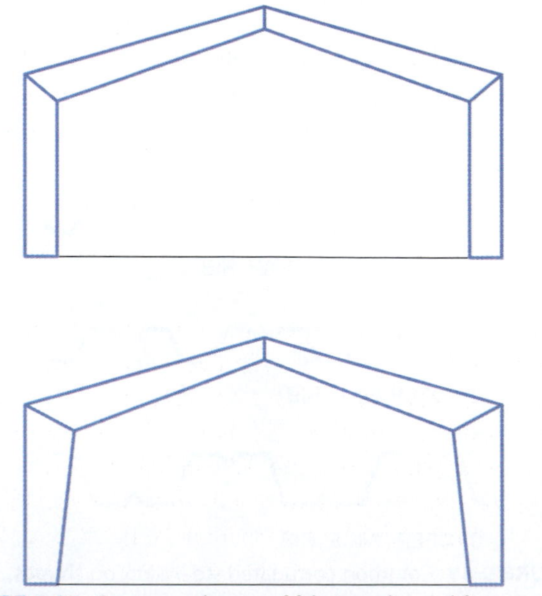

FIGURE 9-21 Common shapes of fabricated metal frames.

FIGURE 9-23 Steel girts are installed between the frame to support metal siding. *Courtesy Gisela Smith.*

Metal siding is screwed to the girts to complete the wall. The roof is made with steel purlins spaced at 24", 32", 48", or 60" (600, 800, 1200, 1500 mm) o.c. Purlins provide support for the roofing material. The spacing of the purlins depends on the sheet metal used to complete the roof. Both the roof and wall frames are reinforced with steel cable X-bracing to resist lateral loads.

Representation on Drawings

The engineering team working with the fabricator will complete the structural drawing required to build the structure. This includes drawings showing exact locations of all framing members and connections similar to those shown in Figure 9-24. Figure 9-25 shows an example of a section for a steel structure.

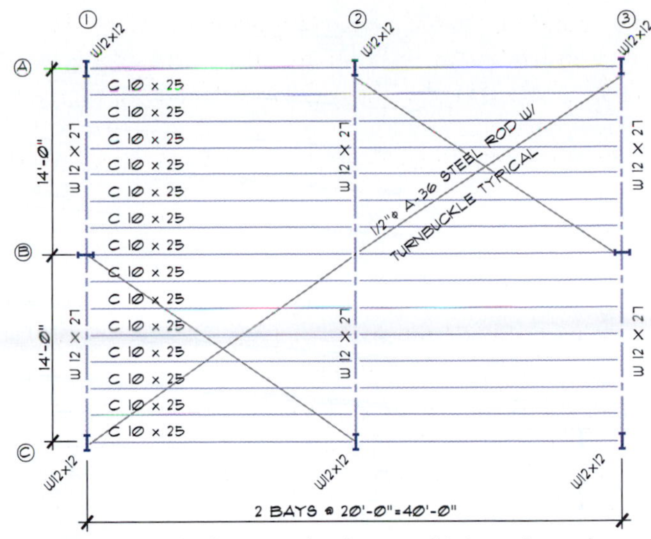

FIGURE 9-24 The framing plan for a prefabricated metal building.

MULTILEVEL STEEL FRAMES

Because of their ability to span large distances, steel is used to frame large sports and assembly arenas, exhibition centers, and multilevel structures. Multilevel structures are built in tiers consisting of vertical columns, horizontal girders, and intermediate support beams.

Multilevel structures framed with steel are similar to the elements of post-and-beam timber construction.

Framing Plans

As with other methods of construction, plans are required to show the location and size of each column, girder, beam, and joist. Columns are usually W-shaped but M and S shapes are also used. Steel columns are shown on the framing plan by representing the proper shape, the location from center to center, and a specification for the size and strength to be used. Steel columns are typically placed in two-story sections. Although smaller columns can be used on upper levels because of decreased loads to be supported, column size is typically kept uniform to make splicing easier. Splices are usually kept about 24" (600 mm) above a floor level so that the connection can be made without interfering with the girder-to-column connection. The load to be supported and the method used for lateral bracing will affect how the columns are attached. With W shapes, a **steel plate** is often added to the outer edge of each column flange to provide stiffness at the connection. On rectangular shapes, the columns can be directly attached with no side plate. Tube steel (TS) can be used for columns for heights below three stories depending on the loads to be supported. Steel studs are used to frame between vertical supports but are not load-bearing when used in conjunction with steel columns.

Girders are also normally formed from W shapes because of the ease of making connections. Intermediate beams may be W, S, M, or C shapes. Depending on the loads to be supported, girders can be built up to increase the bearing capacity. A built-up girder can be composed of standard shapes or fabricated from steel plates. Figure 9-26 shows common methods of built-up girders based on American Institute of Steel Construction (AISC) standards. On the framing and floor plans, girders are usually represented by a single line placed between, but not touching, the columns that will be used for support. The size and spacing of the girders are typically specified. Steel joists and decking covered with lightweight concrete can be used to span between intermediate beams to form floor and roof levels. Reinforced concrete slabs are also used to span between intermediate support beams and will be discussed in Chapter 11. An example of a steel framing plan can be seen in Figure 9-27.

Detail Drawings

Details are required to show each intersection of structural members and the connection method of various materials to the steel frame. Details showing the connections of structural materials will be referenced on the framing plan and sections. Welded joints and bolts are used almost exclusively for steel frames. Bolting and welding methods and symbols were introduced in Chapter 6. Common types of connections can be seen in Figure 9-28.

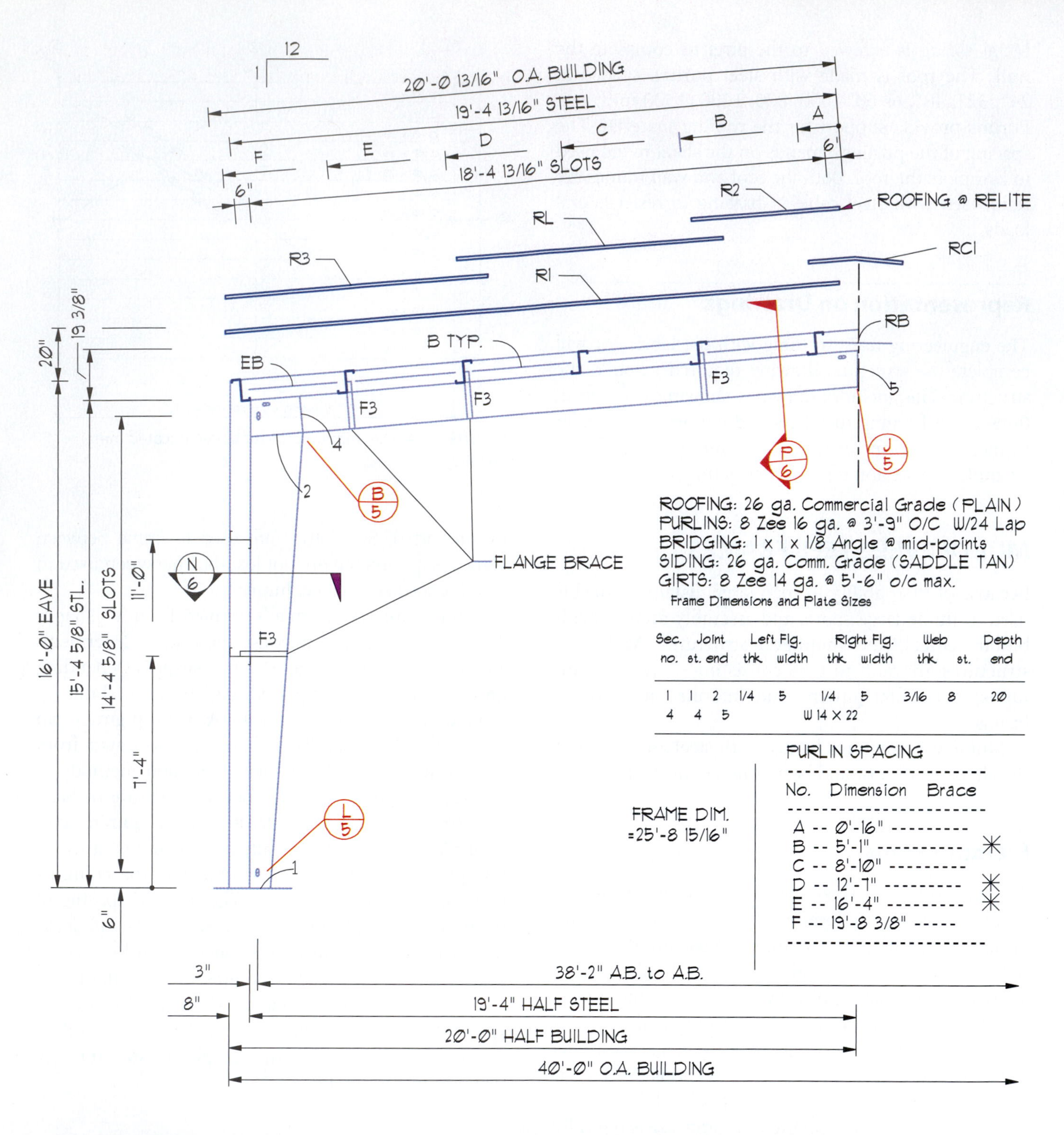
12
1
20'-Ø 13/16" O.A. BUILDING
19'-4 13/16" STEEL
A
B
C
D
E
F
6'
18'-4 13/16" SLOTS
6"
R2
ROOFING @ RELITE
RL
RCI
R3
RI
RB
19 3/8"
20"
B TYP.
EB
F3
F3
F3
5
4
2
P
6
J
5
B
5
FLANGE BRACE
N
6
11'-Ø"
F3
16'-Ø" EAVE
15'-4 5/8" STL.
14'-4 5/8" SLOTS
7'-4"
L
5
1
6"
ROOFING: 26 ga. Commercial Grade (PLAIN)
PURLINS: 8 Zee 16 ga. @ 3'-9" O/C W/24 Lap
BRIDGING: 1 x 1 x 1/8 Angle @ mid-points
SIDING: 26 ga. Comm. Grade (SADDLE TAN)
GIRTS: 8 Zee 14 ga. @ 5'-6" o/c max.
Frame Dimensions and Plate Sizes
Sec. no. | Joint st. end | Left Flg. thk width | Right Flg. thk width | Web thk | Depth st. end
1 1 2 1/4 5 1/4 5 3/16 8 20
4 4 5 W 14 X 22
PURLIN SPACING
No. Dimension Brace
A -- Ø'-16"
B -- 5'-1" ✱
C -- 8'-10"
D -- 12'-7" ✱
E -- 16'-4" ✱
F -- 19'-8 3/8"
FRAME DIM.
=25'-8 15/16"
3"
38'-2" A.B. to A.B.
8"
19'-4" HALF STEEL
20'-Ø" HALF BUILDING
40'-Ø" O.A. BUILDING
RIGID FRAME SECTION @ G.L. 2 & 3 (4 - HALF FRAMES REQ.)
SCALE : 1/4" = 1'-Ø"

FIGURE 9-25 A rigid frame seen in section.

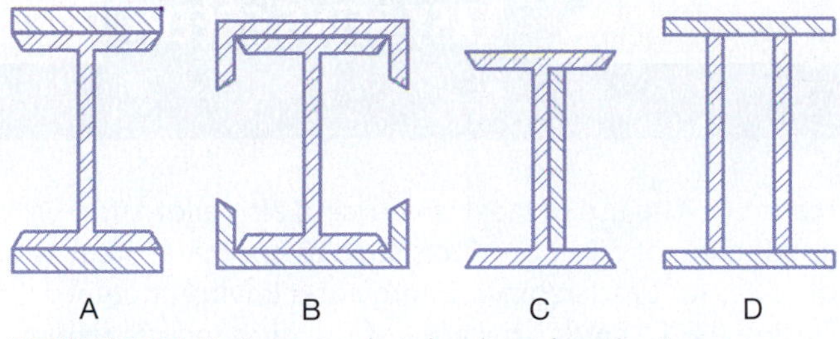

FIGURE 9-26 Built-up steel girders based on the American Institute of Steel Construction recommendations.

Exterior Coverings

Materials such as glass, stone, marble, granite, brick, and precast concrete panels are used to face steel structures. Units can be assembled at the job site or premanufactured and shipped to the site. Panels typically have steel connection plates that can be welded to the steel frame. Lightweight insulated sheet metal panels are also used for an exterior covering. Tempered glass panels set in metal frames are a major exterior covering. Details for connecting the exterior coverings to the frame must also be provided. Details similar to Figure 9-29 are typically referenced on the floor plan and elevations.

Trusses

In addition to the steel trusses already mentioned to support floor and roof loads, steel can be used to form trusses that act as a girder, such as the girder trusses in Figure 9-1. When steel shapes are combined to form a truss, CAD drafters working with the engineering team will draw the details to specify the shapes to be used and the types of connections. Figure 9-30 shows an example of the detail required for the trusses shown in Figure 9-1.

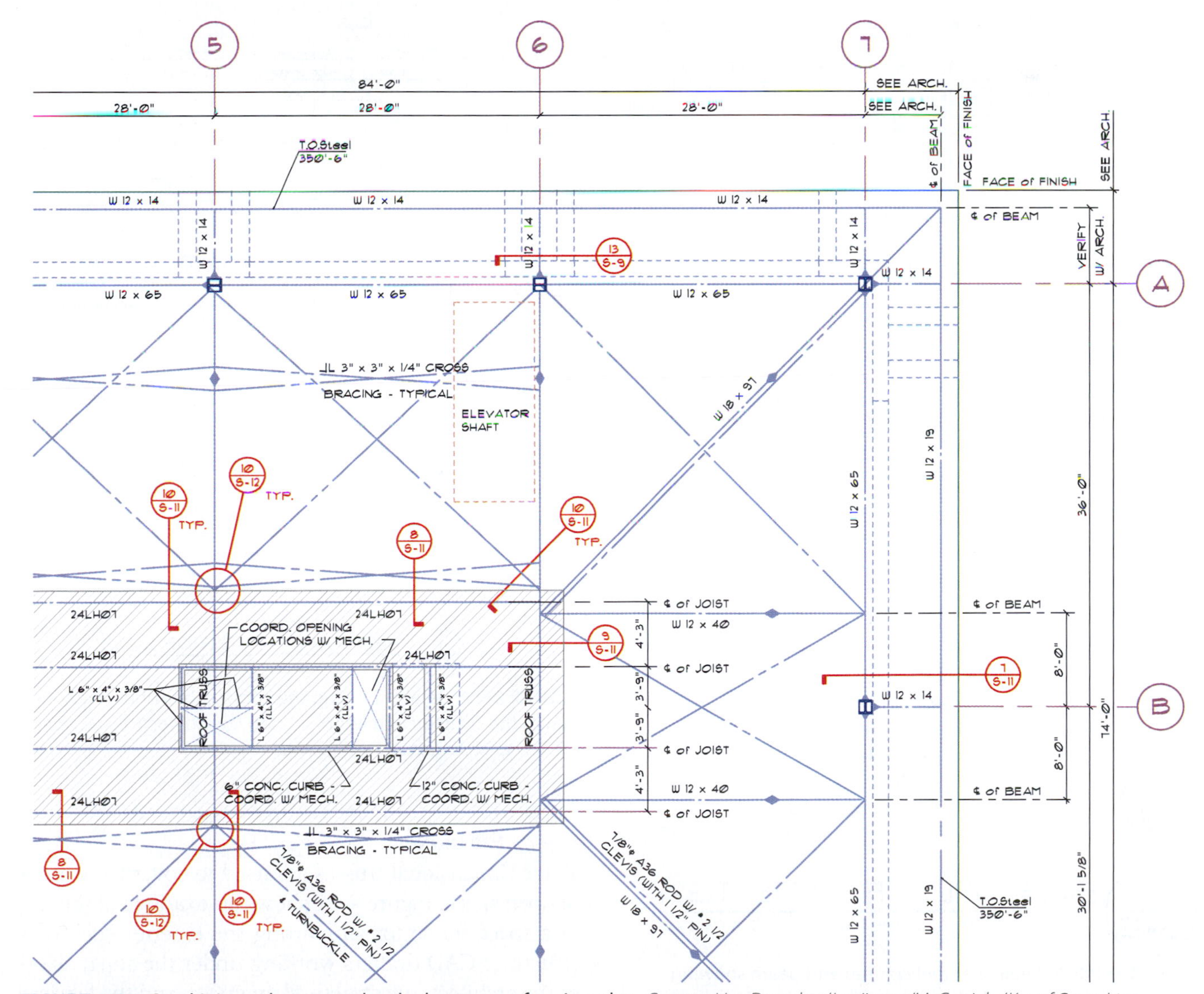

FIGURE 9-27 Steel joist, columns, and rods shown on a framing plan. *Courtesy Van Domelen/Looijenga/McGarrigle/Knauf Consulting Engineers.*

EASY BEING GREEN

Steel is considered a sustainable product because it is completely recyclable, and when compared to other materials, requires relatively low amounts of energy to produce. The steel industry has made immense efforts in the last few decades to limit environmental pollution that used to be associated with its production. Energy consumption and carbon dioxide emissions have been greatly reduced during the production process. But the main sustainable qualities of steel come from its compliance with two key concepts introduced in Chapter 1, which were to reduce and recycle. Steel-framed construction greatly reduces the support members required when compared to other building materials. The use of steel reinforcing also allows for a reduction of other products such as concrete. Steel qualifies for Leadership in Energy and Environmental Design (LEED) credits because of the high recycle rate of steel building products. According to the Steel Recycling Institute, steel is the most widely recycled material in North America. This is largely because it is economically advantageous to do so. It is cheaper to recycle steel than to mine iron ore and then go through the production process to form new steel. Steel does not lose any of its inherent physical properties during the recycling process, and has drastically reduced energy and material requirements than refinement from iron ore.

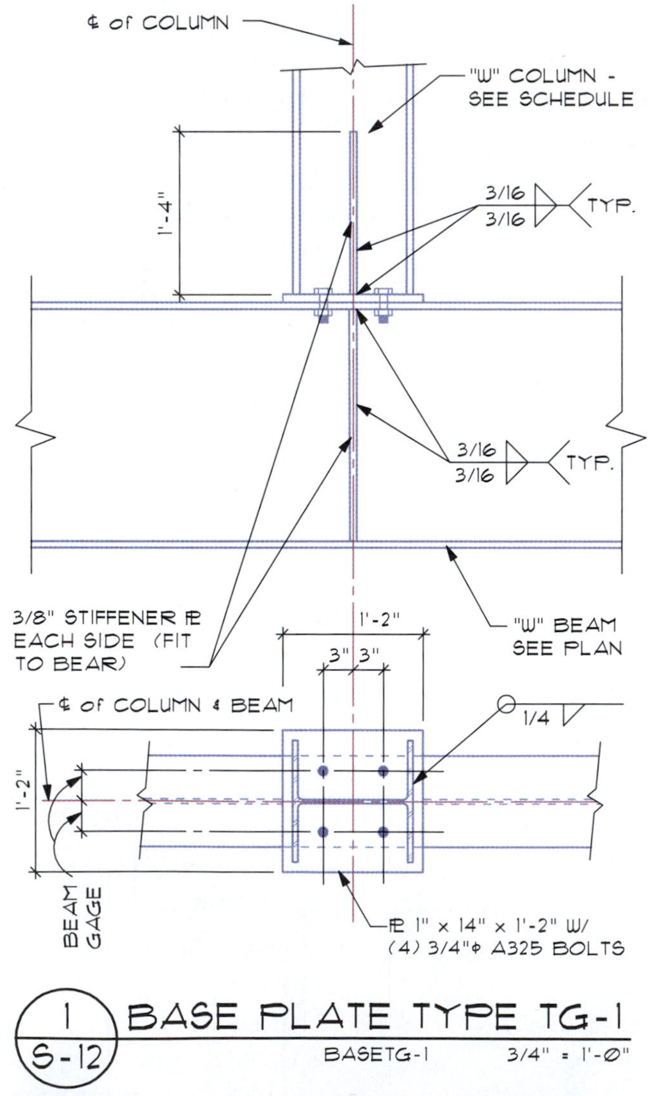

FIGURE 9-28 A detail of a steel column and beam showing both welded and bolted connections. *Courtesy Van Domelen/Looijenga/McGarrigle/Knauf Consulting Engineers.*

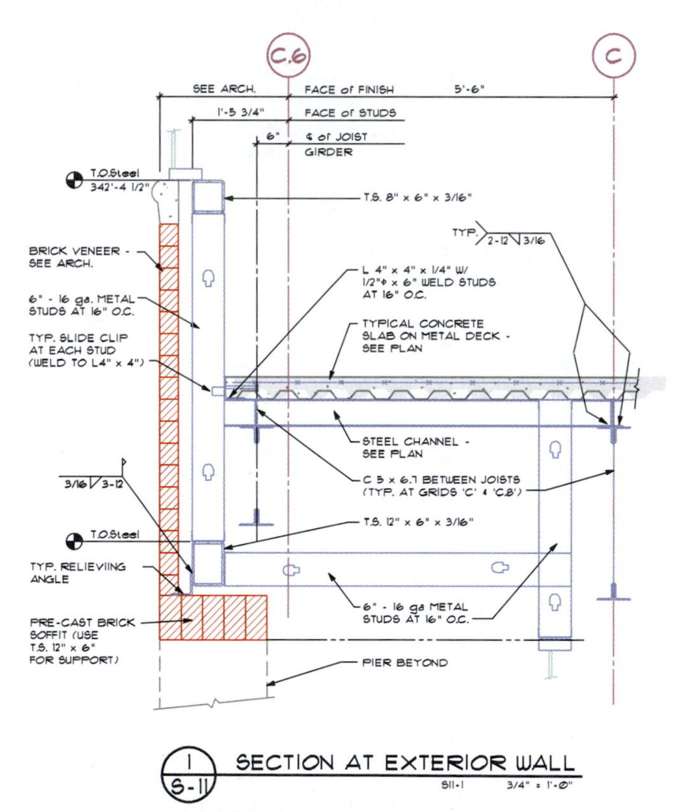

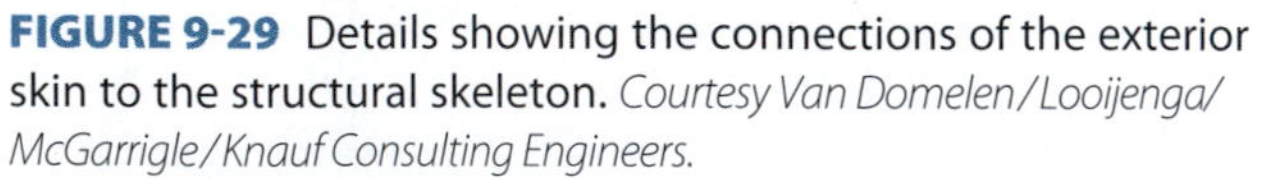

FIGURE 9-29 Details showing the connections of the exterior skin to the structural skeleton. *Courtesy Van Domelen/Looijenga/McGarrigle/Knauf Consulting Engineers.*

Space Frames

Three-dimensional trusses are used to cover large areas of open space. Figure 9-31 shows an example of the use of a space frame for supporting the exterior shell of a structure. CAD drafters working under the supervision of the architect, the structural engineer, and the fabricator will be involved in the detailing of space frames.

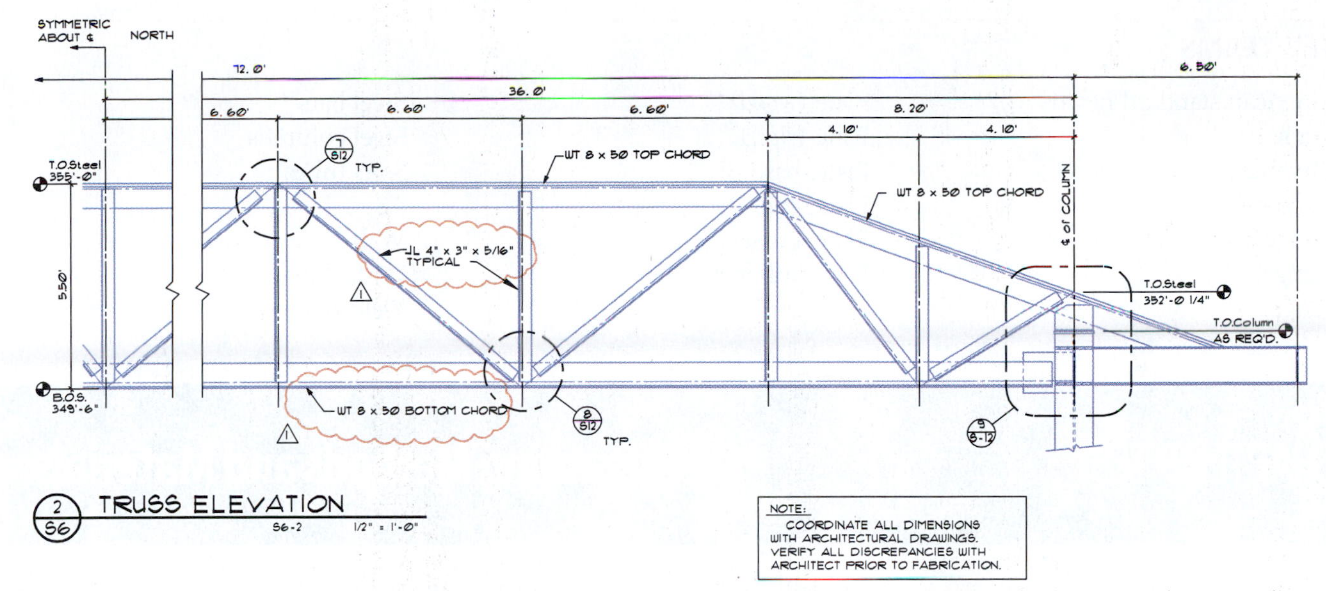

FIGURE 9-30 Steel trusses are often used to form a beam for long spans with smaller trusses set perpendicular to provide intermediate support similar to a panelized framing system used with wood members. *Courtesy Van Domelen/Looijenga/McGarrigle/Knauf Consulting Engineers.*

FIGURE 9-31 Three-dimensional trusses referred to as space frames are formed from lightweight metal. *Courtesy MERO Structures, Inc.*

ADDITIONAL READING

The following Web sites can be used as a resource to help you keep current with changes in building materials.

ADDRESS	COMPANY/ ORGANIZATION
www.aluminum.org	The Aluminum Association, Inc.
www.aisc.org	American Institute of Steel Construction
www.steel.org	American Iron and Steel Institute
www.astm.org	American Society for Testing and Materials
www.cssbi.ca	Canadian Sheet Steel Building Institute
www.cfsei.com	Cold-Formed Steel Engineers Institute
www.copper.org	Copper Development Association
www.worldsteel.org	International Iron and Steel Institute
www.lgsea.com	Light Gauge Steel Engineers Association
www.mbma.com	Metal Building Manufacturers Association
www.metalroofing.com	Metal Roofing Alliance
www.steelframingalliance.com	North American Steel Framing Alliance
www.ssina.com	Specialty Steel Industry of North America
www.sdi.org	Steel Deck Institute
www.steelframing.org	Steel Framing Alliance
www.steeljoist.org	Steel Joist Institute
www.steellinks.com	Steel links search engine
www.recycle-steel.org	Steel Recycling Institute
www.steelroofing.com	Steel Roofing
www.ssma.com	Steel Stud Manufacturers Association
www.unimast.com	Unimast, Incorporated (lightweight steel framing solutions)
www.ussteel.com	United States Steel

KEY TERMS

American standard beams
Angle
Channel
Ductile
Elastic range
Flange
Ledgers
Plate *(wood)*
Plate *(steel)*
Plate, base
Plate, end
Plate, gusset
Plate, stiffener
Plate, top
Steel bars
Steel columns
Steel tubes
Tee
Web
Wide flange
Yield point

CHAPTER 9 TEST Steel Framing Methods and Materials

Answer the following questions with short complete statements. Type the chapter title, question number, and a short complete statement for each question using a word processor. Some answers may require the use of vendor catalogs or seeking out local suppliers.

Question 9-1 What are the two major elements in steel?

Question 9-2 What element must be controlled to keep steel from being brittle?

Question 9-3 If steel is in its elastic stage, is it safe to use and how will it react to a load?

Question 9-4 What happens to steel as it is stretched past its yield point?

Question 9-5 What organization governs the designations of structural steel?

Question 9-6 What are the five most common steel designations used in the construction industry?

Question 9-7 Describe the most common type of steel used in the construction industry in terms of its ASTM designation, yield point, and its maximum carbon content for all shapes.

Question 9-8 List nine common shapes of steel used in construction.

Question 9-9 What does the ultimate tensile strength of steel represent?

Question 9-10 What are the components, cross-sectional shape, and the minimum size and weight of a W shape?

Question 9-11 List the area and size descriptions of a W16 × 50 in decimal inches.

Question 9-12 What information should be given when a channel is specified on a drawing?

Question 9-13 List the three common shapes of steel tubes and their range in sizes.

Question 9-14 What is the difference in size between a bar and a plate?

Question 9-15 List common alloys used throughout the construction of a structure.

Question 9-16 What material is required to specify steel studs on a drawing?

Question 9-17 What is the common depth of steel joists?

Question 9-18 What information about steel joists must be specified on a drawing?

Question 9-19 What are the major components of the walls and roof of a metal-framed structure?

Question 9-20 Use the Internet to list and visit five Web sites related to steel framing. Explain how these sites are related to your class projects.

DRAWING PROBLEMS

Skeletons of most of these details can be accessed from the student CD. Use the drawings on the CD to complete the required views. Use scale factors suitable for plotting each detail at a scale of 3/4" = 1'–0" minimum. Label all material with generic notes that can be altered with each use. Select all sizes based on material found in *Sweets Catalogs*, *Sweet Spot*, vendor catalogs, or Web sites. Save each detail as a wblock. Make the following assumptions to complete details 9-1 through 9-5 unless told otherwise by your instructor.

- **Assume all floor joists to be approximately 10" deep.**
- **Assume all wall studs to be approximately 4" wide.**
- **Place a note with each detail to reference the framing plan for all joists, stud and beam sizes, and spacings.**

Problem 9-1 Draw a side and end view showing a steel stud wall supporting floor joists. Use a runner track that matches the depth of the joists and second runner track that matches the studs.

Problem 9-2 Show a generic end view of the joist/foundation intersection. Show and specify a runner track attached to the foundation attached to an 8" wide concrete stem wall with an L-shaped foundation clip anchor.

- Attach the anchor to the stem wall with 1/2" diameter kwik bolt with 2 1/4" minimum embedment. Attach the anchor to the runner track using (6) 5/8" #10-16 (super-tite) or equal screws.

- Show floor joists w/a web stiffener at the joist/stem wall intersection.
- Provide 3/4" plywood sheathing screwed to the joists.

Problem 9-3 Show a view of floor reinforcement where floor joists rest on a steel wide-flange beam. Show and specify the following materials:

- Attach the floor to the support beam using # 91 anchor clips at each side of each joist.
- Use a 36" long joist stiffener centered over the support beam.
- Attach stiffener to joists with 1/4" –14 x 3/4" long screws, 2" o.c. placed 1 1/2" up/down along joist/joist stiffener.

Problem 9-4 Draw a side and end view to show typical wall reinforcement. Provide a horizontal runner track on one side of the wall studs. Notch the track flanges to lap studs. Attach the runner track to the studs using (2) # 1/4" dia. × 3/4" screws at each stud. Provide 1/2" gypsum board over each side of the studs. Do not show the top or bottom of the studs.

Problem 9-5 Draw the side and end view showing a 7' high opening over a typical window or door in a 10' high wall. Specify the following materials:

- 4" wide studs with a top track.
- Use back-to-back joists attached w/(2) # 10 x 5/8" screws @ 12"o.c. over the opening.
- Provide a stiffener plate at each end of the joists used as a header over the opening.
- Provide nested jamb studs at each end of the header.
- Uses cripple studs @ 16" o.c. below the header with runner tracks at the top/bottom of the studs.
- Attach the runner track at the bottom of the studs to the jamb studs. Note on the drawing that the runner is to be 8" longer than the opening. Clip the flanges of the track 4" from each end bent at the clipped flange to attach to the nested joist studs with (4) #10-16 × 5/16.

Problem 9-6 Draw the end view of each typical steel shape with appropriate hatching and save each shape as a wblock, named for the specific shape.

Guidelines for problems 9-7 through 9-15. Unless your instructor gives other instructions, complete the following details using the guidelines presented in "Drawing Criteria for Completing Details" of Chapter 6. Skeletons of most of these details can be accessed from the student CD. Adjust all scale factors as required. Use these drawings as a base to complete the assignments:

- **Show all required views to describe each connection.**
- **Draw each detail at a scale of 1" = 1'-0" unless noted.**
- **Provide a plan view of each plate that contains bolts to show the bolt to column relationship. If bolts do not fit in their specified location, submit a sketch showing your proposed bolt location to your engineer prior to changing the engineer's design.**

Problem 9-7 Show a plan view showing two steel columns in a 2 × 6 stud wall. Show the columns as 36" from center to center. Specify TS 5 × 5 × 3/8" columns of ASTM A500, grade B steel. Cover the exterior side of the wall with 7/8" exterior plaster over 5/8" metal lath with 5/8" Type X gyp board on exterior side of the stud cavity containing the steel columns.

Problem 9-8 Show a W12 × 22 with a 3 × 6 DFL #1 top plate with 1" minimum overhang on each side with a 5/8" diameter 4" stud anchor at 32" o.c. welded to top flange with a 3/16" field fillet weld. Hang 2 × 12 DFL #1 floor joists at 12" o.c. from the top plate with an appropriate hanger for supporting 1300 lbs. Cover the bottom of the floor and beam with 7/8" exterior stucco over 3/8" metal lath using connections similar to Figure 12.18. Provide 1 1/2" lightweight concrete flooring over 3/4" T & G plywood subfloor.

Problem 9-9 Show a W12 × 30 supporting steel joist. Weld a 925SJ16 floor joist at 16" o.c. to the top flange with a 1/8" × 3" field fillet weld at each side. Provide a web stiffener plate where the joists rest on the beam and weld with fillet weld. Cover the bottom of the floor and beam with 7/8" exterior stucco over 3.4# metal lath. Provide 1 1/2" lightweight concrete flooring over 3/4" T & G plywood subfloor.

Problem 9-10 Show a W14 × 120 supporting TJL24 joists at 32" o.c. support on a 3 × 6 DFPT sill w/5/8" Ø studs @ 32" o.c. welded to the top flange with a 1/8" field fillet weld.

Problem 9-11 Use the attached sketch to complete the required drawing. Use a TS 4 × 4 × 3/16" column to support a W16 × 31 beam with a 1/4" end stiffener plate welded with a 3/16" fillet weld. Rest beam on a 5 1/2" × 11" × 1/2" steel cap plate with (2)-7/8" diameter bolts to beam and weld plate to column with 3/16" fillet weld, all around. Provide TS 4 × 4 × 3/16" chevron (diagonal brace) welded to 14 × 11 × 3/8" gusset plate welded to column and cap plate with 3/16" fillet weld at all contact points.

- Provide a plan view showing the relationship of the bolts, the column, and the gusset plate.
- Bolt 3 x 4 DFL #1 top plate with a 5/8"Ø stud anchor at 32" o.c. staggered welded to top flange with a 3/16" field fillet weld.

Set a 6 × 12 DFL #1 beam level with the top plate and support beam with a 5 × 7 × 7 ga. steel bucket connection with a 5/8" diameter M.B. 1 1/2" down and in from edge. Weld bucket to end plate w/1/8" fillet all around. Tie beam to plate with MST27 strap by Simpson Strong-Tie Company.

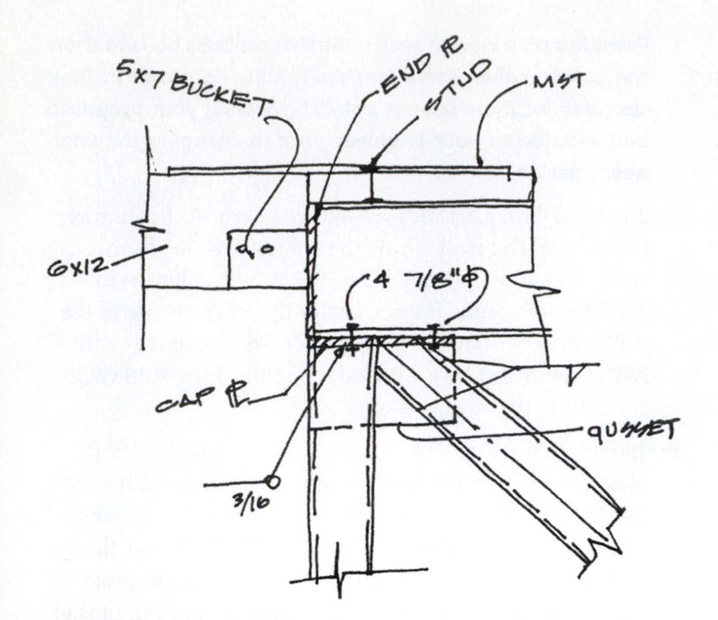

Problem 9-12 Use the attached sketch and a scale of 1 1/2" = 1'–0" to complete the required drawing. Use a TS 5" × 5" × 3/8" column to support a W18 × 76 beam bolted to 11" × 10" × 1 1/4" steel cap plate w/ (4) 5/8" Ø machine bolts to beam and weld plate to column with 3/16" fillet weld-all-around. Butt W18 × 97 beam into left side and W18 × 60 on right. Support each with (2)-4 × 3 1/2 × 12 × 1/4" angles welded to W18 × 76 with 1/4 × 11 1/2" fillet weld and (4)-3/4"Ø machine bolts through 4" leg through WF. Set bolts 1 1/2" from angle edge @ 3" o.c. Set the top bolt 4 1/2" down from top flange.

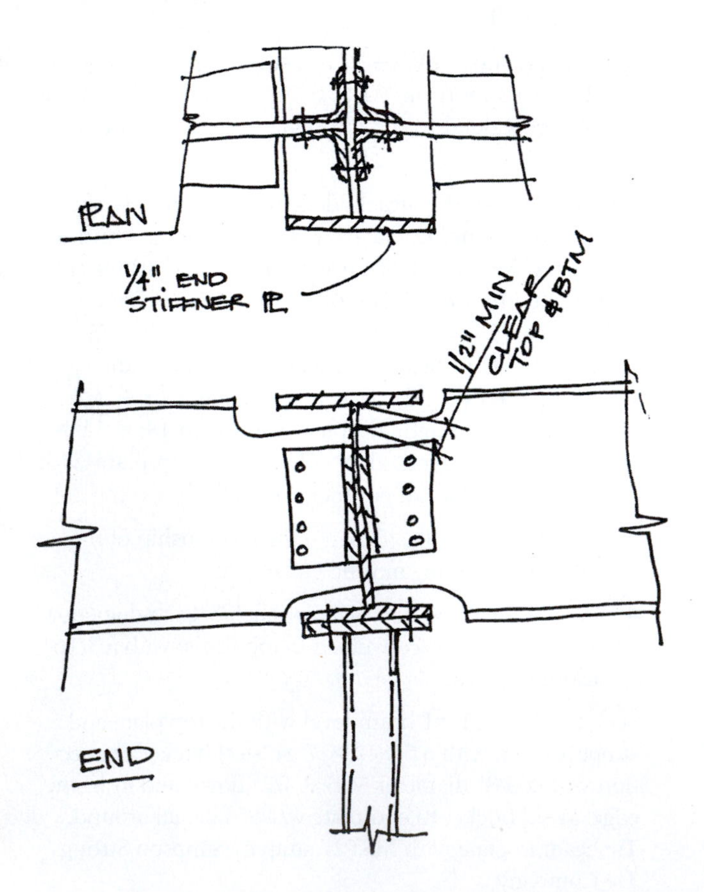

Use a scale of 3/4" = 1'–0" to draw the following details. Base nailing on International Building Code (IBC) standards unless your instructor tells you otherwise. Specify all material based on common local practice. Show and specify all connecting materials. Show a front and side view or a side and top view to describe each connection.

Problem 9-13 A 6" × 6" × 3/8" steel column will be supported by a 3/4" × 10" × 10" base plate. Set on a 1" mortar base. Use a 3/16" fillet weld-all-around to form the connection. The plate will be attached to a concrete footing using (4)-3/4" dia. × 12" A.B. laid out on a 7" grid that is laid out diagonally to the column.

Problem 9-14 Show a 3 × 6 DFL #1 sill supported on a W12 × 30. Use 3/4"Ø stud anchors at 36" o.c. staggered 1 1/2" from plate edge. Use a 1/8" fillet, field weld all around. Use 2"Ø washers. Provide 2 × 12 DFL #1 floor joist @ 16" o.c. to the plate. Provide solid blocking between the floor joist, and provide 3/4" std. grade plywood nailed with 10d @ 4"/8" o.c. to support 1 1/2" lightweight concrete over 55 # felt.

Problem 9-15 Show a 6" × 6" × 3/16" steel column with a Simpson Co. HL46 angle welded on two opposing sides with a 3/16" fillet weld-all-around in the field. Support a 5 1/8 × 15 glu-lam beam on each angle w/ 1/4" clear between the column and the beam. Bolt the beam to the angle using lag screws per the manufacturer's size recommendations. Use a Simpson Co. MST 27 strap on each side of the beams, bolted to each beam and welded to the steel column with a 1/8" × 3" long fillet field weld. Place the strap so that minimum recommended sizes for bolt placement in the strap will be met for wood and steel. Review Chapter 6 for minimum clearances required for placement.

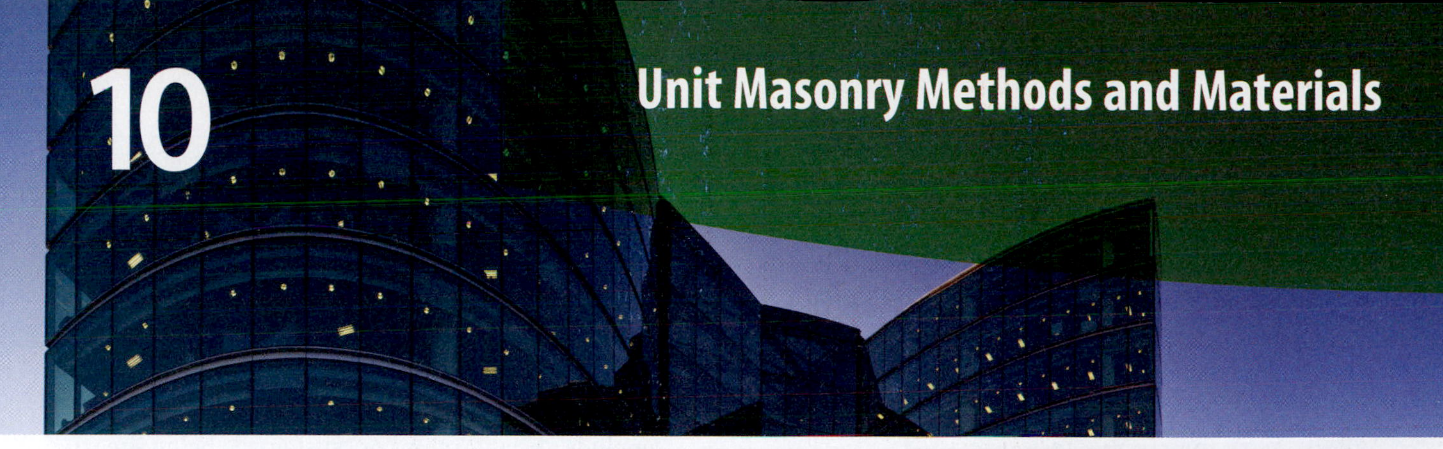

10 Unit Masonry Methods and Materials

Masonry products have been in use for thousands of years because of their durability and aesthetic qualities. They provide an environmentally friendly use of clay, water, and other natural resources used in their construction, as well as excellent life-cycle cost analysis. **Masonry construction** consists of setting building materials in a bed of **mortar.** Brick, stone, glass, and concrete masonry units (CMUs) are the materials commonly used in masonry construction.

This chapter examines methods of construction typically used with brick and concrete masonry units. Topics to be explored include:

- Brick construction.
- Concrete masonry construction.
- Steel reinforcement.
- Mortar and grout.

BRICK CONSTRUCTION

Bricks are made in various colors and textures by pressing different types of clay, shale, and a combination of oxides in a mold of a desired shape and size. Brick is usually produced in red, brown, and gray tones. Common surfaces include smooth, water or sand struck, scored, wire-cut, combed, and roughened. Bricks can also be finished with a ceramic glaze—providing a polished finished in any color. The International Building Code (IBC) relies on information provided by the American Society for Testing and Materials (ASTM) for most technical information related to clay and shale masonry units. Key standards include:

ASTM C34	Structural clay load-bearing wile tile
ASTM C56	Structural clay nonload-bearing wile tile
ASTM C62	Building brick—solid masonry units made from clay or shale
ASTM C126	Ceramic-glazed structural clay facing tile
ASTM C212	Structural clay facing tile
ASTM C216	Facing brick—solid masonry units made from clay or shale
ASTM C652	Hollow brick—hollow masonry units made from clay or shale
ASTM C1088	Solid units of thin veneer brick
ASTM C1405	Glazed brick—single-fired solid brick units

Brick Types

Bricks are produced in solid, cored, and hollow units. Cored units are considered to be solid if a minimum of 75% of the cross-sectional area is solid. A brick is considered to be hollow if it has at least 25%, but not greater than 60% void areas. Vertical cores are placed in bricks to reduce the weight of the brick.

Common or *building brick* is most widely used in the construction industry. Face brick is produced to standards that indicate units in specific sizes, textures, and colors. Other types of brick used in construction include the following:

- *Adobe brick* is made from a mixture of natural clay and straw that is placed in molds and dried in the sun. These units require protection from rain and subsurface moisture unless a moisture-proofing agent is added.
- *Back-up brick* is inferior units used in applications where they can't be seen, for example, behind face brick.
- *Economy brick* is a brick with nominal dimensions of 4" × 4" × 8" (100 × 100 × 200 mm).
- *Engineered brick* is a brick with nominal dimensions of 4" × 3.2" × 8" (100 × 80 × 200 mm).
- *Fire brick* is brick with a high resistance to high temperature, used for the facing material of the firebox of a fireplace.
- *Hollow bricks* are masonry units of clay or shale that are cored in excess of 25% of the gross cross-sectional area.

- *Jumbo brick* is a generic term indicating a brick larger in size than the standard. Some producers use this term to describe oversized brick of specific dimensions.
- *Norman brick* is a brick whose nominal dimensions are 4" × 2 2/3" × 12" (100 × 66 × 300 mm).
- *Paving brick* is masonry units with a hard, dense surface used for floor applications.
- *Roman brick* is brick whose nominal dimensions are 4" × 2" × 12" (100 × 50 × 300 mm).

The type of brick to be used is normally indicated by the drafter on the floor and elevations as well as in written specifications. The surface of the brick to be exposed is also often specified. Each surface of a unit has a specific name, as shown in Figure 10-1. The surface name is often referred to in written specifications to determine which surface will be displayed or cut. For some applications, the mason must cut units to meet the design criteria. Common methods of cutting brick are illustrated in Figure 10-2.

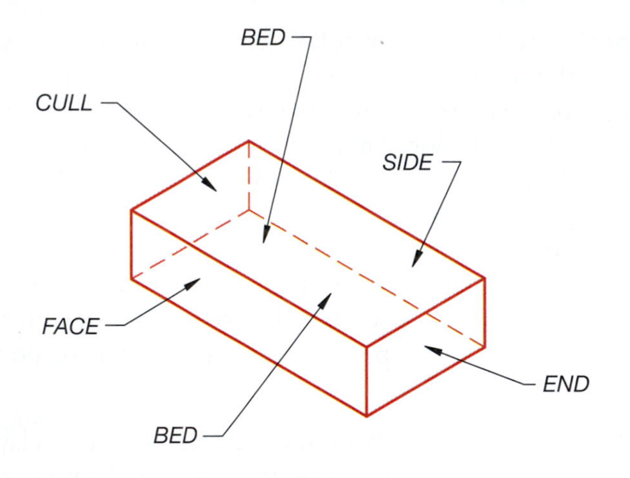

FIGURE 10-1 Each surface of a brick is named to indicate how it is to be positioned.

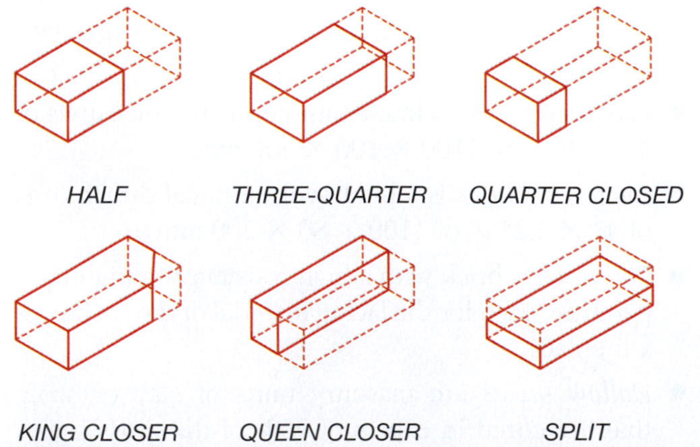

FIGURE 10-2 Brick can be cut into various shapes based on the pattern used to position each unit.

Brick Quality

Strength, appearance, and durability determine the quality of brick. The quality is specified in the written specifications. Common brick is classified into three grades, determined by the weather conditions it will be required to withstand.

- Grade *SW* is for severe weather conditions, which include heavy rains or sustained below-freezing conditions.
- Grade *MW* is for moderate weather conditions for use in areas with moderate rainfall and some freezing conditions.
- Grade *NW* is for use in areas with negligible weathering from minimal rainfall and above-freezing temperatures.

Face brick is rated for durability of exposure and is available in grades SW and MW. It is also identified by the ASTM specifications, which dictate the range allowed in size, texture, color, and structural quality.

- Grade *FBA* brick is nonuniform in size, color, and texture.
- Grade *FBS* brick allows for variations in mechanical perfection with a wide range of color variation.
- Grade *FBX* has the highest degree of mechanical perfection and is the most controlled for color variation.

Hollow brick is classified by SW or MW as well as by factors that affect its appearance.

- Graded *HBA* is nonuniform in size, color, and texture.
- Grade *HBS* is more controlled than HBA brick but allows for some size variation and a wide color range.
- Grade *HBX* has the highest degree of mechanical perfection with the least variation in size and the smallest range in color.

Brick Sizes

To effectively control the cost in using brick and to minimize labor, structures using masonry units should conform to the size of the unit being used. Bricks vary in size not only because of the variety of bricks made, but also because of shrinkage in the drying process.

Modular bricks were developed to regulate the coursing dimensions. An economy-8 modular brick is 3 1/2" high and 7 1/2" long laid with 1/2" joints. It will produce courses exactly 4" high and 8" long. Brick sizes are controlled by the *Standard Guide for Modular Construction of Clay and Concrete Masonry Units* (ASTM document E835/E835M). This document establishes metric dimensions for clay and concrete products based on the basic building module of 100 mm. Many common brick sizes are within a millimeter or two of metric module sizes and nearly all can fit within a 100 mm module vertically by using 10 mm joint widths. Verify the exact size of masonry units and their availability with local suppliers. Common nominal sizes of standard bricks are listed in Table 10-1.

Brick Shapes and Placement

Common shapes of brick are shown in Figure 10-3. One of the most popular features of brick is the wide variety of positions and patterns that can be created. These patterns are achieved by placing bricks in various positions relative to one another. The position in which the brick is placed will alter what the brick course is called. Figure 10-4 illustrates common brick positions and their names. Figure 10-5 shows examples of how these patterns would be represented. Common patterns for laying masonry units are illustrated in Figure 10-6. The brick pattern is represented and specified on the elevation.

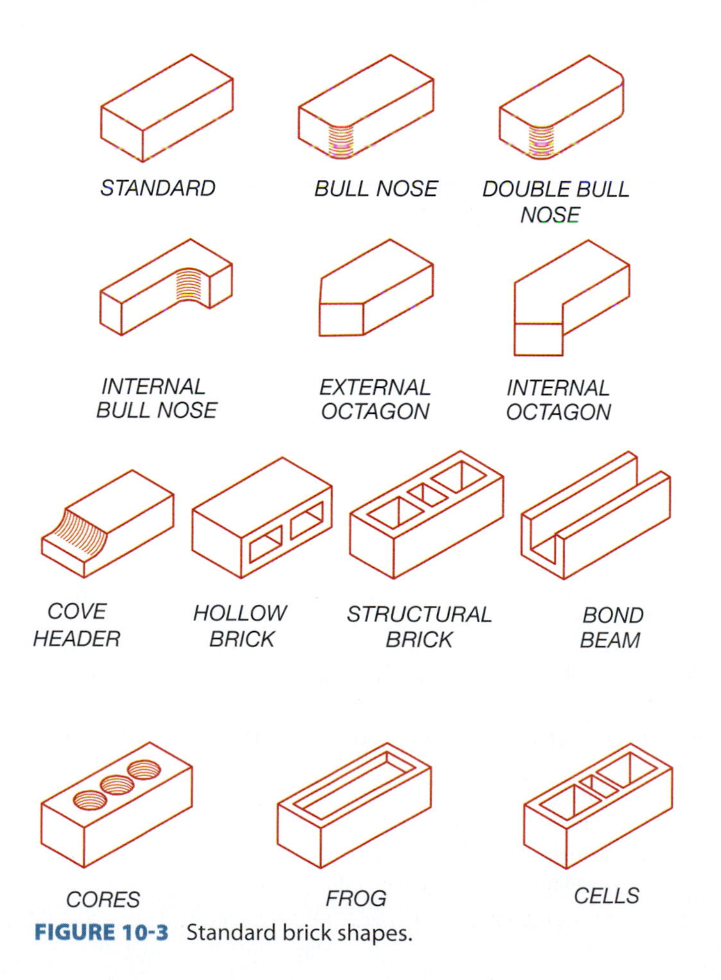

FIGURE 10-3 Standard brick shapes.

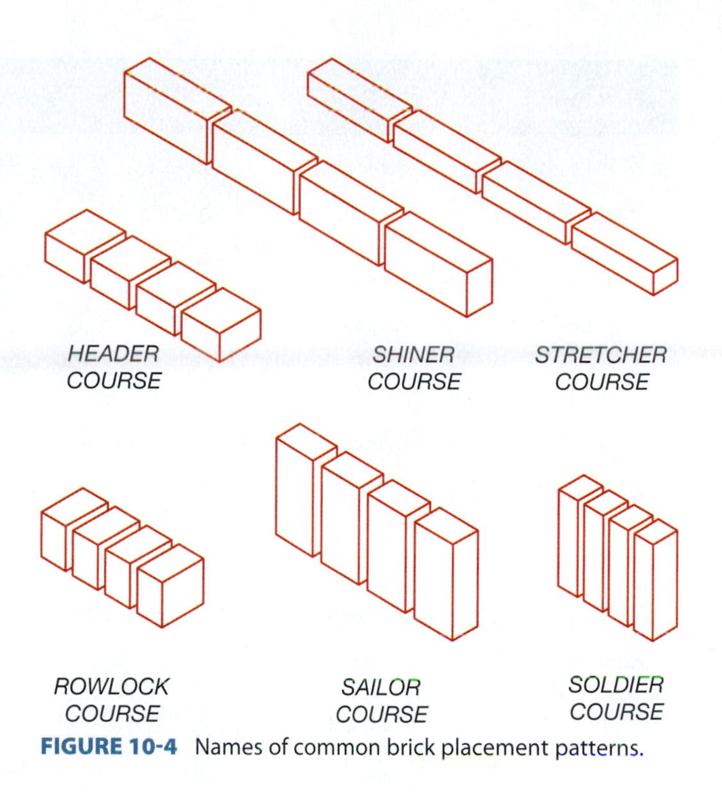

FIGURE 10-4 Names of common brick placement patterns.

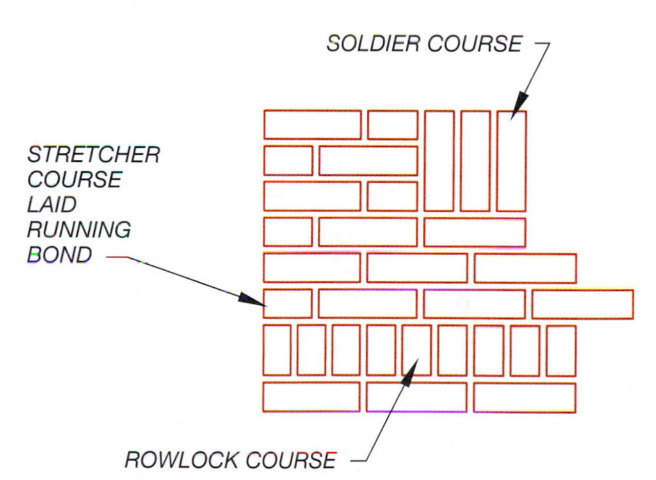

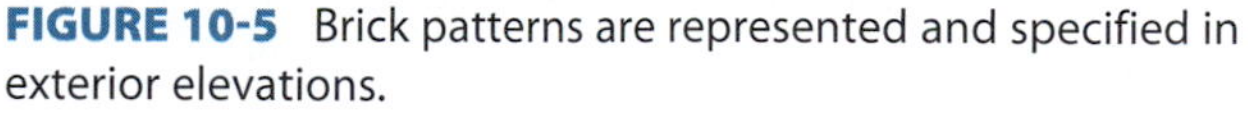
FIGURE 10-5 Brick patterns are represented and specified in exterior elevations.

Wall Construction

Bricks can be placed in any of the positions shown in Figures 10-4 and 10-6 to form a wall, as well as a variety of bonds and patterns. Masonry units are laid in rows called **courses** and in vertical planes called **wythes**. Figure 10-7 shows an example of single-wythe construction with the temporary supports still in place. Single-wythe walls are made with structural

Unit Name		Size (in.)	Actual Size (mm)	Actual Metric Size (mm)	Modular Metric Vertical Coursing
Modular	width	3 1/2"	89	90	3:8"
Metric modular	height	2 1/4"	57	57	3:200 mm
	length	7 1/2"	190	190	
Engineer	width	3 1/2"	89	90	5:16"
Modular	height	2 3/4"	70	70	5:400 mm
	length	7 1/2"	190	190	
Roman	width	3 1/2"	89	90	4:8"
	height	1 5/8"	41	40	4:200 mm
	length	11 1/2"	292	290	
Norman	width	3 1/2"	89	90	3:8"
Metric Norman	height	2 1/4"	57	57	3:200 mm
	length	11 1/2"	292	290	
Engineer	width	3 1/2"	89	90	5:16"
Norman	height	2 3/4"	70	70	5:400 mm
	length	11 1/2"	292	290	
Utility	width	3 1/2"	89	90	2:8
Metric Jumbo	height	3 1/2"	89	90	5:400 mm
	length	11 1/2"	292	290	
Standard	width	3 5/8"	92	90	3:16"
	height	2 1/4"	57	57	3:200 mm
	length	8"	203	200	
*Engineer	width	3 5/8"	92	5:16"	
Standard	height	21 3/16"	71	70	5:400 mm
	length	8"	203		
*King	width	3"	76		5:16"
	height	2 3/4"	70	70	5:400 mm
	length	9 5/8"	245		

Note: Bricks identified with an * are not a modular unit in either inch-pound or the metric system. Vertical coursings are based on 9 to 11 mm mortar joint, depending on the brick height.

TABLE 10-1 Common Nominal Masonry Unit Sizes *(Courtesy Construction Metrication)*

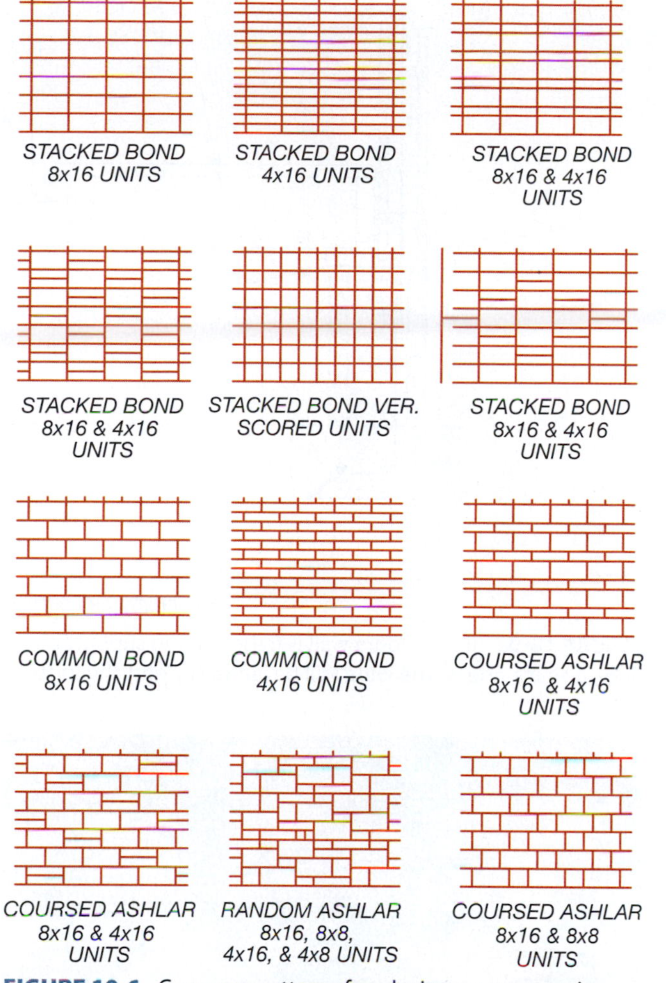

FIGURE 10-6 Common patterns for placing masonry units.

FIGURE 10-7 A single-wythe masonry wall can be constructed of structural brick with steel reinforcing. Masonry units over openings are temporarily supported by wood members until the grout can cure and achieve its full strength. *Courtesy Sara Jefferis.*

bricks (see Figure 10-3) with steel reinforcing placed per local codes to resist lateral loads. Brick walls normally consist of two wythes, with a 2" to 3" air space between each wythe. This type of construction is referred to as a **cavity wall.** Rigid foam insulation can be placed against the inner wythe to increase the R-value of the wall. If the space is filled with grout, it is called a *grouted cavity* wall. Reinforcing steel can be placed in the air space between wythes and then surrounded by solid grout.

If the wythes are connected, the wall is referred to as a **solid wall.** A **bond** is the connection between the wythes to add stability to the wall so that the entire assembly acts as a single structural unit. It can be made by either wire **ties** or by masonry members. The Flemish and English bonds of Figure 10-8 are the most common methods of bonding two wythes together. The Flemish bond consists of alternating headers and stretchers in each course. (A course of masonry is one unit in height.) An English bond consists of alternating courses of headers and stretchers. In each case the header spans between wythes to keep the wall from separating. The bonding course is usually placed at every sixth course.

The cavity between wythes is typically 2" (50 mm) wide and creates approximately a 10" (250 mm) wide wall with masonry exposed on both the exterior and interior surfaces. The air space between wythes provides an effective barrier to moisture penetration to the interior wall. Weep holes must be provided in the lower course of the exterior wythe to allow moisture to escape. Rigid insulation can be applied to the interior wythe to increase the insulation value of the air space in cold climates. Care must be taken to keep the insulation from touching the exterior wythe so that moisture is not transferred to the interior. Metal ties are embedded in mortar joints at approximately 16" (400 mm) to tie each pair of the wythes together.

If a masonry wall is to be used to support a floor, a space one wythe wide will be left to support the joist similar to Figure 10-9a. For a single-wythe wall, the structural member can be supported on a full wythe, and a queen closer is used to create a cavity for the structural member (see Figure 10-9b). Joists are usually required to be strapped to the wall so that the wall and floor will move together under lateral stress. The strap will be specified on the framing plan and the details. The end of the joist must be cut on an angle. This cut is called a *fire cut* and will be specified in the details. If the floor joists were to be damaged by fire, the fire cut will allow the floor joist to fall out of the wall, without destroying the wall. A fire cut is shown in Figure 10-10.

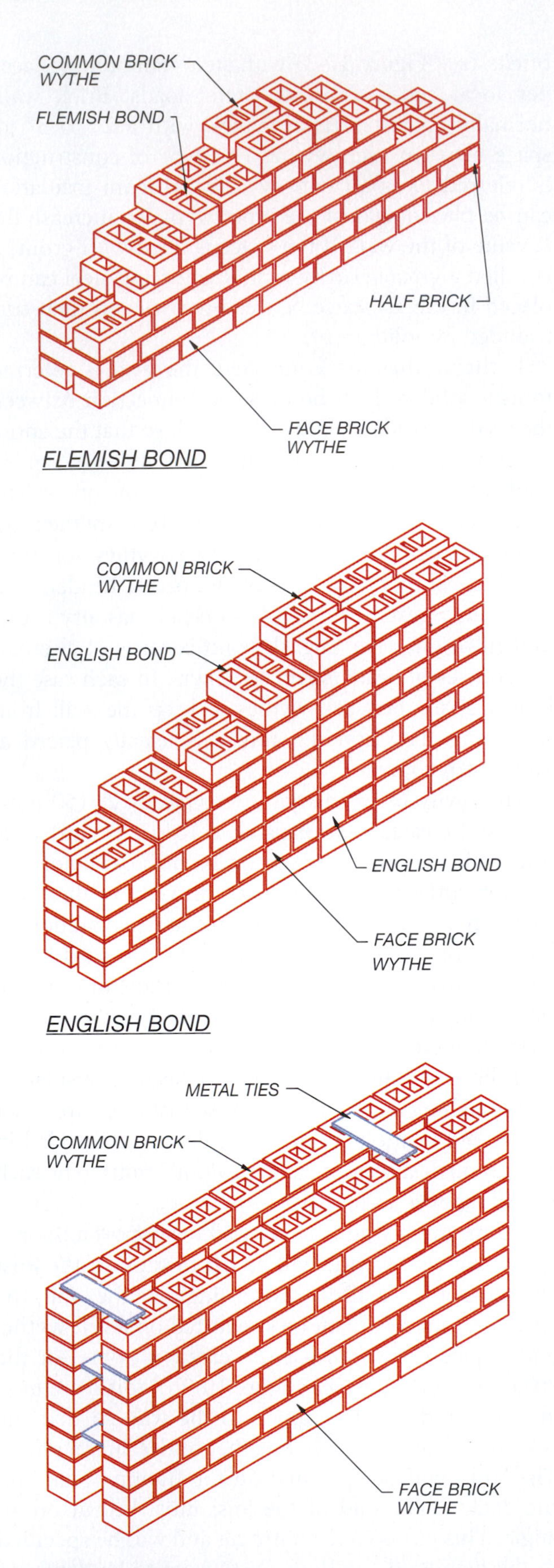

FIGURE 10-8 Brick walls can be reinforced by using metal ties or by placing brick bonds to connect each wythe. Common bonds include the Flemish and English.

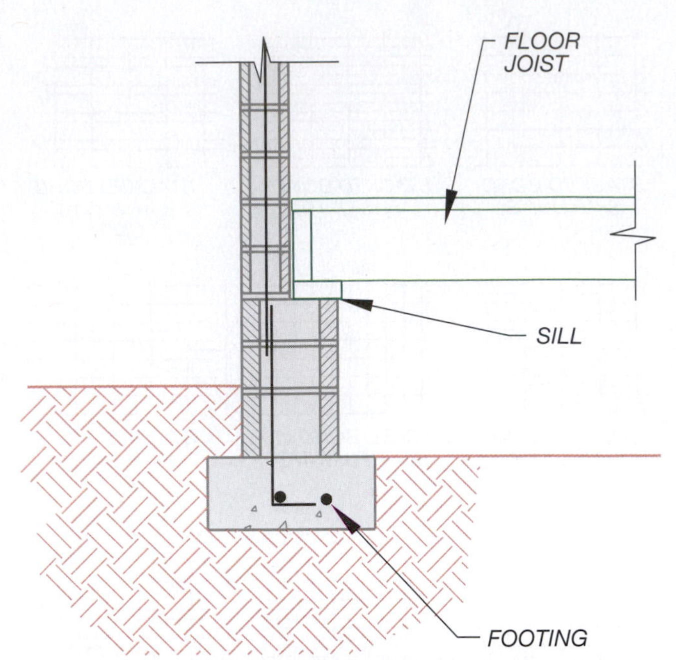

FIGURE 10-9a If a masonry wall is to be used to support a floor, a space one wythe wide will be left to support the joists.

FIGURE 10-9b Floor or roof framing members can be supported on a full brick, whereas a half-brick is used to hide the end of the structural member. *Courtesy Tereasa Jefferis.*

Brick is very porous and absorbs moisture easily. Some method must be provided to protect the end of the joist from absorbing moisture from the masonry. The end of the joist is wrapped with 55# felt and set in a 1/2" (13 mm) air space. The interior of a masonry wall also must be protected from moisture. A layer of hot asphaltic emulsion can be applied to the inner side of the interior wythe and a furring strip can be attached to the wall. In addition to supporting sheet rock or plaster, the space between the furring strips can be used to hold batt or rigid insulation.

An alternative method of attaching a floor to a brick wall is with the use of a pressure-treated **ledger** bolted to the wall with the bolts tied to the wall reinforcing. The joists are connected to the ledger with metal hangers. Figure 10-11 shows an example of the use of a ledger. When a roof-framing system is to be supported on masonry, a pressure-treated plate is usually bolted to the brick in a manner similar to the way a plate is attached to a concrete foundation.

Figure 10-12 shows how roof members are typically attached to masonry.

Brick Joints

The joints between each course of brick must be specified on the construction drawings. Common methods of finishing the mortar in joints are illustrated in Figure 10-13. In addition to the type of joint being specified for decorative purposes, joints must also be used to relieve stress in the wall. Walls longer than 200' (60,960 mm) or walls in a building having two or more wings must have expansion joints. An **expansion joint** is a seam placed in a wall to relieve cracking caused by expansion or contraction. Rather than causing random cracks throughout the brick, the stress is relieved in the expansion joint. Two common methods of constructing an expansion joint in brick walls are shown in Figure 10-14.

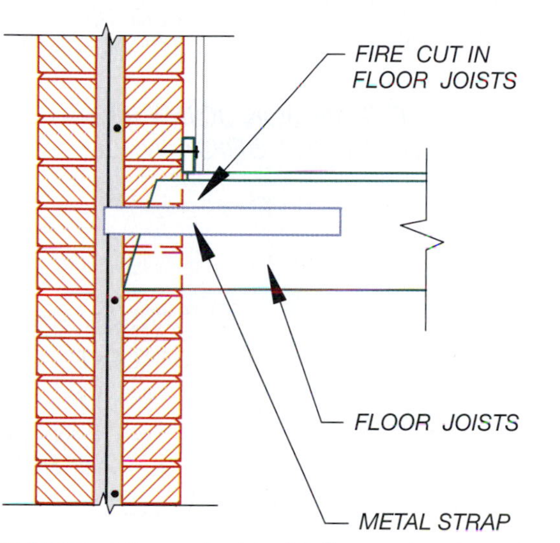

FIGURE 10-10 A fire cut is placed in floor and roof members to prevent wall damage in case of a fire.

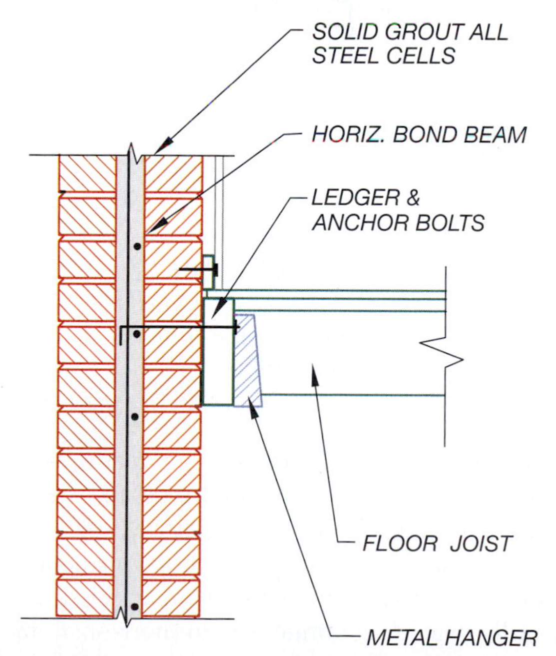

FIGURE 10-11 A double-wythe brick wall with the floor system supported by a wood ledger that is bolted to the wall.

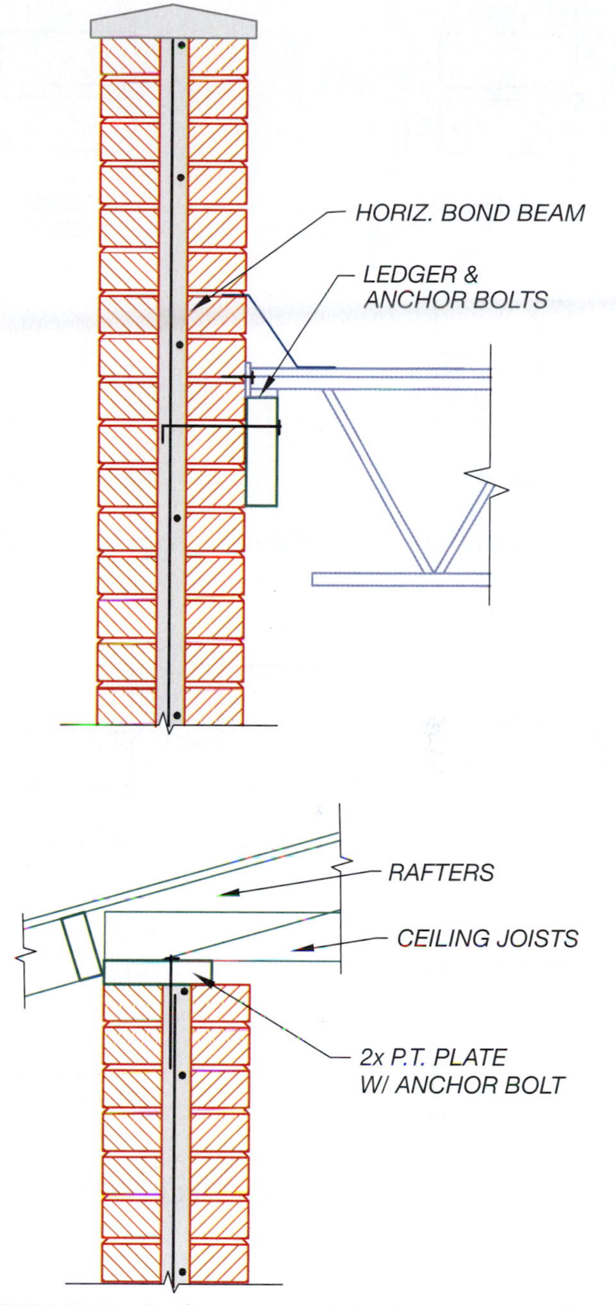

FIGURE 10-12 Roof trusses are attached to a masonry wall using a ledger or by resting on a pressure-treated plate.

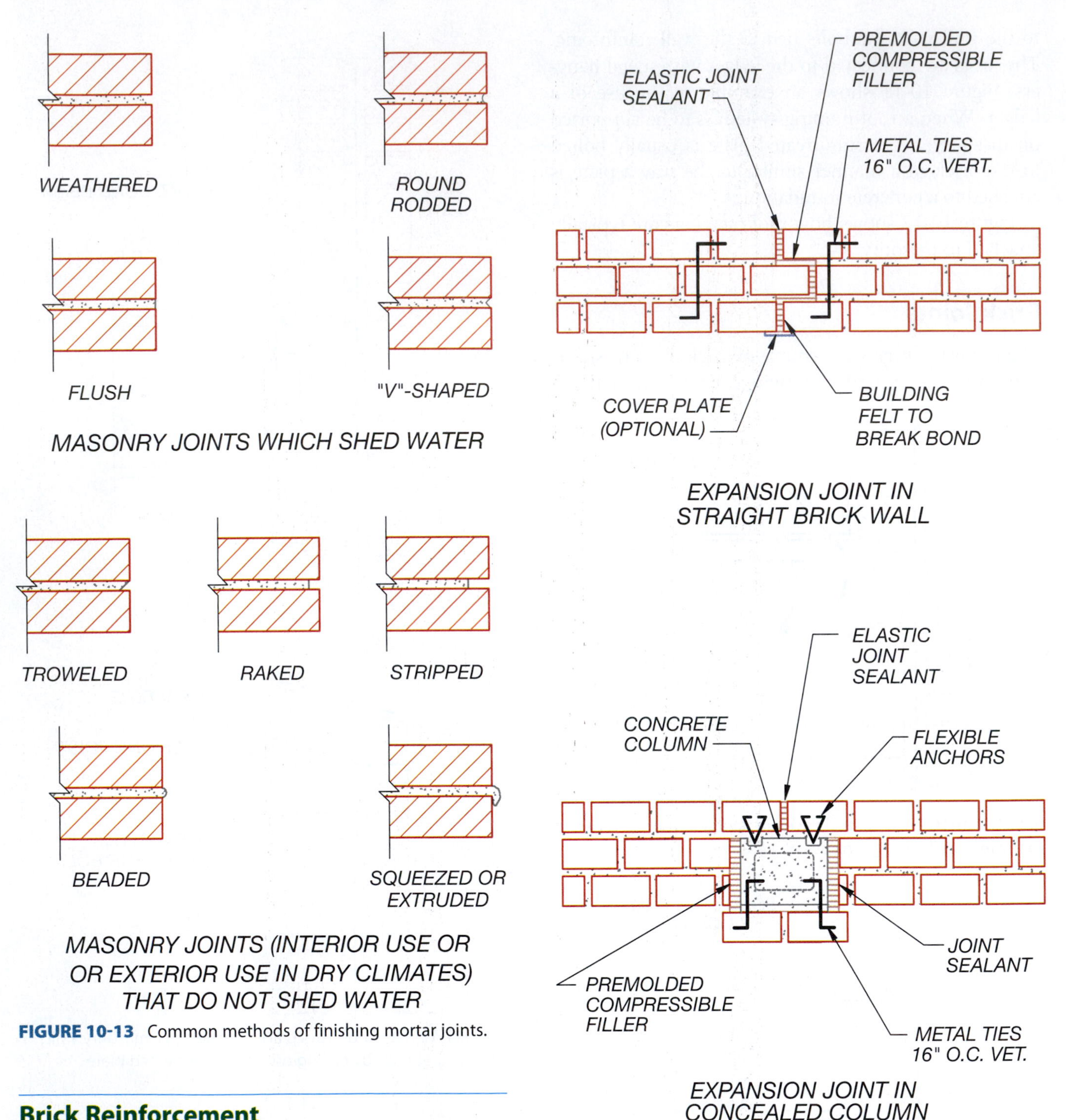

FIGURE 10-13 Common methods of finishing mortar joints.

FIGURE 10-14 Two methods of constructing expansion joints to relieve cracking in masonry walls.

Brick Reinforcement

Load-bearing masonry walls must be reinforced. The architect or engineer will determine the size and spacing of the **rebar** to be used based on the loads to be supported and the stress from wind or seismic loads. Rebar specifications will be discussed later in this chapter.

Most brick walls are reinforced with a wire reinforcement pattern similar to those shown in Figure 10-15. The IBC requires wire that meets ASTM-A82 specifications. Depending on the manufacturer, wire is usually 8 or 9 gauge or 3/16" diameter. Welded wire reinforcement and wire ties do not change size for metric use. The engineer will specify the required type and size based on the vertical loads to be supported, and the seismic and wind loads will affect the placement. Wire reinforcement similar to that shown in Figure 10-15 is usually placed continuously in horizontal mortar joints at six-course intervals. The top course and the first two courses above and below any wall opening should

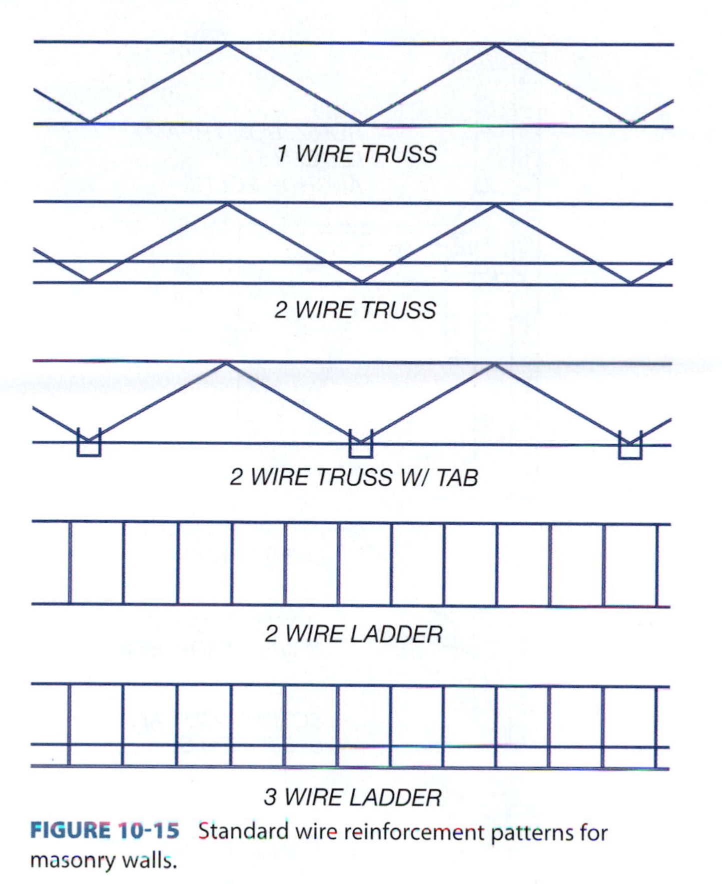

FIGURE 10-15 Standard wire reinforcement patterns for masonry walls.

be reinforced in all walls. The reinforcement should extend 24" (600 mm) minimum beyond each side of the opening. The cavity space between wythes is filled with grout and normally reinforced with vertical and horizontal steel. Steel reinforcement will be introduced later in this chapter.

Brick Veneer Construction

Using **brick veneer** over a wood, steel, or concrete masonry load-bearing wall allows the amount of brick required to be cut in half, greatly reducing the cost of construction. Care must be taken to protect the wood or steel frame from moisture in the masonry. Brick is installed over a 1" (25 mm) air space and a 15# layer of felt is applied to the framing. The veneer is attached to the framing with 26-gauge metal ties at 24" (600 mm) o.c. along each stud. Figure 10-16 shows an example of how masonry veneer is attached to a wood- or steel-frame wall. Masonry can also be attached to a steel frame as shown in Figure 10-17. When the flange is parallel to the masonry, straps measuring approximately 2" × 7" × 1/8" (50 × 180 × 3 mm) are used to bond the masonry to the steel. When the web is parallel to the masonry, 16-gauge straps are typically used to connect the masonry to the steel frame.

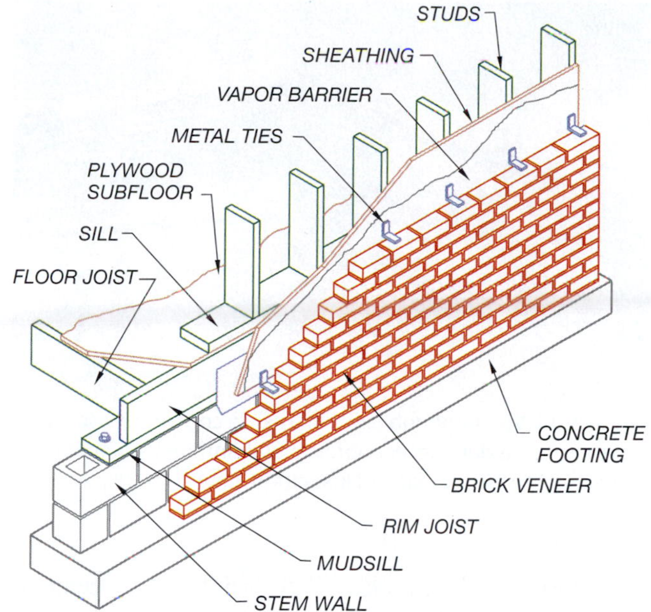

FIGURE 10-16 Brick is attached to wood or steel stud walls by the use of metal ties placed on each stud.

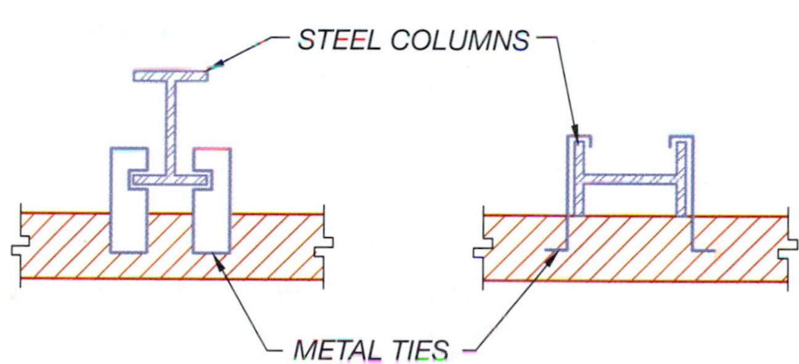

FIGURE 10-17 Brick can be attached to a steel skeleton by the use of metal ties that attach to the steel column.

CONCRETE MASONRY CONSTRUCTION

Concrete masonry units (CMUs) similar to Figure 10-18 provide a durable, economical building material with excellent structural and insulation values. The IBC relies on information provided by the ASTM for most technical information related to CMUs. Key standards include:

ASTM C55	Concrete brick
ASTM C73	Calcium silicate face brick
ASTM C90	Load-bearing concrete masonry units
ASTM C744	Prefaced concrete and calcium silicate masonry units

CMU walls are usually single-wythe, but they may be combined with a wythe of decorative stone or brick. Figure 10-19 shows standard wall construction. The IBC does not allow concrete block walls to exceed 35' (10,700 mm) in height between diaphragms. CMUs can

FIGURE 10-18 Although labor intensive, concrete masonry units provide a durable, economical building material that provides excellent structural and insulation values. *Courtesy Janice Jefferis.*

be waterproofed with clear waterproof sealers, cement-based paints can be used as the exterior finish, or the walls can be covered with stucco. Pressure-treated wood furring strips can be attached to the interior side of the block to support sheet rock, or the interior surface can be left exposed. Concrete blocks are also used for below-grade construction, which is discussed in Chapter 24.

Grades

Four classifications are used to define concrete blocks for construction. These include hollow, load-bearing (ASTM C90), solid load-bearing (ASTM C145), and nonload-bearing blocks (ASTM C129) that can be either solid or hollow blocks. Solid blocks must be 75% solid material in any cross-sectional plane. Blocks are also classified by their weight as normal, medium, and lightweight blocks. The weight of the block is affected by the type of aggregate used to form the unit. Normal aggregates such as crushed rock and gravel produce a block weight of between 40 and 50 lbs (18 and 23 kg) for an 8" × 8" × 16" block. Lightweight aggregates include coal cinders, shale, clay, volcanic cinders, pumice, and vermiculite. Use of a lightweight aggregate will produce approximately a 50% savings in weight. Walls made of CMUs are able to qualify for up to a four-hour fire rating. The type of aggregate used in the block and the thickness of the block dictate the fire rating. Blocks made with pumice in a minimum of 6" (150 mm) wide units produce a four-hour rating.

Common Sizes

CMUs come in a wide variety of patterns and shapes. In the United States, the standard module for concrete

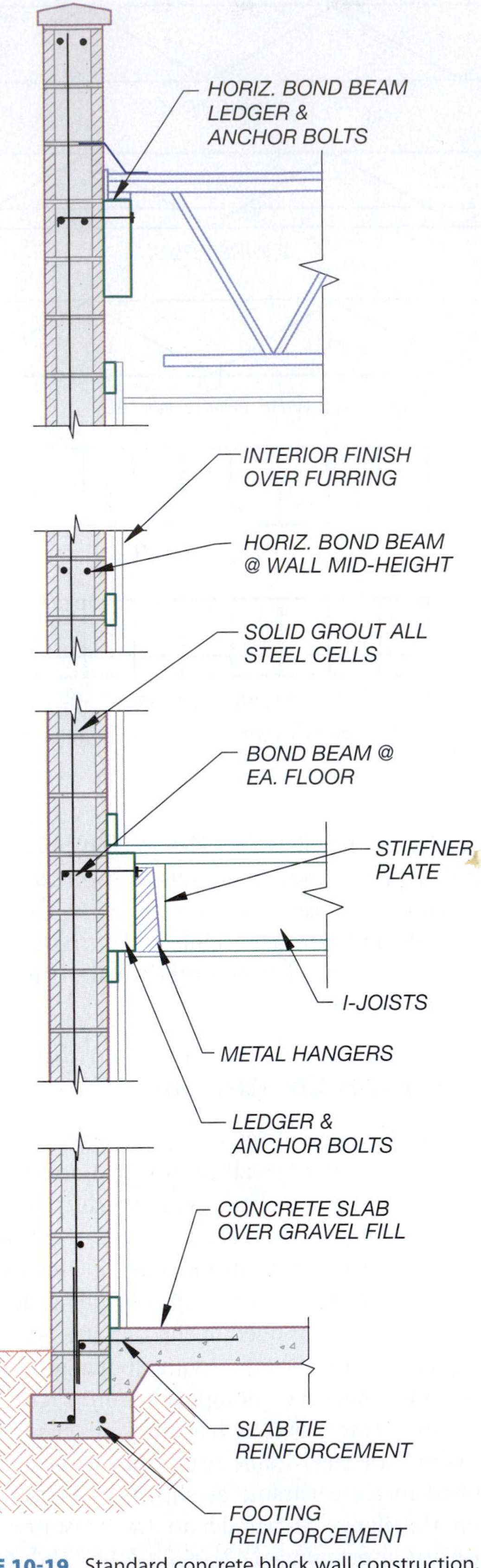

FIGURE 10-19 Standard concrete block wall construction. Walls are reinforced with steel rebar that is placed in solid grout to strengthen the wall. Extra steel referred to as a *bond beam* is typically placed where each floor or roof intersects the wall, at the mid-height between each floor, and at the top of the wall.

masonry units is 4 inches. Standard inch-pound blocks are manufactured to 4", 6", 8", 10", and 12" nominal widths, 4" and 8" heights, and 8" and 16" lengths. The most commonly used sizes are 8" and 16" blocks with 8 × 8 × 16 the industry standard. Each dimension of a concrete block is actually 3/8" smaller to allow for a 3/8" mortar joint in each direction.

Metric Block Sizes

The size of metric block will depend on whether the block size is based on soft or hard conversion. Soft conversion requires no physical change in product size. An 8 × 8 × 16 would be relabeled as a 203 × 203 × 406. This relabeling of inch-pound block is referred to as a *soft conversion*. A designer working with a 4" design module would now be working with a design module of 101.6.

With *hard metric conversion*, masonry blocks are manufactured to metric specifications based on a 100 mm module. Hard metric conversion requires the use of new molds to produce blocks that are 200 × 200 × 400 mm. This requires block manufacturers to purchase new block molds, maintain dual inventories, and develop ways of distinguishing between similar looking metric and inch-pound blocks during storage and shipping. Metric (hard) blocks are manufactured to nominal dimensions of 100, 200, 250, and 300 mm widths, 100 and 200 mm heights, and 200 and 400 mm lengths. Actual dimensions are 10 mm smaller than the nominal size to allow for the vertical and horizontal mortar joints. A designer would now be working with a design module of 100.

Metric Layout

The designer and the CAD drafter must take care in planning and drawing a CMU structure. Severe problems will result if the hard conversion metric blocks are used in a building designed with soft metric dimensions and vise versa. With soft conversion, a 4" module will be 101.6 mm or 1.6% larger than a 100 mm block. This may seem unimportant, but the difference of 1.6 mm (approximately 1/16") will accumulate to become 3.2 mm (1/8") in 8 inches, 6.4 mm (1/4") in 16" and 19.2 mm (3/4") in 48 inches. This size difference will be huge in a wall of several hundred feet. Because concrete block is too big to fit within the standard 100 mm module without changing all block sizes, federally funded block structures can be drawn in either inch/pound or metric sizes. The U.S. Congress allows the use of either inch-pound blocks or metric blocks for federal construction projects. The National Concrete Masonry Association makes the following recommendations for those involved in the design of a metric structure:

1. The architect should verify the availability of the size and type of building components before establishing if the design is to be based on soft or hard conversions.
2. Hard conversions should be used only when all components are readily available in metric modular sizes from multiple suppliers.
3. Soft metric design should be used when hard metric components are not readily available.
4. Soft metric structures should be designed in inch-pound units and then all dimensions should be converted to their metric equivalents.
5. If hard metric design methods are to be used, special consideration should be given to the use of custom 7.5" (190.5 mm) high units and the placement of openings.

Figure 10-20 shows the difference in layouts between hard and soft conversions for horizontal and vertical coursings.

Modular Block Layout

To minimize cutting and labor cost, concrete block structures should be kept to their modular size. When using the standard 8" × 8" × 16" (200 × 200 × 400 mm) CMU, structures that are an even number of feet long should have dimensions that end in 0" or 8". For instance, walls that are 24'–0" or 32'–8" long are each modular. Walls that are an odd number of feet should end with 4". Walls that are 3'–4", 9'–4", and 15'–4" are modular, but a wall that is 15'–8" requires blocks to be cut.

Common Shapes

In addition to the exterior patterns, concrete blocks also come in a variety of shapes, as shown in Figure 10-21. Individual blocks are manufactured with two or three cores. Solid blocks are also manufactured. Two-core concrete blocks reduce heat conduction by approximately 4%, have more space for mechanical pipes and conduits, and weigh about 4 lbs (1.8 kg) less than a three-core block. These cores also allow for the placement of steel reinforcement bars, although steel-reinforcing mesh—similar to the mesh used for brick—can also be used. Reinforcement is discussed later in this chapter.

Concrete blocks come in a variety of surface finishes. A standard block has a smooth but porous finish. The texture of the block can be modified by changing the

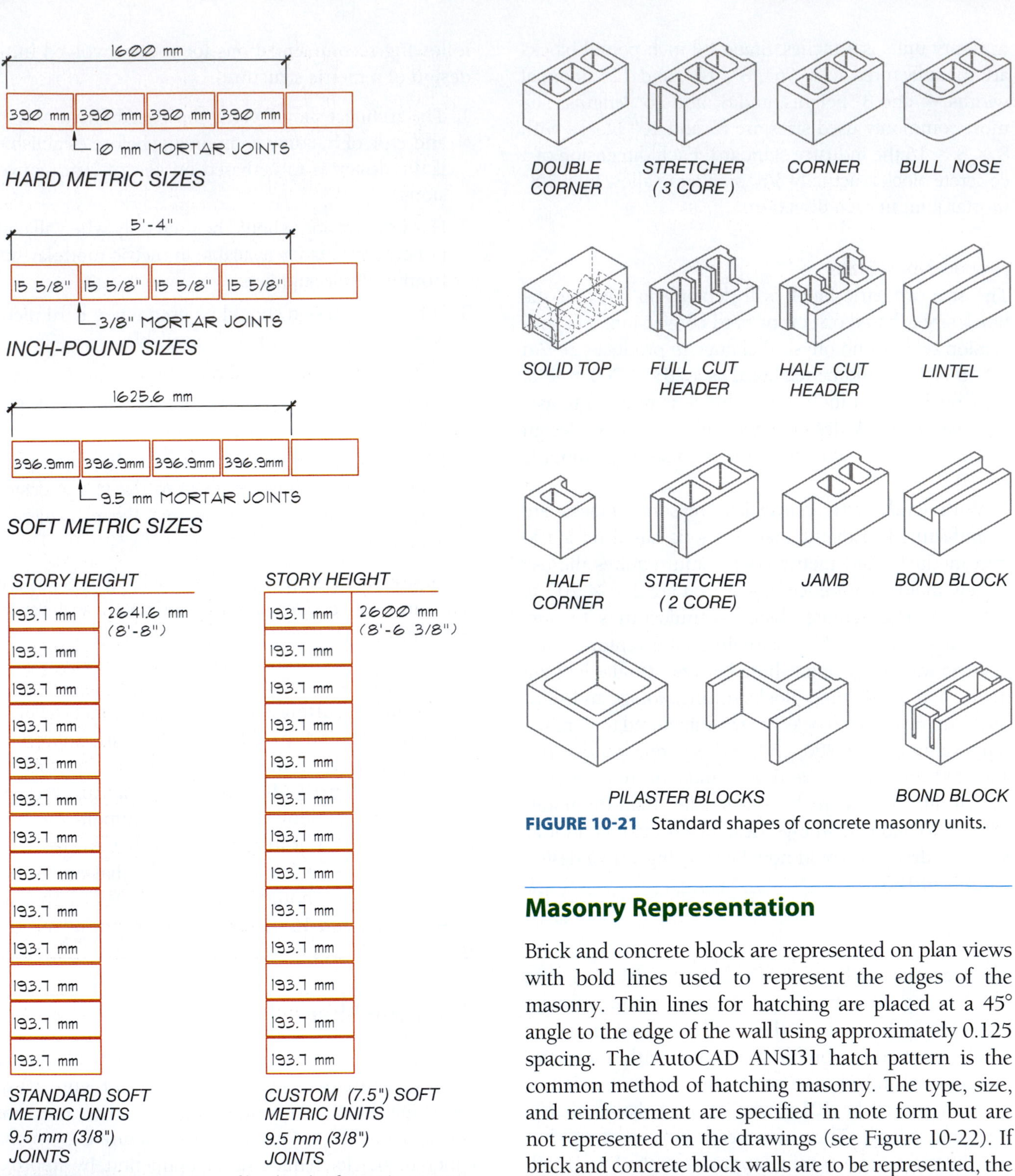

FIGURE 10-20 A comparison of hard and soft metric conversions for horizontal and vertical coursings.

FIGURE 10-21 Standard shapes of concrete masonry units.

coarseness of the aggregate used to make the block. *Grid face* or *scored face* blocks are made with seams cut into the block, so that they resemble 8" × 8" × 8" (200 × 200 × 200 mm) blocks rather than 16" (400 mm) blocks. Seams are also available at 4" (100 mm) o.c. Blocks are also available with a raised or recessed geometric shape pressed into the face.

Masonry Representation

Brick and concrete block are represented on plan views with bold lines used to represent the edges of the masonry. Thin lines for hatching are placed at a 45° angle to the edge of the wall using approximately 0.125 spacing. The AutoCAD ANSI31 hatch pattern is the common method of hatching masonry. The type, size, and reinforcement are specified in note form but are not represented on the drawings (see Figure 10-22). If brick and concrete block walls are to be represented, the hatch patterns are placed at opposing angles.

The size and location of concrete block walls are dimensioned on the floor plan. Concrete block is dimensioned from edge to edge of walls, as shown in Figure 10-23. Openings in the wall are also located by dimensioning to the edge. The scale of the drawing affects the drawing method used to represent masonry shown in sections and details. At scales under 1/2" = 1'-0", the wall is usually drawn just as in plan view. At larger scales, details typically reflect the cells of the

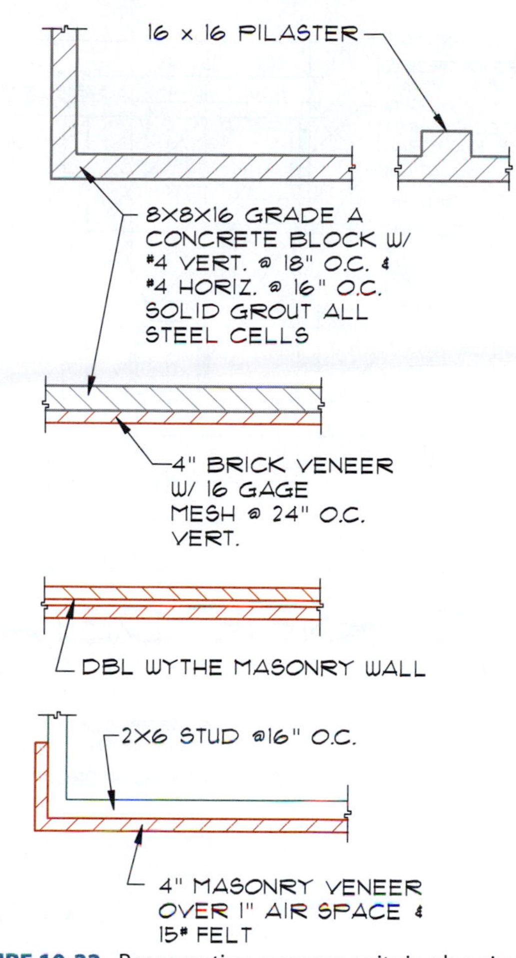

FIGURE 10-22 Representing masonry units in plan view.

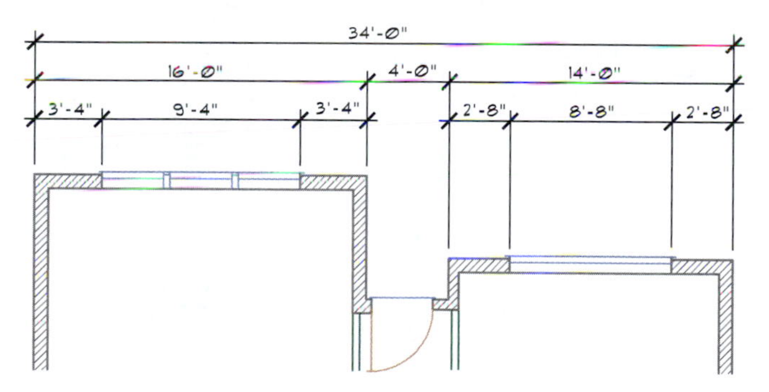

CONCRETE MASONRY UNITS

FIGURE 10-23 Structural brick and concrete masonry units are dimensioned from exterior face to exterior face. Brick veneer is not dimensioned in plan view. The drafter should take great care to maintain modular sizes as masonry walls are dimensioned.

block. Figure 10-24 shows methods of representing concrete block in cross section.

Figure 10-25 shows an example of a section and the notes that usually accompany concrete block construction.

When CMUs have to support loads from a beam, a pilaster is placed in the wall to carry the loads. A

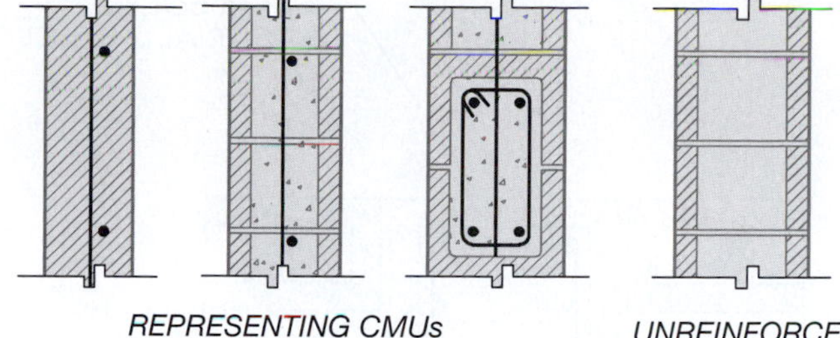

REPRESENTING CMUs IN SECTION & DETAIL

UNREINFORCED MASONRY

FIGURE 10-24 Methods of representing CMUs in sections and details.

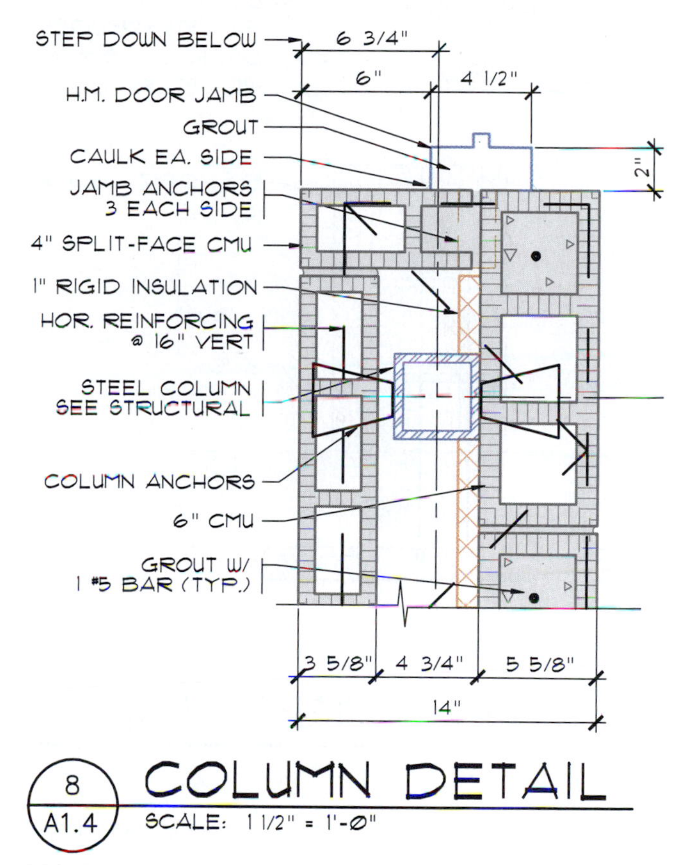

FIGURE 10-25 Representing brick and CMUs in detail. *Courtesy G. Williamson Archer A.I.A., Archer & Archer P.A.*

pilaster is a thickened area of wall used to support gravity loads or to provide lateral support to the wall when the wall is to span long distances. Figure 10-26 shows a detail showing the size and thickness of a pilaster. Figure 10-22 shows how a pilaster is represented in plan view.

Details similar to Figure 10-27 are required to show reinforcement connections at each intersection and where different materials are joined to the blocks. This will require the drafter to draw wall-to-footing, wall-to-slab, wall-to-wall, wall-opening, and wall-to-roof details. When a wood floor or roof is to be supported by the block, a ledger is bolted to the block. Holes are punched in the block to allow an anchor bolt to

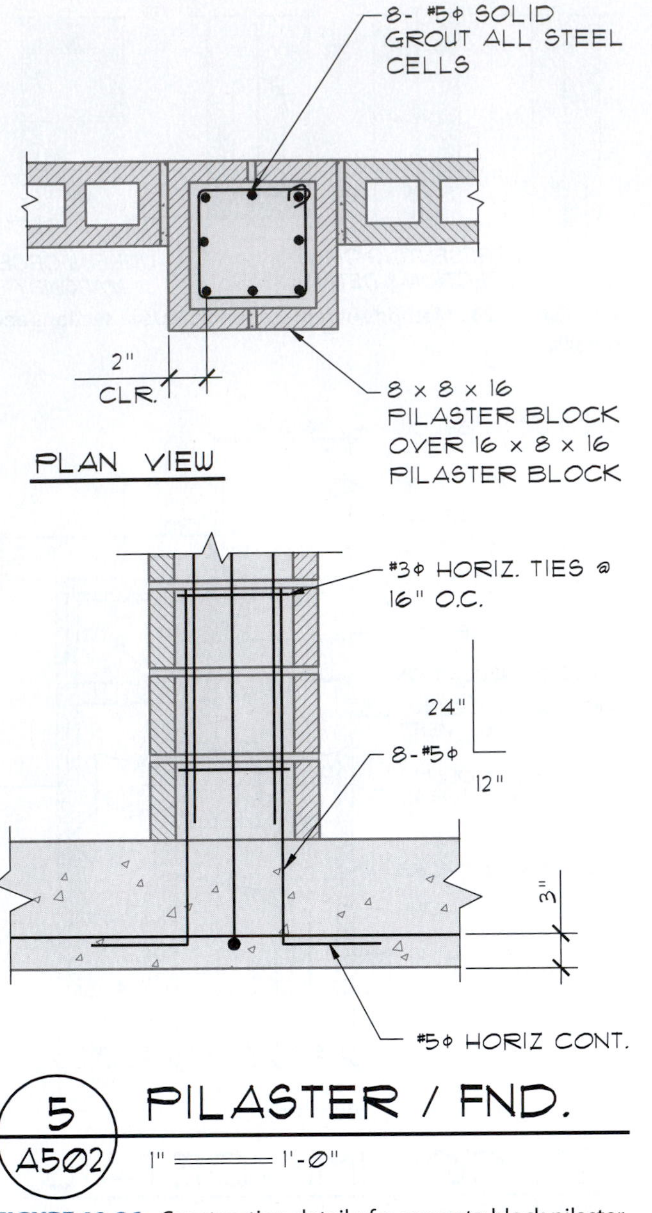

FIGURE 10-26 Construction detail of a concrete block pilaster.

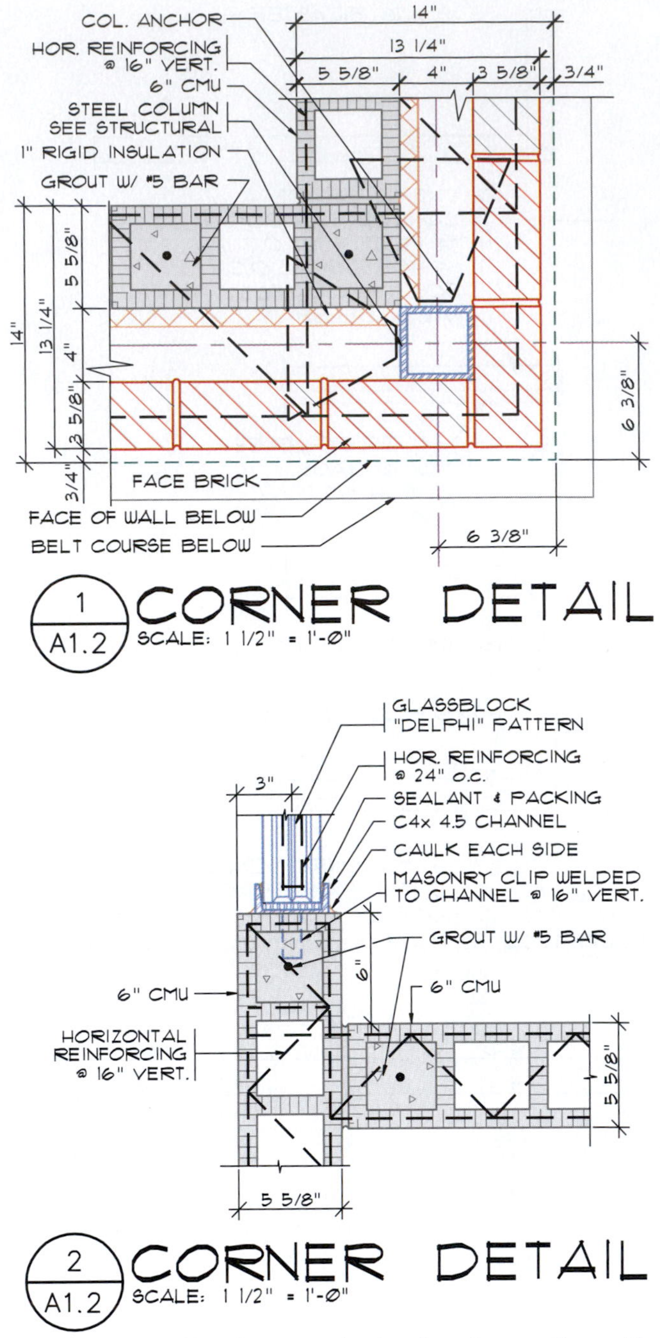

FIGURE 10-27 Details are required to show intersections with each material to be used as well as intersections of floor and roof systems. *Courtesy G. Williamson Archer A.I.A., Archer & Archer P.A.*

penetrate into the cell, where it is wired to the wall reinforcing. The block is reinforced and the bolt and rebar are held in place by filling the entire cell with grout. Figures 10-28a and 10-28b show details for wood and steel floor connections to a block wall. Care must be taken to tie the ledger bolts into the wall reinforcement, as well as securely tying the floor joist to the ledger. Figure 10-28c shows a ledger attached to a CMU wall with beam hangers in place.

The same procedure must be followed when a roof is tied into a wall. In addition, methods of keeping the intersection waterproof must be implemented. The roofing is usually extended up the wall over a cant strip that is covered with metal flashing, which sheds water from the joint. Figure 10-29 shows examples of how the roof intersections are represented in detail. Figure 10-30 shows examples of common mortar wall caps that are used to protect the top of the masonry from moisture.

STEEL REINFORCEMENT

Masonry units are excellent for resisting forces from compression, but they are poor for resisting forces in tension. Steel is excellent for resisting forces of tension, but it tends to buckle under forces of compression. The combination of these two materials forms an excellent unit for resisting great loads from lateral and vertical

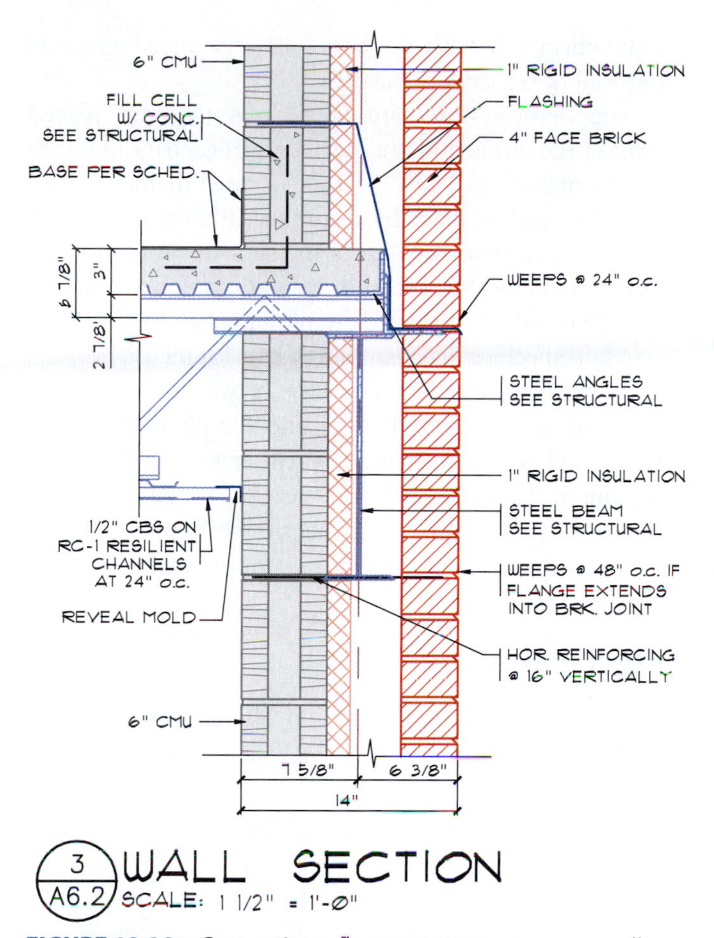

FIGURE 10-28a Connecting a floor system to a masonry wall. *Courtesy G. Williamson Archer A.I.A., Archer & Archer P.A.*

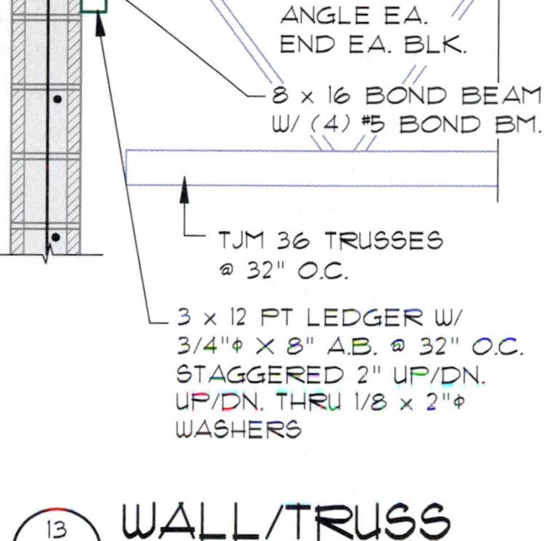

FIGURE 10-28b Connecting a floor system to a masonry wall using a pressure-treated ledger.

FIGURE 10-28c A pressure-treated ledger attached to a CMU wall. Metal hangers have been attached in preparation for installation of floor beams. *Courtesy Janice Jefferis.*

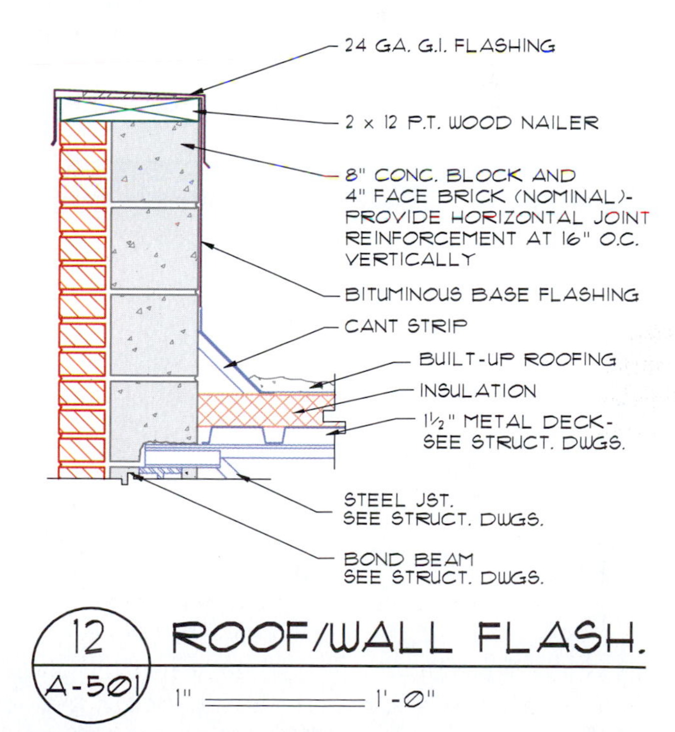

FIGURE 10-29 The method used to waterproof the wall/roof intersection must be represented in detail as well as the written specifications.

forces. Reinforced masonry structures are stable because the masonry, steel, grout, and mortar bond together so effectively. Loads that create compression on the masonry are transferred to the steel. By careful placement of the steel, the forces will result in tension on the steel and be safely resisted.

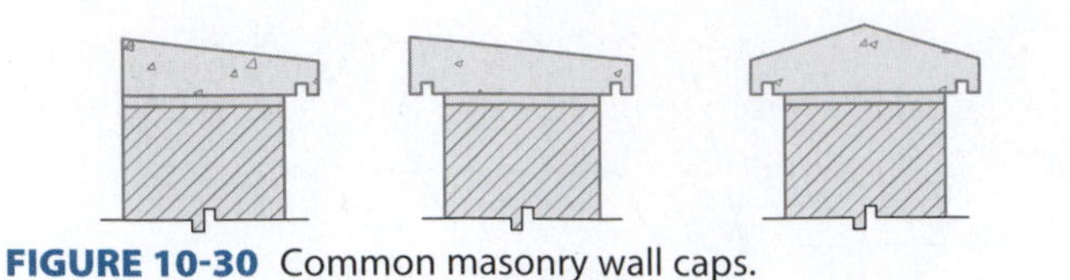

FIGURE 10-30 Common masonry wall caps.

Reinforcing Bars

Placing steel in the masonry creates a wall known as *reinforced masonry construction*. An intersecting horizontal and vertical grid of concrete reinforced with steel runs throughout a concrete block wall, forms the frame for a structure. Although wire mesh similar to that used with brick can be used in concrete block, steel reinforcing bars are the most common method of reinforcing concrete block. The IBC sets specific guidelines for the size and spacing of rebar based on the seismic zone of the construction site. Reinforcement is specified by the engineer throughout the calculations and sketches based on the building code and the Portland Cement Association (PCA) and the Concrete Reinforcing Steel Institute (CRSI) guidelines, and it must be accurately represented by the drafter.

Rebar Sizes and Shapes

Reinforcing bars can be either smooth or deformed. Smooth bars are primarily used for joints in floor slabs. Because the bars are smooth, the concrete of the slab does not bond easily to the bar. This allows the concrete to expand and contract. Bars used for reinforcing masonry walls are usually deformed similar to those shown in Figure 10-31 so that the concrete will bond more effectively with the bar. This prevents slippage as the wall flexes under pressure.

The **deformations** are small ribs that are placed around the surface of the bar. Deformed bars range in size from 3/8" to 2 1/4" (10 to 57 mm) diameter. Bars are referenced on plans by a number rather than a size. A number represents the size of the steel in approximately 1/8" increments. Steel reinforcing bars conform to ASTM A615M standards for metric reinforcing bars. This ASTM standard specifies that reinforcing bars for metric projects are to be listed by their soft metric conversion sizes. This standard also governs the size and the physical characteristics of rebar. Steel may be either grade 40, with a minimum yield strength of 40 ksi (275 MPa), or grade 60, with a minimum yield strength of 60 ksi (415 MPa). Common bar sizes of steel are shown in Table 10-2.

Steel Placement

Reinforcing steel is required in all masonry walls in seismic zones C, D, E, and F. Wall reinforcing is held

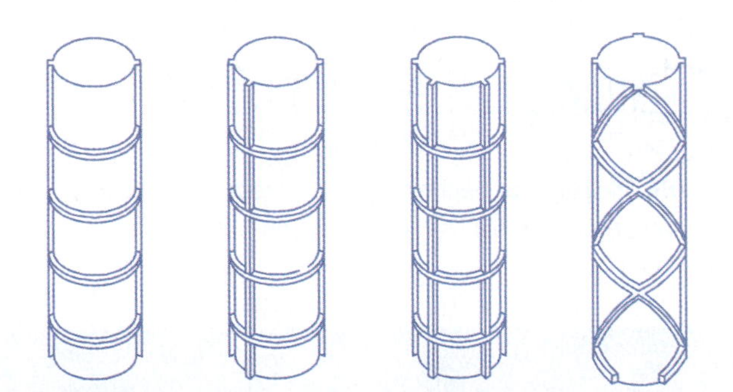

FIGURE 10-31 Common deformed steel reinforcing patterns.

Traditional Bar Size	Nominal Diameter in Inches	Nominal Diameter in Millimeters	Soft Metric Size
# 3	0.375	9.5	# 10
# 4	0.500	12.7	# 13
# 5	0.625	15.9	# 16
# 6	0.750	19.1	# 19
# 7	0.875	22.2	# 22
# 8	1.000	25.4	# 25
# 9	1.128	28.7	# 29
# 10	1.270	32.3	# 32
# 11	1.410	35.8	# 36
# 14	1.693	43.0	# 43
# 18	2.257	57.3	# 57

TABLE 10-2 Common Bar Sizes and Dimensions of Steel

in place with wire ties and by filling each masonry cell containing steel with grout. When exact locations within the cell are required by design, the steel is wired into position, so that it can't move as grout is added to the cell. The placement of steel varies with each application, but steel reinforcement is placed on the side of the wall that is in tension.

Because an aboveground wall will receive pressure from each side, masonry walls have the steel centered in the wall cavity. For cantilevered retaining walls, steel is placed near the soil side of the wall. For full-height retaining walls anchored at the top and bottom, the steel is placed near the side that is opposite the soil. The location of steel in relation to the edge of the wall must be specified in the wall details.

Vertical Reinforcing

Figure 10-32 shows the vertical steel placed in a CMU column. Vertical reinforcing is required in walls at 48" (1200 mm) o.c. with a maximum spacing of 24" (600 mm) if stacked bond masonry is used. IBC requires a vertical piece of steel to be placed within 16" from each end of a wall. Vertical steel can be represented in details similar to Figures 10-19, 10-24, or 10-26. Additional vertical reinforcing is usually required at the edges of openings in walls to serve the same function as a post in wood construction. For small loads, two vertical bars in the same cell may be sufficient. As loads increase, the size of the vertical bond and the number of bars used will increase. Horizontal ties are normally added when four or more vertical bars are required to keep the bars from separating. Figure 10-33 shows an example of a reinforced doorjamb.

FIGURE 10-32 Vertical steel placed in a CMU column with horizontal ties placed to keep the vertical steel from separating. *Courtesy Tom Worcester.*

Horizontal Reinforcing

Figure 10-34 shows the horizontal wall steel extending from a joint in a masonry wall. Horizontal reinforcement is approximately 16" (400 mm) o.c., but the exact spacing depends on the seismic or wind loads to be resisted and is designed by the engineer for each specific use. In addition to the horizontal steel required by the building code, extra steel is placed at the midpoint of a wall between each floor level (see Figures 10-19 and 10-28b). Generally, two bars are placed at the midpoint

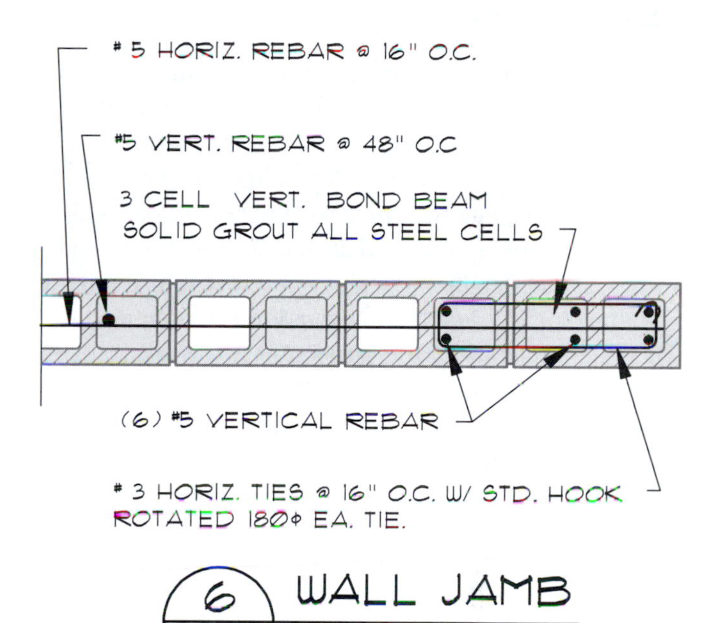

FIGURE 10-33 In addition to the required wall steel, concrete block beside wall openings requires special reinforcement. Vertical reinforcing beside the opening is held in place by horizontal ties.

FIGURE 10-34 Steel reinforcing can be seen extending from this masonry wall. *Courtesy Richard Schmitke.*

of each wall level, forming a reinforced area referred to as a **bond beam.** Extra reinforcing is also added where each floor or roof level intersects the wall. A 16" deep bond beam with two pieces of steel at the top and bottom is a standard method of constructing the bond beam at floor and roof levels. An 8" deep bond beam with two pieces of steel is also placed at the top of concrete block walls. Another common location for a bond beam is over the openings for doors and windows. Figure 10-35 shows a detail representing the reinforcing required over a door opening. The exact depth, quantity of rebar, and ties will be specified by the engineer based on the load to be supported and the span of the bond beam.

Steel Overlap

Because of the limits of construction, steel must often be lapped to achieve the desired height or length. In-line steel bars are lapped and wired together so that individual bars will act as one. Depending on the loads to be resisted, the engineer may require **laps** to be welded. The amount of lap required at steel intersections can be found in the details or in the written specifications of the drawings. Where the information is placed depends on the complexity of the project. In a typical wall section, the lap is not represented. The intersection of a wall at the foundation, for example, is a typical place where steel is lapped. The lap is shown in these details because the reinforcing steel must consist of two separate bars. Figures 10-26 and 10-36 show details with the lapped steel. The L-shaped bar extends from the footing into the wall, and the second piece of steel for the wall is tied to the foundation steel to form a secure bond

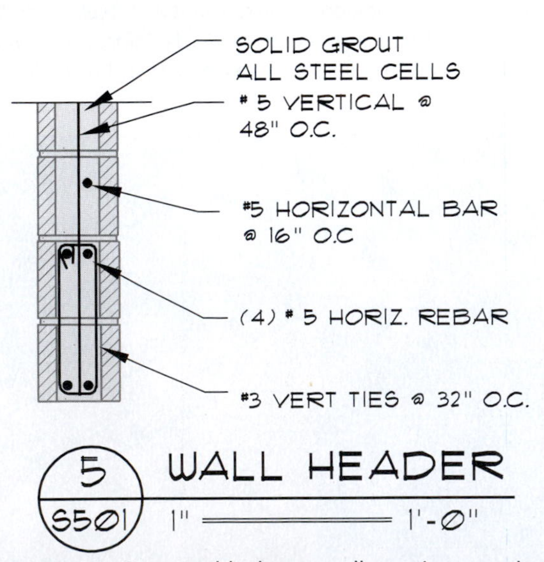

FIGURE 10-35 Concrete block over wall openings requires special reinforcement. Above an opening, horizontal steel is held together with vertical ties.

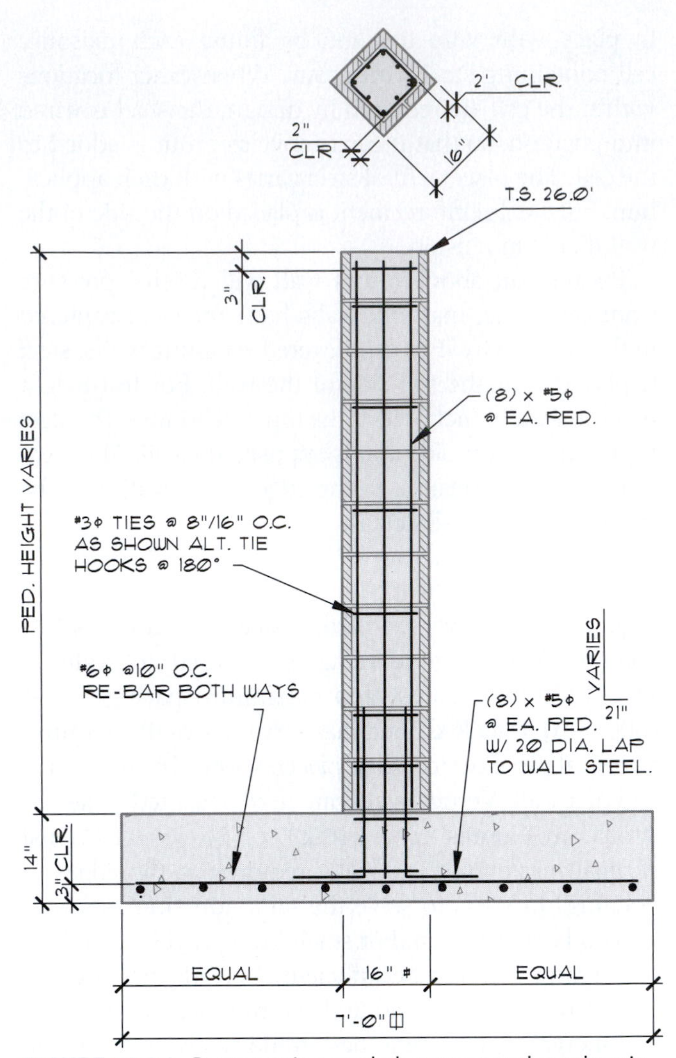

FIGURE 10-36 Representing steel placement and overlaps in details.

between the wall and the foundation. The engineer usually places a note in the calculations indicating the desired overlap of steel to be specified by the drafter. If the engineer specifies the use of #5 bar with a 36-diameter lap, the drafter would need to show the steel overlapping 23" (0.065 diameter × 36 [required lap] = 23") on the drawing. Laps of steel based on diameter are listed in Table 10-3.

Steel Bends and Hooks

A *bend* allows steel reinforcement to be continuous at a corner. A **hook** ties intersecting members to each other. For example, the horizontal steel in a footing may be hooked around the vertical steel in an intersecting footing to ensure that the horizontal can't move in a lateral direction. Unless the engineer has provided a detail, hooks and bends should be detailed to match the minimum standards set by the Concrete Reinforcing Steel Institute (CRSI). Figure 10-37 shows the minimum

Number of Diameters	SIZE OF BAR								
	#3	#4	#5	#6	#7	#8	#9	#10	#11
20	6	10	13	15	18	20	23	26	29
22	8	11	14	17	20	22	25	28	32
24	9	12	15	18	21	24	28	31	34
30	11	15	19	23	27	30	34	39	43
32	12	16	20	24	28	32	37	41	46
36	14	18	23	27	32	36	41	46	51
40	15	20	25	30	35	40	46	51	57
48	18	24	30	36	42	48	55	61	68

TABLE 10-3 Inches of Lap Corresponding to the Number of Bar Diameters

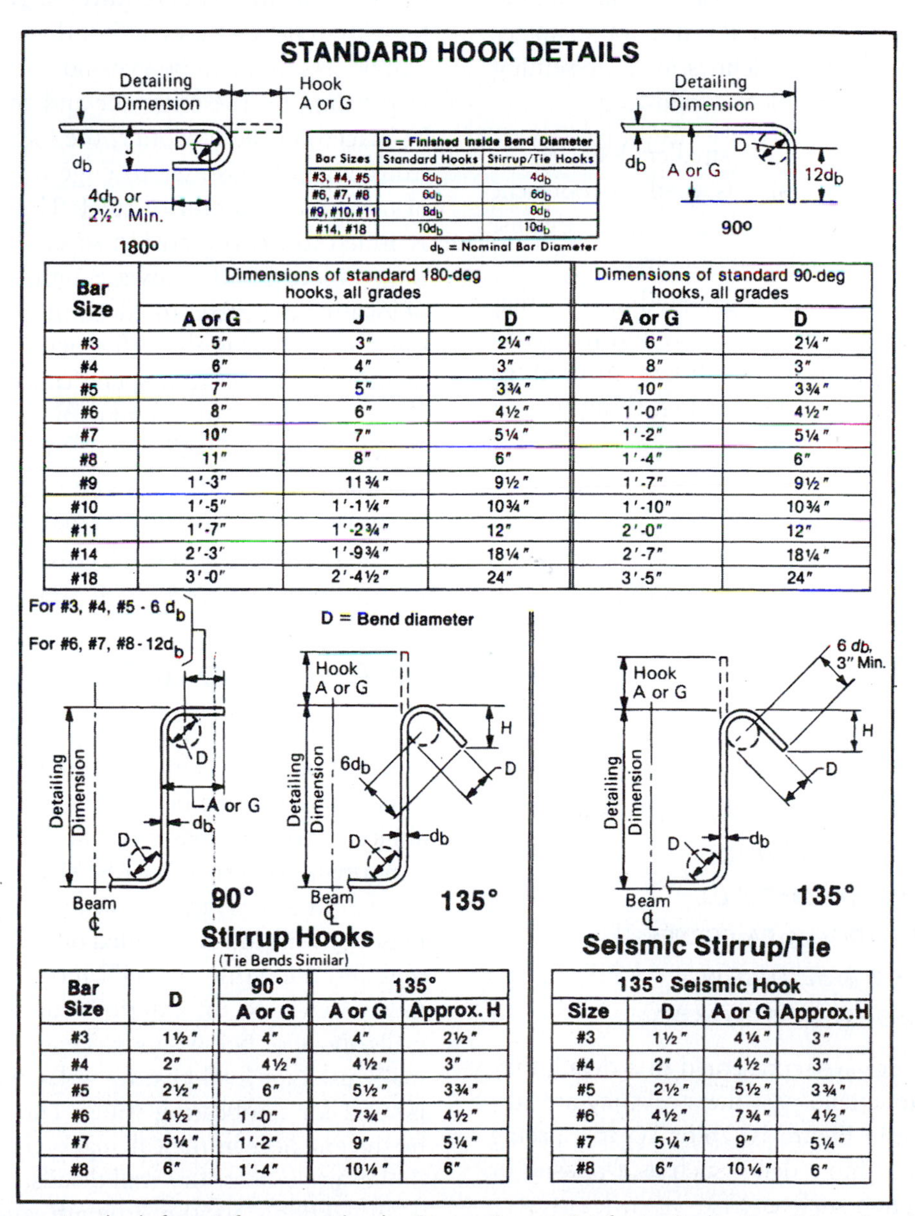

Bar Sizes	D = Finished Inside Bend Diameter: Standard Hooks	Stirrup/Tie Hooks
#3, #4, #5	6db	4db
#6, #7, #8	6db	6db
#9, #10, #11	8db	8db
#14, #18	10db	10db

Bar Size	Dimensions of standard 180-deg hooks, all grades			Dimensions of standard 90-deg hooks, all grades	
	A or G	J	D	A or G	D
#3	5"	3"	2¼"	6"	2¼"
#4	6"	4"	3"	8"	3"
#5	7"	5"	3¾"	10"	3¾"
#6	8"	6"	4½"	1'-0"	4½"
#7	10"	7"	5¼"	1'-2"	5¼"
#8	11"	8"	6"	1'-4"	6"
#9	1'-3"	11¾"	9½"	1'-7"	9½"
#10	1'-5"	1'-1¼"	10¾"	1'-10"	10¾"
#11	1'-7"	1'-2¾"	12"	2'-0"	12"
#14	2'-3"	1'-9¾"	18¼"	2'-7"	18¼"
#18	3'-0"	2'-4½"	24"	3'-5"	24"

Bar Size	D	90°	135°	
		A or G	A or G	Approx. H
#3	1½"	4"	4"	2½"
#4	2"	4½"	4½"	3"
#5	2½"	6"	5½"	3¾"
#6	4½"	1'-0"	7¾"	4½"
#7	5¼"	1'-2"	9"	5¼"
#8	6"	1'-4"	10¼"	6"

135° Seismic Hook			
Size	D	A or G	Approx. H
#3	1½"	4¼"	3"
#4	2"	4½"	3"
#5	2½"	5½"	3¾"
#6	4½"	7¾"	4½"
#7	5¼"	9"	5¼"
#8	6"	10¼"	6"

FIGURE 10-37 Minimum standards for reinforcement hooks. *Courtesy Concrete Reinforcing Steel Institute, Schaumberg, IL.*

standards for hooks and bends. The location and length of the hook is normally placed on the details and in the specifications.

Rebar Representation

The CAD technician needs to specify the quantity of bars, the bar size by number, the direction the steel runs, and the grade on the drawings where steel is referenced. On drawings such as the framing plan, walls are drawn, but the steel is not drawn. Steel specifications are generally included in the wall reference as shown in Figure 10-22. When shown in section or detail, steel is represented by a bold line, which can be solid or dashed depending on office practice. A solid circle represents steel when it is shown in end view. The AutoCAD DONUT command is effective for drawing rebar in end view. Make sure the FILL command is ON, and that the interior diameter value has been set to zero. Guidelines for representing reinforcing steel in detail include the following:

- On drawings plotted at a scale smaller than 3/4" = 1'–0", a constant symbol size is used to represent various steel bar sizes. With larger scaled details, variations in bar sizes can be represented.
- Steel is often drawn at an exaggerated size so that it can be seen more clearly. Rebar should not be so small that it blends with the hatch pattern used for the grout, or so large that it is the first thing seen in the detail.

Figures 10-24 through 10-28c show common examples of representing steel in detail. The direction the steel is to be laid is specified to add clarity to the drawing. Steel direction is specified using the terms vertical, horizontal, or diagonal. Occasionally the letters *E.W.* or **E/W** are used to indicate that the steel is running each way. This might include the horizontal and vertical steel placed using uniform spacing in a wall, or steel placed in each direction in a horizontal mat in a foundation. Notes that are placed on the drawing might include:

#5 Ø VERT @ 48" O.C.
#5 Ø HORIZ @ 32" O.C.
(3) - #5 Ø DIAG. @ EA. CORNER
#5 Ø @ 32" O.C. E.W.
(8) - #5 Ø E.W. 3" UP FROM BTM.

When hooks or ties are represented in a detail, they should be shown to extend past the steel they are surrounding, as shown in Figure 10-26 or 10-36. Spacing of ties often varies within a detail such as a pilaster or column. A specification such as

#3 TIES @ 8/16" O.C.

indicates that the ties are to be a #3 diameter with spacings of 8" and 16". The pattern is then shown in a detail similar to Figure 10-36. Notice that the three bars at the top and bottom are placed at 8" spacings and the bars in the middle of the column are spaced at 16" o.c.

Locating Steel

Dimensions are required to show the location of the steel from the edge of the masonry. The engineer may provide a note in the calculations such as

(7)- #5 HORIZ. @ 3" UP/DN.

Although this note could be placed on the drawings exactly as is, a better method is shown in Figure 10-36. The quantity and size of the steel are specified in the detail, but the locations are placed using separate dimensions. This requires slightly more work on the part of the technician, but the visual specifications will be less likely to be overlooked or misunderstood. Depending on the engineer and the type of stress to be placed on the masonry, the location may be given from edge of masonry to edge of steel or from edge of masonry to center of steel. The term *clear* or *CLR* is added to the dimension to describe the steel location. Figure 10-36 shows examples of each method of locating concrete. In addition to the information in the details, steel is also referenced in the written specifications. The written specifications detail the grade and strength requirements for general areas of the structure such as walls, floors, columns, or retaining walls. Figure 10-38 shows an example of steel specifications that accompany a small office structure. See Chapter 11 for additional types of steel reinforcing used with other types of concrete construction.

MORTAR AND GROUT

Mortar used to bond masonry products and steel is composed of portland cement, sand, lime, and clean water. Other materials can be added to the mix to increase its ability to bond to masonry units and steel. Mortar mix is normally governed by ASTM C144. The strength of the mortar is of critical importance to the strength of the masonry wall and is specified on the architectural drawings and the written specifications. Types M and S are most often used for exterior walls and are suitable for walls above or below grade. Type N or S mortar is normally used for load-bearing walls, while type O mortar is used for nonbearing walls. The mortar is specified by the engineer in the calculations and specified by the drafter in the masonry wall details.

In addition to the strength and makeup of the mortar, the type of joint is usually specified. Joints used

REINFORCING

1. ALL REINFORCING STEEL TO BE ASTM A615 GRADE 60, EXCEPT TIES, STIRRUPS & DOWELS TO MASONRY TO BE GRADE 40. W. W. F. SHALL CONFIRM TO ASTM A185 AND SHALL BE 6 x 6-W1.4XW1.4 WWF MATS.

2. FABRICATE AND INSTALL REINF. STEEL ACCORDING TO THE "MANUAL OF STANDARD PRACTICE FOR DETAILING REINFORCED CONCRETE STRUCTURES" ACI STANDARD 315.

3. PROVIDE 2'-0" x 2'-0" CORNER BARS TO MATCH HORIZ. REINFORCING IN POURED IN-PLACE WALLS & FTGS. @ ALL CORNERS & INTERSECTIONS.

4. SPLICES IN WALL REINFORCING SHALL BE LAPPED 30 DIAMETERS (2'-0" MINIMUM AND SHALL BE STAGGERED AT LEAST 4' AT ALTERNATE BARS.

5. ALL OPENINGS SMALLER THAN 30" x 30" THAT DISRUPT REINFORCING SHALL HAVE AN AMOUNT OF REINFORCING EQUAL TO THE AMOUNT DISRUPTED PLACED ON BOTH SIDES OF OPENING & EXTENDING 2'-0" EA. SIDE OF OPENING.

6. PROVIDE THE FOLLOWING REINFORCING AROUND WALL OPENINGS LARGER THAN 30" x 30":
   A. (2) #5 OVER OPENING x OPENING WIDTH PLUS 2'-0" EACH SIDE.
   B. (2) #5 UNDER OPENING x OPENING WIDTH PLUS 2'-0" EACH SIDE.
   C. (2) #5 EACH SIDE OF OPENING x FULL STORY HEIGHT.
   D. PROVIDE 90 DEGREE HOOK FOR BARS AT OPENINGS IF REQUIRED EXTENSION PAST OPENING CANNOT BE OBTAINED.

7. PROVIDE (2) #4 CONTINUOUS BARS AT TOP AND BOTTOM AND AT DISCONTINUOUS ENDS OF ALL WALLS.

8. PROVIDE (2) #5 X OPENING DIMENSION PLUS 2'-0" EACH SIDE AROUND ALL EDGES OF OPENINGS LARGER THAN 15" x15" IN STRUCTURAL SLABS AND PLACE (1) #4 x 4'-0" AT 45 DEGREES TO EACH CORNER.

9. PROVIDE DOWELS FROM FOOTINGS TO MATCH ALL VERTICAL WALL, PILASTER, AND COLUMN REINFORCING (POURED-IN-PLACE COLUMNS & WALLS). LAP 30 DIAMETERS OR 2'-0" MINIMUM.

10 ALTERNATE ENDS OF BARS 12" IN STRUCTURAL SLABS WHENEVER POSSIBLE.

11 ALL WALL REINFORCING TO BE PLACED IN CENTER OF WALL UNLESS SHOWN OTHERWISE ON THE DRAWINGS.

CONCRETE BLOCK

1. DESIGN F'M = 1500 PSI FOR SOLID GROUTED WALLS, 1350 PSI FOR WALLS WITH REINFORCED CELLS ONLY GROUTED.

2. ALL CMU UNITS TO BE GRADE N TYPE I LIGHTWEIGHT UNITS PER ASTM C90 DRY TO INTERMEDIATE HUMIDITY CONDITION PER TABLE NO. 1.

3. MORTAR TO BE PER IBC TABLE 2103.7(1) TYPE S OR PER MANUF. RECOMMENDATION TO REACH CORRECT STRENGTH. GROUT TO HAVE MINIMUM STRENGTH TO MEET DESIGN STRENGTH AND MADE WITH 3/8" MINUS ACCURATE.

4. CONTRACTOR TO PREPARE AND TEST 3 GROUTED PRISMS & 3 UNGROUTED PRISMS PER EVERY 5000 SQUARE FEET OF WALL, DURING CONSTRUCTION.

5. ALL WORK SHALL CONFORM TO SECTION 2103 THROUGH SECTION 2109 OF THE LATEST INTERNATIONAL BUILDING CODE.

6. REINFORCING FOR MASONRY TO BE ASTM A615 GR. 60 PLACED IN CENTERS OF CELLS AS FOLLOWS (UNLESS OTHERWISE INDICATED):
   A. VERTICAL: 1- #5 @ 48" O.C. PLUS 2- #5'S FULL HEIGHT EACH SIDE OF OPENINGS, UNLESS SHOWN OTHERWISE ON DRAWINGS.
   B. HORIZ.: 2- #4'S @ 48" O.C. (FIRST BOND BEAM 48" FROM GROUND FLOOR) PLUS 2- #4'S AT TOP OF WALL AND AT EACH INTERMEDIATE FLOOR LEVEL.
   C. LINTELS: LESS THAN 4'-0" WIDE-(2)- #4'S IN BTM OF 8"DEEP LINTEL. 4'-0" TO 6'-0" WIDE-(2)- #4'S IN BOTTOM OF 16" DEEP LINTEL.
   D. CORNERS AND INTERSECTIONS: (1)- 24" x 24" CORNER BAR AT EACH BOND BEAM SAME SIZE AS HORIZONTAL REINFORCING.

7. GROUT ALL CELLS FULL @ EXTERIOR OR LOAD-BEARING WALLS. GROUT ALL CELLS THAT CONTAIN REINFORCING OR EMBEDDED ITEMS.

8. ELECTRICAL BOXES, CONDUIT AND PLUMBING SHALL NOT BE PLACED IN ANY CELL THAT CONTAINS REINFORCING.

FIGURE 10-38 Written specifications for steel reinforcing, and concrete block for a multilevel office structure are often placed with the structural drawings. *Courtesy Van Domelen/Looijenga/McGarrigle/Knauf Consulting Engineers.*

with masonry units are similar to the joints used with brick construction and are shown in Figure 10-13.

Grout is composed of the same components as mortar, but the grading is slightly different. Grout is specified as fine or coarse and governed by ASTM C33. Specifications are placed on the drawings or in the project manual to control the proportions of elements in grout.

ADDITIONAL READING

The following Web sites can be used as a resource to help you keep current with changes in unit masonry.

ADDRESS	COMPANY/ ORGANIZATION
www.ambrico.com	American Brick Company
www.concrete.org	American Concrete Institute
www.astm.org	ASTM International
www.bia.org	Brick Industry Association
www.confast.com	Concrete Fastening Systems, Inc.
www.crsi.org	Concrete Reinforcing Steel Institute
www.nibs.org/MetricNews	Construction Metrication
www.masonryresearch.org	Council for Masonry Research
www.generalshale.com	General Shale Brick
www.imiweb.org	The International Masonry Institute
www.maconline.org	Masonry Advisory Council
www.masonrydetails.com	MasonryDetails.com
www.masonryinstitute.org	Masonry Institute of America
www.masonrysociety.org	The Masonry Society
www.masonrystandards.org	Masonry Standards Joint Committee
www.ncma.org	National Concrete Masonry Association
www.lime.org	National Lime Association
www.portcement.org	Portland Cement Association
www.rmmi.org	Rocky Mountain Masonry Institute
www.brick-wscpa.org	Western States Clay Products Association

EASY BEING GREEN

Clay, adobe, and concrete masonry units are key components of green building and sustainable design. Brick is made primarily from clay and shale, two of the most abundant natural resources. In addition to using readily available resources, brick manufacturing incorporates many sustainable practices of its own and has captured great production efficiencies that reduce its environmental impact. Brick also contributes to sustainable design through its long life span, energy efficiency, durability, recycled content, local availability, acoustic insulation, low construction waste, and potential for reuse.

Adobe is a natural building material common to the U.S. Southwest. Made from soil that has suitable sand and clay content and then air-dried in the sun, adobe bricks typically have extremely low greenhouse gas emissions and embodied energy. Most commercially available adobe bricks are "stabilized" with cement or asphalt additives, but adobe bricks are very commonly made on-site without stabilizers.

Like many conventional products, CMUs can be a green product by the manner in which they are produced, by the amount of material used, or the efficiency they provide to the finished structure. Some blocks earn Leadership in Energy and Environmental Design (LEED) green points by using post-industrial recycled content such as fly ash or ground blast-furnace slag to produce the blocks. CMUs are also available with specially designed expanded polystyrene insulation inserts that reduce the use of natural materials, and increase the energy performance of the block. Other concrete products provide superior energy performance by using innovative web designs. Many CMUs feature reduced material use by way of a finished exterior face such as split-faced block, or other decorative patterns that allow fewer materials to be used in the production process.

KEY TERMS

Bond
Bond beam
Brick
Brick veneer
Cavity wall
Concrete masonry units (CMUs)
Course
Deformations
Expansion joint
E/W
Hook
Lap
Ledger
Masonry construction
Mortar
Rebar
Solid wall
Tie
Wythe, double
Wythe, single

CHAPTER 10 TEST Unit Masonry Methods and Materials

QUESTIONS

Answer the following questions with short complete statements. Type the chapter title, question number, and a short complete statement for each question using a word processor. Some answers may require the use of vendor catalogs or seeking out local suppliers. If math is required to answer a question, show your work.

Question 10-1 A structure is to be 33'–8" wide, and covered with baby Roman brick. Is the structure modular? Explain your answer.

Question 10-2 Explain the difference between a course and a wythe.

Question 10-3 A wall is to be built of Roman bricks, approximately 12'–0" high. How high must the wall be to be modular?

Question 10-4 List two common methods of bonding wythes together.

Question 10-5 List the two most common methods of masonry bonds.

Question 10-6 What minimum code is wire reinforcement required to meet?

Question 10-7 Use local vendor catalogs to find a suitable connector for attaching masonry veneer to steel stud framing.

Question 10-8 List four classifications of CMUs.

Question 10-9 What factors dictate the fire rating of a block wall?

Question 10-10 List the approximate spacing of concrete block reinforcement.

Question 10-11 Explain the guidelines for the length of modular walls.

Question 10-12 What is the name of the piece of wood that attaches to the side of concrete block walls to support wood framing? To the top of concrete block?

Question 10-13 Explain how beams are supported with masonry walls.

Question 10-14 What do the numbers of rebar correspond to?

Question 10-15 What would be the benefit of having smooth rebar?

DRAWING PROBLEMS

Unless your instructor gives other instructions, complete the following details using the guidelines presented in "Drawing Criteria for Completing Details," in Chapter 6 and save them as wblocks. Skeletons of most of these details can be accessed from the student CD. Adjust all scale factors as required. Use these drawings as a base to complete the assignments:

- **Unless noted, all wall construction is to be 8" × 8" × 16" CMU.**
- **Reinforce all CMU walls with #5 rebar at 16" o.c. horizontal and 48" o.c. vertical unless otherwise noted.**
- **Specify in each detail that all steel cells are to be solid grouted, or provide a general note to cover all details assigned to the page.**
- **Show a bond beam (2 - #5); at the mid-height of each wall; at the top of all walls (4-#5); and at the intersection of all roof and floor construction to the CMU wall (4 - #5). Unless noted, specify all steel to be centered and all steel cells to be solid grouted.**
- **Where L-shaped reinforcing is specified, assume a continuous bar of equal diameter to the L is to be placed at the bend in the L to keep the L from pulling away from the footing or wall.**
- **For details showing masonry at the foundation level, provide 3/16" (4.8 mm) diameter weep holes at a maximum spacing of 33" (838 mm) o.c.**
- **For details showing masonry at the roof level, provide a cant strip similar to Figure 10-29, and a masonry wall cap similar to Figure 10-30.**
- **Show and specify all connecting materials base nailing on IBC standards unless your instructor tells you otherwise. Specify all material based on common local practice.**
- **Assemble the details required by your instructor into an appropriate template sheet for plotting.**

Problem 10-1 Draw a detail to represent a brick veneer wall on an 18" × 8" concrete footing for a two-level apartment project. Reinforce the footing with (1)-#5, 3" up from the bottom. Form the stem wall with CMUs. Use a 2 × 6 DFPT sill supporting 2 × 8 floor joist, supporting 3/4" ply floor sheathing, and 2 × 6 studs at 16" o.c. Cover the wall with brick veneer over 1/2" wafer board and Tyvek. Specify the appropriate connection of the brick to the wall.

Problem 10-2 Show standard trusses with a 24" overhang, and a 4/12 pitch supported by a 2 × 6 DFPT plate w/ 5/8" A.B. @ 36" o.c. Support the trusses on a reinforced CMU wall.

Problem 10-3 Show an 8" × 16" wide footing 18" into the grade to support CMUs. Show a 5" concrete slab at grade level on gravel fill. Thicken the slab to 10" × 10" at the edge and show a #5 × 24" × 12" L @ 48" o.c. from wall to slab. Reinforce the footing with (2)-#5 continuous, 3" up from the bottom of the footing. Show a #5 × 12" long L projecting from the footing into the wall. Extend the footing steel far enough into the wall to get a 24-diameter lap. See Table 11-1 for minimum concrete coverage and placement of foundation steel.

Problem 10-4 Show an 8" × 16" bond beam for a door header in an 8" block wall. Use (4)-#6 horizontal bars continuous with #3 ties at 24" o.c. Provide #5 vertical at 48" o.c. and #5 @ 24" o.c. horizontal.

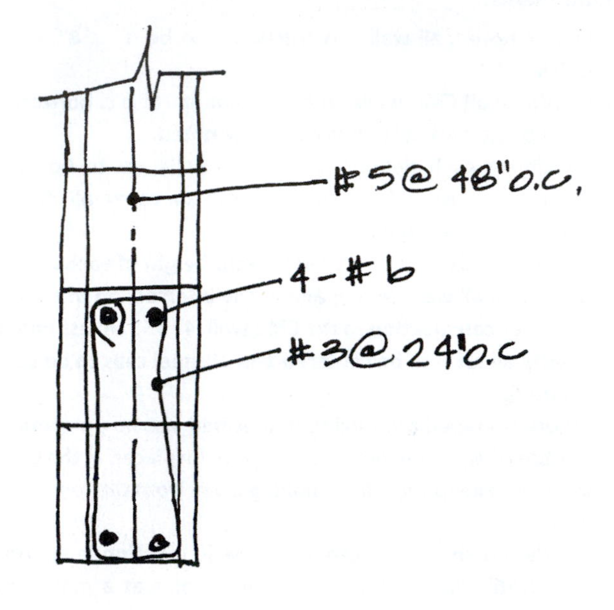

Problem 10-5 Show a plan view of a wall opening. Reinforce the last three cells of the wall with (2)-#5 vertical rebar. Tie with #3 ties at 24" o.c. Specify horizontal steel to be #5 at 16" o.c. with standard hook around the #5 at 24" o.c. vertical steel.

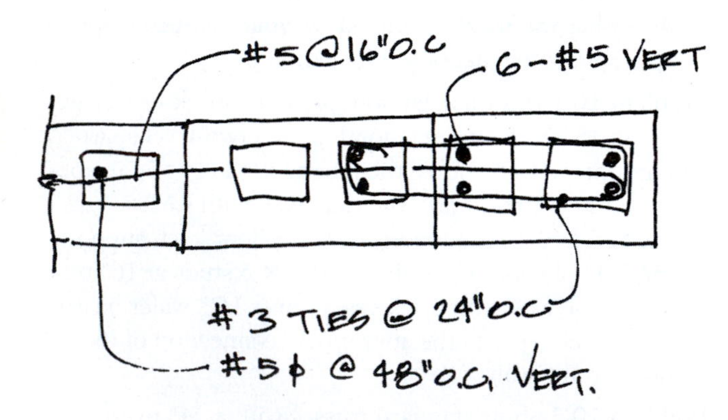

Problem 10-6 Show the intersection of a wood floor and a block wall formed from 8" × 8" × 16" CMUs. Use a 3 × 12 DFPT ledger with 3/4" diameter ×12" anchor bolts @ 32" o.c. staggered 2" up/dn, supporting 2 × 12 DFL #1 joists @ 16" o.c. with appropriate metal hangers. Reinforce the wall with #5 @ 32" o.c. vertical and #5 @ 16" o.c. horizontal. Provide an 8" × 16" deep bond beam w/ (4)-#5 @ ledger. Bolt a 3" high × 4" wide × 3/8" plate above floor joists with 3/4" diameter machine bolt at 48" o.c. and connect an 18-gauge Simpson strap capable of resisting 1000 # floor loads below 3/4" plywood sheathing. Weld the strap to the wall plate with 1/8" fillet field weld.

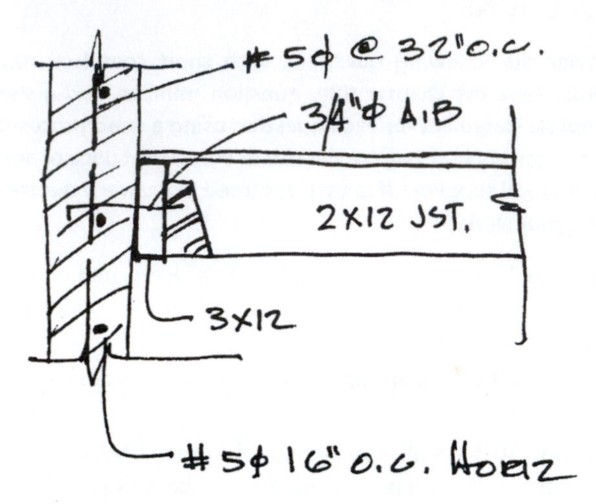

Problem 10-7 Draw a similar detail to problem 10-6 with the joists parallel to the wall at 16" o.c. Use 5/8" diameter anchor bolts at 48" o.c. staggered 2" up/dn. In addition to the minimum wall steel, provide an 8 × 8" deep bond beam with (2)-#5 @ ledger. Provide solid blocking at 48" o.c. for 48" out from wall. Bolt a 3" high × 4" wide × 3/8" plate above floor joist with 1/2" diameter × 8" anchor bolts at 48" o.c. and connect 1 1/2" × 48" × 3/16" strap below 3/4" plywood sheathing. Weld the strap to the wall plate with 1/8" fillet field weld and nail the strap to blocks with (48)-10d nails.

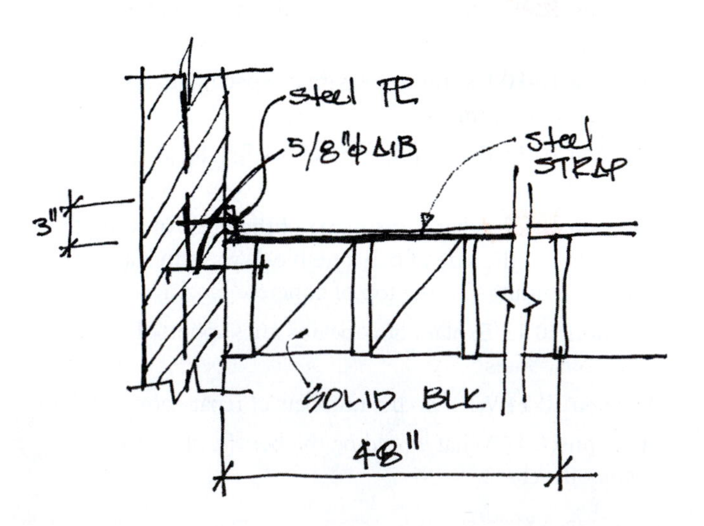

Problem 10-8 Show a plan view of a 16" × 16" pilaster. Reinforce with (8)-#5 vertical 2 1/2" clear of outside face. Tie with #4 horizontal at 16" o.c. Show typical wall reinforcement on each side of the pilaster,

with the horizontal steel required for the wall passing through the pilaster.

Problem 10-9 Show a side view of the pilaster drawn in problem 10-8. Use an appropriate GLB beam seat to hold a 6 3/4 × 15 fb 2400 glu-lam beam and support 9000# @ 375 psi minimum. Weld the seat to wall reinforcement with 1/8" fillet. Specify beam bolting based on the manufacturer of the seat.

Problem 10-10 Show a 3 × 12 DFPT ledger with 3/4" dia. A.B. @ 32" o.c. staggered 2" up/dn supporting TJW 24 open-web truss joist @ 32" o.c. Reinforce wall w/ #5 @ 32" o.c. vertical and #5 @ 16" o.c. horizontal. Reinforce the wall at the ledger with (4)-#5 in an 8" × 16" bond beam. Bolt a 4" high × 6" wide × 1/2" plate above trusses with 1/2" × 8" diameter A.B. @ 64" o.c. and connect Simpson Company MST27 @ 64" o.c. Connect the strap to the plate with 1/8" fillet field weld. Show a 30" high parapet above the roof trusses and cap with a double-sloping masonry cap. Show a 6" wide cant strip covered with 24-ga. flashing and built-up roofing.

Problem 10-11 Draw a CMU wall with brick veneer resting on a 20" wide × 12" footing that extends 18" minimum into natural grade. Provide (3)-#5 continuous 3" up from the bottom of the footing and a #5 × 24 × 6" bend @ 48" o.c. Extend to the steel with a minimum of 20 diameter lap. Reinforce the wall with #5 at 48" o.c. vertical and #5 at 24" o.c. horizontal steel. Provide a 4" concrete slab over 4" gravel fill. Reinforce the slab with WWM. Attach the brick veneer to the CMU with metal ties at 24" o.c. each direction.

Problem 10-12 Use the attached sketch to draw a section view showing an 8" wide × 48" maximum high retaining wall that will be on each side of a recessed loading dock. The retaining wall ranges in height from 16" to 48", depending on its position in the loading dock. Support the wall on a 12" deep footing that tapers from 30" to 54" (as the wall gets higher, the footing gets wider). Show the footing extending from 8" to 12" on the soil side. Place a 4" diameter French drain at the base of the wall placed in an 8" × 24" × 3/4" gravel bed. Extend the wall 6" min above the finish grade. Show a 5" thick slab reinforced with W12/12 × 4/4 mesh 3" from top surface over 4" gravel fill. Reinforce the footing with grade 40, #5 at 8" o.c. each way 3" up/down. Show a grade 60, #5 at 10" o.c. × 24" L extending from the footing, with a 40Ø lap to the vertical wall steel. Reinforce the wall with grade 60, #5 at 16" o.c. each way, 2" clear of wall.

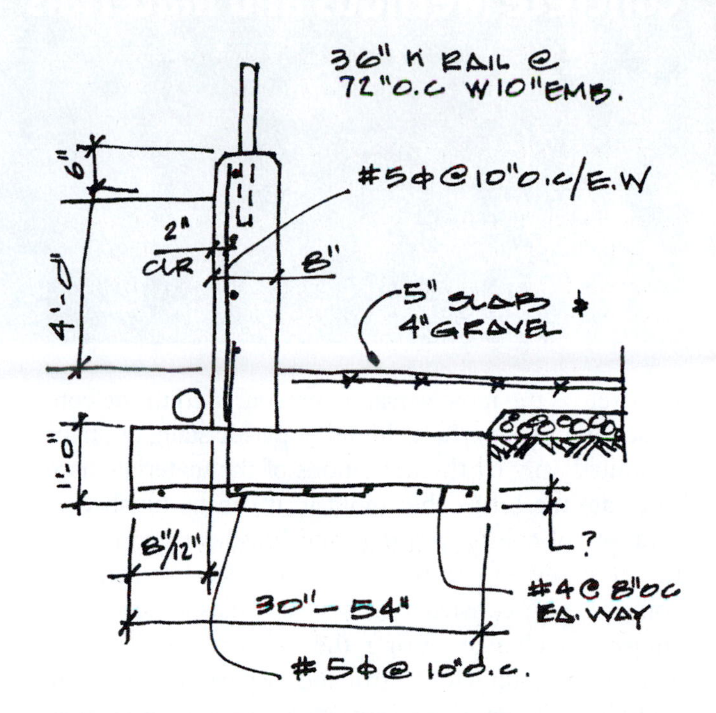

Problem 10-13 Edit detail 10-2 to show a 6" wide wall supporting roof trusses with a 4/12 pitch. Cover the exterior side of the CMU wall with brick veneer with ties at 24" o.c. each way. Provide 1× furring strips on the interior side of the wall to support 1/2" gypsum board.

Problems 10-14a and 10-14b Edit detail 10-10 to show the intersection of a floor to a CMU wall. Reinforce the wall with #5 at 48" o.c. vertical and #5 at 24" o.c. horizontal. Reinforce the wall at the ledger with (4)-#5 in an 8 × 16 bond beam. Attach a 3 × 12 DFPT ledger placed with the top of the ledger set at 10'–6" above the finished floor level. Attach the ledger to the wall with 3/4" Ø × 12" anchor bolts at 24" o.c. staggered 2" up/down. Use the ledger to support TJM 28 open-web floor joists at 32" o.c. Attach the trusses to the ledger as per manufacturers specifications. Bolt a 4" high × 6" wide × 1/2" plate above the trusses with 1/2" diameter anchor bolts at 64" o.c. and connect Simpson Company MST36 @ 64" o.c.

Edit this detail to show the use of Simpson HD2A on each side of trusses at 96" o.c. Drill the truss at the job site to allow for (2)-5/8" diameter bolts to pass through the truss. The bolts are to be a minimum of 16" from the end of the truss chord. Provide 5/8" anchor bolts for each anchor to wall in addition to ledger bolting.

Problem 10-15 Combine details 10-3 and 10-10 to create a partial wall section. Set the top of ledger at 12"–0" above the finish floor. Provide (2)-#5 at 48" o.c. horizontal and (2)- #5 horizontal at the wall top.

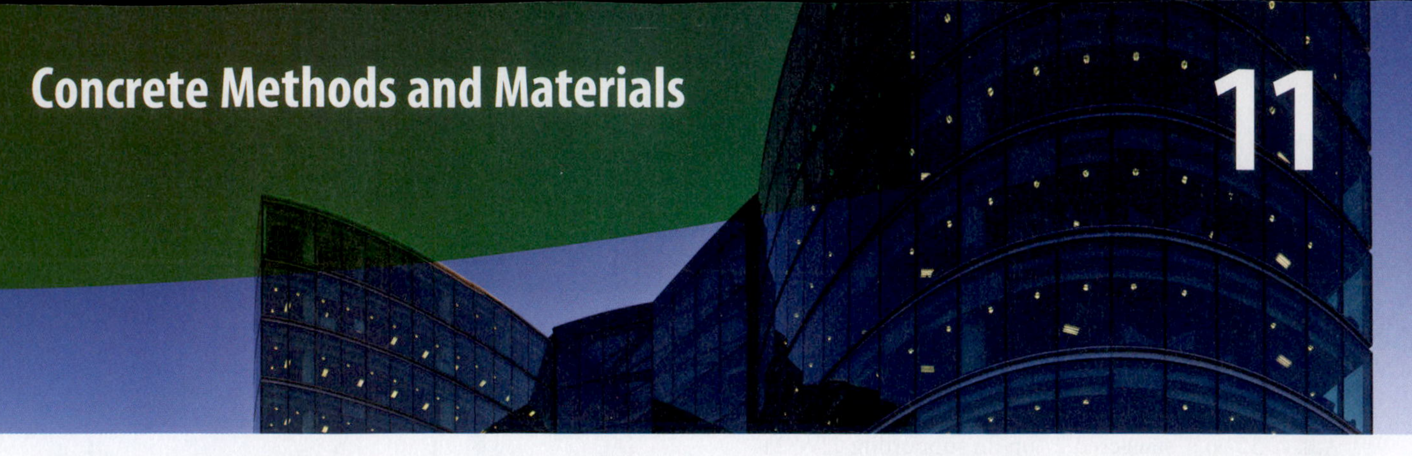

Concrete Methods and Materials

11

Concrete is the most versatile material used in the construction industry. In its liquid or plastic state, its shape is limited only by the limitations of the materials used to create the forms that mold it. It can be finished in a variety of colors, textures, and finishes to withstand the most severe weather conditions. Because concrete comes to the construction site in a plastic state, it is subject to change during the construction process. This requires on-site control and inspections to ensure that all of the engineer's specifications have been met. Engineering specifications often include information on how to mix, transport, prepare forms, place, reinforce, finish, and cure concrete so that all design values can be achieved. More so than with any other material, the CAD drafter must pay strict attention to detailing concrete to ensure the building materials will provide a strong, durable construction method. This chapter will introduce:

- The physical properties of concrete.
- Types of concrete construction.
- Reinforcement placement and specifications.
- Floor systems.
- Walls.
- Beams.
- Columns.
- Joints in concrete.
- Conduit.

PHYSICAL PROPERTIES OF CONCRETE

Concrete can be greatly altered depending on the materials and proportions of materials used to make the concrete. The methods of pouring, setting, and curing will be affected by the weather at the site where concrete is poured, the temperature, the rate of placement, and the size of the area to be poured. All of these factors can affect the appearance, strength, and weight of the finished concrete. Most of the qualities of concrete are governed by ACI 318 (American Concrete Institute) standard, which the engineer will specify to govern the quality best suited to the project. Although the contractor is responsible for verifying that the standards of the mix and pouring procedures are met, knowledge of concrete and the pouring process will help the drafter better understand the contents of the specifications and details being provided.

Concrete Materials

Concrete is a mixture of portland cement, aggregate, miscellaneous additives called admixtures, and water, and it is cast while in a plastic or liquid state (see Figure 11-1). The written specifications in the drawings and in the project manual specify the proportions of the mixture and the conditions under which they can be used.

Cement

Portland cement is a mixture of lime, silica sand, iron oxide, and alumina. When mixed with clean water, this mixture will harden and set into the form in which it is molded. Portland cement is usually required to conform to ASTM C150, which is the standard specification for portland cement. Five types of portland cement are used in the construction industry.

- *Type I* is referred to as standard or normal cement. Common uses include beams, columns, and floor slabs.
- *Type II* cement is used in localities where a strong resistance to sulfate is not required. It is used primarily for members that require a lower heat of hydration, such as large footings or foundations. The hydration process will be discussed as the water mixture is considered. Type II cement has a lower early strength than Type I cement. A cement designated Type I/II meets the specifications for both Type I and Type II and may be used where either is specified.
- *Type III* cement achieves a high early strength, which continues to accelerate as it cures. Type III cement has the same strength after one day that Type I cement has after three days. After 28 days, Type I and III cements will have achieved about the same

FIGURE 11-1 Construction methods, reinforcement, and placement must all be considered as the drawings that detail concrete structures are completed. *Courtesy Constructionphotographs.com.*

strength. Type III cement is often used for precast concrete and for cold-weather pours because of its high heat hydration. This quality also makes Type III cement unsuitable for use with large pour members.

- *Type IV* cement is used for massive pour construction and is generally considered low-heat cement. Type IV cement is excellent where cracking must be kept to a minimum.
- *Type V* cement is used in areas that will be exposed to water and high amounts of sulfate content.

Water

The amount of water is an important element in making concrete. The quantity to be used must be specified on the working drawings. **Potable water** that is drinking-water quality and free of oils or excessive amounts of acid or alkali must be used. Although water from a municipal water supply is usually adequate, water must conform to ASTM C94 *Standard Specification for Ready-Mixed Concrete*. In addition to controlling the water in the mix, the engineer may provide specifications for the soaking of the forms prior to use, as well as methods of keeping the concrete damp during the early stages of the curing process.

When water is added to cement, a chemical reaction known as *hydration* takes place, causing a bonding of the molecules. As the mixture begins to harden, it releases heat in a process known as **heat of hydration.** Methods of controlling the rate of evaporation of water from the exposed surfaces, and hydration within the concrete, are often specified by the engineer and must be included in the written specification for the job. Control of the amount of water affects not only the hydration but also the workability of the concrete. Excess water is held to a minimum to keep the concrete from becoming porous. This keeps cracks from developing, in addition to limiting voids in the concrete.

Aggregates

Although the aggregates do not affect the chemical process that causes the cement to harden, they are controlled in the specifications because they play an important role in the quality of concrete. Sand, gravel, and crushed rock are used for the **aggregates** in the concrete mix. Aggregates should be rounded, insoluble, clean, and free of dust or other substances so that the cement paste can bond to the aggregate. It is also important that the aggregate not contain elements that will chemically react when added to water or cement. Aggregate materials are divided into fine and coarse aggregates based on the size of the material used. Fine material is usually sand, but rocks small enough to pass through a 1/4" (6 mm) sieve are also used. Anything that will not pass through a No. 4 sieve is considered coarse aggregate.

The aggregate for most construction is between 3/4" and 1 1/2" (20 and 40 mm) washed gravel or crushed stone conforming to ACI 318–3.3. The ACI does not allow the size of the aggregate to exceed 1/5 of the pour width between forms; 1/3 of the slab depth; or 3/4 of the clear rebar spacing. The aggregate can be altered, affecting the weight of the mixture. Aggregate will be discussed further in the "Lightweight Concrete" section of this chapter. The drafter needs to note the ingredients, proportions, or the size of the aggregate specified by the engineer on the drawings or in the written specifications.

Admixtures

Any substance other than cement, water, or aggregates that is added to concrete to affect its workability, accelerate its setting time, or alter its hardness is known as an **admixture.** Admixtures are listed in the written specifications for concrete drawings and are used to increase early strength, increase ultimate strength, accelerate or retard setting time, or increase workability. ACI 318-3.6 governs chemical admixtures. Common admixtures include hydrated lime, calcium chloride, and kaolin. When calcium chloride is added to the water, concrete with a fast setting time is produced. This concrete is known as **high-early-strength concrete.** A high degree of strength is obtained quickly, allowing for protection from rain or freezing conditions. High-early-strength concrete allows for rapid removal of forms that can be reused again, further reducing the cost of construction. The engineer or contractor determines the admixtures required for the mix, based on expected weather conditions at the time of the pour, and the set speed.

Air-Entrained Concrete

Air can be injected into the water before it is added to the mixture, forming microscopic air bubbles in the mix. Air-entrained concrete meeting ASTM 360 and C33 is effective for resisting deterioration that results from contraction and expansion caused by freezing and thawing. Adding air to the cement mixture also improves its workability and lowers the tendency for mixture separation. Air-entrained concrete is identified as Types IA, IIA, and IIIA and is used in ways similar to its non-air-containing counterpart. The air bubbles that form in the hardened concrete provide space for expansion of moisture during freezing conditions.

Mixture Ratios

The ratio of the concrete mix is very important to the setting, curing, and ultimate strength of the concrete. A common ratio for concrete is 1:2:4, which represents 1 part of cement, 2 parts of sand, and 4 parts of gravel. Other common mixes are 1:3.75:5, and 1:2.5:3. Concrete is usually specified by its batch weight, giving a ratio of water to cement (w/c) stated in pounds. If a batch of concrete contains 200 pounds of water and 500 pounds of cement, the w/c would be:

$$200/500 = 0.40 \text{ (lb of water per lb cement)}$$

The ACI limits the water/cement ratio based on the desired strength and the air content. The engineer will specify the exact proportions of water to cement to be used to achieve the highest strength and maintain a workable mixture. If too much water is added, the strength and durability of the concrete will be compromised. The strength of concrete is measured in terms of its ultimate compressive strength in pounds per square inch (psi) or megapascals (MPa) and represented by the symbol F'c. A concrete with a compressive strength of 2500 psi is referred to as 2500 pound concrete. Because of the variations in producing concrete, the working stress of concrete is used for design purposes. The design strength is established by the engineer and is based on a portion of the ultimate strength and expressed as a fraction of F'c. Concrete is considered to reach its ultimate compression strength at 28 days after pouring based on ASTM C192, but concrete actually continues to get harder at a very slow rate. Forms are typically removed after approximately seven days.

Concrete Testing

To determine if concrete meets the requirements of the engineer, testing can be done on the slump, compressive strength, and air content of the concrete. A *slump test* determines the consistency and workability of concrete in its plastic state. This test can be performed at the mixing site or at the job site, with results specified by ASTM C143.

The ultimate compressive unit strength of the concrete (F'c) is tested by taking samples of the concrete in its plastic state. Concrete samples are placed in 6" diameter × 12" long molds. The concrete is allowed to harden and is tested at various stages of the curing process. Cores are generally tested at days 3, 7, and 28, to see the amount of compression that can be resisted. If the test results fall below the design values intended for the concrete, the engineer will need to determine if the pour is suitable for design conditions. Careful planning by the engineer and quality control at the mixing site typically provide concrete well above the minimum design value.

TYPES OF CONCRETE CONSTRUCTION

Common methods of concrete construction include plain, reinforced, prestressed, precast, prefabricated, lightweight, thin shell, and pneumatically applied concrete. With each method, the engineer will provide calculations and sketches for the drafter to complete the finished drawing.

Plain Concrete

Concrete that contains no reinforcement, or contains less reinforcement than the minimum requirements of ACI 318, is considered **plain concrete.** Plain concrete must have a minimum compressive strength of 2500 psi (17.24 MPa) and is used primarily for foundations and slabs of office structures. Unreinforced walls are required by code to be a minimum of 5 1/2" (140 mm) thick, and slabs must be a minimum of 3 1/2" (90 mm) thick. Walls must be connected at each floor and roof level. The engineer will determine the exact thickness of the concrete member based on the amount of loads to be resisted.

Reinforced Concrete

Concrete is extremely strong in resisting loads that cause compression, but very weak at resisting forces that cause tension. **Reinforced concrete** contains steel reinforcement to increase the tensile strength of the concrete. Typically the concrete reinforcement is rebar (introduced in Chapter 10). The bond of the concrete to the surface of the reinforcing bars and the resistance provided by the deformations make the concrete and steel act as one material. Figure 11-2 shows an example of reinforced concrete construction. **Welded wire fabric** is also used to reinforce concrete and to carry stress from tension. Methods of reinforcement specific to concrete construction will be introduced later in this chapter.

Prestressed Concrete

Most concrete members are stressed by loads that cause bending. To resist the forces of bending, members are frequently prestressed. Thin horizontal members such as beams and slabs are most likely to be prestressed to resist tension in the member. In reinforced construction, the reinforcing steel carries all of the forces of tension. In prestressed construction, the entire concrete member is effective at resisting the forces of tension. Prestressed members are used for pilings, columns, wall panels, floor slabs, and roof panels. The stressing is applied by placing high-strength steel reinforcing bars or cables that have been stretched. The bars can be stretched prior to the concrete being poured in a process called **pretensioning**. Steel can also be stretched after the concrete has been poured and hardened around the steel for a process called **post-tensioning**.

In pretensioning, high-strength stranded steel wires are stretched into position before the concrete is poured. After the concrete is poured and allowed to harden, the tension on the wires is released. As the steel contracts and attempts to regain its original shape, the wires conduct compressive forces throughout the concrete. The compression in the concrete helps prevent cracking from deflection and allows the size of the member to be reduced when compared to nonstressed members.

In the post-tensioning process, large-diameter steel rods or cables are placed in conduit that is embedded within the concrete, with one end of the steel anchored into the concrete. After the concrete has hardened, the steel is stretched by hydraulic jacks (Figure 11-3) and then attached to the concrete so that the steel remains under tension. Post-tensioned reinforcing is often used for elevated slabs for parking garages; residential or commercial buildings; and residential foundations, walls, and columns. The use of post-tensioned reinforcing can result in thinner concrete sections, longer spans between supports, stiffer walls to resist lateral loads, and stiffer foundations to resist the effects of shrinking and

FIGURE 11-2 The forming of reinforced concrete columns for a structure. Reinforced concrete contains steel reinforcement, referred to as rebar, to resist forces that cause tension in the concrete. *Courtesy Dick Schmitke.*

FIGURE 11-3 Concrete can be reinforced by placing tension on steel cables within a concrete member. When the cables are tensioned after the concrete has cured, the process is referred to as post-tensioning. *Courtesy Portland Cement Association.*

swelling soils. Concrete that is put into compression has the added advantage of creating slabs and walls that have fewer visible cracks that can allow the passage of moisture and termites.

Precast Concrete

Precast concrete members are cast and cured in an off-site location and then transported to the job site. Figure 11-4 shows an example of a precast floor panel being positioned. When a large number of identical units are to be built, time, money, and space at the job site are saved when members are cast off-site. Members can be built ahead of time and shipped as needed to the construction site. Beams, columns, and small wall panels are usually precast. Panels are normally welded to the building frame at the top and bottom of the panel, allowing multilevel structures to be built one floor at a time. Figure 11-5 shows the connection of precast concrete panels attached to the structural steel frame.

FIGURE 11-5 Precast concrete panels can be welded to a steel skeleton. *Courtesy Sam Griggs.*

Prefabricated Concrete Units

Prefabricated concrete units are formed at the job site and then lifted into place. This form of construction is also known as tilt-up construction in many parts of the country. Tilt-up construction is usually used for large single-level structures. Large industrial complexes or retail sales outlets represent ideal uses of concrete tilt-up structures. Wall panels are cast in a horizontal position over the floor slab or in a bed of sand. Once they have cured, they are lifted into place by a crane similar to the panels in Figure 11-6.

FIGURE 11-4 Precast concrete members can be cast and tensioned off-site and then positioned as needed. *Courtesy Portland Cement Association.*

FIGURE 11-6 Concrete wall panels can be cast in a horizontal position over the floor slab and lifted into place by a crane once they have cured. *Courtesy Gisela Smith.*

Panels can be attached to precast columns, welded to adjacent panels similar to Figure 11-7, or attached to the structural skeleton. Panel construction requires a plan view similar to Figure 11-8 showing the location of each panel; elevations similar to Figure 11-9, showing the location of each opening and placement of the reinforcing steel; and details to show typical connections such as panel-to-panel, panel-to-foundation, slab-to-panel, and roof-to-panel. An example of the intersection of two beams to a wall can be seen in Figure 11-10. Precast panels will be discussed further in the section on methods of wall construction later in this chapter.

Lightweight Concrete

Concrete weighs approximately 150 pounds per cubic foot (2400 kg/m³). The weight can be lessened depending on the aggregate used or chemicals that are injected into the mix. Lightweight concrete is classified in three groups according to the unit weight per cubic foot:

- The lightest concrete has a unit weight of between 20 and 70 lb/ft³ (320 and 1120 kg/m³) and is classed as *insulating lightweight concrete*. Because its compressive strength is usually below 1000 psi (6.9 MPa), it is not used as a structural material, but as a protective covering for fireproofing. It is frequently poured over a wood or steel decking to form a floor.
- *Structural lightweight concrete* has a weight of up to 115 lb/ft³ (1840 kg/m³) with a compressive strength of 2000 psi (13.8 MPa).
- *Semi-lightweight concrete* has a unit weight of from 115 to 130 lb/ft³ (1840 to 2080 kg/m³). Its ultimate strength is comparable to that of normal concrete.

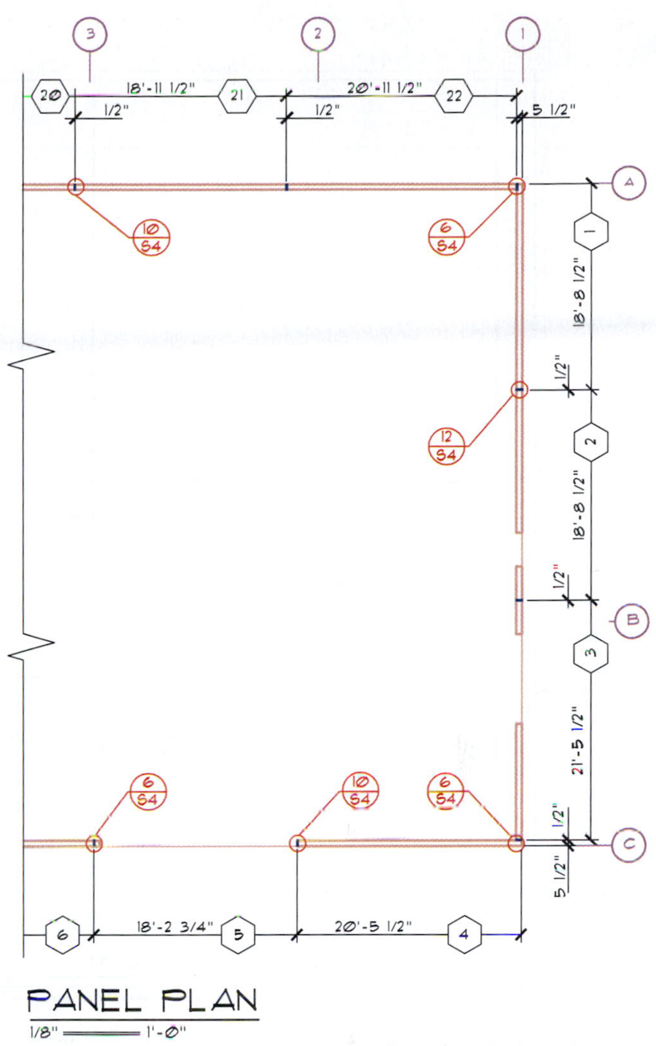

FIGURE 11-8 Precast members are often identified on the floor plan or a separate panel plan. The numbers in the hexagons refer to a specific concrete panel shown in a drawing similar to Figure 11-9. *Courtesy Ginger M. Smith, Kenneth D. Smith Architect & Associates, Inc.*

FIGURE 11-7 Concrete panels are connected to each other by the use of steel plates that are set into the concrete and welded to each other. *Courtesy Zachary Jefferis.*

Thin Shell

Thin-shell or eggshell construction can be used to form 3D structures made of one or more curved slabs or folded plates. Concrete as thin as 2 1/2" (65 mm) thick with a compressive strength of 3000 psi (20.7 MPa) can be used to form the shell. The strength of the shell is achieved by the shape of the concrete into which it is molded. In addition to the material saved in the roof, less material is needed for supporting walls, columns, and foundations due to the reduced weight of the roof. Arcs or barrels, hyperbolic paraboloid domes, and folded plates are common shapes used for thin-plate construction because of their inherent strengths; but there are many variations that can be achieved with these basic shapes.

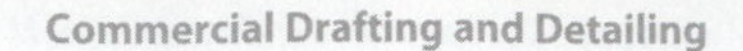

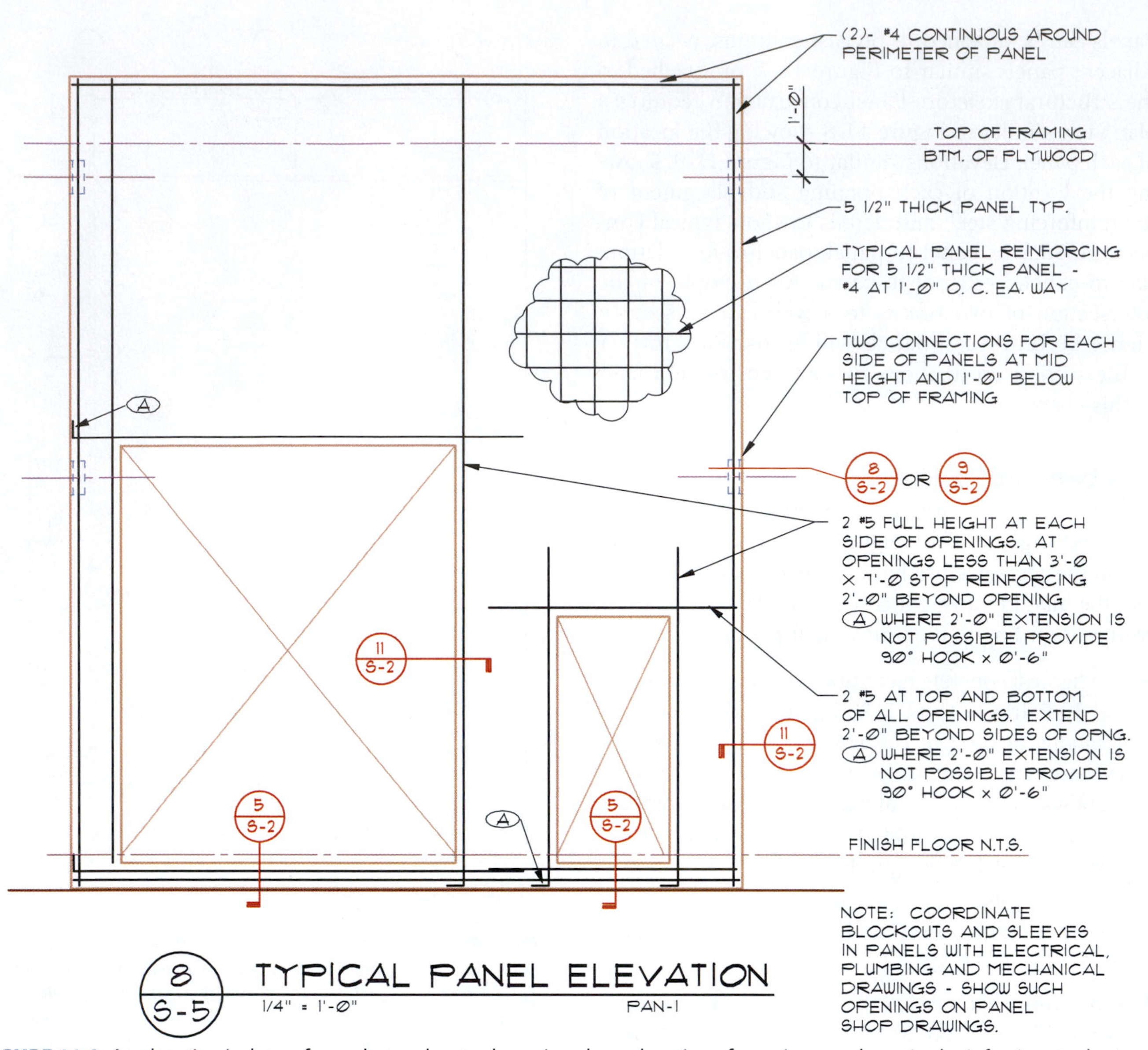

FIGURE 11-9 An elevation is drawn for each member to show size, shape, location of openings, and required reinforcing steel. *Courtesy Van Domelen/Looijenga/McGarrigle/Knauf Consulting Engineers.*

Shotcrete

Portland cement or plaster that is applied with a compressed air gun is considered shotcrete or pneumatic concrete. Shotcrete is extremely dense and strong. It tends to have a high resistance to weathering and is an excellent waterproofing material because of its low absorption rate.

REINFORCEMENT PLACEMENT AND SPECIFICATIONS

As with masonry products, deformed steel bars and welded wire mesh are the major reinforcing materials of concrete construction. Concrete reinforcement patterns and specifications are usually more complicated than masonry construction.

Reinforcing Bars

Steel bars having the same characteristics as those used with masonry are used to reinforce concrete construction, although the CAD technician will need to be mindful of different configurations, spacings, and clearances. Steel reinforcing is required to meet the ASTM standards listed in the International Building Code (IBC). Nondeformed bars can be used for spiral reinforcing of columns and must conform to ASTM standards. Structural steel shapes, steel pipes, and steel tubing are also used for concrete reinforcement and must meet the specifications of ASTM required by the code. The engineer will select the standard to be met, and the drafter will place the required standards on the concrete drawings and written specifications. Figure 11-11 shows the reinforcing notes that

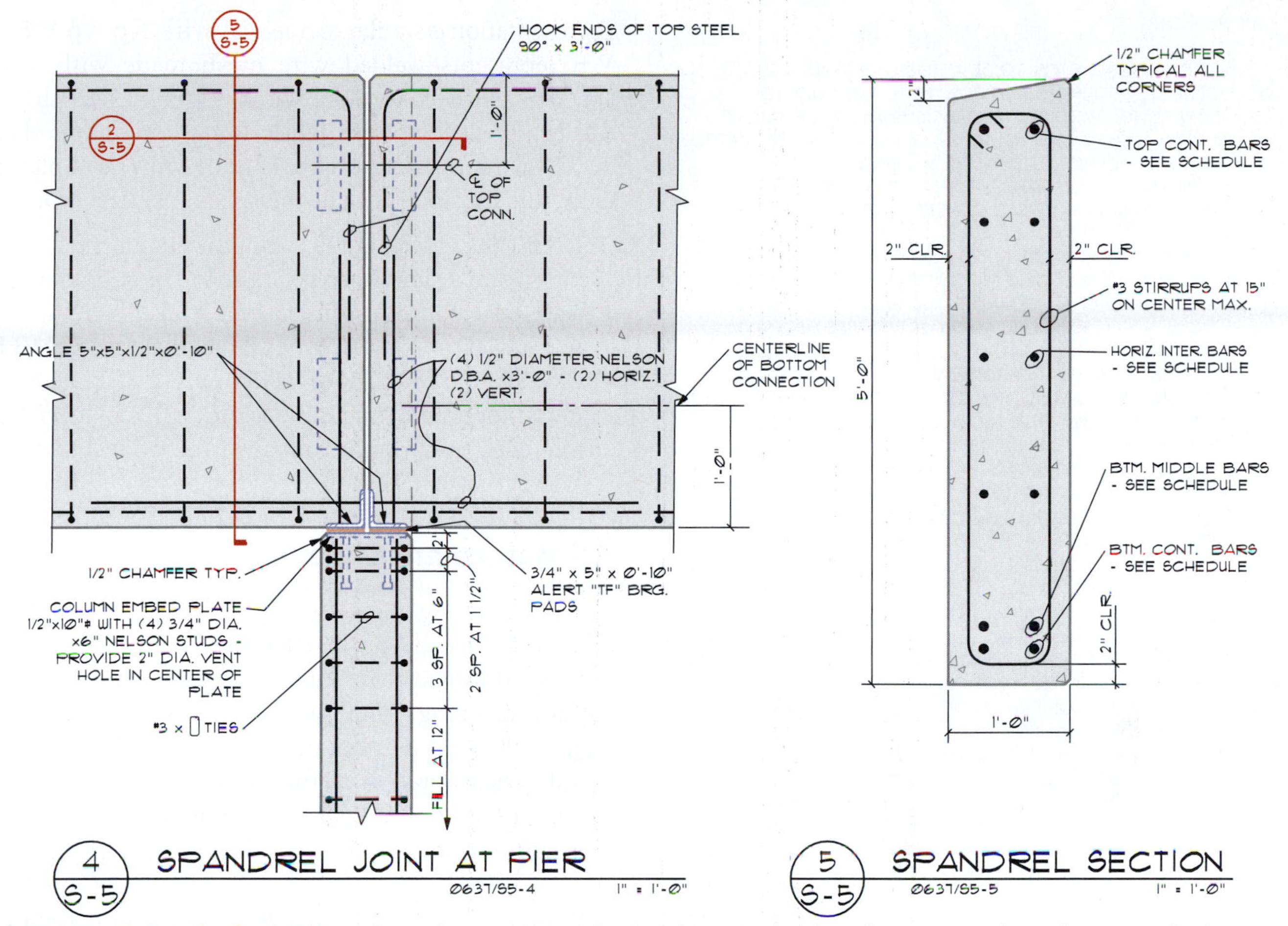

FIGURE 11-10 Details of connections for each concrete member are drawn to detail reinforcement and attachment methods.
Courtesy Van Domelen/Looijenga/McGarrigle/Knauf Consulting Engineers.

supplement the drawings for the tilt-up structure used throughout this text.

Unlike masonry construction, reinforcement in poured concrete also requires the use of **bolsters** and **chairs** to maintain a uniform positioning. Each is made from steel or plastic wire, but protective coatings or a galvanized finish can be applied to steel wires. The finish is often specified in the written specifications that accompany the drawings. Figure 11-12a shows several common types of supports that are designed to place the steel in the desired finished position. Figure 11-12b shows their placement. *Bolsters* come in lengths up to 10' (3000 mm) and are designed to hold the lower steel-reinforcing mat in position. *Chairs* are designed to hold upper-level pieces of steel in position within a slab. These members are not specified on the structural drawings but are referenced on fabrication drawings. A minimum of 1" (25 mm) is required by the IBC/ACI 318-7.6 between parallel layers of reinforcing, and each layer must be staggered so parallel members do not align. The spacing of steel mats must be specified on the structural drawings using methods similar to Figure 11-13.

Welded Wire Fabric

Wire fabric, or mesh as it is sometimes called, is laid in perpendicular patterns and is used for reinforcement of floor, wall, and roof slabs. Wire meshes are required to meet ASTM standards of the IBC or of those specified by the engineer. Patterns are made in one-way rectangular and two-way square patterns.

One-way mesh is made of heavy longitudinal wires that are spaced from 2" to 16" (50 to 400 mm) apart. A lighter tie wire is spaced from 1" to 18" (25 to 460 mm) apart and welded at each intersection. Common wire fabric dimensions can be seen in Figure 11-14. The specification of 6 × 18-W8 × D4 represents *welded wire fabric* with a spacing of 6" × 18" using No. 8-gauge smooth wire for the longitudinal wires with No. 4-gauge tie wires. The letter *D* can be substituted for the *W* to represent deformed wire. The wire mesh is represented by a thick line and specified in detail as seen in Figure 11-13. A distance from either the top or bottom of the slab must be provided. Two-way wire fabric is arranged in a square pattern with equal sized wires used in each direction. The designation and

REINFORCING:

1. ALL REINFORCING STEEL TO BE ASTM A615 GRADE 60 EXCEPT TIES, STIRRUPS AND DOWELS TO MASONRY TO BE GRADE 40. WELDED WIRE FABRIC SHALL CONFORM TO ASTM A-185 AND SHALL BE 6x6-W1.4 x W1.4 WWF MATS.
2. FABRICATE AND INSTALL REINFORCING STEEL ACCORDING TO THE "MANUAL OF STANDARD PRACTICE FOR DETAILING REINFORCED CONCRETE STRUCTURES" ACI STANDARD 315.
3. PROVIDE 2'-0" x 2'-0" CORNER BARS TO MATCH HORIZONTAL REINFORCING IN POURED-IN-PLACE WALLS AND FOOTINGS AT ALL CORNERS AND INTERSECTIONS.
4. SPLICES IN WALL REINFORCING SHALL BE LAPPED 30 DIAMETERS (2'-0" MINIMUM) AND SHALL BE STAGGERED AT LEAST 4'-0" AT ALTERNATIVE BARS.
5. ALL OPENINGS SMALLER THAN 30" x 30" THAT DISRUPT REINFORCING SHALL HAVE AN AMOUNT OF REINFORCING EQUAL TO THE AMOUNT DISRUPTED PLACED BOTH SIDES OF OPENING AND EXTENDING 2'-0" EACH SIDE OF OPENING.
6. PROVIDE THE FOLLOWING REINFORCING AROUND WALL OPENINGS LARGER THAN 30" x 30":
   A. (2) #5 OVER OPENING x OPENING WIDTH PLUS 2'-0" EACH SIDE.
   B. (2) #5 UNDER OPENING x OPENING WIDTH PLUS 2'-0" EACH SIDE.
   C. (2) #5 EACH SIDE OF OPENING x FULL STORY HEIGHT.
   D. PROVIDE 90 DEGREE HOOK FOR BARS AT OPENINGS IF REQUIRED EXTENSION PAST OPENING CANNOT BE OBTAINED.
7. PROVIDE (2) #4 CONTINUOUS BARS AT TOP AND BOTTOM AND AT DISCONTINUOUS END OF ALL WALLS.
8. PROVIDE (2) #5 x OPENING DIMENSION PLUS 2'-0" EACH SIDE AROUND ALL EDGES OF OPENINGS LARGER THAN 15" x 15" IN STRUCTURAL SLABS AND PLACE (1) #4 x 4'-0" AT 45 DEGREES TO EACH CORNER.
9. PROVIDE DOWELS FROM FOOTING TO MATCH ALL VERTICAL WALL, PILASTER, AND COLUMN REINFORCING (POURED-IN-PLACE COLUMNS AND WALLS). LAP 30 DIAMETERS OR 2'-0" MINIMUM.
10. ALTERNATE ENDS OF BARS 12" IN STRUCTURAL SLABS WHENEVER POSSIBLE.
11. ALL WALL REINFORCING TO BE PLACED IN CENTER OF WALL UNLESS SHOWN OTHERWISE ON THE DRAWINGS.

FIGURE 11-11 Many building departments require reinforcement specifications to be placed with the structural drawings rather than in the project manual. *Courtesy H.D.N. Architects A.I.A.*

representation is similar to one-way wire. A 6 × 6-W8 × W8 represents welded wire mesh made with a 6" spacing with No. 8-gauge wires. Figure 11-15 shows the use of welded wire fabric for a foundation slab. WWM is represented by a thick line with an X placed along the line to represent the spacing of the grid.

> *Note:* When locating the mesh, the engineer will provide a dimension from either the top or bottom of the slab. Don't alter this dimension. It may be tempting to change a dimension for a 5" slab that reads 2" CLR of TOP, to be 3" CLR of BTM because of space limitation for placing the dimensions. The engineer has referenced the dimension from the critical surface. Don't change it.

Reinforcement Coverage

Steel reinforcing bars are typically placed as close as possible to the outer edge of the member being reinforced to resist the forces of tension and to resist surface cracking. The distance between the outer edge of the steel and the outside edge of the concrete member is referred to as the *cover*. The engineer will specify exact requirements for **coverage**, based on the requirements for weather, fire protection, and the space required to develop a bond between the concrete and the steel to develop adequate bending resistance within the component being detailed. The drafter should specify the distance, as shown in Figure 11-16. Depending on the experience level of the drafter, the engineer may not always provide a specification for coverage, intending the minimum recommended coverage to be followed. These minimum standards can be seen in Table 11-1.

Concrete Exposure	Minimum Cover (inches /mm)
Concrete cast against and permanently exposed to earth	3/75
Concrete exposed to earth or weather:	
No. 5 bars, W31 or D31 wire and smaller	1 1/2/38
No. 6 through No. 18 bars	2/50
Concrete not exposed to weather or in contact with ground slabs, walls, and joists:	
No. 11 bars and smaller	3/4/20
No. 14 and No. 18 bars	1 1/2/38
Beams and columns:	
Primary reinforcement, ties, stirrups, and spirals	1 1/2/38
Shells, folded plate members:	
No. 5 bar, W31 or D31 wire and smaller	3/4/20
No. 6 bars and larger	1/2/13

TABLE 11-1 Minimum Concrete Coverage Recommendations

SYMBOL	BAR SUPPORT ILLUSTRATION	TYPE OF SUPPORT	STANDARD SIZES
SB	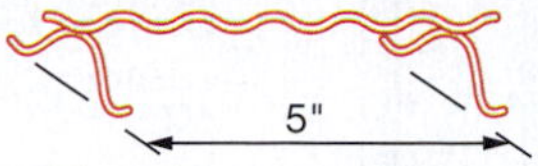 5"	Slab Bolster	3/4, 1, 1 1/2, and 2 inch heights in 5 ft. and 10 ft. lengths
SBU*	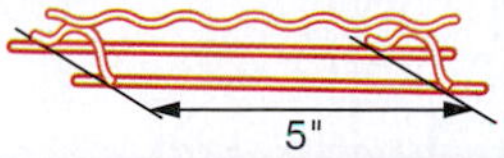 5"	Slab Bolster Upper	Same as SB
BB	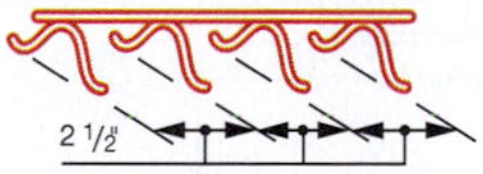 2 1/2"	Beam Bolster	1, 1 1/2, 2; over 2" to 5" heights in increments of 1/4" in lengths of 5 ft.
BBU*	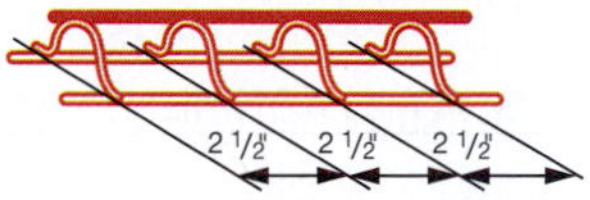 2 1/2" 2 1/2" 2 1/2"	Beam Bolster Upper	Same as BB
BC	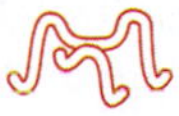	Individual Bar Chair	3/4, 1; 1 1/2, and 1 3/4" heights
JC	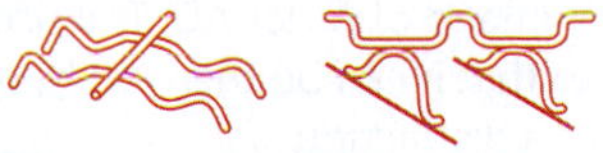	Joist Chair	4, 5, and 6 inch widths and 3/4, 1, and 1 1/2 inch heights
HC		Individual High Chair	2 to 15 inch heights in increments of 1/4 in.
HCM*	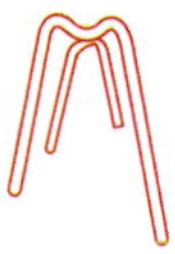	High Chair for Metal Deck	2 to 15 inch heights in increments of 1/4 in.
CHC	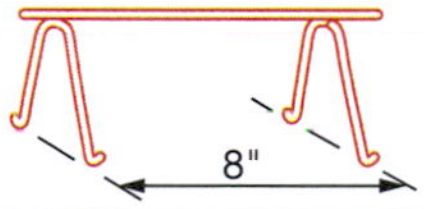 8"	Continuous High Chair	Same as HC in 5 foot and 10 foot lengths
CHCU*	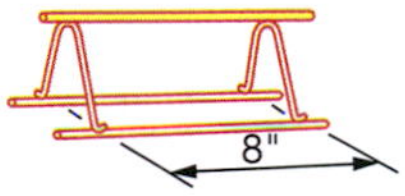 8"	Continuous High Chair Upper	Same as CHC
CHCM*	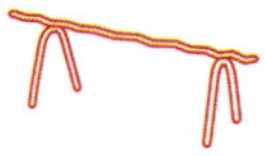	Continuous High Chair for Metal Deck	Up to 5 inch heights in increments of 1/4 in.
JCU**	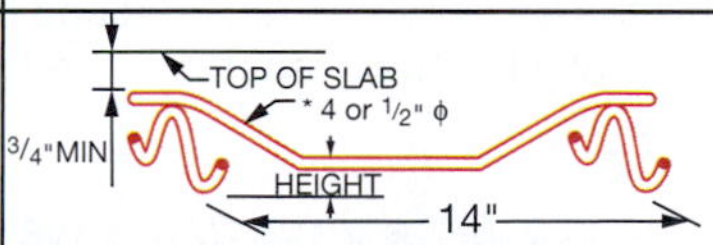 TOP OF SLAB; * 4 or 1/2" ϕ; 3/4"MIN; HEIGHT; 14"	Joist Chair Upper	14" span. heights —1" through + 3 1/2" vary in 1/4 increments

* Available in Class 3 only, except on special order.

* * Available in Class 3 only, with upturned or end bearing legs.

FIGURE 11-12a Typical supports used to hold reinforcing in the correct position relative to each other, and the edge of the concrete.
Courtesy Concrete Reinforcing Steel Institute, Schaumberg, IL.

FIGURE 11-12b Reinforcing bars are tied to other rebar with thin pieces of wire. The groups of rebar (mats) are lifted above what will become the face of the panel by the use of plastic chairs. *Courtesy LeRoy Cook.*

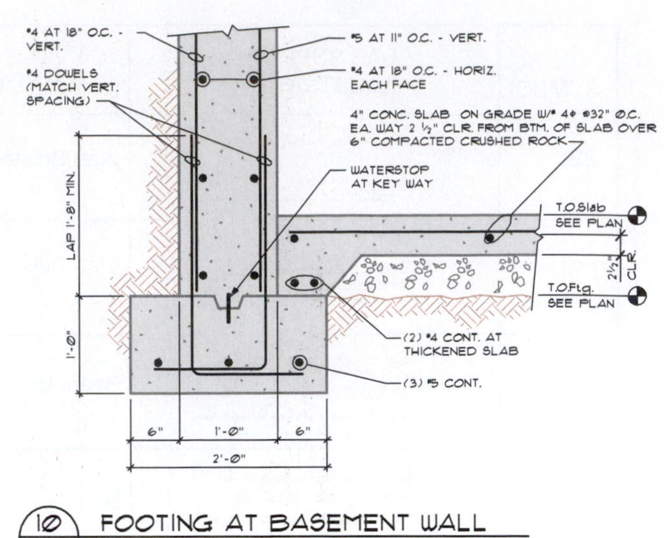

FIGURE 11-13 The spacing of steel mats must be specified on the structural drawings. Steel that is parallel to the cutting plane is shown as a thick line. Steel that is perpendicular to the cutting plane is represented using a solid circle placed using the DONUT command of AutoCAD. The size of the rebar should be enhanced so that it can be seen easily, but not to the point that it becomes a distraction. *Courtesy Van Domelen/Looijenga/McGarrigle/Knauf Consulting Engineers.*

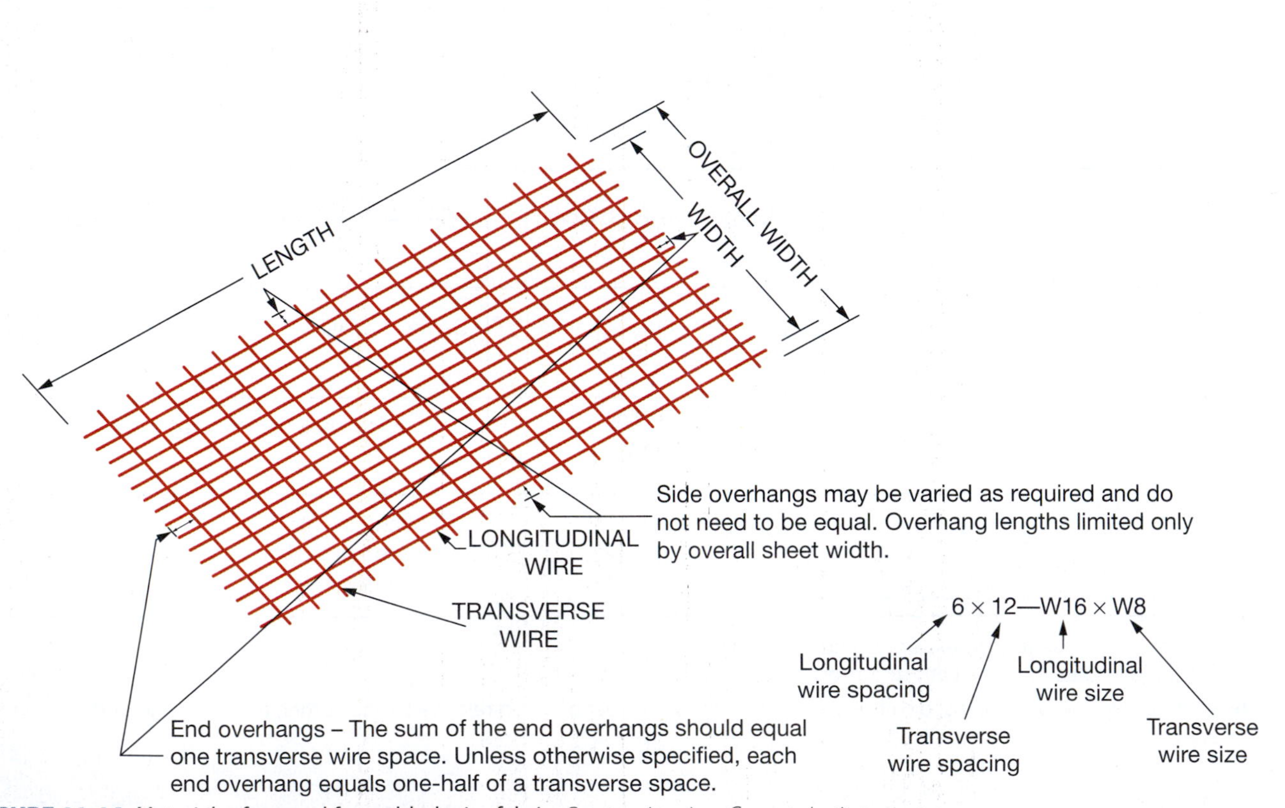

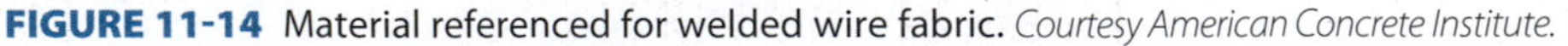

FIGURE 11-14 Material referenced for welded wire fabric. *Courtesy American Concrete Institute.*

FIGURE 11-15 Welded wire fabric used to reinforce a concrete slab. The mesh is resting on plastic chairs to provide the proper elevation from the bottom of the slab. *Courtesy Constructionphotographs.com.*

FLOOR SYSTEMS

Concrete floor systems are typically constructed as cast-in-place, one- and two-way reinforced slab systems, or as lift systems.

Cast-in-Place Floors

Cast-in-place floor systems can be formed over a wood or steel decking. Wood is a common forming material for cast-in-place reinforced slabs. Concrete is placed over wood decking for the floor system for many business, retail sales, or apartment structures. The concrete is typically 1 1/2" (40 mm) thick. Steel decking like the decking in Figure 11-17 (introduced in Chapter 9) is a common method of supporting concrete floor slabs for occupancies that must support heavier floor loads. Ribbed decking is welded to steel trusses or beams that provide the support for the floor system. Ribs with a depth of 1 1/2", 3", or 4 1/2" (40, 80, or 115 mm) are most often used for floor slabs. Figure 11-18 shows an example of a poured-in-place floor system.

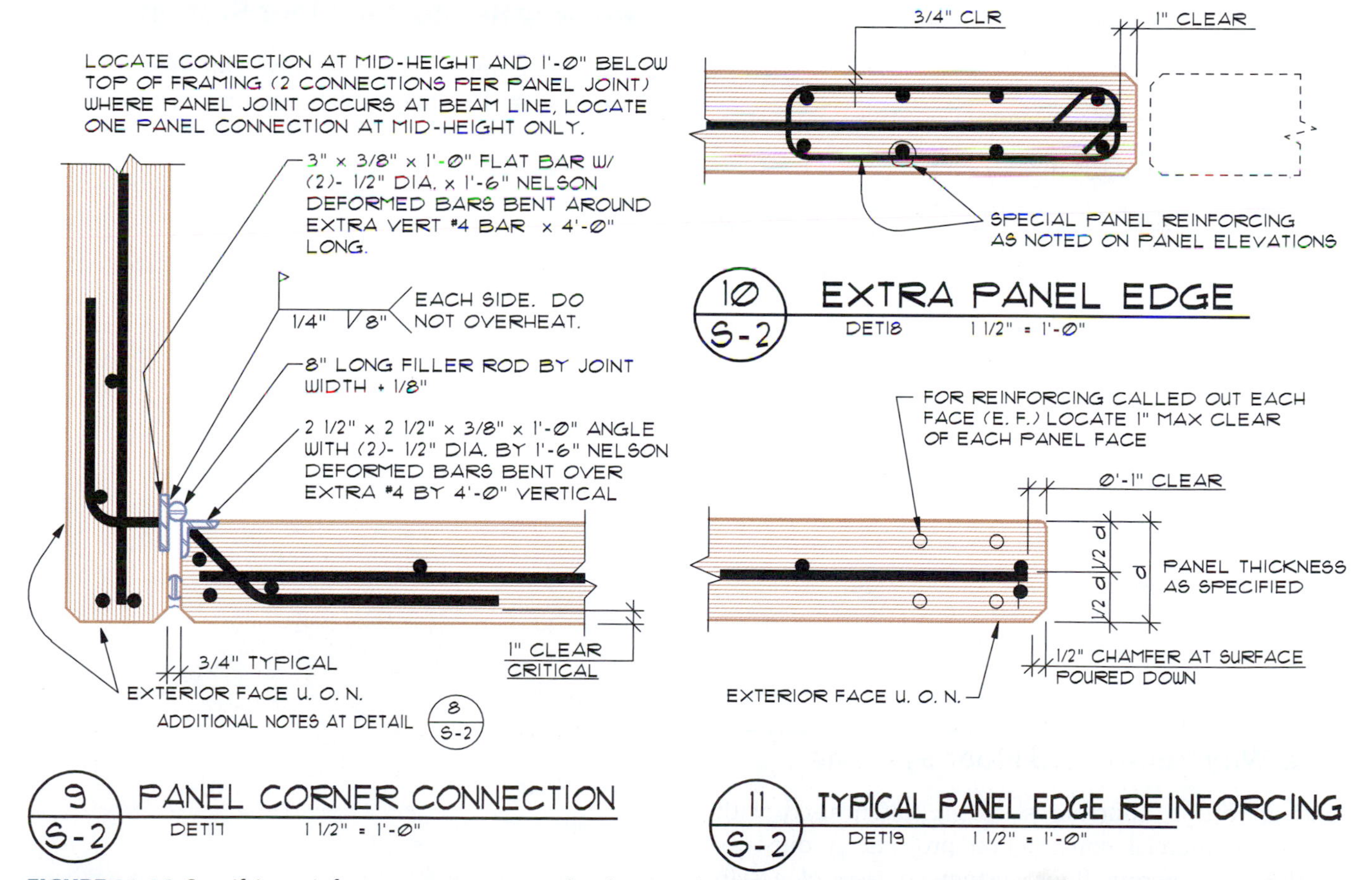

FIGURE 11-16 Specifying reinforcement coverage in details. Key information to be provided includes the size, spacing, and placement relative to the edge of the concrete. The method used to hold the reinforcement in place is referenced in the project manual. *Courtesy KPFF Consulting Engineers.*

FIGURE 11-17 A common method of supporting concrete floor slabs for occupancies that must support heavier floor loads is the use of ribbed steel decking ranging in depth from 1 1/2" to 4 1/2" (40 to 115 mm) welded to beams or trusses. *Courtesy Lee Gleason.*

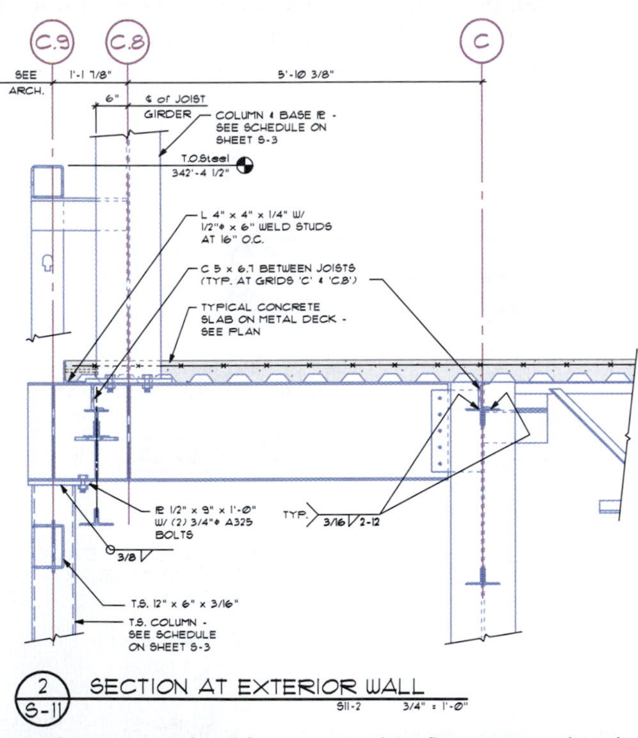

FIGURE 11-18 A detail for a cast-in-place floor system placed over metal decking. *Courtesy Van Domelen/Looijenga/McGarrigle/Knauf Consulting Engineers.*

One-Way Reinforced Floor Systems

Self-supporting concrete slabs are often used with larger commercial construction projects. A **one-way reinforced concrete floor system** consists of a slab supported by parallel reinforced concrete beams supported on reinforced-concrete columns. This system of construction is normally considered feasible for slabs spanning from 10' to 35' (3000 to 10 500 mm). Figure 11-19 shows an example of a one-way floor system. Intermediate beams placed at right angles to the main support beams can also be used for longer spans or heavier loads. A versatile system called a *one-way joist* or *one-way ribbed floor* uses narrow beams at close repetitive spacing. Joists are typically limited to a minimum width of 4" (100 mm) with a maximum spacing of 30" (750 mm) clear. The depth of the slab can range from 1 1/2" to 3" (38 to 75 mm) depending on the span with the minimum depth of the joist required to be three times the depth of the slab. The floor slab and support beams are poured to make a monolithic structure. *Monolithic* is used to describe the process of pouring several components in one pour so that all units act as one. The system is usually constructed using metal forms to construct voids between the joists. The system also can be constructed using precast, pretensioned floor units. Figure 11-20 shows examples of common precast floor shapes. Figure 11-21 shows an example of a precast floor panel, and Figure 11-22 shows an example of a detail drawn to show support for floor panels.

Two-Way Reinforced Floor Systems

As the name implies, the **two-way reinforced concrete floor system** has major slab reinforcement running

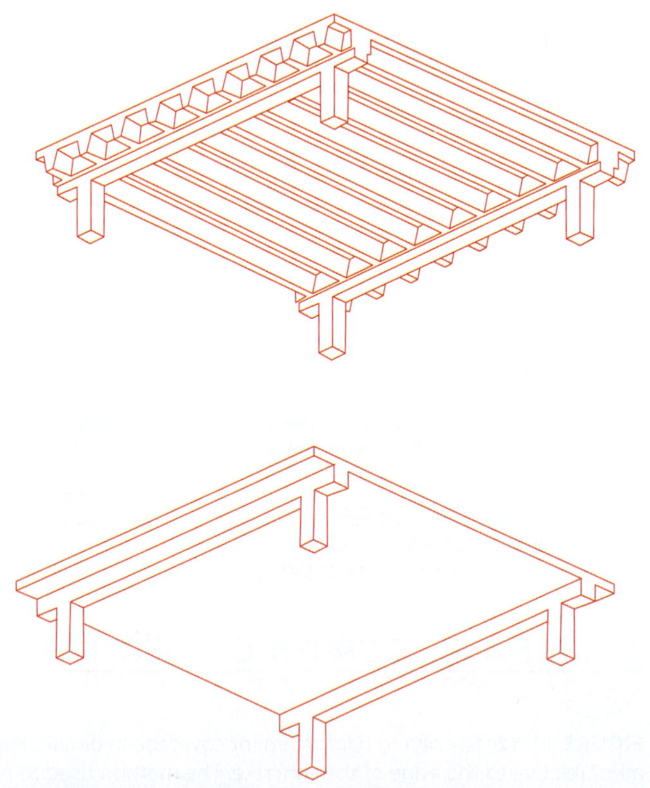

FIGURE 11-19 One-way slab and one-way joist floor systems.

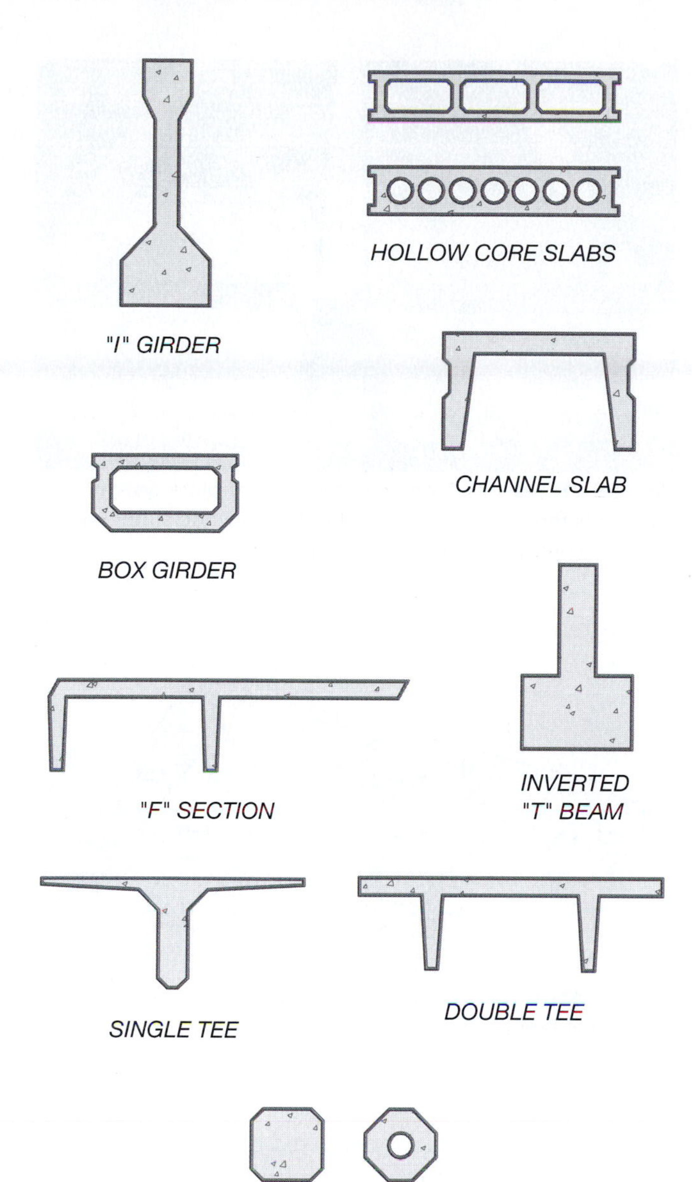

FIGURE 11-20 Common precast shapes.

FIGURE 11-21 Precast concrete hollow-core slab units. *Courtesy Portland Cement Association.*

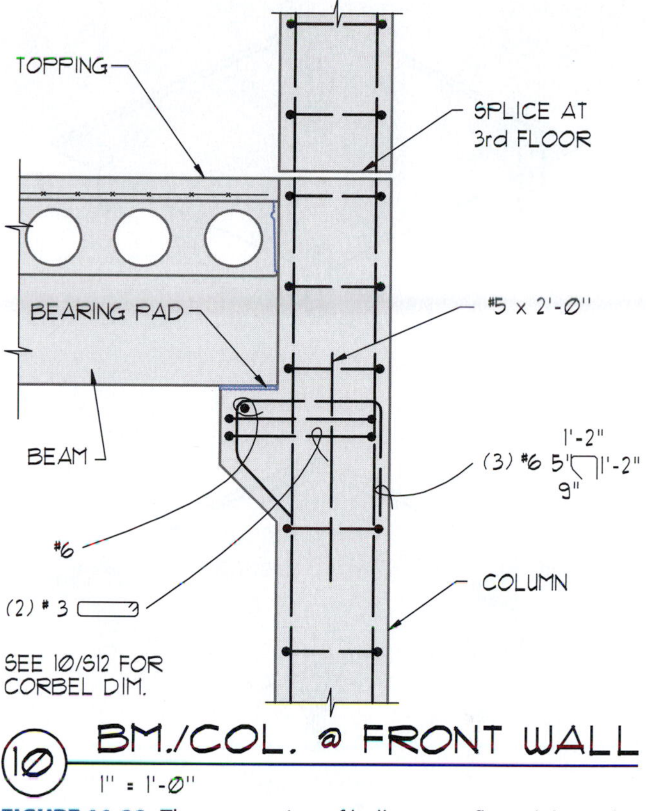

FIGURE 11-22 The connection of hollow-core floor slabs with supporting columns. *Courtesy H.D.N. Architects A.I.A.*

in two directions. Figure 11-23 shows three common methods of forming a two-way system. With each system, columns are usually placed in square patterns. In its simplest form (at top in Figure 11-23), a solid slab is placed over the support columns. This method of construction would be suitable for floors with light loads such as apartment or small office structures. A flat plate with internal two-way reinforcing (at middle in Figure 11-23) can be placed over the columns to support the slab loads and better distribute lateral loads from the floor into the column. The plate allows the overall depth of the floor to remain thin, while still providing ample area to resist shear forces. Shear is created in the floor slab as the gravity loads try to push the slab downward, forcing the post to punch a hole in the slab. The plate can be further reinforced by placing a crown at the top of the column. The crown allows necessary depth to help transfer lateral loads from the slab into the column while still keeping the slab thickness to a minimal depth.

As loads increase, beams are placed below the slab and supported on reinforced columns at the intersection of the beams (at bottom in Figure 11-23). Beams can either be precast or formed in monolithic construction. A fourth method of two-way reinforcement is called a *waffle-flat-plate* system. A two-way

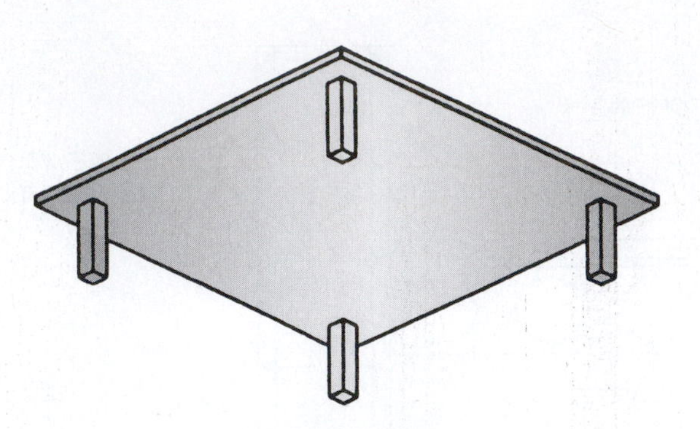

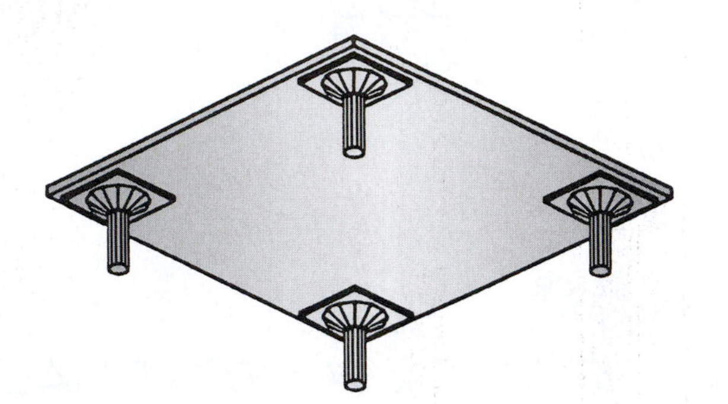

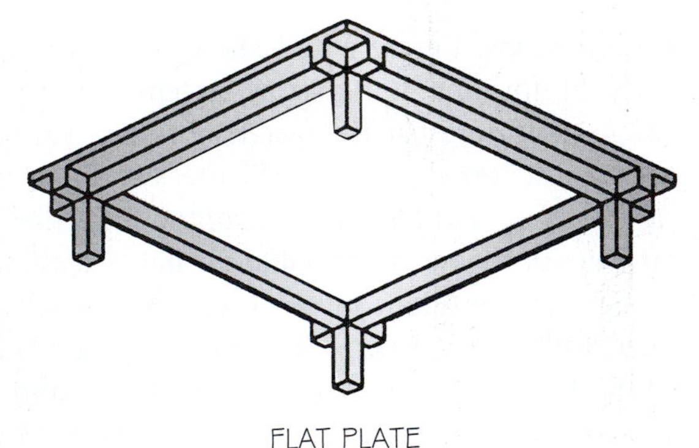
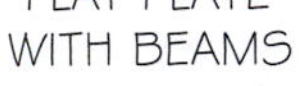

FIGURE 11-23 Three standard methods of constructing two-way floor systems include flat slab, flat slab with drop panels, and flat plate with beams.

grid of narrow beams is placed below the slab, with a void formed by metal pans placed below the slab. Figures 11-24a and 11-24b show examples of a waffle-flat-plate slab system and typical section. Common spacing of the cores ranges from 20" to 30" (500 to 750 mm) with a joist width of 4", 5", and 6" (100, 125, or 150 mm) typical. The joist depth ranges in 2" (50 mm) increments from 6" to 14" (150 to 350 mm) depending on the load, span, and manufacturer of the metal pan. Most heavy metal pans can be removed after the concrete has cured—allowing reuse. Lightweight

FIGURE 11-24a The forming of a waffle-flat-plate system. Fiberglass forms are placed over temporary wood supports until the concrete is poured and cured.

FIGURE 11-24b A waffle-flat-plate system is a common method of forming a two-way floor system. *Courtesy Portland Cement Association.*

metal pans can be left in place and then covered with a layer of metal lath to support a finish coat of plaster.

Lift Slabs

With the lift-slab method of construction, floor slabs for a multilevel structure are poured on the ground, one above another. Support columns are fabricated prior to pouring the slabs, and the slabs, ranging in thickness from 6" to 10" (150 to 250 mm) deep, are formed around the columns. The depth is determined by the spacing of supports and the loads to be supported. Once the slabs have cured, the slabs are lifted into place by hydraulic jacks and fastened to supporting columns. This system is used primarily for what are classed as lightweight loads, encountered in large-scale office structures, apartments, or dormitories.

Walls

Concrete walls can either be poured in the vertical or horizontal position. Walls taller than 10' (3000 mm), which can't easily be formed with plywood, are typically poured in a horizontal position and lifted into place.

Poured-in-Place Concrete Walls

Poured-in-place walls are formed by placing plywood forms in the desired shape, as shown in Figures 11-25a and 11-25b. Figure 11.25c shows an example of a detail used to explain cast-in-place concrete. Reinforcing is wired into position when one side of the form has been installed. Once the reinforcing has been placed in its proper location, the forms can be completed. Metal ties are used to keep the forms in their proper alignment. Some types of ties remain in the wall; others are removed and leave a small V-shaped indentation in the wall. The indentation can either be left exposed or hidden with mortar. Notes in the written specifications will indicate how tie holes will be treated after tie removal. Methods for placement of concrete are also typically provided. Figure 11-26 shows how poured walls are represented in plan views.

FIGURE 11-25b A poured-in-place wall after the forms are removed. *Courtesy Aaron Jefferis.*

FIGURE 11-25a Poured-in-place walls are formed by placing plywood forms in the desired shape. *Courtesy Megan Jefferis.*

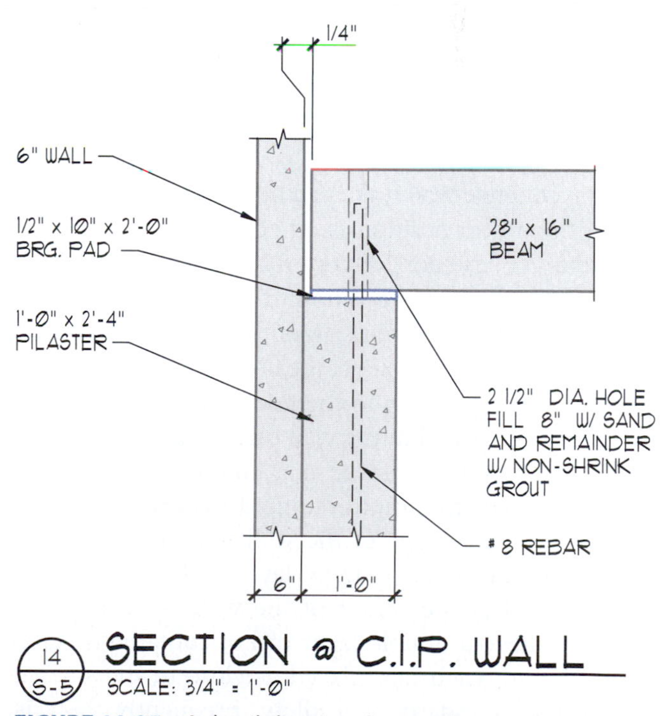

FIGURE 11-25c A detail showing the attachment of a precast beam to a cast-in-place wall.

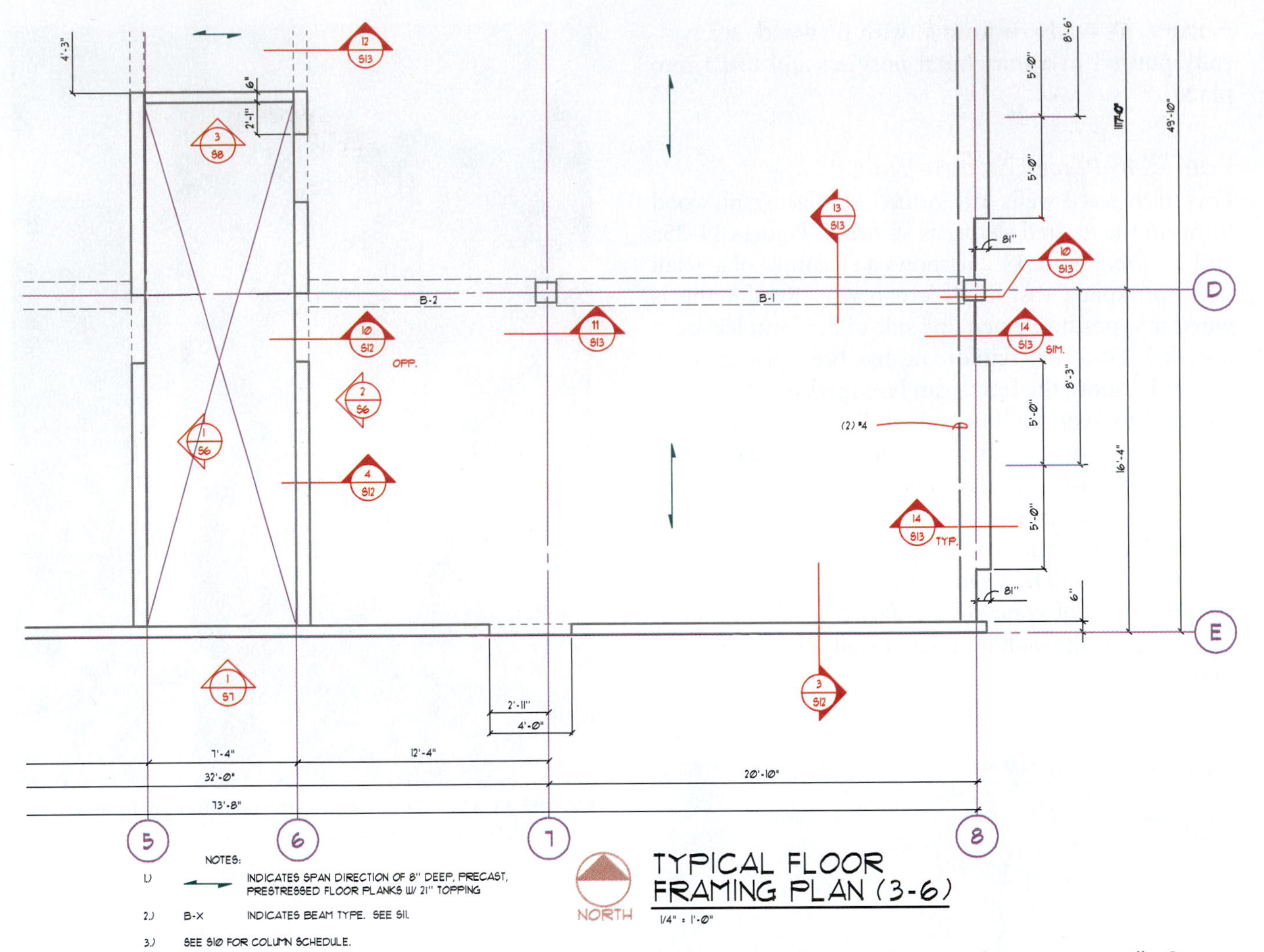

FIGURE 11-26 A floor plan for a multilevel structure framed with stacked concrete columns and cast-in-place concrete walls. *Courtesy KPFF Consulting Engineers.*

Tilt-Up Walls

Casting concrete wall units in a horizontal position and lifting them into position is an economical method of concrete construction. Figure 11-27 shows a precast wall being lifted into position. The casting surface is usually the concrete floor slab of the structure. A liquid bond breaker is sprayed over the floor slab to prevent the wall slab from bonding to the floor. Door and window openings are framed in the wall prior to pouring.

Concrete walls must be a minimum width of 4" (100 mm). The minimum required steel reinforcing is based on a percentage of the wall thickness. Thinner walls normally have a single layer of horizontal and vertical steel at the center of the wall. Thicker walls normally have a double layer of horizontal and vertical steel with each mat placed as close to the surface as coverage standards will allow. Frequently there is a horizontal bar referred to as a *chord bar*, placed near the roof diaphragm. The chord bars in each panel are joined together by welding a transfer angle from panel to panel, as shown in Figure 11-28. These angles tie all the panels together to resist horizontal forces from the roof diaphragm.

The engineer determines the required wall thickness and steel reinforcement, which the drafter must reflect on the floor framing plans similar to Figure 11-8, and wall details. Precast concrete construction requires details showing the reinforcement of panels similar to Figure 11-9, connection of panels to each other, placement and reinforcement of openings similar to Figure 11-10, anchors for connecting members, as well as fabrication and erection methods.

Figure 11-29a shows an example of a fabrication drawing for placing panel attachments for lifting. Figure 11-29b shows steel placed in the wall to resist the stress that the wall will encounter as it is lifted into place, as well as the stresses that must be resisted over the life of the structure.

FIGURE 11-27 Precast concrete walls are poured on top of the floor slab, lifted into place, and welded to adjoining panels. *Courtesy Constructionphotographs.com.*

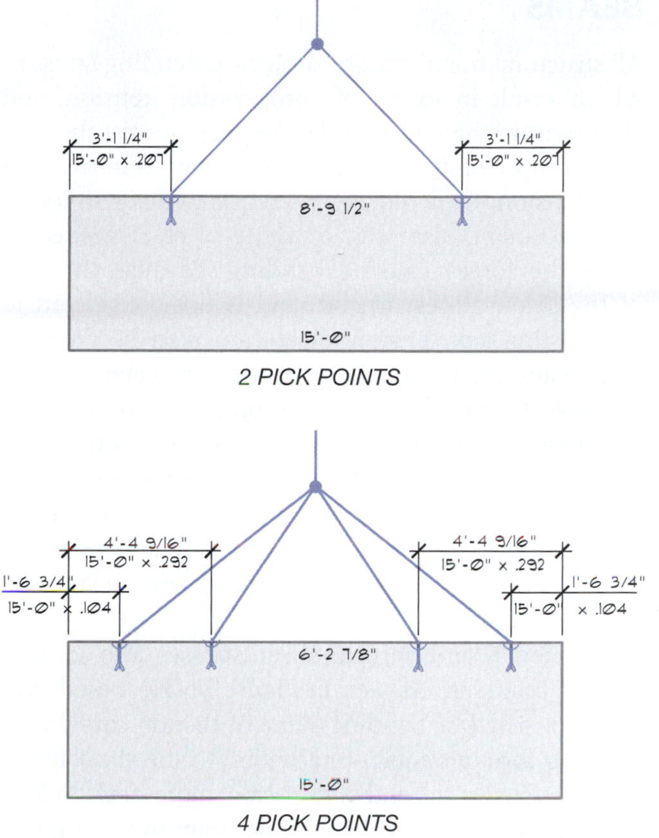

FIGURE 11-29a Details provided by the concrete fabricator are required to explain each step of the forming and moving process. This detail shows the connection points for lifting two specific types of concrete panels. *Courtesy Jim Ellsworth.*

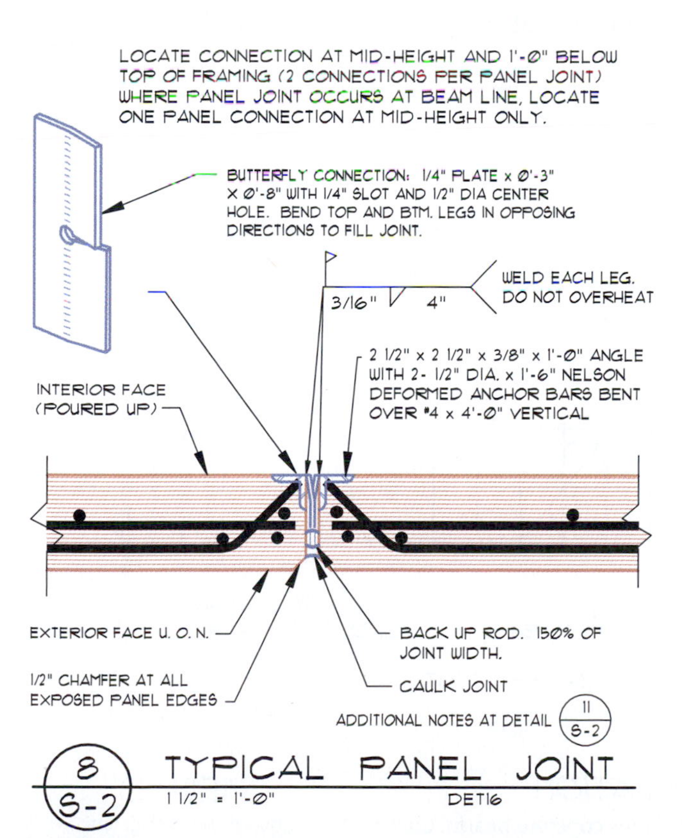

FIGURE 11-28 A connection detail for wall panels. *Courtesy Van Domelen/Looijenga/McGarrigle/Knauf Consulting Engineers.*

FIGURE 11-29b Extra steel is placed by each wall opening to resist the stress that the wall will encounter as it is lifted into place as well as the stresses that must be resisted over the life of the structure. *Courtesy LeRoy Cook.*

BEAMS

All structural members are subject to bending stresses, which result in forces of compression, tension, and shear within the member. The surface nearest the load is in compression and the surface away from the load is in tension. Concrete beams require details to locate the placement, size, and quantity of steel needed to resist the forces causing bending. Because the side of the beam closest to the load is not considered to be resisting any tension, steel can often be omitted. The beam shape can be altered from rectangular to T-shape to increase the compression surface and decrease the tension surface. If a beam is continuous over a support, the surface in tension will change to the upper surface over the support. Bars from the lower surface are often bent on an angle to reinforce the top surface of the beam as it passes over a support, as seen in Figure 11-30.

Horizontal and vertical shear stresses also tend to make a beam fail. As seen in Figure 11-30, these forces can be resisted by bending some of the top bars downward, to form an angle. In addition to the shear forces being resisted at the end of the beam, more steel is added to the bottom of the center of the beam to resist forces causing bending. Another method of resisting the shear forces is with the use of *stirrups*, which are U-shaped rebar, hung from the compression bars around the tension bars.

COLUMNS

Columns must withstand heavy compression loads. As the height of the column increases, stress from rotation and bending must also be resisted by the column. To reduce the size of the concrete, and reduce cracking, steel reinforcement is added to a column near the surface. In its simplest form, a poured concrete column can be square, rectangular, or round, with steel reinforcement placed in an arrangement similar to the reinforcement patterns in Figure 11-31.

Vertical steel is placed near the surface of the column and then tied with horizontal ties at approximately 8" to 12" (200 to 300 mm) intervals to restrict spreading of the steel. Figure 11-32 shows common methods used for column reinforcement.

Circular Columns

Circular columns are formed in premanufactured fiber tubes. Because of the difficulty of making circular ties to reinforce the vertical steel in the field, circular columns are reinforced with continuous spiral hoops that wrap around the vertical steel. These hoops can be shipped to a job site or bent at the site with special equipment. Spiral reinforcing steel is often used for square columns (see Figure 11-31) because it is better than individual ties at resisting loads. Spiral reinforcing for cast-in-place concrete is

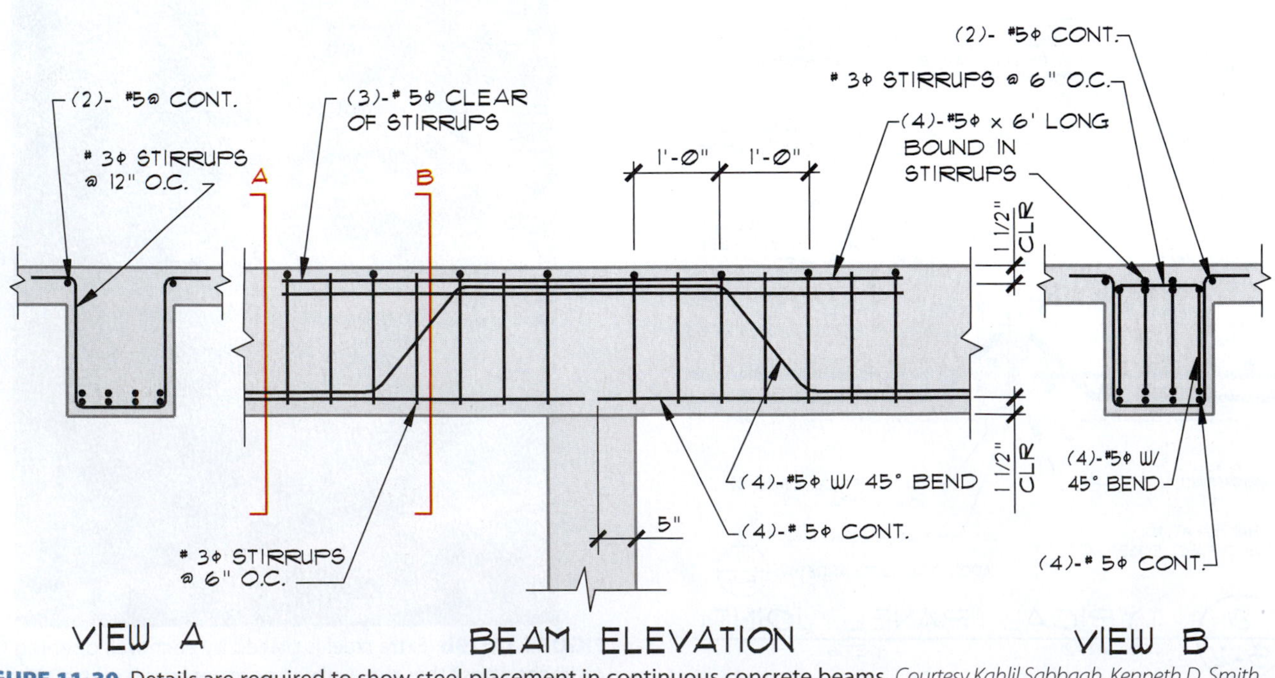

FIGURE 11-30 Details are required to show steel placement in continuous concrete beams. *Courtesy Kahlil Sabbagh, Kenneth D. Smith Architect & Associates, Inc.*

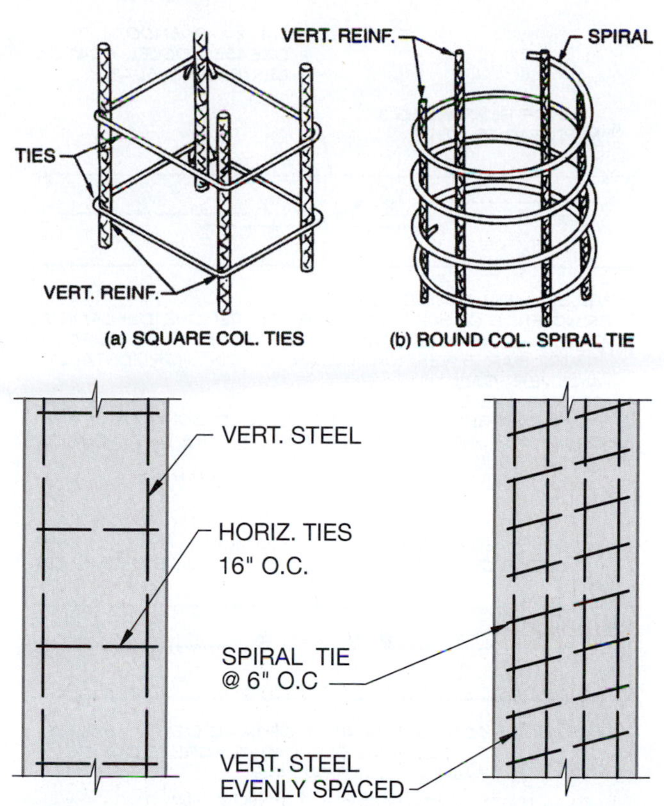

FIGURE 11-31 Steel reinforcement is made from ties placed perpendicular to, or in a spiral pattern around the main reinforcing steel.

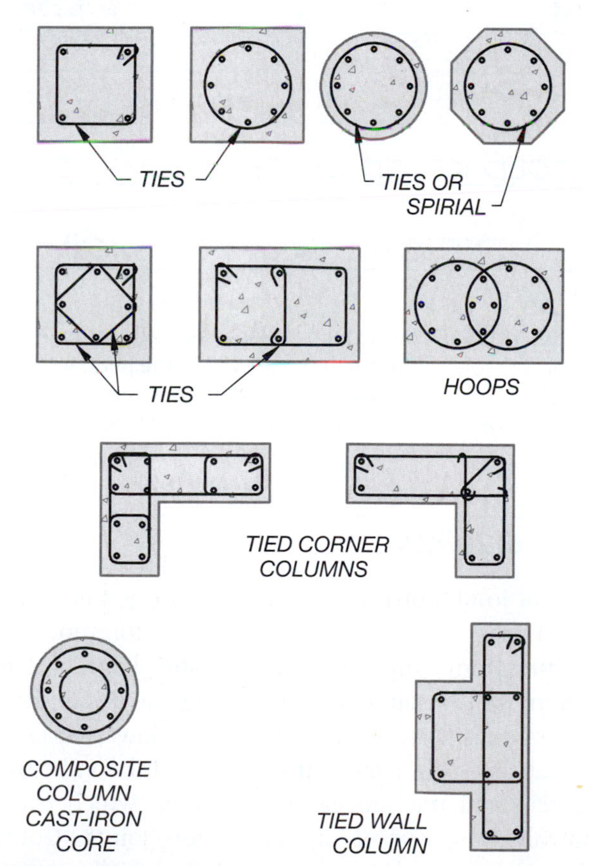

FIGURE 11-32 Standard placements of steel reinforcement for irregularly shaped columns.

required by the IBC/ACI 318-7.6 to be a minimum of No. 3 bar with a minimum spacing of 1" (25 mm) and a maximum spacing of 3" (75 mm). Columns with an irregular shape are used throughout the construction industry. Details will need to be drawn of each column to clarify the placement of reinforcement and ties.

Bundled Bars

Small bars of steel are often bundled in groups to form larger members that can be used to reinforce concrete as seen in Figure 11-33. When steel is wired or welded together, the bundle of rebar acts as one unit. The IBC/ACI 318-7.7.4 limits bundle bars to the following:

- Four pieces of equal-sized rebar per bundle.
- Laps in individual bundles in flexural members must terminate at different points with a minimum of 40 db (bar diameters) stagger.
- Concrete-beam bundles are limited to the use of No. 11 or smaller bars.

Composite Columns

As the need to resist greater loads is increased, columns are reinforced with solid steel shapes. A **composite column** uses a steel tube filled with concrete in the center of the column, surrounded with reinforced concrete around the outer surface (at left in Figure 11-34). The metal core is not allowed to account for more than 20% of the total area of the column. A minimum of 3" (75 mm) of clearance between the core and the spiral steel must be provided.

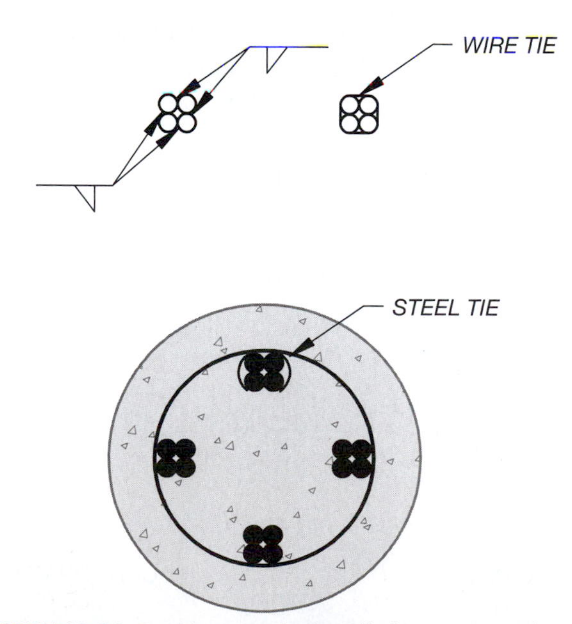

FIGURE 11-33 Steel bars can be bundled in groups to form larger members by the use of wire ties or welding.

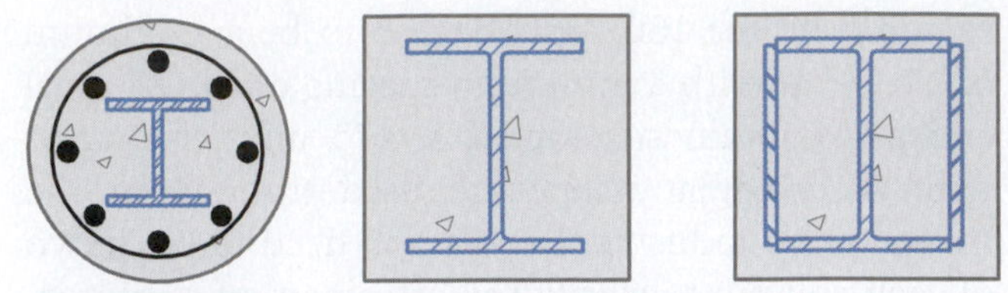

FIGURE 11-34 Placement of steel reinforcement for composite columns.

Combination Columns

A **combination column** is a column with a steel shape such as a W encased in wire mesh and surrounded in concrete. Common types of composite columns can be seen in Figure 11-34. The mesh must be a minimum of 4 × 8/W1.4 × W1.4 with the 8" wires parallel to the axis of the column. Typically 2 1/2" (65 mm) of concrete is required to surround the column. Both composite and combination columns are used for the lower floors of multilevel structures. The steel is used as part of the concrete support system rather than just reinforcing the concrete.

JOINTS IN CONCRETE

Concrete tends to shrink approximately 0.66" per 100' (16.7 mm/30 500 mm) as the moisture in the mix hydrates and the concrete hardens. Concrete also expands and shrinks throughout its life based on the temperature and the moisture in the supporting soil. This can cause the floor slab and other concrete members to crack. Control, construction, and **isolation joints** are placed in concrete floor and wall assemblies to help control possible cracking. Each can be seen in Figure 11-35. The engineer will specify the type of joint to be used as well as where it will be located, and how it will be constructed. Joint locations in slabs are specified on the slab-on-grade and foundation plans. These drawings will be explored in Chapter 24. Details similar to Figure 11-36 show the type of joint to be used and are typically placed with or near the slab plan. Joints in walls are normally specified on the elevations or on panel elevations for tilt-up structures.

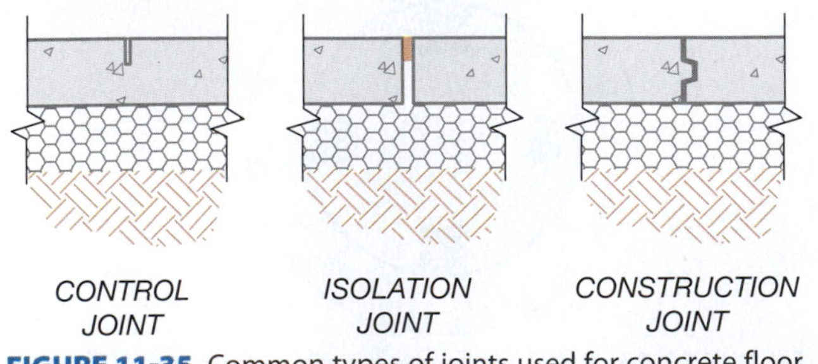

FIGURE 11-35 Common types of joints used for concrete floor slabs and for other structural concrete members.

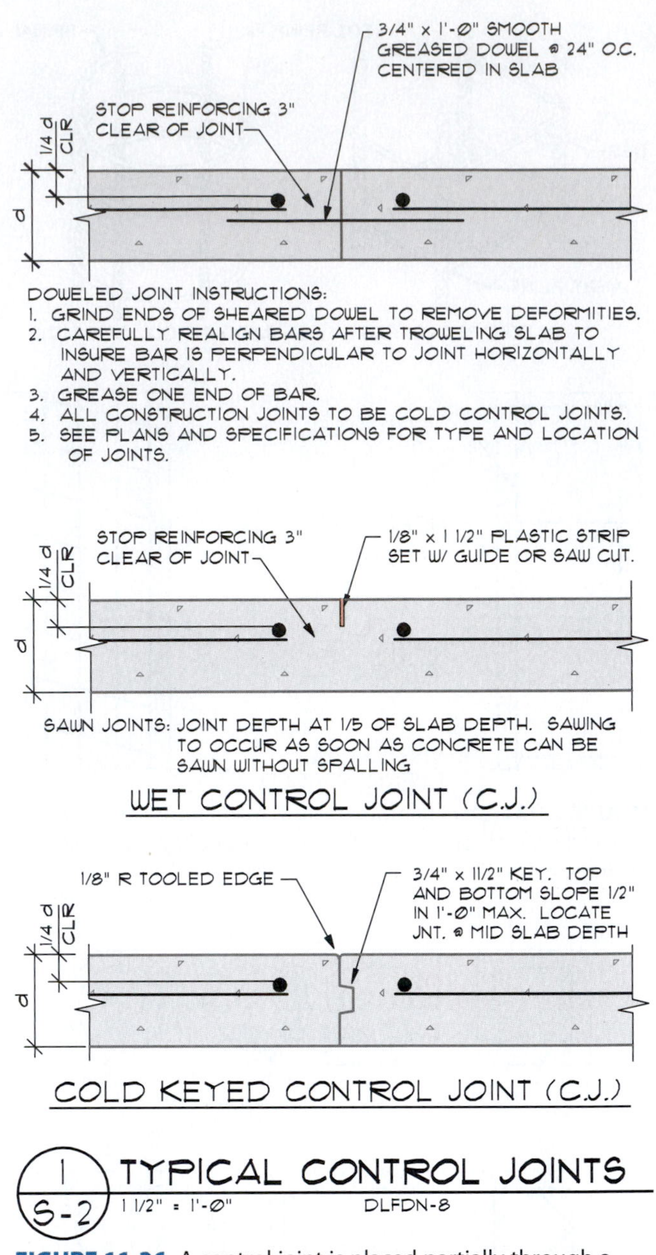

FIGURE 11-36 A control joint is placed partially through a slab or wall to allow stress to be relieved in the joint. An isolation joint is placed completely through a concrete member to allow for expansion and contraction. *Courtesy Michael & Kuhns Architects, P.C.*

Control Joints

Control joints, or *contraction joints*, are joints that are placed in concrete members to allow for movement resulting from temperature change, shrinkage, or deformation. As the slab contracts during the initial drying process, tensile stress is created in the lower surface of the slab as it rubs against the soil. The friction between the slab and the soil causes cracking, which can be controlled by *control* or contraction joints. Control joints are sawn, formed, or tooled partially through the concrete. By placing the joint partially through a slab or wall, structural stability and water tightness are maintained. The joint weakens the surface of the concrete slightly, causing any cracks that develop from movement to occur along the control joint. These joints do not prevent cracking, but they do control where the cracks will develop in the slab.

Control joints are usually one-third of the slab depth. The American Concrete Institute (ACI) suggests control joints be placed a distance in feet equal to about 2 1/2 times the slab depth. For a 5" (125 mm) slab, joints would be placed at approximately 12.5' (3800 mm) intervals. The engineer will provide specific locations based on the soil conditions at the job site. The location of control joints should be represented by a line and specified by a note referenced to the joint. The spacing and method of placement can be specified in the general notes placed with the foundation plan for simple projects, on the slab-on-grade plan if required, or in the specifications within the project manual.

Control joints are also placed in concrete walls. The roughness of the cracks that develop and the reinforcing steel keep the wall from moving independently once cracks occur. The joint is usually about one-third the depth of the wall or structural member. The engineer determines the size of joints and where they are to be placed, based on the size of the member and the loads that it will be subjected to. Joints are placed at a maximum of 20' (6000 mm) intervals in walls unless the wall contains multiple openings. Other guidelines based on the American Concrete Institute are shown in Table 11-2.

Joints are typically provided within 10' to 15' (3000 to 4500 mm) of building corners. For multistory structures, joints are often centered above an opening. To induce cracking in the control joint, one-half of the horizontal reinforcement is often stopped at the joint. Documentation by the engineer will specify any special treatment of steel reinforcing as well as the location of each joint. Although the engineer may specify the location of joints in a note within the structural calculations, the CAD technician should physically represent the joint on the drawings rather than merely placing a note.

Wall Height Recommended	Vertical Control Joint Spacing
2–8 ft (610–2440 mm)	3 × the wall height
8–12 ft (2440–3660 mm)	2 × the wall height
>12 ft (3660 mm)	equal to the wall height

TABLE 11-2 Recommended Joint Spacing

Isolation Joints

Isolation or *expansion joints* are used to separate a floor slab or wall panel from adjacent structural members so that stress cannot be transferred from one member to another and cause cracking. These joints will fully penetrate through the member and allow for expansion of the concrete caused by moisture or temperature. Isolation joints are typically between 1/4" to 1/2" (6 to 13 mm) wide and are placed at intersections of members such as walls to slabs, or at stress points within a slab. Isolation joints should be drawn and specified using methods similar to those used to represent control joints. Figure 11-37 shows a partial slab-on-grade plan with isolation joints placed between slabs. Isolation joints are also provided around each column base so that forces causing the column to rotate will not affect the slab. This joint also allows for settling of the column footing due to normal shrinkage.

Care must be taken when representing mesh in details that show the intersections of slabs. As a general rule, smooth steel dowels are often placed across the joint to allow lateral movement of the joint and to prevent out-of-plane movement. No mesh should be allowed to cross the joint because the mesh would restrict expansion and contraction. Figures 11-38a and 11-38b show a slab floor joint placed around a concrete pedestal. Material used for the filler at the joint is also specified in the joint detail. The exterior side is usually caulked with a weather-resistant caulking that will remain flexible. The caulking is placed over a flexible gasket material to further seal the seam.

Construction Joints

Joints provided to facilitate the construction process are referred to as **construction joints**, or as *keyed* or *pour joints*. Construction joints are located where one day's placement will end, or when slab construction must be interrupted long enough that new concrete will not bond with old concrete. The joint is made to provide support between the two slabs. The key is a beveled edge that is about one-fifth of the slab thickness in depth, and one-tenth of the slab thickness in width. The method used to form the joint must be specified with information about other joint types.

Because the crew placing the concrete determines the location, these joints are not shown on the drawings. Based on the specifications of the engineer, the drafter does need to show the reinforcement that will extend into the next phase of work. Construction joints in floors, beams, and girders are restricted to the middle

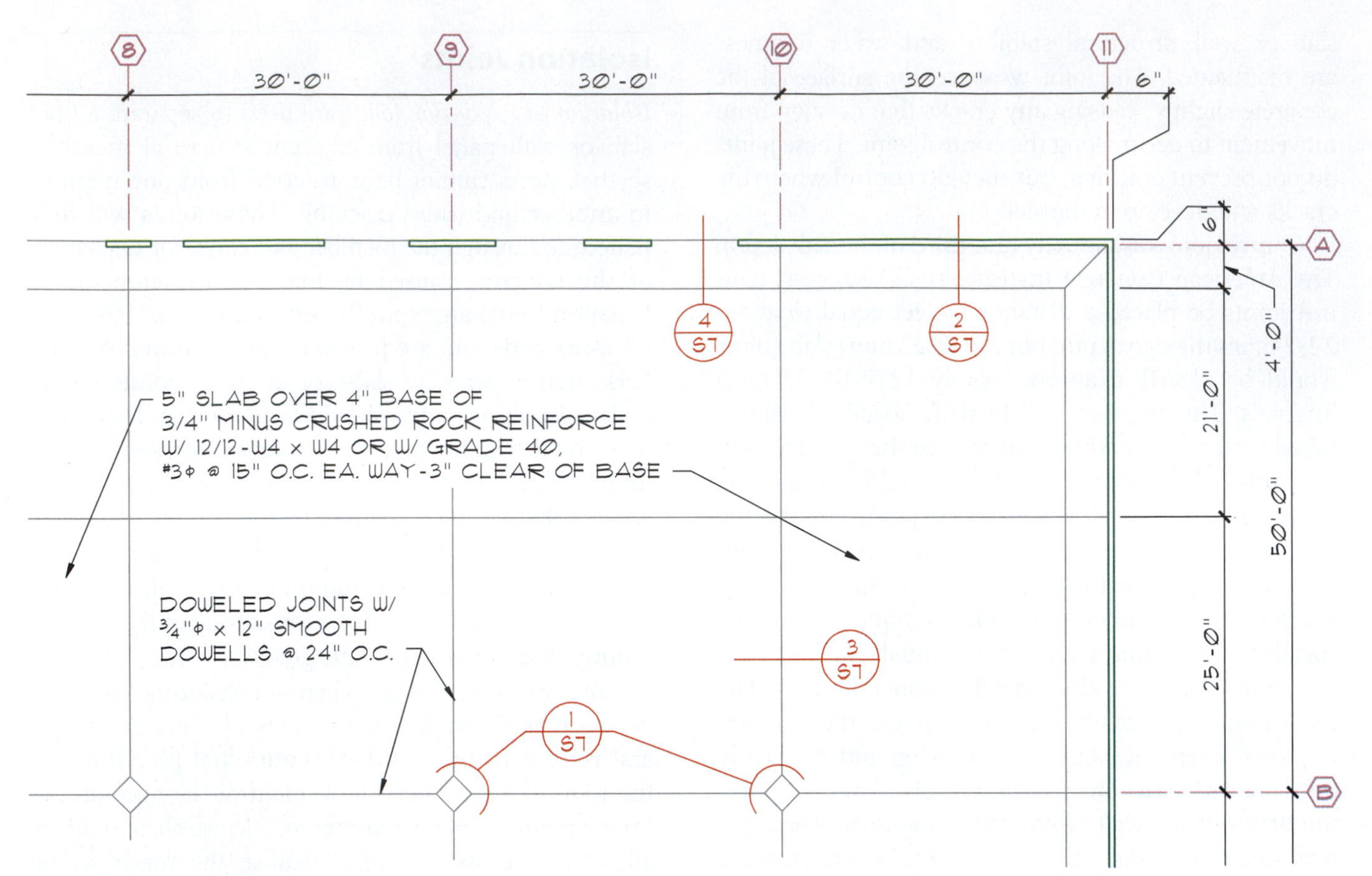

FIGURE 11-37 Representing joints on a concrete slab-on-grade plan. *Courtesy David Ambler, Kenneth D. Smith Architect & Associates, Inc.*

FIGURE 11-38a The concrete slab is poured over the pier that provides support for the steel column. Gravel fill will be placed over the pier and a small slab will be poured to provide the finished flooring. The new slab will be isolated from the floor slab so that any movement of the column will not damage the floor slab. *Courtesy Sandy Clark, Divine Designs & Associates, Inc.*

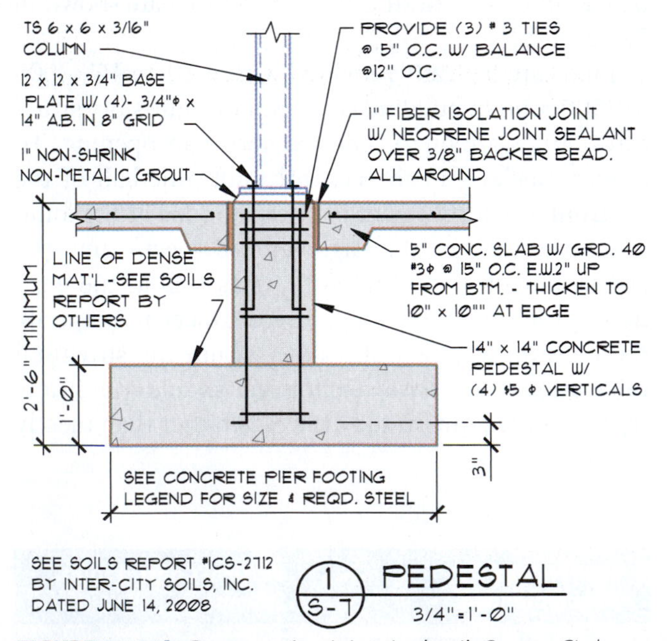

FIGURE 11-38b Representing joints in detail. *Courtesy Gisela Smith, Kenneth D. Smith Architect & Associates, Inc.*

EASY BEING GREEN

Concrete is an environmentally friendly product in all stages of its life span, from raw material production to demolition. Key benefits of concrete products include resource efficiency, life span, thermal mass, and afterlife. The predominant raw material for the cement in concrete is limestone, the most abundant mineral on earth. Concrete can also be made with fly ash, slag cement, and silica fume, all waste by-products from power plants, steel mills, and other manufacturing facilities.

Concrete is ordered and placed as needed and does not need to be trimmed or cut after installation. Wash water is frequently recycled using trucks equipped with devices that collect wash water and return it to the drum where it can be returned to the ready-mix concrete plant for recycling. Extra concrete is often returned to the ready-mix plant where it is recycled or used to make jersey barriers or retaining wall blocks; or it can be washed to recycle the coarse aggregate.

Precast structural and architectural concrete products are friendly to the environment because they have optimized geometries, and they reduce environmental damage on the construction site compared with pouring concrete on-site. Waste from overage is also eliminated through precasting at a plant. Through aeration, the weight of precast concrete components is reduced by up to one-third while improving its insulation value. Some panels are cast with integral foam insulation.

Insulating concrete forms (ICFs) provide a labor-efficient means of making insulated poured-concrete walls, floors, and roof decks. ICFs are permanent forms that are left in place after the concrete has cured. Most of these products are made from expanded polystyrene (EPS) foam produced with a non-ozone-depleting blowing agent. Some ICFs are made from a composite of wood waste or EPS beads and portland cement. The environmental advantages of ICF walls include higher R-values, and their use can result in reduced concrete content compared with conventionally formed concrete walls.

Once concrete is made, it provides durable, long-lasting structures that will not rust, rot, or burn. Life spans for concrete building products can be double or triple those of other common building materials. Concrete structures are not only long-lasting; they are efficient. Structures built with concrete walls, foundations, and floors are highly energy efficient because of their ability to absorb and retain heat. With careful planning, occupants can significantly cut their heating and cooling bills and install smaller-capacity HVAC equipment.

Once a concrete structure has served its purpose, the usefulness of the concrete can continue. After a concrete structure has served its original purpose, the concrete can be crushed and recycled into aggregate for use in new concrete, or used as backfill. Recycling concrete from demolition projects can result in considerable savings because it saves the costs of transporting concrete to the landfill and eliminates the cost of disposal. Crushed concrete may be reused as an aggregate in new portland cement or any other structural layer. Generally it is combined with a virgin aggregate when used in new concrete. However, recycled concrete is more often used as aggregate in a sub-base layer.

third of the span. Horizontal construction joints are usually placed in vertical members for each floor level. Vertical joints are usually located at corners, pilaster, or column edges, or on other elements where they will be hidden.

CONDUIT

In addition to the structural considerations of planning and drawing the floor slabs and walls, provision must also be made for providing conduits for the passage of electrical, plumbing, and heating, ventilation, air-conditioning (HVAC) ducts. Plastic or fiber conduit is fastened to the reinforcing material to allow for passage of electrical wiring and plumbing lines. Copper and aluminum should not be connected to reinforcement because of electrolysis. The engineer is responsible for determining the size and location of all penetrations into concrete members, and the drafter details and specifies the required material. Care must be taken to avoid using aluminum conduit or piping in concrete unless the member has been coated to prevent a reaction with either the concrete or steel. The IBC/ ACI 318-6.3 limits the material, size, and placement of the conduit. Restrictions include the following:

- Conduits or pipes in a column can't displace more than 4% of the surface area of the column.
- The outer diameter of the conduit can't be greater than one-third the thickness of the member it passes through without the approval of the building department.

- Conduit can't be spaced closer than three conduit diameters or widths on center.
- The conduit must be of a size that will not require the cutting of the reinforcement. Conduit can be considered as replacing structurally in compression the displaced concrete under the following conditions:
  - The conduit must not be exposed to rusting or other deterioration.
  - The conduit is uncoated or galvanized iron or steel with a minimum thickness of schedule 40 steel.
- The conduit has a nominal interior diameter of 2" (50 mm) maximum with a minimum spacing of 3 diameters on center.

When placed in below-grade concrete or concrete that will be exposed to weather conditions, the conduit must have a minimum of 1 1/2" (40 mm) of concrete cover. Conduit not exposed to either condition requires only a 3/4" (19 mm) cover. Whenever possible, electrical, plumbing, and HVAC lines and equipment are placed in the space below the floor slabs for a suspended ceiling.

ADDITIONAL READING

The following Web sites can be used as a resource to help you keep current with changes in concrete materials.

ADDRESS	COMPANY/ ORGANIZATION
www.aci-int.org	American Concrete Institute
www.archprecast.org	Architectural Precast Association
www.concretenetwork.net	Concrete Information and Resources
www.cmpc.org	Concrete Masonry Promotions Council
www.concretenetwork.com	ConcreteNetwork.com
www.crsi.org	Concrete Reinforcing Steel Institute
www.concretethinker.com	Concrete Thinking for a Sustainable World
www.iccsafe.org	Concrete Manual
www.ncma.org	National Concrete Masonry Society
www.precast.org	National Precast Concrete Association
www.nrmca.org	National Ready Mixed Concrete Association
www.cement.org	Portland Cement Association
www.post-tensioning.org	Post-Tensioning Institute
www.pci.org	Precast/Prestressed Concrete Institute
www.rrc-info.org.uk	Reinforced Concrete Council
www.soils.org	Soil Science Society of America
www.tilt-up.org	Tilt-Up Concrete Association
www.wirereinforcementinstitute.org	Wire Reinforcement Institute

KEY TERMS

Admixture
Aggregate
Bolsters
Bundled bars
Chairs
Combination column
Composite column
Concrete, air-entrained
Concrete, high-early-strength
Concrete, lightweight
Concrete, plain
Concrete, post-tensioning
Concrete, precast
Concrete, prefabricated
Concrete, prestressed
Concrete, pretensioning
Concrete, reinforced
Coverage
Heat of hydration
Joint, construction
Joint, control
Joint, isolation
One-way reinforced concrete floor systems
Portland cement
Potable water
Two-way reinforced concrete floor systems
Welded wire fabric

CHAPTER 11 TEST Concrete Methods and Materials

QUESTIONS

Answer the following questions with short complete statements. Type the chapter title, question number, and a short complete statement for each question using a word processor. Some answers may require the use of vendor catalogs or seeking out local suppliers. If math is required to answer a question, show your work.

Question 11-1 List six areas that are often addressed by concrete specifications to ensure that all design values are met.

Question 11-2 What standard controls portland cement?

Question 11-3 What Type of cement should be specified if an early high strength is needed?

Question 11-4 What type of cement should be used for large pour members requiring low heat of hydration?

Question 11-5 What is Type IV cement used for?

Question 11-6 Define "potable water."

Question 11-7 List elements that should be kept from water to be used for concrete.

Question 11-8 What two qualities of concrete will the water affect?

Question 11-9 What size aggregate is used for most construction projects?

Question 11-10 List three guidelines required by the IBC for sizing aggregate.

Question 11-11 List two methods of describing the contents of a concrete mixture.

Question 11-12 What does the water/cement ratio affect?

Question 11-13 When will concrete mixed to ASTM C192 standards reach its ultimate strength?

Question 11-14 List and briefly describe the eight major areas of concrete construction.

Question 11-15 What does ASTM A-36 cover?

Question 11-16 What is the minimum spacing between parallel strands of rebar?

Question 11-17 How are perpendicular strands of rebar held in place?

Question 11-18 What holds the upper layer of a double-layered rebar mesh?

Question 11-19 Describe the note: W6 ×12 10/10.

Question 11-20 Define "monolithic."

Question 11-21 What keeps tilt-up wall panels from sticking to the floor slab?

Question 11-22 Describe general steel requirements for tilt-up walls thicker than 10 inches.

Question 11-23 Describe two methods of reinforcing round columns.

Question 11-24 Describe the difference between a composite and a combination column.

Question 11-25 List and describe three types of concrete joints.

DRAWING PROBLEMS

Unless your instructor gives you other instructions, complete the following details using the guidelines presented in "Drawing Criteria for Completing Details," in Chapter 6 and the additional guidelines listed below. Skeletons of most of these details required for this chapter can be accessed from the student CD.

ADDITIONAL DRAWING GUIDELINES

- **Unless noted, all wall construction is to be poured concrete.**
- **Unless noted, specify all walls to be 6" thick 4000 psi concrete with grade 60, #5 bars at 10" o.c. each way.**
- **Note that all steel is to extend to be within 2 inches of wall edges. Steel is to be 2" clear of exterior wall surface. Place unspecified steel per minimum specifications listed in this chapter.**

Problem 11-1 Use the following sketch to draw a section view showing TJW/36 open-web trusses at 32" o.c. intersecting a tilt-up concrete wall. Wall steel will be shown in other details, so no steel will be shown in this detail. Support the trusses on a 3 × 12 DFPT ledger bolted with 3/4"Ø × 8" anchor bolts placed at 48" o.c. through 1/8" × 2"Ø washers staggered 3" up/down. Provide 2 × 4 solid blocking between the trusses at the ledger and anchor each block with a Simpson Co. A-35 at each end of each block.

Provide a 3/4" × 4" (high) × 3" steel plate bolted to the wall with 5/8" × 4" M.B. at each third truss. Weld a MST27 strap to plate with 1/8" fillet field weld. Nail the strap to the truss as per the manufacturer's specifications. Cover the roof with 5/8" plywood sheathing. Use the information presented in problem 10-10 to waterproof the roof/wall intersection.

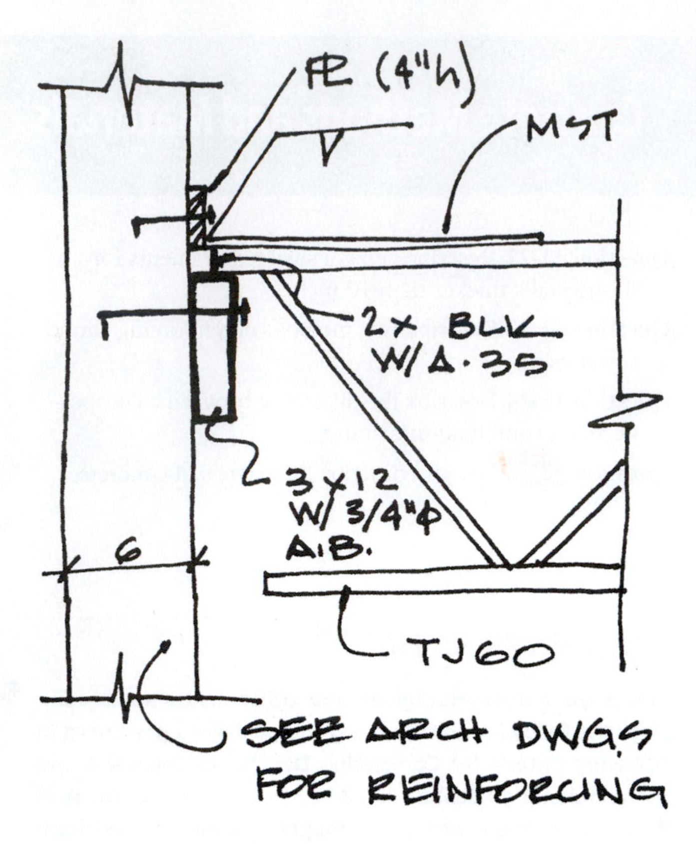

Problem 11-2 Use the following sketch to draw a section view showing a wall panel intersecting a 30" × 12" deep concrete footing that extends 18" into the natural grade. No steel needs to be drawn in the wall and footing. Reference the architectural drawings for steel location.

Show the base of the footing extending 18" into the natural grade. Show a 3" × 3" × 9" deep sleeve at 48" o.c. Support the wall on 1" nonshrink grout with 7/8" diameter × 14" structural connector by Richmond or equal centered in wall and footing.

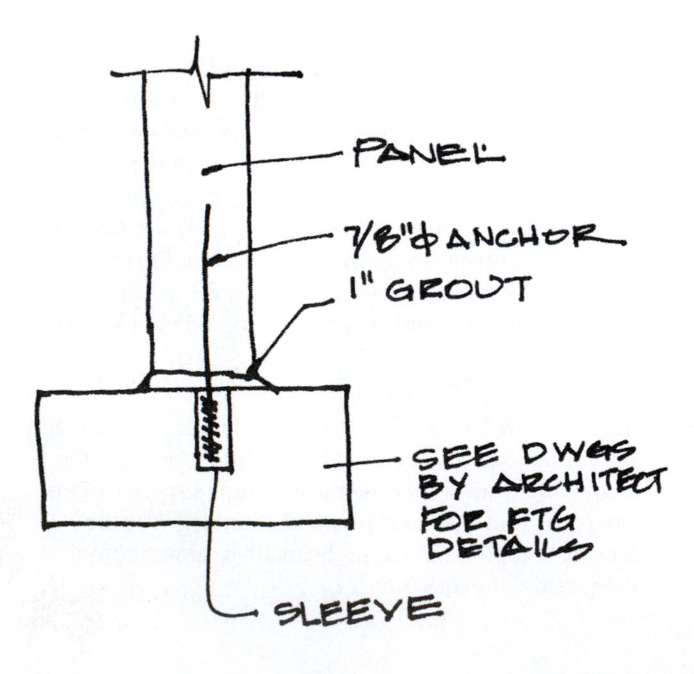

Problem 11-3 Use the attached sketch and a scale of 1/4" = 1'–0" to draw an elevation of a concrete wall panel. Show a portion of the wall detailing the following steel placement for the interior panel centered in the wall using:

- Grade 60, #5 @ 10" o.c. vertical.
- Grade 40 #4 bars @ 12" o.c. horizontal.
- Extend all wall steel to within 2" clear of panel edge.

Place grid symbols as indicated.

Draw an additional detail of each panel opening using a scale of 1/2" = 1'–0" to locate and represent steel. Reference these details on the original elevation. Reinforce each opening with additional steel placed 1" clear of the interior face and 1 1/2" clear of doorway opening each side:

- (3)-grade 60, #6 @ 8" o.c. vertical bars beside each opening.
- (3)-grade 60, #4 @ 10" o.c. horizontal above and below each door.
- Grade 40, #4 @12" o.c. × 36" vertical above and below door.
- Reinforce each corner with (2)-grade 60, #5 × 36" diagonal bars at 8" o.c. Place the first diagonal bar at the intersection of the vertical and horizontal reinforcing.

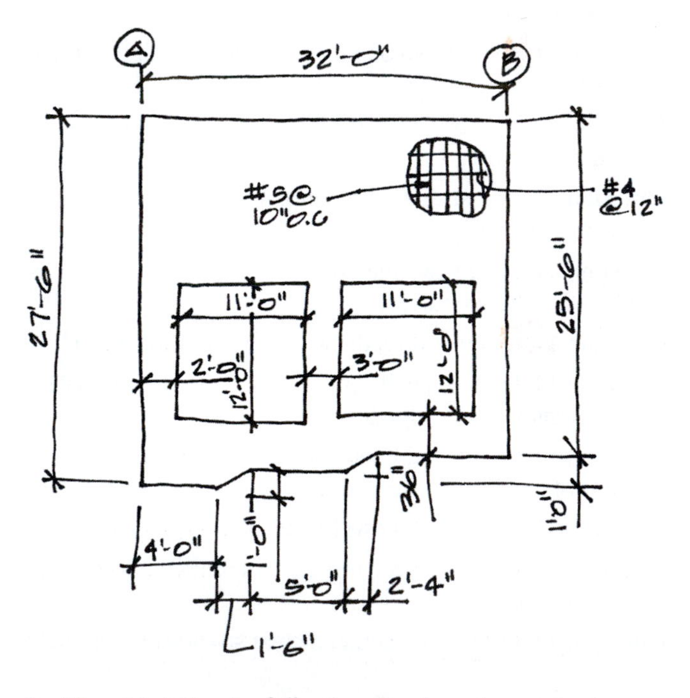

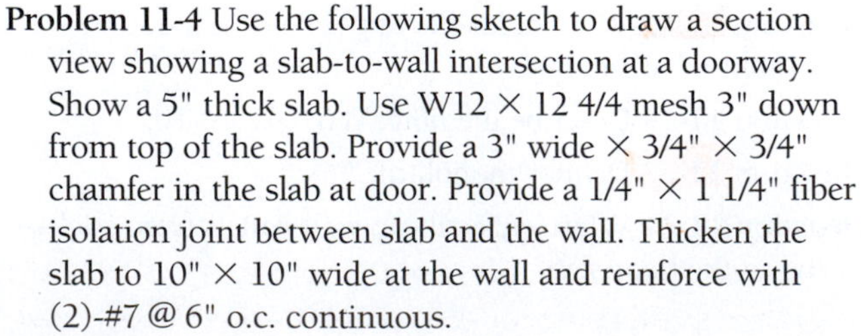

Problem 11-4 Use the following sketch to draw a section view showing a slab-to-wall intersection at a doorway. Show a 5" thick slab. Use W12 × 12 4/4 mesh 3" down from top of the slab. Provide a 3" wide × 3/4" × 3/4" chamfer in the slab at door. Provide a 1/4" × 1 1/4" fiber isolation joint between slab and the wall. Thicken the slab to 10" × 10" wide at the wall and reinforce with (2)-#7 @ 6" o.c. continuous.

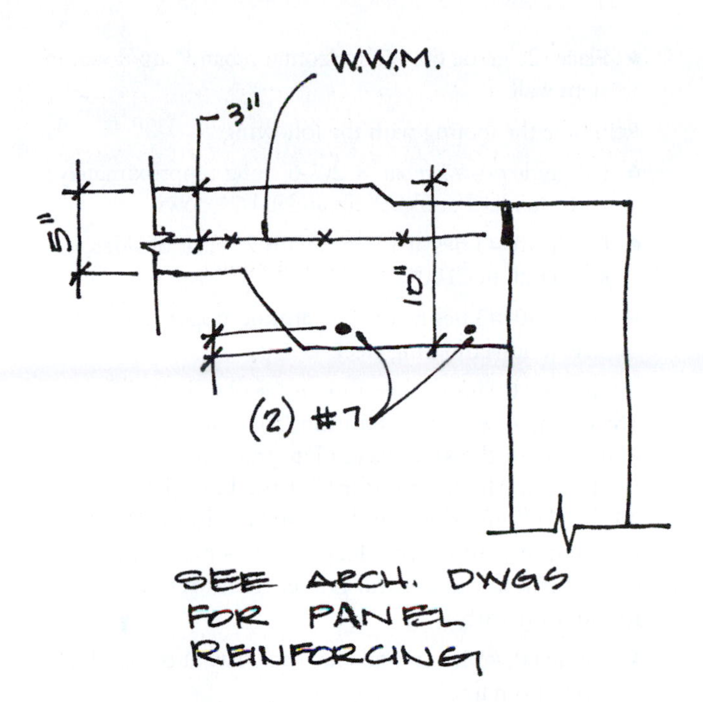

Problem 11-5 Use the following sketch to draw a section view showing an 8" wide × 48" maximum high retaining wall. Support on a 14" deep footing that tapers from 30" to 54". Show the footing extending 8" to 12" at soil side, with a 4" diameter French drain in 8" × 24" × 3/4" gravel bed. Extend the wall 6" min. above the finish grade. Show a 5" thick slab reinforced with W12/12 × 4/4 mesh 3" from top surface over 4" gravel fill. Reinforce the footing with grade 40, #4 @ 8" o.c. each way 3" up/dn. Show a grade 60, #5 @ 10" o.c. × 24" L into footing, with a 40Ø lap to the vertical wall steel. Reinforce the wall with grade 60, #5 @10" o.c. each way, 2" clear of wall.

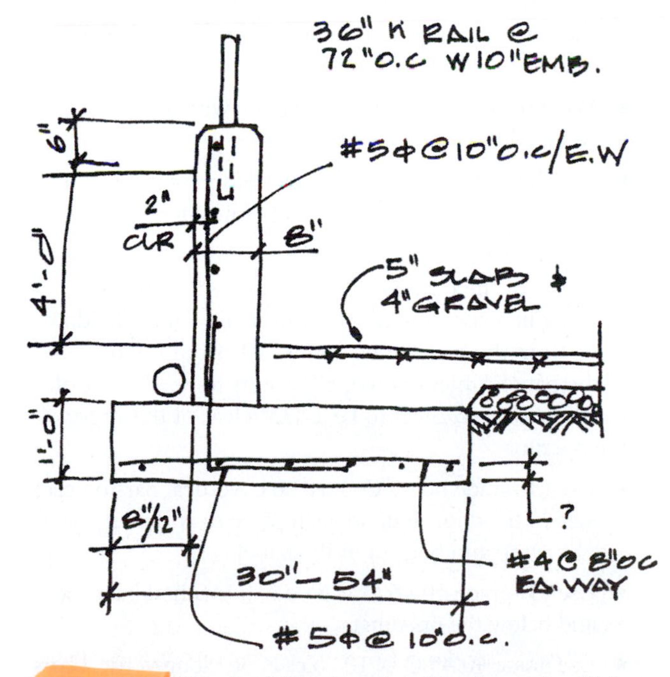

Problem 11-6 Use the following sketch and a scale of 1" = 1'–0" to draw a section view showing a 14" × 14" × 60" high pedestal (from top of footing/top of pedestal). Show a 5" thick slab with W12 × 12/4 × 4 mesh 3" from top surface with a 1/4" × 1 1/4" fiber filled isolation joint filled with a neoprene backer bead level with the top of the pedestal. Support the pedestal on a 14" × 6'–6" × 6'–6" concrete footing. Reinforce the footing with:

- Grade 40, #5 @12" o.c. each way 3" up from bottom.
- Grade 40, #4 @ 8" o.c. each way 2" down from top.
- Extend the (8)-grade 60, #5 × 6" leg in the footing into the pedestal and lap with the pedestal steel with a 20 diameter lap.

Reinforce the pedestal with:

- (8)-grade 60, #5 vertical 1 1/2" clear of face.
- Tie vertical pedestal steel with #3 @ 1 1/2" up/dn. @ 5/10" o.c. per sketch.
- Provide (4)-3/4" dia. × 16" A.B. on an 8" pattern with a 2 1/2" minimum projection.

Show a plan view of the pedestal and footing and indicate that the pedestal is rotated 45° to the footing.

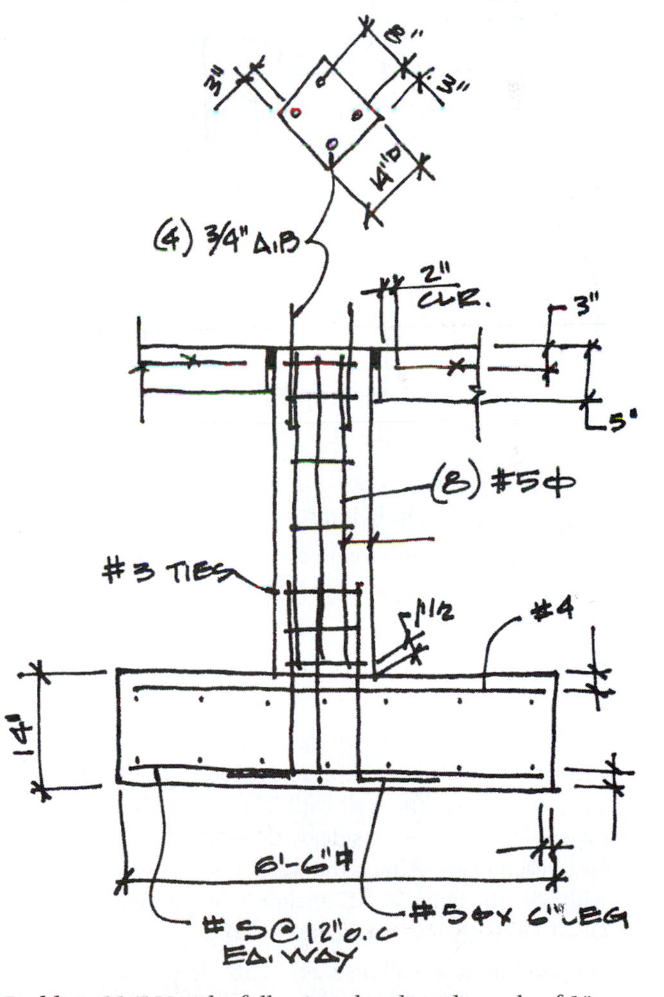

Problem 11-7 Use the following sketch and a scale of 1" = 1'–0" to draw a section and plan view showing the intersection of wall panels. Use (2)-3/4"ø × 2 1/2" coil inserts through 1" diameter holes in an 11"h × 10"w × 1/2" plate. Inserts to be 1 1/2" minimum from plate edge and 3" minimum from concrete edge. Attach the edge of

the plate to 3" × 13" × 1/2" plate embedded into concrete panel with 3" minimum edge clearance. Provide (2)-3/4"ø × 3" studs at 8" o.c. centered and welded to back of the 13" plate with 1/8" fillet all around weld. Provide 1 1/2" plate overlap and weld with 1/4" fillet weld. Place the connectors at 60" o.c. along each wall joint. Seal the exterior side of the wall with neoprene sealant and 3/4" backer bead.

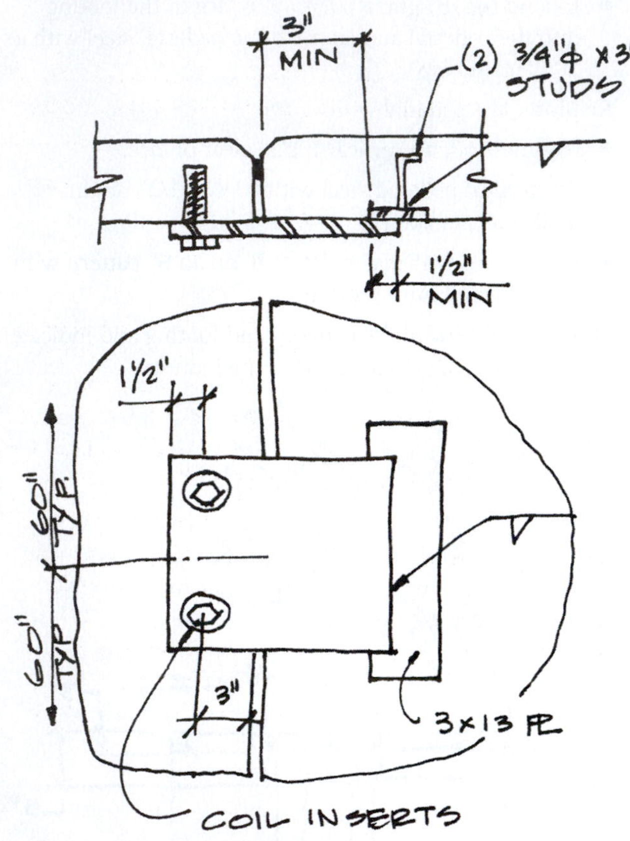

Problem 11-8 Draw a section view of an 18" wide × 36" deep concrete beam. Reinforce the beam with:

- (6)-grade 60, #8 horizontal 3" up from bottom.
- (4)-#6 horizontal bars 2" down from top.

Wrap the horizontal steel with #4 ties at 12" o.c. and stagger each lap 180°. Ties to be 1 1/2" minimum clear of the beam face.

Problem 11-9 Draw a section view showing a wall framed with 2 × 8 studs at 16" o.c. resting on 7 1/2" × 12" deep stem wall. Cover the wall with 2 layers of 5/8" type X gypsum board on each side, with joints laid perpendicular to each other. Attach the wall sill to the stem wall with 3/4" diameter × 12" anchor bolts through a 2 × 8 DFPT sill with 2"Ø washers at 48" o.c.

Support the stem wall on a 24" × 36" deep concrete footing. Provide a 5" concrete slab at each side of the stem wall that is 6" below the top of the stem wall. Thicken the slab as required to rest on the footing. Reinforce the slab with:

- W12 × 12/ 4 × 4 mesh 3" from top surface.
- Provide a 1/4" × 1 1/4" fiber filled isolation joint with a neoprene backer bead.
- Place (2)-grade 60, #5 horizontal rebar 3" up/down in stem wall.

Reinforce the footing with the following:

- (6)-grade 60, #5 rebar × 20'–6" long (approximately) with standard hook, 3" up and 1 1/2" down.
- Grade 40, #3 ties at 16" o.c. with 2" clear of sides, stagger all ties 180°.
- Grade 40, #3 ties at 12" o.c. around upper stem wall steel to lower footing steel.

Problem 11-10 Draw a section and a plan view of a 16" diameter concrete column. Also provide a plan view of the steel base plate that supports the steel T.S. at the foundation. Set the base plate at 45° to the foundation and the steel column. The foundation is to be 16" deep × 36" square and extend 24" in the natural grade. Reinforce the foundation with:

- Grade 60, #6 horizontal bars each way at 8" o.c. 1 1/2" down from top.
- Provide grade 60, #7 bars each way at 6" o.c. 3" up from the bottom.
- Provide (8)-#6 × 18" leg in the foundation extending into the concrete column with a 20 diameter lap to vertical steel.

A concrete column is to be centered on the foundation. Reinforce the column with:

- (8)-grade 60, #6 vertical bars evenly spaced.
- Wrap the vertical steel with a grade 40, #4 spiral tie at 6" o.c. 2" clear of exterior face.
- Provide a T.S. 6" × 6" × 3/8" column centered in the concrete supported on an 11" × 11" × 1/2" base plate.
- Weld the column to the base plate with a 3/16" fillet weld, all around.
- Attach the base plate to the foundation with (4)-3/4" × 12" anchor bolts 1 1/2" clear of plate edge.

Problem 11-11 Use a suitable scale to draw an enlarged elevation of the reinforcing steel for an 8' × 8' door opening in a concrete tilt-up wall panel. Place the door 2'–0" from the panel edge and 4'–10" from the bottom of the panel. Unless noted, all steel is to be 1" clear of the exterior face. Steel to be 1 1/2" clear of the edges of the opening.

- Use (3)-grade 60 #5 Ø @ 10" o.c. vertical bars on each side of the door. Extend each piece to within 2" clear of the top and bottom of the panel.
- Use (3)-grade 60 #5 Ø @ 8" o.c. horizontal bars above and below the opening.
- Use grade 40 #5 Ø @ 10" o.c. × 36" long vertical bars above and below the opening.
- Provide (2)-grade 40, #5 × 48" diagonal rebar 2" clear of each corner, and 10" o.c.

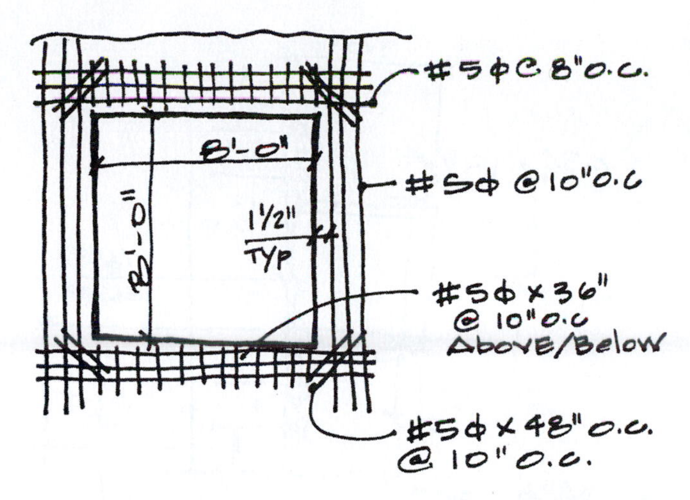

Problem 11-12 Draw a section using one of the listed sizes and reinforcement patterns that can be used to show six different widths. Each footing is to be 14" deep with rebar that is 3" up from the bottom of the footing and 2" clear of each edge. All steel is to be ASTM A615 grade 40. Dimension the width of the footing with a letter to represent each of the following sizes. Create a table to represent the following sizes and reinforcing:

Width	Reinforcing
20" × 20"	(3)-#5 Ø rebar each way
22" × 22"	(3)-#5 Ø rebar each way
24" × 24"	(4)-#5 Ø rebar each way
32" × 32"	(5)-#5 Ø rebar each way
34" × 34"	(5)-#5 Ø rebar each way
36" × 36"	(6)-#5 Ø rebar each way

Problem 11-13 Draw a section view of a joint in a 5" thick concrete slab over 4" gravel fill. Provide 12 × 12/6 × 6 WWM centered in the slab. Provide a 1/4" fiber isolation slab joint with neoprene joint sealant over a 3/8" backer bead. Keep the mesh 2" minimum clear of the slab joint. Reinforce the joint with #6 Ø × 18" smooth rebar with a paper shield placed at 18" o.c. placed 2" clear of the bottom of the slab.

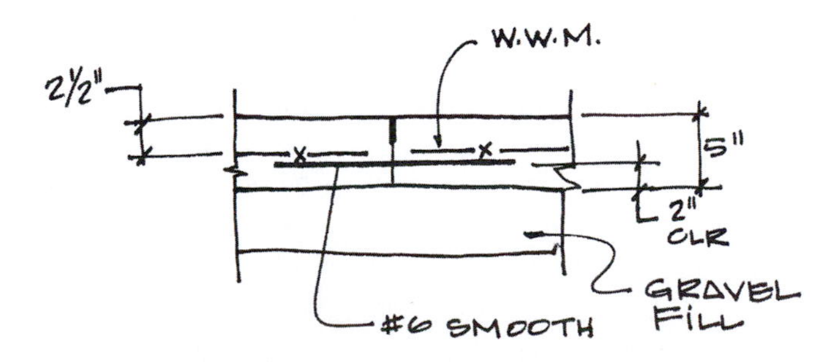

Problem 11-14 Draw a section view of a perimeter joint in a 5" thick slab that is 48" from a 6" exterior concrete tilt-up wall. Place the slab over 4" gravel fill. Provide 12 × 12/6 × 6 WWM centered in the slab. Provide a 1/4" fiber isolation slab joint with neoprene joint sealant over a 3/8" backer bead. Keep the mesh 2" minimum clear of the slab joint. Reinforce the joint with #6 Ø × 36" smooth rebar with a paper shield placed at 15" o.c. placed 2" clear of the bottom of the slab. Tie the slab to the wall with a 3/4" Ø × 24" long coil rods at 45" with 3" wall penetration and 2" clear of the bottom of the slab.

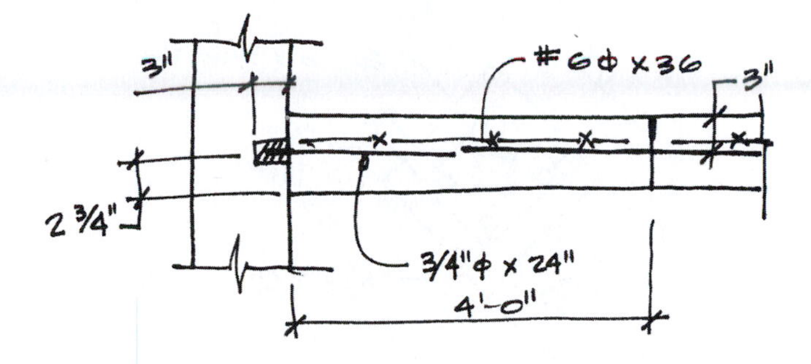

Problem 11-15 Use the attached sketch to draw plan and elevation views of a concrete tilt-up panel corner joint. Use a 4" × 15" × 3/8" steel plate inserted flush into a panel 3 1/2" from the panel joint. Attach (2)-3/4" Ø stud anchors at 9" o.c. centered on the plate and attached with 1/8" fillet weld, all around into the panel.

Provide (2)-3/4"Ø × 2 3/4" long coil inserts at 9" o.c. Use a 4" × 8" × 10"h × 3/8" angle with (2)-1" Ø holes, 1 1/2" from each plate edge to allow connection to the coil inserts. Lap this plate over the recessed 4" × 15" × 3/8" plate with a 1 1/2" minimum lap and weld at the job site with 3/16" fillet weld. Provide connection plates 1'–0" from top and bottom of each panel and at 6'–0" o.c. maximum along panel joint.

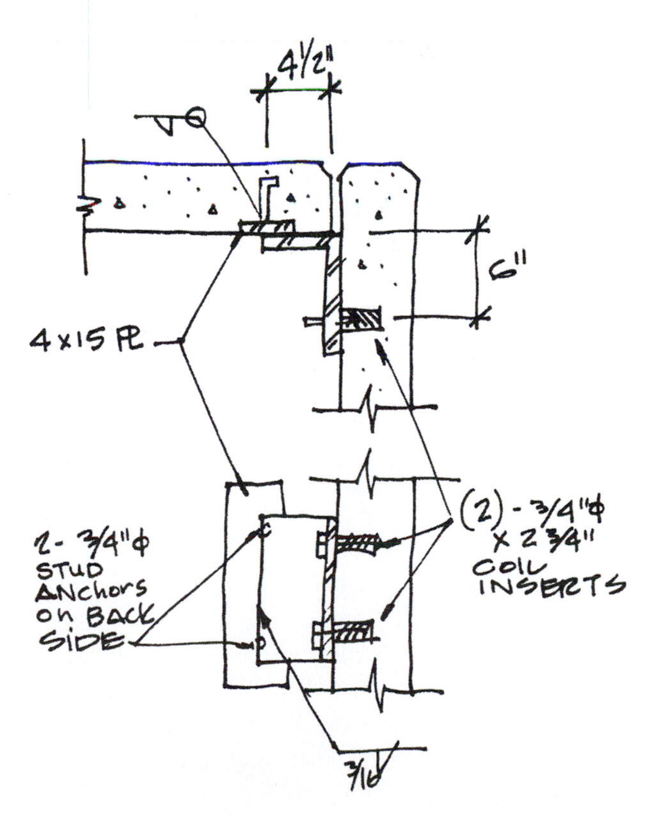

Problem 11-16 Draw a plan and section view of a 6" × 6" × 3/16" T.S. column resting on a 16" square concrete pedestal. Support the column on a 12" × 12" × 1/2" base plate on 1" dry pack with (4)-3/4" Ø × 15" A.B. in a 9" grid. Bolts to have a 3" projection through the slab. Attach the T.S. column to the base plate with 3/16" fillet weld, all around.

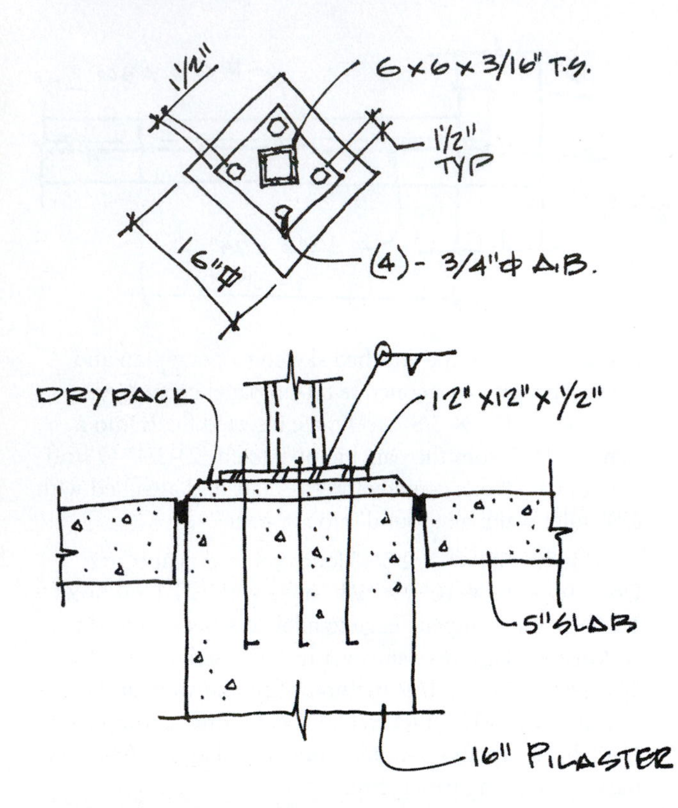

Problem 11-17 Use the following sketch to draw a plan and section view of a 6 3/4" × 43 1/2" laminated beam intersecting a concrete tilt-up wall. Support the beam with (2)-9" × 10" × 15 1/2" × 5/16" angles. Bolt the angles to the wall with (6)-3/4" Ø × 4 1/8" taper bolts at each angle and use (2)-3/4" Ø × 9" A307 bolts with standard washers through the beam. Provide a 7/8" bearing plate welded to angles with 5/16" fillet welds.

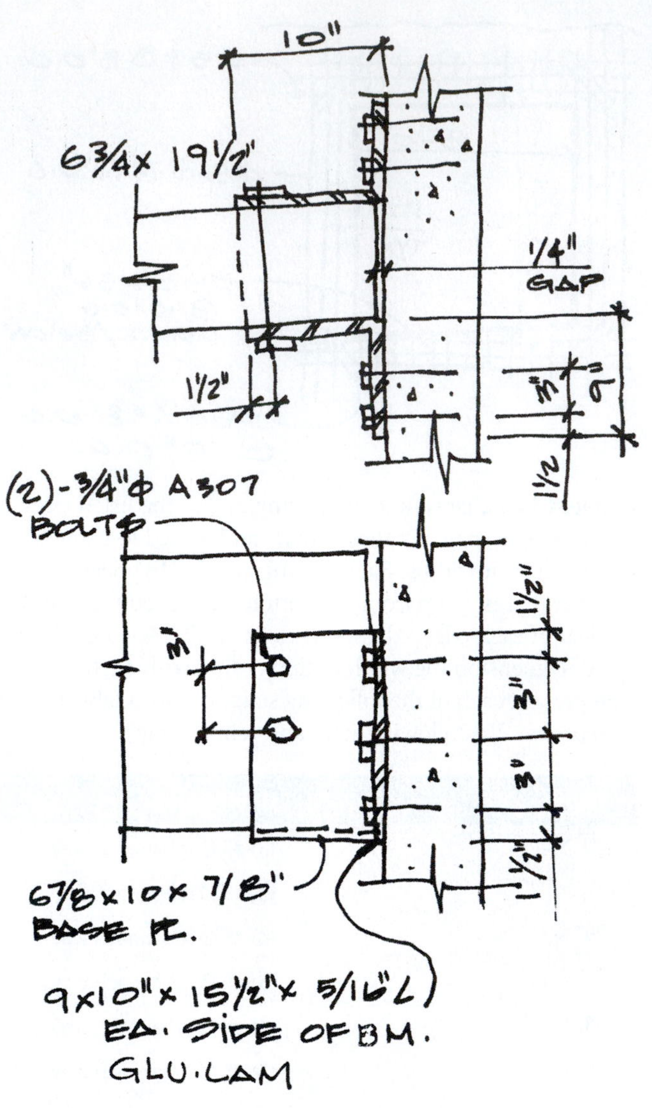

Problem 11-18 Edit problem 10-6 to be suitable for a 6" wide tilt-up wall.

Problem 11-19 Edit problem 10-7 to be suitable for a 6" wide tilt-up wall.

Problem 11-20 Edit problem 10-10 to be suitable for a 6" tilt-up wall.

12 Fire-Resistive Construction

The International Building Code (IBC) has many requirements for incorporating fire-resistive construction in a portion of a building. Shafts, hallways or corridors, parapets, exterior walls, and structural components have to be of fire-resistive construction under some conditions. Sometimes entire buildings have to be built of fire-resistive construction. Some buildings must not only be fire-resistive, but they must also be built entirely of noncombustible materials such as steel or concrete.

The most common methods of fire protection are covering wood or metal members with gypsum board or plaster, encasing metal members in concrete, and spraying metal beams and posts with fire-resistant materials. This chapter will explore:

- Objectives of fire protection.
- Methods of protecting materials including:
  - Walls.
  - Wall openings.
  - Columns.
  - Parapet walls.
  - Floor and ceiling assemblies.
  - Shafts.
  - Roof and ceiling assemblies.
  - Heavy timber.
  - Fire-stops.

OBJECTIVES OF FIRE PROTECTION

The primary reasons that the code requires fire protection is to protect human lives and property (the building). The code attempts to do this by:

- Protecting structural members from damage by fire and therefore preventing a structural collapse.
- Containing the fire and preventing it from spreading to other portions of the structure or to other structures.
- Limiting the amount of fuel available to the fire.

Fire Resistance

Fire resistance is measured in time. If an assembly (see assembly below) is able to resist fire for an hour, it is a 1-hour assembly. The code generally discusses 1-, 1 1/2-, 2-, 3-, and 4-hour assemblies.

Fire-Resistive Assembly

An example of a **fire-resistive assembly** is shown in Figure 12-1. This particular assembly is for an interior metal stud wall and is rated for 2-hour fire protection. The wall has two layers of gypsum board on each side, which must be of a certain type and thickness and must be installed in a specific way. Detailed instructions for installation, such as direction of layers and method of attachment, and so on, are usually called out in the specifications (see Chapter 14). Metal studs used in an assembly must also meet certain requirements of width, gauge, and spacing. All of the parts must be in place (for example, gypsum board must be on both sides of the studs) to be a fire-resistive assembly. Figure 12-2 shows two examples of a 1-hour floor/ceiling assembly.

Product manufacturers hire testing labs to perform tests on assemblies using their products. The results of these tests are published by organizations such as National Fire Protection Association and Underwriters' Laboratories. A list of these assemblies is available in Chapter 7 of the IBC, from the Gypsum Association, and from other building-related organizations. If manufactured joists are to be used on a project, the joist manufacturer can often supply the architectural team with the necessary report to draft details and specifications for the assembly. Building codes usually contain the necessary information for detailing a lumber assembly.

METHODS OF PROTECTING STRUCTURAL MEMBERS

The method of protecting an assembly varies depending on its location in the building, its location relative to the property lines, and the construction

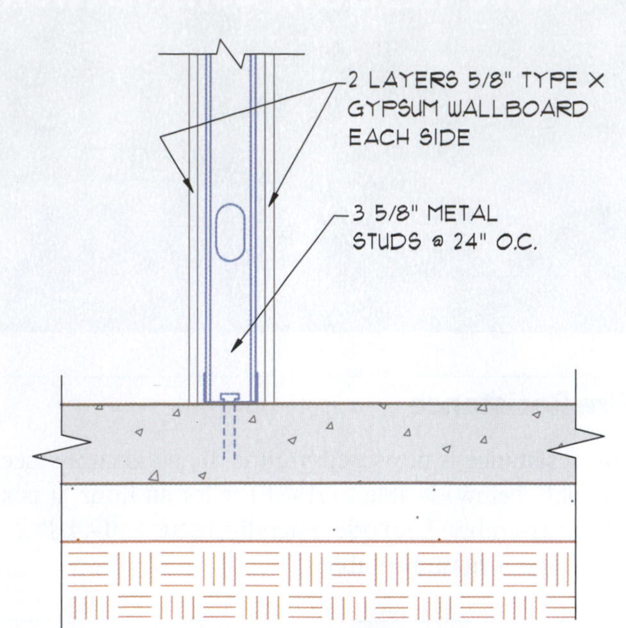

FIGURE 12-1 A fire-resistive assembly for an interior metal stud wall rated for 2-hour fire protection. *Courtesy Robin Smith, Kenneth D. Smith Architect & Associates, Inc.*

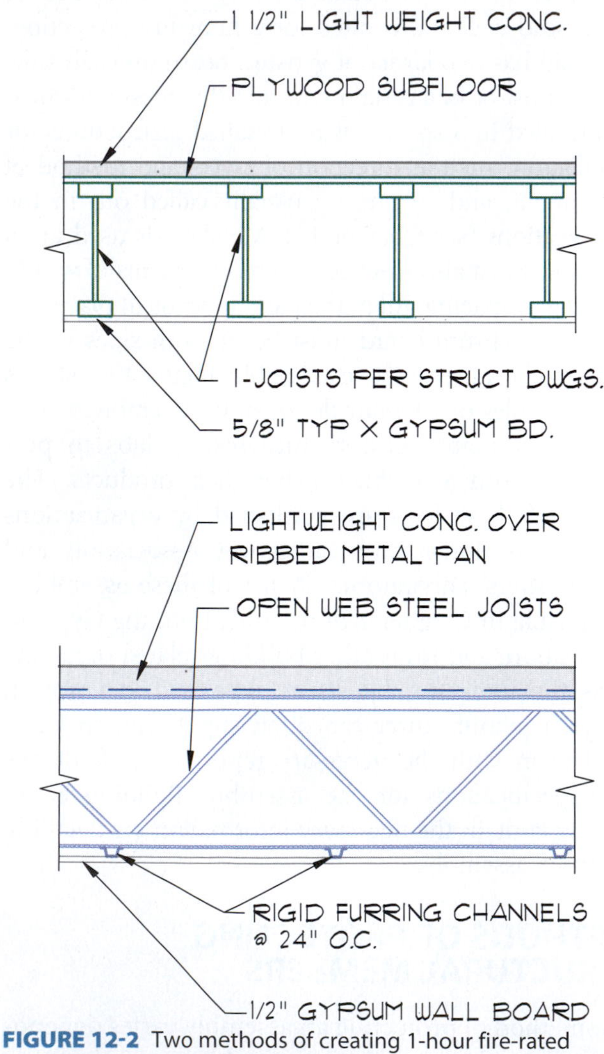

FIGURE 12-2 Two methods of creating 1-hour fire-rated floor assemblies. *Courtesy Jamie Smith, Kenneth D. Smith Architect & Associates, Inc.*

material that is being protected. The code lists assembly requirements for walls, floor and ceiling assemblies, roof and ceiling assemblies, columns, shafts, corridors, and wall openings. The code also lists requirements for penetrations through each of these assembly types.

Protective Wall Materials

The material used to form the wall will affect the fire resistance of the assembly. The code addresses concrete, masonry, wood, and steel assemblies. The location of the wall will also affect the construction.

Concrete

Concrete is a fire-resistive material; therefore, concrete walls are automatically fire-resistive. The amount of fire resistance provided by a concrete wall is a function of the type of concrete used and the thickness of the wall, as shown in Figure 12-3. In this chart the four columns on the right show the required thickness of concrete necessary for 1-, 2-, 3-, or 4-hour fire resistance. Five types of concrete and masonry products are listed. It is important to know what kind of concrete is to be used in the area where the project is to be constructed so that the proper thickness of concrete can be specified.

Masonry

Masonry is also a naturally fire-resistive material and its resistance depends on its thickness and the type of masonry used, as shown in Figures 12-4 and 12-5. Figure 12-4 deals with hollow masonry units. The table refers to minimum equivalent thickness of these units. As stated in the notes below the chart, the equivalent thickness is the average thickness of the solid material in the unit. This number can be obtained from the concrete block manufacturer.

Fire ratings are listed from 0.50 or one-half hour to four hours. The chart lists the minimum equivalent thicknesses required to achieve these fire ratings for five different types of material that hollow concrete masonry units can be made of. Figure 12-5 shows the fire resistance of other types of masonry units.

Wood and Metal

With the exception of heavy timber that was discussed in Chapter 7, wood is not considered a fire-resistive material. Steel by itself is never considered to be fire-resistive because it weakens and softens under relatively

RATED FIRE-RESISTANCE PERIODS FOR VARIOUS WALLS AND PARTITIONS [a,o,p]

MATERIAL	ITEM NUMBER	CONSTRUCTION	MINIMUM FINISHED THICKNESS FACE-TO-FACE[b] (Inches)			
			4 hour	3 hour	2 hour	1 hour
1. Brick of clay or shale	1-1.1	Solid brick of clay or shale[c]	6	4.9	3.8	2.7
	1-1.2	Hollow brick, not filled.	5.0	4.3	3.4	2.3
	1-1.3	Hollow brick unit wall, grout or filled with perlite vermiculite or expanded shale aggregate.	6.6	5.5	4.4	3.0
	1-2.1	4" nominal thick units at least 75 percent solid backed with a hat-shaped metal furring channel 3/4" thick formed from 0.021" sheet metal attached to the brick wall on 24" centers with approved fasteners, and 1/2" Type X gypsum wallboard[e] attached to the metal furring strips with 1"-long Type S screws spaced 8" on center.	—	—	5[d]	—
2. Combination of clay brick and load-bearing hollow clay tile	2-1.1	4" solid brick and 4" tile (at least 40 percent solid).	—	8	—	—
	2-1.2	4" solid brick and 8" tile (at least 40 percent solid).	12	—	—	—
3. Concrete masonry units	3-1.1[f,g]	Expanded slag or pumice.	4.7	4.0	3.2	2.1
	3-1.2[f,g]	Expanded clay, shale or slate.	5.1	4.4	3.6	2.6
	3-1.3[f]	Limestone, cinders or air-cooled slag.	5.9	5.0	4.0	2.7
	3-1.4[f,g]	Calcareous or siliceous gravel.	6.2	5.3	4.2	2.8
4. Solid concrete[h,i]	4-1.1	Siliceous aggregate concrete.	7.0	6.2	5.0	3.5
		Carbonate aggregate concrete.	6.6	5.7	4.6	3.2
		Sand-lightweight concrete.	5.4	4.6	3.8	2.7
		Lightweight concrete.	5.1	4.4	3.6	2.5
5. Glazed or unglazed facing tile, nonload-bearing	5-1.1	One 2" unit cored 15 percent maximum and one 4" unit cored 25 percent maximum with 3/4" mortar-filled collar joint. Unit positions reversed in alternate courses.	—	6 3/8	—	—
	5-1.2	One 2" unit cored 15 percent maximum and one 4" unit cored percent maximum with 3/4" mortar-filled collar joint. Unit positions side with 3/4" gypsum plaster. Two wythes tied together every fourth course with No. 22 gage corrugated metal ties.	—	6 3/8	—	—
	5-1.3	One unit with three cells in wall thickness, cored 29 percent maximum.	—	—	6	—
	5-1.4	One 2" unit cored 22 percent maximum and one 4" unit cored 41 percent maximum with 1/4" mortar-filled collar joint. Two wythes tied together every third course with 0.030-inch (No. 22 galvanized sheet steel gage) corrugated metal ties.	—	—	6	—
	5-1.5	One 4" unit cored 25 percent maximum with 3/4" gypsum plaster on one side.	—	—	4 3/4	—
	5-1.6	One 4" unit with two cells in wall thickness, cored 22 percent maximum.	—	—	—	4

FIGURE 12-3 The amount of fire resistance produced by a concrete wall is a function of the wall thickness and the type of concrete used. This partial table [720.1.(2)] contains footnotes that are important to its use. *Courtesy 2009* International Building Code, *copyright © 2009. Washington, DC. International Code Council. Reproduced with permission. All rights reserved. www.iccsafe.org.*

TABLE 721.3.2
MINIMUM EQUIVALENT THICKNESS (inches) OF BEARING OR NONBEARING CONCRETE MASONRY WALLS[a,b,c,d]

TYPE OF AGGREGATE	FIRE-RESISTANCE RATING (hours)														
	1/2	3/4	1	1 1/4	1 1/2	1 3/4	2	2 1/4	2 1/2	2 3/4	3	3 1/4	3 1/2	3 3/4	4
Pumice or expanded slag	1.5	1.9	2.1	2.5	2.7	3.0	3.2	3.4	3.6	3.8	4.0	4.2	4.4	4.5	4.7
Expanded shale, clay or slate	1.8	2.2	2.6	2.9	3.3	3.4	3.6	3.8	4.0	4.2	4.4	4.6	4.8	4.9	5.1
Limestone, cinders or unexpanded slag	1.9	2.3	2.7	3.1	3.4	3.7	4.0	4.3	4.5	4.8	5.0	5.2	5.5	5.7	5.9
Calcareous or siliceous gravel	2.0	2.4	2.8	3.2	3.6	3.9	4.2	4.5	4.8	5.0	5.3	5.5	5.8	6.0	6.2

For SI: 1 inch = 25.4 mm.

a. Values between those shown in the table can be determined by direct interpolation.

b. Where combustible members are framed into the wall, the thickness of solid material between the end of each member and the opposite face of the wall, or between members set in from opposite sides, shall not be less than 93 percent of the thickness shown in the table.

c. Requirements of ASTM C 55, ASTM C 73, ASTM C 90 or ASTM C 744 shall apply.

d. Minimum required equivalent thickness corresponding to the hourly fire-resistance rating for units with a combination of aggregate shall be determined by linear interpolation based on the percent by volume of each aggregate used in manufacture.

FIGURE 12-4 Table 721.3.2 of the *International Building Code* can be used to determine the fire resistance for hollow masonry units. *Courtesy 2009* International Building Code, *copyright © 2009. Washington, DC. International Code Council. Reproduced with permission. All rights reserved. www.iccsafe.org.*

TABLE 721.4.1(1)
FIRE-RESISTANCE PERIODS OF CLAY MASONRY WALLS

MATERIAL TYPE	MINIMUM REQUIRED EQUIVALENT THICKNESS FOR FIRE RESISTANCE[a, b, c] (inches)			
	1 hour	2 hour	3 hour	4 hour
Solid brick of clay or shale[d]	2.7	3.8	4.9	6.0
Hollow brick or tile of clay or shale, unfilled	2.3	3.4	4.3	5.0
Hollow brick or tile of clay or shale, grouted or filled with materials specified in Section 721.4.1.1.3	3.0	4.4	5.5	6.6

For SI: 1 inch = 25.4 mm.

a. Equivalent thickness as determined from Section 721.4.1.1.

b. Calculated fire resistance between the hourly increments listed shall be determined by linear interpolation.

c. Where combustible members are framed in the wall, the thickness of solid material between the end of each member and the opposite face of the wall, or between members set in from opposite sides, shall not be less than 93 percent of the thickness shown.

d. For units in which the net cross-sectional area of cored brick in any plane parallel to the surface containing the cores is at least 75 percent of the gross cross-sectional area measured in the same plane.

FIGURE 12-5 Table 721.4.1(1) of the *International Building Code* can be used to determine the fire resistance for masonry walls. *Courtesy 2009* International Building Code, *copyright © 2009. Washington, DC. International Code Council. Reproduced with permission. All rights reserved. www.iccsafe.org.*

low temperatures. When these materials are used for studs in a wall, any necessary fire resistance is achieved by covering the studs with a fire-resistive material such as type X gypsum board. There are many variations to fire assemblies for steel and wood stud walls. Good sources for these assemblies are the code, the Gypsum Association's *Fire Resistance Design Manual*, and *The Plaster Resource Manual*.

Wall Openings

Fire-resistance rated walls and structural members are built for two primary reasons. The first is to prevent structural collapse that would occur if the building were to burn down. The second is to prevent the spread of fire into other areas of the building or to other buildings. When the purpose of the fire-resistant assembly is to prevent the spread of fire and smoke, openings in that wall must receive special treatment.

If openings were to remain open during a fire, the purpose of the wall would be defeated by smoke and fire pouring through the opening. If a window or door in the opening were to be readily destroyed by heat or if it were easy to break, the purpose of the wall would also be defeated. For this reason, doors and windows installed in firewalls to contain fire and smoke must meet certain requirements. Door and window assemblies must carry labels stating that they have been tested and have passed fire and/or smoke tests. Different firewall ratings and types require different ratings for openings. These assemblies are usually labeled as 20-minute, 3/4-hour, 1-hour, 1 1/2-hour, 2-hour, 3-hour, or 4-hour assemblies. A fire-rated door assembly consists of a fire-rated and labeled door and frame, proper hinges, and a closing device. The closing device will either be self-closing or automatically closing, as required by the code.

Frequently, a fire resistant separation has an opening that needs to remain open all of the time. The opening at a counter between a manufacturing F-2 or Storage S-2 from offices is one example of this type of condition. The code defines the cafeteria as an assembly area, and as such there must be a fire-separation wall between the cafeteria and the kitchen, which is a very likely place for a fire to start. Figure 12-6 shows such an opening. In the case of a fire, a sensing device would cause a fire door to automatically roll down and meet the countertop to close off the opening and prevent the spread of smoke and fire. This is an example of an automatic closing device. Swinging doors are frequently equipped with self-closing devices.

In all cases, the amount of opening that is allowed in a fire-rated wall is limited by area. In some cases, such as an exterior wall that is close to the property line, openings are prohibited altogether. An example of a fire-rated wall with limited openings is a **firewall**, a wall that separates portions of a building to form separate buildings. These firewalls are allowed to have a maximum size of 120 square feet in a non-sprinklered building. The aggregate width of all openings shall not exceed 25% of the length of the wall in each story.

Columns

Columns are normally wrapped with gypsum board, plaster, or sprayed with a fire-protective coating. Figures 12-7 and 12-8 show protected columns. The amount of protection required for a steel column depends on the weight of the steel member. In general, two layers of gypsum board provide 2-hour protection for a lightweight column and 3-hour protection for a heavy column (Figure 12-8). The minimum weight required for the steel column will be listed in the assembly specifications from the testing lab. Coordination with the structural engineer is necessary to achieve the necessary protection in the most economical manner.

Another consideration for fire protection at columns is that they are frequently subject to impact from cars, people, carts, forklifts, and other moving objects. The material that provides the fire protection for columns will frequently need to be protected from such impact. This can be accomplished in several ways: Barriers can be placed between the column and the traffic, the fire-resistive material can be covered by another material such as sheet metal, or the column can be set upon a concrete pedestal.

FIGURE 12-6 Fire-rated doors with a closing device that is activated by heat or smoke are used at openings that need to remain open for normal business operations.

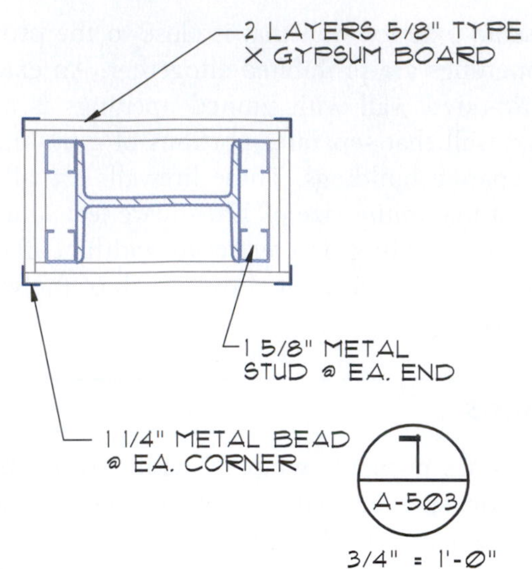

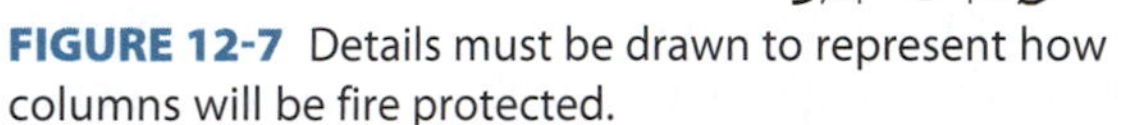

FIGURE 12-7 Details must be drawn to represent how columns will be fire protected.

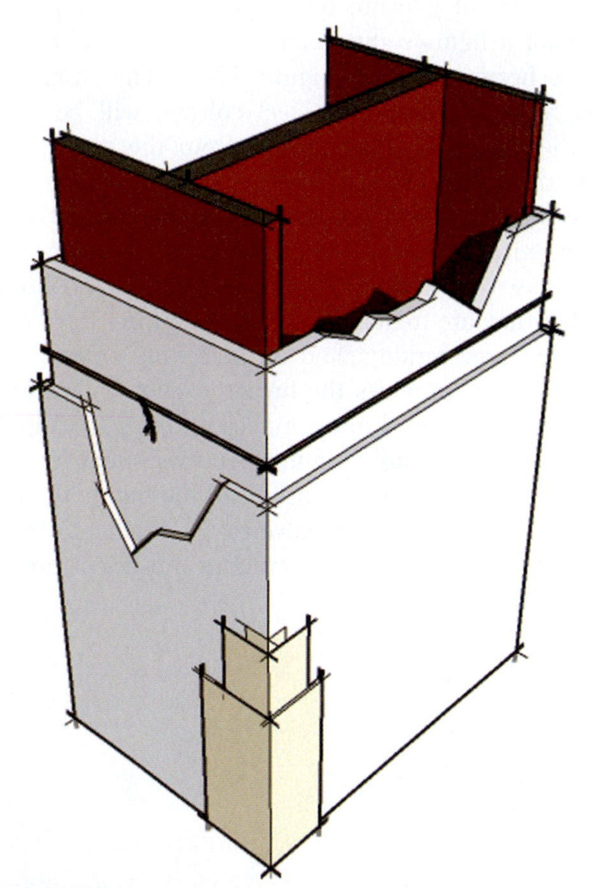

FIGURE 12-8 One or more layers of Type X gypsum board, or sprayed-on fire-protective coating can be used to provide fire protection for steel columns.

Corridors

Corridors are frequently required to be fire rated in order to protect the occupants of a building while they exit. All corridors serving H occupancies are required to be 1-hour fire rated. Corridors serving A, B, E, F, M, S, and U occupancies are required to be 1-hour fire rated if the building is not sprinklered and the corridor serves more than 30 occupants. R occupancies with more than 10 occupants are required to have a 1/2-hour fire-rated corridor. I-1 and I-3 occupancies are required to have a 1-hour corridor, whereas I-2 and I-4 occupancies are not. When the corridor is required to be fire rated, walls are required to be of 1-hour construction. Corridor ceilings are required to be constructed, as they would be for a 1-hour floor/ceiling assembly. The doors opening onto these fire-rated corridors are required to be 20-minute assemblies.

The basic concept here is to give the occupants of a building an exit path that is safe from fire and smoke. This requires that the doors be tightly sealed and that any penetrations such as air-conditioning ducts be properly treated. Air-conditioning ducts serving these corridors will normally be required to have fire and smoke dampers installed in them. A **fire and smoke damper** is a device that shuts off the air duct when fire or smoke is detected so that smoke is not pumped into the corridor from other areas of the building. Figures 12-9 through 12-11 illustrate basic principles

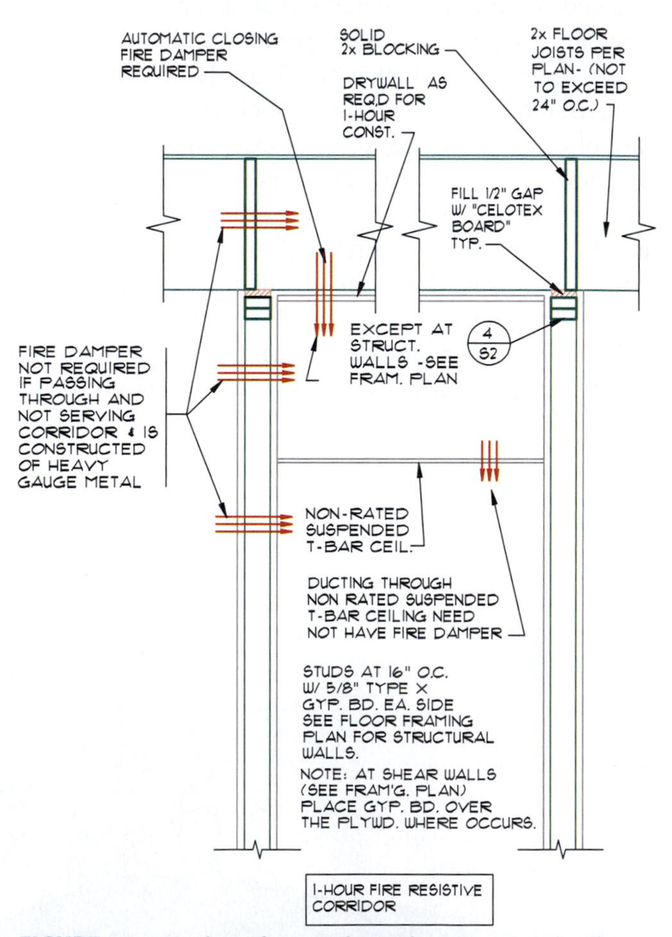

FIGURE 12-9 A 1-hour fire-rated corridor constructed with wood walls and 5/8" Type X gypsum board. *Courtesy Dean K. Smith, Kenneth D. Smith Architects & Associates, Inc.*

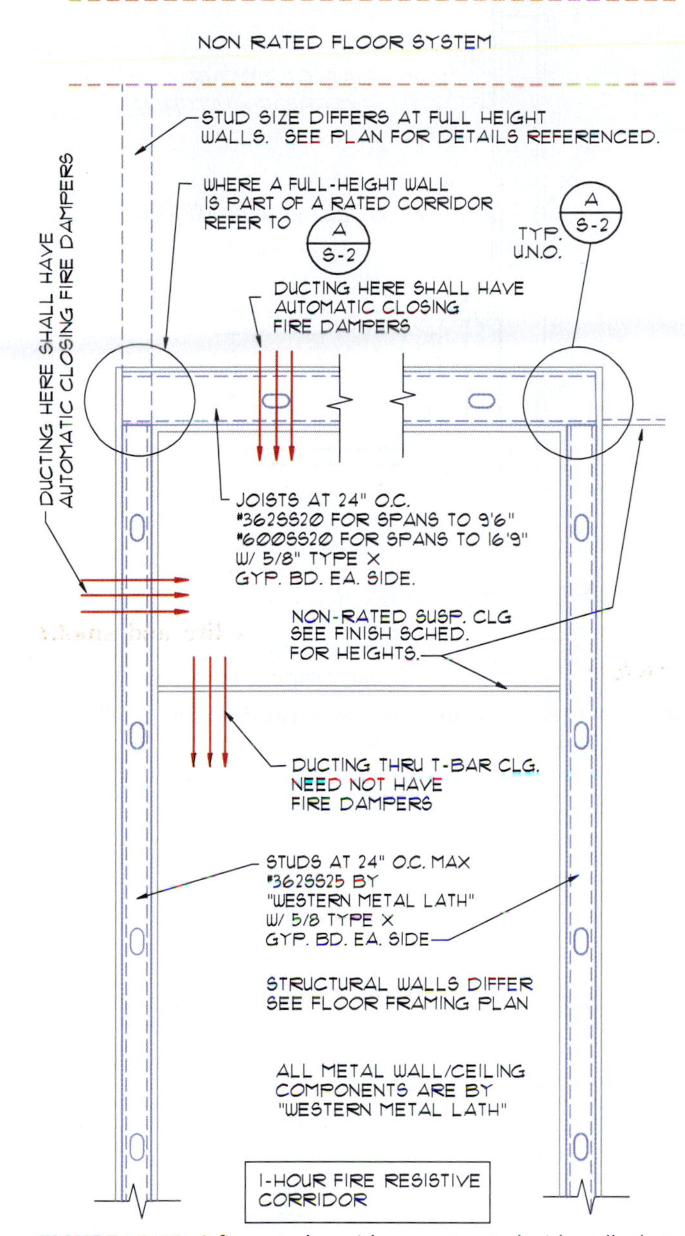

FIGURE 12-10 A fire-rated corridor constructed with walls that do not extend to the underside of the floor joists. *Courtesy Dean K. Smith, Kenneth D. Smith Architects & Associates, Inc.*

in designing fire-rated corridors. All three examples are for buildings that can be built of combustible construction and are not otherwise fire-rated. Buildings that are required to be fire-rated or that are required to be constructed of noncombustible construction will require slightly different details.

Figure 12-9 shows a section of a typical fire-rated corridor. The walls (fire partitions) are constructed as typical 1-hour walls with 2 × 4 studs and 5/8" type X gypsum board. These walls extend all the way up to the underside of solid wood floor joists and require solid blocking to extend from the top of the walls to the underside of the floor sheathing. The undersides of the

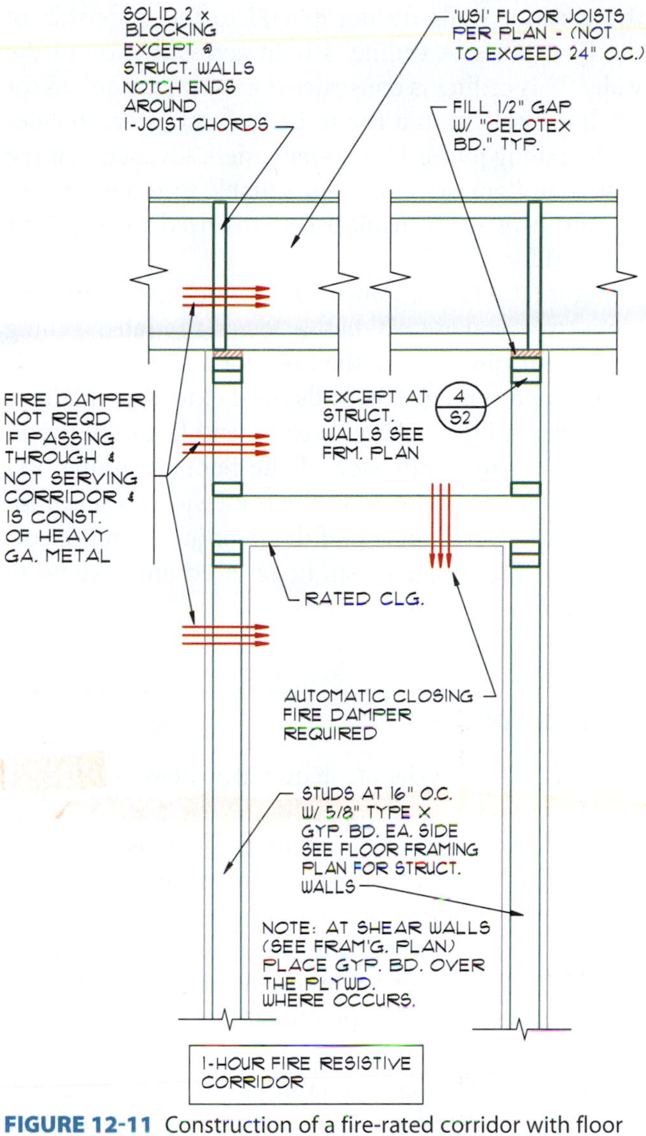

FIGURE 12-11 Construction of a fire-rated corridor with floor joists that are not suitable for receiving a ceiling. *Courtesy Dean K. Smith, Kenneth D. Smith Architects & Associates, Inc.*

floor joists have gypsum board installed as if they were a part of a 1-hour floor/ceiling assembly. Fireproofing methods may be slightly different if the floor joists are trusses instead of solid wood members. A 1-hour surface is not required on the upper side of the floor joists. A non-fire-rated ceiling may be installed below the fire-rated ceiling if it is noncombustible or made of fire-retardant treated wood. Ducts that serve the corridor and penetrate the fire-rated wall or ceiling must have fire and smoke dampers. Ducts that penetrate the optional ceiling do not require fire and smoke dampers. In each of the systems shown in Figures 12-9 and 12-11, ducts that do not serve the corridor may pass through the fire-rated assemblies if they are constructed of a minimum of 26-gauge metal.

Another way to construct a fire-rated corridor is to build a tunnel as shown in Figure 12-10. In this case the corridor walls do not extend to the underside of the floor joists. A ceiling is built across the top of the walls. This ceiling is constructed exactly as required for a 1-hour wall, with a fire-resistive finish on both sides of the ceiling joists. This system offers advantages if the structural floor system is not suitable to receive a ceiling finish or if mechanical systems need to pass over the corridor.

Figure 12-11 combines some elements of the previous two systems. It has a lower fire-rated ceiling surface for cases where the floor joists are not suitable to receive a ceiling. The walls must extend to the floor joists and solid blocking must extend from the top of the wall to the underside of the floor sheathing. On the outside of the corridor, the fire-resistive wall finish extends to the underside of the floor joists. In the corridor, the fire-resistive wall finish need only extend to the fire-resistive ceiling.

Parapet Walls

Many fire-rated walls are required to have a parapet, a portion of the wall extending above the roof. The parapet is required to be as fire-resistive as the lower portion of the wall. It is required to extend at least 30" above the roof. If the roof slopes at a rate greater than 2 in 12 toward the parapet, the parapet will be required to extend a minimum of 30" above the highest point of the roof within the distance that fire protection of openings would be required.

Exterior walls that are required to be fire protected are often required to have a parapet, but there are many exceptions to this rule such as when the wall terminates into a 2-hour roof assembly or when the wall is far enough away from the property line that protected openings are not required. Firewalls are usually required to have a parapet. Occupancy separation walls, also referred to as fire barriers, do not require parapets.

Figures 12-12 and 12-13 are fire-resistive parapet details. Take note of the continuation of the fire-resistive material. In Figure 12-12, a solid wood ledger forms part of the fire-resistive materials. In Figure 12-13, the parapet wall sits on top of the roof sheathing. This is a design that is frequently avoided, because the parapet has to be braced to prevent it from being pushed over by wind or other forces. Figure 12-14 shows a concrete wall with a parapet. This detail could easily be altered to apply to masonry. Note that for all required parapets if the roofing material is combustible, it must stop 18" minimum from the top of the wall and a noncombustible material such as sheet

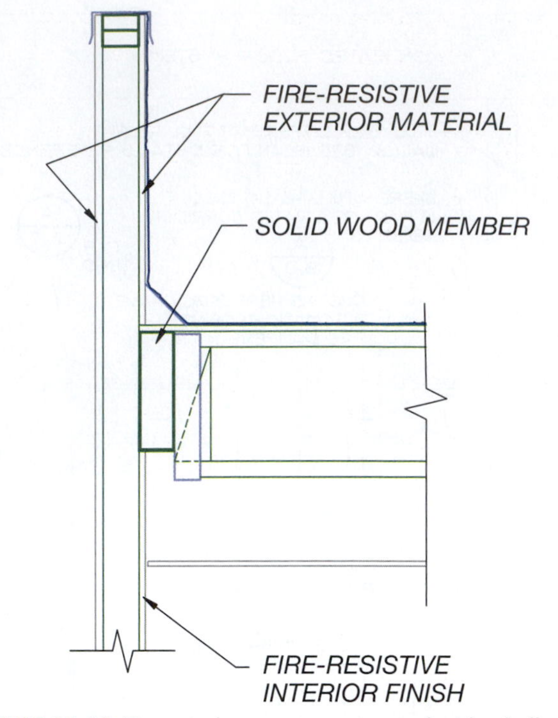

FIGURE 12-12 Fire-rated parapet constructed with a balloon-framed wall.

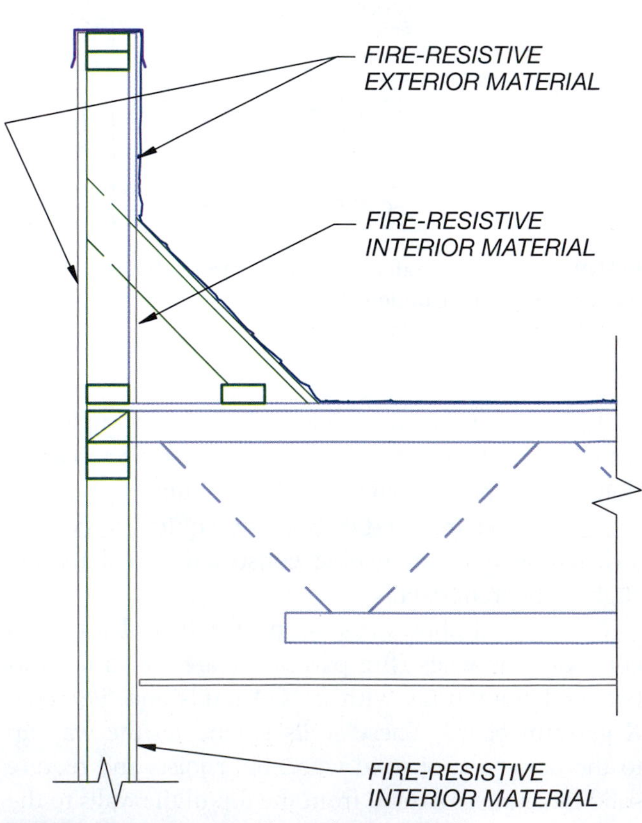

FIGURE 12-13 Fire-rated braced parapet wall.

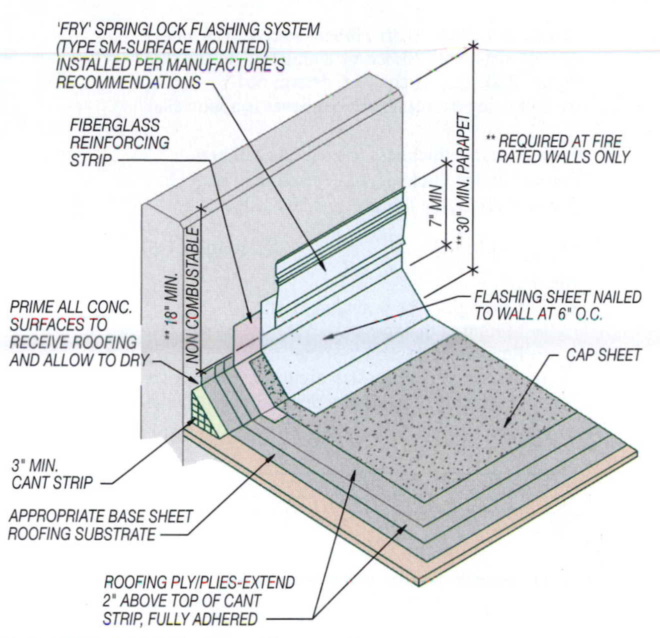

FIGURE 12-14 Concrete fire-rated parapet.

metal must be added in this portion of the wall. In the case of a concrete or masonry wall it would not be necessary to add anything to the upper 18" of the wall for fire protection.

Floor/Ceiling Assemblies

Floors and ceiling assemblies are typically constructed from concrete, wood, or metal.

Concrete

A structural concrete floor can be a floor/ceiling fire assembly if it is the proper thickness. Figure 12-15 is a table from the IBC that shows required thicknesses for these assemblies. If the concrete is something other than a slab with a uniform thickness, its equivalent thickness can be calculated. A method of calculating equivalent thickness is shown in Figure 12-16. The concrete provides the necessary protection for both the floor side and ceiling side of the assembly. A suspended ceiling is frequently placed below the concrete to conceal utilities such as air-conditioning ducts, plumbing, and electrical lines. In most cases the suspended ceiling is not considered part of the fire assembly.

Wood and Metal

If wood or metal is used for the structural system, it must be protected by other materials to achieve a fire rating. The upper drawing in Figure 12-2 is an example of protection provided for a wood floor assembly. The floor assembly is built using factory-built wood I-joists for structural support. A thin layer of lightweight concrete is poured over a plywood subfloor. The concrete serves as protection on the floor side of the assembly. In the lower example, open-web steel joists are used and the concrete is poured over fluted metal sheathing.

Both floor assemblies shown in Figure 12-2 provide fire protection on the ceiling side of the structural members with a layer of Type X gypsum wallboard. When I-joists are used to support the floor, the gypsum board

VALUES OF $R_n^{0.59}$ FOR USE IN EQUATION 7-4

TYPE OF MATERIAL	THICKNESS OF MATERIAL (inches)											
	1 1/2	2	2 1/2	3	3 1/2	4	4 1/2	5	5 1/2	6	6 1/2	7
Siliceous aggregate concrete	5.3	6.5	8.1	9.5	11.3	13.0	14.9	16.9	18.8	20.7	22.8	25.1
Carbonate aggregate concrete	5.5	7.1	8.9	10.4	12.0	14.0	16.2	18.1	20.3	21.9	24.7	27.2[c]
Sand-lightweight concrete	6.5	8.2	10.5	12.8	15.5	18.1	20.7	23.3	26.0[c]	Note c	Note c	Note c
Lightweight concrete	6.6	8.8	11.2	13.7	16.5	19.1	21.9	24.7	27.8[c]	Note c	Note c	Note c
Insulating concrete[a]	9.3	13.3	16.6	18.3	23.1	26.5[c]	Note c	Note c	Note c	Note c	Note c	Note c
Air space[b]	–	–	–	–	–	–	–	–	–	–	–	–

For SI: 1 inch = 25.4 mm, 1 pound per cubic foot = 16.02 kg/m³.

a. Dry unit weight of 35 pcf or less and consisting of cellular, perlite, or vermiculite concrete.
b. The $R_n^{0.59}$ value for one 1/2" to 3 1/2" air space is 3.3. The $R_n^{0.59}$ value for two 1/2" to 3 1/2" air spaces is 6.7.
c. The fire-resistance rating for this thickness exceeds 4 hours.

FIGURE 12-15 The type of concrete and the thickness affects the fire resistance of a floor assembly based on Table 721.2.1.2(1) of the IBC. *Courtesy 2009* International Building Code, *copyright © 2009. Washington, DC. International Code Council. Reproduced with permission. All rights reserved. www.iccsafe.org.*

721.2.2.1
MINIMUM SLAB THICKNESS (inches)

CONCRETE TYPE	FIRE-RESISTANCE RATING (hour)				
	1	1 1/2	2	3	4
Siliceous	3.5	4.3	5.0	6.2	7.0
Carbonate	3.2	4.0	4.6	5.7	6.6
Sand-lightweight	2.7	3.3	3.8	4.6	5.4
Lightweight	2.5	3.1	3.6	4.4	5.1

For SI: 1 inch = 25.4 mm.

720.2.2.1.2 Slabs with sloping soffits. The thickness of slabs with sloping soffits (see Figure 720.2.2.1.2) shall be determined at a distance $2t$ or 6 inches (152 mm), whichever is less, from the point of minimum thickness, where t is the minimum thickness.

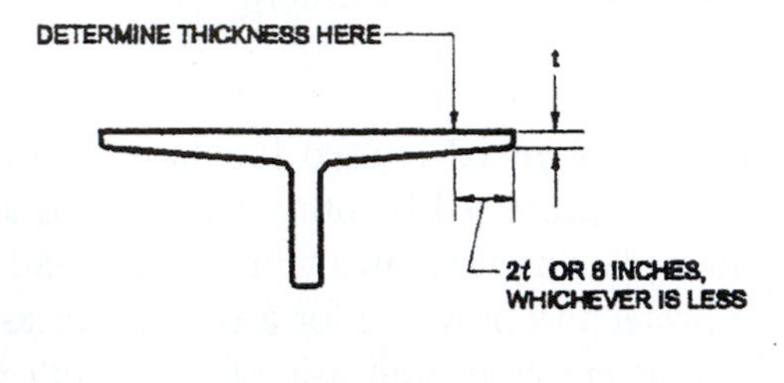

For SI: 1 inch = 25.4 mm.

FIGURE 720.2.2.1.2
DETERMINATION OF SLAB THICKNESS FOR SLOPING SOFFITS

720.2.2.1.3 Slabs with ribbed soffits. The thickness of slabs with ribbed or undulating soffits (see Figure 720.2.2.1.3) shall be determined by one of the following expressions, whichever is applicable:

For $s \geq 4t$, the thickness to be used shall be t
For $s \leq 2t$, the thickness to be used shall be t_e
For $4t > s > 2t$, the thickness to be used shall be

$$t + \left(\frac{4t}{s} - 1\right)\left(t_e - t\right) \quad \textbf{(Equation 7-5)}$$

where:

s = Spacing of ribs or undulations.
t = Minimum thickness.
t_e = Equivalent thickness of the slab calculated as the net area of the slab divided by the width, in which the maximum thickness used in the calculation shall not exceed $2t$.

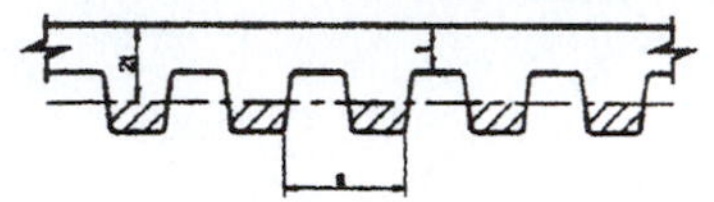

NEGLECT SHADED AREA IN CALCULATION OF EQUIVALENT THICKNESS

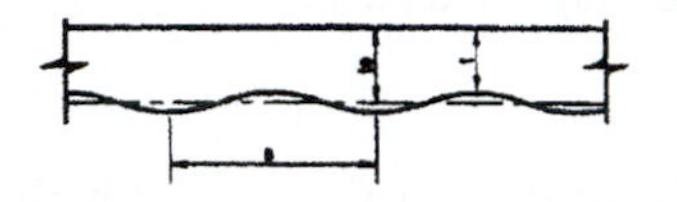

For SI: 1 inch = 25.4 mm.

FIGURE 720.2.2.1.3
SLABS WITH RIBBED OR UNDULATING SOFFITS

FIGURE 12-16 Calculation method for fire-rating equivalent thickness of ribbed concrete based on Table 721.2.2.1 of the IBC.

is attached directly to the bottom chord of the joists. When open-web steel joists are used, the gypsum board is fastened to metal furring channels that are wire tied to the bottom chord of the truss. Both details require coordination with the structural engineer to ensure that the trusses meet depth and spacing requirements for the selected fire assembly. The type of lightweight concrete used will vary depending on whether the concrete will be the wear surface (exposed) or not. Another factor for selecting the type of lightweight concrete is the type of activity taking place on the floor. Office floors do not receive the same wear and tear that an industrial area would. An industrial area would require stronger lightweight or regular concrete with a tougher wear surface than an office area.

Both details in Figure 12-2 will depend on additional information in the specifications for proper gypsum board application such as the direction that each layer of gypsum wallboard is to be laid, and the type and spacing of drywall screws for each layer. Other materials that could be used effectively to provide fire protection are plaster or a suspended ceiling system. When using steel joists, the ceiling is sometimes omitted. When this is done, a fire-protective coating is sprayed on the joists and the underside of the deck. Three basic types of coatings are used to protect structural steel: The least expensive types of fireproofing are cementitious and sprayed mineral fibers, which are rough and somewhat uneven and meant to be hidden. The third type is intumescent paint. This material looks like ordinary paint but when it is exposed to heat, it swells or expands to form a thick protective layer. This type of fireproofing is expensive and is rarely used unless the structural steel is to be left exposed to make an architectural statement.

Shafts

Vertical shafts extending through floors can play a key part in aiding the spread of fire. A shaft can act like a chimney, pulling smoke and fire from floor to floor. For this reason, the construction of any shaft that penetrates through more than one floor is highly regulated by the model codes. Openings into shafts that extend through more than one floor are required to have fire and smoke dampers, devices that close off the opening when fire or smoke is detected. The shaft itself can be constructed like a wall. Optionally there are numerous assemblies available specifically for shafts. The Gypsum Association's *Fire Resistance Manual* lists several of them.

Roof/Ceiling Assemblies

When working with the roof, a floor/ceiling assembly is selected and the materials above the rafters are replaced with appropriate roofing materials, as shown in Figure 12-17. The roofing materials are required to be fire retardant. A fire-retardant roof is categorized by the code as being either a Class A or Class B roof assembly. Class A roof assemblies are capable of passing a test for severe fire exposure. Class B roof assemblies are capable of passing a test for moderate fire exposure. Most conditions for a nonresidential building can be satisfied with a Class B roof. The typical built-up roof shown in Figure 18-15 could be Class A or B depending on its exact makeup. Other typical Class A roofing materials are concrete, slate, clay, and roofing materials listed by their manufacturer as passing ASTM E 108 for severe fire exposure. Other Class B roofing materials include metal sheets, metal tiles, and roofing materials listed by their manufacturer as passing ASTM E 108 for moderate fire exposure.

Beams

Metal beams that project outside the floor/ceiling or roof/ceiling assemblies are required to have their own fire protection. Steel beams are normally wrapped with gypsum board or plaster, or sprayed with a fire-protective coating. Figure 12-18 shows an example of a beam that extends below the rated ceiling and needs its own fire protection. Wood beams are frequently large enough to be considered heavy timber and would not have to be protected. Concrete beams are

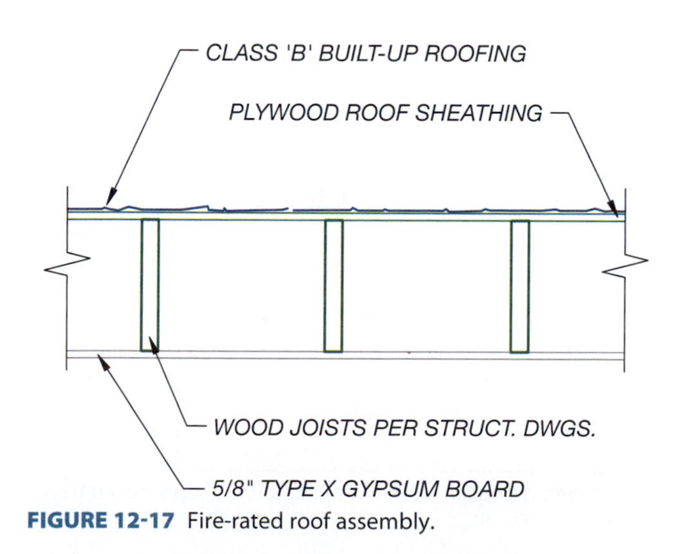

FIGURE 12-17 Fire-rated roof assembly.

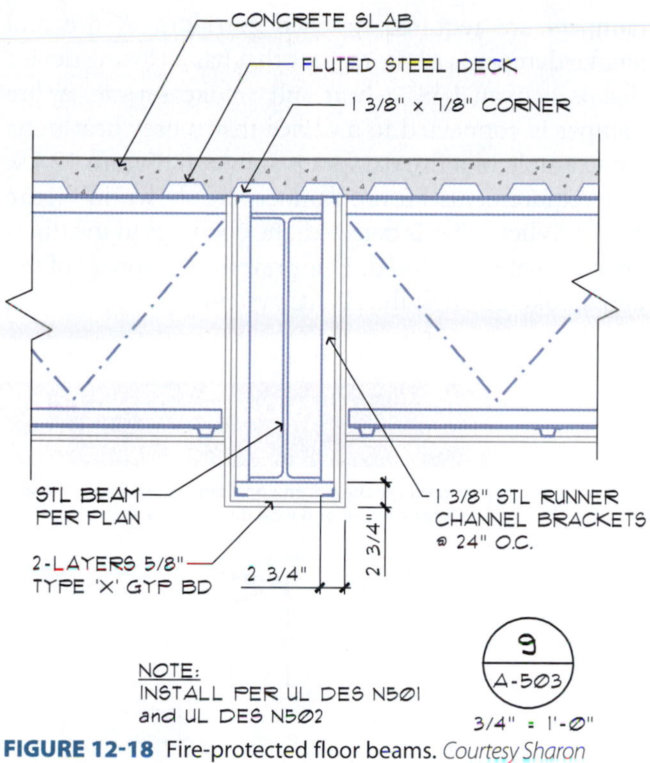

FIGURE 12-18 Fire-protected floor beams. *Courtesy Sharon Chambliss, Divine Designs & Associates, Inc.*

considered to be fire protected if the reinforcing steel is adequately covered.

Penetrations

Fire-resistive surfaces will normally have to be penetrated to allow for the installation of building equipment such as electrical, HVAC, and plumbing. These penetrations must be properly dealt with in order to preserve the integrity of the fire-resistive system.

Plumbing Plumbing penetrations can be minimized by keeping plumbing off of fire-rated walls. If this is not possible, metal piping will probably have to be used in lieu of plastic. The pipe will not normally completely fill the hole it goes through. The area between the pipe and the wall material will have to be filled with an approved fire-stop material.

HVAC Ductwork for heating and air-conditioning frequently has to penetrate firewalls and ceilings, as shown in Figures 12-9, 12-10, and 12-11. The duct penetration itself is not usually a problem if it is constructed of the proper gauge of metal. It is the air register (outlet) or return air (inlet) register that presents the problem. These openings into the ductwork normally remain open at all times. Fire and/or smoke dampers are required at these openings. **Fire dampers** are required in walls that separate occupancies. Fire and smoke dampers are required in rated corridors. A fire and smoke damper is an air register that has a closing device that is activated by a heat and smoke sensor. A fire damper is connected to a device that senses heat only. An example of a fire damper is shown in Figure 12-19. The requirements for fire dampers are shown in Figure 12-20. When a fire is detected, the openings to the ducts are automatically closed. This prevents the spread of fire and smoke through the HVAC system.

Electrical Electrical fixtures create more wall and ceiling penetrations than any other building system. These penetrations of wall or floor/ceiling assemblies normally require special consideration and detailing. Figures 12-21 and 12-22 illustrate some of these considerations when nonmetallic wiring is used. These examples apply to wiring for televisions, phones, computers, and sound systems, as well as electrical wiring. Wiring passing through walls framed with wood,

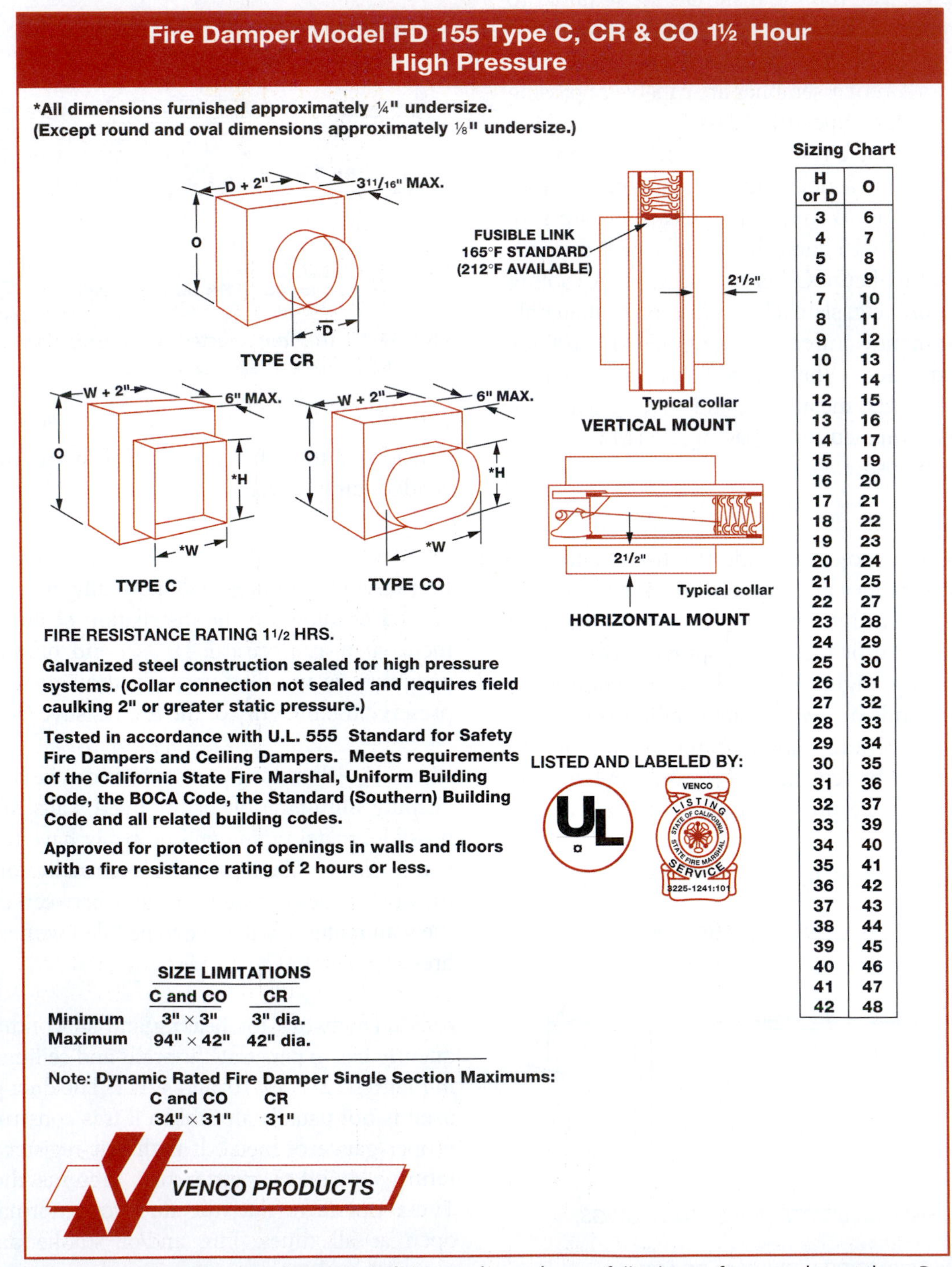

	C and CO	CR
Minimum	3" × 3"	3" dia.
Maximum	94" × 42"	42" dia.

C and CO	CR
34" × 31"	31"

Sizing Chart

H or D	O
3	6
4	7
5	8
6	9
7	10
8	11
9	12
10	13
11	14
12	15
13	16
14	17
15	19
16	20
17	21
18	22
19	23
20	24
21	25
22	27
23	28
24	29
25	30
26	31
27	32
28	33
29	34
30	35
31	36
32	37
33	39
34	40
35	41
36	42
37	43
38	44
39	45
40	46
41	47
42	48

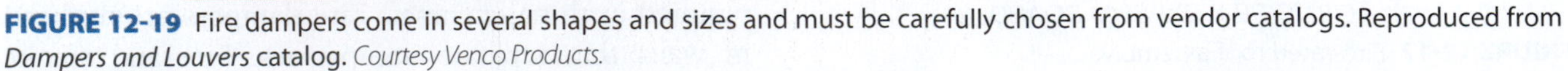

FIGURE 12-19 Fire dampers come in several shapes and sizes and must be carefully chosen from vendor catalogs. Reproduced from *Dampers and Louvers* catalog. *Courtesy Venco Products.*

TABLE 716.3.2.1
FIRE DAMPER RATING

TYPE OF PENETRATION	MINIMUM DAMPER RATING (hour)
Less than 3-hour fire-resistance-rated assemblies	1.5
3-hour or greater fire-resistance-rated assemblies	3

FIGURE 12-20 Fire dampers have different ratings based on the ratings of the assembly they are to penetrate. *Courtesy 2009* International Building Code, *copyright © 2009. Washington, DC. International Code Council. Reproduced with permission. All rights reserved. www.iccsafe.org.*

engineered lumber, or steel studs should be placed in a metal conduit as shown in a Figure 12-21. Note that the metal conduit extends 5' beyond the firewall in each direction to reduce the likelihood of fire entering and passing through the conduit. Electrical convenience outlets that are recessed into firewalls also require special detailing, as shown in Figures 12-21 and 12-22.

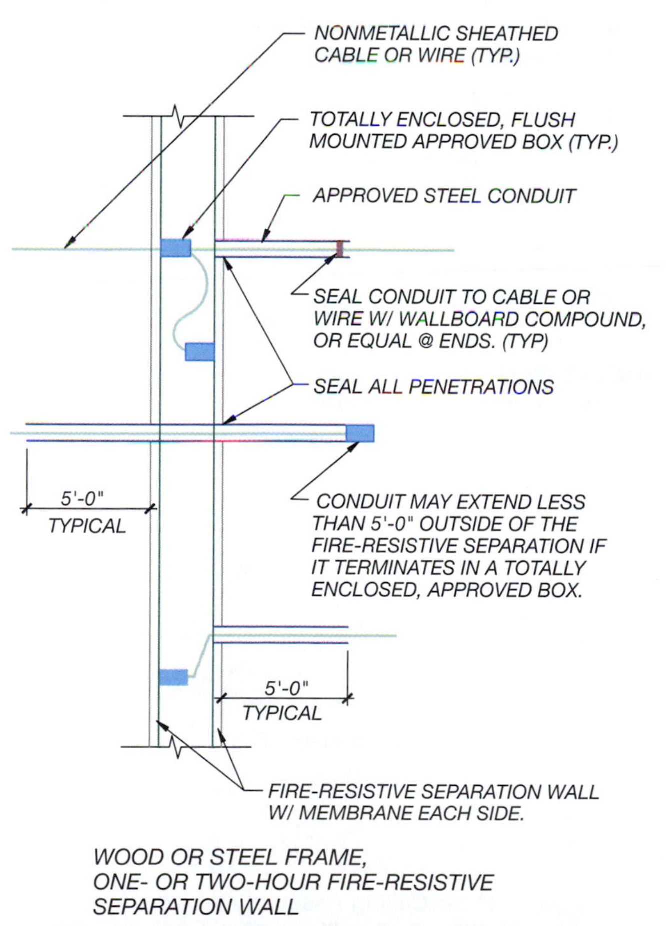

FIGURE 12-21 Electrical penetrations in fire-resistive walls must be carefully considered based on their location to the wall. *Courtesy Julie Searls, Divine Designs & Associates, Inc.*

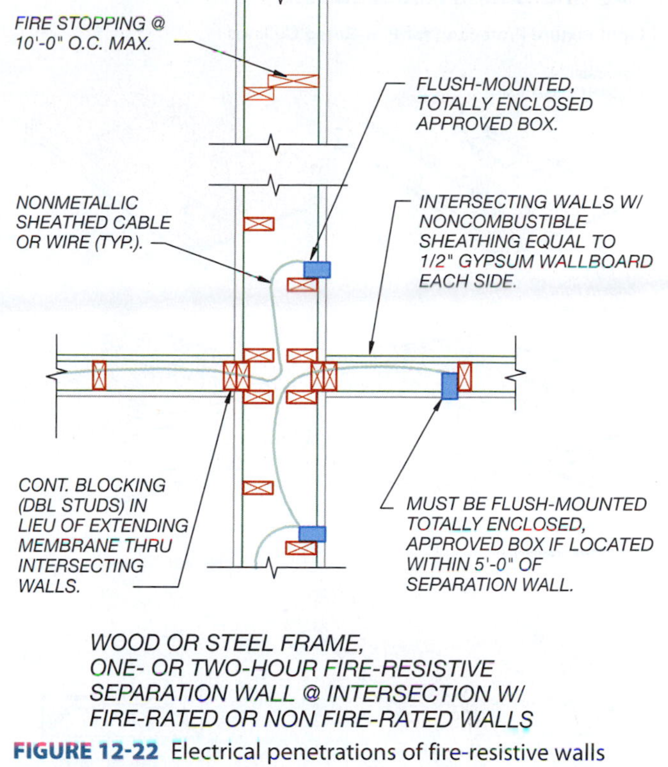

FIGURE 12-22 Electrical penetrations of fire-resistive walls must be carefully detailed in the architectural drawings. This type of installation can also be used with steel-framed walls if the membranes are continuous through the intersecting walls. *Courtesy Terrel Broiles, Divine Designs & Associates, Inc.*

One of the most common electrical penetrations of a fire-resistive assembly is a drop-in light fixture in a fire-rated suspended ceiling system. Figure 12-23 illustrates methods of preserving the integrity of the fire-rated ceiling at lighting fixtures in a suspended ceiling. The methods illustrated in this figure have been carefully tested by an approved testing laboratory.

Heavy Timber

Under some conditions, the code will allow heavy timber construction to be substituted for fire-rated construction. Thick and deep wood beams, thick post, and thick floor or roof sheathing can be left unprotected under this condition. The code describes the minimum dimensions required for a wood member to be considered heavy timber (see Chapter 7). The minimum required size varies for a vertical member (post) and a horizontal member (beam). Thick wood members will char on the outside but not burn through for an extended period of time. In a fire, the outside of the post or beam will be charred and become structurally useless. The inner portion of the wood will remain sound and have adequate strength to keep the structure from collapsing.

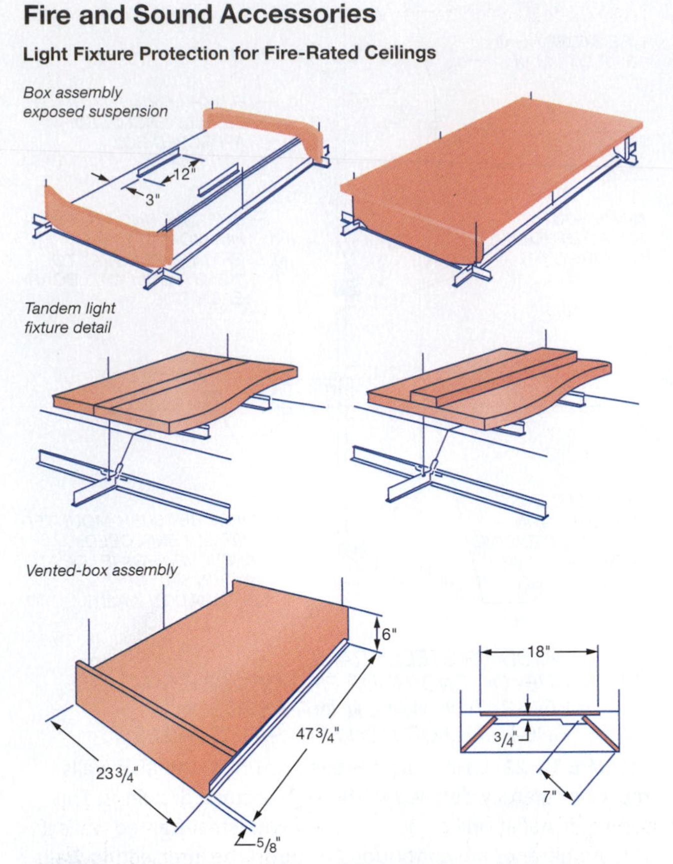

FIGURE 12-23 Fire-rated assemblies need special treatment at penetrations. This figure shows some ways to maintain a fire rating when a light fixture penetrates a fire-rated suspended-ceiling system. *Courtesy United States Gypsum Company.*

Fire and Draft Stops

Long, concealed spaces such as attics or crawl spaces can aid the spread of fire, especially in combustible construction. For this reason the code requires these concealed spaces, both horizontal and vertical, to be blocked off at certain points and intervals. Figure 12-24 shows a typical **draft stop** at a floor ceiling space. The draft stops are intended to prevent fire from spreading from one part of the structural system to another.

Spaces between the ceiling and the floor above in a non-sprinklered building of combustible construction need to be divided into controlled areas by draft stops. An attic must be divided up in a similar manner. Allowable areas for the space between the ceiling and the floor in nonresidential occupancies are limited to a maximum of 1000 square feet; attics have a limit of 3000 square feet. Figure 12-25 shows an example of a draft stop above a suspended ceiling system. If the floor or attic space is properly sprinklered, draft stops are not required.

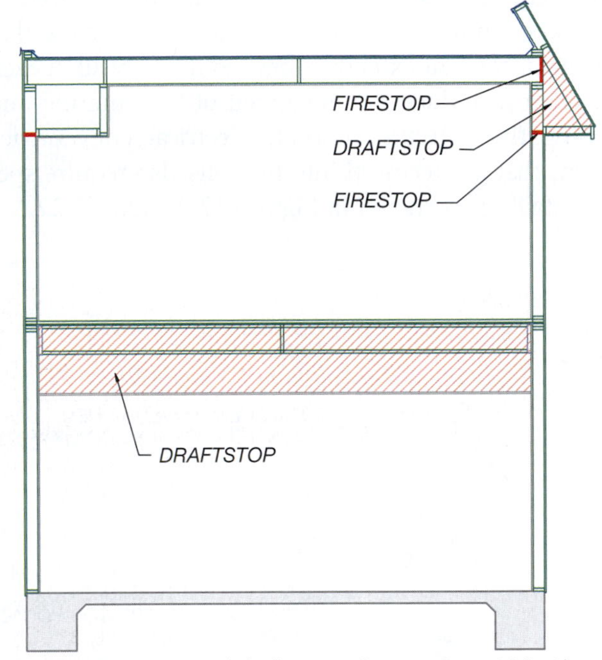

FIGURE 12-24 Fire and draft stops. *Courtesy Ginger M. Smith, Kenneth D. Smith Architects & Associates, Inc.*

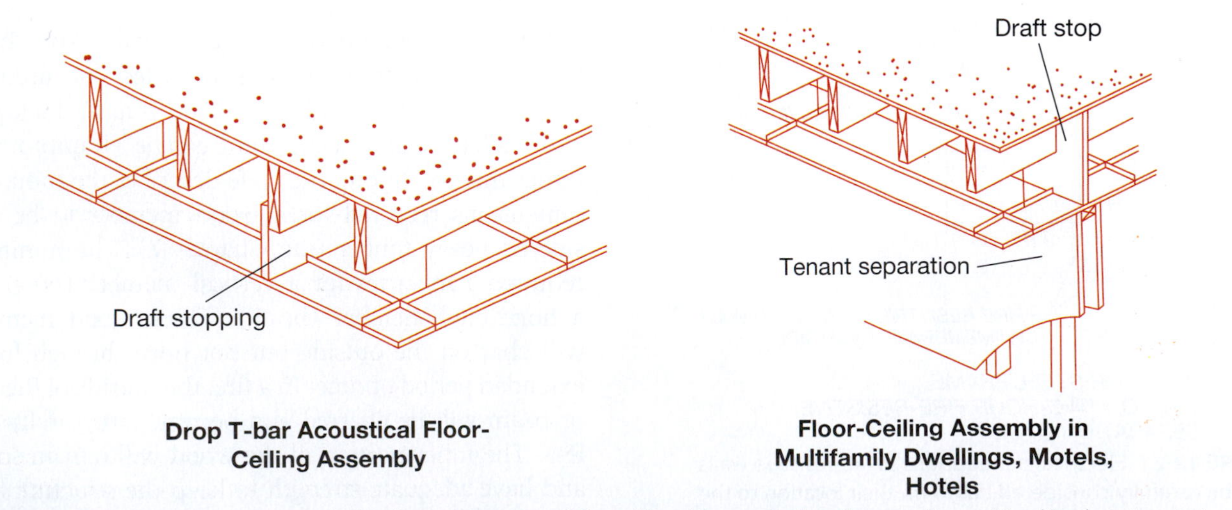

FIGURE 12-25 Draft stops at suspended ceilings. Reproduced from *Wood Frame Design. Courtesy Western Wood Products Association.*

ADDITIONAL READING

The *International Building Code and Commentary Volume I and II* have excellent illustrations and interpretations of the code that is available at www.iccsafe.org. The following Web sites can be used as a resource to help you find additional information.

ADDRESS	COMPANY/ ORGANIZATION
www.astm.org	ASTM International
www.pacificfirelab.com	Chilworth Pacific Fire
www.contegointernational.com	Contego International
www.gypsum.org	Gypsum Association (The entire *Fire Resistance Design Manual* is available on-line at this Web site.)
www.firefree.com	International Fire Resistant Systems, Inc.
www.paintooritect.com	International Fireproof Technology, Inc.
www.nailor.com	Nailor Industries
www.nfpa.org	National Fire Protection Association
www.ul.com	Underwriters' Laboratories

KEY TERMS

Draft stop
Fire damper
Fire resistance
Fire-resistive assembly
Firewall
Parapet
Smoke damper

CHAPTER 12 TEST

Fire-Resistive Construction

QUESTIONS

Answer the following questions with short complete statements. Type the chapter title, question number, and a short complete statement for each question using a word processor.

Question 12-1 What is fire-resistive construction trying to protect?

Question 12-2 Describe three ways that fire protection of a structural member is achieved.

Question 12-3 How thick must a solid brick wall be in order to be rated as a 4-hour firewall?

Question 12-4 What special consideration would be given to the fire protection of a steel column in a parking garage?

Question 12-5 What is a fire damper?

Question 12-6 What are the two primary purposes of a firewall?

Question 12-7 What is the primary purpose of a 20-minute fire door?

Question 12-8 What function does the parapet serve on a firewall?

Question 12-9 Wood burns and metal does not; but, in a fire-resistive building, an 8" × 8" wood post might not have to be protected from fire while an 8" steel column would. Explain why.

Question 12-10 Use the drawing on the student CD in the RESOURCE MATERIAL/CHAPTER 12 file to identify the six figures from this chapter that are referenced in the building section.

DRAWING PROBLEMS

Use the drawing guidelines and criteria presented in Chapters 2 and 6 to complete the following drawings.

Problem 12-1 Draw and label a parapet detail for a concrete block wall located 1' from a property line. Use orthographic projection and include appropriate callouts on the detail, assuming that there will be specifications accompanying the working drawings. Make sure that the surface materials within the top 18" of the parapet are noncombustible.

Problem 12-2 Using Figure 12-7 as a guide, draw and label a detail showing a W 18 × 40 steel column with 2-hour fire protection. Use and specify the following materials:

- Use a base layer of 5/8" Type X gypsum board applied around the column and held in place with paper masking tape.

- Provide a second layer of 5/8" Type X gypsum board applied over first coat, held in place with paper masking tape.
- Provide a face layer of either:
    - No. 24 MSG galvanized steel column covering consisting of two L-shaped sections with snap-lock sheet steel joints.
    - No. 22 MSG galvanized steel column covers consisting of two L-shaped sections with lap joints fastened with No. 8 × 1/2" sheet metal screws at 12" o.c.

Problem 12-3 Use figure 12-7 as a guide to draw and label a detail showing a 6" × 6" × 3/8" steel column with 1-hour fire protection. Use and specify the same material options that are used to protect the steel in problem 12-5.

Problem 12-4 Use the drawing on the student CD to complete a detail that will supplement the structural drawings, and show a 1-hour wood stud exterior wall within 12" of a property line. In addition to noting standard construction materials, specify the following minimum fire-protection materials:

- A 2 × 4 stud parapet wall. Extend the parapet wall 30" minimum above the finished roofing materials.
- Protect the parapet wall that extends above the finished roofing with 3/8" plywood and exterior stucco to match the exterior.
- Cap the top of the parapet wall with (2) 2 × 4 top plates and flash the top of the wall with 26 g.i. coping.
- Cover the exterior sides of the wall with 7/8" cement plaster w/ metal lath attached to studs with 6d common nails at 7" o.c. driven to 1" minimum penetration and bent over. Plaster mix to be 1:4 for scratch coat, and 1:5 for brown coat. By volume cement to sand per IBC Table 720.1(2) item #15-1.2 & 1.3. Provide a stucco screed where the stucco meets the finished roofing.
- Provide 3/8" plywood sheathing per structural drawings over both sides of 2 × 6 studs with (2) 2 × 6 studs with 10' high plate supporting TJM30 @ 24" o.c. (show trusses parallel to wall).
- Provide 2 × nailer where roof sheathing intersects the wall.
- Cover the interior side of the wall with 5/8" Type X gypsum board that extends to the bottom of the roof sheathing.
- Provide wall insulation appropriate for your area.

Problem 12-5 Use problem 9-7 as a base and the drawing on the student CD to show an exterior wall that contains (2) 5" × 5" × 3/8" T.S. columns that are approximately 45" apart (outside to outside). Frame the wall with 2 × 6 studs at 16" o.c. with one stud on the outside edge of each column so that the total wall length is 48" from exterior face of stud to exterior face of stud. Wrap the wall with (2)-layers of 5/8" Type X gypsum board covered with 7/8" exterior plaster applied as per problem 12-4.

Section 3

Preparing Architectural and Civil Drawings

13 Structural Considerations Affecting Design

Buildings should stand up. They should remain standing regardless of the forces applied to them. Not many would argue with these statements. There have been a few exceptions to this principle over the years, such as the houses of Japan that were made of bamboo and paper. They were designed to be lightweight so that when they fell over, due to the shaking of an earthquake, no one was hurt; the occupants just put the pieces back together. This is not a practical approach for today's factories or office buildings. This chapter will explore:

- Common loads that affect structures.
- How loads are distributed throughout structures.
- The process engineers follow to determine the loads to be resisted.
- The effects of seismic and wind forces on a structure.
- General design consideration for horizontal forces.

LOADS AFFECTING STRUCTURES

Buildings must be designed to meet several types of loads including:

- Dead loads
- Live loads
- Environmental loads

Some buildings need to resist **dynamic loads**, which are loads imposed by a moving object such as a car. The information in this chapter is intended to broaden your knowledge of how loads affect the structures you will be drawing. The chapter is not intended to make you an engineer. It is not essential that CAD technicians have an understanding of loads, but it is very helpful. If the CAD drafter can visualize how a load affects the building and can follow that **load's path** from its origin to the foundation, it will help the technician visualize conditions that need to be detailed on the plans. The way that building components are put together may have a major impact on how the structure behaves. As an example, hanging a load off the side of a column instead of directly on top of the column would introduce a tendency to bend, referred to as bending moment, into the column that it might not have been designed for. CAD technicians can better serve as a safeguard for the engineer and the designer with an understanding of how loads are distributed into the foundation.

Dead Loads

The weights of building materials such as rafters, roofing materials, floor materials, and walls are referred to as **dead loads**. Dead loads are determined by the actual weight of the materials from which the building is constructed. Asphalt roof shingles and the layers of felt under them have a given weight. The weight of air-conditioning units can be readily found in the manufacturer's literature. Plywood, framing lumber, glass, and gypsum board all have a given weight. The weights of these materials are readily available from manufacturer's literature or reference books, similar to the table seen in Figure 13-1. These materials are frequently combined into commonly used assemblies as shown in Figure 13-2. Loads from the materials are considered to be in place for the life of the building.

Live Loads

Buildings need to resist changing vertical loads from people, furniture, machinery, and other similar features, which are called **live loads**. Live loads are the loads determined by the use and occupancy of the building and are therefore much less exacting than dead loads. They are not static, but rather, constantly changing. Live loads are spelled out by the building code. At the floor, live loads are based on occupancy type (see Chapter 4) as shown in Figures 4-4a and 4-4b. Live loads are approximations. The live load on a roof may assume that a worker is carrying a load of roofing material across it. The live load on a ceiling may assume that someone is crawling across it to fix an electrical connection. People may be packed tightly on a balcony, watching a performance below. Building codes do not

WEIGHTS OF BUILDING MATERIALS			
Material	**PSF**	**Material**	**PSF**
Ceilings		**Plywood Sheathing**	
Acoustical tile	1.0	3/8"	1.1
Channel suspended	1.0	1/2"	1.5
1/2" gypsum board	2.2	5/8"	1.8
5/8" gypsum board	2.8	3/4"	2.2
		1"	3.0
Decking		1 1/8"	3.3
2"	4.4		
		Roofing	
Floor Finishes		Asbestos, corrugated 1/4"	3.0
Asphalt tile	2.0	15 lb felt	0.85
Brick pavers	10.0	3 ply and gravel	5.5
Ceramic tile 3/4"	10.0	4 ply and gravel	6.0
Concrete (lightweight per in.)	10.0	5 ply and gravel	6.5
Concrete (reinforced per in.)	12.0		
Hardwood 1"	4.0	**Shingles**	
Linoleum	2.0	Asphalt	2.0
Marble	30.0	Asbestos cement	4.0
Subfloor per in. of depth	3.0	Book tile 2"	12.0
Quarry tile 3/8"	5.0	Book tile 3"	20.0
Terrazzo 1"	13.0	Cement tile	16.0
Vinly asbestos tile	1.3	Clay tile	14.0
Wood floor joists 12" O.C. 16" O.C.		Fiberglass	0.5
2 × 66.0	5.0	Ludowici	10.0
2 × 86.0	6.0	Roman	12.0
2 × 107.0	6.0	Slate 1/4"	10.0
2 × 128.0	7.0	Spanish	19.0
		Wood	3.0
Insulation			
Rigid 1"	1.5	**Walls**	
Fiberboard 1"	2.0	4" glass block	18.0
Foam board per in.	0.2	Glass 1/4" plate	3.3
Poured in place	2.0	Window (glass, frame, sash)	8.0
4" batts	1.7	Glazed tile	18.0
6" batts	2.5	Gypsum board 1/2"	2.2
10" batts	4.5	Gypsum board 5/8"	2.8
		Marble	15.0

FIGURE 13-1 Standard weights of common building materials.

(continued)

WEIGHTS OF BUILDING MATERIALS			
Material	**PSF**	**Material**	**PSF**
Masonry 4" thick unless noted		Limestone	55.0
Brick	38.0	Terra-cotta tile	25.0
		Stone	55.0
Concrete block		Plaster 1"	8.0
4" hollow	30.0	Plaster 1" w/wood lath	10.0
6" hollow	42.0	Porcelain enamel, steel	3.0
6" hollow	55.0	Stucco 7/8"	10.0
Hollow clay tile	18.0	For more exact weights, see manufacturers' specifications.	
Hollow gypsum tile	13.0		

FIGURE 13-1 *(continued)* Standard weights of common building materials.

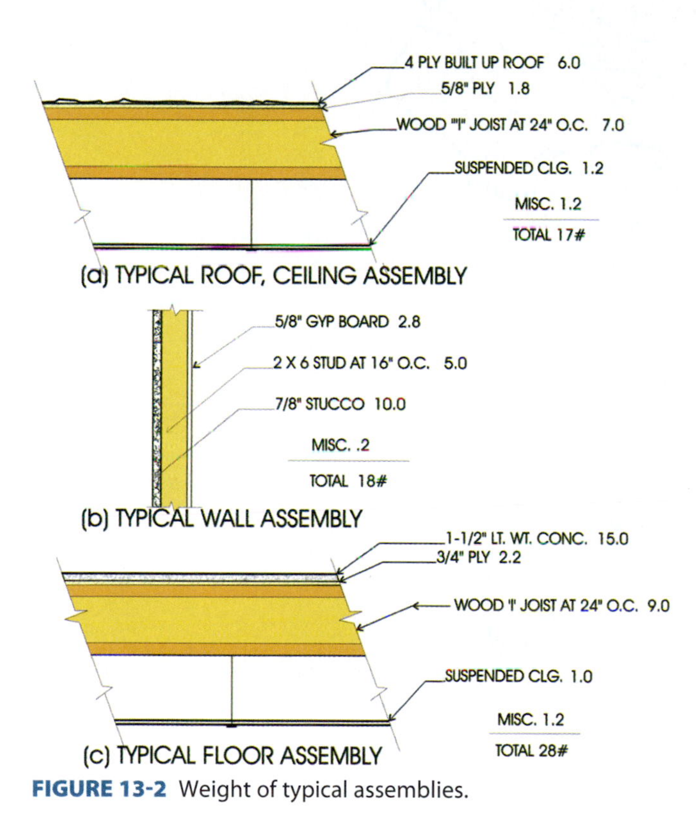

FIGURE 13-2 Weight of typical assemblies.

try to anticipate every conceivable live load condition. A laborer working on the roof, loading a wheelbarrow full to the brim with heavy material may cause a failure. The engineer and architect need to anticipate unusual loads that may occur in the building they are designing, such as a room full of heavy file cabinets. They may need to design for loads beyond those called for in the code. Live loads are a judgment call by the code and by the professional.

Environmental Loads

Buildings also need to resist vertical environmental loads from snow and rain, and be able to support vertical loads from their own weight. Buildings must be able to resist **horizontal environmental loads** from wind, soil, flooding, and earthquakes. Six basic environmental loads are addressed in the code. They are snow loads, rain loads, wind loads, soil lateral loads, flood loads, and earthquake loads. Snow loads and rain loads act vertically on the building. The other environmental loads act laterally or horizontally on the building.

Snow Loads

Snow creates a live load. Snow loads vary greatly across the country. Some areas experience heavy snowfall, whereas others don't have any snow. A typical snow load map is shown in Figure 13-3. The snowfall in one area may be substantially different from the snowfall a few miles down the road. The shape of a structure may cause increased localized loading from snowdrifts. Specific design loads for snow should be obtained from the local building department.

Rain Loads

There are conditions where rainwater will be allowed to accumulate on the roof of a building. If this condition exists then the structure must be designed to support the weight of the water. Roofs that are very flat (less than 1/4" per foot) are assumed to pond water. Because of this, their structure must be designed to support water. When roofs need roof drains, they must also have

FIGURE 13-3 A ground snow load chart and the local building department must be consulted to determine design snow loads.

Courtesy 2009 International Building Code, *copyright © 2009. Washington, DC. International Code Council. Reproduced with permission. All rights reserved. www.iccsafe.org.*

overflow drains located 2" above the roof as shown in Figure 18-30. These overflow drains are provided to drain water off the roof if the primary drain is clogged. The roof structure must be designed to support the water that would accumulate if the primary drain were clogged.

Earth Pressure

Retaining walls are designed to resist the lateral force from soil piled up against the wall. As seen in Figure 13-4, restrained or full-height retaining walls, which are tied to a diaphragm at their top, act like a beam spanning between the ground and the floor or roof of the building. The floor or **roof diaphragm** must be designed to resist the loads that the retaining wall will transfer to it. As seen in Figure 13-5, partial-height retaining walls, which are not connected to a diaphragm at their top, resist loads by cantilevering off their footings. The footing transfers the horizontal load from the earth they are retaining into the soil below. The further the footing extends from the wall the less the pressure is on the soil below.

The earth produces a relatively **static load** against the wall. It is assumed to act like a fluid with a weight

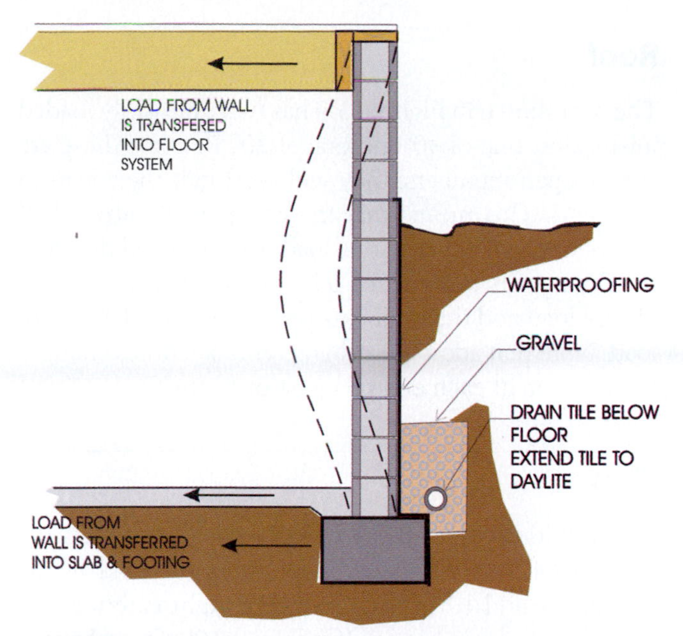

FIGURE 13-4 Restraining walls act like a beam spanning between each floor level. The lateral load of the soil causes the "beam" to bend in the middle.

FIGURE 13-5 Retaining walls resist loads by cantilevering from the footing. Forces from the exposed side of the wall are not considered because they are much smaller than the forces caused by the pressure of the soil on the wall.

of about one-third its actual weight, or about 30 pounds per cubic foot (4.71 Kn/m^3) but it is subject to change. Soil is heavier when it is wet. Different soil types absorb water differently. In order to minimize the effects of moisture, efforts are made to remove the water. This is done by installing a drain line behind the wall, as shown in Figure 13-4, or by letting the water flow through weep holes at the base of the wall, shown in Figure 13-6. The pressure also changes when another load, such as a car in Figure 13-7, or the footing for a building is placed on top of the earth, as in Figure 13-8. This type of loading is called a *surcharge*.

Load Design Considerations

Buildings are not designed to withstand every conceivable force. If they were, they would not be affordable. Buildings are designed to resist all dead loads and all

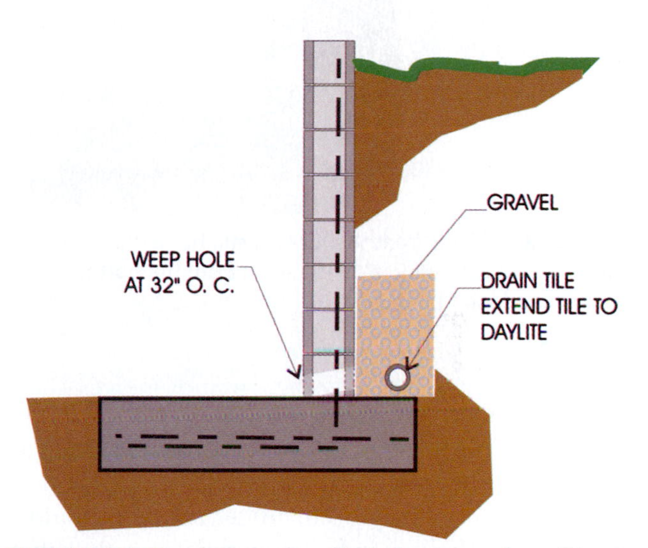

FIGURE 13-6 To minimize the effect of water on the soil load, a weep hole is placed in the bottom of the retaining wall.

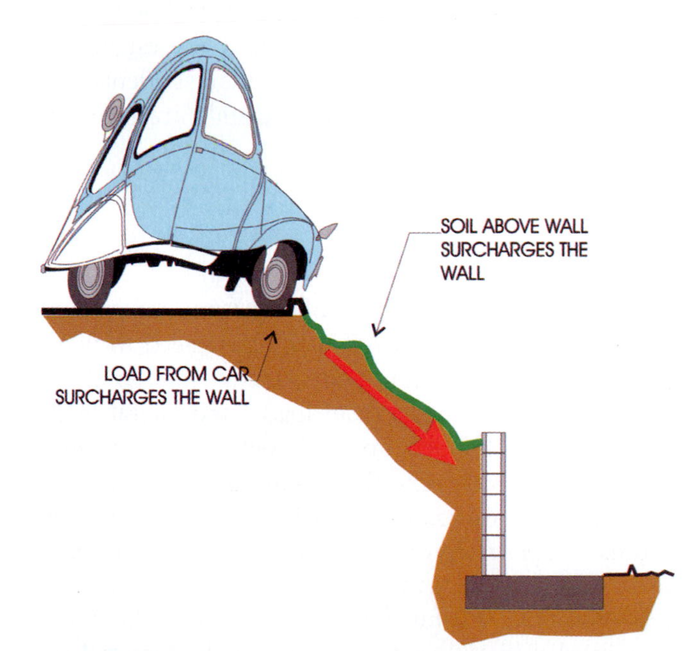

FIGURE 13-7 The pressure resisted by a wall is altered when a load is supported by the soil. The load is transferred through the soil to the wall, causing a condition referred to as a surcharge.

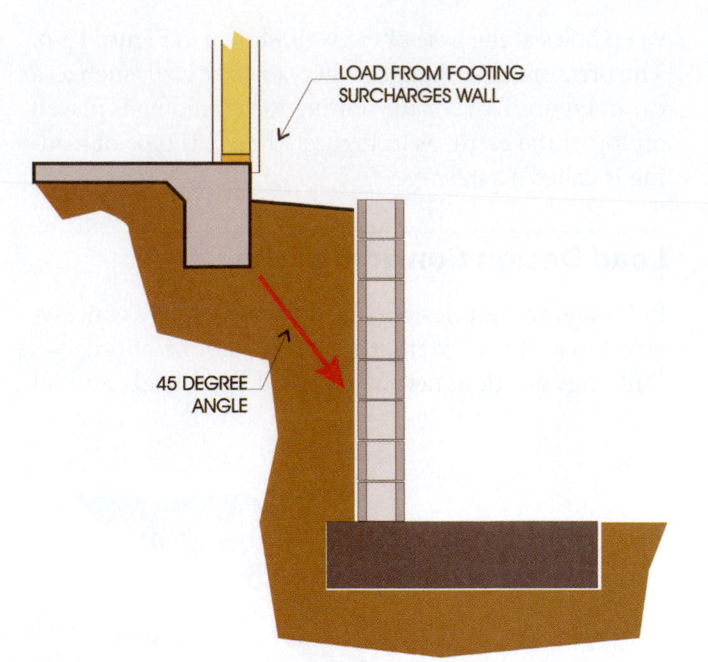

FIGURE 13-8 Loads from the footing surcharge the wall below a point extending at a 45° angle from the bottom of the building footing.

anticipated live loads and dynamic loads. They are designed to resist most anticipated lateral forces with no damage. They should perform well, but some nonstructural damage may occur under severe conditions such as very high winds or moderate earthquakes. Under the most severe storm or earthquake conditions, the building might have structural problems beyond repair, but it should not fail in such a manner as to cause loss of life. Details should be designed to be *ductile*, that is, to give rather than fail suddenly and explosively. Examples of ductile details are given in other chapters of this book.

LOAD DISTRIBUTION

The contractor builds the building from the ground up. The engineer and CAD technician need to work in the opposite direction. Starting from the top of the building, the live loads, dead loads, and lateral loads are collected. Members are designed to support those loads. These members then transfer their loads to other members until the load is transferred into the ultimate supporting member—the earth. The primary reason for the structural portion of plans is to illustrate this transfer. Load transfers are depicted on the drawings in plan views, sections, and details. Figure 13-9 shows a simple two-story structure that will be examined throughout this chapter. You will follow the loads depicted in this figure, from the roof to the soil.

Roof

The roof shown in Figure 13-9 has two uniformly loaded joist spans, one of 40' and one of 20'. Because these are simple-span members, they will send half their load to each end. This means that the center wall carries half the total load (30'), and this load is transferred down to the floor joists below. The left exterior wall carries 20' of roof load and the right exterior wall carries 10' of roof load. Note that adding overhangs and balconies changes the reaction at each end of a joist or beam.

Floor

At the middle floor, the left exterior wall supports the 20' of roof load discussed above, plus the load of the wall above and 10' of floor load. The right exterior wall is a bit more complicated. It supports 10' of roof load as discussed above, 20' of floor load and the wall above. The right wall also receives a point load from the center wall. The point load consists of the weight of the wall and the roof it supports. Because this wall falls at the center of the 40' floor joist span, the floor joists transfer half the weight of the point load to the right wall and half to the center support on the lower floor.

The center support at the lower floor for the floor above is a beam and three columns. Just like the right exterior wall, the beam receives loads from the center wall above, half the wall load and half the roof load, supported by the wall. It also receives half the left floor joist load (10') and half the right joist load (20').

Foundation

The continuous left footing receives the weight of the left exterior wall, plus all of the loads supported by the wall. The continuous right footing does the same. It receives all the loads the right exterior wall supports, plus the load of the right wall. The center support is not continuous and does not have a continuous footing under it. The beam transfers the loads from above and its own weight to the post. These posts then transfer their loads to pier footings. The weight of the footing will need to be considered along with the other loads to arrive at a total load to be supported by the soil.

Soil

All loads are eventually transferred into the earth or the soil. To determine the footing size, the engineer will need to know the bearing value of the soil. The building code provides a table giving allowable bearing

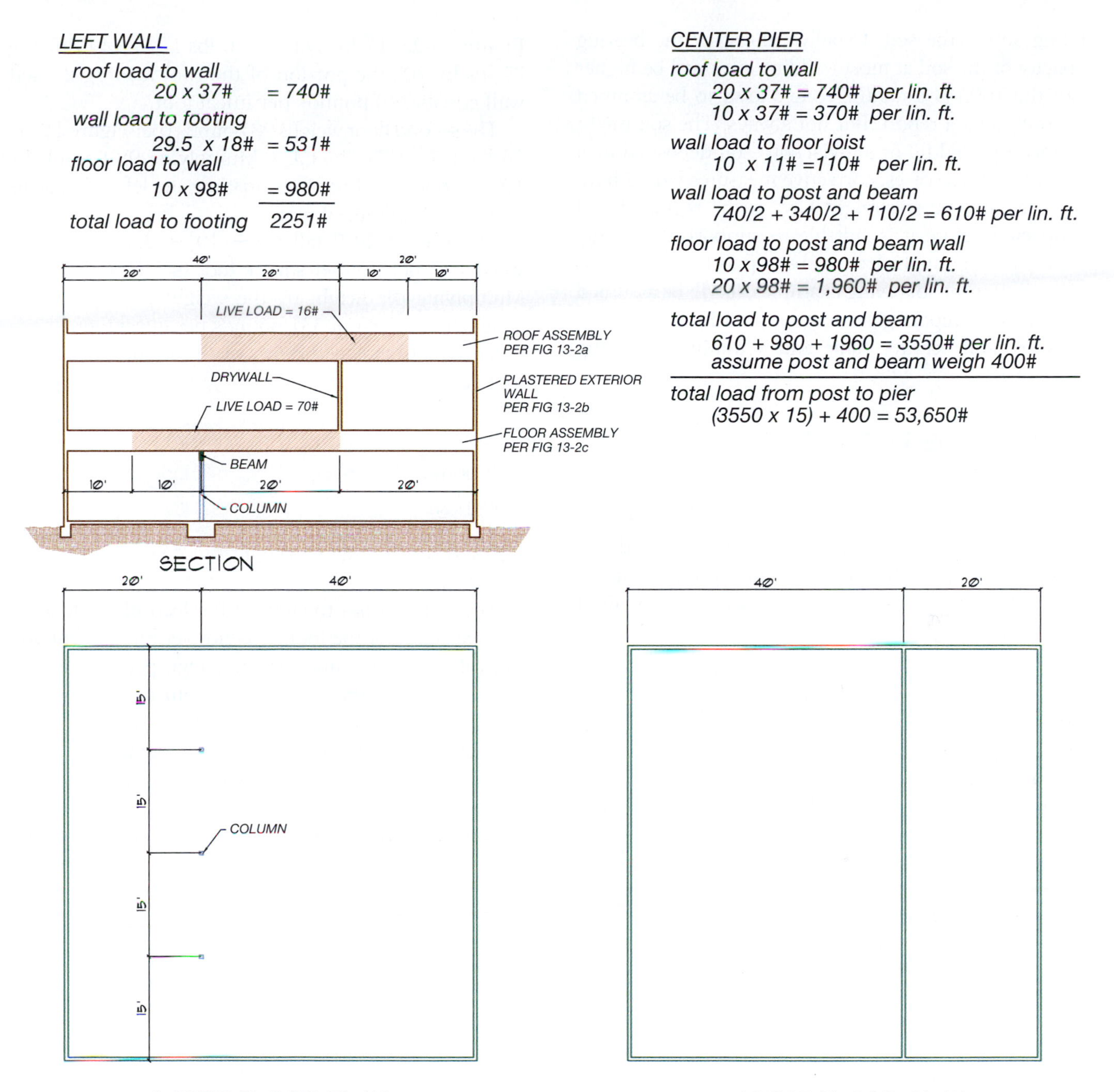

FIGURE 13-9 Transferring loads to the soil. The calculations shown in this figure will be used to determine the loads on the left and center footings.

values in pounds per square foot for various soil types. These values can be used without a soils report, but this is seldom the wise thing to do unless you are supporting very light loads. A soils investigation and perhaps a geological investigation should be done. To make a soils report, a civil engineer must do a soils investigation at the building site. The engineer first visits the site to obtain soil samples at various depths and locations on the site. These soil samples are then taken to a soils lab, tested, and classified. After this is done, the soils engineer will prepare a soils report. A good soils report will contain, among other things, soil bearing values and construction design recommendations. There will be recommendations for minimum footing width, depth, and reinforcing; slab thickness, underlayment, and reinforcing; and paving materials. The report will also give equivalent fluid pressure, and friction coefficient values for designing retaining walls, expansive soils information, and grading recommendations. The **equivalent fluid pressure** is a simplified way to measure the force that the earth will exert on the wall. The **friction coefficient** value measures the resistance of the footing sliding across the soil. Usually values for the bearing capacity of the soil at most building sites will be higher than the 1000 pounds the code allows to be assumed without using a report, but not always. The soil might be uncompacted fill or alluvial soil (soil deposited by a stream). In these cases, corrective measures would have to be taken. The site might have to be regraded and compacted before it is suitable to support a building, or the building might have to be supported by piles or caissons. For the following examples, it will be assumed that the soils report gives a value of 1500 psf for the allowable soil bearing value and a minimum footing depth of 18".

DETERMINING LOADS

The balance of this chapter will provide an introduction to compiling loads. Before the actual loads are examined, the technician must again be reminded that determining the loads is the responsibility of the engineer. The CAD technician's role is to carefully specify all of the engineer's calculations.

Footing Calculations—Left Wall

The loads for the left wall transfer straight down, so they are relatively simple to calculate. The footing will be supporting roof loads, floor loads, and wall loads. The roof weighs 37 pounds per square foot (per Figure 13-2a, 17 lbs D.L. + 20 lbs L.L.). Multiplying 37 lbs by 20', the portion of the roof loading the left wall equals 740 pounds per lineal foot.

The second floor weighs 98 pounds (per Figure 13-2c, 28 lbs D.L. + 70 lbs L.L.). Multiplying 98 pounds by 10', the portion of the floor loading the left wall equals 980 pounds per lineal foot.

The wall is 29.5' tall (6 + 10 + 3.5 + 10) and weighs 18 pounds per square foot (see Figure 13-2b). Multiplying the height by the weight per square foot results in a weight of 531 pounds per lineal foot.

The total load to the top of the footing per lineal foot is:

Roof load	740 lbs
Floor load	980 lbs
Wall load	531 lbs
Total load	2251 lbs

The soil also has to support the load of the footing. The exact size of the footing is not yet known, but you can calculate how much it will weigh per square foot. If this weight is then subtracted from the soil bearing value, the weight of the footing can then be ignored for the rest of the calculation. Assume the footing will be 18" thick, and the weight of the concrete is 150 pounds per cubic foot. Using Figure 13-10, it can be seen that the footing will weigh 225 pounds (1.5 × 150) per square

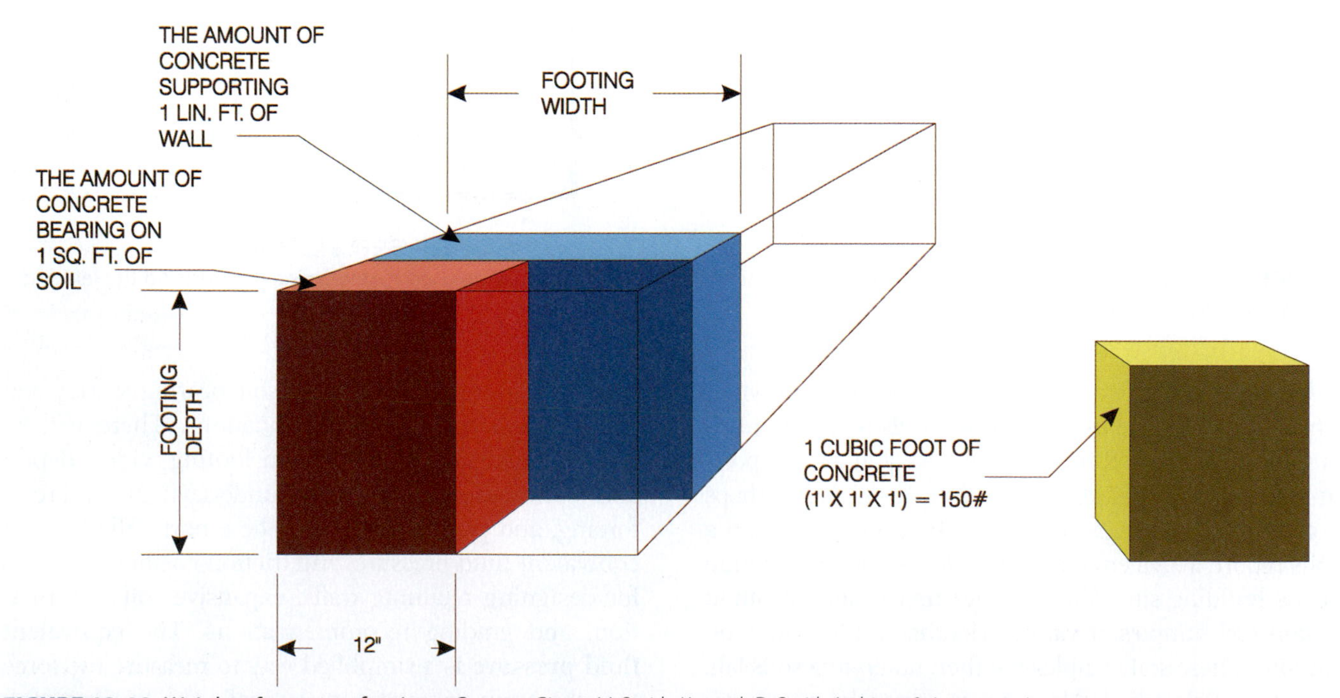

FIGURE 13-10 Weight of concrete footings. *Courtesy Ginger M. Smith, Kenneth D. Smith Architect & Associates, Inc.*

foot. This weight of the concrete can be subtracted from the soil bearing value, leaving a value of 1275 psf (1500 – 225). Dividing the 2251-pound load to the top of the footing by the remaining soil bearing value of 1275 pounds results in a required footing width of 1.77' or 1'–9 1/4". Because this would not be a convenient size to trench, a 2' wide footing would probably be specified for this condition.

Footing Calculation—Center Support

The loads for this support are not as simple to calculate because the loads are not stacked directly above the pier.

The roof will transfer 1110 pounds {37 lbs × [(40/2) + (20/2)]} to the interior stud wall. The center wall weighs 110 pounds (11 lbs × 10'). The wall transfers its weight and the weight of the roof load, a total of 1220 pounds, to the floor joists.

Floor joists on the right side of the center support beam transfer half the load from the wall above to the right exterior wall and half, 610 pounds, to the center beam. The floor itself weighs 2940 pounds {98 lbs × [(20/2) + (40/2)]}. This means that the floor joists transfer 3550 pounds per lineal foot to the center beam. Assuming that the post and beam weigh 400 pounds, 53,650 pounds [400 + (3550 × 15)] will be transferred to the pier footing.

Next, subtract the weight per square foot of the pier from the soil bearing value. As shown in the previous example, the weight of an 18" deep footing is 225 pounds per square foot. Subtracting this value from the allowable soil bearing value of 1500 pounds leaves an adjusted value of 1275 pounds per square foot. Dividing the load of 53,650 pounds by the adjusted soil bearing value gives a value of 42.08 square feet. Using the square root of this number will determine the size of a square pier. This results in a pier size of 6.48 feet square. A pier measuring 6.5 feet square would likely be specified for this condition.

THE EFFECTS OF SEISMIC AND WIND FORCES ON A STRUCTURE

So far only vertical loads have been explored. Forces that push sideways are somewhat less obvious. These sideway forces are referred to as **lateral forces**. Every building experiences lateral forces. The two most common sources of lateral forces are seismic forces that result from earthquakes and wind. Buildings must be constructed to resist these forces as well as the vertical forces discussed in earlier in this chapter. The building code contains maps like those in Figures 13-11 and 13-12 to classify the intensity of wind and earthquake forces that are likely to occur in various parts of the country. Rather than use this map, engineers frequently use the U.S. Geological Survey (USGS) Web site, which lets them obtain information on a specific zip code area. The code also contains minimum design requirements to resist these forces.

Earthquakes

Earthquakes happen without warning and last for a very short period of time, perhaps 10 to 20 seconds. A shift in the Earth's crust sets off shock waves that shake and rattle buildings. Most earthquakes in North America occur in California and Alaska, but they have occurred in other parts of the continent. The code contains maps, similar to the snow load and wind speed maps, showing the expected ground motion caused by an earthquake (see Figure 13-11). The shock wave from an earthquake causes a building to move laterally and vertically. It is assumed that the weight of the building and normal design for vertical forces will take care of the vertical motion, so vertical forces are generally ignored in seismic design. The lateral movement is very dramatic. It is not a constant or uniform force. It shakes the building back and forth with different intensities. The type of structure (solid masonry, moment-resisting frame, and so on) and the nature of the soil under the building are major factors in the effect that an earthquake will have on a building.

Buildings are not designed to be earthquake proof. They are designed to be earthquake resistant. Protecting the people inside the building is more important than protecting the building itself. It is better for a building to twist and bend in a severe earthquake than it is for it to suddenly break apart. With this in mind, engineers try to design connections that will bend or stretch rather than snap. This type of connection is called a ductile connection. Ductile connections give the building a better chance of remaining standing during and after an earthquake and give the building's occupants a better chance of evacuating. The building may be as useless as if it had collapsed, but fewer lives will be lost.

Wind Loads

Every portion of the world has wind, and wind patterns are far more predictable than earthquakes. There are

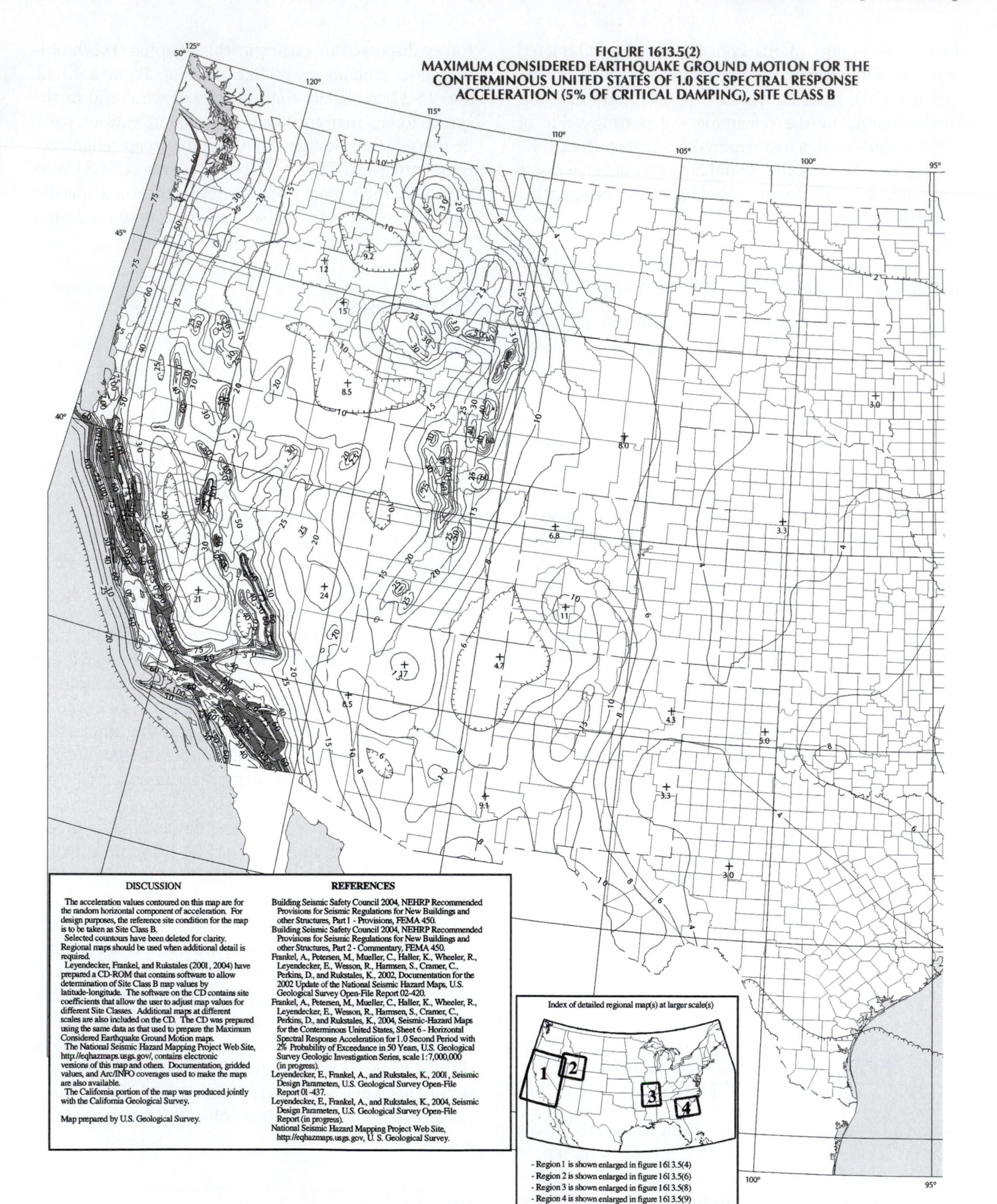

FIGURE 13-11 A map showing the maximum considered earthquake motion for the western portion of the United States. *Courtesy 2009* International Building Code, *copyright © 2009. Washington, DC. International Code Council. Reproduced with permission. All rights reserved. www.iccsafe.org.*

warning systems in place for high-wind conditions. Even so, hurricanes and tornadoes are very destructive forces. Because destructive wind events happen for a longer period of time and over a larger portion of North America, they cause a much greater loss of life and property than earthquakes. Figure 13-12 is a partial wind speed map of the United States. It has wind contour lines ranging from 90 to 150 miles per hour.

Wind causes both horizontal and vertical movement that must be considered by the engineer. The vertical forces are called *uplift*. As seen in Figure 13-13, uplift is generally caused in one of two ways. Wind blowing into a structure through an opening increases

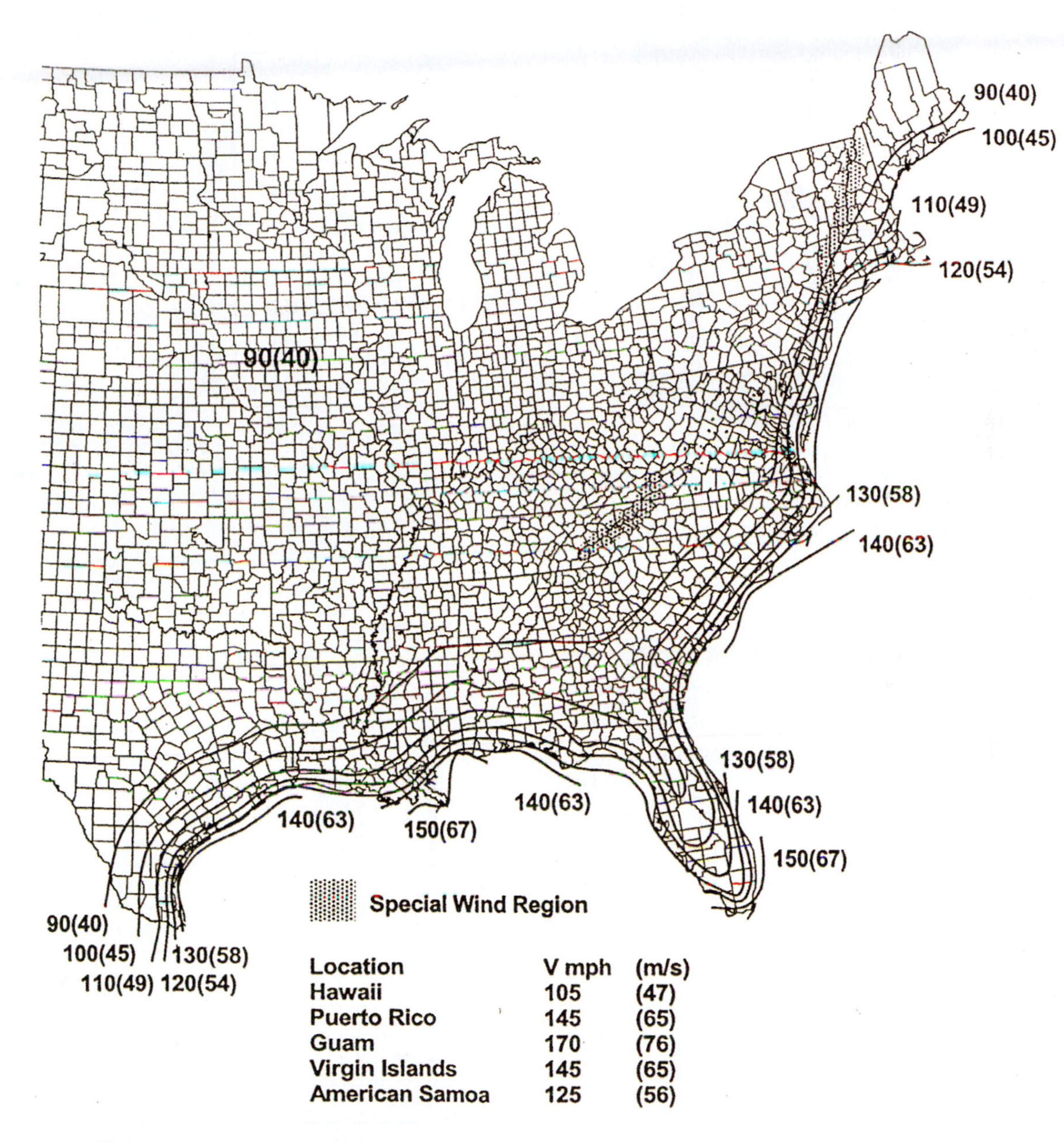

Notes:

1. **Values are nominal design 3-second gust wind speeds in miles per hour (m/s) at 33 ft (10 m) above ground for Exposure C category.**
2. **Linear Interpolation between wind contours is permitted.**
3. **Islands and coastal areas outside the last contour shall use the last wind speed contour of the coastal area.**
4. **Mountainous terrain, gorges, ocean promontories, and special wind regions shall be examined for unusual wind conditions.**

FIGURE 13-12 A map showing the basic wind speeds for the eastern portion of the United States. *Courtesy 2009* International Building Code, *copyright © 2009. Washington, DC. International Code Council. Reproduced with permission. All rights reserved. www.iccsafe.org.*

FIGURE 13-13 Uplift is caused by wind blowing into the structure and attempting to lift the structure, or by creating a negative pressure as it blows over the structure attempting to suck the roof off the structure. *Courtesy Jamie Smith, Kenneth D. Smith Architect & Associates, Inc.*

the pressure and tries to lift the roof. The wind blowing across the top of a structure creates a negative pressure that tries to lift or suck the roof off its supports. Both these vertical forces require that the roof be firmly anchored to the supporting elements. Figure 13-14 shows uplift connections for wall/floor members that are designed to transfer the uplift force into the footing. Horizontal wind forces are treated in very much the same manner as seismic forces. Wind speeds can reach as high as 200 miles per hour in a hurricane and 600 miles per hour in a tornado. As with earthquakes, buildings are not designed to withstand the worst wind event without damage but are designed to protect human life.

GENERAL DESIGN CONSIDERATIONS FOR HORIZONTAL FORCES

Figure 13-15a shows an open-frame stud wall with a lateral force pushing on it as indicated by the arrows. If the joints between the top plate and the studs are not stiff enough, the wall will try to fold into a parallelogram as shown in Figure 13-15b. If the wall is stiff enough to resist these forces but is not anchored well

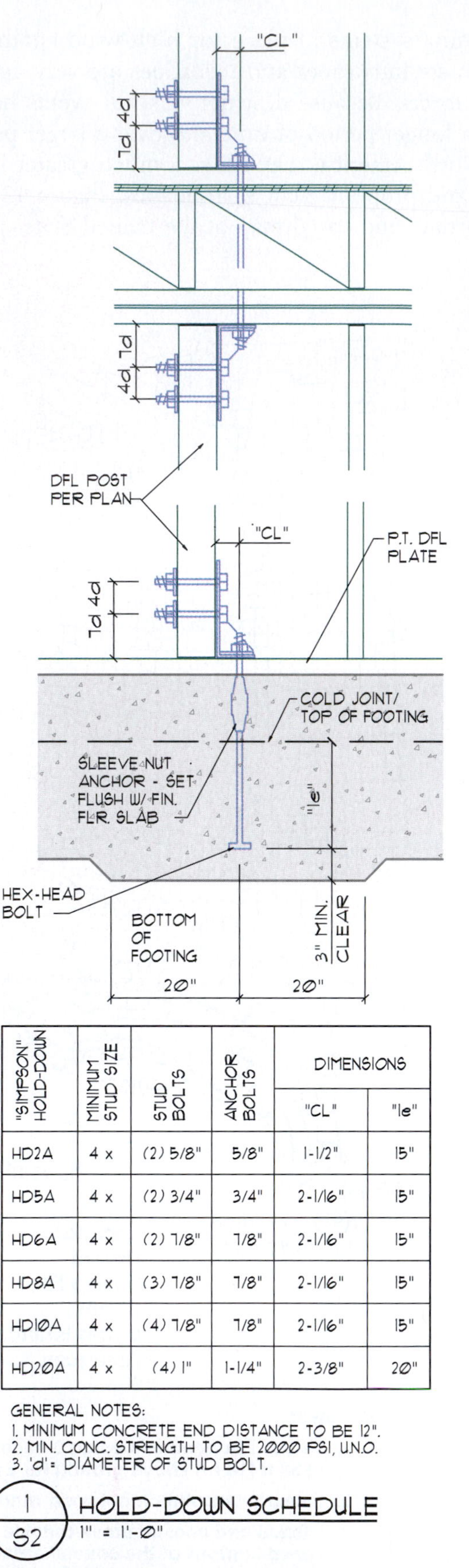

"SIMPSON" HOLD-DOWN	MINIMUM STUD SIZE	STUD BOLTS	ANCHOR BOLTS	DIMENSIONS	
				"CL"	"le"
HD2A	4 x	(2) 5/8"	5/8"	1-1/2"	15"
HD5A	4 x	(2) 3/4"	3/4"	2-1/16"	15"
HD6A	4 x	(2) 7/8"	7/8"	2-1/16"	15"
HD8A	4 x	(3) 7/8"	7/8"	2-1/16"	15"
HD10A	4 x	(4) 7/8"	7/8"	2-1/16"	15"
HD20A	4 x	(4) 1"	1-1/4"	2-3/8"	20"

GENERAL NOTES:
1. MINIMUM CONCRETE END DISTANCE TO BE 12".
2. MIN. CONC. STRENGTH TO BE 2000 PSI, U.N.O.
3. 'd'= DIAMETER OF STUD BOLT.

1 / S2 HOLD-DOWN SCHEDULE
1" = 1'-0"

FIGURE 13-14 Hold-down connections for wall/floor members that are designed to transfer the forces of uplift into the foundation. *Courtesy Dean K. Smith, Kenneth D. Smith Architect & Associates, Inc.*

to the foundation, it will tend to slide along the foundation or will try to lift up and turn over, as shown in Figure 13-15c.

The wall can be stiffened by applying a rigid sheathing such as the plywood shown in Figure 13-16a or by adding diagonal straps or wood let-in braces as shown in Figure 13-16b. A wall with rigid sheathing is called a **shear wall.** Figure 13-17 shows a typical wood framed plywood shear wall. An alternative material could be used for the wall, such as concrete block that makes a rigid solid wall as shown in Figure 13-16c, or a steel rigid frame as shown in Figure 13-16d. Figure 13-18 shows a shear wall made of steel studs and straps. Two connections the CAD technician would need to complete to explain the wall construction are shown in Figure 13-19.

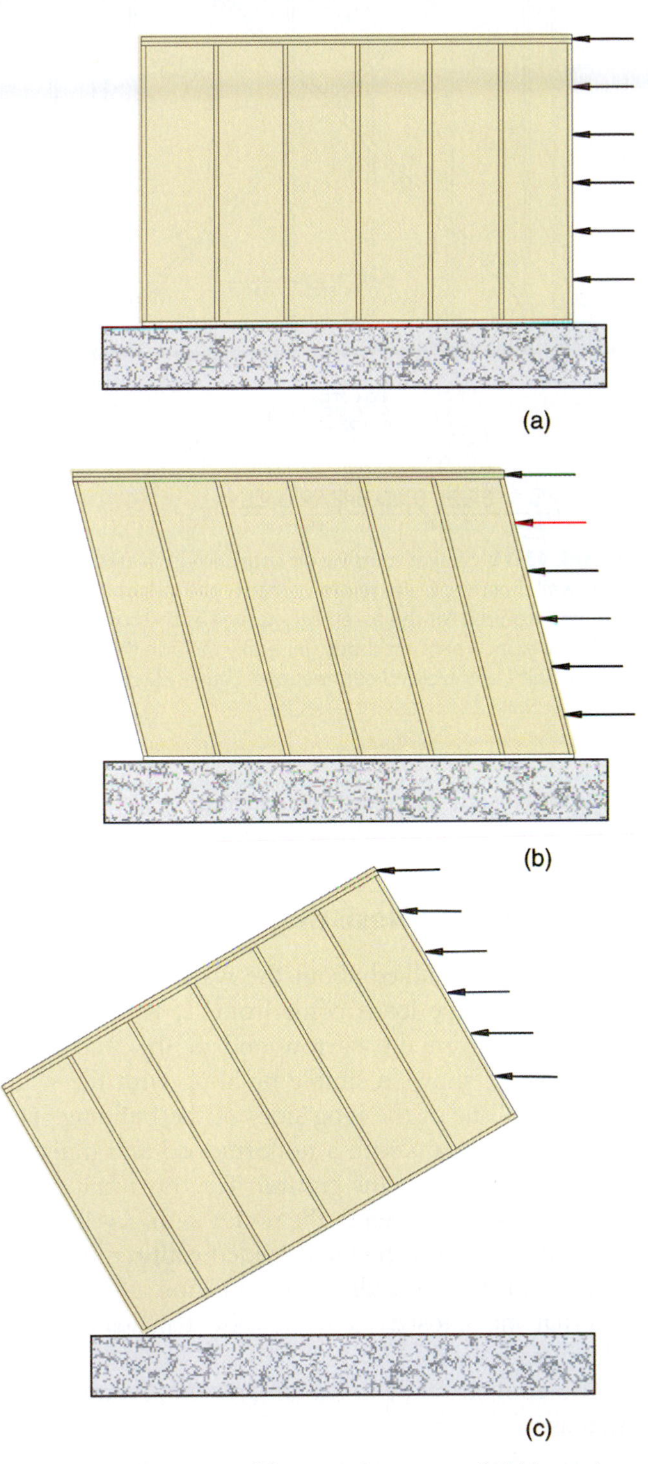

FIGURE 13-15 (a) Lateral forces acting on a wall. (b) Lateral forces will try to flatten or turn the wall into a parallelogram, or (c) lift and rotate the wall. *Courtesy Robin Smith, Kenneth D. Smith Architect & Associates, Inc.*

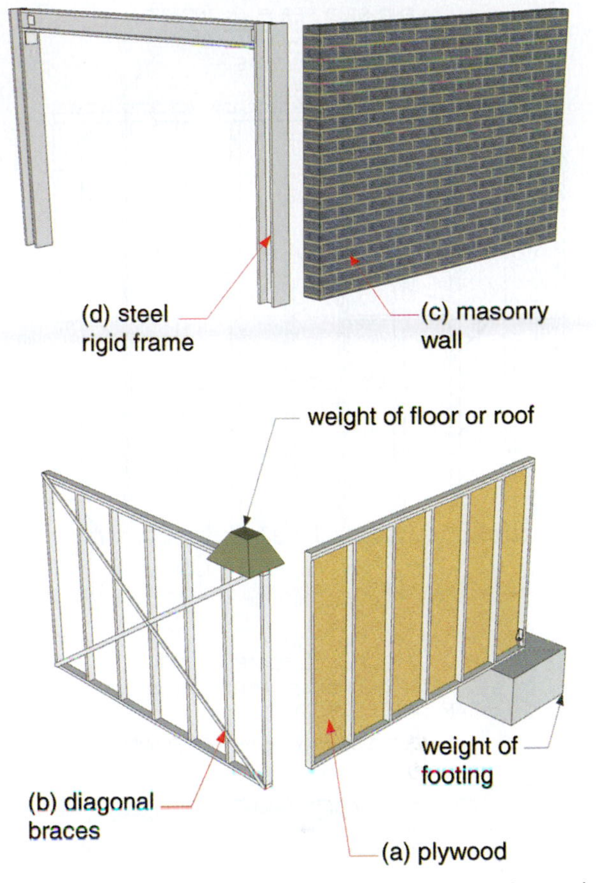

FIGURE 13-16 Shear wall examples. Shear walls can be made of many materials including (a) studs and plywood, (b) studs with diagonal strapping, or (c) masonry. A steel rigid frame (d) serves the same function as a shear wall. Shear walls depend on weight from the foundation below or roof and floor loads from above to keep them from lifting and rotating.

Stiffening the wall is only a part of the solution. The wall must also be tied to the foundation with anchor bolts or similar devices to prevent sliding, as shown in Figures 13-17 and 13-18. Shear walls also need adequate weight at the end to prevent overturning. The footing itself may be heavy enough to prevent overturning, or a pier may need to be added to the end of the wall, as shown in Figure 13-16a. Special bolting, called hold-downs, that are shown in Figures 7-4, 13-17, 13-18, and 13-19, may be needed on the end of the wall to firmly tie the wall to the footing. The end of a wall may also be held down by a load being applied to the top

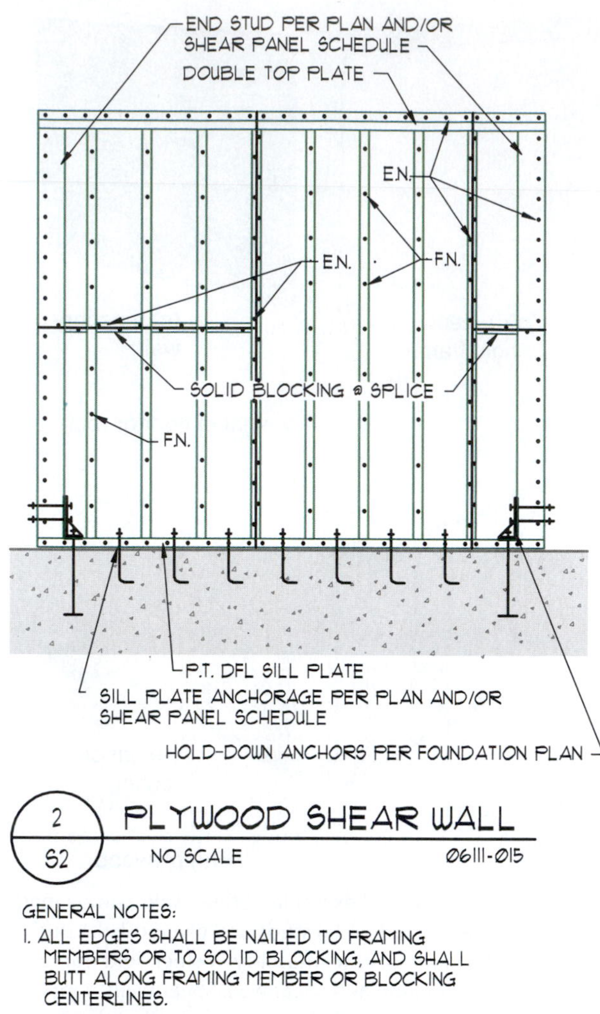

GENERAL NOTES:
1. ALL EDGES SHALL BE NAILED TO FRAMING MEMBERS OR TO SOLID BLOCKING, AND SHALL BUTT ALONG FRAMING MEMBER OR BLOCKING CENTERLINES.
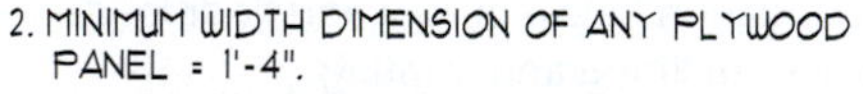
2. MINIMUM WIDTH DIMENSION OF ANY PLYWOOD PANEL = 1'-4".
3. NO PENETRATIONS SHALL BE ALLOWED THROUGH ANY SHEAR WALL WITHOUT THE EXPRESSED WRITTEN CONSENT OF ENGINEER UNLESS SPECIFICALLY SHOWN ON THESE PLANS.

FIGURE 13-17 A plywood shear wall detail must show and specify the size and type of the wall reinforcement as well as any extra framing members that must be provided, and the methods to be used to attach each feature. Attachment methods generally include gluing, nailing, and bolting. *Courtesy Dean K. Smith, Kenneth D. Smith Architect & Associates, Inc.*

of the wall such as a header, shown in Figure 13-16b. Figures 13-15 and 13-16 show the lateral force acting in one direction. This is never the actual case. Wind loads or seismic loads can come from any direction. For engineering design purposes, the lateral forces are calculated as if they act perpendicular to the building's surfaces. The building must be designed to resist lateral forces from all sides. Figure 13-20 shows a tall, thin, and very rigid wall that resists a large lateral force. The wall has hold-down bolts on each side of it and is anchored to a footing that projects beyond the wall on each side. This extended footing would have extra reinforcing steel encased in the concrete and is called a *grade beam*.

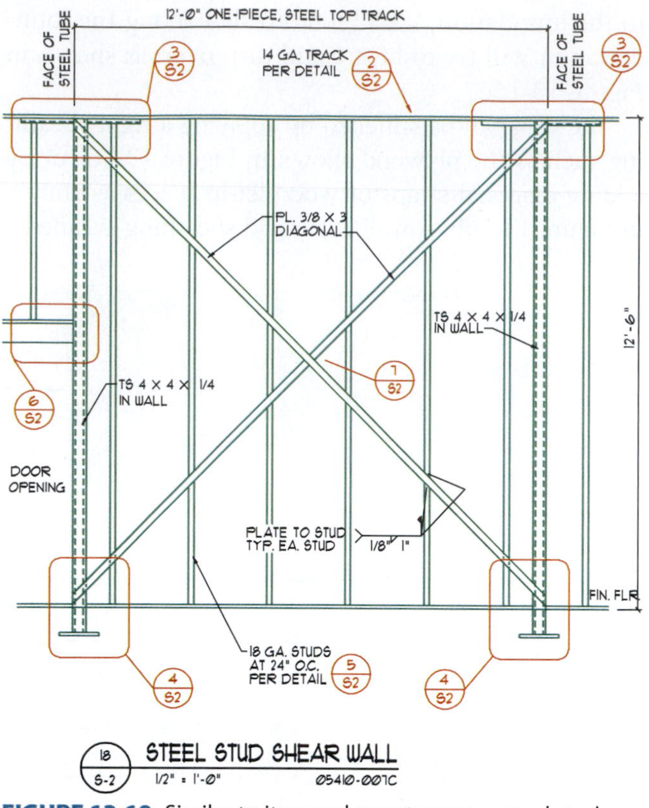

FIGURE 13-18 Similar to its wood counterpart, a steel stud shear wall must show the reinforcement method, each framing member required for the assembly, as well as the connection methods. Attachment methods generally include the use of screws, bolts, and welded connections. *Courtesy Dean K. Smith, Kenneth D. Smith Architect & Associates, Inc.*

The grade beam will resist overturning by changing the leverage point and by adding weight.

Collecting and Resisting Loads

So far, we have talked about the wall resisting loads, but where do the loads come from? It isn't from the wind blowing on the narrow end of the wall itself. Figure 13-21 shows a simple building with the wind blowing on one of the long sides of the building. The wall acts as a beam with a uniform load and transfers half the wind load to the ground. The other half of the wind load is transferred to the roof. Figure 13-22 is the same building with the wind-loaded wall removed for clarity. The arrows indicate the portion of the wind load that was transferred to the roof. This load pushes sideways on the roof.

The roof is made rigid in one of several ways, such as:

- A layer of plywood nailed to the joists.
- Metal decking welded to open-web steel joists.
- Cable X bracing attached to metal rigid frames.

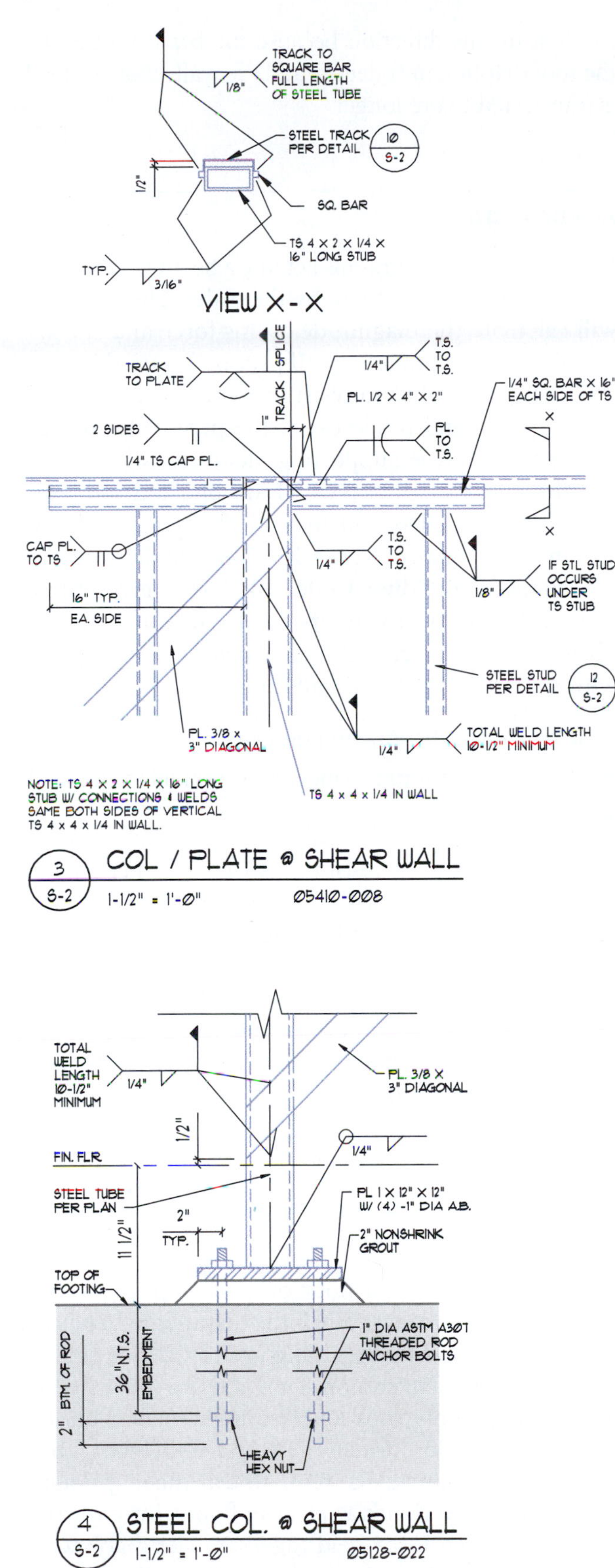

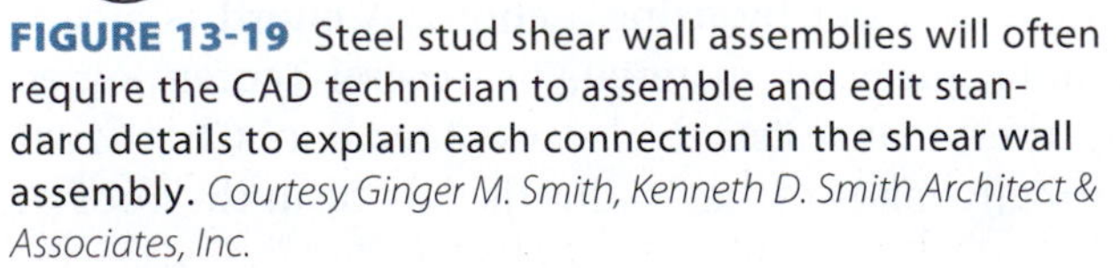

FIGURE 13-19 Steel stud shear wall assemblies will often require the CAD technician to assemble and edit standard details to explain each connection in the shear wall assembly. *Courtesy Ginger M. Smith, Kenneth D. Smith Architect & Associates, Inc.*

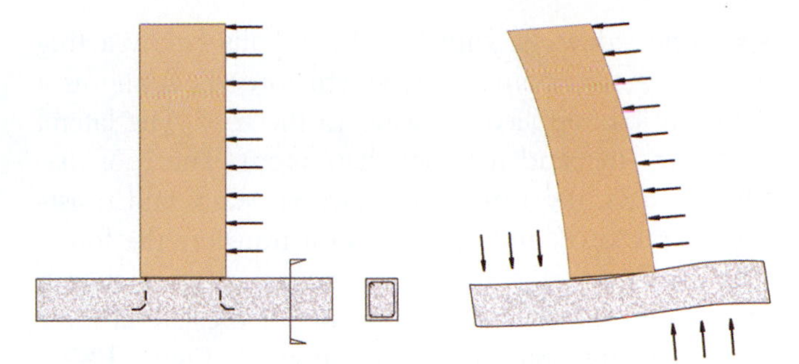

FIGURE 13-20 Lateral forces transfer from the wall to the footing and apply pressure to the soil.

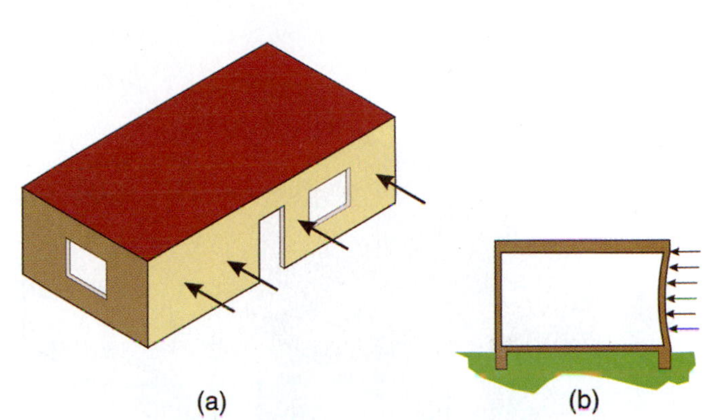

FIGURE 13-21 Lateral forces put pressure on the walls and try to bend them.

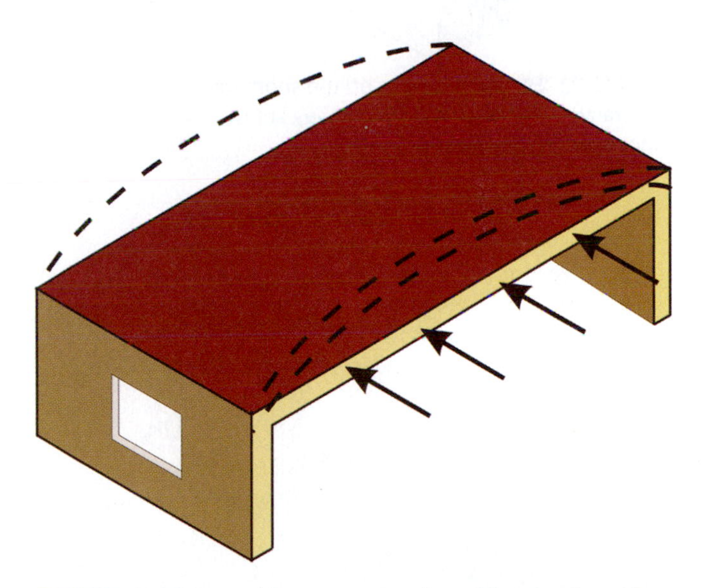

FIGURE 13-22 Lateral forces are transferred from walls to the roof causing the roof diaphragm to bend. The roof diaphragm transfers the forces back to the perpendicular walls and then to the footings.

or by another method similar to these. A rigid roof surface is called a *roof diaphragm* and is discussed in more detail later in this chapter. It is very similar to a shear wall, except that it is a horizontal surface instead of a vertical surface. A roof diaphragm acts like a beam spanning sideways, with the shear walls below acting as the reaction points. The lateral force is acting as a uniform load applied sideways to the roof. The lateral forces try to bend the roof diaphragm. The roof diaphragm acts like a uniformly loaded beam, and resists the lateral force, and in doing so it transfers the forces into the shear walls. It has a horizontal reaction at each end that is transferred to the walls. If the lateral force were to push on the narrow wall shown in Figure 13-23, it would try to bend the roof diaphragm in the other direction, as shown in Figure 13-24. The loads are easier to resist in this direction because the beam formed by the roof diaphragm is deeper and the walls that the load is transferred to are longer.

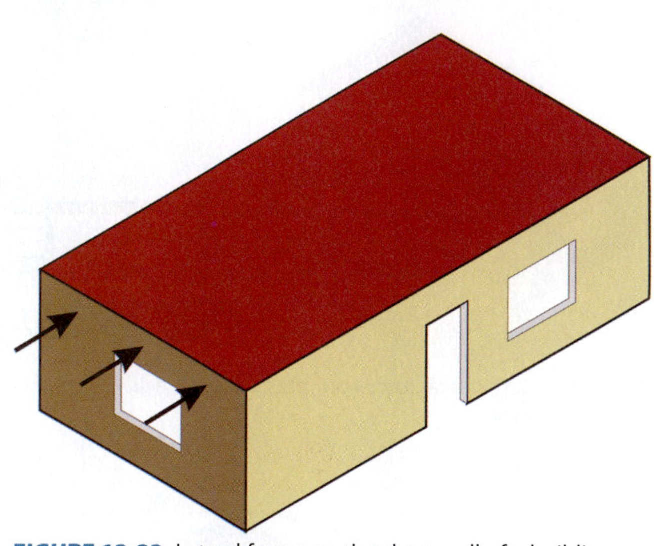

FIGURE 13-23 Lateral forces on the short wall of a building are generally easier to resist because they are transferred into the long walls.

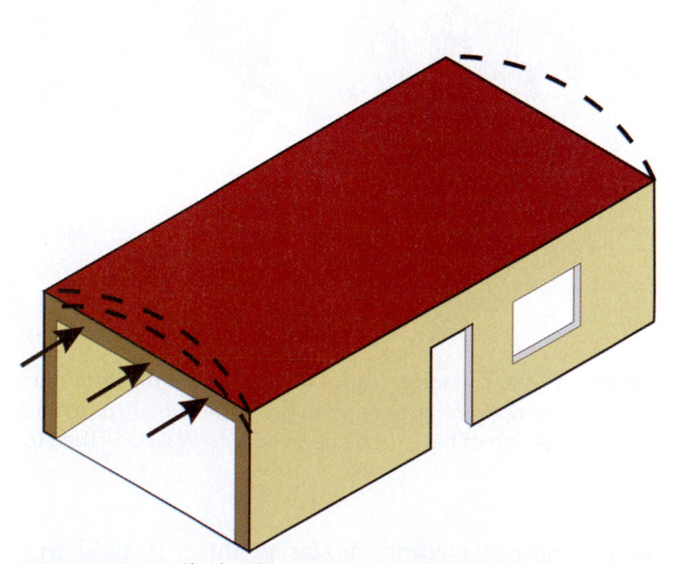

FIGURE 13-24 The loads are easier to resist in this direction because the beam formed by the roof diaphragm is deeper than that shown in Figure 13-22 and the wall that the load is transferred to is longer.

Diaphragms

Floor and roof diaphragms act like a beam to resist the forces transferred to them by the walls. The engineer will calculate the magnitude of this force and specify a system to resist that force. If plywood is used for the sheathing, its ability to resist the forces is dependent on the grade and thickness of the plywood, the size and spacing of the supporting members, whether or not the edges of the plywood are blocked, the pattern the plywood is laid in, and the type, size, and spacing of the nails. Figure 13-25 is a table from the International Building Code (IBC) that gives these values. According to this table, 15/32" Structural 1 plywood laid per case 3 and nailed to 3 × 6 joists with a single row of 10d nails at

- 4" o.c. at the diaphragm boundary
- 4" o.c. at continuous panel edges parallel to the load
- 6" o.c. at all other panel edges

would be capable of resisting 480 pounds per square foot. Figure 13-26 shows common placement patterns for placing plywood over framing members.

Diaphragm values for metal decks are available from their manufacturer. Metal sheathing is either welded or screwed to its supporting member. Frequently, metal decks have concrete poured in them to add rigidity. Concrete floors and roofs make excellent diaphragms.

Chord Forces

As discussed earlier, a floor or roof diaphragm acts like a horizontal beam. When the diaphragm bends as shown by the dotted lines in Figure 13-22, one side of the diaphragm is in compression and the other side is in tension. The plywood and framing members on the compression side normally need no reinforcing, but tension on the far side is trying to tear the diaphragm apart, and the individual pieces of sheathing on the diaphragm need to be held together. The diaphragm is held together by using a chord. A **chord** is a continuous piece of material, or several pieces joined together, that resists the forces of tension. Chords are normally made of steel. Chords may be in the roof or

TABLE 2306.2.1(1)
ALLOWABLE SHEAR (POUNDS PER FOOT) FOR WOOD STRUCTURAL PANEL DIAPHRAGMS WITH FRAMING OF DOUGLAS FIR-LARCH, OR SOUTHERN PINE[a] FOR WIND OR SEISMIC LOADING[h]

PANEL GRADE	COMMON NAIL SIZE OR STAPLE[f] LENGTH AND GAGE	MINIMUM FASTENER PENETRATION IN FRAMING (inches)	MINIMUM NOMINAL PANEL THICKNESS (inch)	MINIMUM NOMINAL WIDTH OF FRAMING MEMBERS AT ADJOINING PANEL EDGES AND BOUNDARIES[g] (inches)	BLOCKED DIAPHRAGMS				UNBLOCKED DIAPHRAGMS	
					Fastener spacing (inches) at diaphragm boundaries (all cases) at continuous panel edges parallel to load (Cases 3, 4), and at all panel edges (Cases 5, 6)[b]				Fasteners spaced 6″ max. at supported edges[b]	
					6	4	$2^1/_2$[c]	2[c]		
					Fastener spacing (inches) at other panel edges (Cases 1, 2, 3 and 4)[b]				Case 1 (No unblocked edges or continuous joints parallel to load)	All other configurations (Cases 2, 3, 4, 5 and 6)
					6	6	4	3		
Structural I grades	8d ($2^1/_2$″ × 0.131)	$1^3/_8$	$^3/_8$	2	270	360	530	600	240	180
				3	300	400	600	675	265	200
	$1^1/_2$ 16 Gage	1		2	175	235	350	400	155	115
				3	200	265	395	450	175	130
	10d[d] (3″ × 0.148″)	$1^1/_2$	$^{15}/_{32}$	2	320	425	640	730	285	215
				3	360	480	720	820	320	240
	$1^1/_2$ 16 Gage	1		2	175	235	350	400	155	120
				3	200	265	395	450	175	130
Sheathing, single floor and other grades covered in DOC PS 1 and PS 2	6d[e] (2″ × 0.113)	$1^1/_4$	$^3/_8$	2	185	250	375	420	165	125
				3	210	280	420	475	185	140
				2	240	320	480	545	215	160
				3	270	360	540	610	240	180
	$1^1/_2$ 16 Gage	1		2	160	210	315	360	140	105
				3	180	235	355	400	160	120
	8d ($2^1/_2$″× 0.131″)	$1^3/_8$	$^7/_{16}$	2	255	340	505	575	230	170
				3	285	380	570	645	255	190
	$1^1/_2$ 16 Gage	1		2	165	225	335	380	150	110
				3	190	250	375	425	165	125
	8d ($2^1/_2$″ × 0.131″)	$1^3/_8$	$^{15}/_{32}$	2	270	360	530	600	240	180
				3	300	400	600	675	265	200
	10d[d] (3″ × 0.148″)	$1^1/_2$		2	290	385	575	655	255	190
				3	325	430	650	735	290	215
	$1^1/_2$ 16 Gage	1		2	160	210	315	360	140	105
				3	180	235	355	405	160	120
	10d[d] (3″ × 0.148″)	$1^1/_2$	$^{19}/_{32}$	2	320	425	640	730	285	215
				3	360	480	720	820	320	240
	$1^3/_4$ 16 Gage	1		2	175	235	350	400	155	115
				3	200	265	395	450	175	130

FIGURE 13-25 Allowable stress for wood floor and roof diaphragms. *Courtesy 2009 International Building Code, copyright © 2009. Washington, DC. International Code Council. Reproduced with permission. All rights reserved. www.iccsafe.org.*

TABLE 2306.2.1(1)—continued
ALLOWABLE SHEAR (POUNDS PER FOOT) FOR WOOD STRUCTURAL PANEL DIAPHRAGMS WITH FRAMING OF DOUGLAS FIR-LARCH, OR SOUTHERN PINE[a] FOR WIND OR SEISMIC LOADING[h]

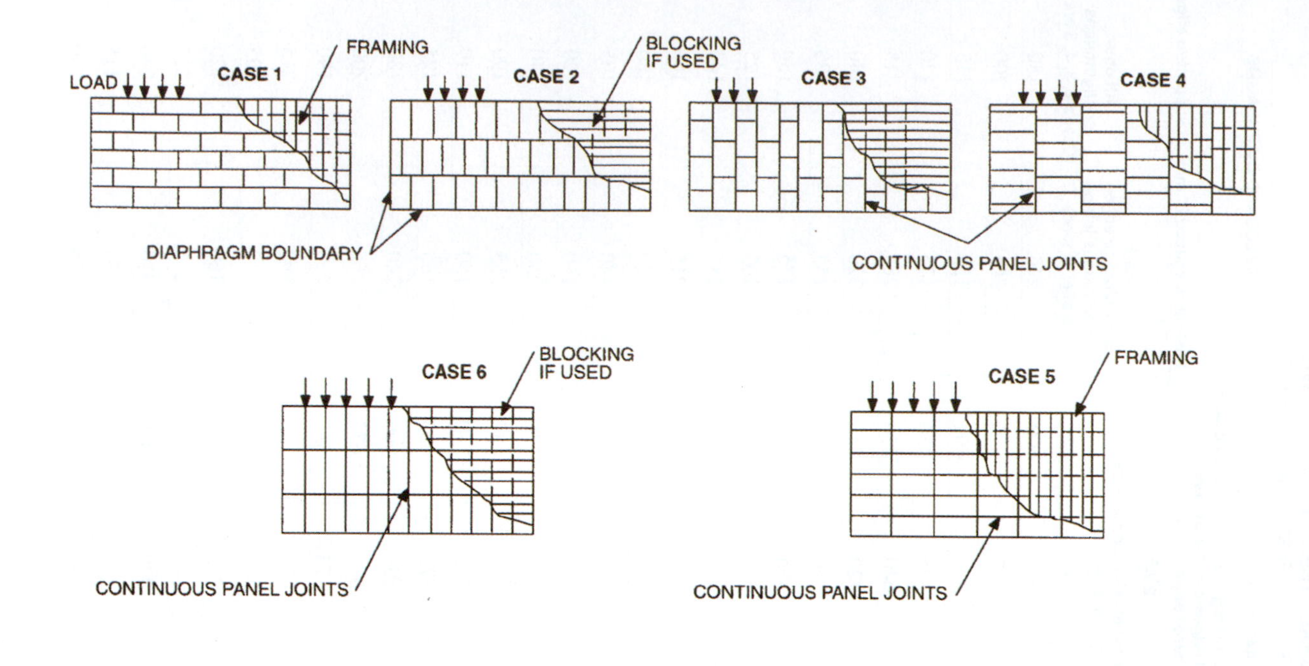

For SI: 1 inch = 25.4 mm, 1 pound per foot = 14.5939 N/m.

a. For framing of other species: (1) Find specific gravity for species of lumber in AF&PA NDS. (2) For staples find shear value from table above for Structural I panels (regardless of actual grade) and multiply value by 0.82 for species with specific gravity of 0.42 or greater, or 0.65 for all other species. (3) For nails find shear value from table above for nail size for actual grade and multiply value by the following adjustment factor: Specific Gravity Adjustment Factor = [1-(0.5 - SG)], where SG = Specific Gravity of the framing lumber. This adjustment factor shall not be greater than 1.

b. Space fasteners maximum 12 inches o.c. along intermediate framing members (6 inches o.c. where supports are spaced 48 inches o.c.).

c. Framing at adjoining panel edges shall be 3 inches nominal or wider, and nails at all panel edges shall be staggered where panel edge nailing is specified at $2\frac{1}{2}$ inches o.c. or less.

d. Framing at adjoining panel edges shall be 3 inches nominal or wider, and nails at all panel edges shall be staggered where both of the following conditions are met: (1) 10d nails having penetration into framing of more than $1\frac{1}{2}$ inches and (2) panel edge nailing is specified at 3 inches o.c. or less.

e. 8d is recommended minimum for roofs due to negative pressures of high winds.

f. Staples shall have a minimum crown width of $\frac{7}{16}$ inch and shall be installed with their crowns parallel to the long dimension of the framing members.

g. The minimum nominal width of framing members not located at boundaries or adjoining panel edges shall be 2 inches.

h. For shear loads of normal or permanent load duration as defined by the AF&PA NDS, the values in the table above shall be multiplied by 0.63 or 0.56, respectively.

FIGURE 13-26 Sheathing and framing configurations. *Courtesy 2009* International Building Code, *copyright © 2009. Washington, DC. International Code Council. Reproduced with permission. All rights reserved. www.iccsafe.org.*

in the wall that the roof attaches to. Figure 13-27 shows a chord that is made of reinforcing steel cast into a tilt-up wall panel. The walls are stood up and a piece of angle iron is welded to the chord bars to join them together so that they act as one continuous piece. In a concrete block structure, the bond beam at the top of the wall shown in Figure 13-28 acts as the chord. The reinforcing steel in the upper bond beam is lapped in the wall so that it acts as one continuous piece. Figure 13-29 shows a block wall with two #5 bars acting as a chord where the floor ledger attaches to the wall. This figure also shows the edge nailing of the plywood that is used to transfer the diaphragm loads into the ledger and the bolting required to transfer the diaphragm loads from the ledger to the wall.

Shear Walls

Shear walls may be made of many different materials. Common materials are concrete, concrete block, brick, and plywood placed over a wood or lightweight steel frame. Shear walls are tied to the floor or roof diaphragm above. They transfer the loads from this diaphragm to the foundation. In a multistory building, shear walls transfer loads to shear walls below and then into the

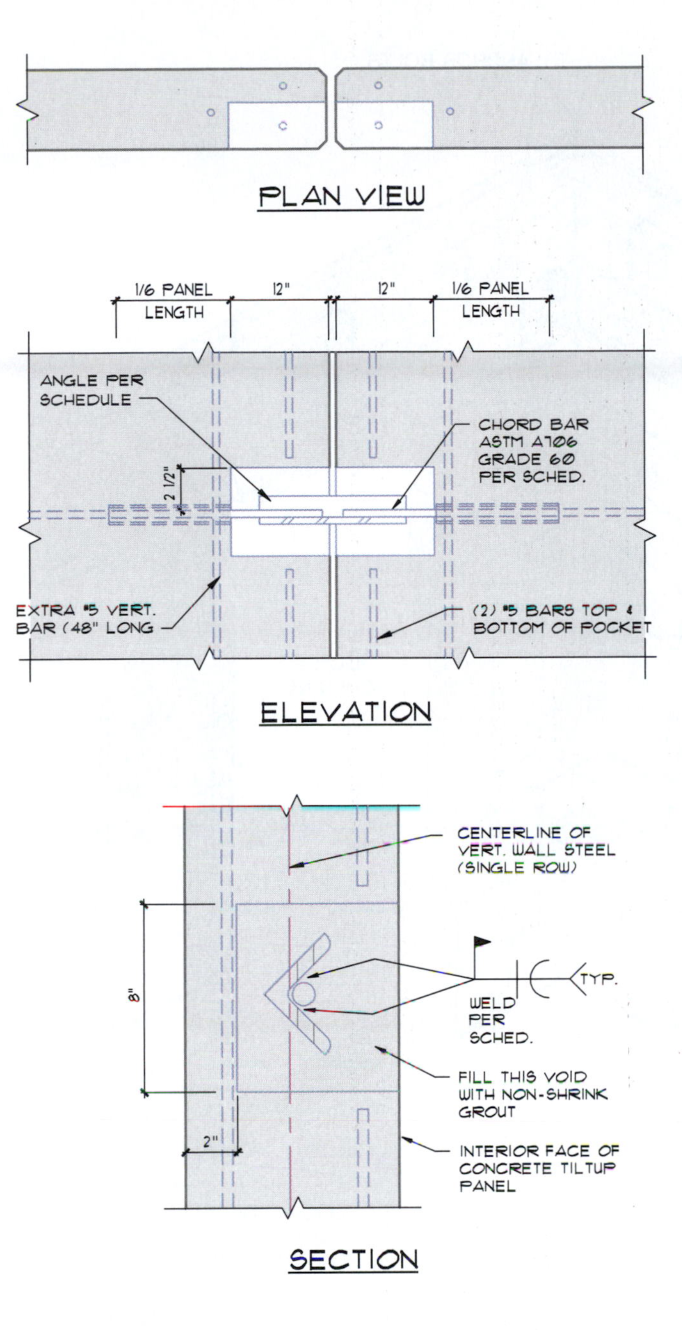

MARK	CHORD BAR	ANGLE	WELD LENGTH
1	#6	∠ 2 1/2 x 2 1/2 x 1/4 x 1'-0"	4"
1A	#7	∠ 2 1/2 x 2 1/2 x 1/4 x 1'-2"	4 1/2"
1B	#8	∠ 2 1/2 x 2 1/2 x 3/8 x 1'-3"	5"
1C	#9	∠ 2 1/2 x 2 1/2 x 3/8 x 1'-3"	5 1/2"
1D	#10	∠ 3 x 3 x 1/2 x 1'-6"	6 1/2"

FIGURE 13-27 A chord in a tilt-up wall. A chord keeps the edges of diaphragms from tearing apart when the diaphragm tries to bend. *Courtesy Dean K. Smith, Kenneth D. Smith Architect & Associates, Inc.*

foundation. It is important that the wall be adequately tied to the diaphragm above and the footing below. Lateral loads on the shear wall try to lift one end of the wall and footing, while pushing down on the other end of the wall and footing, as shown in Figure 13-20. The footing must be sized to support the lateral loads from the shear wall as well as the normal vertical loads. Frequently, these walls will be extended to the floor sheathing or roof sheathing, as shown in Figure 13-30. This allows both the plywood floor diaphragm and the plywood shear panel to be nailed to the top plate to transfer lateral loads from the diaphragm to the shear wall.

Concrete

Tilt-up concrete walls often have additional steel added at each end. A void is left at the bottom of the panel that is also open to the inside face of the panel. A steel bar is left exposed in this pocket. Reinforcing steel is placed in the footing that projects above the top of the footing and up into the void in the panel. The steel in the footing must be placed very precisely so that it closely aligns with the steel in the panel when the panel is tilted up and put in the proper location. A piece of angle iron is placed in the void and welded to the reinforcing steel projecting from the footing and to the reinforcing steel that was left exposed in the panel, as shown in Figure 13-31. By connecting the reinforcing steel in the panel to the reinforcing steel in the footing, the lateral or overturning forces in the panel will be transferred to the footing.

Concrete Block

Concrete block shear walls have vertical steel placed closer together than a typical CMU wall and all of the cells may be required to be solid grouted. All the vertical steel must be connected to steel extending from the footing. The steel from the footing laps the steel in the wall and the concrete grout transfers the loads from one bar to the other.

Wood Shear Walls

Plywood is nailed to one or both sides of the studs that form the shear wall. If the plywood is nailed to both sides of a wall, the studs need to be thicker than the normal 2" (50 mm) nominal size so that the nails penetrating from both sides do not split the stud. Figure 13-32 shows values for particleboard shear walls for different thickness and nailing conditions. The code has similar tables for plywood and plaster.

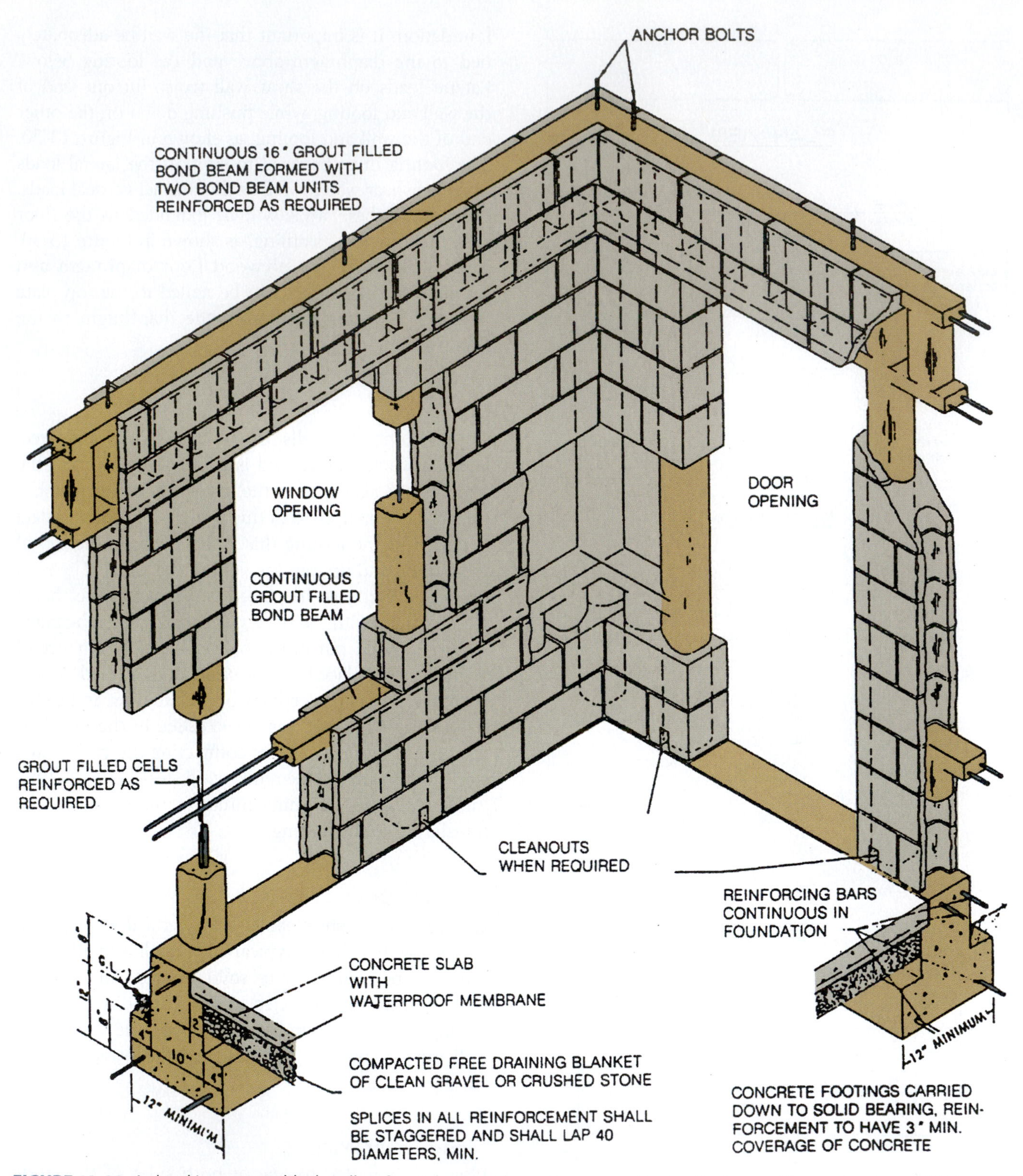

FIGURE 13-28 A chord in concrete block walls is formed using rebar set in mortar. *Courtesy the Concrete Masonry Association of California and Nevada.*

A typical hold-down detail that might be used at each end of a shear wall is shown in Figure 13-14. A manufactured bracket such as a Simpson Strong-Tie hold-down anchor with tested values is attached to the stud and into the foundation. The bolt is cast into the footing as the concrete is poured, and then the bracket is attached to the post as the framing is completed.

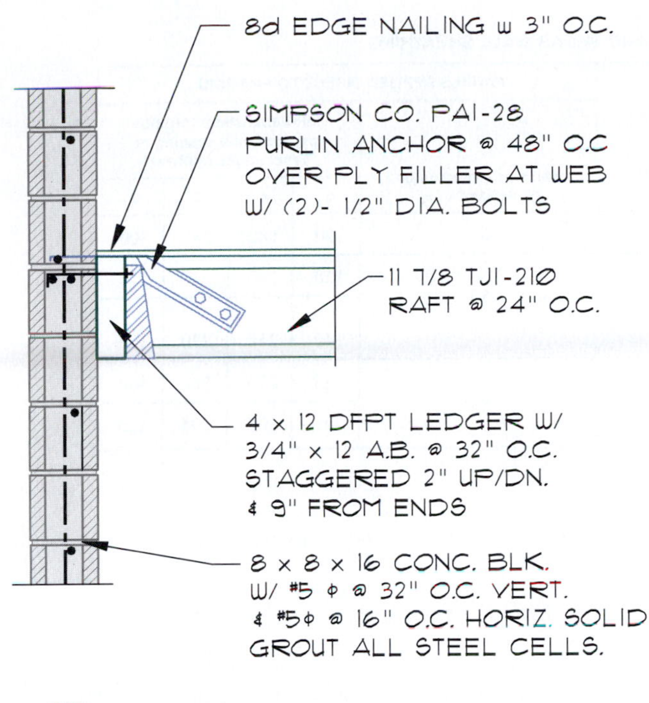

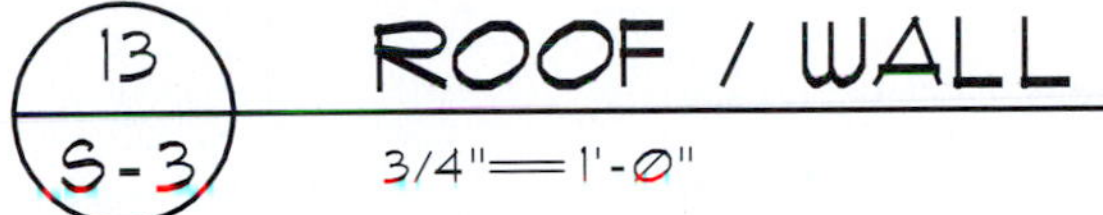

FIGURE 13-29 Diaphragm transfer to block wall. *Courtesy Gisela Smith, Kenneth D. Smith Architect & Associates, Inc.*

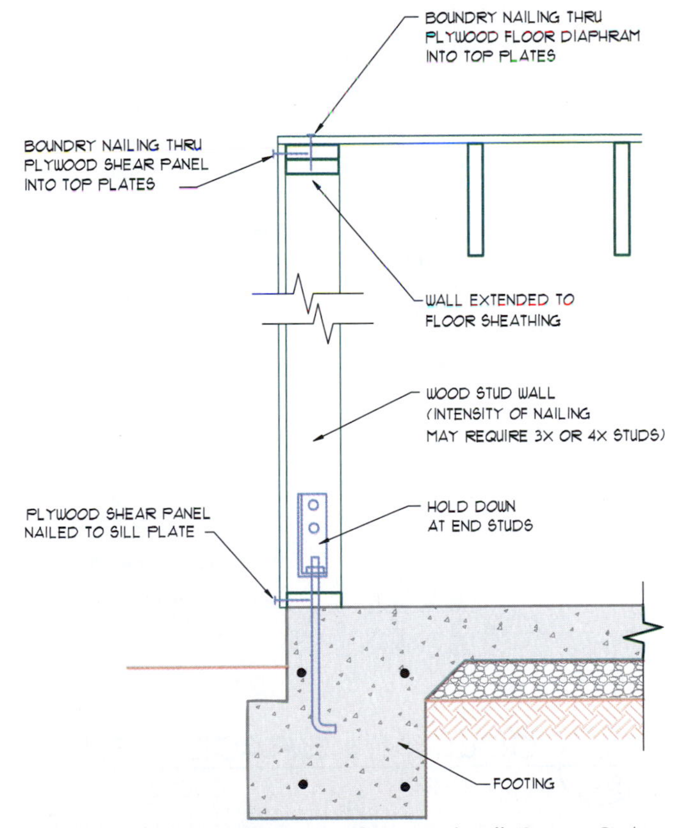

FIGURE 13-30 Diaphragm transfer to stud wall. *Courtesy Gisela Smith, Kenneth D. Smith Architect & Associates, Inc.*

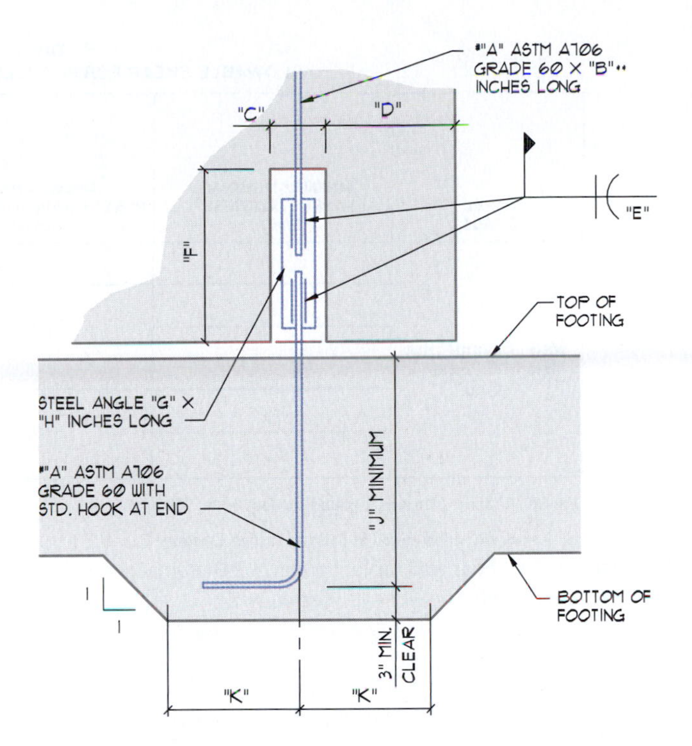

HOLD-DOWN SCHEDULE

MARK	"A"	"B"**	"C"	"D"*	"E"	"F"	"G"	"H"	"J"	"K"
8A	6	60	5	6	4	16	2-1/2" X 2-1/2" X 1/4"	12	15	18
8B	7	72	5	6	4	18	2-1/2" X 2-1/2" X 1/4"	14	20	20
8C	8	84	5	6	5-1/2	19	2-1/2" X 2-1/2" X 3/8"	15	26	22
8D	9	96	5	6	6	19	2-1/2" X 2-1/2" X 3/8"	15	32	25
8E	10	108	6	6	6-1/2	22	3" X 3" X 1/2"	18	38	28
8F										
8G										

* NOTE: IN THE CASE THAT AN ABUTTING CONC. PANEL THICKER THAN 7" WOULD HINDER ACCESS TO THE POCKET "D" SHALL BE ADJUSTED TO EQUAL THE WIDTH OF THE ABUTTING PANEL SO AS TO AVOID CONFLICT.

** NOTE: SEE PANEL ELEVATION DRAWING WHERE A LONGER "B" MAY BE REQUIRED.

FIGURE 13-31 Hold-down to foundation detail, tilt-up. *Courtesy Dean K. Smith, Kenneth D. Smith Architect & Associates, Inc.*

Braced Frames

Braced frames made of steel are frequently used instead of shear walls as forces increase. They consist of vertical, horizontal, and diagonal members forming triangles. Figure 13-33 shows two examples of braced frames. Figure 13-34 shows typical braced frame details.

TABLE 2306.5
ALLOWABLE SHEAR FOR PARTICLEBOARD SHEAR WALL SHEATHING

PANEL GRADE	MINIMUM NOMINAL PANEL THICKNESS (inch)	MINIMUM NAIL PENETRATION IN FRAMING (inches)	PANELS APPLIED DIRECT TO FRAMING: Nail size (common or galvanized box)	Allowable shear (pounds per foot) nail spacing at panel edges (inches)[a]: 6	4	3	2
M-S "Exterior Glue" and M-2 "Exterior Glue"	3/8	1 1/2	6d	120	180	230	300
	3/8	1 1/2	8d	130	190	240	315
	1/2			140	210	270	350
	1/2	1 5/8	10d	185	275	360	460
	5/8			200	305	395	520

For SI: 1 inch = 25.4 mm, 1 pound per foot = 14.5939 N/m.

a. Values are not permitted in Seismic Design Category D, E or F.

FIGURE 13-32 Shear wall table. *Courtesy 2009* International Building Code, *copyright © 2009. Washington, DC. International Code Council. Reproduced with permission. All rights reserved. www.iccsafe.org.*

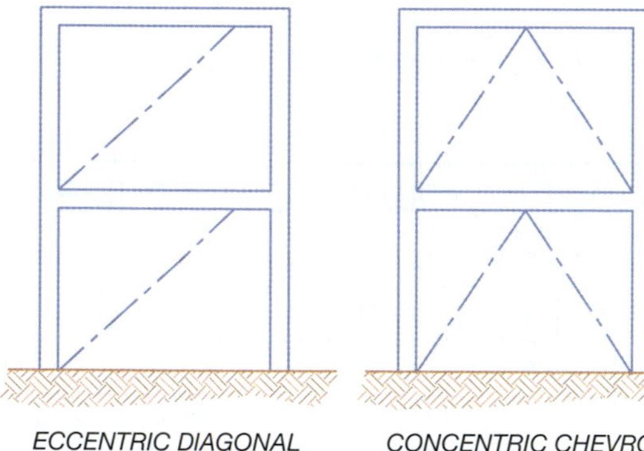

FIGURE 13-33 Braced frame construction features either eccentric or concentric bracing.

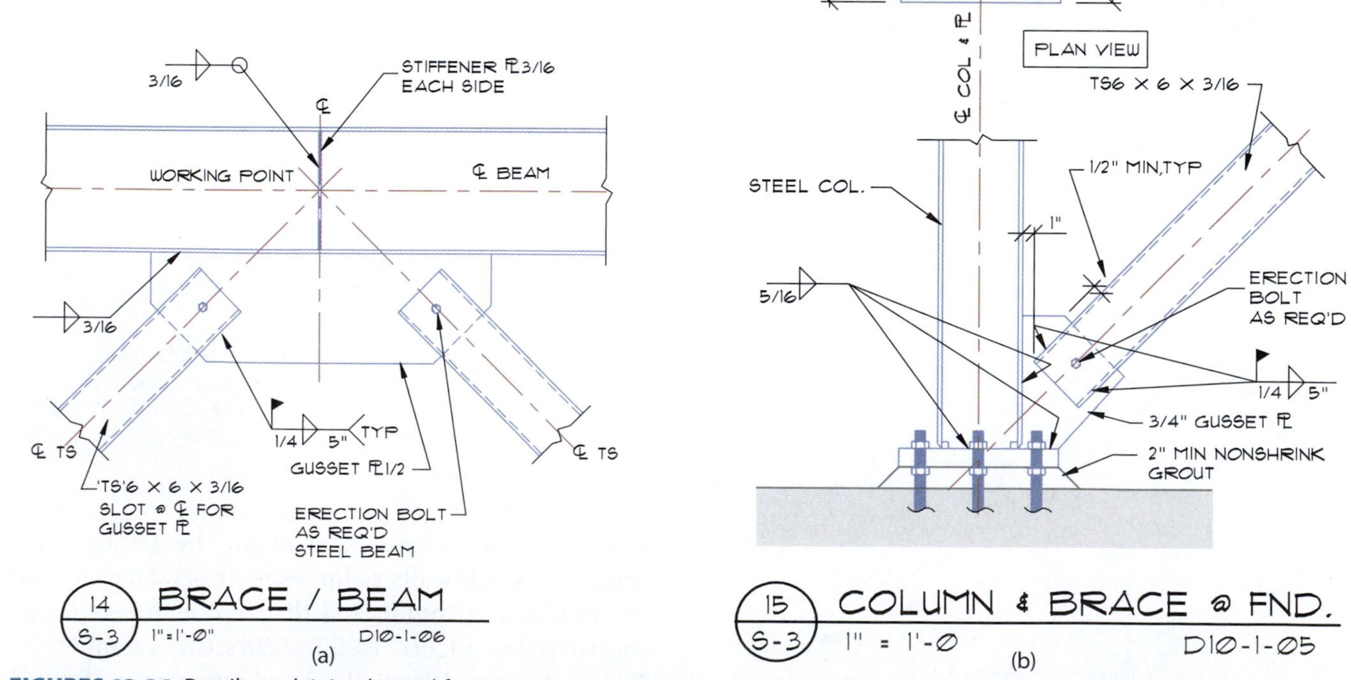

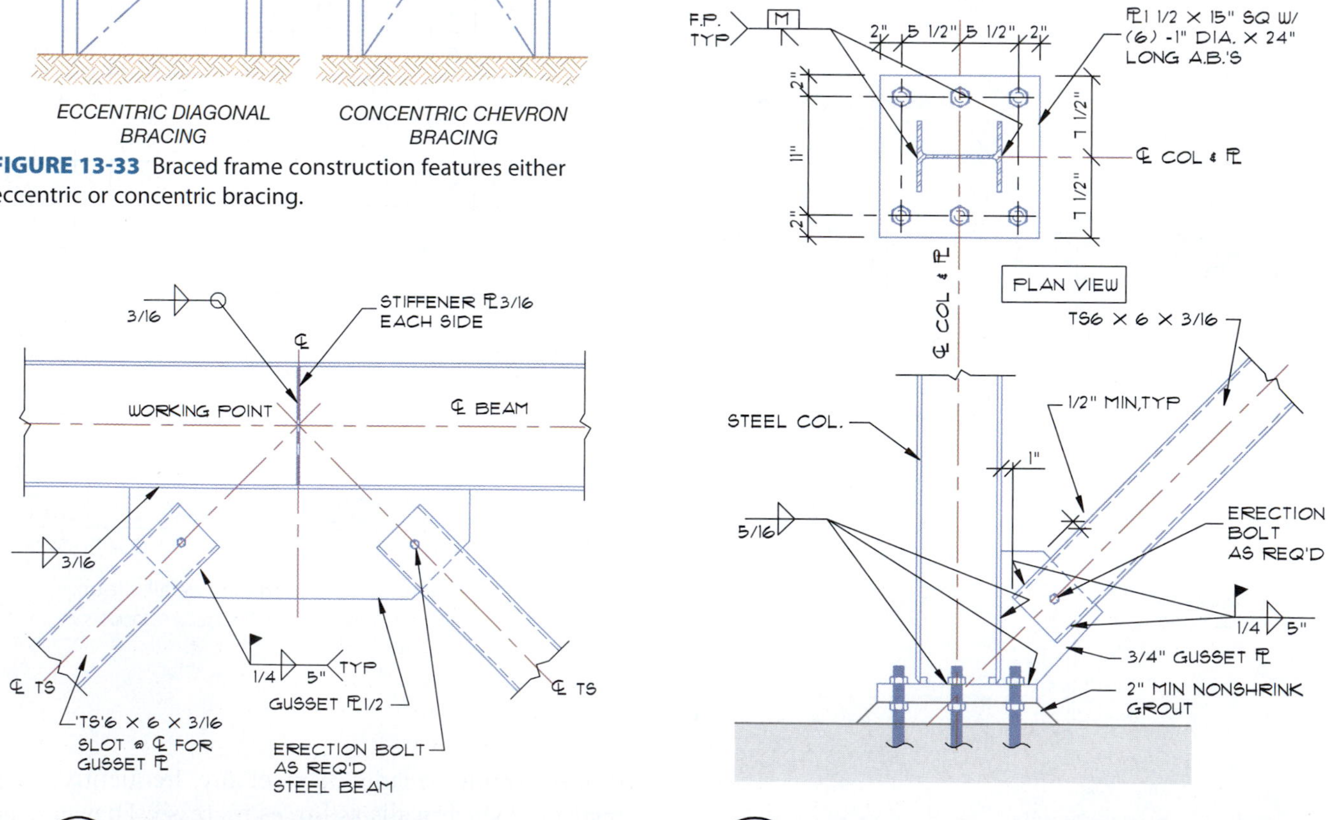

FIGURES 13-34 Details explaining braced frame connections must be drawn including (a) brace-to-beam, and (b) brace-to-column connections. *Courtesy Dean K. Smith, Kenneth D. Smith Architect & Associates, Inc.*

Moment Frames

Moment frames are made of steel members and depend on a very rigid connection of the horizontal members to the vertical members. Figure 13-35 illustrates a connection that a CAD technician would be required to draw to represent a bolted connection. A similar connection can be accomplished with rivets or by welding. When connected, the members all act as one. Lateral forces induce bending in the horizontal and vertical members, as illustrated in Figure 13-36.

FIGURE 13-36 Vertical and lateral forces acting on a moment frame.

Drags

When buildings have irregular shapes, as shown in Figure 13-37, there is frequently a need to transfer or drag the forces from a floor or roof diaphragm into the shear wall. This can be done through a beam or joist. A **drag** is a beam that is attached to the edge of a diaphragm to transfer lateral loads from the diaphragm into a shear wall or braced frame. The beam must then be fastened to the shear wall in a secure manner. Figure 13-38 is a typical detail showing the drag and the attachment to the shear wall. The bracket has steel rebar welded to it and extending back into the concrete wall to transfer the loads. Bolts through the bracket and the beam transfer the load to the beam. The roof diaphragm is securely fastened to the beam with nails.

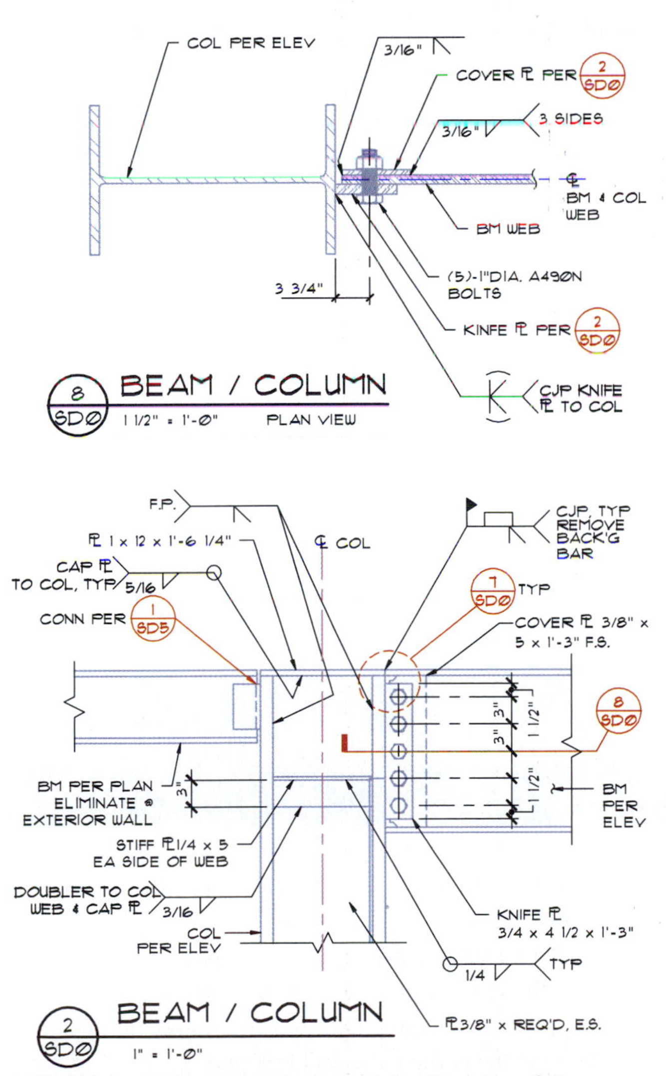

FIGURE 13-35 Moment connection details are carefully designed to be ductile. If the connection fails under extreme conditions, it is designed to give way gradually, not suddenly.
Courtesy Ginger M. Smith, Kenneth D. Smith Architect & Associates, Inc.

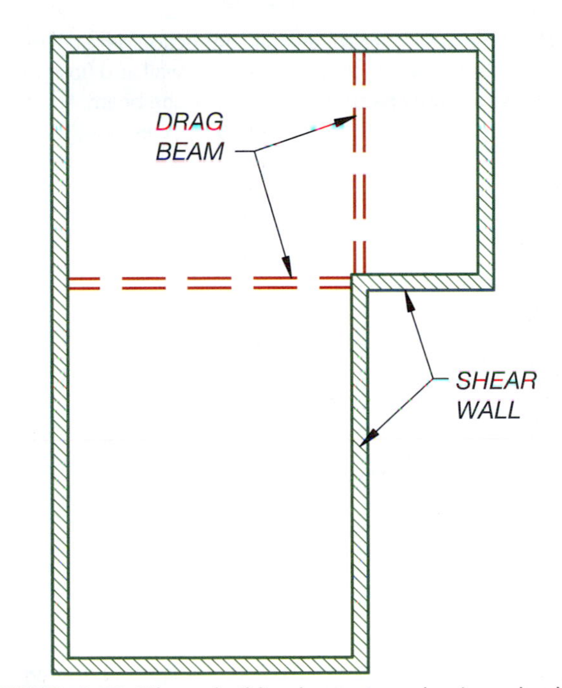

FIGURE 13-37 When a building has an irregular shape, loads may need to be transferred into shear walls by use of a drag beam.

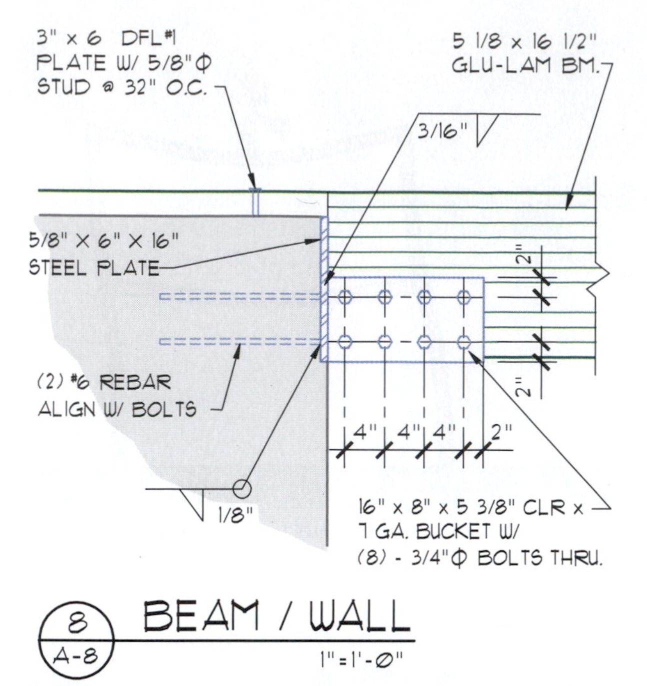

FIGURE 13-38 A drag to shear wall connection detail. The beam must be firmly attached to the shear wall and floor or roof diaphragm must be firmly attached to the beam. *Courtesy Ginger M. Smith, Kenneth D. Smith Architect & Associates, Inc.*

ADDITIONAL READING

Two additional resources include *The IBC Handbook-Structural provisions* and *Soils, Earthwork, and Foundations: A Practical Approach*. Both can be obtained at www.iccsafe.org. In addition, the following Web sites can be used as a resource to help obtain additional information about structural issues.

ADDRESS	COMPANY/ ORGANIZATION
www.asdipsoft.com	Advanced Design International Programs
www.acec.org/programs/case.htm	Council of American Structural Engineers
www.scitation.aip.org	*Journal of Structural Engineering*
www.gostructural.com	SDS/2 Design Data
www.strucalc.com	StruCalc Design Software
www.seaint.prg	The Structural Engineers Association—International

KEY TERMS

Braced frames
Chord
Drag
Equivalent fluid pressure
Friction coefficient
Lateral forces
Load's path
Loads, dead
Loads, dynamic
Loads, horizontal environmental
Loads, live
Loads, static
Moment frames
Retaining walls
Roof diaphragm
Shear wall

CHAPTER 13 TEST Structural Considerations Affecting Design

QUESTIONS

Answer the following questions with short complete statements. Type the chapter title, question number, and a short complete statement for each question using a word processor. If math is required to answer a question, show your work.

Question 13-1 What type of load does a desk produce on a floor?

Question 13-2 What is the wind speed in Detroit?

Question 13-3 What live load is produced by the occupants of a restaurant?

Question 13-4 What term is used to define a load placed on top of the soil and transferred into a retaining wall?

Question 13-5 How much weight is transferred to the top of the footing at the right wall in Figure 13-9?

Question 13-6 What is the actual psf transferred to the soil for the left wall of the structure examined in this chapter?

Question 13-7 What is the significance of triangles in braced frames?

Question 13-8 What causes seismic forces? In what direction do they try to move a structure?

Question 13-9 What does a hold-down do?

Question 13-10 How can 640 pounds per foot be resisted if 15/32" Structural 1 plywood is used for a roof diaphragm?

Question 13-11 What is the difference between a case 1 and a case 5 roof diaphragm?

Question 13-12 Why are weep holes provided at the bottom of retaining walls?

Question 13-13 What is a drag?

Question 13-14 What is the importance of a chord?

Question 13-15 What is a moment frame?

Project Manuals and Written Specifications

14

Construction documents consist of the data needed to build a structure and are used to clearly convey the intentions of the owner, the architectural team, and each consultant involved with the project. Written specifications provide a method of supplementing the working drawings regarding the quality of materials to be used and labor to be supplied. The drafter needs to remember that both the drawings and the written specifications are legal documents that must be very carefully researched and prepared. Both may someday be used in a court of law as a standard for quality for a specific structure. To help you gain an understanding of the written documents that support the drawings, this chapter will explore:

- The goal of specifications.
- Specification placement.
- Using the Construction Specification Institute (CSI) *MasterFormat.*
- Types of specifications.
- Writing specifications.

EXPLORING WRITTEN SPECIFICATIONS

The construction drawings and written specifications must be compatible so that the project can be accurately bid and built. They must be clear in order to avoid misinterpretation. The drawings visually define the relationships between materials, products, and systems within the structure by showing the location and size of each element. Drawings should show the following:

- Location of materials, fixtures, and equipment.
- Sizes, quantities, and the physical relationship between elements.
- Dimensions to pinpoint the size and location of each material and component.
- The size, type, finish, and hardware associated with rooms, windows, and doors.

Written specifications express the requirements in words. Specifications describe the following:

- The quality and type of material.
- Required gauge, size, or capacity of material and equipment.
- The quality of workmanship.
- Methods of fabrication.
- Methods of installation.
- Test and code requirements.

Specifications provide information regarding the quality of materials and workmanship, methods of installation, the desired performance to be achieved at completion, and how performance is to be measured. The drafter must understand the role of each type of specification to fully understand notations that are placed on the drawings. This includes knowledge of placement, formats, and types of specifications.

SPECIFICATION PLACEMENT

The specifications for a project are either found in a **project manual** or placed on one or more sheets of the construction drawing. The American Institute of Architects (AIA) recommends that the documentation for every project contain a project manual. In order to reduce cost, many owners request the architect to substitute less detailed specifications into the drawings than are found in a project manual. Projects that are funded by federal, state, or municipal money are required to have a project manual.

Specifications within the Construction Drawings

Many owners of small, privately funded projects opt for specifications that are less specific than those found in a project manual. These specs are placed on one or more sheets of the construction drawings. Many engineering firms place the written specifications for structural material within the drawings to meet building department requirements and to establish minimum standards for construction materials. Examples of the drawing specifications that are part of the steel structure used throughout this text are shown in Figure 14-1. See the student CD in the RESOURCE MATERIAL/CHAPTER 14 file for the complete specification.

Project Manual

The term *specs* has often been used by contractors and their subcontractors to refer to the written specifications that are used to supplement the *plans*. The terms *project manual* and *drawings* have become more representative of the contents of each aspect of the construction documents. The *project manual* contains many documents related to a specific construction project in addition to the written specifications. The manual is produced by the combined efforts of the owner, the architect, and a representative for each of the consulting engineers involved with the project. Several portions are developed under the supervision of attorneys. A project manual often contains the following separate

GENERAL STRUCTURAL NOTES

CODE REQUIREMENTS:

Conform to the IBC 2006 edition, as amended by the state of Oregon.

TEMPORARY CONDITIONS:

The contractor shall be responsible for structural stability during construction. The structure shown on the drawings has been designed for stability under the final configuration only.

EXISTING CONDITIONS:

All existing conditions, dimensions, and elevations are to be field verified. The contractor shall notify the architect of any significant discrepancies from that shown on the drawing.

DESIGN CRITERIA:

Design was based on the strength and deflection criteria of the 2006 IBC. In addition to the dead loads, the following loads and allowables were used for design, with live loads reduced per IBC.

Roof: - 25 PSF L.L.(snow drift as shown on plans.

Floors: Retail - 75 PSF L.L.

Allowable soil bearing pressure: 2000, 2500 PSF (per soils report) Retaining walls: 40 PCF (equivalent fluid pressure)

Wind: 80 MPH - Exposure B

Earthquake design is based on the following:

Z = 0.30, I = 1.0, C = 2.75, Rw = 6

Design and detailing based on criteria for SEISMIC ZONE C.

SUBMITTALS:

Shop drawings shall be submitted to the architect prior to fabrication and construction for all structural items including the following:

- Concrete mix designs, concrete and masonry reinforcement, embedded steel items, structural steel, glued laminated members, and prefabricated wood joist.
- If the shop drawings differ from or add to the design of the structural drawings, they shall bear the seal and signature of a structural engineer, registered in the state of Oregon. Any changes to the structural drawings are the subject of review and acceptance of the architect.
- Design drawings, shop drawings, and calculations for the design and fabrication of items that are designed by others including: Prefabricated wood joist, skylights, window wall and all other glazing systems, and seismic restraints for mechanical, plumbing, electrical, equipment, machinery, and associated piping shall bear the seal and signature of a structural engineer, registered in the state of Oregon and shall be submitted to the architect prior to fabrication.
- Calculations are to be included for all connections to the structure, considering localized effects. Design is to be based on the requirements of the 2000 IBC for the following:
  - Earthquake zone C
  - Wind zone 80 MPH,
  - Exposure B
- Field engineered details, developed by the contractor, that differ from or add to the structural drawings shall bear the seal and signature of a structural engineer registered in the state of Oregon and shall be submitted to the architect prior to construction.

FIGURE 14-1 Structural specifications governing major components of the skeleton of a structure may be placed in the project manual or on the structural drawings. *Courtesy KPFF Consulting Engineers.*

documents that cover a specific portion of the construction process:

- Invitation to bid.
- Instructions to bidders.
- Bid form.
- Bond form.
- Form of agreement.
- General conditions of the contract for construction.
- Supplementary conditions.
- Specifications.
- Addenda.

Each document is a legal contract in itself, which as a whole constitutes the project manual. Whenever possible, AIA forms for each document should be used because they are well known to the construction industry. These documents have been tested and interpreted in the courts and are widely understood by the construction industry.

Invitation to Bid

An **invitation to bid** is a summary of the bidding and construction procedures for a project. It is usually about one page in length and is used to advise potential bidders about the existence of a project. If a contractor has already been secured for the project prior to bidding, an invitation is not a part of the project manual. On privately funded projects, based on the request of the owner, the architect may mail the invitation to a few selected contractors known to have a proven track record of success with this type of project. On publicly funded projects, laws require the invitation to be published in a prescribed form in a newspaper.

Instruction to Bidders

The purpose of the **instruction to bidders** is to tell each bidder the format to be used for the reviewing of bids. The AIA standard form provides information about the bidding process and procedure including the following:

- Definitions.
- Bidder's representations.
- Bidding documents.
- Bidding procedures.
- Consideration of bids.
- Post-bid information.
- Performance bond and payment bond.
- Form of agreement between owner and contractor.

Bid Forms

The purpose of a **bid form** is to provide a uniform presentation of the cost associated with the construction project. The bid form provides blank lines where the bidder is to insert the price for specific portions of the project. Using a uniform bid form limits bidders in the use of *exclusions* and *substitutions*, which often make it unclear what costs the bid actually covers.

Bond Forms

A **bond** is a legal document, which assures that the contractor will provide the goods or services represented in the contract agreement. Three types of bonds used in the construction industry include a bid, performance, and a labor and materials payment bond. A *bid bond* provides assurance that a contractor will sign the contract if the firm is awarded the bid. A *performance bond* is posted by the winning bidder as a guarantee that the firm will complete the job and not become defunct during the process. The *labor and material payment bond* is posted by the winning bidder as a guarantee that all bills for material and labor used for this project will be paid by the contractor.

Form of Agreement

The **Form of Agreement** is a legal contract that includes the who, what, how much, and when of the project. These aspects of the project are usually represented in the following five elements of the agreement:

- Identification of the parties involved.
- Statement of the work to be performed.
- Statement of the consideration.
- Time of performance.
- Signatures of the parties involved.

General Conditions of the Contract for Construction

This document specifies the relationship of each party who signed the contract and how the contract will be administered. The AIA format of general conditions includes the following 14 categories overseeing the contract:

- General provisions.
- Owner.
- Contractor.
- Administration of the contract.
- Subcontractors.
- Construction by owner or by separate contractors.
- Changes in work.
- Time.
- Payments and completion.
- Protection of persons and property.
- Insurance and bonds.

- Uncovering and correction of work.
- Miscellaneous provisions.
- Termination or suspension of the contract.

Supplementary Conditions

The general conditions are used to address elements that are common to all construction. The supplementary or special conditions are used to address the elements that are found only within a specific project. The **supplementary conditions** are written to address portions of the project that are special to that specific project that do not fit into the preceding section of the contract.

Specifications

Although only one of many documents, the *written specifications* constitutes the bulk of a project manual; it is this portion of the project manual that affects the drafter. Depending on the size of the project and the firm, an experienced drafter may be required to alter or update specifications. Every drafter involved with the project needs to be aware of the contents of the written specifications so that the drawings and the specs can be coordinated. The following sections of this chapter address this aspect of the project manual.

Addenda

Although not always used, **addenda** can be used to amend the contract during the bidding process but prior to the awarding of the contract. Changes issued after the contract has been awarded are referred to as *change notices* or *change orders* and are not considered in this chapter.

CSI *MASTERFORMAT*

Several formats have been used to write specifications during the last 50 years. An alphabetical listing of materials, listings based on the order of use, and hundreds of personal styles were common prior to the development of the standard for today's construction industry. First published in 1963, most construction specifications are now written according to the 2004 ***MasterFormat* system** published by the Construction Specification Institute (CSI) in the United States, and by Construction Specifications Canada (CSC) in Canada. This system consists of five groups, five major subgroups, and fifty divisions. The two major groups include *Procurement and Contract Requirements (00)* and the *Specifications* groups. The five major subgroups and their divisions include: General requirements (Division 01), Facility Construction (Divisions 02-19), Facility Services (Divisions 20-29), Site and Infrastructure (Divisions 30-39), and Process Equipment (Divisions 40-49). These listings can be used for production, distribution, filing, and retrieval of construction documents. Each major division of the *MasterFormat* is related to a major grouping of the construction process. The major listings and their divisions for *MasterFormat* include the following:

Procurement and Contracting Requirements Group

Division 00	Procurement and contracting requirements

Specifications Group

Division 01	General requirements

Facility Construction Subgroup

Division 02	Existing conditions
Division 03	Concrete
Division 04	Masonry
Division 05	Metals
Division 06	Wood, plastics, and composites
Division 07	Thermal and moisture protection
Division 08	Openings
Division 09	Finishes
Division 10	Specialties
Division 11	Equipment
Division 12	Furnishings
Division 13	Special construction
Division 14	Conveying equipment
Divisions 15 through 19	Reserved for future expansion

Facility Service Subgroup

Division 20	Reserved
Division 21	Fire suppression
Division 22	Plumbing
Division 23	Heating, ventilating, and air-conditioning
Division 24	Reserved
Division 25	Integrated automation
Division 26	Electrical
Division 27	Communications
Division 28	Electronic safety and security
Division 29	Reserved

Site and Infrastructure Subgroup

Division 30	Reserved
Division 31	Earthwork
Division 32	Exterior improvements

Division 33	Utilities
Division 34	Transportation
Division 35	Waterway and marine construction
Division 36 through 39	Reserved for future expansion

Process Equipment Subgroup

Division 40	Process integration
Division 41	Material processing and handling equipment
Division 42	Process heating, cooling, and drying equipment
Division 43	Process gas and liquid handling, purification, and storage equipment
Division 44	Pollution control equipment
Division 45	Industry-specific manufacturing equipment
Division 46	Reserved
Division 47	Reserved
Division 48	Electrical power generation
Division 49	Reserved

Notice that numbers are placed in order, but many unassigned numbers are available for individual projects. These same divisions are used in the *Sweets Catalogs* as well as most other vendor material. Each division is in turn divided into numbered sections represented by six-digit numbers that are explained in Table 14-1. The numbers consist of three pairs of numbers, with each pair defining a level of specificity. The first pair of numbers represents the division number; in this case division 03 is Concrete. The second pair of numbers (11) represents second-level headings, in this case Concrete Forming. The third pair of numbers represents third-level headings, in this case the 13 represents Structural Cast-in-Place Concrete Forming. An optional fourth pair of numbers (.19) can be used if greater specificity is required. The number 03 11 13.19 represents Falsework. Additional pairs of numbers and letters can be added for user-defined numbers. Figure 14-2 shows the listings for Section 03 11 00. See the CSI *MasterFormat* Web site for a complete listing of each division. See the student CD in the RESOURCE MATERIAL/CHAPTER 14 file for the complete specification.

Parts of a Division

In addition to the five subgroups, and the fifty divisions, specifications can be further divided into three subsections that deal with general, materials, and execution.

Part One

1.02 Part One, *General* deals with the scope of a section by describing work, related definitions, quality control, submittals, shop drawings, guarantees, and warranties. Common components of the

DIVISION 03—CONCRETE	
03 00 00	**Concrete**
03 01 00	Maintenance of Concrete
03 05 00	Common Work Results for Concrete
03 06 00	Schedules for Concrete
03 08 00	Commissioning Concrete
03 10 00	**Concrete Forming and Accessories**
03 11 00	Concrete Forming
03 11 13	Structural Cast-In- Place Concrete Forming
03 11 13.13	Concrete Slip Forming
03 11 13.16	Concrete Shoring
03 11 13.19	Falsework
03 11 23	Permanent Stair Forming
03 15 00	Concrete Accessories

FIGURE 14-2 Division and subdivision listing for Division 3 of the CSI *MasterFormat*. Note that Section 03 11 has an expanded listing to show the third-level heading. See the CSI Web site for a complete listing of all headings and their subheads.

Level	Sample CSI Number	Meaning
1	**03** 11 13	Division number representing major construction division (e.g., Concrete)
2	03 **11** 13	Second-level heading representing major subsections (e.g., Concrete Forming)
3	03 11 **13**	Third-level heading representing minor sub-subsections (e.g., Structural Cast-in-Place Concrete Forming)
4	03 11 13.**19**	Optional modifier (e.g., Falsework)
L5	03 11 13.19.**12ABC**	User-defined modifiers

TABLE 14-1 CSI Numbering Format

General section include the following: 1.01 Summary References

1.03 Definitions
1.04 System descriptions
1.05 Submittals
1.06 Quality assurance
1.07 Delivery, storage, and handling
1.08 Project and conditions
1.09 Sequencing and scheduling
1.10 Warranty
1.11 Maintenance

Part Two

The Materials section describes materials, products, and equipment that are to be used. Common components of the Materials section include the following:

2.01 Manufacturers
2.02 Materials
2.03 Manufactured units
2.04 Equipment
2.05 Components
2.06 Accessories
2.07 Mixes
2.08 Fabrication
2.09 Source quality control

Part Three

Execution describes the method that is to be used to install materials or products specified in Part Two, and how work is to be performed. Common components of the Execution section include the following:

3.01 Examination
3.02 Preparation
3.03 Erection/Installation/Application
3.04 Field quality control
3.05 Adjusting
3.06 Cleaning
3.07 Demonstration
3.08 Protection
3.09 Schedules

Figure 14-3 shows a partial listing for the three-part specification for Exterior Insulation and Finish System (EIFS) for the structure shown in Chapters 16 and 17. See the student CD in the RESOURCE MATERIAL/ CHAPTER 14 file for the complete specification. Keep in mind that the following are the specifications for only one component of the structure. Each component of the structure is listed in the project manual using the three-part system. Components are placed within the manual according to their CSI number. Notice that at the start of each page of the specifications, the job number, project name, CSI section number, the component name, and the page number of the section are listed.

TYPES OF SPECIFICATIONS

Regardless of where the specifications are contained, six common methods are used to present technical specifications. These methods are cash allowance, descriptive, open, performance, proprietary, and reference specifications. A construction document can contain each type of specification and should not be limited to one specific type.

Cash Allowance Specifications

Cash allowance specifications are used when the information regarding quality or quantity has not been determined. These specifications can be found in a project manual or within the working drawings. The cash allowance temporarily replaces drawings for a specified portion of work and instructs all bidders to plan for a specified amount of money to cover that work. This money is then dispersed under the architect's direction when the work is completed.

This type of specification often handles finish items such as hardware, carpeting, and lighting fixtures. Although exact quantities can be determined from the drawings, the owner may wish to determine quality at a later date in the building process.

Cash allowances are also used because the owner may not know who will occupy the structure. For example, a structure can be built for the intended use of medical professionals. Because there is such diversity in the profession, individual occupants will have specific needs. These needs can best be met by cash allowances. When cash allowances are given, information should include the dollar amount of the allowance, installation methods, and the method of measuring cost to be applied against the allowance. When work covered by the allowance is complete, savings can be credited to the contractor or owner. If allowance costs are exceeded, the contractor is usually entitled to charge additional fees. An example of a cash allowance specification is shown in the example below:

Allow $6500.00 for wall brackets and hanging incandescent fixtures. All other fixtures indicated on drawings to be included in bid.

8712 CCC Wilsonville Section 07240

Page 1

EXTERIOR INSULATION AND FINISH SYSTEM

Part 1—General

1.1 WORK INCLUDED

A. Provide field applied exterior insulation and finish system as indicated on Drawings and specified herein.

1.2 SYSTEM DESCRIPTION

A. Structural Requirements: Deflection of exterior wall substrate shall not exceed L/240 or L/360.

B. Substrate Requirements: Substrates shall be as indicated on the Drawings.

1.3 QUALITY ASSURANCE

A. Manufacturer's Qualifications:

1. Manufacturer shall have successful performance history over at least five (5) years in the local geographic area.
2. Manufacturer shall have established ongoing design and specification assistance to Architects and Engineers.
3. Manufacturer shall have established ongoing contractor applicator training programs for at least five (5) years in local geographic area.
4. Manufacturer shall have established warranty program and a five (5) warranty on this Project.
5. Manufacturer shall have full-scale fire test reports and documentation of IBC compliance.
6. Manufacturer shall have resident Field Technical Service personnel.
7. Manufacturer shall attend job site preapplication conference and provide inspection of work in progress and completed work.
8. Manufacturer shall have local inventory of products.

B. Applicator's Qualifications:

1. Applicator shall be licensed, bonded, and insured.
2. Applicator shall have successful performance history with Exterior Insulation and Finish Systems over at least five (5) years in the local geographic area.
3. Applicator shall be trained and approved by manufacturer for at least five (5) years.
4. Applicator shall have an established ongoing training program for workmen, including manufacturer's training.

FIGURE 14-3 Written specifications that are several hundred pages long are a major component of a project manual. The specification contained on the following pages is the specification for *one* component. Each fixture, unit, and system in the structure must be specified. *Courtesy Ron Lee, Architects Barrentine, Bates & Lee, A.I.A.*

Descriptive Specifications

Descriptive specifications are the most detailed type of specification and are used when the architect assumes total responsibility for the performance of a system. Descriptive specifications are exclusively used within a project manual and describe the method of assembly, physical properties, installation, arrangement, or any other category of information needed to fulfill the manufacturer's requirements, or owner's demands for a specific system. The specification for EIFS in Figure 14-3 is an example of a descriptive specification.

Performance Specifications

Specifications that define the desired results of a product or system are referred to as **performance specifications.** Using this type of specification, the exact makeup of individual components is not described, allowing suppliers maximum input. Performance specifications describe the conditions or parameters a product or system must function in, as well as the method of measurement to be used to judge compliance. An example of a performance specification can be seen in Figure 14-4. See the student CD in the RESOURCE MATERIAL/CHAPTER 14 file for the complete specification.

Proprietary Specifications

Specifications that call for materials by their trade name and model name are referred to as **proprietary specifications.** Individual products, specific pieces of equipment, or entire systems can be referenced using this method. Material provided by the manufacturer's literature forms the basis for the specification that can be found in both project manuals and drawing specifications.

SECTION 07 90 00—CAULKING

Part 1—General

1.01 SUMMARY

A. The purpose of the caulking work is to provide a positive barrier against penetration of air and moisture at the joints between items where caulking is essential to continued integrity of the barrier.

B. Comply with governing codes and regulations.

1.02 MATERIALS

A. Caulking material shall be properly selected to resist the type of abuse it will receive. It shall have expansion and contraction characteristics suitable to the installed conditions. Use only material that is best suited to the installation and is so recommended by the caulking material manufacturer.

B. Provide sealants in colors suitable for the location. Colors are to be selected from the manufacturer's standards.

1.03 INSTALLATION

A. Beginning work means acceptance of substrates.

B. Install caulking in strict accordance with the manufacture's recommendations. Caulking bead shall be true to line and surface, of uniform width and smooth, solid and slightly below the surface or adjacent materials.

C. Cure and protect sealants as directed by the manufacturer. Replace or restore damaged sealants. Clean adjacent surfaces.

D. Caulking is to be protected from the work of other trades during and after installation.

FIGURE 14-4 A performance specification describes the conditions or parameters a product or system must function in.

Proprietary specifications may be written in either a closed or open form. *Closed* or *sole source* specifications require that only one specific manufacturer be allowed to supply the desired product. Closed specifications are often used on renovations where a specific product is required to match existing conditions. An example of a proprietary specification can be seen in the specification 3.4 B for finish hardware in the example below:

3.4 MANUFACTURERS

A. Hinges: Stanley (Specification base), McKinney, Hagger, or accepted substitute.

B. Locks and Latchsets: Russwin, no other substitutions.

C. Exit Devices: Von Duprin or accepted substitute.

Courtesy Architects Barrentine, Bates & Lee, A.I.A.

Open or *equal* bid specifications allow for substitutions of products that are deemed to be equal by the architect. With an open specification, the architect may name several acceptable products, allowing the contractor or owner to choose based on availability or price. Typically, the specification names a minimum of three products that could be used when in a project manual. Although the names of three products are listed, the qualifications for only one of the three are listed. When placed within the drawings, the phrase "or equal" or "accepted substitute" is sometimes used to allow the contractor to submit names of products judged to meet the project requirements.

Reference Specifications

A minimum standard for quality or performance can be established by using a **reference specification**. Typically used in conjunction with other types of specifications, reference specifications refer to minimum levels of quality established by recognized testing authorities such as UL (Underwriters Laboratories), ASTM (American Society for Testing and Materials), ANSI (American National Standards Institute), ACI (American Concrete Institute), and ICBO (International Conference of Building Officials). Specifications should include a date to ensure the latest standard is used and should also reference which portions of the standard are to be applied. An example of a reference specification can be seen in the specification for custom casework in the example below:

Part 2 Products

2.1 MATERIALS

A. **Lumber:** Transparent finished wood AWI Premium grade as indicated in Section 100 in AWI Quality Standards.

B. **Plywood:** APA, PS 1 for softwood. PS 1-71 and Industry Standard, I.S1 for hardwood 5 ply minimum.

C. **Particleboard:** Mat-formed wood particleboard, CS 236-66, Type 1-B-2 except 2-B-2 for sink countertops and similar moisture exposed work, 45-pound density, 3/8" to 1 1/4" thick, Duraflake by Willamette Industries.

D. **Fasteners:** Nails, staples, and screws to comply with Section 400 in AWI Quality Standards.

Courtesy Architects Barrentine, Bates & Lee, A.I.A.

WRITING SPECIFICATIONS

Specifications are written by an architect, engineer, or professional writer with expertise in legal and architectural matters. Rarely is a drafter involved with the actual writing of specifications. Many smaller offices have master sets of specifications, which can be edited for similar projects. These *masters* can serve as a core set of specifications that are edited for each project, saving hours of labor. A senior drafter is often required to update the master specs to meet the requirements of the current project. A wide variety of software programs are also available to aid in writing specifications. Programs typically offer several options and allow the user to pick the options that meet the needs of the project.

The Relationship of Drawings to Specifications

A key to good construction documentation is to remember the goal of each portion of the documentation. Drawings present pictures of the structure and give the size, form, location, and arrangement of various components by use of lines, symbols, and text.

Specifications provide written descriptions of type and quality of materials; labor and components; and methods of testing, fabricating, and installing components within the structure. Each form should complement the other without repeating information.

- Drawings should be used to depict the physical relationship.
- Specifications should be used to show the quality and types of material in the structure.

Duplication between the drawings and specs—unless it is exact word for word—can lead to contradiction, misunderstanding, and a court date. Word-for-word duplication is redundant and harmful to productivity. Because many privately funded projects do not have a project manual, duplication throughout the drawings is often used to avoid misunderstandings.

The drawings and the specs should be developed together to achieve a balanced relationship. At the onset of a project, someone on the architectural team is usually assigned the task of keeping a checklist for the project. The checklist generally includes a schedule of what drawings are required, and what will appear on each drawing, related schedules, or what needs to appear in the specifications. Material that is added to the preliminary spec list can then be noted on the plans by the drafter in a shortened note form. For instance, a wall can be described in a detail or section as an 8" concrete wall. The project manual would contain the specifications for the type of concrete, how it is to be mixed, when it can be poured, and numerous other features about quality. It is imperative that the drafter be well informed about material placed in the project manual so that materials can be adequately described. Figure 14-5 shows a detail of the EIFS that depends on the specifications seen in Figure 14-3. Structural drawings done by an engineering team tend to be more redundant in callouts to meet the demands of the building department. Figure 14-6 shows an example of a steel detail that is placed with the specifications. The written specifications can be seen in Figure 14-1.

Writing Metric Symbols and Names

Dimensions and sizes found in the specifications should never be expressed in dual units. The same units found throughout the drawings should be expressed throughout the specifications to reduce dimensioning time, lessen the chance of conversion errors, and eliminate confusion. If units are to be expressed in metric, all sizes should be listed in millimeters except for large distances associated with the site drawings. Notations describing area should be expressed in square meters,

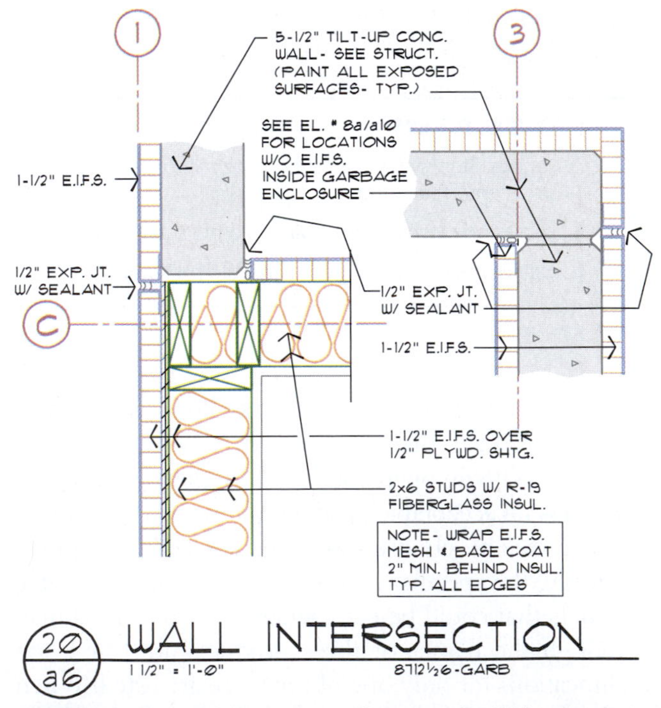

FIGURE 14-5 Drawings and specifications should be developed together so that information is not duplicated. This detail relies on the information contained in Figure 14-3. *Courtesy Architects Barrentine, Bates & Lee, A.I.A.*

EASY BEING GREEN

By some estimates, approximately 20% of the greenhouse gas emissions that contribute to global warming are attributed to the building sector. This is due to a combination of factors including energy usage, the continued creation of impervious surfaces, and construction debris generation. Green, sustainable building techniques, incorporating products that are friendly to the environment coupled with the right kind of maintenance, can help reduce this trend and ultimately contribute to the creation of environmentally, financially, and socially sustainable communities.

When writing specifications, select products that not only meet the needs of the project, but that are Leadership in Energy and Environmental Design (LEED) certified. Use the tools provided by the Life Cycle Cost Analysis (LCCA) to select products and materials that offer the most cost-effective design alternatives for the project while providing the best possible alternatives for the project and for those who will use and be impacted by the project.

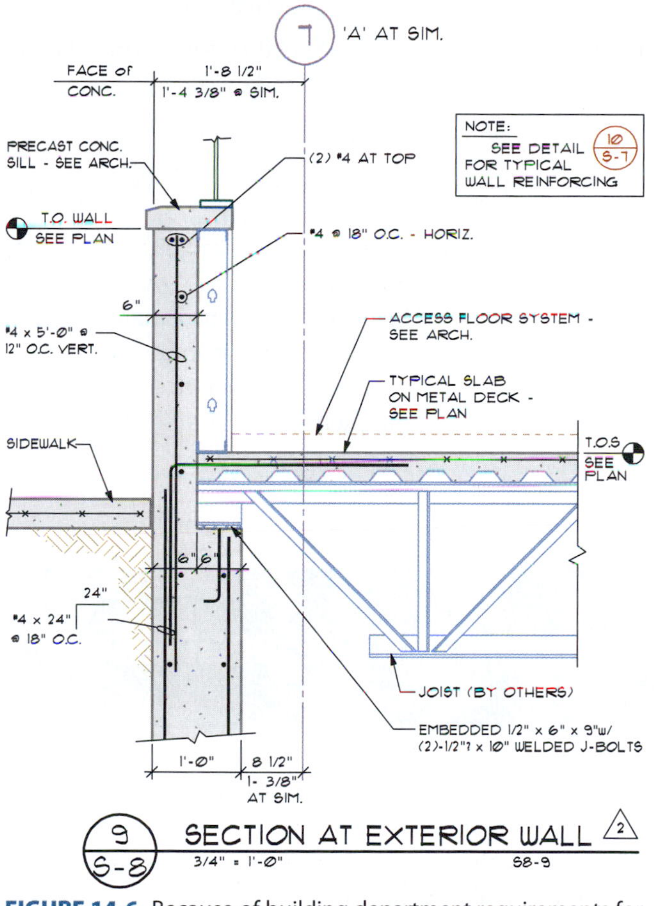

FIGURE 14-6 Because of building department requirements for drawings, structural drawings often duplicate information contained in written specifications. The drafter must carefully coordinate both parts of the project. *Courtesy KPFF Consulting Engineers.*

and expressions of volume should be expressed in cubic meters. Fluid measurements should be expressed as liters. Loads that have traditionally been expressed as pounds per square foot (psf) should be expressed as kilograms per square meter (kg/m^2) or kilonewtons per square meter (kN/m^2).

Kilograms per square meter should be used for expressing floor loads because most live and dead loads are measured in kilograms. Many engineers use kilonewtons per square meter (kN/m^2) or their equivalent, megapascals (MPa), for structural calculations.

- Additional guidelines for expressing metric units within specifications can be found on the student CD in Appendix B.

Additional information on metric conversions can be obtained from the following:

Construction Metrication Council (This council is no longer active; however, findings and previous newsletters are available through NIBS.)
National Institute of Building Sciences (NIBS)
1090 Vermont Ave. NW
Suite 700
Washington, DC 20005-4905
1-202-289-7800
Web site: nibs@nibs.org

American Institute of Architects
AIA Bookstore
1735 New York Ave. NW
Washington, DC 20006
1-800-242-3847
Web site: www.aia.org

American National Standards Institute, Inc.
25 W. 43rd St.
New York, NY 10036
1-212-642-4900
Web site: www.ansi.org

Construction Specifications Institute
99 Canal Center Plaza
Suite 300
Alexandria, VA 22314-1791
1-800-689-2900
Web site: www.csinet.org

ADDITIONAL READING

The following Web sites can be used as a resource to help you keep current with changes in writing specifications.

ADDRESS	COMPANY/ ORGANIZATION
www.aiaonline.com	American Institute of Architects
www.ansi.org	American National Standards Institute
www.buildinggreen.com	BuildingGreen.com (green products listed by CSI divisions)
www.nibs.org	Construction Metrication Council
www.csc-dcc.ca	Construction Specifications Canada
www.csinet.org	Construction Specifications Institute
www.reedfirstsource.com	First Source
www.greenformat.com	GreenFormat
www.masterformat.com	MasterFormat
www.pcea.org	Professional Construction Estimators Association of America, Inc.
www.scip.com	Specifications Consultants in Independent Practice

KEY TERMS

Addenda
Bid form
Bond
Form of Agreement
General Conditions of the Contract
Instruction to bidders
Invitation to bid
MasterFormat system
Project manual
Specifications
Specifications, cash allowance
Specifications, descriptive
Specifications, performance
Specifications, proprietary
Specifications, reference
Supplementary conditions

CHAPTER 14 TEST Project Manuals and Written Specifications

QUESTIONS

Answer the following questions with short complete statements. Type the chapter title, question number, and a short complete statement for each question using a word processor. Some answers may require the use of vendor catalogs or seeking out local suppliers.

Question 14-1 What are the two common locations for specifications and what are the major differences between the two systems?

Question 14-2 What areas of information should drawings describe?

Question 14-3 Why is duplication in the drawings and specifications a possible problem?

Question 14-4 What types of information should the specifications include?

Question 14-5 List common documents that are typically part of the project manual.

Question 14-6 Name and describe the common format for writing specifications within a manual.

Question 14-7 How are the major divisions of the CSI system broken down?

Question 14-8 Describe the three major portions of a specification written to meet the CSI format.

Question 14-9 Describe the CAD technician's role in the writing of specifications.

Question 14-10 List and describe the five major methods of writing a specification.

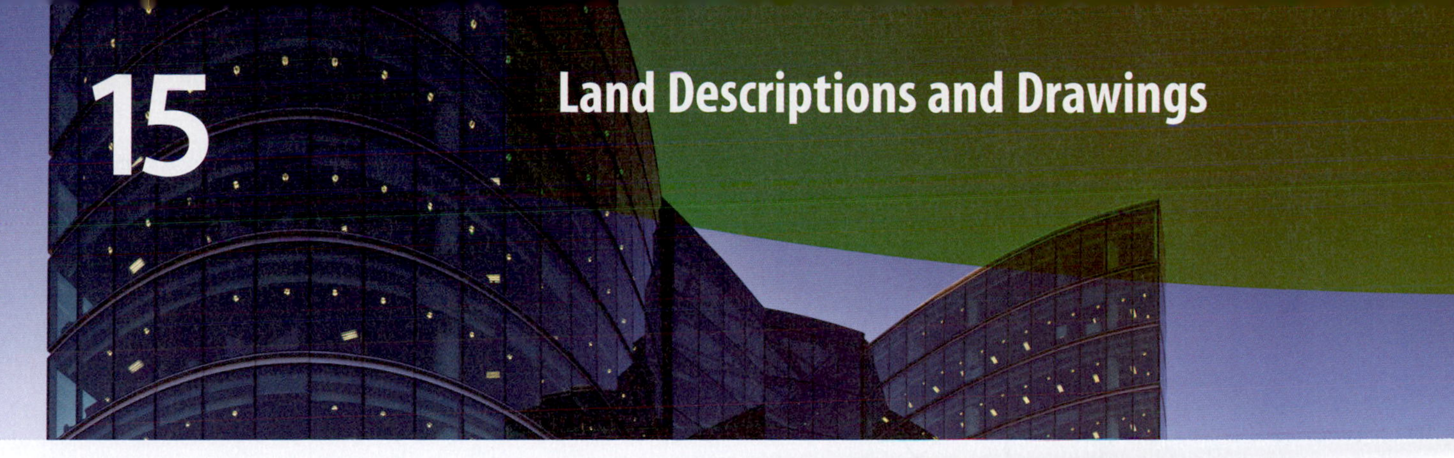

15 Land Descriptions and Drawings

Every project to be built requires one or more drawings and a legal description to describe the construction site. This chapter will explore the use of:

- Legal descriptions.
- Common types of land drawings including a vicinity map, site plans, and grading plans.
- Common drawing steps required to complete a site plan and site-related drawings.

The **site plans** that will be started in problems 15-3 through 15-6 will serve as the building sites for the projects that will be started in Chapter 16 and completed in the remaining chapters. Even if you are not required to draw a site plan for your class project, remember that there may be information on the site plan that will be needed to complete future drawings.

LEGAL DESCRIPTIONS

For legal purposes, each piece of land is outlined by a description of the property known as a **legal description.** The legal description may be given in several forms, such as a **metes and bounds system**, a *rectangular system*, or a *lot and block system*. The type of description to be used depends on the contour of the area to be described or the requirements of the municipality reviewing the plans.

Metes and Bounds

A metes and bounds description is also referred to as a long description. A copy of the description can be obtained from a title company or the assessor's office. A complete description can be added to the site plan or attached to the drawings on a separate sheet of paper. This system provides a written description of the property in terms of measurements of distance and angles of direction from a known starting point. The known starting point is referred to as the *true point of beginning* in a legal description. The true point of beginning is usually marked by a steel rod or by a bench mark established by the U.S. Geological Society, (USGS). This is referred to as a *monument*.

Metes

The metes are measured in feet, yards, rods, or surveyor's chains. A rod is equal to 16.5 feet (5000 mm) or 5.5 yards. A chain is equal to 66 feet (20,100 mm) or 22 yards. Directions are given from a monument such as a bench mark established by the USGS or an iron rod set from a previous survey. This location is known as the point of beginning, and may be several hundred feet away from the property to be described. Directions are given from the point of beginning to a specified point on the perimeter of the property to be described. Directions are then given to outline the property with all distances set in units of feet expressed in one-hundredth of a foot rather than the traditional feet and inches normally associated with construction. Metric units should be expressed in either millimeters or meters. Centimeters are not used on construction drawings.

Bounds

The **boundaries** of property are described by **bearings**, which are angles referenced to a quadrant on a compass. Figure 15-1 shows the four possible quadrants and descriptions of lines within a quadrant. Bearings are always described by starting at north or south and turning to the east or west. Bearings are expressed in angles, minutes, and seconds. Each quadrant of the compass contains 90°, each degree contains 60 minutes, and each minute can be divided into 60 seconds. A degree is represented by the ° symbol, a minute is represented by the ' symbol, and seconds are represented by the " symbol. Bearings are used to describe the angular location of a property line. Some properties are also defined by their location to the centerline of major streets. When property abuts a body of water such as a creek or river, a boundary in angles may not be given and the property boundary is defined by the center point of the body of water. A metes and bounds legal description would resemble the description shown in the following example:

> A tract of land situated in the Southeast quarter of the Northeast quarter of Section 17, T3S, R1W of the

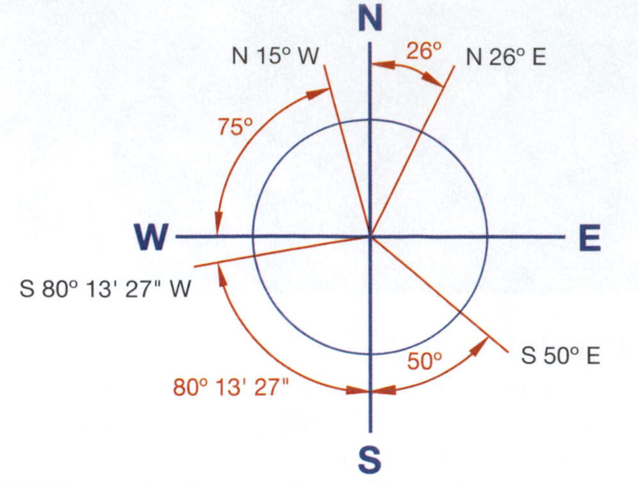

FIGURE 15-1 Bearings are referenced by quadrants on a compass that begin by looking either north or south and turning to look either east or west.

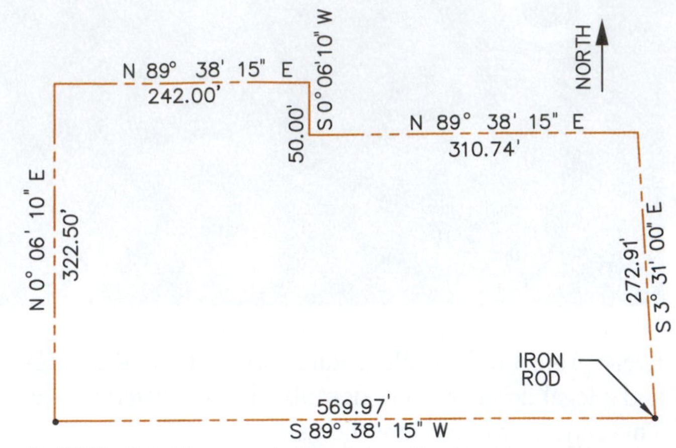

FIGURE 15-2 The parcel of land described by the metes and bounds legal description listed above.

Willamette Meridian, Clackamas County, Oregon, being more particularly described as follows, to wit. Beginning at the 5.8 inch iron rod at the Southwest corner of the Southeast quarter of the Northeast quarter of said Section 17; thence North 0°06'10" East along the West line of the Southeast quarter of the Northeast quarter, 322.50'; thence leaving said West line North 89°38'15" East 242.00 feet; thence South 0°06'10" West parallel with said West line of the Southeast quarter of the Northeast quarter, 50.00 feet, thence North 89°38'15" East 310.74 feet to the Westerly right of way line of Bell Road No. 113; thence South 3°31' East along said Westerly right of way line, 272.91 feet to a 5/8 inch iron rod; thence leaving said Westerly right of way line, South 89°38'15" West 569.97 feet to the point of beginning.

This description would be listed on legal documents describing the property as well as the site plan. Figure 15-2 shows the land described by the legal description. A short description of the property should be copied exactly onto the site.

Rectangular Systems

Many areas of the country refer to land based on its location by latitude and longitude. Parallels of latitude and meridians of longitude were used by the U.S. Bureau of Land Management in states that were originally defined as public land states. As the land was divided, large parcels of land were defined by what are known as *basic reference lines*. There are 31 pairs of standard lines in the continental United States and three in Alaska. Principal meridians and base lines can be seen in Figure 15-3. These divisions of land are described as the *great land surveys*. As the initial division of land was started, the first six principal meridians were numbered. The last numbered meridian passes through Nebraska, Kansas, and Oklahoma. All subsequent meridians are defined by local names. The great land surveys were further divided by surveys that define land by townships and sections.

Townships

Baselines and meridians are divided into six-mile square parcels of land called **townships.** Each township is numbered based on its location to the principal meridian and base line. Horizontal tiers are numbered based on their position above or below the base line.

Vertical tiers are defined by their position east or west of the principal meridian. Township positions can be seen in Figure 15-4. The township highlighted in Figure 15-4 would be described as *Township No. 2 North, Range 3 West* because it is in the second tier north of the base line and in the third row west of the principal meridian. The name of the principal meridian would then be listed. This information is abbreviated as T2NR3W on the site plan.

Sections

Land within townships is broken down into one-mile square parcels known as **sections.** Sections are assigned numbers from 1 to 36 (as shown in Figure 15-5) beginning in the northeast corner of the township. Each section is further broken down into quarter sections. Each section contains 640 acres or 43 560 square feet. Quarter sections are broken down as seen in Figure 15-6 using divisions of quarters, and quarters

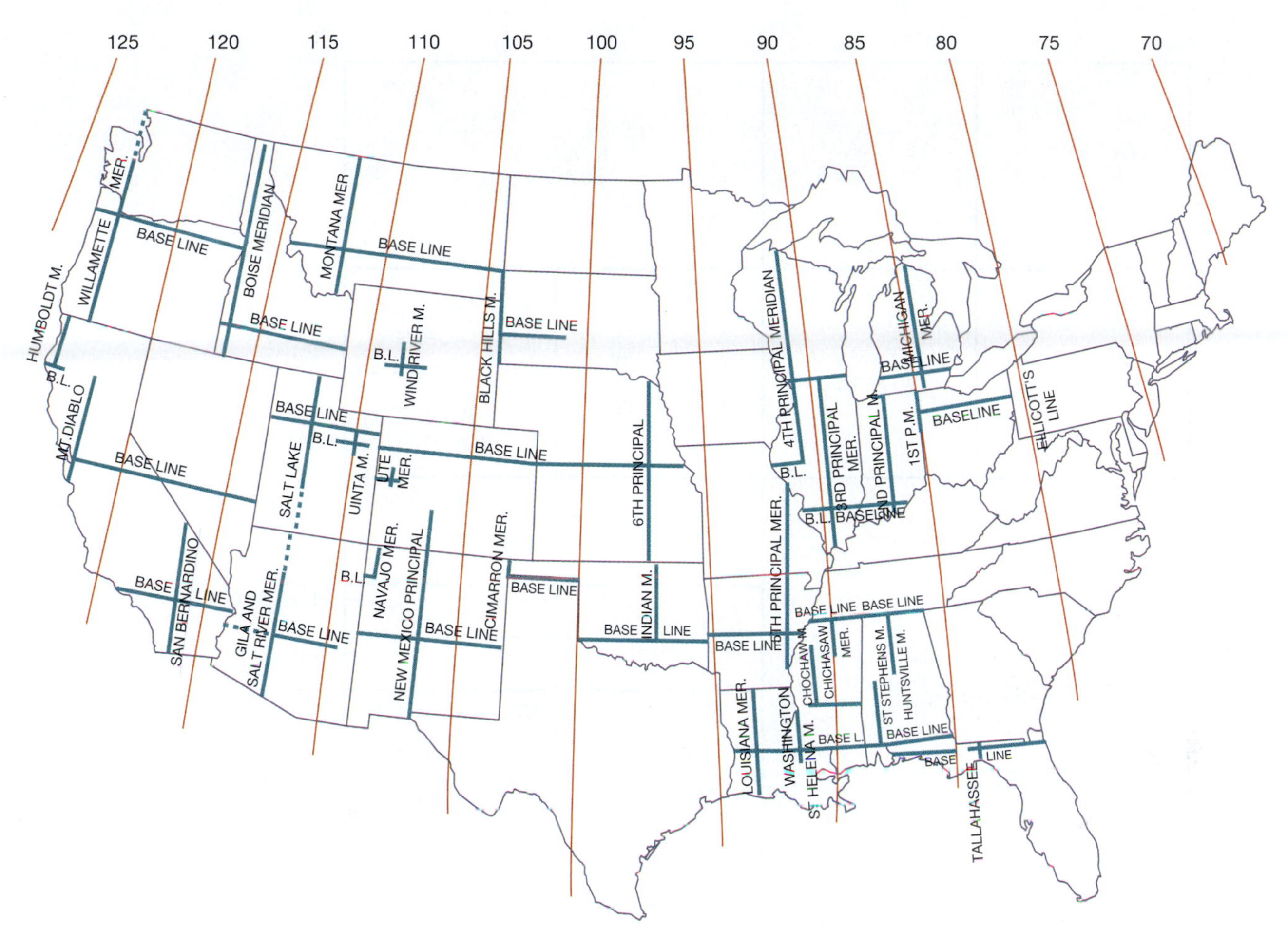

FIGURE 15-3 Principal meridians and basic reference lines are used to divide land in the continental United States.

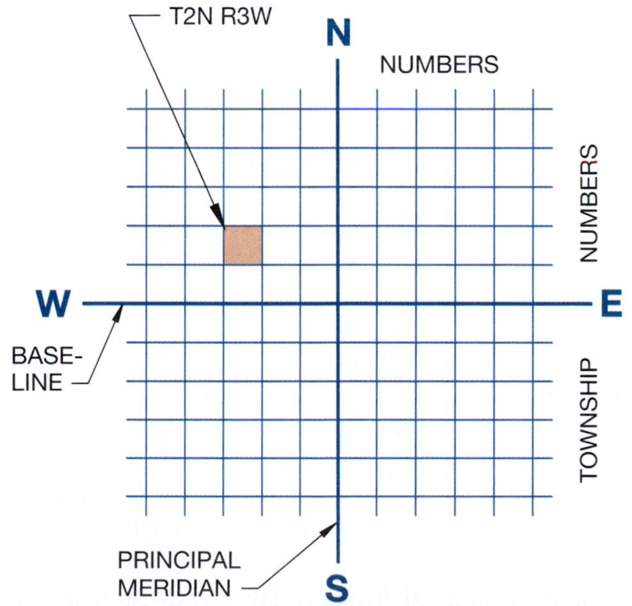

FIGURE 15-4 A township is a six-mile square portion of land that is defined by its position to a principal meridian and a baseline. The indicated township is referred to as T2NR3W because it is in the second tier north of the baseline and three columns west of the principal meridian.

or halves of quarters. These small segments can be further broken down into quarters or halves again.

Specifying the Legal Description

The legal description, based on the rectangular system, lists the portion of the land to be developed described by its position within the section, the section, the

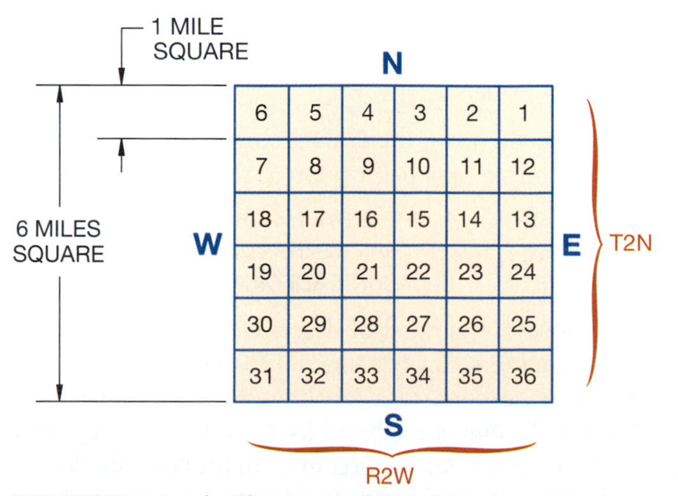

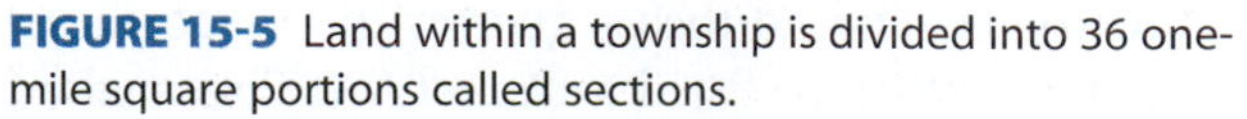

FIGURE 15-5 Land within a township is divided into 36 one-mile square portions called sections.

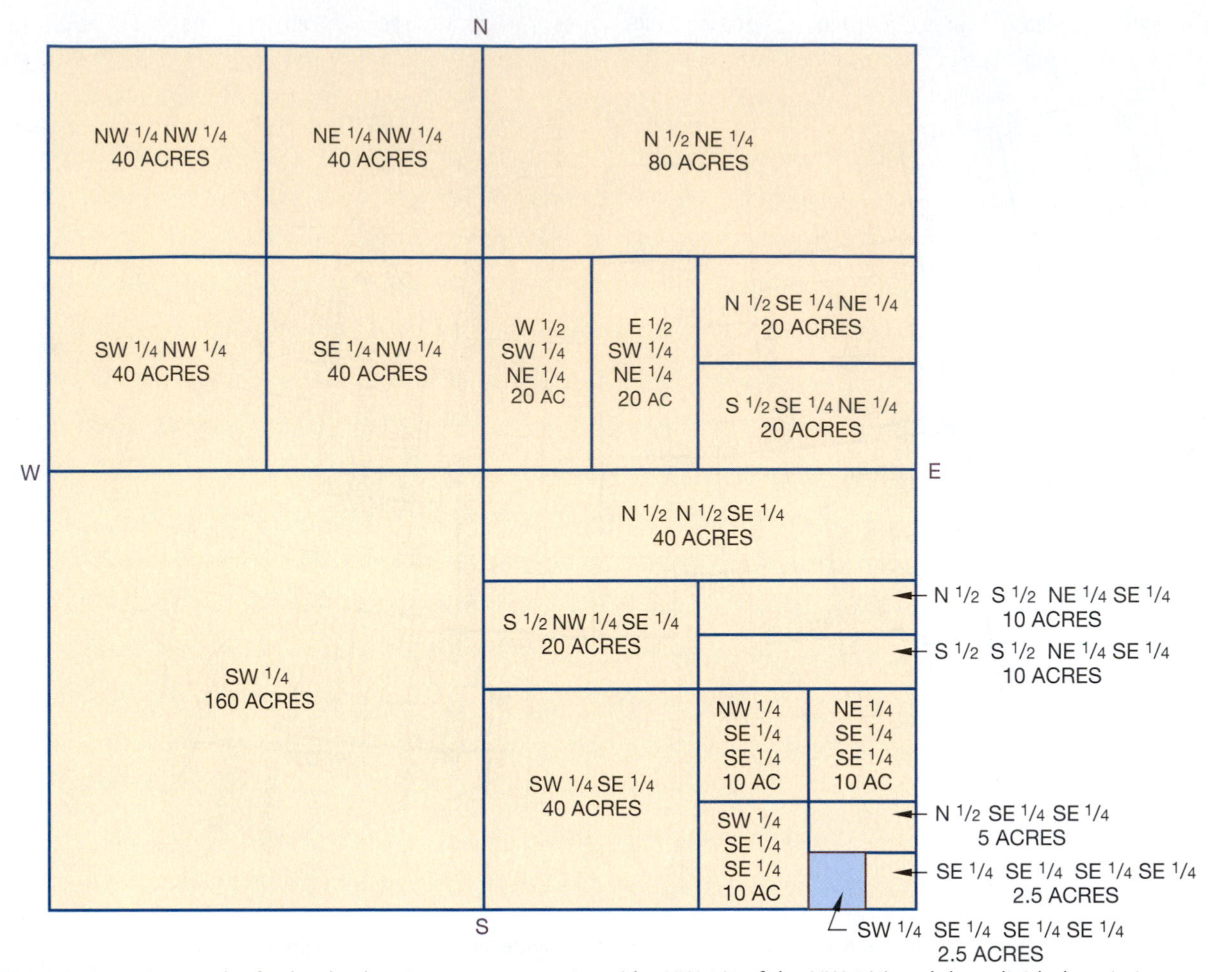

FIGURE 15-6 A section can be further broken into quarter sections (the NW 1/4 of the NW 1/4) and then divided again into a quarter of a quarter section (the NW 1/4 of the SE 1/4 of the SE 1/4) and then divided one more time into quarters (the SE 1/4 of the SE 1/4 of the SE 1/4 of the SE 1/4).

township, and the principal meridian. A typical legal description resembles the following description:

> The Southeast one quarter, of the Southwest one quarter of the Southwest one quarter of Section 31, Township No. 2 North, Range 3 West of the San Bernardino meridian, City of El Cajon, in the state of California.

On the site plan, the CAD technician often abbreviates this into a legal description as follows:

THE SE 1/4, OF THE SW 1/4 OF THE SW 1/4,
S31, T2N, R3W, SAN BERNARDINO MERIDIAN,
EL CAJON, CALIFORNIA.

Because the method of describing quarters of quarters of quarters can become confusing, many municipalities have gone to a labeling system using letters. Quarter sections are labeled as A, B, C, or D. Quarter sections are further divided into quarters by the letters A, B, C, or D. The northeast quarter of the northeast quarter would then be listed as Parcel AA. This method of land description works especially well in areas where the land contour is fairly flat. In areas with irregular land shapes, the rectangular land description can be linked with a partial metes and bounds description to accurately locate the property.

Lot and Block

This system of describing land is usually found within incorporated cities. Most states require the filing of a subdivision map that defines individual lot size and shape as land is being divided. Each lot is defined on a subdivision map by a length and angle of each property line as well as a legal description. On older subdivision maps, land is divided into areas based on neighborhood, called *subdivisions*. The subdivision is then divided into blocks based on street layout. The block is further divided into lots. On newer maps, most municipalities assign a number to a parcel of land that corresponds to either a tax account or a map number. Lots can be irregularly shaped or rectangular. The shape of the lot is often based on the contour of the surrounding land and the layout of streets. A typical legal description for a lot and block description would resemble the following description:

LOTS NO. 1, 2, AND 3 OF MAP #17643
OF THE CITY OF BONITA, COUNTY
OF SAN DIEGO, CALIFORNIA.

LAND DRAWINGS

The most common drawing for describing property is the site plan. The site plan is started in conjunction with the floor plan, although it probably will not be finished until other architectural drawings are completed. With the property boundaries drawn, the preliminary design for the floor plan can be inserted into the site plan. Preliminary designs for access and parking can then be determined, and adjustments to both the site and floor plans can be made.

The size of the project and the complexity of the site will dictate the drawings required to describe site-related construction, who will develop the needed drawings, and when they will be developed. In addition to the site plan, a demolition plan, topography, grading, landscape, sprinkler, freshwater, and sewer plan may be required to describe the alterations to be made to the site. On simple construction projects, all of these plans can be combined into one site plan. This plan would typically be the first sheet of the architectural drawings and labeled A-1. For most projects, the site-related drawings precede the architectural drawings and are listed as civil drawings (C-1 or L-1).

The size and complexity of the project will determine who will draw the project. On most projects, each of the site-related drawings will be prepared by a CAD drafter working for a civil engineer, surveyor, or landscape architect and then given to the architect to be incorporated into the working drawings. On simple projects, a CAD technician working for the architect or engineer may complete the drawings under the supervision of a landscape architect or a civil engineer.

Vicinity Map

A **vicinity map** is used to show the area surrounding the construction site. It is placed near the site plan and is used to show major access routes to the site. This could include major streets, freeway on and off ramps, suggested routes for large delivery trucks, and rail routes. The drafter should prepare a vicinity map showing an area that reflects the size of the project. If building components primarily come from the surrounding area, the map needs to reflect only the immediate construction area. If materials come from several different cities or states, the area of the vicinity map should be expanded to aid drivers who may not be familiar with the area. Figure 15-7 shows an example of a vicinity map.

COMMON SITE PLAN ELEMENTS

The *site plan* for a commercial project is the basis for all other site-related drawings. It shows the layout and size of the property, the outline of the structure to be built, north arrow, ground and finish floor elevations, setbacks, parking and access information, and information about utilities. The results of engineering studies and soils reports related to the site will be summarized on the site plan.

Common Linetypes

The shape of the construction site is drawn based on information provided by the legal description or subdivision map. Common lines found on site-related drawings and the AutoCAD linetype used to represent them include:

- Property lines—PHANTOM2 or PHANTOMX2.
- Outlines of the building footprint—CONTINUOUS.

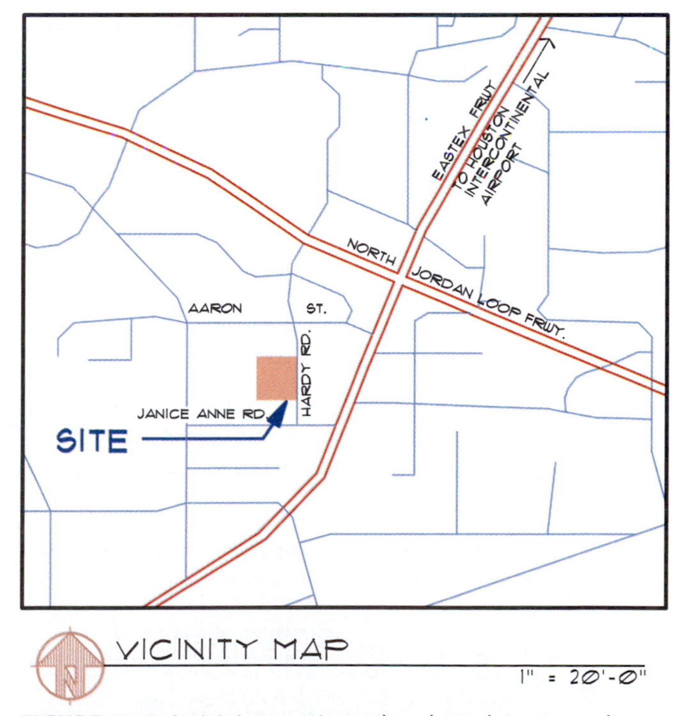

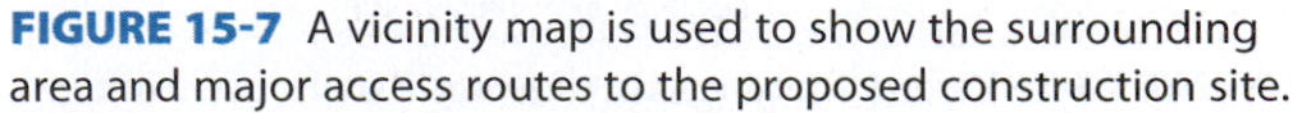

FIGURE 15-7 A vicinity map is used to show the surrounding area and major access routes to the proposed construction site.

- Centerlines of access roads or easements—CENTER, CENTER2, or CENTERX2.
- Edges of easements and building setbacks—DASHED, DASHED2, DASHEDX2, HIDDEN, HIDDEN2, or HIDDENX2.
- Utility lines—a dashed or hidden pattern. Each line is labeled with a G (gas), S (sewer), W (water), P (power), or T (communications) to denote its usage.

Specifications must also be provided to distinguish between existing utilities and those that the contractor is expected to provide. The location of each utility referenced to the property should also be provided. Figure 15-8 shows a partial site plan for a small commercial structure and the appropriate linetypes. Sidewalks, patios, stairs, and parking outlines are typically represented by a continuous linetype, with each item properly labeled.

Representing the Structure

The structure to be built must be accurately represented and easily distinguishable from the property and utilities. Common methods of representing a structure include the use of a polyline to define the perimeter and a hatch pattern such as ANSI31 or ANSI37 to further highlight the structure. This method can be seen in Figure 15-8. Many firms attach the floor into the site plan using the XREF command. This offers the advantage of an up-to-date site plan each time the drawing is opened. This can be especially important if the footprint of the structure is altered or door locations are moved. If the project consists of more than one structure, either a building number or name should be used to identify each building. The method used to identify each building on the site plan should correspond to the title placed with the floor, framing, and

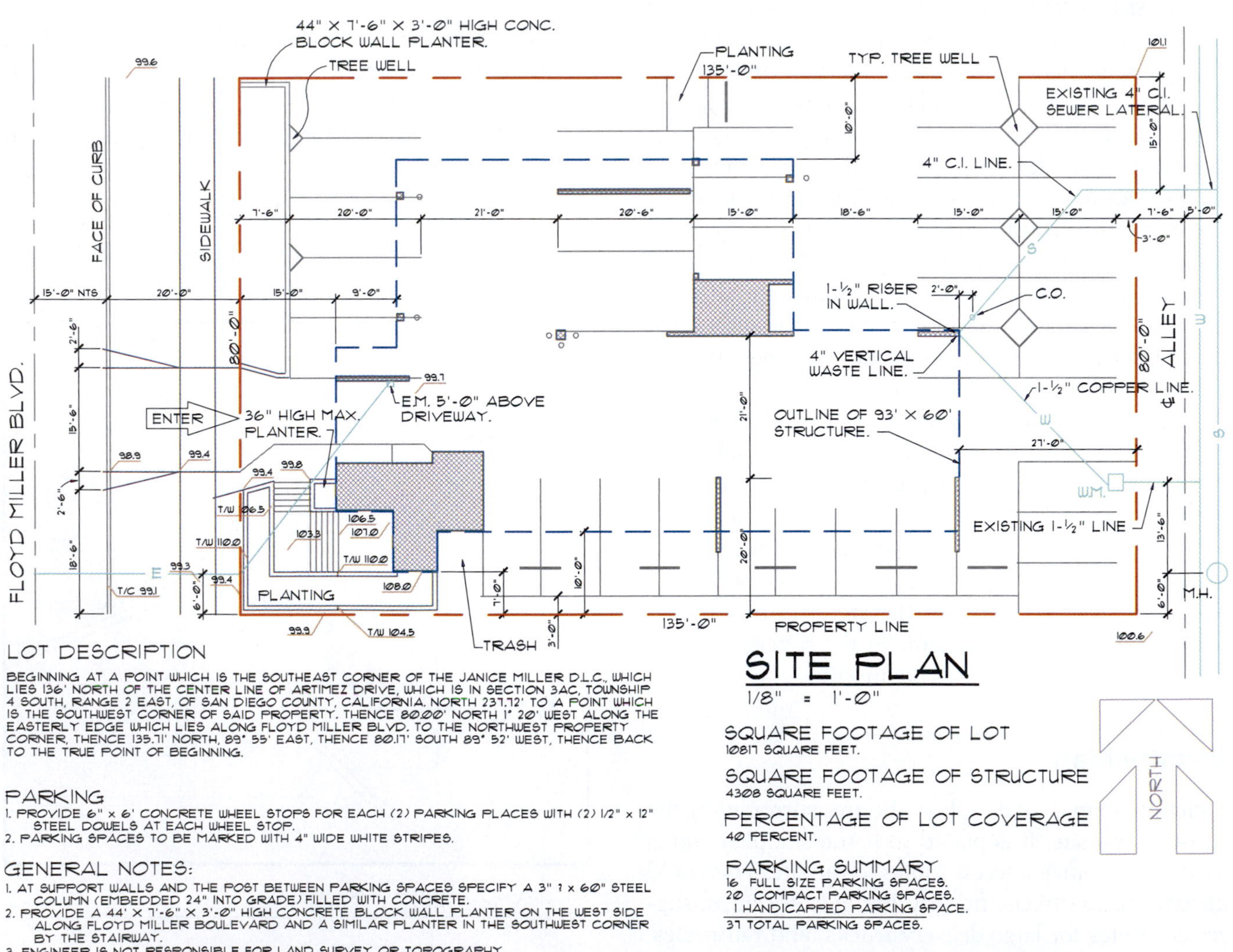

44" X 7'-6" X 3'-0" HIGH CONC. BLOCK WALL PLANTER.
TREE WELL
PLANTING
135'-0"
TYP. TREE WELL
EXISTING 4" C.I. SEWER LATERAL.
4" C.I. LINE.
FACE OF CURB
SIDEWALK
1-½" RISER IN WALL.
C.O.
4" VERTICAL WASTE LINE.
1-½" COPPER LINE.
ALLEY
FLOYD MILLER BLVD.
ENTER
36" HIGH MAX. PLANTER.
E.M. 5'-0" ABOVE DRIVEWAY.
OUTLINE OF 93' X 60' STRUCTURE.
W.M.
EXISTING 1-½" LINE
M.H.
PLANTING
TRASH
PROPERTY LINE
LOT DESCRIPTION
BEGINNING AT A POINT WHICH IS THE SOUTHEAST CORNER OF THE JANICE MILLER D.L.C., WHICH LIES 136' NORTH OF THE CENTER LINE OF ARTIMEZ DRIVE, WHICH IS IN SECTION 3AC, TOWNSHIP 4 SOUTH, RANGE 2 EAST, OF SAN DIEGO COUNTY, CALIFORNIA. NORTH 237.72' TO A POINT WHICH IS THE SOUTHWEST CORNER OF SAID PROPERTY. THENCE 80.00' NORTH 1° 20' WEST ALONG THE EASTERLY EDGE WHICH LIES ALONG FLOYD MILLER BLVD. TO THE NORTHWEST PROPERTY CORNER, THENCE 135.71' NORTH, 89° 55' EAST, THENCE 80.17' SOUTH 89° 52' WEST, THENCE BACK TO THE TRUE POINT OF BEGINNING.
PARKING
1. PROVIDE 6" x 6' CONCRETE WHEEL STOPS FOR EACH (2) PARKING PLACES WITH (2) 1/2" x 12" STEEL DOWELS AT EACH WHEEL STOP.
2. PARKING SPACES TO BE MARKED WITH 4" WIDE WHITE STRIPES.
GENERAL NOTES:
1. AT SUPPORT WALLS AND THE POST BETWEEN PARKING SPACES SPECIFY A 3" ? x 60" STEEL COLUMN (EMBEDDED 24" INTO GRADE) FILLED WITH CONCRETE.
2. PROVIDE A 44' X 7'-6" X 3'-0" HIGH CONCRETE BLOCK WALL PLANTER ON THE WEST SIDE ALONG FLOYD MILLER BOULEVARD AND A SIMILAR PLANTER IN THE SOUTHWEST CORNER BY THE STAIRWAY.
3. ENGINEER IS NOT RESPONSIBLE FOR LAND SURVEY OR TOPOGRAPHY.
4. EXTERIOR SIGNS TO COMPLY WITH COUNTY SIGN ORDINANCE.
5. ALL PLANTERS TO BE CONSTRUCTED WITH A SEPARATE PERMIT.
SITE PLAN
1/8" = 1'-0"
SQUARE FOOTAGE OF LOT
10817 SQUARE FEET.
SQUARE FOOTAGE OF STRUCTURE
4308 SQUARE FEET.
PERCENTAGE OF LOT COVERAGE
40 PERCENT.
PARKING SUMMARY
16 FULL SIZE PARKING SPACES.
20 COMPACT PARKING SPACES.
1 HANDICAPPED PARKING SPACE.
37 TOTAL PARKING SPACES.
NORTH

FIGURE 15-8 Varied linetypes and line widths are used to represent materials on a site plan.

foundation plans. The use of the structure should also be specified within the outline of the structure with titles such as PROPOSED 2-LEVEL OFFICE STRUCTURE. Consideration must also be given by the drafter to accurately distinguish between portions of the structure that are in contact with the ground and those that are supported on columns.

Commercial projects often include structures that are to be demolished to make way for the new project. These structures must be accurately located and distinguished from new construction. A separate demolition plan similar to Figure 15-9 may also be used to supplement the site plan.

Parking Information

Information about access, areas to be paved, parking spaces, ramps, and walkways must be shown on the site plan. Because requirements vary from city to city, the drafter must verify requirements for each project. Continuous lines are used to represent each parking and paving component with the exception of the centerline of access roads. Many offices hatch concrete walkways with a random pattern of dots so they can be easily distinguished from asphalt paving areas. In addition to each parking space being represented, parking areas should be represented using a method

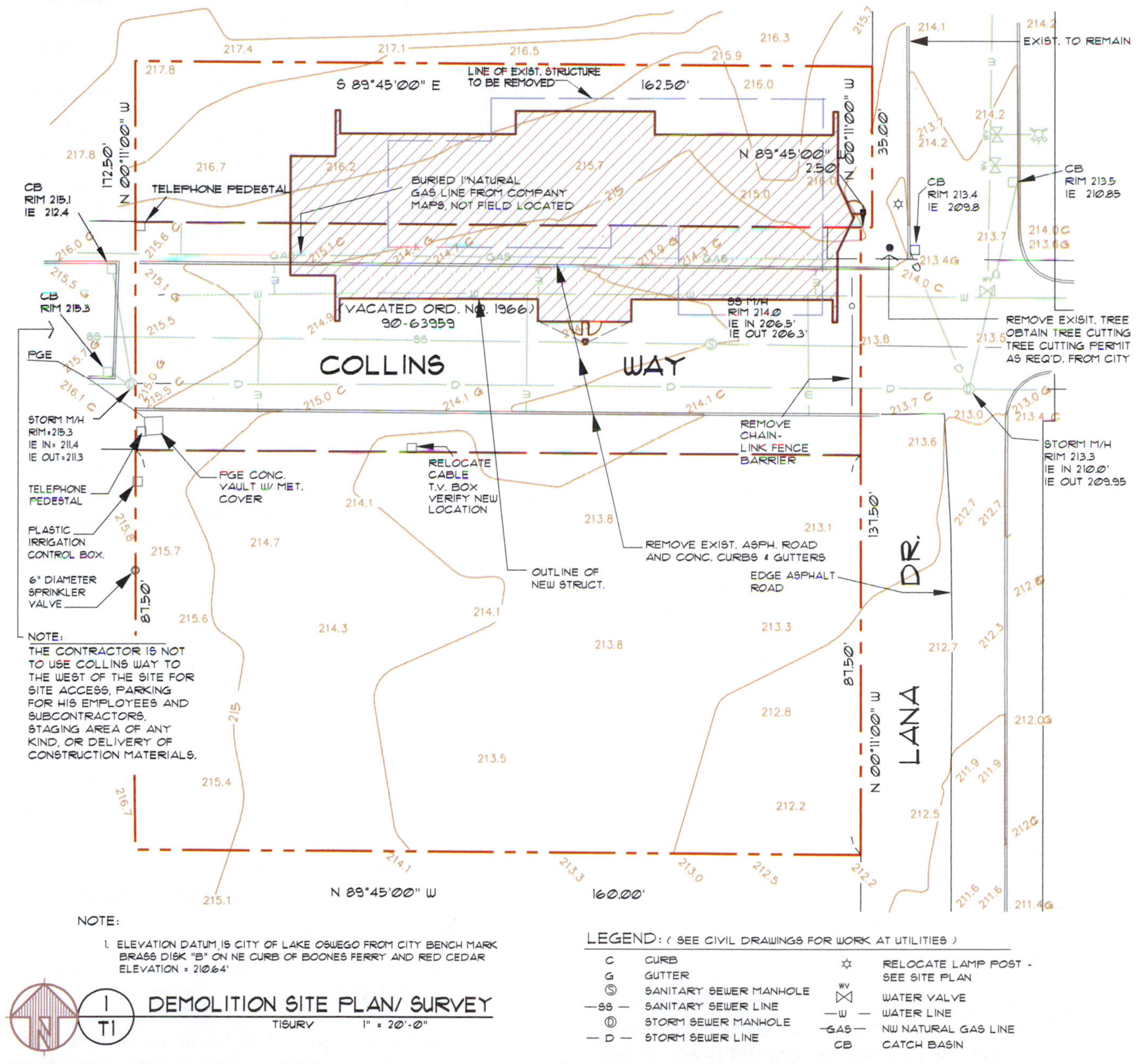

FIGURE 15-9 When existing structures or other features must be demolished to make way for a proposed project, they can be indicated on a site plan or a separate demolition plan will be provided. *Courtesy Architects Barrentine, Bates & Lee, A.I.A.*

similar to Figure 15-10. Many municipalities require a parking schedule to be part of the site plan. A parking schedule can be used to specify the number, type, and size of each category of parking space. Parking spaces are specified as full, compact, handicapped, or van handicapped. Common sizes are:

Full space: 9' × 20' (2740 × 6100 mm)

Compact: 8' × 16' (2440 × 4875 mm)

ADA-approved handicapped space: 14' × 19' (4270 × 5790 mm)

ADA-approved handicapped van space: 16' × 19' (4875 × 5790 mm) minimum

It is important to remember that there is a wide variance in sizes depending on the municipality and the direction of entry into the space. Figure 15-11 shows an example of common alternatives for parking based on the angle of entry. Perpendicular spaces can be shorter than spaces that require parallel parking. Spaces that are placed on an angle require different lengths, widths, and driveway size as the entry angle is varied. Parking spaces next to obstructions such as raised planters or building supports should have added width to ease access and allow the minimum required width to be maintained.

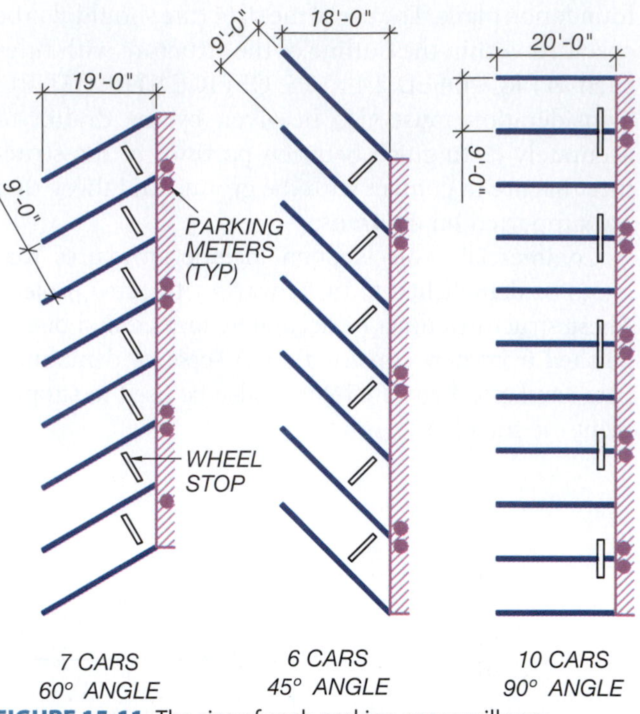

FIGURE 15-11 The size of each parking space will vary depending on the angle of approach. Each city has its own parking requirements, which the drafter should verify prior to starting the site plan.

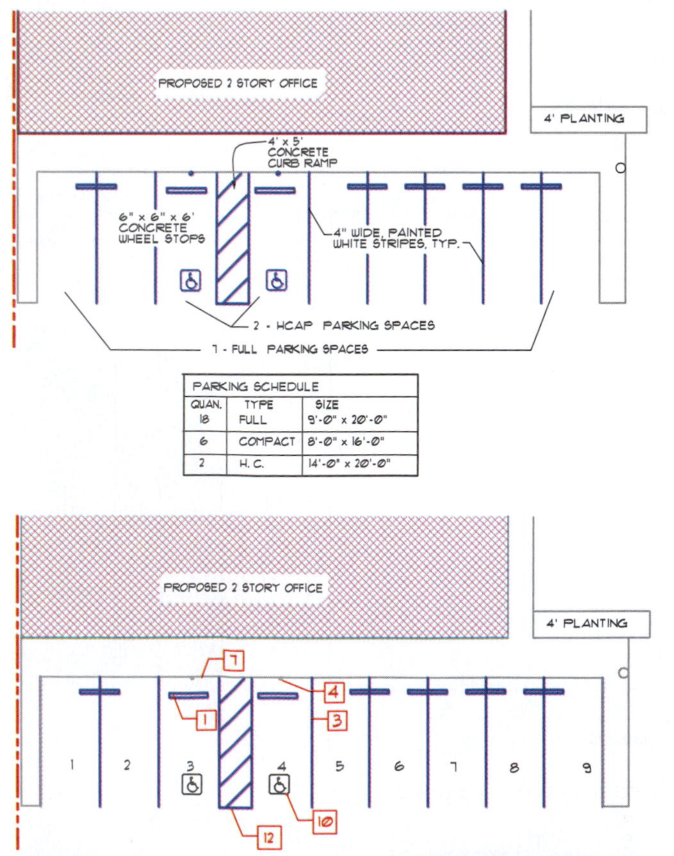

QUAN.	TYPE	SIZE
18	FULL	9'-0" x 20'-0"
6	COMPACT	8'-0" x 16'-0"
2	H. C.	14'-0" x 20'-0"

FIGURE 15-10 Parking spaces must be clearly defined on the site plan by either referencing each space or by referencing each group of spaces. The use of a parking schedule and keyed notes can greatly aid in keeping the drawing from becoming cluttered.

Building supports often require a permanent protective device to be installed for protection from vehicle damage. The addition of steel columns filled with concrete near each wood column is a common method of protecting wood columns from damage caused by careless drivers. These columns must be represented and specified on the site plan. Many offices have a standard detail in a symbols library that can be inserted into the site plan. Wheel stops for individual spaces are also required to be drawn and specified. Six-foot (1830 mm) long wheel stops are typically used for spaces that are 90° to the access drive. Stops are normally placed 24" (600 mm) from the front-end of the stall and straddle the dividing line between spaces so that one stop is shared by two spaces. Building supports and wheel stops can be represented as shown in Figures 15-8 and 15-10.

Site Dimensions

Dimensions must be placed to locate each item represented on the site plan. Three types of dimensions are often found on a site plan, including:

- **Land descriptions**: Land sizes are represented in feet and hundredths of a foot using notations such as 100.50' or in meters (30.6 m). Property line dimensions are placed parallel to the property line with no use of dimension or extension lines.
- **Building descriptions**: Overall sizes of structures are often placed parallel to the side of the structure and specified in feet and inches using notations such as 75'–4" or 22.8 m for meters. These overall dimensions are also usually placed without the use of dimension lines.
- **Site components:** Objects such as light poles, signs, catch basins, parking boundaries, and the structure are located using dimension and extension lines.

Objects such as sidewalks or planters can often be described in a note rather than by dimensions. Figure 15-12 shows examples of each type of dimension. Notice the symbol used to designate each property

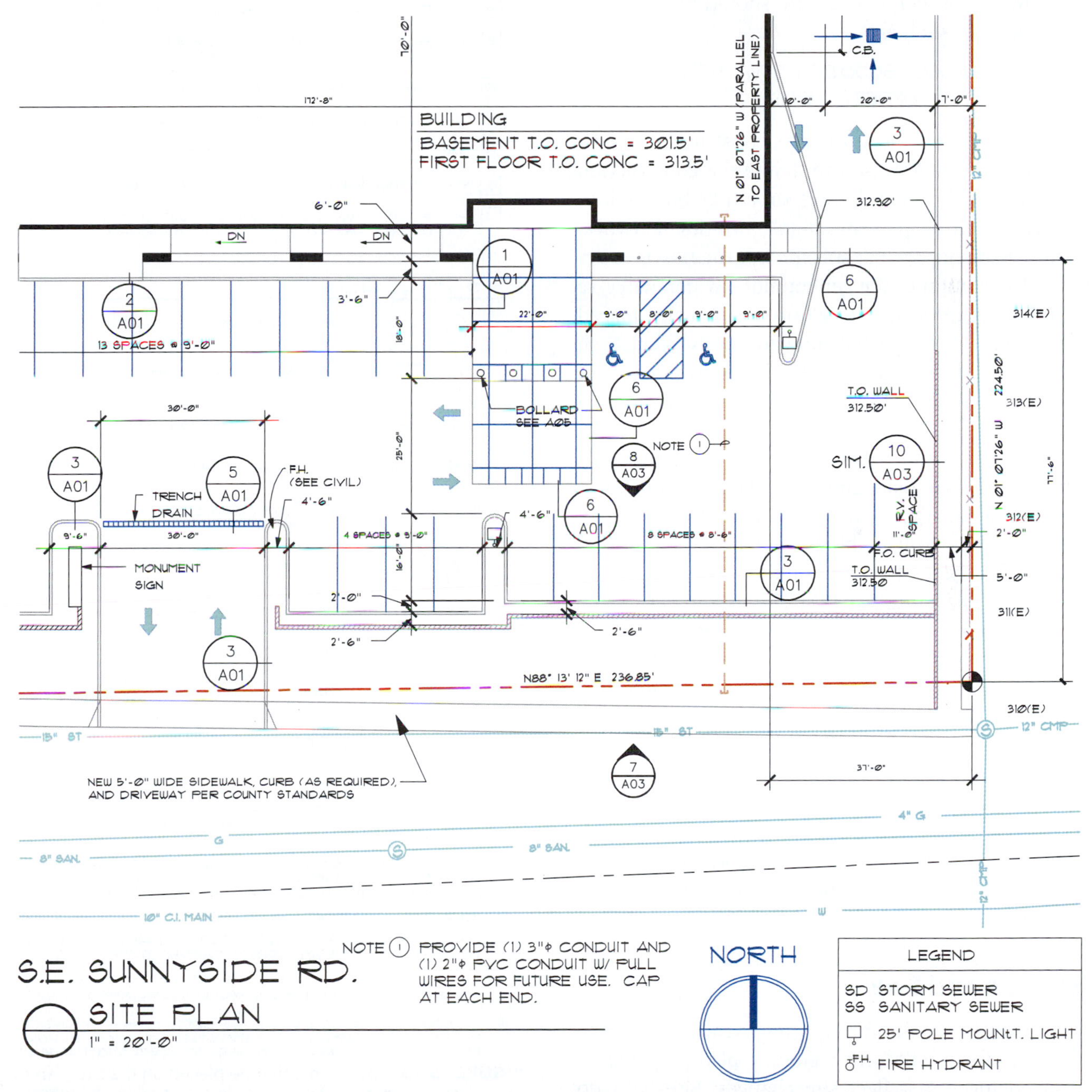

FIGURE 15-12 Site dimensions are expressed in engineering units and placed parallel to the property line they describe. Overall sizes of the structure are placed parallel to the side of the structure they describe and expressed in feet and inches. Both of these types of dimensions are placed without the use of dimension and extension lines. All signs, parking boundaries, light poles, and drains should be dimensioned using extension and dimension lines with text expressed in architectural units. *Courtesy Peck, Smiley, Ettlin Architects.*

corner. The symbol is often omitted from rectangular sites but is very helpful in locating small changes of angle on irregularly shaped lots.

Annotation

In addition to drawing and locating information, notes must be placed on the site plan to completely specify required construction. General annotation on a site drawing should be placed using the guidelines given in Chapter 3. Text describing the names of streets, and describing the proposed structure should be treated as titles. Notations such as

A PROPOSED 3000 SQ. FT. TWO-STORY
OFFICE COMPLEX

are used to define the proposed construction. Local and general notes are used to define the construction to be completed. General notes should be broken into categories such as parking, signs, utilities, flatwork, and miscellaneous so that specific information is easier to find. In addition to the general note, which will give the full specification, a local note is usually placed on the plan with a partial note to clarify what each symbol represents. Figure 15-13 shows an example of common notes and symbols that are often placed on a site plan.

Notice that these notes are referenced to specific portions of the site plan, similar to the site plan shown in Figure 15-10. Abbreviations exclusive to site-related drawings are also used. Abbreviations common to site-related drawings can be found on the student CD in the RESOURCE MATERIAL/CHAPTER 15 folder. Many plans include common abbreviations used throughout the drawings on a title page. In addition to information used to describe the construction site, general information is also placed on the site plan. General information might include a table of contents, a list of consultants, and information about the overall construction project. This would include information about the occupancy, construction type, and building areas per floor. An alternative to placing this general information on the site plan is to provide a title sheet that includes the vicinity map.

Elevation and Swale

Major changes of contour requiring excavation of large quantities of soil will have a grading plan prepared by a civil engineer to reflect site changes. Sites that do not require extensive excavation often reflect finished grade **elevations** on the site plan. Four common methods of denoting elevation can be seen in Figure 15-14.

KEY NOTES (□ SYMBOL)

1 — EDGE OF EXIST. PAVING- PROVIDE NEW ASPH. PAVING AND BASE PER CITY STANDARDS AND TRANSITION FLUSH W/ EXIST. PAVING VERIFY GRADE ELEVATIONS
2 — 4" ⌀STEEL PIPE BOLLARD- PAINT TO MATCH BLDG. TRIM (5/A1)
3 — 4" WIDE PAINTED STRIPES- WHITE COLOR
4 — HCAP PARKING SIGN- TYP. (6/A1)
5 — NEW 4'-6" WIDE CONC. WALK (2/A1) (3/A1) SIM.
6 — NEW ASPHALT PAVING OVER COMPACTED GRAVEL BASE
7 — NEW CONC. ISLAND (WALK) PROVIDE CONTROL JOINTS (CJ) AT MAX 5'-0" O.C. ALIGN W/ WINDOW MULLION AS NOTED ON PLANS W/ EXPANSION JOINTS (EJ) AT MAX. 25'-0" O.C. UNLESS NOTED OTHERWISE ON PLANS (2/A1)
8 — OPENING THRU CURB FOR DRAINAGE SEE CIVIL & GRADING PLANS
9 — REMOVE EXIST. AC PAVING AND CONC. CURBS & GUTTERS, ETC. (SEE DEMO. SITE PLAN / SURVEY FOR ADDITIONAL NOTES)
10 — PAINTED HCAP. PARKING SYMBOL- BLUE COLOR (4/A1)
11 — NEW DRIVEWAY & CURB CUT PER CITY STANDARDS
12 — NEW CONC. CURB, GUTTER AND SIDEWALK PER CITY STANDARDS
13 — "COMPACT" PAINTED ON PARKING STALL- WHITE COLOR
14 — RED COLOR PAINTED CURB (TOP & SIDE) & WHITE COLOR LETTERS STATING "NO PARKING FIRE LANE" ON CURB FACE.

SITE LEGEND

— EXISTING MAN HOLE TO REMAIN
— NEW LIGHT POLE AND BASE (KIM # 1A-EKG401-150HPS 120 -BLE OR APPROVED SUBSTITUTE MOUNT @16' ABOVE LIGHTED SURFACE (4/A2)
— EXIST. LIGHT POLE AND BASE TO REMAIN
CB. — EXISTING CATCH BASIN TO REMAIN
— NEW VEHICLE DIRECTION ARROW PAVEMENT MARKING
— NEW PAINTED HANDICAP PARKING SYMBOL (4/A1)
AD. — NEW AREA DRAIN SEE CIVIL DRAWING C-
CJ — CONTROL JOINT SEE DETAIL (2/A1)
EJ — EXPANSION JOINT SEE DETAIL (2/A1)

SITE PLAN NOTES

1. COORDINATE ALL WORK UNDER THIS CONTRACT WITH EXISTING UTILITIES AND STREET WORK ON AND OFF THE SITE. MAKE ALL ADJUSTMENTS REQUIRED FOR CONNECTIONS OR ALTERATIONS CAUSED BY WORK UNDER THIS CONTRACT.
2. PATCH ALL PAVING (ASPHALT, CONC.,ETC.) WHERE UTILITIES ARE ADDED OR MOVED. SAWCUT EDGES WHERE EXISTING PAVING IS REMOVED. PATCH IN A SMOOTH, EVEN MANNER AS DIRECTED BY ARCHITECT.
3. SEE SHEET #A2 FOR NEW AND EXISTING GRADES.
4. COORDINATE ALL STREET WORK WITH THE CITY OF LAKE OSWEGO ALONG S.W. LANA DRIVE- ALL STREET WORK TO CONFORM TO APPLICABLE CITY STANDARDS.
5. COORDINATE ALL WORK IN THIS CONTRACT WITH WORK DONE BY OTHERS.
6. THE CONTRACTOR IS NOT TO USE COLLINS WAY TO THE WEST OF THE SITE FOR SITE ACCESS, PARKING FOR HIS EMPLOYEES AND SUBCONTRACTORS, STAGING AREA OF ANY KIND, OR DELIVERY OF CONSTRUCTION MATERIALS.

FIGURE 15-13 Notes can either be placed on the site plan as seen in Figure 15-8 or referenced to a list of general notes.

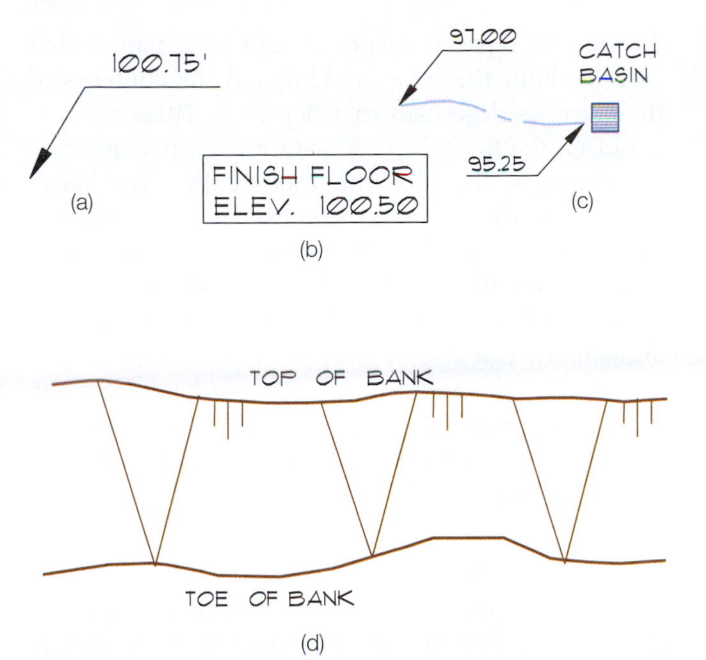

FIGURE 15-14 Four methods of referencing ground elevation include: (a) an elevation symbol, (b) a note to describe the elevation, (c) a swale line with elevation symbols, and (d) a bank indicator to represent the top and toe of a bank.

The elevation of ground level is indicated on the site plan with a note similar to

FINISH FLOOR ELEVATION 100.75'.

A symbol similar to a leader line is used to indicate the elevation of each property corner as well as other important features. The elevation of the specified location is indicated above the leader line. Comparisons of elevations can also be indicated by indicators such as TW (top of wall), which are placed above the symbol or BF (bottom of footing), which is placed below the symbol.

A third method of showing minor change of elevation is with the use of a swale indicator. A **swale** is a small valley used to divert water away from a structure. The slope of the swale is dependent on the surface material being drained. A minimum slope of 2% should be provided for dirt, and a slope of between 1% and 2% is used for asphalt or concrete paving areas.

The fourth method of showing a change of contour is with a bank indicator. A **bank indicator** is represented by a V placed between lines that indicate the top and toe of the bank. The bottom of the V is placed at the bottom of the bank. Three short lines are normally placed between the Vs to indicate the top of the bank.

Drainage

Once the structure and pavement areas have been determined, a swale for drainage can be indicated on a site plan. On the simplest of projects, swales to divert rainwater from the building perimeter can be shown on the site plan. On complex projects, the drainage system may be shown on a separate drainage plan. An underground concrete box called a **catch basin** is often placed at the low point of swales to divert water from the site. Catch basins act as a funnel to collect water from parking areas and channel it into the storm system. A metal cover that is level with the ground surface, called a drainage grate, covers a catch basin. The grate allows water to flow into the catch basin without allowing anyone to fall in. Water flows through the grate into the catch basin, and it is then funneled into pipes that connect to public storm sewers. The pipe that connects the construction site to the public sewer pipe is referred to as a *lateral*. The location of the lateral should be indicated on the site plan and be noted as *existing* or *new*.

A catch basin is placed at low areas of the structure, such as loading docks where water will naturally collect. The engineer designs the system and determines the required change of elevation to ensure proper runoff. Elevation markers referred to as spot grades should be established on the plan to specify the finished grading to ensure runoff. Markers at the corners of paved areas and at each drain should be indicated. The elevation, size, type, and location of all grates and drainage lines should also be specified on either the site or drainage plan. The grate elevation is shown using a symbol, as shown in Figure 15-14a. If a drainage plan is drawn, the elevation of the inside surface of the bottom of the drainage pipe is also specified. This elevation is known as an invert elevation. Either the minimum slope required or specific elevations along the drainage pipe are specified on a drainage plan.

Site Plan Development

Figure 15-15 shows an example of a completed site plan for a very simple site. Figure 15-16 shows a site plan that will have supplemental drawings. In its original state, the site plan also included the notes seen in Figure 15-13 and the details seen in Figure 15-17.

Figure 15-10 shows the grading information for the site. Other related drawings are introduced throughout this chapter. Because the site plan serves as a basis for so many other drawings, it is important to clearly understand what additional site drawings are required and who will be completing them. Storing all site-related drawings in one file, using external referencing, or using a separate drawing file for each drawing are the most common methods used to develop a site plan. See Chapter 3 for a review of methods for assembling drawings and for guidelines to assigning layer prefixes and titles.

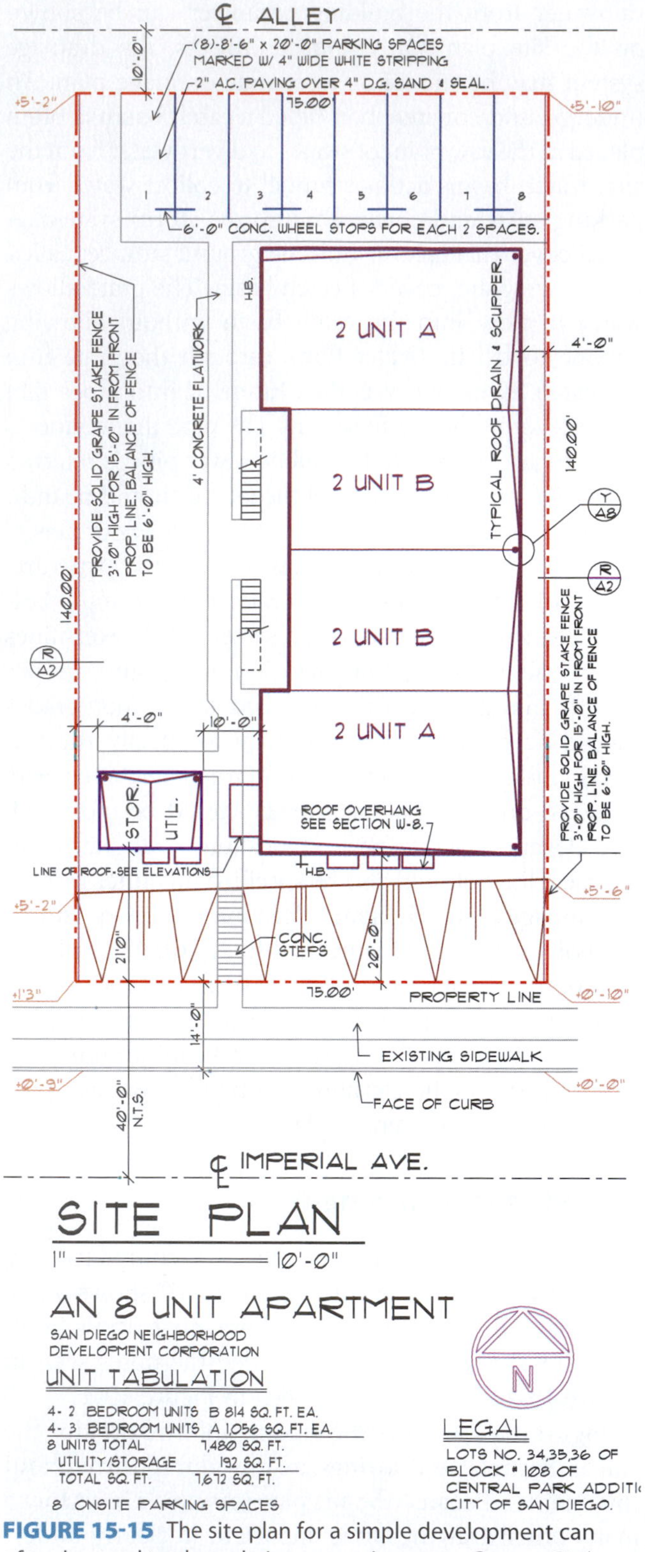

FIGURE 15-15 The site plan for a simple development can often be used to show drainage requirements. *Courtesy Gisela Smith, Kenneth D. Smith Architect & Associates, Inc.*

One Drawing File

On small-scale projects, if each of the site drawings is to be prepared by one firm, all site-related drawings can be stored within one drawing file by placing each drawing on separate layers. Assign layer names that describe both the base drawing and the contents of the layer, as described in Chapter 3. Titles such as C-BLDG, C-PROP, or C-ANNO will clearly describe the contents. To plot the topography plan, basic items from the site plan such as the building and the property and all TOPO layers should be set to THAW, with all other layers such as site annotation set to FREEZE. The use of separate layouts in one drawing file can also be used to store multiple drawings in one file. Separate layouts can be created for the site plan, topography plan, and landscape plan in one drawing, allowing easy selection of the desired drawing for plotting.

External Reference

Site-related drawings are an excellent example of drawings that can be referenced to a base drawing. A CAD technician working for the architectural team can draw the basic SITE plan information in a base drawing that will be reflected on all other site drawings. Electronic copies of this drawing file can be given to other consulting firms who will develop the landscape, sprinkler, and grading drawings. Using external referencing allows each firm to have a current drawing file as a base, while each firm progresses with its work independently.

Setting Site Plan Parameters

Offices usually have stock template drawings containing common site-related linetypes, dimension and text variables, and symbols. As a student, you may need to develop your own template drawings for site plans. Site plans are often plotted at a scale factor such as 1" = 10' (1:120) or 1" = 20' (1:240). Preferred metric scales for civil drawings are as follows:

- 1:100 (close to 1/8" = 1' − 0")
- 1:200 (close to 1/16" = 1' − 0")
- 1:500 (close to 1" = 40' − 0")
- 1:1000 (close to 1" = 80' − 0")

This requires a template drawing to be developed reflecting engineering values rather than architectural values. This can be done by making a wblock of a title sheet developed in an introductory CAD class. Parameters such as UNITS, LIMITS, SNAP aspect, and GRID sizes can then be set to meet the specific needs of the site plan. Common layers can also be set up to divide site information from other information that will be stored with the site file.

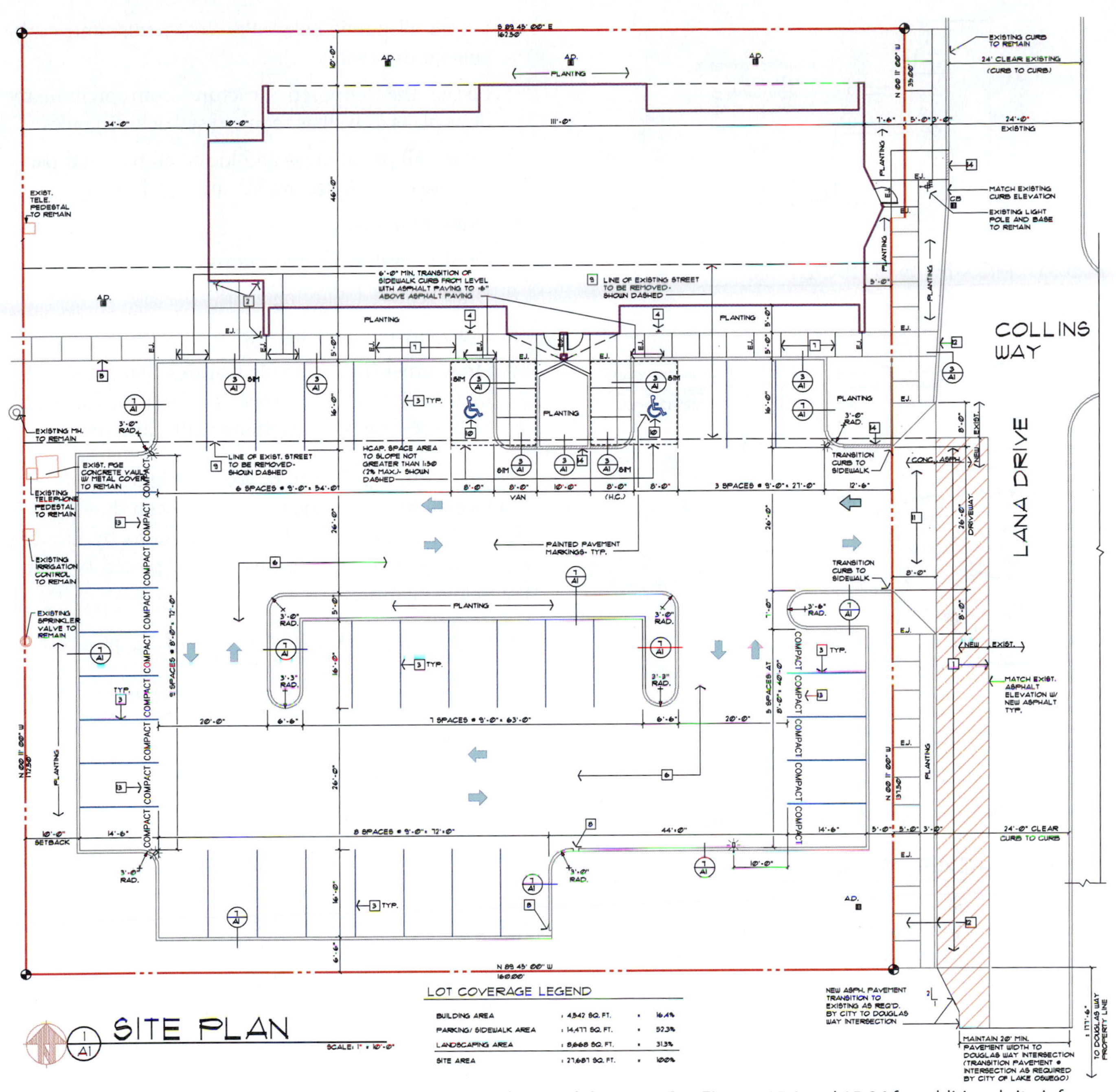

FIGURE 15-16 A site plan for a structure that will have supplemental drawings. See Figures 15-9 and 15-26 for additional site information. *Courtesy Architects Barrentine, Bates & Lee, A.I.A.*

Common Site Plan Details

The details required for a site plan vary widely from office to office and will vary based on the complexity of the construction. Common areas requiring details on a site plan include lighting supports, sign supports, parking barriers, parking painting details, sidewalks, curbs, and decorative walls. Figure 15-17 shows an example of common site details. Each is considered a standard detail and could be stored as a wblock in a file labeled as \PROTO\SITE\ (dwg file name) that is edited for specific job requirements. The engineer or project coordinator will generally note for the drafter the details that should be inserted into the site plan, and the drafter is expected to compile and edit the details to match project requirements.

COMPLETING A SITE PLAN

The software that is used to create the site drawings will greatly affect how the drawing is created. The following discussion assumes the use of AutoCAD or "AutoCAD vanilla" as many professionals refer to the software. AutoCAD works well for simple projects or in the hands

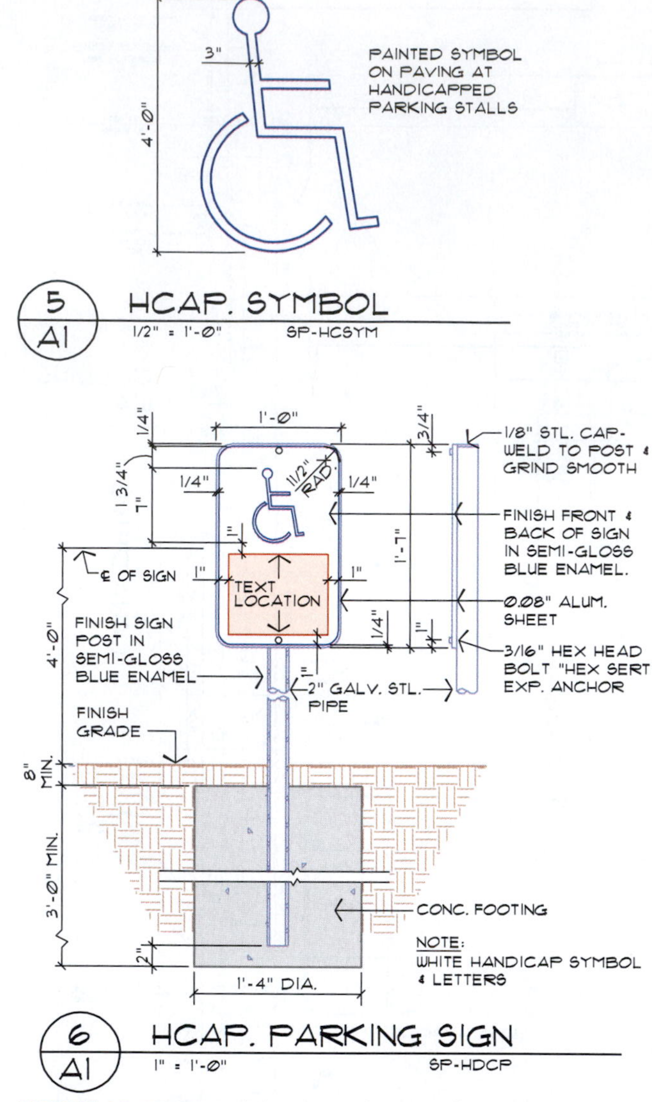

FIGURE 15-17 Typical details can be stored as wblocks and inserted into a drawing as needed. *Courtesy Architects Barrentine, Bates & Lee, A.I.A.*

of skilled CAD users. But there is a better way. The use of AutoCAD Civil 3D or AutoCAD Land Desktop can greatly increase the productivity of a skilled CAD operator. AutoCAD works with drawing geometry; AutoCAD Civil 3D is a model builder. As the site plan is created using Civil 3D, information is stored that will automate the development of the plan, profiles, and data management for site-related projects.

If you're creating drawings using AutoCAD Civil 3D the general steps will be the same, but the procedure will be much easier because of the automation provided by the software. The site plan can be completed using AutoCAD by following these steps:

1. Working from a plat map or metes and bounds legal description, lay out the shape of the site.
2. Establish the center of all access roads.
3. Locate all public sidewalks, easements, and curb-cuts for driveways.
4. Locate the proposed structure from preliminary floor plans as well as required setback distances.
5. Define all paved areas and locate all required parking spaces, drainage swales, and catch basins.
6. Draw all utilities.
7. Draw all walkways and ramps.
8. Draw required planting areas.
9. Draw any required signage.
10. Draw other human-made features that are specific to this site, such as trash enclosures, furniture, benches, flagpoles, retaining walls, and fencing.

Steps 1 through 10 can be seen in Figure 15-18.

Once all human-made items have been drawn, each should be dimensioned. Place dimensions by locating major items first. Dimensions can be placed by using the following order:

11. Dimension and list bearings of each property surface.
12. Dimension the location of the structure to the property.
13. Dimension the minimum required yard setbacks.
14. Dimension the footprint of the structure.
15. Dimension the locations of streets, public sidewalks, and easements.
16. Dimension all paved areas as well as individual parking spaces and catch basins.
17. Dimension all utility locations where they enter the site.
18. Dimension all other miscellaneous human-made features.

Steps 11 through 18 can be seen in Figure 15-19.

With all features drawn and located, add any special symbols necessary to describe the material being used. Typically, this would include hatching concrete walkways with a dot pattern. Other symbols to be added might include property corner markers, a north arrow, elevation markers, catch basins, fire hydrants, or clean-outs.

The next stage of completing the site plan is to provide notations to define all human-made features that have been added to the site. Care should be taken to distinguish between existing material and material that is to be provided. Occasionally, existing material to be removed from the site must also

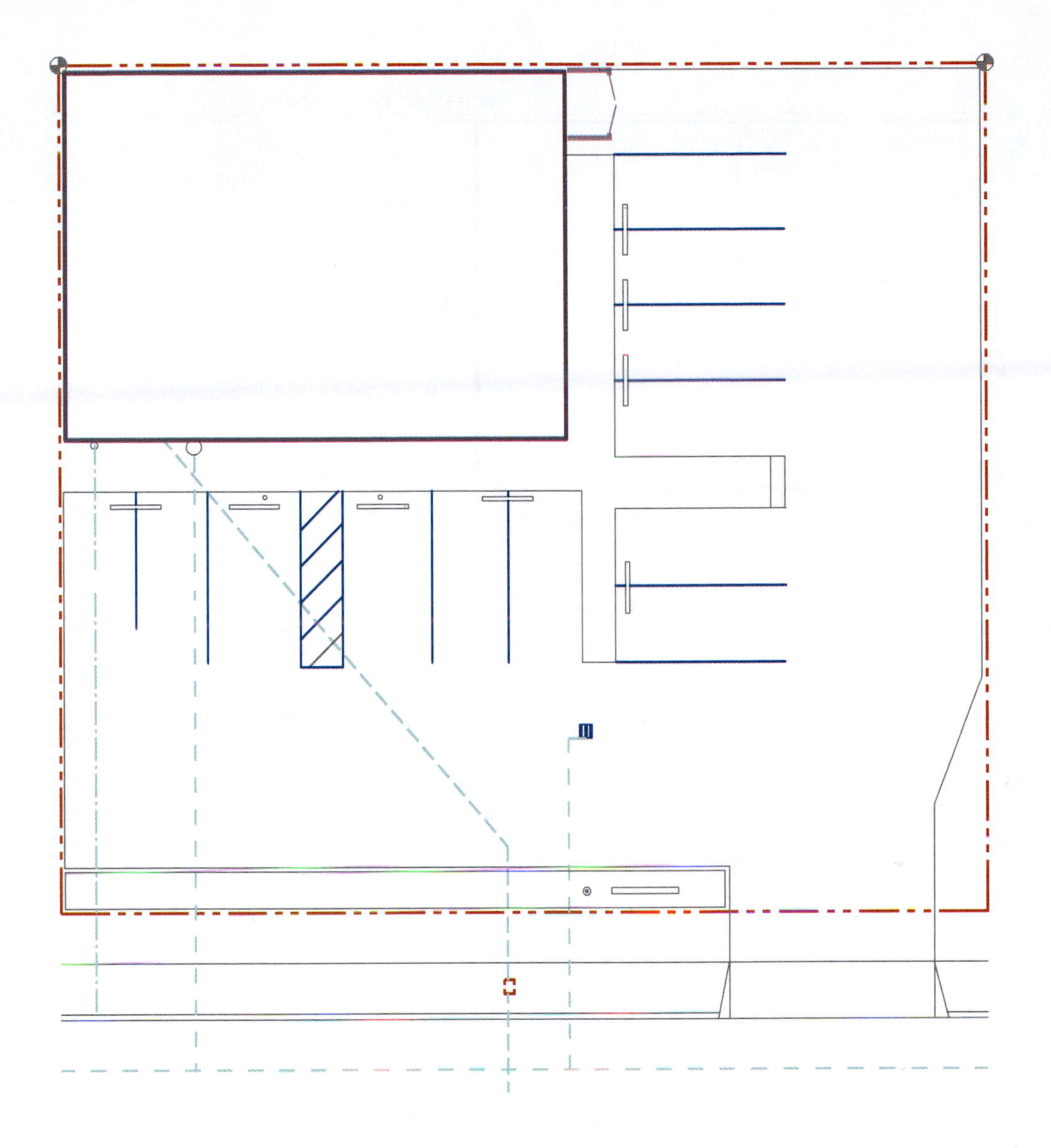

FIGURE 15-18 Steps 1 through 10 can be used to establish the basic parameters of the site plan.

be specified. Notes that should be included on a site plan are as follows:

19. Specify proposed building use and square footage.
20. Include finished floor elevations.
21. Include the legal description.
22. Specify all streets, public walkways, curbs, and driveways.
23. Describe all utilities.
24. Describe all paved areas and specific parking areas.
25. Describe specific features that have been added to the site, including furniture, flagpoles, retaining walls, fencing, and planting areas.
26. Add details to show construction of human-made items.
27. Add general notes to describe typical construction requirements.
28. Specify a drawing title and scale.

Figure 15-20 shows a completed site plan.

RELATED SITE DRAWINGS

In addition to the site and demolition plans, several related drawings could accompany a set of working drawings. Related drawings include a topography plan, grading plan, paving plan, site utility plan, landscaping plan, and sprinkler plan. A topography plan shows

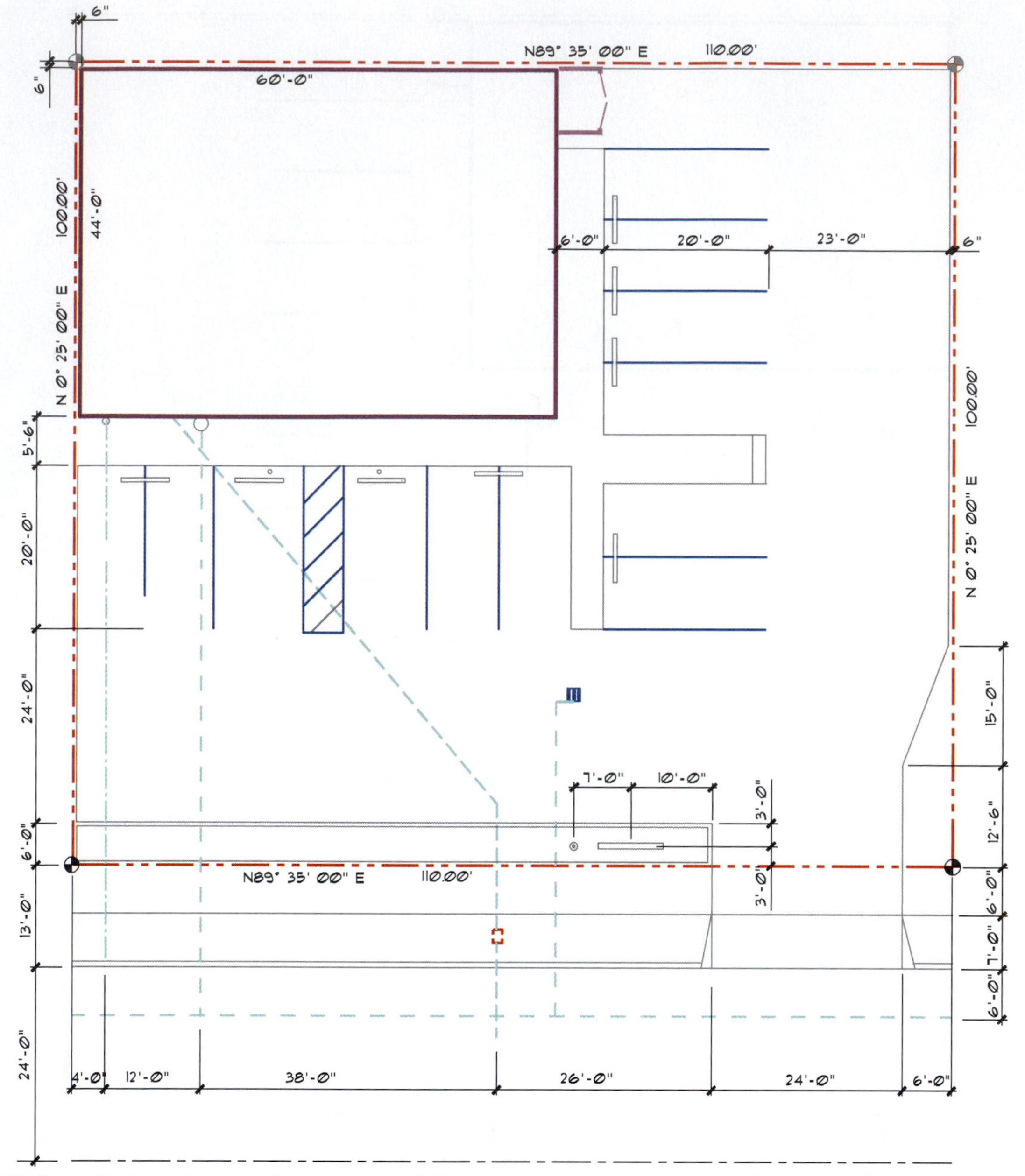

FIGURE 15-19 Steps 11 through 18 can be used as guidelines to locate parking and other human-made components as well as providing their locations.

the existing contour of the construction site. This plan is normally prepared by a licensed surveyor based on notes from a field survey or by a civil engineer. It is based on existing municipal drawings describing the site, on measurements taken by the surveyor, or by aerial photographic methods. Once the shape of the site is prepared, the results of the field survey are translated onto the drawing by a drafter working for a civil engineer. A grading plan is used to show the finished configuration of the building site. A grading plan for major projects is designed and completed under the direction of a civil engineer. A landscaping plan similar to Figure 15-21 is completed by drafters working for a landscape architect. On small projects, drafters working for the architectural team may complete the project under the supervision of the project architect. In arid climates, a sprinkler plan similar to Figure 15-22 will be required to maintain landscaping. The same team that provides the landscape plan usually completes the sprinkler drawing. Because the landscape and sprinkler drawings usually fall under the supervision of the landscape architect, procedures for creating these drawings will not be presented in this chapter.

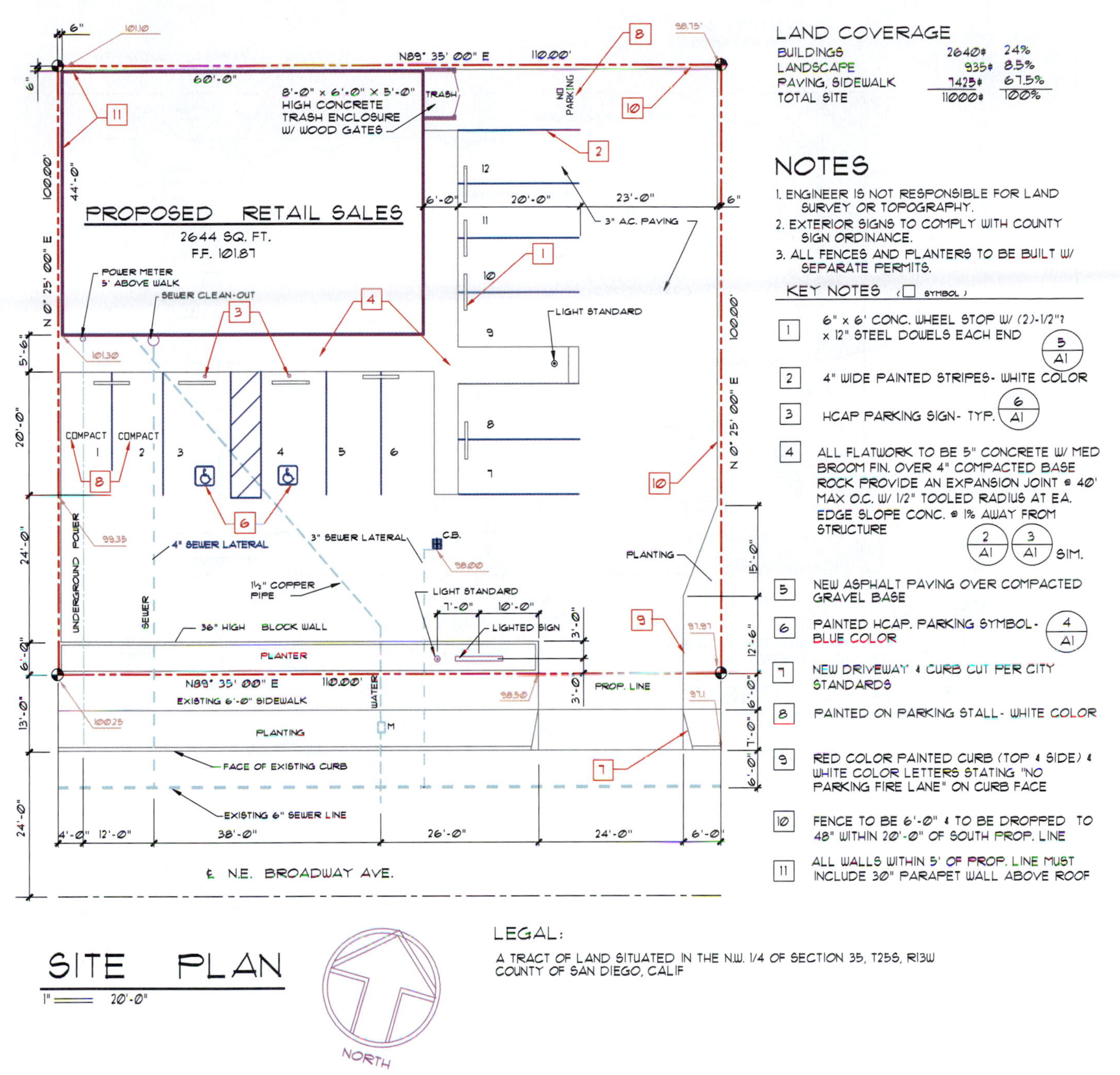

FIGURE 15-20 The completed site plan includes notes and dimensions to describe all human-made features of the project.

Topography Plans

Figure 15-23 shows an example of a sketch provided to reflect the existing surface elevations. A grid is drawn to represent each known elevation location. Notice the six elevations in the northwest corner of the survey. Between grid 69.41 and 76.44, seven contour lines are required to represent the change of elevation. It can be assumed the contours fall at an even spacing because the surveyor did not change the distance between grids to reflect a rise or depression. Six contour lines are required between grids 69.74 and 75.58, and eight contour lines are required to reflect the change in elevation between grids 69.41 and 77.49.

Once the distance between grids has been divided into the required divisions, points of equal elevation can be connected, as seen in Figure 15-24a. The use of a polyline aids in finishing the drawing. As the distance between contour lines decreases, the land becomes steeper. As the distance between contour lines increases, the land becomes flatter. The topography plan must be completed before accurate estimates of soil excavation or movement can be planned.

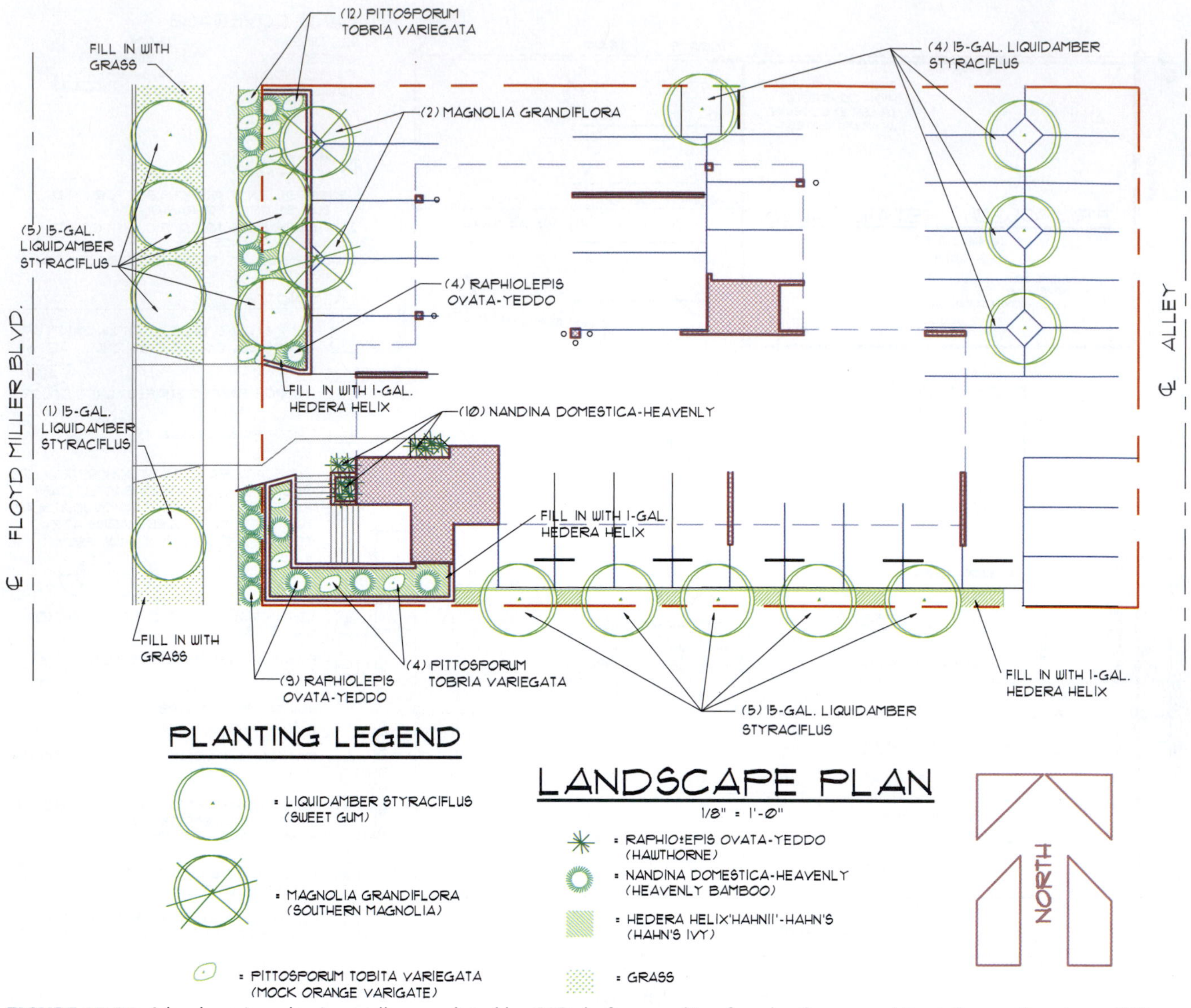

FIGURE 15-21 A landscaping plan is usually completed by CAD drafters working for a landscape architect. On small projects, CAD technicians working for the architectural team may complete the project under the supervision of the project architect.

Contour Lines and Symbols

Once the known elevations have been converted to contour lines, the lines can be curved and altered to provide clarity. Unlike the real contour of the site, in the initial layout stage, the lines run from point to point and have distinct directional change at each known point. These angular contour lines can be curved using the Fit Curve option of the PEDIT command. PEDIT leaves each vertex in its exact location but changes the straight line to a curved line. The Width option of the PEDIT command can also be used to alter the width of the contour lines. Most offices use varied lineweights to help distinguish between contours. Depending on the spacing of the contours, lines representing every 5' or 10' (1500/3000 mm) are usually highlighted and labeled as shown in Figure 15-24b.

Grading Plans

The grading plan shows the finished configuration of the building site. On simple plans, the finished grades will be placed over the existing grades, or the proposed grades can be shown separately from the existing grades. Figure 15-25 shows an example of a simple grading plan. On simple projects, a drafter working for the architectural team may complete the grading plan. A drafter working for the civil engineer translates preliminary designs by the architectural team and the topography drawings into the finished drawings. The grading plan shows the finished contour lines, areas of cut and fill, building footprint, and the outline of parking areas, walkways, patios and steps, catch basins, and drainage provisions. Figure 15-26 shows a portion of a grading plan.

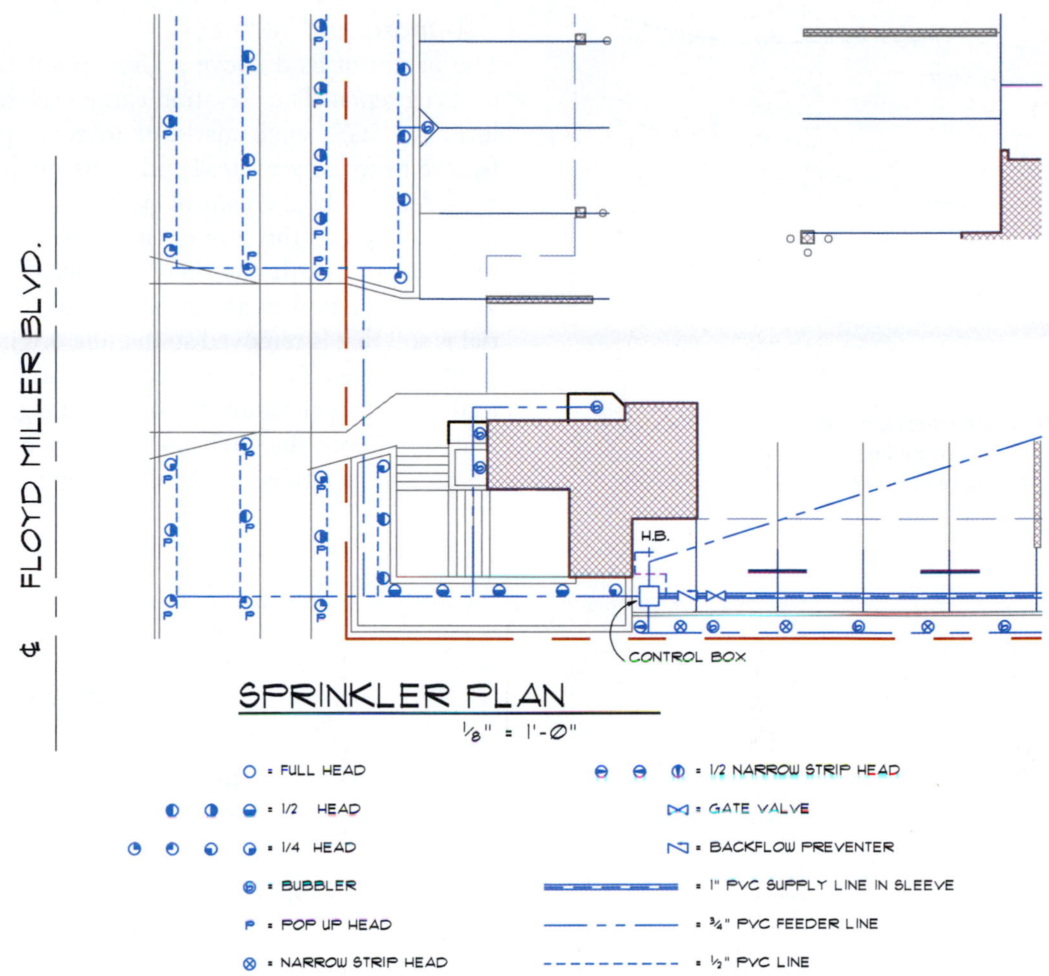

FIGURE 15-22 In arid climates, a sprinkler plan will be required to maintain landscaping. The same team that provides the landscape plan usually completes the sprinkler drawing. *Courtesy Matthew Jefferis.*

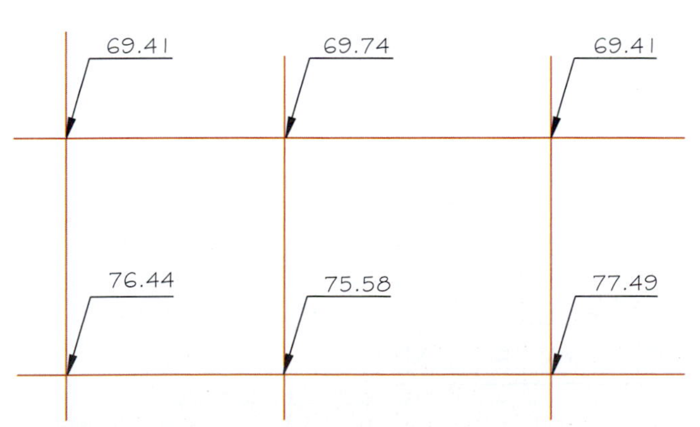

FIGURE 15-23 Existing topography is determined from spot grades determined by the survey team. These elevations can then be placed in a drawing provided by the civil engineer that lists the known elevation at specific points at the site.

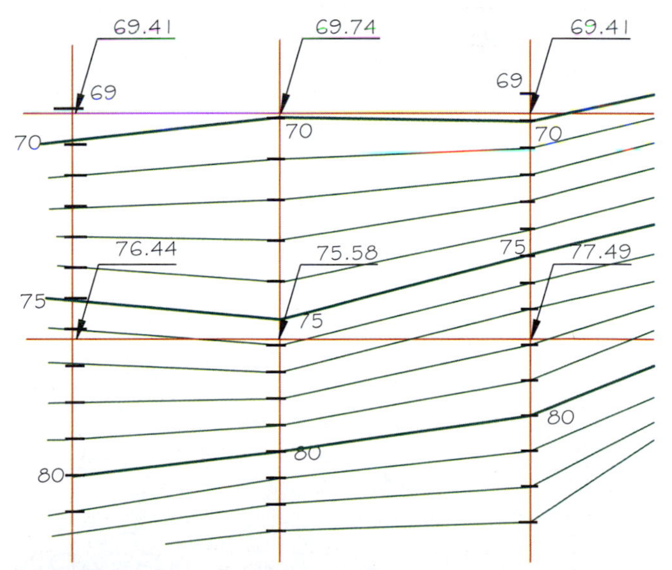

FIGURE 15-24a Once known elevations are located, lines representing a specific elevation can be placed. Grades can be determined by estimating the rise or fall between two known points. Because lines for grades 70 through 76 occur in the upper left grid, the distance between the two points is divided to represent each elevation.

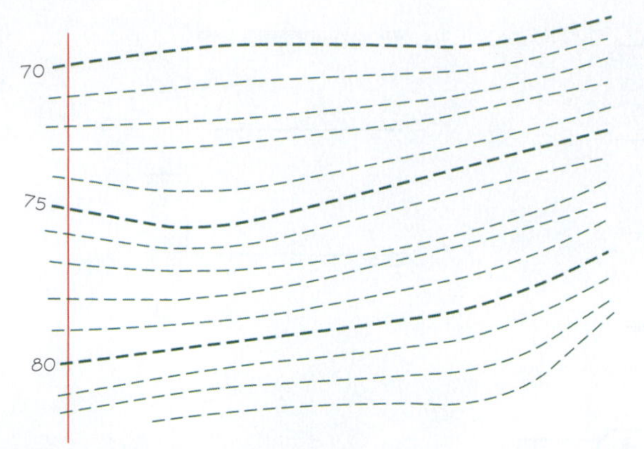

FIGURE 15-24b Dashed lines are used to represent existing contours. The angular contour lines of Figure 15-24a were curved using the Fit Curve option of the PEDIT command.

Common Land Terms

The height of land above a given point is referred to as its *elevation*. The elevation can be referenced to sea level, a USGS bench mark, or to some predominant feature near the job site. Land in its unaltered state is considered to be the *natural grade*. **Finish grade** refers to the shape of the ground once all excavation and movement of earth has been completed. Earth that is moved is referred to as cut or fill material. **Cut** material is soil that is removed so that the original elevation can be lowered. **Fill** material refers to soil that is added to the existing elevation to raise the height. The point that represents the division between cut and fill is referred to as **daylight**. The civil engineer determines

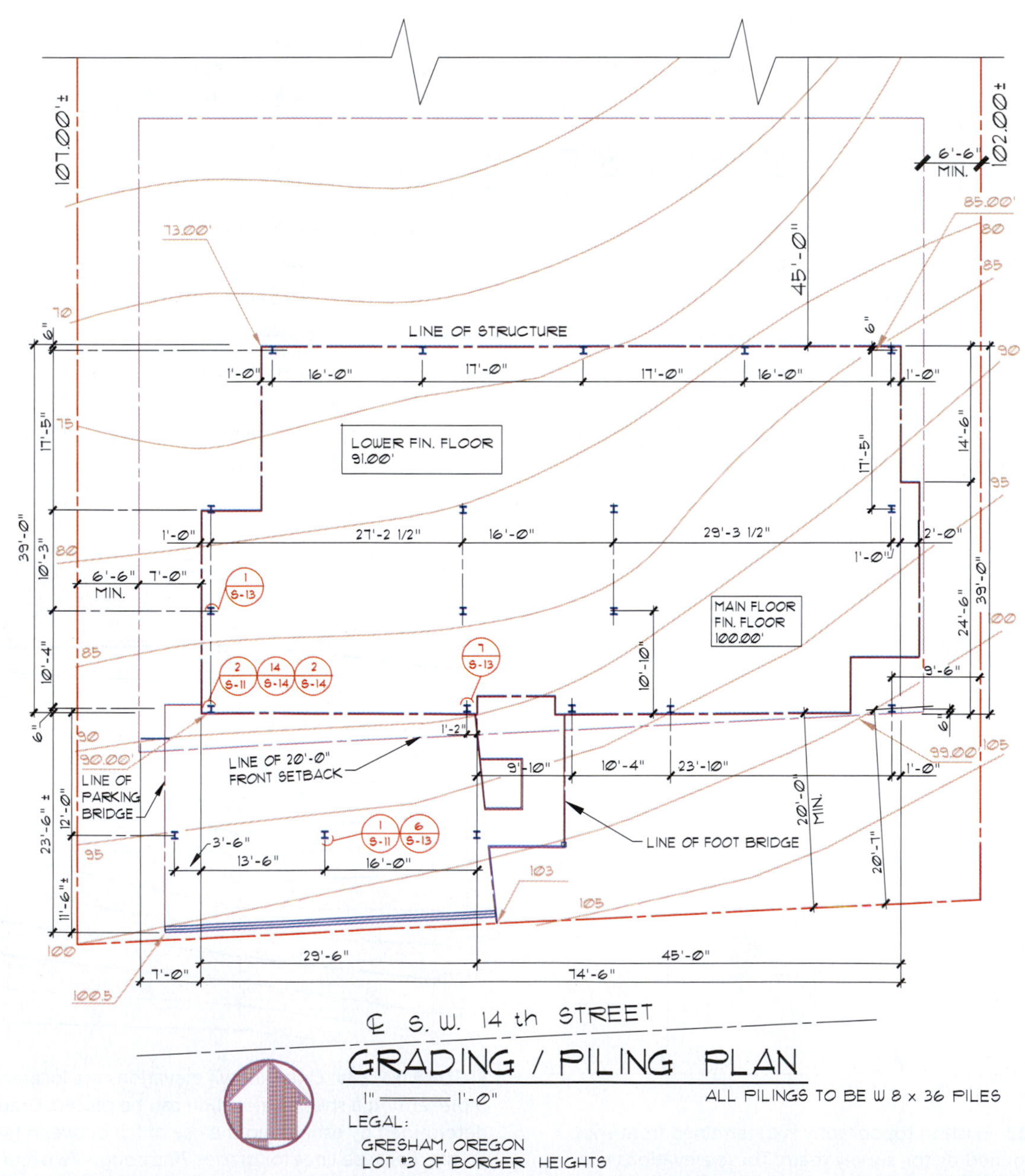

FIGURE 15-25 A simple grading plan showing the existing grades, spot elevations, piling locations, and the finish floor level of each portion of the hillside office.

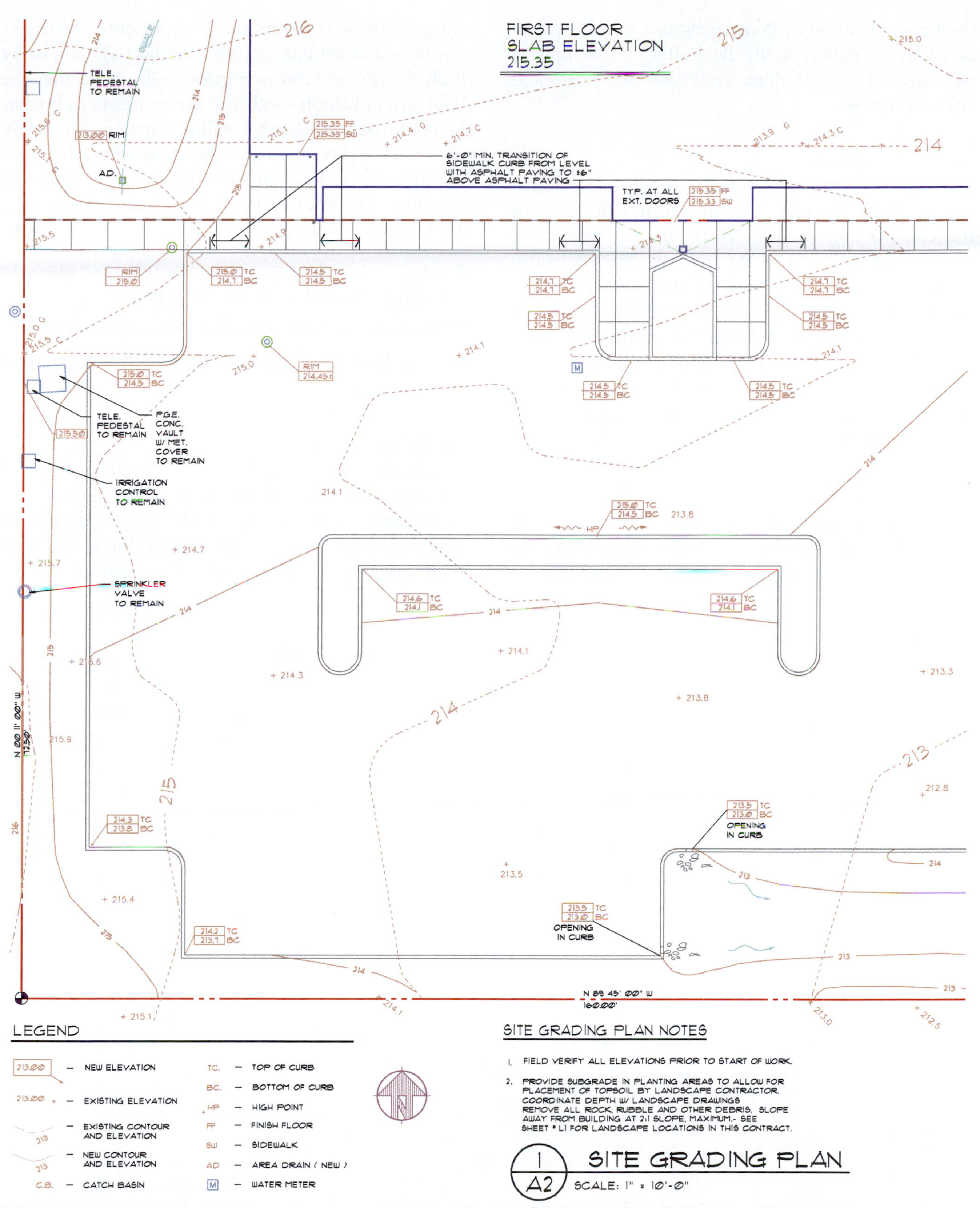

FIGURE 15-26 The grading plan is usually prepared by CAD drafters working for a civil engineer. The architectural team completes the design based on the work of the civil engineer. *Courtesy Van Domelen/Looijenga/McGarrigle/Knauf Consulting Engineers.*

the finish elevation of floor levels and major areas of concern. It is then typically the drafter's job to indicate the extent of cut and fill material based on the desired angle of repose.

Angle of Repose

The angle of the cut or fill bank that is created is referred to as the **angle of repose.** The municipality that governs the construction project determines the maximum angle of repose based on the soil type. For fill banks, a common angle of repose is often set at a 2:1 angle. For every two units of horizontal measurement, one vertical unit of elevation change can occur. A common angle for cut banks is 1.5:1. The engineer specifies variations in the angle of repose that will be allowed near footings or retaining walls to minimize loads that must be supported.

Representing Contours

If the new and existing contours are to be combined on one plan, care must be taken to distinguish between contours. Thin dashed lines often represent existing contours. New contours can be represented by thickened continuous lines, as seen in Figure 15-25, or by dashed lines, as seen in Figure 15-26. Note that the bank symbol can be added to the contours to further highlight new banks that will be created. The angle of the new bank should also be specified on the grading plan. In addition to the lines used to represent contours, key information that must be placed on a grading plan includes:

- Critical spot elevations to assure proper drainage to catch basins and other discharge points.
- Sizes of new drainage facilities with control grades.
- Spot elevations at the corners and points adjacent to the building entrances.

Grading Plan Details

Details are often required to show the construction of drainage materials and retaining walls. Figure 15-27 shows an example of each. Both types of details are wblocks, which would be edited by the drafter, based on project requirements specified by the engineer.

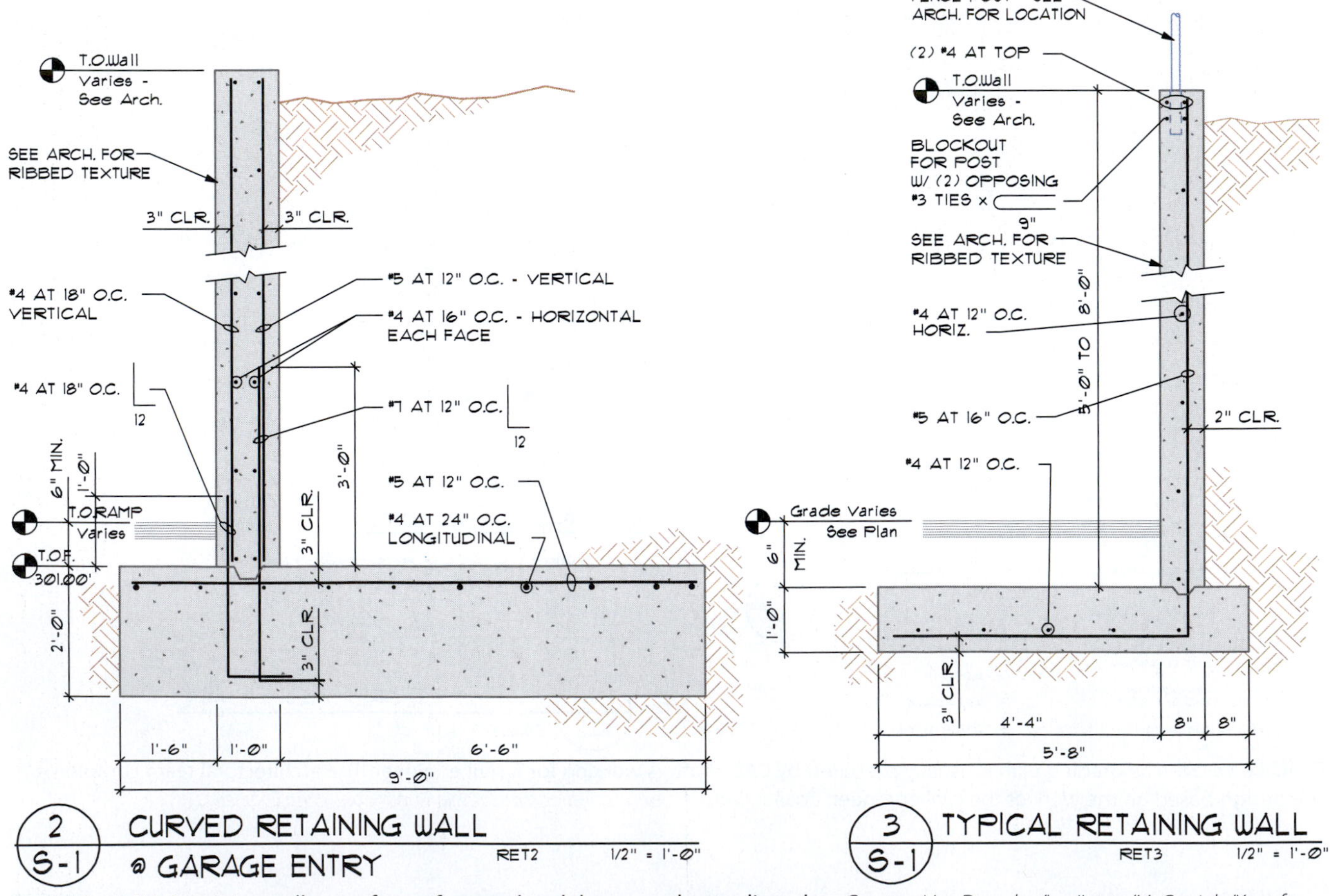

FIGURE 15-27 Retaining walls are often referenced and drawn on the grading plan. *Courtesy Van Domelen/Looijenga/McGarrigle/Knauf Consulting Engineers.*

EASY BEING GREEN

Site selection and development is a key area of green construction. The LEED rating system awards credits for sustainable sites in several areas. Areas of site-related work and relevant CSI division numbers for future research include:

- Erosion and Sedimentation Control—This area of site development addresses methods of construction in ecologically sensitive areas including:
  - 07—Thermal and moisture protection
  - 31—Earthwork
  - 32—Exterior improvements
  - 33—Utilities
- Urban Redevelopment—The efficiency of transportation to and from a structure can greatly affect the environment. This area of the LEED rating system allows credit for the transportation considerations of a project that influence the location and various land-use features on the measure of energy use. The primary focus of this credit is on energy usage and the associated environmental impacts of energy use such as pollution. Measures to reduce transportation energy use can have very significant ancillary benefits relating to water runoff, urban heat island mitigation, and habitat protection, while creating more vibrant, livable communities.
- Brownfield Redevelopment—An important strategy for protecting prime agricultural land, open space, and woodlands against the forces of sprawl is to build within existing urban areas on previously disturbed sites. Many abandoned sites suffer the scars of use in a time when environmental controls were unknown or unheeded. Their soil and groundwater may be contaminated with toxic heavy metals, petrochemicals, and soil gases. Such gases include methane from decomposing organic matter or landfills, hydrocarbons from underground fuel spills, pesticides that have been used around buildings, and other volatile organic compounds (VOCs). A LEED credit can be gained by recycling, or reusing these abandoned structures or sites.
- Alternative Transportation—Studies by *Environmental Building News* have shown that an average commercial office building in the United States built to American Society of Heating, Refrigerating, and Air-Conditioning Engineers (ASHRAE) 90.1-2004 energy standards uses more than twice as much energy in getting workers to and from work as the building uses in its operation. A LEED credit is awarded for a wide range of strategies regarding land-use planning in relation to transportation, which can significantly reduce the transportation energy intensity of a building.
  - See CSI Section 12—Furnishings for products related to this category.
- Reduced Site Disturbance—This area of the LEED rating system provides credit for designing and building structures that work with landscapes rather than against nature. This area of the standard deals with the protection of trees during site work, restoration of wetlands, selection of environmentally responsible landscape materials, and infiltration strategies for storm water management. Relevant areas of CSI for research related to this area include:
  - 07—Thermal and moisture protection
  - 31—Earthwork
  - 32—Exterior improvements
- Storm Water Management—The environmental impact of oil and other chemicals that collect in parking lots can be severe. In many developed areas, storm water runoff is the single largest source of water pollution. For most municipalities, that means these chemicals will usually make their way into streams, rivers, and lakes. This section offers LEED credit for how effectively runoff is controlled and cleaned. Key areas of CSI for research related to this area include:
  - 07—Thermal and moisture protection
  - 10—Specialties
  - 22—Plumbing
  - 31—Earthwork
  - 32—Exterior improvements
  - 33—Utilities
- Landscape and Exterior Design to Reduce Heat Islands—This area of LEED offers credits for methods of reducing the heating and cooling loads placed on a structure. Key areas considered in the standards include products that add to the ability of a structure to reflect heat, and products that aid in keeping the heat they produce. Products range from reflective paints to green roofs (see Chapter 18). Relevant areas of CSI for research related to this area include:
  - 06—Wood, plastics, and composites
  - 07—Thermal and moisture protection
  - 10—Specialties
  - 32—Exterior improvements

(continued)

- Light Pollution Reduction—Although you may think that light pollution only affects your ability to look up and see the night sky, light pollution is wreaking havoc in certain natural systems such as sea turtle nesting in Florida and migrating birds and sycamore trees in urban parks. Not just related to animal habitat, there are some who suggest that human health is affected by lack of darkness. This area of the LEED standard provides credits for dealing with light pollution because light pollution is really often a matter of poor planning and wasted energy. Through proper planning and usage, additional green benefits can be gained by reducing the amount of electricity that must be generated. Relevant areas of CSI for research related to this area include:
  - 01—General requirements
  - 26—Electrical

ADDITIONAL READING

The following Web sites can be used as a resource to help you keep current with changes in professional design careers.

ADDRESS	COMPANY/ ORGANIZATION
www.oars97.com	AbTech Industries (Clean water solutions)
www.itsrecycled.com	American Recycled Plastics, Inc. (Landscaping products)
www.clivusmultrum.com	Clivus Multrum, Inc. (Water conservation)
www.contechstormwater.com	Contech Stormwater Solutions
www.curbappealmaterials.com	Curb Appeal Materials (Landscaping products)
www.ernstseed.com	Ernst Conservation Seeds
www.hancor.com	Hancor (Storm water control)
www.weathertrak.com	HydroPoint Data Systems (Water management)
www.invisiblestructures.com	Invisible Structures, Inc. (Storm water management)
www.nedia.com	Nedia Enterprises (Erosion control)
www.nesea.org	Northeast Sustainable Energy Association

KEY TERMS

Angle of repose
Bank indicator
Bearings
Boundaries
Catch basin
Cut
Daylight
Elevations
Fill
Finish grade
Legal description
Legal description, metes and bounds
Metes
Sections
Site plans
Swale
Townships
Vicinity map

CHAPTER 15 TEST Land Descriptions and Drawings

QUESTIONS

Answer the following questions with short complete statements. Type the chapter title, question number, and a short complete statement for each question using a word processor. Some answers may require the use of vendor catalogs or seeking out local suppliers. If math is required to answer a question, show your work.

Question 15-1 What is a monument as it relates to a site drawing?

Question 15-2 List four units of measurement that might be referred to in a legal description.

Question 15-3 What two directions are used to commence a bearing?

Question 15-4 What would the following designation represent on a site plan? 100.67'S37° 30'45"W

Question 15-5 What would the following designation represent on a site plan? S28AA, T3S, R1E

Question 15-6 List three components of a legal description commonly used for incorporated areas.

Question 15-7 List two common sources of site drawings.

Question 15-8 List and describe three methods of organizing site-related files for a small commercial project.

Question 15-9 Explain the difference between a topography plan and a grading plan.

Question 15-10 What is the purpose of a vicinity map?

Question 15-11 Sketch examples of linetypes to represent the following materials: property line, centerline, easement, sewer, water line, parking stop, and curb edge.

Question 15-12 List and describe two common hatch patterns that can be used to highlight a building.

Question 15-13 List the minimum size requirements for a compact, full, and handicapped parking space that is perpendicular to the access route.

Question 15-14 List two common methods of removing rainwater from a site.

Question 15-15 Give an example of how the top of a wall with an elevation of 100.67' resting on a footing supported by soil with an elevation of 95.75' would be specified on a site plan.

Question 15-16 Sketch a bank indicator representing a slope.

Question 15-17 How are existing and new grades typically represented on a plan?

Question 15-18 Describe the process of changing field sketches to a topography map.

Question 15-19 What is an angle of repose?

Question 15-20 Give the common proportions of cut and fill banks.

Question 15-21 Verify with a local zoning department the minimum size requirement for a compact, full, and handicapped parking space that is parallel to the access route.

Question 15-22 Determine with the local zoning agency the required back-out space for perpendicular parking spaces.

Question 15-23 Determine with the local zoning department the legal description of three prominent structures in your area.

Question 15-24 Obtain a land map and legal description from a local title company of the property your school is built on.

Question 15-25 Interview a local landscape architect, civil engineer, and landscape contractor to assess the drafting opportunities in your area and to obtain examples of the types of drawings local professionals expect drafters to draw.

Question 15-26 Take a minimum of 15 photographs representing the installation or completion of work specified on a site, grading, landscape, and sprinkler plan.

DRAWING PROBLEMS

Note: The skeleton of each drawing problem can be found on the student CD in the DRAWING PROBLEMS/CHAPTER 15 folder. Use these drawings as a base to complete the assignment.

***These base drawings are to be used as a guide only.* Keep in mind that the technician who started these drawings was fired. It is your responsibility to ensure that all objects in a drawing meet the minimum company standards.**

Minimum Drawing Standards

Unless your instructor provides other instructions, draw the site plan that corresponds to the floor plan that will be drawn in Chapter 16, using appropriate line and scale factors for plotting at the designated scale. Use appropriate drawing methods to represent and label all material. Provide the following minimum materials for each site plan:

- Unless a specific scale is provided for a project, establish drawing parameters to allow for the plotting of the drawing at the largest scale possible on a D size template.

- Specify all material based on common local practice.
- Use separate layers for each major feature, text, and dimensions. Create a layer titled MISC for random features.
- Provide dimensions to locate all features.
- Hatch materials such as landscaping, walkways, and the proposed structure with the appropriate hatch pattern.
- Use an appropriate architectural text font to label and dimension each drawing as needed. Refer to *Sweets Catalogs*, vendor catalogs, or the Internet to research needed sizes and specifications.
- Keep all text 3/4" minimum from the object being described, and use an appropriate architectural style leader line to reference the text to the drawing.
- Provide a detail marker, with a drawing title, scale, and problem number below each problem.
- Provide and number the specified parking spaces. Areas in front of all overhead doors (coordinate with the floor plan in the next chapter) to be marked "loading zone, no parking."
- Insert the standard notes from the student CD/RESOURCE MATERIAL/CHAPTER 15 into your site plan. Unless your instructor gives you other instructions, edit the notes to meet your specific project and your area of the country.
- Assume all utilities to be in the street unless noted at the following distances from the property line:
  - Storm sewer 3'
  - Sewer 6'
  - Water 15'
  - Telephone 18'
  - Specify a clean-out 24" from structure in the sewer line
  - Specify that all utilities are to be underground
- Establish swales to provide drainage to catch basins that drain to the storm sewer.
- Assume 24'-wide curbcuts with streets to be 40' from property line to the centerline with a 5'-wide sidewalk located 3' from the property line.
- Provide 14'-high light standards to light all public parking areas.
- Use the site and the legal description provided with each project. Create a vicinity map for your project assuming that it will be built at the current location of your school. If a legal description is not provided, call a title company in your area and order a plat map for a local business.

The site drawings for projects 15-3 through 15-6 are related to the projects drawn in the balance of Sections 3 and 4, and will require referencing sketches in other chapters to gain additional information. Use the following assumptions to complete the details found in each chapter in Section 2, 3, and 4. These "assumptions" should be considered as company standards.

> *Warning:* Submitting drawings that do not meet the following minimum standards may earn you the wrath of your supervisor, slow your progress in achieving a pay raise, or lead to your dismissal. Remember these are minimum standards. Your drawings should exceed them.

Correcting Drawing Errors

As you progress through the projects contained in Chapters 15 through 24, you will find that some portions of the drawings do not match things that have already been drawn. Each project has errors that will need to be solved. The errors are placed to force you to think in addition to draw. Most of the errors are so obvious that you will have no trouble finding them. If you think you have found an error, do not make changes to the drawings until you have discussed the problem and possible solutions with your engineer (your instructor).

Order of Drawing Precedence

As a CAD drafter, you will need to sort through conflicting information. It is your responsibility to coordinate the material that you draw with other drawings for your selected project. To sort through conflicts, use the following order of precedence:

1. Written changes given to you by the engineer (your instructor) as change orders.
2. Verbal changes given to you by the engineer (your instructor).
3. The engineer's calculations (written instruction in the text).
4. Sketches provided by the engineer (sketches provided in each of the following chapters).
5. Lecture notes and sketches provided by your instructor.
6. Sketches that you make to solve problems; verify alternatives with your instructor.

Problem 15-1 Vicinity Map. Draw a vicinity map of your school showing major access routes and important landmarks within a five-mile radius.

Problem 15-2 Legal Descriptions. Use the following description to draw and label a site plan that could be plotted at a scale of 1/8" = 1'–0".

Beginning at a point that is the NE corner of the G.M. Smith D.L.C., which lies 250 feet north of the centerline of N.E.122 street which is in Section 1ab, Township 6 North, Range 1 West of Los Angeles County, California thence South 225.00 feet to a point which is the Southwest corner of said property, thence North 85.00 feet 1°25'30" West along the easterly edge which lies along N.E. Sweeney Drive (25' to street centerline) to the Northwest property corner, thence 140.25 feet South 89°15'00" East, thence South 85.25 feet 1°25'30" East, thence westerly along the North boundary of N.E. 122 street to the true point of beginning.

Problem 15-3 Site Plan. Use the attached sketch to draw a site plan to be plotted at a scale of 1" = 10'–0". The lot is 120' × 110' and is Lot 4 and 5 of Arrowhead Estates, Rapid City, South Dakota.

The project fronts onto Sioux Parkway with 36' from the property line to centerline of the roadway. Tee water to the property 75' east of the westerly property line from a main line 30' south of the property line. Provide a water shutoff valve in the garage of each unit. Access city sewer line, which is 25' south of the property with a new line 25' east of the westerly property line. Provide a clean-out in the lower planting strip of each unit. Provide a continuous drainage grate across each driveway. Set each drain 3" below the finish floor level at each door. Provide a 7.5' wide easement along the northerly property line for a telephone easement. Provide each unit with a fenced yard. Dimension each unit as needed and specify all grades. Interpolate as needed to determine the grades for each corner of the structure.

Problem 15-4 Site Plan. Use the attached drawing to complete the site plan. Provide a minimum of 13 parking spaces including (1)-handicapped space, and a maximum of (3)-8 × 16 compact spaces. Provide an 8 × 8 × 6 foot high enclosed trash area with double wood gates.

Use the following legal description to complete the project:

Beginning at a point, which is the Southeast corner of the Stark D.L.C., which is in section 34AA, Township 5 South, Range 2 East of San Diego County, California. East 13.72 chains to a point, which is the Southwest corner of the said property. Thence 120.00' North 16° 20' West to the Northwest corner; thence 125.00' North 73° 35' East; thence 120.00' South 16° 25' East; Thence +/– 125'.00 back to the true point of beginning.

Problem 15-5 Site Plan. Use the attached drawing to complete the required site plan. Coordinate the parking areas between door openings and ensure that no overhead doors will be blocked by public parking. Provide curved corners on all planting curbs. Locate each catch basin centered in each driveway.

Problem 15-6 Site Plan. Use the attached drawing to complete the required site plan. Coordinate the parking areas between door openings and ensure that no overhead doors will be blocked by public parking. Provide curved corners on all planting curbs. Locate each catch basin centered in each driveway. The elevation marked 14.90' on the east property line is 41' from the north property line. Show grade lines for each 1' grade interval.

Problem 15-7 Topography. Use the attached drawing to lay out the represented grades. Draw the site plan using a scale of 1" = 10'–0". Show grades at 1' intervals and highlight contours every 5'.

Floor Plan Components, Symbols, and Development

16

The floor plan forms the core of the architectural working drawings and is typically the first drawing created as the drawing set is being developed. Information on the floor plan will impact nearly every other drawing in the architectural and structural drawings. A **floor plan** is created by passing a horizontal cutting plane through the walls at approximately five feet above the floor level and removing the upper portion of the structure. This process is repeated for each level of a structure to reveal the locations of all prominent items for that specific level. The floor plan also serves as a basis for the drawing of the ceiling, framing, mechanical, plumbing, and electrical plans. Key elements shown on the floor plan include the walls, doors, windows, and cabinets. This chapter will explore:

- Floor plan development.
- Floor plan components including walls, windows, doors, openings, plumbing symbols, and interior equipment.
- Placing dimensions and the changes required when dimensioning light-framing, timber, masonry, and concrete construction.
- Placing drawing annotations including door, window, and finish schedules.
- Drawing symbols including north arrows; grade, section, detail, and elevation markers; and wall construction symbols.
- Steps for completing a floor plan.
- Floor plan related drawings including space plans and reflected ceiling plans.
- Layering.

FLOOR PLAN DEVELOPMENT

Initially, the floor plan is developed as a schematic drawing by the architect to meet the design needs of the client. This plan is initially sketched on either graph or sketch paper to define the arrangement of basic shapes. Figure 16-1 shows an example of a preliminary floor plan. Once the best possible solution is determined, the plan is refined and drawn to scale. Depending on the office structure, and the skill of the drafter, this may be the first involvement of a drafter in the development of the floor plan. Generally, a senior drafter or designer who will head the project draws the preliminary floor plans for presentation to the client, planning commissions, and review boards. Figure 16-2 shows an example of the preliminary drawing of the floor plan. During this stage of development, the floor plan and other preliminary drawings are given to other consulting firms for preliminary input on structural, mechanical, and plumbing considerations. Keep in mind that as the preliminary drawings are prepared, the floor plan is developed in conjunction with the elevations, site and roof plans, and sections.

Although the preliminary floor plan may be drawn first, as each of the other preliminary drawings is developed, changes may be required for the floor plan.

Once input from the owner, review boards, and other consultants has been compiled, the floor plan to be used for the construction drawings can be started. This plan is usually completed by a drafter or a team of drafters, depending on the size and complexity of the structure to be drawn. The floor plan will become a reference for all other architectural drawings. Figure 16-3 shows a completed floor plan.

Drawing Scales

The size and complexity of the project and the paper size to be used dictate the scale to be used to draw the floor plan. The paper size usually remains standard within an office, with all projects typically drawn on 24" × 36" or 30" × 42" material. The architect generally sets the scale of the floor plan for an inexperienced drafter. On small structures, a scale of 1/4" = 1'–0" is used. On large simple structures, scales of 1/8" = 1'–0" (1:50) or 1/16" = 1'–0" (1:200) are common. Scales of 3/16" = 1'–0", 3/32" = 1'–0", or 1/16" = 1'–0" are used for large structures with detailed areas of the plan enlarged to scales such as 1/4" = 1'–0", 3/8" = 1'–0", or 1/2" = 1'–0". Because of the few choices of metric scales, large metric structures typically must be divided into zones and spread over two or more drawing sheets.

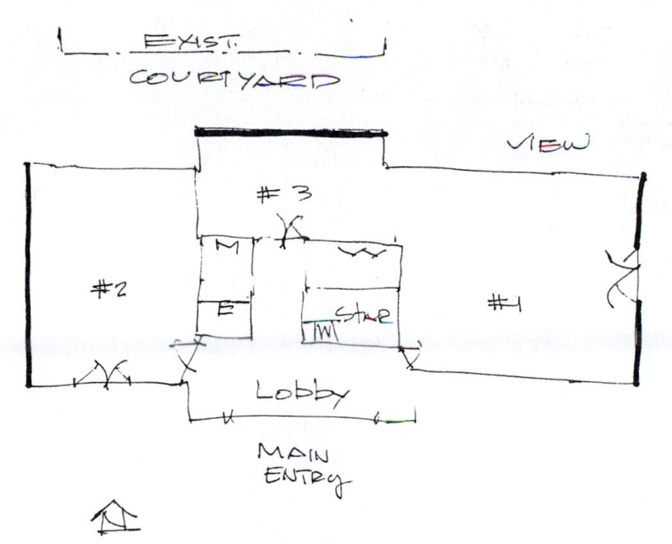

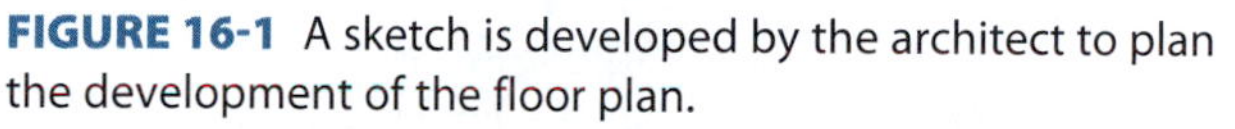
FIGURE 16-1 A sketch is developed by the architect to plan the development of the floor plan.

Scales of 1:50 or 1:20 can be used for enlarged metric plans. Examples of enlarged floor plans can be seen later in this chapter. In selecting a scale, the floor plan must be drawn at a scale that is large enough to reflect necessary details of the structure. For large projects, the floor plan is often drawn in segments with the use of a key plan and match lines.

Key Plans and Match Lines

A **key plan** is a very small-scale floor plan that shows the overall footprint of the structure, with specific areas of the structure divided into zones. A key plan is used when a structure is too large to fit on one sheet or if a specific portion of the structure is to be enlarged to show details that are too complex for the scale used to show the balance of the structure. The key plan can be placed anywhere on the drawing sheet that space allows, but should always be placed in the same location throughout the drawing set. Figure 16-4 shows a key plan for an office structure. Breaking the floor plan into zones allows a larger scale to be used than would be possible if the structure was drawn as a whole unit.

When the entire floor plan is not shown on one sheet, a *match line* must be provided to show how the portions of the floor plan relate to each other. Figure 16-5 shows Zone B of the structure shown in Figure 16-4. When the drawings are assembled, care should be taken to always use the same order of placing zoned drawings within the drawing set. On multilevel drawings, if the main floor level is divided into east and west zones, all levels should be divided into east and west zones.

FLOOR PLAN COMPONENTS

One of the key roles of a floor plan is to show the location and materials used to construct the walls, windows, doors, cabinets, and the dimensions and special symbols. Although the symbols are similar to those used in residential drafting, a wider variety of wall materials must be represented on commercial construction drawings.

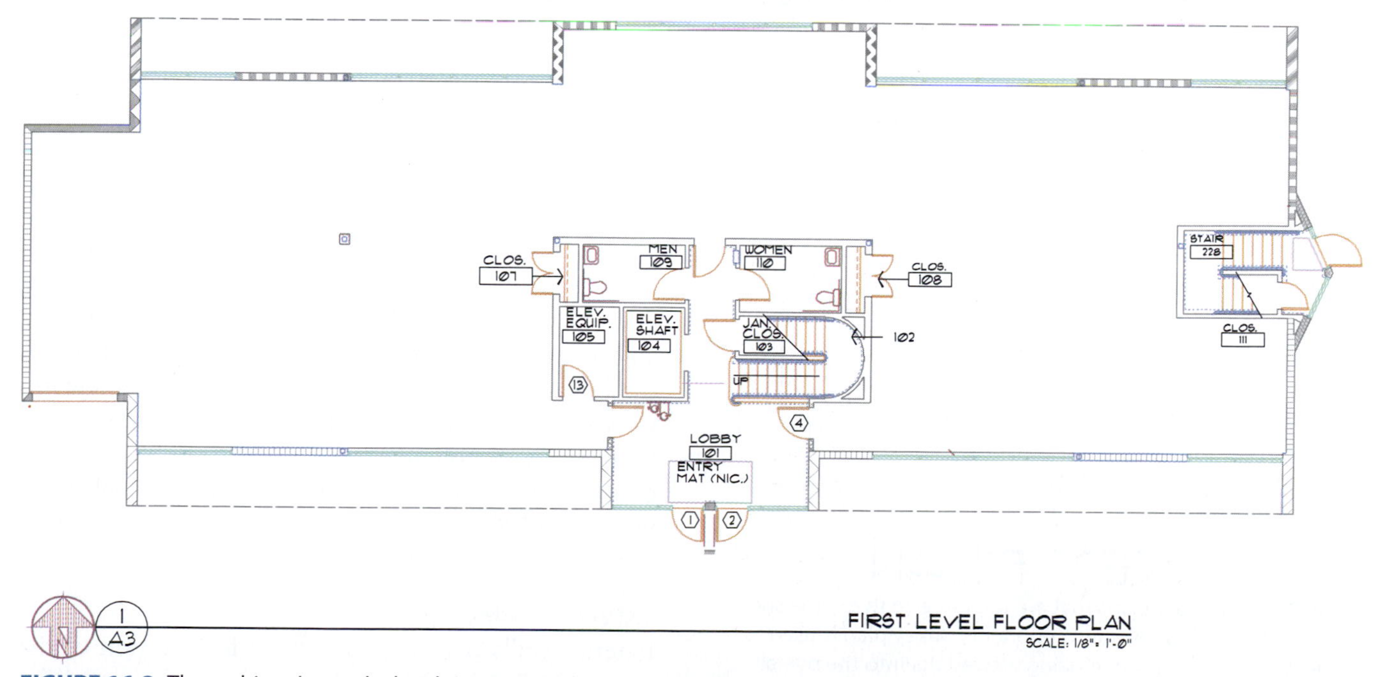

FIGURE 16-2 The architect's rough sketch is converted to a preliminary floor plan by a project manager or senior drafter. *Courtesy Architects Barrentine, Bates & Lee, A.I.A.*

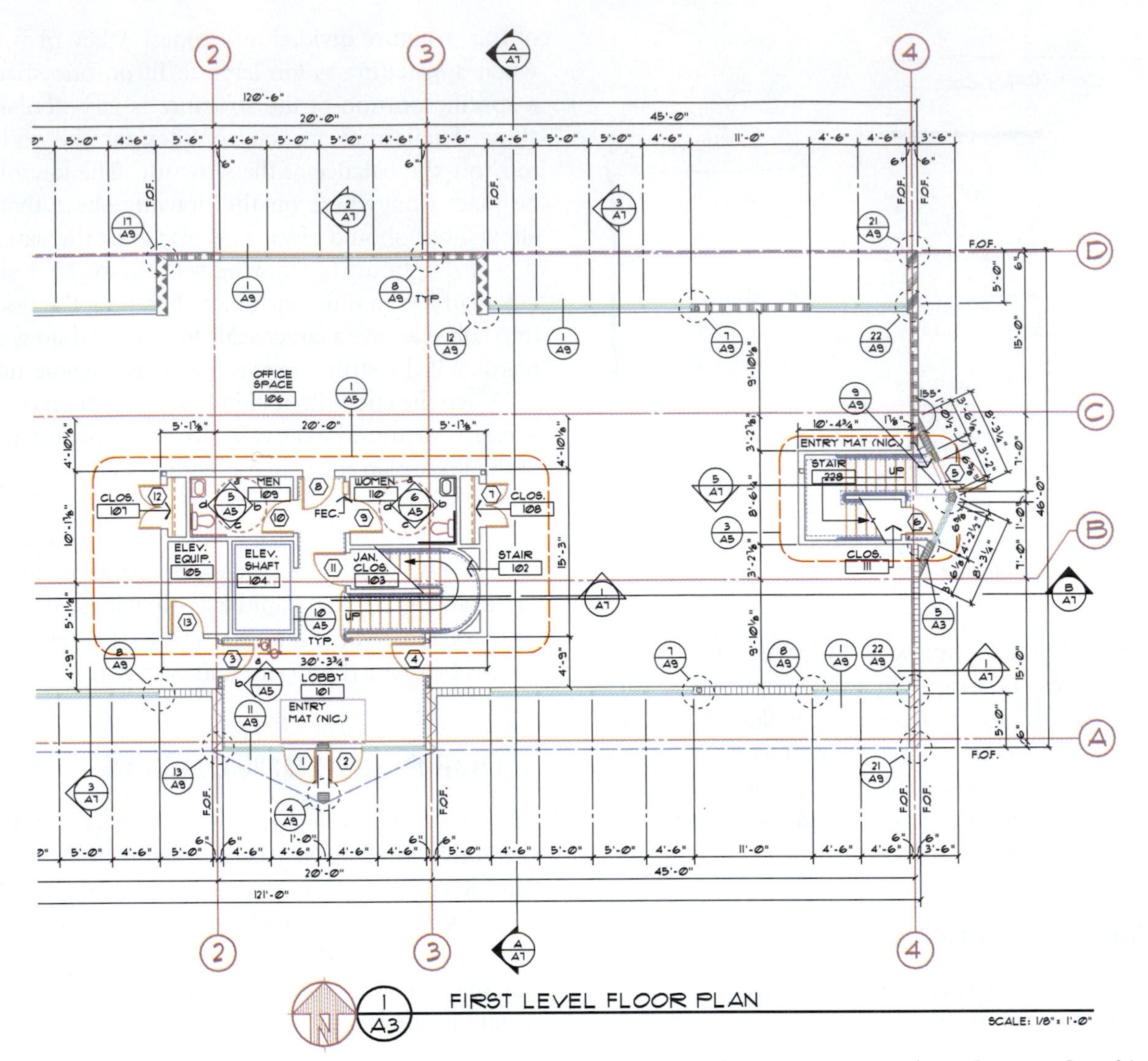

FIGURE 16-3 A completed floor plan often requires the efforts of several CAD technicians. *Courtesy Architects Barrentine, Bates & Lee, A.I.A.*

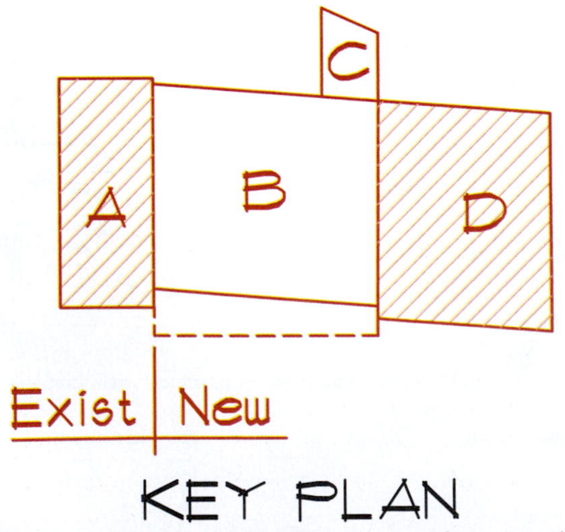

FIGURE 16-4 Floor plans that are too large to fit on one sheet can be broken into several components. A key plan is used to indicate where the portion being viewed fits into the overall plan. The floor plan representing portion B can be found in Figure 16-5. *Courtesy Chris DiLoreto, DiLoreto Architects, LLC.*

Walls

Walls are the vertical members of a structure and can be used to meet each of the three goals of a building component. Although walls are typically vertical to accommodate windows and doors, International Building Code (IBC) defines walls as any portion of the structure angled at 60° or greater to be considered a wall. Walls can be constructed of single units forming a framework, as with wood and steel studs, or may be monolithic units, as with concrete walls. Walls are classified as either *exterior* or *interior* and *bearing* or *nonbearing* walls. The IBC also refers to shear, faced, parapet, fire, party, and retaining walls.

Exterior Walls

Exterior walls help form the basis for the skin of the structure. The architectural team draws the wall locations to meet the needs of the client. The structural team

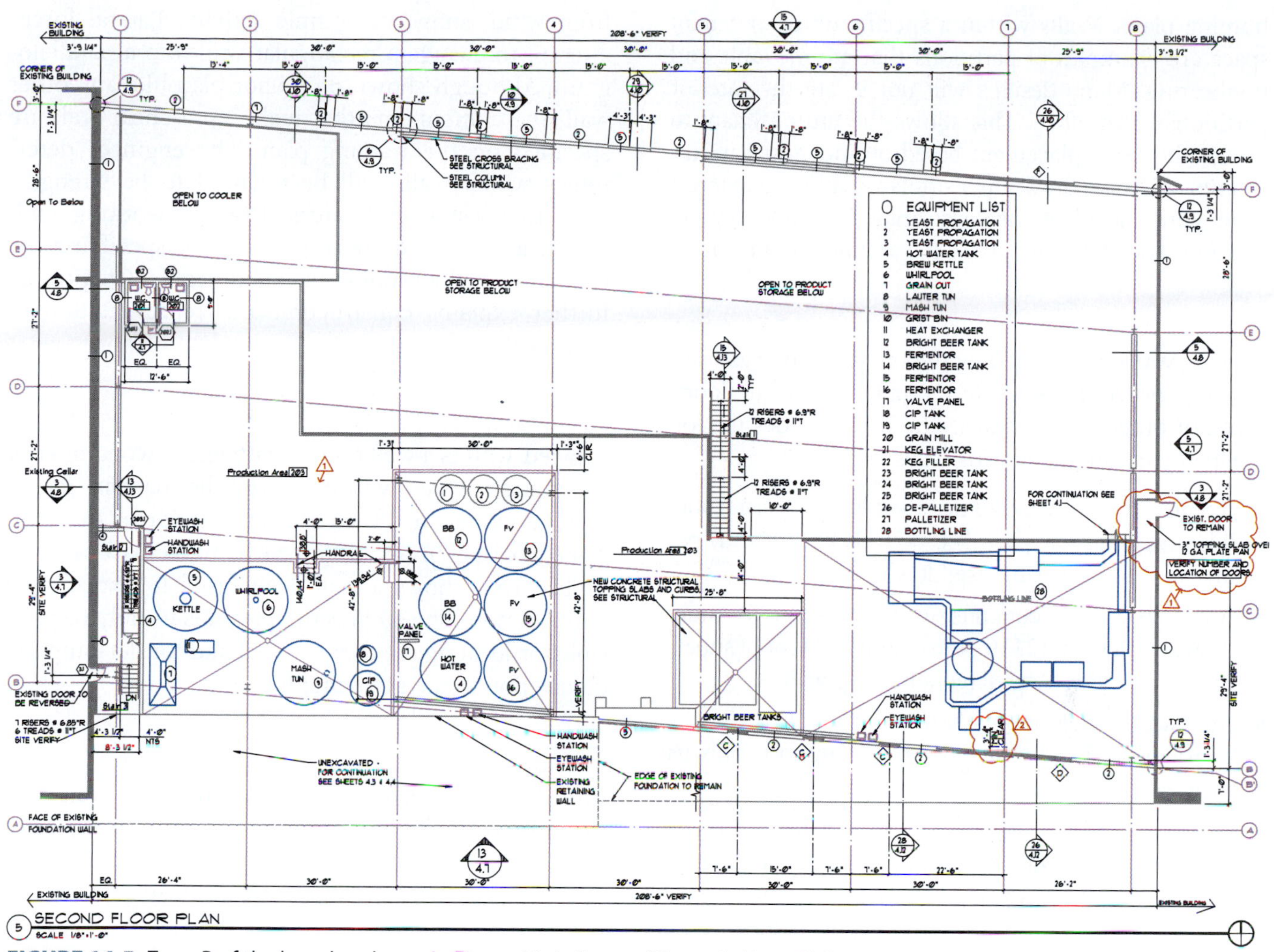

FIGURE 16-5 Zone B of the key plan shown in Figure 16-4. *Courtesy DiLoreto Architects, LLC.*

is responsible for detailing the construction of these walls. These walls may be required to:

- Restrict heat loss in winter but allow air flow in summer.
- Contain interior humidity, while repelling exterior humidity.
- Allow for view and access to the outdoors, or totally restrict access for security reasons.
- Welcome guests and inhabitants while keeping intruders out.

Exterior walls will be the recipient of external forces from wind and rain and must be capable of transferring this force to the shell of the structure or to the lateral bracing system. Although it is the architect's responsibility to design an exterior covering to protect the structure, the CAD technician should have an understanding of these forces to understand how the structure resists the forces. An understanding of the forces that affect the building will provide meaning to the drawings that the drafter is required to complete. Exterior walls are typically permanent in nature, but they may be either bearing or nonbearing. The building codes consider walls that separate the structure from interior courtyards as exterior walls. Exterior walls are represented on all of the plan views of the project, and are specified on the floor plan. The framing plan shows specific measures that must be taken to build each wall and connect it to the frame of the structure. Framing plans are discussed in Chapter 22.

Interior Walls

Interior walls, or **partitions** as they are sometimes called, are used to define space. They may be bearing or nonbearing, fixed or removable. Walls that form hallways, enclose elevator shafts, surround bathrooms, or surround a specific tenant space in an office complex are examples of *fixed partitions*. Fixed walls and load-bearing partitions will be represented on each plan view. Their finish will be specified on the floor plan and materials used for their construction will be specified on the framing plans. Walls within a specific office or tenant space are examples of partitions that are movable and nonbearing. Many designs will not locate the interior partitions of an office. This allows the future tenant to design the exact placement based on individual needs. Interior walls are drawn in a similar manner to exterior walls constructed of the same material, and they are generally the responsibility of the architectural team.

Bearing Walls

Bearing walls are defined by the building code as walls that support compression loads transferred from floors, roofs, or other bearing walls. The IBC provides the following definitions of bearing walls:

- A wall made of wood or steel studs is a bearing wall if it supports more than 100 pounds (1459 N/m) per lineal foot of superimposed load.
- Any masonry or concrete wall that supports more than 200 pounds (2119 N/m) per lineal foot of superimposed load is considered a bearing wall.
- Any wall that is more than one story in height is considered a bearing wall, even if it supports only its own weight.

Bearing walls are shown by the architectural and structural teams on the floor and framing plans.

The structural team is usually responsible for detailing the construction of the bearing walls. Chapters 13 and 22 discuss distribution of loads to walls and weights of common building materials. Chapters 19 and 23 discuss methods of drawing wall sections to show the construction of bearing walls.

Nonbearing Walls

One-story walls that support only their own weight, or that support less than the acceptable lower limits of bearing walls, are considered **nonbearing walls.** Exterior nonbearing walls are often referred to as **curtain walls.** A load-bearing post or column may be contained in a nonbearing wall without requiring the wall to support any loads other than its own. Nonbearing walls are not shown on the framing plan. A drafter working with the architectural team usually completes details showing the construction of nonbearing walls.

Shear Walls

In addition to supporting loads that are transferred down through the structure to the foundation, **shear walls** are used to resist lateral (horizontal) forces transferred through the structure. These forces may come from wind, snow, or seismic activity. Lateral forces attempt to change a rectangular wall into a parallelogram. Although shown on a floor plan like any other wall, the components that make up a shear wall are specified on the framing plan. The engineer determines which walls will be required to be strengthened to resist lateral forces. Drafters working with the structural team are responsible for specifying the materials on the framing plan and details. Chapter 22 further explains construction and representation of shear walls.

Faced Walls

A **faced wall** is a wall framed with either wood or steel studs and covered with masonry. The masonry facing may be either on the interior or exterior of the wall. The two faces must be joined to move as one unit. Figure 16-6 shows a wall detail used to explain the construction. The architectural team is usually responsible for locating the face material and for detailing the connection to the structural frame.

Parapet Walls

Introduced in Chapter 12, a **parapet** is a wall that extends above the roofline. A parapet wall is typically used to hide mechanical equipment located on low-pitch roofs (as seen in Figure 16-7), or to provide fire

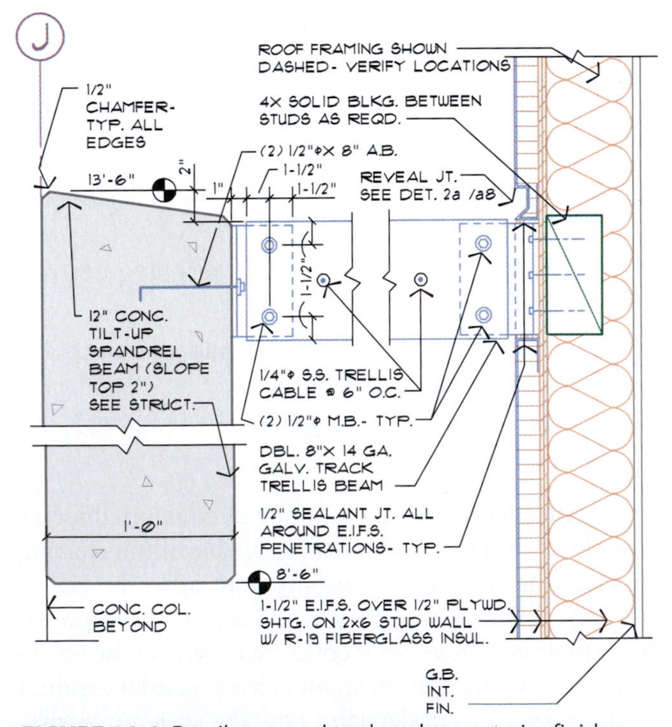

FIGURE 16-6 Details are used to show how exterior finishes are attached to the skeleton. *Courtesy Architects Barrentine, Bates & Lee, A.I.A.*

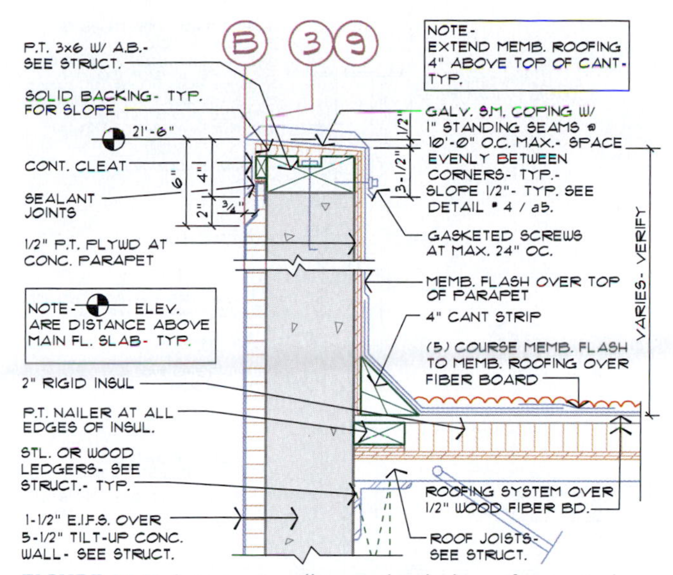

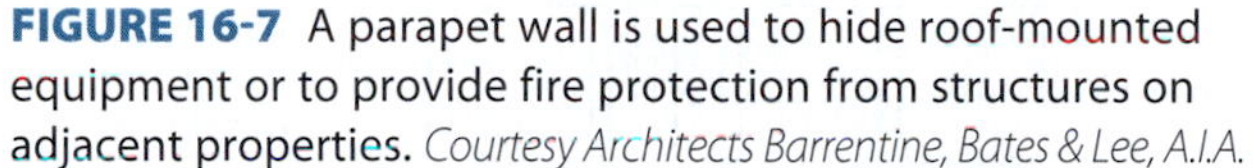
FIGURE 16-7 A parapet wall is used to hide roof-mounted equipment or to provide fire protection from structures on adjacent properties. *Courtesy Architects Barrentine, Bates & Lee, A.I.A.*

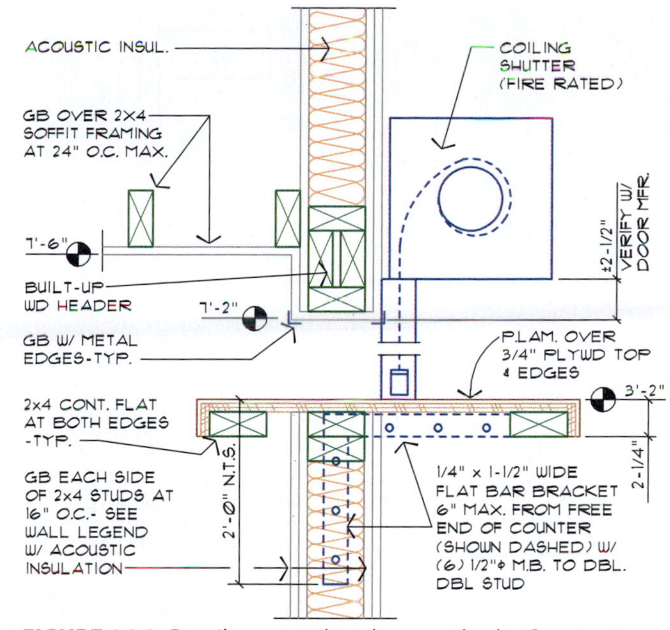

FIGURE 16-8 Details are used to show methods of protecting a structure from the spread of smoke and fire. *Courtesy Architects Barrentine, Bates & Lee, A.I.A.*

protection to the roof from other structures and vice versa. Locations are determined during the design stage by the architectural team. CAD technicians working with the architectural team are responsible for locating parapet walls on the roof and roof framing plans. Details showing the construction of parapet walls are also usually part of the architectural drawings.

Fire Walls

Chapters 4 and 12 introduce occupancies and fire construction requirements. A **fire wall** is a wall that has a specific fire rating and can resist the spread of fire for a specific amount of time due to the materials that were used to construct the wall. For example, a wood stud wall covered with a layer of 5/8" (16 mm) type X gypsum board on each side has a 1-hour fire resistance rating. A wall covered with two layers of 5/8" (16 mm) type X gypsum board on each side has a 2-hour fire rating. A steel beam protected with 2" (50 mm) of carbonate or lightweight concrete has a 4-hour fire rating. Table 720.1.1 of the IBC provides a comprehensive listing of common methods of protecting structural materials with various building materials. The Gypsum Association also has a comprehensive listing of methods of protecting wood and steel members from fire. Materials used to achieve a specific fire rating are typically specified on the floor plan by the architectural team. Details similar to Figure 16-8 are used to specify the fireproofing methods.

Party Walls

The wall used to separate adjoining residential units of an apartment or condominium project is referred to as a **party wall**. Generally, a party wall is constructed of two rows of parallel studs that are arranged in such a manner that sounds from one side of the wall will not transfer to the other side. The IBC defines a party wall as any wall located on a lot line between adjacent buildings that is used for joint service between the two buildings. Any such wall is required to have 2-hour minimum fire protection. Party walls are represented on each plan view, and specified on the floor plan. Except for the thickness, they will resemble any other wood or steel framed wall. Figure 16-9a shows common arrangements for placing the studs used to form a party wall. Figure 16-9b shows a typical section for a 2-hour fire-rated party wall.

Retaining Walls

A retaining wall is a masonry wall designed to resist lateral displacement of soil or other materials. Retaining walls are shown on the floor and framing plans if they extend above the floor level, but they are always represented on the foundation plan. Wall locations can be determined by the architectural team or a civil engineer. Details showing the construction of the wall are grouped with other concrete details by the engineering team. Figure 16-10 shows an example of an internal retaining wall completed by an architectural firm.

Representing Walls

The drafter is required to specify and distinguish each type of wall based on office standards for the construction crew. Walls are shown in plan view by pairs of bold

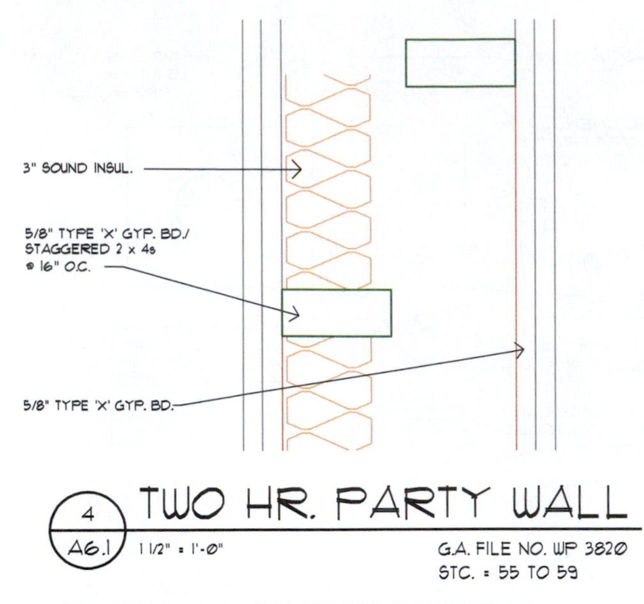

CONSTRUCTION TYPE: GYPSUM WALLBOARD, WOOD STUDS

BASE LAYER 5/8" TYP X GYP. WALLBOARD OR VENEER BASE APPLIED AT RIGHT ANGLES TO EACH SIDE OF DOUBLE ROW OF 2x4 WOOD STUDS 16" O.C. ON SEPARATE PLATES 1" APART WITH 6d COATED NAILS, 1 7/8" LONG, 0.085" SHANK, 1/4" HEADS, 24" O.C. FACE LAYER 5/8" TYPE X GYPSUM WALLBOARD OR VENEER BASE APPLIED AT RIGHT ANGLES TO EACH SIDE OF STUDS OVER BASE LAYER WITH 8d COATED NAILS, 2 3/8" LONG, 0.100" SHANK, 1/4" HEADS, 8" O.C. STAGGER JOINTS 16" O.C. EACH LAYER AND SIDE. SOUND TESTED USING 3 1/2" GLASS FIBER 0.75 PCF STAPLED TO STUDS IN STUD SPACES ON ONE SIDE AND WITH NAILS FOR BASE LAYER SPACED 6" O.C. (LB)

FIGURE 16-9a A party wall is constructed of two rows of parallel studs that are arranged in such a manner that sounds from one side of the wall will not transfer to the other side.

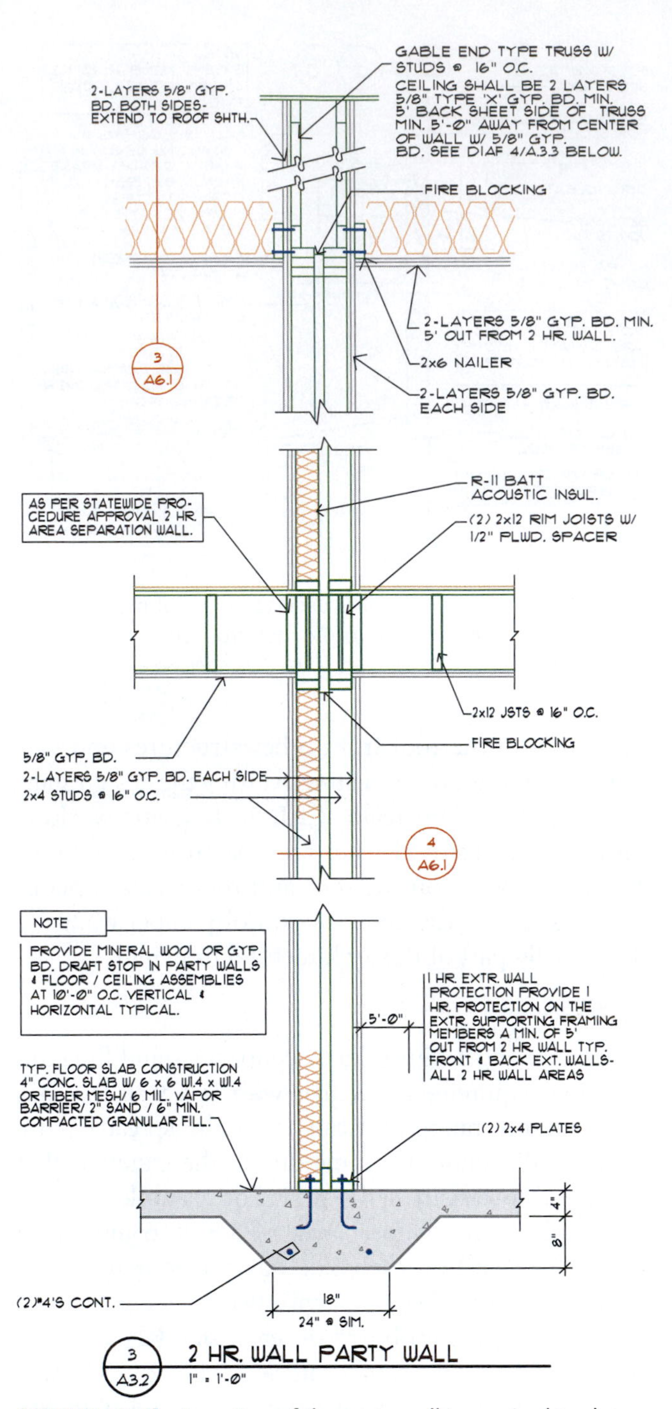

FIGURE 16-9b A section of the party wall is required to show methods of construction. *Courtesy Scott R. Beck, Architect.*

parallel lines. The distance between the lines is determined by the type of material being represented.

Common materials used for walls were introduced in Chapters 7 through 11. Several types of walls can be seen on the plan shown in Figure 16-11. A wall legend similar to Figure 16-12 is usually provided to show what each pattern represents. If demolition or remodeling is required to be represented on the floor plan, additional symbols must also be used to distinguish between material to be removed and new construction. A separate floor plan similar to Figure 16-13 is typically used to represent material that must be used.

Windows

The drawing scale to be used and the type of structure affect the method used to represent the windows on a floor plan. Thin lines are used to represent the glass, sill, and jamb. Multifamily residential projects often use window symbols similar to those used with residential construction. Common symbols are shown in Figure 16-14. Many commercial projects use fixed glass set in an aluminum frame, referred to as **storefront windows.** Storefront windows can be represented in plan view as seen in Figure 16-15. The scale that is used to represent the floor plan alters the exact appearance of storefront windows. Figure 16-16 shows three common methods used to represent storefront windows, based on the scale of the floor plan.

In addition to representing the window in plan view, specifications and details are required to indicate the size, material, and installation of all windows. The

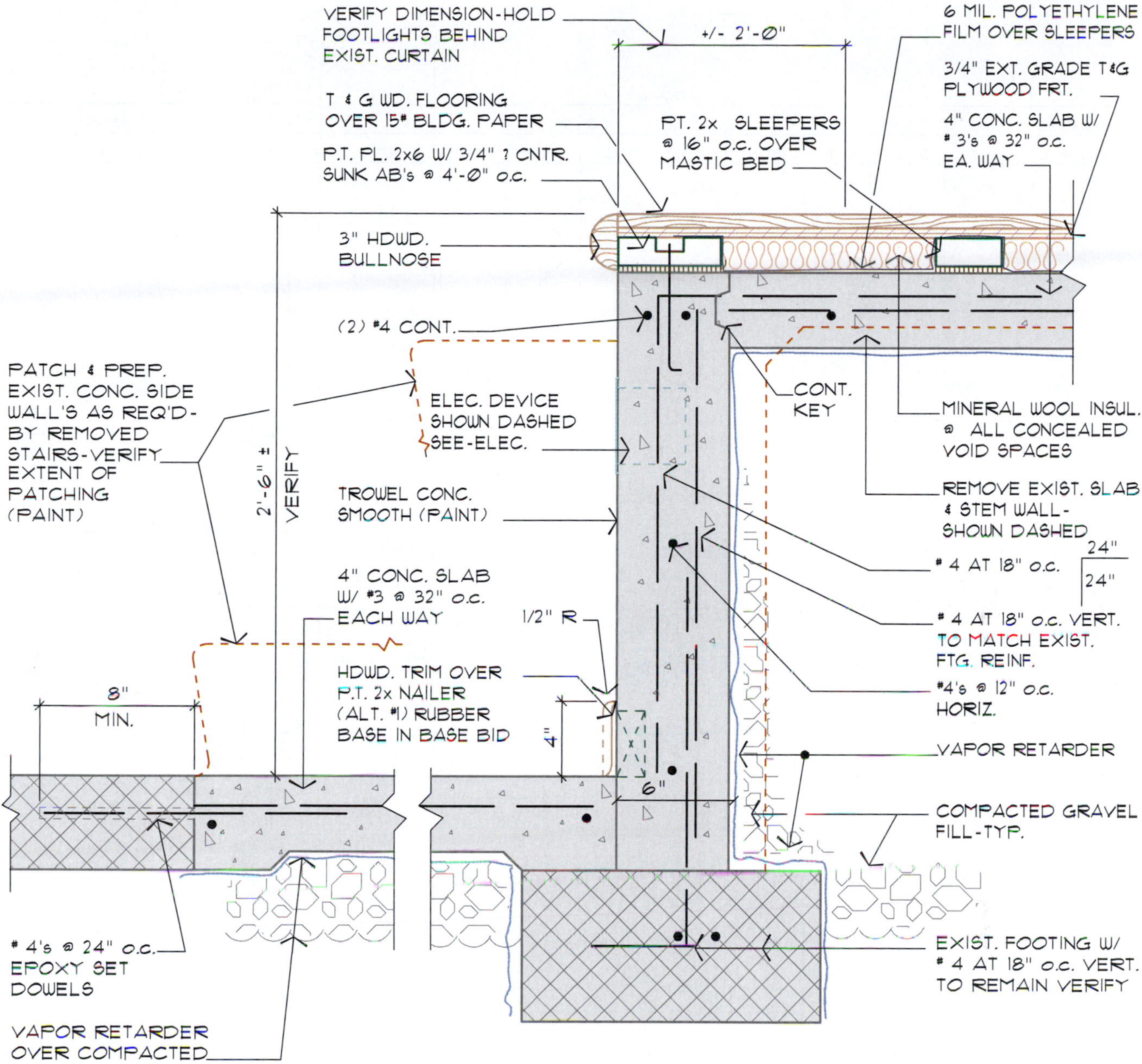

FIGURE 16-10 Retaining wall details are used to show how soil loads are resisted. Drafters working for the architect, structural engineer, or civil engineer can complete the details depending on where the wall is located. Walls that are part of the structure are generally detailed by the architectural team based on the designs of the structural engineer. *Courtesy Architects Barrentine, Bates & Lee, A.I.A.*

size and type of the window are usually represented in a window schedule. A circle is placed by each window on the floor plan, with a letter or number placed in the symbol to reference the window to the schedule. The use of schedules is discussed later in this chapter. Window details are provided to represent how the window jamb, sill, and head attach to the wall. Figure 16-17 shows an example of a typical window detail referenced to the floor plan. Detail symbols are discussed later in this chapter.

Doors

Many of the same types of door symbols used on residential drawings are also used on commercial drawings. Common door symbols can be seen in Figure 16-18. Swinging doors are normally shown open 90° to the wall they are in. A thin line, thick line, or pairs of thin lines can be used to represent the door. The line that represents the door swing is drawn with a thin line. For doors that swing in two directions, the secondary

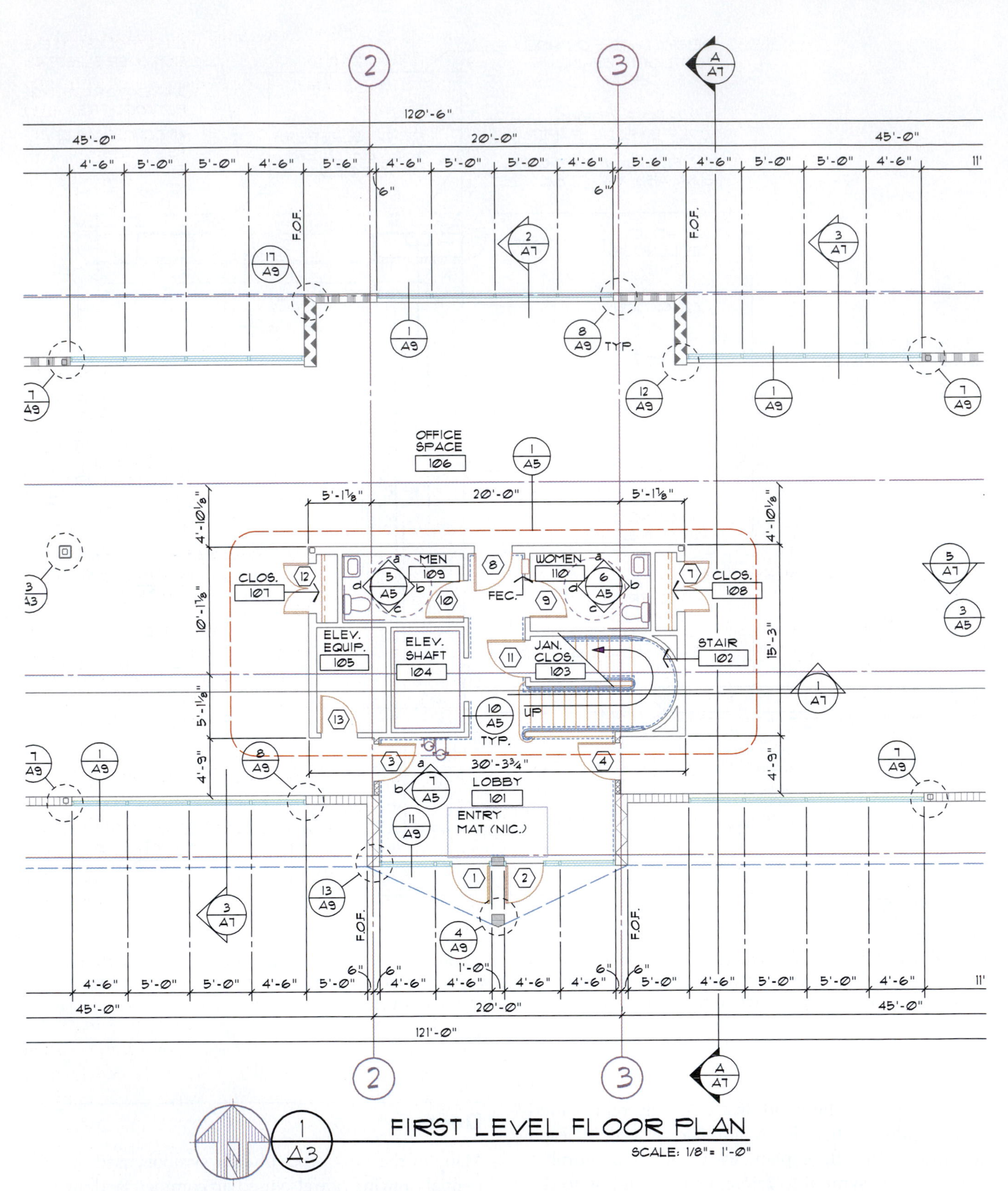

FIGURE 16-11 The floor plan can be used to represent different wall material. *Courtesy Architects Barrentine, Bates & Lee, A.I.A.*

WALL LEGEND

EXTERIOR WALLS

1. EXT. INSULATION & FINISH SYSTEM (EIFS.) OVER 1/2" PLYWD. WALL SHTG. OVER 2 x 6 STUDS AT 16" O.C. W/ R-19 FOIL FACE FIBERGLASS INSUL. & 5/8" GYP. BOARD INTERIOR FINISH (CONTINUE DOUBLE LAYER OF FINISH FROM INTERIOR WALLS AT OFFSETS IN EXTERIOR WALLS)

2. EXTERIOR INSULATION & FINISH SYSTEM (EIFS.) OVER 6" TILT-UP CONC. WALLS W/ PAINTED EXPOSED CONC. ON GARBAGE ENCLOSURE SIDE ONLY- (8'-10" HIGH WALL)

3. EXTERIOR INSULATION & FINISH SYSTEM (EIFS.) OVER 6" TILT-UP CONC. WALLS W/ SEALED OR PAINTED INTERIOR FINISH WALLS W/ SEALED OR PAINTED INTERIOR FINISH.

4. 12" TILT- UP CONC. COLUMNS- PAINT ALL EXPOSED SURFACES.

INTERIOR WALLS

5. 2x4 STUDS AT 16" OC. W/ 3" ACOUSTIC INSUL. IN STUD CAVITY G.B. ON ONE SIDE (LEAVE AIR SPACE BETWEEN STUDS AND /OR TILT- UP CONC. WALL) EXTEND FROM FLOOR LINE UP TO ROOF FRAMING (ONE HOUR FIRE RATED CONSTRUCTION)

6. 6" TILT- UP CONC. WALL W/ EXTERIOR INSUL. & FINISH SYSTEM (AT EXTERIOR) ABOVE ALL LOW ROOF AREAS. SEAL OR PAINT ALL EXPOSED CONC. SURFACES

7. 2x4 STUDS AT 16" OC. W/ 3" ACOUSTIC INSUL.- (2) LAYERS OF G.B. ON BOTH SIDES EXTEND FROM FLOOR LINE UP TO ROOF FRAMING (END SECOND LAYER OF G.B. 6" ABOVE CEILING LINE- 2x6 STUDS AS NOTED ON FLOOR PLAN (ONE-HOUR) FIRE RATED CONSTRUCTION AT LOBBY (15) & CORRIDORS)

8. 2x4 STUDS AT 16" OC. W/ 3" ACOUSTIC INSUL.- G.B. ON BOTH SIDES OF WALL EXTEND FROM FLOOR LINE UP TO FLOOR FRAMING, UNLESS OTHERWISE NOTED (ONE HOUR FIRE RATED AS NOTED ON PLANS).

9. 2x4 STUDS AT 16" OC. W/ 3" ACOUSTIC INSUL.- 1/2" RESELEINT CHANNELS AT 16" O.C. W/ G.B. ON ONE SIDE & (2) LAYERS OF G.B. ON OPPOSITE SIDE-EXTEND FROM FLOOR LINE UP TO 6" ABOVE CEILING LINE.

10. METAL HAT CHANNELS AT 16" OC. FURRING OVER CONC. WALL OR CONC. CONDUIT ENCASEMENT W/ G.B. FINISH- EXTEND FROM FLOOR LINE UP TO FLOOR FRAMING

11. 2x4 STUDS AT 16" OC. W/ 3" ACOUSTIC INSUL. & (2) LAYERS OF G.B. ON ONE SIDE (LEAVE AIR SPACE BTWN STUDS) EXTEND FROM FLOOR LINE UP TO ROOF FRAMING (ONE HOUR FIRE RATED CONSTRUCTION)

WALL LEGEND NOTES

1. USE TYPE 'X' G.B. ON ALL CORRIDOR WALLS & ALL WALLS SHOWN TO BE FIRE RATED
2. COORDINATE WALL FRAMING W/ PLUMBING & MECH. ROUGH-INS (URINAL, WC. CARRIERS & DUCT PENETRATIONS)
3. WALLS SHOWN BACK TO BACK (IE. PLUMBING WALLS) TO BE FINISHED ONLY ON OUTSIDE (EXPOSED) FACE OF STUDS AS SCHEDULED IN WALL LEGEND)
4. FIRETAPE ONLY ALL GYPSUM BOARD SURFACES ABOVE CEILINGS AT CONCEALED SPACES - TYP.
5. SEE STRUCTURAL DRAWINGS FOR SHEAR WALL LOCATIONS & CONC. REINFORCING REQUIREMENTS - TYP.
6. PROVIDE ACOUSTIC SEALANT AT ALL WALL EDGES, INTERSECTIONS & PENETRATIONS- TYP.
7. HOLD ALL ACOUSTIC INSULATION AT CENTER OF STUD CAVITIES -TYP.

FIGURE 16-12 Wall legends are often used to explain the symbols used on a floor plan. *Courtesy Michael & Kuhns Architects, P.C.*

direction is often shown using thin dashed lines. As with windows, doors are also represented in schedules, specifications, and details. Hexagons with either a letter or a number are also used to key doors on the floor plan to the door schedule. The same symbol used to key windows to the schedule should never be used for doors. If numbers are used to represent windows, letters should be used to key doors to the door schedule. In structures with a large number of rooms, the door may also be referenced to a specific room.

Openings

In addition to doors and windows, relites, transoms, skylights, archways, cased openings, and pass-throughs must be represented and specified, and detail symbols must be placed on the floor plan. *Relites,* or sidelights as they are often called, are windows that are mounted beside a door. They are drawn as a window on the floor plan, but they may be referenced with the door symbol or given a separate symbol and treated as a window. A *transom* is a window placed above a doorway. They are not drawn on the floor plan but are referenced with a separate symbol placed beside the door symbol.

Skylights are often drawn as a square or rectangle on the floor plan with either a thin dashed or continuous line depending on office preference. On a simple plan, the specifications to describe the skylight can be placed on the floor plan. On complicated plans, skylights are referenced to a schedule that explains the size, type, and manufacturer. Details that show how the skylight is installed are referenced on the roof plan. An *archway* is a wall opening that usually has no trim. A *cased opening* is an opening in a wall that has trim similar to a doorjamb but has no trim for door stops. A *pass-through* is an opening in a wall that does not go all the way to the floor. It may be trimmed or untrimmed.

Floors

The floor system provides the horizontal surface for each level of the structure. Some occupancies such as theaters or assembly halls have a sloping floor to enhance viewing or sound reproduction. In multilevel structures, the floor of an upper space is also part of the ceiling of the lower level. The architectural team determines the shape and limits of the floor. The structural team designs the necessary supports and the material to construct the floor system. If the floor is carefully connected to supporting walls, the floor system can also be used to resist lateral loads from the walls. The shape and levels of the floor are shown and specified on the floor plan.

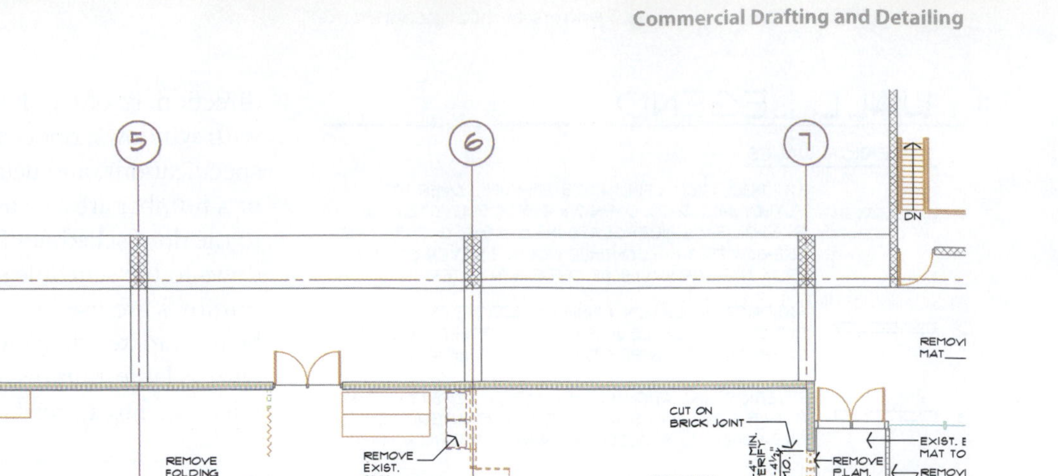

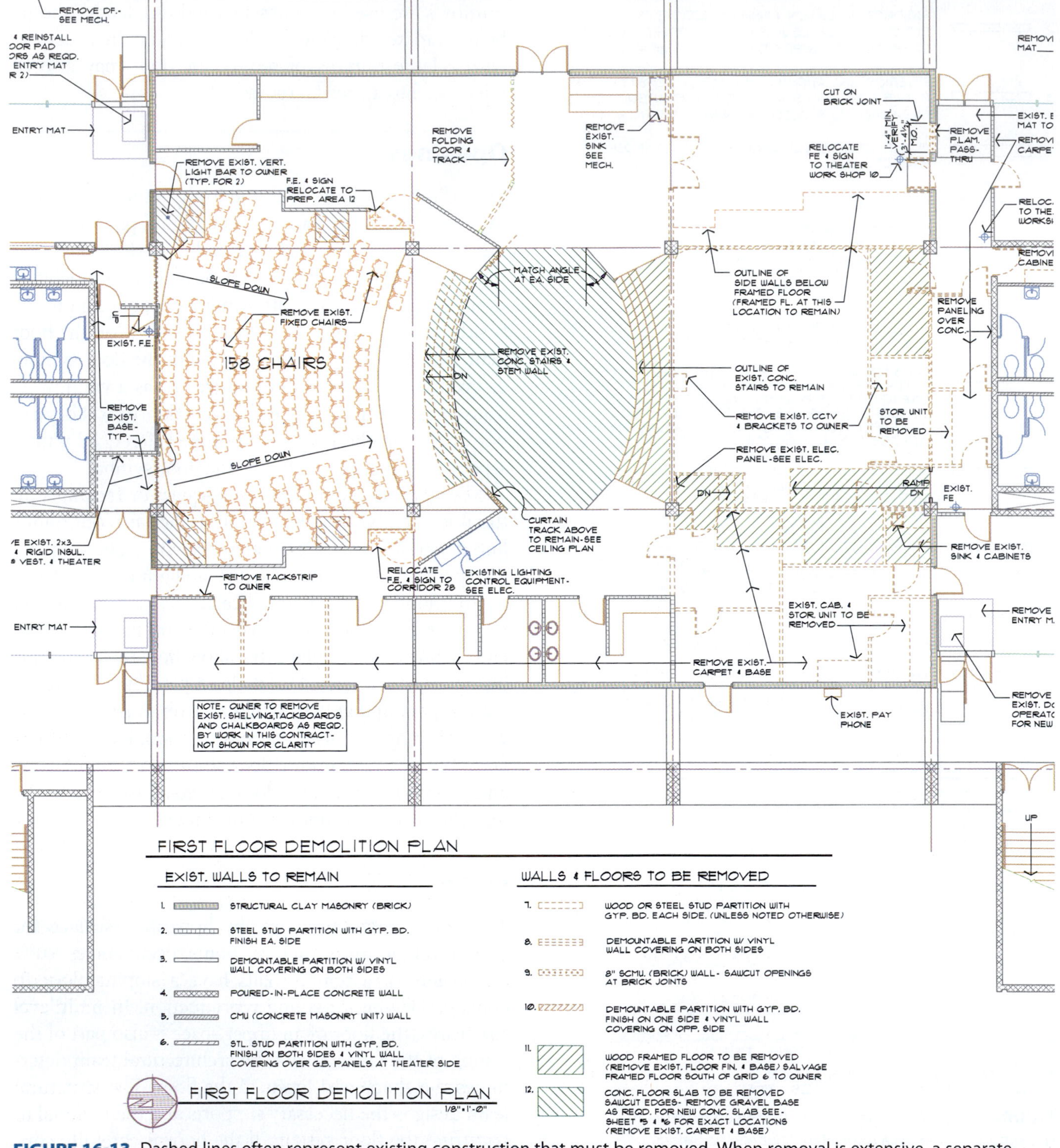

4
5
6
7
DN
REMOVE DF.-
SEE MECH.
ENTRY MAT
REMOVE EXIST. VERT.
LIGHT BAR TO OWNER
(TYP. FOR 2)
F.E. & SIGN
RELOCATE TO
PREP. AREA 12
REMOVE
FOLDING
DOOR &
TRACK
REMOVE
EXIST.
SINK
SEE
MECH.
CUT ON
BRICK JOINT
RELOCATE
FE & SIGN
TO THEATER
WORK SHOP 10
REMOVE
PLAM.
PASS-
THRU
MATCH ANGLE
AT EA SIDE
OUTLINE OF
SIDE WALLS BELOW
FRAMED FLOOR
(FRAMED FL. AT THIS
LOCATION TO REMAIN)
SLOPE DOWN
REMOVE EXIST.
FIXED CHAIRS
EXIST. F.E.
158 CHAIRS
REMOVE EXIST.
CONC. STAIRS &
STEM WALL
DN
OUTLINE OF
EXIST. CONC.
STAIRS TO REMAIN
REMOVE
EXIST.
BASE-
TYP.
REMOVE EXIST. CCTV
& BRACKETS TO OWNER
STOR. UNIT
TO BE
REMOVED
REMOVE EXIST. ELEC.
PANEL-SEE ELEC.
SLOPE DOWN
RAMP
DN
EXIST.
FE
CURTAIN
TRACK ABOVE
TO REMAIN-SEE
CEILING PLAN
REMOVE EXIST.
SINK & CABINETS
REMOVE TACKSTRIP
TO OWNER
RELOCATE
F.E. & SIGN TO
CORRIDOR 28
EXISTING LIGHTING
CONTROL EQUIPMENT-
SEE ELEC.
EXIST. CAB. &
STOR. UNIT TO BE
REMOVED
REMOVE
ENTRY M.
REMOVE EXIST.
CARPET & BASE
NOTE- OWNER TO REMOVE
EXIST. SHELVING,TACKBOARDS
AND CHALKBOARDS AS REQD.
BY WORK IN THIS CONTRACT-
NOT SHOWN FOR CLARITY
EXIST. PAY
PHONE
UP
FIRST FLOOR DEMOLITION PLAN
EXIST. WALLS TO REMAIN
1. STRUCTURAL CLAY MASONRY (BRICK)
2. STEEL STUD PARTITION WITH GYP. BD. FINISH EA. SIDE
3. DEMOUNTABLE PARTITION W/ VINYL WALL COVERING ON BOTH SIDES
4. POURED-IN-PLACE CONCRETE WALL
5. CMU (CONCRETE MASONRY UNIT) WALL
6. STL. STUD PARTITION WITH GYP. BD. FINISH ON BOTH SIDES & VINYL WALL COVERING OVER G.B. PANELS AT THEATER SIDE
WALLS & FLOORS TO BE REMOVED
7. WOOD OR STEEL STUD PARTITION WITH GYP. BD. EACH SIDE. (UNLESS NOTED OTHERWISE)
8. DEMOUNTABLE PARTITION W/ VINYL WALL COVERING ON BOTH SIDES
9. 8" SCMU. (BRICK) WALL- SAWCUT OPENINGS AT BRICK JOINTS
10. DEMOUNTABLE PARTITION WITH GYP. BD. FINISH ON ONE SIDE & VINYL WALL COVERING ON OPP. SIDE
11. WOOD FRAMED FLOOR TO BE REMOVED (REMOVE EXIST. FLOOR FIN. & BASE) SALVAGE FRAMED FLOOR SOUTH OF GRID 6 TO OWNER
12. CONC. FLOOR SLAB TO BE REMOVED SAWCUT EDGES- REMOVE GRAVEL BASE AS REQD. FOR NEW CONC. SLAB SEE-SHEET *5 & *6 FOR EXACT LOCATIONS (REMOVE EXIST. CARPET & BASE)
FIRST FLOOR DEMOLITION PLAN
1/8"=1'-0"

FIGURE 16-13 Dashed lines often represent existing construction that must be removed. When removal is extensive, a separate demolition plan is provided. *Courtesy Architects Barrentine, Bates & Lee, A.I.A.*

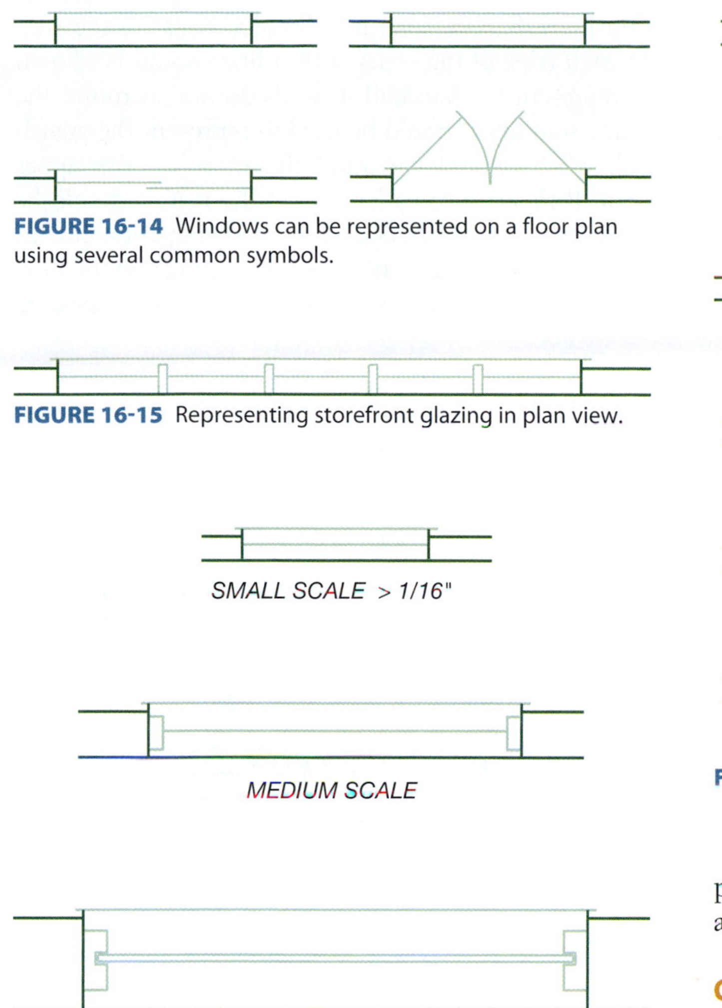

FIGURE 16-14 Windows can be represented on a floor plan using several common symbols.

FIGURE 16-15 Representing storefront glazing in plan view.

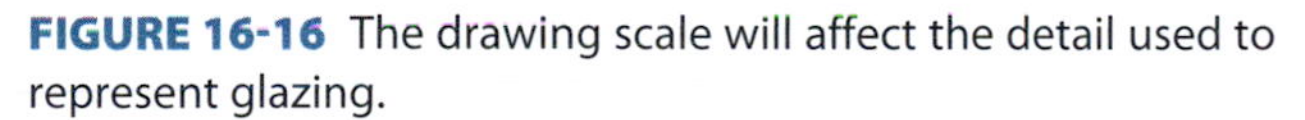

FIGURE 16-16 The drawing scale will affect the detail used to represent glazing.

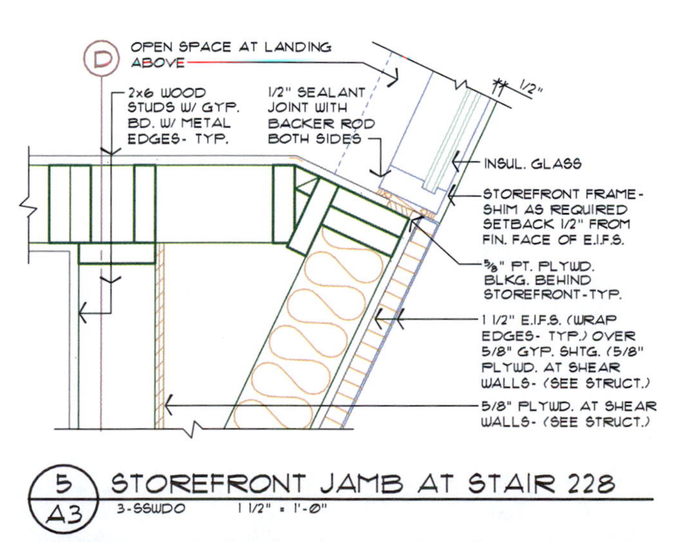

FIGURE 16-17 Window details explain how the window and finish materials are to be placed and are referenced to the floor plan. *Courtesy Architects Barrentine, Bates & Lee, A.I.A.*

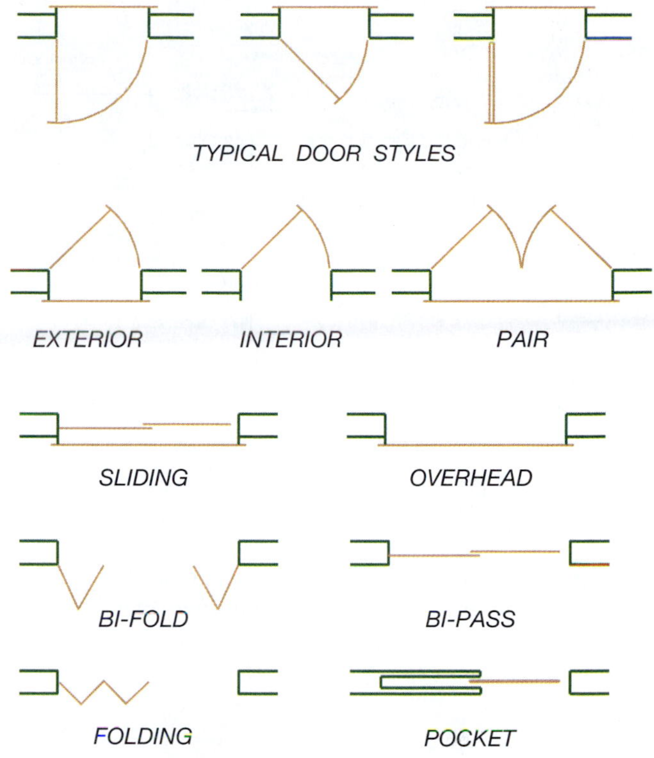

FIGURE 16-18 Representing doors in plan view.

Construction of the floor is shown on the framing plans and in the sections and details. Each can be created as a wblock and inserted as needed.

Changes in Floor Elevation

Steps, stairs, and ramps are common methods of changing elevation that must be shown on the floor plan. Each is represented on the floor plan by thin lines. *Steps* represent the change in elevation between two or more levels of the same floor. A *ramp* is an inclined floor that connects two different floor elevations on the same floor level. An elevated stage within a room would be an example of where steps or a ramp would be used to travel from one level to another. In addition to representing the steps, specifications are placed on the plan to indicate the number and maximum rise and minimum run of the steps. Ramps or inclined floors are generally represented on the floor plan, as shown in Figure 16-19. Ramps should include elevations of each floor level serviced by the ramp or provide the rise and run in slope of the ramp.

Stairs represent the change in elevation between two or more different floors and require the use of two or more floor plans to show the complete stair. Because stairs connect two different floor levels, the lower portion of the stair is shown on the lower floor and the upper portion of the stair is shown on the upper floor. Figure 16-20 shows how a straight stair run can be represented on a floor plan. Each step is represented by

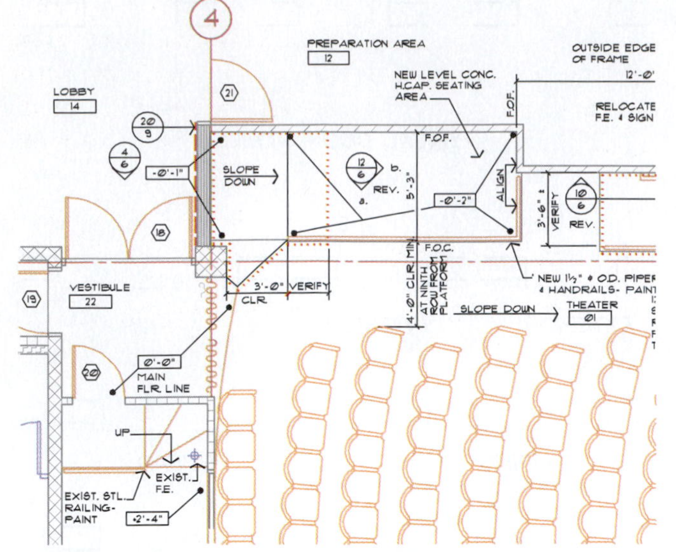

FIGURE 16-19 Ramps and sloping floors and their heights should be represented and specified on the floor plan. *Courtesy Architects Barrentine, Bates & Lee, A.I.A.*

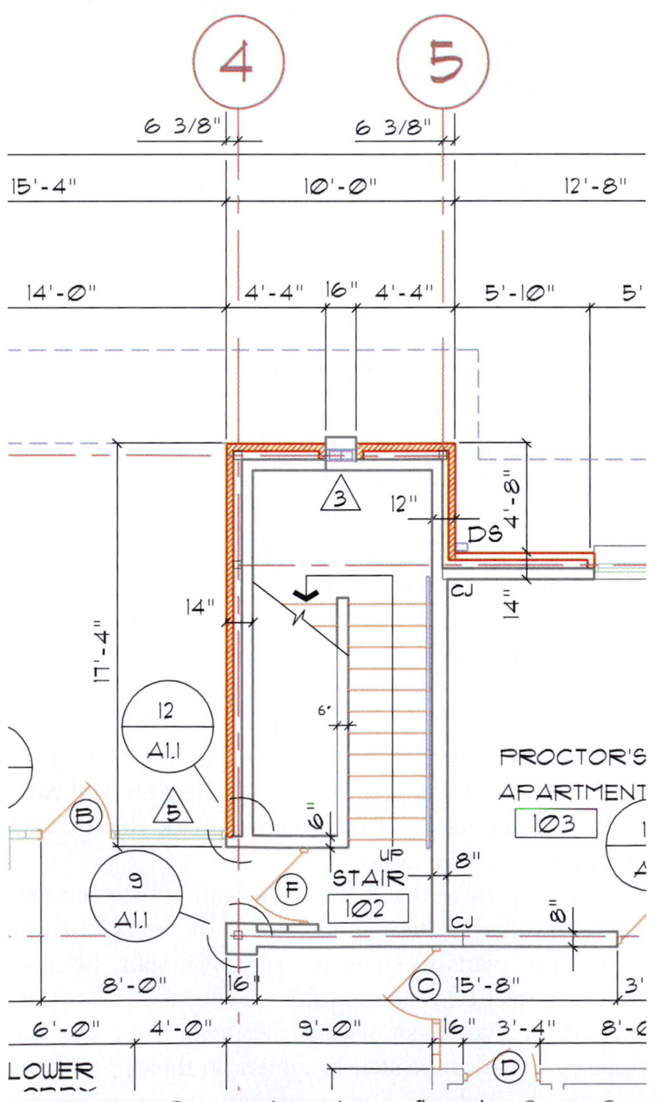

FIGURE 16-20 Representing stairs on a floor plan. *Courtesy G. Williamson Archer A.I.A., Archer & Archer P.A.*

thin lines that are spaced a distance equal to the run. At each edge of the stairs, a thin line should be drawn to represent the handrail. If walls do not surround the stairs, thin lines should be used to represent the guardrail, which would surround the stairs on the upper floor. If the stair has a landing in the run, it should be represented on each floor plan showing the stair, as seen in Figure 16-21. Showing the landing and the next few steps establishes a visual reference point in placing

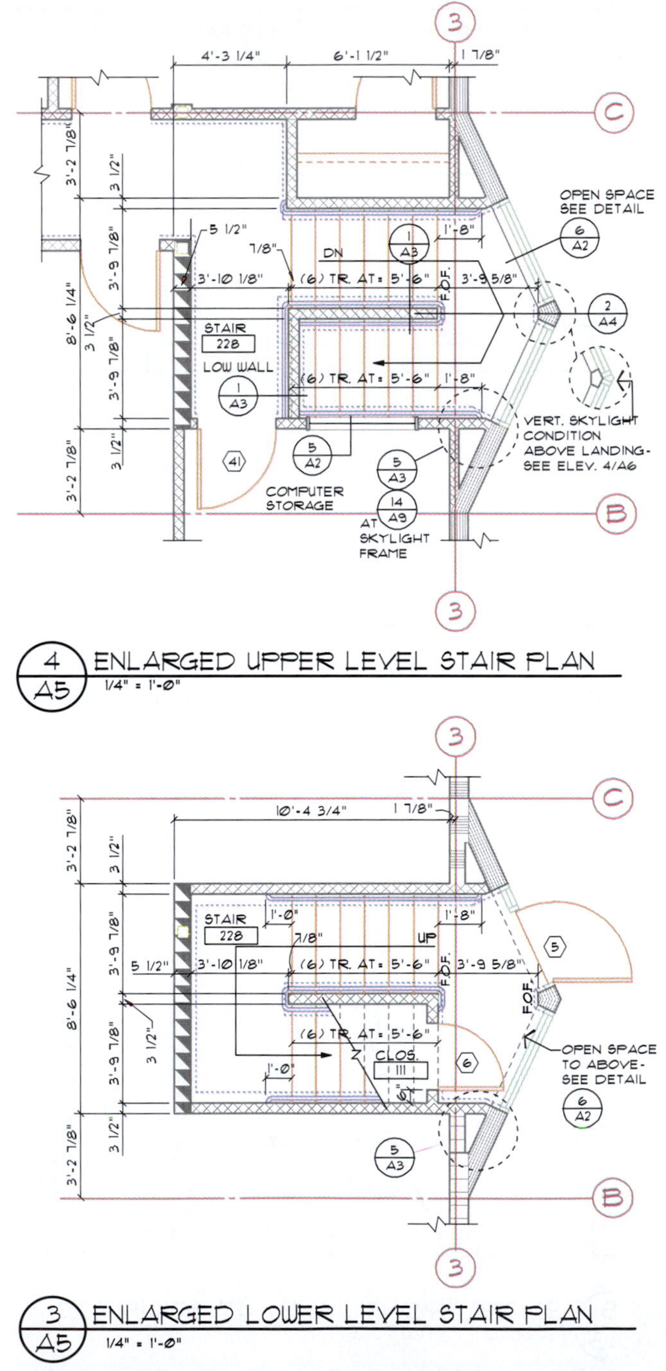

FIGURE 16-21 If a stair has a landing between floors, it must be represented on each floor plan level that the stair serves. *Courtesy Architects Barrentine, Bates & Lee, A.I.A.*

the landing. On structures with more than three levels of stacked stairs, the landing shown on each floor plan represents the landing that is above that floor, except for the uppermost floor plan. A stair for a four-level structure can be seen in Figure 16-22.

Plumbing Symbols

Plumbing symbols are added to the floor plan by the use of blocks. The type of structure greatly influences the plumbing fixtures that must be represented on the floor plan. An apartment may show common items such as tubs, showers, toilets, lavatories, sinks, spas, and pools. A warehouse may require drinking fountains, lavatories, toilets, urinals, and equipment specific to a particular occupancy. Figure 16-23 shows plumbing symbols that may be represented on the floor plan. Although the symbols are self-explanatory, many firms label each fixture on the floor plan and show each fixture in the interior elevations. Figure 16-24 shows a bathroom for an office structure. Methods used to draw interior elevations will be discussed in Chapter 20. The cabinet elevations that would accompany the plan can be seen in Figure 20-8.

Interior Equipment and Furnishings

The complexity of the structure, the drawing scale, and the amount and type of cabinets affect where interior equipment and furnishings are shown. Cabinets and other interior furnishing supplied by the contractor can be shown on a floor plan, a space plan, or a fixture schedule. Common materials that are represented as the interior furnishings include the following:

- Built-in cabinets.
- Base cabinets and countertops, upper cabinets.
- Shelves, display racks, and display counters.

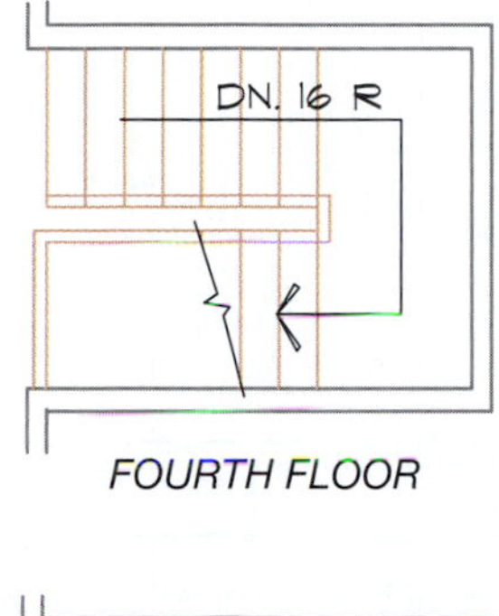

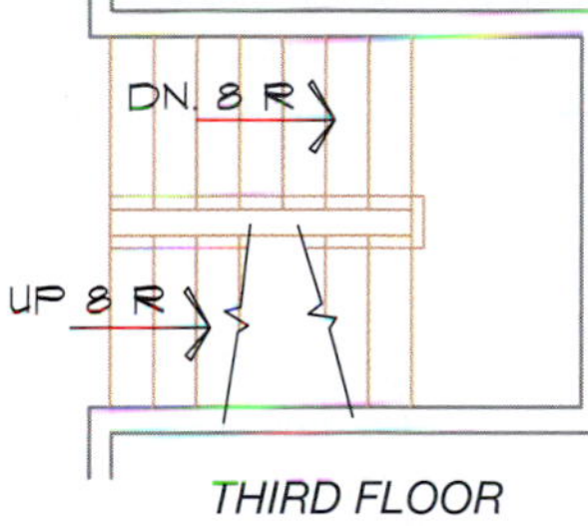

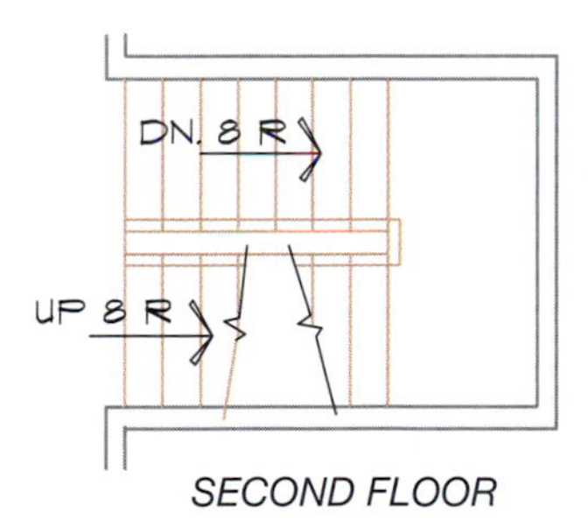

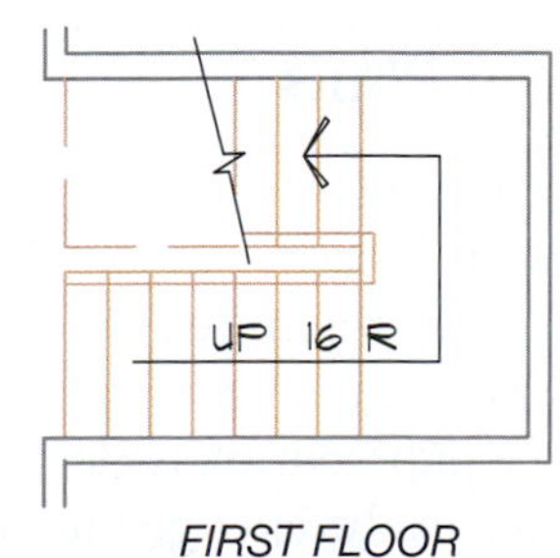

FIGURE 16-22 Stairs for a multilevel structure must represent each run direction.

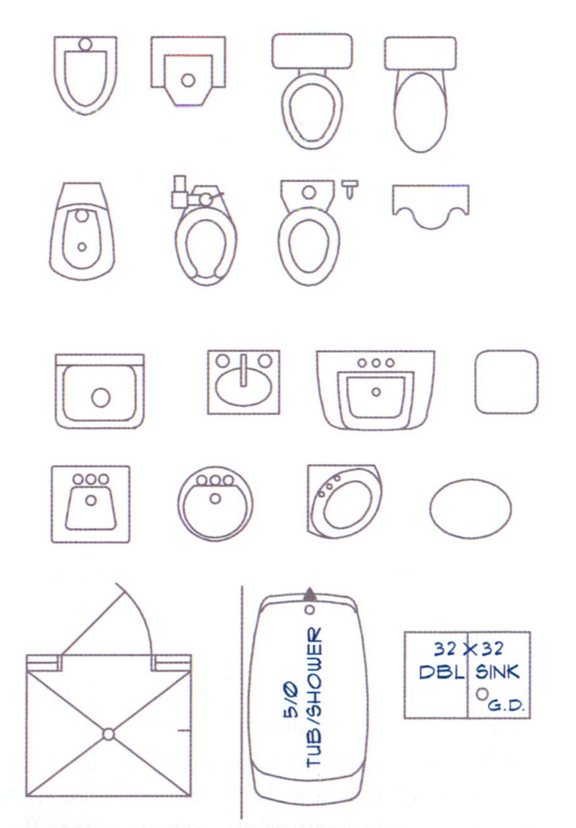

FIGURE 16-23 A wide range of plumbing fixtures may need to be represented on the floor plan, depending on the occupancy. Common symbols are available from third-party vendors.

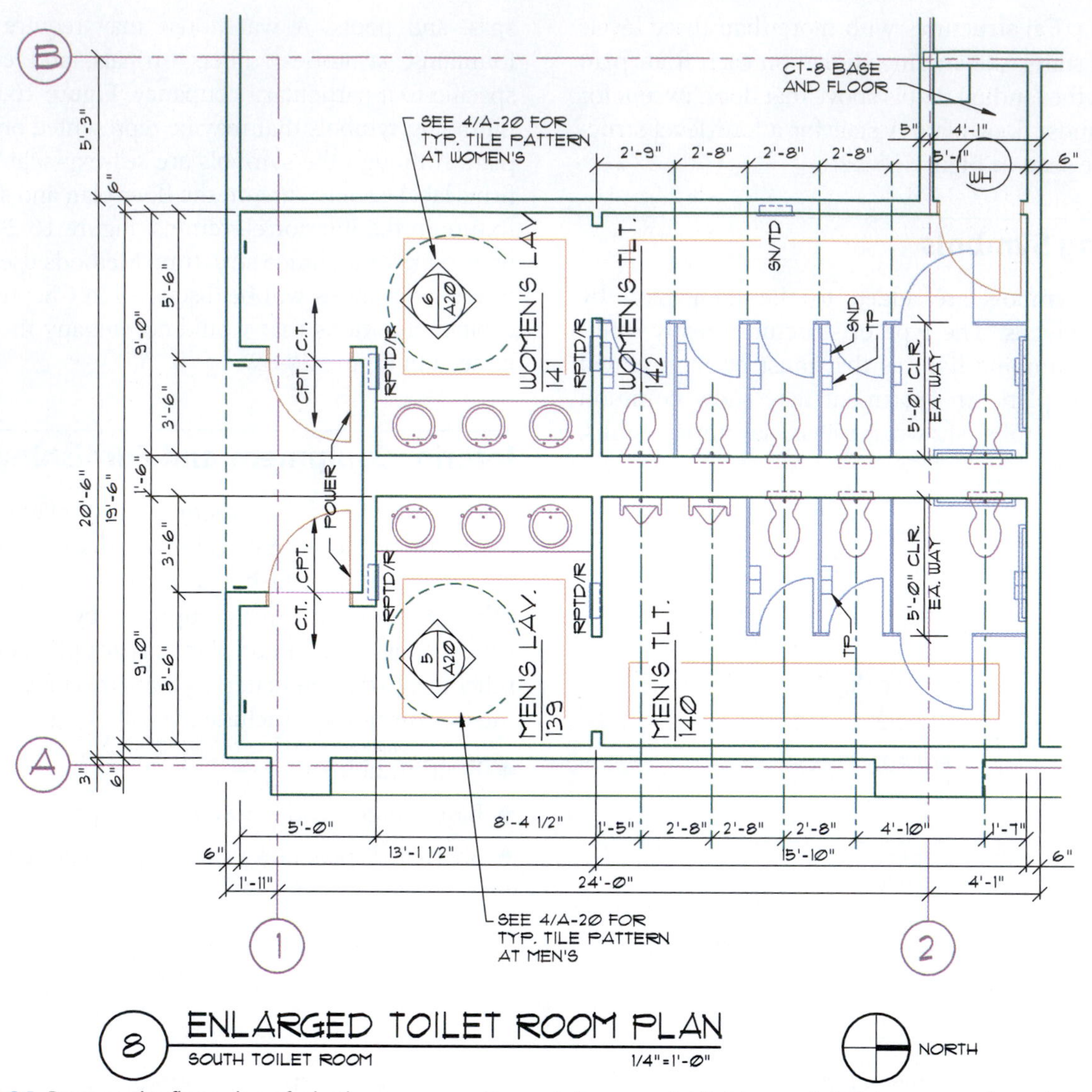

FIGURE 16-24 Because the floor plan of a bathroom generally must show special finishes and all fixtures, an enlarged plan of the area is often provided. *Courtesy Michael & Kuhns Architects, P.C.*

- Mechanical equipment such as water heaters, washers and dryers, cooking equipment, heating or cooling units, condensers, and compressors.
- Electrical equipment such as cash registers, communications equipment, or computer workstations.
- Built-in furnishings such as seating or benches.
- Special equipment specific to the occupancy such as churches, schools, restaurants, medical facilities, malls, or theaters.

Blocks for some features are available from third-party vendors, but most blocks must be developed by the drafter and stored for future use. Thin lines are used to represent the item being drawn. The method and the detail used to represent each item are affected by the scale of the floor plan. The complexity used to show a feature can be reduced if the item is manufactured and must be installed rather than fabricated according to the working drawings. Features that must be installed are often represented on an enlarged floor plan. Features that must be fabricated at the job site must be explained by a detail, which must be referenced on the floor plan.

Enlarged Floor Plans

Enlarged scale floor plans are used to provide support information to specific areas of the main floor plan. Stairwells, toilet rooms, elevator shafts, kitchens, mechanical and electrical rooms, and rooms that feature large amounts of equipment are typically represented on a floor plan as well as the enlarged floor plan. All items mentioned in the preceding section that are shown on the floor plan can be represented on an enlarged floor plan. Although features are shown on both plans, the annotation to describe features should be placed only on the enlarged plan. Information that should be duplicated on both the small and the enlarged floor plans includes room name and numbers, partition types, and column grids. To indicate that an enlarged plan has been drawn, a dashed line similar to Figure 16-11 is placed on the floor plan around the room or area to be described. Figures 16-21 and 16-24 show examples of enlarged floor plans.

PLACING DIMENSIONS

Dimensions are used on the floor plan to locate walls, windows, doors, and interior features. Not all windows, doors, and interior furnishings have dimension lines to locate them if they can be located by their position to a known wall. The material being represented will affect the method of placing dimensions. Refer to Chapter 3 for a review of basic dimensioning concepts and placement guidelines. In addition to those guidelines, consideration should be given to the effect the material has on the dimension style.

Light-Frame Dimensions

Light-frame structures constructed of wood or steel studs are usually dimensioned by following the four dimension groupings presented in Chapter 3. Dimensions are placed on each side of the structure as follows:

- The outer line defines the overall limits of the structure.
- The next line of dimensions is used to describe changes in shape or major jogs. The changes are dimensioned from the exterior face of the material used to build one wall to the exterior face of the next wall. The extension line is sometimes labeled with text parallel to the extension line that reads F.O.S. or F.S., representing the face of stud.
- The third line of dimensions represents the distance from the edge of exterior walls to the center (edge for some architectural offices) of interior walls.
- Used to locate openings, the fourth line of dimensions extends from a wall to the center of the openings.

Figure 16-25 shows dimensioning methods for a light-frame structure.

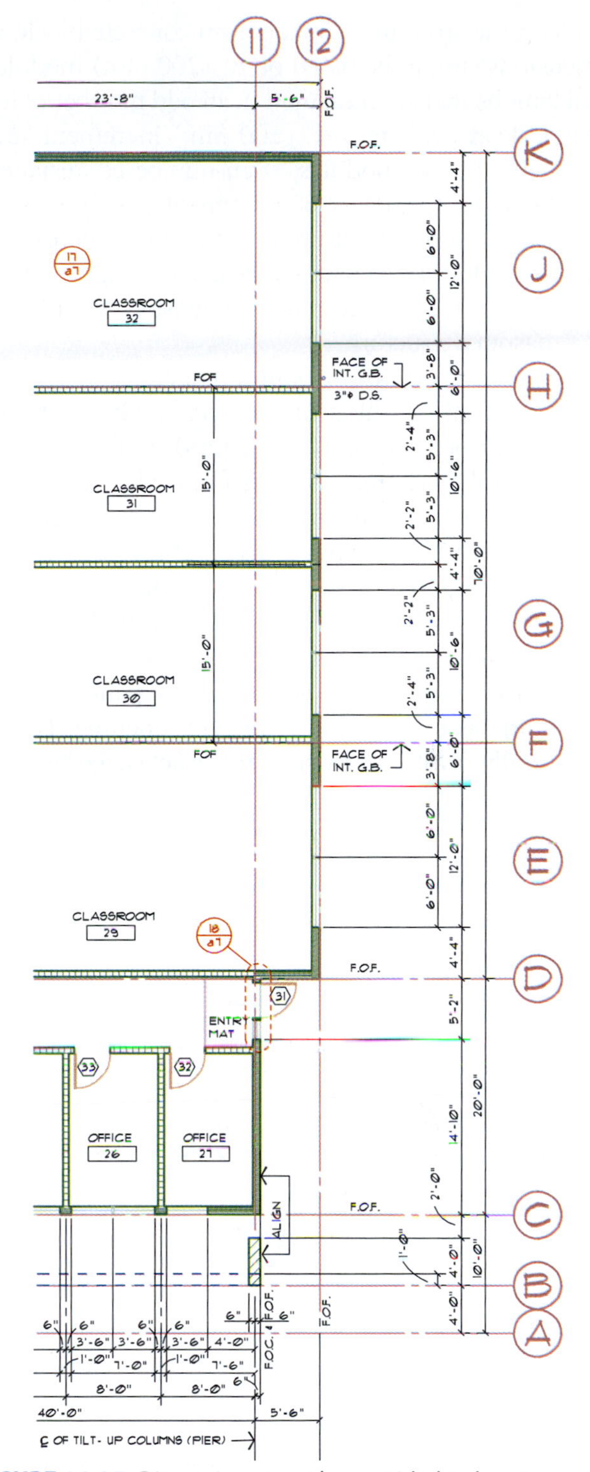

FIGURE 16-25 Dimensions must be provided to locate every feature. Each successive line of dimensions must add up to the dimension on the line above it. *Courtesy Architects Barrentine, Bates & Lee, A.I.A.*

Masonry Dimensions

Structures made of concrete block, poured concrete, or tilt-up concrete are dimensioned using line placement methods similar to those in light-frame construction. Differences include the following:

- Masonry walls are always dimensioned to edge and never to center.
- Openings for doors and windows are also dimensioned from edge rather than from center.

When the structure is made from concrete block, all dimensions should be based on 8" (200 mm) modules. Wall lengths that are measured in an odd number of feet should always end in a 4" (100 mm) increment such as 15'–4". To be modular, even-numbered distances will always end in 0" or 8" increments, such as 8'–0" or 10'–8". Interior light-frame walls are dimensioned to their centers as previously described. Figure 16-26 shows an example of common dimensioning practices for a masonry structure.

Structures made from poured concrete are dimensioned in similar fashion to concrete block but do not need to conform to the 8" (200 mm) module. Tilt-up concrete walls are dimensioned from edge to edge of each panel or from one grid point to the next grid on the architectural drawings. Figure 16-27 shows how concrete tilt-up walls can be dimensioned. Notice that openings in the walls *are not* located on the floor plan. The drawings for a tilt-up structure will include an elevation of each panel showing exact panel sizes as well as the size and location of each opening. Details are typically provided to define the exact space between panels.

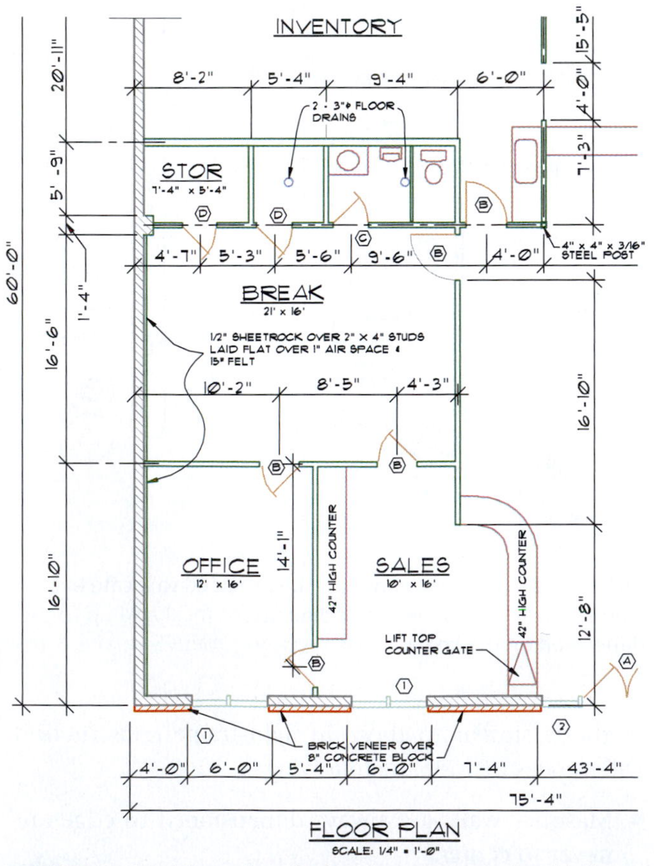

FIGURE 16-26 Dimensions for locating masonry should always be in modular sizes unless the project manager provides other instructions. *Courtesy Wil Warner.*

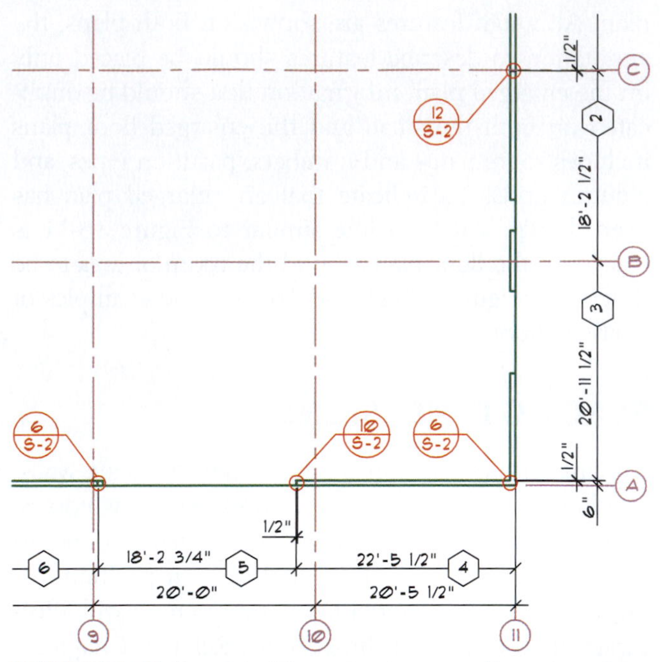

FIGURE 16-27 Concrete tilt-up structures are dimensioned from panel joint to panel joint or from grid to grid. *Courtesy Dean M. Smith, Kenneth D. Smith Architect & Associates, Inc.*

Steel and Timber Dimensions

In addition to the four types of dimensions used with other materials, structures framed with steel or timber require dimensions to determine the grid placement of the vertical columns. The grid shows the location of each vertical column and is placed after the overall dimensions have been placed. Once the structural shell has been erected, other dimensions can be used to locate walls and openings in the skeleton. Walls are typically referenced back to the grid line of the structural material. Figure 16-28 shows a floor plan for a structure framed of steel.

PLACING DRAWING ANNOTATION

The use of local notes to specify materials on the floor plan was introduced in Chapter 3 and has been mentioned throughout this chapter. Material that must be identified by notation should be identified after dimensions are placed so that the text does not interfere with dimension placement. Text can be referenced to the item being described by the use of a leader line as seen in Figure 16-29. The leader line should be kept as close to the feature as possible, without crowding the text or other features. If a note is lengthy, a shortened version can be placed on the drawing to identify the feature, with the complete specification given in a general note.

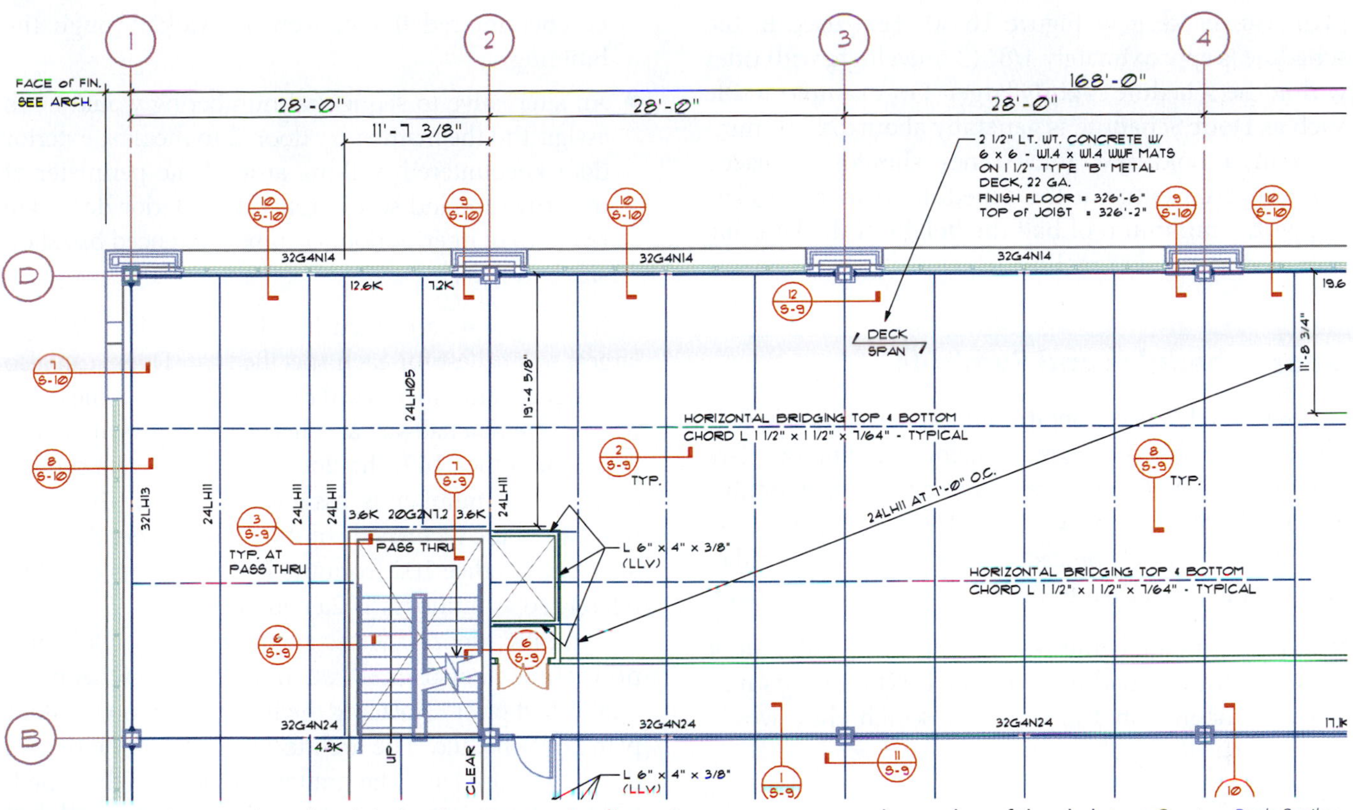

FIGURE 16-28 Structural steel and timber framing require dimensions to represent each member of the skeleton. *Courtesy Peck, Smiley, Ettlin Architects.*

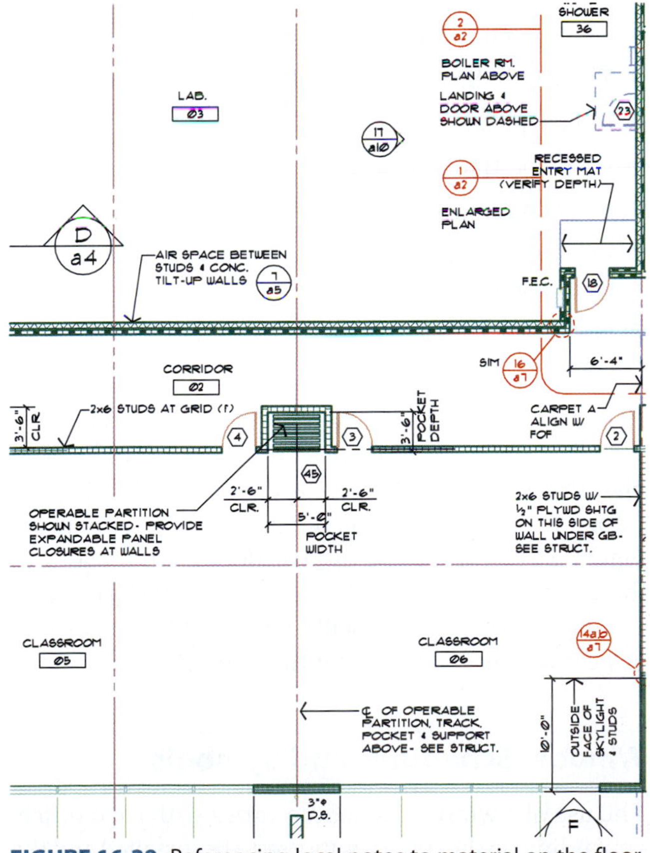

FIGURE 16-29 Referencing local notes to material on the floor plan. *Courtesy Architects Barrentine, Bates & Lee, A.I.A.*

Many professionals place general notes on the floor plan to ensure the compliance of the code that governs construction and to clarify all construction. These general notes can be placed beside the floor plan or on a separate specification page that is included with the working drawings. When a material is specified in a general note, a shortened version of the note should still be given on the floor plan to denote the accurate location of the specified material. A useful order for placing text on a drawing would be:

1. Room titles
2. Component identification
3. General notes
4. Drawing titles

Schedule Notations

Schedules are used on most architectural and structural drawings to reduce the information that must be placed in and around the drawing. Schedules add clarity to a drawing by removing notations and placing the notations either at the edge of the floor plan or on another sheet. Common schedules associated with floor plans include door, window, finish, hardware, and appliance schedules. Symbols used on the floor plans can be seen in Figure 16-30. Text used in the schedule is approximately 1/8" (3 mm) high, with titles within the schedule slightly larger. For example, a title such as **Door Schedule** is generally about 1/4" (6 mm) high in a bold text font. Lines should be placed between each entry in the schedule to increase clarity, with a minimum of half the height of the lettering provided to avoid crowding.

Door Schedules and Symbols

A door is referenced on the floor plan to a schedule by the use of a hexagon containing a number (see Figure 16-30a). The text representing the door symbol is typically 1/8" (3 mm) high in a 1/4" (6 mm) hexagon. Depending on office procedure, the number may either be sequential or related to the room containing the door.

- With sequential numbering, the main entry door is usually assigned the number 1. Other doors may be numbered in the order in which they would be encountered if you were to walk through the building.
- An alternative to sequential numbering would be to assign 1 to the main entry door, 2 to the next exterior door encountered walking around the perimeter of the structure and so on. Once all exterior doors are referenced, interior doors can be referenced based on size and type.

Interior doors are often listed in the schedule from largest to smallest for each specific type. Doors can also be assigned a number based on the location of the door. If you were in a hallway and needed to represent a door going into office 102, the door would receive a symbol of 102. The number is assigned based on the room number on the secure side of the door. If more than one door enters office 102 from the hall, the doors could be referenced as numbers 102a and 102b.

Typically, doors are assigned a door symbol on a print of the floor plan. A rough draft of the schedule is established prior to placing the information on the floor plan and schedule. The architect will assign the type of doors to be used and the drafter will develop the schedule based on company policy. Although material will vary depending on the occupancy and complexity of a structure, a schedule will normally contain information related to:

- The reference number.
- The width, height, and thickness.
- Door type and UL rating.
- Door and frame material.
- Door and frame finish.
- References for details showing the jamb, header, and threshold construction.
- Notations to clarify door operation, location, or construction.

Figure 16-31 shows a portion of the door schedule, the notes, and door drawings for the doors on the floor plan that will be developed at the end of the chapter. Doors need to be detailed using details similar to Figure 16-32 to show how the door frame relates to the exterior finish and the skeleton of the structure. Details showing door installation should be placed on or near the sheet containing the floor plan.

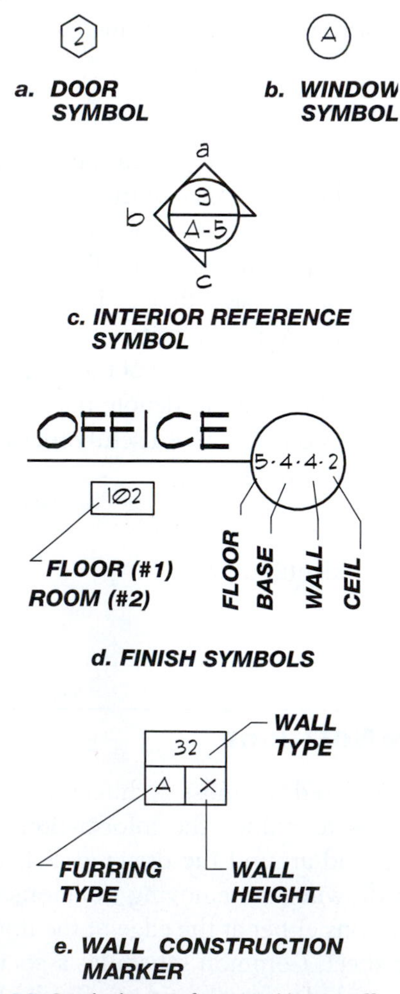

FIGURE 16-30 Symbols are often used by an office to explain materials referenced to schedules.

Window Schedules and Symbols

The use of a window schedule varies with each office, depending on the type of window to be installed and the amount of detailing provided. On the structure used

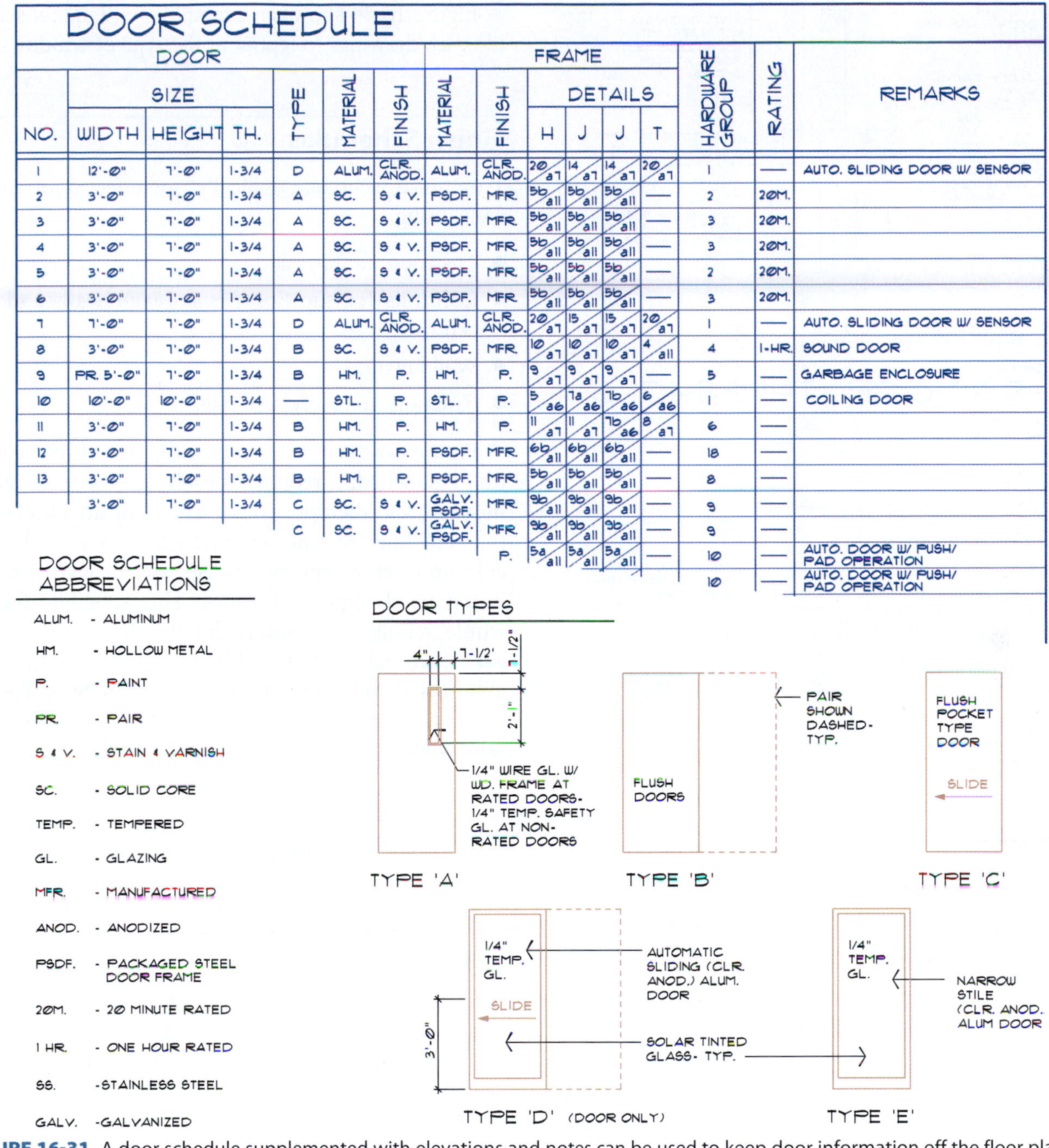

DOOR SCHEDULE

NO.	DOOR SIZE WIDTH	DOOR SIZE HEIGHT	DOOR SIZE TH.	DOOR TYPE	DOOR MATERIAL	DOOR FINISH	FRAME MATERIAL	FRAME FINISH	FRAME DETAILS H	FRAME DETAILS J	FRAME DETAILS J	FRAME DETAILS T	HARDWARE GROUP	RATING	REMARKS
1	12'-0"	7'-0"	1-3/4	D	ALUM.	CLR. ANOD.	ALUM.	CLR. ANOD.	20/a7	14/a7	14/a7	20/a7	1	—	AUTO. SLIDING DOOR W/ SENSOR
2	3'-0"	7'-0"	1-3/4	A	SC.	S & V.	PSDF.	MFR.	5b/all	5b/all	5b/all	—	2	20M.	
3	3'-0"	7'-0"	1-3/4	A	SC.	S & V.	PSDF.	MFR.	5b/all	5b/all	5b/all	—	3	20M.	
4	3'-0"	7'-0"	1-3/4	A	SC.	S & V.	PSDF.	MFR.	5b/all	5b/all	5b/all	—	3	20M.	
5	3'-0"	7'-0"	1-3/4	A	SC.	S & V.	PSDF.	MFR.	5b/all	5b/all	5b/all	—	2	20M.	
6	3'-0"	7'-0"	1-3/4	A	SC.	S & V.	PSDF.	MFR.	5b/all	5b/all	5b/all	—	3	20M.	
7	7'-0"	7'-0"	1-3/4	D	ALUM.	CLR. ANOD.	ALUM.	CLR. ANOD.	20/a7	15/a7	15/a7	20/a7	1	—	AUTO. SLIDING DOOR W/ SENSOR
8	3'-0"	7'-0"	1-3/4	B	SC.	S & V.	PSDF.	MFR.	10/a7	10/a7	10/a7	4/all	4	1-HR.	SOUND DOOR
9	PR. 5'-0"	7'-0"	1-3/4	B	HM.	P.	HM.	P.	9/a7	9/a7	9/a7	—	5	—	GARBAGE ENCLOSURE
10	10'-0"	10'-0"	1-3/4	—	STL.	P.	STL.	P.	5/a6	7a/a6	7b/a6	6/a6	1	—	COILING DOOR
11	3'-0"	7'-0"	1-3/4	B	HM.	P.	HM.	P.	11/a7	11/a7	7b/a6	8/a7	6	—	
12	3'-0"	7'-0"	1-3/4	B	HM.	P.	PSDF.	MFR.	6b/all	6b/all	6b/all	—	18	—	
13	3'-0"	7'-0"	1-3/4	B	HM.	P.	PSDF.	MFR.	5b/all	5b/all	5b/all	—	8	—	
	3'-0"	7'-0"	1-3/4	C	SC.	S & V.	GALV. PSDF.	MFR.	9b/all	9b/all	9b/all	—	9	—	
				C	SC.	S & V.	GALV. PSDF.	MFR.	9b/all	9b/all	9b/all	—	9	—	
								P.	5a/all	5a/all	5a/all	—	10	—	AUTO. DOOR W/ PUSH/ PAD OPERATION
													10	—	AUTO. DOOR W/ PUSH/ PAD OPERATION

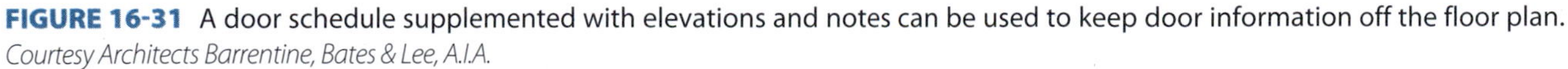

FIGURE 16-31 A door schedule supplemented with elevations and notes can be used to keep door information off the floor plan.
Courtesy Architects Barrentine, Bates & Lee, A.I.A.

throughout this chapter, no window schedule is provided. Sizes would be determined based on the dimensions provided on the floor plan (see Figure 16-25); the exterior elevations; and jamb, header, and sill details. Figure 16-33 shows a portion of the exterior elevations that references the height and construction details. If a window schedule is used, it should include:

- The reference letter.
- The width and height.
- Type of window.
- Type of glass to be used.
- Frame material and finish.
- Fire rating.
- References for details showing the jamb, header, horizontal and vertical mullions, and sill construction.
- Notations to clarify operation, location, or construction.

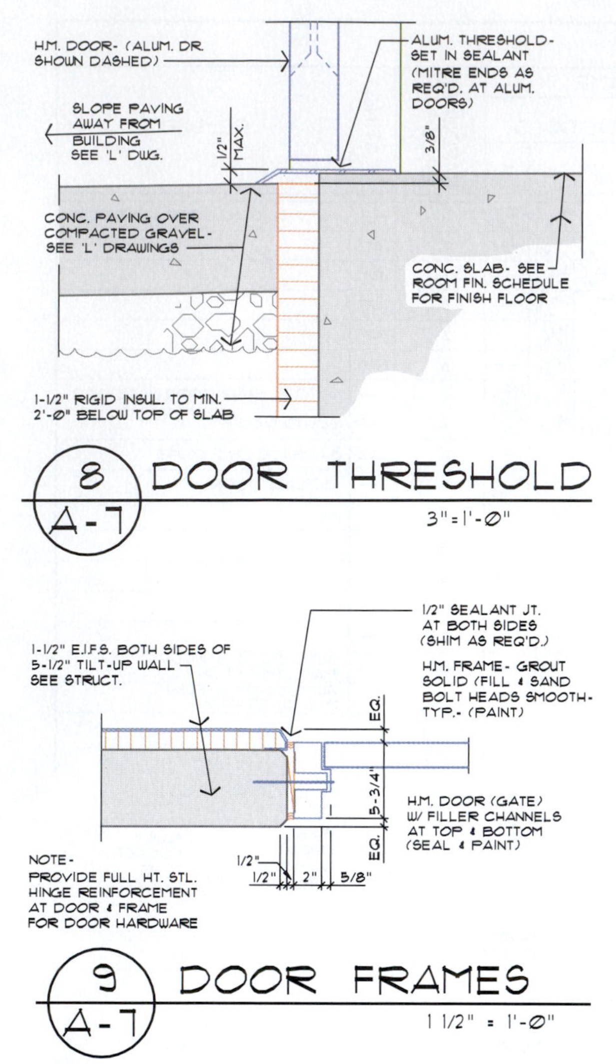

FIGURE 16-32 Details showing door information must be referenced to the floor plan. *Courtesy Architects Barrentine, Bates & Lee, A.I.A.*

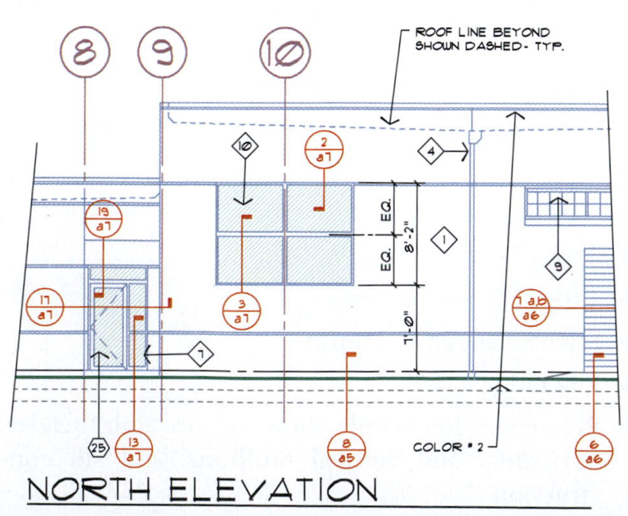

FIGURE 16-33 Information for windows for a simple structure can often be shown on the elevations rather than the floor plan. *Courtesy Architects Barrentine, Bates & Lee, A.I.A.*

Figure 16-34 shows an example of a window schedule with drawings to represent the type of windows.

Finish Schedules

The finish schedule typically includes information concerning:

- The room number and name.
- Columns for the floor, base, each wall, and the ceiling.
- Wall height.
- Notations to clarify materials.

Figure 16-35 shows an example of a finish schedule. This schedule represents each room with a specific number, and then provides a legend of finishes below the schedule. Another common format is to have a column to represent each finish, and provide a check in a box of required finishes. This schedule format would require six columns for the floor to represent each material. Notice in Figure 16-35 that the wall columns are referenced to *N*, *S*, *E*, and *W*. Several walls also have notations for two types of materials on the same wall.

Interior Reference Symbols

Interior reference markers like those in Figure 16-30c are used to reference interior elevations to the floor plan. A 3/8" to 1/2" (9 to 13 mm) diameter circle is used to identify each detail. The circle with an arrow is placed in the room to indicate which direction the marker is referencing. Four arrows can be placed around the circle to represent each wall of the room. If an arrow is not placed on a specific side of a circle, this indicates that no elevation is to be drawn for that wall. The page number where the detail is drawn is placed in the circle. Each arrow is numbered to represent a specific elevation. Interior elevations will be discussed in Chapter 20.

Interior Finish Symbols

To supplement the interior elevations, the use of finish symbols like the symbol in Figure 16-30d referenced to a finish schedule can be used to specify interior finishes. When several options for construction material must be specified, finish symbols can be used to add clarity. No matter what the symbol, it is typically placed near the room title or room number. Within the symbol, references for the floor, base, wall, and ceiling material should be given. This information is then specified in a finish schedule.

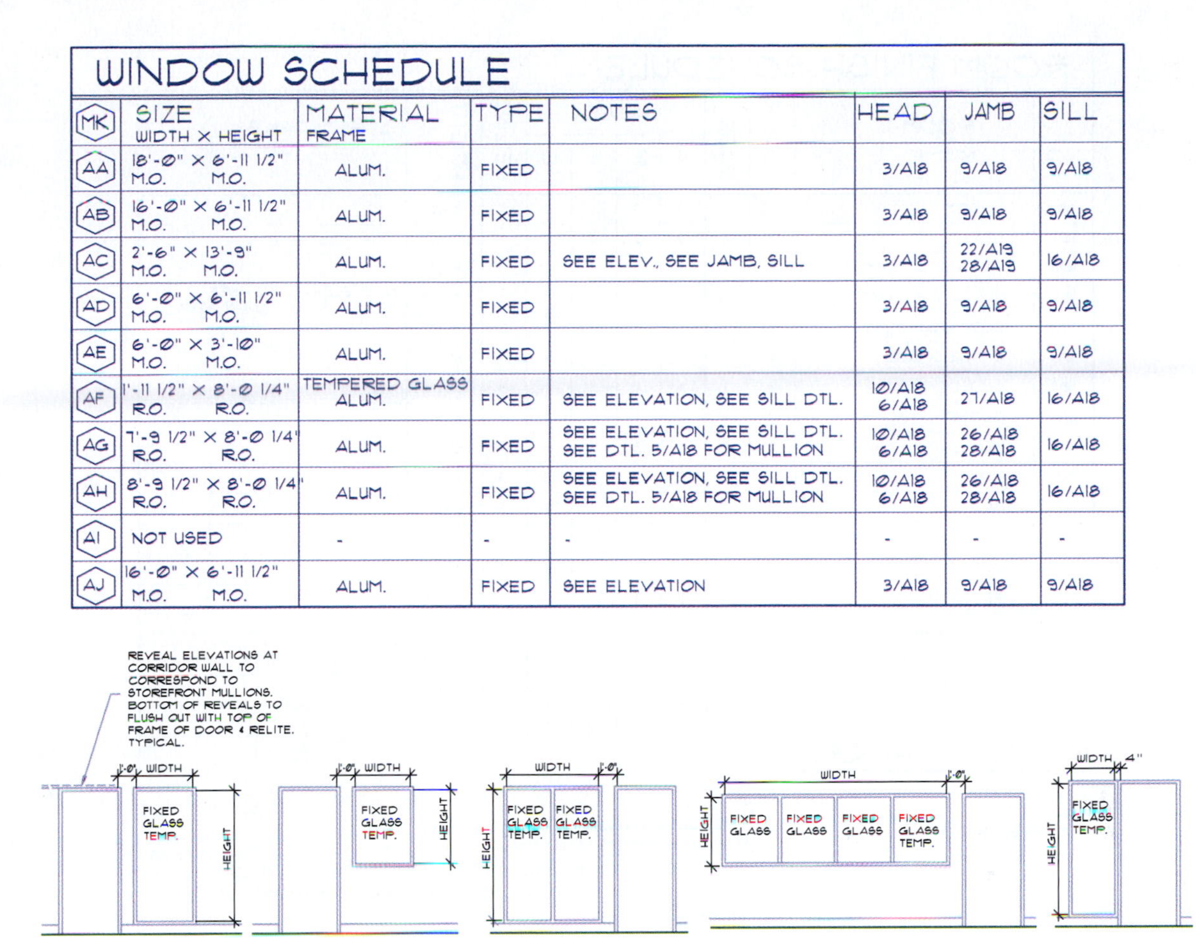

WINDOW SCHEDULE

MK	SIZE WIDTH X HEIGHT	MATERIAL FRAME	TYPE	NOTES	HEAD	JAMB	SILL
AA	18'-0" X 6'-11 1/2" M.O. M.O.	ALUM.	FIXED		3/A18	9/A18	9/A18
AB	16'-0" X 6'-11 1/2" M.O. M.O.	ALUM.	FIXED		3/A18	9/A18	9/A18
AC	2'-6" X 13'-9" M.O. M.O.	ALUM.	FIXED	SEE ELEV., SEE JAMB, SILL	3/A18	22/A19 28/A19	16/A18
AD	6'-0" X 6'-11 1/2" M.O. M.O.	ALUM.	FIXED		3/A18	9/A18	9/A18
AE	6'-0" X 3'-10" M.O. M.O.	ALUM.	FIXED		3/A18	9/A18	9/A18
AF	1'-11 1/2" X 8'-0 1/4" R.O. R.O.	TEMPERED GLASS ALUM.	FIXED	SEE ELEVATION, SEE SILL DTL.	10/A18 6/A18	27/A18	16/A18
AG	7'-9 1/2" X 8'-0 1/4" R.O. R.O.	ALUM.	FIXED	SEE ELEVATION, SEE SILL DTL. SEE DTL. 5/A18 FOR MULLION	10/A18 6/A18	26/A18 28/A18	16/A18
AH	8'-9 1/2" X 8'-0 1/4" R.O. R.O.	ALUM.	FIXED	SEE ELEVATION, SEE SILL DTL. SEE DTL. 5/A18 FOR MULLION	10/A18 6/A18	26/A18 28/A18	16/A18
AI	NOT USED	-	-	-	-	-	-
AJ	16'-0" X 6'-11 1/2" M.O. M.O.	ALUM.	FIXED	SEE ELEVATION	3/A18	9/A18	9/A18

FIGURE 16-34 A window schedule is used to keep the floor plan from becoming cluttered. *Courtesy Michael & Kuhns Architects, P.C.*

DRAWING SYMBOLS

Symbols are added to drawings to add clarity, eliminate notations, and to cross reference materials from one drawing to another. Symbols that must be shown on a floor plan include the north arrow, grid markers, section markers, detail markers, cabinet reference symbols, elevation symbols, and finish symbols. Because there are many variations to symbols used on the floor plan, a legend should be included with the floor plan.

North Arrow

A north arrow should be placed on the floor plan, as well as every other plan view to help orient the structure to the site. The north arrow should be simple and is typically placed in a circle measuring between 3/4" and 1" (20 and 25 mm) diameter. The symbol is often placed near the drawing title. North is placed at the top of the page if possible, although it can be rotated to the right side of the drawing sheet because of project and sheet size limitations. The symbol can be used to designate true north, magnetic north, and plan north. True north points to the North Pole, whereas magnetic north is a compass point deviating slightly from true north. Magnetic north is rarely indicated on plans, and should be shown relative to true north when represented. Plan north provides a reference to north that has been rotated to be parallel to the grid used to define the structure. True north can be indicated if the orientation is skewed by use of a north reference arrow. Examples of each are shown in Figure 16-36a.

Grid Markers

Grids are used to provide reference points established at a uniform distance throughout the project. Grid markers are used to reference each major change in shape of the structure, as well as structural columns, load-bearing walls, shear walls, and other structural elements that lie along a selected grid size. The size of the grid will vary with each project based on the module used to design the structure. Grid lines are also used as the basis for locating dimensions. A 3/8" to 1/2" (9 to 13 mm) diameter circle is typically used to represent the grid and column lines of a structure. A medium-width centerline extends from the circle through the drawing. Either 1/8" (3 mm) high text or numbers are used to

ROOM FINISH SCHEDULE

NO.	ROOM NAME	FLOOR	BASE	WALLS N	E	S	W	CEILING	HEIGHT	REMARKS
101	LOBBY	2/5	2	2	2	2/4	2	1	9'-0"	
102	STAIR	1	4	2	2	2	2	1	21'-0"	
103	JANITOR CLOSET	3	1	1	1	1	1	3	9'-0"	
104	ELEVATOR SHAFT	4	—	7	7	7	7	5	29'-0"	CAB ELEVATIONS- SEE INTERIOR ELEVATION 9/A5.
105	ELEV. EQUIPMENT	4	1	1	1	1	1	3	9'-0"	
106	OFFICE SPACE	4	1	1/4	1	1/4	1	1	9'-0"	
107	CLOSET	4	1	1	1	1	1	3	9'-0"	
108	CLOSET	4	1	1	1	1	1	3	9'-0"	
109	MEN	5	2	3	3	3	3	3	9'-0"	
110	WOMEN	5	2	3	3	3	3	3	9'-0"	
111	CLOSET	3	1	1	1	1	1	3	9'-0"	
201	LOBBY	1	4	—	2	—	2	1	9'-0"	
202	RECEPTION	1	4	2	2	2/4	2	1/4	9'-0"	
203	JANITOR CLOSET	3	1	1	1	1	1	3	9'-0"	
204	CONFERENCE	1	4	1	1	1	1	1	9'-0"	
	CLOSET	1	4	1	1	1	1	3	9'-0"	
				1	1	1	1	3	9'-0"	
									9'-0"	

ROOM FINISH KEY

FLOORS
1. CARPET (NIC.)
2. ENTRY MAT (NIC.)
3. SHEET VINYL (NIC.)
4. SEALED CONCRETE
5. CERAMIC TILE (NIC.)
6. HARDWOOD (NIC.)

BASE
1. 4" COVED RUBBER BASE (NIC)
2. CERAMIC TILE (NIC.)
3. HDWD. SKIRT (NIC.)
4. 4" CARPET (NIC.)

NOTES
— NO WORK REQUIRED

WALLS
1. GYPSUM BOARD (PAINT- NIC.)
2. GYPSUM BOARD W/ REVEALS (PAINT- NIC.)
3. CERAMIC TILE WAINSCOT- 2x2 THINSET OVER W.R. G.B. (PAINT- NIC.)
4. STOREFRONT SYSTEM
5. E.I.F.S.
6. METAL RAILING (PAINT)
7. GYPSUM BOARD- ROUGH TAPED

CEILINGS
1. 2X4 SUSP. ACOUSTIC TILE (NON-RATED)
2. SUSPENDED TYPE "X" GYPSUM EXT. SOFFIT BOARD (1 HR. FIRE RATED)
3. SUSPENDED GYP. BD. (NON RATED)
4. SKYLIGHT SYSTEM
5. GYPSUM BOARD- ROUGH TAPED

FIGURE 16-35 A finish schedule is used to group information relating to interior finishes and keep the floor plan uncluttered. *Courtesy Architects Barrentine, Bates & Lee, A.I.A.*

define each grid. To place vertical grid symbols, follow these guidelines:

- Place the grid symbol at the top end of the grid lines.
- Use sequential numbers to define the grid with 1 on the left side of the grid.
- Use numbers such as 3.5 to define items that are placed at the midpoint of grids 3 and 4.

Use the following guidelines for placing horizontal grid symbols:

- Place grid markers on the right end of the grid line.
- Use sequential letters, starting with A at the bottom of the grid, progressing through the alphabet for grids moving toward the top of the grid.
- To avoid confusion with the numbers 1 or 0, do not use the letters I or O.
- Use a letter followed by a number, such as B.5, to represent a column that lies midway between grids.

Figure 16-36b shows an example of grid reference symbols. Figure 16-28 and all of the other floor plans in this chapter show the use of grid markers. Grids should remain the same as each type of plan view is developed and for multiple levels of a structure. One exception to this guideline is a building with walls that step back from each other as the building rises in height. For example, on the architectural drawings, as the height progresses and the size of the building decreases, some grids will be deleted. A lower floor that had grids A through H may require only grids C through H on the upper levels due to decreased width. On the structural

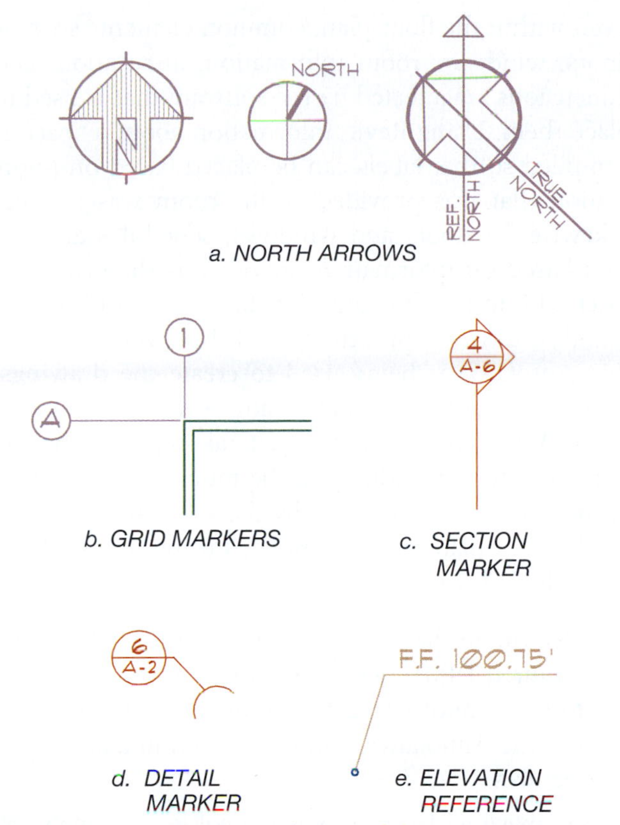

FIGURE 16-36 Standard symbols used throughout floor plans.

drawings for the same project, each grid from lower floors must be shown on each level, even if no columns or walls are in that portion of the grid.

The placement of grids will depend on the major building material for the project. Common grid locations based on material include:

- For structures framed using wood, composite, or steel studs, grids should be placed to align with the exterior face of exterior walls, and through the center of interior partitions, posts, or columns.
- For buildings framed using steel columns with curtain walls, grids are typically placed to represent the center of the column.
- For buildings constructed of concrete masonry units (CMUs) or poured concrete, grids should be placed to align with the exterior face of exterior walls, and through the center of interior partitions and columns.

The grid lines for poured concrete may be placed on the interior side of the exterior walls if the majority of the interior features will correspond better to the grid lines.

Section Markers

Section markers and section lines are used to indicate where each building and wall section is cut. A 5/8" (16 mm) diameter circle with 1/8" (3 mm) text should be used to identify the section. A bold phantom line is used to indicate where the cutting plane has passed through the structure. An arrow is placed around the circle to indicate the viewing direction. As seen in Figure 16-36c, the circle is divided into two portions, with a letter representing the section placed over the page number. Chapters 19 and 23 will discuss methods for drawing sections.

Detail Markers

Detail markers are used to reference construction details to the floor plan. A 5/8" (16 mm) diameter circle with 1/8" (3 mm) text should be used to identify each detail. A thin line is used to reference where the detail occurs. The circle referencing the detail is divided into two portions, with a number representing the detail placed over the page number. Figure 16-36d shows an example of a detail marker and the detail it represents.

Elevation Symbols

An elevation symbol is used to represent changes in height of a specific material. On plan views, the symbol consists of the elevation placed above a horizontal line, which is connected to the material by an inclined line with an arrow terminator. An elevation symbol can be seen in Figure 16-36e. Elevation symbols are used on floor, framing, site, and foundation plans, and on exterior elevations. Figure 16-19 shows examples of elevations referenced to a floor plan. Chapters 17, 19, and 23 introduce elevation symbols used to specify vertical locations on elevations, sections, and details.

Wall Construction Symbols

A final symbol that may be shown on the symbol legend is a symbol for wall construction. When required, this symbol will reference several details of wall construction to a specific wall represented on a floor plan. Figure 16-30e shows one option for creating a wall construction symbol. This particular symbol references several wall details, which are shown in Figure 16-37.

COMPLETING A FLOOR PLAN

The software that is used to create the floor plan will greatly affect how the drawing is created. The following discussion assumes the use of AutoCAD (or "AutoCAD vanilla" as many professionals refer to the software). AutoCAD works well for simple projects or in the hands of skilled CAD users. But there is a

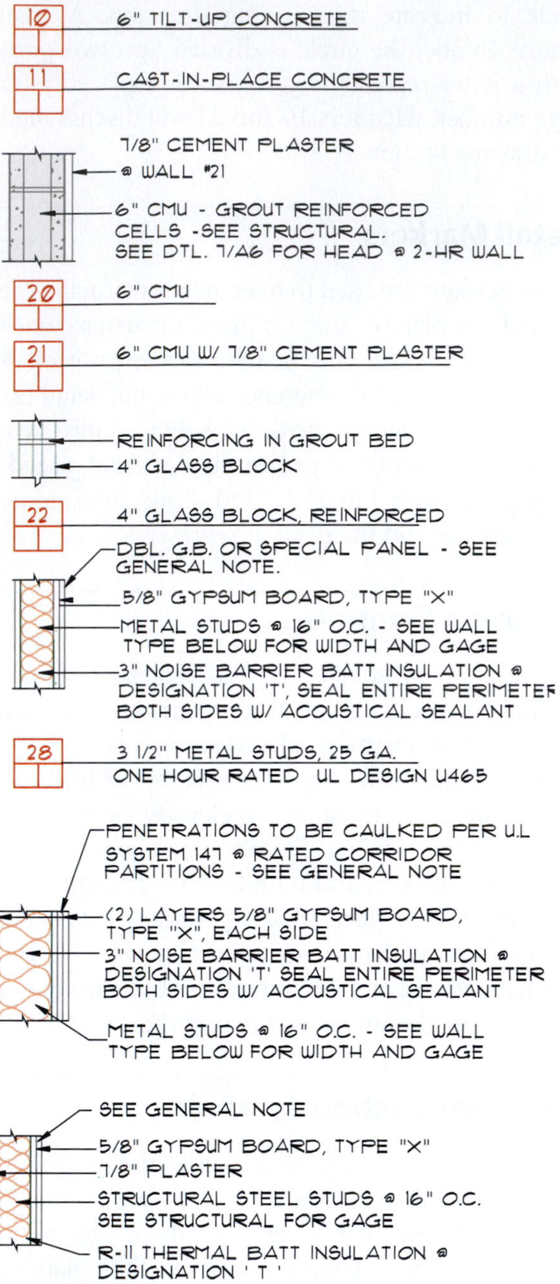

FIGURE 16-37 Symbols referencing wall construction should be placed on or near the floor plan. *Courtesy Michael & Kuhns Architects, P.C.*

better way. The use of AutoCAD Revit can greatly increase the productivity of a skilled CAD operator. AutoCAD works with drawing geometry; AutoCAD Revit is a parametric model builder. As the floor plan is created using Revit, information is stored that will automate the creation of stairs, elevations, sections, the roof plan, and other key elements of a drawing set. Even within the floor plan, common elements such as doors, windows, room information, annotation, and dimensions are affected by the software that is used to place them. Using Revit, information about a space is compiled so that labels can be placed based on information that was provided as the room was created. Likewise for doors and windows, schedules are created based on information provided as the symbol is inserted into the drawing. For the balance of Sections 3 and 4, drawing procedures will be given assuming that AutoCAD is being used to create the drawings. If you're creating drawings using AutoCAD Revit or AutoCAD Architecture, the general steps will be the same, but the procedure will be much easier because of the automation provided by the software.

The floor plan can be completed using AutoCAD by using the following steps:

1. Determine the scale to be used and establish the required DIMVARS, LTSCALE, and text scale factors. If AutoCAD 2008 or newer is to be used, use the Annotative feature to determine all scaling features.
2. Establish all layers and assign colors and linetypes. Save these settings in a template for future projects.
3. Draw all exterior walls.
4. Draw all interior walls.
5. Draw all openings. Include windows, doors, archways, skylights, and pass-through openings.
6. Draw all required stairs, steps, and ramps.
7. Draw all built-ins.
8. Draw all required plumbing symbols.
9. Draw all required equipment and interior furnishings.
10. Draw all required hatch patterns.

The floor plan should now resemble the floor plan shown in Figure 16-38. With all required items drawn, the dimensions can be completed using the following steps:

11. Draw all grid lines.
12. Place all overall dimensions.
13. Dimension all major jogs.
14. Dimension all walls that intersect an exterior wall.
15. Dimension all openings in the exterior walls.
16. Dimension all interior walls.

The floor plan with completed dimensions will now resemble Figure 16-39. The plan can be completed by adding required symbols and written specifications.

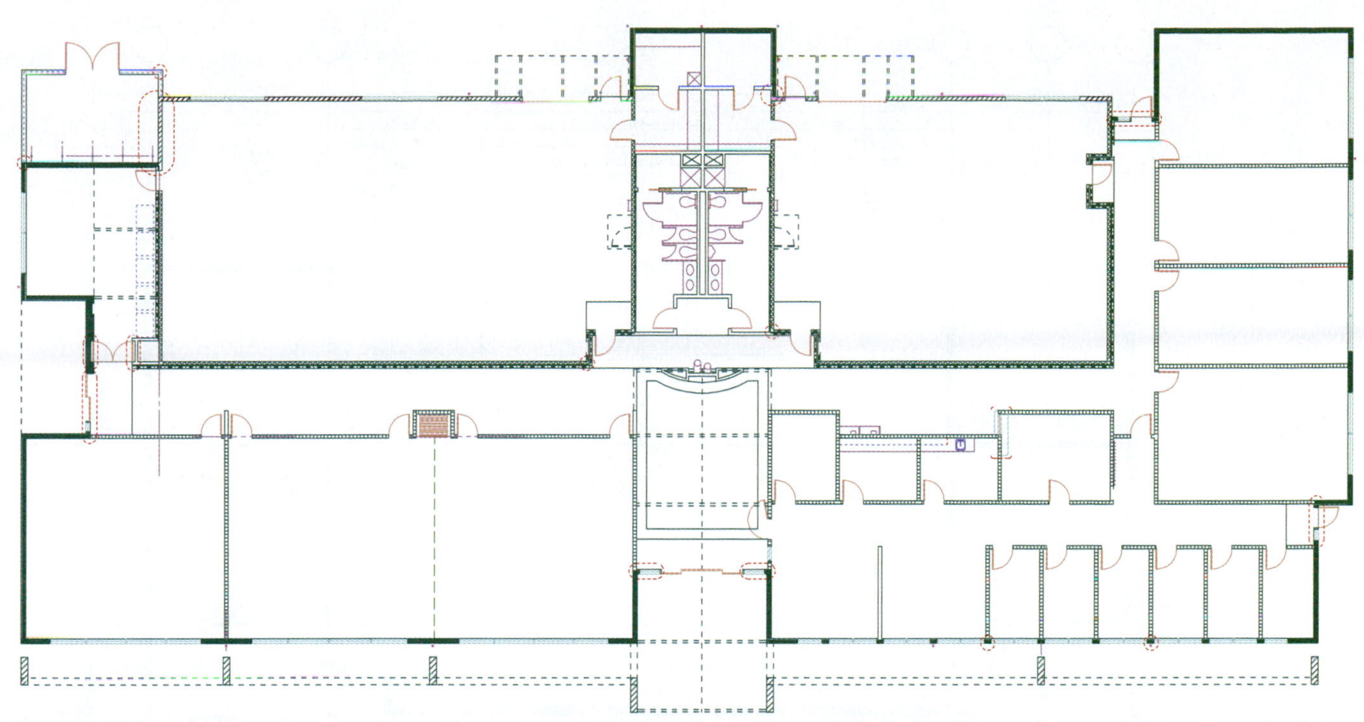

FIGURE 16-38 The floor plan showing walls, openings, cabinets and built-ins, and plumbing fixtures. *Courtesy Architects Barrentine, Bates & Lee, A.I.A.*

17. Place a title and scale below the drawing.
18. Draw a north arrow.
19. Provide identification text in each grid symbol.
20. Provide room names and numbers.
21. Provide all door and window symbols.
22. Locate anticipated section and detail markers. Additional symbols can be completed as the sections and details are drawn.
23. Locate partition type and interior elevation symbols.
24. Specify all elevation changes. Include changes in floors required for various materials.
25. Draw all required finish and wall construction symbols.
26. Draw a schedule to reference all symbols used on the floor plan.
27. Show and provide text to identify all features that are not represented in a schedule such as the termination of floor materials within a room, plumbing fixtures, fire equipment, floor drains, built-in casework, shelving lockers, benches, kitchen casework and equipment; openings in the floor such as shafts for dumbwaiters, mechanical shafts, electrical shafts, and HVAC shafts; atria, stairs, and escalators; overhead features such as balconies, skylights, beams, and roof overhangs.
28. List all general notes required to explain the floor plan.
29. Provide all material necessary to complete the title block.

The complete floor plan will now resemble the plan shown in Figure 16-40.

SPACE PLANS

When large amounts of interior furnishing must be represented on the floor plan, the plan can often become unclear. Many firms represent interior furnishing on a separate plan called a *space plan*. This plan uses the walls and openings shown on the floor plan as a base, but does not show the other material typically associated with a floor plan. The space plan shows all specialty items such as interior built-ins as well as movable furnishings such as shelving, display cabinets, book racks, sales counters, display cases, and seating. Equipment specific to the occupancy such as dental chairs, turnstiles, or examination tables is also represented on space plans. Figure 16-41 shows an example of a space plan for a small theater.

REFLECTED CEILING PLANS

A **reflected ceiling plan** shows the ceiling of a specific level of a structure. The floor plan is created by the cutting plane passing through the walls five feet above the

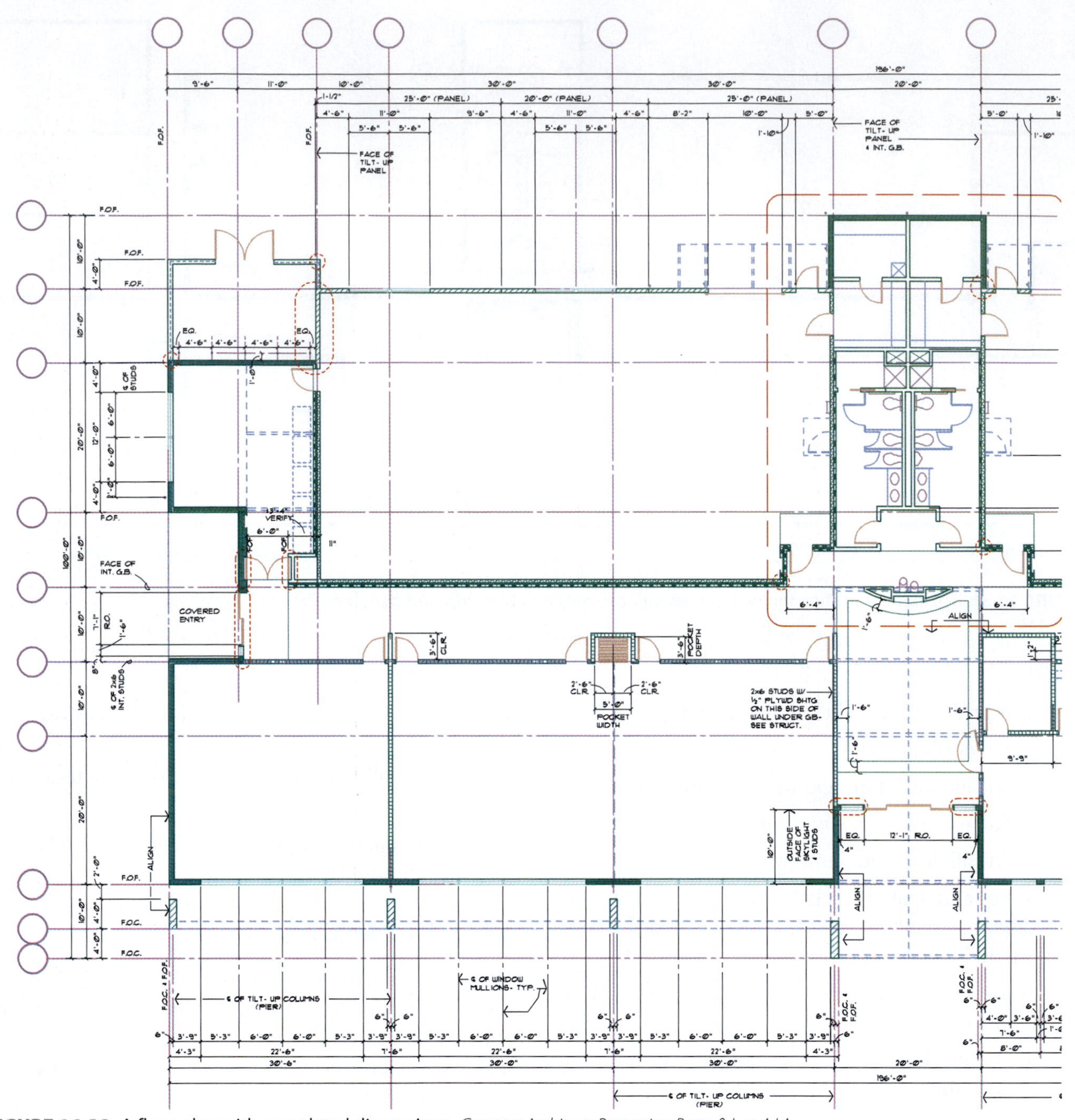

FIGURE 16-39 A floor plan with completed dimensions. *Courtesy Architects Barrentine, Bates & Lee, A.I.A.*

floor, allowing the viewer to see the floor. The ceiling plan is a mirror image of the ceiling, which is the source of the drawing title. Acoustical ceilings supported by wires are typically used on commercial structures to hide the structural material. These ceilings are often referred to as *T-bar* ceilings, because of the lightweight metal frames that support the ceiling tiles. A reflected ceiling plan shows the location, starting point, and type of material used to form the ceiling. The plan may also show the location of light panels and ceiling-mounted heat registers. These features may also be shown on a separate plan supplied by the mechanical engineer. Other features that should be shown on a reflected ceiling plan include:

- References to ceiling-related details such as ceiling edge conditions, expansion control joints, seismic joints, and lighting coves.
- Ceiling material indications including the extent of materials if more than one material is used.
- Exit lights, light fixtures, sprinkler heads, supply and return grills, smoke detectors, speakers, and related dimensions to explain these features.

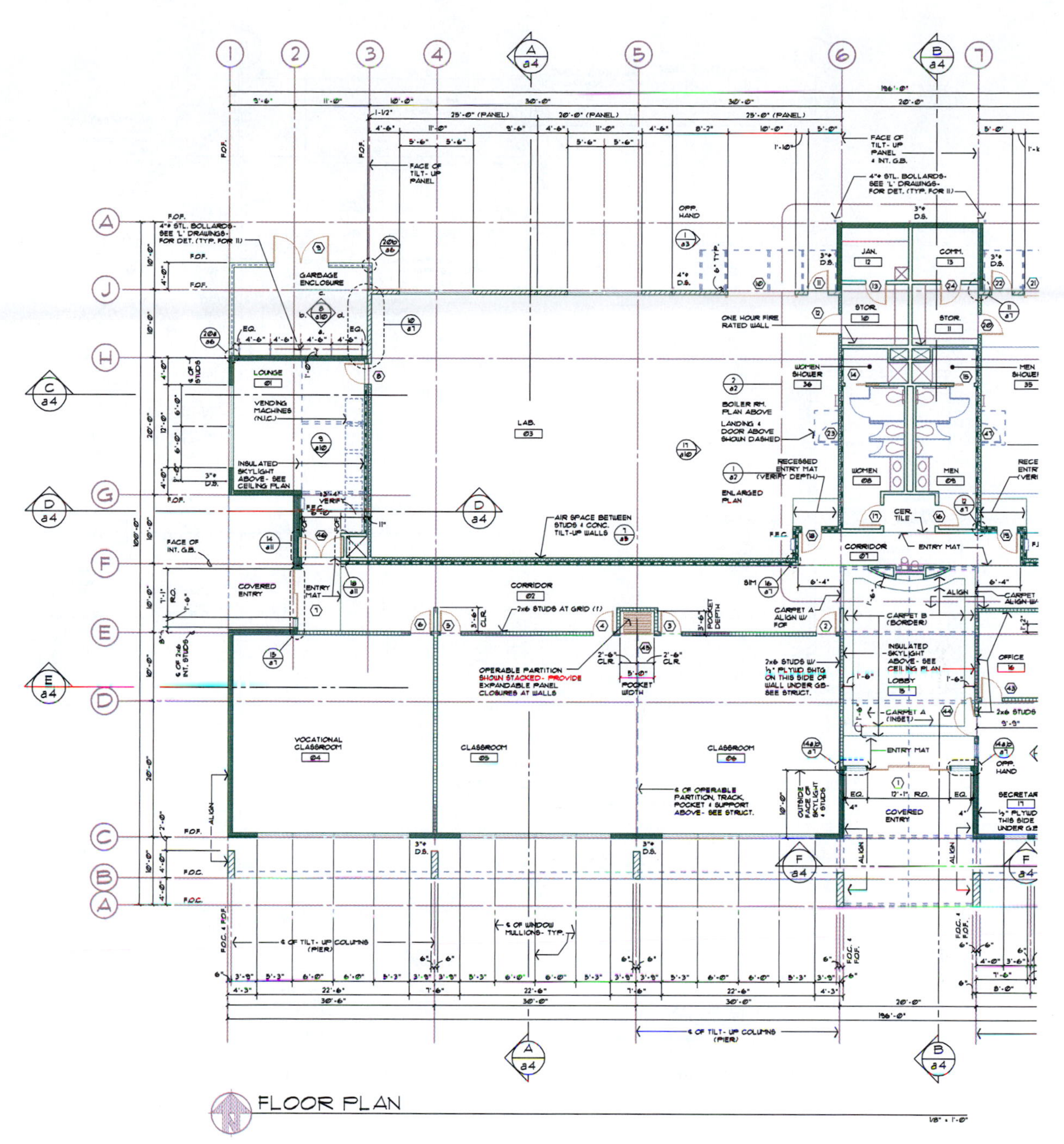

FIGURE 16-40 A completed floor plan with notations and symbols. *Courtesy Architects Barrentine, Bates & Lee, A.I.A.*

- Ceiling access panels to meet the needs of the mechanical and electrical trades.
- Skylights and roof hatches with related dimensions.
- Plenum barriers.
- Elements located above the ceiling that require specific locations such as fire-rated horizontal enclosures, catwalks, disappearing stairs, and HVAC equipment.

Below are some final guidelines for drawing reflected ceiling plans.

- Do not show door swings or openings unless the opening extends to the ceiling.
- Do not reference building or wall sections on this plan.

Figure 16-42 shows an example of a reflected ceiling plan and examples of the symbol legend and notes that explain construction.

LAYERING

Because the floor plan is used as a base for so many different drawings, careful thought should be given to layering as the drawing file is being established. Walls, windows, doors, built-in cabinets, and plumbing

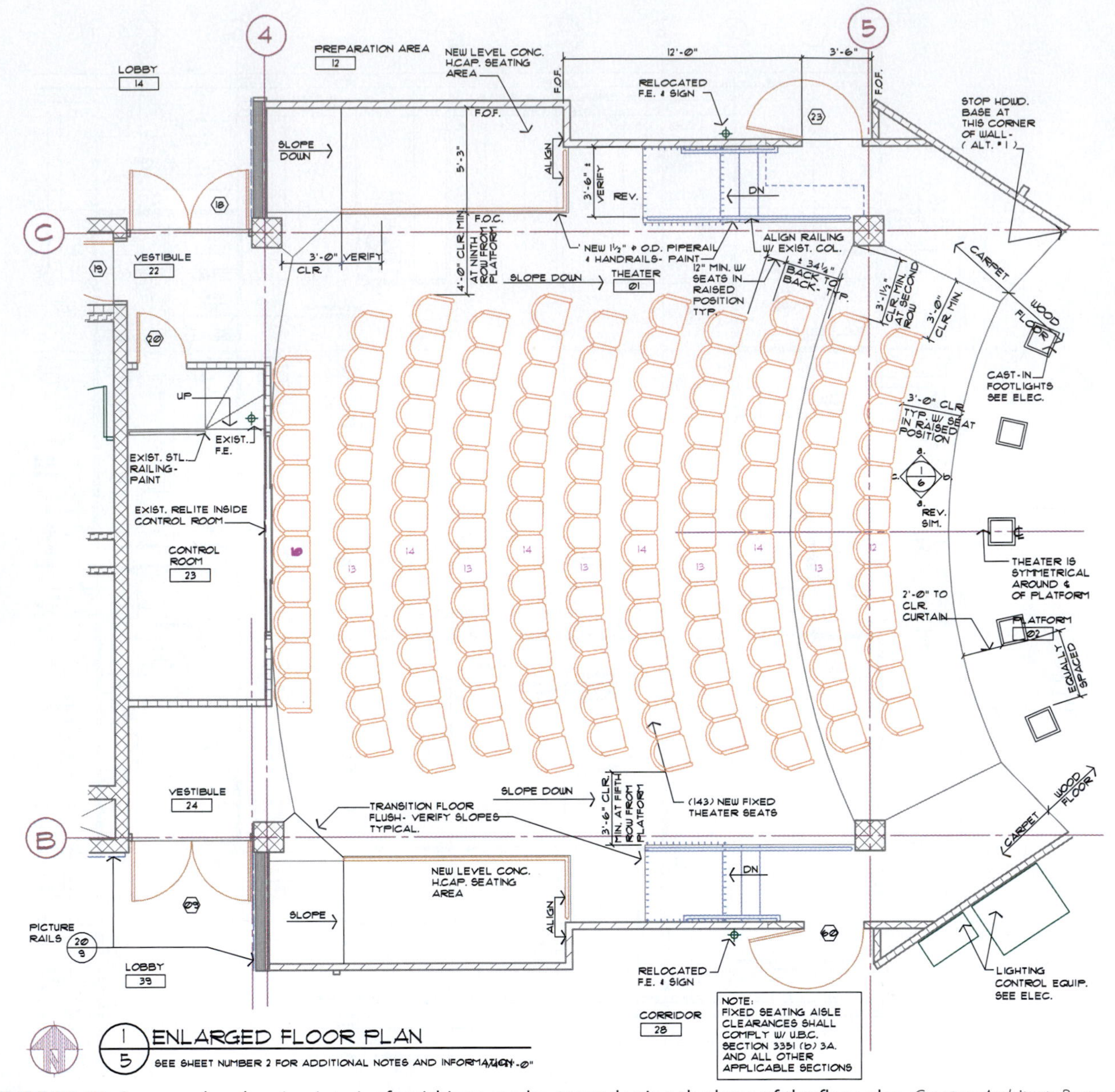

FIGURE 16-41 A space plan showing interior furnishings can be created using the base of the floor plan. *Courtesy Architects Barrentine, Bates & Lee, A.I.A.*

symbols are represented on other drawings that use the floor plan as a drawing base. All other material must be placed on a layer that can be frozen so that it will not be displayed. Chapter 3 introduced the American Institute of Architects (AIA) formats for defining layers. A partial list of recommended layer names is provided on the student CD in the APPENDIX folder. These names should be used to represent walls, doors, glazing, floor information, ceiling information, interior and exterior elevations, details, and reflected ceiling plan information.

Because of the wide variety of structures that will be drawn, each office usually establishes its own variation of the AIA system for naming layers. On complex structures, instead of having one layer for all walls, layers may be established for full-height, partial-height, curtain, and movable walls. Each of the eight major areas of the floor plan should be further divided as the floor plan grows more complicated. Added layers might include layers for furniture, area calculations, and occupancy information.

In addition to layer names that define information within the drawing, specific layers that describe the drawing should also be established. This includes layers to contain information that will be placed in the title block, dimensions, symbols, hatching symbols, schedules, and enlarged floor plans. The AIA and the National CAD Standard (NCS) have a complete list of modifiers available, which should be consulted as drawing standards are established.

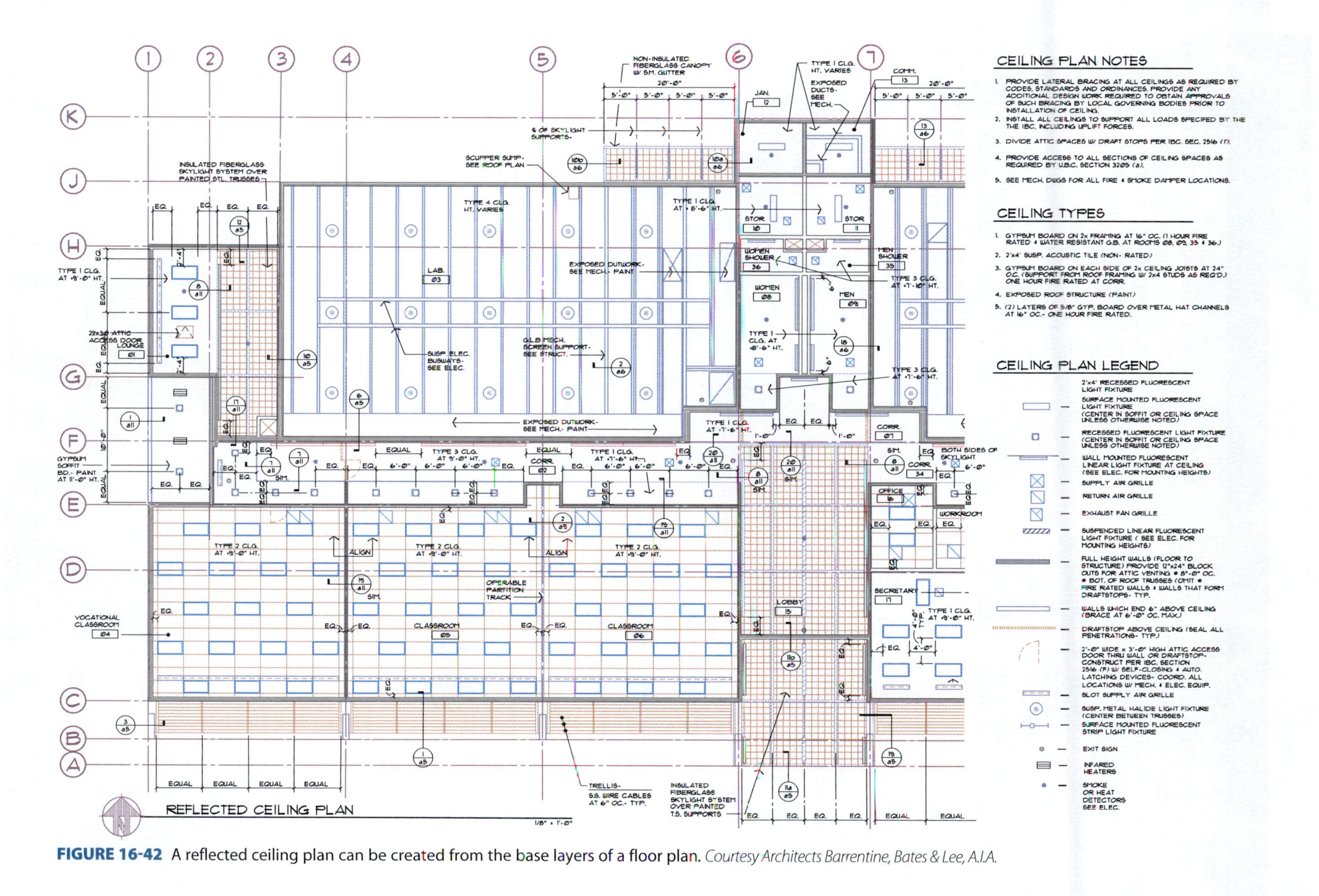

FIGURE 16-42 A reflected ceiling plan can be created from the base layers of a floor plan. *Courtesy Architects Barrentine, Bates & Lee, A.I.A.*

EASY BEING GREEN

Key elements shown on a floor plan include walls, windows, doors, openings, plumbing fixtures, cabinets, and finishing materials. These are also key areas to be considered when planning for sustainable construction. Energy credits from five Leadership in Energy and Environmental Design (LEED) categories can be represented on the floor plan. Following each of these five categories of LEED credits is a list of CSI sections that contain materials, fixtures, and equipment related to each credit. As you read through the five categories, you'll notice that many of the CSI listings are found in more than one credit category. That's important to realize because the factors that make a structure green are woven through all facets of the project. The floor plan and the related plans such as the mechanical, plumbing, and electrical plans are used to specify information related to gaining LEED credits. Information to be shown on the other related floor plans includes credits for:

- Innovation and Design Process—This LEED category represents two credits that rate the design process of a structure including the certification, standards, and codes that regulate green construction.
- Water efficiency—dealing with innovative waste water technologies and water reduction. Relevant areas of CSI for research related to this area include:
  - 11 – Equipment
  - 22 – Plumbing
  - 32 – Exterior improvements
  - 33 – Utilities
  - 44 – Pollution control equipment
- Indoor Environmental Quality—This far-reaching LEED category covers two prerequisites and eight credits. Prerequisites include minimum indoor air quality (IAQ) performance and the control of tobacco and other chemical pollutants. Eight LEED credits are given for the areas of interior finishing materials, furnishing, plumbing fixtures, HVAC equipment, and electronic safety and security equipment. Relevant areas of CSI for research related to this area include:
  - 01 – General requirements
  - 09 – Finishes
  - 11 – Equipment
  - 12 – Furnishings
  - 22 – Plumbing
  - 23 – Heating, ventilating, and air-conditioning
  - 28 – Electronic safety and security
  - 32 – Exterior improvements
  - 33 – Utilities
  - 44 – Pollution control equipment
- Energy and Atmosphere—This LEED category represents three prerequisites and six credits that rate the energy efficiency of a structure. Prerequisites include energy and atmosphere, minimum energy performance, and chlorofluorocarbon (CFC) reduction in HVAC & R equipment. Credits are given for optimizing energy performance, renewable energy, eliminating HCHCs and halons, green power, and management and verification methods. The scope of this category can be seen by the number of relevant areas of research related to this credit. CSI sections and subjects include:
  - 01 – General requirements
  - 03 – Concrete
  - 04 – Masonry
  - 05 – Metals
  - 06 – Woods, plastics, and composites
  - 07 – Thermal and moisture protection
  - 08 – Openings
  - 09 – Finishes
  - 10 – Specialties
  - 11 – Equipment
  - 12 – Furnishings
  - 13 – Special construction
  - 22 – Plumbing
  - 23 – Heating, ventilating, and air-conditioning
  - 26 – Electrical
  - 27 – Communications
  - 28 – Electronic safety and security
  - 32 – Exterior improvements
  - 33 – Utilities
  - 35 – Waterway and marine construction
  - 42 – Process heating, cooling, and drying equipment
  - 44 – Pollution control equipment
- Material and Resources—This LEED category represents one prerequisite and seven credits that rate the energy efficiency of a structure. The prerequisite for this section relates to the storage and collection of recyclables. Credits are available for the area of reuse, construction waste management, resource reuse, recycle content, use of local and regional materials, and certified wood. Relevant areas of CSI for research related to this area include:
  - 01 – General requirements
  - 02 – Existing conditions
  - 03 – Concrete
  - 04 – Masonry

(continued)

- 05 – Metals
- 06 – Woods, plastics, and composites
- 07 – Thermal and moisture protection
- 08 – Openings
- 09 – Finishes
- 10 – Specialties
- 11 – Equipment
- 12 – Furnishings
- 23 – Heating, ventilating, and air-conditioning
- 26 – Electrical
- 32 – Exterior improvements
- 33 – Utilities
- 34 – Transportation
- 35 – Waterway and marine construction

ADDITIONAL READING

The list of links to floor-related information is too lengthy to include here. Rather than listing individual sites, use your favorite search engine to research products from the CSI categories listed in the previous section. Other key areas to research include wall materials, window materials, glazing, doors, skylights, plumbing fixtures, cabinets, and furnishing.

KEY TERMS

Floor plan
Key plan
Partitions
Reflected ceiling plan
Storefront windows
Walls, bearing
Walls, curtain
Walls, faced
Walls, fire
Walls, nonbearing
Walls, parapet
Walls, party
Walls, shear

CHAPTER 16 TEST

Floor Plan Components, Symbols, and Development

QUESTIONS

Answer the following questions with short complete statements. Type the chapter title, question number, and a short complete statement for each question using a word processor. Some answers may require the use of vendor catalogs or seeking out local suppliers. If math is required to answer a question, show your work.

Question 16-1 List three types of floor plans, explain their origin, and state which plan a drafter is most likely to work on.

Question 16-2 List five common drawing scales for a floor plan.

Question 16-3 List three common drawing scales appropriate for drawing an enlarged floor plan.

Question 16-4 A floor plan for a large office structure needs to be drawn but it will not fit on the specified vellum size. Without reducing the scale, how can the plan be completed?

Question 16-5 List two drawing symbols that must be added to the floor plan described in question 16-4.

Question 16-6 List five categories of information that must be shown on the floor plan.

Question 16-7 List five categories of exterior dimensions.

Question 16-8 Describe two common types of notes found on a floor plan.

Question 16-9 List the dimension commands that would be useful for dimensioning a rectangular building.

Question 16-10 Describe the relationship between a floor plan and a reflected ceiling plan.

Question 16-11 Use graph paper and make a sketch of the floor plan of your drafting lab showing the dimensions of all walls and openings.

Question 16-12 Use graph paper and make a sketch of the floor plan of your drafting lab and show the layout of all furniture and equipment.

Question 16-13 Use graph paper and make a sketch of the floor plan of your drafting lab and show a reflected ceiling plan of your drafting lab.

Question 16-14 Contact architectural offices or contractors in your city and collect blueprints of five structures. Compare differences in drawing techniques between these drawings.

Question 16-15 Visit five construction sites, and with the permission of the superintendent, photograph the construction of wood, concrete block, concrete tilt-up, poured concrete, steel studs, and steel-frame construction.

DRAWING PROBLEMS

Use the reference material from preceding chapters, local codes, and vendor catalogs to complete one of the following projects. Unless other instructions are given by your instructor, draw the floor plan that corresponds to the site plan that was drawn in Chapter 15. Skeletons of the plans can be accessed from the student CD in the DRAWING PROJECTS/CHAPTER 16 folder. Use these drawings as a base to complete the assignment with the following assumptions:

- Use appropriate symbols, linetypes, dimensioning methods, and notations to complete the drawing.
- Determine any unspecified sizes based on material or practical requirements.
- Unless specified, select a scale appropriate to plotting on D-size material and determine LTSCALE and DIMVARS or use the Annotative feature of newer releases of AutoCAD.

Note: The sketches in this chapter are to be used as a guide only. As you progress through the floor plan you will find that some portions of the drawings do not match things that have been drawn on the site plan. Each project has errors that will need to be solved. The errors are placed to force you to think in addition to draw. Most of the errors are so obvious that you will have no trouble finding them. If you think you have found an error, do not make changes to the drawings until you have discussed the problem and possible solutions with your engineer (your instructor).

Information is provided on drawings in Chapters 17 through 24 relating to each project that might be needed to make decisions regarding the floor plan. Chapter 22 will be especially helpful in completing the basic layout of the floor plan. You will act as the project manager and will be required to make decisions about how to complete the project. Any information not provided must be researched and determined by you unless your instructor (the project architect and engineer) provides other instructions.

RESOLVING DRAWING CONFLICTS

When conflicting information is found, information from preceding chapters should take precedence. If conflicting information is found, use the following order of precedence:

1. **Written changes given to you by the engineer (your instructor) as change orders.**
2. **Verbal changes given to you by the engineer (your instructor).**
3. **The engineer's calculations (written instruction in the text).**
4. **Sketches provided by the engineer (sketches provided in each of the following chapters).**
5. **Lecture notes and sketches provided by your instructor.**
6. **Sketches that you make to solve problems; verify alternatives with your instructor.**

GENERAL DIRECTIONS

Use the attached drawings, the drawings provided on the student CD, and the information provided with other related problems in Chapters 15 through 24 to complete the required building floor plan for the project that was started in Chapter 15.

Warning: Submitting drawings that do not meet the following minimum standards may earn you the wrath of your supervisor, slow your progress in achieving a pay raise, or lead to your dismissal. Remember these are minimum standards. Your drawings should exceed them.

- Draw the building floor plan at the largest scale possible to fit on D-size material.
- Establish grids based on the NCS guidelines presented in the reading assignment for individual projects.
- Adjust all scale, lineweights, text, and dimension factors as required for the selected scale.
- Specify all material based on common local practice.
- Provide access that complies with the information provided in Chapters 4 and 5.
- Use separate layers for each major drawing component based on the recommended AIA guidelines presented on the student CD.
- Use dimensions to locate features as required by common practice.
- Provide annotation to specify all materials.
- Assume all trusses are to be provided Weyerhaeuser. Verify all truss sizes from tables found on their Web site.

- Hatch materials with the appropriate hatch pattern when appropriate.
- Use an appropriate architectural text font to label and dimension each drawing as needed.
- Refer to *Sweets Catalogs*, vendor catalogs, or the Internet to research needed sizes and specifications.
- Use schedules when possible to keep the drawing from being cluttered.
- Provide section and elevation markers for the related drawings from Chapters 17 through 23.
- Provide a drawing title, scale, and problem number below each detail.

Problem 16-1 Four-Unit Condominium—Wood or Metal Framing, Trussed Roof. Use the drawing and the DRAWING PROJECTS/ CHAPTER 16/16.1.PDF file on the student CD to complete a building floor plan for each level. The building is to contain four units that match the layout shown on the site plan. Each end unit is to have a bay window with center units to have flat glass. Place all required annotation in one unit, electrical information in another unit, and place all required unit dimensions in another unit. A separate unit is to be completed at a later time to be used for showing all framing material, which will be discussed in Chapter 22. Specify all door and window sizes using a schedule.

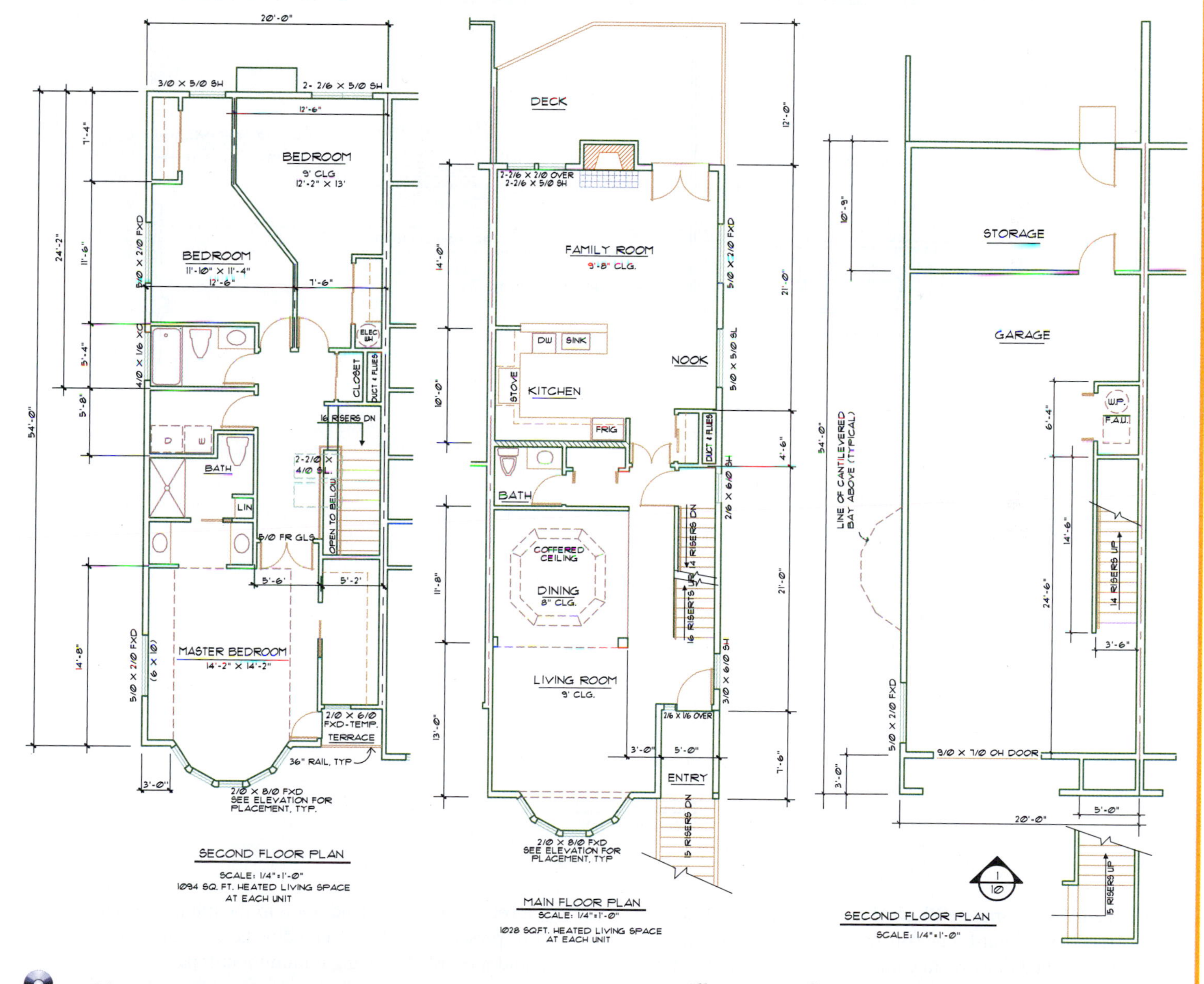

Problem 16-2 Retail Sales: Concrete Block, Wood Frame, Truss Roof. In addition to the minimum standards presented in the general directions, use the following drawings to complete the floor plan for the retail sales outlet using the following criteria:

- All exterior walls are 8" × 8" × 16" CMUs.
  - The south and east walls are to have a 4" brick veneer.
  - The west wall should have an attractive block or pattern because it will be visible.

- The north wall will be hidden by an existing structure on adjoining property.
- The interior side of masonry walls, except those in the inventory and parts department, and all interior walls are to be covered with sheet rock.
- Interior walls are to be 2× studs.
- Panel the office and sales areas.
- All windows are to be 84" high storefront.
- Exterior doors to be 3' × 7' double swinging tempered glass with a solar tint.
- Provide a sign over each pair of exterior doors to note that they are to remain unlocked during business hours. Use 32" wide doors throughout interior.

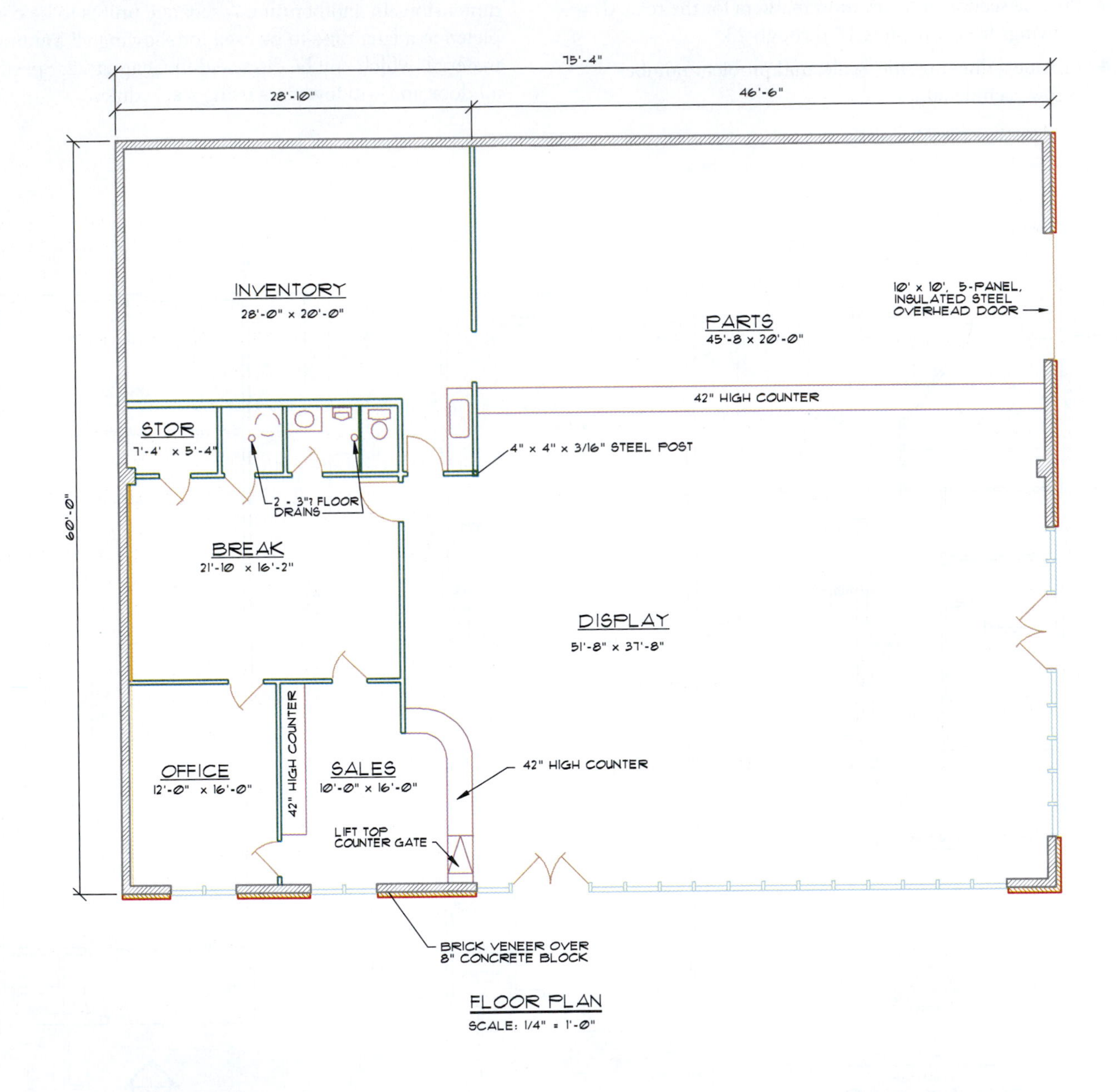

Problem 16-3 Concrete Block Warehouse. Material: concrete block, light-frame construction, panelized roof. Use the attached drawing to complete the floor plan for the concrete block warehouse to be used by a factory direct furniture company. Determine the occupancies of each portion of the structure and any required fire ratings. No matter the rating, a 1-hour fire wall is required at grid 5 to meet future plans. In addition to the minimum standards presented in the general directions, your solution should meet the following minimum criteria:

- Use 32' wide grids (typically located to the inside of the CMU walls) referenced to meet NCS guidelines.
- Construct the warehouse and office walls using 8" CMUs. Represent the pilasters that are specified on the

framing plan. Reference the pilaster details in Chapter 22 for the required size.

- Provide for two rows of 6" × 6" × 3/16" steel columns running north/south.
  - Provide one column on each side of the fire wall for a total of 10 columns.
- Provide a 1-hour fire wall framed with 2 × 6 studs between the display and storage area at grid 5. Specify the doors located in this wall to be 1-hour rated.
- Draw an enlarged floor plan of the sales/staff area. Show a sales counter in the sales area, as well as room for some floor display and public access.
  - Install counters in the staff room to allow for a stove, refrigerator, and double sink.
  - Provide two bathrooms near the sales/display area with one that is open to the public. Provide a 3" diameter floor drain for each bath.
- In the sales/display area, halls, bathrooms, and staff room, provide for a suspended ceiling that is 9' above the floor. Verify that a minimum distance of 18" will be provided above the suspended ceiling to the bottom of any structural members to provide space for HVAC ductwork.
- Provide a unisex bathroom in the storage area with its own water heater, lavatory, toilet, and urinal.

Mezzanine

- Plan for a mezzanine floor system to be placed above the display area. Assume 12' from finish floor of the display area to finish floor of the mezzanine.
- The west portion of the upper floor will be used as office space and will have a suspended ceiling at 8'.
- Provide 84" high storefront windows on the south and west sides that align with the windows on the lower floor.
- Provide 84" high × 15' ± storefront windows in the northwest corner of the mezzanine.
- Provide one unisex bathroom (sink and toilet only) for staff only. Place the bathroom so that the piping can share the same wall with the bath below in the storage area. The east portion of the upper floor will be used for lightweight storage with a minimum height of 9' to framing members.
- Research and specify an 8' × 8' service elevator capable of lifting 4000 pounds that will not interfere with the roof framing. Submit copies of vendor material showing requirements for the elevator to your instructor prior to completing the floor plan.
- Design two sets of stairs to provide access to the mezzanine. The stair serving the northeast stair will require a landing to serve a 3' × 7' 1-hour fire door.

Alternate Office/Staff Layout

- Design and draw the bathrooms somewhere in the staff area, providing all information for their installation. Specify all other interior information for the office/staff areas to be provided by the tenant. If your instructor allows this option to be used:
  - Do not provide a slab in the office area when the foundation plan is completed.
  - Do not show any interior walls or doors unless they are required for the construction of the bathrooms.
  - Specify all doors and windows as indicated in the exterior walls.
- Provide for a suspended ceiling in the display, sales, and break rooms 9' above the floor. Verify that a minimum distance of 18" will be provided above the suspended ceiling to the bottom of any structural members to provide space for HVAC ductwork.
- Provide a unisex bathroom in the storage area with its own water heater, lavatory, toilet, and urinal.

Door and Glazing Requirements

Provide a pair of glass doors to provide access from the exterior into the sales area, with as much glass as code allows on the south and west walls. Provide windows or skylights for the break room. Provide the following openings while maintaining modular sizes.

- Provide a pair of 3'–0" × 7'–0" steel doors to provide access from the sales area into the display area, and provide a large area of glass on the south and west sides of the sales office.
- Provide approximately 10' wide × 84" high storefront glass on the west and south walls of the display areas.
- Provide (1)-3'–0" exterior steel door on the north wall of the display area.
- Provide (3)-8' overhead doors in the loading dock that is 48" below the storage/display finish floor and (4)-11' × 11' overhead steel roll-up doors and (1)-3' door in the west wall of the storage area.
- Provide (1)-8' × 8' roll-up door and (1)-3' door, both with 1-hour ratings to provide access from the display to the storage area.
- Provide (4)-22" × 48" skylights per each grid over the storage area.

Loading Dock

- Provide a loading dock on the north side of the building, centered on the (3) service doors. The dock is to be 48" below the storage/display finish floor. Parking on the site plan will need to be coordinated with the loading dock.
  - Indicate a 36" high rail made of 1 1/2" Ø steel where the wall exceeds 30" above the slab in the loading dock.

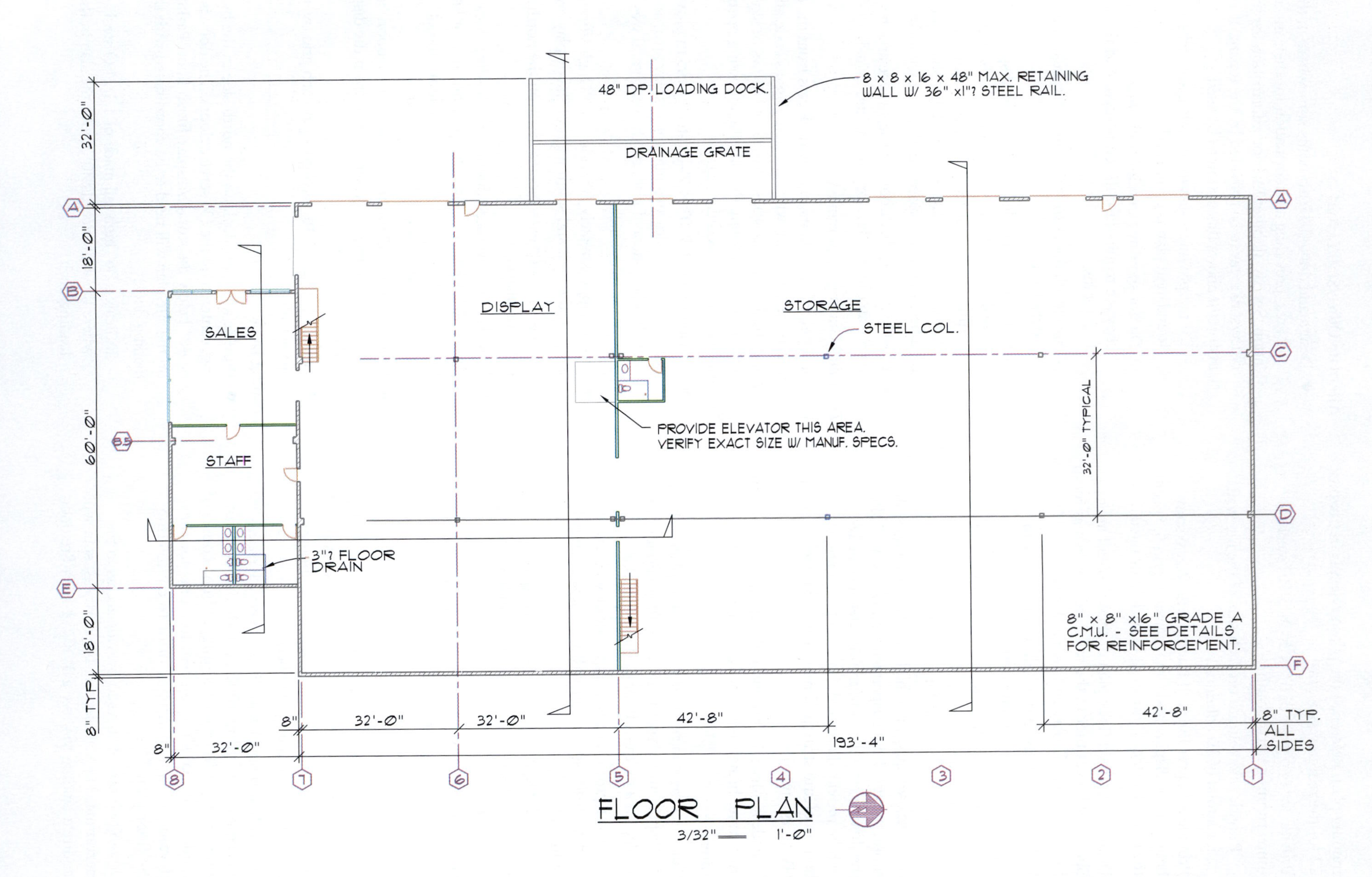
48" DP. LOADING DOCK.
8 x 8 x 16 x 48" MAX. RETAINING WALL W/ 36" x1"? STEEL RAIL.
DRAINAGE GRATE
DISPLAY
STORAGE
STEEL COL.
SALES
STAFF
PROVIDE ELEVATOR THIS AREA. VERIFY EXACT SIZE W/ MANUF. SPECS.
32'-0" TYPICAL
3"? FLOOR DRAIN
8" x 8" x16" GRADE A C.M.U. - SEE DETAILS FOR REINFORCEMENT.
32'-0"
18'-0"
60'-0"
18'-0"
8" TYP.
8"
32'-0"
32'-0"
42'-8"
193'-4"
42'-8"
8" TYP. ALL SIDES
FLOOR PLAN
3/32" = 1'-0"

Problem 16-4 Tilt-Up Warehouse. Material: concrete tilt-up, truss roof. Use the attached drawing to complete the floor plan for the concrete tilt-up warehouse to be used by a cabinet manufacturer. Part of the structure will be used by the sales staff for display, and for assembly and storage of units to be delivered to the job site. Although there will be no hazardous manufacturing on-site, varnish and other types of stains will be applied to cabinets on-site in the northeast portion of the storage area. Determine the occupancies of each portion of the structure and any required fire ratings. No matter the rating, a 1-hour fire wall is required at grid F to meet future plans. In addition to the minimum standards presented in the general directions, your solution should meet the following minimum criteria:

- Draw a floor plan showing all walls, openings, and columns at a scale of 3/32" = 1'–0".
- Construct the exterior walls of the warehouse and office using 6" thick concrete panels.
  - The east/west walls of the warehouse are to be constructed of four panels that are approximately 24.3' wide and extend the full width of the building (97').
  - The north/south wall joints to match grids.
  - Dimension the location of all openings in the concrete walls on separate panel elevations that will be completed later using information in Chapter 23.
- Lay out the warehouse using 24' grids from west/east.
- Assume all interior wood walls will be 2 × 4 stud walls. Use 2 × 6 studs for the fire wall at grid F.
- Provide a row of 6" × 6" × 3/16" TS columns to be located along grid 3 at the following locations:
  - Column 1: 12' to the west of grid J.
  - Column 2: @ grid H.
  - Column 3: 12" to the east of grid F.
  - Column 4: 12" to the west of grid F.
  - Column 5: 24' to the west of grid F.
  - Column 6: 24' to the west of grid F, 24' north of grid 3.
- Place a wall running e/w 32' north of the south side of the storage area to divide the storage and display areas. Cover the display side of the wall with 1/2" gypsum board.
- Provide a unisex bathroom in the shop area with its own water heater, lavatory, toilet, and urinal.

Sales/Office Area

Design a 48' × 70' office structure on the west end of the structure centered on grid 3. Use this area for a sales area on the northeast side with a sales counter between the reception area placed in the northwest corner as well as room for some floor display and public access.

- Provide two offices on the west side, two bathrooms, and a break room in the balance of this area. Bathrooms should have easy access to the display/storage area and the break room and must be available for public access.
- Provide for a gas water heater. All heating and cooling units will be roof-mounted.
- Assume the use of a suspended ceiling to be at 9' above the floor.
  - Verify that a minimum distance of 18" will be provided above the suspended ceiling to the bottom of any structural members to provide space for HVAC ductwork.
- Draw an enlarged floor plan of the office area at a scale of 1/4" = 1'–0".
  - Install counters in the break room to allow for a stove, refrigerator, and double sink.
  - Provide two bathrooms near the sales/display area. One must be open to the public. Provide a 3" diameter drain for each bath.
  - One of the bathrooms should have easy access to the break room.
- The display area is to have a 9'–0" suspended ceiling.
- The lower storage area will be used for storage of all chemical materials.

Alternate Office/Staff Layout

Design and draw the bathrooms somewhere in the staff area, providing all information for their installation. Specify all other interior information for the office/staff areas to be provided by the tenant. If your instructor allows this option to be used:

- Do not provide a slab in the office area when the foundation plan is completed.
- Do not show any interior walls or doors unless they are required for the construction of the bathrooms.
- Specify all doors and windows as indicated in the exterior walls.

Mezzanine

A mezzanine floor system is to be placed above the display/storage areas. Assume 12' from finish floor of the display area to finish floor of the mezzanine.

- The area over the lower storage area will be used for storage of lightweight assembled units.
- The north portion of the upper floor will be used as office space and will have a suspended ceiling at 8'.
  - Verify that a minimum distance of 18" will be provided above the suspended ceiling to the bottom of any structural members to provide space for HVAC ductwork.
- Provide one unisex bathroom (sink and toilet only) for staff only that will be above the bathroom located in the display area.
- The south portion of the upper floor will be used for lightweight storage.

- Research and specify an 8' × 8' service elevator capable of lifting 4000 pounds that will not interfere with the roof framing. Submit copies of vendor material showing requirements for the elevator to your instructor prior to completing the floor plan.
- Provide at least one 42" wide stairway to this area. Verify with local codes to see if more are required.

Doors and Glazing Requirements

- Provide (2)-8'–0" × 8'–0" overhead doors in display area and (3)-11'–0" × 12'–0" overhead doors in loading dock.
- Provide an 8'–0" roll-up door, and a 3'–0" metal door in the tilt-up wall between the display and the sales areas.
- Provide (1) 3'–0" metal door between the storage area and the break rooms.
- Provide (1) 8' metal roll-up door in the wall at grid F centered in bay and another door on the south section of wall F from storage area to storage area.
- Provide (1) 3' × 7' personnel door near each roll-up door in the storage areas.
- Provide 7'–0" high storefront glass in the north and west sides of the display area.
    - Provide for 24" minimum width panel support beside each opening or limit openings to 10'–0" maximum width at panel edge if glass is to extend to the corner.
- Provide a pair of 3'–0" × 7'–0" glass entry doors from the exterior into the sales area.
- Provide 84" high storefront glazing along the north and west walls of the sales area.
- Storefront windows in each office would be nice but not required.
- At the mezzanine level, provide 84" high storefront windows along the north and west sides. Windows on the north wall to extend to align with the east edge of the west door on the lower floor.
- Provide (4)-22" × 48" skylights per each grid over the shop and upper-level office areas.

Loading Dock

- Provide a loading dock on the north side of the building, centered on the storage area.
    - The three doors must be centered on the dock.
    - The 3' door should not be in the loading dock.
    - Indicate a 36" high rail made of 1 1/2" Ø steel where wall extends 30" above slab in the loading dock.

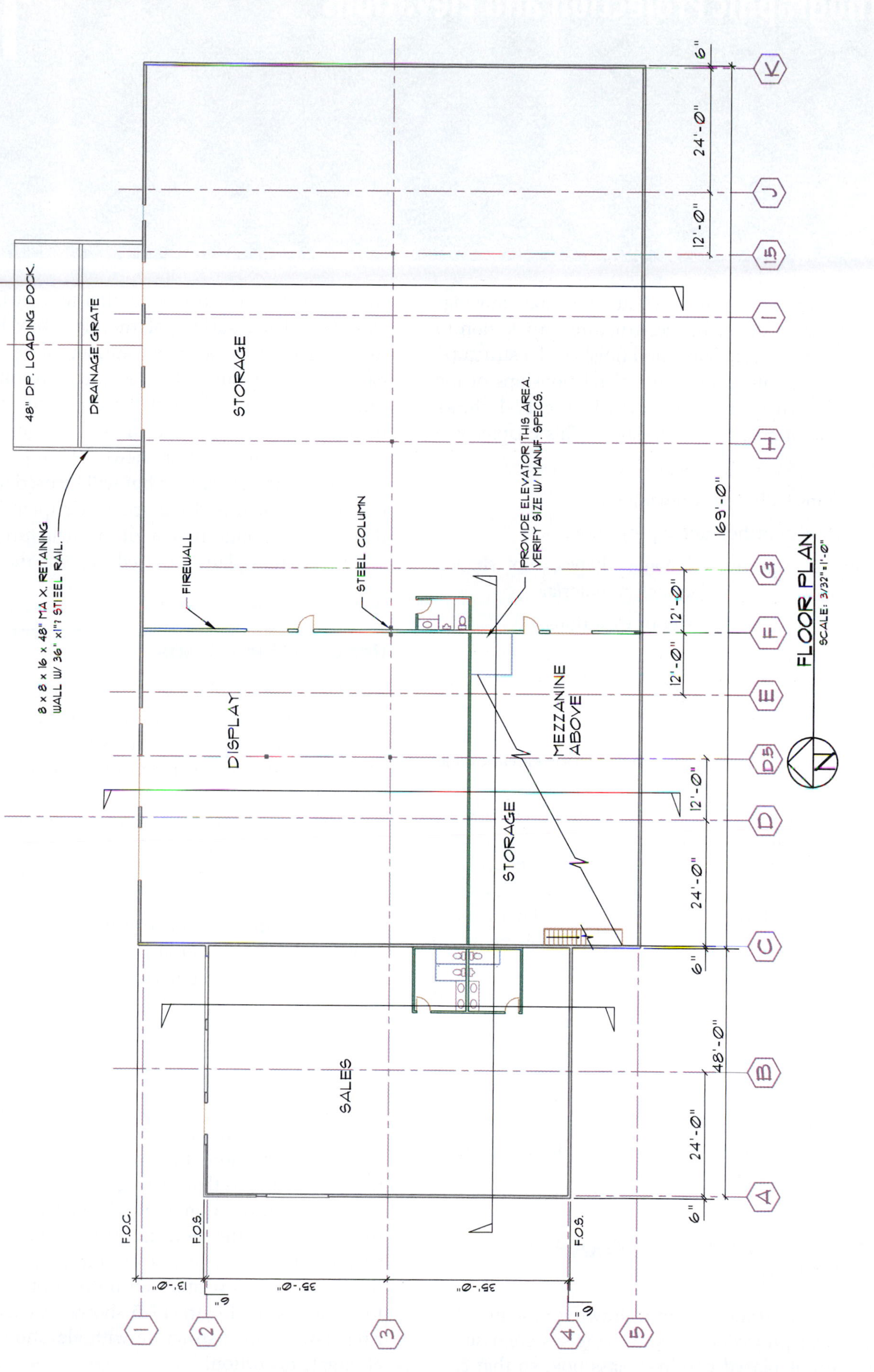
48" DP. LOADING DOCK
DRAINAGE GRATE
8 x 8 x 16 x 48" MAX. RETAINING WALL W/ 36" x1"? STEEL RAIL.
STORAGE
FIREWALL
STEEL COLUMN
PROVIDE ELEVATOR THIS AREA. VERIFY SIZE W/ MANUF. SPECS.
DISPLAY
STORAGE
MEZZANINE ABOVE
SALES
FLOOR PLAN
SCALE: 3/32"=1'-0"
169'-0"
24'-0"
12'-0"
12'-0"
12'-0"
12'-0"
24'-0"
48'-0"
24'-0"
6"
F.O.C.
F.O.S.
F.O.S.
13'-0"
35'-0"
35'-0"
K
J
I.5
I
H
G
F
E
D.5
D
C
B
A
1
2
3
4
5
N

Orthographic Projection and Elevations

17

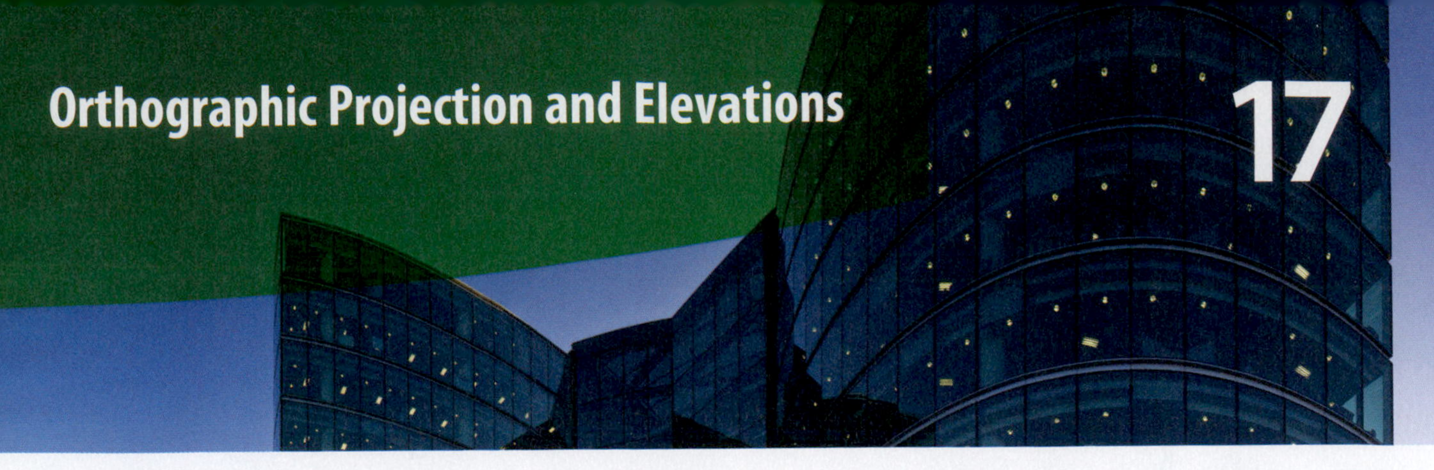

Elevations are the portions of architectural drawings that show the exterior of the structure. In addition to showing the exterior shapes and finishes of a structure, these drawings also show vertical relationships of the material shown in the plan views. Figure 17-1 shows an example of an exterior elevation. This chapter will explore:

- Elevations within the drawing set.
- Principles of orthographic projection.
- Drawing projection of major shapes, roof shapes, grades, openings, and common materials.
- Steps for completing exterior elevations.

ELEVATIONS WITHIN THE DRAWING SET

The architectural team draws the exterior elevations as part of the preliminary designs. In conjunction with the floor and site plans, the elevations are initially developed as sketches to investigate the shapes, materials, and relationship of the structure to the site. The front and one side elevation are typically drawn, but other elevations may be provided for complex structures. Major materials used to protect the structure from the elements are represented and briefly noted. Shades and shadows, landscaping, vehicles, and people may be shown to add a sense of reality. Figure 17-2 shows an example of a preliminary elevation. Although a drafter may not draw the preliminary elevations, these drawings will be useful later as a reference, when drawing the elevations that will be a part of the working drawings. Elevations are generally placed after the floor plans within the architectural drawings on sheets labeled A-2.

PRINCIPLES OF ORTHOGRAPHIC PROJECTION

Elevations are drawn using a drawing system called *orthographic projection*. This drawing system assumes a structure is placed within a glass box, so that each surface of the structure is parallel to a surface of the glass box. Each surface of the glass box is referred to as a *picture plane*. Each surface of the structure parallel to a picture plane is then presumed to be projected to the surface of the glass box. The box is then unfolded to produce the front, right side, rear, left side, top, and bottom views. The top view is not shown with the elevations but will be used to form the roof plan, which is discussed in Chapter 18. Rather than having a bottom view of the structure, the floor and foundation plans are used to show the outline of the structure.

Required Elevations

An elevation of each side of the structure is required to accurately represent construction. Although each parallel surface of the structure is projected to a picture plane, the distance of surfaces from the picture plane is very difficult to represent. An elevation of each surface is required to represent changes in shape and varied distance from the picture plane. With the comparison between two or more elevations, a structure can be clearly explained.

A minimum of four elevations are required to explain exterior materials for a structure. When a structure has an irregular shape such as a U shape, parts of the building are hidden in the basic four elevations. An elevation of each surface should be drawn, as shown in Figure 17-3. If a structure has walls that are not at 90° to each other, as shown in Figure 17-4, these walls will not be parallel to a picture plane and will be distorted when shown in an elevation. To eliminate this distortion, the structure is rotated within the "glass box" so that each surface of the structure is seen in true projection. Only the portion of the structure parallel to the viewing plane is shown, with the balance of the structure seen in another view. Match lines are used to relate each portion of the elevation to related parts. Figure 17-5 shows an example of an irregularly shaped structure with elevations drawn to eliminate distortion.

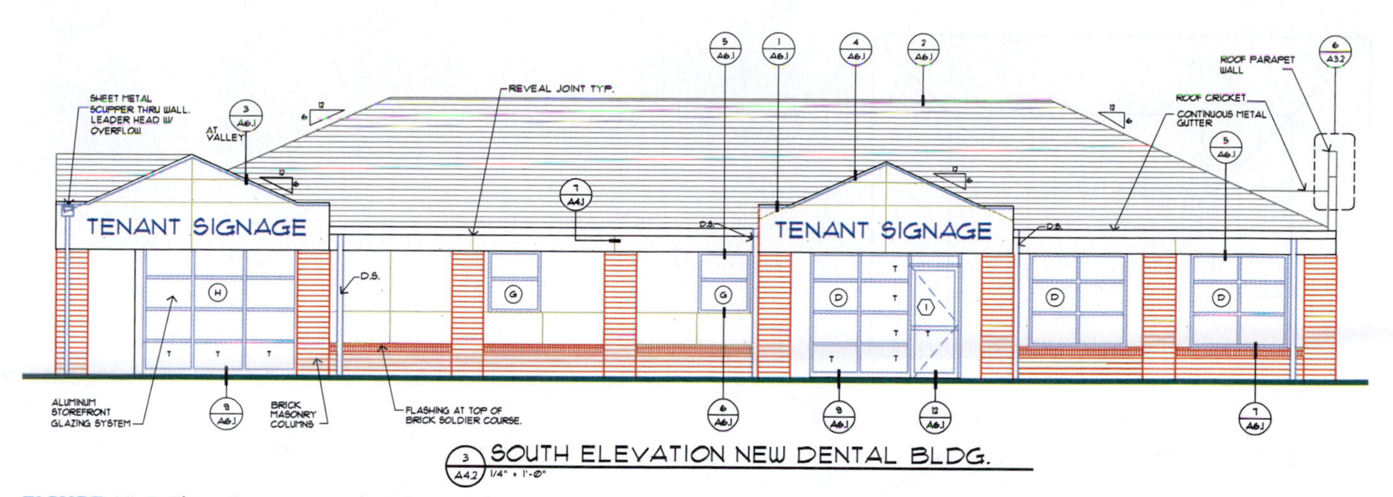

FIGURE 17-1 Elevations are used to show the materials used to protect the skeleton from the elements, and to show vertical relationships. *Courtesy Scott R. Beck, Architect.*

FIGURE 17-2 A preliminary drawing created by the project manager is often given to a drafter to help in the creation of an elevation. *Courtesy Scott R. Beck, Architect.*

Elevation Scales

Elevations are typically drawn at the same scale as the floor plan. Matching the scales allows the elevations to be drawn by using the floor plan or plans to project horizontal distances for walls, openings, and roof projections. Some offices provide elevations at a smaller scale to show the basic shape of the structure, with an enlarged set of elevations to provide details for construction. Figure 17-6 shows a small-scale elevation.

DRAWING PROJECTION

Exterior elevations can be easily constructed by using the floor, roof, and grading plans, as well as the preliminary sections, to project needed information. If the plan views are combined in one drawing file, required layers can be thawed for viewing to produce a base drawing from which to project the elevations. A wblock of the required layers should be created so that the floor plan is not accidentally damaged. This wblock can be copied and rotated to allow each surface to be projected. Figure 17-7 shows an example of how elevations can be projected from a wblock using AutoCAD.

Drawing Major Building Shapes

With the wblock created to project walls and openings, lines should be established to represent each finish floor, the upper and lower limits of the windows and doors,

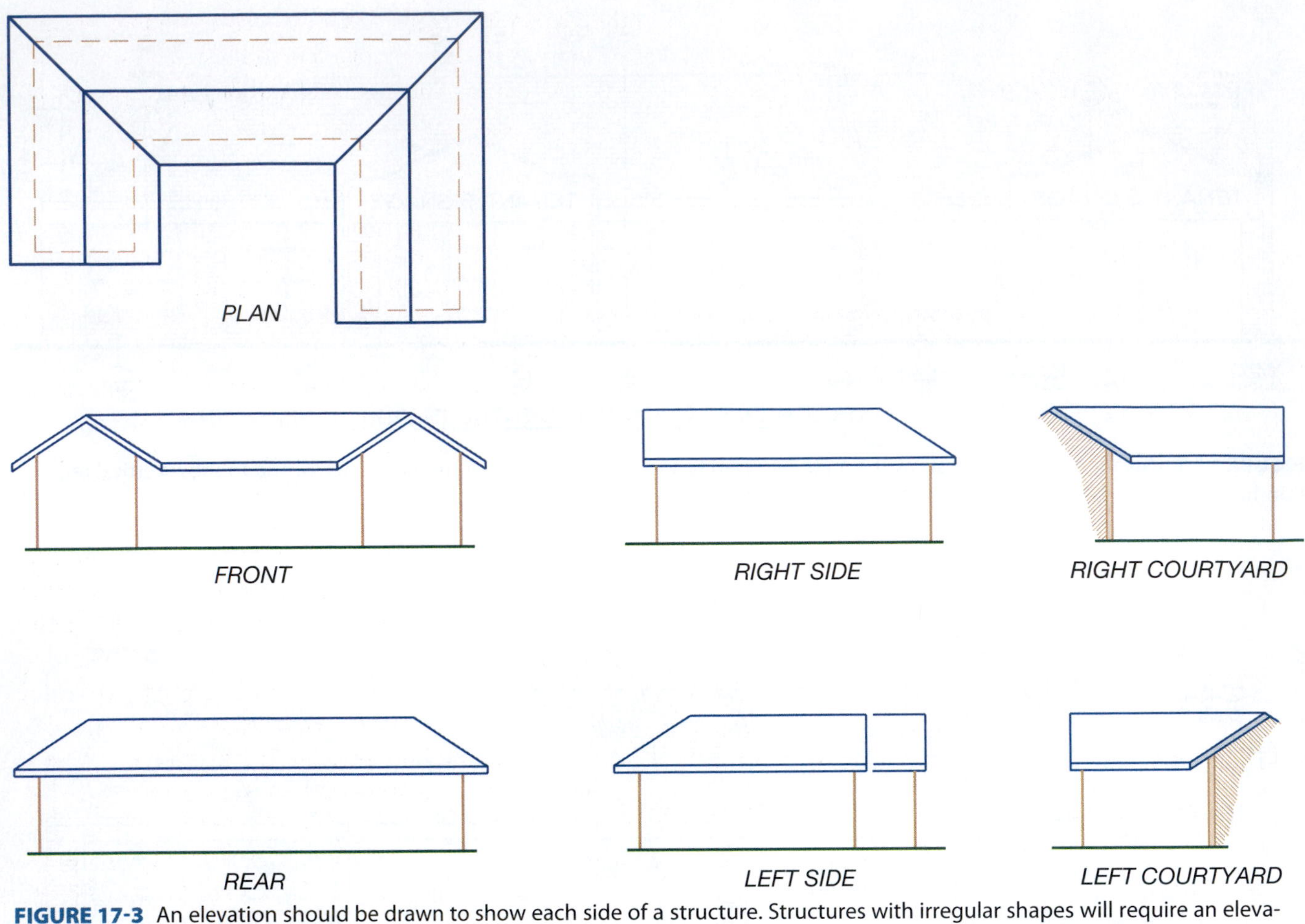

FIGURE 17-3 An elevation should be drawn to show each side of a structure. Structures with irregular shapes will require an elevation of each hidden surface.

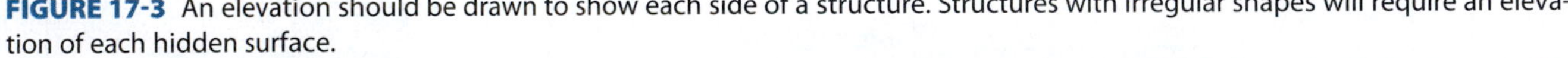

the ceiling, roof, and the top of any rails, balconies, or walls. The elevation should be projected by drawing surfaces closest to the picture plane first and then drawing each receding surface. Drawing all surfaces prior to attempting to show any openings will aid in the projection. Figure 17-8 shows the projection for major shapes of an elevation.

Projecting Roof Shapes

The shape of the roof must be determined from preliminary elevations and sections. For structures that have shapes other than a flat roof, the roof plan must be drawn before the elevations can be completed. The overhangs of the roof should be included in the wblock used to project the elevations. The roof shape is established by projecting the limits of overhangs from the floor plan down to the elevations. The pitch of the roof and the height of bearing walls can be determined from preliminary sections. Figure 17-9 shows the projection of the roof to the base drawing.

Drawing Grades

As the elevations are started, major shapes of a structure can be established from the finished floor line. To complete the elevations, the structure must be referenced to the site by showing the finished grade elevations. Most architects use a bold line to represent the finished grade. Many firms place a small break in the ground line at each surface change and use a progressively smaller line to represent increasing distance of the surface to the picture plane. For relatively flat sites, the grade can be determined based on the type of foundation to be used. For structures with concrete slabs, the finished floor must be a minimum of 8" (200 mm) above the finished grade. For sloping sites, the ground needs to be established by using the grading plan. The grading plan shows the finished floor elevation and spot grades for each corner of the structure. By starting at the floor line, the drafter can measure down the required distance to establish the ground line specified on the grading plan. Once grades are determined,

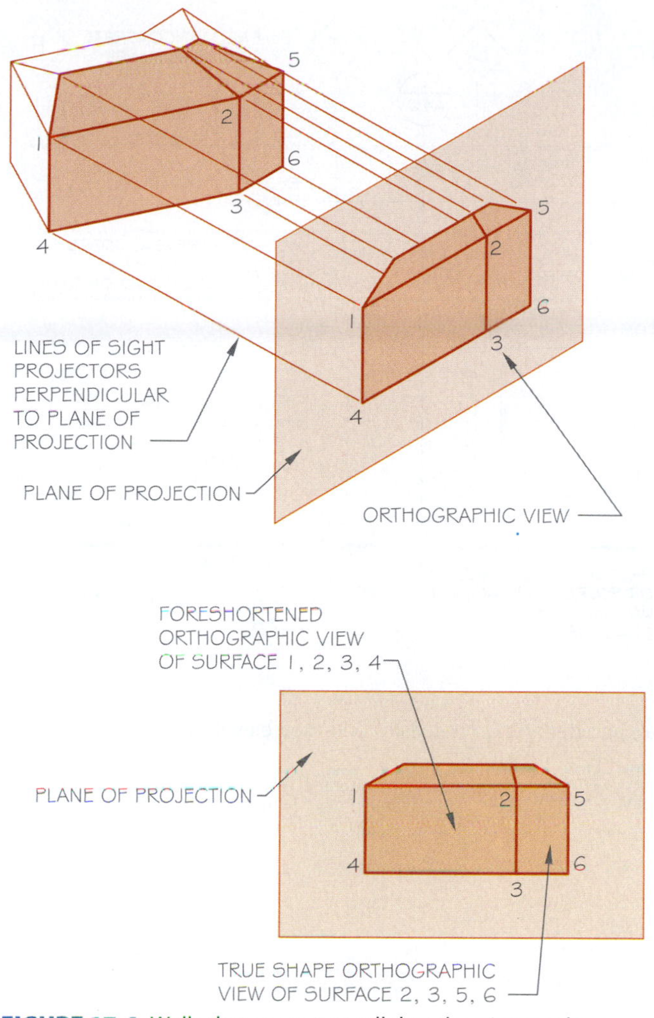

FIGURE 17-4 Walls that are not parallel to the picture plane will be distorted.

the outline of exterior footings can be drawn using a dashed-line pattern. Figure 17-10 shows a grading plan for a structure and the resulting ground line and footings on the elevation.

Drawing Windows, Doors, and Skylights

Once each major shape has been projected, doors and windows can be drawn matching the specifications from the floor plan or window schedule. Typically, each window is inserted as a block from a reference library that has already been established. If you must create your own blocks, create a block of each type of window to be used and store the block on a disk in a directory labeled \PROTO\ELEV\WIN####. Text describing the size or type of window, such as sldg, fix, or awn can be inserted in place of the ####. The size can be altered by stretching the block once it has been inserted into the drawing and exploded. This results in slight distortions.

Figure 17-11 shows examples of several styles of window blocks for use on commercial drawings. The exact location of the block can be determined by projecting the opening from the floor plan.

Door locations can be projected from the floor plan in the same manner as a window. Door styles should be confirmed with the door schedule. As door blocks are inserted, be sure that the sill of the door aligns with the finish floor. Figure 17-12 shows a sample of the wide variety of doors available from third-party vendors.

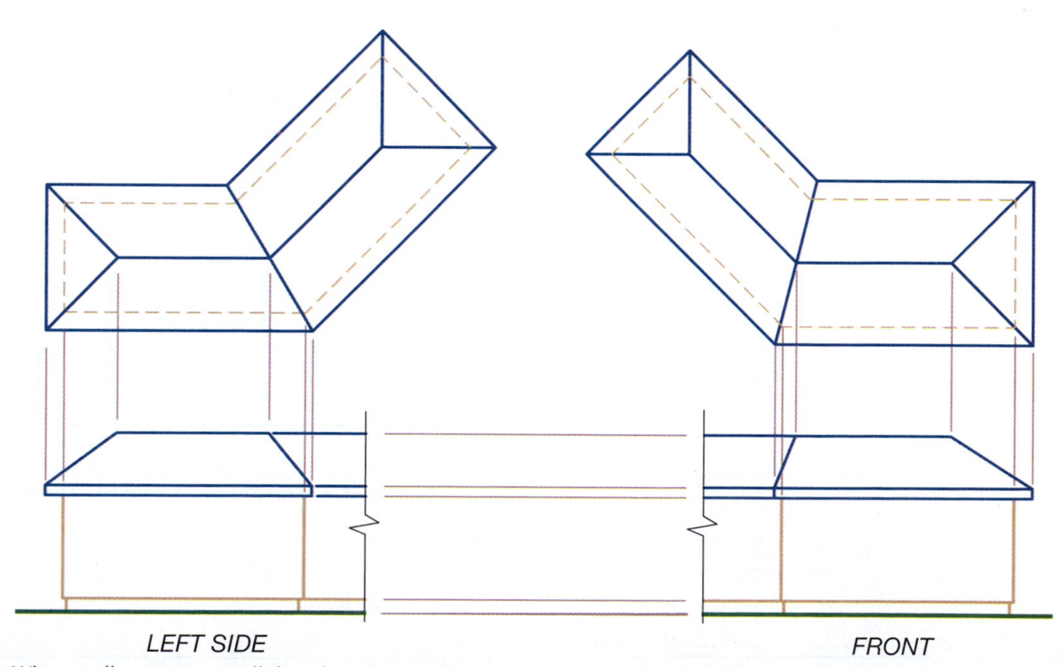
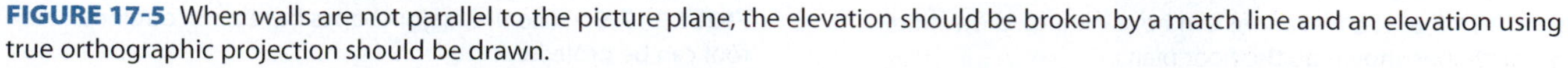

FIGURE 17-5 When walls are not parallel to the picture plane, the elevation should be broken by a match line and an elevation using true orthographic projection should be drawn.

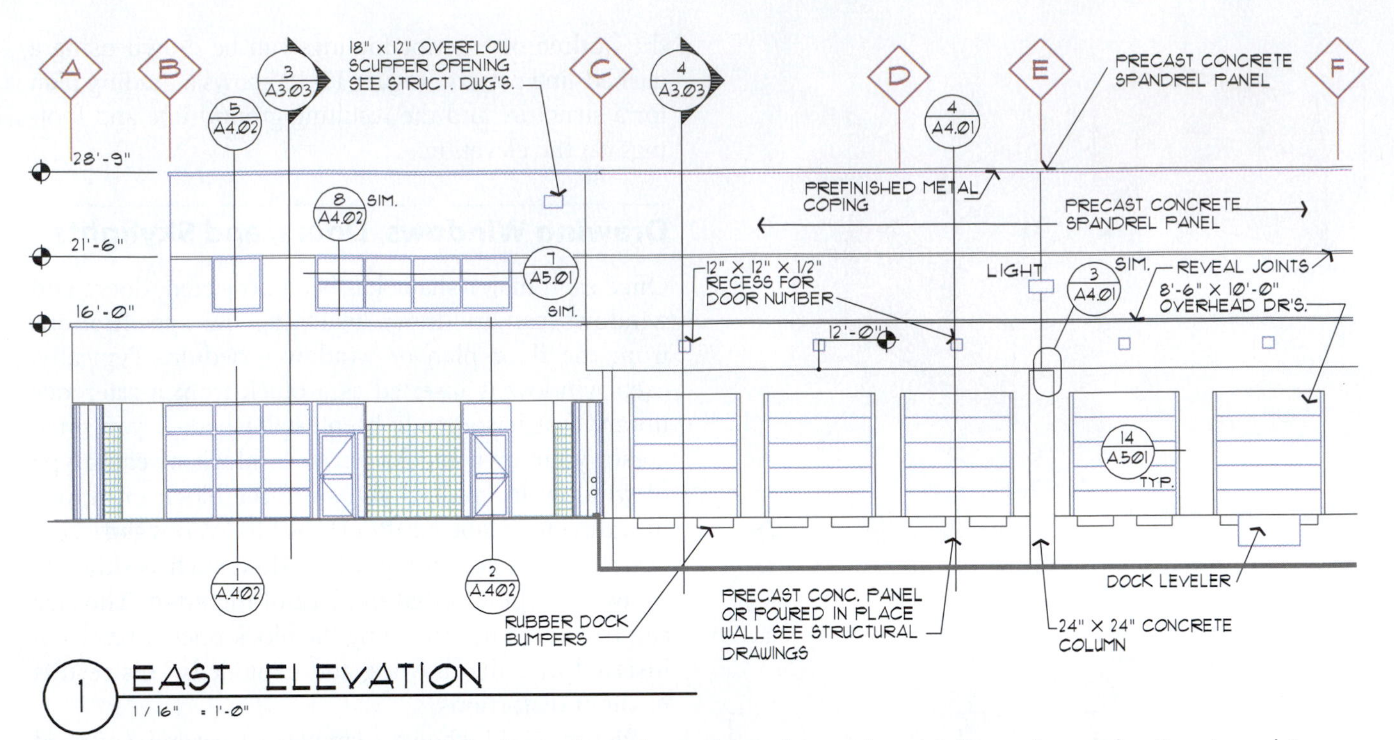

FIGURE 17-6 Elevations drawn at small scale do not show all materials and depend on details or enlarged elevations to explain material placement. *Courtesy Architects Barrentine, Bates & Lee, A.I.A.*

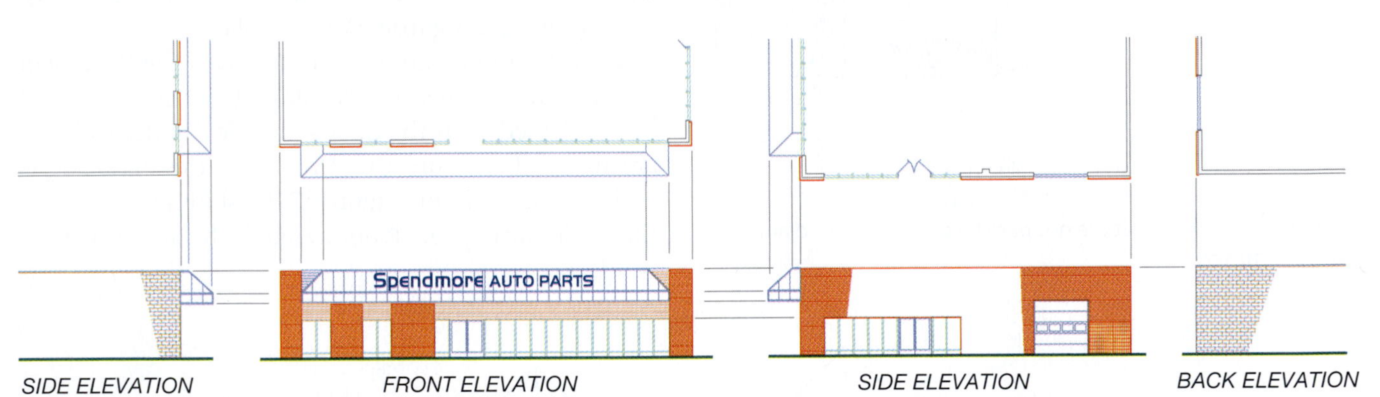

FIGURE 17-7 With a wblock created from the floor plan, the elevations can be projected easily. Place elevations side by side to project horizontally. Once the elevations are completed, they can be moved into other positions for plotting.

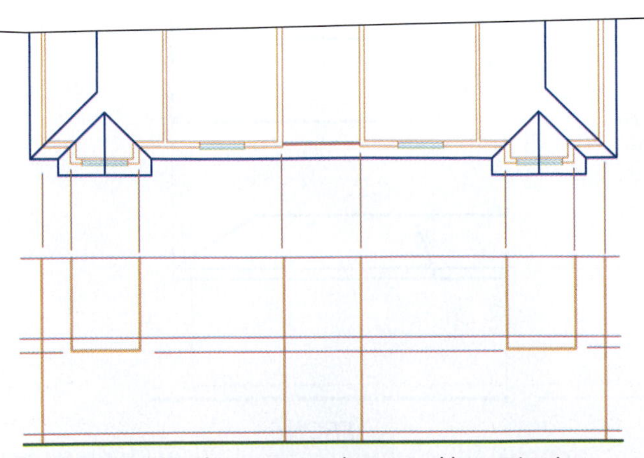

FIGURE 17-8 The elevations can be started by projecting major shapes shown on the floor plans.

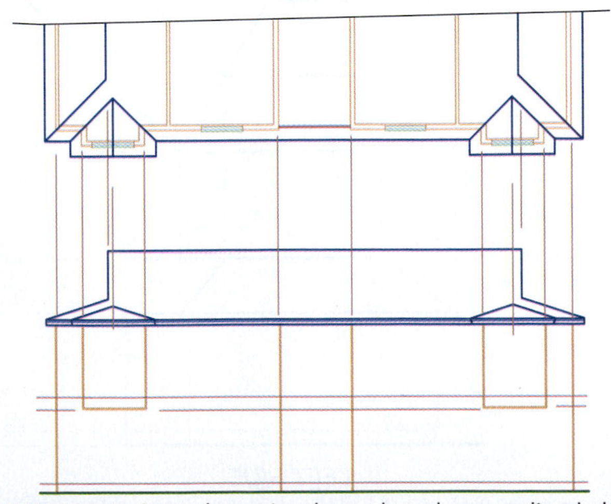

FIGURE 17-9 Once the major shapes have been outlined, the roof can be projected.

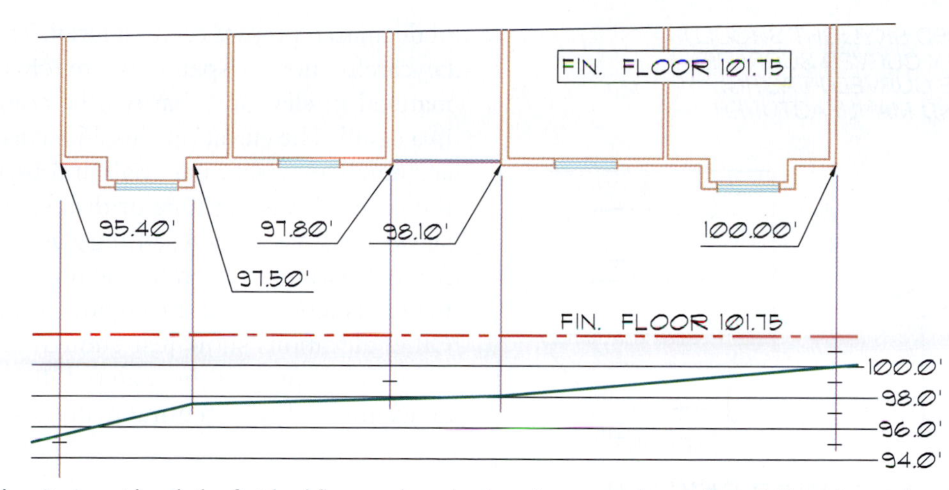

FIGURE 17-10 If the site is not level, the finished floor and grade elevations must be projected. Elevations taken from the site or grading plan can be used to establish the finish grade by using the finished floor line as a base and offsetting the required distance for each point.

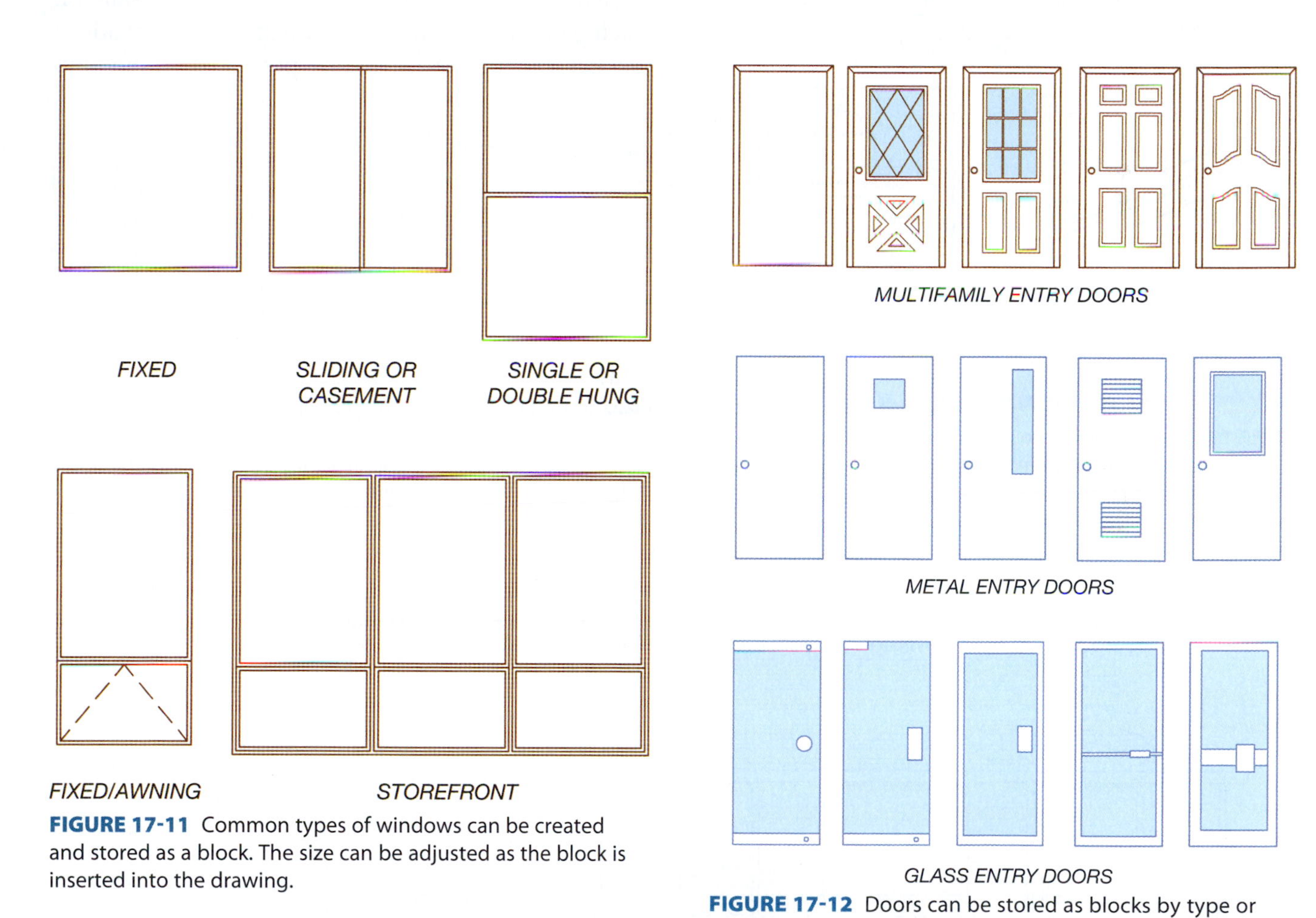

FIGURE 17-11 Common types of windows can be created and stored as a block. The size can be adjusted as the block is inserted into the drawing.

FIGURE 17-12 Doors can be stored as blocks by type or material.

When an opening must be projected in an inclined wall or roof surface, the opening will not be seen in its true shape because it is not parallel to the picture plane. Inclined materials must be projected from an elevation or section where the inclined surface shows as a line rather than a plane. Figure 17-13 shows an example of projecting a skylight on an inclined roof.

Material Representation

The method of representing materials varies depending on the scale of the drawing, the type of material, and which elevation is being represented. As the scale of the drawing increases, so should the amount

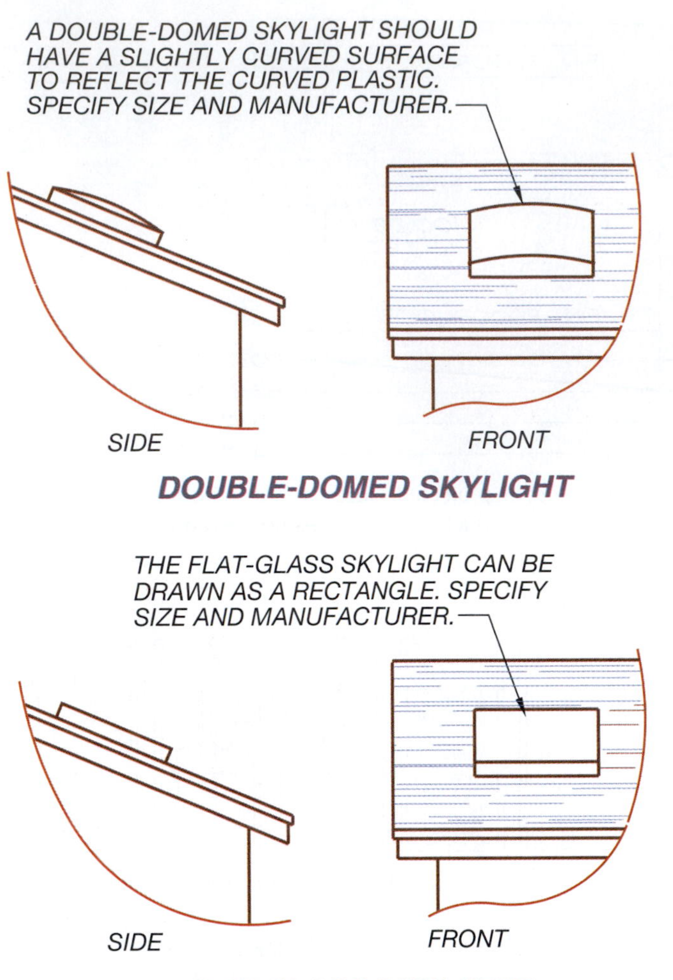
A DOUBLE-DOMED SKYLIGHT SHOULD HAVE A SLIGHTLY CURVED SURFACE TO REFLECT THE CURVED PLASTIC. SPECIFY SIZE AND MANUFACTURER.
SIDE
FRONT
DOUBLE-DOMED SKYLIGHT
THE FLAT-GLASS SKYLIGHT CAN BE DRAWN AS A RECTANGLE. SPECIFY SIZE AND MANUFACTURER.
SIDE
FRONT
FLAT-GLASS SKYLIGHT

FIGURE 17-13 Materials to be represented on an inclined surface must be projected from the surface where the material shows its true size.

of detail to represent each material. The drafter should be careful not to spend too much time showing a material in elevation that will be completely detailed in a detail. The elevation should have enough detail to accurately represent materials and surfaces, but not to the point of wasting time or disk space. Figure 17-14 shows methods of representing glass in elevations. The front elevation is generally the only elevation that has materials represented throughout the entire elevation. Other elevations show just enough of the material to accurately represent the material. Figure 17-15 shows an example of representing materials in elevations.

Using AutoCAD Hatch Patterns

Materials are often represented by using a hatch pattern created by AutoCAD or a third-party vendor. Common hatch patterns used to represent materials in elevations can be seen in the elevations presented throughout

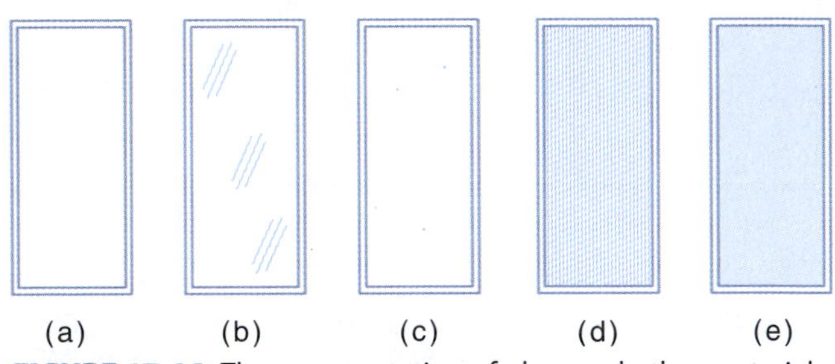
(a)
(b)
(c)
(d)
(e)

FIGURE 17-14 The representation of glass and other materials should be influenced by the drawing scale. Methods used to represent glass include: (a) clear, (b) simple representation, (c) dot, (d) solid line, and (e) solid fill.

FIGURE 17-15 Representing materials in elevation. Roofing and masonry can be represented using AutoCAD hatch patterns. The siding can be represented using the ARRAY or COPY commands.

this chapter. Drafters typically encounter problems in scaling the hatch pattern and defining the extents of the pattern placement when working with the HATCH command. If AutoCAD 2008 or newer is used, the Annotative feature of the HATCH command can be used to control the pattern scale. For older versions of AutoCAD, inserting a hatch pattern is often a matter of trial and error in picking the scale pattern. For a pattern to be used accurately, it should fit the scale of the drawing. Figure 17-16 shows an elevation with a variety of hatch patterns that have been applied with good use of scaling.

FIGURE 17-16 Materials can be created in AutoCAD by using hatch patterns, inserting blocks, or by using the ARRAY and OFFSET commands. *Courtesy H.D.N. Architects, A.I.A.*

Dimensioning

Chapter 3 introduced basic concepts to consider as structures are dimensioned. The use of dimensions varies for each office and depends on the complexity of the structure. Typically, the height from floor to ceiling is indicated on the elevations. Other dimensions that may be specified on the elevations include the following:

- Floor to floor for multilevel structures.
- Floor to rails at balconies or aboveground decks.
- Window height and size.
- Height and width exterior finishes.
- Roof pitches.

Figure 17-17 shows an example of a fully dimensioned elevation.

Elevation Symbols

Elevations often use symbols to keep the drawing uncluttered. Common symbols include the following:

- Finished floor lines.
- Match lines.
- Grid line markers.
- Elevation symbols.
- Section markers.
- Detail markers.

Floor and ceiling lines can be represented by a long-short-long pattern. If a horizontal siding is used, the floor line should be represented with a thick line so that it can be easily distinguished from the siding.

If the elevation must be split because of length, a match line should be placed in the same location that was used on the floor plan. Grid lines should be placed on the elevation to match those used on the floor plan. Because the elevation is seen from the exterior of the structure, and the floor plan is seen when looking down on the structure (as shown in Figure 17-18), the grid lines will be opposite on the elevations to what they are on the floor plan.

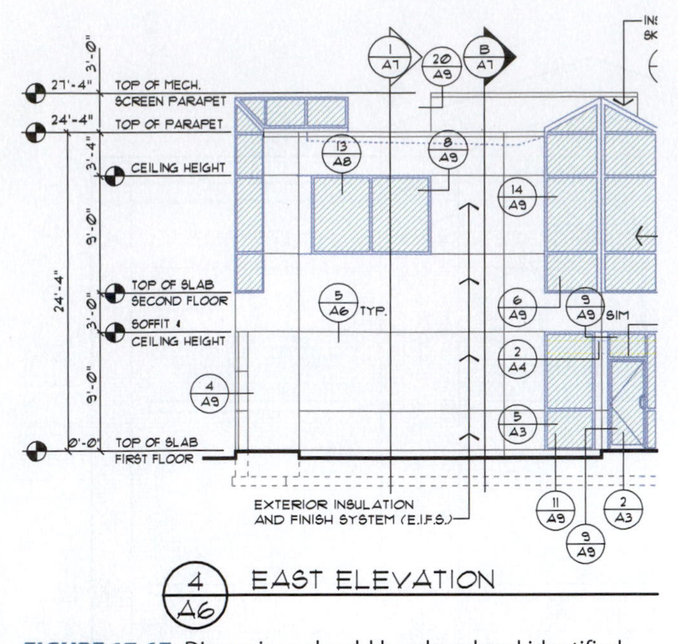

FIGURE 17-17 Dimensions should be placed and identified to clearly specify all vertical relationships. *Courtesy Architects Barrentine, Bates & Lee, A.I.A.*

Heights that are specified or dimensioned on the elevation are usually highlighted by an elevation symbol, as shown in Figure 17-19. Section markers like those used on the floor plan are placed on the elevations. The same symbol and location should be identical to that of the floor plan. Detail markers to represent exterior treatment, window or door construction, and special construction are also referenced on the elevation. Detail symbols are like those used on the floor plan. Figure 17-19 shows an example of the use of each of these symbols.

Elevation Notations

Each material represented on the elevation should be specified by a note or detail using the procedures described in Chapter 3. Notes can be either a local note referenced to the material by a leader line, or a general note. When general notes are used, a short reference should be specified on the elevation. Many offices use notes with a reference symbol, which can be placed in the elevation to maintain clarity in the drawing. Notes should be carefully worded to avoid discrepancies with the specifications.

Layering

Using the guidelines presented in Chapter 3, the materials represented on the elevations should be separated by layers. Most offices use titles that start with A-ELEV and are followed by an appropriate modifier to describe layer contents. Modifiers that describe the contents or use vary for each office. Suggested modifiers based on National CAD Standards (NCS) are listed on the student CD in Appendix C. Other common methods of naming layers include names based on the type of line to be used or the weight of the line when plotted. Layer names such as FINE, LIGHT, MED, HEAVY, DASH, FLOOR, or CONTINUOUS, describe

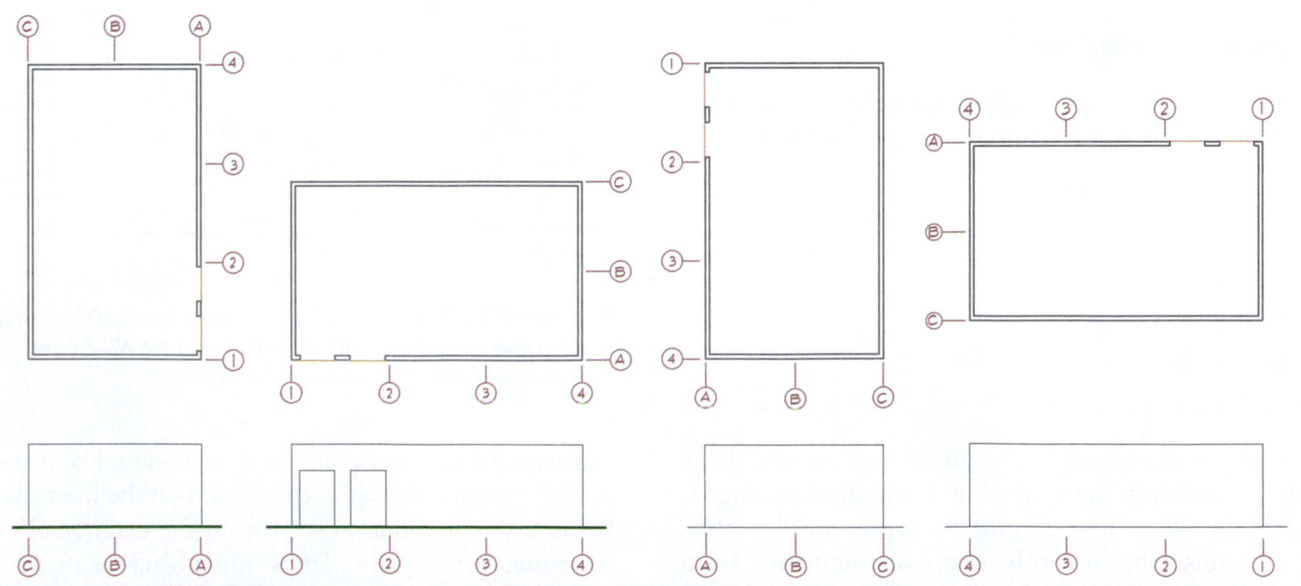

FIGURE 17-18 Although most drafters understand the basics of orthographic projection, the elevations might be drawn backward if the floor plan is not properly rotated.

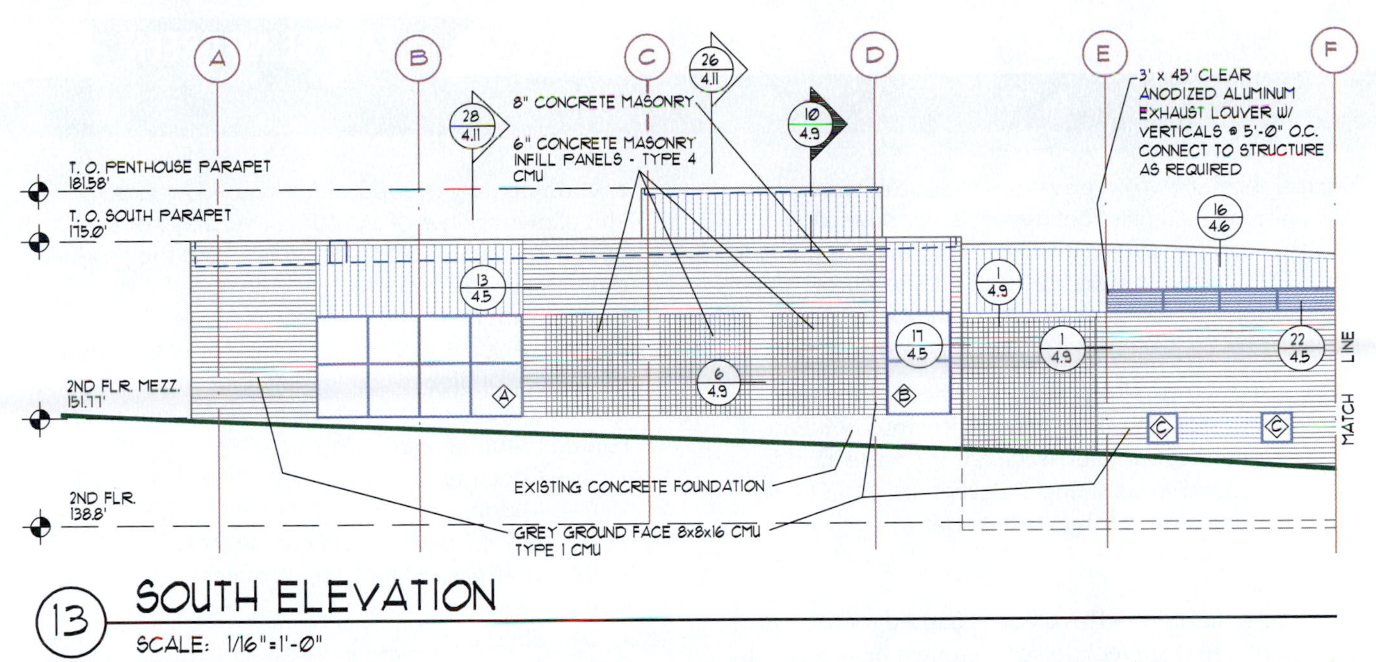

FIGURE 17-19 Annotation is provided by the use of local notes. Dimensions can be expressed using traditional dimension methods or by the use of elevation symbols. *Courtesy Architects Barrentine, Bates & Lee, A.I.A.*

the contents in general terms but may not be specific enough to adequately separate material for plotting.

COMPLETING ELEVATIONS

The elements that make up elevations have been presented throughout this chapter. These components will be combined to complete the front elevation that matches the structure presented throughout Chapter 16. The software that is used to create the floor plan will greatly affect how the drawing is created. The following discussion assumes the use of AutoCAD. If you're creating elevations using AutoCAD Revit or AutoCAD Architecture, the general steps will be the same, but the procedure will be much easier because of the automation provided by the software.

1. Freeze all material on the floor plan except for layers that contain the walls, exterior openings, roof outlines, and grades. Make a wblock of this drawing and save it as ELEV BASE.
2. Close the current FLOOR file and open the ELEV BASE file.
3. Establish layers for dimensions, text grids, outlines, roofs, doors, and windows with a prefix of A-ELEV.
4. Make three additional copies of the FLOOR wblock and rotate them for projection for each elevation similar to the layout presented in Figure 17-7.
5. Project lines from the floor plans to represent the limits of each exterior wall.
6. Draw a line to represent the finish floor line for each level.
7. Establish lines to represent the tops of all walls, rails, and the bottoms of all footings and below-grade construction.
8. Use preliminary drawings to establish the roof slope and draw the required inclined roof pitches.
9. Draw all gutters, downspouts, and scuppers.
10. Draw all skylights, roof-mounted equipment, equipment screens, ladders, and penthouses that project above the parapet walls.
11. Represent all building identification graphics and surface-mounted signage.

The drawing should now resemble Figure 17-20.

12. Establish lines to represent the top and bottom of all openings.
13. Project the locations of all exterior openings from the floor plans.
14. Use appropriate layers to insert blocks for all doors and windows.
15. Locate all dock bumpers.
16. Use appropriate layers to provide hatch patterns to represent the type and extent of all exterior materials.

EASY BEING GREEN

Key elements shown on elevations include exterior wall finishing materials, roof materials, windows, doors, skylights, and finishing materials. These products primarily earn LEED credits from the Materials and Resources division. Information to be shown on the exterior elevations includes credits from the following categories:

- Resource Reuse—This credit is earned for the use of materials that are salvaged from previous building sites for reuse. Common materials for reuse include concrete and wood siding. Relevant areas of CSI for research related to this area include:
  - 03 – Concrete
  - 09 – Finishes
- Recycled Content—This credit is earned for the use of materials that are recycled or contain a high amount of recycled materials. Relevant areas of CSI for research related to this area include:
  - 04 – Masonry
  - 06 – Woods, plastics, and composites
  - 07 – Thermal and moisture protection
  - 08 – Openings
  - 09 – Finishes
- Local/Regional Materials—This credit is earned for the use of materials that are produced near the construction site to minimize damage to the environment by transportation-related factors. This allows the use of a product that may not be usually be considered to be "green," but the product does receive credit for reducing damage from transportation.
- Rapidly Renewable Materials—This credit is earned for the use of biobased materials, biocomposites, and biofibers. Areas of CSI that will be useful for research related to this area include:
  - 03 – Concrete
  - 04 – Masonry
  - 06 – Woods, plastics, and composites
  - 07 – Thermal and moisture protection
  - 08 – Openings
  - 09 – Finishes
  - 10 – Specialties
- Certified Wood—This credit is earned for the use of materials such as lumber products originating in certified forests. Areas of CSI that will be useful for research related to this area include:
  - 01 – General requirements
  - 03 – Concrete
  - 06 – Woods, plastics, and composites
  - 07 – Thermal and moisture protection
  - 08 – Openings
  - 09 – Finishes

17. Draw any site-related elements such as retaining walls or loading docks.
18. Draw all light fixtures.
19. Represent all control and expansion joints for concrete masonry structures.

The elevation should now be completely drawn and should resemble Figure 17-21. Although only one elevation is shown, all elevations should be drawn as each stage is completed. The elevations can be completed by using the following steps:

20. Provide all grid and match lines and their notations.
21. Label all floor, rail, ceiling, and rooflines.
22. Dimension for all floor-to-floor, floor-to-rails, and window heights so that your drawing resembles Figure 17-22.
23. Provide notations to specify all exterior materials that were drawn in steps 6 through 19.
24. Provide all required general notes.
25. Provide building section markers and reference markers and notations for all required details. Detail markers are added as the details are drawn.
26. Provide a title and scale.

The completed front elevations should now resemble Figure 17-23. The completed elevations for the entire structure are shown in Figure 17-24.

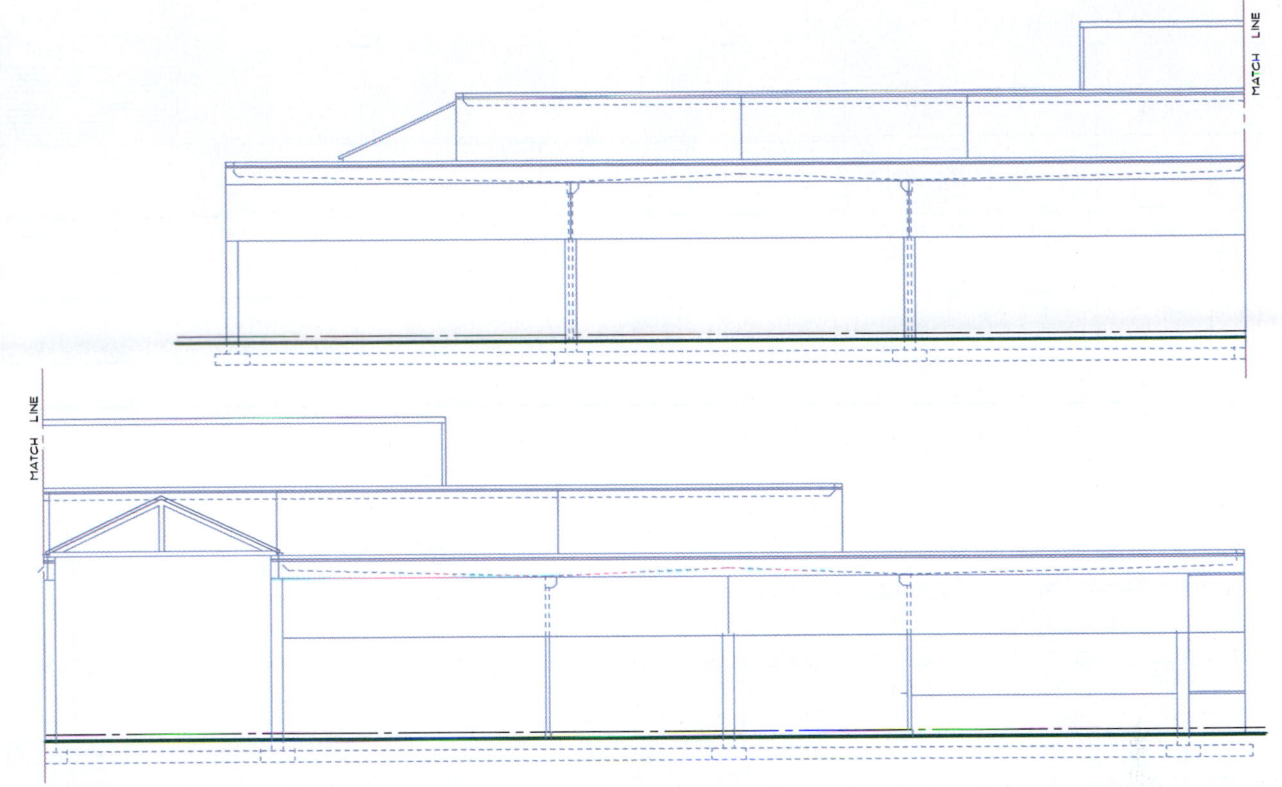

FIGURE 17-20 The structure presented in Chapter 16 is used to project each of the major shapes. Although only one elevation is being presented, a drafter would be required to represent each surface. Figure 17-24 shows the entire structure. *Courtesy Architects Barrentine, Bates & Lee, A.I.A.*

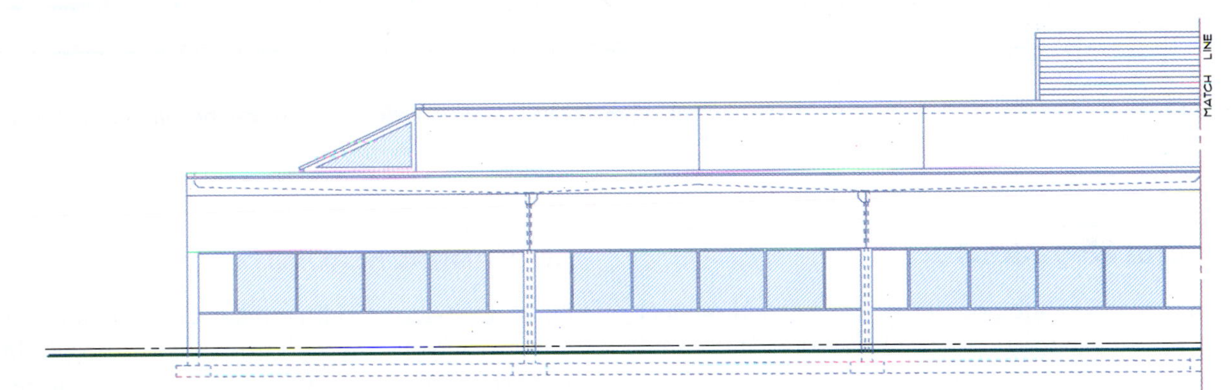

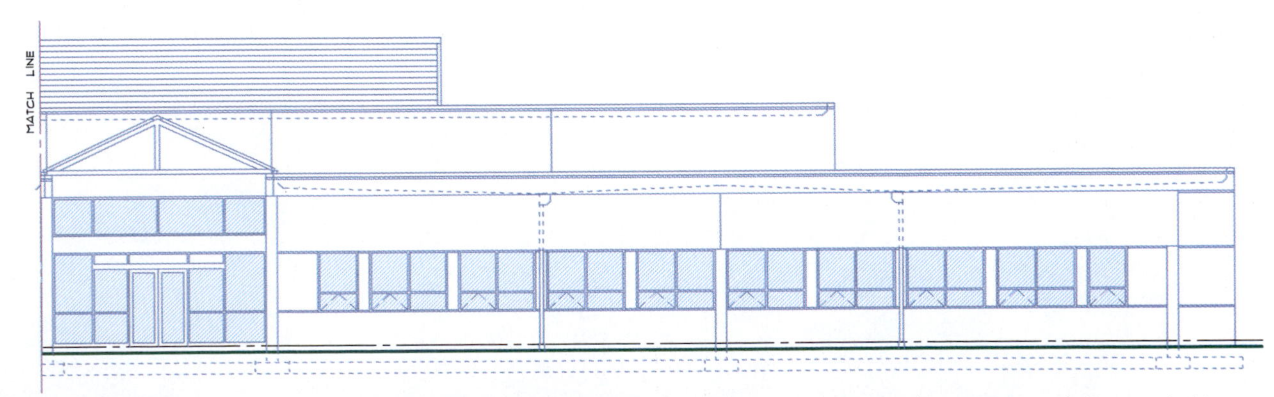

FIGURE 17-21 Once major shapes have been created, windows, doors, and trim patterns can be represented. *Courtesy Architects Barrentine, Bates & Lee, A.I.A.*

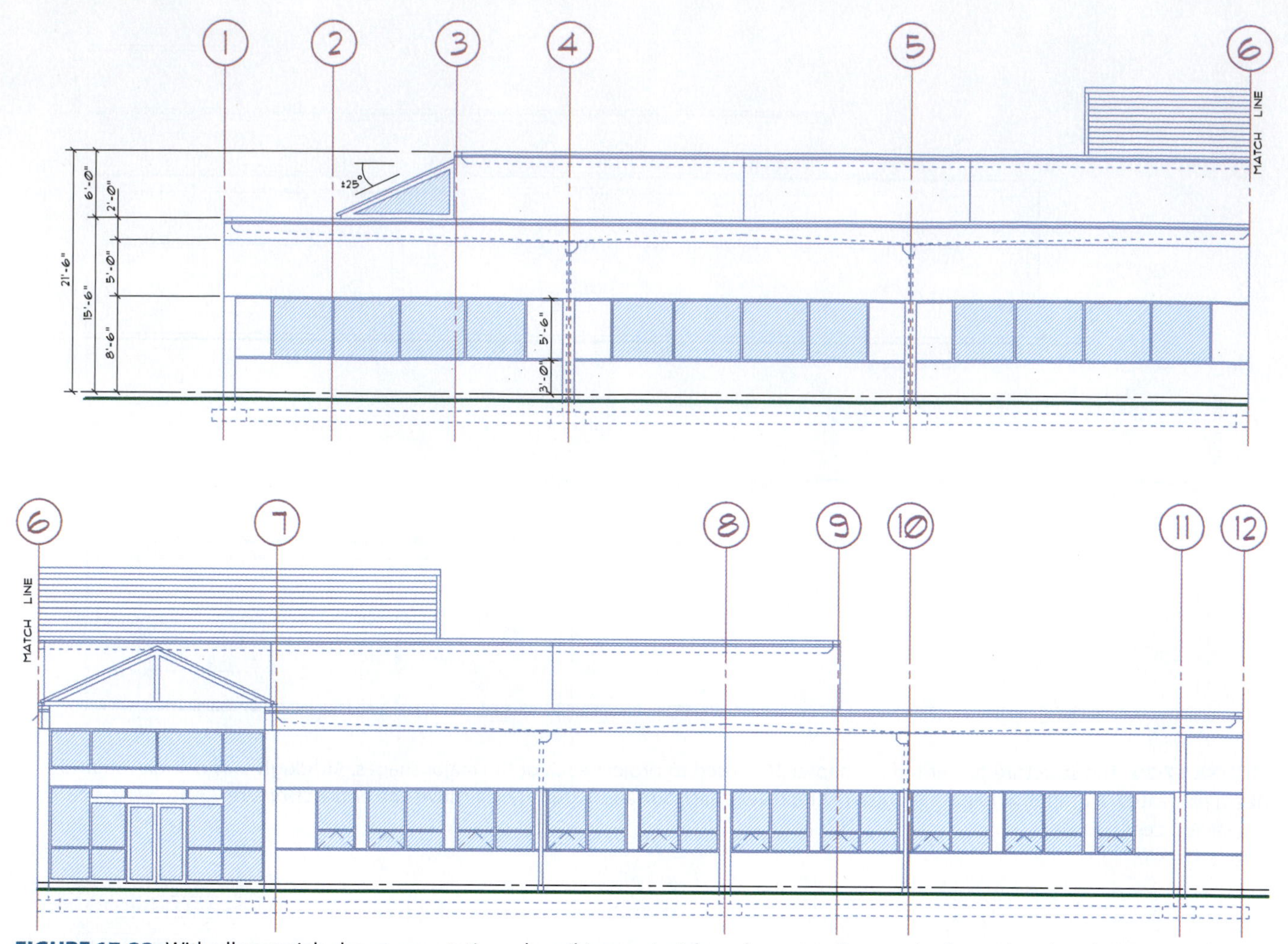

FIGURE 17-22 With all materials drawn, annotations describing materials and construction methods can be placed. *Courtesy Architects Barrentine, Bates & Lee, A.I.A.*

Structural Elevations

Structures made of site-poured concrete, wood or steel frame, or tilt-up concrete panels often require structural elevations as well as the exterior elevations. Figure 17-25 shows an example of the structural elevation that corresponds to Figure 16-5. Various types of structural elevations and guidelines for completing them are covered in Chapter 23.

ADDITIONAL READING

Links to exterior siding related information are too numerous to list here. Rather than listing individual Web sites, use your favorite search engine to research products from the CSI categories listed in wall materials, window materials, glazing, doors, siding, skylights, plumbing fixtures, cabinets, and furnishing.

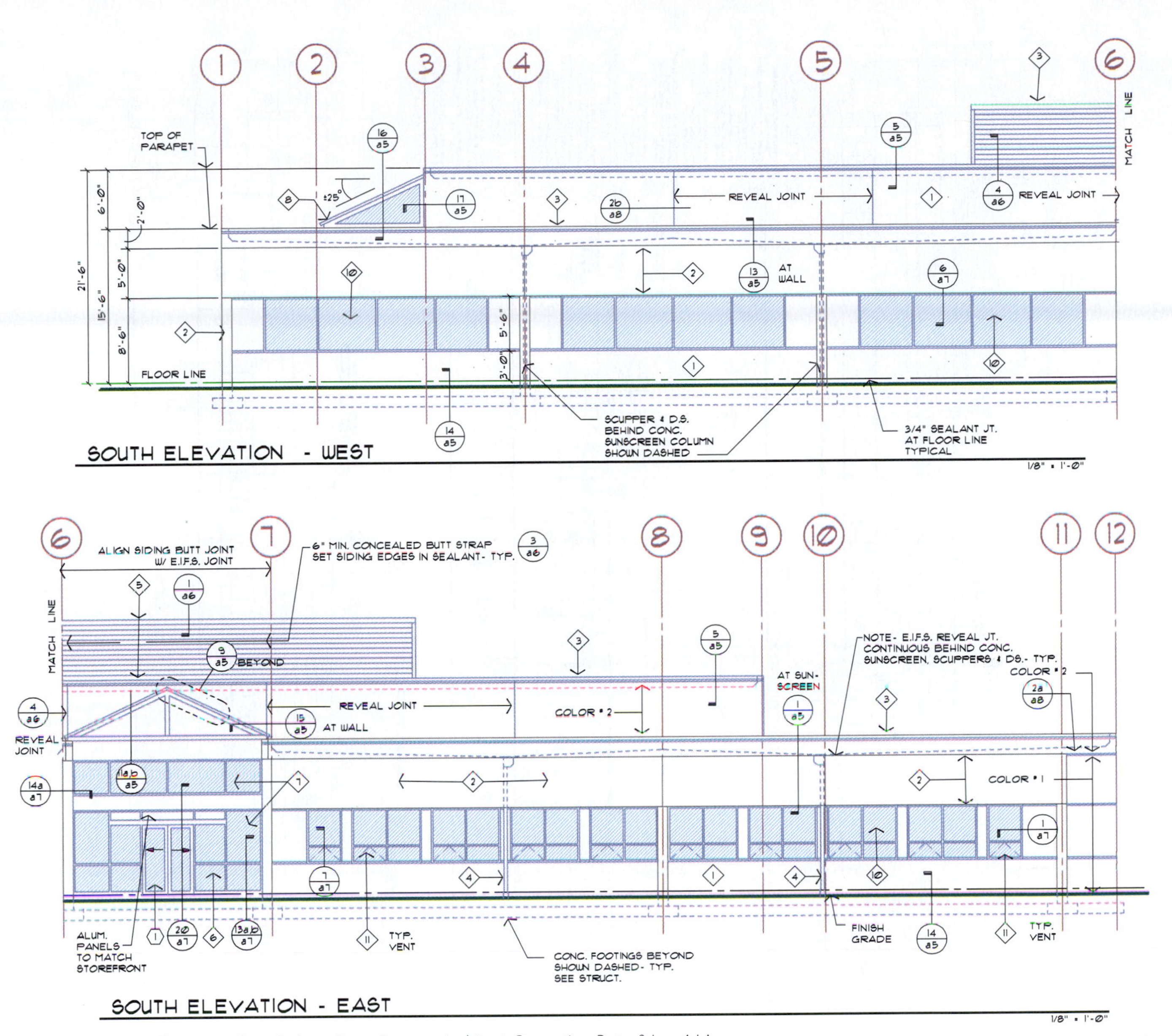

FIGURE 17-23 The completed elevation. *Courtesy Architects Barrentine, Bates & Lee, A.I.A.*

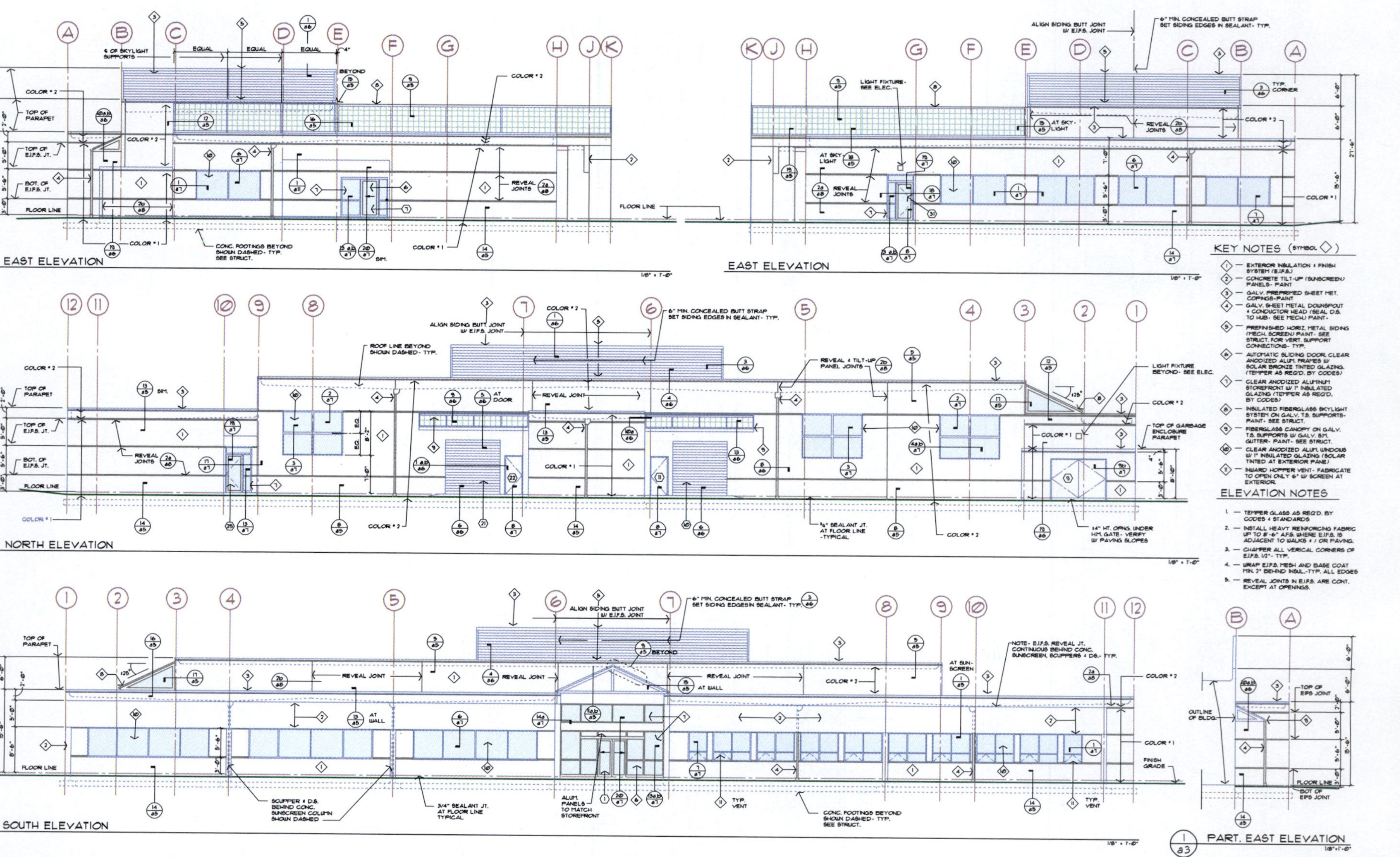
EAST ELEVATION
EAST ELEVATION
NORTH ELEVATION
SOUTH ELEVATION
PART. EAST ELEVATION
KEY NOTES (SYMBOL ◇)
1 — EXTERIOR INSULATION & FINISH SYSTEM (E.I.F.S.)
2 — CONCRETE TILT-UP (SUNSCREEN) PANELS- PAINT
3 — GALV. PREPRIMED SHEET MET. COPINGS-PAINT
4 — GALV. SHEET METAL DOWNSPOUT & CONDUCTOR HEAD (SEAL D.S. TO HUB- SEE MECH.) PAINT-
5 — PREFINISHED HORIZ. METAL SIDING (MECH. SCREEN) PAINT- SEE STRUCT. FOR VERT. SUPPORT CONNECTIONS- TYP.
6 — AUTOMATIC SLIDING DOOR. CLEAR ANODIZED ALUM. FRAMES W/ SOLAR BRONZE TINTED GLAZING. (TEMPER AS REQ'D. BY CODES)
7 — CLEAR ANODIZED ALUMINUM STOREFRONT W/ 1" INSULATED GLAZING (TEMPER AS REQ'D. BY CODES)
8 — INSULATED FIBERGLASS SKYLIGHT SYSTEM ON GALV. T.S. SUPPORTS- PAINT- SEE STRUCT.
9 — FIBERGLASS CANOPY ON GALV. T.S. SUPPORTS W/ GALV. S.M. GUTTER- PAINT- SEE STRUCT.
10 — CLEAR ANODIZED ALUM. WINDOWS W/ 1" INSULATED GLAZING (SOLAR TINTED AT EXTERIOR PANE)
11 — INWARD HOPPER VENT- FABRICATE TO OPEN ONLY 6" W/ SCREEN AT EXTERIOR
ELEVATION NOTES
1. — TEMPER GLASS AS REQ'D. BY CODES & STANDARDS
2. — INSTALL HEAVY REINFORCING FABRIC UP TO 8'-6" A.F.S. WHERE E.I.F.S. IS ADJACENT TO WALKS & / OR PAVING.
3. — CHAMFER ALL VERICAL CORNERS OF E.I.F.S. 1/2"- TYP.
4. — WRAP E.I.F.S. MESH AND BASE COAT MIN. 2" BEHIND INSUL.-TYP. ALL EDGES
5. — REVEAL JOINTS IN E.I.F.S. ARE CONT. EXCEPT AT OPENINGS.

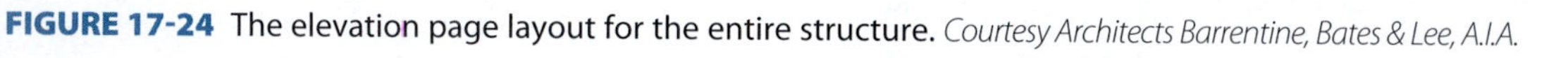

FIGURE 17-24 The elevation page layout for the entire structure. *Courtesy Architects Barrentine, Bates & Lee, A.I.A.*

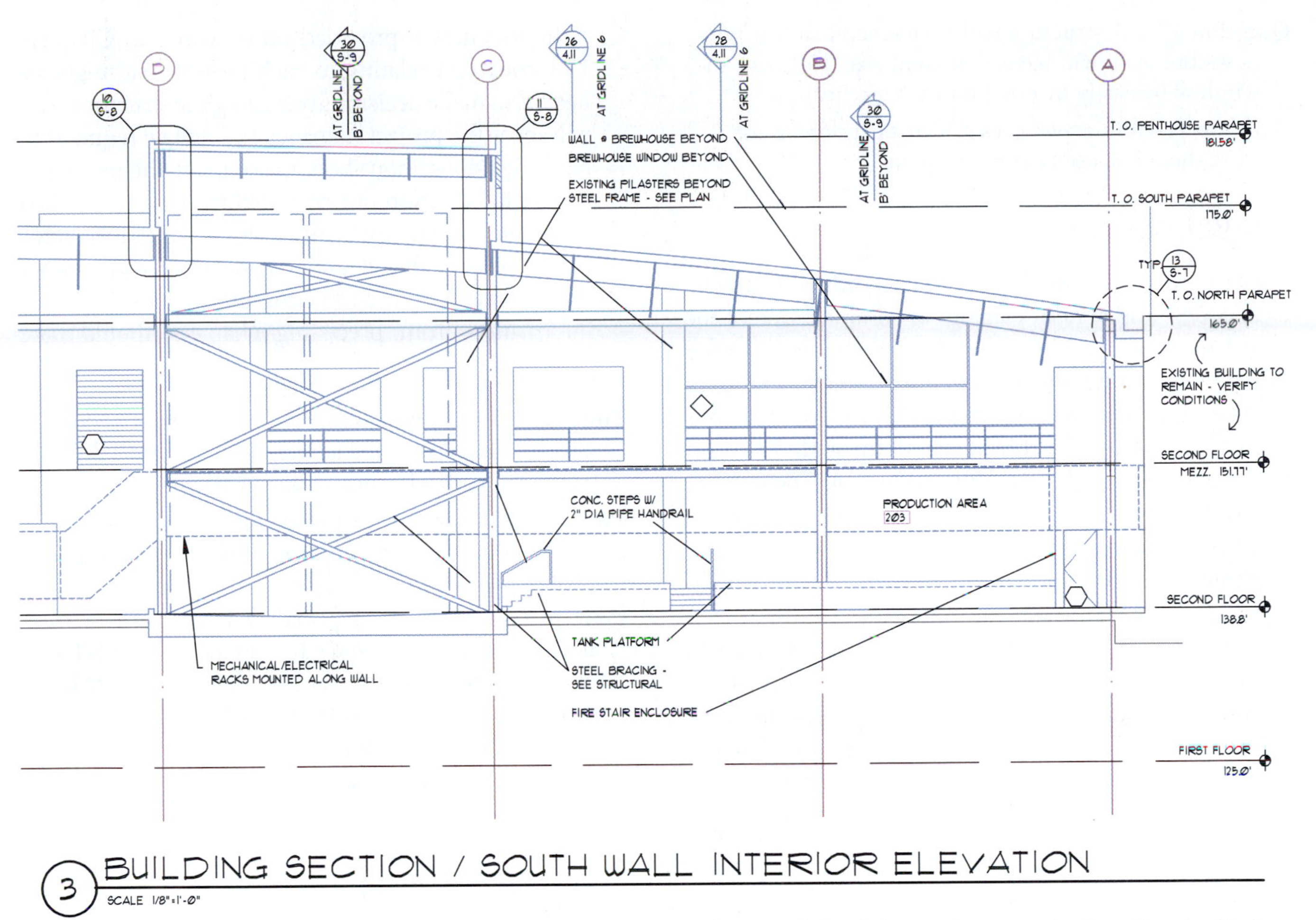

FIGURE 17-25 Structural elevations are drawn to show how major components of the skeleton relate to each other. Chapter 23 discusses the relationship of structural elevations to the other structural drawings. *Courtesy DiLoreto Architects, LLC.*

CHAPTER 17 TEST Orthographic Projection and Elevations

QUESTIONS

Answer the following questions with short complete statements. Type the chapter title, question number, and a short complete statement for each question using a word processor. Some answers may require the use of vendor catalogs or seeking out local suppliers.

Question 17-1 An elevation is too long to fit on the paper that is being used. How can this elevation be drawn?

Question 17-2 What process is used to draw elevations?

Question 17-3 List the drawings that may be needed to draw exterior elevations.

Question 17-4 Describe how irregularly shaped structures should be handled for exterior elevations.

Question 17-5 Describe the differences in material representation for the front and other elevations.

Question 17-6 What materials are typically dimensioned on elevations?

Question 17-7 Describe symbols that are typically placed on elevations.

Question 17-8 List the complete layer names for the following materials: concrete footings, grid numbers and letters, window glazing, detail and section markers, and roof materials.

Question 17-9 A structure will have several sliding windows but each will have a different size. How can the window be easily inserted for each application?

Question 17-10 In drawing a structure on a sloping site, what should be used as the base point?

DRAWING PROBLEMS

Use the reference material from preceding chapters, local codes, and vendor catalogs to complete one of the following projects. Unless your instructor gives other instructions, draw the elevations that correspond to the floor plan that was drawn in Chapter 16. Skeletons of the elevations for problems 17-3 through 17-6 can be accessed from the student CD in the DRAWING PROBLEMS/CHAPTER 17 folder. Use these drawings as a base to complete the assignment. Use appropriate symbols, linetypes, dimensioning methods, and notations to complete the drawing. Determine any unspecified sizes based on the material presented in Sections 3 and 4. Unless specified, select a scale appropriate to plotting on D-size material that matches the scale of the floor plan drawn in Chapter 16. Determine LTSCALE, DIMVARS, and all other drawing factors as needed.

> *Note: The sketches in this chapter are to be used as a guide only*. As you progress through the elevations you will find that some portions of the drawings do not match things that have been drawn on the floor plan. If you think you have found an error, do not make changes to the drawings until you have discussed the problem and possible solutions with your engineer (your instructor).

Information is provided on drawings in Chapters 15 through 24 relating to each project that might be needed to make decisions regarding the structure. You will act as the project manager and will be required to make decisions about how to complete the project.

Any information not provided must be researched and determined by you unless your instructor (the project architect and engineer) provides other instructions. When conflicting information is found, information from preceding chapters should take precedence.

Problem 17-1 Draw blocks to represent a plain glass door, a pair of glass doors, and a solid metal personnel door. Save each block with an appropriate name.

Problem 17-2 Draw a block to represent a sliding, single-hung, and fixed window. Save each block with an appropriate name.

Problem 17-3 Using the site plan drawn in Chapter 15, the floor plan drawn in Chapter 16, and the drawing below of an exterior and an interior unit, draw the required elevations. Use the following materials:

- 300# composition shingles
- Horizontal siding with a 6" exposure over 1/2" wafer board and Tyvek
- 1 × 3 corner trim
- 1 × 6 fascias and barge rafters

Provide notes to specify all exterior materials, and provide grid lines to match the floor plan. Alter exterior materials and window sizes and shapes to provide an alternative front elevation using a Spanish style, and another front elevation providing a very traditional style.

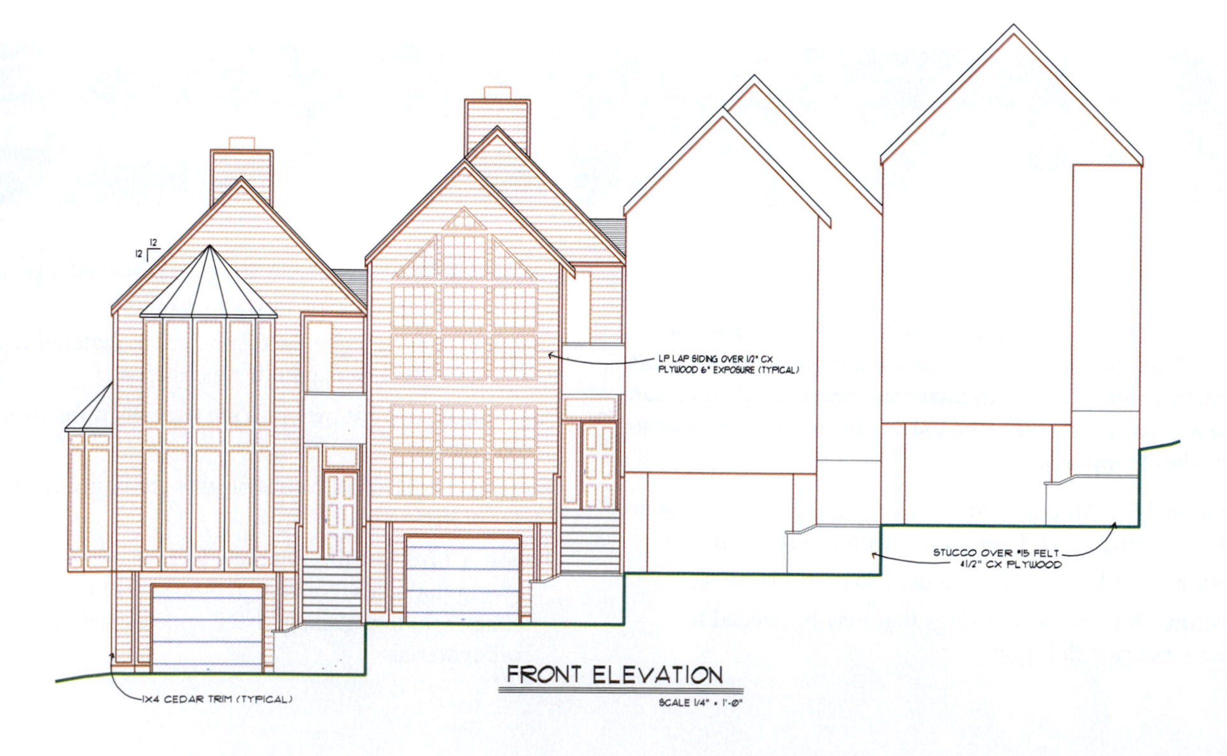

Problem 17-4 Draw the required elevations to represent the retail sales outlet. In addition to showing the 8" × 8" × 16" grade A concrete blocks, show a brick veneer over the south and east walls. Use 26 gauge metal roofing installed as per manufacturer's specifications.

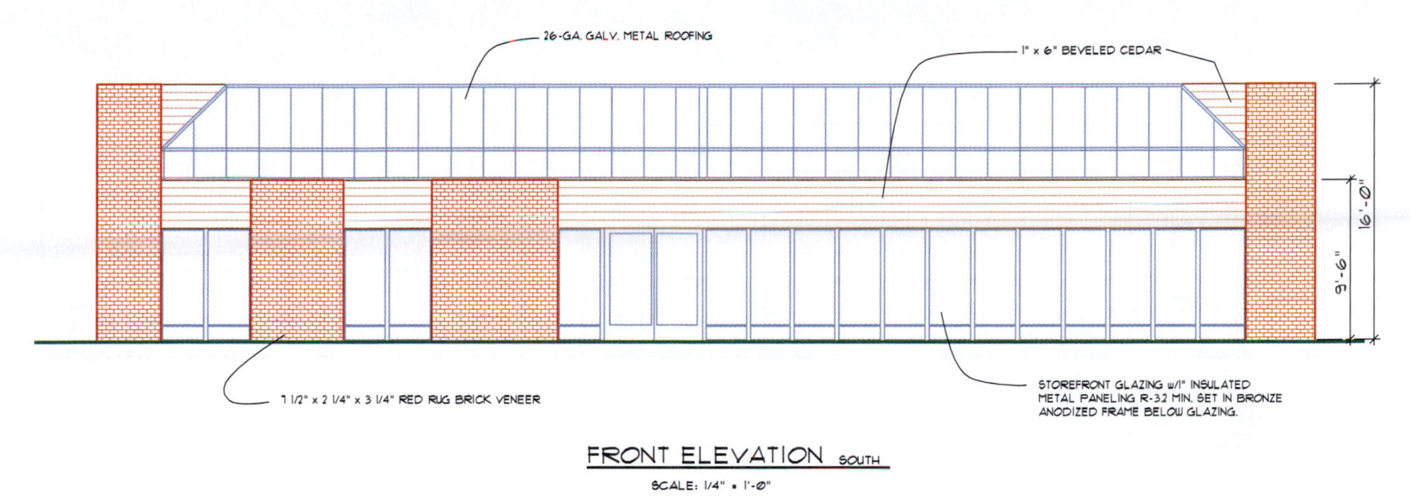

Problem 17-5 Use the floor and site plan to draw each of the elevations for this structure. Use horizontal cedar siding with a 6" exposure over 1/2" wafer board at the roof awning. Cover the south and west office walls with EIFS. Design a decorative pattern to enhance the west elevation of the warehouse. A sign displaying the company name and logo should also be prominent on the north wall.

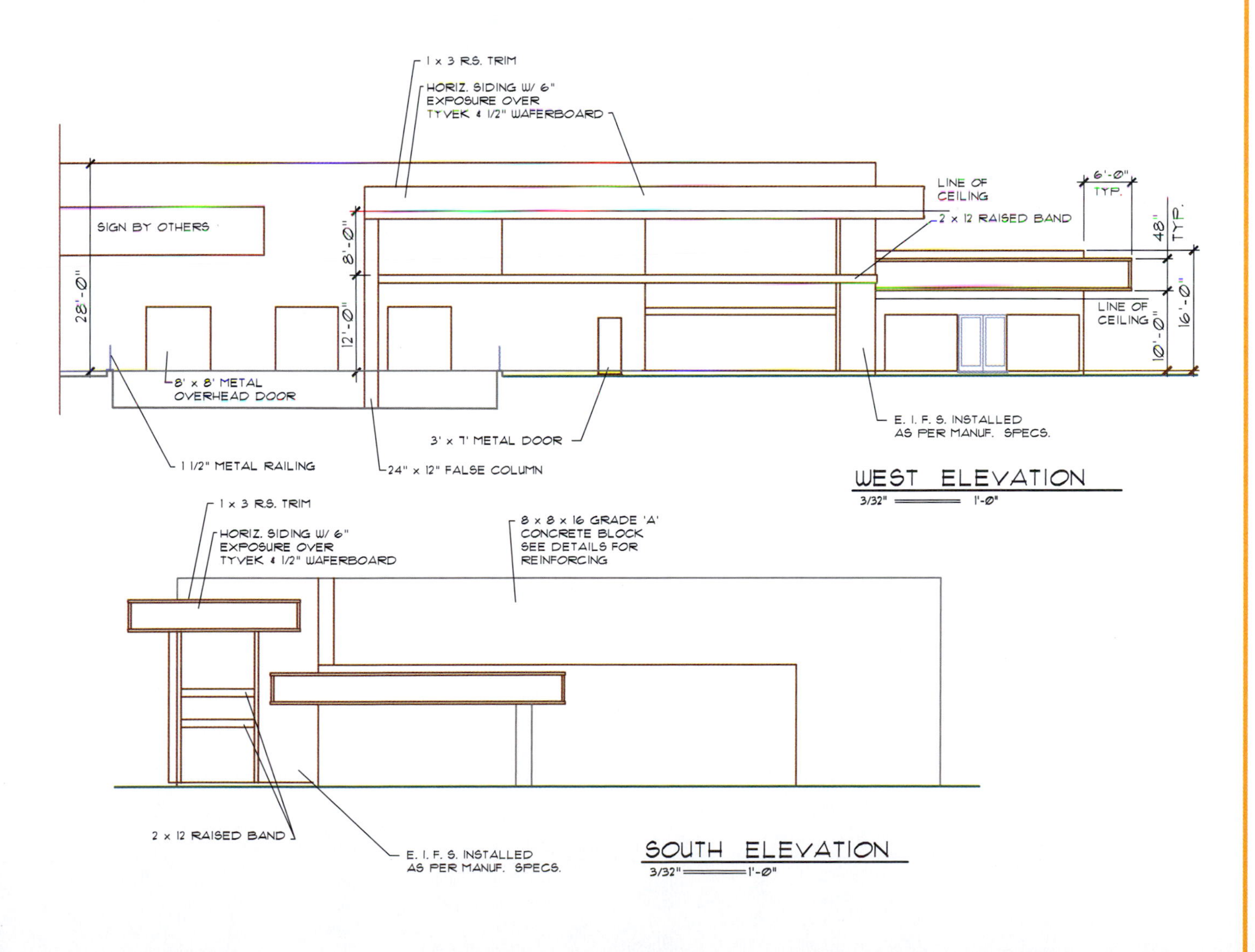

Problem 17-6 Use the floor and site plan to draw each of the elevations for this structure. Use 26 gauge (ga.) metal roofing for the projected roof. Use a combination of a grooved inset into each panel and a painted band on all walls to minimize the building height.

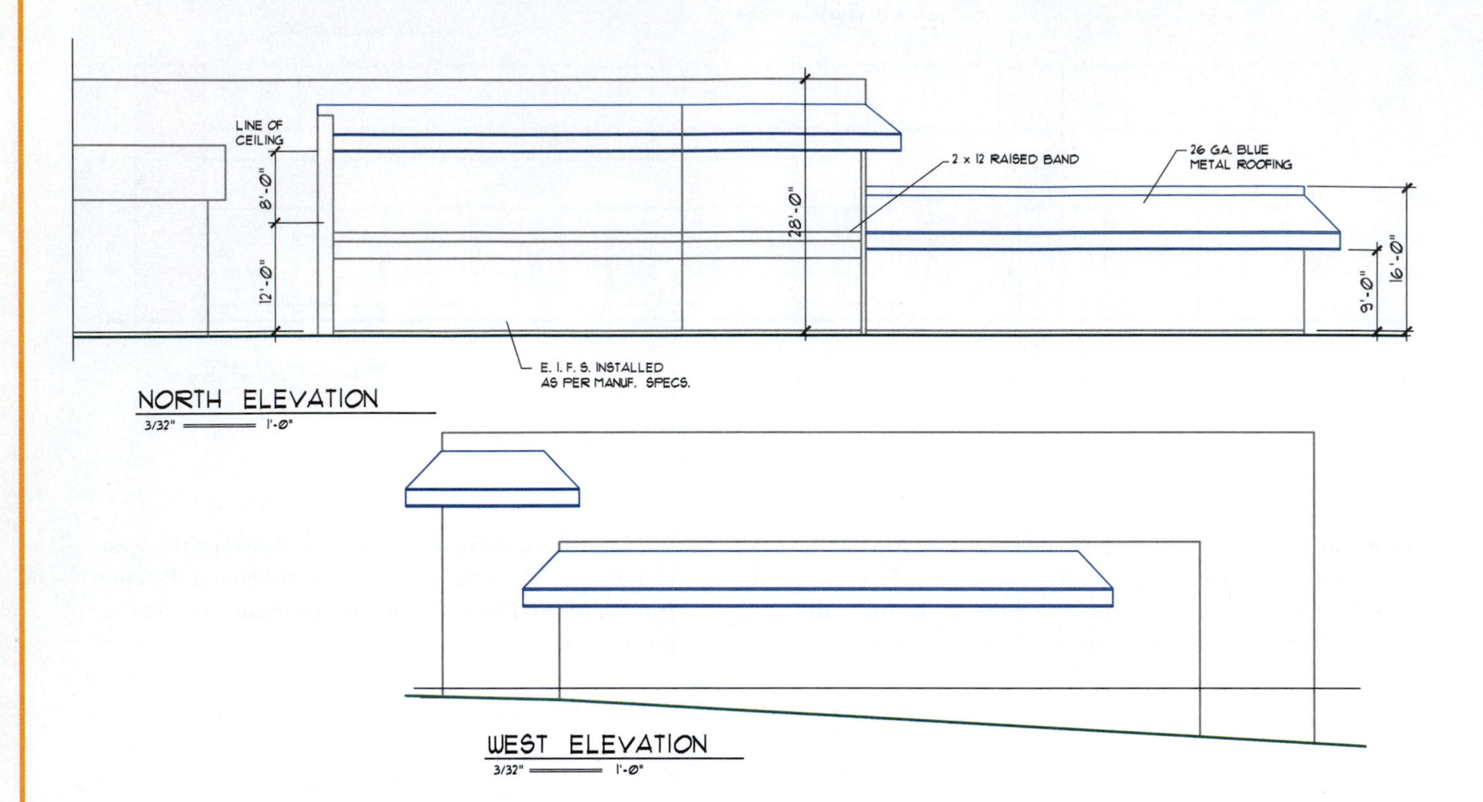

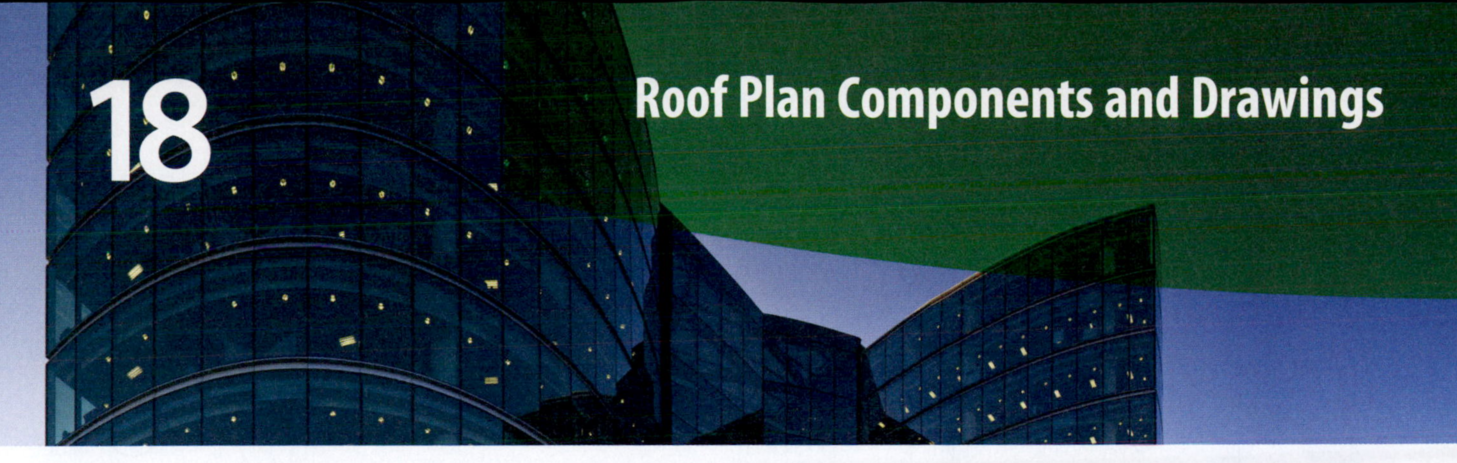

18 Roof Plan Components and Drawings

The roof plan is used to show the shape, the roofing materials, the direction of slope, and the drainage method of each roof surface. Although many commercial structures have a roof with a slight slope, each of the shapes traditionally associated with light-frame construction can also be used with smaller commercial structures. This chapter will explore:

- Roof pitch and its effect on the roof shape.
- Roof materials.
- Common roof components.
- Methods of completing a low-sloped roof plan.
- Methods of completing a high-sloped roof plan.

The roof plan is the part of the architectural drawings developed to show a top view of the structure in orthographic projection. An example of a roof plan is shown in Figure 18-1. On high-sloped roofs, the roof plan is developed with other preliminary drawings to help the clients visualize how the structure will look. On low-sloped roofs, a roof plan may not be started until the working drawings are started. On simple structures, the plan may be drawn at a small scale such as 1/8" = 1'–0" (1:100) or smaller. On more complicated structures, the roof plan is drawn at the same size as the floor plan, and divided into sections by match lines that match the floor plan and elevations. The construction of the roof is represented on a roof framing plan similar to Figure 18-2. The roof framing plan is part of the structural drawings, which will be discussed in Chapter 22.

ROOF PITCH

Pitch is a ratio of the vertical rise to the horizontal run used to describe the angle of the roof slope. Pitch is described in units comparing horizontal run to the vertical rise. The slope may be measured in inches, millimeters, feet, or any other unit as long as the same unit is used for each measurement. The slope will be shown on the sections and elevations. The direction and intersections between roof surfaces must be shown on the roof plan, but the pitch is not represented. The angle equivalents for representing each pitch can be seen in Table 18-1.

Low-Sloped Roofs

Low-sloped roofs are divided into two categories by the IBC. The IBC requires a minimum of 1/4"/12" (2% slope) for drainage. Manufacturers of roofing material typically consider low-sloped roofs to be those with less than a 2:12 pitch (17% slope). A second category of material can be used for roofs with pitches that range from 2 1/2:12 (21% slope) to 4:12 (33% slope). Because of the size of many commercial structures, a flat, low-sloped roof is the only practical shape that can be used economically. The method of framing the roof will depend on the size of the structure and the arrangement of space. For a structure

Roof	Pitch Angle
1:12	4°30'
2:12	9°30'
3:12	14°0'
4:12	18°30'
5:12	22°30'
6:12	26°30'
7:12	30°0'
8:12	33°45'
9:12	37°0'
10:12	40°0'
11:12	42°30'
12:12	45°0'

TABLE 18-1 Common Angles for Drawing Roof Pitches

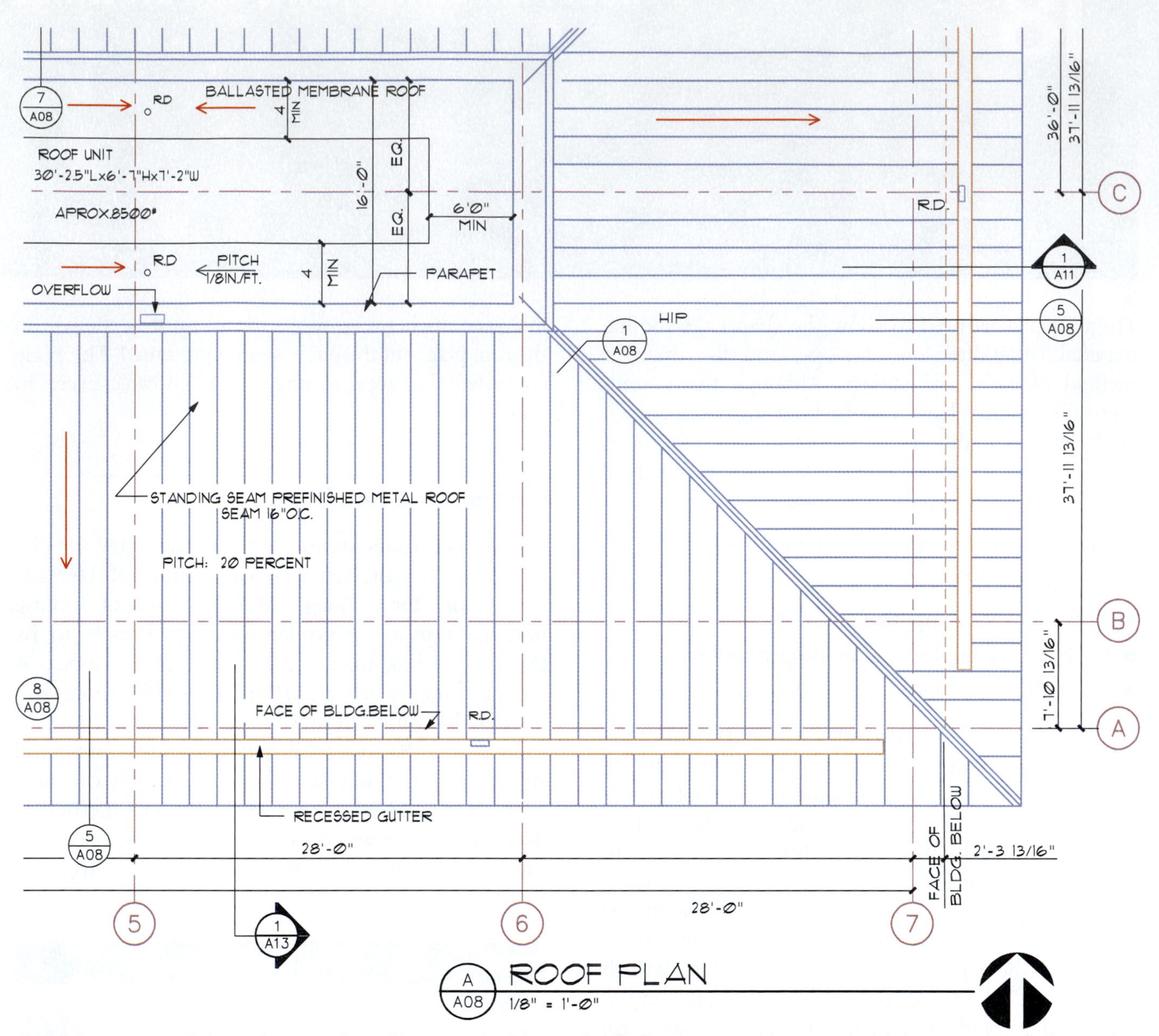

FIGURE 18-1 A roof plan is drawn to show the contours of the top level of a structure and is considered the top view in orthographic projection. *Courtesy Peck, Smiley, Ettlin Architects.*

composed of a single space, a single-slope shed roof similar to Figure 18-3 can be used. In locations that do not receive much rain, any water collected on the roof may be allowed to drain off the low edge of the roof directly to the ground. In moister regions, water will be diverted to one or more collection points and then removed from the roof by a series of drains. Water removal methods are discussed later in this chapter.

Many low-sloped roofs are surrounded by *parapet walls*, which are exterior walls that extend above the roof. Parapet walls are provided for low-sloped roofs to hide mechanical equipment from public view. Depending on the location of the structure relative to the property line, building codes require a minimum of a 30" (750 mm) high parapet wall regardless of the pitch. This is to protect the roof from a fire on adjoining property, or vice versa. The parapet wall location and height must be specified on the roof plan. Wall construction is specified in details similar to Figure 18-4, which must be referenced to the roof plan. The parapet wall creates a dam effect, trapping water on the low side of the roof. Lightweight metal called *flashing* is used to protect each wall/roof intersection from water seepage. To protect from water penetration when two different materials meet, flashing is also used on the tops of parapet walls and at each penetration of the roof. The type

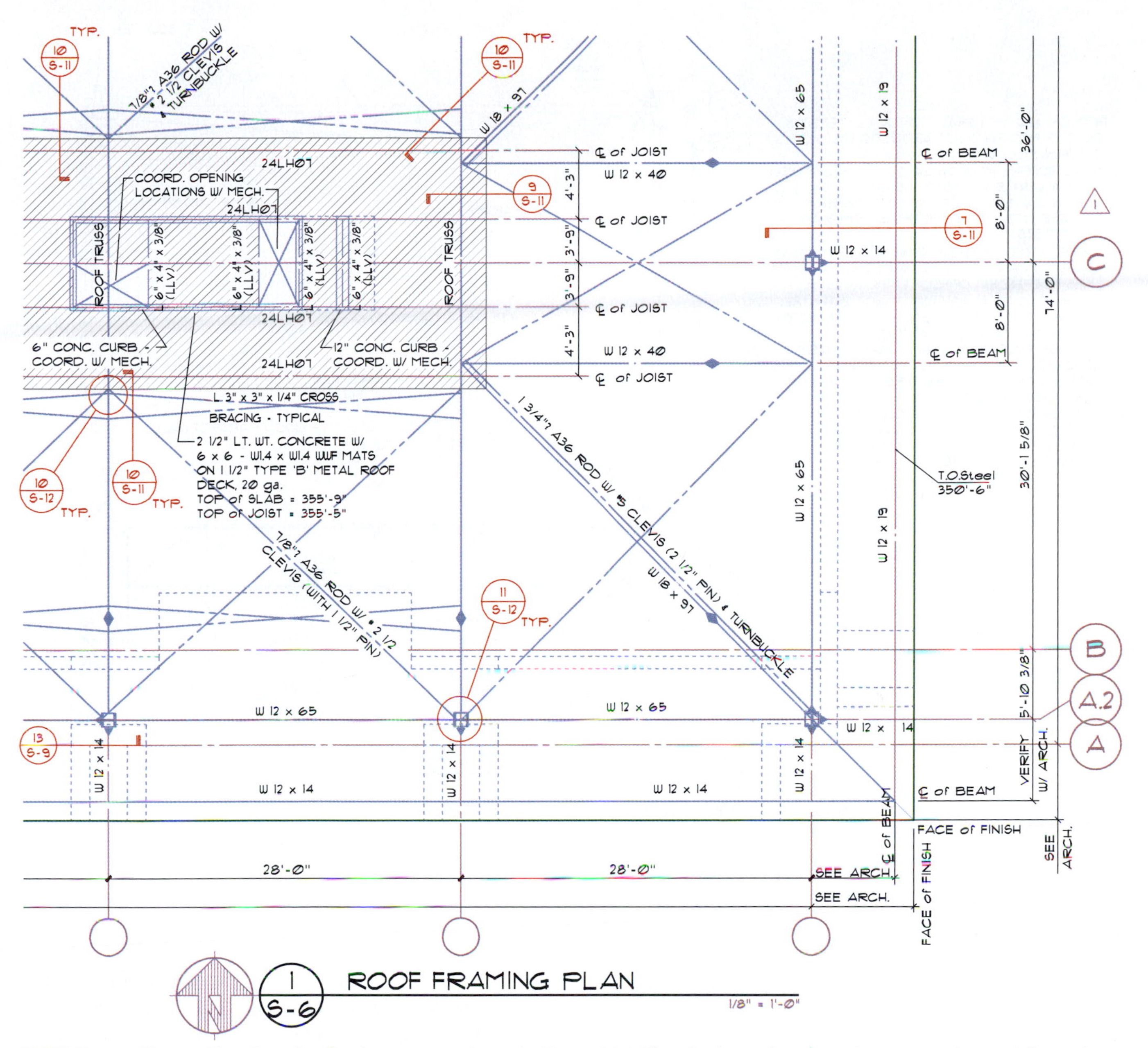

FIGURE 18-2 The roof framing plan for the structure shown in Figure 18-1. The plan is used to show the structural materials used to construct the roof. *Courtesy Van Domelen/Looijenga/McGarrigle/Knauf Consulting Engineers.*

and gauge of the metal and the type of lap are specified on the roof details.

A **cant strip** is a small block of wood or other material that is placed below the roofing material where the parapet wall intersects the roof. The cant strip provides support to the flashing and roofing. As shown in Figure 18-4, the cant is cut to provide an inclined surface to slope water away from the intersection. The cant strip is not always specified on the roof plan, but is generally referenced in details similar to Figure 18-4. As the length of the structure increases, a single direction slope can still be used, but additional measures are required to distribute rainwater. As the distance between roof drains is increased, the cant strip is replaced by a cricket. A **cricket** or *saddle* is a small "fake roof" built to divert water. The cricket is built over the roof on its low side to divert water to roof drains. It can be constructed of plywood or sheet metal depending on the size. Figures 18-3 and 18-5 show an example of a roof framed with crickets. Smaller crickets are also framed on the upper edge of any roof penetrations such as skylights and chimneys. An alternative to framing crickets is to alter the slope of the entire roof by using a framing system typically referred to as a butterfly roof. With a single-sloped roof, framing members hang from a ledger that is

1. PROVIDE 400# PEA GRAVEL OVER 4 LAYERS MIN. BUILT-UP ROOFING W/ HOT ASPHALTIC EMULSION BETWEEN. EA. COURSE.
2. METAL ROOFING TO BE 26 GA. GALV. STEEL PANELS WITH GALLERY BLUE FACTORY BAKED FINISH.
3. ROOF AND OVERFLOW DRAINS TO BE GENERAL PURPOSE TYPE WITH NON-FERROUS DOMES AND 4" DIA. OUTLETS.
4. OVERFLOW DRAINS TO BE SET WITH INLET 2" ABOVE DRAIN INLET AND SHALL BE CONNECTED TO DRAIN LINES INDEPENDENT OF ROOF LINES.
5. SCUPPERS TO BE 4" HIGH AND 7" WIDE WITH 4" RECT. CORRUGATED D.S. PROVIDE A 6" x 9" CONDUCTOR HEAD AT THE TOP OF DOWNSPOUTS.
6. FOUR-TON, ROOF-MOUNTED HEAT PUMP FURNISHED & INSTALLED BY HVAC CONTR. COORDINATE INSTALLATION OF DUCTWORK THROUGH JOIST WITH TRUSS-JOIST INSTALLER.
7. FOUR-TON, ROOF-MOUNTED AIR CONDITIONER FURNISHED & INSTALLED BY HVAC CONTRACTOR. COORDINATE INSTALLATION OF DUCTWORK THROUGH JOIST WTIH JOIST-INSTALLER.
8. INSERT AND ROOF CURBING FOR HVAC TO BE FURNISHED & INSTALLED BY GENERAL CONTRACTOR.

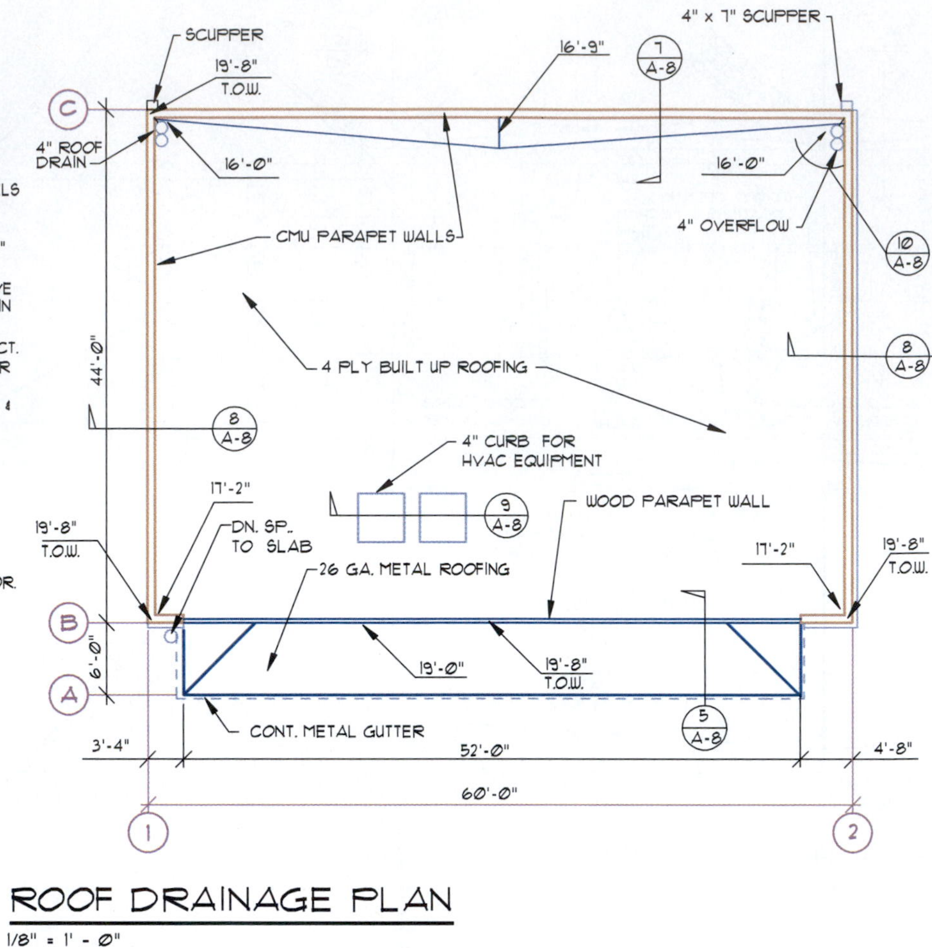

FIGURE 18-3 Simple structures often use a single-slope roof to direct water off the roof.

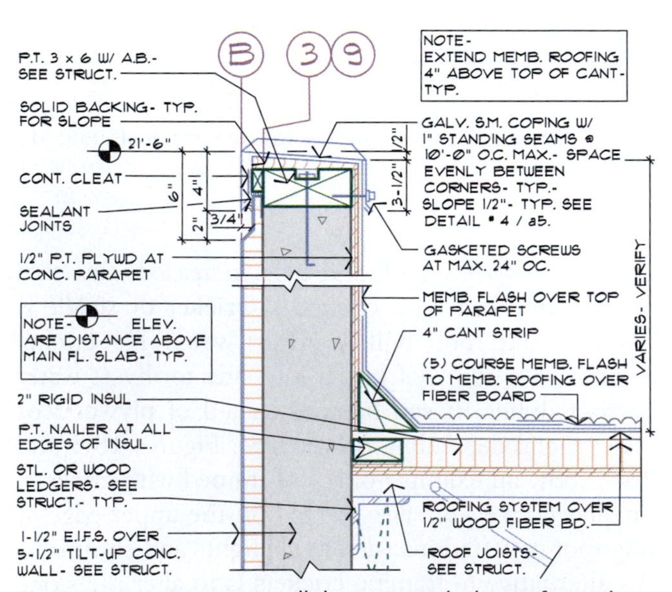

FIGURE 18-4 A parapet wall that surrounds the roof must be referenced on the roof plan with construction methods showing the intersection in details, which are referenced to the roof plan. *Courtesy Architects Barrentine, Bates & Lee, A.I.A.*

attached to the exterior wall. The ledger is parallel to the floor. With a butterfly roof, the ledger is set on an incline, which provides additional slope to what is normally considered a flat roof. Figure 18-6 shows the outline of a butterfly roof in elevations. Figure 18-7 shows an example of a butterfly roof plan.

High-Sloped Roofs

High-sloped roofs are generally considered to be any roof with greater than a 4:12 pitch (33% slope). Many of the roof shapes found in residential construction are also used in light construction and small commercial uses. The most common high-sloped roof shapes include mansard, gable, hip, Dutch hip, and gambrel roofs.

Mansard

A **mansard roof** is an inclined roof that is used to cover one or more full stories of a structure. A partial mansard

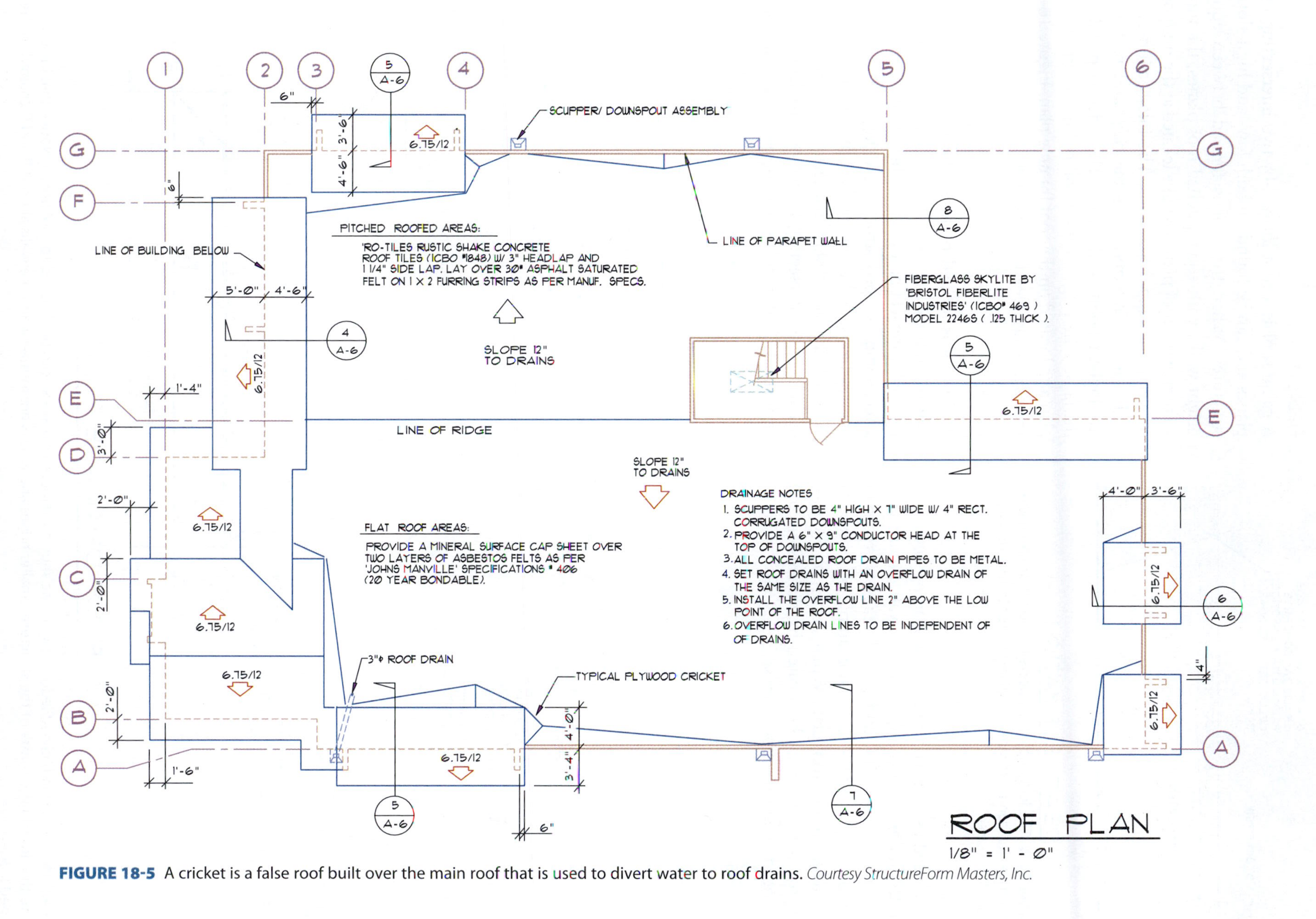

SCUPPER/ DOWNSPOUT ASSEMBLY
LINE OF PARAPET WALL
LINE OF BUILDING BELOW
PITCHED ROOFED AREAS:
'RO-TILES RUSTIC SHAKE CONCRETE ROOF TILES (ICBO #1848) W/ 3" HEADLAP AND 1 1/4" SIDE LAP. LAY OVER 30# ASPHALT SATURATED FELT ON 1 X 2 FURRING STRIPS AS PER MANUF. SPECS.
SLOPE 12" TO DRAINS
FIBERGLASS SKYLITE BY 'BRISTOL FIBERLITE INDUSTRIES' (ICBO# 469) MODEL 22465 (.125 THICK).
LINE OF RIDGE
SLOPE 12" TO DRAINS
FLAT ROOF AREAS:
PROVIDE A MINERAL SURFACE CAP SHEET OVER TWO LAYERS OF ASBESTOS FELTS AS PER 'JOHNS MANVILLE' SPECIFICATIONS # 406 (20 YEAR BONDABLE).
DRAINAGE NOTES
1. SCUPPERS TO BE 4" HIGH X 7" WIDE W/ 4" RECT. CORRUGATED DOWNSPOUTS.
2. PROVIDE A 6" X 9" CONDUCTOR HEAD AT THE TOP OF DOWNSPOUTS.
3. ALL CONCEALED ROOF DRAIN PIPES TO BE METAL.
4. SET ROOF DRAINS WITH AN OVERFLOW DRAIN OF THE SAME SIZE AS THE DRAIN.
5. INSTALL THE OVERFLOW LINE 2" ABOVE THE LOW POINT OF THE ROOF.
6. OVERFLOW DRAIN LINES TO BE INDEPENDENT OF OF DRAINS.
3"ϕ ROOF DRAIN
TYPICAL PLYWOOD CRICKET
6.75/12
ROOF PLAN
1/8" = 1' - 0"

FIGURE 18-5 A cricket is a false roof built over the main roof that is used to divert water to roof drains. *Courtesy StructureForm Masters, Inc.*

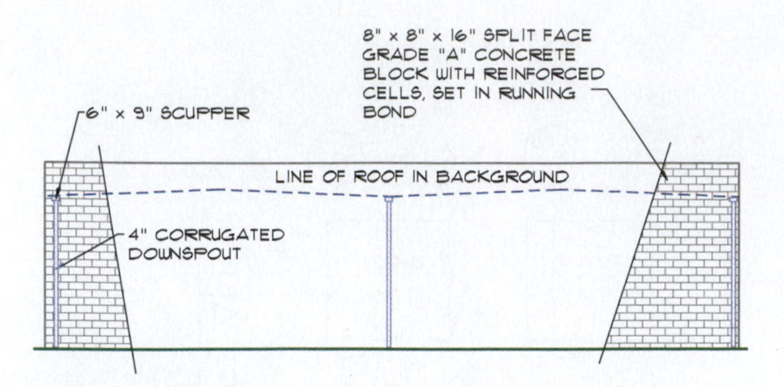
8" x 8" x 16" SPLIT FACE GRADE "A" CONCRETE BLOCK WITH REINFORCED CELLS, SET IN RUNNING BOND
6" x 9" SCUPPER
LINE OF ROOF IN BACKGROUND
4" CORRUGATED DOWNSPOUT
REAR ELEVATION NORTH

FIGURE 18-6 The ledger that supports the roof framing members at the wall can be inclined to divert water to roof drains. *Courtesy Wil Warner.*

can be used as a visual screen, as a shading device, or as a decoration to break up a parapet wall. Figure 18-8 shows examples of full and partial mansard roofs in elevation. Figure 18-9 shows an example of representing a mansard on a roof plan. When shown on the roof plan, the mansard must be located by dimension and referenced to construction details.

Gable

A **gable roof** is a roof formed by two intersecting roof planes and can be found on both low- and high-sloped structures. When the roof is formed between equal-height walls, with equal-pitched roof planes, the intersection of the roof planes will be located at the midpoint between the supporting walls. The intersection of the roof planes is referred to as the *ridge* and represents the highest point of the roof. The ridge is parallel to the floor. The ridge location will be shown on the roof plan and located on the roof framing plan. Figure 18-10 shows an example of a gable roof.

Hip

A **hip roof** is formed by four intersecting roof planes. When a hip roof is formed on a square structure, the roof will intersect at a point. When a hip is framed on a rectangular-shaped structure, a ridge will be formed halfway between the long walls of the structure. A *hip* is an inclined roof member that forms an external corner between two intersecting planes of a roof. A *valley* is an inclined roof member that forms an internal corner

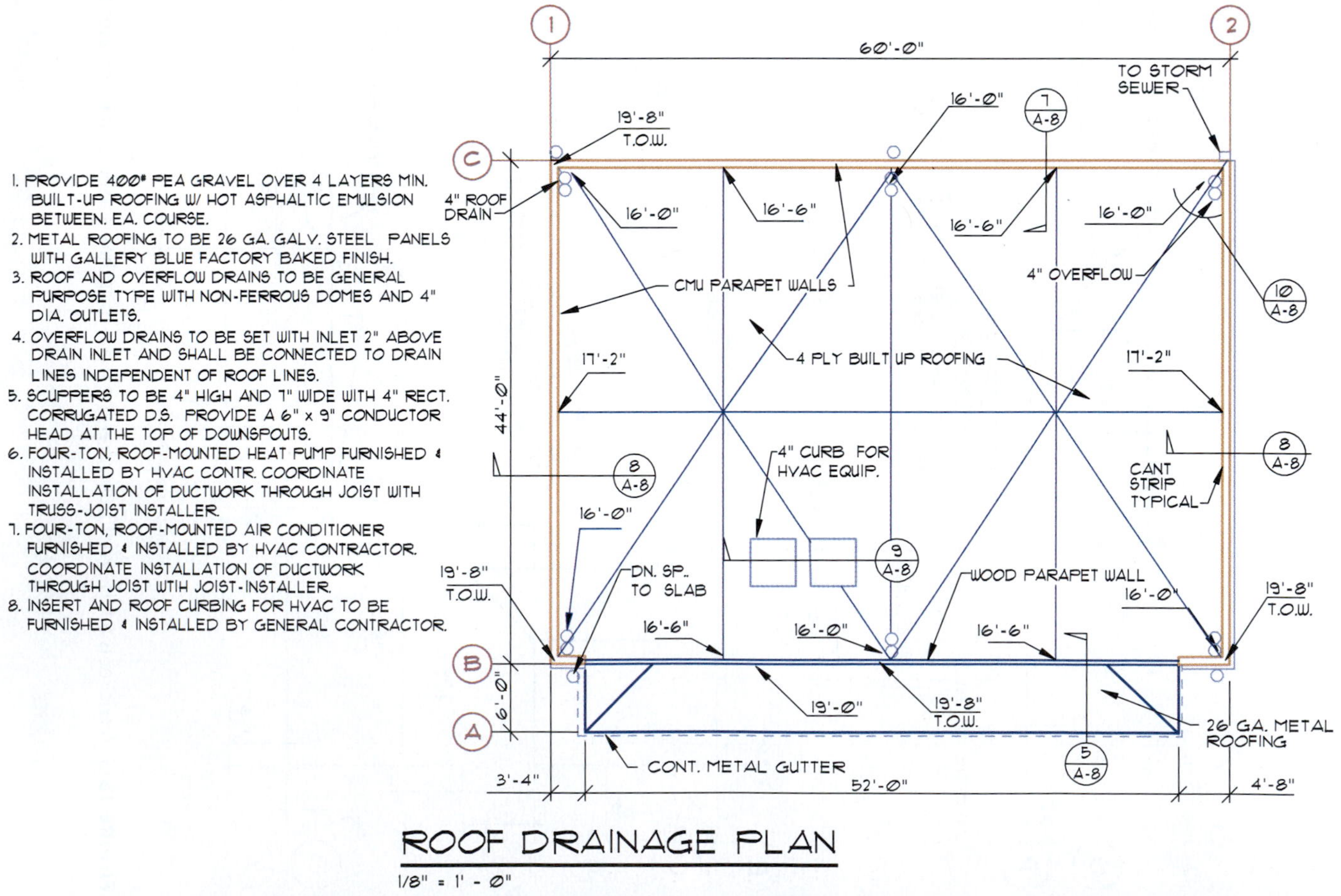
1. PROVIDE 400# PEA GRAVEL OVER 4 LAYERS MIN. BUILT-UP ROOFING W/ HOT ASPHALTIC EMULSION BETWEEN. EA. COURSE.
2. METAL ROOFING TO BE 26 GA. GALV. STEEL PANELS WITH GALLERY BLUE FACTORY BAKED FINISH.
3. ROOF AND OVERFLOW DRAINS TO BE GENERAL PURPOSE TYPE WITH NON-FERROUS DOMES AND 4" DIA. OUTLETS.
4. OVERFLOW DRAINS TO BE SET WITH INLET 2" ABOVE DRAIN INLET AND SHALL BE CONNECTED TO DRAIN LINES INDEPENDENT OF ROOF LINES.
5. SCUPPERS TO BE 4" HIGH AND 7" WIDE WITH 4" RECT. CORRUGATED D.S. PROVIDE A 6" x 9" CONDUCTOR HEAD AT THE TOP OF DOWNSPOUTS.
6. FOUR-TON, ROOF-MOUNTED HEAT PUMP FURNISHED & INSTALLED BY HVAC CONTR. COORDINATE INSTALLATION OF DUCTWORK THROUGH JOIST WITH TRUSS-JOIST INSTALLER.
7. FOUR-TON, ROOF-MOUNTED AIR CONDITIONER FURNISHED & INSTALLED BY HVAC CONTRACTOR. COORDINATE INSTALLATION OF DUCTWORK THROUGH JOIST WTIH JOIST-INSTALLER.
8. INSERT AND ROOF CURBING FOR HVAC TO BE FURNISHED & INSTALLED BY GENERAL CONTRACTOR.
60'-0"
TO STORM SEWER
19'-8" T.O.W.
4" ROOF DRAIN
16'-0"
16'-6"
CMU PARAPET WALLS
4" OVERFLOW
4 PLY BUILT UP ROOFING
17'-2"
44'-0"
4" CURB FOR HVAC EQUIP.
CANT STRIP TYPICAL
DN. SP. TO SLAB
WOOD PARAPET WALL
19'-0"
19'-8" T.O.W.
6'-0"
CONT. METAL GUTTER
26 GA. METAL ROOFING
3'-4"
52'-0"
4'-8"
ROOF DRAINAGE PLAN
1/8" = 1' - 0"

FIGURE 18-7 By using inclined ledgers like those shown in Figure 18-6, water can be diverted to roof drains. Notice that in addition to the ridge in the center of the structure, smaller ridges and valleys are formed along the walls at grids B and C created by the inclined ledgers.

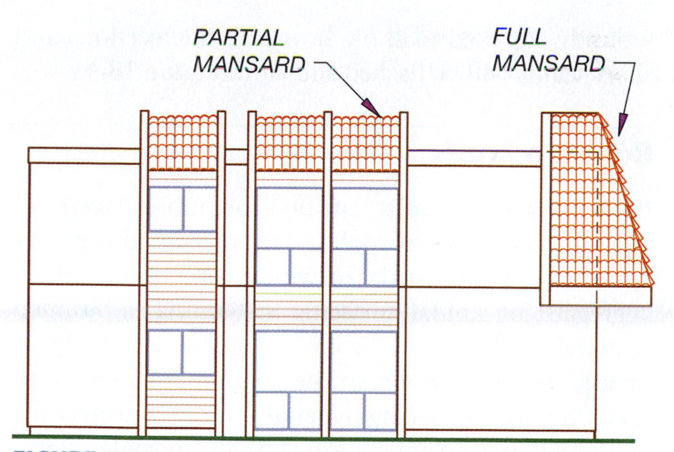

FIGURE 18-8 A partial mansard roof can be used on low-sloped roofs to hide mechanical equipment or to provide shade for windows. A full mansard can also be used to hide roof-mounted materials or to visually reduce the height of a structure.

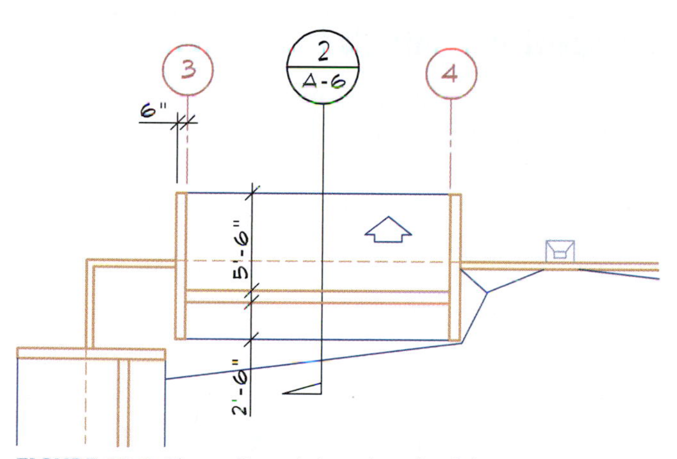

FIGURE 18-9 The walls and sloped roofs of the mansard must be reflected on the roof plan.

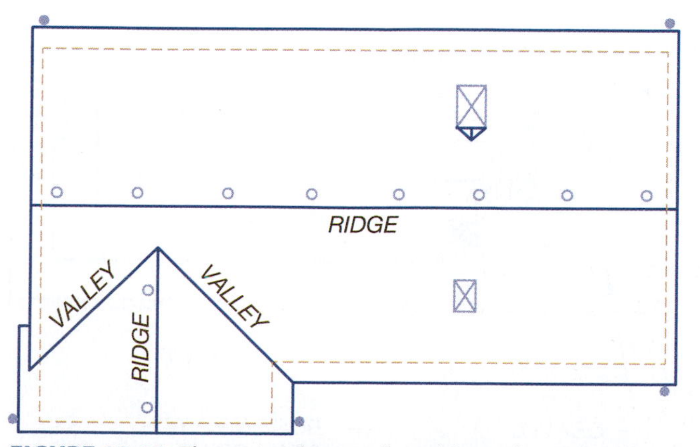

FIGURE 18-10 The major shapes of a gable roof include the ridge, valley, and hips. The walls perpendicular to each ridge are referred to as gable end walls.

between two intersecting planes of a roof. Both hips and valleys are common to high-sloped roofs and are shown in Figure 18-11.

Dutch Hip

A **Dutch hip**, sometimes referred to as a *Dutch gable roof*, is a combination of a gable and a hip roof. A wall is placed parallel to the short support wall at the roof level. This wall is used to form a break in the roof plan, as shown in Figure 18-12. The roof pitch and desired effect determine the location of the wall. Figure 18-13 shows a Dutch hip represented on a roof plan.

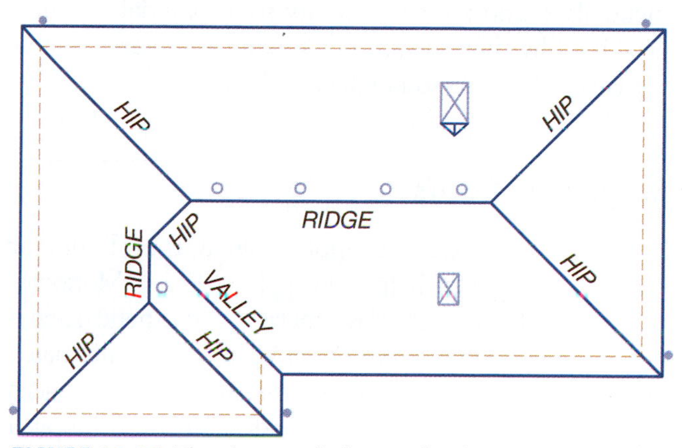

FIGURE 18-11 The shapes of a hip roof include ridges, hips, and valleys. A hip roof provides a protective overhang on all sides of the structure.

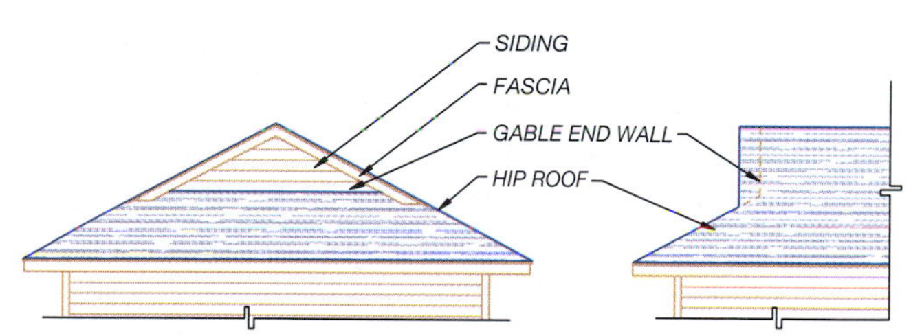

FIGURE 18-12 A Dutch hip roof combines features of a hip and gable roof. The plane of the hip is terminated before it reaches the ridge, and a gable wall is built at the roof level.

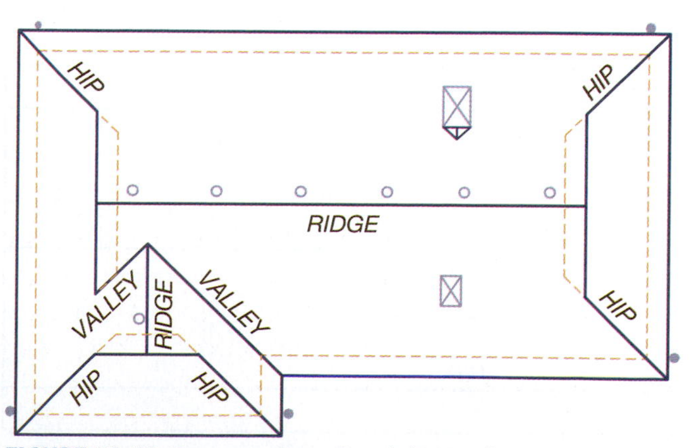

FIGURE 18-13 Representing a Dutch hip roof on a roof plan.

Gambrel

A **gambrel roof** is formed by two different roof levels. The upper portion is a gable roof, and the lower portion is a steep-shed roof. The ridge and the limits of each roof plane must be shown on the roof plan. Figure 18-14 shows a gambrel roof plan.

ROOF MATERIALS

The type of material used for weather protection must be specified on the roof plan. Common roofing materials include single-ply and built-up roofs, shingles, metal slate, and tile roofs. Many roof materials are also described by their weight per square. A **square** of roofing is equal to 100 square feet (9.3 m^2).

Single-Ply Roofs

Single-ply roofs can be applied as a thin liquid or sheet made from Ethylene Propylene Diene Monomer (EPDM), which is an elastomeric or synthetic rubber material. Polyvinyl chloride (PVC), chlorosulfonated polyethylene (CSPE) and polymer-modified bitumens are also used. These materials are designed to be applied to roof decks with a minimum pitch of 1/4:12 (2% slope). In liquid form, these materials can be applied directly to the roof decking with a roller or sprayer to conform to irregular-shaped roofs.

Single-sheet roofs are rolled out and bonded together to form one large sheet. The sheet can be bonded to the roof deck by mechanical fasteners. Some applications are not attached to the roof deck and are held in place by gravel material that is placed over the roofing material to provide ballast. A typical specification for a single-ply roof would specify the material, the application method, and the aggregate size. Many firms also specify a reference for the material to be installed according to the manufacturer's specification. Details are also typically supplied to show how roof intersections and sheet seams will be flashed similar to Figure 18-4.

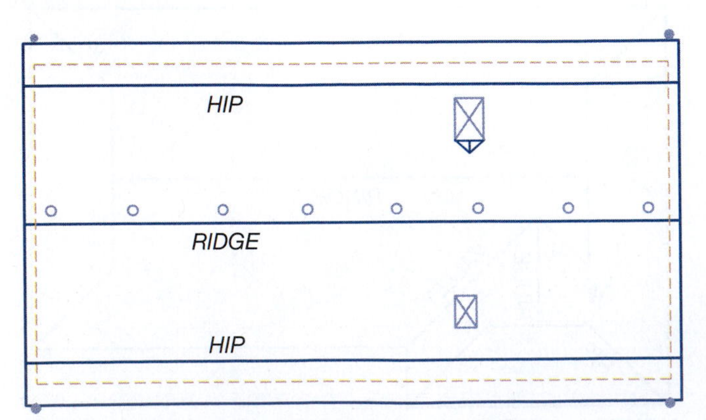

FIGURE 18-14 Representing a gambrel roof plan.

Built-Up Roofs

Built-up roofs consist of two or more layers of bituminous-saturated roofing felt, cemented together with bitumen and surfaced with a cap sheet, mineral aggregate, or similar surfacing material. The bitumen material used to bond the layers together is usually tar or asphalt. In addition to the roofing felts and bitumens, a gravel surfacing material is used to protect the exposed surface from abrasions. Built-up roof systems can be applied over any type of roof deck and are suitable for low- or high-sloped roofs. Generally, they are to be applied to roofs with a minimum pitch of 1/4:12 (2% slope) but less than a 2:12 pitch (17% slope). Figure 18-15 shows the process for applying a built-up roof system.

Composition Shingles

Composition or asphalt shingles come in a variety of colors and patterns. Standard shingles come in a 12" × 36" (300 × 900 mm) size and are divided into three tabs, with a weight based on the square footage

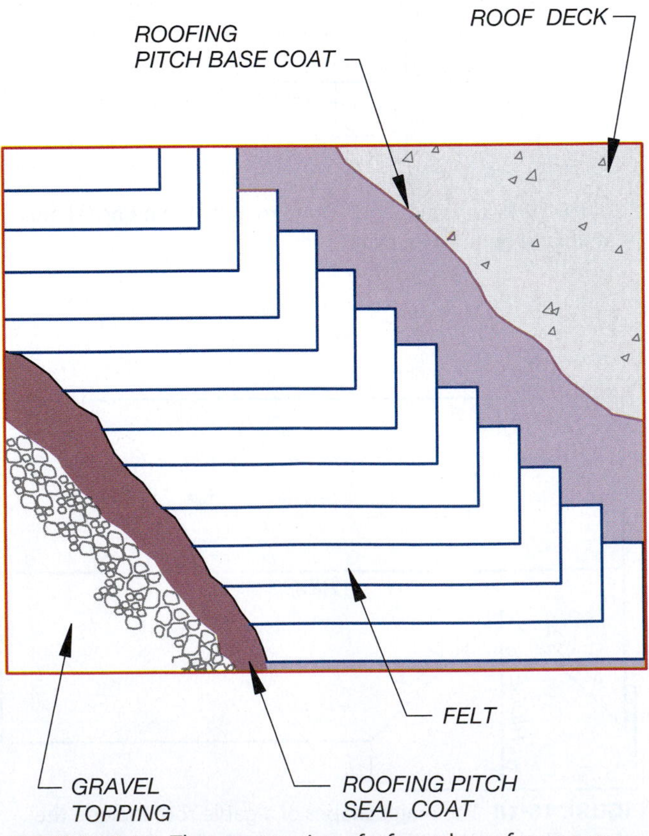

FIGURE 18-15 The construction of a four-ply roof over a concrete roof deck.

weight. Three-tab composition shingles have a weight of 235 pounds per square and are installed over 15# felt. Shingles divided into random widths and thicknesses with a weight of 300 pounds per square are also available. Composition shingles are designed for use on roofs with a minimum pitch of 2:12 (17% slope). Underlayment requirements vary depending on the roof pitch and the design wind speed.

Wood Shingles

Wood shingles are applied to roofs with a 3:12 pitch (25% slope) or greater over 1 × 4 or 1 × 6 (25 × 100 or 25 × 150) spaced sheathing and 15# felt. In areas of the country where the average daily temperature in January is 25°F (–4°C) or less, wood shingles must be applied over special underlayment. Wood shingles laid in cold climates must be installed over two layers of underlayment cemented together or over a self-adhering polymer modified bitumen sheet laid over plywood sheathing for the first 24" (610 mm) from the eave edge. An additional layer of 15# felt is placed between each row of shingles. Wood shakes are thicker and have a rougher texture than shingles. Wood shakes are applied in a similar manner as a shingle but must be applied to roofs with a 4:12 pitch (33% slope) or greater. Both wood shakes and shingles can be chemically treated for fire resistance. Shingles made from masonite are also available. Masonite shingles simulate wood shakes and provide a fire-resistant material.

Seamed Metal

A sheet metal roof made of copper, zinc alloy, or galvanized or coated stainless steel is a common alternative for a commercial roof when a 3:12 or greater pitch (25% slope) is used. A pitch as low as 1/4:12 (2% slope) can be used, depending on the type of seam used to join the metal sheets. Figure 18-16 shows common seam methods. When specified on the roof plan, the metal gauge, panel size, material, and seam pattern must be specified.

Corrugated Metal

Corrugated steel sheets can be used for either siding or roofing material. Metal sheets in either rounded or angular bent patterns similar to Figure 18-17 are available and can be installed on roofs with a pitch of 3:12 (25% slope) or greater. In addition to steel, panels are also available in aluminum, galvanized steel, fiberglass, and corrugated structural glass. When specified, the manufacturer, material, the panel size and weight, required lap, finish, and fastening method must be placed on the plans. Details of installation are also referenced to the roof plan.

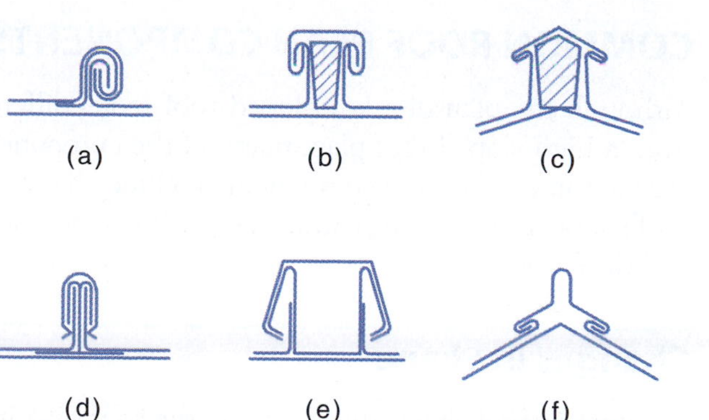

FIGURE 18-16 Examples of metal seams include: (a) double lock standing seam, (b) wood batten seam, (c) wood batten ridge seam, (d) prefabricated standing seam, (e) prefabricated batten seam, and (f) ridge seam.

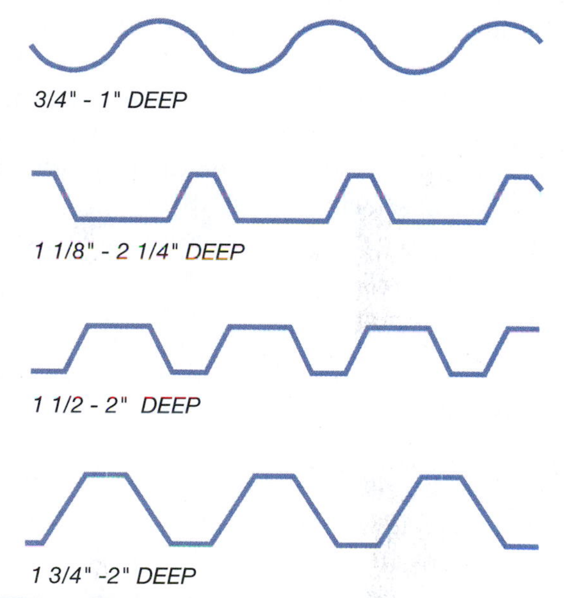

FIGURE 18-17 Standard corrugated and ribbed roofing patterns.

Tile

Tile roofing can be clay, concrete, or metal units that overlap to form the weather protection. Tiles come in a variety of patterns, colors, and shapes and are usually installed on roofs having a pitch of 3:12 or greater. Tile can be installed over either solid or spaced sheathing. Attachment methods will vary depending on the pitch and the maximum basic wind speed. On the roof plan, the manufacturer, style, color, weight, and application method must be specified.

COMMON ROOF PLAN COMPONENTS

Although the plan of a low-sloped roof looks different from a high-sloped roof plan, many of the components are the same. Common components include changes in roof shape, roof openings, drainage methods, notations, and dimensions.

Changes in Shape

Changes in roof shape are shown on the roof plan with continuous object lines. A ridge, hip, and valley are the most common changes of roof shape.

Locating Ridges

The ridge between equal-pitched planes formed from equal-height walls can be located by determining the midpoint between the two supporting walls. When the walls are an unequal height, or a sawtooth roof is formed with unequal roof pitches, a section will need to be drawn so that the ridge location can be projected in the roof plan, as shown in Figure 18-18.

When two gabled structures intersect each other, the intersection of the ridges needs to be determined. With equal pitches over each portion of the structure, the portion that is widest will have the highest ridge. The ridge over the narrower portion of the structure will intersect the highest gable at a point where both roofs are of equal height. This point can be determined by drawing a line to represent the valley formed between the two planes. The valley will always be formed at an angle that is one-half the angle formed between the two intersecting walls. Figure 18-19 shows the intersection of two roofs of unequal widths and the merging of the ridge lines.

Representing Valleys

When one gable or hip roof intersects another gable roof, a valley is created. If the two intersecting roof planes are at an equal pitch, the line representing the valley is drawn on the roof plan at an angle that is one-half the angle formed by the intersecting support walls.

For perpendicular support walls, the valley is represented by a line drawn at a 45° angle on the roof plan, as shown in Figure 18-11. When the intersecting roof planes are at unequal pitches, sections must be drawn to determine where the valley will be created. Figure 18-20 shows an example of determining the intersection of two ridges with different heights and pitches. When the intersecting roof planes are not perpendicular to each other and are over portions of the structure with different widths, the ridges, valleys, and hips must be determined using sections, as seen in Figure 18-21.

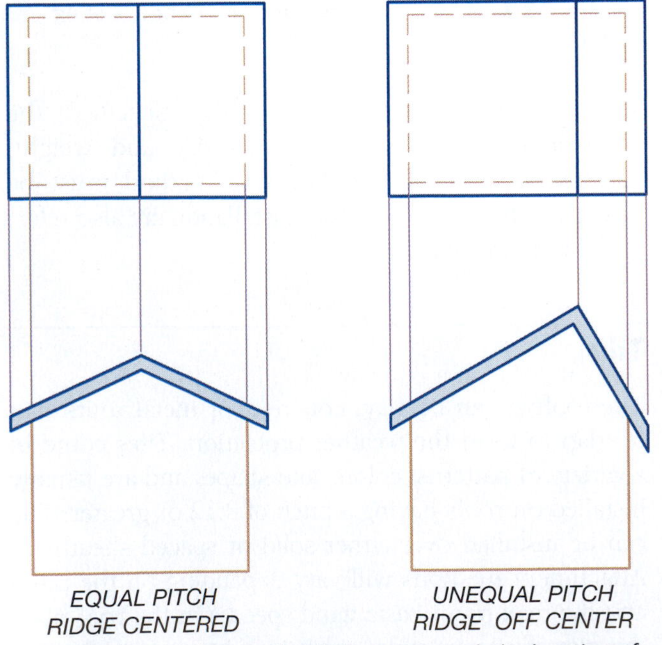

FIGURE 18-18 When roof pitches are unequal, the location of the ridge on the roof plan must be projected from sections.

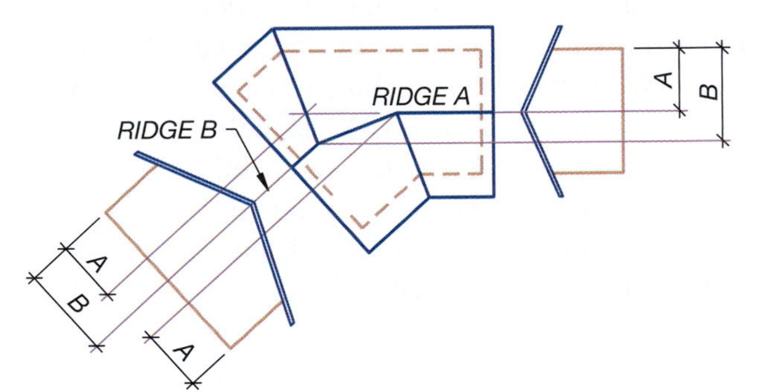

FIGURE 18-19 The intersection of two roofs that are not perpendicular to each other. The hip and valley will be parallel to each other and will be formed at an angle that is equal to one-half the angle formed between the two intersecting support walls.

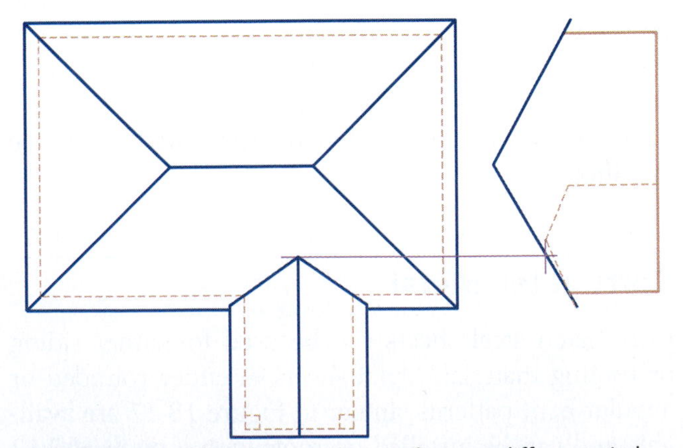

FIGURE 18-20 When intersecting roofs are at different pitches, the ridge intersection and valleys must be projected from an elevation or section.

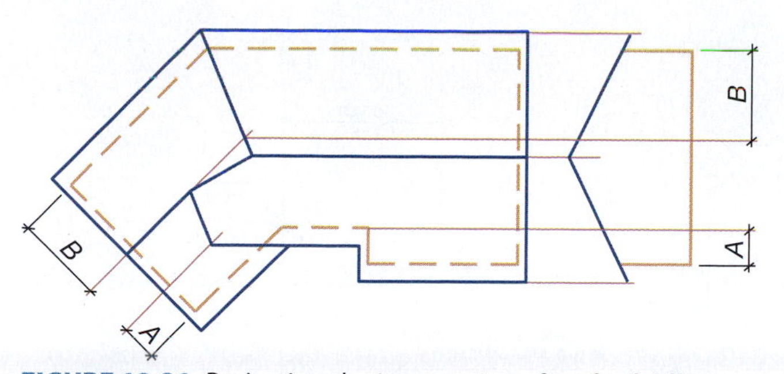

FIGURE 18-21 Projecting the intersection of roofs of different widths.

The intersection of two gables that are supported by walls with different heights must often be represented on a roof plan. Using the roof pitch, the drafter must determine the distance the lower roof will require to rise to an equal height on the upper roof. Figure 18-22 shows the intersection of an entry portico roof framed with 12' (3700 mm) high support walls, the main roof framed with 10' (3000 mm) high walls, and each roof framed with a pitch of 6:12 (50% slope).

Roof Openings

Each penetration of the roof is usually shown on the roof plan. Major openings shown on the roof plan include skylights, chimneys, vent pipes, and openings for HVAC ducts.

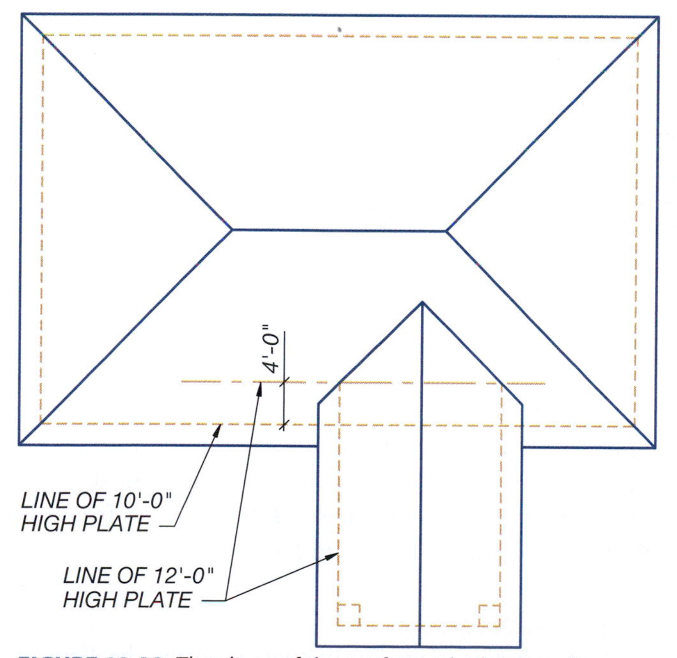

FIGURE 18-22 The slope of the roof must be known when drawing the intersection of two different roofs supported on walls of differing heights. With a 6:12 pitch, a horizontal distance of 48" (1200 mm) is required to reach the starting height of the upper roof. By offsetting the 10' (3000 mm) wall 48" (1200 mm), the valley between the two roofs can be located from the point where the 12' (3600 mm) high wall intersects this 48" (1200 mm) offset.

Skylights

Individual skylights are shown on the roof plan by a rectangle that represents the size of the skylights. On inclined roofs, the true size of the skylight will be altered because of the roof angle. An X is sometimes placed in the skylight to accent its location. Figure 18-23 shows an example of how skylights are represented and specified on a roof plan. On plans with a large number of skylights, the size and type of each skylight can be shown in a schedule similar to Figure 18-24. Large skylight units need to be represented by a method that distinguishes the skylight from the roof. The hatch pattern used to represent the skylight varies for each office, but it must be represented in a symbol schedule attached to the roof plan. In addition to showing the location, the method of protecting the opening from water must also be shown. The waterproofing method is shown in details similar to Figure 18-25, which are referenced on the roof plan.

Chimneys

A fireplace is common in many multifamily units and business occupancies such as restaurants and small offices. Chimneys are represented on the roof plan according to their type. Masonry and metal chimneys in wood chases are represented as a rectangle with an X through it. Metal chimneys are represented by a circle, which represents the required diameter of the chimney. Figure 18-11 shows a method of representing chimneys.

Plumbing and Mechanical Penetrations

Because of the risk of leaking, plumbing vents are shown on many low-pitched roofs. Circles or squares, depending on their shape, represent vents and exhaust flues. Their location is determined by their location on the floor plan and is generally not located by dimensions on the roof plan. Details such as those in Figure 18-26 are referenced to the roof plan to show waterproofing methods. Because openings for mechanical equipment must be shown on the roof framing drawing, they may or may not be shown on the roof plan. The location of equipment is generally shown and specified in relation to surrounding walls or screens on the roof plan. Both the screens and curbs for mounting the equipment should be referenced on the roof plan as shown in Figure 18-27.

Roof Drainage

Once the shape of the roof is indicated, the next priority is to show how water will be diverted from the surface.

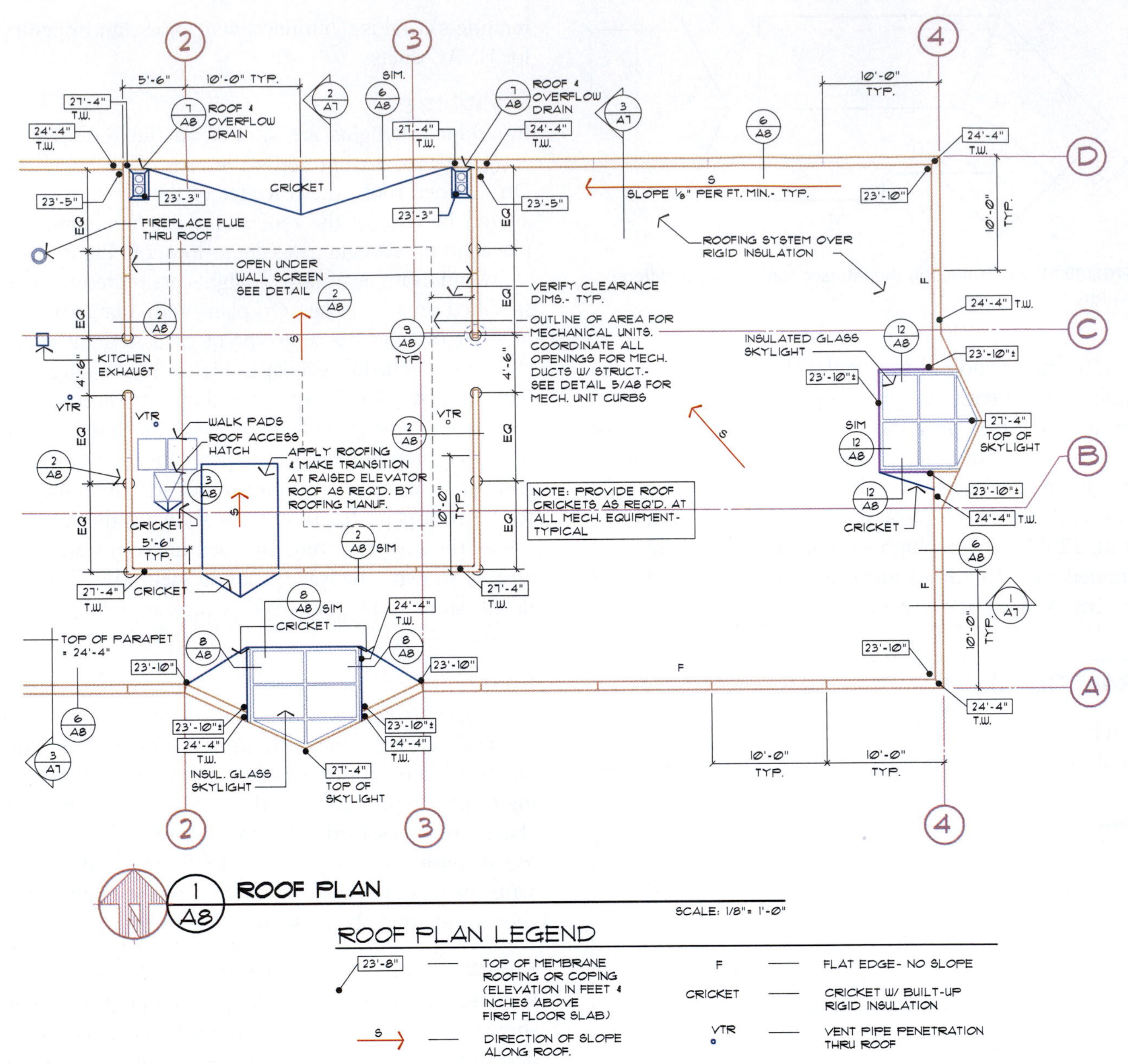

FIGURE 18-23 A completed roof plan for a single-slope roof. *Courtesy Architects Barrentine, Bates & Lee, A.I.A.*

SKYLIGHT SCHEDULE

MARK	LOCATION	PANEL DIMENSION	INSIDE CURB DIMENSION	FINISH DRYWALL DIMEN.
SL-1	CORRIDOR	5'-0" X 5'-0"	4'-8 1/2" X 4'-8 1/2"	4'-7 1/4" X 4'-7 1/4"
SL-2	PRODUCTION	5'-0" X 5'-0"	4'-8 1/2" X 4'-8 1/2"	---------------
SL-3	DROP-OFF	5'-0" X 16'-0"	4'-8 1/2" X 15'-8 1/2"	---------------
SL-4	RETAIL	5'-0" X 16'-0"	4'-8 1/2" X 15'-8 1/2"	---------------
SL-5	LUNCH ROOM	5'-0" X 12'-0"	4'-8 1/2" X 11'-8 1/2"	---------------
SL-6	OFFICES	4'-4" X 6'-4"	4'-1 1/4" X 6'-1 1/4"	4'-0" X 6'-0"
SL-7	OFFICES	5'-0" X 5'-0"	4'-8 1/2" X 4'-8 1/2"	4'-7 1/4" X 4'-7 1/4"
SL-8	CORRIDOR	5'-0" X 16'-0"	4'-8 1/2" X 15'-8 1/2"	4'-1 1/4" X 15'-7 1/4"

FIGURE 18-24 On plans with a large number of skylights, a schedule can be used to improve clarity. *Courtesy Michael & Kuhns Architects, P.C.*

In order to outline drainage, the slope indicators, elevation markers, drains, and overflow drains must all be shown to indicate how water will be removed from the roof. Each is shown in Figure 18-23.

Slope Indicators

A *slope indicator* is an arrow that indicates the flow of water on the roof. Arrows are generally placed so that they point away from the ridge to the low points of the roof.

Elevation Markers

Once the slope has been indicated, the elevation of the roof must be shown. Elevations for the roof are

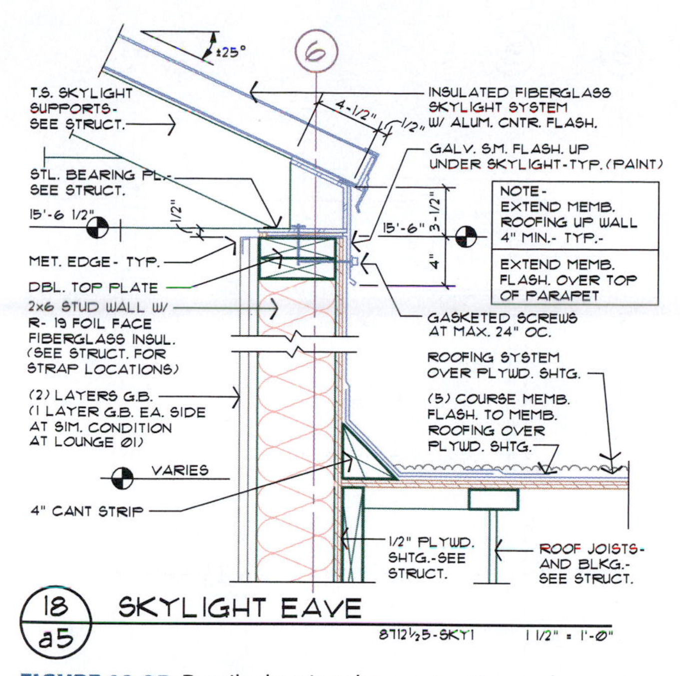

FIGURE 18-25 Details showing the construction and mounting of skylights must be referenced on a low-sloped roof plan. *Courtesy Architects Barrentine, Bates & Lee, A.I.A.*

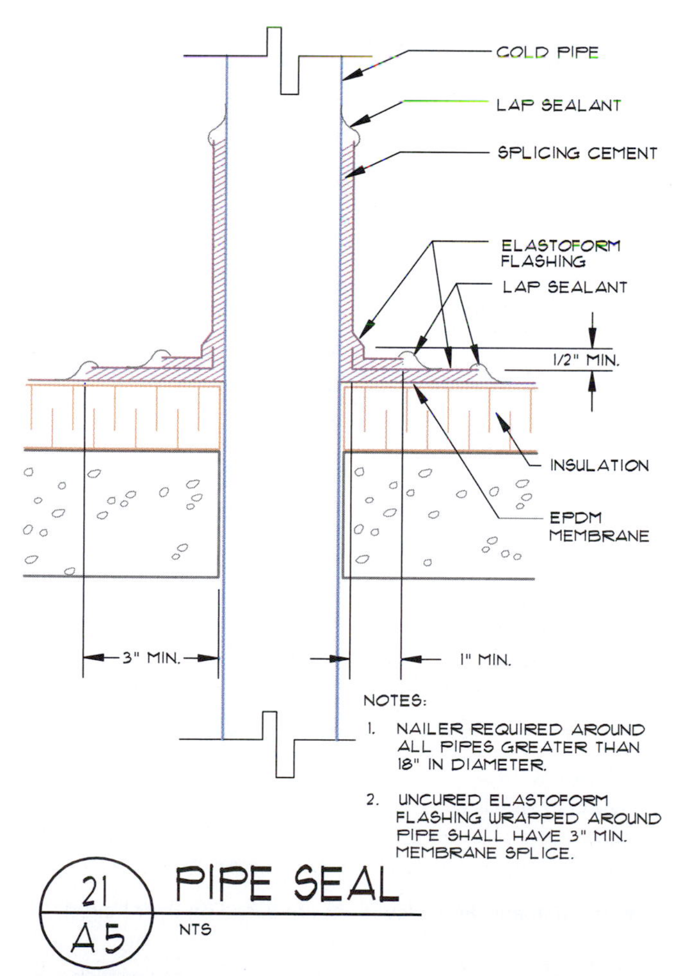

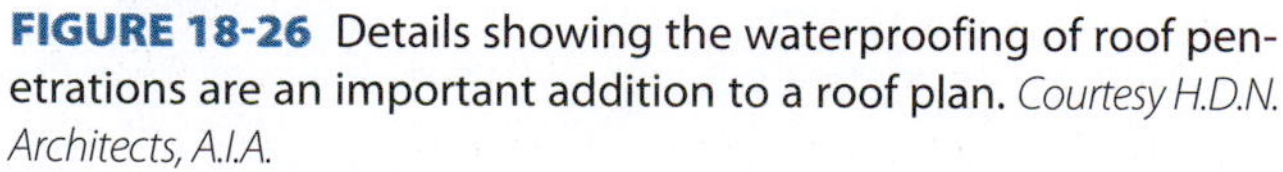

FIGURE 18-26 Details showing the waterproofing of roof penetrations are an important addition to a roof plan. *Courtesy H.D.N. Architects, A.I.A.*

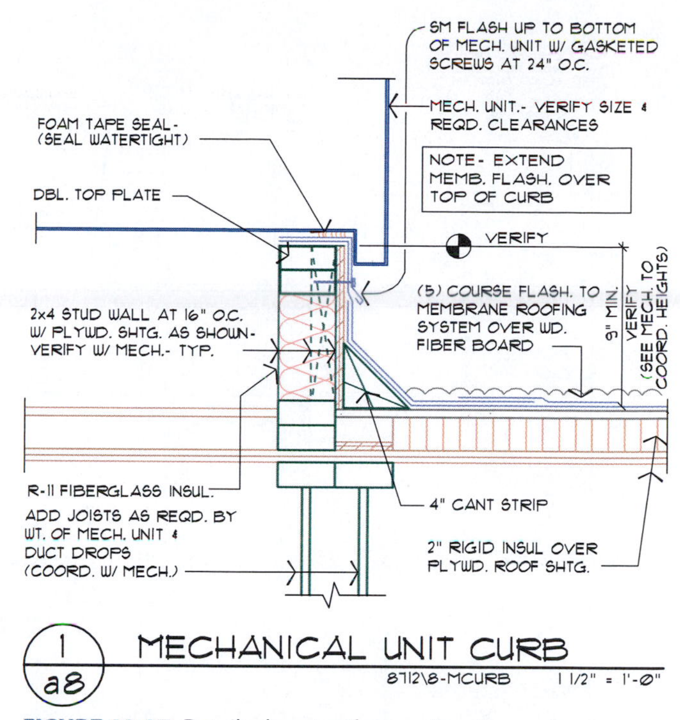

FIGURE 18-27 Details showing the construction and waterproofing of curbs for mechanical equipment must be referenced to the roof plan. *Courtesy Architects Barrentine, Bates & Lee, A.I.A.*

represented with symbols similar to those used on the floor and site plans. Elevations for the roof are typically given from either the finished floor level or from the height of the ridge. The base point for the elevations should be indicated on the roof plan in either a legend or in general notes. Figures 18-23 and 18-28 show examples of elevations that are based on the finished floor. When elevations are given from the ridge, the height of the finished roof at the ridge is usually referred to as elevation *00*. All other points are expressed as heights relative to the ridge. Points lower than the ridge will have negative numbers, and points above the ridge are represented by positive numbers. Heights are determined by multiplying the distance by the desired slope. For instance, with the minimum roof pitch, a ridge 30'–0" from a wall will require 7.5" (30' × 0.25") of fall to properly shed water. Figure 18-29 shows an example of elevations based on the ridge height.

Gutters

The roof system is designed to direct water from the ridge downward. For sloped roofs built in areas of the country that receive large amounts of rain, gutters are placed on the low edge of the roof to collect the water. The gutter has a slight slope to direct water to a downspout. The downspout transfers the water from the roof level to the grade level. Depending on the area of the country and the rain to be dispersed, gutters may be connected to a dry well or to a storm

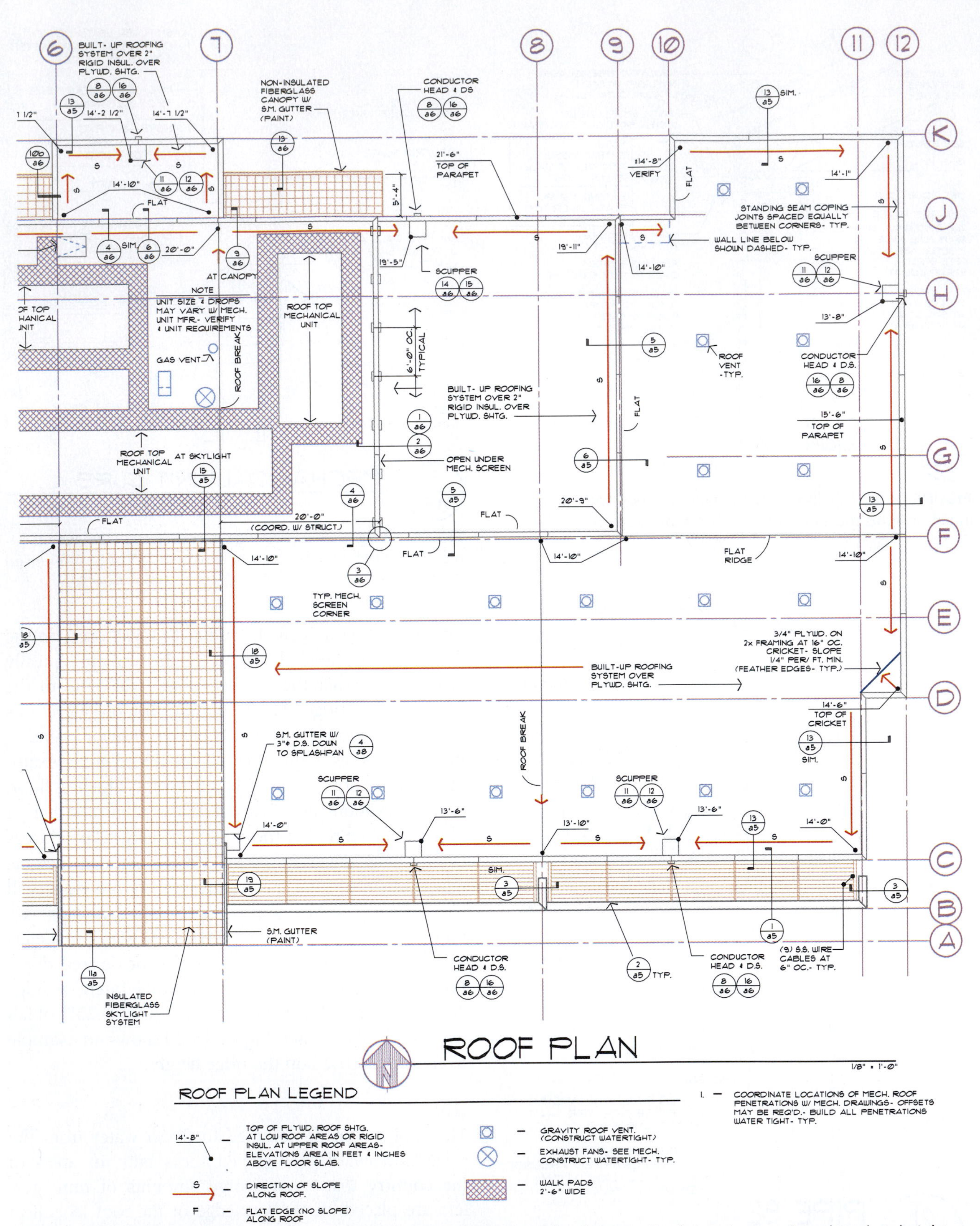

FIGURE 18-28 A completed roof plan for a shed roof with multiple slopes for drainage. Roof elevations are based on their height above the finish floor level. *Courtesy Architects Barrentine, Bates & Lee, A.I.A.*

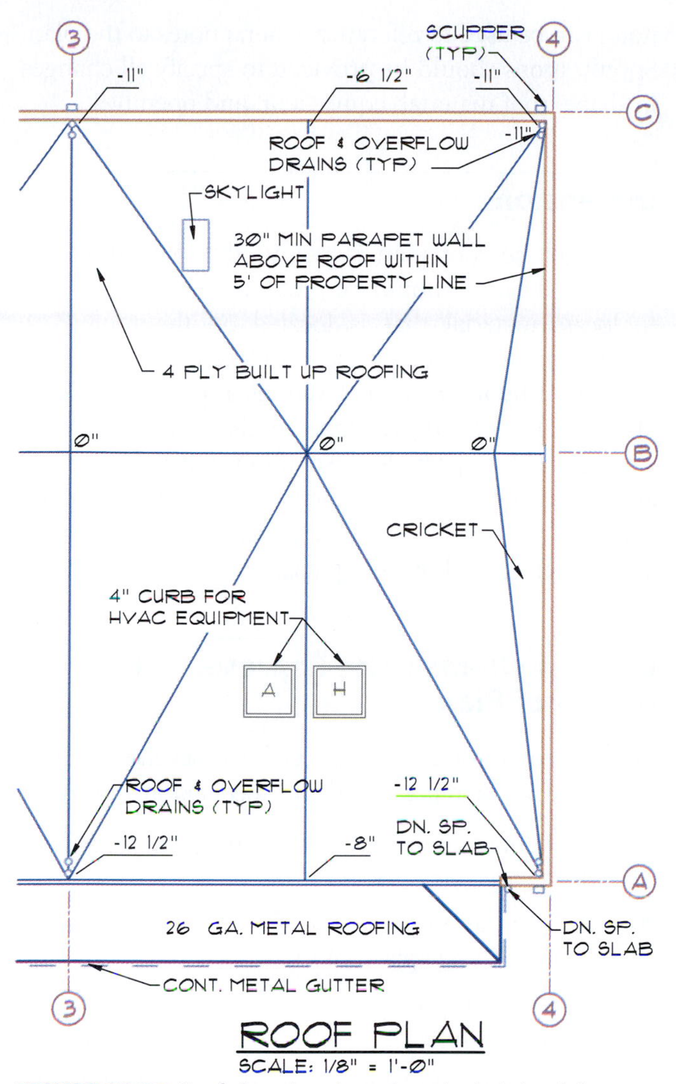

FIGURE 18-29 Roof elevations based on the height of the ridge (grid B) of the roof. Heights shown on the walls at grids A and C represent a height that is below the ridge line.

sewer. The location of the gutter is specified on the roof plan but not drawn. The location, size, and material of the downspouts should be indicated on the roof plan. Construction of the gutter and downspouts is shown in details similar to Figure 18-30.

Overflow Drains and Scuppers

Low-sloped roofs that are enclosed by a parapet wall trap water on the low side of the roof. *Drains* are placed in the roof to remove water from the roof wherever the water cannot run over the edge of the roof. A cricket is used to direct water to the drains in a method similar to that shown in Figure 18-5. Where drains are required, an overflow drain must also be provided to drain water from the roof in case the drain becomes blocked. The overflow drain must be the same size and be installed 2" (50 mm) above the roof drain. The drain directs water to a downspout, which in turn transports water to the

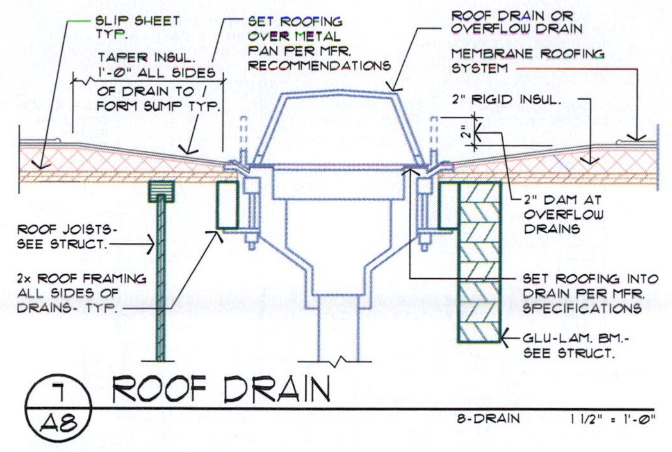

FIGURE 18-30 A detail showing the mounting of a roof drain.
Courtesy Architects Barrentine, Bates & Lee, A.I.A.

waste disposal system. The overflow drain is connected to a separate drain that often connects to a hole in the wall. Water flowing from this drain indicates that the drainage system requires maintenance.

An opening in the parapet wall is an alternative to providing drains and overflow drains. A funnel-like collector called a *scupper* is placed on the outside of the hole to funnel water from the roof to a downspout. A scupper can be represented on the roof plan as shown in Figure 18-7 and would be shown in details similar to those in Figure 18-31.

Diverters

For structures built in areas of the country that receive only small amounts of rain, water is often allowed to flow over the edge of the roof directly to the ground. To keep water from dripping onto walkways at entries, a metal strip called a *diverter* is placed on the roof. The diverter is placed so that it will direct water on the roof to an area where no one will be splashed. A bed of gravel is generally placed on the ground below the low point of the diverter to decrease the chance of erosion. A diverter can be represented on the roof plan as shown in Figure 18-32.

Annotation

As with other drawings, annotation must be placed on the roof plan to specify the material that has been drawn. Much of the material that will need to be specified has been presented throughout this chapter. Where possible, local notes can be used to identify each material. If space does not allow for the entire specification to be placed in or near the drawing, an abbreviated portion of the note should be placed by the object, and the complete note should be provided in a listing of general notes. To add clarity to the drawing, keyed

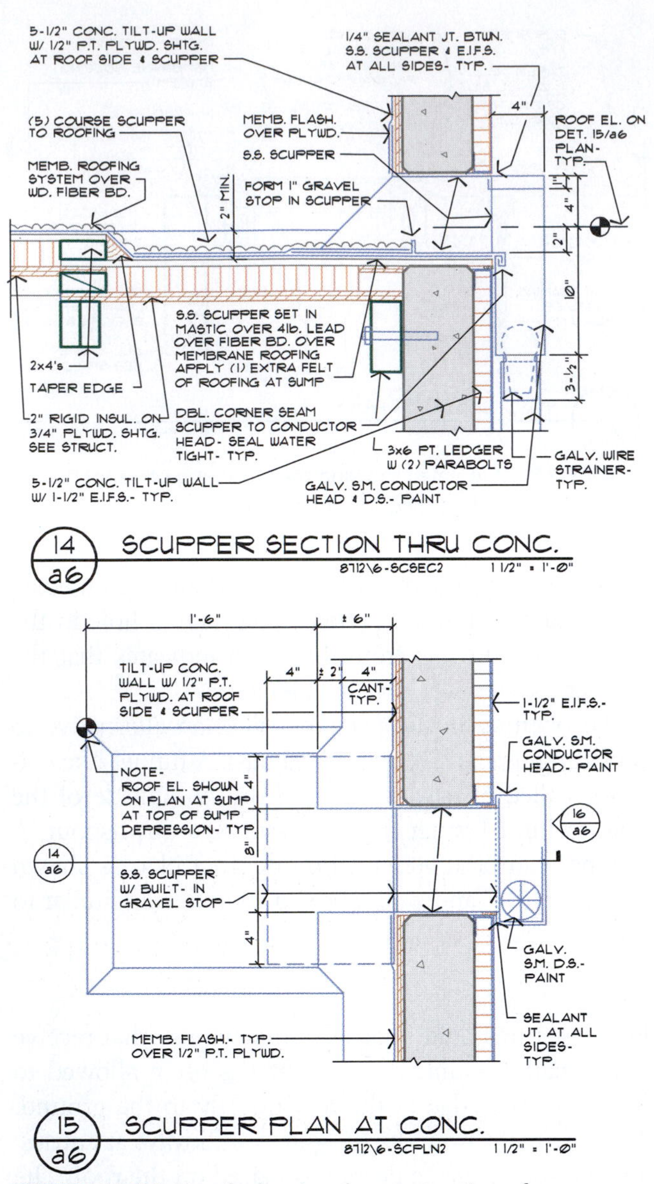

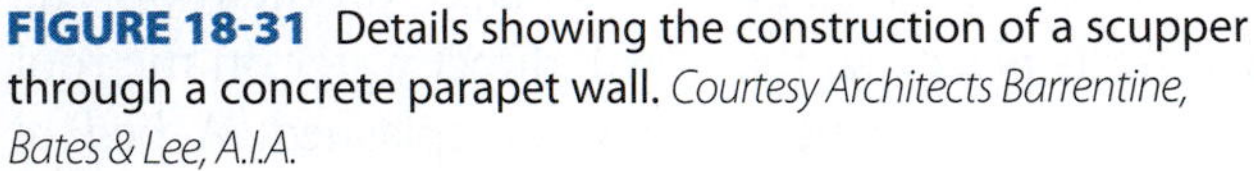

FIGURE 18-31 Details showing the construction of a scupper through a concrete parapet wall. *Courtesy Architects Barrentine, Bates & Lee, A.I.A.*

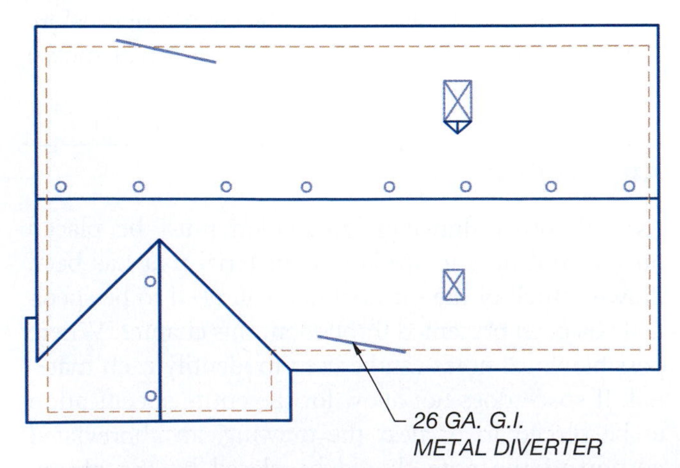

FIGURE 18-32 In areas with minute amounts of rainfall, a metal diverter can be installed on the roof to drain water away from doorways.

notes can be used to reference general notes to the plan. Specifications should be provided to specify all changes in shape, roof material, equipment, and openings.

Dimensions

Although the roof plan is not used to construct the roof, dimensions should be provided on the roof plan to locate material that is located on the roof. Base dimensions should be provided to locate each grid, just as with the floor plan and elevations. Equipment should also be dimensioned. On sloped roofs, overhang dimensions should be provided. Anything that is located on another plan is often not dimensioned on the roof plan. Figures 18-5, 18-7, 18-23, and 18-28 show examples of dimensions placed on roof plans.

Common Items to be Represented on a Roof Plan

When preparing to complete a roof plan, several common features must be represented and specified including:

- The extent and location of each change in roof shape and roof crickets.
- The extent and direction of fall to each roof drain, downspout, and splash block.
- Overflow drains and scuppers.
- Penthouse and required walking surfaces, changes in roof material, and roofing control and seismic joints.
- Roof-mounted features such as antennas, lightning arresters, and HVAC equipment. Coordinate the material on the roof plan with supplemental drawings such as mechanical and electrical drawings.
- Roof penetrations, skylights, and roof-mounted wall screens.
- Roof access and ladders.
- Reference to details for any roof material listed above.

COMPLETING A LOW-SLOPED ROOF PLAN

The convenience store shown in Chapter 16 will be used to show the process for drawing a low-sloped roof. The architect's sketch shown in Figure 18-33 will be used as a guide to complete the roof plan. Use layers based on the NCS found in Appendix C on the student CD to complete the drawing.

1. Using the sketch, draw the exterior walls.
2. Draw the outline of the mansard roof.

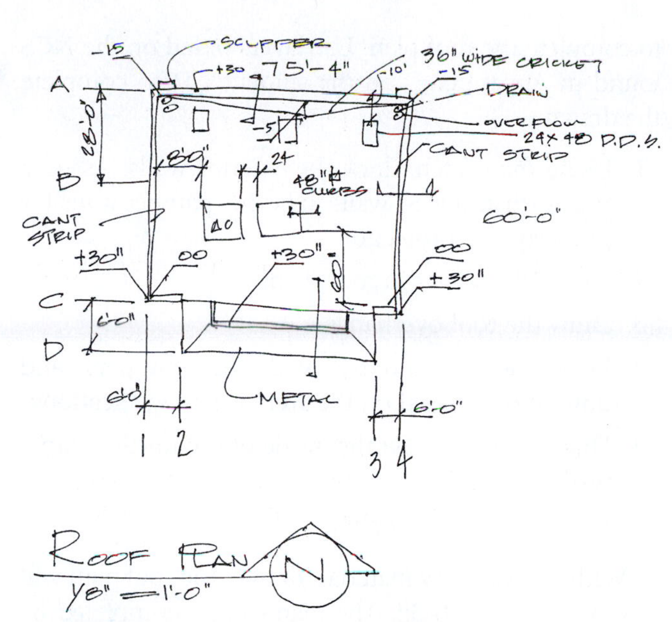

FIGURE 18-33 A sketch developed by the architect or project manager is used by the drafter to develop a roof plan.

3. Draw the drain, overflow drains, and scuppers based on the architect's specifications.
4. Draw the cricket on the inside of the north parapet wall.
5. Draw required 48" × 48" × 4" high curbs for HVAC equipment.
6. Draw required skylights.

With all necessary materials drawn, the roof plan will resemble Figure 18-34. The plan can be completed by using the following steps:

7. Draw all grid lines.
8. Provide overall dimensions.
9. Dimension the mansard.
10. Dimension all openings.
11. Dimension each skylight.

The roof plan is now dimensioned and will resemble Figure 18-35. The drawing can be completed by adding the required notations:

12. Place the title and scale below the drawing.
13. Specify all materials that can be grouped as general notes.
14. Draw the north arrow.
15. Provide identification text for each grid symbol.
16. Locate and specify all elevations.
17. Specify all changes in roof shape.
18. Specify all drainage devices.
19. Specify all roof-mounted equipment.
20. Specify all roof openings.
21. Draw detail markers for anticipated details.

The completed roof plan is shown in Figure 18-3.

A similar structure with a ridge at grid B and a butterfly roof is shown in Figure 18-36. This structure can be drawn using steps similar to those used to draw the single-sloped roof. Once the walls and mansard have been drawn, the ridge should be located. Because the architect has specified that three sets of drains are to be located on grids A and C, the roof can be divided into four quadrants to locate the butterfly.

The roof pitch for points 3 and 5 should be determined prior to determining the height at the drains.

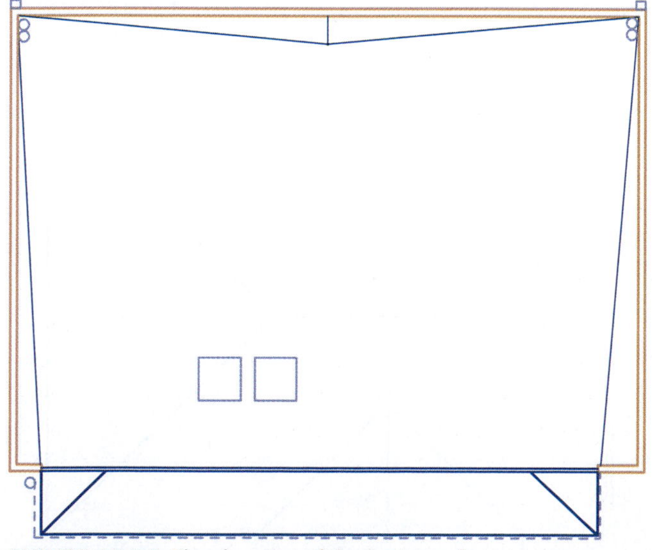

FIGURE 18-34 The drawing of the basic roof components based on the sketch in Figure 18-33.

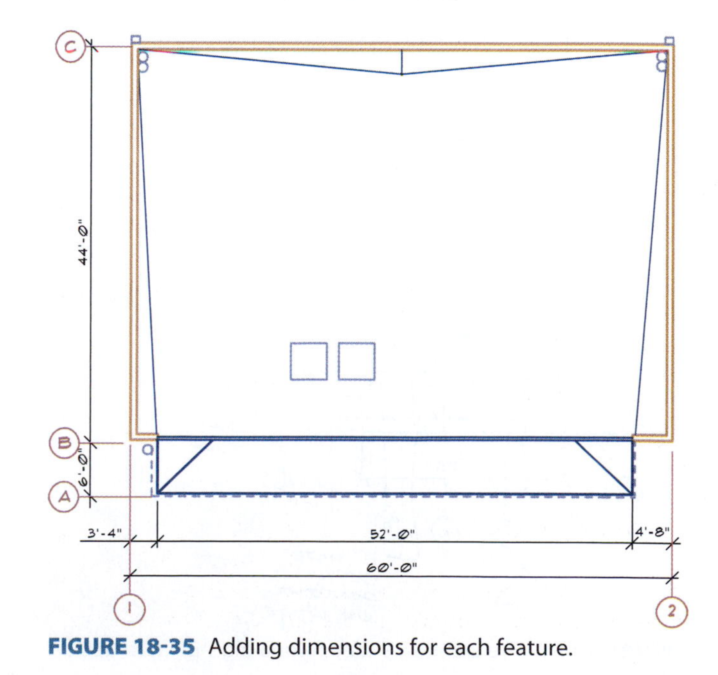

FIGURE 18-35 Adding dimensions for each feature.

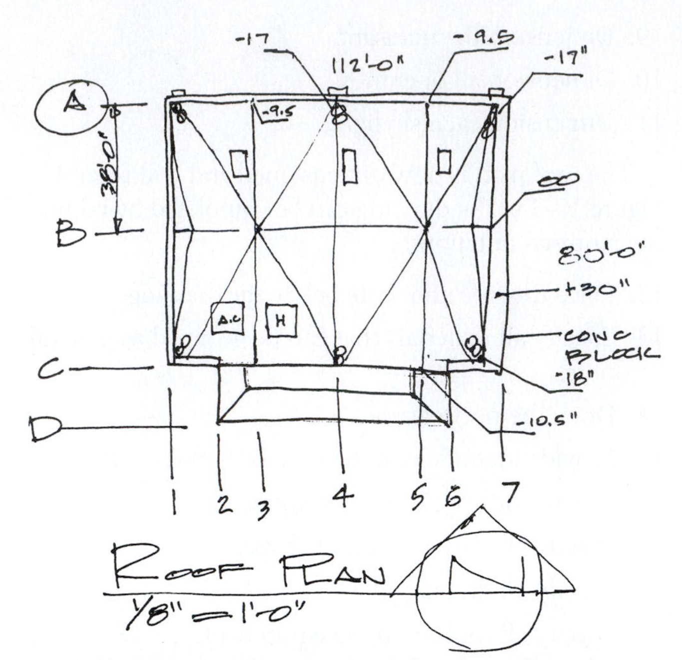

FIGURE 18-36 The architect's sketch for a butterfly roof.

Using a pitch of 1/4:12", the slope from B-3 to C-3 requires a 10" drop. The distance from C-3 to C-4 is 28 feet, which would require an additional drop of 7". Using the same materials that were used for the structure in Figure 18-4, the completed roof plan would resemble the structure in Figure 18-7.

COMPLETING A HIGH-SLOPED ROOF PLAN

A light-framed retail outlet will be used to show the process for drawing a high-sloped roof. The architect's sketch shown in Figure 18-37 will be used as a guide to complete the roof plan. Use layers based on the NCS found in Appendix C on the student CD to complete the drawing.

1. Using the sketch, draw the exterior walls. Assume the north and west walls to be 30" parapet walls for property line protection.
2. Draw all changes in roof shape.
3. Draw the roof overhangs.
4. Draw the drain, overflow drains, scuppers, and downspouts based on the architect's specifications.
5. Draw the cricket on the inside of the north parapet wall.
6. Draw required skylights.

With all necessary materials drawn, the roof plan will resemble Figure 18-38. The plan can be completed by using the following steps:

7. Draw all grid lines.
8. Provide overall dimensions.
9. Dimension all overhangs.
10. Dimension all openings.
11. Dimension the skylights.
12. Place the title and scale below the drawing.
13. Draw the north arrow.
14. Provide identification text for each grid symbol.
15. Locate and specify all elevations.
16. Specify all changes in roof shape.
17. Specify all drainage devices.
18. Specify any roof-mounted equipment.
19. Specify all roof openings.
20. Draw detail markers for anticipated details.

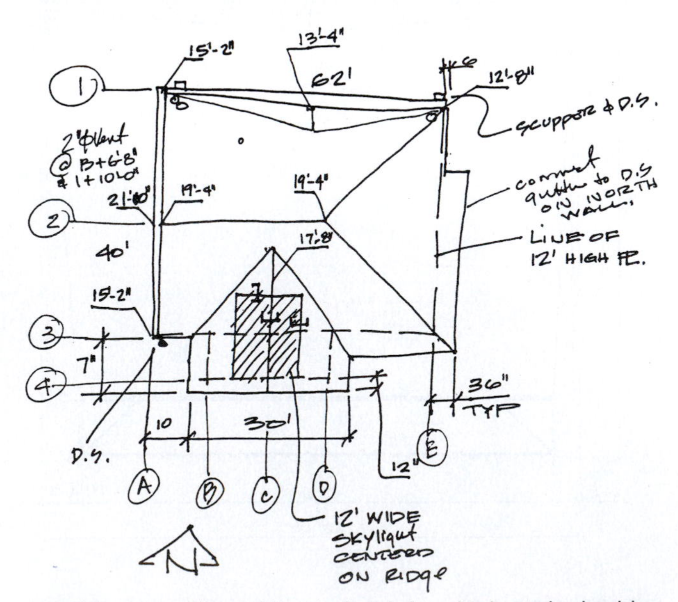

FIGURE 18-37 The architect's sketch for a high-pitched gable roof.

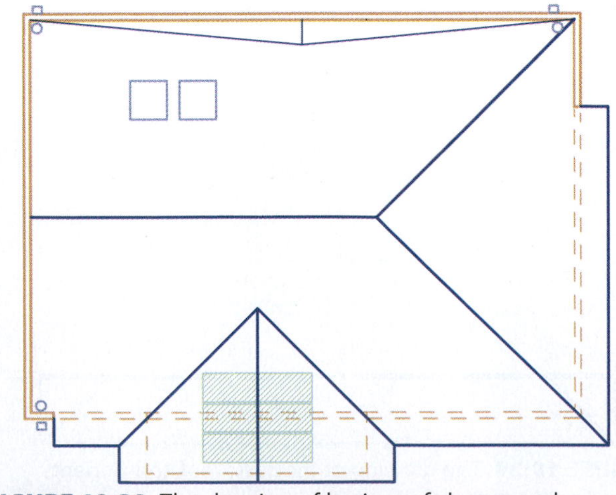

FIGURE 18-38 The drawing of basic roof shapes and components.

EASY BEING GREEN

As you think of energy efficient roof systems, you might quickly think of gardens of flowers or freshly mown lawns. This image is typically referred to as a green roof. The term *green roof* is generally used to describe a roof system consisting of some type of vegetation growing in soil, planted over a waterproof membrane, a root barrier, and drainage and irrigation systems. This type of roof system has become popular in many applications such as municipal buildings, including industrial facilities, offices, and other commercial property. In Europe, they are widely used for their storm water management and energy savings potential, as well as their aesthetic benefits. Although popular, this green roof system is far from the only type of sustainable roof system.

Two other categories of roofing are available to the design team to gain Leadership in Energy and Environmental Design (LEED) credits. Cool roofs and sustainable roofs can be used to make a structure more earth friendly. A cool roof is a roof that features highly reflective materials. It can be something as simple as finishing the roof by painting it with a coating of light-colored water sealant. In the same way that white clothing helps keep you cool in the summertime, white roofs reflect sunlight and heat. More high-tech cool roofs feature special light-colored, reflective membranes that reflect heat from the roof surface to keep the building cool.

Although not as flashy as a garden roof, asphalt, clay, and composition shingles can be just as green if produced and installed to meet certain criteria. To be considered sustainable, a roofing system must meet five key criteria in the areas of energy, environment, endurance, economics, and engineering.

- Energy—High-performance roofing materials are available that can be a powerful asset in reducing energy consumption and forming an energy efficient roof. When used with appropriate insulation on low-sloped or flat roofs, high-emissivity products can:
  - Reduce building energy consumption by up to 40%.
  - Improve insulation performance to reduce winter heat loss and summer heat gain.
  - Preserve the efficiency of rooftop air-conditioning and potentially reduce HVAC capacity requirements.
  - Decrease the effects of urban heat islands and related urban air pollution.
- Environment—A roof system is generally considered to be friendly to the environment if it is designed, constructed, maintained, rehabilitated, and demolished with an emphasis throughout its life cycle on using natural resources efficiently and preserving the global environment. This would include roofing materials that are produced, applied, and reused using methods that comply with LEED guidelines.
- Endurance—The amount of time a product can be used before being replaced is a key feature in its LEED certification. Endurance for a roofing component is typically measured in terms of reliability, water absorption, wind and fire resistance, maintenance, and repair. No matter how green a roof is, it still has to protect the building for years in all types of weather.
- Economics—While some types of roofing may have lower initial costs, the true costs of a roofing system are measured over its total life cycle. These include maintenance and repair costs, energy savings, and tear-off and disposal costs. The total cost of a product through its entire life cycle must be considered rather than just the initial cost during building construction to gain LEED credits. This is a very important criterion for building owners: High-performance roofing systems must be economical if they are to become viable, real-world options.
- Engineering—Smart, coordinated engineering is not only the essential enabler for the other E's of high-performance roofing materials, it is the key to whole-building design. The roofing products must be considered in relation to all other products that are used to form the exterior shell of the structure.

As you work with the design team to meet the goals of the five E's, key elements shown on a roof plan include finishing roof materials, skylights, drainage systems, and mechanical equipment. Each product earns LEED credits primarily from the Materials and Resources division. Information to be shown on the roof plan includes credits primarily from CSI Section 07—Thermal and moisture protection. Specific areas to research include:

- 07 31 13 – Asphalt shingles
- 07 31 16 – Metal shingles
- 07 31 19 – Mineral fiber cement shingles
- 07 31 26 – Slate shingles
- 07 31 29 – Wood shingles and shakes
- 07 31 33 – Plastic and rubber shingles
- 07 32 13 – Clay roof tiles
- 07 34 00 – Building integrated photovoltaic roofing
- 07 41 00 – Roof panels

The completed roof plan is shown in Figure 18-39.

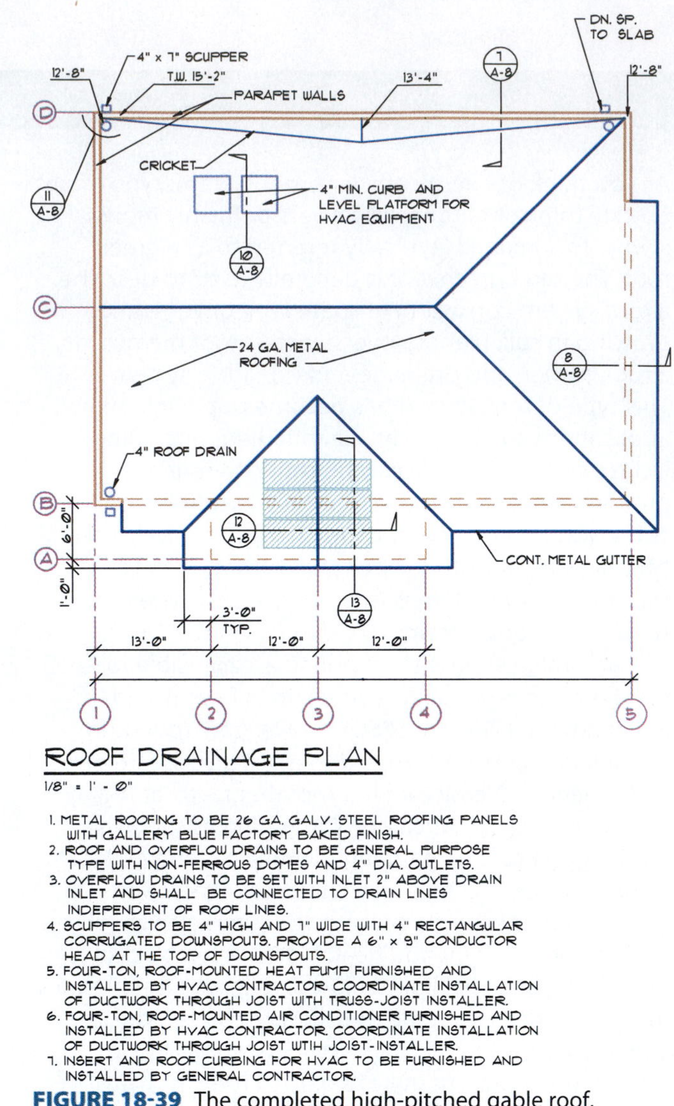
DN. SP. TO SLAB
4" x 7" SCUPPER
T.W. 15'-2"
12'-8"
13'-4"
12'-8"
PARAPET WALLS
CRICKET
4" MIN. CURB AND LEVEL PLATFORM FOR HVAC EQUIPMENT
24 GA. METAL ROOFING
4" ROOF DRAIN
CONT. METAL GUTTER
6'-0"
1'-0"
3'-0" TYP.
13'-0"
12'-0"
12'-0"
ROOF DRAINAGE PLAN
1/8" = 1' - 0"
1. METAL ROOFING TO BE 26 GA. GALV. STEEL ROOFING PANELS WITH GALLERY BLUE FACTORY BAKED FINISH.
2. ROOF AND OVERFLOW DRAINS TO BE GENERAL PURPOSE TYPE WITH NON-FERROUS DOMES AND 4" DIA. OUTLETS.
3. OVERFLOW DRAINS TO BE SET WITH INLET 2" ABOVE DRAIN INLET AND SHALL BE CONNECTED TO DRAIN LINES INDEPENDENT OF ROOF LINES.
4. SCUPPERS TO BE 4" HIGH AND 7" WIDE WITH 4" RECTANGULAR CORRUGATED DOWNSPOUTS. PROVIDE A 6" x 9" CONDUCTOR HEAD AT THE TOP OF DOWNSPOUTS.
5. FOUR-TON, ROOF-MOUNTED HEAT PUMP FURNISHED AND INSTALLED BY HVAC CONTRACTOR. COORDINATE INSTALLATION OF DUCTWORK THROUGH JOIST WITH TRUSS-JOIST INSTALLER.
6. FOUR-TON, ROOF-MOUNTED AIR CONDITIONER FURNISHED AND INSTALLED BY HVAC CONTRACTOR. COORDINATE INSTALLATION OF DUCTWORK THROUGH JOIST WTIH JOIST-INSTALLER.
7. INSERT AND ROOF CURBING FOR HVAC TO BE FURNISHED AND INSTALLED BY GENERAL CONTRACTOR.

FIGURE 18-39 The completed high-pitched gable roof.

ADDITIONAL READING

Links to roof-related information are too numerous to list here. Rather than listing individual Web sites, use your favorite search engine to research products from the CSI categories listed above as well as the following headings: roofing materials, skylights, green roofs, green roofing, green roofing construction, green roofing design, and green roofing technology. A few helpful sites as you develop your own database of links include:

ADDRESS	COMPANY/ ORGANIZATION
www.asla.org/land/050205/greenroofcentral.html	American Society of Landscape Architects
www.epa.gov/hiri/strategies/greenroofs.html	Environmental Protection Agency
www.greengridroofs.com	Green Grid Roofs
www.greenroofs.com	GreenRoofs.Com (Green Roof Council Newsletter)
www.greenroofs.org	Green Roofs for Healthy Cities
www.greenroof.org	ZinCo

KEY TERMS

Cant strip
Cricket
Dutch hip
Gable roof
Gambrel roof
High-sloped roofs
Hip roof
Low-sloped roofs
Mansard roof
Pitch
Square

CHAPTER 18 TEST Roof Plan Components and Drawings

QUESTIONS

Answer the following questions with short complete statements. Type the chapter title, question number, and a short complete statement for each question using a word processor. Some answers may require the use of vendor catalogs or seeking out local suppliers.

Question 18-1 Why are flat roofs used for a majority of commercial projects?

Question 18-2 List two reasons for using parapet walls.

Question 18-3 At what angle does the building code consider a roof a *steep roof*?

Question 18-4 What is the difference between a cricket and a cant strip?

Question 18-5 What is typically used to bind together layers of a built-up roof?

Question 18-6 List the weight per square foot for the two common types of composition shingles.

Question 18-7 Describe how the ridge of a sawtooth roof is located.

Question 18-8 Describe the angle used to represent a valley between two intersecting portions of a structure when represented on a roof plan.

Question 18-9 How should skylights be specified on a roof plan?

Question 18-10 List two methods of determining elevations on a roof plan.

DRAWING PROBLEMS

Use the reference material from preceding chapters, local codes, and vendor catalogs to complete one of the following projects.

Unless your instructor provides other instructions, draw the roof plan that corresponds to the floor plan that was drawn in Chapter 16. Skeletons of the roof plan for problems 18-1 through 18-4 can be accessed from the student CD in the DRAWING PROBLEMS/CHAPTER 18 folder.

Use these drawings as a base to complete the assignment. Use appropriate symbols, linetypes, dimensioning methods, and notations to complete the drawing. Determine any unspecified sizes based on material or practical requirements. Unless specified, select a scale appropriate to plotting on D-size material and determine LTSCALE and DIMVARS.

Information is provided on drawings in Chapters 15 through 24 relating to each project that might be needed to make decisions regarding the roof plan. You will act as the project manager and will be required to make decisions about how to complete the project. Any information not provided must be researched and determined by you unless your instructor (the project architect and engineer) provides other instructions. When conflicting information is found, information from preceding chapters should take precedence.

Minimum Standards

Your completed drawings should meet or exceed the following minimum standards unless noted:

- Create the necessary layers to keep the roof information stacked above, but separated from, the floor plan. Freeze all material not required by the roof plan.
- Note that all downspouts are to be connected to a storm sewer connection separate from site drainage.
- Roof and overflow drains are to be general-purpose type with nonferrous domes and 4" drains. Overflow drains to be set 2" above drain inlet and shall be connected to drain lines independent of roof drains.
- Scuppers to be 4" high and 7" wide with 4" rectangular corrugated downspouts. Provide a 6" × 9" conductor head at the top of each downspout.
- Flat roofs to have 400 pounds of pea gravel over a 4-ply built-up roof.
- Locate the following elevations for problems 18-2, 18-3, and 18-4:
  - Each corner of the parapet wall.
  - Each end of the ridge.
  - Each corner of the roof.
  - The roof at each drain location.
- Inclined roofs to have 26-ga. metal roofing.

Problem 18-1 Draw a roof plan for the four-unit complex using the following parameters:

- The roof is to be built at a pitch of 12:12 and will be covered with 300# composition shingles.
- For the two interior units, provide a 4' minimum covered entry deck.

- The gable end walls are to have a 12" overhang.
- Exterior units will have a 24" overhang. All units will have a 6" overhang as it projects over lower units.
- Form a 24" high maximum cricket over the roof between units to divert water to scuppers and downspouts.
- Provide five scuppers and downspouts on each gable end wall. Provide a continuous gutter on the outer walls of each outer unit and connect to a downspout.
- Provide a 12:12 roof over each bay with a continuous gutter connected to a downspout located where it will not be seen.

Problem 18-2 Draw a roof plan using the following parameters:

- Provide a double-shed roof with the ridge formed by a beam running east/west centered over the 4" × 4" steel column.
- Establish the ridge 13'–0" above the finished floor.
- Provide a slope of 1/4:12 with all elevations listed from the finish floor.
- Place three scuppers and downspouts on the front and rear and establish cant strips to provide drainage to each roof drain.
- Provide continuous gutters on the mansard surrounding the main roof.
    - Project the mansard 6' from the edge of the block wall.
- Provide a 4" curb for roof-mounted heating and air-conditioning equipment.
    - Curb to be installed by general contractor, based on sizes supplied by the HVAC supplier.
- Show (4)-22" × 48" double-domed skylights equally spaced over the parts and inventory areas and one centered over the break area.

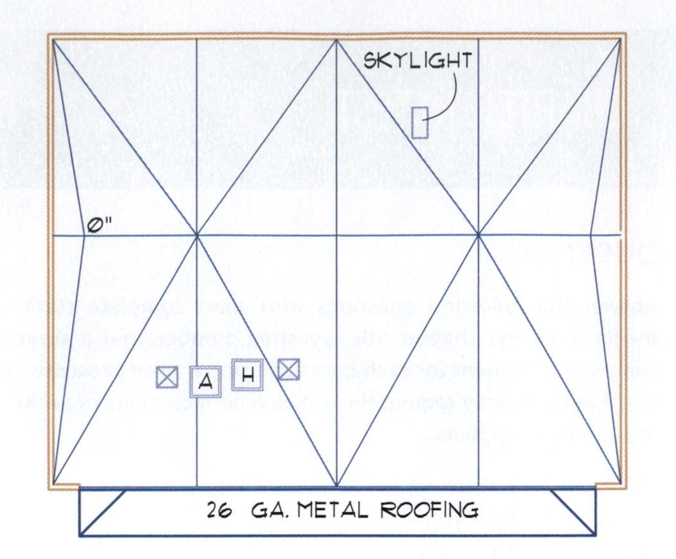

Problem 18-3 Design a roof plan for the structure started in problem 16-3. Coordinate this plan with the framing plan and adjust all heights to provide the minimum ceiling heights specified earlier. Use the following parameters to complete the drawings:

- Provide a ridge along the beam placed over the west row of columns (grid C) from 1 to 7.
    - Use a double-shed system over the warehouse and provide a slope with a minimum of 1/4:12 to grids A and F.
- Provide four drains on each side of the warehouse.
- Use a double-shed roof over the office area sloping from a ridge beam placed at grid C.5 between grids 7 and 8.
    - Establish the warehouse ridge to be 30" below the parapet wall but maintain minimum ceiling heights specified with the floor plan.
- Place a drain at grid B dependent on glazing placement, and a second drain at grid E7.5.
- Provide (4)-22" × 48" skylights per each grid over the mezzanine.

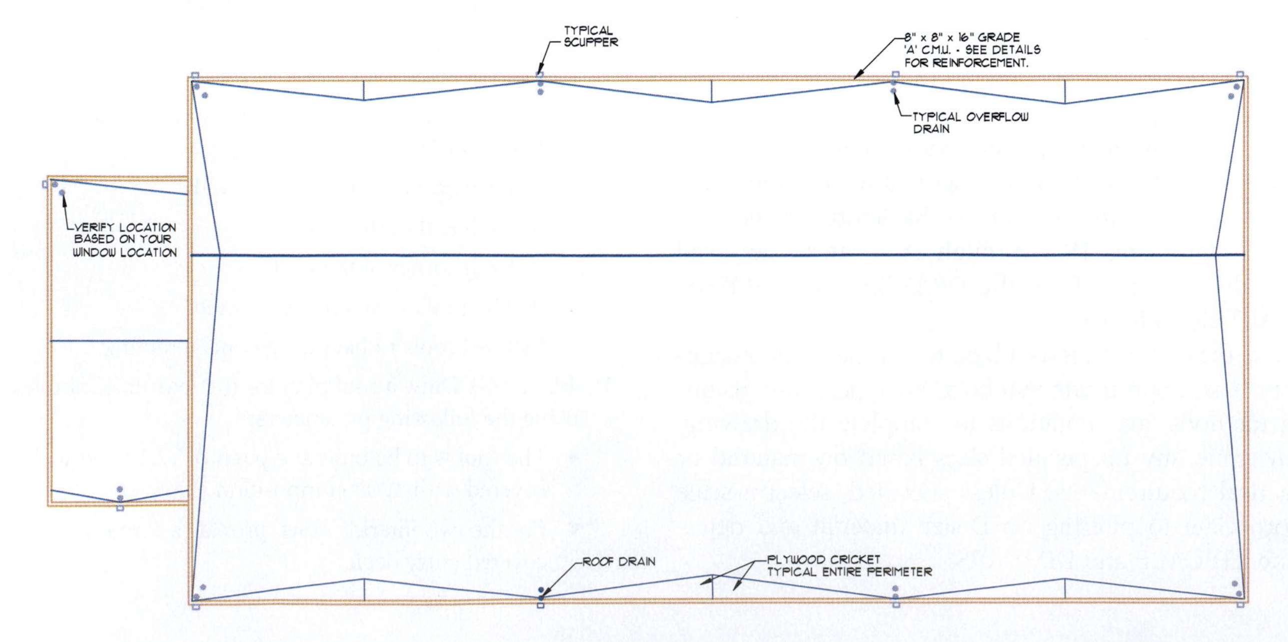

Problem 18-4 Serve as the project manager and design a butterfly roof plan for this structure. Coordinate this plan with the framing plan and adjust all heights to provide the minimum ceiling heights specified earlier. Use the following parameters to complete the drawings:

- Provide a ridge over the warehouse at grid 3 from C to K.
  - Provide five drains on each side of the warehouse. (Divide the warehouse roof between grids 1C-K and 5C-K into four equal parts. Place a drain at each of the three points that were just located as well as in each corner.)
    - Locate the midpoint between each drain along grids 1 and 5. These points will establish the high point of the ledger. Provide a minimum slope of 1/4:12 from the ridge to ledger high points on grids 1 and 5.
    - Provide an additional minimum slope of 1/4:12 from the ledger high points to each of the drains along grids 1 and 5. This will place the drains at a low point and cause water to flow from the ridge to the drain.
- Provide a ridge over the office area at grid 3 from grids A to C.
  - Provide a minimum slope of 1/4:12 from the ridge to grids 2 and 4.
  - Provide one drain at grid 2 placed dependent on glazing placement, and one drain at grid B-4.
  - Provide a minimum slope of 1/4:12 along the ledger to the drains. Use a layout similar to the one over the warehouse.
- Establish each ridge so that it is 30" below the parapet wall. Maintain the minimum ceiling heights specified with the floor plan.
- Label the following elevations:
  - Each corner of the parapet wall.
  - Each end of the ridge.
  - Each corner of the roof.
  - The roof at each high ledger point.
  - The roof at each drain location.
- Provide (4)-22" × 48" skylights per each grid over the mezzanine.

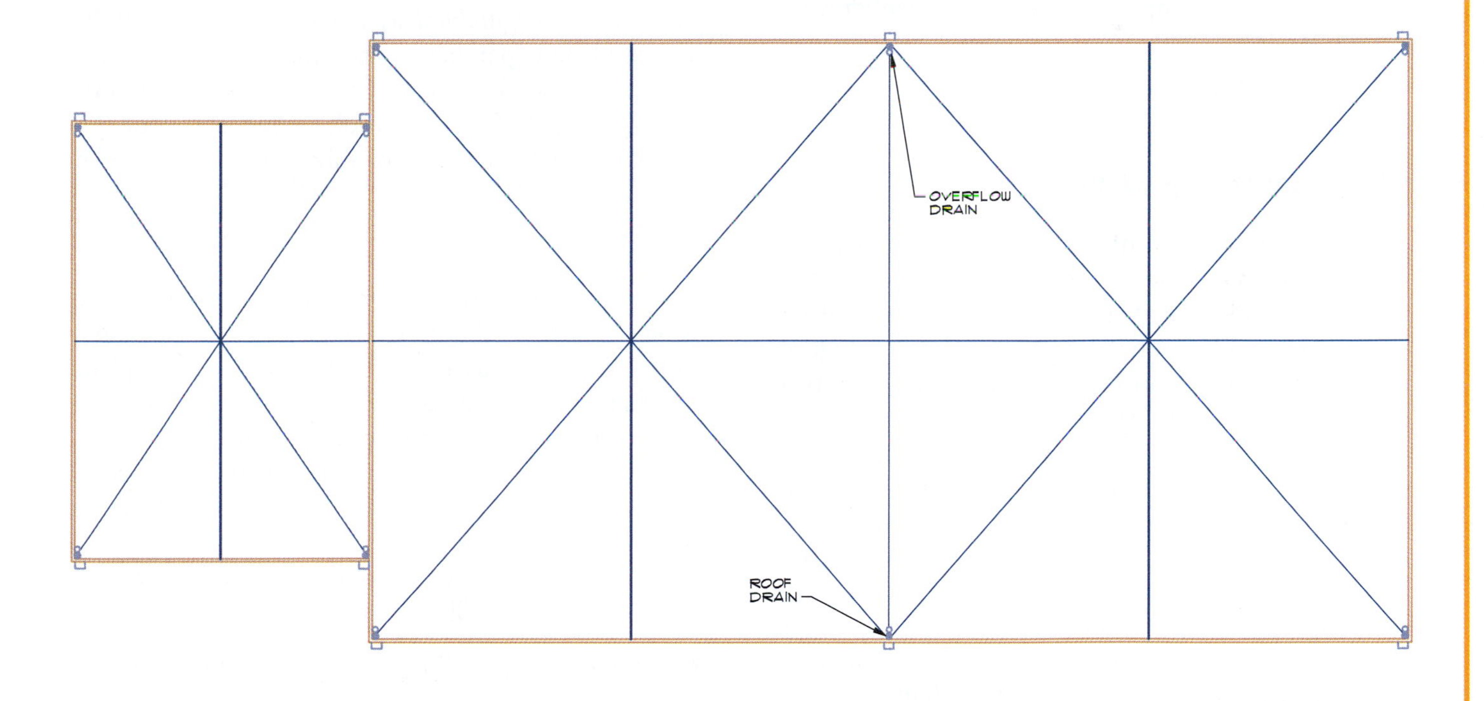

Drawing Sections

19

Building sections are drawn to show the vertical relationships of materials represented on each of the plan views and the exterior elevations. Exterior elevations show the material that will be used to shield the structure from the weather. Sections show the materials used to construct the walls, floors, ceilings, and roof components, and their vertical relationships to each other. This chapter will explore:

- Section origination.
- Types of sections.
- Section development.
- The major drawing considerations when drawing sections.
- Steps for completing a section.

SECTION ORIGINATION

Building sections are the result of passing a viewing plane through a structure to reveal the construction methods being used. Material that is in front of the viewing plane cannot be seen. Material that is behind the viewing plane is projected to the plane and reproduced in the section. Another method of visualizing a section is to think of the viewing plane as a giant saw that slices vertically through a structure and divides it into two portions. One portion is removed to allow viewing of the portion that remains. Using the saw analogy, the viewing plane is referred to as a *cutting plane*. The location of the cutting plane is shown on the floor plan using symbols similar to those shown in Figure 19-1. Notice that with each symbol, an arrow is included to indicate which portion of the structure is being viewed. Figure 19-2 shows an example of a floor plan with section markers. Figure 19-3 shows the section that is specified by the section marker 1/A12. Notice that material the cutting plane passes through on the floor plan is represented on the section. Material in the background such as doors, windows, and cabinets is also usually shown.

TYPES OF SECTIONS

Full sections, partial sections, and details may be used to represent a structure. Each has a specific use in representing how walls are constructed and how they intersect with floor and roof systems.

Full or Building Sections

Full sections are the views that result from passing the cutting plane through the entire structure. Full sections are meant to give an overall view of a specific area of the structure and may be either longitudinal or transverse. A *longitudinal section* is produced when the cutting plane is parallel to the long axis of the structure. It is generally perpendicular to most structural materials used to frame the roof, ceiling, and floor systems. Because the framing members are perpendicular to the cutting plane, they are seen as if they had been cut. Figure 19-4 shows an example of a longitudinal section.

A **transverse section** is produced when the cutting plane is parallel to the short axis of the structure, and it is often referred to as a cross section. The cutting plane for a transverse section is usually parallel to the materials used to frame the roof, ceiling, and floor systems and generally shows the shape of the structure better. Figure 19-5 shows a transverse section for a structure formed from precast concrete.

No matter the type of section, a building section should include the following components:

- Columns and grid lines to match the lines used on the floor plan.
- Other building section references that intersect the building section.
- Room names and numbers of the areas cut by the cutting plane.
- Floor-to-floor dimensions. (The National CAD Standards [NCS] recommends against the use of elevations.)
- Finish elevations.

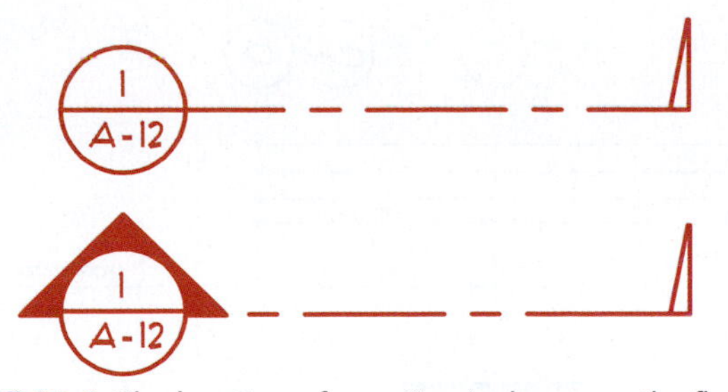

FIGURE 19-1 The location of a section is shown on the floor and framing plans by use of a cutting plane. The upper letter represents a specific section. The bottom number represents the page where the section can be found.

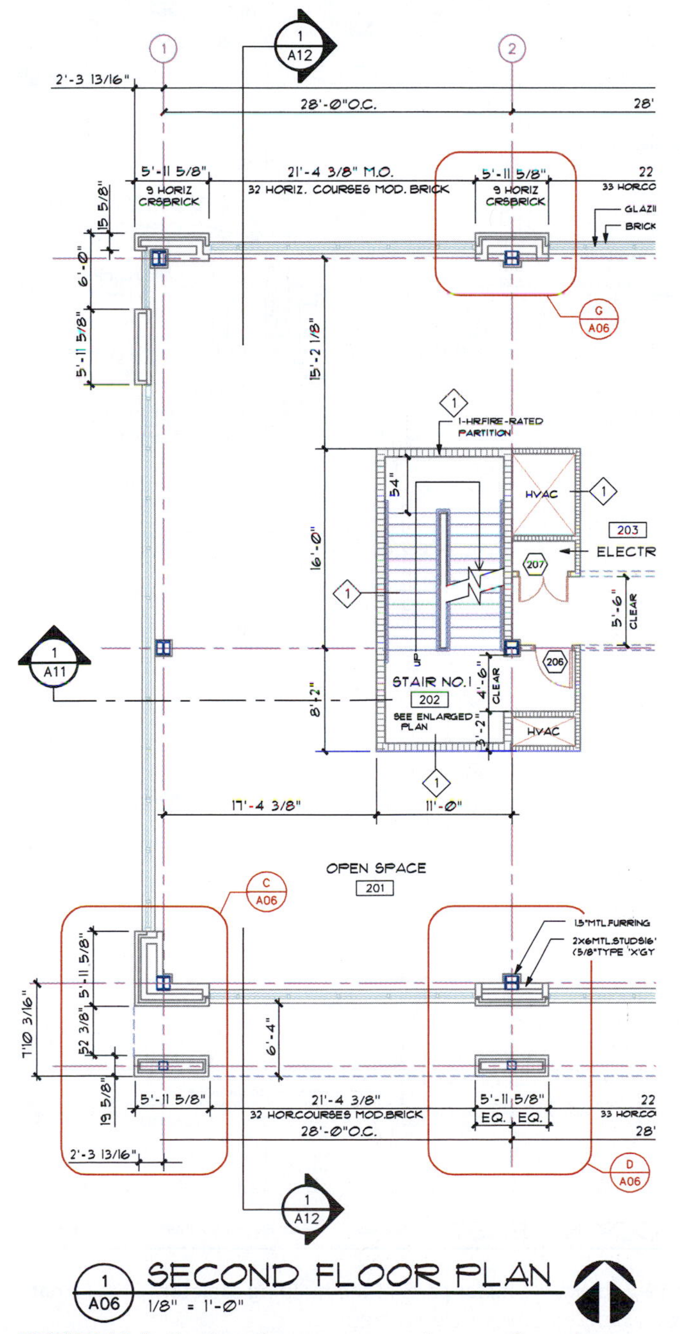

FIGURE 19-2 Cutting plane representation on a floor plan.
Courtesy Peck, Smiley, Ettlin Architects.

- Ceilings and partitions that are cut by the cutting plane.
- Major materials, symbols, and lists of abbreviations.

Partial Sections

A **partial section** is a section that does not go completely through the structure and can be used to show construction materials that are not seen in other sections. Figure 19-6 shows an example of a partial section that is referenced to the section shown in Figure 19-4. Partial sections are used to show only a specific area of the structure, whereas other sections are used to define the balance of the structure.

A partial section can also be used to show construction materials of a specific wall. When a section is used to show one specific wall, it is also referred to as a *wall section*. Wall sections can be referenced to the floor plan or to other sections. Figure 19-7 shows a wall section. Wall sections are drawn at a larger scale than a full section and are drawn to provide information on one specific type of wall.

Detailed or Wall Sections

Full and partial sections are often drawn at a scale that cannot adequately display the needed detail of all materials. *Detailed sections* are drawn to provide clarity for a small area, such as the intersection of the floor system to a wall. Sections, wall sections, and details can be thought of as different stages of the ZOOM command in AutoCAD. With each zoom, a smaller area is seen, but each material in the display increases in size. Details provide more specific information about a smaller area than seen in other types of sections. Figure 19-8 shows an example of a detail that is referenced to the section shown in Figure 19-6.

Wall sections should include the following components by both drawing representation and written specifications:

- Exterior wall types at a scale that allows the full height of the wall to be represented without the use of break lines, whenever possible.
- Exterior and interior finishes and materials.
- Finish grades.
- Floor levels and floor-to-floor dimensions.
- Profiles of built-in equipment.

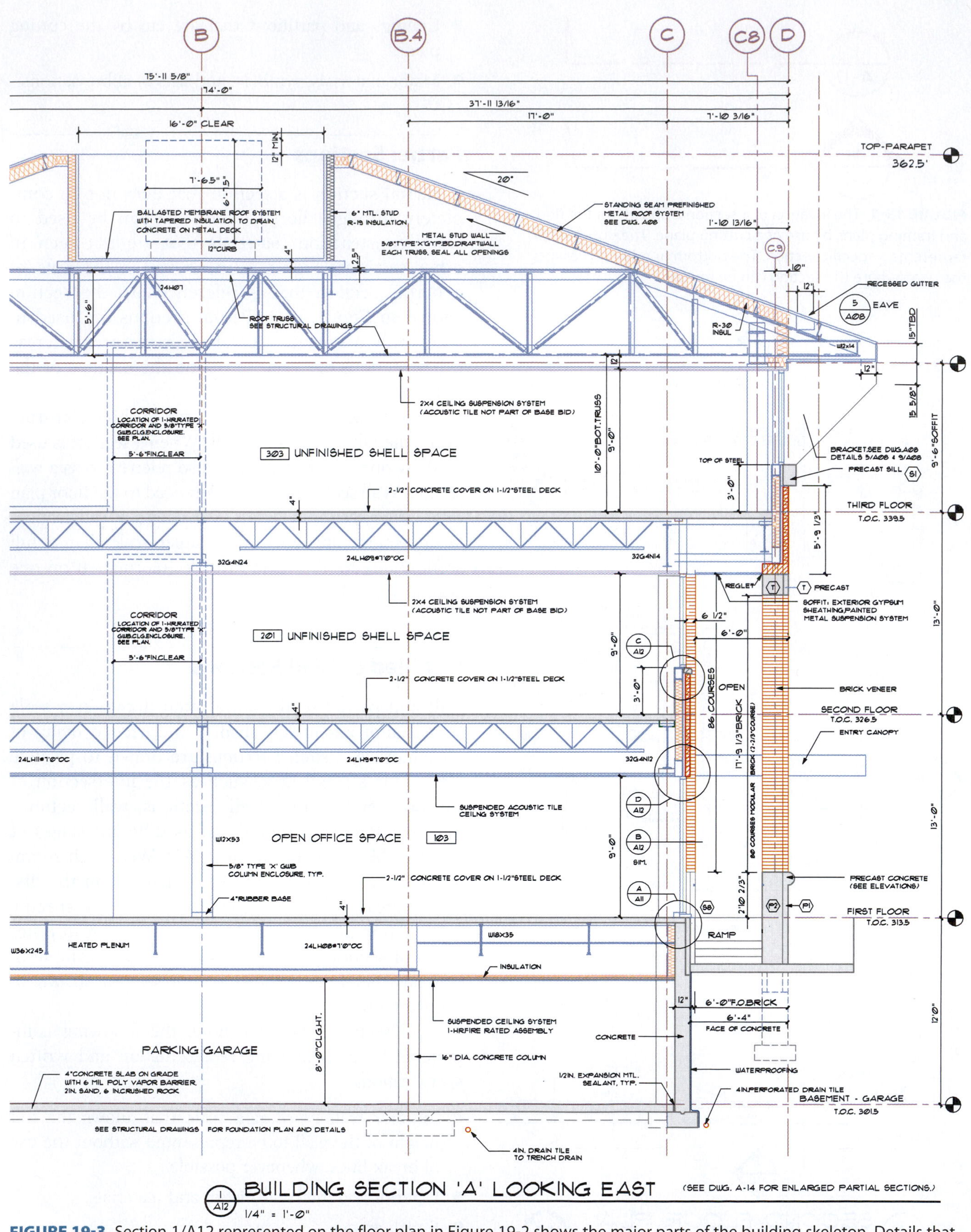

FIGURE 19-3 Section 1/A12 represented on the floor plan in Figure 19-2 shows the major parts of the building skeleton. Details that explain construction are referenced to the section and the floor plan. *Courtesy Peck, Smiley, Ettlin Architects.*

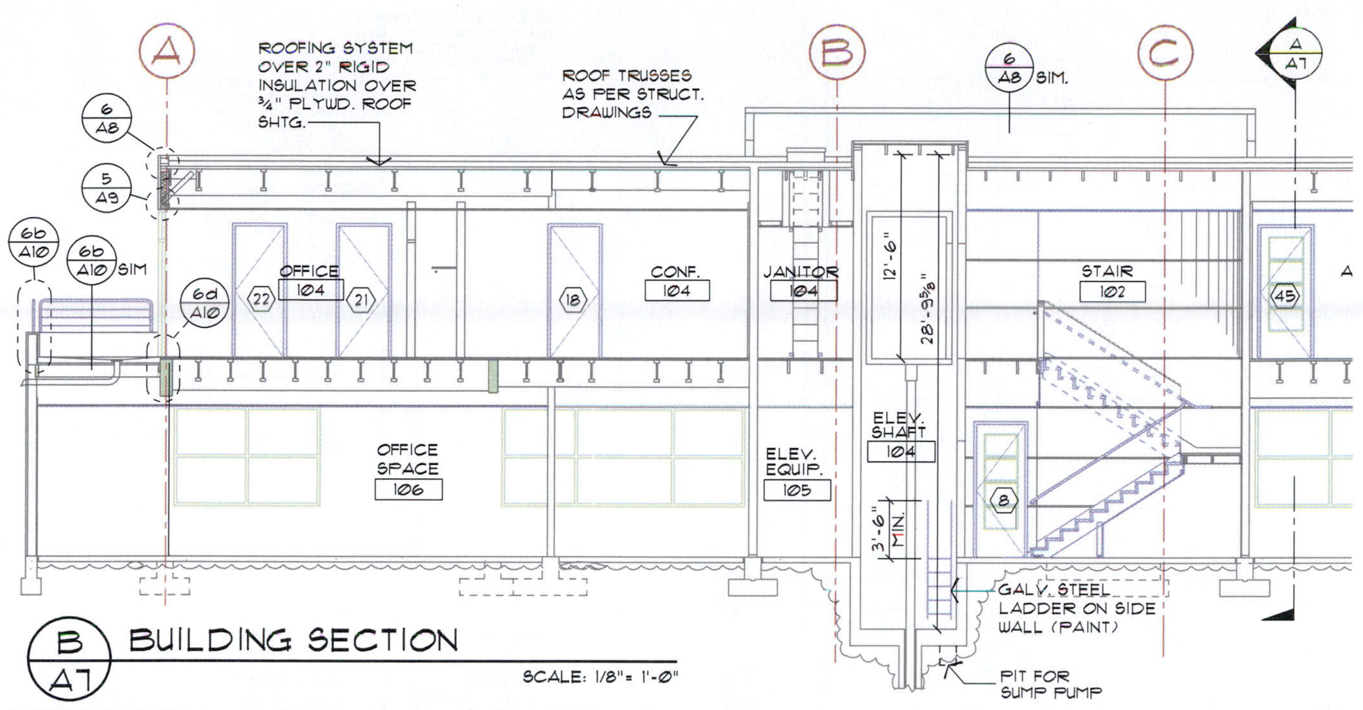

FIGURE 19-4 This longitudinal section, specified on the floor plan in Figure 16-3, was produced by passing the cutting plane parallel to the long axis of the structure. *Courtesy Architects Barrentine, Bates & Lee, A.I.A.*

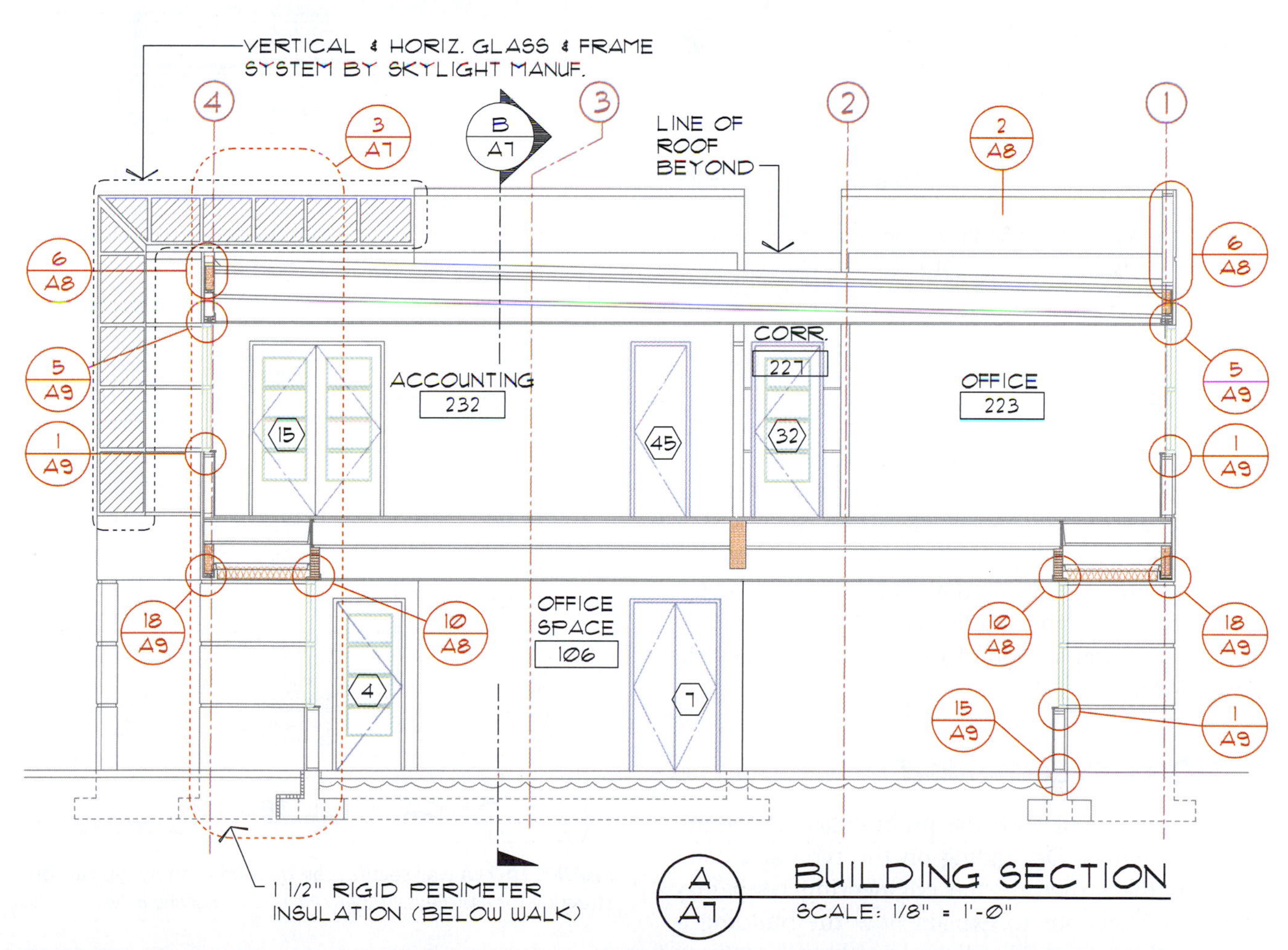

FIGURE 19-5 This transverse section, specified on the floor plan in Figures 16-3 and 16-11, was produced by passing the cutting plane parallel to the short axis of the structure. *Courtesy Architects Barrentine, Bates & Lee, A.I.A.*

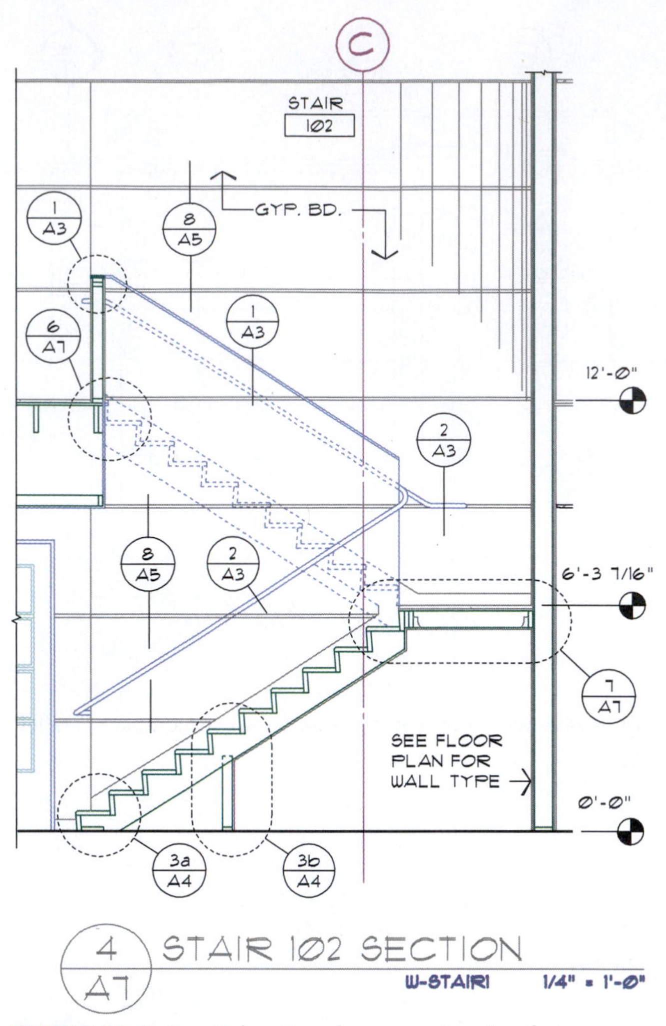

FIGURE 19-6 A partial section shows construction features of one specific area. This section is an enlargement of the stair shown in Figure 19-4. *Courtesy Architects Barrentine, Bates & Lee, A.I.A.*

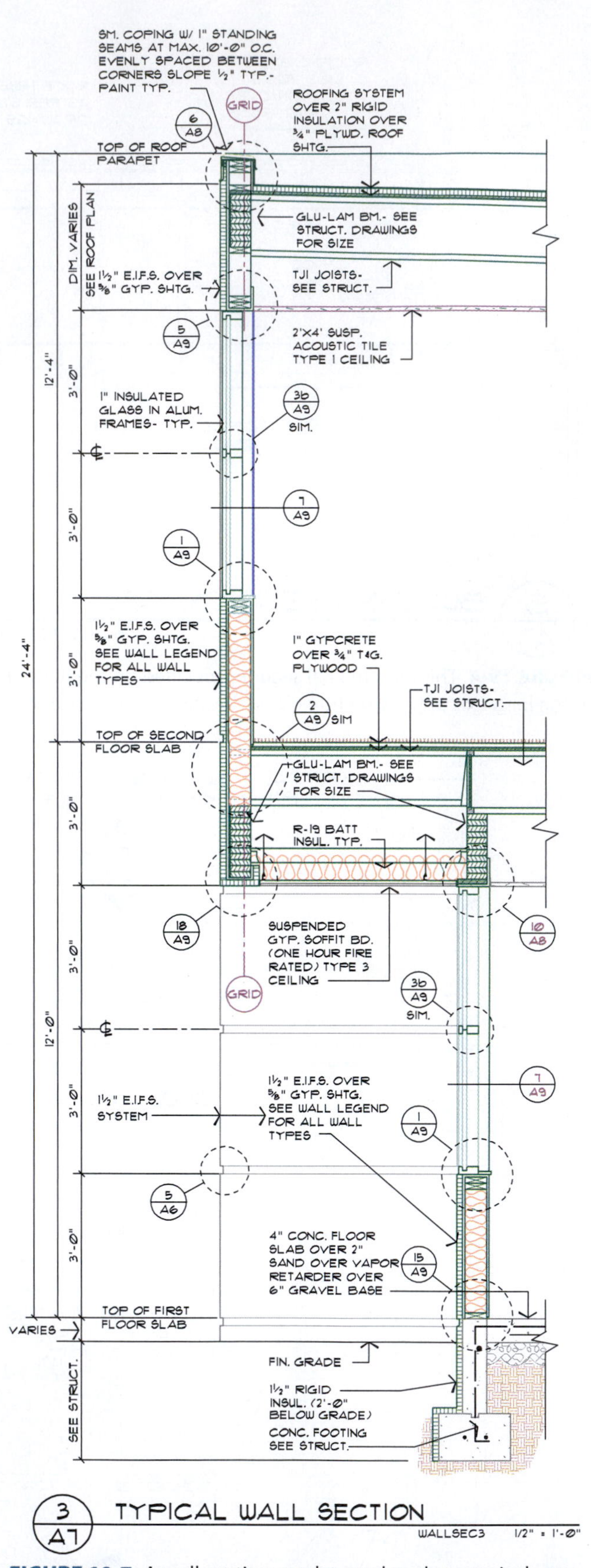

FIGURE 19-7 A wall section can be used to show typical construction for a project. *Courtesy Architects Barrentine, Bates & Lee, A.I.A.*

Section/Elevations

Although not as common as other types of sections, a section can be combined with elevations to explain construction of a structure. Section/elevations can be either full or partial views. Figure 17-25 shows an example of a structural elevation that superimposes structural material over the exterior elevation. Figure 19-9 shows a structural section of the same area shown in Figure 17-25.

SECTION DEVELOPMENT

Sections are part of both the preliminary and architectural drawings. Depending on the type and complexity of the structure, the engineering team may also provide sections to explain how the building is constructed. Chapter 23 will explore the use of structural sections.

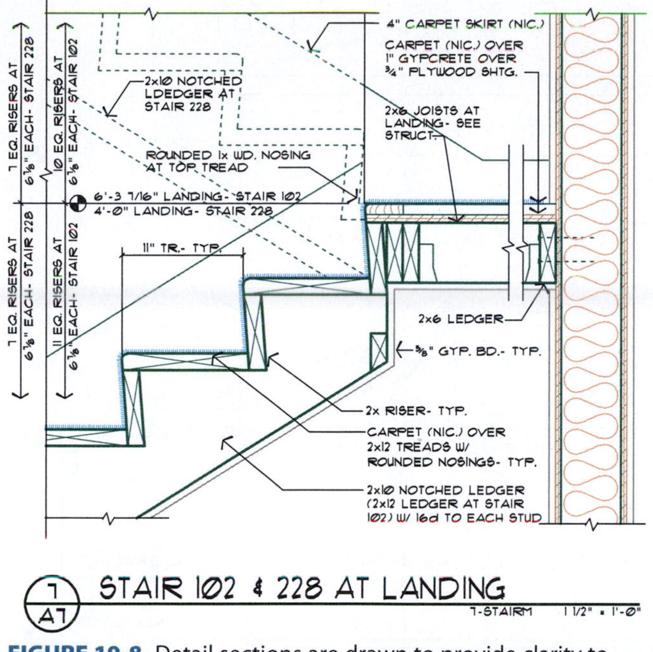

FIGURE 19-8 Detail sections are drawn to provide clarity to small areas. The detail is referenced to the section shown in Figure 19-6. *Courtesy Architects Barrentine, Bates & Lee, A.I.A.*

Preliminary Drawings

A section is one of the preliminary drawings developed by the design team to represent the typical shape of the structure to the client. Once the creative options of the structure have been explored in the preliminary elevations, the architect generally develops sections to explore possible construction techniques. A senior drafter may occasionally develop the section using rough sketches developed by the architect. In the design stage, the section is not intended to accurately represent materials, but to show the shape of the structure, major areas of openings such as windows and skylights, and the vertical relationships of the structure. Walls, floors, and roof systems are represented by solid black lines, which represent thickness rather than specific construction methods. Materials in the background are represented by thin lines, or may not be shown at all. The distances between floor and ceiling, or overall heights, can be shown to represent heights critical to the design. Elevation symbols can be used in place of dimensions to specify heights. Figure 19-10 shows an example of a preliminary section.

To help the print reader relate to the section, room names are specified with large text, and important materials are represented with smaller text. A title and scale should be placed below the drawing. The actual size of the text will vary depending on the size and complexity of the structure. People, plants, automobiles, or other items that will help to identify the use are also placed in the section. Sun angles or shading can be placed to help define how the structure will relate to the site and environment.

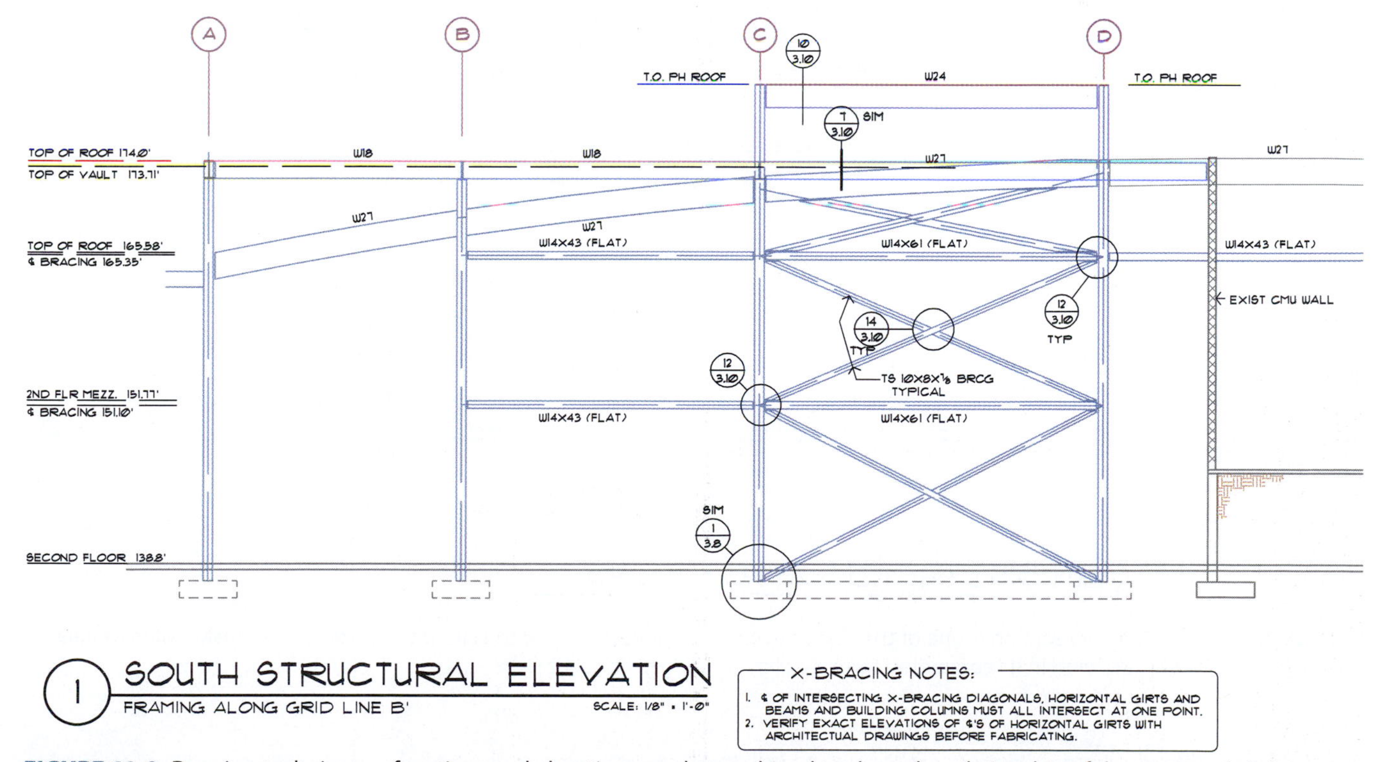

FIGURE 19-9 Drawing techniques of sections and elevations can be combined to show the relationship of the structural skeleton to other materials. See Figure 17-25 to compare the elevation for the same area. *Courtesy Charles J. Conlee P.E., Conlee Engineers, Inc.*

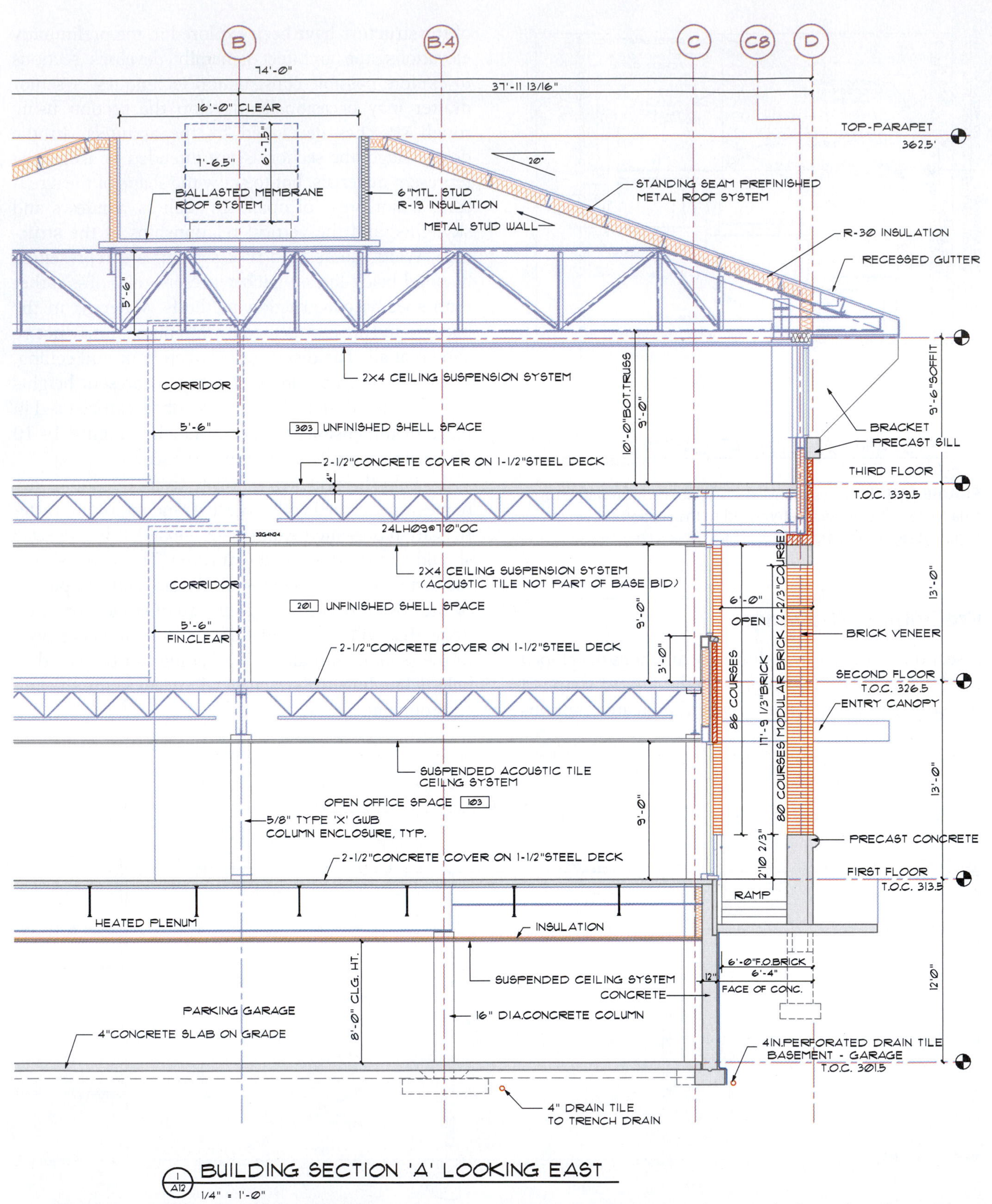

FIGURE 19-10 A preliminary section is one of the first drawings completed to show the client the vertical relationship of materials. Figure 19-3 shows this drawing in its completed form. *Courtesy Peck, Smiley, Ettlin Architects.*

Sections in the Architectural Drawings

A CAD drafter will often convert the preliminary section into the drawing that will be part of the working drawings. The preliminary section and other sketches developed by the design team, as well as a print of the floor plan with the desired cutting planes, are usually given to the technician to indicate which sections should be completed. For experienced CAD technicians, the architect will depend on the drafting team to determine the required sections. Junior drafters will typically not be involved with the initial drawing of the sections but will usually be introduced to sections by adding information to the drawing base from check prints.

Section Scales

The number of sections to be drawn is determined by the complexity of the structure. The size of each section is determined by the complexity and material used for construction. Full sections are generally drawn at the same size as the floor plan. Wall sections are usually drawn at a scale of 3/8" = 1'–0" (1:20) or larger. A scale of 1/2" = 1'–0" (1:20) or larger is used to represent materials shown in detail.

Details should be drawn at a scale that will be sufficient to show the detail of all materials in the area being represented. Generally, a scale of 3/4" = 1'–0" (1:10) or larger should be used. Common scales used for details and their closest metric counterpart include 1" = 1'–0", 1 1/2" = 1'–0", and 3" = 1'–0" (1:5). The scale used for wall sections and details is often determined by blocks, which are on file in the office library. Figure 19-11 shows an example of a party wall that is stored as a block and then inserted into the drawing set. Once inserted into the drawing base, the section can be edited to meet specific conditions of the job.

MAJOR SECTION DRAWING COMPONENTS

Whether the sections are part of the preliminary, architectural, or structural drawings, consideration needs to be given to several key areas when creating the drawings. The CAD technician will need to consider how materials are to be represented, how to place dimensions, how and which symbols are appropriate for the drawing, how text is placed, and if and how blocks can be used.

Material Representation

The method used to represent each material will vary depending on the scale that is used to create the section.

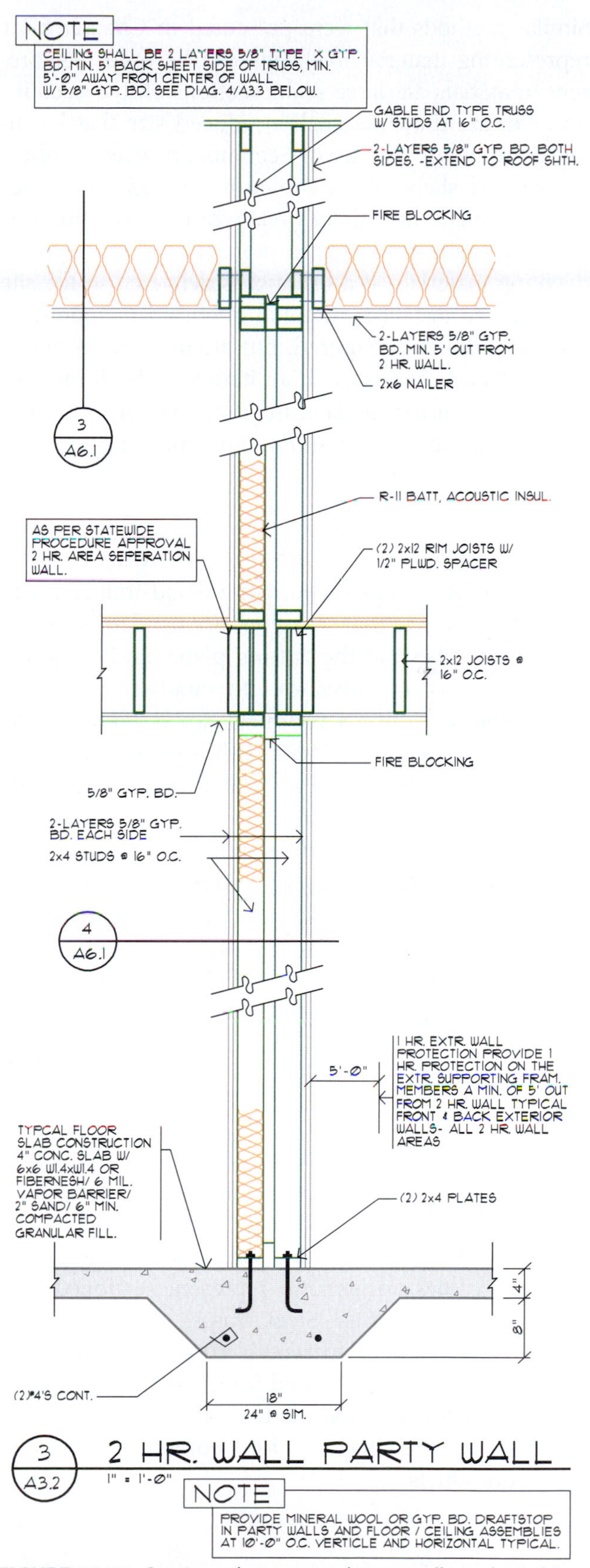

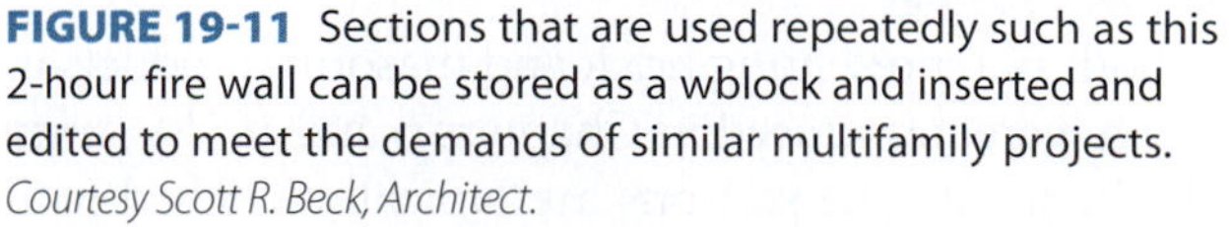

FIGURE 19-11 Sections that are used repeatedly such as this 2-hour fire wall can be stored as a wblock and inserted and edited to meet the demands of similar multifamily projects. *Courtesy Scott R. Beck, Architect.*

Similar methods that were presented in Chapter 6 for representing material in details can be used to represent materials in large-scale sections. Most sections however are drawn at such a reduced size that hatching patterns are not used. Sections are used to show the general shape of the structure, and as a reference map to explain how details relate to the structure. It is also important not to spend more time than necessary detailing materials. If a product is delivered to the site ready to be installed, minimal attention to representing the product is required, but attention to installing the product is necessary. If an item in a detail must be constructed at the job site, the drawings must provide enough information for all the different trades depending on the drawings.

Wood, Timber, and Engineered Products

Figure 19-12 shows a section for a wood-framed medical facility. Notice that thin lines are used to represent studs placed beyond the cutting plane, and the plates are represented by polylines. On small-scale sections, the lumber and timber products can be drawn using their nominal size. The sizes of thin materials such as plywood are exaggerated so that they can be clearly represented. Trusses perpendicular to the cutting plane can be represented by polylines showing the shapes of the truss, similar to those in Figure 19-4. When parallel to the cutting plane, trusses can be represented by a thin line, which represents the chords and webs, similar to Figure 19-3. See Figure 6-22 for methods of representing other common wood members in sections.

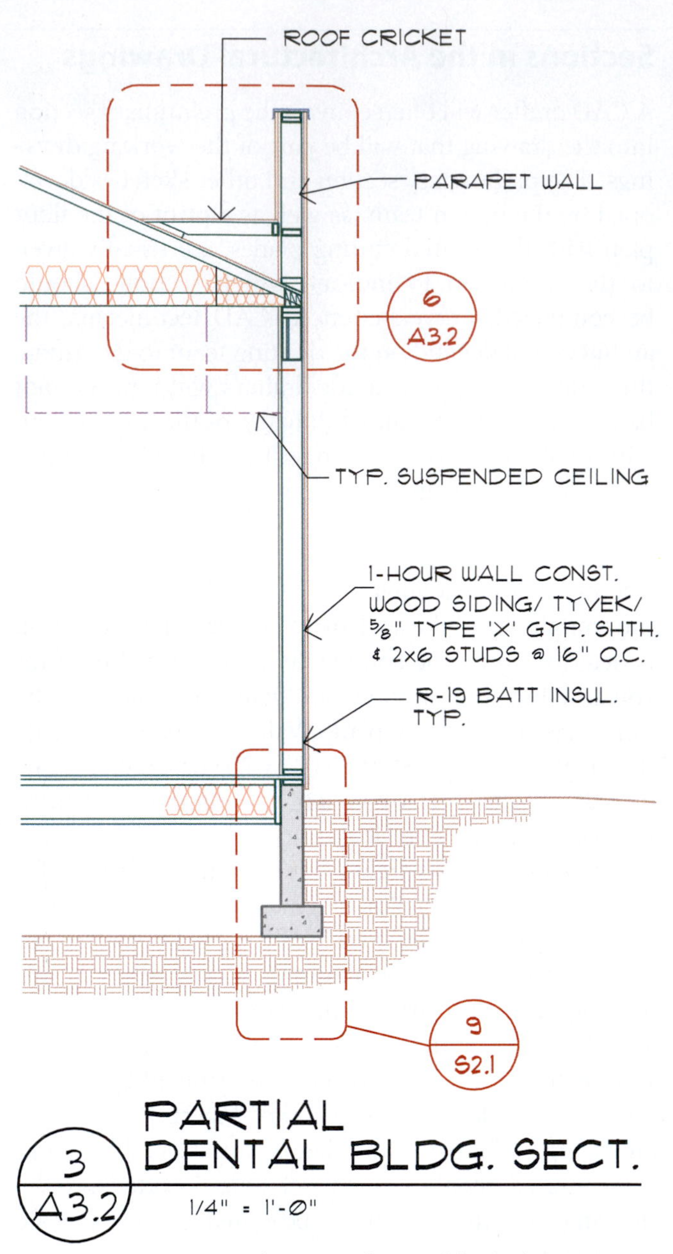

FIGURE 19-12 A partial section of a wood-framed structure. *Courtesy Scott R. Beck, Architect.*

Steel

The size of the drawing will affect how sectioned steel members are represented. At small scale, steel members are generally represented by a solid, thick polyline that represents the desired shape. As the scale increases, pairs of parallel lines are used to represent sectioned steel structural components. Steel trusses are represented in a manner similar to wood trusses. Figure 19-13 shows a wall section formed with steel products. See Figure 6-22 for common methods of representing steel components, and Chapter 9 for a complete description of steel products.

Unit Masonry

Methods of representing brick and masonry products in section were introduced in Chapters 6 and 10. Masonry units in small-scale sections are typically hatched with diagonal lines with no attempt to represent individual units or cavities. As the scale of the section increases, individual units, as well as cavities within the unit, and grouting between the units are represented. Individual hatch patterns are used to differentiate between the masonry unit and the grout. Steel reinforcing can be represented by either a hidden or continuous polyline. Figure 19-14 shows an example of a wall section, representing unit masonry and brick veneer.

Concrete

Chapters 6 and 11 introduced the use of concrete and methods for representing concrete in section views. Thick lines are used to represent the edges of poured-in-place or precast concrete members.

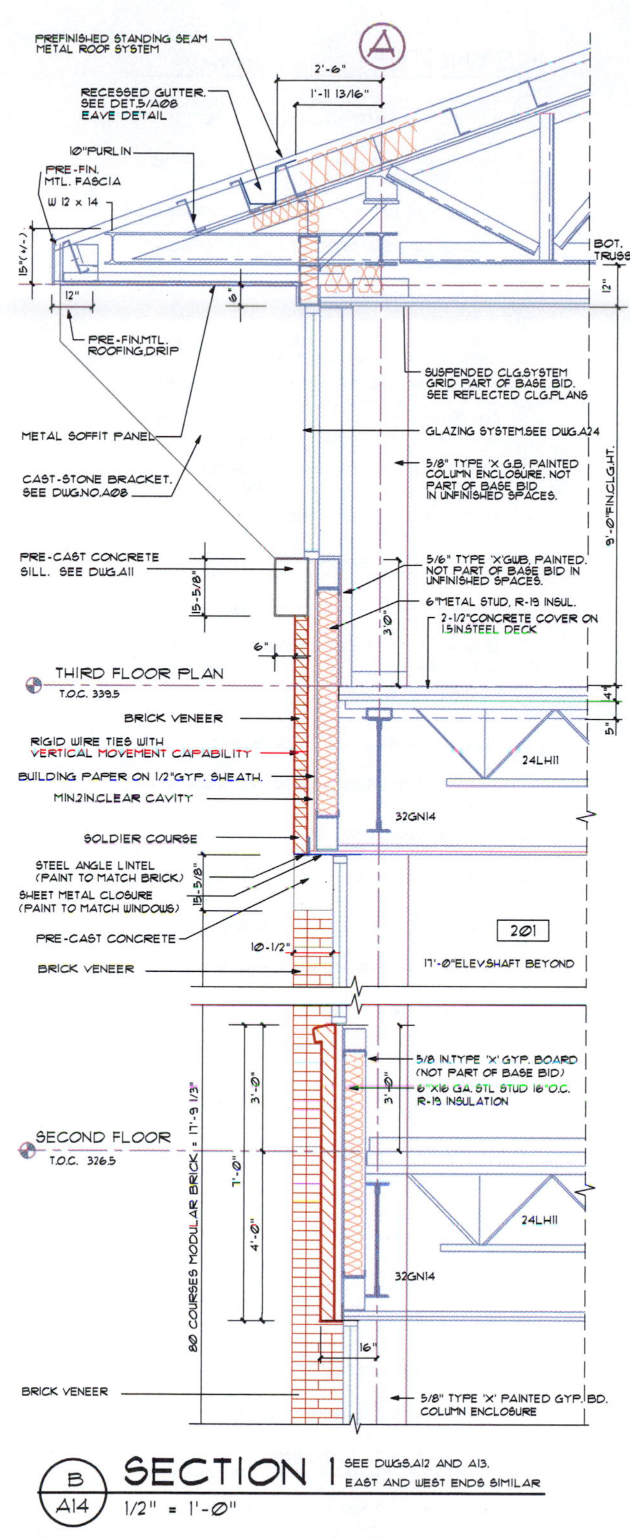

FIGURE 19-13 Steel materials seen in section. *Courtesy Peck, Smiley, Ettlin Architects.*

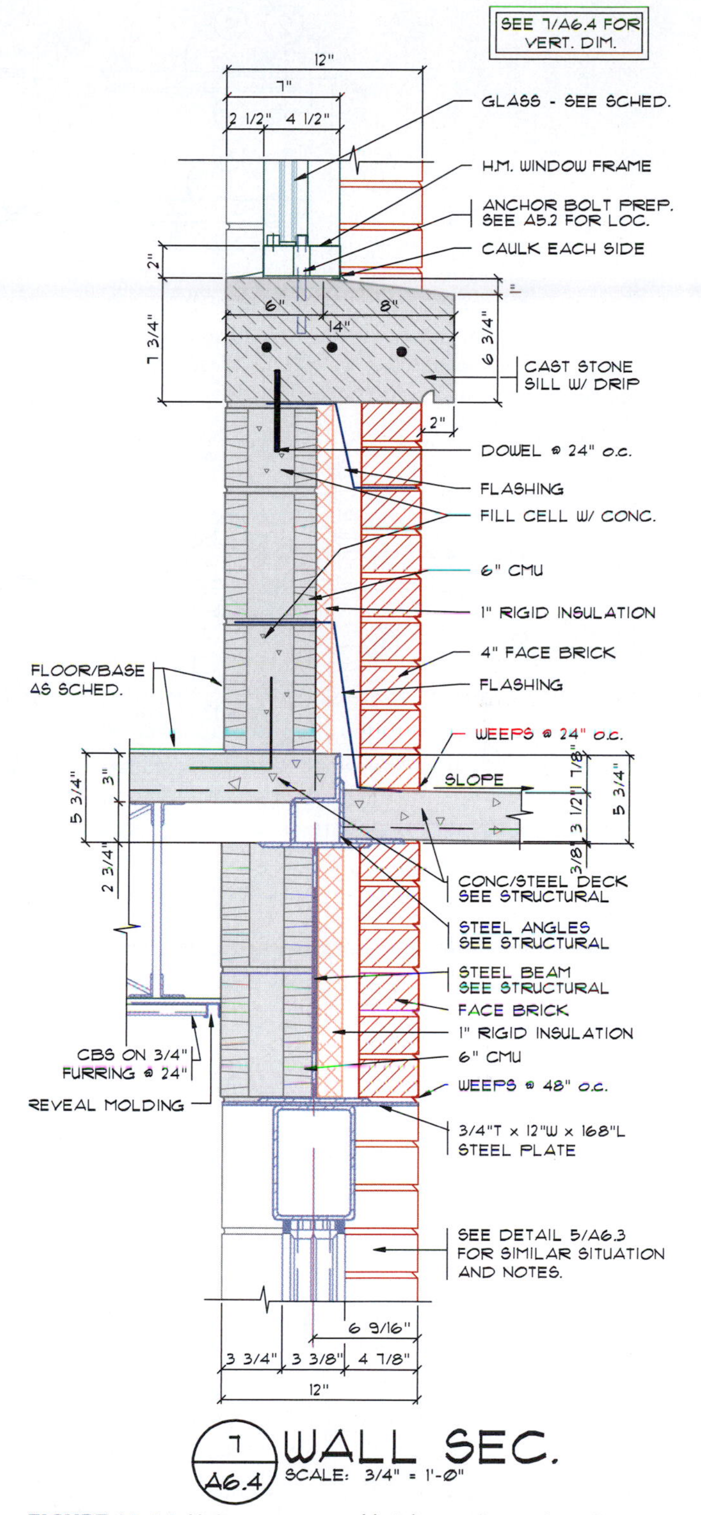

FIGURE 19-14 Unit masonry and brick seen in section. *Courtesy G. Williamson Archer, A.I.A., Archer & Archer P.A.*

A hatch pattern consisting of dots and small triangles is used to further define the concrete from other materials. Figure 19-15 shows an example of a wall section for a precast structure. Because of the complexity of concrete construction, the section depends on many details to show construction of each concrete member. Large-scale sections and details such as Figure 19-16 are usually part of the structural drawings and show individual cavities, grouting, and caulking materials.

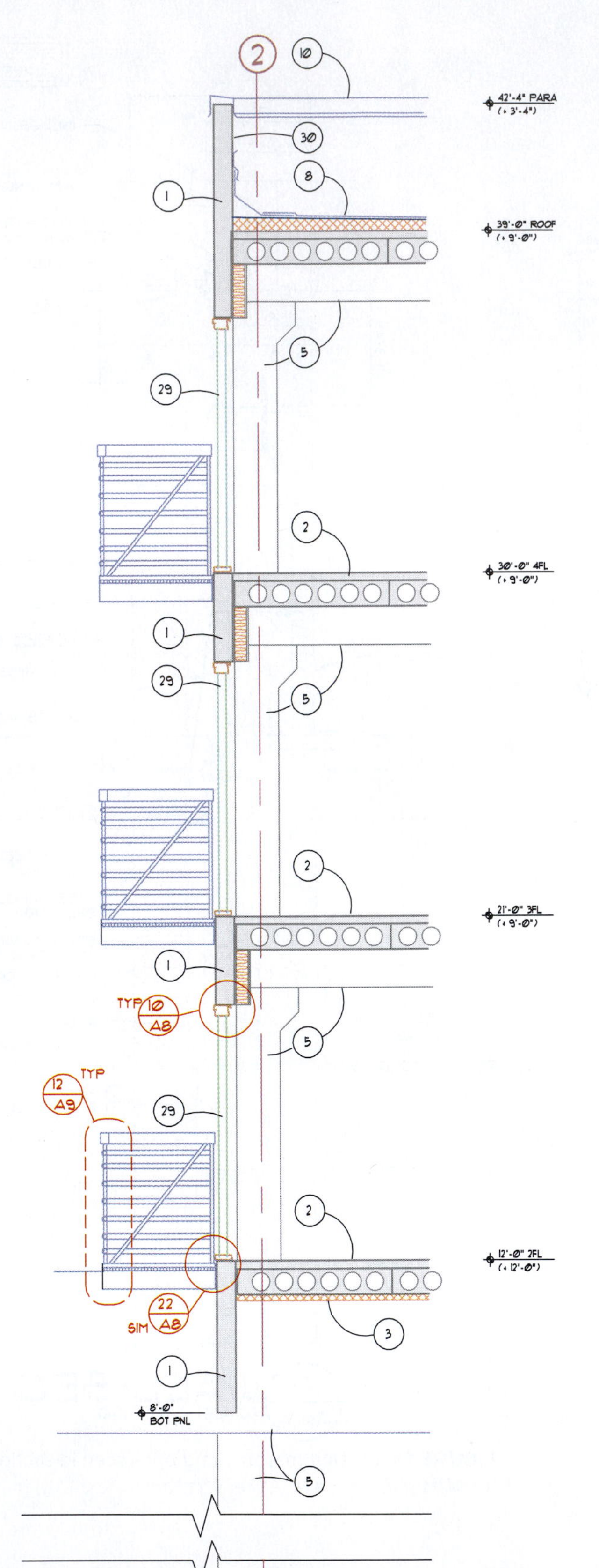

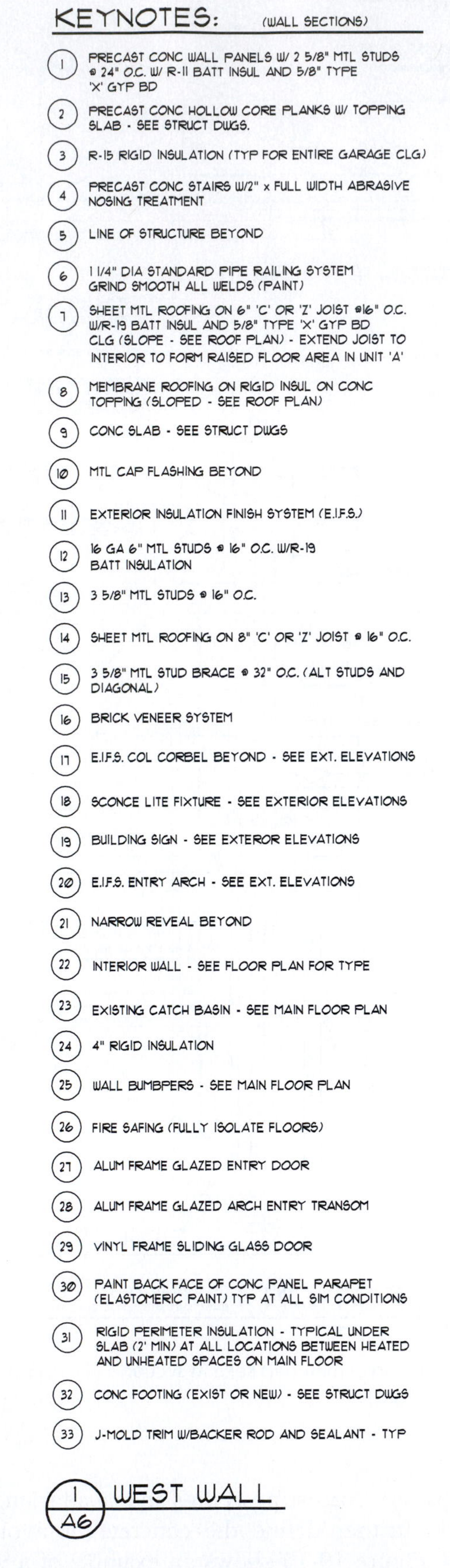

KEYNOTES: (WALL SECTIONS)
1 PRECAST CONC WALL PANELS W/ 2 5/8" MTL STUDS @ 24" O.C. W/ R-11 BATT INSUL AND 5/8" TYPE 'X' GYP BD
2 PRECAST CONC HOLLOW CORE PLANKS W/ TOPPING SLAB - SEE STRUCT DWGS.
3 R-15 RIGID INSULATION (TYP FOR ENTIRE GARAGE CLG)
4 PRECAST CONC STAIRS W/2" x FULL WIDTH ABRASIVE NOSING TREATMENT
5 LINE OF STRUCTURE BEYOND
6 1 1/4" DIA STANDARD PIPE RAILING SYSTEM GRIND SMOOTH ALL WELDS (PAINT)
7 SHEET MTL ROOFING ON 6" 'C' OR 'Z' JOIST @16" O.C. W/R-19 BATT INSUL AND 5/8" TYPE 'X' GYP BD CLG (SLOPE - SEE ROOF PLAN) - EXTEND JOIST TO INTERIOR TO FORM RAISED FLOOR AREA IN UNIT 'A'
8 MEMBRANE ROOFING ON RIGID INSUL ON CONC TOPPING (SLOPED - SEE ROOF PLAN)
9 CONC SLAB - SEE STRUCT DWGS
10 MTL CAP FLASHING BEYOND
11 EXTERIOR INSULATION FINISH SYSTEM (E.I.F.S.)
12 16 GA 6" MTL STUDS @ 16" O.C. W/R-19 BATT INSULATION
13 3 5/8" MTL STUDS @ 16" O.C.
14 SHEET MTL ROOFING ON 8" 'C' OR 'Z' JOIST @ 16" O.C.
15 3 5/8" MTL STUD BRACE @ 32" O.C. (ALT STUDS AND DIAGONAL)
16 BRICK VENEER SYSTEM
17 E.I.F.S. COL CORBEL BEYOND - SEE EXT. ELEVATIONS
18 SCONCE LITE FIXTURE - SEE EXTERIOR ELEVATIONS
19 BUILDING SIGN - SEE EXTEROR ELEVATIONS
20 E.I.F.S. ENTRY ARCH - SEE EXT. ELEVATIONS
21 NARROW REVEAL BEYOND
22 INTERIOR WALL - SEE FLOOR PLAN FOR TYPE
23 EXISTING CATCH BASIN - SEE MAIN FLOOR PLAN
24 4" RIGID INSULATION
25 WALL BUMPERS - SEE MAIN FLOOR PLAN
26 FIRE SAFING (FULLY ISOLATE FLOORS)
27 ALUM FRAME GLAZED ENTRY DOOR
28 ALUM FRAME GLAZED ARCH ENTRY TRANSOM
29 VINYL FRAME SLIDING GLASS DOOR
30 PAINT BACK FACE OF CONC PANEL PARAPET (ELASTOMERIC PAINT) TYP AT ALL SIM CONDITIONS
31 RIGID PERIMETER INSULATION - TYPICAL UNDER SLAB (2' MIN) AT ALL LOCATIONS BETWEEN HEATED AND UNHEATED SPACES ON MAIN FLOOR
32 CONC FOOTING (EXIST OR NEW) - SEE STRUCT DWGS
33 J-MOLD TRIM W/BACKER ROD AND SEALANT - TYP
1
A6
WEST WALL

FIGURE 19-15 Precast concrete panels in section with keyed notes. *Courtesy H.D.N. Architects, A.I.A.*

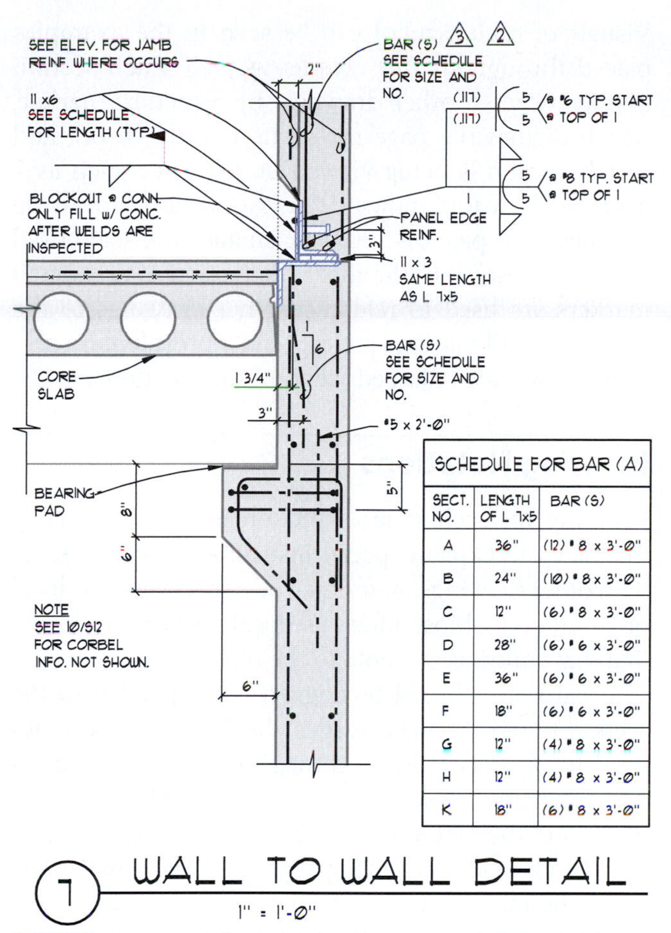

SCHEDULE FOR BAR (A)

SECT. NO.	LENGTH OF L 7x5	BAR (S)
A	36"	(12) #8 x 3'-0"
B	24"	(10) #8 x 3'-0"
C	12"	(6) #8 x 3'-0"
D	28"	(6) #6 x 3'-0"
E	36"	(6) #6 x 3'-0"
F	18"	(6) #6 x 3'-0"
G	12"	(4) #8 x 3'-0"
H	12"	(4) #8 x 3'-0"
K	18"	(6) #8 x 3'-0"

FIGURE 19-16 Details drawn by the structural team supplement the sections drawn by the architectural team. *Courtesy Van Domelen/Looijenga/McGarrigle/Knauf Consulting Engineers.*

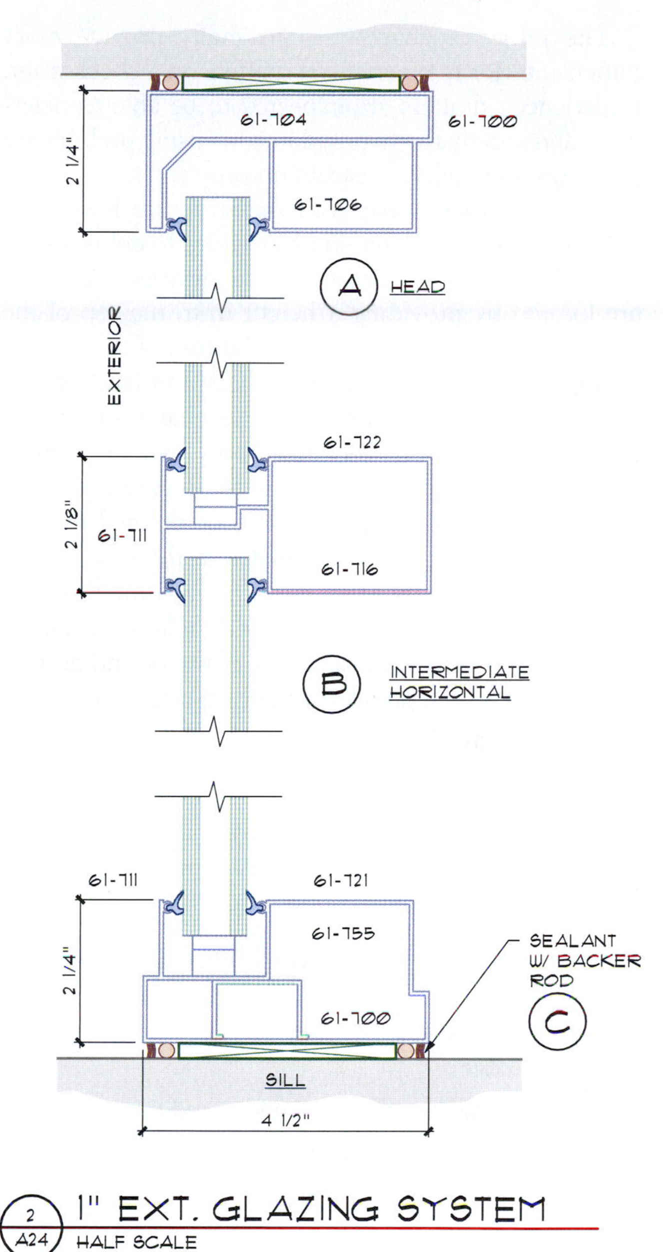

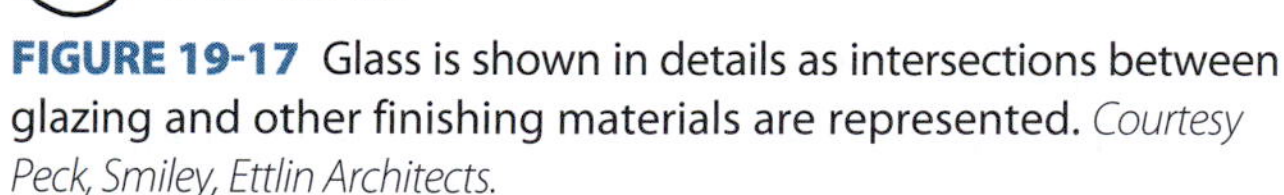

FIGURE 19-17 Glass is shown in details as intersections between glazing and other finishing materials are represented. *Courtesy Peck, Smiley, Ettlin Architects.*

Glass

A single thin line or pairs of thin lines are used to represent glass in section drawings. Little attention is given to intersections between the glazing and window frames in section drawings. Details similar to Figure 19-17 are used to describe head, jamb, and sill construction.

Placing Dimensions on Sections

Dimensions are an important element of full, partial, and wall sections. Both vertical and horizontal dimensions may be placed on sections, whereas partial sections and details generally show only vertical dimensions.

Vertical Dimensions

Vertical heights in section views are represented using the methods that were introduced in Chapters 3 and 6 for placing dimensions on details. Figure 19-7 shows an example of using standard dimension placement. Figures 19-3 and 19-6 show another common method of specifying elevations above a specific surface. For wood-frame structures, vertical dimensions are generally given from the bottom of the sole plate to the top of the top plate. This dimension also provides the height from the top of the floor sheathing to the bottom of the framing member used to frame the next level. A common alternative is to provide a height from the top of the floor sheathing to the top of the next level of floor sheathing. Other common vertical exterior dimensions include the following:

- Steel decking: from the top of decking.
- Steel stud walls: from plate to top of channel.
- Structural steel: to top of steel member.
- Masonry units: to top of unit with distance and number of courses provided.
- Concrete slab: from top of slab or panel.

The project manager will generally provide exact dimensions for inexperienced drafters on a check print. Experienced drafters are expected to be able to determine and specify the required heights using preliminary drawings, or similar construction projects.

Once the major shapes of the structure have been defined, dimensions should be provided to define openings, floor changes, or protective devices. Openings are located by providing a height from the top of the floor decking or sheathing to the bottom of the header. Changes in floor height and the height of landings are dimensioned in a similar manner as changes in height between floor levels. Inclined floors can be defined using a slope indicator, similar to the method of defining roof pitch. The slope can also be defined by providing a vertical dimension to define the total height difference between floor levels. Other common interior dimensions that should be provided include the height of railings, partial walls, balconies, planters, and decorative screens. When possible, interior dimensions should be grouped together.

Horizontal Dimensions

The use of horizontal dimensions on full sections varies greatly depending on each office. When provided, horizontal dimensions should be placed so that:

- They are located from grid lines to the desired member.
- Exterior wood and concrete members are referenced to their edge.
- Interior wood members are referenced to a centerline.
- Interior concrete members are referenced to an edge.
- Steel members are referenced to their centers.

The distance for roof overhangs and balcony projections may also be placed on sections. Figure 19-3 shows the use of horizontal dimensions on a full section.

Drawing Symbols

Sections use symbols that match those of the floor, roof, and elevation drawings to reference material. Symbols that might be found on the section include the following:

- Grid markers.
- Elevation markers.
- Section markers.
- Detail markers.
- Room names and numbers.

As the skills for drawing details were introduced in Chapter 6, examples of each symbol were given. Visuals of each symbol can be seen in the examples placed throughout this chapter as well. Each section is referenced to other drawings by a section marker, which defines the page the section is drawn on, and which section is being viewed. A reference such as 1 over A–500 would indicate that the section is drawing number 1 on page A–500. The smaller the scale used to draw the section, the more likely section and detail markers are used to reference other drawings to the section. Detail markers are especially prevalent on sections to provide enlarged views of intersections.

Drawing Notations

Lettering on each type of section is used to specify material and explain special insulation procedures. As with other drawings, notes may be either placed as local or keyed notes. Most offices use local notes with a leader line that connects the note to the material.

Local notes should be aligned to be parallel to the section to aid the print reader. On full sections, notes need to be placed neatly throughout the entire drawing. Wherever possible, notes should be placed on the exterior of the building. For wall sections, aligned notes can greatly add to drawing neatness. The smaller the scale, the more generic the notes tend to be on a section. For instance, on a full section, roofing that might be specified as:

MEMB. ROOFING SYSTEM OVER RIGID
INSUL. OVER PLYWD. SHTG.

would be referenced by complete notes for the roofing, insulation, and plywood in the roofing details. Related notes should be grouped together within the same area of a section. The contents of notes will also be affected by the project manual, which is discussed in Chapter 14. As with other types of drawings mentioned in previous chapters, the Construction Specifications Institute (CSI) *MasterFormat* reference number can be used on a section to reference products. The roofing membrane specified in the note above would resemble:

ROOFING MEMBRANE - 07 52 19

The rigid insulation and plywood sheathing specified in the note above would require additional similar notations.

Blocks

Software such as AutoCAD Revit or AutoCAD Architecture can be a great aid in developing sections from information provided as the floor plan was created. With these programs, sections are generated automatically from plan views using the same colors, layers, and linetypes of the plan views. The software, based on the material selected, automatically assigns hatch patterns. Annotation can also be assigned based on common assemblies created using CSI guidelines. If a section is to be created using AutoCAD, careful planning can greatly reduce the CAD technician's job as sections are prepared. By planning and sketching the required sections, repetitive features can be drawn once and copied to other sections. Because of the similarities of many structures, common intersections can be drawn and stored as a wblock. These wblocks can be inserted into a drawing with all required notations, symbols, and dimensions, greatly reducing drawing time. Assembling and connecting drawing blocks can be an efficient method of creating a section, rather than drawing the section line by line. Figure 19-18 shows an example of a footing and roof detail for a concrete block structure. These details can be inserted into a drawing base and moved the proper distance to reflect the distance from floor to roofing as seen in Figure 19-19. Once a wall section has been created, the MIRROR command can be used to create the opposite wall, which can then be moved the appropriate distance to represent the total width of the building. Blocks of details showing a truss to a laminated beam and a column to footing were inserted at the center and connected to each other. The section can then be completed by stretching the slab from one detail to another, and by drawing the truss. The completed section, shown in Figure 19-20, required very little actual drawing because of the use of wblocks. For students, access to block libraries is limited. For that reason, the following explanation of drawing sections will assume no use of blocks.

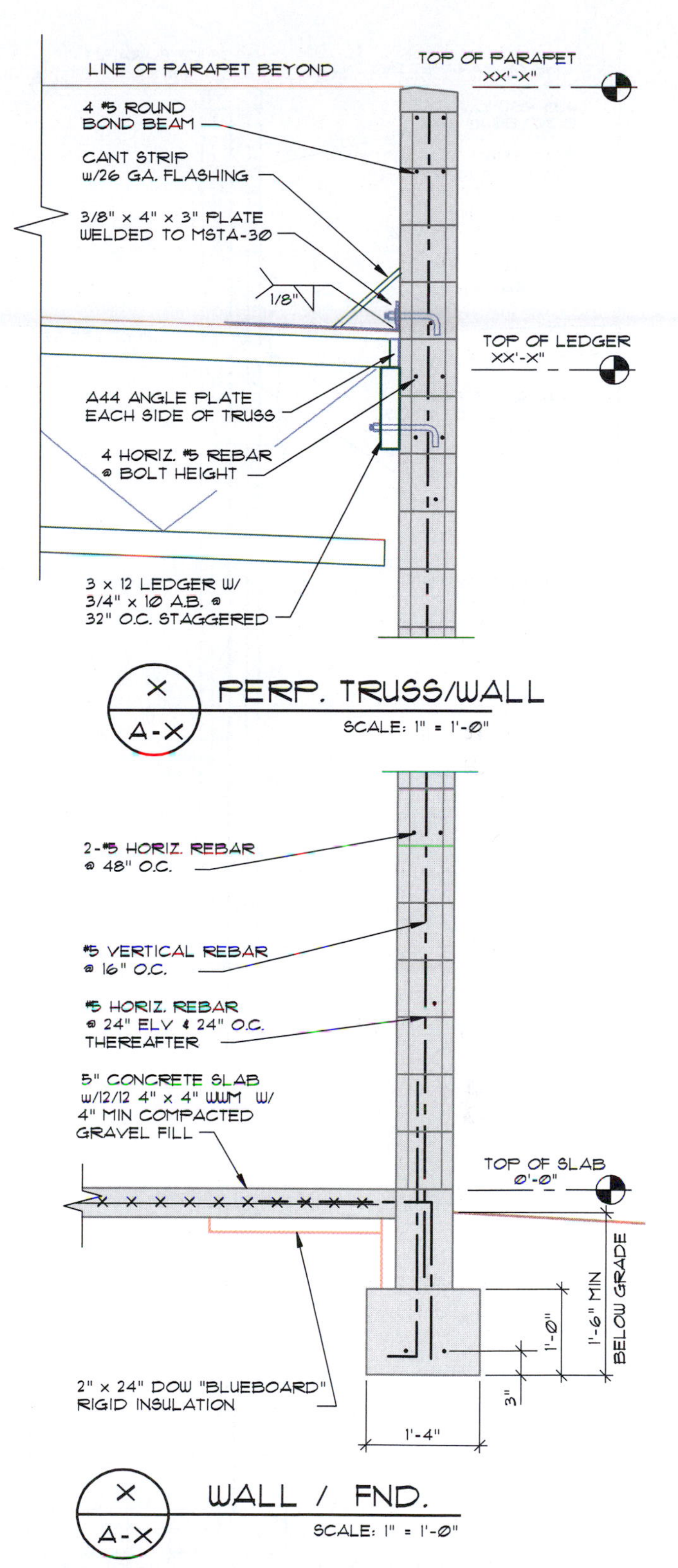

FIGURE 19-18 Stock details such as these footing and truss details can be joined together to form a partial wall section.

COMPLETING A SECTION

A CAD technician will need a floor plan, foundation plan, grading plan, roof plan, exterior elevations, and preliminary sections to complete the building sections. For relatively flat lots or if spot grades are indicated on the elevations, the grading plan is not required.

1. Place a cutting plane on the floor plan to determine the view to be created. The following guidelines will be used to draw section A/A-7 for the structure represented in Figures 16-3 and 16-11.
2. Create a separate block of the floor plan, which can be used to project the section. *Do not use the floor plan.*
3. Establish horizontal lines to represent the finished floor, ceiling, and roof levels.
4. Project lines representing each wall cut by the cutting plane from the floor plan into the drawing area.
5. Project lines to represent each column and beam cut by the cutting plane from the floor plan into the drawing area.
6. Using the foundation and grading plans, establish the foundation locations.
7. Establish a line to represent the finish grade location.
8. Using elevations on the roof or exterior elevations, establish the height of all exterior walls.

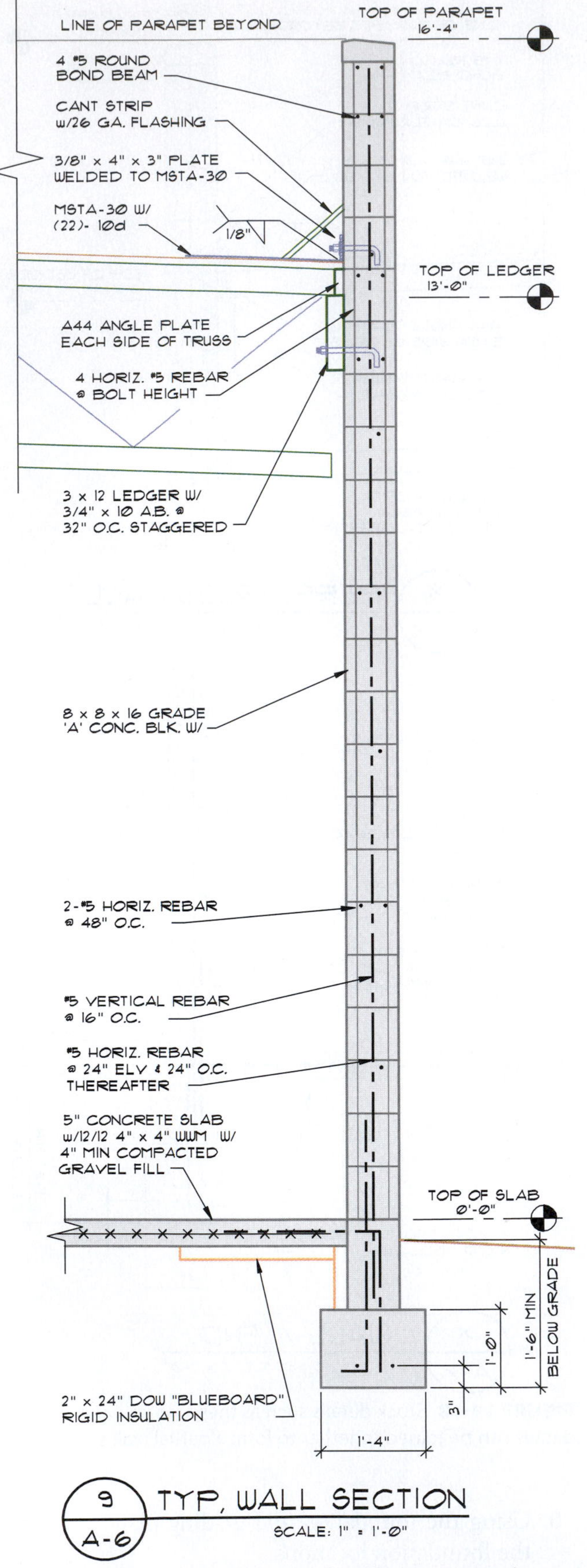

FIGURE 19-19 A wall section completed by joining stock details.

The projection of the section should now resemble Figure 19-21. The section can be completed using the following steps:

9. Trim all projection lines to form the outline of the section.
10. Use the elevations on the roof plan to establish the pitch of the trusses.
11. Use vendor catalogs to determine the depth of the trusses, and draw the roof trusses.
12. Draw the finished roofing system above the trusses.
13. Draw all beams and ledgers required to support the trusses.

Figure 19-22 shows the development of the section. Once the base of the section is complete, detail can now be added to the drawing using the following steps:

14. Draw any required ceilings.
15. Show any interior walls.
16. Show any false ceilings.
17. Show all material used to complete the roof and floor systems.

The entire section cut by the cutting plane has now been drawn and should resemble Figure 19-23. The following steps can be used to complete the drawing:

18. Show materials that lie beyond the cutting plane.
19. Provide grid lines and specifications.
20. Provide horizontal reference lines and elevations.
21. Show detail markers for areas to be enlarged. Although detail specifications are provided in this example, they can't usually be provided until the entire drawing is drawn.
22. Show all room names and numbers.
23. Place all required notes based on the architect's specifications.
24. Place a title and scale below the drawing.

The completed section should now resemble Figure 19-5.

Layer Guidelines for Sections and Details

Separation of material by layer and color can greatly aid in the development of a section. A prefix of A-SECT is recommended by the NCS guidelines. Modifiers can be used to define materials. Many firms divide modifiers by materials such as CONC, CBLK, BRK, STL, WD, or PTWD. See Chapter 3 for layer naming guidelines and Appendix C on the student CD for a list of common layer names used on sections.

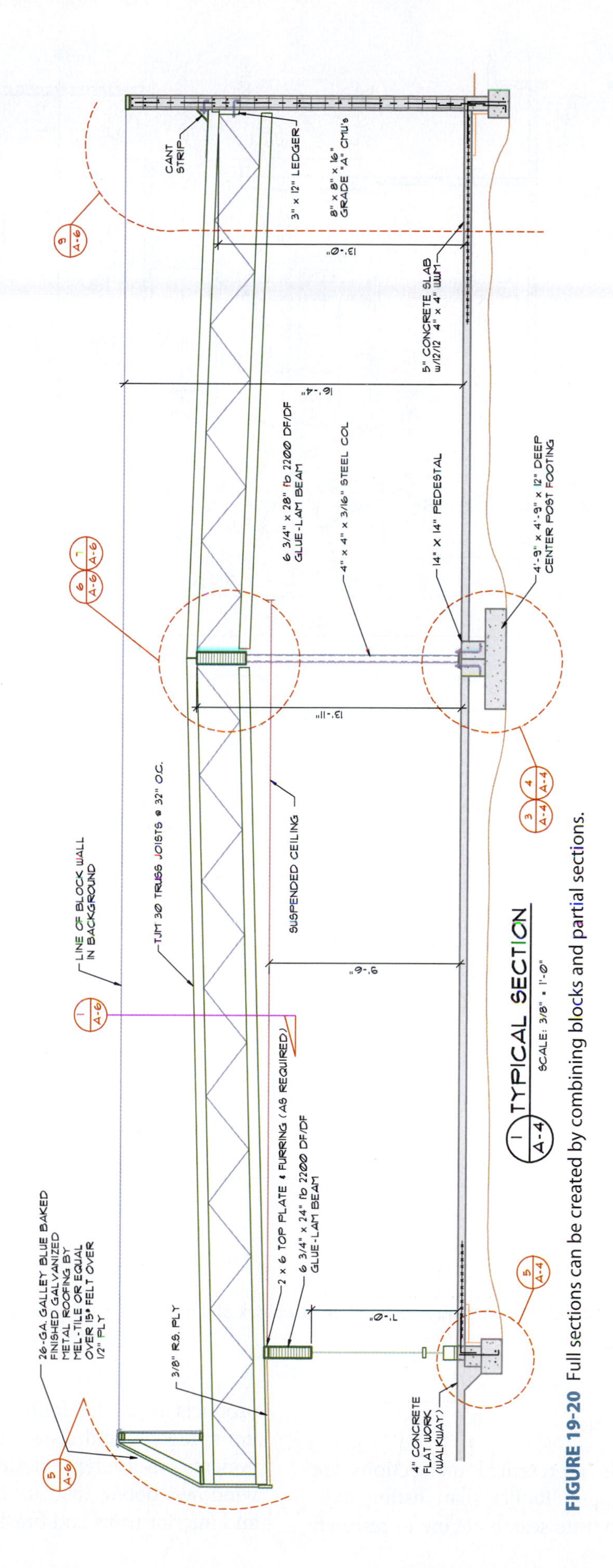

FIGURE 19-20 Full sections can be created by combining blocks and partial sections.

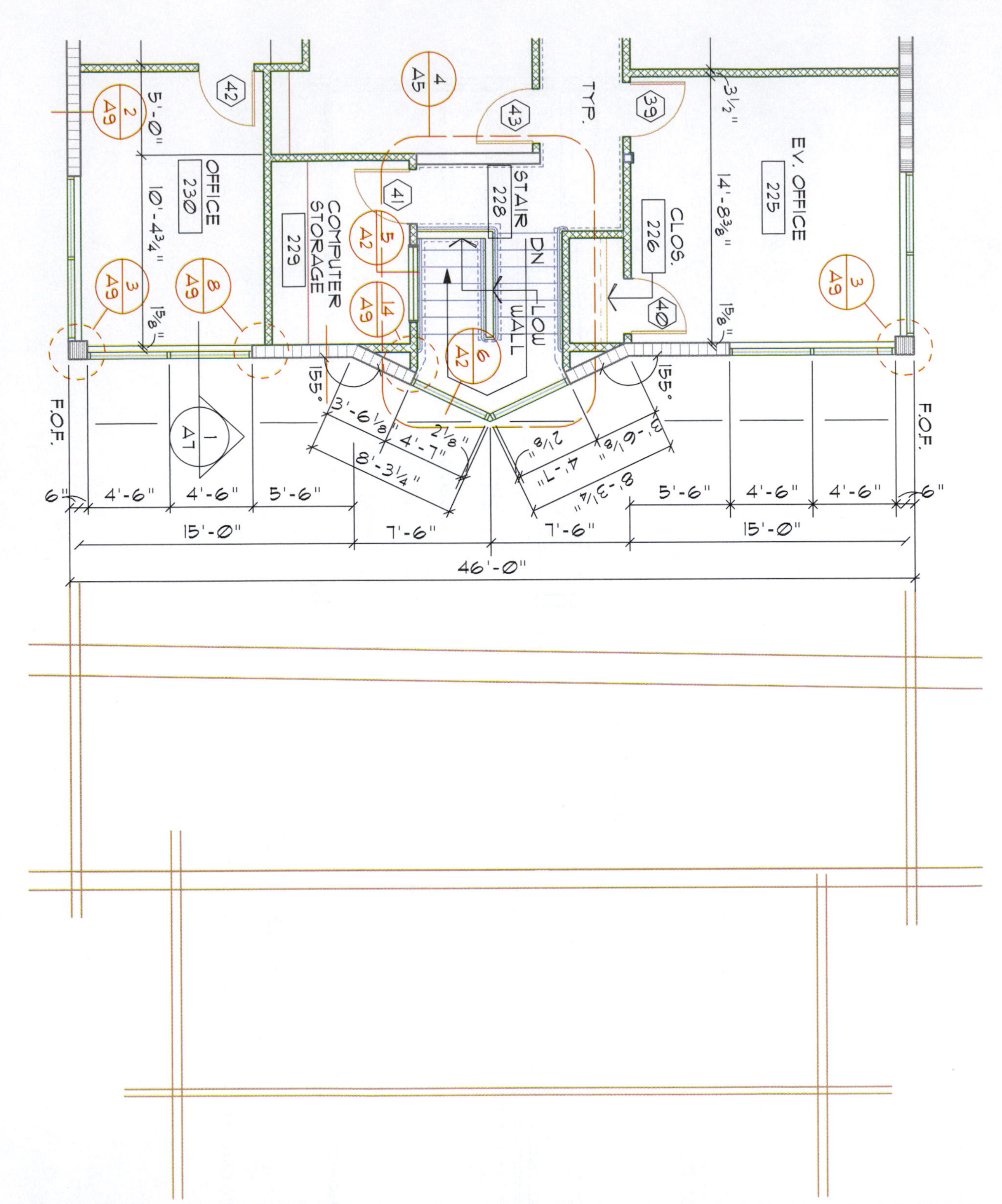

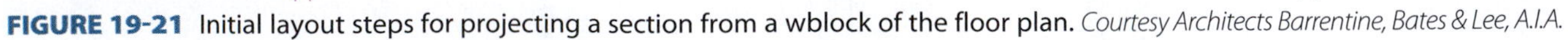

FIGURE 19-21 Initial layout steps for projecting a section from a wblock of the floor plan. *Courtesy Architects Barrentine, Bates & Lee, A.I.A.*

ADDITIONAL READING

Links for the materials represented on sections are too numerous to list here. Rather than listing individual sites, use your favorite search engine to research products from the following CSI categories: foundations, footers, and slabs; structural systems and components; sheathing; exterior trim and finish; roofing; windows; doors; insulation; flooring; floor coverings; and interior trims and finishes.

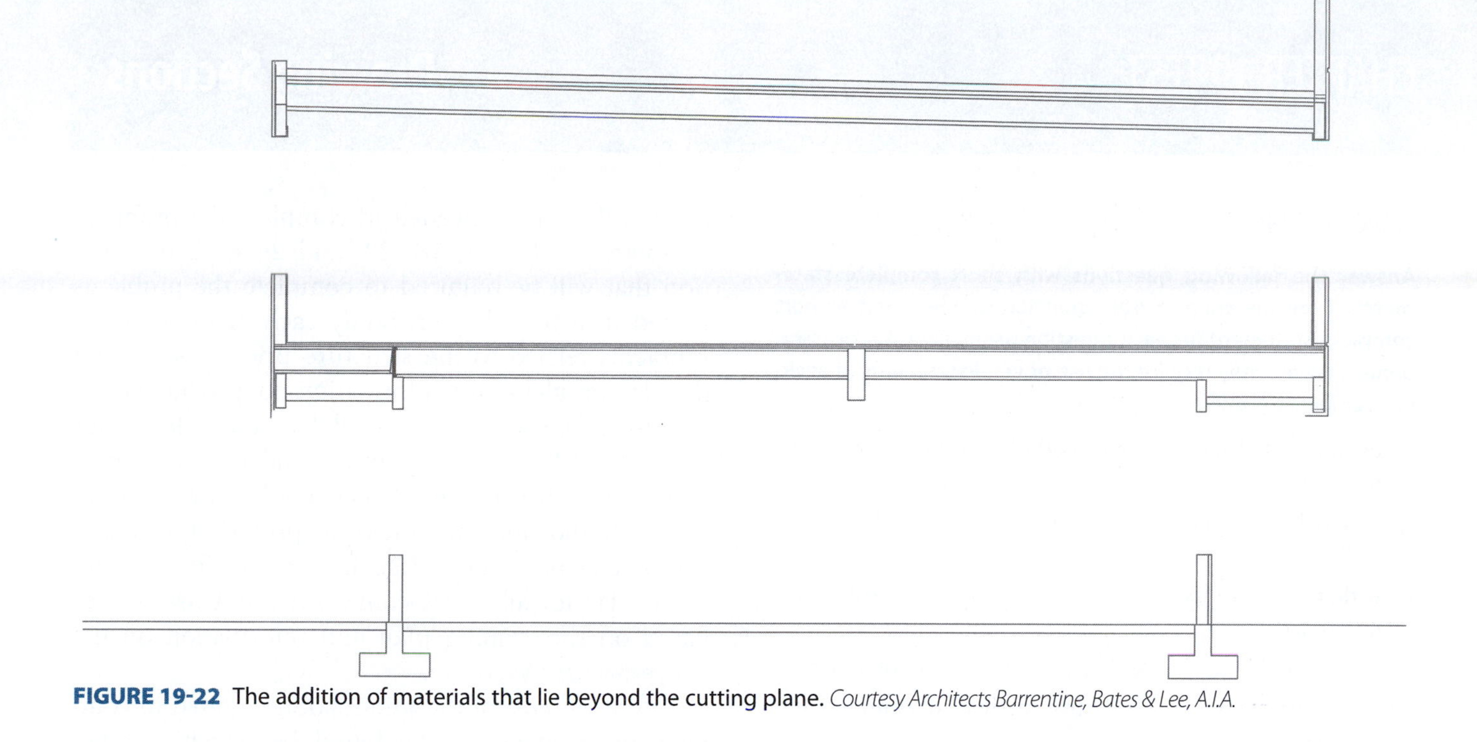

FIGURE 19-22 The addition of materials that lie beyond the cutting plane. *Courtesy Architects Barrentine, Bates & Lee, A.I.A.*

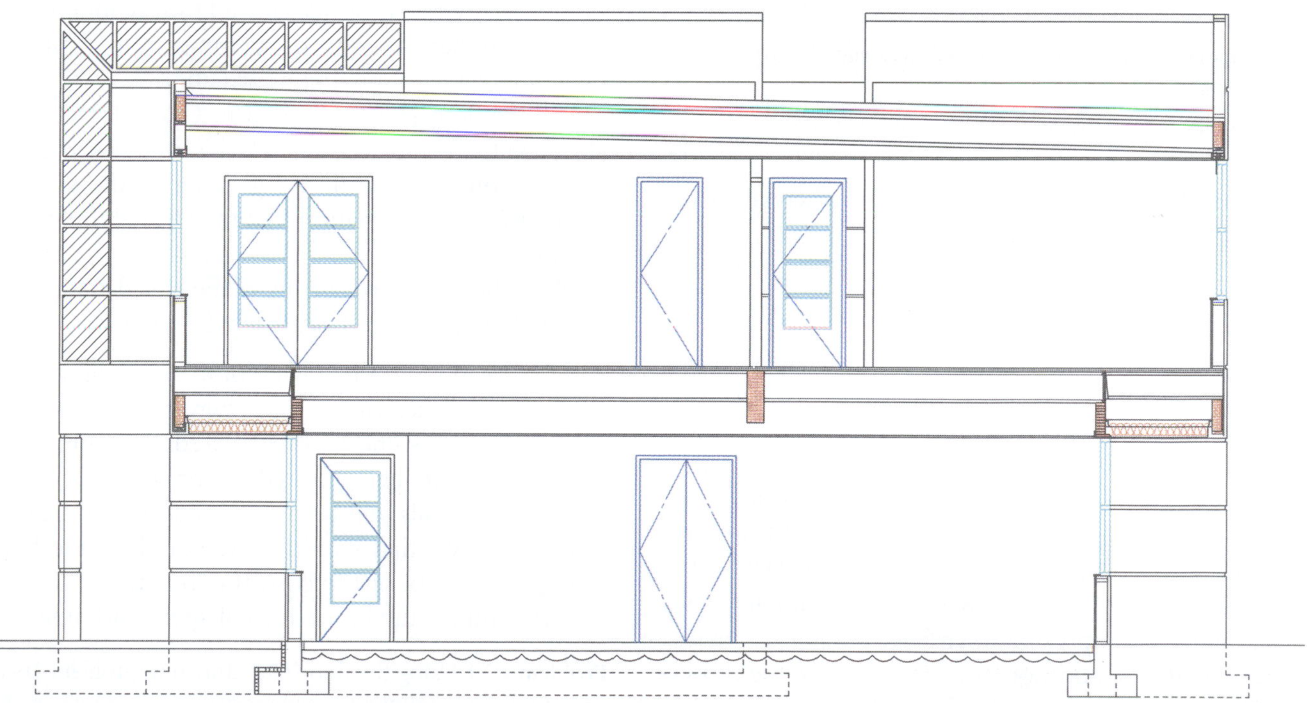

FIGURE 19-23 Drawing the major shapes of the structure and projection of all ceilings and detailing of each floor and opening. *Courtesy Architects Barrentine, Bates & Lee, A.I.A.*

KEY TERMS

Partial section

Transverse section

CHAPTER 19 TEST Drawing Sections

QUESTIONS

Answer the following questions with short complete statements. Type the chapter title, question number, and a short complete statement for each question using a word processor. Some answers may require the use of vendor catalogs or seeking out local suppliers.

Question 19-1 What drawings might be required to draw a full section?

Question 19-2 Explain the difference between a full and a wall section.

Question 19-3 List two types of full sections and explain the differences.

Question 19-4 What relates the section to the floor plan?

Question 19-5 Briefly describe two methods of completing a section.

Question 19-6 List guidelines for selecting the scale for a section, wall section, and a detail.

Question 19-7 Describe two common methods of describing vertical heights.

Question 19-8 List common symbols associated with sections.

Question 19-9 Explain variations in notes between a section and a detail.

Question 19-10 Explain the various roles a drafter will play in the development of sections.

DRAWING PROBLEMS

Use the reference material from preceding chapters, local codes, and vendor catalogs to complete one of the following projects. Unless your instructor provides other instructions, draw the sections that correspond to the floor plan that was drawn in Chapter 16. Skeletons for problems 19-1 through 19-4 can be accessed from the student CD in the DRAWING PROJECTS/CHAPTER 19 folder. Use these drawings as a base to complete the assignment. Use appropriate symbols, linetypes, dimensioning methods, and notations to complete the drawing. Determine any unspecified sizes based on material or practical requirements. Unless specified, select a scale appropriate to plotting on D-size material and determine LTSCALE and DIMVARS.

You will act as the project manager and will be required to make decisions about how to complete the project. By now you're far enough into this structure to realize that no one drawing problem contains all of the information needed to complete the problem. Chapters 15, 16, 17, 18, 22, and 24 have information that will be required to complete the problems provided in this chapter. Study each of the drawing problems related to the structure prior to starting a drawing problem. Any information not provided must be researched and determined by you unless your instructor (the project architect and engineer) provides other instructions. When conflicting information is found, information from preceding chapters should take precedence. Use the Web site for Simpson Strong-Tie for all metal connectors, and use trusses based on the framing plan and information on the Weyerhaeuser Web site.

Although most of the information required to complete the sections can be found by searching other related problems, some of the drawings will require you to make a decision on how you would solve a particular problem. Make several sketches of possible solutions based on examples found throughout this text and submit them to your instructor prior to completing each drawing with missing information.

Use the floor plans provided in Chapter 16 to complete the following sections. Decide on an appropriate scale to complete the drawing. Refer to the details in Chapter 23 for information that may be helpful in completing the sections.

Place the necessary cutting plane and detail bubbles on the floor plan to indicate each drawing.

Note: Keep in mind as you complete these drawings that the finished drawing set may contain more sections and in some cases hundreds of more details than you are being asked to draw. Some of the required details will be introduced in later chapters. Other details that would be required for the building department or the construction crew will not be drawn due to the limitations of space and time.

Problem 19-1 Using the floor plan drawn in problem 16-1, complete the following drawings:

- A section cutting through each unit, showing the elevation change for each floor level at a scale of 1/4" = 1'–0".
- A wall section of a party wall at a scale of 1/2" = 1'–0".
- A detail of each retaining wall at a scale of 3/4" = 1'–0".

Minimum Project Standards

Your completed drawings should meet or exceed the following minimum standards unless noted:

- Dimension each major material and provide unit dimensions, floor-to-floor heights, and elevations.
- Use the partial section and details to completely specify required materials.
- The lower level of each unit is to have a 4" concrete slab and 8'–0" high walls; the main and upper floors are to have 9'–0" ceilings.
- Frame the roof using standard roof trusses. Use standard/scissor trusses to provide a vaulted ceiling over the master bedroom.
- Use 2 × 6 studs at 16" o.c. for exterior walls and 2 × 4 studs at 16" o.c. for interior walls.
- Use Figure 19-11 as a guide, and design a fire wall that reflects the change in elevation for each unit based on local codes for your area. Use solid blocking in the upper wall of the party wall opposite the top plate for the lower wall.
- Select and specify TJIs or open-web trusses suitable to span from party wall to party wall.
- Support the exterior wall at the upper end of the project on an 8" wide × 8' high concrete retaining wall supported on a 12" × 16" wide footing. Reinforce the wall with #5 bars at 18" o.c. E.W. steel to be 2" clear of interior side of wall.

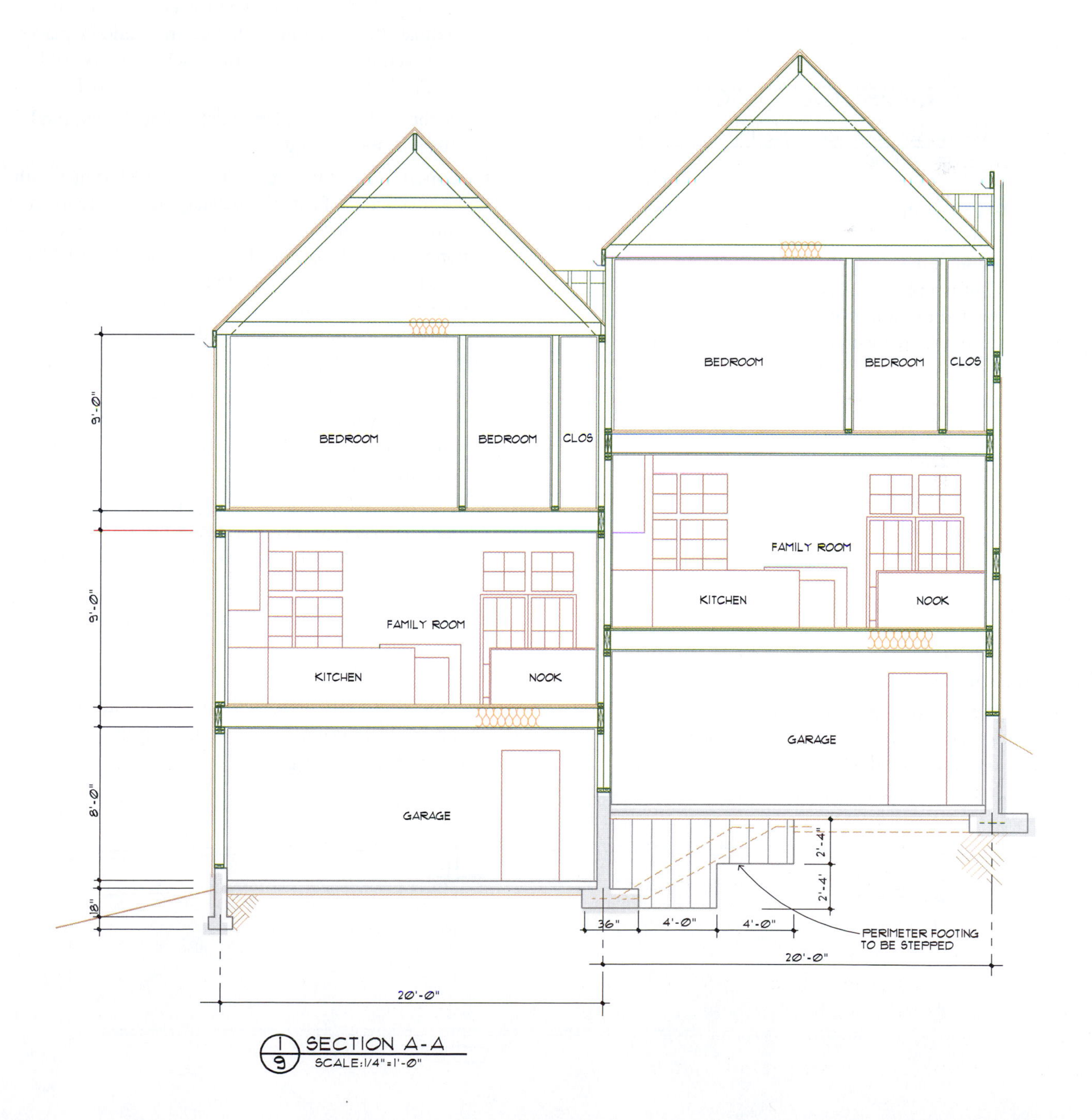

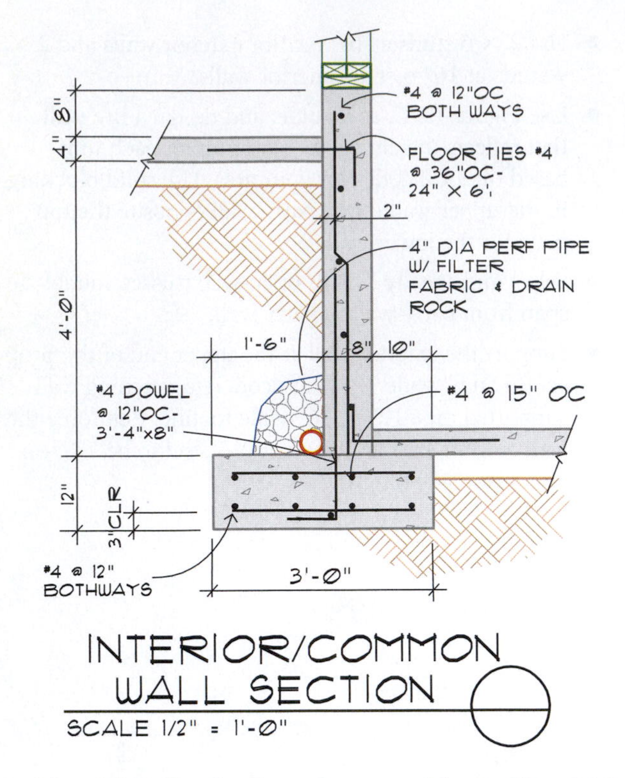

Problem 19-2a Use the floor plan created for Problem 16-2, the drawing above, and Figure 19-20 as a guide to draw a full section of the structure at a scale of 3/8" = 1'–0". Run the cutting plane north/south through the cantilevered roof and the storefront glass of the display area. Reference and draw the indicated details. Specify the following heights:

Top of storefront glass 8'–0"

Suspended ceiling 9'–6"

Top of 2 × 6 stud wall 11'–0"

Top of ledger at north wall 12'–0"

Top of masonry wall 16'–0" minimum

Minimum Project Standards

Your completed drawings should meet or exceed the following minimum standards unless noted:

- Block walls to be 8 × 8 × 16 grade A concrete blocks with #5 vertical at 36" o.c. and #5 at 24" o.c. horizontally.
- Provide (2)-#5 bond beams at 48" o.c. Use (4)-#5 at top of wall. Solid grout all steel cells.
- Support walls as per the foundation drawings in Chapter 24.
- Use (2)-#5 rebars in an 8" × 8" bond beam at ledger height.
    - Provide (2)-#5 continuous rebar in the footing 2" clear of the bottom and extend a #5 × 36" L from the footing to the wall with a 6" toe.
    - Space footing steel to match the wall steel.
- Provide a slab as per foundation plan.
- Use an 8 3/4" × 28 1/2" glu-lam for the ridge. Determine the location of the ridge based on the location of the steel column and pilaster on the floor plan.
- Although the location of the pilaster *must* remain modular, the centerline of the beam should be placed at an even number of feet from each wall if possible. *Edit the dimensions on the floor plan as required.*
- See the roof framing plan to determine the required truss size and spacing.
- Support the roof trusses on 3 × 12 DFPT ledger with 3/4" diameter A.B. at 24" o.c. staggered 3" up/down with 2" diameter washers. Use (4)-#5 rebars in 8 × 16 bond beam at ledger height. Provide a 3 × 12 DFPT ledger for the walls parallel to the trusses. Attach to the wall with 5/8" diameter × 10" A.B. at 36" o.c. staggered 3" up/down with 2" diameter washers.
    - Use 3 × 4 blocking between trusses with an appropriate "A" clip from block to ledger.
    - Use an MST strap capable of resisting 2700# of tension stress. Place the strap over the top chord of trusses at 8'–0" o.c.
- Weld the strap to a 3" × 4" × 1/4" steel plate with 1/8" fillet weld topside only. Weld (2)-3/4" × 6" studs to the backside of the plate and weld with 1/8" fillet welds all around.
- Use 4× ripped cant strip to support 26 gauge flashing at all wall/roof intersections. Use the following drawings to complete the following details.

Use the materials specified in the minimum standards, and the materials specified in Chapters 22 and 23 to complete this drawing.

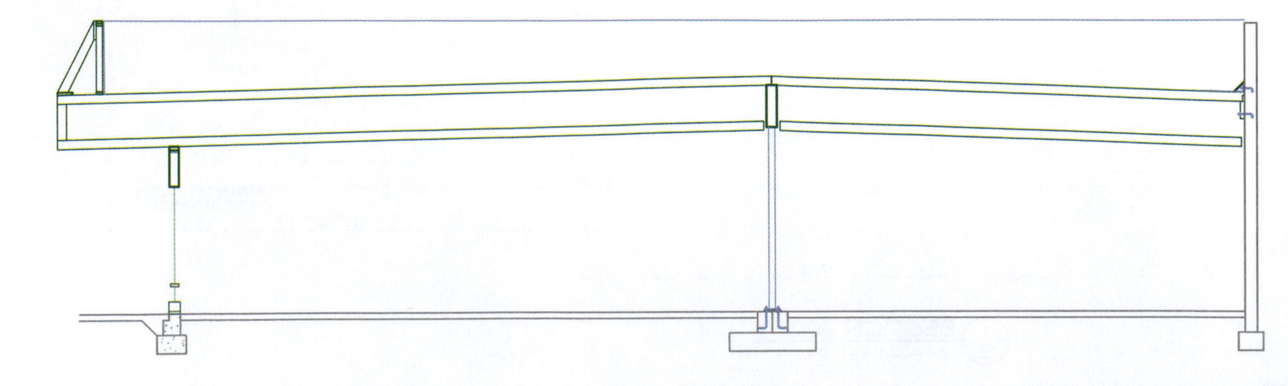

Problem 19-2b Draw and label an enlarged wall section (bearing wall) showing the roof/wall intersection and wall construction. See Chapters 22 and 23 for additional information that may be required to complete this drawing.

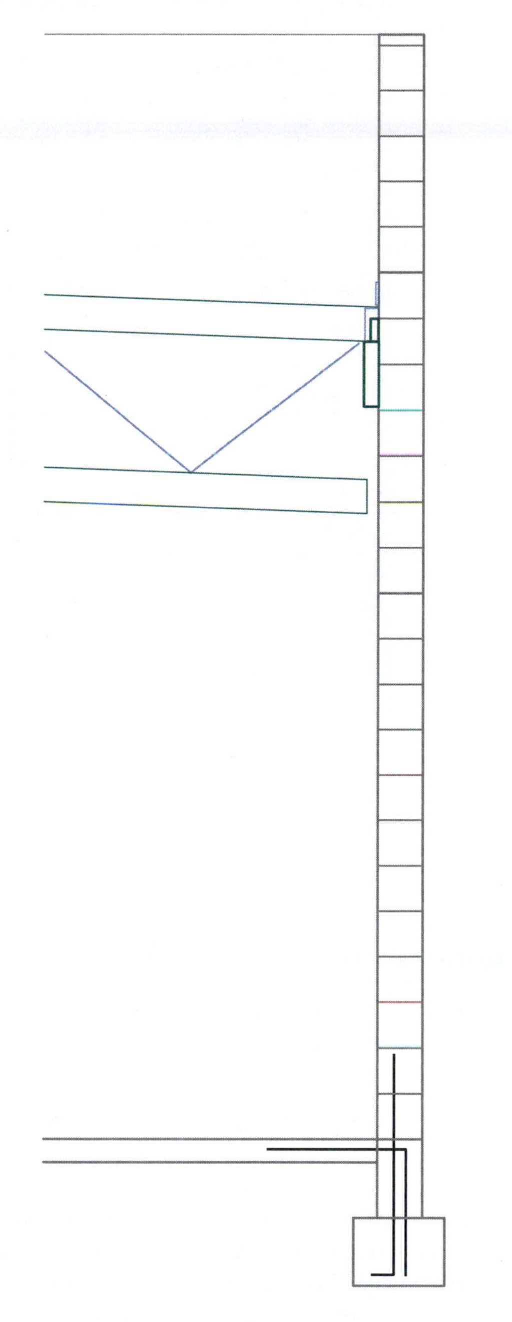

Problems 19-3a through 19-3d Use the floor plan from problem 16-3 to complete the required drawings. Use the parameters outlined below and on the drawings that follow to complete the required sections. Use the framing plans in Chapter 22 to determine all beam sizes.

Minimum Project Standards

Your completed sections should meet or exceed the following minimum standards unless noted:

- Roofing to be 4-ply built-up asphaltic fiberglass class A roofing over 5/8" 42/20 interior APA plywood with exterior glue roof sheathing laid perpendicular to purlins. Stagger the sheathing seams at each purlin. Nail with 10d common nails @ 4" o.c. at all panel edges and blocked areas, @ 6" o.c. at all supported panel edges @ unblocked areas, and at 12" o.c. at all intermediate supports.
    - Use MST 27 straps at 8'–0" o.c. to tie the trusses to the wall ledger.
- Walls to be 8" × 8" × 16" grade A concrete blocks reinforced with #5 verticals at 32" o.c. and #5 horizontal @ 24" o.c. with (2)-#5 @ 48" o.c.
    - Provide 8" × 16" bond beam with (4)-#5 at each roof and floor ledger, at midpoints between floor, and at top of all walls.
    - Provide an 8" × 16" bond beam over openings in the walls up to 10' wide and support with (4)-#5 horizontal bars with #3 ties @ 24" o.c. Provide an 8" × 24" bond beam over openings in the walls greater than 10' wide and support with (6)-#5 horizontal bars with #3 ties at 24" o.c.
- Support the walls of the office and warehouse as per the specifications in Chapter 24. All footings to extend 18" into natural grade.
- Provide a 4" thick slab for the office area and a 5" thick slab in the warehouse.

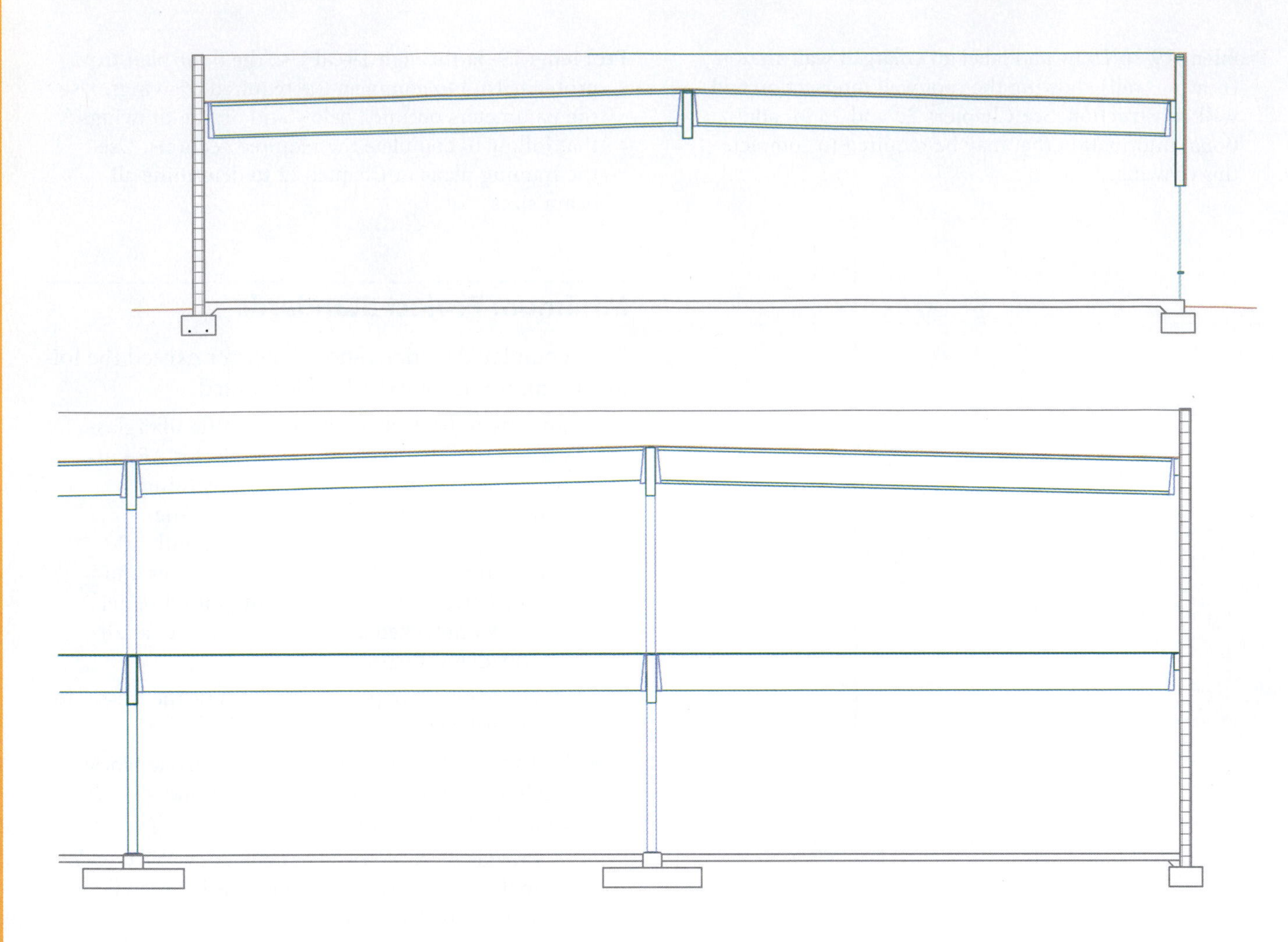

Problem 19-3a Use the architectural and structural drawings in Chapter 16 through 24 to complete the required section through the office area.

Problem 19-3b Use attached drawings and the balance of the architectural and structural drawings to complete the required section through the warehouse and mezzanine area.

Problem 19-3c Use the drawings provided for problem 19-3b and the balance of the architectural and structural drawings to complete the required section through the warehouse with no mezzanine.

Problem 19-3d Use the architectural and structural drawings to complete the required longitudinal section through the office, warehouse, and mezzanine.

Problems 19-4a through 19-4d Use the floor plan from problem 16-4 along with the following drawings to complete the required sections. Use the framing plans in Chapter 22 to determine all beam sizes.

Minimum Project Standards

Your completed sections should meet or exceed the following minimum standards unless noted:

- Roofing to be 4-ply built-up asphaltic fiberglass class A roofing over 5/8" 42/20 interior APA plywood with exterior glue roof sheathing laid perpendicular to trusses. Stagger seams at each truss. Nail with 10d common nails at 4" o.c. at all panel edges and blocked areas, at 6" o.c. at all supported panel edges at unblocked areas and at 12" o.c. at all intermediate supports.
    - Use MST 27 straps @ 8'–0" o.c. to tie the trusses to the wall ledger, and to tie the blocking (nonbearing walls at diaphragm) to ledger. Align straps at ridge with wall straps (see framing plans).
- All concrete walls are to be 6" thick with steel reinforcement as per the panel elevations in Chapter 23. Use trusses based on the sketches provided with the framing plans.
- Provide a concrete slab for the office and warehouse areas based on the foundation drawings.

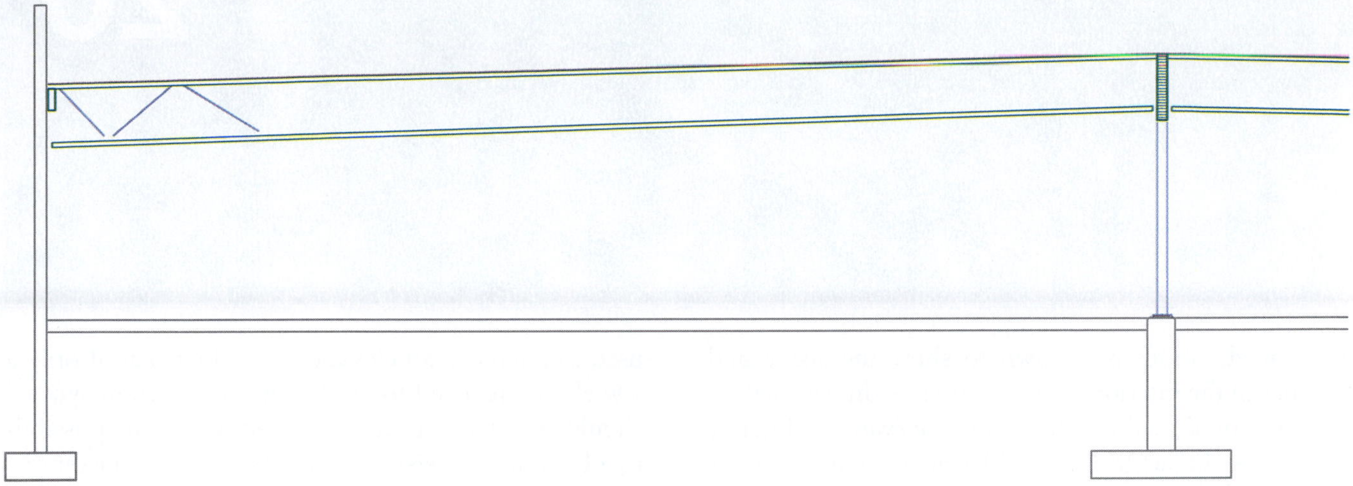

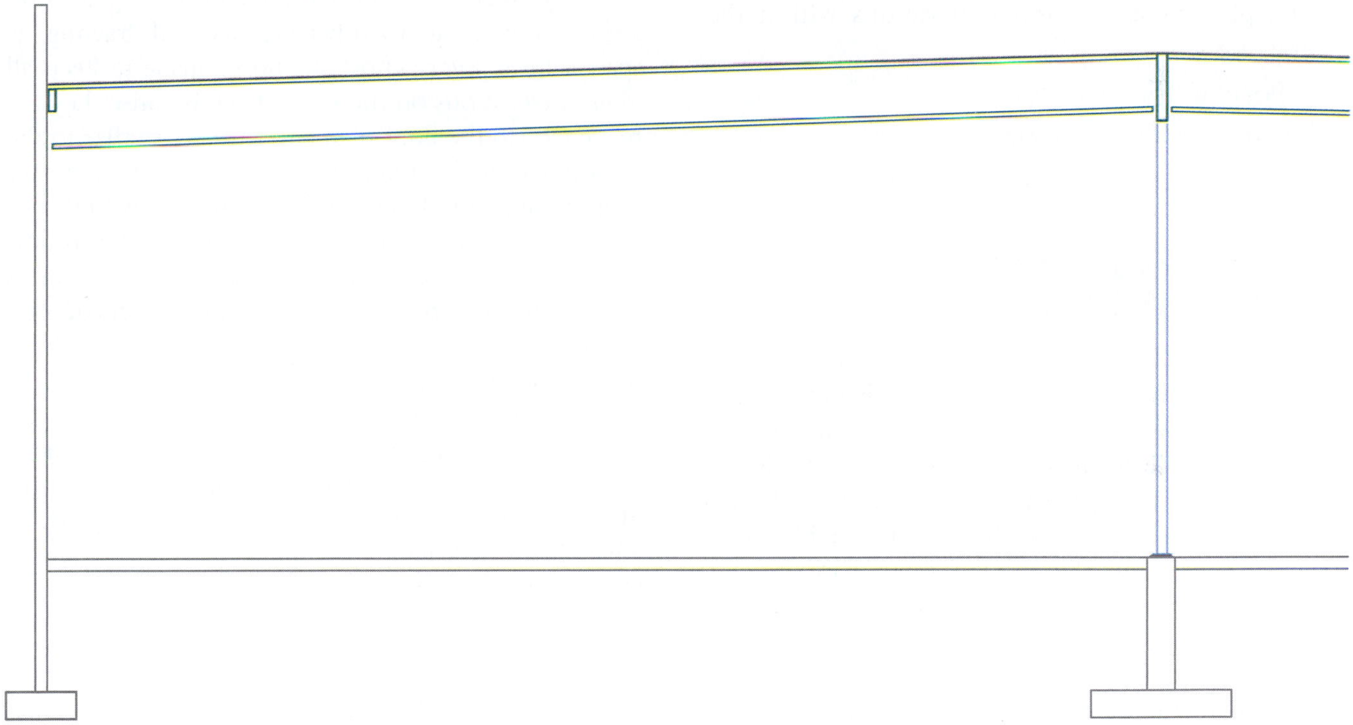

Problem 19-4a Use the balance of the architectural and structural drawings to complete the required section through the office area.

Problem 19-4b Use the attached drawing and the balance of the architectural and structural drawings to complete the required partial section through the warehouse and mezzanine area.

Problem 19-4c Use the drawings provided for problem 22-4b and the balance of the architectural and structural drawings to complete the required partial section through the warehouse with no mezzanine.

Problem 19-4d Use the architectural and structural drawings to complete a longitudinal section through the office, warehouse, and mezzanine.

Interior Elevations

20

Interior elevations are drawn to show the shape and finishes of the interior of a structure in a similar method that was used to draw the exterior elevations. Interior elevations similar to Figure 20-1 are drawn whenever a wall has features that are built into the wall. The complexity of the structure and office practice affect the method used to present materials in elevation. This chapter will explore:

- The placement of interior elevations within the drawing set.
- Types of interior drawings.
- Minimum drawing content.
- The drawing sequence for interior elevations.

ELEVATIONS WITHIN THE DRAWING SET

Except for a few features that are critical to the design of a structure, interior elevations are generally not a part of the preliminary elevations. Features such as an elaborate fireplace or intricate entry that is related to the design would be included in the preliminary design. Interior elevations are usually drawn after all other architectural drawings have been started. The floor plan, exterior elevations, and sections do not have to be complete before the interior elevations are started, but the basic arrangement of materials must be fairly well-established before the interior elevations can be started. Once basic shapes have been defined and the architectural team is progressing with the working drawings, the design team can now plan the arrangements of interior features by making preliminary drawings. Once approved by the client, these drawings can be completed by the CAD technician.

Referencing Interior Elevations

Interior elevations are based on the position of materials that are shown on the floor plan. Elevations are referenced to the floor plan by the use of a title or a symbol. Titles such as North Lobby Elevation can be used to reference an elevation to a floor plan if only a few elevations need to be drawn. An elevation symbol should be placed on the floor plan where any possibility of confusion exists. Figure 20-2 shows an example of an elevation symbol represented on a floor plan and Figure 20-3 shows the corresponding elevation. Although interior elevations are referenced to the floor plan, their location in the architectural drawings will vary greatly. Interior elevations can be placed on a sheet containing any other architectural drawing, as space allows. Every effort should be made to keep all interior elevations on the same sheet. Because they are referenced to the floor plan and they are similar to the exterior elevations, interior elevations for a very simple project are often placed on one of these drawings. For projects such as a commercial kitchen, or other rooms with multiple interior features, a separate sheet is required to show the interior elevations. Sheets of interior elevations are often placed immediately behind the exterior elevations (the 200 numbered sheets) or with the enlarged drawings (the 400 numbered sheets) in the drawing set. Regardless of where they are placed within the drawings, the location of interior elevations, like other drawings, should be clearly indicated in the table of contents, which is contained on the title page. See Chapter 2 for a review of title pages and drawing placement.

Drawing Projection

Interior elevations are similar to the exterior elevations in that they are drawn using orthographic projections based on a viewing plane. In theory, a viewing plane is placed on the floor plan parallel to each wall containing material that must be explained. Objects are then projected onto the plane to create a two-dimensional view of the wall. The location of the viewing plane is indicated on the floor plan by using the reference symbol. Objects parallel to the viewing plane are projected to the viewing plane just as with exterior elevations. In drawing the interior elevations, in addition to using the floor plan to determine the shape, the sections can be used to determine interior

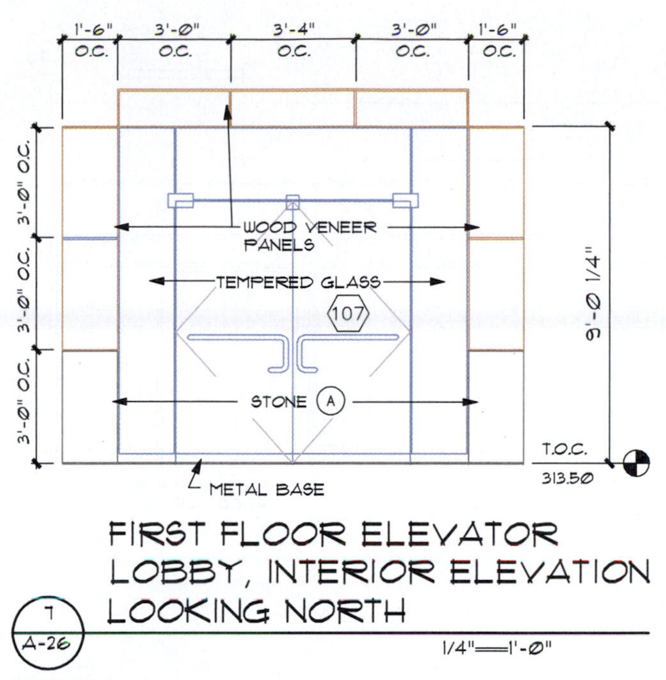

FIGURE 20-1 Interior elevations are drawn whenever a wall has features that must be built or installed as part of the construction contract. *Courtesy Ned Peck, Peck, Smiley, Ettlin Architects.*

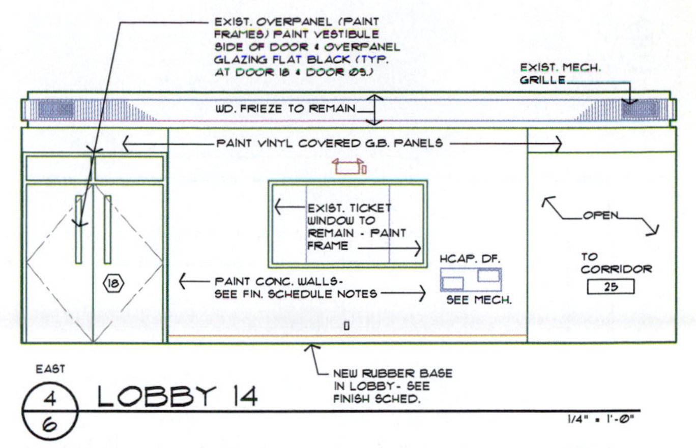

FIGURE 20-3 The elevation referenced to the floor plan shown in Figure 20-2. *Courtesy Architects Barrentine, Bates & Lee, A.I.A.*

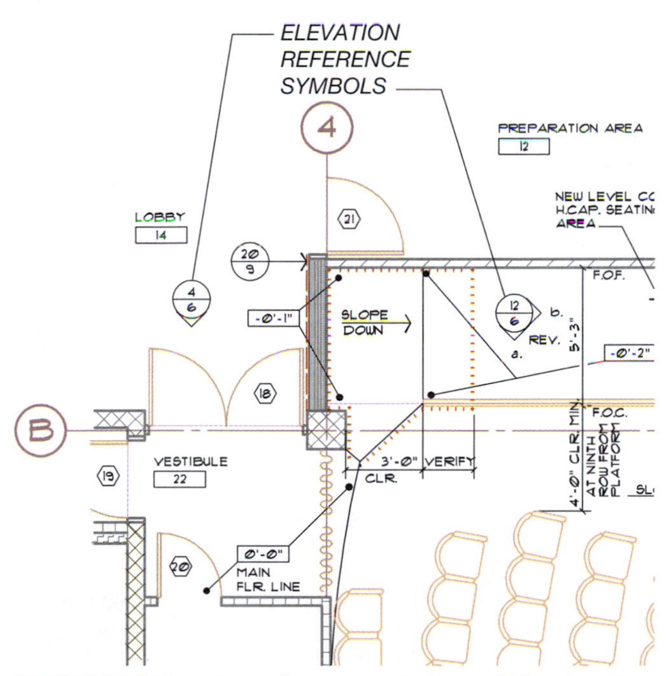

FIGURE 20-2 Elevation reference symbols should be placed to define the elevations to be drawn and their location. *Courtesy Architects Barrentine, Bates & Lee, A.I.A.*

heights. Base cabinets are often 36" (900 mm) high, with upper cabinets typically placed 18", 24", or 30" (450, 600, or 750 mm) above the lower counters. Because cabinet placement and sizes are so dependent on the equipment that will be installed in the cabinets, the drafter should rely on vendor catalogs and drawing standards such as *Time Saver Standards* and *Architectural Graphic Standards*.

TYPES OF ELEVATIONS

The material shown in an interior elevation will vary depending on the complexity of the structure. Generally, elevations can be grouped into the categories of wall elevations and cabinet elevations.

Wall Elevations

Wall elevations can show an entire wall or a partial wall. All materials that are built into or mounted on the wall should be shown. This includes showing special finishes as well as items such as automated teller machines, fire extinguishers, drinking fountains, and phone stalls, similar to the elevation shown in Figure 20-4. The elevation becomes a source to reference details similar to those shown in Figure 20-5, which can be used to explain construction of the enclosures as well as mounting of equipment. Wall elevations are also used for defining the limits of interior finishes. Figure 20-6 shows an example of the lobby of a two-story structure and the various materials used to cover the structural frame.

Cabinet Elevations

Perhaps the most common use of interior elevations is to show the cabinets, appliances, grab bars, and plumbing fixtures used in bathrooms and kitchens. Nearly every structure drawn contains at least one bathroom

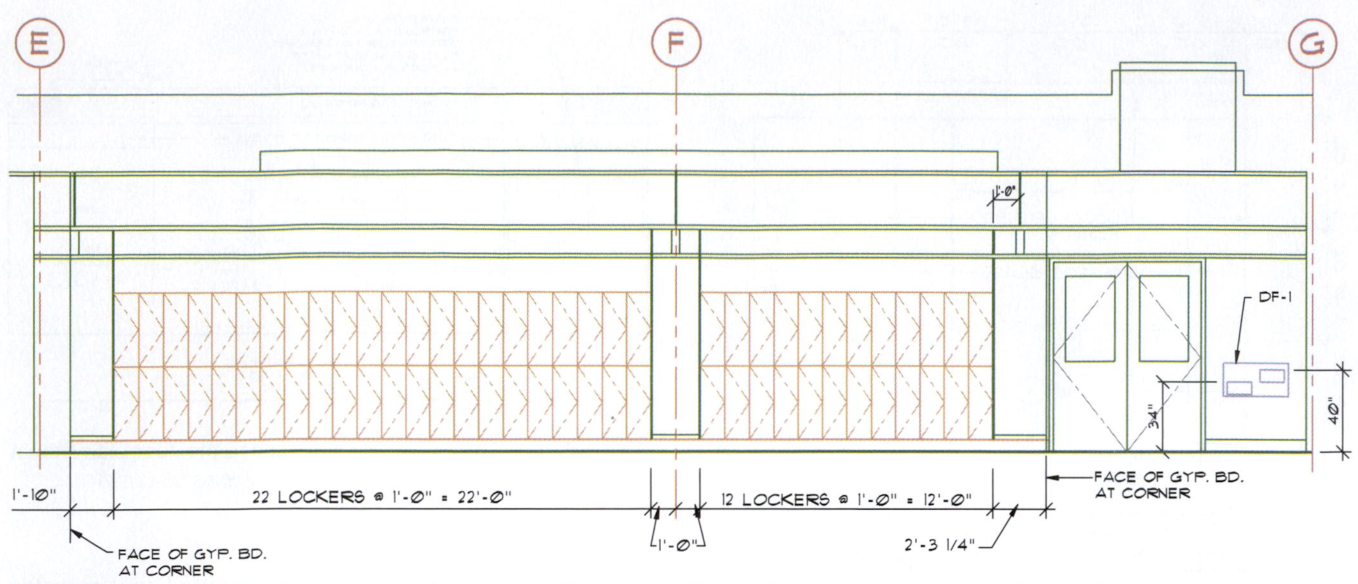

FIGURE 20-4 An interior elevation is used to show the location of all special equipment. *Courtesy Michael & Kuhns Architects, P.C.*

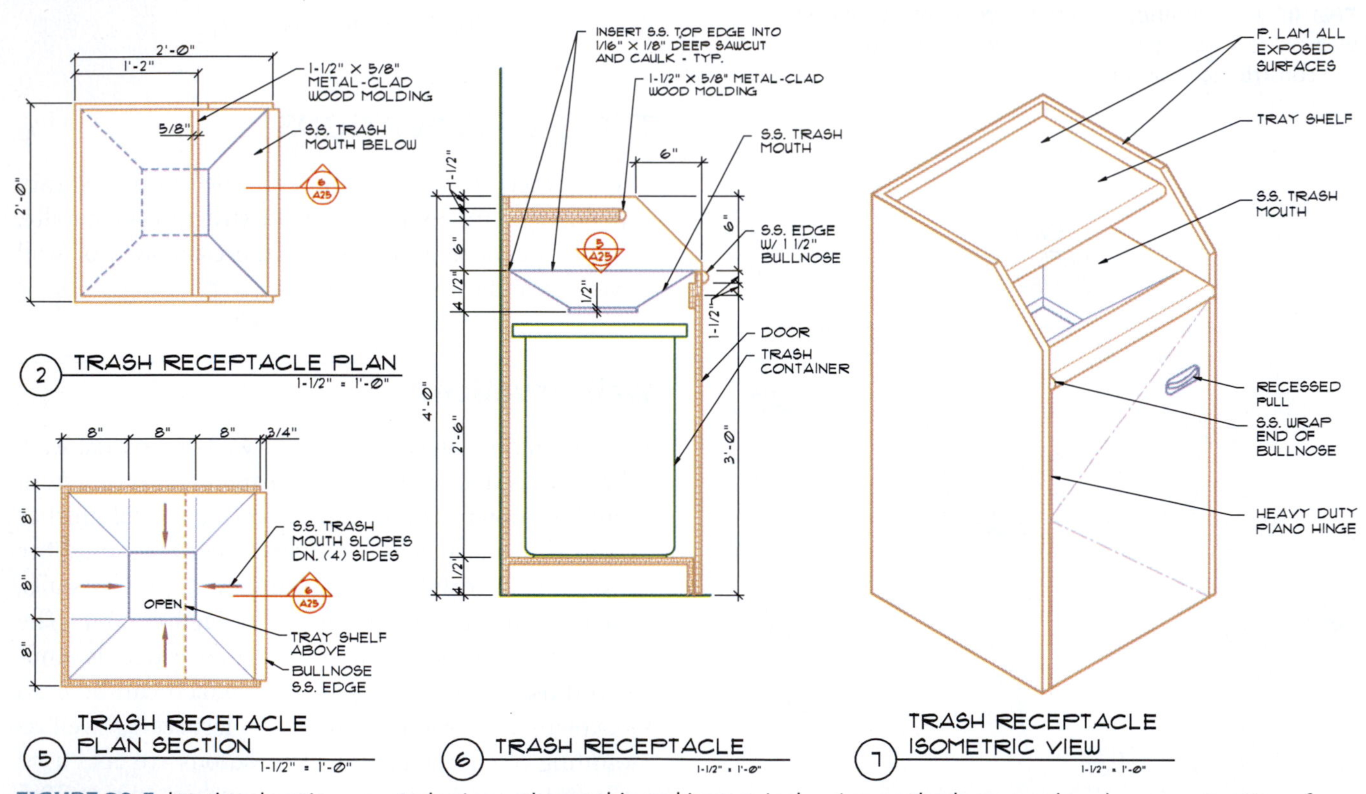

FIGURE 20-5 Interior elevations created using orthographic and isometric drawing methods are used to show construction of specialized cabinets. *Courtesy Michael & Kuhns Architects, P.C.*

that requires elevations. Figure 20-7 shows the elevation for a simple washroom and standard details that accompany the elevation to ensure minimum allowances to meet both the model code and Americans with Disabilities Act (ADA) standards. Figure 20-8 shows the elevation of each wall of the bathroom shown in Figure 16-24.

For multifamily occupancies, a CAD technician working with the architectural team draws an elevation of each wall showing kitchen and bathroom cabinets and fixtures. Drawings must show each fixture and all door and drawer locations of the adjoining cabinets. Figure 20-9 shows the bath and kitchen elevations for a unit in a multilevel apartment

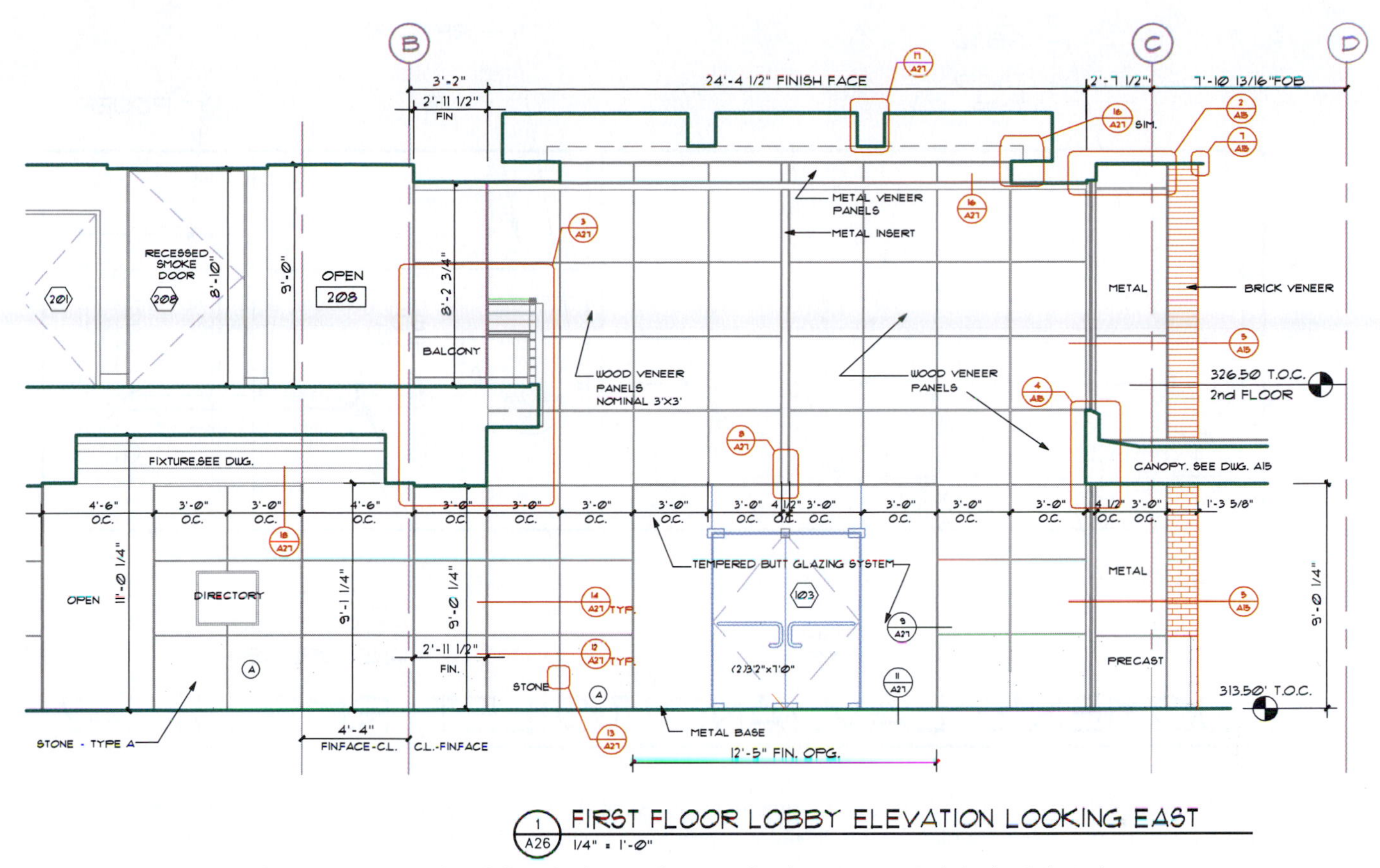

FIGURE 20-6 Interior elevations are used to define the limits of interior finishes. *Courtesy Peck, Smiley, Ettlin Architects.*

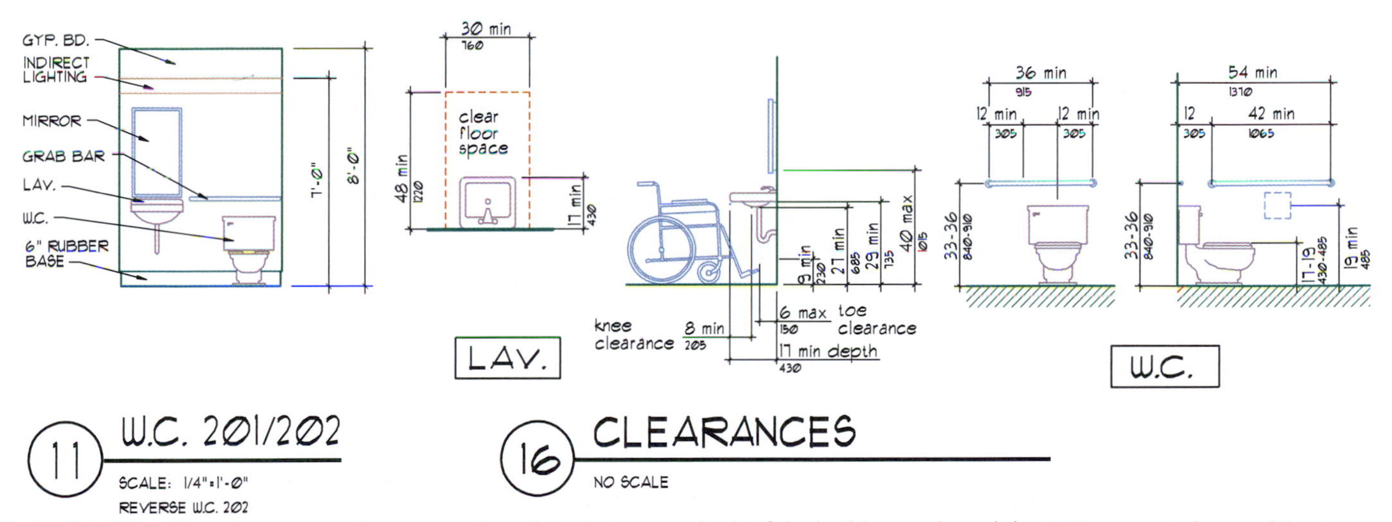

FIGURE 20-7 Elevations are used to ensure that all minimum standards of the building code and the ADA are met. *Courtesy DiLoreto Architects, LLC.*

complex. Elevations for each unit are provided. When a cooking area is used to serve the public, the supplier of the kitchen equipment generally develops the elevations for the kitchen equipment. Figure 20-10 shows the floor plan developed for a commercial kitchen. In addition to the cabinet drawings, many plans include drawings that show specialty equipment to be used in the kitchen. CAD drafters working for a consulting firm specializing in interior fixtures draw these drawings. Several of the specialty drawings can be found in Chapter 2. Figure 20-11 shows the interior elevations for this kitchen. Generally, drawings for specialty equipment are placed separate from the interior drawings and are placed in a section of the drawings titled Food Service (FS). The location of these drawings is listed in the table of contents.

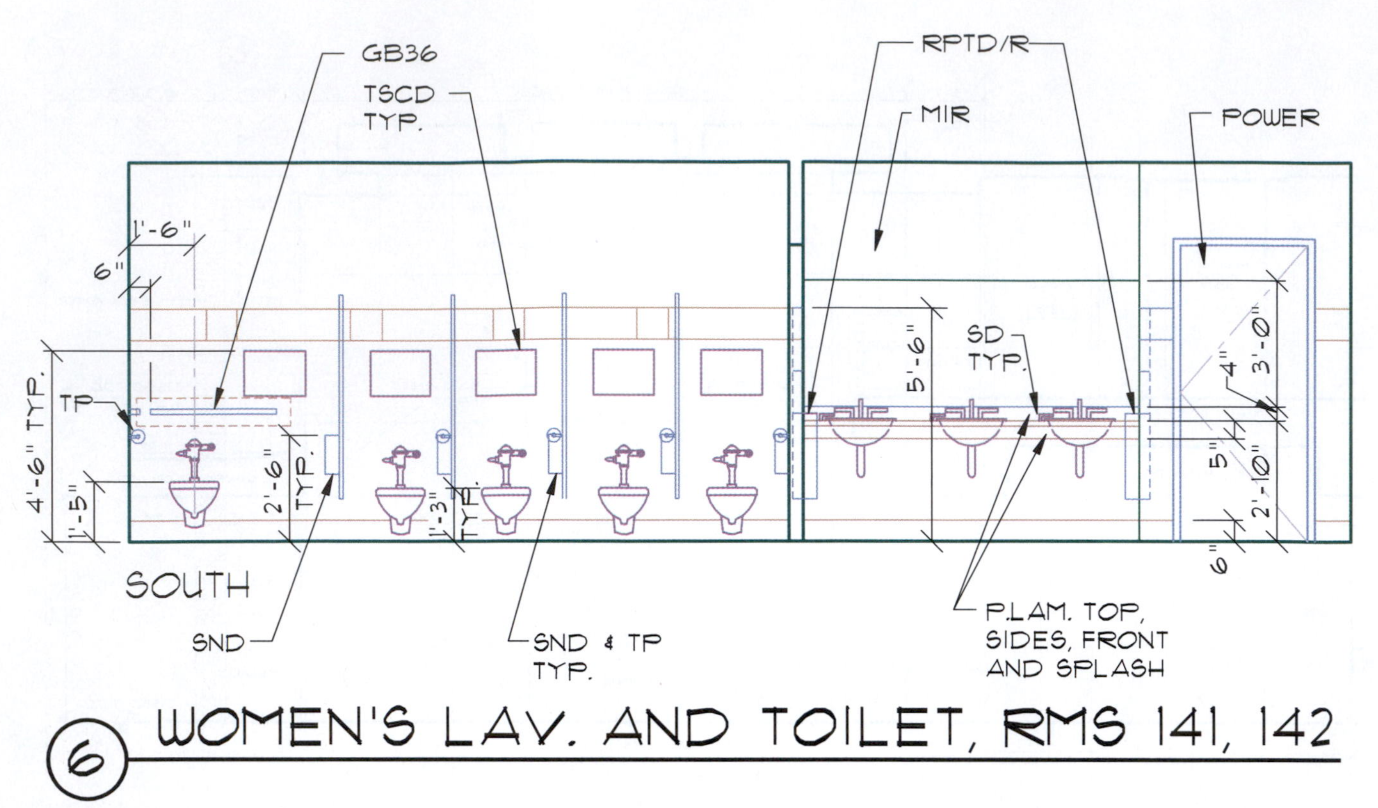

FIGURE 20-8 Because of the large amount of fixtures, equipment, and special finishes, an elevation for a bathroom can become quite complex. The corresponding floor plan for this elevation can be seen in Figure 16-24. *Courtesy Architects Barrentine, Bates & Lee, A.I.A.*

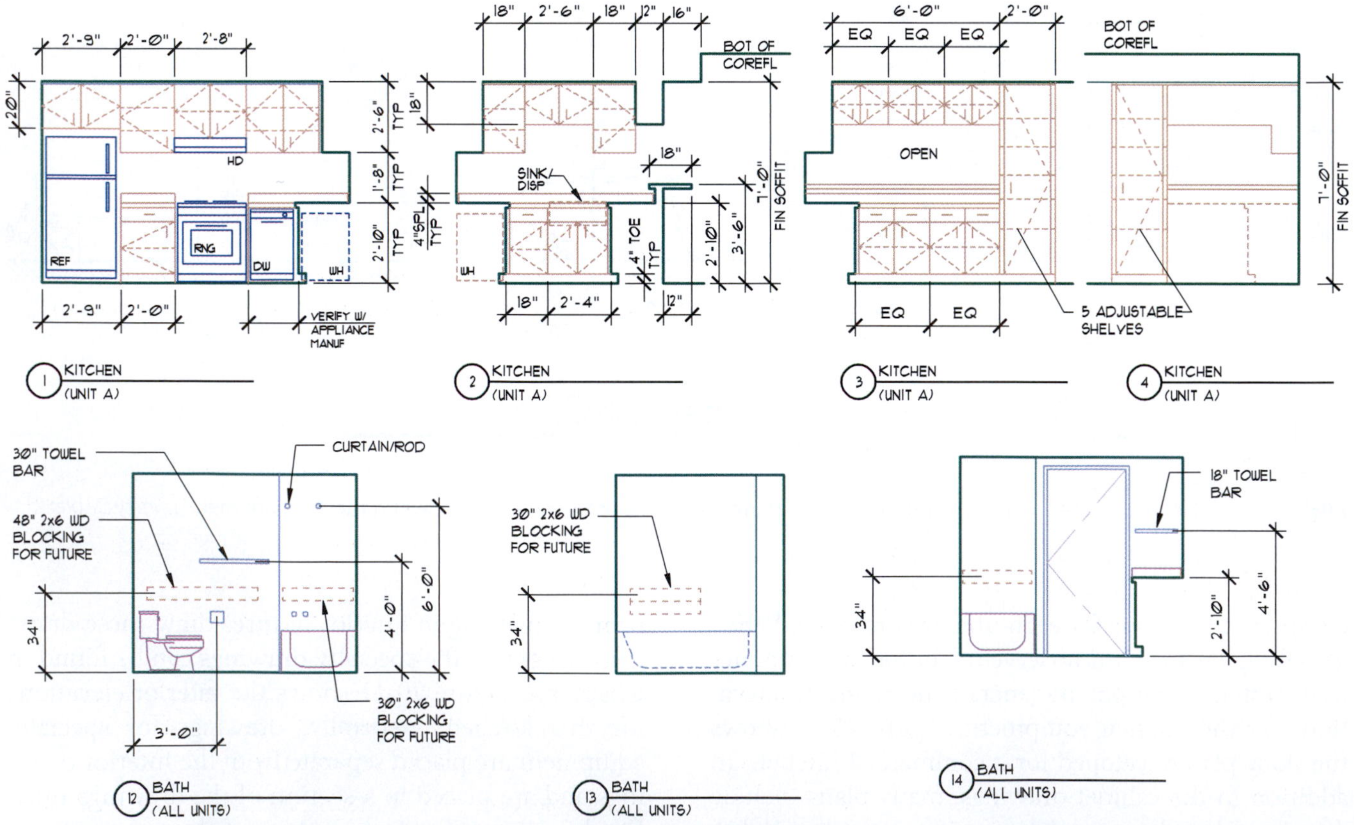

FIGURE 20-9 The elevations for a multifamily complex will often resemble the elevations drawn for a single-family residence. *Courtesy H.D.N. Architects, A.I.A.*

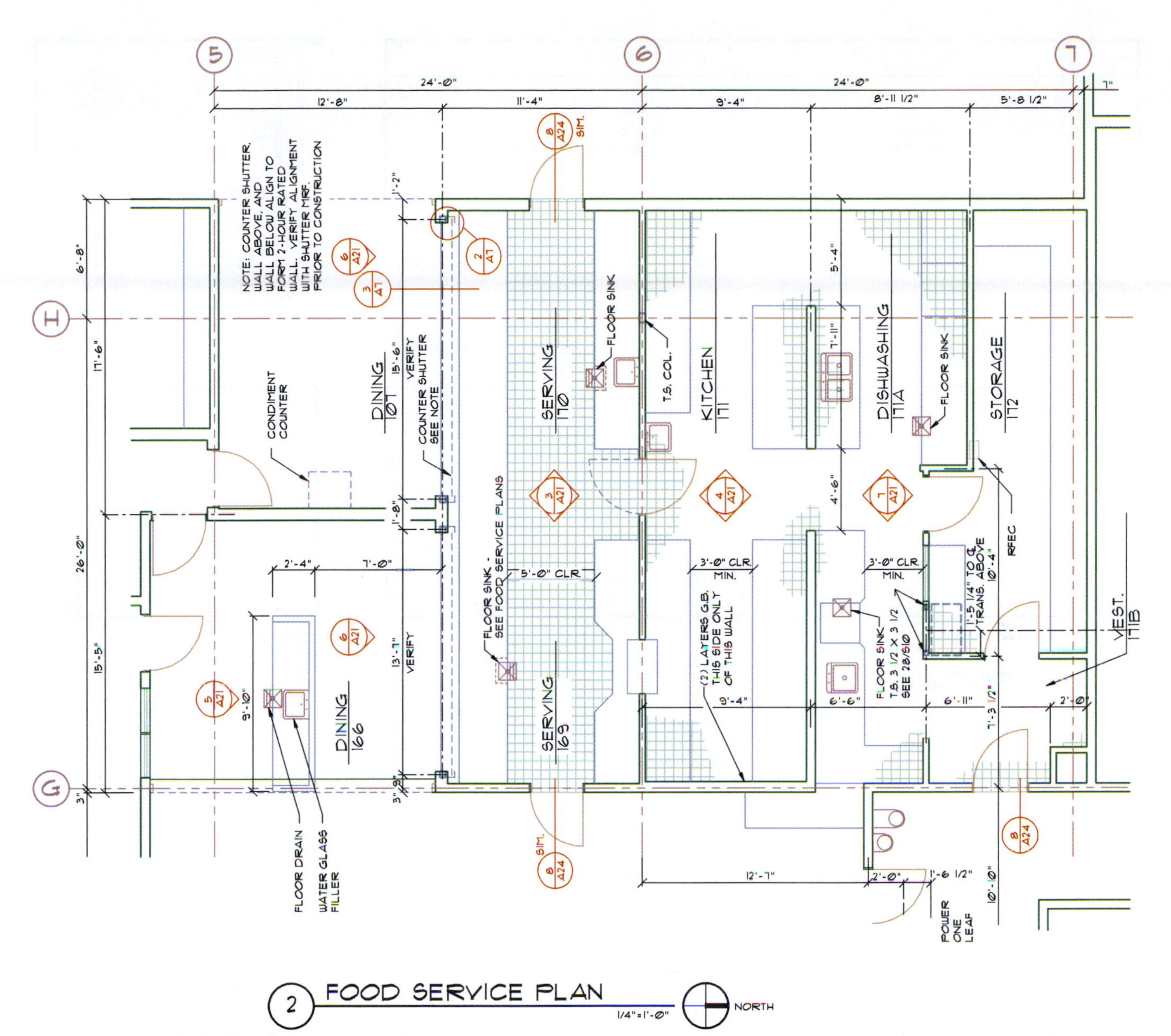

FIGURE 20-10 The kitchen for a commercial complex is far more detailed than that of a multifamily project and requires additional drawings from subcontractors. The corresponding floor plan for this kitchen can be seen in Figure 2-5. Use the grids to help coordinate the two plan views. *Courtesy Michael & Kuhns Architects, P.C.*

Cabinet Sections

A third type of interior drawing that may be provided is an interior section. An interior section is used to show construction methods and materials of interior built-ins. Cabinet sections should be provided when the built-in is fabricated at the job site. When cabinets are built off-site, the cabinet supplier often develops the drawings required for cabinet construction, and submits them to the architectural team for approval. When required, the drafter must be careful to clearly define each material that will be used. Figure 20-12 shows a cabinet section and common methods of representing thin materials.

MINIMUM DRAWING CONTENTS

No matter the type of interior drawings that are provided, several basic types of information should be provided. Based on the National CAD Standards (NCS), minimum drawing contents include:

- Access panels for all features specified on the electrical or mechanical plans.
- Changes in wall materials. If more than one material is to be placed on a wall, the extent of the material should be drawn and dimensioned.
- Dimensions—all vertical specifications, and any horizontal dimensions that are not represented on plan views.

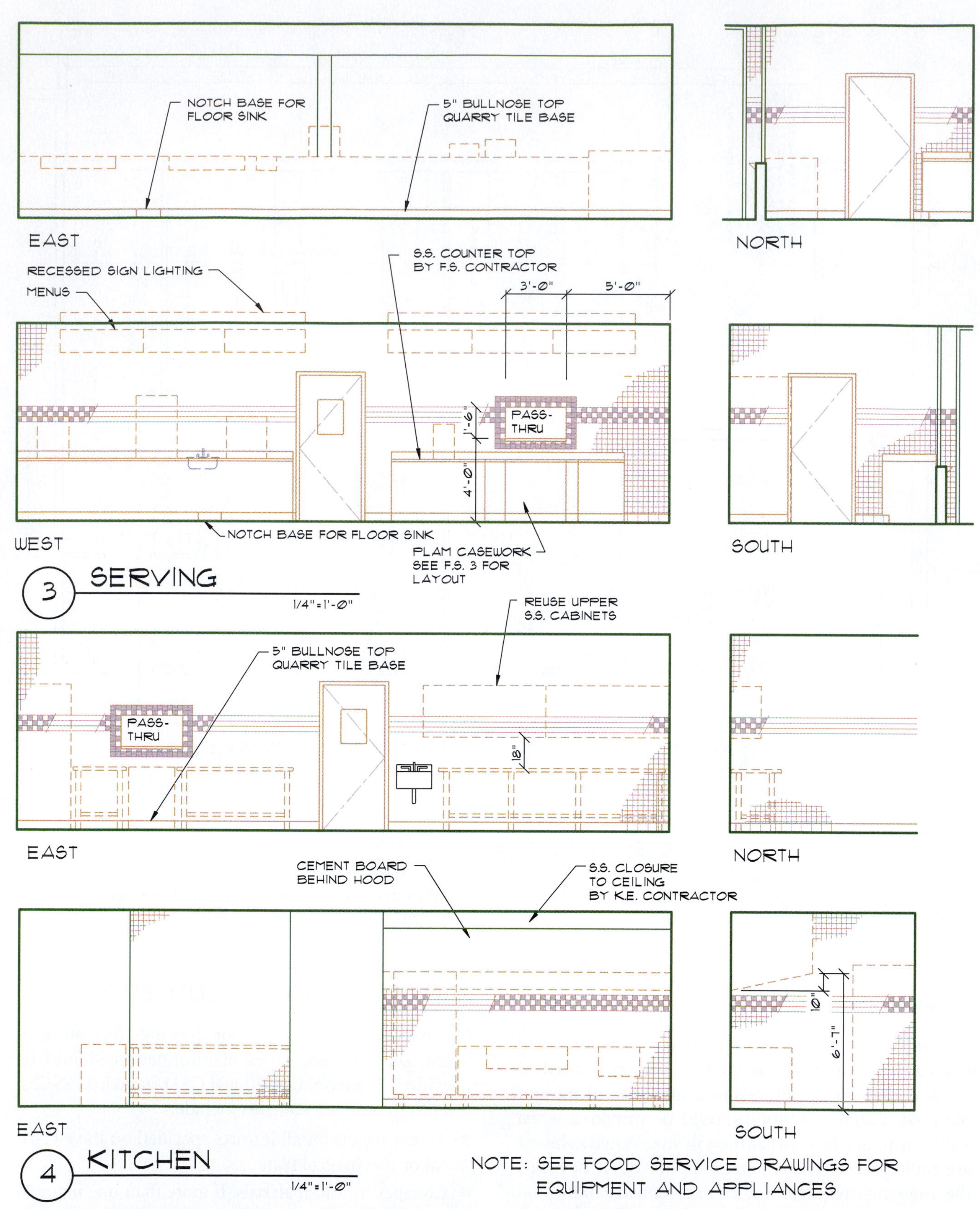

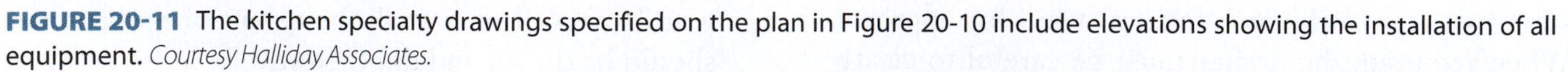

NOTE: SEE FOOD SERVICE DRAWINGS FOR EQUIPMENT AND APPLIANCES

FIGURE 20-11 The kitchen specialty drawings specified on the plan in Figure 20-10 include elevations showing the installation of all equipment. *Courtesy Halliday Associates.*

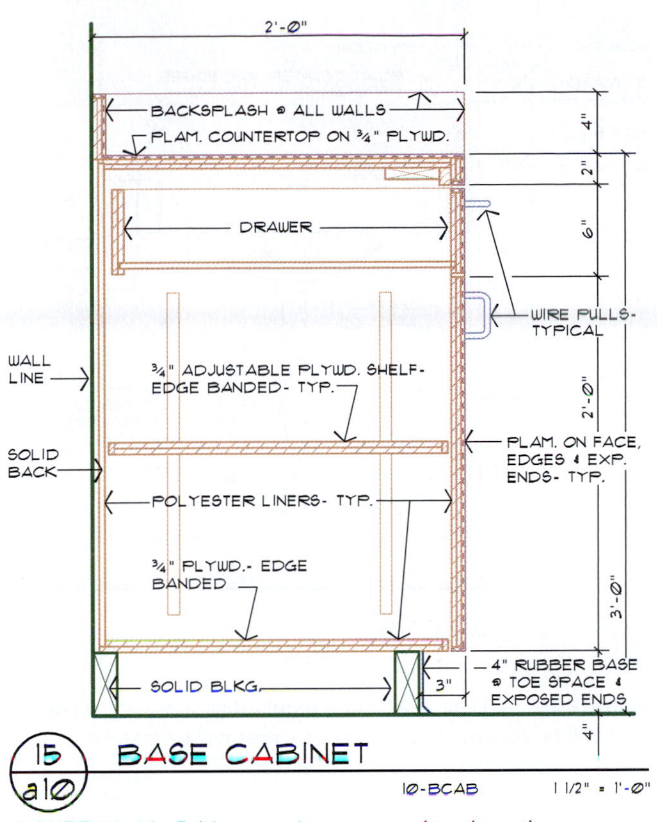

FIGURE 20-12 Cabinet sections are used to show the construction of specialty equipment. *Courtesy Architects Barrentine, Bates & Lee, A.I.A.*

- Door heights and locations.
- Grid lines.
- Large pipe and duct penetrations.
- Mounting heights and locations of all equipment and fixtures.
- Outlets such as data, power, telephone, and other miscellaneous outlets that will affect casework, equipment, or placement of other key interior features.
- Coordination of each interior drawing with interior and equipment drawings.

Choosing a Drawing Scale

The complexity of the materials to be drawn determines the scale used to complete the cabinet elevations and details. A scale of 1/4" = 1'–0" (1:50) is often used for elevations, but both smaller and larger scales can be found on professional examples. In choosing the drawing scale, the technician must be able to show sufficient detail to meet the demands of the drawings. If the drawings are to be used to represent premanufactured materials, a drawing using a small scale showing little detail can be used. If the drawings are used to construct the item, a scale sufficient to show all detail must be used. Scales for construction elevations are often drawn at a scale of 3/4" or 1/2" = 1'–0" (1:20). Details are usually drawn at a scale of 1" = 1'–0" or larger (1:10).

Material Representation

The symbols that are used to represent materials in sections and details are also used to represent material drawn in cabinet details. Continuous lines represent materials that are represented in elevation, with varied thickness. Thick lines usually represent the floor, walls, and ceiling, which define the view to be seen. On L-shaped cabinets, the line that represents the drawing limits will follow the contour of the cabinets when a cabinet is shown in end view. Figure 20-13 shows the outline of a room with

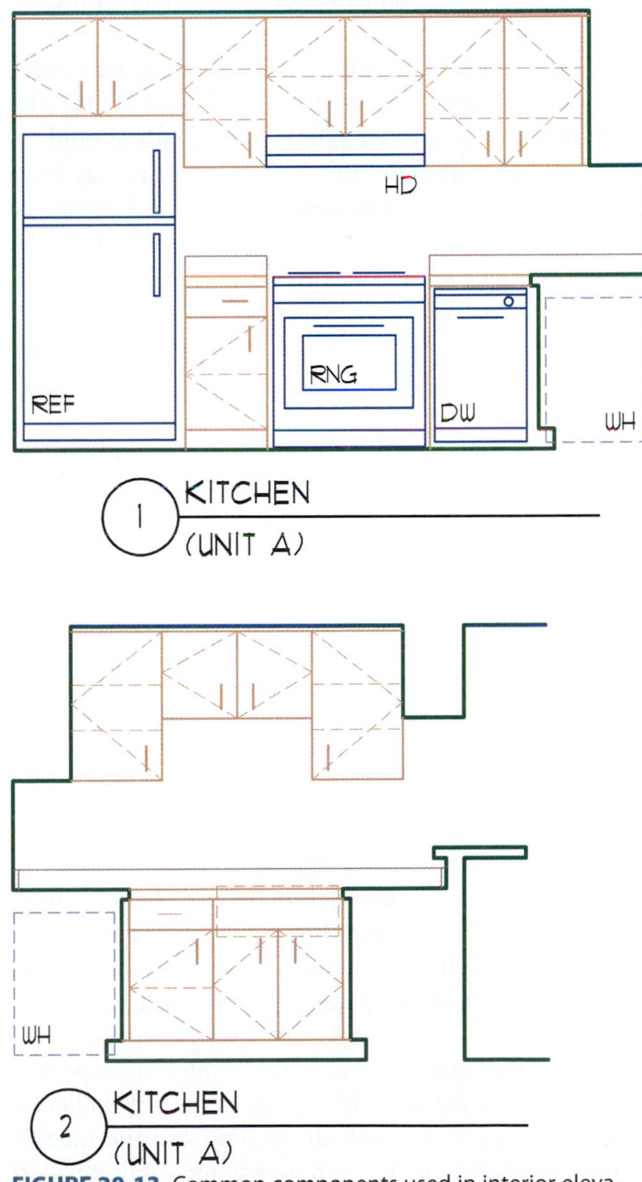

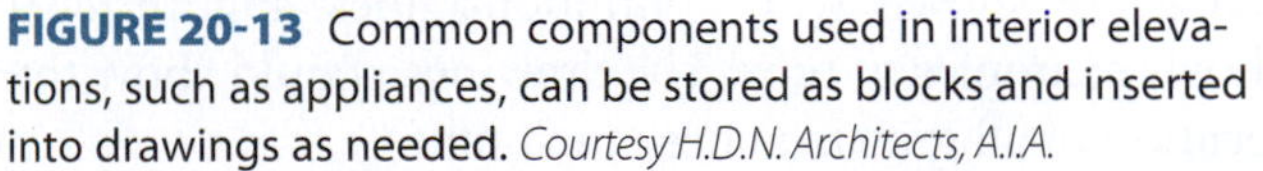

FIGURE 20-13 Common components used in interior elevations, such as appliances, can be stored as blocks and inserted into drawings as needed. *Courtesy H.D.N. Architects, A.I.A.*

L-shaped cabinets. Common features represented in the elevations, such as the range, dishwasher, and refrigerator, are usually inserted as blocks.

Cabinetry

The outline of cabinets and the lines representing doors, drawers, and exposed shelves are thin, continuous lines. A single thin, dashed line is used to represent shelves that are hidden behind doors. On larger scale drawings, pairs of continuous lines are used to represent shelves. Knobs and handles can be represented by a short, thick line or a circle, to show the approximate style of hardware. Hinges are not usually shown, but door swings are represented by continuous or dashed lines in the shape of a V, drawn through the door. The point of the V represents the hinged side of the door. Figure 20-13 shows an elevation of cabinets with door swings represented.

When drawn in section, doors, drawers, and hardware can be represented as shown in Figure 20-12. The location of partitions for public toilet stalls should be represented on the elevations. Thin continuous lines can be used to represent walls that are perpendicular to the viewing plane. Doors can be omitted to show the location of toilets, or they can be represented by thin dashed lines to show the direction of swing. Figure 20-8 shows the location of premanufactured toilet partitions.

Plumbing Fixtures

Because of the repetitive nature of elevations, plumbing symbols should be represented by user or third-party blocks. Cabinet-mounted lavatories and sinks can be represented by hidden or continuous lines to define the outline of the unit. Fixtures built into a counter can be drawn using the same methods as for a cabinet-mounted unit. Wall-mounted and pedestal lavatories are drawn using thin continuous lines to represent the outline and some detailing of the unit. Figures 20-7, 20-8, 20-9, and 20-13 show examples of sinks and lavatories in elevation. Figure 20-14 shows a detail of a counter-mounted sink.

Thin lines are used to represent water closets and urinals, showing the basic components of the fixture. Blocks should be developed to show both a front and side view of each unit. Figure 20-15 shows common methods of representing water closets and urinals. A thin continuous line should be used to represent the outlines of tubs and showers, with the interior shape represented by dashed lines. When the unit is perpendicular to the viewing plane, both the outline and the interior shape are shown with continuous lines. Figure 20-9 shows examples of tubs. The drawings should show the location of all grab bars near each unit.

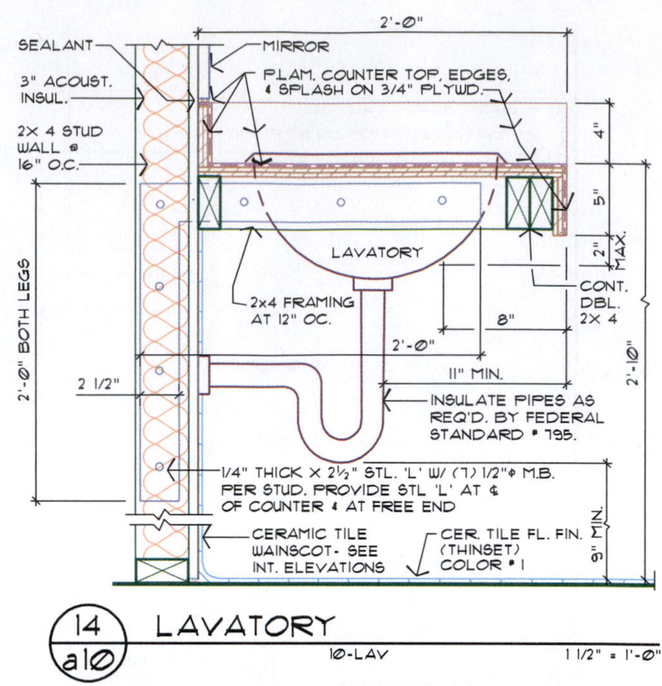

FIGURE 20-14 Details are often required to show how cabinets will be constructed and how fixtures will be installed. *Courtesy Architects Barrentine, Bates & Lee, A.I.A.*

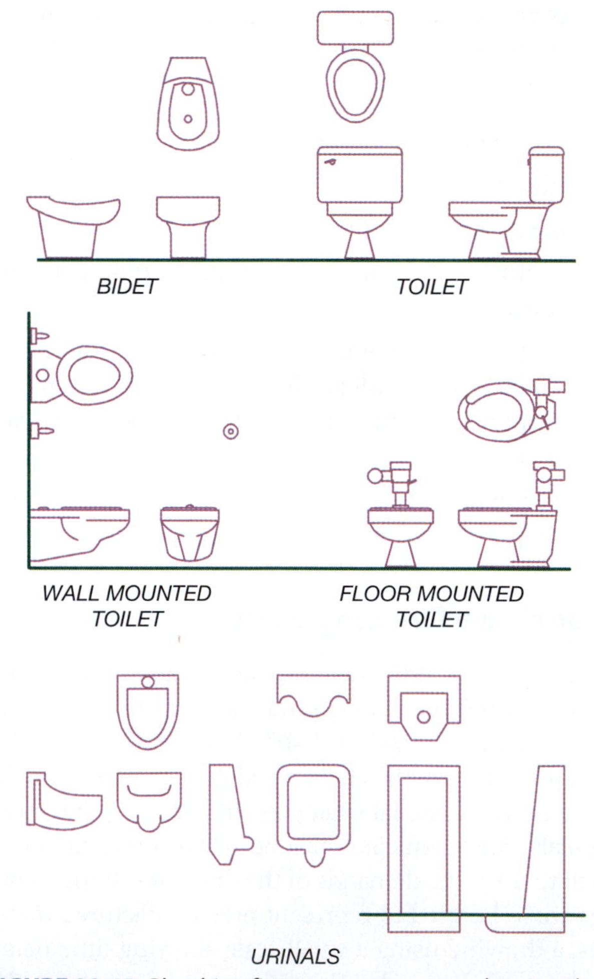

FIGURE 20-15 Plumbing fixtures common to architectural drawings. *Courtesy Berol Rapidesign.*

Appliances

Kitchen appliances are represented by a box that outlines the shape or by a detailed representation. The detailed representation should not be drawn to represent an exact model, but a generic representation. Figures 20-9 and 20-13 show cabinet elevations with a stove and a refrigerator. These appliances can be easily recognized but are not of a specific unit. In creating blocks, show doors or drawers that open and the general location of control features, but avoid spending time adding features that do not aid in installation.

Special Equipment

The occupancy of the structure dictates what equipment is required. Equipment for fire protection, drinking fountains, telephone stalls, mirrors, paper towel dispensers, access ladders, and building directories should all be represented on the elevations. Vendor specification can be an excellent source for obtaining information for drawing the needed feature. Many equipment suppliers will provide disk copies of their products, similar to the blocks shown in Figure 20-16.

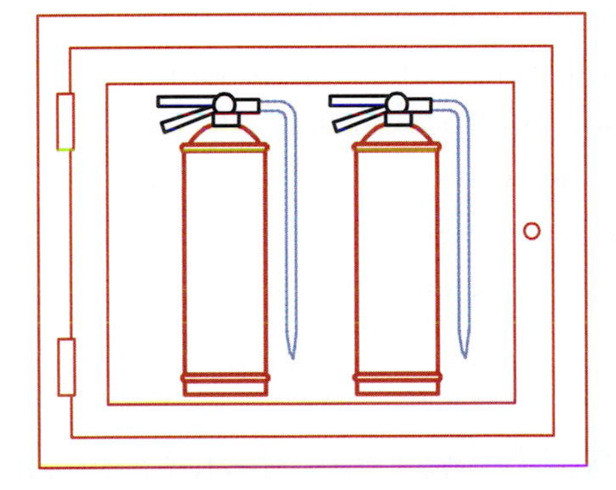

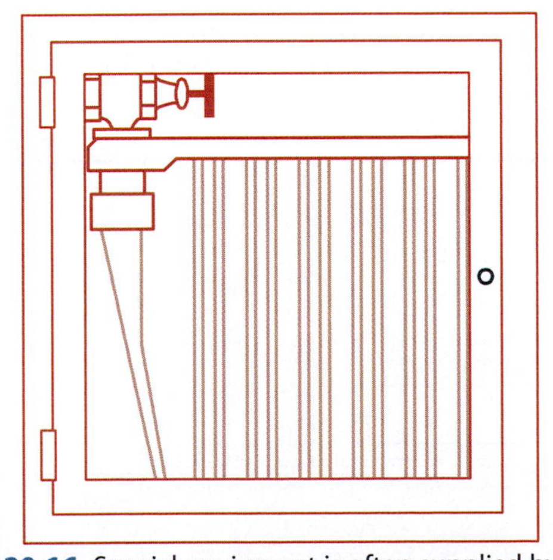

FIGURE 20-16 Special equipment is often supplied by vendors or in libraries available from third-party suppliers. *Courtesy Berol Rapidesign.*

Drawing Notations

The materials represented in elevation require both written and dimensional clarification. Notations must be added to each elevation to clearly define objects that have been drawn. As with sections, the larger the drawing scale, the more detailed the specifications should be. Information to be specified on the elevation is provided by the architect's preliminary drawings and check prints, and by consultation with *Time Saver Standards* or *Architectural Graphics Standards*. Vendor specifications should also be consulted to determine complete specifications. In addition to notations specifying each material, many cabinet elevations contain complete specifications to provide instructions for mounting and installation of each material to be used.

Vertical dimensions should be provided to locate the limits of all counter heights, cabinets, special equipment, and soffits that are not specified on the sections. Horizontal dimensions are used to locate special equipment from grid lines. Horizontal dimensions are also used to locate divisions in cabinets as shown in Figure 20-17. Cabinets and specialty items are generally referenced to a grid line or to a wall that is referenced to a grid. Although no numbers have been given for the dimensions shown in Figure 20-17, the cabinetmaker will determine the exact size required to make four equally spaced units, based on the finished dimensions of the construction site. When the architect does not specify dimensions in the preliminary drawings,

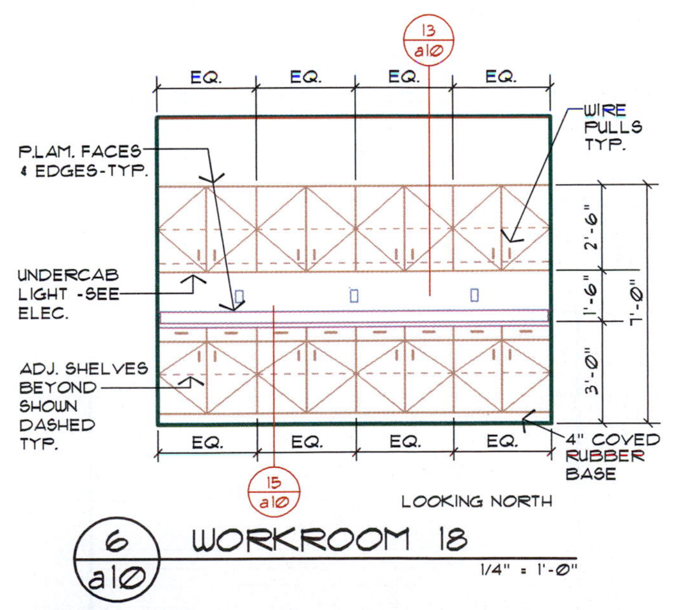

FIGURE 20-17 Vertical dimensions must be supplied for all material or equipment that will be installed. Horizontal dimensions can be omitted if they have been placed on a plan view. *Courtesy Architects Barrentine, Bates & Lee, A.I.A.*

the CAD drafter should provide dimensions based on common practice, building and ADA codes, and supplier recommendations.

DRAWING SEQUENCE

Figure 20-18 shows a portion of the floor plan for two rooms from the educational complex shown in Figures 16-38 and 16-40. This unit is used for completing the cabinets for these rooms. Assume that they are 36" base cabinets, with the bottom of the upper cabinets set at 4'–6" and the top at 7'–0" above the floor. Common steps in completing an interior elevation include:

1. Establish a viewing plane on the floor plan to indicate the needed elevations.
2. Use a block of the floor plan to project the outline of all walls that enclose the elevations.

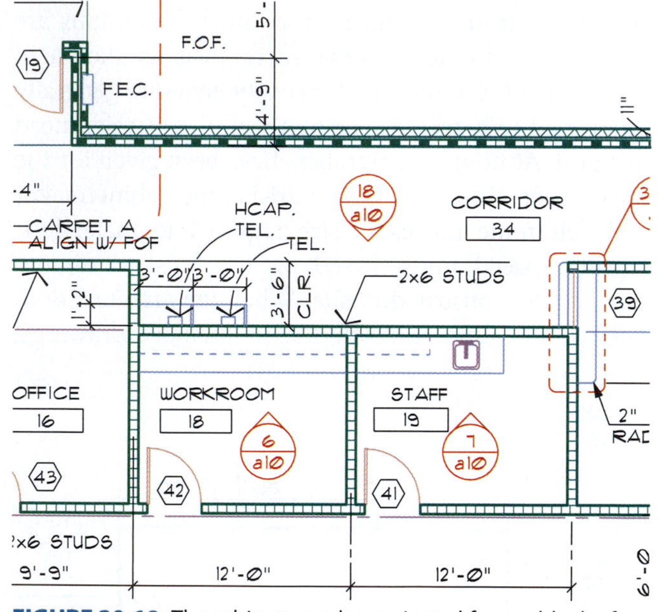

FIGURE 20-18 The cabinets can be projected from a block of the floor plan, or drawn by following the dimensions on the plan.

3. Establish grid lines on the elevations.
4. Establish a line to represent the finish floor.
5. Use the building sections to determine the heights of all ceilings and openings that are parallel to the viewing plane.
6. Draw the outline of each cabinet shape.

The initial layout should resemble Figure 20-19.

7. Insert blocks to represent each appliance.
8. Draw doors and shelves in the upper cabinets.
9. Draw doors and swings, drawers, toe kicks, and backsplash required for the lower cabinets.
10. Draw all equipment such as sinks, breadboards, and fixtures.
11. Draw all required interior finishes.

The cabinet elevations should now resemble Figure 20-20.

12. Draw and label each grid line.
13. Provide horizontal dimensions to locate cabinets to each grid line.
14. Place all horizontal dimensions, locating the edge of cabinets and door locations.
15. Place all vertical dimensions, locating all the heights of all cabinets and equipment.
16. Place notes to specify all materials and equipment.
17. Reference any required details.
18. Provide titles and drawing scales.

The completed elevations are shown in Figure 20-21.

Layer Guidelines for Interior Elevations

When cabinet drawings are placed on a separate sheet, the drawing file is generally named I-ELEV-####. The owner's name or the job number should be substituted

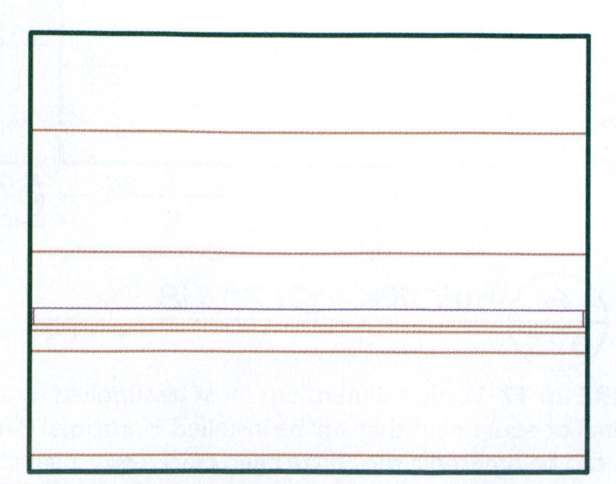

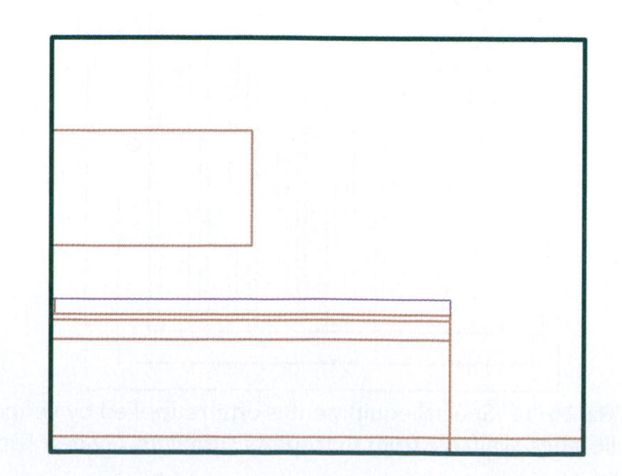

FIGURE 20-19 The initial layout of interior elevations.

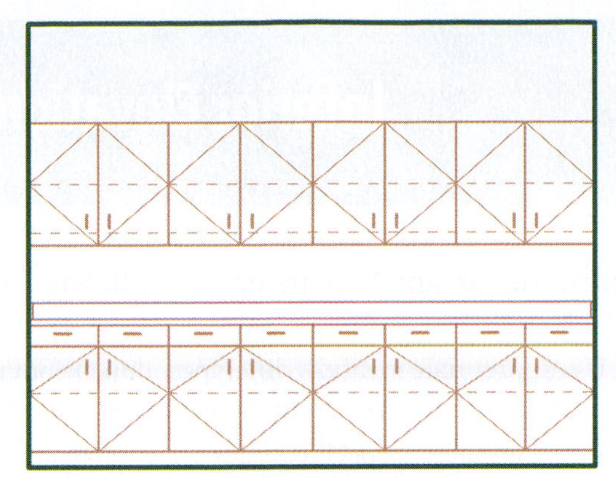

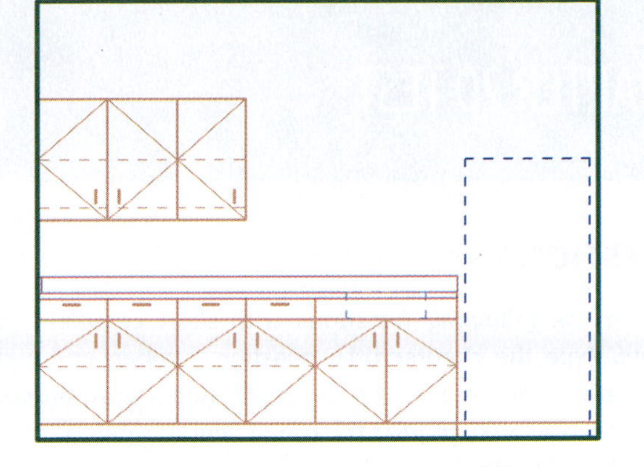

FIGURE 20-20 Layout of the major cabinet components.

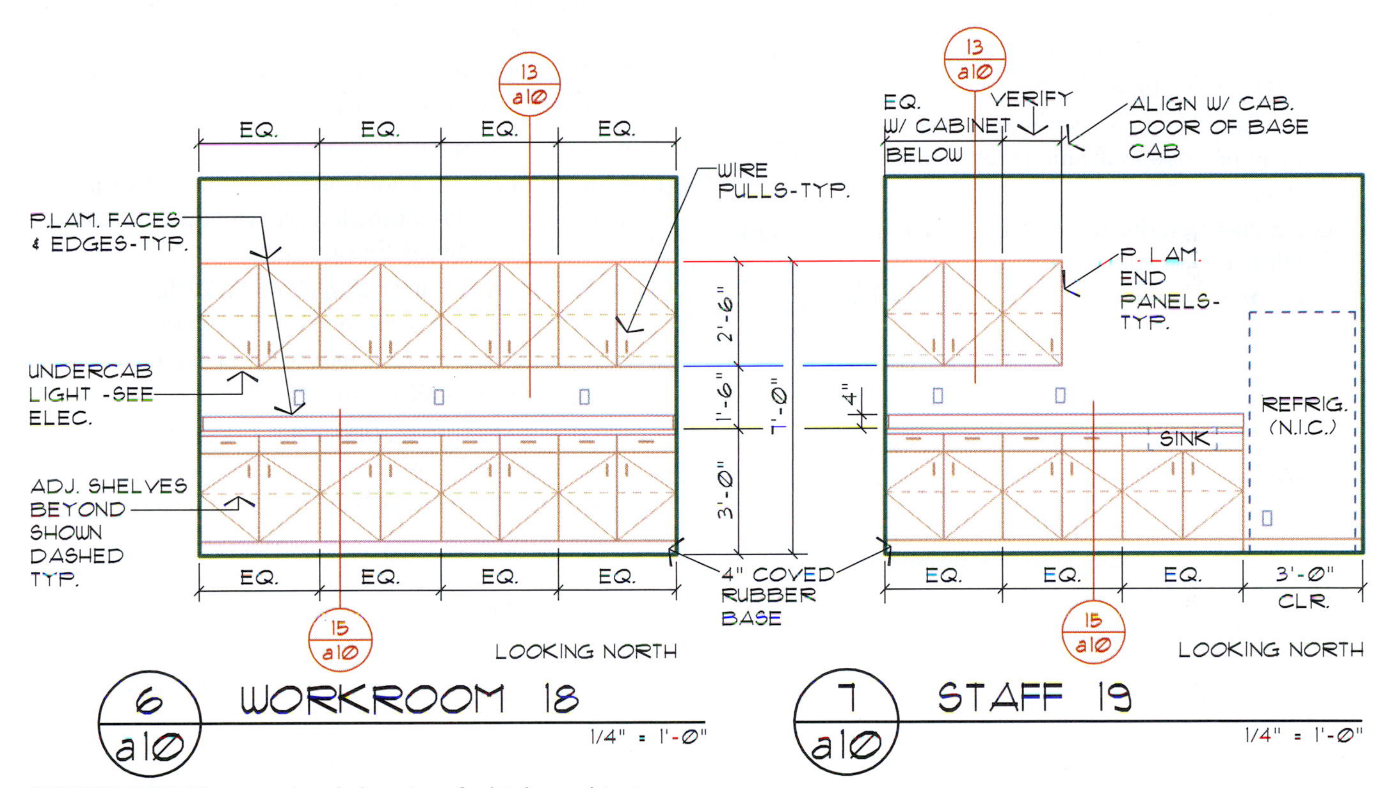

FIGURE 20-21 The completed elevations for kitchen cabinets.

for ####. Cabinet elevations should be placed on layers with a prefix of A-ELEV if they are being placed in drawing files that contain other drawings. Modifiers such as APPL, CABS, CNTR, EQIP, HDWR, or PLMB can be used to describe layers such as appliances, cabinets, counters, equipment, hardware, and plumbing fixtures. Cabinets that are placed in their own .dwg file can have a wider variety of file names. Many professional firms label interior elevations based on linetype or usage. Layer titles such as FINE, HEAVY, and LIGHT can be used to define linetypes. Titles such as BORDER, TEXT, and DIMEN can be used to describe both the drawing contents and general drawing information. Suggested layer names based on NCS can be found on the student CD in Appendix C.

ADDITIONAL READING

Links for the materials represented on the interior elevations are too numerous to list here. Rather than listing individual sites, use your favorite search engine to research products from the following Construction Specifications Institute (CSI) categories: interior finish and trim, paints and coatings, caulks and adhesives, plumbing, lighting, electrical, appliances, furniture, and furnishings.

CHAPTER 20 TEST Interior Elevations

QUESTIONS

Answer the following questions with short complete statements. Type the chapter title, question number, and a short complete statement for each question using a word processor. Some answers may require the use of vendor catalogs or seeking out local suppliers.

Question 20-1 What drawings are required to draw the interior elevations?

Question 20-2 Describe placement of the interior elevations within the drawing set.

Question 20-3 List three common sources for cabinet and equipment specifications required for cabinet drawings.

Question 20-4 Describe two methods of referencing cabinet drawings to other drawings.

Question 20-5 Describe the selection of scales for cabinet drawings.

Question 20-6 To what are horizontal dimensions on cabinet elevations referenced?

Question 20-7 What is the common height of a base and upper cabinet?

Question 20-8 What is the goal of a cabinet section?

Question 20-9 List two methods of showing the direction of door swings.

Question 20-10 List three types of interior drawings.

DRAWING PROBLEMS

Use the reference material from preceding chapters and vendor catalogs to complete one of the following projects. Unless your instructor provides other instructions, draw the elevations that correspond to the floor plan that was drawn in Chapter 16. Because of the wide range of design latitude within each problem, no skeletons for the problems are provided on the student CD. You will act as the project manager and will be required to make decisions about how to complete the project. *Study each of the drawing problems related to the structure prior to starting a drawing problem.* Any information not provided must be researched and determined by you unless your instructor (the project architect and engineer) provides other instructions. Develop the base drawing to complete the assignment using the following minimum criteria:

- Use appropriate symbols, linetypes, dimensioning methods, and notations to complete the drawing.
- Determine any unspecified sizes based on material or practical requirements.
- Unless specified, select a scale appropriate to plotting on D-size material and determine LTSCALE and DIMVARS.
- Place the necessary viewing plane symbols on the floor plan to indicate each drawing.
- Use a 3" toe space unless noted.

Problem 20-1a Use the floor plan from problem 16-1 to draw the elevations for the kitchen. Use the following guidelines to complete the drawings:

- Drop the ceiling to 8'–0" high in the kitchen and indicate the limits of the lowered ceiling on the floor plan.
- Draw the base cabinets with a 36" high ceramic tile counter with a full splash.
- Draw the bar behind the sink as 42" high and expand the width to 15" wide.
- Provide 18" clearance between the countertop and the upper cabinets.
- Amend the floor plan to show the upper cabinets to be over the stove and provide a range hood with 30" clearance to the cabinet.
- Provide a bank of drawers on the nook side of the sink.
- Provide an area for tray storage on the bathroom side of the stove with the balance of the base cabinets to have 4" deep drawers over doors.
- Provide a breadboard over a drawer between the refrigerator and the stove.

Problem 20-1b Design three different elevations for the fireplace in the family room. Use entirely different styles and materials, and include the doors and windows on each side for each elevation.

Problem 20-1c Draw an elevation of the master bathroom cabinets including the linen closet.

The linen closet is to have shelves above (2)-12" deep drawers. The vanity countertop is to be 32" high plastic laminate with a 4" backsplash. Provide a mirror from the backsplash to 6'–0" above the floor.

Problem 20-2 Use the floor plan created for problem 16-2 to design and draw the elevation for an oak cabinet with shelves along the north wall of the office that meets the following parameters:

- The lower 36" is to have 4" deep drawers over doors.
- Shelves are to be on rollers.
- Part of the lower cabinet should include rollout drawers (behind the doors) suitable for legal-size filing.
- The upper portion should include shelves for books and a stereo system.

Problem 20-3 Use the floor plan in problem 16-3 and design and draw an elevation of the cabinets in the break room.

Problem 20-4 Design a 24" deep storage cabinet from floor to ceiling along the north wall of the lower storage area that meets the following parameters:

- The 20 feet closest to the warehouse area will be used to store flammable materials and should be lockable.
- Ten feet of the cabinet should be divided into lockable storage units for employees.
- The balance of the area will be used for general storage and should have a base cabinet with enclosed shelves, a plastic laminate countertop at 36" high with a 4" backsplash, and upper shelves behind doors.

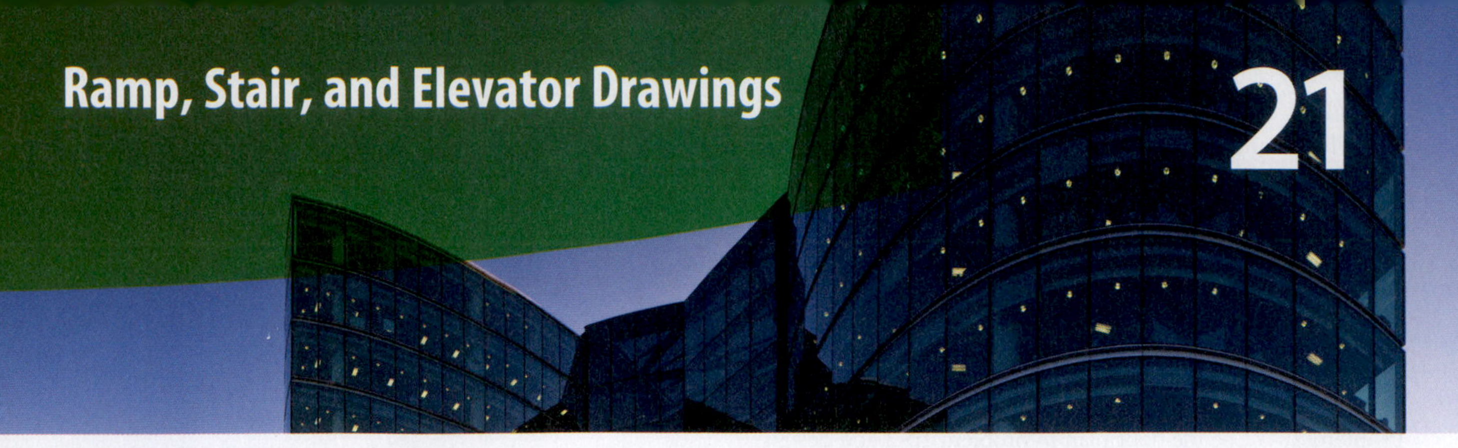

Ramp, Stair, and Elevator Drawings

21

Changes in floor elevation are accomplished by using a ramp, stairs, elevator, escalators, or a combination of each. Although the architect is responsible for the design, a CAD technician should be familiar with the code requirements and vendor drawings associated with each method of changing floor levels. This chapter will introduce the drawings required for:

- Ramps.
- Stairs.
- Major stair materials.
- Drawing representation.
- Elevator drawings.
- Escalator drawings.

RAMPS

A ramp is an inclined surface that connects two different floor levels. Design and construction are regulated by both the building code and Americans with Disabilities Act (ADA) requirements. (See Chapter 5 for a review of ADA requirements.) Chapter 16 introduced ramp representation on the floor plan. This chapter addresses code requirements, methods of representation on drawings, and common sections and details that a drafter is required to work with.

Code Requirements

Ramps are classified by International Building Code (IBC) based on usage. Ramps that are used in an aisle for an exit from a structure must also meet the exit requirements that were presented in Chapter 12. This chapter focuses on code requirements for ramps that are used to achieve a change in floor levels.

Minimum Ramp Sizes

The IBC requires ramps to have a minimum width equal to the required width for the corridor and a minimum clear width of 36" (900 mm) for the ramp and clear width between the handrails. Handrails are allowed to project into the required ramp width a maximum of 4 1/2" (114 mm) on each side provided the minimum clear width of 36" (914 mm) is maintained. Any decorative trim or other features may project a maximum of 1 1/2" (40 mm) on each side. The rise of ramps is restricted to 30" (762 mm) total with a maximum slope of 1 vertical unit per each 8 horizontal units, or a 12.5% slope. Ramps used as a means of egress are restricted to a maximum slope of 1 vertical unit per each 12 horizontal units, or an 8% slope. The cross-slope of a ramp, measured perpendicular to the direction of travel, cannot be greater than 1 vertical unit per each 48 horizontal units (2% slope). If a door swings over an egress ramp, the door when opened in any position cannot reduce the width of the ramp to less than 42" (1067 mm). A minimum of 80" (2032 mm) headroom must be provided above a ramp.

Landings

Except in non-accessible Group R-2 and R-3 units, a landing with a length of 60" (1525 mm) measured in the direction of the ramp is required at the top and bottom, points of turning, doors, and the entrance and exit of all ramps. Group R-2 and R-3 units can have a landing with a minimum length of 36" (914 mm). The landing width must be as wide as the widest ramp that connects to the ramp. The slope of a landing cannot be greater than 1 vertical unit per each 48 horizontal units (2% slope).

Railings, Handrails, and Curbs

With the exception of ramps in aisles that serve seating, ramps with a rise greater than 6 inches (152 mm) must have a handrail on both sides of the ramp. The handrail must be between 34" and 38" (864 and 965 mm) in height. The rail must extend horizontally 12" (305 mm) past the top and bottom of the ramp. For round rails, the handgrip portion of the rail must be between 1 1/4" and 2" (32 and 51 mm) wide and must be smooth with no sharp corners. Rails must project 1 1/2" (38 mm) clear from walls to provide hand space. To prevent dropping off the ramp edge, a curb or barrier must be provided at the floor or ground surface of the ramp and landing. The curb must be constructed to prevent the passage of a 4" diameter sphere. Guards must be provided in accordance with IBC section 1013.

Ramp Representation

The width and length of a ramp can be represented on the floor plan using thin lines. The angle can be represented by listing either the slope or the elevation of each end, as seen in Figure 21-1. The slope and construction method must be represented by means of a section or detail that is referenced to the floor plan. Figure 21-2 shows an example of a ramp detail.

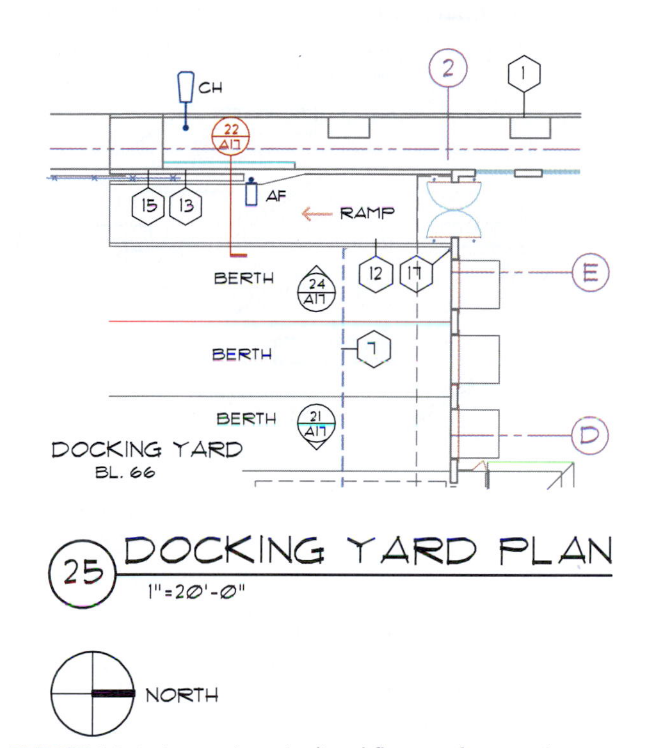

FIGURE 21-1 A ramp is an inclined floor surface used to connect two different floor elevations that are often represented on the floor and site plans. *Courtesy Michael & Kuhns Architects, P.C.*

STAIRS

Stairs are a primary method of changing floor levels in a multilevel structure. To effectively work with the architect's stair drawings, the CAD drafter must be familiar with common terms, layout methods, and code requirements.

Stair Terminology

The exact terminology for stair construction may vary in different parts of the country. Figure 21-3 shows each of the common components of a stair that must be represented by a drafter. The layout of a stair is dictated by the rise and run.

- The **rise** is the vertical distance from one floor level to the next.
- The **run** is the total length required to form the stairs.
- The total rise is made up of individual **risers**, which are the vertical portion of a step. The height of a riser is measured from one step to the next.
- The **tread** is the horizontal portion of a step. The length of a tread is measured from the face of the riser at the rear of the tread to the nosing at the front-end of a tread.
- The *nose* is the front edge of a step. Nosing is material that is added to the nose of the tread for decoration and to provide a slip-resistant surface. The type and design of nosing varies depending on the material used to build the stair. Nosings usually project approximately 1 1/8" (28 mm) past the riser.

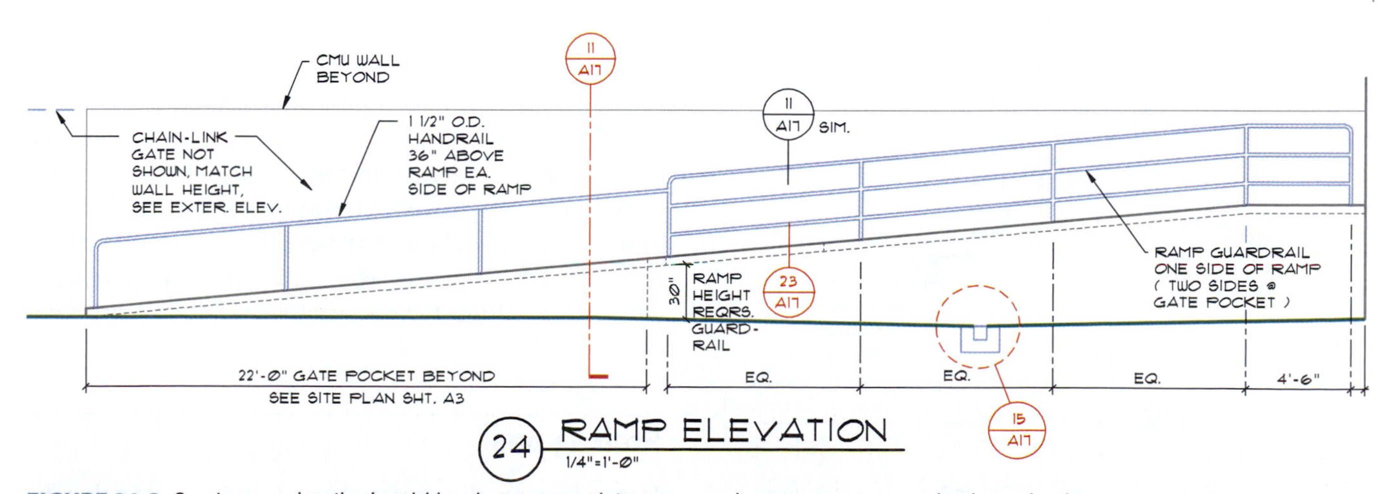

FIGURE 21-2 Sections or details should be drawn to explain ramp angle, construction methods, and railings. *Courtesy Michael & Kuhns Architects, P.C.*

- A **stringer** provides support for the tread and riser. In some areas of the country, it is referred to as a *carriage* or *stair jack*. The stringer may be placed below or beside the risers.
- The *headroom* for stairs is the vertical distance above the stair measured from the tread nosing to any obstruction above the stair.

Figure 21-4 shows the relationship of the stringer to the tread for various materials. The stringer terminates at a floor level or at an intermediate platform called a *landing*. Landings are inserted into long stair runs to provide users with a resting point, or as a means of changing the direction of the run. In multilevel structures, stair landings are often stacked above each other.

To provide stability to the user, a handrail must extend 12" (300 mm) past the upper riser. The lower end of the rail must continue to slope for the depth of one tread beyond the bottom riser. The number of handrails required is based on the width of the stairs and is discussed later in this chapter. The handrail on the open side of a stair terminates at a newel post.

- A *newel post* is a large, fancy post used to provide support at the end of a railing. The handrail is connected to the open side of the stair by a vertical *baluster*.
- A **guardrail** is the railing that is placed around stairs to protect people from falling from the floor into the stairwell.

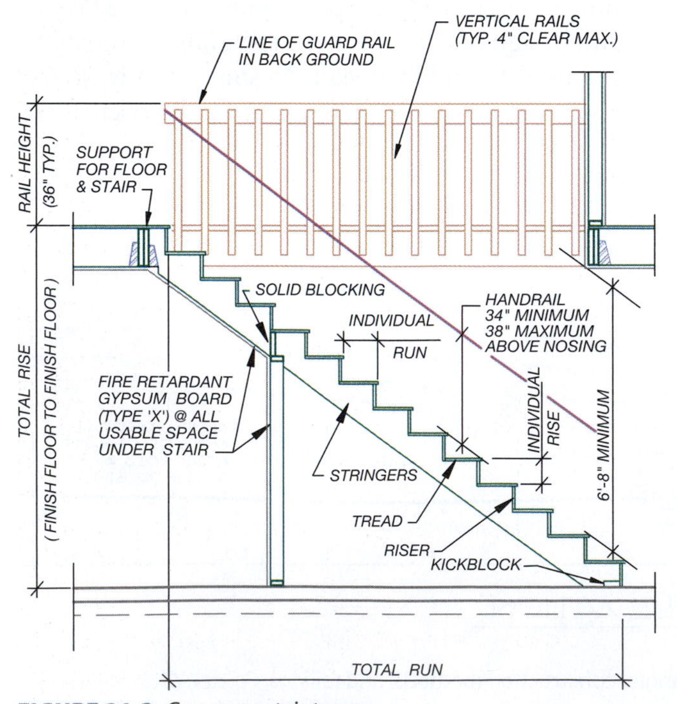

FIGURE 21-3 Common stair terms.

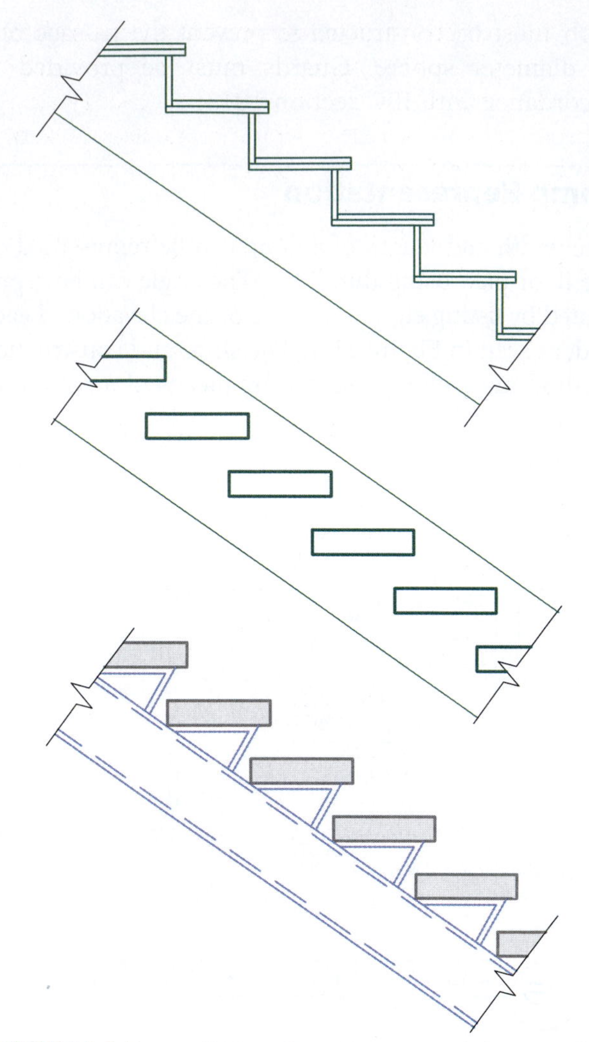

FIGURE 21-4 Depending on the occupancy and the type of construction, stair treads may be directly supported by the stringer, placed between the stringers, or be above the stringer and supported by metal brackets.

- A **stairwell** is the vertical shaft where the stairs are to be placed. The design of the baluster connection to the stair and floor systems is of great importance for the safety of the stair users.

Depending on the height of the structure, the occupancy, and fire-resistive construction used, stairs may need to be separated from the balance of the structure by an enclosed stairwell. To control the spread of smoke and flames throughout a structure, access to stairwells is restricted by doors that must have a minimum 1-hour fire rating. Review Chapter 12 for fireproofing of stairwells.

Stair Types

Depending on the material used, stairs may either be open or enclosed. Stairs framed with wood, steel, or precast concrete risers on a steel stringer can be framed as open stairs with no riser. Each of these materials can also be used to form enclosed stairs. Stairways that serve as the only access to another floor cannot be open because of ADA restrictions. The method of framing is not represented in plan view but is shown in sections and details. In addition to the type of riser construction, the configuration of the run can also be altered to meet design needs.

- Straight stair runs are the most economical to build. If space for the run is limited, a landing can be inserted into the run and the direction of the run can be altered. L- and U-shaped stairs are two common variations to a straight run.
- An L-shaped stair contains a 90° bend in the total run. The change in direction occurs at a landing, which may be placed in any portion of the run. A U-shaped stair causes the user to make a 180° turn at a landing. The landing can be placed at any point of the run, but it is usually placed at the midpoint so that each run is of equal length.
- A curved or circular stairwell can be used to enhance the design of an entry. *Curved stairways* are formed by using one or more radii to form the stair. Curved stairs have wedge-shaped treads.
- A *spiral stair* is a stair that has a circular form in plan view with treads that radiate about a central support column. Spiral stairs can be used only within an individual unit or an apartment that serves an area less than 250 square feet (23 m^2) and serves less than five occupants.

Figure 21-5 shows an example of how each type of stair is represented on a floor plan.

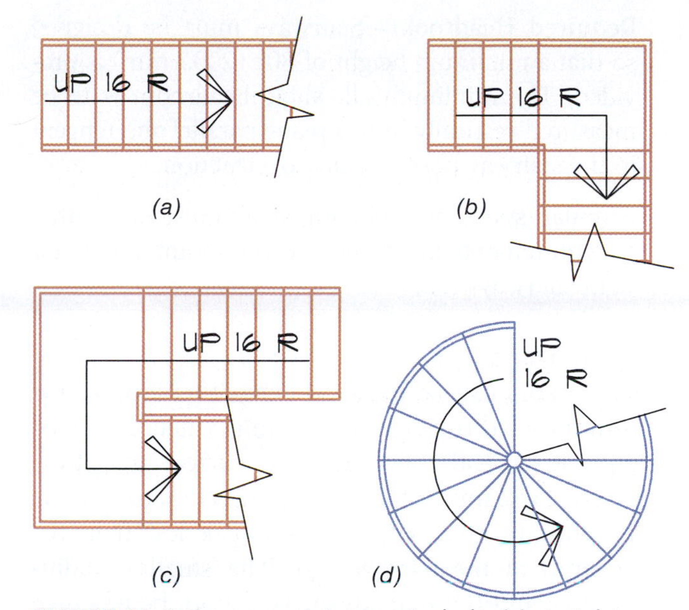

FIGURE 21-5 Common stair types include: (a) straight, (b) L-shaped with a landing at any point of the run, (c) U-shaped with a landing at any point of the run, and (d) spiral. The occupancy load and the type of construction limit the use of spiral stairs.

Minimum Code Design Requirements

Any stairs having two or more risers (except stairs used exclusively to service mechanical equipment) are regulated by the IBC and the ADA. A drafter must be familiar with these codes so that drawings and details are accurate.

Required Sizes for Straight-Run Stairs

The IBC specifies minimum sizes for the width, height, individual rise and run, and the headroom for stair construction.

Required Width—Stairways serving an occupant load of less than 50 can be a minimum of 36" (914 mm) wide. Stairs serving 50 or more occupants must be a minimum of 44" (1118 mm) wide.

Required Riser Height—4" (102 mm) minimum and 7" (178 mm) maximum measured vertically between the leading edges of adjacent treads. For dwelling units in Group R-2 and R-3 occupancies, and for Group U occupancies that are accessory to a Group R-3 occupancy, or accessory to individual units in Group R-2 occupancies, the maximum rise can be increased to 7 3/4" (197 mm).

Maximum Stair Rise—A landing must be provided for every 12' (3658 mm) of vertical rise.

Required Tread Depth—11" (279 mm) minimum measured horizontally between the vertical planes of the foremost projection of adjacent treads and at right angles to the tread's leading edge. For dwelling units in Group R-2 and R-3 occupancies, and for Group U occupancies that are accessory to a Group R-3 occupancy, or accessory to individual units in Group R-2 occupancies, the minimum tread depth can be reduced to 10" (254 mm).

Landing—A landing equal in length and width to the width of the stair must be provided at the top and bottom of a stairway. If a landing is provided in the stair span, the length of the landing measured in the direction of travel must be equal to the width of the stair run but is not required to exceed 48" (1219 mm).

Tread and Landing Slope—The walking surface of a tread or ramp cannot be greater than 1 vertical unit per each 48 horizontal units (2% slope) in any direction.

Required Headroom—Stairways must be designed so that a minimum height of 80" (2032 mm) is provided. The minimum allowable headroom is to be measured vertically from a plane parallel and tangent to the stairway nosing to any obstruction.

Circular, spiral, and winding stairs must meet other code requirements in addition to the requirements for straight stairways.

Curved Stairs

Curved stairs can be used for an exit as long as the minimum tread run is 11" (279 mm) when measured at a point 12" (305 mm) from the narrow end of the tread (6" [152 mm] for dwelling units in Group R-2). The minimum tread depth shall not be less than 10" (254 mm) at the narrow end. The smallest radius used to lay out the stairway curve cannot be less than twice the width of the stair. Curved stairs that do not have a uniform radius are considered *winders* by the building code.

Winders

Winders can be used as a means of egress only within dwelling units unless they comply with the dimensional criteria and requirements of curved stairs. When used, a required width for the run of 11" (279 mm) is provided at a point 12" (305 mm) from the narrowest portion of the stair. Winders also must be a minimum of 6" (152 mm) at the narrowest portion of the tread. Required tread sizes are shown in Figure 21-6.

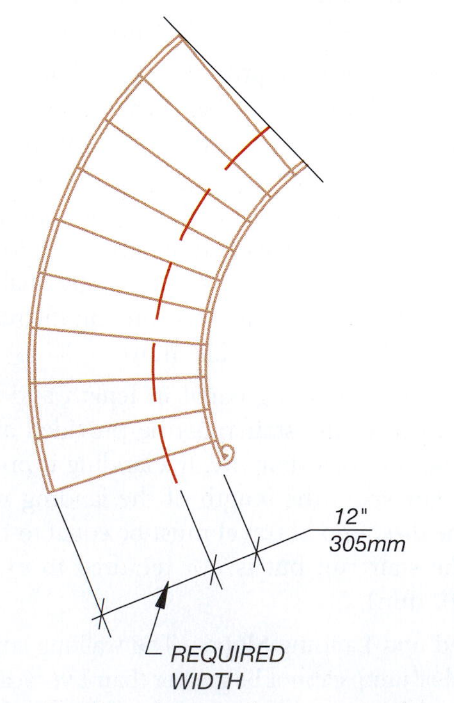

FIGURE 21-6 Tread requirements for a curved stair.

Spiral Stairs

The tread of a spiral stair must provide a clear walking area of 26" (660 mm) from the outer edge of the supporting column to the inner edge of the handrail. A minimum run of 7 1/2" (191 mm) must be provided at a point 12" (305 mm) from the narrowest portion of the tread. The maximum rise for spiral stairs cannot exceed 9 1/2" (241 mm) and minimum headroom of 78" (1981 mm) must be provided. Figure 21-7 shows the minimum tread requirements for a spiral stair.

Rail Design

The placement of handrails and guardrails must be considered as stair drawings are completed.

Handrails

Handrails must be provided on each side of the stair run. Common exceptions include stairs in aisles used to serve seating areas, stairways within dwelling units, and spiral stairs. An intermediate handrail must be provided for wide stairs so that no portion of the width required for egress is more than 30" (762 mm) of a handrail. Handrails must be between 34" and 38" (864 and 965 mm) high above the nosing of treads or landings. Handrails are required to extend horizontally 12" (305 mm) past the top riser and continue to slope for the depth of one tread beyond the bottom riser. The guidelines used for placing handrails by ramps also apply for stairs.

Guardrails

Guardrails are required at non-enclosed floor areas that are 30" (762 mm) above the ground or another floor level. Guardrails are not required on the loading side of loading docks, on the auditorium side of stages, or around service pits that are not accessible to

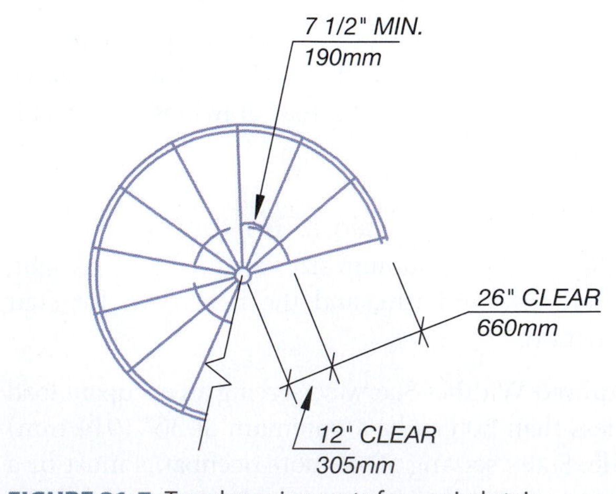

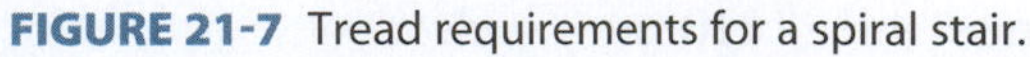
FIGURE 21-7 Tread requirements for a spiral stair.

the public. Guardrails must be 42" (1067 mm) high, except for rails serving occupancies in Group R-3, and within individual dwelling units in group R-2. Guards on the open side of stairways in these two occupancies are allowed to be between 34" to 38" (864 to 965 mm) high. Intermediate railings in a guardrail must be designed so that a 4" (102 mm) diameter sphere cannot pass through any openings up to a height of 34" (864 mm). Openings in the railing above 34" to 42" (864 to 1067 mm) can be designed using a 4 3/8" (111 mm) sphere. The triangular opening formed by the riser, tread, and the bottom rail at the open side of the stairway must be designed so that a 6" (533 mm) diameter sphere cannot pass through the opening. Railings for industrial occupancies that are not open to the public can be designed so that a 21" (533 mm) sphere will not pass through the railing.

Lateral Load Design for Railings Railings must be able to withstand the pressure of people pushing or pulling on the rail. The IBC requires rails to resist a lateral load of 50 plf (0.73 kN/m^2) applied in any direction. Rails are also required to resist a single concentrated load of 200 pounds (0.89 kN) applied from any direction to any portion of the railing. For Group I-3, F, H, and S occupancies that are not accessible to the general public and have an occupant load of less than 50, the minimum load the rail is required to resist can be reduced to 20 plf (0.29 kN/m^2). For guardrails in one- and two-family dwellings, the rail is required to resist only the concentrated load. These loads must be transferred down through balusters into the floor or stair system. Both the connectors used to attach balusters and the material they are connected to must be strong enough to resist the loads to be supported and to resist the tendency to twist as the loads are applied. Figure 21-8 shows an example of a detail used to show the connection of a railing to the floor system.

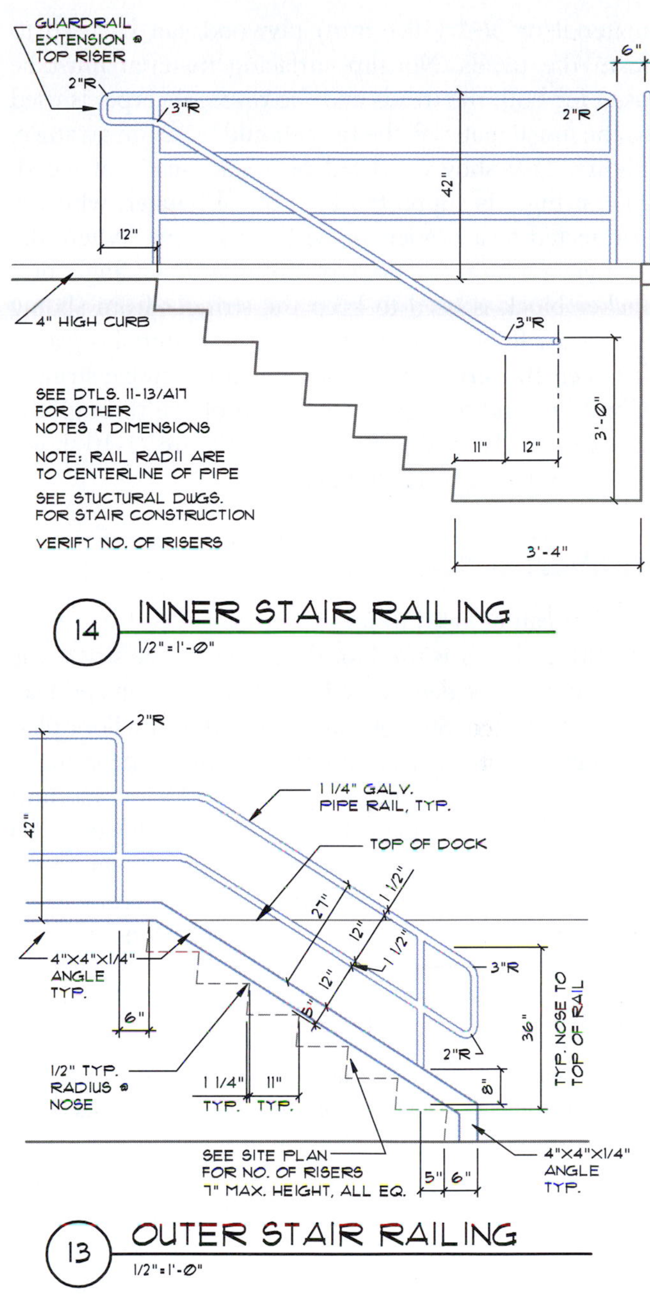

FIGURE 21-8 Details must be drawn to show how the handrail and guardrail will be constructed. This railing is designed for use in an industrial area that is not open to public use. *Courtesy Scott R. Beck, Architect.*

Determining Required Rise and Run

The total rise and run of a stair is based on the legal requirements for the rise and run of each tread. If the floor-to-floor height is 9'–6" and a rise of 7" is used, 17 risers will be required. This is determined by dividing 114" (the total height) by 7" (the maximum rise). Although the answer is 16.28, this must be rounded up to 17 risers. The actual riser height of 6.70" can be determined by dividing 114 inches by 17. The number of treads is always one less than the number of risers. This stair would require 17 risers and 16 treads. If a run of 11 inches is used, the total run would be 176 inches (16 treads × 11").

STAIR MATERIAL

The occupancy and type of construction dictate the material used to form the stairs. Wood, timber, steel, concrete, or a combination of steel and concrete is used to form the stairs.

Wood

Wood stairs are made by using 2 × 12 or 2 × 14 (50 × 300 or 50 × 350) stringers. One-inch (25 mm) tread material or 3/4" (19.1 mm) plywood can be used to form the treads. Nonslip surfacing material must be used for both the treads and the risers. If carpet is used as the finish material, the riser should be set on an angle. Figure 21-9 shows a detail of a stair made of wood. The stringer is supported by a metal hanger, which is connected to a header in the floor system. Where the stringer rests on the floor platform, a metal angle or a kicker block is used to keep the stringer from sliding across the floor. A block of 2× (50×) material is placed between the stringers at midspan to minimize stringer vibrations and to reduce the spread of fire through the chase formed between the stringers. Figure 21-10 shows an example of a wood-framed stair.

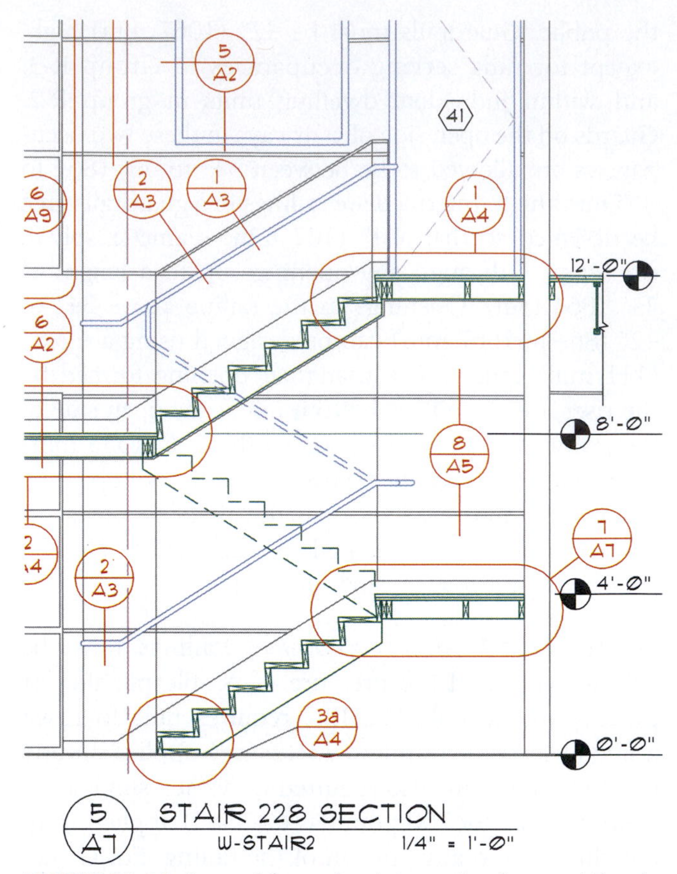

FIGURE 21-10 A wood-framed stair in section for a multilevel office complex. *Courtesy Architects Barrentine, Bates & Lee, A.I.A.*

Timber

Timber can be used to form open stairs. A 3 × 12 (75 × 300) or larger is used for the stringer. The size of the stringer must be determined based on the span and load to be supported. Stringers are attached to the floor platform by use of a metal angle at each end of the stringer. Treads are generally made of 2" or 3" (50 or 75 mm) material and are attached to the stringer by use of a metal angle or let into the stringer. Figure 21-11 shows examples of each method of attachment. Figure 21-12 shows an example of a stair using timber construction.

Steel

Steel stairs are found in many types of commercial and light industrial construction because of the need for noncombustible construction. Steel stairs can be built similar to wood stairs, using a channel as the stringer and steel plates for the risers and treads. Stringers are attached to each floor system by the use of a steel angle. Handrails are generally made of 1 1/4" diameter metal pipe and are supported on metal uprights called *balusters*. Balusters are welded to the channel stringer. Figure 21-13 shows a portion of a steel stair. As shown in Figure 21-14, steel treads can be made in several configurations. For exterior applications, plates forming a grating or made from extruded metal are used to shed water. Metal is also used to form a pan to support concrete steps. The combination of a metal frame and concrete steps is quite common in many

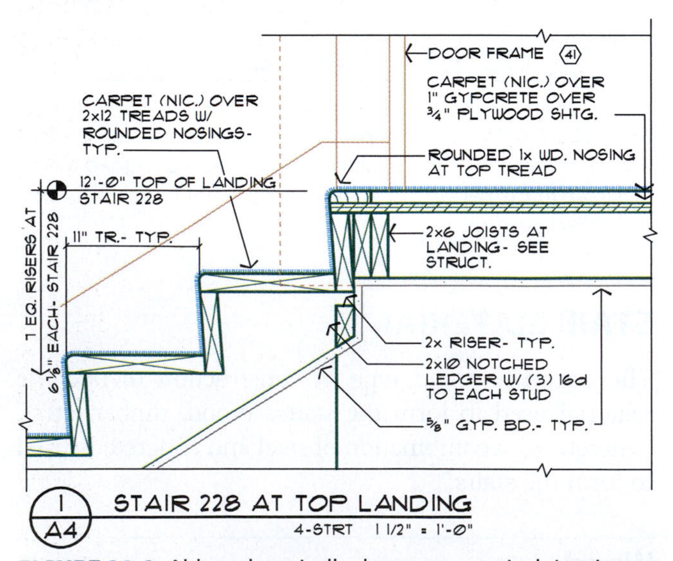

FIGURE 21-9 Although typically drawn as a vertical riser in a small-scale section, a riser covered with carpet is generally inclined and must be represented in detail. *Courtesy Architects Barrentine, Bates & Lee, A.I.A.*

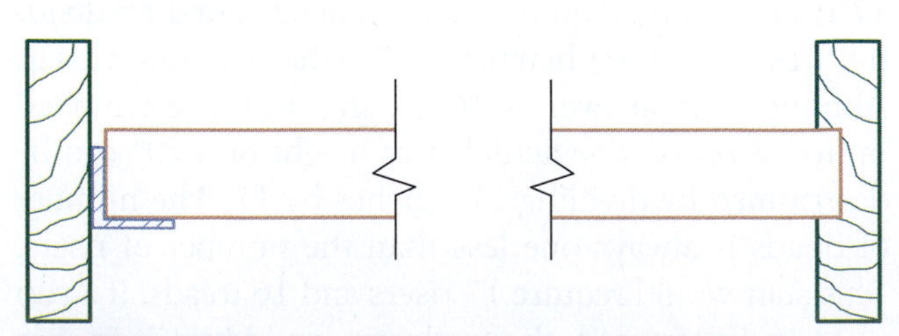

FIGURE 21-11 Timber treads may be supported by a metal angle or let into the stringer.

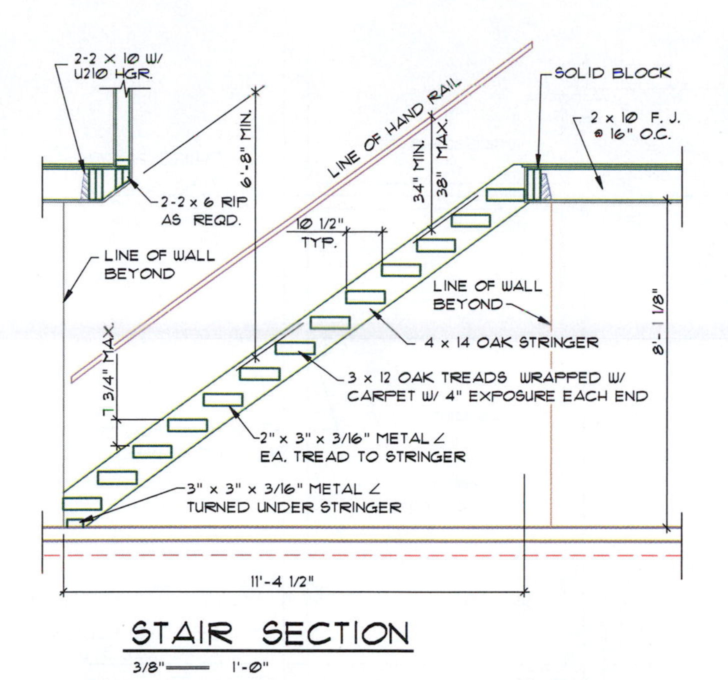

FIGURE 21-12 A stair section for a timber stair in a Group R-3 occupancy.

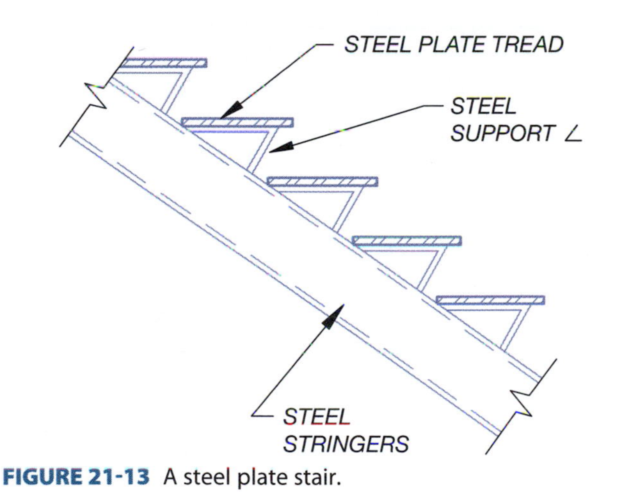

FIGURE 21-13 A steel plate stair.

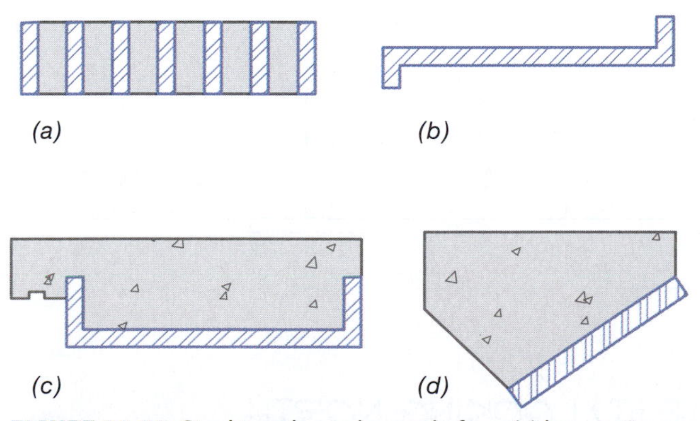

FIGURE 21-14 Steel treads can be made from (a) bar grating, (b) steel plates with a textured surface, (c) concrete treads supported by a steel pan, and (d) precast concrete treads with a steel mounting plate.

office structures. Figure 21-15 shows an example of a stair framed with steel and concrete. Figure 21-16 shows examples of details required to supplement the section.

Concrete

Because of their high cost, concrete stairs are used in only the most restrictive fire rating situations. When used, the stairs are designed as an inclined one-way slab with an irregular upper surface. Concrete stairs can be cast after the wall and floor systems have been cast, or at the same time. Drawings for concrete stairs are generally done by CAD drafters working for the engineering team. Figure 21-17 shows an example of a stair formed using concrete. Reinforcing details can be seen in Figure 21-18. Concrete stairs similar to Figure 21-19 are formed at grade.

DRAWING REPRESENTATION

The CAD technician will generally be required to work with stairs in plan views, sections, and details.

Plan Views

Stairs and the line work used to represent them were first introduced in Chapter 16. The stairway must be shown on the floor plan before details and sections can be started. The sections must be planned however, before the floor plan can be drawn. Using directions from earlier in this chapter, the number of risers and treads must be determined. When drawing a stair that goes to another level on the same floor, the entire run of the stair is shown. Figure 21-20 shows an example of a stair going up to an intermediate floor level. For a stair that extends between two full floors, the lower portion of the stair is shown on the main floor plan, and the upper end is shown on the second floor plan, as shown in Figure 21-21.

In planning the layout of the stairs on the floor plan, the required width, tread length, and number of risers must be known. Using a height of 9'–6" from floor to floor, it is determined that 17 risers and 16 treads are required with an overall run of 176". The layout of the stair on the floor plan can be completed by using the following steps:

1. Draw a line to represent the width of the stair.
2. Start at the upper floor landing, and place a line to represent the ending point. The length of the run can be determined as illustrated in Figure 21-22.

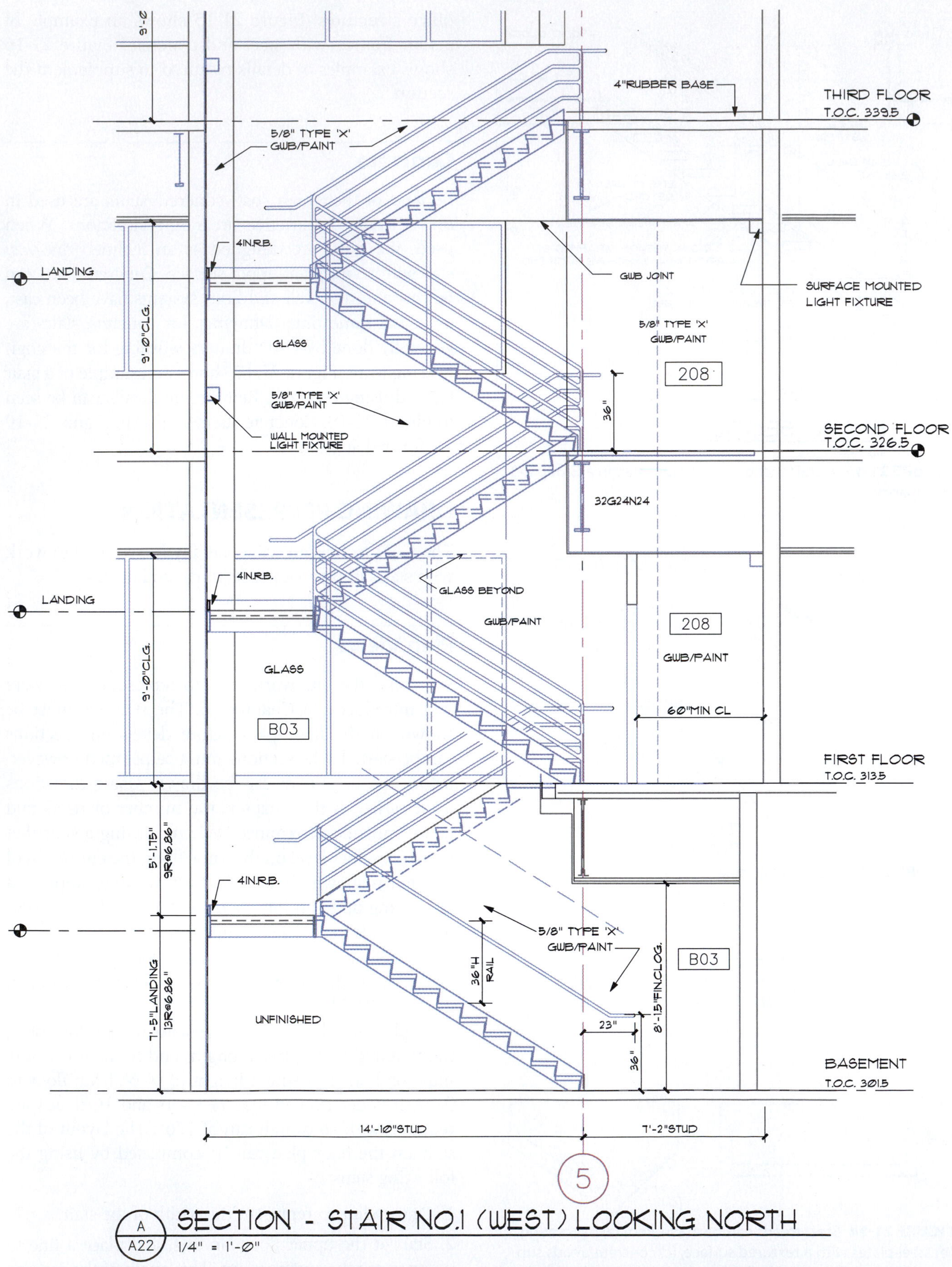

FIGURE 21-15 A stair section for a multilevel steel stair. The plan view is shown in Figure 21-23. *Courtesy Peck, Smiley, Ettlin Architects.*

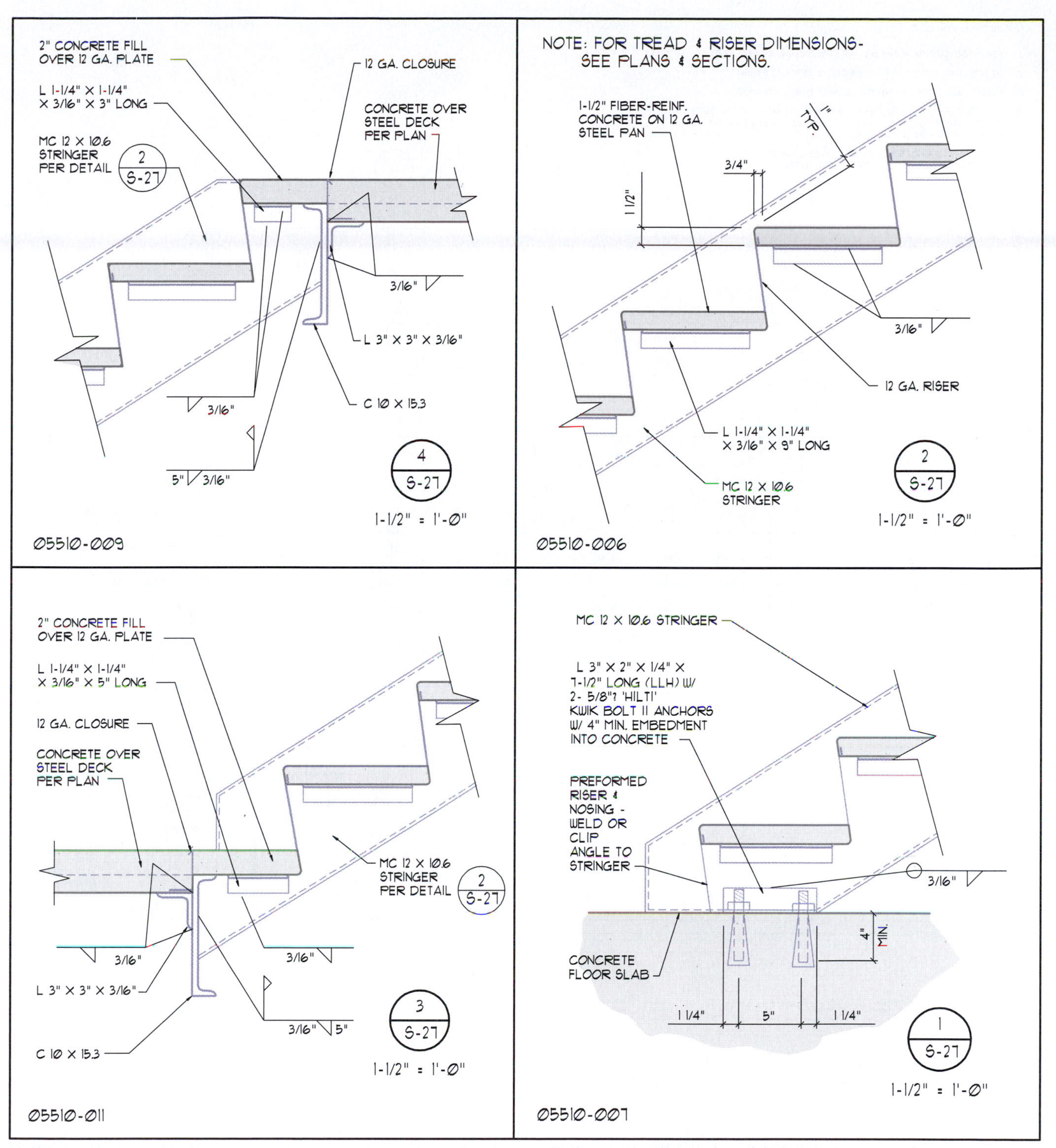

FIGURE 21-16 Details for construction of a steel stair. *Courtesy Dean Smith, Kenneth D. Smith Architect & Associates, Inc.*

3. Use either OFFSET or ARRAY to show enough treads to establish the pattern. The run should be terminated with a short break line at the end of the run.
4. Represent the handrail by a line offset into the stairs approximately 1". There is no need to accurately represent the actual rail offset on the floor plan because it will be included in the stair details.
5. Label the rails, floor above, the direction of travel, and the number of risers per flight. Figure 21-21 shows the completed lower floor plan. A similar process can be used to draw the upper portion of the run on the upper floor plan.

For structures with U-shaped or other shaped stairs, the layout process is similar. Remember, on multilevel stairs the highest and lowest floor levels show only one

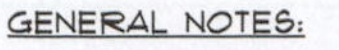

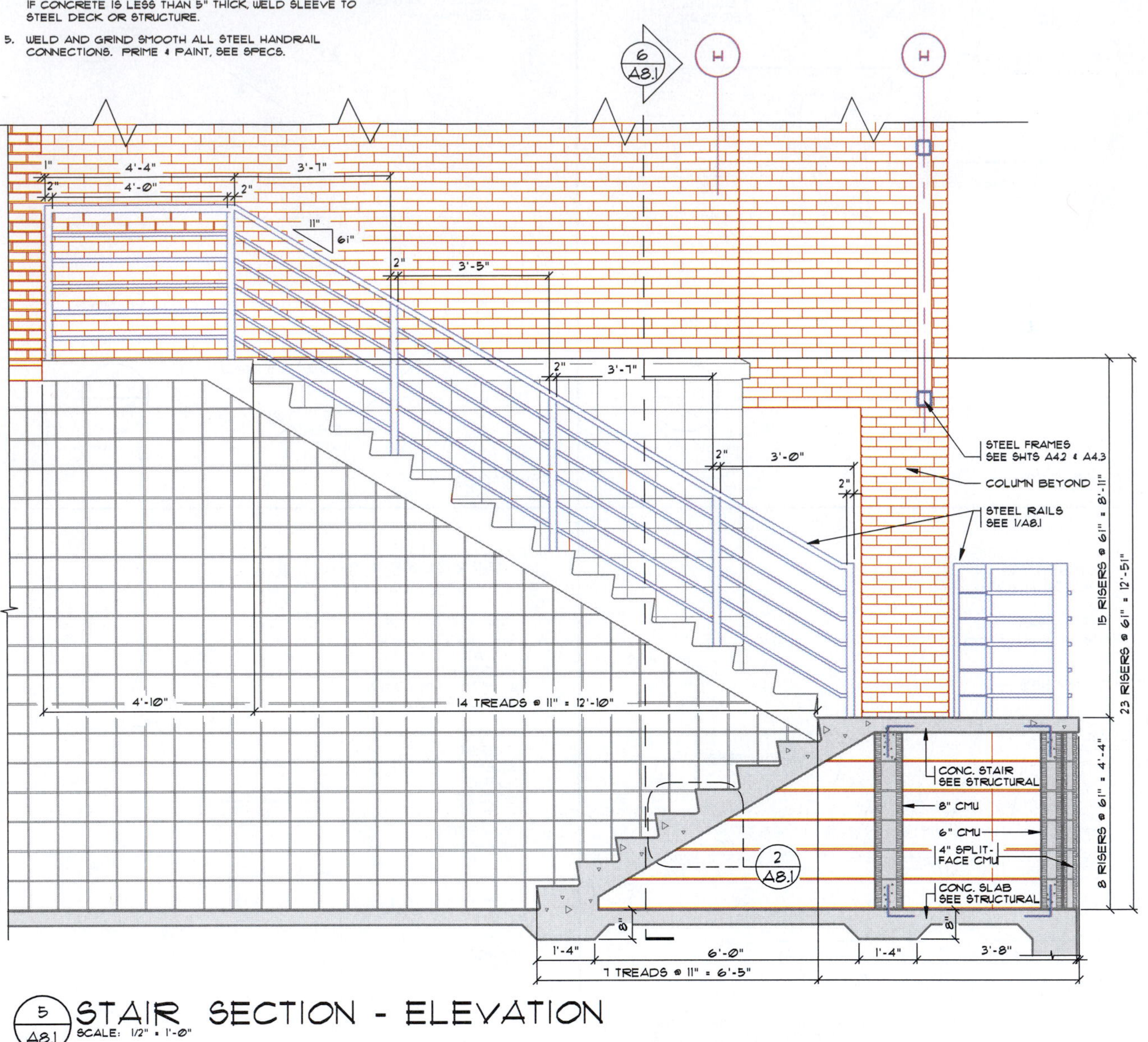

FIGURE 21-17 Sections showing the construction of a concrete stairway. Background information is often shown to help the print reader relate the section to the structure. *Courtesy G. Williamson Archer, A.I.A., Archer & Archer, P.A.*

run. Middle floors show runs going both up and down. Figure 21-23 shows how multilevel stairs would be represented. Notice that the stair going up is always on the same side of the stair and the upper run is opposite the lowest run.

Drawing Stair Sections

Stair sections are typically drawn at a scale of 1/4" = 1'–0", 3/8" = 1'–0", or 1/2" = 1'–0". The space available, the detail to be shown, and the amount of related details to be used affect the selection of the scale. If the stair is to be premanufactured, only drawings required to attach the stair are supplied. Straight steel, concrete, and combination stairways are drawn using similar methods and should represent the following items in section:

- Draw the stair section so that it is referenced to a reference grid whenever possible.
- When drawing enlarged plan views of stairs, place the lowest floor level at the sheet, with each upper level placed vertically above it in order.

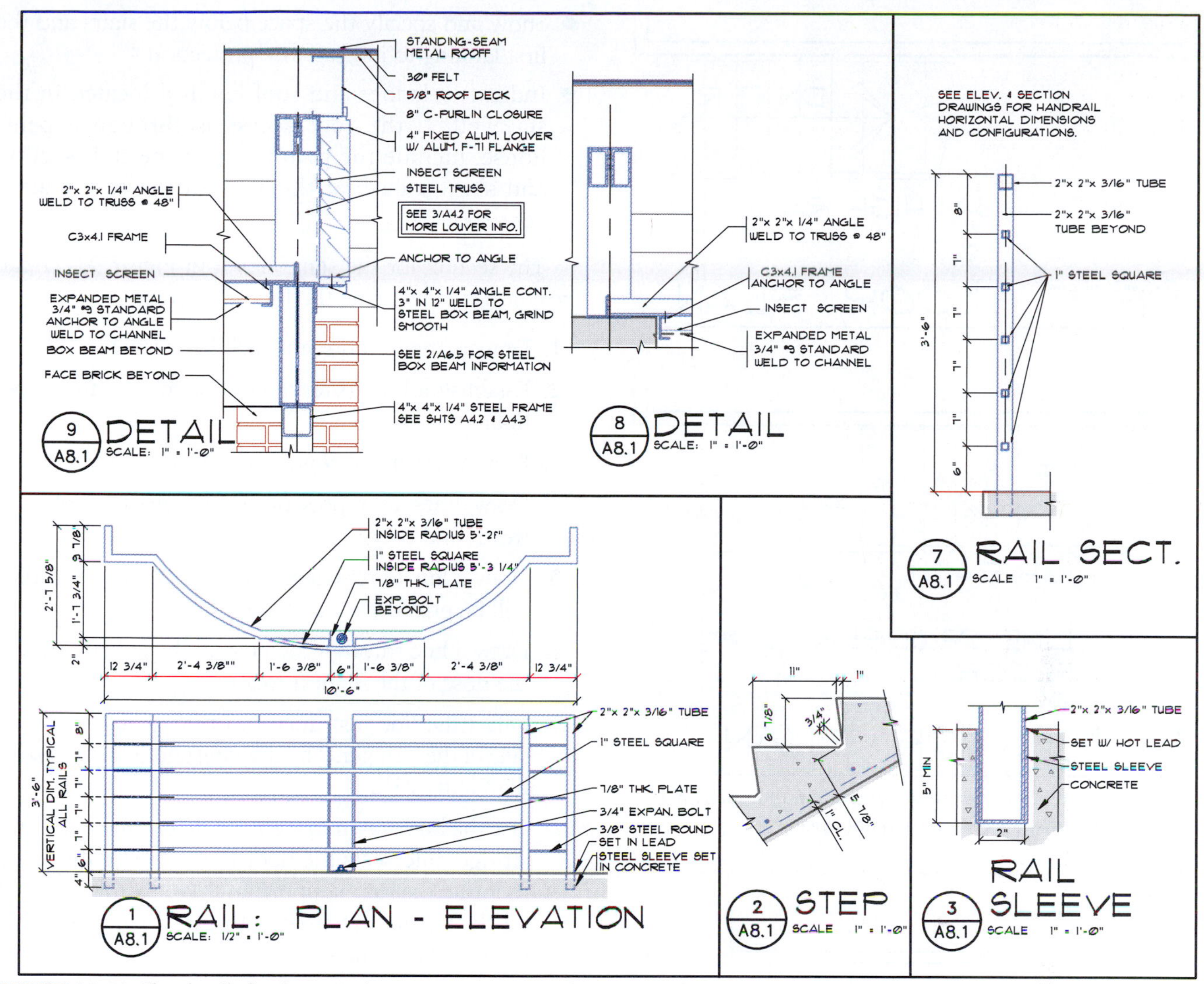

FIGURE 21-18 The details for the stair shown in Figure 21-17. *Courtesy G. Williamson Archer, A.I.A., Archer & Archer, P.A.*

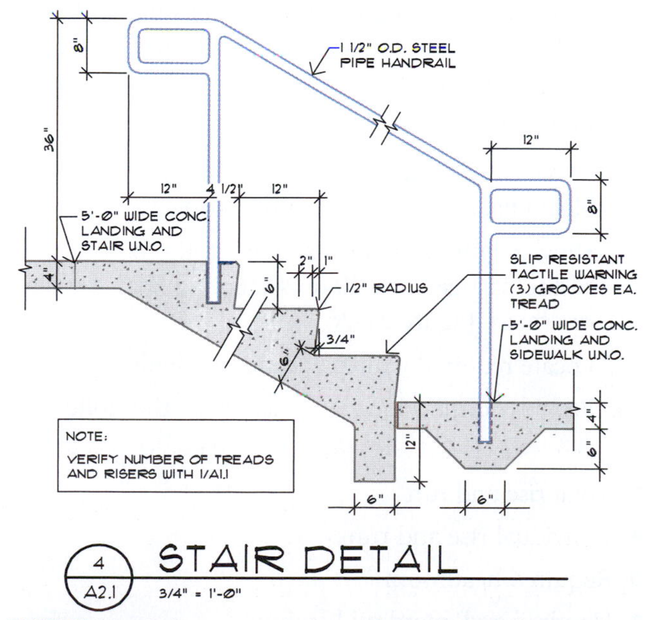

FIGURE 21-19 A concrete stairway formed at grade. *Courtesy Scott R. Beck, Architect.*

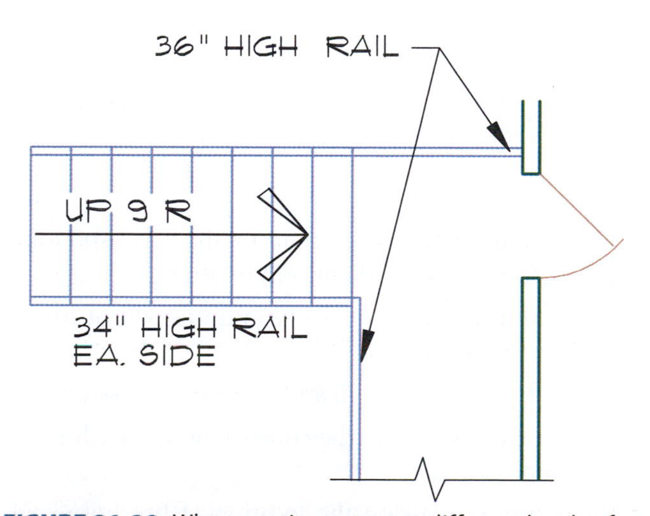

FIGURE 21-20 When a stair serves two different levels of the same floor, both ends of the stair run are shown in the plan view.

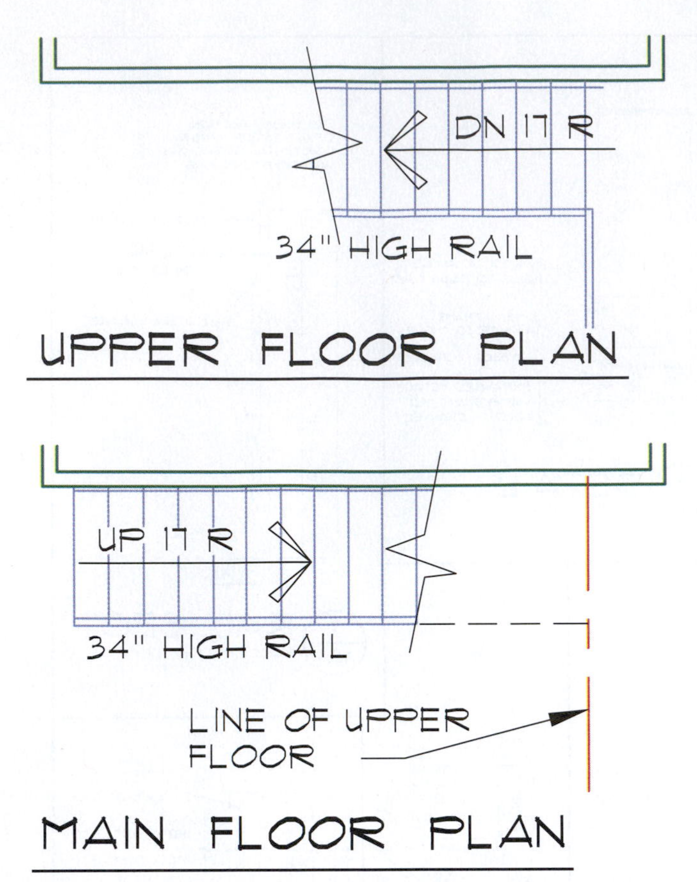

FIGURE 21-21 When a stair serves two different floors, only one end of the run is shown on each floor plan.

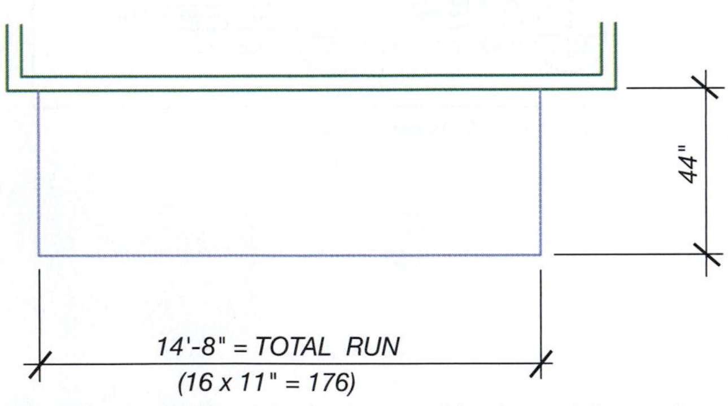

FIGURE 21-22 The stair plan is started by determining and representing the limits and width of the stair.

- Place stair sections adjacent to the corresponding enlarged plan view whenever possible.
- Show and dimension the size and number of treads, risers, and headroom.
- Show details of handrail and guardrail construction.
- Show details showing the construction of each tread, riser, and landing.
- Show and dimension the location of fire hose cabinets that are located in stairwells.
- Show and specify the space below the stairs and the first landing to indicate fire protection.
- Indicate a ladder and roof hatch if located in the stairwell. If the roof access is through a penthouse, include the penthouse in the stair section and show a curb at the access door to the adjacent roof.

The section for the stair shown in Figure 21-21 can be completed by using the following steps:

1. Draw a line to represent each floor level.
2. Establish a line to represent the edge of the upper floor.
3. Establish a line to represent the total run.
4. Draw lines to represent the required number of treads and risers.
5. Draw a line to represent where each tread and riser will be placed.
6. Draw a line through the nose of the lowest tread to the nose of the highest tread.
7. Offset the line just drawn to represent the bottom side of the stringer. The drawing should now resemble Figure 21-24.
8. Use the line drawn in step 7 to determine where the stringer intersects the floor framing. Use this point to locate the stair support. In this example, (2)-2 × 12 floor joists are used with metal hangers to the floor joist and the stringer.
9. Using the grid that was established in step 4, draw a tread and a riser similar to Figure 21-25.
10. Using the COPY command with a running OSNAP of INT, copy the tread and riser to reproduce the required number of treads.
11. Eliminate all layout lines.
12. Provide a 2× kick block at the toe of the stair and blocking at the midspan of the stringer.
13. Show required finish material on the bottom side of the stringer and floor system so that the stair resembles Figure 21-26.
14. Locate required guardrails and handrails.
15. Provide dimensions to represent the following sizes:

- Total rise and run
- Individual rise and runs
- Required headroom
- Handrail and guardrail heights

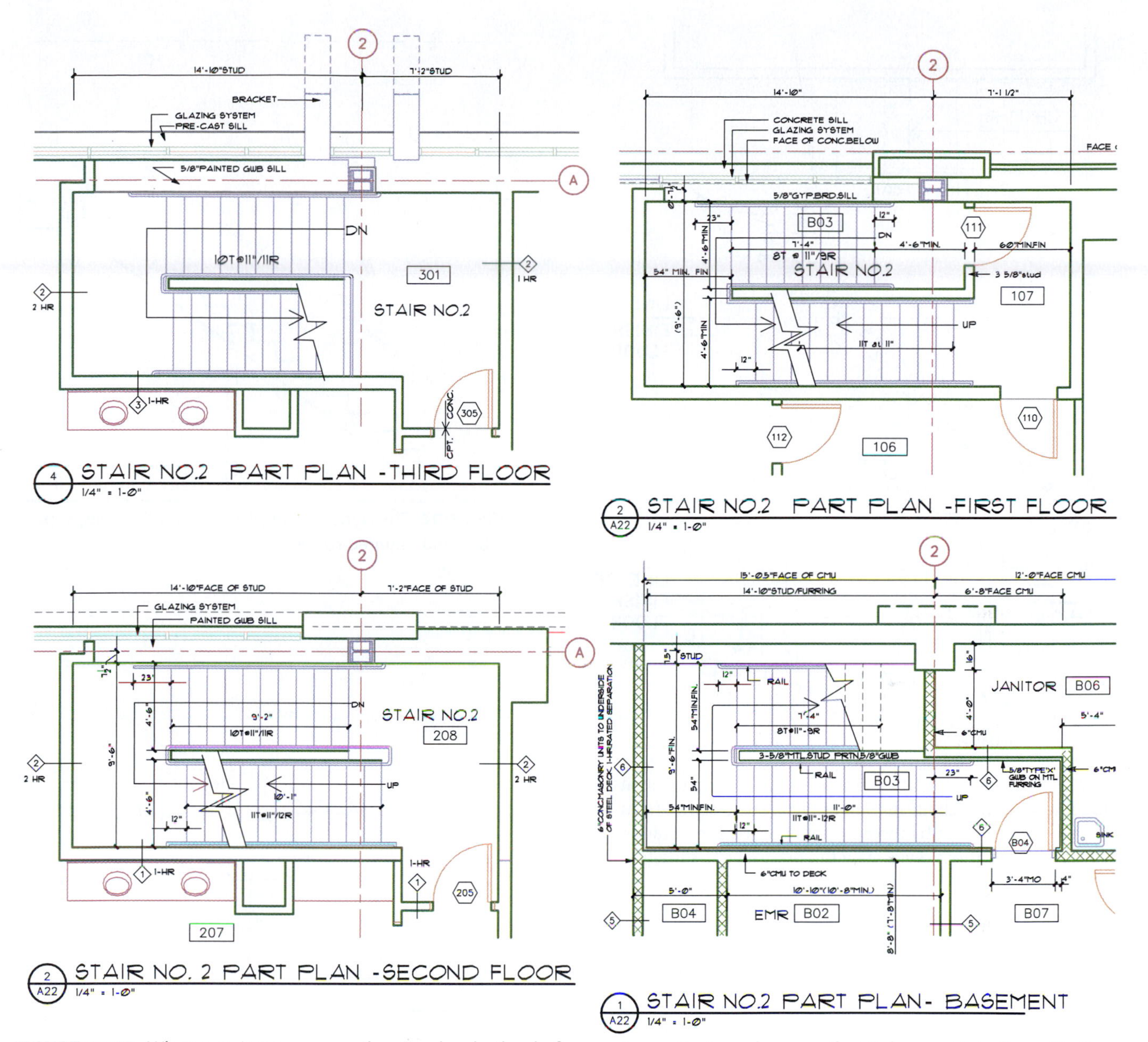

FIGURE 21-23 When a stair serves more than two levels, the drafter must coordinate each run so that it does not interfere with the headroom of other stairs in the run. *Courtesy Peck, Smiley, Ettlin Architects.*

16. Use leaders to label each size and component used to make the stair. Material represented in a detail does not have to be completely specified.

Notice the lack of notes in each of the professional examples throughout this chapter and the dependence on details for a clear explanation of construction.

17. Place a title and scale below the drawing.
18. Place required detail tags to locate related details to the section.

The completed stair section should resemble Figure 21-27.

Drawing Stair Details

Stair details should be drawn at a scale appropriate to the detail that must be shown. Scales ranging from 3/4" = 1'–0" to 3" = 1'–0" (1:20 to 1:5) are typically used for stair details. Material typically shown in detail includes:

- Stringer-to-upper-floor-system connection.
- Stringer-to-lower-floor-system connection.
- Baluster-to-stringer connection.
- Stringer-to-tread connection.

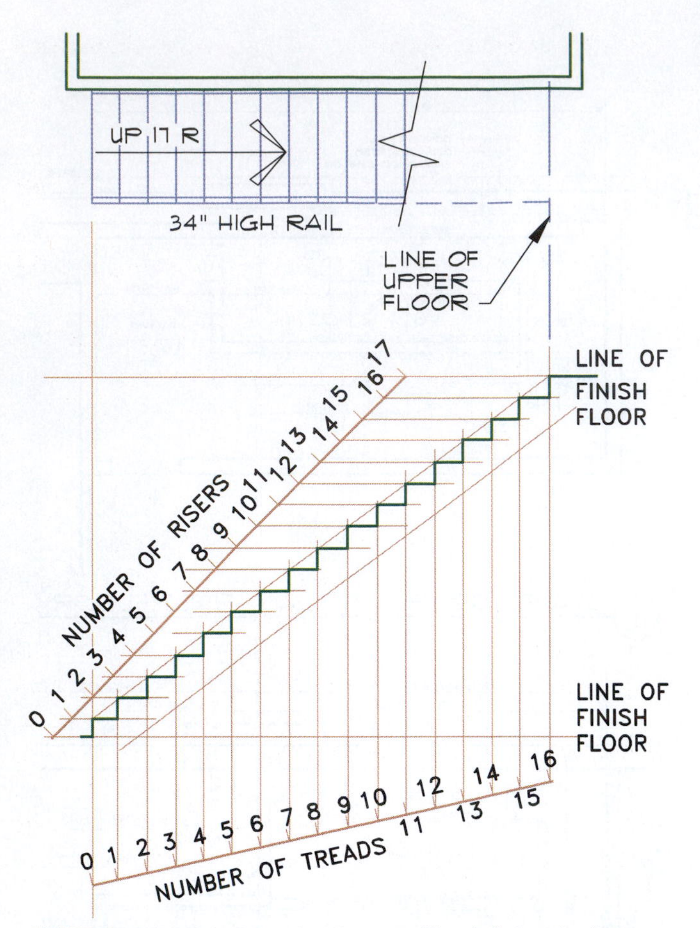

FIGURE 21-24 A stair section is started by defining the limits and representing each finished floor level. The distance from floor to floor is divided by the number of risers. The distance representing the run is divided by the number of treads.

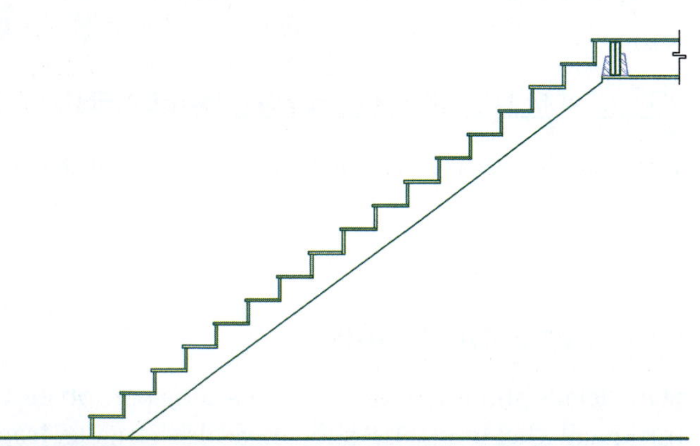

FIGURE 21-25 Once a grid for the stair has been determined, individual treads and risers can be drawn.

Figure 21-18 shows several of the supplemental details for the stair drawn in Figure 21-15. Each is a stock detail edited to meet the specific needs of this stair. Because of the scale used for details, it is important to use varied lineweights and appropriate hatch patterns to accent each material used.

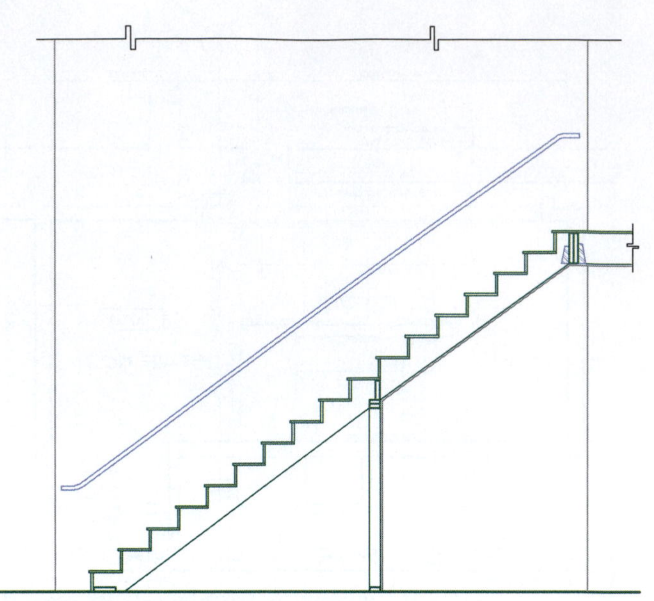

FIGURE 21-26 The layout is completed by adding rails, support walls, and required supports.

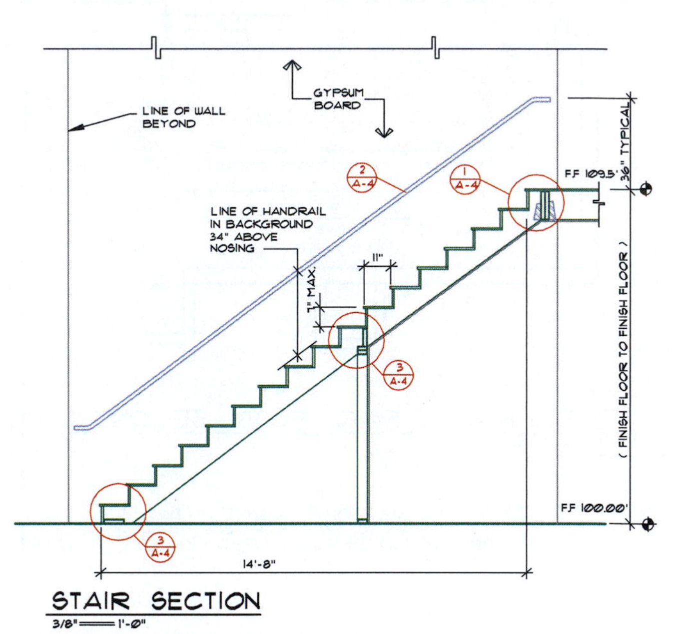

FIGURE 21-27 A completed stair section. The absence of notation is allowed because of the use of four details to further explain construction.

ELEVATOR DRAWINGS

The elevator manufacturer provides the bulk of the drawings required to install elevators. The responsibility of CAD drafters working with the architectural team is to accurately represent the shaft size and location on each plan view based on the design of the architect. Figure 21-28 shows how an elevator can be represented on enlarged floor plans. Sections and

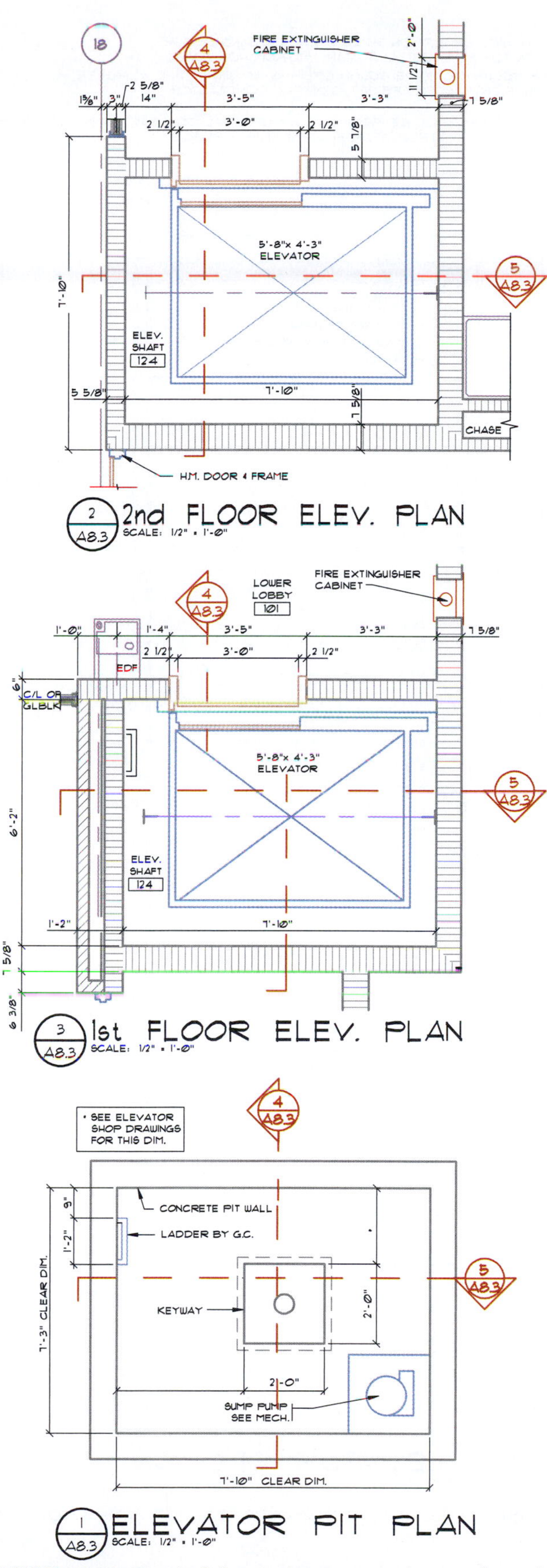

FIGURE 21-28 The shaft for an elevator must be coordinated with the supplier and carefully detailed on each floor. *Courtesy G. Williamson Archer, A.I.A., Archer & Archer, P.A.*

details similar to Figure 21-29 need to be drawn and should represent:

- The door and elevator height.
- Vertical heights between floors around the elevator shaft.
- The depth of the elevator pit, sump pump, and shaft vent, or the height of the elevator loft.
- Enlarged details of the pit ladder, door head and sill, slab edge, pit ladder, and vent placement.
- Wall clearances based on the manufacturer's requirements.

The *elevator pit* is the area below the lowest floor required to provide clearance for the elevator and the shaft mechanism. The *elevator loft* is provided at the upper end of the shaft for the same purpose. The drafter needs to work very closely with specifications supplied by the manufacturer to detail required sizes for the elevator. If the elevator is custom designed, the interior elevations of the car are drawn so that the interior of the elevator matches the building interior. These elevations should be placed on the same sheet as other elevator information and not placed with other interior elevations.

ESCALATOR DRAWINGS

The manufacturer of the escalator provides the bulk of the drawings required to install escalators. The responsibility of CAD drafters working with the architectural team is to accurately represent the width, length, and location on each plan view based on the design of the architect.

ADDITIONAL READING

Links for the materials represented on the stair and ramp drawings are too numerous to list here. Rather than listing individual sites, use your favorite search engine to research products from the following Construction Specifications Institute (CSI) categories: wood, steel, concrete, interior finish and trim; paints and coatings; caulks and adhesives.

NOTES:

1. HOISTWAY ENCLOSURES MUST HAVE SUBSTANTIALLY FLUSH SURFACES ON HOISTWAY SIDE EXCEPT ON SIDES WHERE LANDING OCCURS. ANY SETBACKS, PROJECTIONS, OR BEAMS OF MORE THAN 2" MUST BE BEVELED ON TOP SIDE NOT LESS THAN 75° FROM HORIZONTAL.
2. LEAVE OUT FRONT WALL OF HOISTWAY FULL WIDTH BY 7'4" HIGH, WHERE OPENINGS OCCUR OR LEAVE ROUGH OPENING 1'0" WIDER AND 4" HIGHER SO THAT ENTRANCES CAN BE SET IN PROPER RELATIONSHIP TO GUIDE RAILS. ONCE ENTRANCES ARE SET, GENERAL CONTRACTOR IS TO COMPLETE FRONT WALLS AND GROUT FRAMES AND SILLS.
3. ANY SLEEVES, CUT OUTS, CHASES, AND RECESSES REQUIRED BY ELEVATOR CONTRACTOR SHALL BE PROVIDED BY GENERAL CONTRACTOR IN LOCATIONS DESIGNATED ON ELEVATOR LAYOUT DRAWINGS. GENERAL CONTRACTOR TO FILL VOIDS AFTER ELEVATOR EQUIPMENT IS IN PLACE.
4. PROVIDE VINYL COMPOSITION TILE FOR ELEVATOR CAR FLOOR.

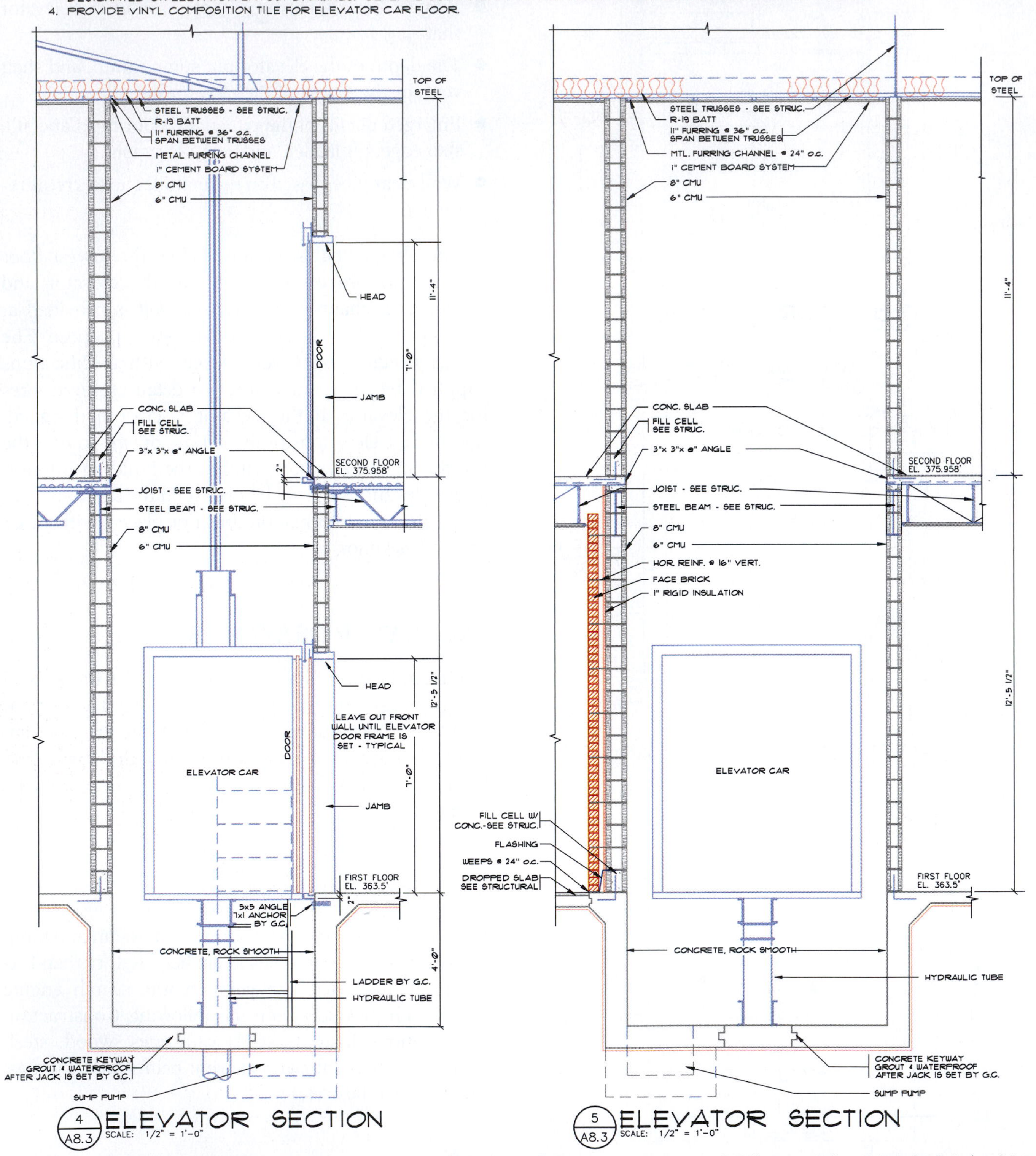

FIGURE 21-29 The sections necessary to supplement the stair shown in Figure 21-28. *Courtesy G. Williamson Archer, A.I.A., Archer & Archer, P.A.*

KEY TERMS

Guardrail
Handrail
Ramp
Rise
Risers
Run
Stairwell
Stringer
Tread

CHAPTER 21 TEST Ramp, Stair, and Elevator Drawings

QUESTIONS

Answer the following questions with short complete statements. Type the chapter title, question number, and a short complete statement for each question using a word processor. Some answers may require the use of vendor catalogs or seeking out local suppliers. If math is required to answer a question, show your work.

Question 21-1 What restrictions govern stairs and ramps?

Question 21-2 When are ramps required to have a handrail?

Question 21-3 Can a handrail be mounted to a bracket that will extend three inches from the support wall?

Question 21-4 What is the maximum riser height allowed by the IBC for a stairway serving an office? An upper floor of an apartment?

Question 21-5 When is a guardrail required?

Question 21-6 What is the required spacing of the balusters of a guardrail?

Question 21-7 What is the minimum width required for a spiral stair tread?

Question 21-8 Determine the total run and the number of risers and treads required for an office structure with 11'–0" from floor to floor.

Question 21-9 List five common materials used for treads.

Question 21-10 Describe two methods of attaching a wood tread to its support.

Question 21-11 Why are the drafter's responsibilities limited when specifying an elevator?

Question 21-12 List details that typically accompany a stair section.

Question 21-13 What elevator drawings is a drafter usually required to work on?

Question 21-14 What information must be specified for an escalator?

Question 21-15 What is the minimum information that should be labeled on a stairway on a floor plan?

DRAWING PROBLEMS

Use the reference material from preceding chapters, local codes, and vendor catalogs to complete one of the following projects.

Unless your instructor provides other instructions, draw the stair sections that correspond to the floor plan that was drawn in Chapter 16. No skeletons for these problems are available on the student CD. You will need to generate the required drawings from scratch using appropriate symbols, linetypes, dimensioning methods, and notations to complete the drawing. Determine any unspecified sizes based on material or practical requirements. Unless specified, select a scale appropriate to plotting on D-size material and determine LTSCALE and DIMVARS.

Although most of the information required to complete a project can be found by searching other related problems, some of the problems will require you to make a decision on how you would solve a particular problem. Make sketches of possible solutions and submit them to your instructor prior to completing each problem.

- Place the necessary cutting plane on the floor plan to indicate each drawing.
- All heights listed can be assumed to be from finish floor to finish floor.
- Save each drawing as a wblock with a name that accurately reflects the contents.

Problem 21-1 Draw a ramp that will connect two different floor levels in a warehouse that are not open to the public. The ramp will be a total of 30" high. Determine the required length. Assume a 5" thick concrete slab, and thicken the intersection of the top and bottom of the ramp to 10" × 20". Reinforce the slab with (3)-#5ø bars continuous throughout the ramp, 2" up from the bottom of the slab. Use (2)-#5 bars at the top and bottom of the ramp (at the thickened areas) 1 1/2" clear of the footing bottom. Use #5 @ 24" o.c. for the length of the ramp.

Problem 21-2 Draw a U-shaped stair to be used in an office structure with 10'–0" from floor to floor. Determine the total required run for each flight and the rise and run for each tread. Use steel stringers, rails, and balusters, and a metal pan to support each riser. Use the drawings in this chapter to determine material sizes unless your instructor provides other instructions.

Problem 21-3 Draw a straight-run stair to be used in an office structure with 12'–0" from floor to floor. Determine the total required run for each flight and rise and run for each tread. Use open wood treads with laminated stringers. Use the examples in this chapter for selecting materials.

Problem 21-4 Draw a section for a stair to be used in a town house unit of an apartment complex. The lower floor will be a concrete slab and the upper floor will be framed with 2 × 12 floor joists with 3/4" plywood and 1 1/2" concrete deck. Assume a total rise of 9'–0". Determine the required number of treads, risers, and the total run. Frame the stair using (3)-2× material for the stringers. Select appropriate material for treads and risers.

Problem 21-5 Draw the required stairs for the structure that was started in Chapter 15. If suitable, use one of the blocks created in problem 1-3 as a base drawing and edit as required. Provide details to show stringer-to-floor, tread-to-stringer, and stringer-to-baluster connections.

Section 4

Preparing Structural Drawings

22 Drawing Framing Plans

In Chapter 2 you were introduced to the drawings that make up a set of construction drawings. The drawings used to build the frame of a structure are the structural drawings. This chapter will explore:

- The drawings generally included with the structural drawings.
- Framing plans including the CAD skills that are required, common features, and representation of materials on a framing plan.
- Steps for drawing the framing plans for floor levels and the roof level.

Understanding the major concepts contained on these drawings and how the drawings are integrated into the entire set of construction drawings is a must for drafting a set of plans.

STRUCTURAL DRAWINGS

The same types of drawings used to present the architectural information can be used to present structural information. These include plan views, elevations, sections, details, schedules, and written specifications. Structural plan views include framing plans of the roof level and each floor level. This chapter discusses each of the framing plans, and subsequent chapters introduce the remaining structural drawings.

The area of the country you are in, the type of building to be erected, and the occupancy of the structure dictate what materials are to be used. Common materials used to form the shell of the structure include sawn lumber, heavy timber, poured concrete, and concrete block are represented on the framing plan. With the exception of concrete block, each of these materials can be used to frame the roof and floor systems represented on the framing plan. Typically, several of these materials may be incorporated into the framework and reflected in the framing plan.

FRAMING PLANS

The base layers of the floor plan are used to create the framing plans. Figure 16-38 shows the base layers of the floor plan that are displayed when starting a framing plan. For a very simple project similar to problem 16-1, framing information can be placed on the floor plan and separated from the architectural information by the use of layers. The framing plan for a simple project can also be started by making a copy of the floor plan to serve as a base for the framing plan so that two completely separate drawings are used to show the floor and framing information. The most common method of developing the framing plan is for the architectural team to supply an electronic copy of the floor plan that contains the basic elements of the floor plan to the structural engineer. CAD drafters working under the supervision of the engineer will use the structural calculations and drawings supplied by the architectural team to complete the project. Structural material can then be attached to the architectural base drawings using the INSERT or XREF commands. If a large complex is being drawn, the plan may be divided into zones and placed on two or more sheets. If zones are used, they must always match those used for the architectural drawings.

Using the mathematical calculations and sketches prepared by the engineering team, CAD technicians draw the framing plans. Figure 22-1 shows a portion of a framing plan. Framing plans are drawn for each level of the structure. For a one-level structure, the material used to frame the roof system is shown on a plan that resembles the roof plan. Because the framing for the first-level floor is shown on the foundation, no framing plan is required. For a multilevel structure, a plan is provided for each level. A three-level structure requires plans for the lower, second, and upper levels. Plans are arranged within the drawing set from the ground level to the roof, reflecting how the structure will be built. The lowest level of the structure is represented on the foundation plan, followed by succeeding floor levels. Framing plans are drawn working

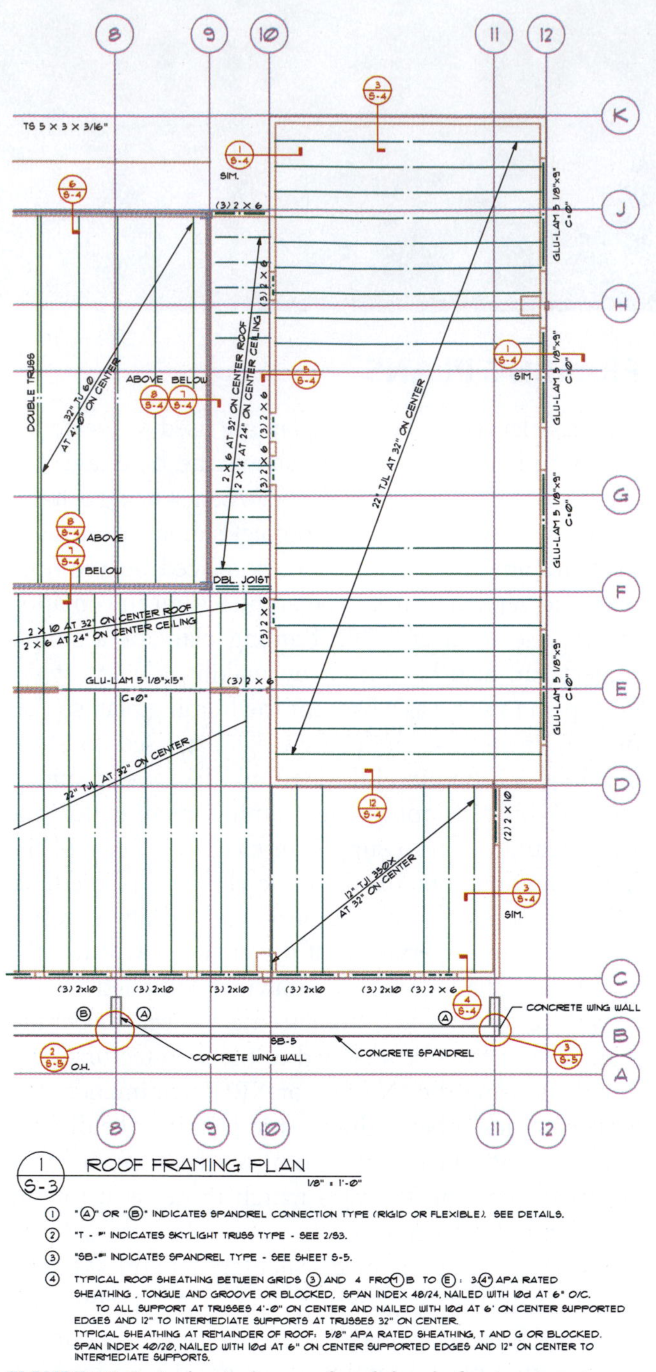

FIGURE 22-1 A plan is drawn of each level of a structure to show all materials needed to construct the skeleton. *Courtesy Van Domelen/Looijenga/McGarrigle/Knauf Consulting Engineers.*

from the top of the structure down to the ground, so that the location for beams and supports is better understood.

The main goal of the framing plan is to represent the location and size of framing members such as beams, joists or trusses, posts and columns, and bracing that resists the stresses applied to the structure. These are the major elements that make up the skeleton of a structure and transfer the weight of vertical loads to the supporting foundation. Framing plans are also used to show the materials that are used to resist the horizontal stress from forces caused by wind, flooding, and seismic activity. Information regarding nailing, bolting, and welding to resist these stresses is also found on the framing plan.

Using CAD Skills for Framing Plans

The framing plan should be used to specify the locations and the materials used to resist the forces of rotation, uplift, and shear. Just as with other drawings, these features can easily be placed using AutoCAD commands such as LAYER, ARRAY, COPY, DIM, TEXT, WBLOCK, and ATTRIBUTE, as well as features such as DesignCenter.

Layers

The complexity of the structure dictates how the framing plan will be arranged. On a simple structure such as a retail sales outlet, the framing plan could be included with other plan views and separated by different layers. With proper planning, information can be divided into layers to define each plan view contained in the drawing file. If the architectural and structural drawings are kept in the same file, the CAD drafter must carefully separate each layer using names based on the National CAD Standards (NCS) guidelines. Recommended names are presented in Appendix C on the student CD. Prefixes such as S-FRAM can be used to define material on the framing plan, with modifiers such as BEAM, JOIS, and ANNO used to further define materials stored on the framing plan.

Larger structures are also more likely to involve several firms to complete the plans. Each firm must have a copy of the base floor plan file to add its material. A separate framing plan can be created by making a block of the walls and columns of the floor plan. Information specific to the framing plan can now be added to this drawing, and the drawing file is then stored as a new drawing file. By adding information to a copy of the floor plan using the external reference (XREF) command, changes can be made to the floor plan, and the framing plan will be updated automatically.

Inserting Blocks and Assigning Attributes

Once the method of creating the framing plan has been determined, information specific to the framing plan can be added. Repetitive information should be created as a block and inserted or referenced into the drawing base. Items such as grids, detail and section markers, and drawing symbols should be created as a block and inserted with ATTRIBUTE to control page and detail numbers. An alternative to inserting common features is to store them in a template on appropriate layers and thaw them as needed.

COMMON FEATURES SHOWN ON A FRAMING PLAN

Regardless of the material used, framing plans have many common features. Because of the large amount of information that needs to be placed on the framing plan, it is important for the CAD technician to develop the framing plan in a logical order.

Bearing Walls and Support Columns

Using the walls and columns drawn for the floor plan is the most efficient way to create the framing plan. This can be done by using either the BLOCK or XREF method. All items shown on the floor plan that are not directly related to forming or supporting the structure should be removed from the drawing to be used to create the framing plan. Figure 22-2 shows the base floor plan used to draw the framing plan for an apartment complex. The right unit shows the information required for the floor plan and the left unit shows the framing plan for the structure.

Locations for Each Beam

Once the drawing base has been prepared, support beams should be drawn. Start with major beams and work to intermediate beams and then to purlins. As beams are labeled, text is typically placed parallel to the member using the methods presented in Section 2 and Figure 22-2. Beams can also be specified using a schedule as shown in Figure 22-3. Notes to specify columns are placed on an angle to help distinguish the post or column from other building materials. Beams and related material should be placed on layers separate from the base material with a layer name such as S FRAM BEAM.

Dimensions

The floor plan contains all the dimensions needed to locate walls and other key architectural features. The framing plan typically contains the dimensions to locate the grids to help the print reader gain an overall sense of the building size without having to go to the architectural drawings, as well as the dimensions to locate each structural component. Figure 22-4 shows an example of dimensions on a framing plan. Methods for dimensioning structural members vary based on the material to be used. These methods are explored later in this chapter after other key framing components have been introduced. Regardless of the material used, dimensions locating structural material should be referenced to a wall, column, or grid that is already located on the floor plan.

Thought should also be given to other drawings that require dimensioning. The overall and grid dimensions on the floor plan will be the same for the framing plan. The majority of dimensions on the lowest framing level are often the same as the dimensions on the foundation plan. Placing the dimensions on several layers can aid in dimensioning the overall project. A layer such as S FRAM DIMN EXT can be used to contain all exterior dimensions that will be needed on all drawings to be dimensioned. A layer such as S FRAM DIMN INT1 could be used for all dimensions that are specific to the first level.

Local Notes

Notes on the framing plan should be placed using the guidelines presented in Chapter 16 for placing notes on a floor plan. Many items on the framing plans will be specified but not drawn. This includes items such as shear panels, hold-down anchors, or metal ties, which can be noted as shown in Figure 22-5. Types of notes specific to each material will be explored later in this chapter. A layer name such as S FRAM ANNO can be used to separate framing text from other material.

Drawing Tags

In addition to showing major materials, framing plans are used as a reference map to coordinate the structural elevations, sections, and details. Tags relating the elevations to the framing plan resemble those introduced in Chapter 16. Detail markers are placed on the framing plan to show the location of details that relate to the framing plan, just as they were on the floor plan. Tags should be placed on a layer separate from other tags, with a title such as S FRAM SYMB. The section and detail tags can be inserted before the detail or section is actually drawn, but the attributes for page and detail numbers cannot be defined until the entire job is near completion. Detail and section tag numbering is explained in Chapter 2.

General Notes

Lengthy notes that specify materials should be placed as a general note to keep the framing plan uncluttered. When this is done, an abbreviated version of the note should be placed on the drawing, and referenced to the full note in the framing notes. An example of this form of notation would be:

Abbreviated note:

5/8" PLY ROOD SHEATH. SEE NOTE 1.

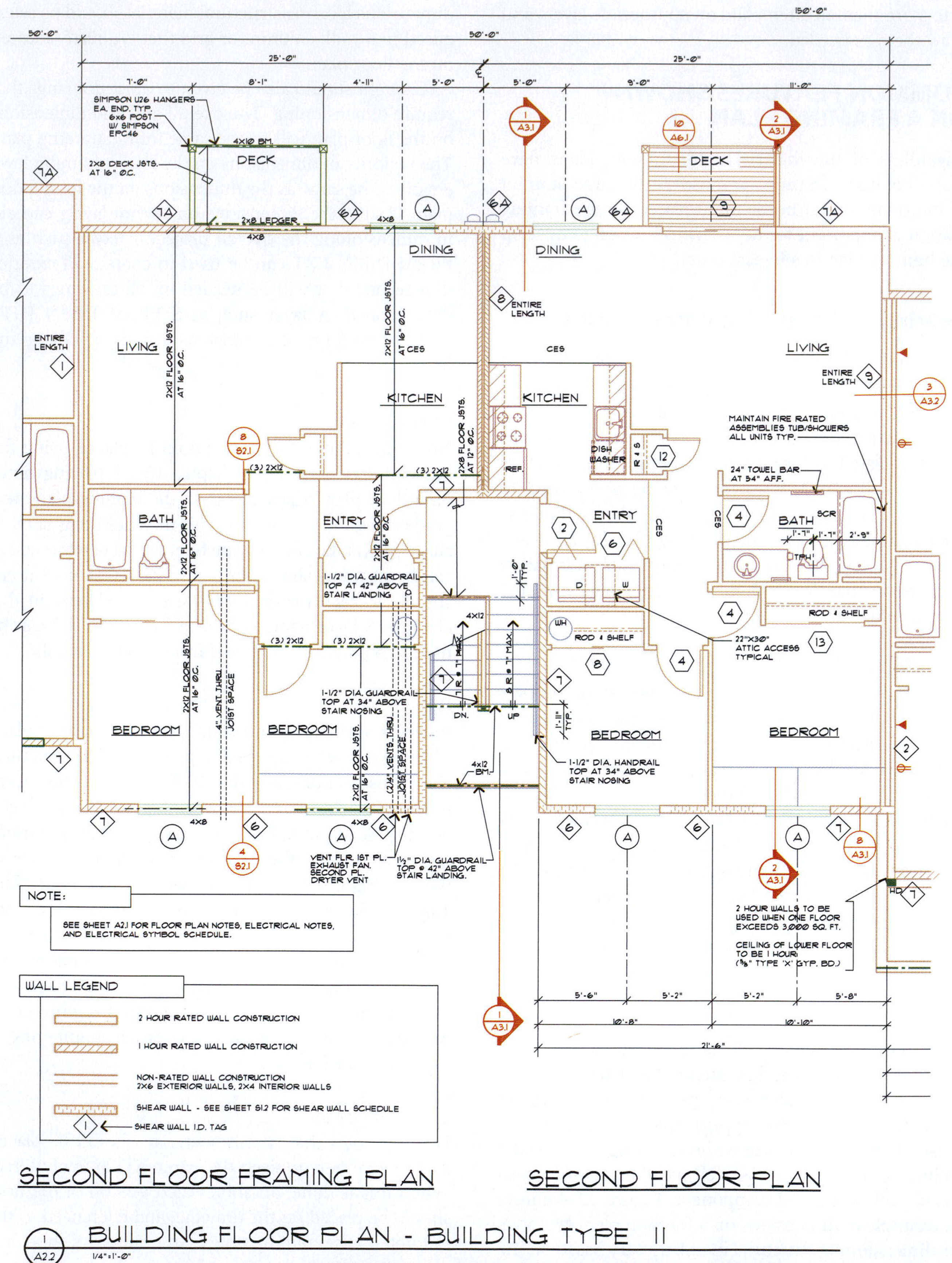

FIGURE 22-2 For simple structures, and for multiunit structures, the framing information can be placed in one unit so that the floor plan remains uncluttered. *Courtesy Scott R. Beck, Architect.*

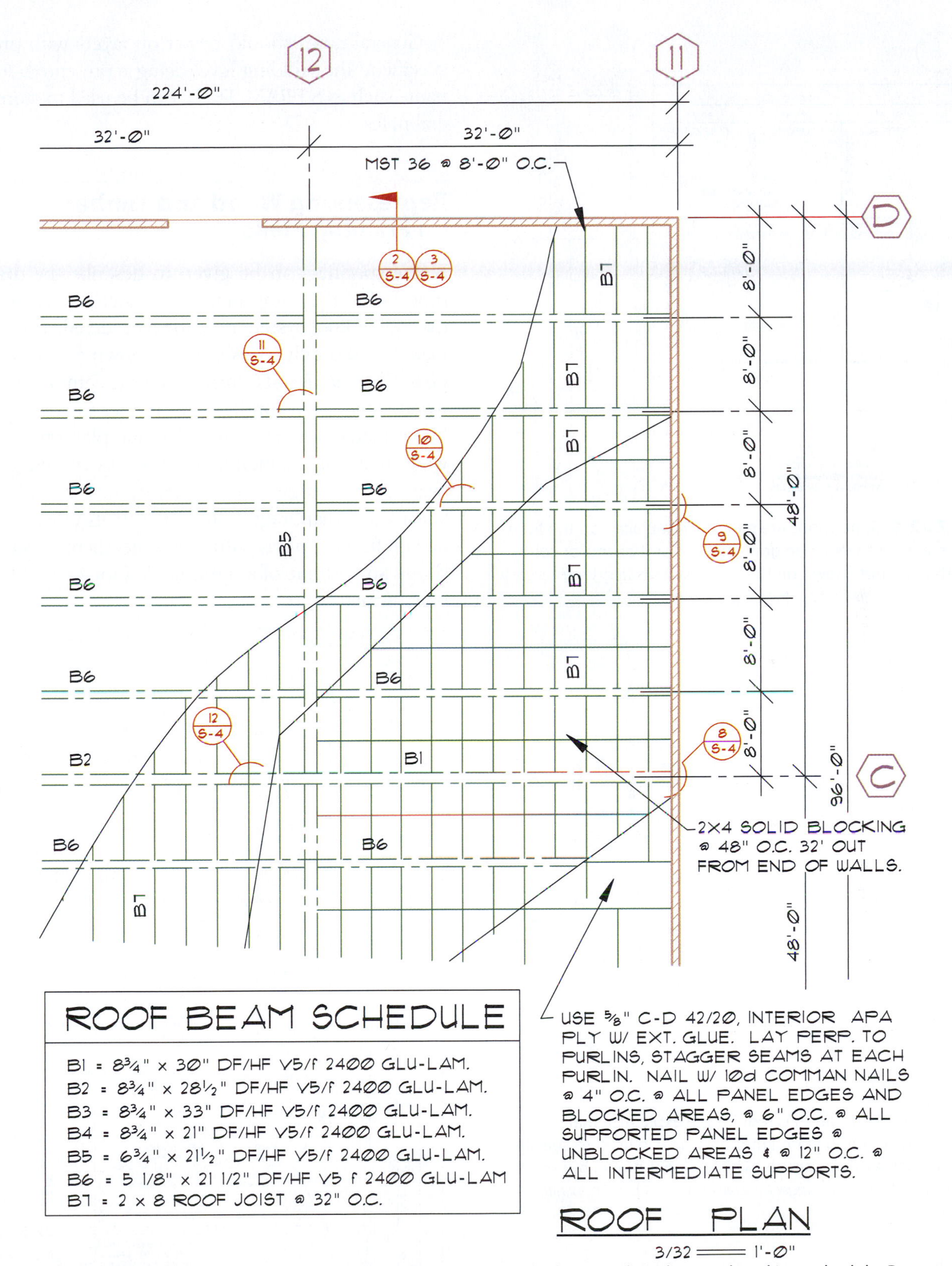

FIGURE 22-3 The specifications for a beam should be written parallel to the beam it describes or placed in a schedule. Because this is a fairly simple structure, information for the laminated beams and trusses has been placed in the same schedule. As structures become more complex, separate schedules for steel, laminated, sawn lumber beams, and trusses should be provided.

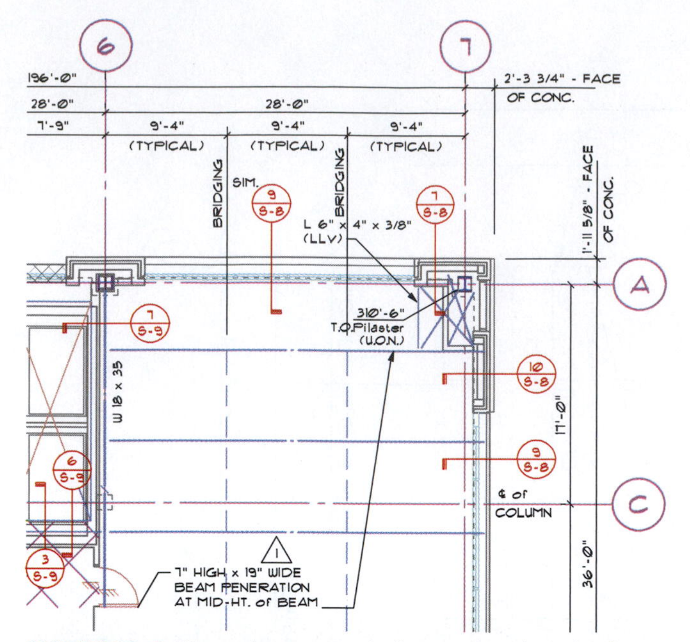

FIGURE 22-4 Dimensions are typically not placed on the framing plan except to describe grid locations and beams or columns that are not shown on the floor plan. *Courtesy Van Domelen/Looijenga/McGarrigle/Knauf Consulting Engineers.*

General notes should be set on layers with prefixes specific to the building level being represented. A layer name such as S FRAM TEXT can be used to store general notes.

Representing Wood and Timber on Framing Plans

Before specifics can be given to describe the drawing process for a framing plan, the material used to form the roof or floor system must be considered. The type of material used influences what is shown on the framing plan. Chapters 7 and 8 introduced the common uses for wood, engineered lumber, and timber in construction. Major materials shown on a framing plan representing wood materials include studs, post, sawn and glu-lam beams, joists, engineered joists, trusses, and plywood. Shear walls, diaphragms, and drag struts must be specified on framing plans with wood members. Figure 22-6 shows an example of a framing plan for a wood-framed office structure.

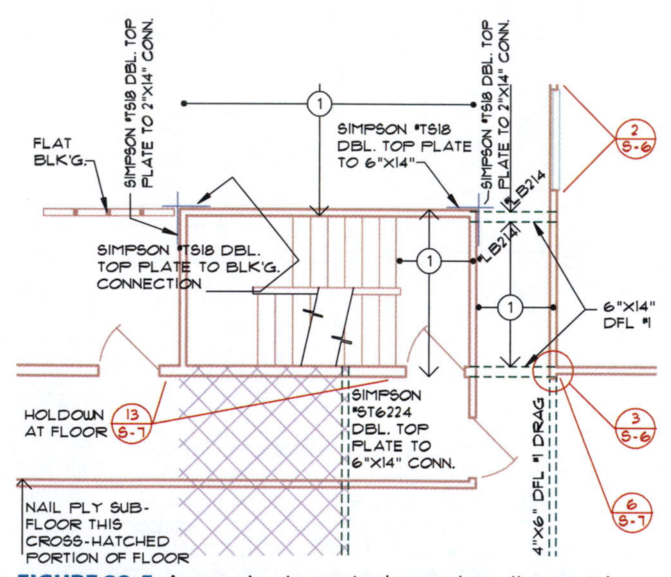

FIGURE 22-5 Annotation is required to explain all materials shown on the framing plan. Annotation is also used to specify materials such as shear panels, which are not drawn but must be referenced. *Courtesy Ginger M. Smith, Kenneth D. Smith Architect & Associates, Inc.*

General note:

ALL ROOF SHEATHING TO BE 5/8" APA 32/16 INTERIOR GRADE WITH EXTERIOR GLUE. LAY FACE GRAIN PERPENDICULAR TO JOIST AND STAGGER ALL JOINTS. USE 10d COMMON NAILS @ 6" O.C. @ BOUNDARY AND EDGES. USE 10D COMMON NAILS @ 10" O.C. @ FIELD UNLESS NOTED.

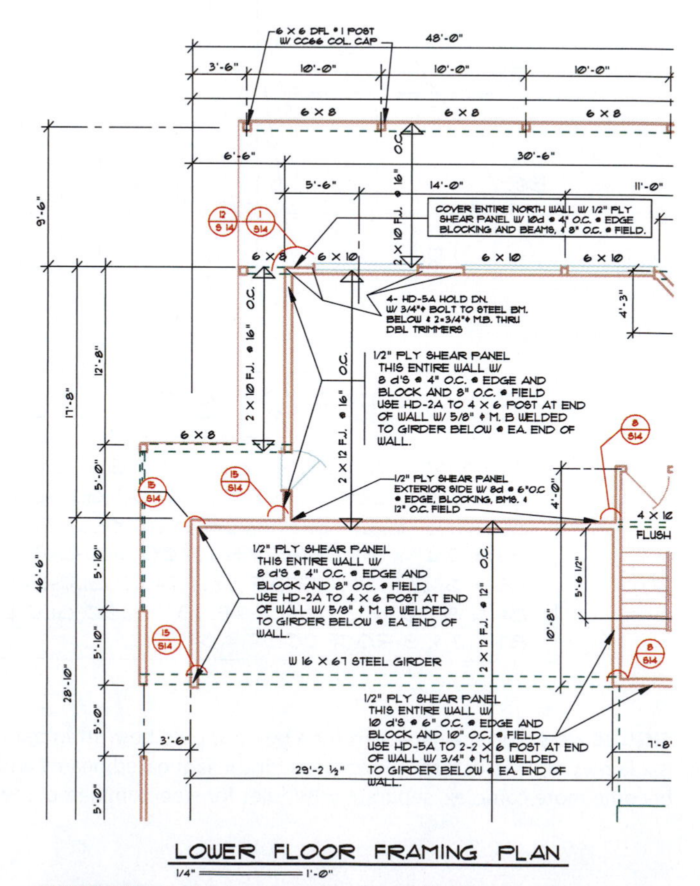

FIGURE 22-6 The framing plan for a hillside wood-framed office structure. Because the back side of the office is 16′ off the ground, walls must be stiffened to resist wracking.

Walls

Stud walls and wood posts appear on the framing plan just as they do on a floor plan. Walls that have special construction such as an extra base or top plate or plywood panels for resisting shear need to be noted and detailed based on the engineer's calculations. Figure 22-6 shows the locations of several different shear walls. Shear walls and metal hangers can be specified by local notes and explained in a detail similar to Figure 22-7.

Shear Walls

Shear walls were introduced in Chapters 13 and 16 as a means of resisting and transferring lateral forces from wind, snow, or seismic activity through the structure. Shown on a floor plan like any other wall, the components used to resist the lateral forces must be specified on the framing plan. The engineer determines which walls will be strengthened to resist lateral forces and CAD drafters are responsible for specifying the materials on the framing plan. The materials used to strengthen the wall are usually not drawn, but are specified in local notes or in a schedule. Special connectors, straps, or hold-down anchors required to reinforce construction are specified in details that are referenced to the framing plan. Figure 22-8a shows an example of how a shear wall can be represented on a framing plan. Figure 22-8b shows the resulting bracing installed in a structure to resist lateral forces.

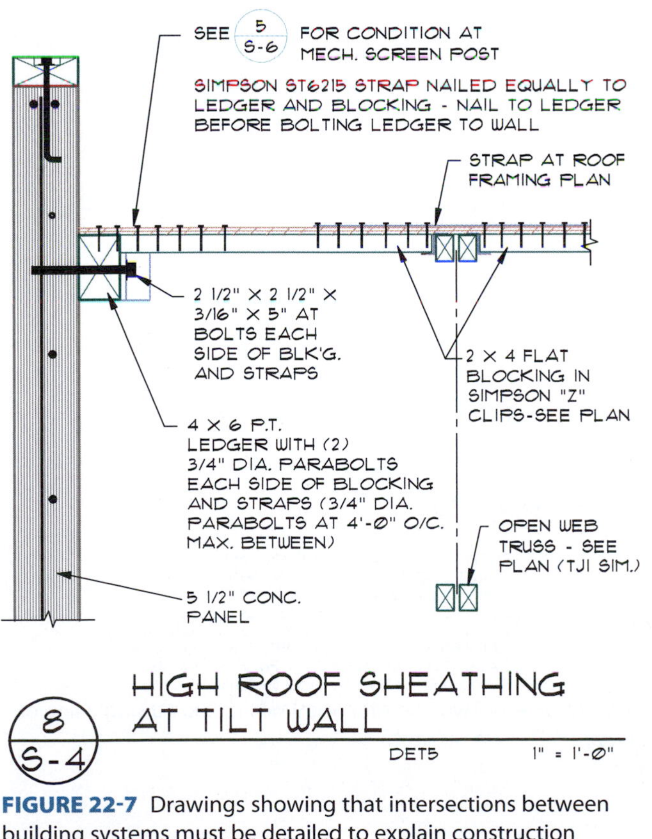

FIGURE 22-7 Drawings showing that intersections between building systems must be detailed to explain construction methods. *Courtesy Dean K. Smith, Kenneth D. Smith Architect & Associates, Inc.*

Beams

Sawn, engineered, and laminated beams are represented by thin dashed lines as shown in Figure 22-9. Beams are located by dimensions from the edge of an exterior wood or masonry wall and from the center of wood interior walls to the center of the beam. When a beam cantilevers past a supporting post, the end of the beam should be dimensioned from the end to the center of the supporting post. The locations of main support columns and beams should be dimensioned from the exterior face of exterior walls or from grid lines. The CAD technician's job is to provide dimensions that define the locations of each beam and column based on the design of the engineer. CAD technicians working for the lumber fabricator or truss manufacturer use the information provided on the framing plan to produce drawings that indicate exactly how the prefabricated material will be constructed, as well as precise measurements for how it will be cut.

Joists and Trusses

Joists or trusses can each be represented on framing plans using two techniques. Figure 22-10 shows a symbol that can be used on simple structures to locate the direction and member to be used. The joist symbol can be created as a block with attributes that can be altered for each application. Many offices show all of the required framing members except where other information cannot be clearly displayed. Figure 22-11 shows a typical method of representing framing members. Details that are required to show how forces will be transferred from the roof or floor system through walls to other areas of the structure must be referenced on the framing plan. Structures framed with wood typically require details showing beam-to-beam, beam-to-wall, joist-to-beam, and joist-to-wall connections. These can be referenced as shown in Figure 22-11, to represent a cutting plane or a view.

Structural Steel Framing Systems

Steel framing can be either a moment frame as shown in Figure 22-12 or a rigid frame as shown in Figure 22-13. Each system is introduced in Chapter 13. Each system has its own unique material to be shown on a framing plan.

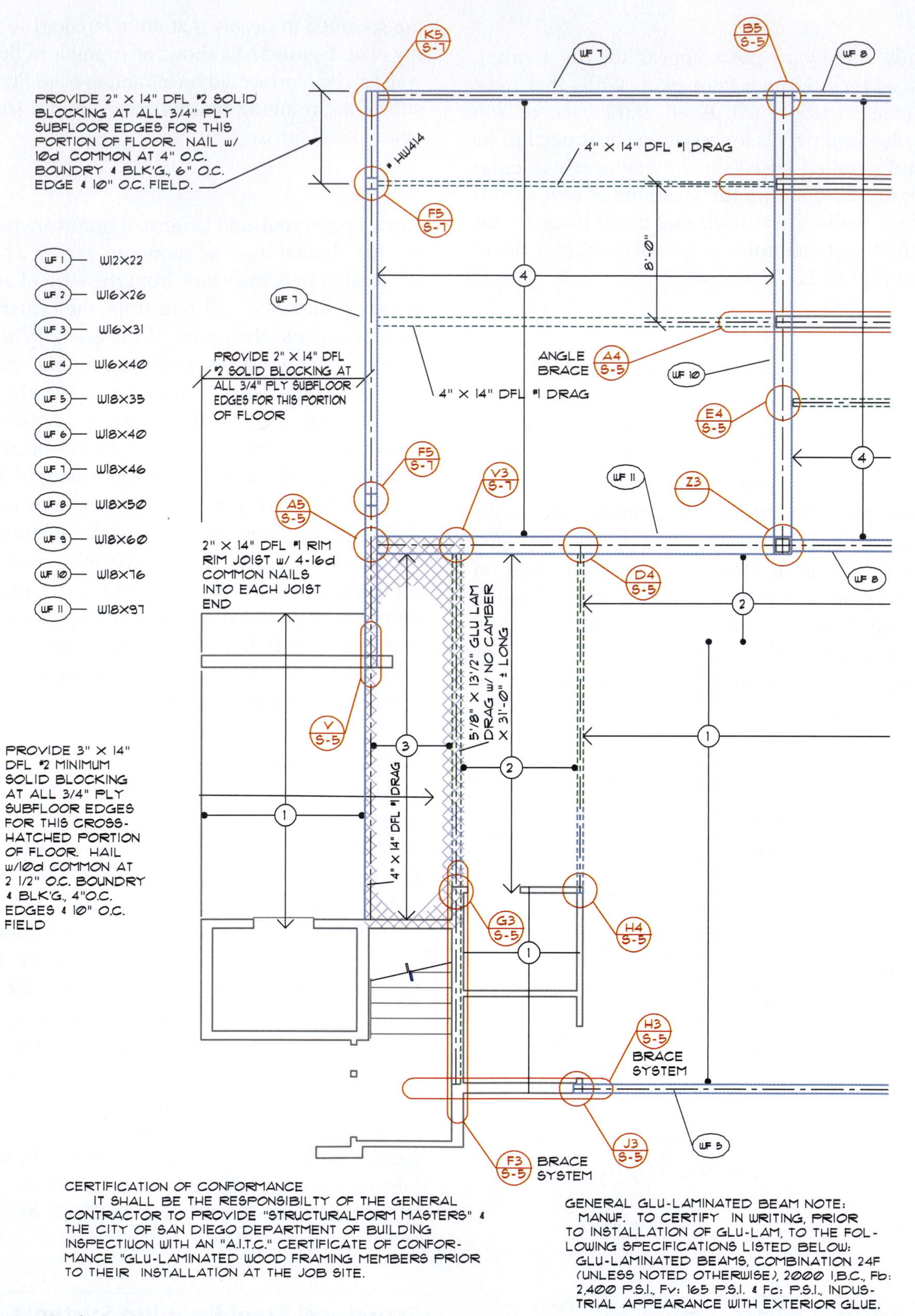

FIGURE 22-8a A wide variation in presentation methods is used to represent beams. Two common methods include a single line or double lines representing the width of the beam. *Courtesy Ginger M. Smith, Kenneth D. Smith Architect & Associates, Inc.*

FIGURE 22-8b Steel beams and bracing from detail F3/S5 specified on the framing plan shown in Figure 22-8a. *Courtesy Janice Jefferis.*

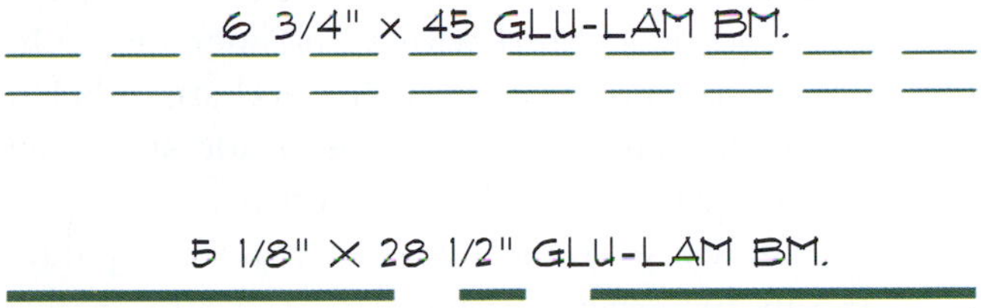

6 3/4" x 45 GLU-LAM BM.

5 1/8" X 28 1/2" GLU-LAM BM.

FIGURE 22-9 A wide variation in presentation methods is used to represent beams. Two common methods include the use of a thick single line, or double lines representing the width of the beam. The complexity of the structure and office practice determine which method is used.

22" TJL @ 32" O.C.

FIGURE 22-10 On simple plans, one symbol can be used to represent the span and spacing of repetitive members.

Moment Frames

Figure 22-14 shows an example of a framing plan using structural steel to form the major support system. Common steel materials were introduced in Chapter 9. The most common shapes to support major loads are typically M-, S-, and W-shaped steel beams.

Rectangular and circular steel columns as well as W shapes are used as vertical supports. Steel angles, tees, and channels are often used to support intermediate loads. Steel cables are used to resist lateral, wind, and seismic loads. Just as with structures framed with wood, framing plans reflecting steel-framed structures are formed using a base drawing that is similar to the floor plan. CAD drafters prepare the plan using the engineer's sketches and calculations, as well as reference manuals published by AISC such as the *Manual*

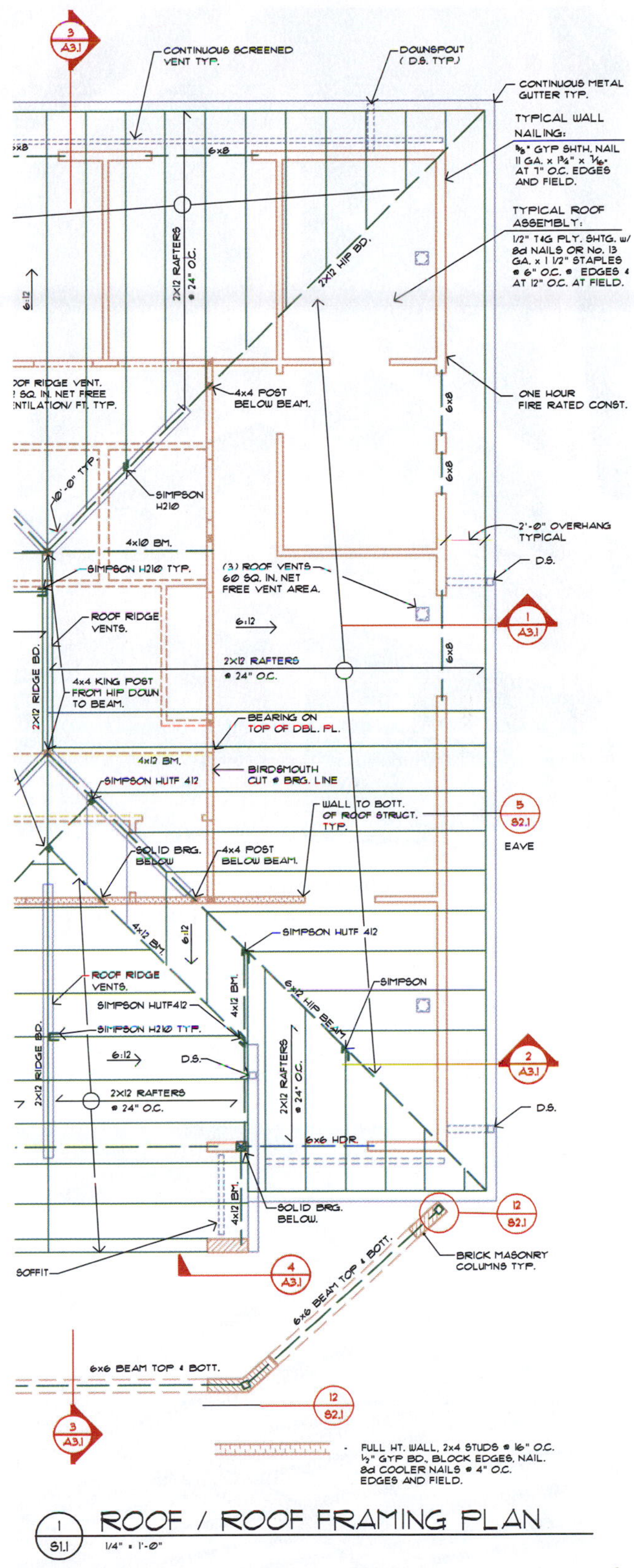

FIGURE 22-11 Most structural drawings show the majority of repetitive members to avoid confusion during the construction process. *Courtesy Scott R. Beck, Architect.*

FIGURE 22-12 Steel is often used to form the skeleton using connections referred to as a moment frame. *Courtesy Mike Jefferis.*

FIGURE 22-13 Many single-story steel industrial buildings are built using a framing system referred to as a rigid frame. *Courtesy David Jefferis.*

of Steel Construction and *Structural Steel Detailing*. The engineer provides sketches and specifications for the selection and location of materials, as would be provided for wood-framed structures. CAD technicians working for the steel fabricator produce drawings providing exact measurements of each piece of steel to be fabricated.

Columns and Beams On the framing plan, steel columns are typically represented by a thick line and are dimensioned from center to center of columns. Exterior columns are dimensioned from the face of the exterior shell to the center of the column. Beams may be represented by centerlines, solid lines, or hidden lines and are located using centerline dimensions from one member to the next. Figure 22-15 shows how steel columns and beams can be specified and dimensioned.

Steel framing members are specified using methods similar to those used with wood members. On complicated plans, framing members are labeled using tables to help the print reader understand the location and quantity required for construction. Steel columns should be represented in a table separate from beams to provide better clarity. Beams of different materials should also be kept in separate beam tables. Depending on the complexity of the structure, a separate plan can be provided for vertical supports and horizontal steel to provide clarity. Figure 22-16 shows a piling plan for the hillside office structure shown in Figure 22-6. Figure 22-17 shows the beam plan for the lower floor of the same structure. Steel framing often requires the elevation of a specific member to be shown on the plan. The height above a specific point, or surface, such as a finish floor level, is noted near the beam or listed in the table specifying the member size.

Decking Steel decking is often used to form a diaphragm in horizontal surfaces. Poured concrete floors placed over the steel deck can also be used to provide rigidity. Steel cables and turnbuckles, steel rods, or steel tubing is used between major supports to resist sheer and rotational stresses. Figure 22-18 shows an example of how the reinforcing steel is specified on a framing plan.

Rigid Frames

Rigid framing or prefab methods were introduced in Chapter 9. Members that make up the frame are typically W shapes, but both S and M shapes are also

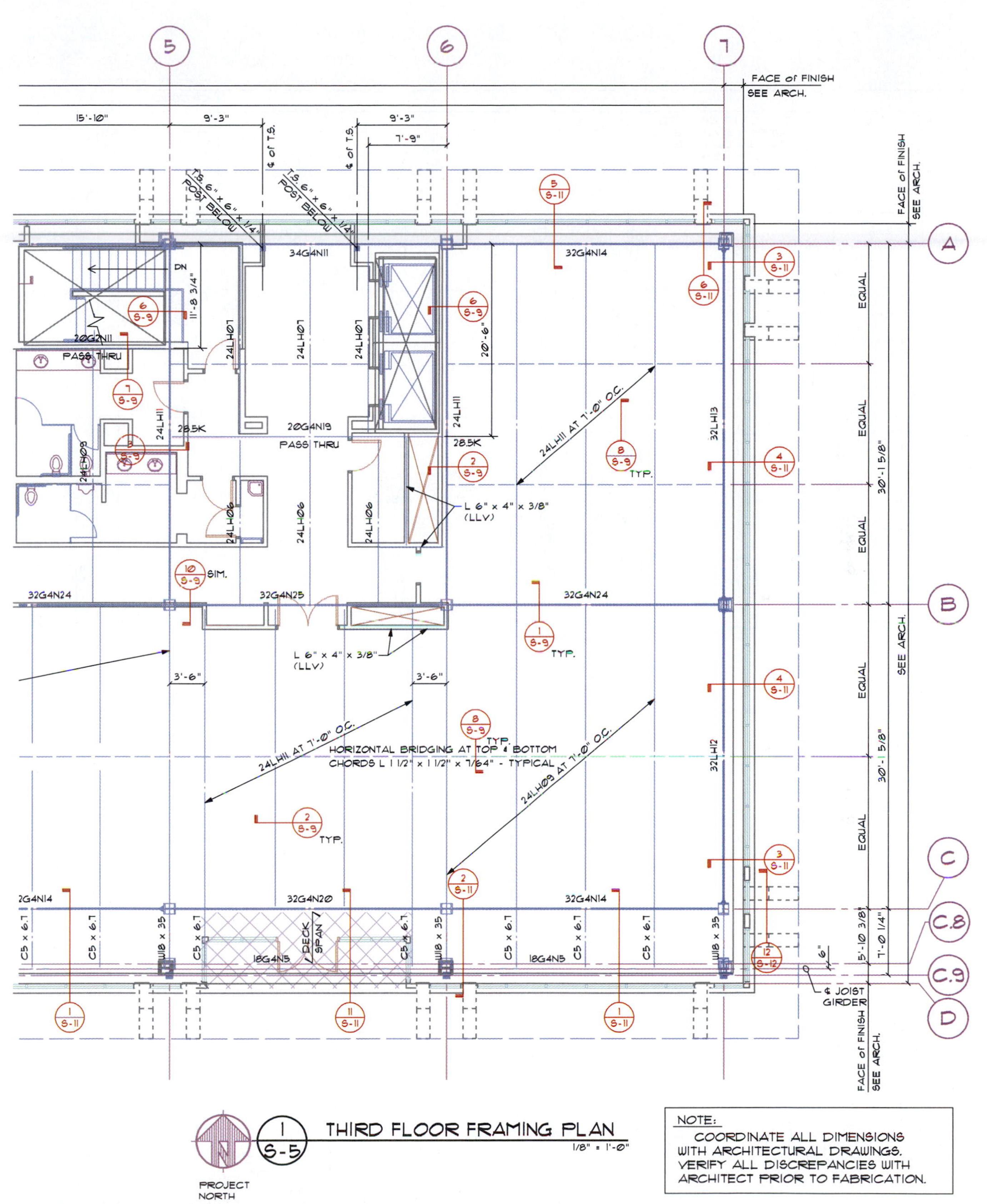

FIGURE 22-14 The framing plan for a steel structure. *Courtesy Van Domelen/Looijenga/McGarrigle/Knauf Consulting Engineers.*

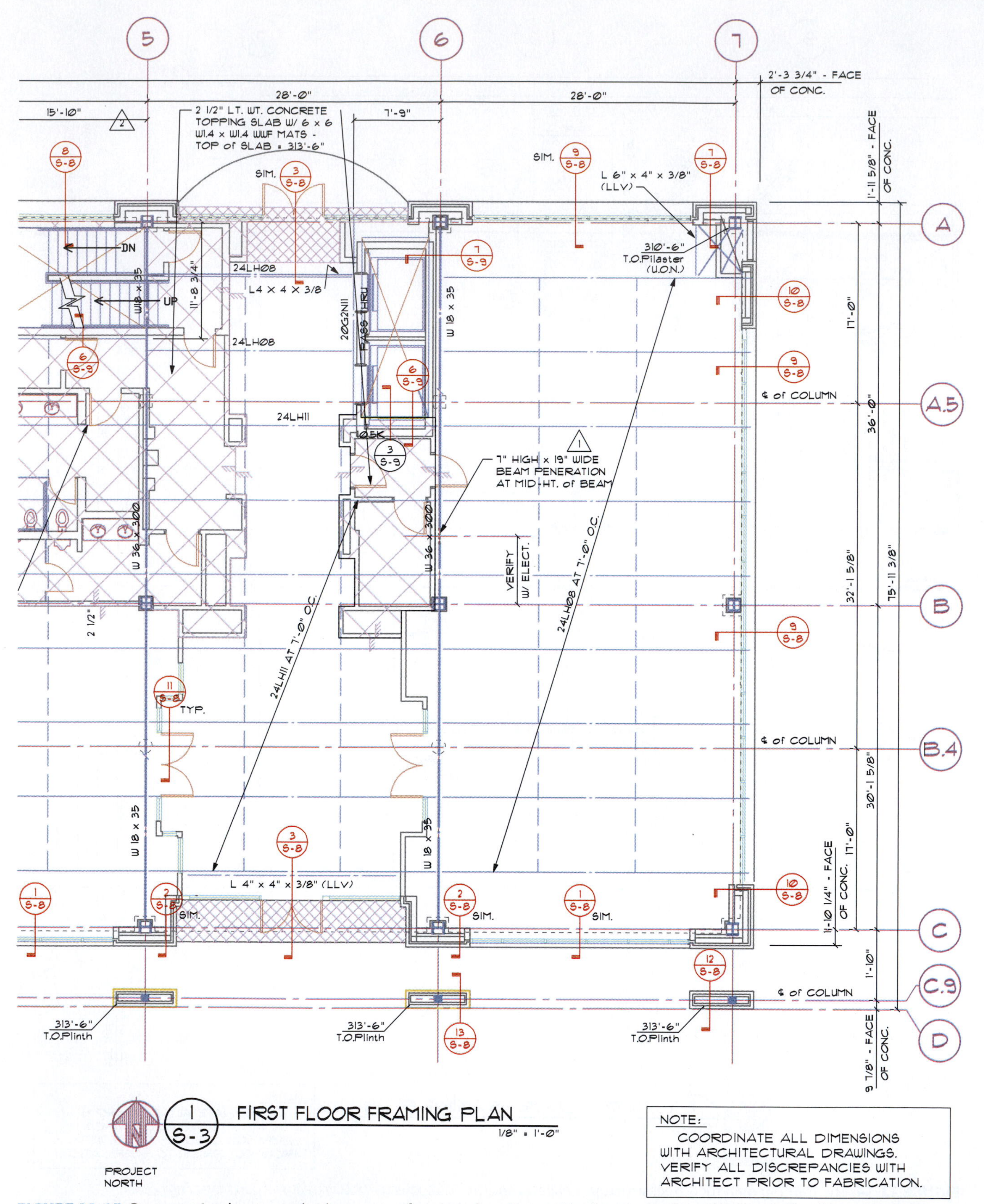

FIGURE 22-15 Representing beams and columns on a framing plan. *Courtesy Van Domelen/Looijenga/McGarrigle/Knauf Consulting Engineers.*

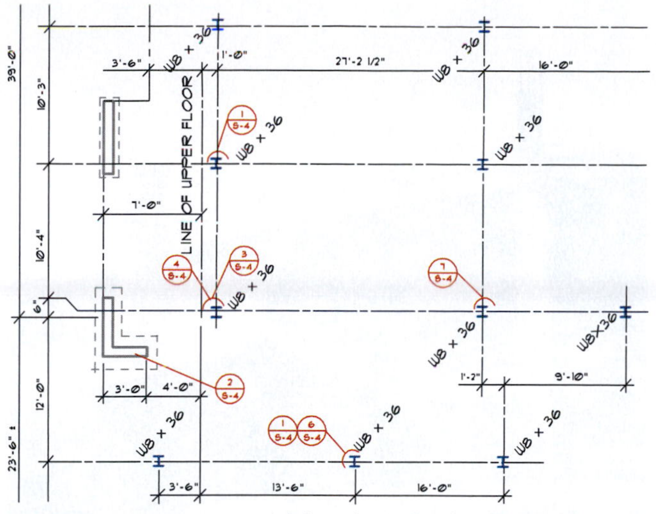

FIGURE 22-16 A portion of the piling plan for the hillside office shown in Figure 22-6.

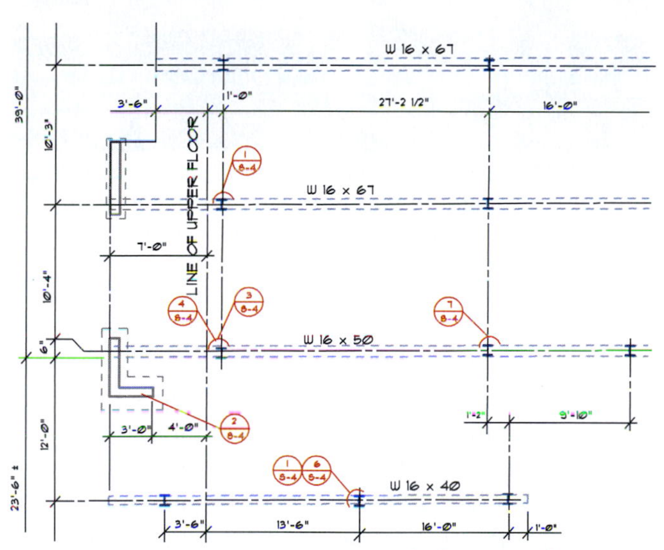

FIGURE 22-17 Because one crew sets the beams and a separate crew drives the pilings, each is placed on a separate plan.

used, depending on the size and spacing of the frame. Horizontal members that span between the frame are primarily channels, but angles and tees are also used. Figure 22-19 shows the framing plan for an industrial building framed with a rigid steel frame. Frame and intermediate members are located based on centerline locations. Support between members is usually developed by the use of steel cables. Framing elevations similar to Figure 22-20 are drawn to show the locations of members used to form the shell supports. Chapter 23 explains how the elevations are drawn. A structure framed using rigid-frame methods is drafted using methods similar to those used to draw a steel-framed structure.

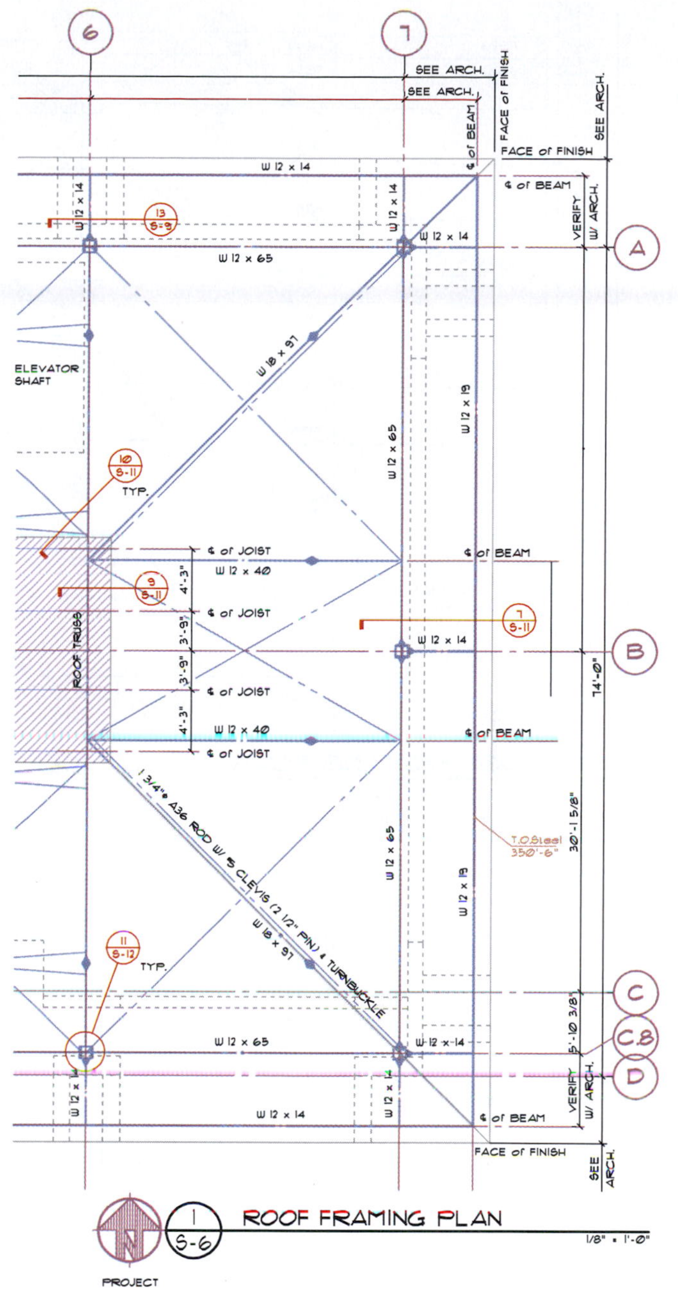

FIGURE 22-18 Representing reinforcing steel on the framing plan. *Courtesy Van Domelen/Looijenga/McGarrigle/Knauf Consulting Engineers.*

Precast Concrete Framing Systems

Precast concrete structures similar to those in Figure 22-21 offer exceptional strength and resistance to seismic stresses as well as a high degree of fire safety. Concrete is also widely used because it can be cast into almost any shape. Concrete structures require drawings to represent the walls, beams, and columns for each specific level. Depending on the complexity of the project, framing plans for concrete structures can be divided into column, beam, wall, and floor and roof plans. CAD

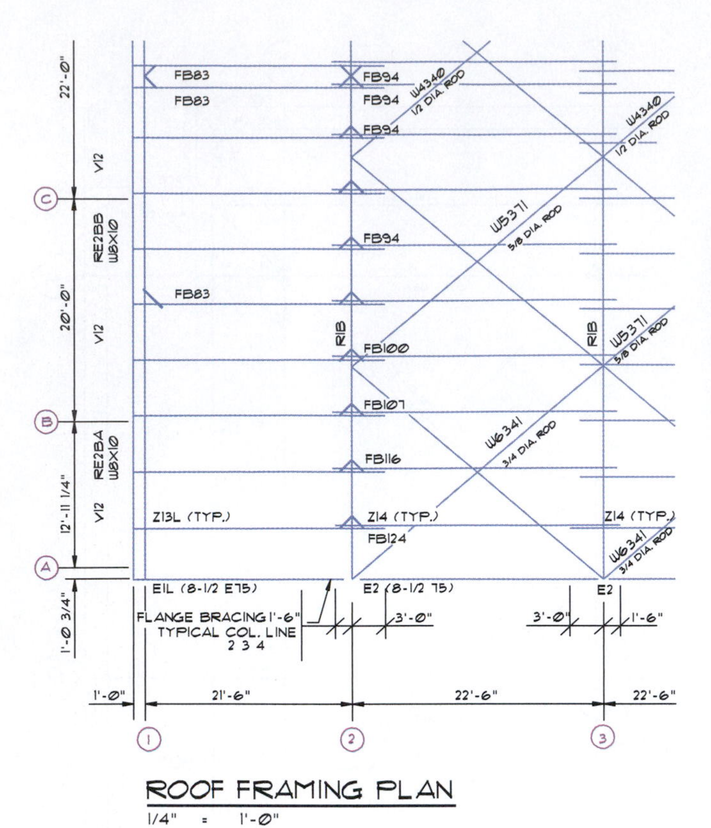

FIGURE 22-19 The roof plan for a rigid-frame structure.

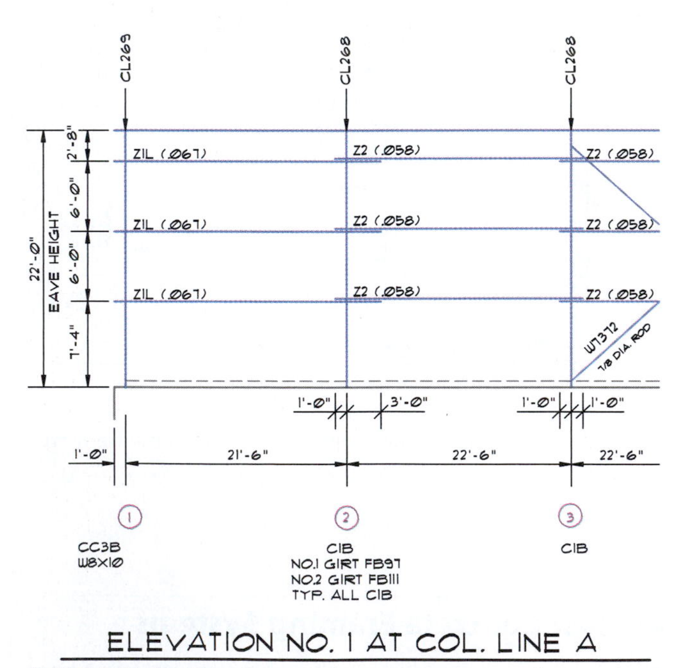

FIGURE 22-20 The elevation for a steel structure shows where each member of the skeleton will be placed.

technicians receive the information they need to draw a concrete framing plan from the architectural drawings and the engineer's sketches and calculations. Walls are located to their edges, and columns are located to their center in a method similar to that used in steel structures.

FIGURE 22-21 Precast and poured-in-place concrete offers exceptional strength and resistance to seismic forces and a high degree of fire safety. These floor and wall assemblies are used to support a multilevel sports facility. *Courtesy Janice Jefferis.*

Concrete Tilt-Up Plans

Forming walls with precast concrete panels, similar to those shown in Figure 22-22, is a common method of construction. The tilt-up plan must show the location of each panel as well as the members that are used to form and support the floor or roof framing. The panel plan is used to reference the location of each panel and resembles Figure 22-23. Panels are dimensioned from center of joint to center of joint, and from edge to edge. Notice in Figure 22-23 that panel 21 has a length of 19'–11 1/2" from center to center, with a distance of 1/2" required at each end between panels. On simple structures, this information can be placed on the floor or roof plans without the use of a separate plan, with specific information for the construction of each panel shown in a panel elevation similar to Figure 22-24.

DRAWING A FRAMING PLAN

Rarely is a CAD technician given the task of drawing a framing plan from its inception. Remember that a drafter is a person who draws the ideas of another in a clear, logical manner. As a new employee in a company,

FIGURE 22-22 Precast concrete panels are poured on the ground and then lifted into position once they have cured. Drawings similar to the elevations shown in Chapter 23 must show the materials used to construct each panel as well as the location of each panel. *Courtesy Janice Jefferis.*

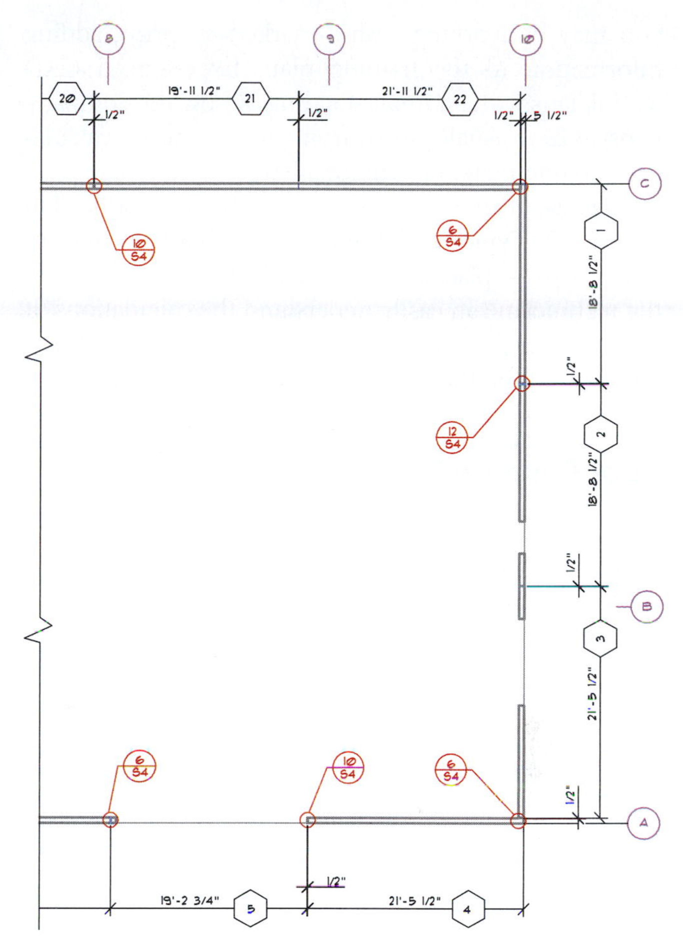
FIGURE 22-23 A panel plan can be used to locate where each panel fits into a structure. Some offices reference elevations to the floor plan. *Courtesy Ginger Smith, Kenneth D. Smith Architect & Associates, Inc.*

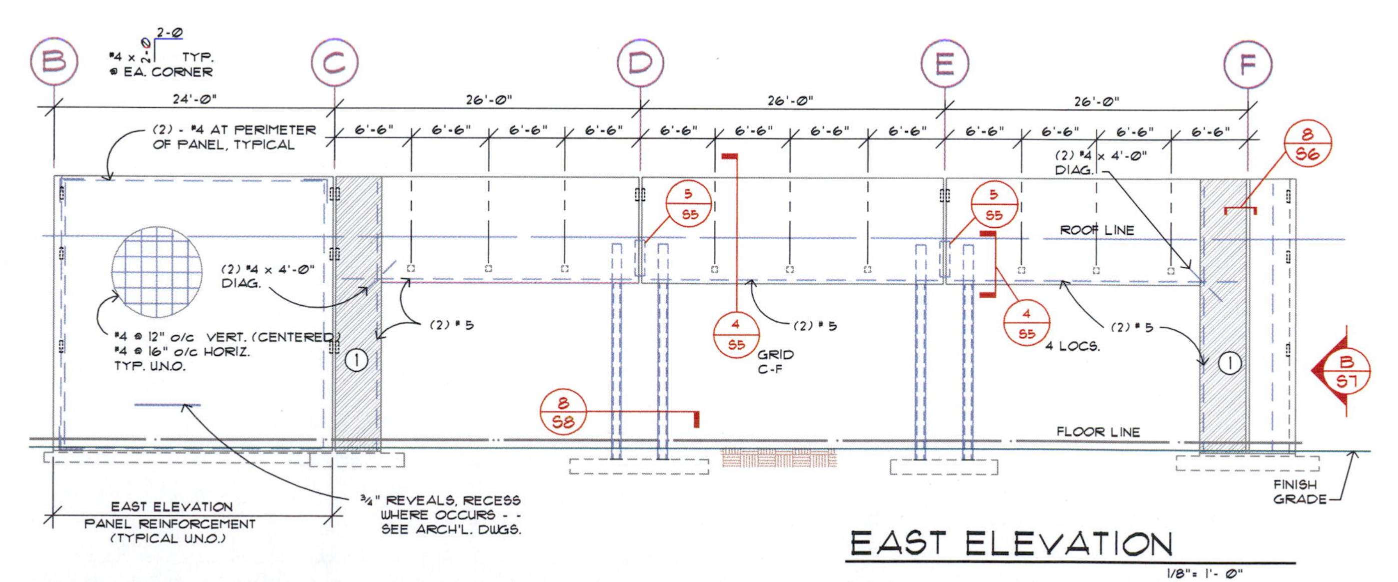

FIGURE 22-24 An elevation of each panel must be created to explain construction methods. An alternative to drawing each panel is to show an elevation of each face of a simple structure. Chapter 23 will present additional information used to explain materials on the framing plan. *Courtesy Charles J. Conlee, P.E., Conlee Engineers, Inc.*

you may be working with a marked-up print, adding information to the framing plan. Experienced CAD technicians with an understanding of how the company is organized usually work from the engineer's calculations to complete the framing plan.

The use of an engineer's calculations is introduced in Chapter 2. Typically the information that the engineer requires to be placed on the plan is highlighted so that the technician can easily understand the calculations. It is the technician's responsibility to place on the drawings everything that is highlighted in the calculations.

Roof Framing Plans

The shape of the roof affects how the drawing is started. Framing plans for a steep-pitched roof can be drawn using the roof plan as a base. Roof plans for structures with a low-sloped roof can use the floor plan for the drawing base.

Drawing Low-Sloped Roof Framing Plans

The structure drawn in Figures 16-38 and 18-28 will be used throughout this example.

1. Use the floor plan base drawing to create the roof framing plan. Freeze all material except the outline of exterior walls, interior bearing walls, and openings in each.

The base plan should now resemble Figure 22-25.

2. Thaw or draw all grid lines and markers.
3. Draw and specify all beams.

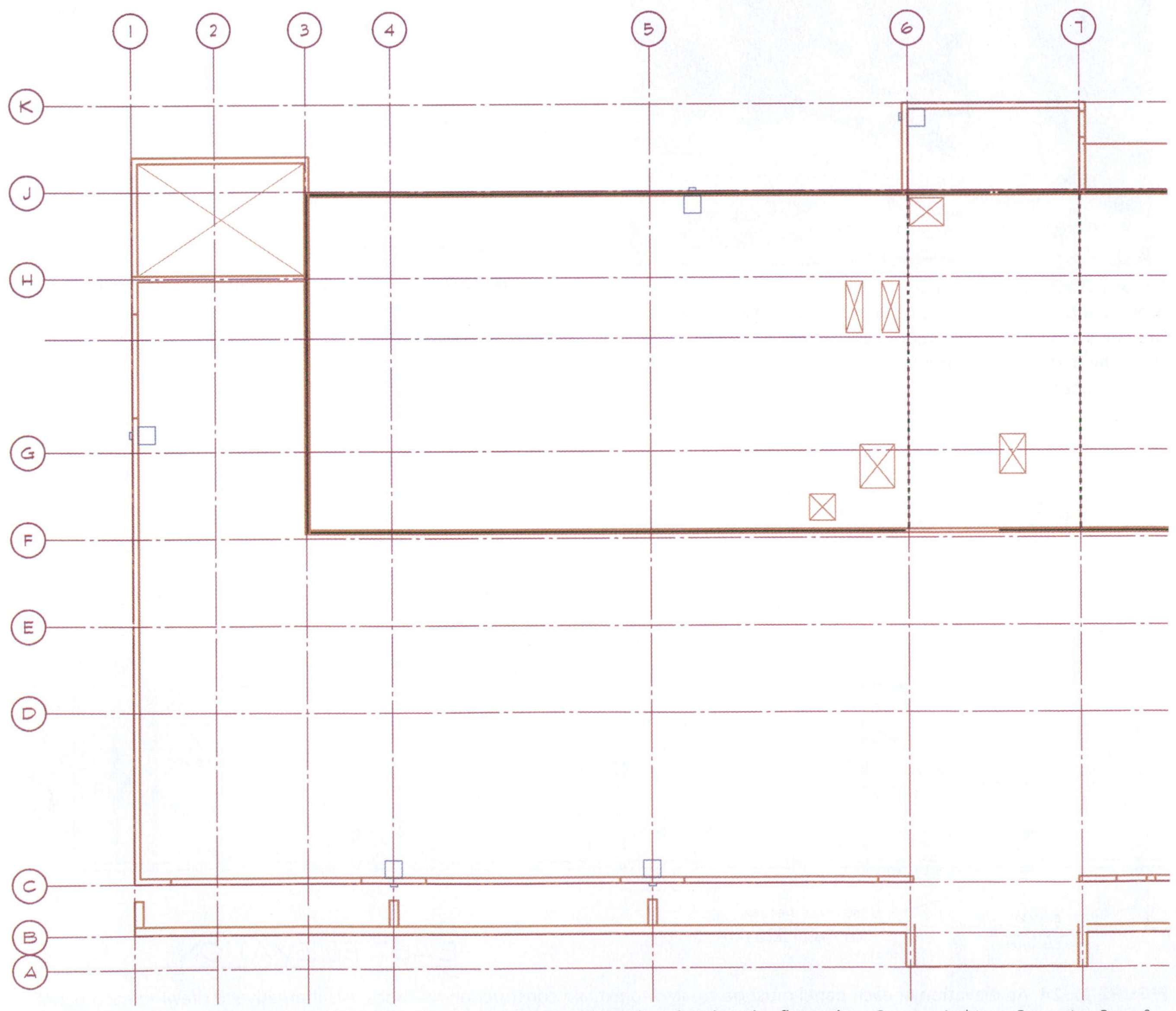

FIGURE 22-25 The roof plan can be started by using the base plan used to develop the floor plan. *Courtesy Architects Barrentine, Bates & Lee, A.I.A.*

4. Draw and specify all openings.
5. Draw and specify the outline of all roof projections.
6. Provide dimensions to locate all beams, openings, and overhangs not dimensioned on the floor or roof plan.

The roof plan should now resemble Figure 22-26.

7. Locate and specify all trusses.
8. Place all local and general notes specified by the engineer.
9. Place all detail and section markers.
10. Place a title and scale below the drawing.

The completed drawing should now resemble Figure 22-27.

Drawing High-Sloped Roof Framing Plans

The plans for a high-sloped roof can be completed using methods similar to those used for a low-sloped roof. The steel structure shown in Chapter 9 and Figures 18-1 and 22-15 will be used as the example.

1. Use the floor plan base drawing to create the roof framing plan. Freeze all material except the outline of exterior walls, interior bearing walls, and columns. This particular plan includes a dashed

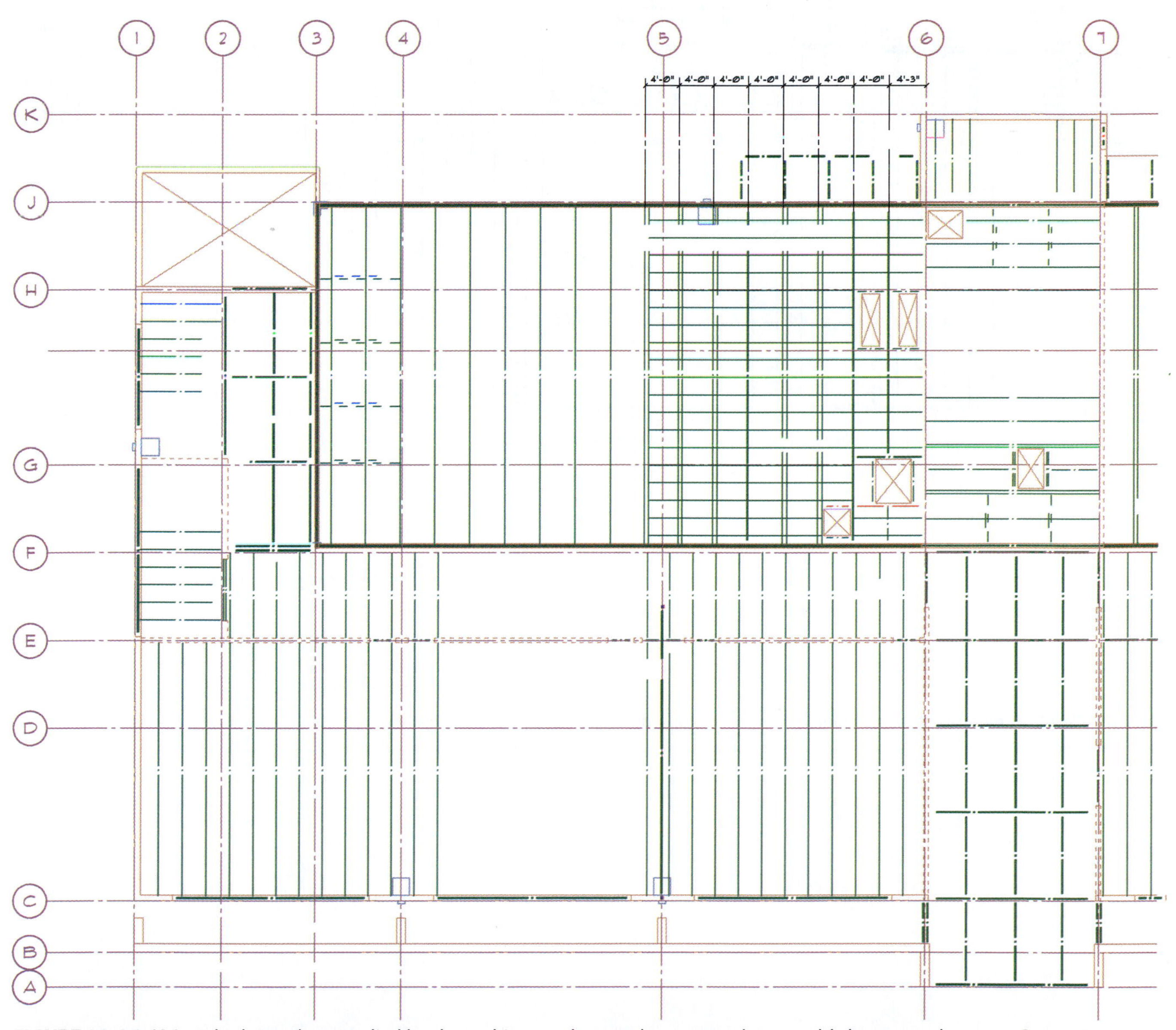

FIGURE 22-26 Using the base plan supplied by the architectural team, the structural team adds beams and trusses. *Courtesy Van Domelen/Looijenga/McGarrigle/Knauf Consulting Engineers.*

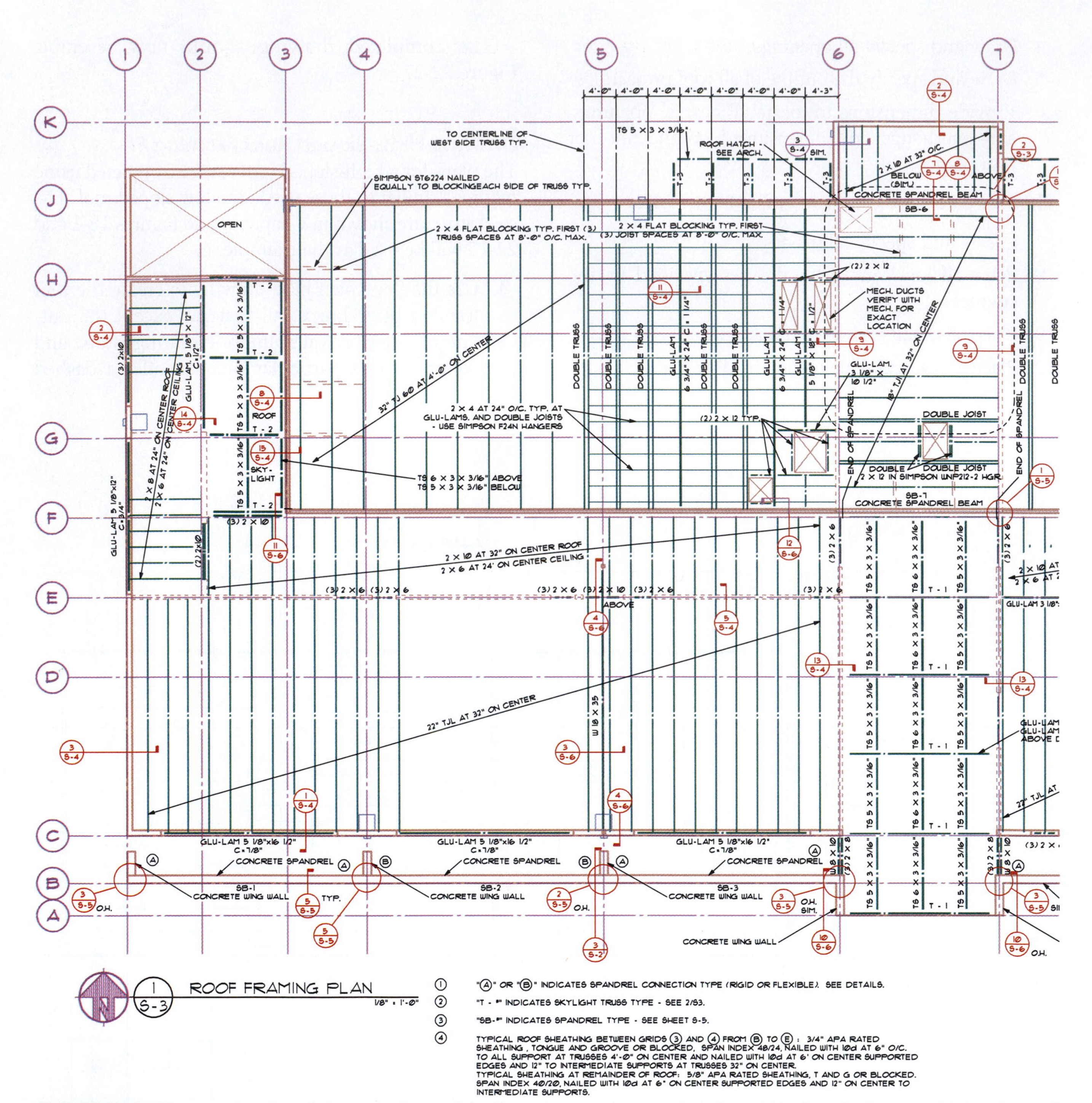

FIGURE 22-27 The completed roof plan with all materials, dimensions, and annotation. *Courtesy Van Domelen/Looijenga/McGarrigle/Knauf Consulting Engineers.*

line to represent supports indicated on the architectural drawings that will support the overhang (see Figure 22-28).

2. Thaw or draw all grid lines and markers.
3. Draw and specify the outline of all overhangs.
4. Draw and specify all changes in roof shape such as ridges, hips, and valleys.

The base plan should now resemble Figure 22-29.

5. Draw and specify all primary beams and their elevations.
6. Draw and specify all openings.
7. Locate and specify all trusses (see Figure 22-30).
8. Locate and specify all bracing and cross bracing.
9. Provide dimensions to locate all beams, openings, and overhangs not dimensioned on the floor or roof plan.
10. Place all local and general notes specified by the engineer.
11. Place all detail and section markers.
12. Place a title and scale below the drawing.

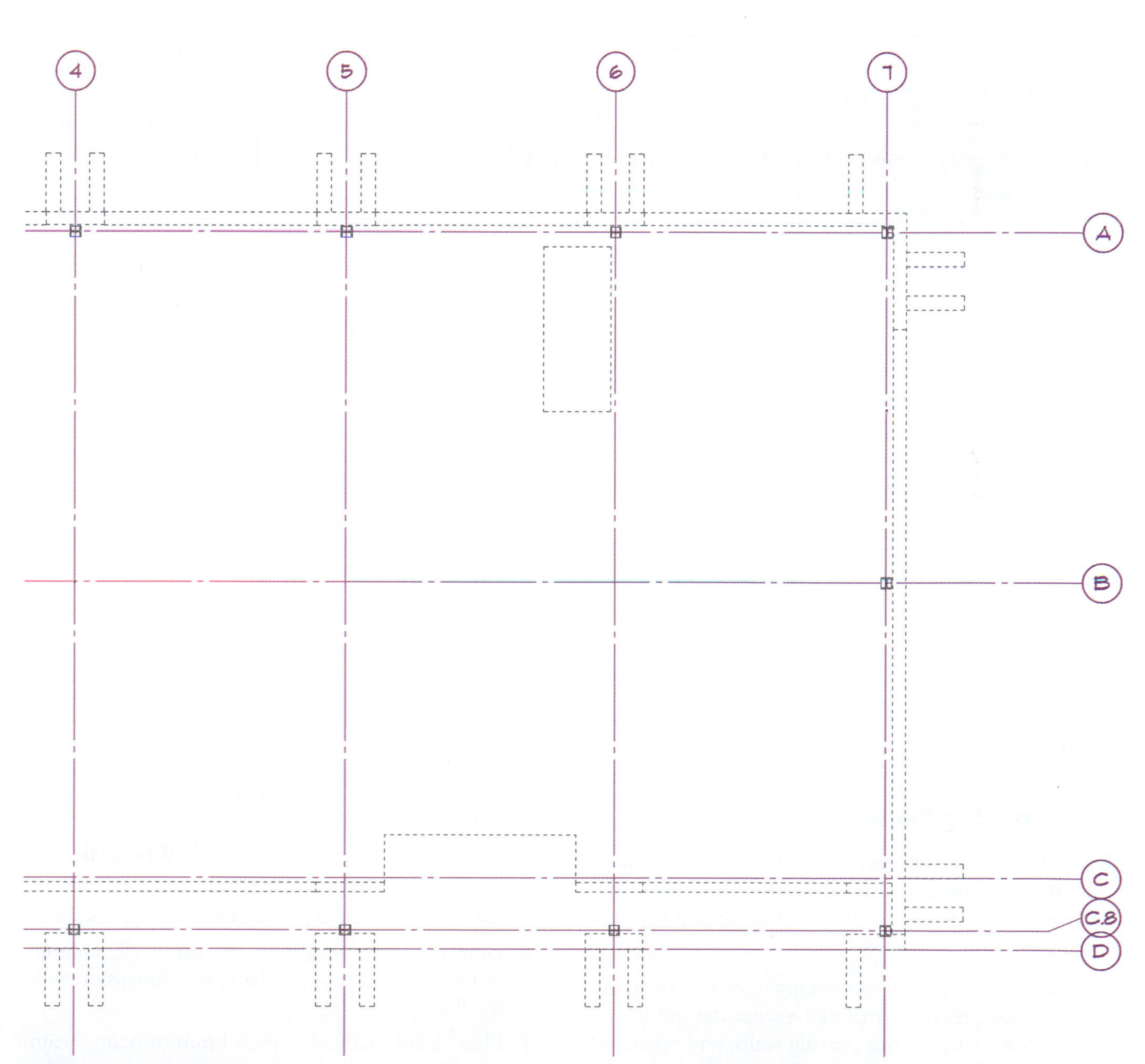

FIGURE 22-28 The base drawings for a high-sloped roof plan. *Courtesy Peck, Smiley, Ettlin Architects.*

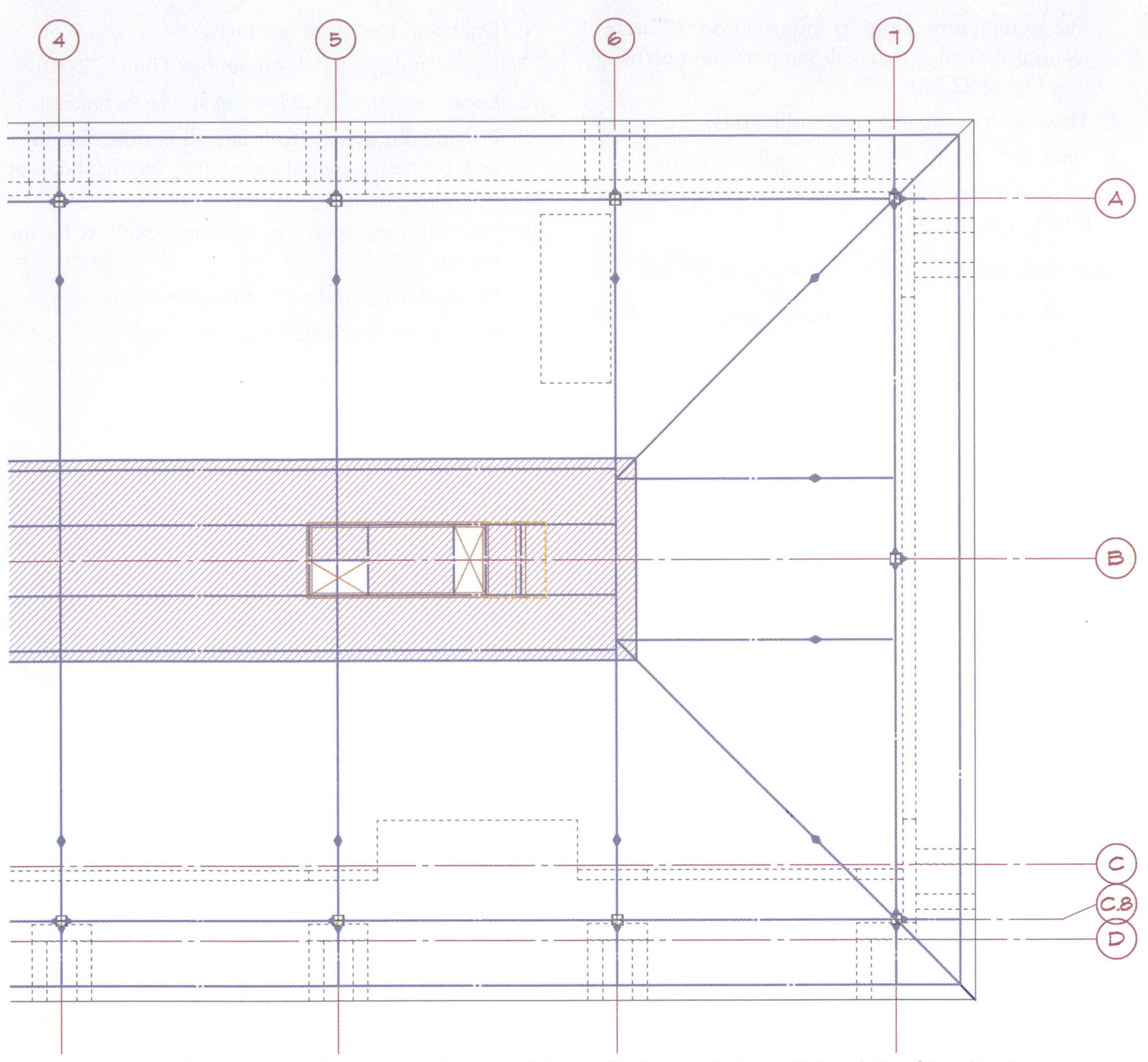

FIGURE 22-29 The layout of major shape changes in the roof. *Courtesy Van Domelen/Looijenga/McGarrigle/Knauf Consulting Engineers.*

The completed drawing should now resemble Figure 22-31.

Floor Framing Plans

A floor framing plan is started using methods similar to those that were used to draw the roof framing plan. The structure shown in Figure 16-3 and 18-23 will be used as the example.

1. Use the floor plan base drawing to create the framing plan. Freeze all material except the outline of exterior walls, interior bearing walls, and openings in each. Thaw or draw all grid lines and markers.

The base plan should now resemble Figure 22-32.

2. Draw and specify interior columns.
3. Draw the centerline locations for the primary and intermediate beams.
4. Draw and specify the outline of all projections of upper floors (Figure 22-33).
5. Label each beam and provide elevation specifications.
6. Dimension the locations of each wall, column, beam, opening, and overhang not dimensioned on the floor plan.
7. Place detail markers at each beam-to-beam, beam-to-wall, or column connection.

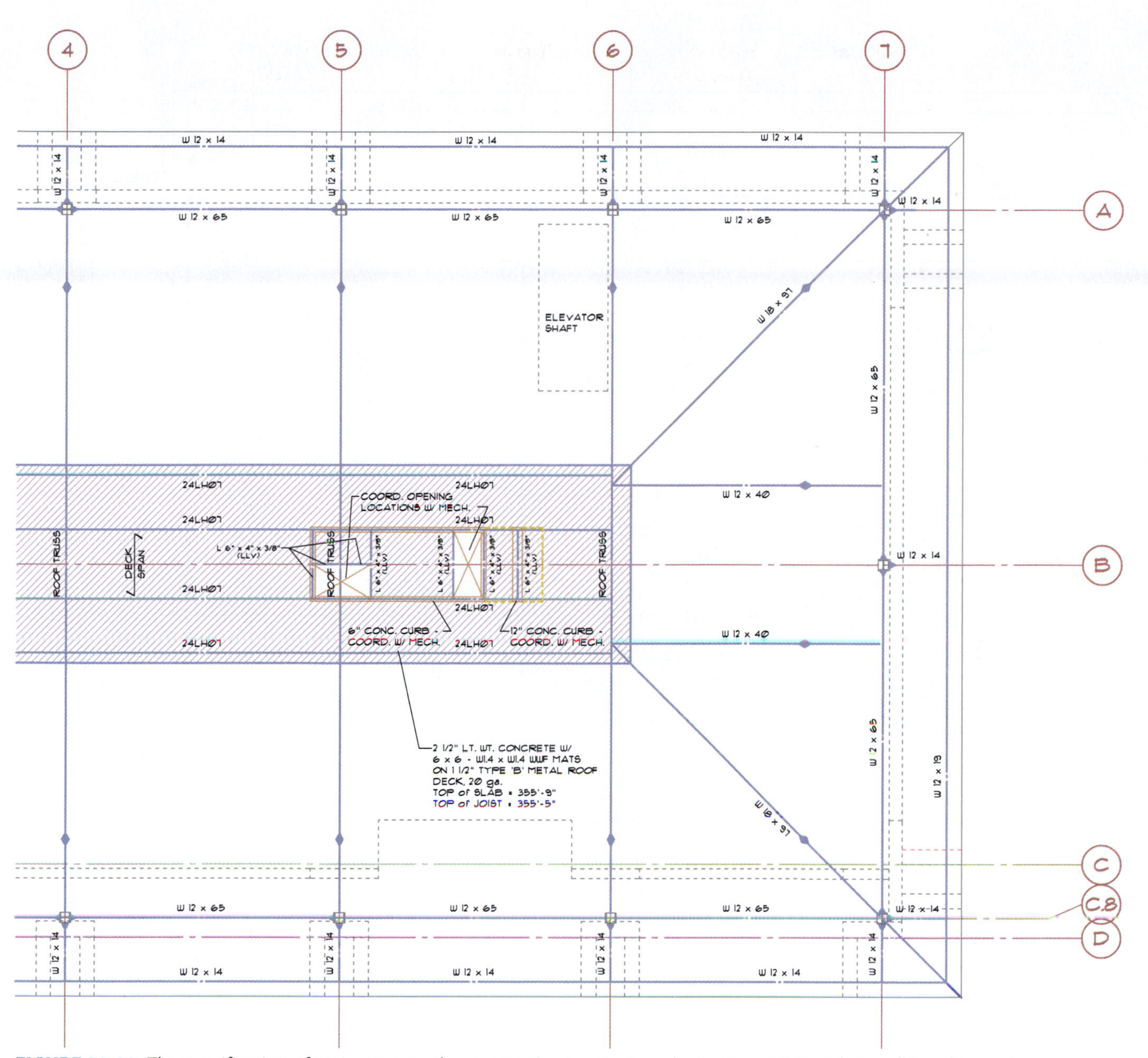

FIGURE 22-30 The specification of major structural supports. *Courtesy Van Domelen/Looijenga/McGarrigle/Knauf Consulting Engineers.*

8. Draw and label each joist or truss span for each area of the structure.
9. Place detail markers at each joist-to-beam and joist-to-wall connection.
10. Locate common section markers.
11. Place all local and general notes specified by the engineer.
12. Place a title and scale below the drawing.

The completed drawing should now resemble Figure 22-34.

Drawing a Framing Plan for Concrete

The drawing procedure for a concrete frame is similar to that used for a steel framing plan. Steps for completing the drawing include the following:

1. Use the floor plan base drawing to create the framing plan. Freeze all material except the outline of exterior walls, interior bearing walls, grids, and openings in each.

The base plan should resemble Figure 22-35.

2. Draw and specify interior columns.

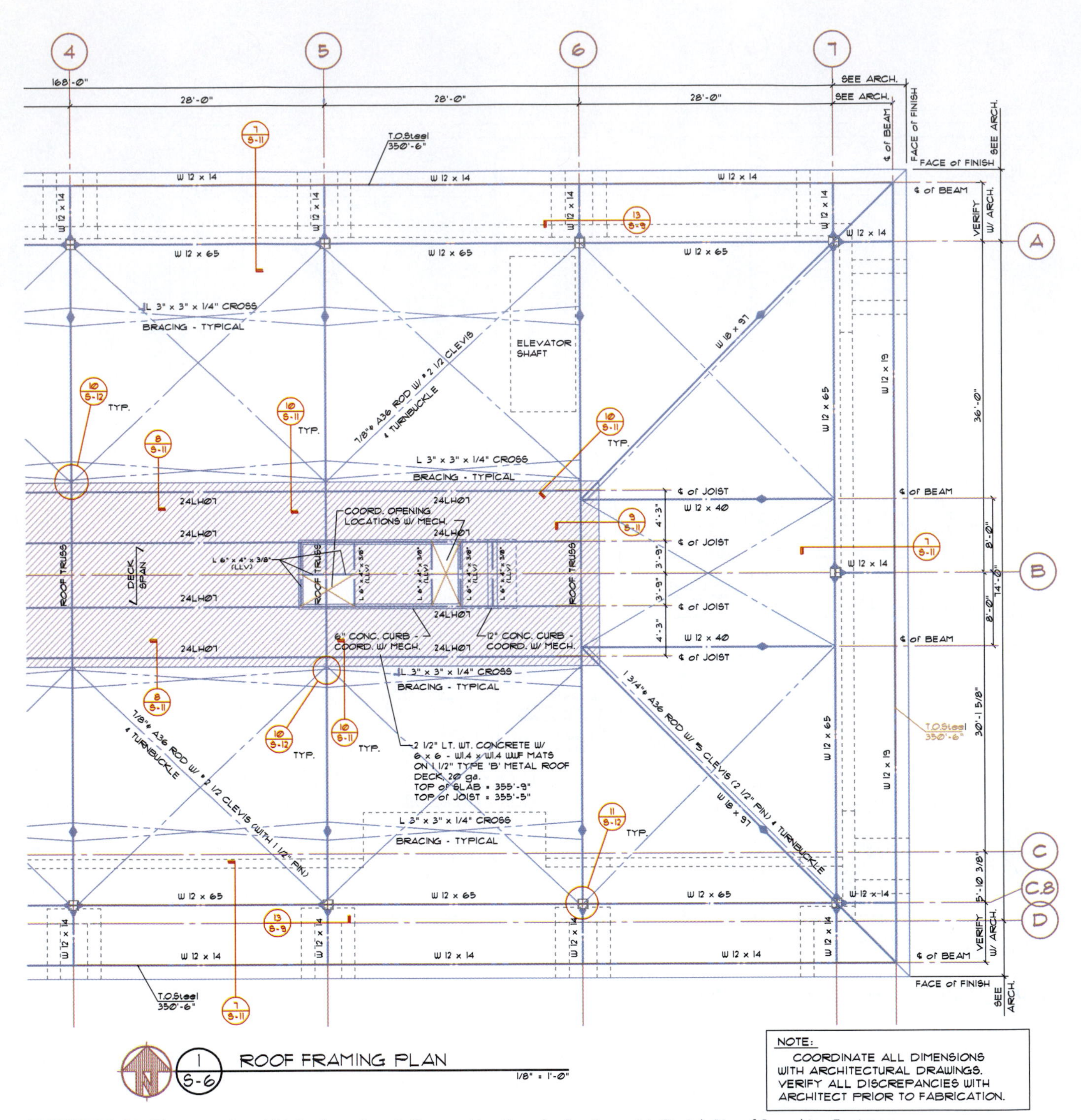

FIGURE 22-31 The completed high-sloped roof. *Courtesy Van Domelen/Looijenga/McGarrigle/Knauf Consulting Engineers.*

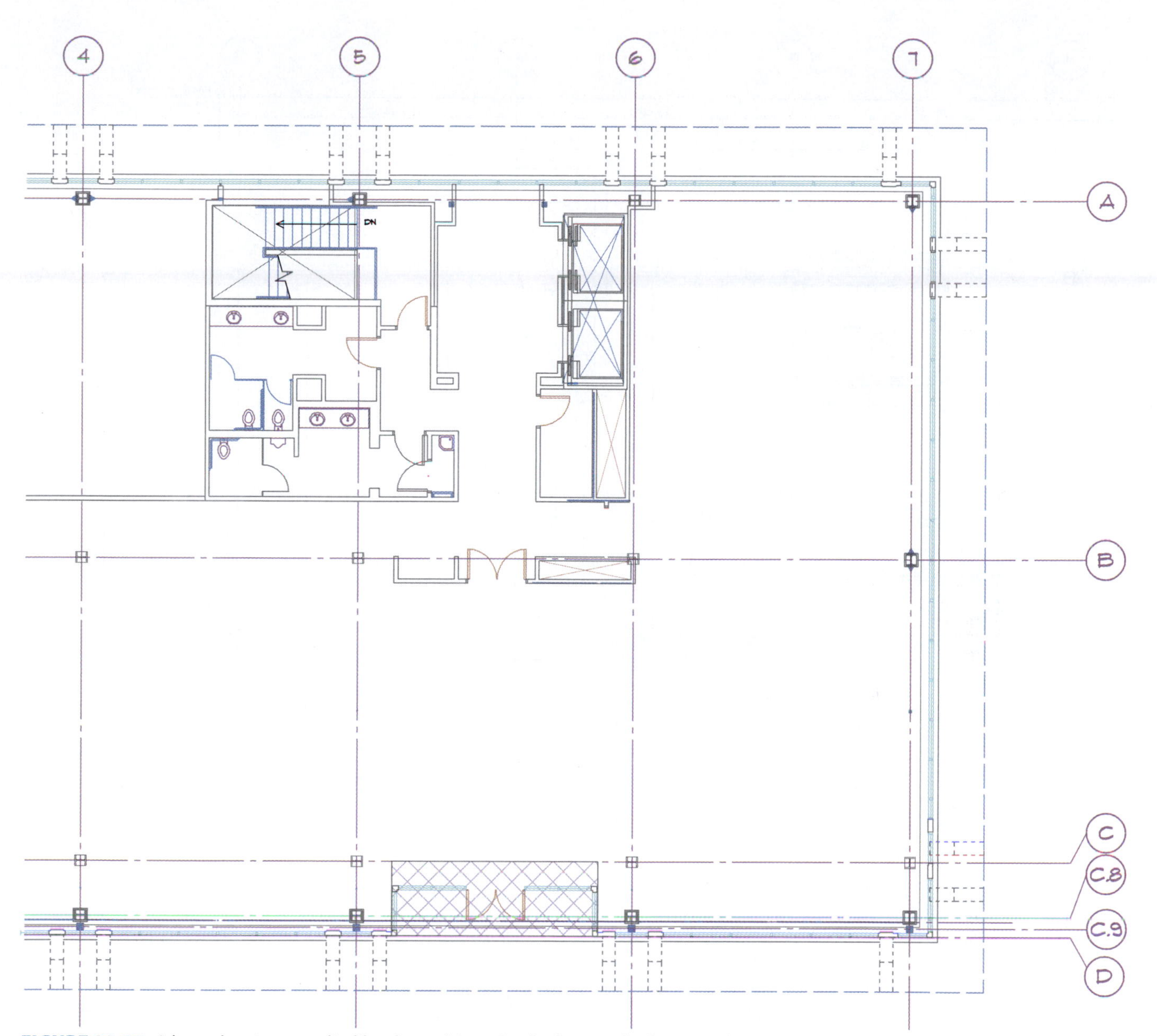

FIGURE 22-32 A base drawing supplied by the architect for the layout of a framing plan. *Courtesy Peck, Smiley, Ettlin Architects.*

3. Draw the centerline locations for each beam not located at a grid line.
4. Draw major primary support beams and then draw intermediate beams.
5. Label each beam and provide elevation specifications.
6. Locate panel indicators.
7. Place detail markers at each beam-to-beam and beam-to-wall or column connection.
8. Dimension the locations of each wall, column, and beam.
9. Locate elevation and section markers.
10. Provide all general notes and grid designations.
11. Place a title and scale below the drawing.

The completed drawing should now resemble Figure 22-36.

Layer Guidelines for Framing Plans

Separation of material by layer and color can greatly aid in the development of framing plans. A prefix of S-FRAM is recommended by the NCS guidelines. Modifiers can be used to define materials. Many firms divide modifiers by materials such as CONC, CBLK, BRK, STL, WD, or PTWD. See Chapter 3 for layer naming guidelines; also refer to Appendix C on the student CD for a list of common layer names used on sections.

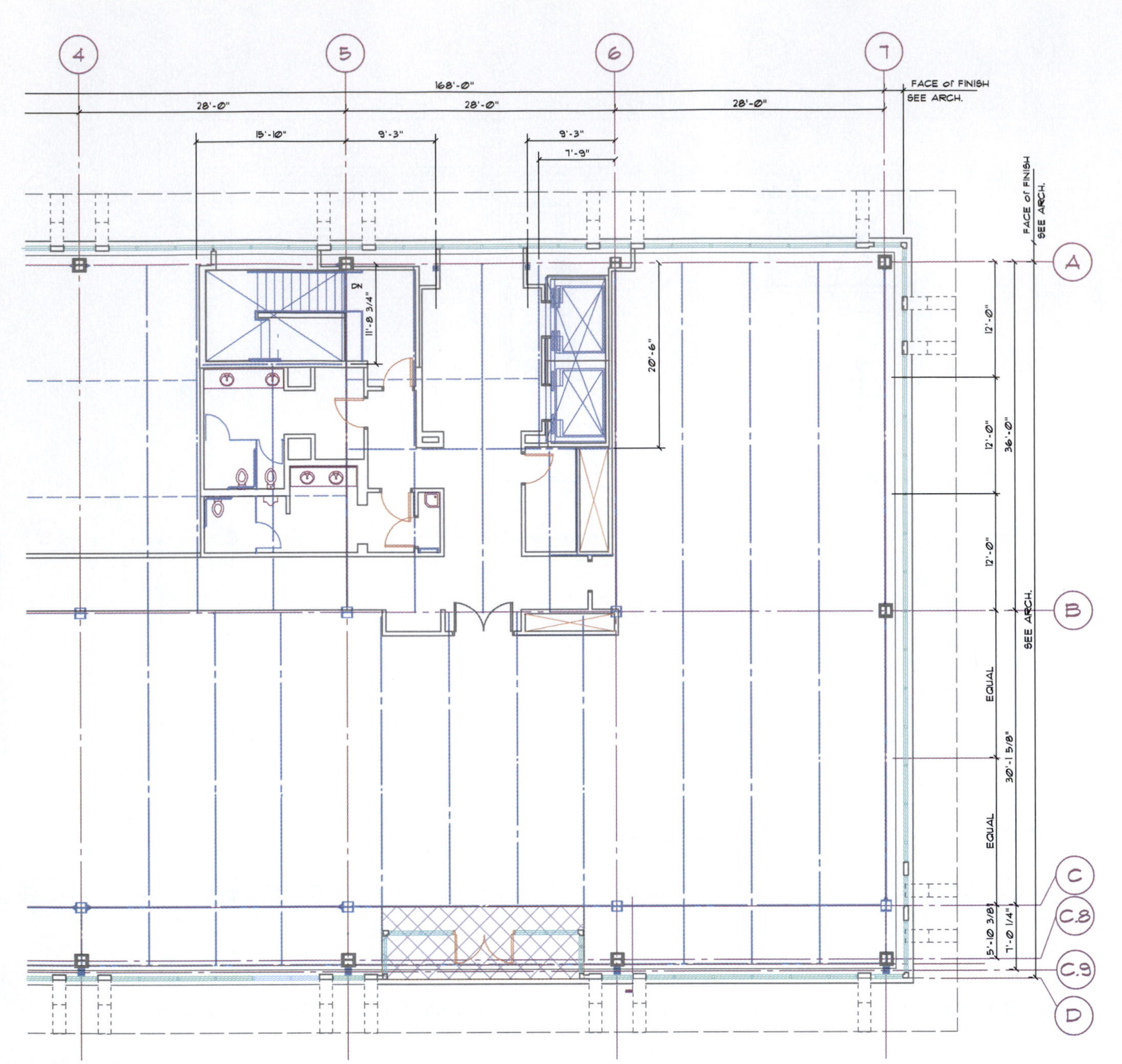

FIGURE 22-33 The layout of major structural materials. *Courtesy Van Domelen/Looijenga/McGarrigle/Knauf Consulting Engineers.*

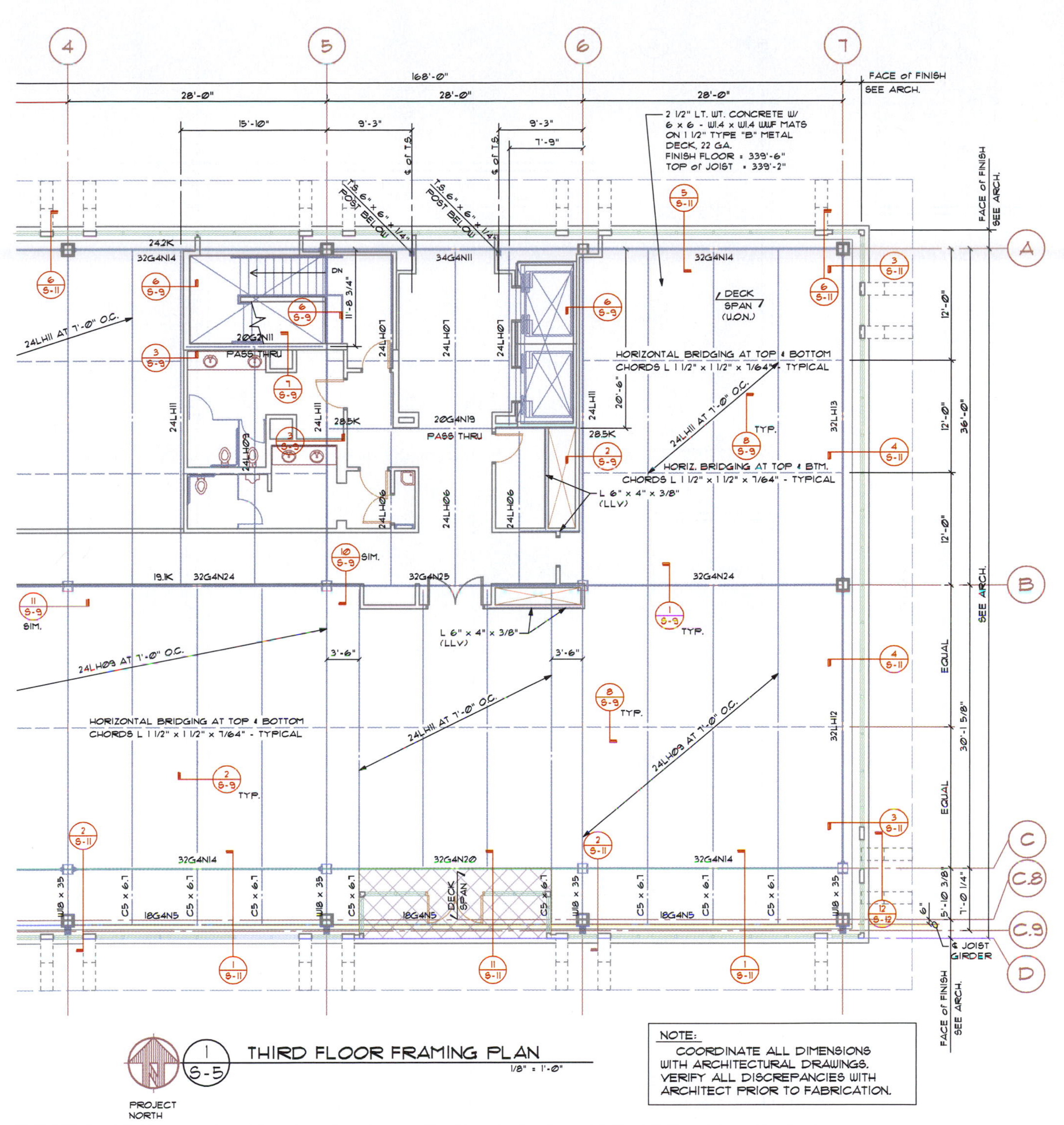

FIGURE 22-34 The completed framing plan. *Courtesy Van Domelen/Looijenga/McGarrigle/Knauf Consulting Engineers.*

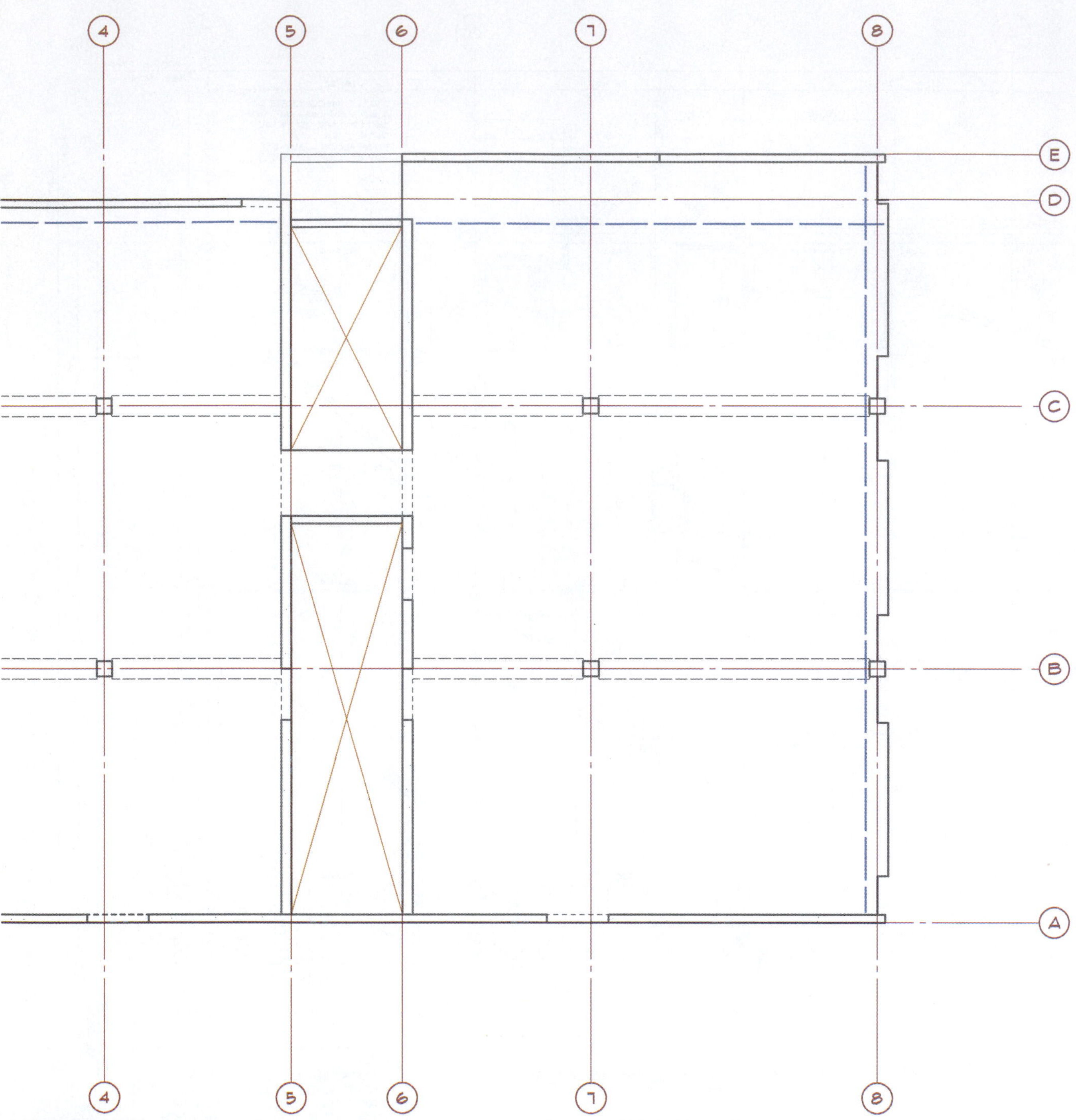

FIGURE 22-35 The base floor plan for a precast concrete structure. *Courtesy H.D.N. Architects A.I.A.*

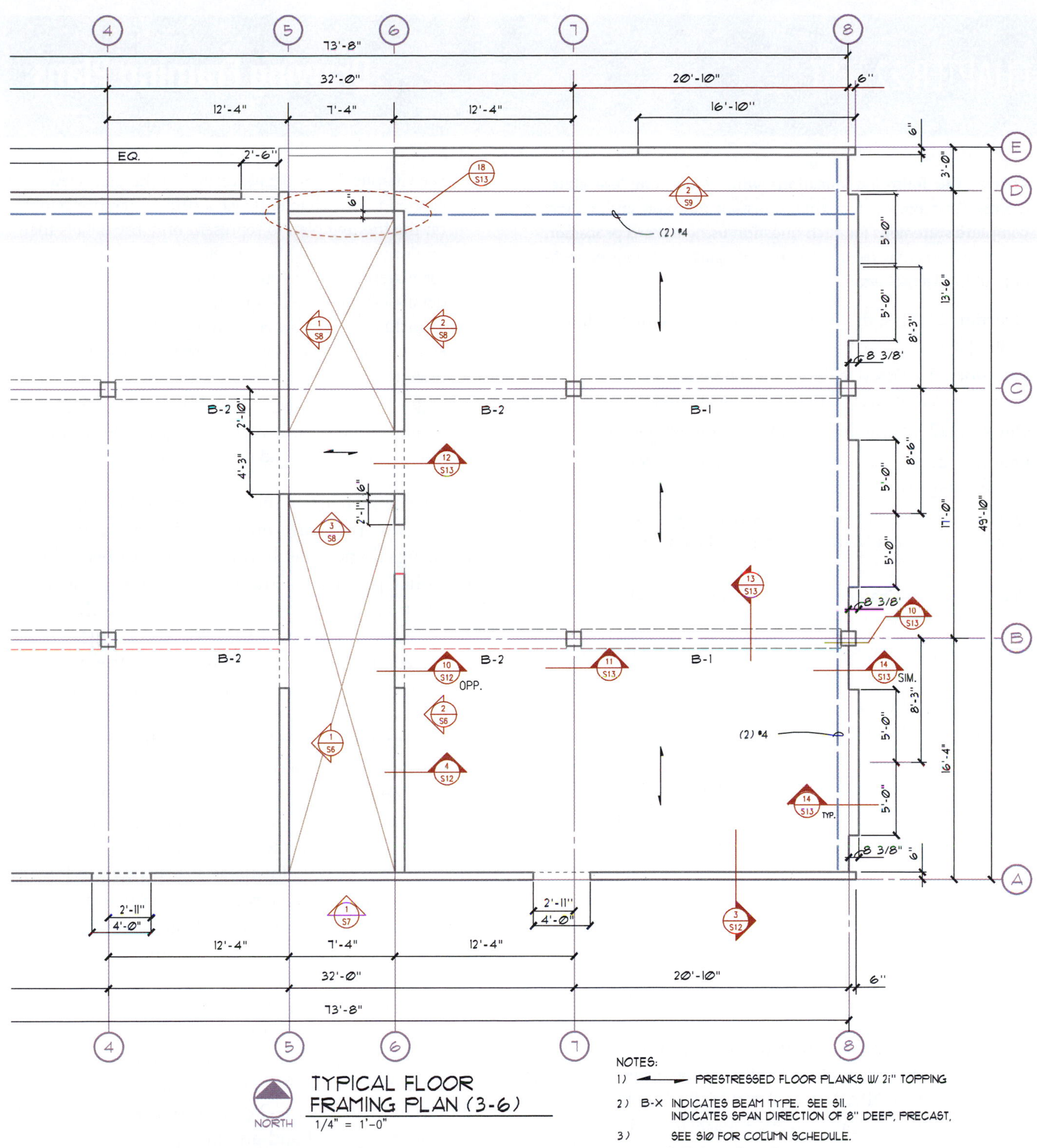

FIGURE 22-36 The completed framing plan for a precast concrete structure. *Courtesy KPFF Consulting Engineers.*

ADDITIONAL READING

Links for the materials represented on the framing plans are too numerous to list here. Rather than listing individual sites, use your favorite search engine to research products from the following Construction Specifications Institute (CSI) categories: wood, steel, concrete, and tilt-up concrete.

CHAPTER 22 TEST Drawing Framing Plans

Answer the following questions with short complete statements. Type the chapter title, question number, and a short complete statement for each question using a word processor. Some answers may require the use of vendor catalogs or seeking out local suppliers.

Question 22-1 List the common scales used to draw a framing plan.

Question 22-2 Describe the major differences between a steel frame and a steel rigid frame structure.

Question 22-3 What is the main goal of a framing plan?

Question 22-4 What is a diaphragm, and how does it affect a framing plan?

Question 22-5 What type of details will typically be referenced on a framing plan for a heavy timber structure?

Question 22-6 What are two common uses for tables on a framing plan for a steel-framed structure?

Question 22-7 How is the elevation of a steel beam typically referenced on a framing plan?

Question 22-8 How are precast concrete panels typically attached to the foundation?

Question 22-9 A concrete component is listed as a 155. What type of structural member would it be?

Question 22-10 List possible layer titles that could be used to divide information on a framing plan.

DRAWING PROBLEMS

Use the reference material from preceding chapters, local codes, and vendor catalogs to complete one of the following projects. Unless your instructor provides other instructions, draw the framing plan that corresponds to the floor plan that was drawn in Chapter 16. Skeletons of the plans and details can be accessed from the student CD/DRAWING PROJECTS /CHAPTER 22 folder. Use these drawings as a base to complete the assignment.

- Use appropriate symbols, linetypes, dimensioning methods, and notations to complete the drawing.
- Create the needed layers to keep major groups of information separated.
- Provide complete dimensions to locate all walls, columns, and beams not represented on the floor plan.
- Determine any unspecified sizes based on material or practical requirements.
- See Chapter 23 for details related to the plan views created in this chapter. Place a detail reference bubble on the framing plan to represent only the details that you or your team will draw. Some generic information for materials such as straps that are specified in the details may need to be specified on the framing plans. Coordinate the information in Chapters 22 and 23 to ensure that all information is placed on the structural drawings.
- Unless specified, select a scale for the framing plan that matches the scale of the floor plan. Determine required LTSCALE and DIMVARS settings.

Note: The sketches in this chapter are to be used as a guide only. You will find that some portions of the drawings do not match things that have been drawn on other portions of the project. Each project has errors that will need to be solved. If you think you have found an error, do not make changes to the drawings until you have discussed the problem and possible solutions with your engineer (your instructor).

Information is provided on drawings in Chapters 16 through 24 relating to each project that might be needed to make decisions regarding the framing plan. You will act as the project manager and will be required to make decisions about how to complete the project. Any information not provided must be researched and determined by you unless your instructor (the project architect and engineer) provides other instructions. When conflicting information is found, information from preceding chapters should take precedence. Use the order of precedence introduced in Chapter 15.

Unless noted, all shear panels are to be 1/2" plywood with 8d nails @ 4" o.c. @ edge and 8d nails @ 8" o.c. in field. Use (2)-2× studs or a 4× post at each end of shear panel and an appropriate Simpson Co. strap or tie to the lower floor or foundation.

Problem 22-1 Framing Plans. Draw the framing plan for each level of the condominium started in problem 16-1. One unit should contain all framing information for the entire floor level.

- Frame the upper level with standard roof trusses.
- Use combination standard/scissor trusses to form a vaulted ceiling over the master bedroom.

- Select and specify appropriate solid-web truss joists to span the width of the unit for each floor system.
- Use (2)-2 × 12 headers for all openings unless noted.
- Use 2 × 8 joists to frame each deck and support with a 5 1/8" × 13 1/2" glu-lam beam at the outer edge of each deck. Support each beam on a 2 × 6 wall between units.
- Use a 5 1/8" × 13 1/2" glu-lam for the header over the garage door.

Use the framing elevations from Chapter 23 as a guide and specify all shear walls and metal straps on each framing plan as well as the elevations and foundation plan.

- The front and rear walls are to be full height shear walls for all units.
- The wall between the kitchen/bathroom on the main floor is also to be a shear wall.
- Place a 5 1/8" × 10 1/2" glu-lam beam in the garage ceiling below the shear wall for all units.

Note: Use the sample calculations posted on the student CD in the REFERENCE MATERIAL folder to specify material that is not specified in the following problems.

Problem 22-2a Framing Plan. Use the floor plan that was created in Chapter 16 as a base to create the framing plan for this structure. Freeze all of the architectural information and create the required layers to add the structural materials. Provide dimensions to locate all walls and openings. Use the information presented in problem 22-2b to complete this drawing and the details to be completed in Chapter 23.

Problem 22-2b Roof Plan. Use the attached drawing to complete the roof framing plan for the structure started in problem 16-2. Coordinate this plan with the details associated with problem 23-2.

- Use 8 3/4" × 28 1/2" glu-lam beams for the east side of the steel column. Use an 8 3/4" × 24" glu-lam beam for the west side of the steel column.
- Use 6 3/4" × 12" beams over the two smaller windows on the south and northeast side, and a 6 3/4" × 24" beam over the opening on the south end of the east side.
- Use a 6 3/4" × 18" beam over the windows on the south end of the east wall.
- Specify each ledger on the framing plan as per specifications given for problems 19-2a, 19-2b, and 23-2a through 23-2j.
- Use TJL65/18 open-web trusses @ 32" o.c. See the Weyerhaeuser Web site (www.ilevel.com).
- Provide 3 × 4 solid blocking between each truss at each bearing point and attach blocks to ledger w/ A34 each end to ledger.
    - Use H3 hurricane ties to each side of each truss.
- Use 5/8" APA group 1 interior ply with exterior glue roof sheathing. Lap plywood seams at 24" o.c.

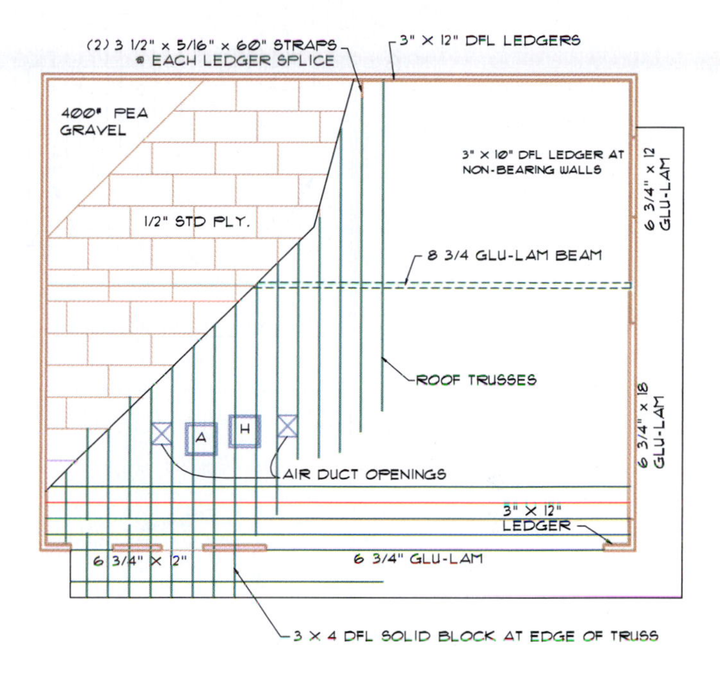

Problem 22-3a Roof Plan. Use the drawing on the next page to complete the roof framing plan for the structure started in problem 16-3. See the details for problem 23-3 for additional information. Use the plywood roof sheathing per problem 22-2.

Roof Beams

All beams to be DF/HF, V-5, Fb 2400 unless noted. Listed spans refer to distance from the column to the indicated cantilever.

Beam 1: 6'–4" into span 2–3	Use 6 3/4" × 34 1/2"
Beam 2:	Use 6 3/4" × 27"
Beam 3: 6'–4" into grid 3	Same size as beam 1
Beam 4:	Same size as beam 2
Beam 5: +8'	Same size as beam 1
Beam 6: Grid B + 30'	Use 6 3/4" × 31 1/2"
Beam 7:	Use TJI HS90/20 truss @ 8'–0" o.c.
Beam 8:	Use 2 × 6 purlins @ 24" o.c.

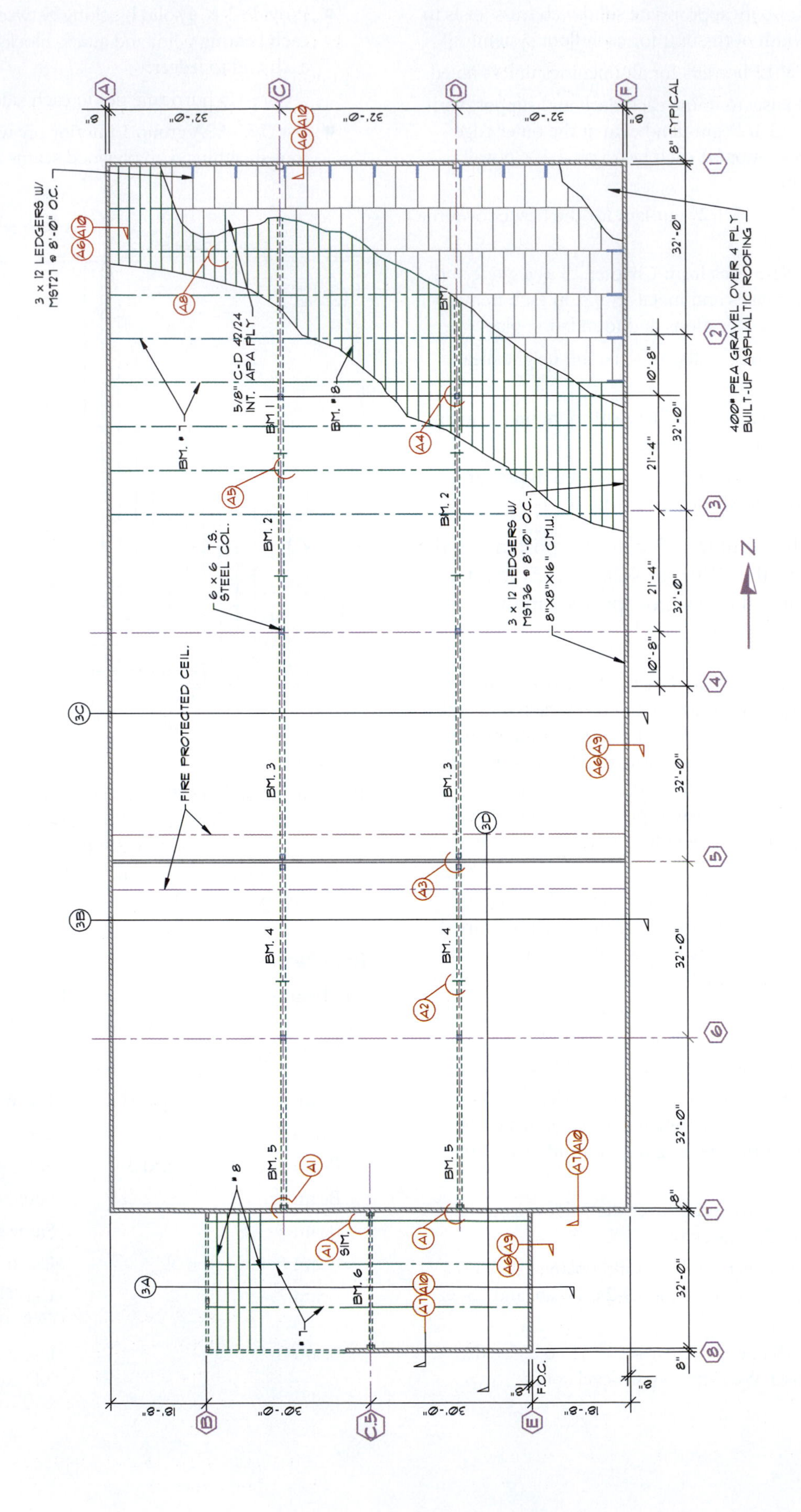
3 x 12 LEDGERS W/
MST27 @ 8'-0" O.C.
5/8" C-D 42/24
INT. APA PLY
400# PEA GRAVEL OVER 4 PLY
BUILT-UP ASPHALTIC ROOFING
8" TYPICAL
6 x 6 T.S.
STEEL COL.
3 x 12 LEDGERS W/
MST36 @ 8'-0" O.C.
8"X8"X16" C.M.U.
FIRE PROTECTED CEIL.
BM. 1
BM. 2
BM. 3
BM. 4
BM. 5
BM. 6
BM. #7
BM. #8
SIM.
F.O.C.

Problem 22-3b Mezzanine Floor Framing. Use the drawings below as a base to develop the framing plan for the mezzanine level started in problem 16-3. Assume the use of 1 1/8" plywood floor sheathing. Develop a beam schedule for framing members. Use DF/HF, V-5, Fb 2400 beams unless noted:

Beam 1:	Span 32'	Use 6 3/4" × 37 1/2"
Beam 2:	Span 32'	Use TJI HS90/28" I-joist at 24" o.c.

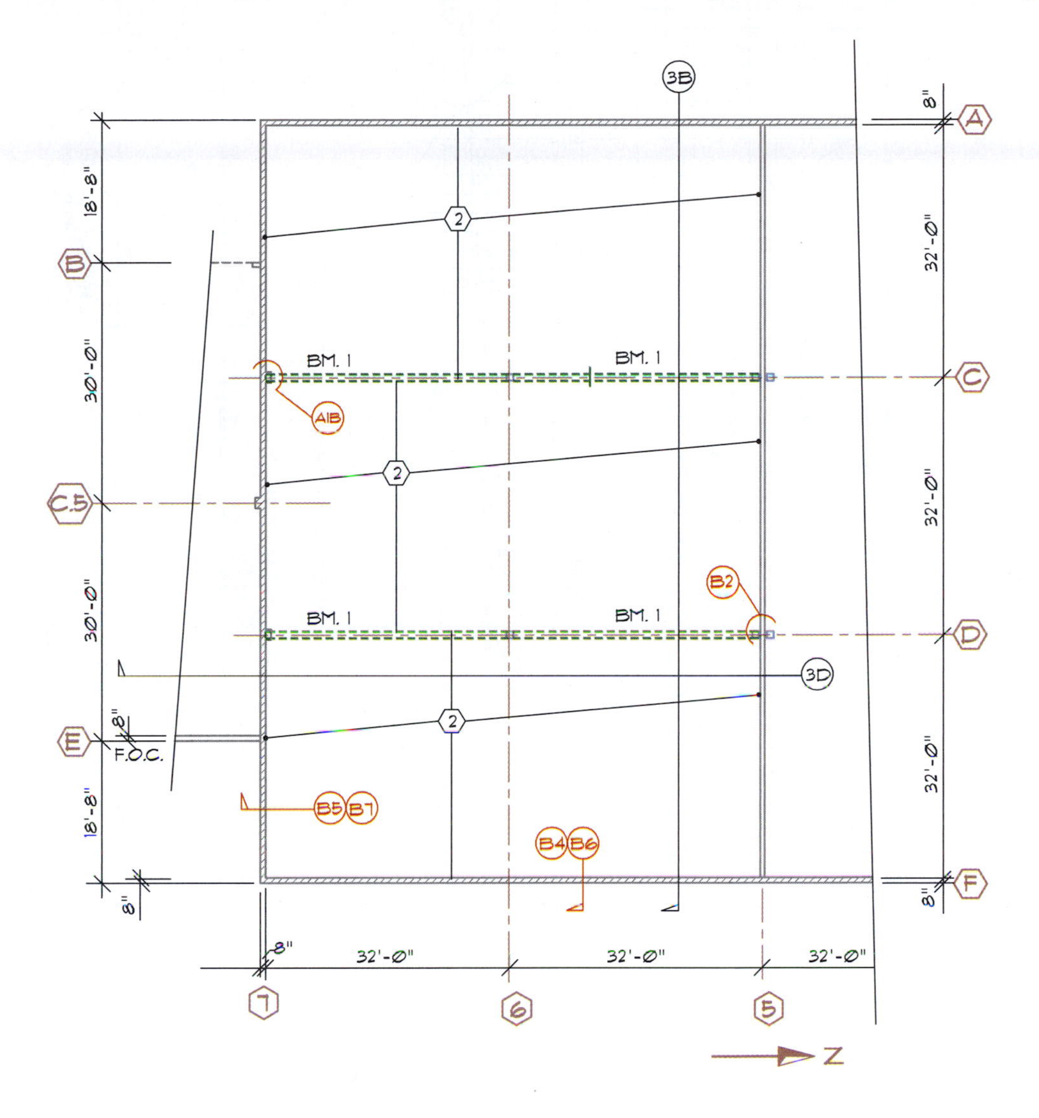

Problem 22-4a Roof Plan. Use the on the next page drawing to complete the roof framing plan for the structure started in problem 16-4. Develop a beam schedule for the roof framing.

Roof Beams

Make the following assumptions unless noted:

All beams to be DF/HF, V-5, Fb 2400 unless noted.

Listed spans assume 36' column spacing plus cantilevers.

Listed spans refer to distance from the column to the indicated cantilever.

Beam 1:	8'–0" east of grid I	Use 6 3/4" ×31 1/2"
Beam 2:	8'–0" east of grid H	Use 6 3/4" × 21"
Beam 3:	8'–0" east of H to F	Same as beam 1
Beam 4:	12'–0" east of D to F	Use 6 3/4" × 24"
Beam 5:	16'–0" east of D to C	Same as beam 1
Beam 6:	48'–0" clear span	Use 6 3/4" × 34 1/2"
Beam 7:	48'–0" clear span	Use TJI HS90/18 trusses @ 32" o.c.
Beam 8:	35'–0" clear span	Use TJI HS90/16 trusses @ 32" o.c.

Roof Blocking

Provide 3 × 4 solid blocking between trusses at each end of the warehouse structure.

- Blocks to extend a minimum of 36' (and modular based on truss locations) from each end wall and to be placed at 48" o.c.
- Bond last truss of blocking pattern w/ (2)-Simpson HD-2 hold-down anchors.

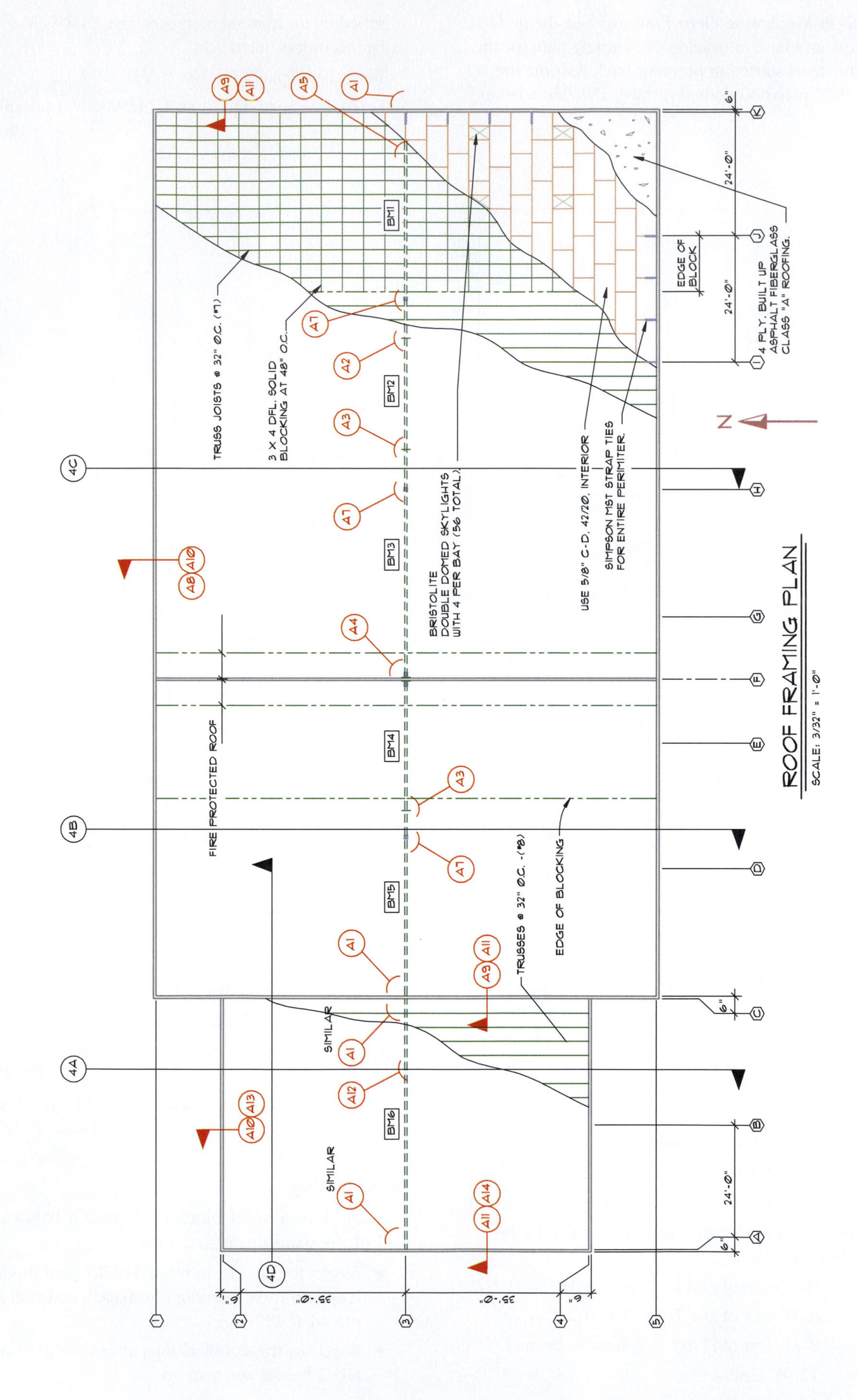
TRUSS JOISTS @ 32" O.C. (*1)
3 X 4 DFL. SOLID BLOCKING AT 48" O.C.
BRISTOLITE DOUBLE DOMED SKYLIGHTS WITH 4 PER BAY (56 TOTAL)
USE 5/8" C-D, 42/20, INTERIOR
SIMPSON MST STRAP TIES FOR ENTIRE PERIMITER.
4 PLY. BUILT UP ASPHALT FIBERGLASS CLASS "A" ROOFING.
EDGE OF BLOCK
FIRE PROTECTED ROOF
TRUSSES @ 32" O.C. -(*8)
EDGE OF BLOCKING
SIMILAR
SIMILAR
BM1
BM2
BM3
BM4
BM5
BM6
24'-0"
24'-0"
24'-0"
6"
35'-0"
35'-0"
N
ROOF FRAMING PLAN
SCALE: 3/32" = 1'-0"

Problem 22-4b Mezzanine Floor Plan. Use the drawings on this page and complete the framing plan for the mezzanine level started in problem 16-4. Develop a beam schedule for framing members. Use DF/HF, V-5, Fb 2400 beams unless noted. Listed spans refer to grids on framing plan.

Mezzanine Floor Framing

Use 1 1/8" floor sheathing.

Beam 1: Span 36' Use 6 3/4" × 37 1/2"
Beam 2: Span 24' Use 6 3/4" × 30"
Beam 3: Span 24' Use TJI HS90/24 @ 32" o.c.
Beam 4: Span 16' Use TJI HS90/24 @ 32" o.c.
Beam 5: Span 32' Use TJI HS90/24 @ 24" o.c.

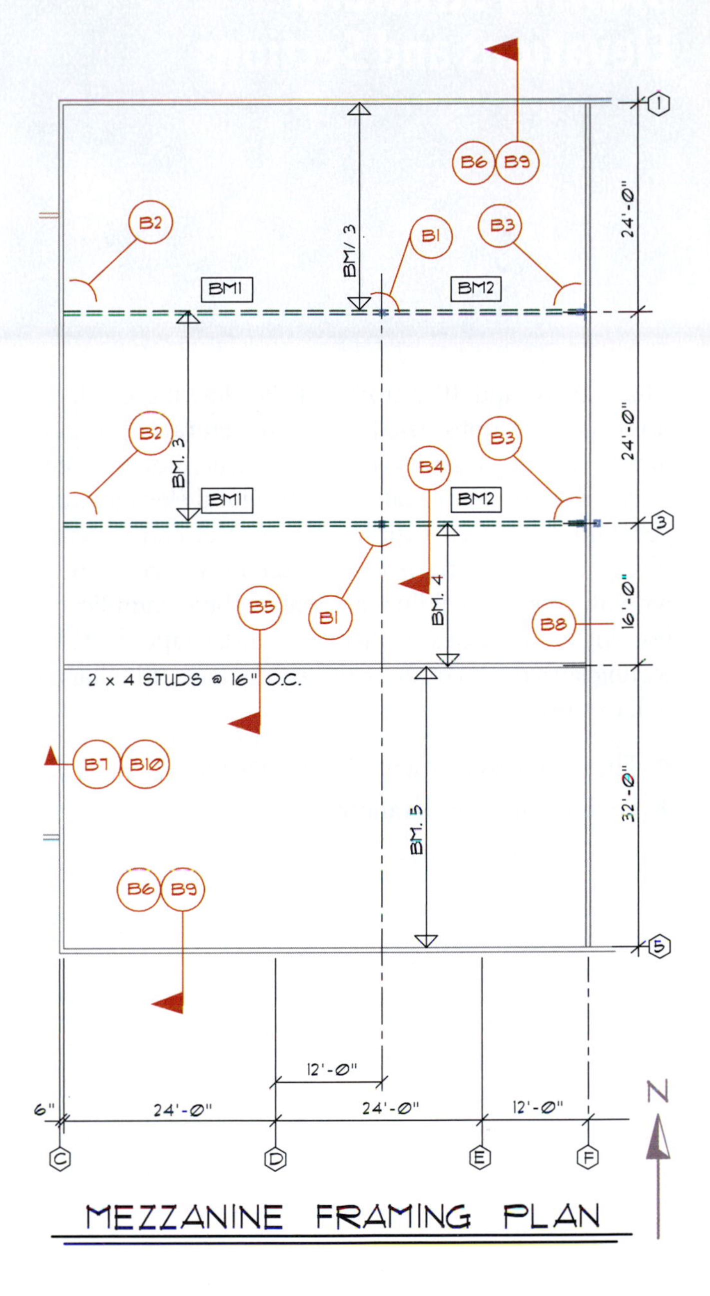

Drawing Structural Elevations and Sections

23

Chapters 16 and 19 introduced the drawing of elevations and sections used in architectural drawings. Similar drawings can also be used in structural drawings to describe structural materials. Structural elevations are generally associated with structural steel and concrete tilt-up construction. Structural sections can be used with any type of building material to show complicated framing intersections. To help you understand the CAD technician's role in creating these drawings, this chapter will explore:

- The purpose of structural steel elevations.
- Poured concrete elevations.
- Concrete tilt-up elevations.
- Key concepts for drawing structural sections, including:
  - Drawing scales.
  - Drawing placement.
  - Representing materials.

STRUCTURAL STEEL ELEVATIONS

Three common types of elevations are used to explain structural steel. Elevations can be used to show the vertical relationship of major framing members shown on the framing plans without showing any architectural information. Figure 23-1 shows an example of the steel skeleton for a warehouse. This type of elevation

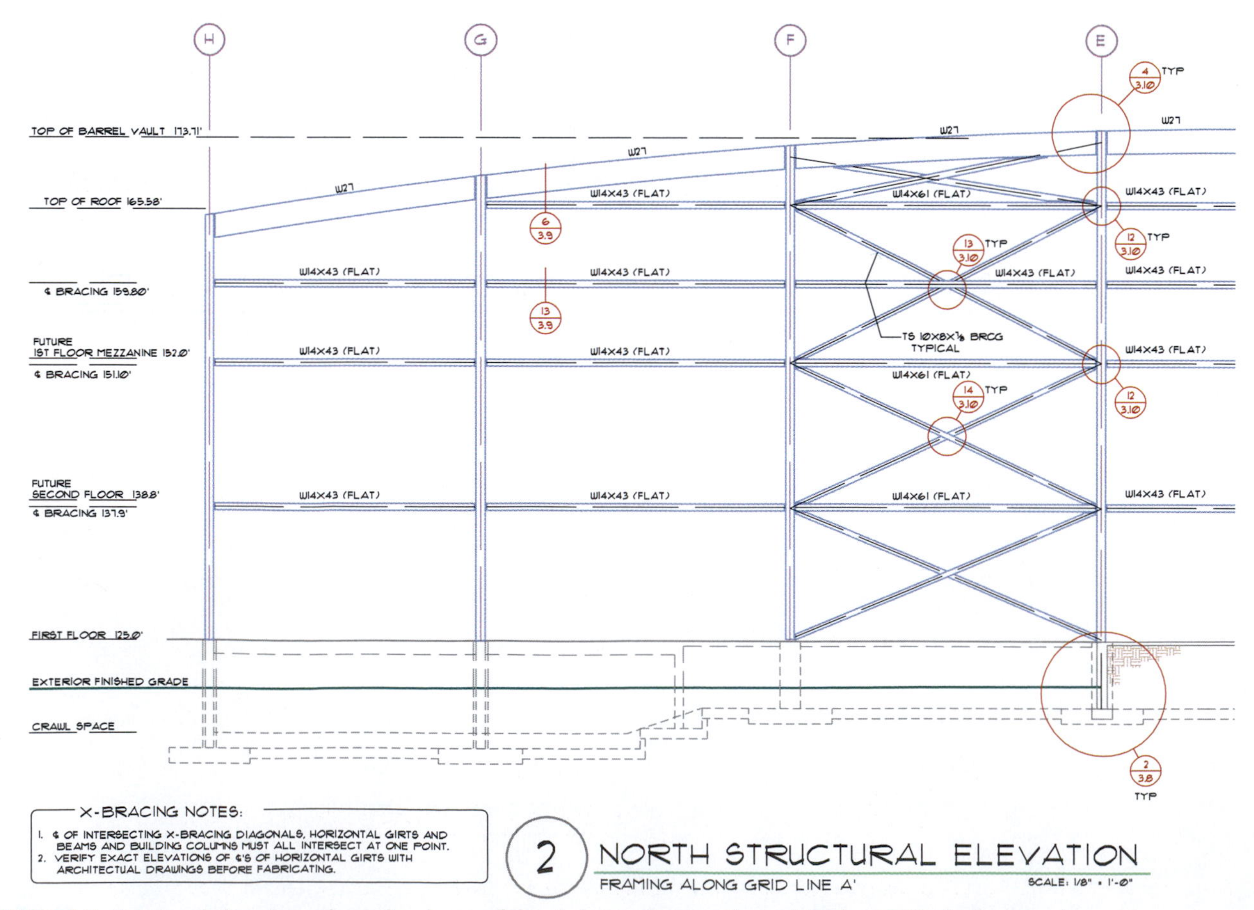

FIGURE 23-1 Structural elevations show the shape of the skeleton. *Courtesy Charles J. Conlee, P.E., Conlee Engineers, Inc.*

provides a vertical reference for materials shown along one specific grid line on the framing plans. Several elevations can be created depending on the complexity of the structure. Details similar to Figure 23-2 are referenced to the elevation to explain construction.

Elevations can also be used to show how the steel framework relates to the architectural members of the structure. Figure 23-3 shows an example of this type of elevation. Notice that this drawing shows similarities to both a structural elevation and an architectural section. Details for structural information are referenced to the elevation.

A third type of steel elevation is used to show major steel components similar to the truss shown in Figure 23-4. This type of elevation is used to fabricate a component off-site. The drawing shows major structural components as well as connection details similar to Figure 23-5.

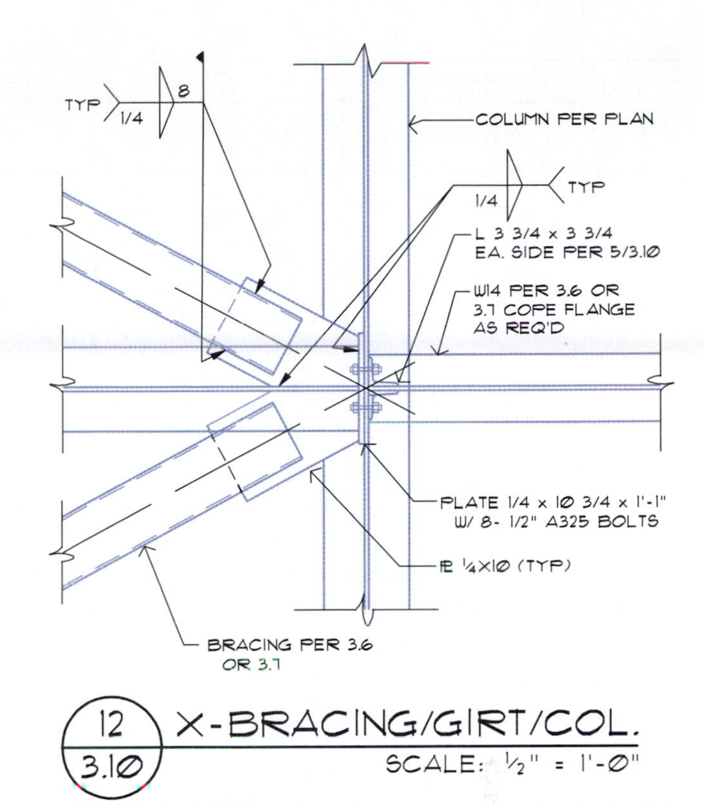

FIGURE 23-2 Details are keyed to structural sections to explain connections. *Courtesy Charles J. Conlee, P.E., Conlee Engineers, Inc.*

POURED CONCRETE ELEVATIONS

The elevations for poured concrete are similar to those used for structural steel. Drawings show the location of structural beams and columns along a specific grid in

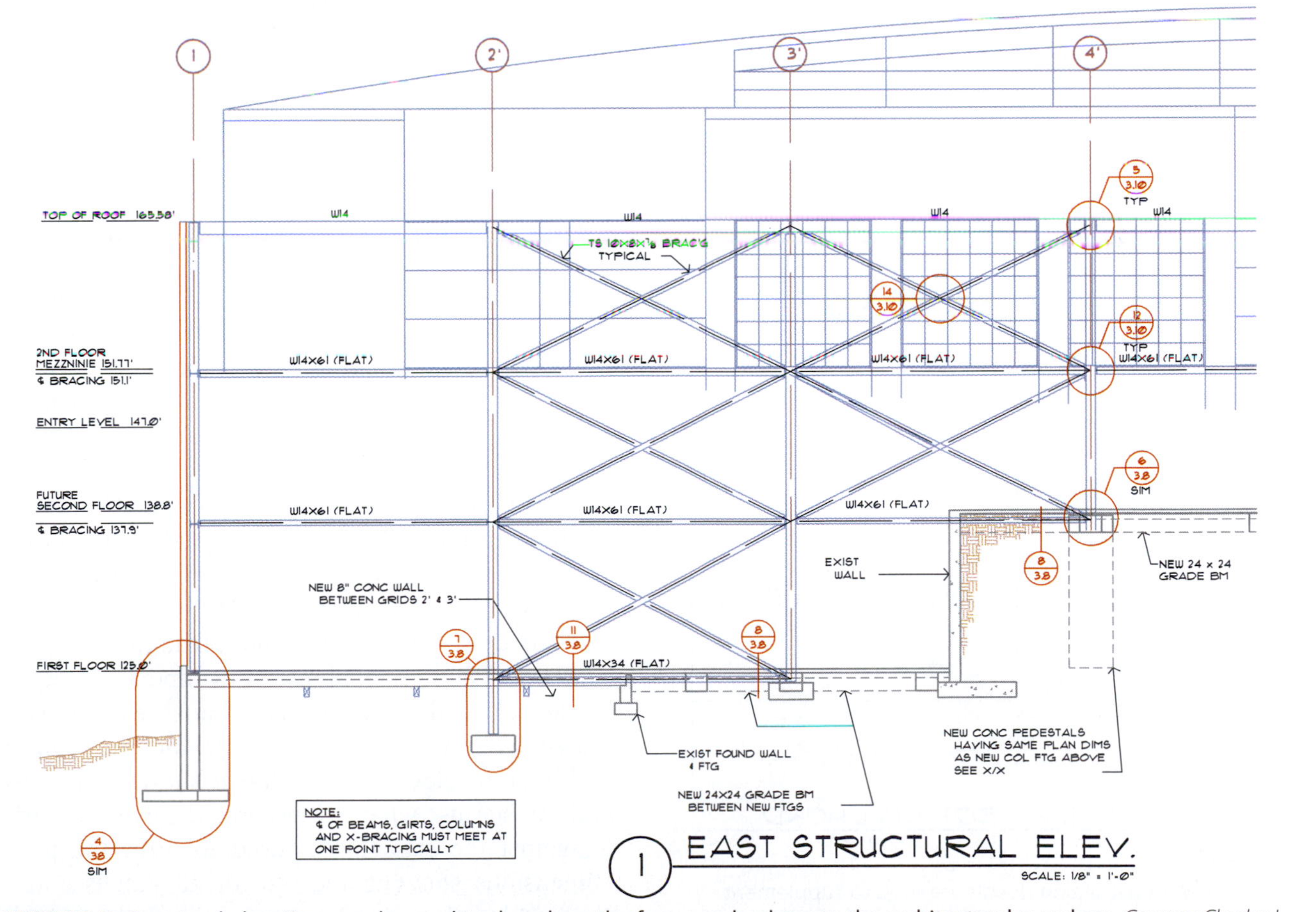

FIGURE 23-3 Structural elevations can be used to show how the framework relates to the architectural members. *Courtesy Charles J. Conlee, P.E., Conlee Engineers, Inc.*

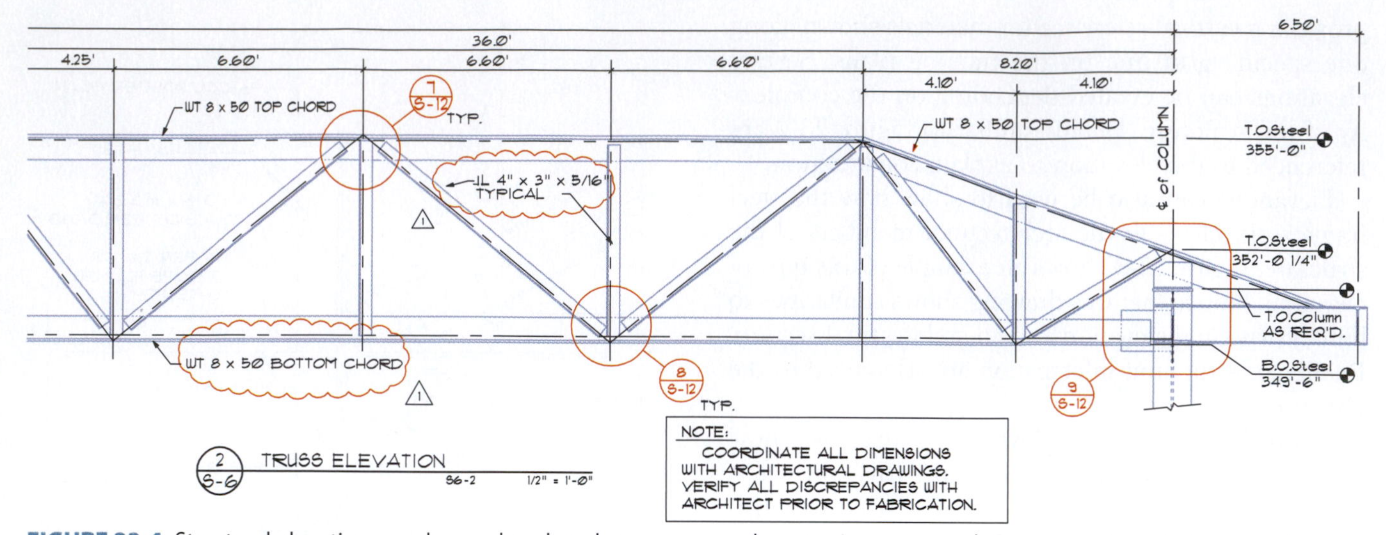

FIGURE 23-4 Structural elevations can be used to show how a structural system is constructed. *Courtesy Van Domelen/Looijenga/McGarrigle/Knauf Consulting Engineers.*

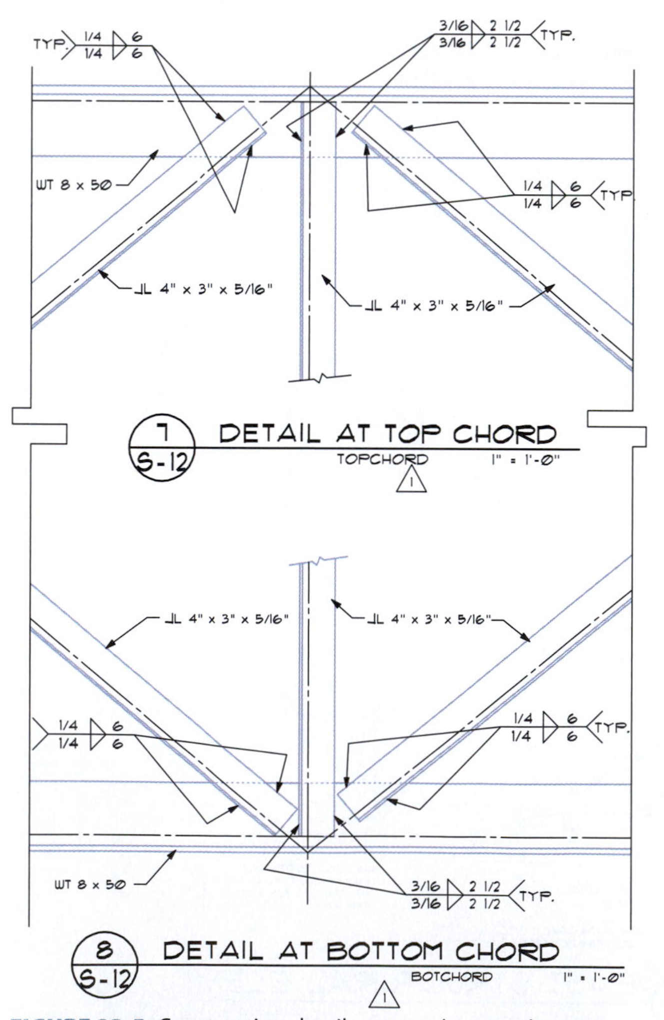

FIGURE 23-5 Construction details are used to supplement structural elevations. *Courtesy Van Domelen/Looijenga/McGarrigle/Knauf Consulting Engineers.*

the structure, as shown in Figure 23-6. In addition to describing distances between members, the elevation also serves as a reference map for listing beam sizes and for connection details similar to Figure 23-7. Elevations can also be more specific and show only specific types of construction, similar to the column elevations shown in Figure 23-8, as well as the size, shape, and reinforcing of specific members.

CONCRETE TILT-UP ELEVATIONS

Elevations for a tilt-up structure are used to show the size, shape, opening locations, and reinforcing for each panel of a structure. Elevations are drawn for tilt-up construction using two common formats. The format used depends on the size and complexity of the structure. Each method is usually referenced on the title page by the name *panel elevation*. One method of showing the construction of panels is to show an entire face of a structure similar to the architectural elevations. A scale of 1/8" or 1/4" = 1'–0" is often used to draw panel elevations. Structural elevations similar to Figure 23-9 show the components required to construct the panels. The panel elevations are drawn as if you are standing inside the building, looking out. Elevations are drawn showing the inside surface, because what will become the exterior surface is against the slab that is used to form the panels. Dimensions showing the size of all panels and the size and locations of all openings should be provided. Because of the small scale, reinforcing steel patterns

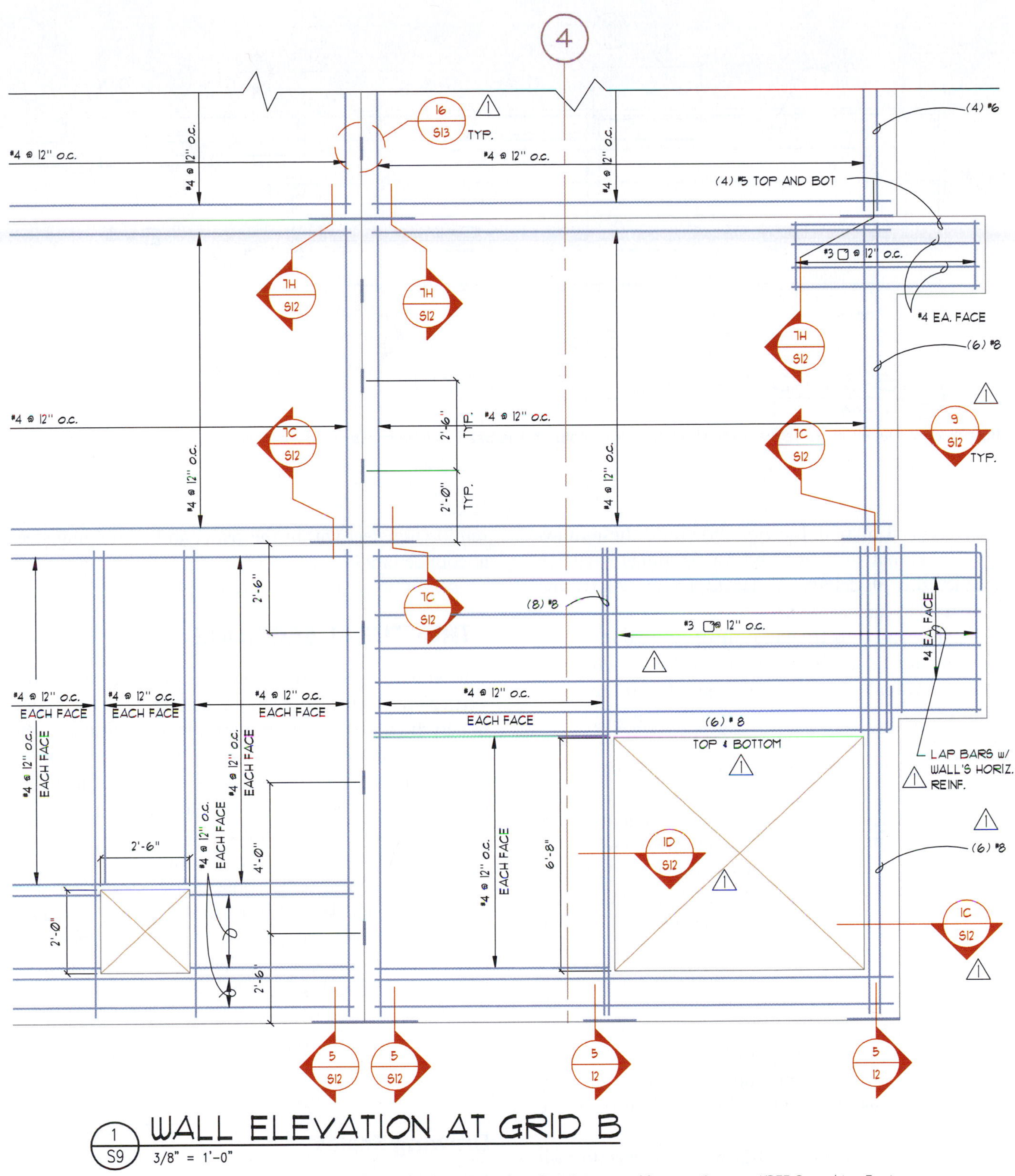

FIGURE 23-6 Elevations for poured concrete show the location of each column and beam. *Courtesy KPFF Consulting Engineers.*

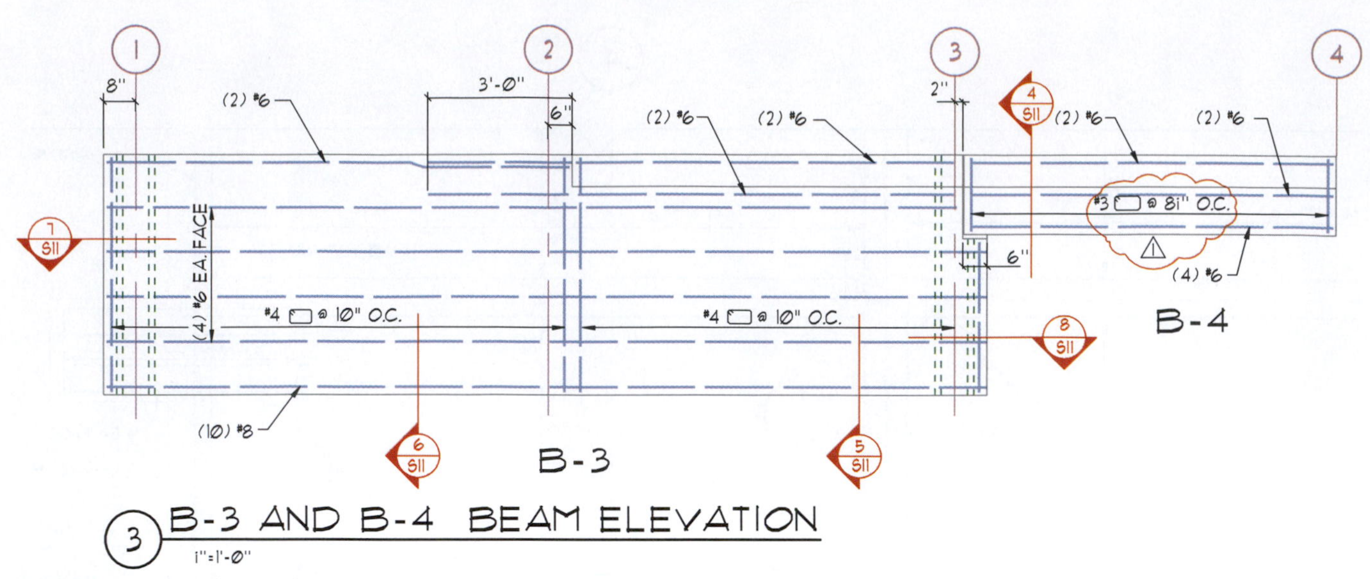

FIGURE 23-7 Details must be drawn to show how each column and beam will be constructed. *Courtesy KPFF Consulting Engineers.*

are referenced by note but not shown on these drawings. Individual steel required for openings is shown in schedules or details similar to Figure 23-10 and referenced to the elevations.

An alternative to drawing all of the panels along one grid line as a whole unit is to draw each panel as an individual panel elevation, similar to the panel seen in Figure 23-11. Using this method, a large-scale elevation is drawn of each panel. The scale depends on the size of the panel and the complexity of the reinforcing pattern to be represented. One elevation can be used to describe several panels if they are exactly the same in every respect. If any variation exists, an elevation of each panel must be drawn. A detail is typically included within each panel elevation to show and specify typical reinforcing steel. Special steel required to reinforce each opening is drawn and specified on each panel elevation. Panels can be referenced to other drawings by referring to the grids located at each end of the panel. Panels can also be referenced to the framing plan by using an elevation reference symbol. This method is similar to the way interior drawings are referenced to the floor plan (see Chapter 16).

Details similar to Figure 23-12 are referenced to each type of panel elevation to show typical corner and joint details. In addition to showing whole panels, elevation of smaller panels, called spandrels are also usually provided. *Spandrels* are concrete panels that are not full height, located over or between areas of glazing. When spandrels are required for a project, an elevation should be provided to specify reinforcing steel. Details similar to Figure 23-13 are typically required to show both reinforcing steel within the panel and restraining steel at connections.

STRUCTURAL SECTIONS

The engineering team draws sections to show the vertical relationship of major material intersections. When drawn, the section would be created using the same guidelines used to create architectural sections. Although referred to as sections on the project title page, structural sections are usually details of a specific area rather than of an entire wall, or a portion of the structure. The structural drawings detail how each component is to be constructed or connected to other materials. The architectural drawings are used to show how each component relates to others in size and location. The same guidelines used to construct architectural details should be used to draft structural details. See Chapters 3 and 6 for a review of linetypes and hatch patterns typically used with sections and details.

Drawing Scales

The scale used to create structural sections and details depends on the intent of the detail. The senior drafter typically provides a sketch and indicates the required scale for a new CAD technician to follow. When determining the size for a detail, the smallest component to be represented should determine the drawing scale. Use a scale that allows the smallest member to be shown clearly. Common scales used for structural

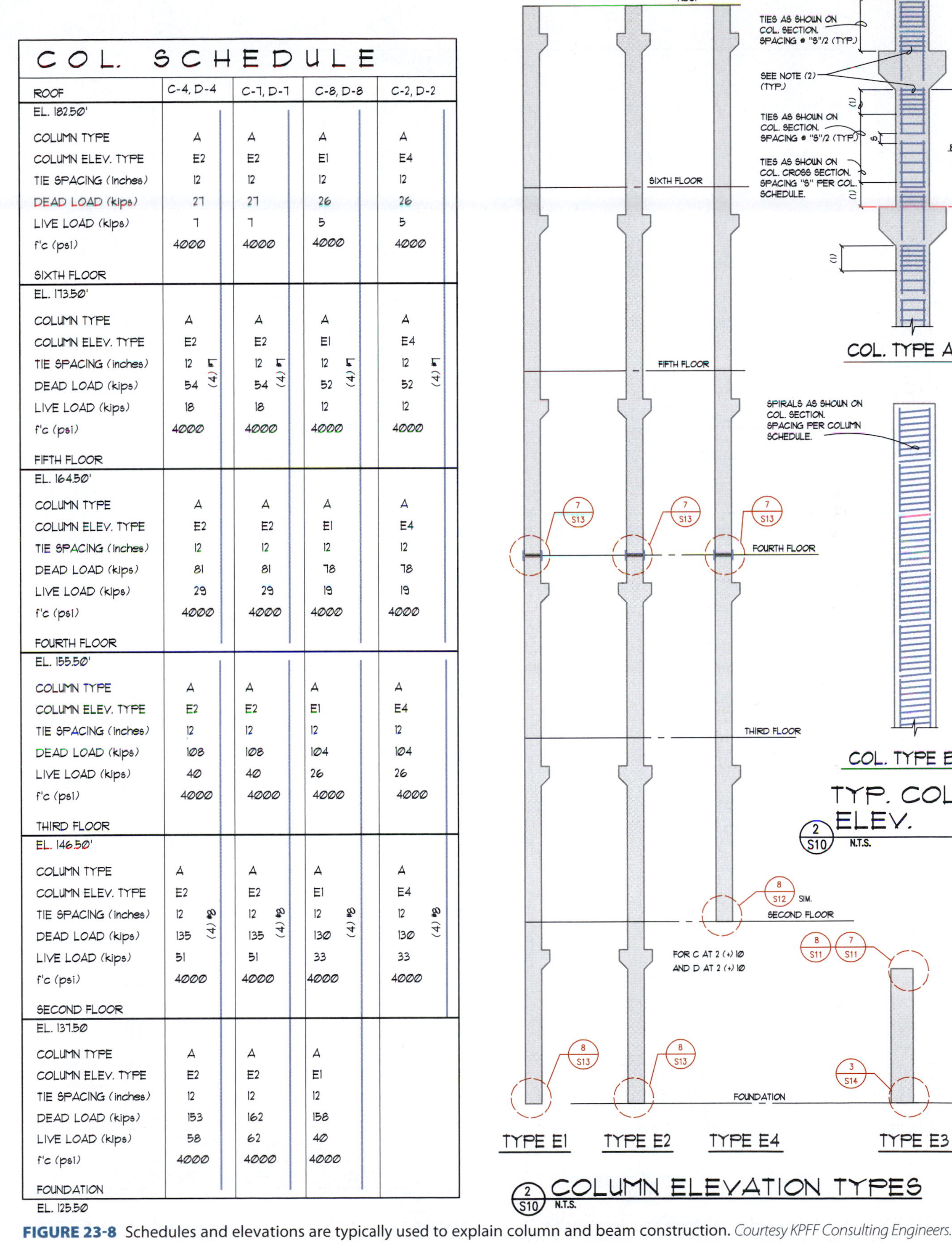

COL. SCHEDULE

ROOF	C-4, D-4	C-7, D-7	C-8, D-8	C-2, D-2
EL. 182.50'				
COLUMN TYPE	A	A	A	A
COLUMN ELEV. TYPE	E2	E2	E1	E4
TIE SPACING (inches)	12	12	12	12
DEAD LOAD (kips)	27	27	26	26
LIVE LOAD (kips)	7	7	5	5
f'c (psi)	4000	4000	4000	4000
SIXTH FLOOR				
EL. 173.50'				
COLUMN TYPE	A	A	A	A
COLUMN ELEV. TYPE	E2	E2	E1	E4
TIE SPACING (inches)	12 (4) #7	12 (4) #7	12 (4) #7	12 (4) #7
DEAD LOAD (kips)	54	54	52	52
LIVE LOAD (kips)	18	18	12	12
f'c (psi)	4000	4000	4000	4000
FIFTH FLOOR				
EL. 164.50'				
COLUMN TYPE	A	A	A	A
COLUMN ELEV. TYPE	E2	E2	E1	E4
TIE SPACING (inches)	12	12	12	12
DEAD LOAD (kips)	81	81	78	78
LIVE LOAD (kips)	29	29	19	19
f'c (psi)	4000	4000	4000	4000
FOURTH FLOOR				
EL. 155.50'				
COLUMN TYPE	A	A	A	A
COLUMN ELEV. TYPE	E2	E2	E1	E4
TIE SPACING (inches)	12	12	12	12
DEAD LOAD (kips)	108	108	104	104
LIVE LOAD (kips)	40	40	26	26
f'c (psi)	4000	4000	4000	4000
THIRD FLOOR				
EL. 146.50'				
COLUMN TYPE	A	A	A	A
COLUMN ELEV. TYPE	E2	E2	E1	E4
TIE SPACING (inches)	12 (4) #8	12 (4) #8	12 (4) #8	12 (4) #8
DEAD LOAD (kips)	135	135	130	130
LIVE LOAD (kips)	51	51	33	33
f'c (psi)	4000	4000	4000	4000
SECOND FLOOR				
EL. 137.50				
COLUMN TYPE	A	A	A	
COLUMN ELEV. TYPE	E2	E2	E1	
TIE SPACING (inches)	12	12	12	
DEAD LOAD (kips)	153	162	158	
LIVE LOAD (kips)	58	62	40	
f'c (psi)	4000	4000	4000	
FOUNDATION				
EL. 125.50				

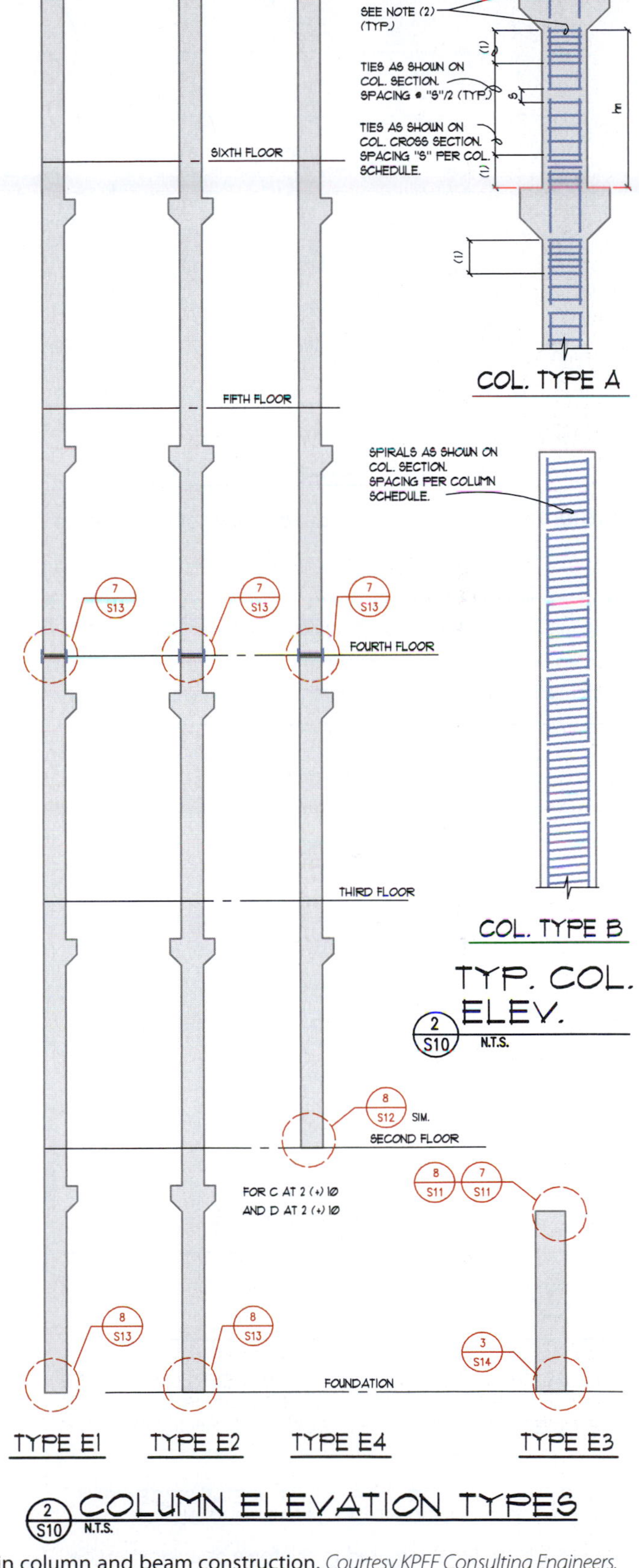

FIGURE 23-8 Schedules and elevations are typically used to explain column and beam construction. *Courtesy KPFF Consulting Engineers.*

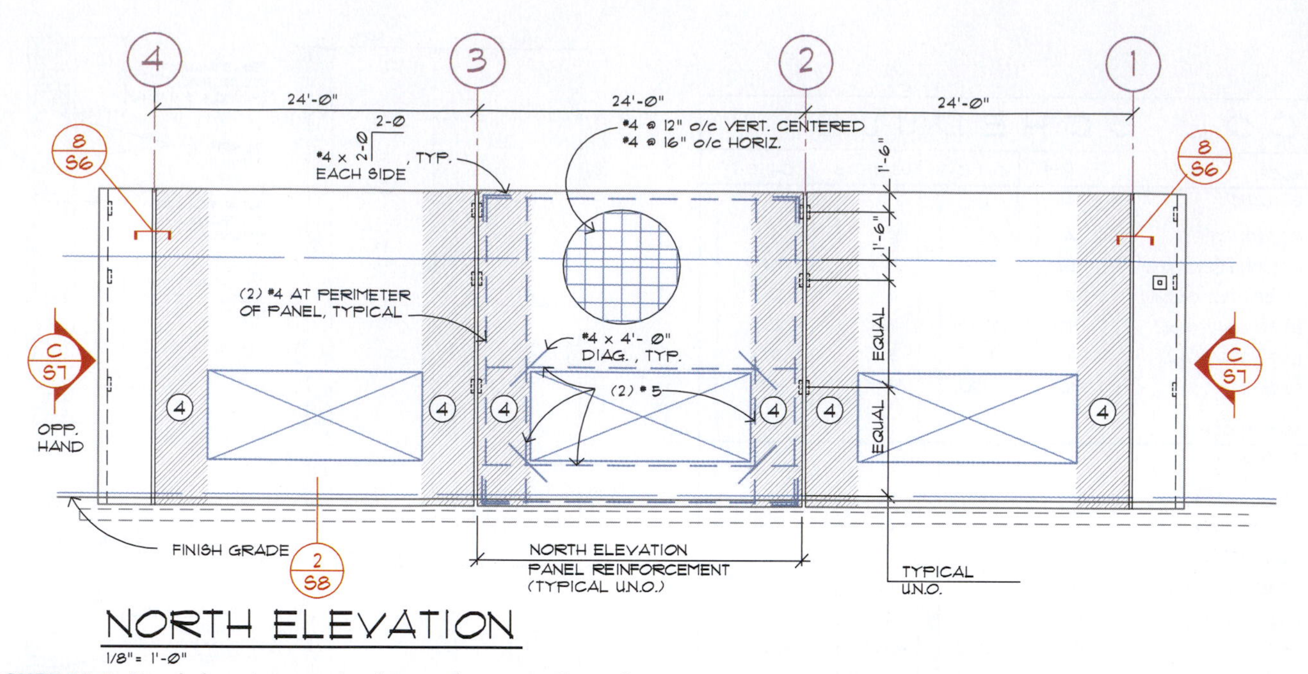

FIGURE 23-9 Panel elevations can be drawn of an entire face of a structure. *Courtesy Bill Berry, Berry-Nordling Engineers, Inc.*

TILT-UP JAMB REINFORCEMENT		
MARK	VERT. REINF. *	DETAIL
1	#4 @ 6" CENTERED	2/S6
2	#4 @ 9" CENTERED	2/S6
3	#4 @ 9" CENTERED	3/S6
4	#4 @ 12" INSIDE FACE #4 @ 8" OUTSIDE FACE	4/S6
5	# 5 @ 8" INSIDE FACE # 5 @ 8" OUTSIDE FACE	5/S6
6	# 5 @ 8" INSIDE FACE # 5 @ 8" OUTSIDE FACE	6/S6
7	(3) - #5 EACH FACE	7/S6

* IN ADDITION PROVIDE (2) # 5 VERTS. AT EDGE OF ALL OPENINGS, (2) #4 VERTS. AT PANEL PERIMETER. SEE PANEL ELEVATIONS, (TYP. U.N.O.)

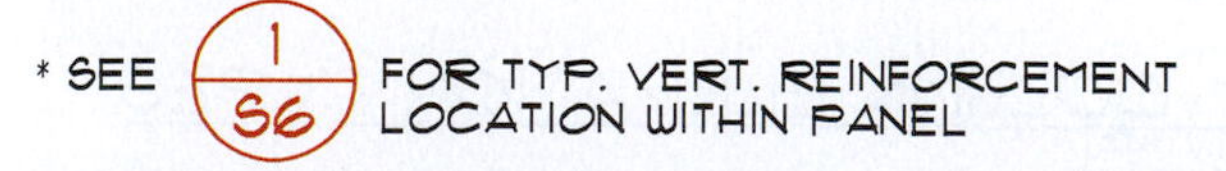

* SEE
1
S6
FOR TYP. VERT. REINFORCEMENT
LOCATION WITHIN PANEL

FIGURE 23-10 Steel required to reinforce openings can be represented in a schedule to keep the drawings uncluttered. *Courtesy Bill Berry, Berry-Nordling Engineers, Inc.*

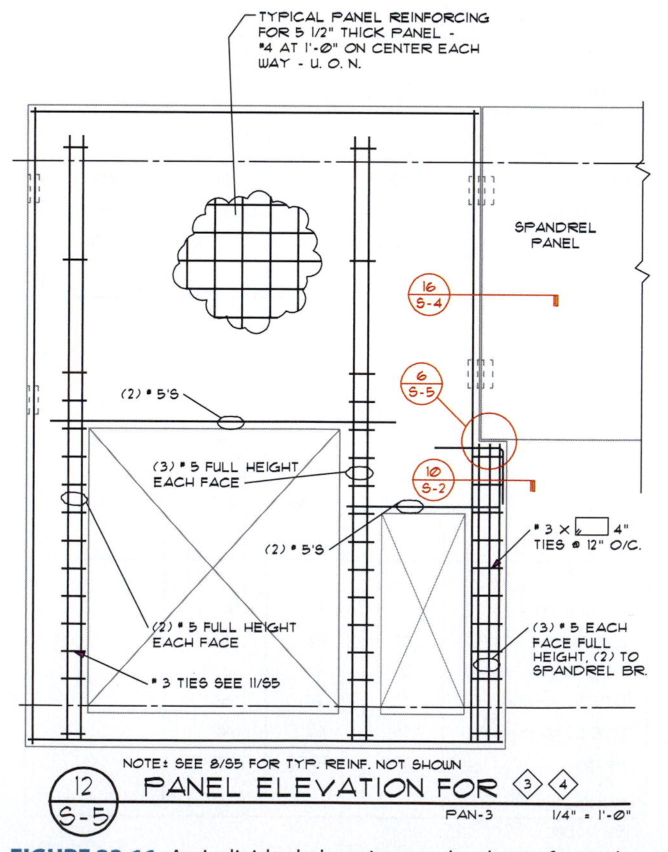

FIGURE 23-11 An individual elevation can be drawn for each panel. *Courtesy Van Domelen/Looijenga/McGarrigle/Knauf Consulting Engineers.*

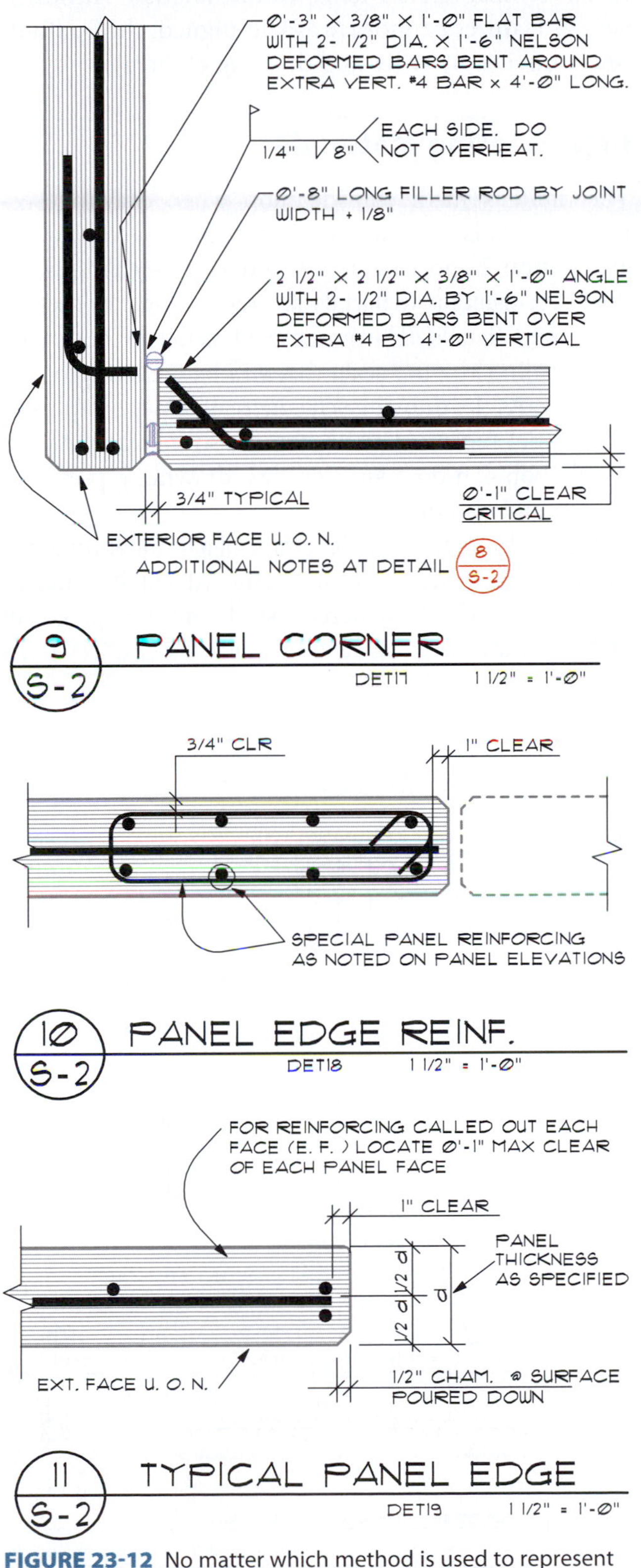

FIGURE 23-12 No matter which method is used to represent the panels, details must be drawn to represent each panel intersection. *Courtesy Van Domelen/Looijenga/McGarrigle/Knauf Consulting Engineers.*

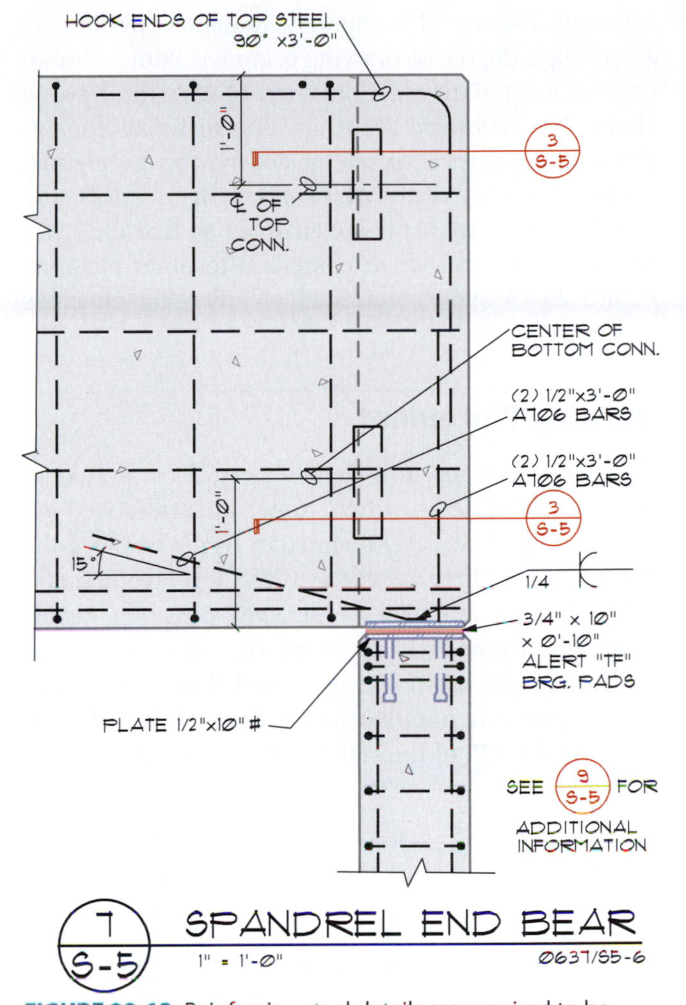

FIGURE 23-13 Reinforcing steel details are required to be drawn for individual panels. *Courtesy Van Domelen/Looijenga/McGarrigle/Knauf Consulting Engineers.*

details and the closest metric counterparts include the following:

In/ft Scale	Metric
3/4" = 1'–0"	1:20
1" = 1'–0"	1:10
1 1/2" = 1'–0"	(no common scale)
3" = 1'–0"	1:5

When several details are to be drawn that relate to the same component, each should be drawn at the same scale if space permits.

When scales smaller than 3/4" = 1'–0" (1:20) are used, structural materials are sometimes drawn at their nominal size of 6" × 12" (150 × 300 mm) rather than their actual size 5 1/2" × 11 1/2" (140 × 292 mm). Structural details drawn at a scale of 3/4" (1:20) and larger should always represent materials with their actual sizes. Review Section 2 for various material sizes.

Although the use of computers enables CAD drafters a very high degree of drawing accuracy, some components of a detail may be drawn out of scale for drawing clarity. In small-scale drawings, thin materials such as plywood, sheet rock, or steel plates may be represented thicker than they really are to add clarity. Care should be taken as components are enlarged so that their size remains consistent when compared to other material. Figure 23-14 shows examples of enlarging materials for clarity.

Drawing Placement

Structural sections and details should be drawn as close as possible to where they are referenced on the drawing. The CAD technician needs to place the details so that they conform to the job layout established by the engineer. Whenever possible, details should be drawn so that common features such as walls, floors, or rooflines are aligned. This is especially true if details are not placed in individual detail boxes. Aligning the top of the slab in each detail adds clarity to the total page layout. When individual detail boxes are used, an attempt should still be made to group details so that similar items can be aligned. Although similar features cannot always be aligned, the box outlines provide an order and drawing clarity.

Representing Materials

The drafter's main consideration when drawing sections or details is to clearly distinguish between materials. Section 2 and Chapters 6 and 19 give examples of how construction materials are represented in section views and details. The amount of detail to be drawn is determined by the scale that will be used to plot the drawing. As the drawing scale increases, the complexity of material representation should also increase. As the detail zooms in on a smaller area, drawing representation should also increase.

Figure 23-15 shows an example of a foundation detail showing six materials. Using standard hatch patterns, the soil, gravel fill, concrete, steel reinforcing, grout, and structural steel can each be quickly identified. Even

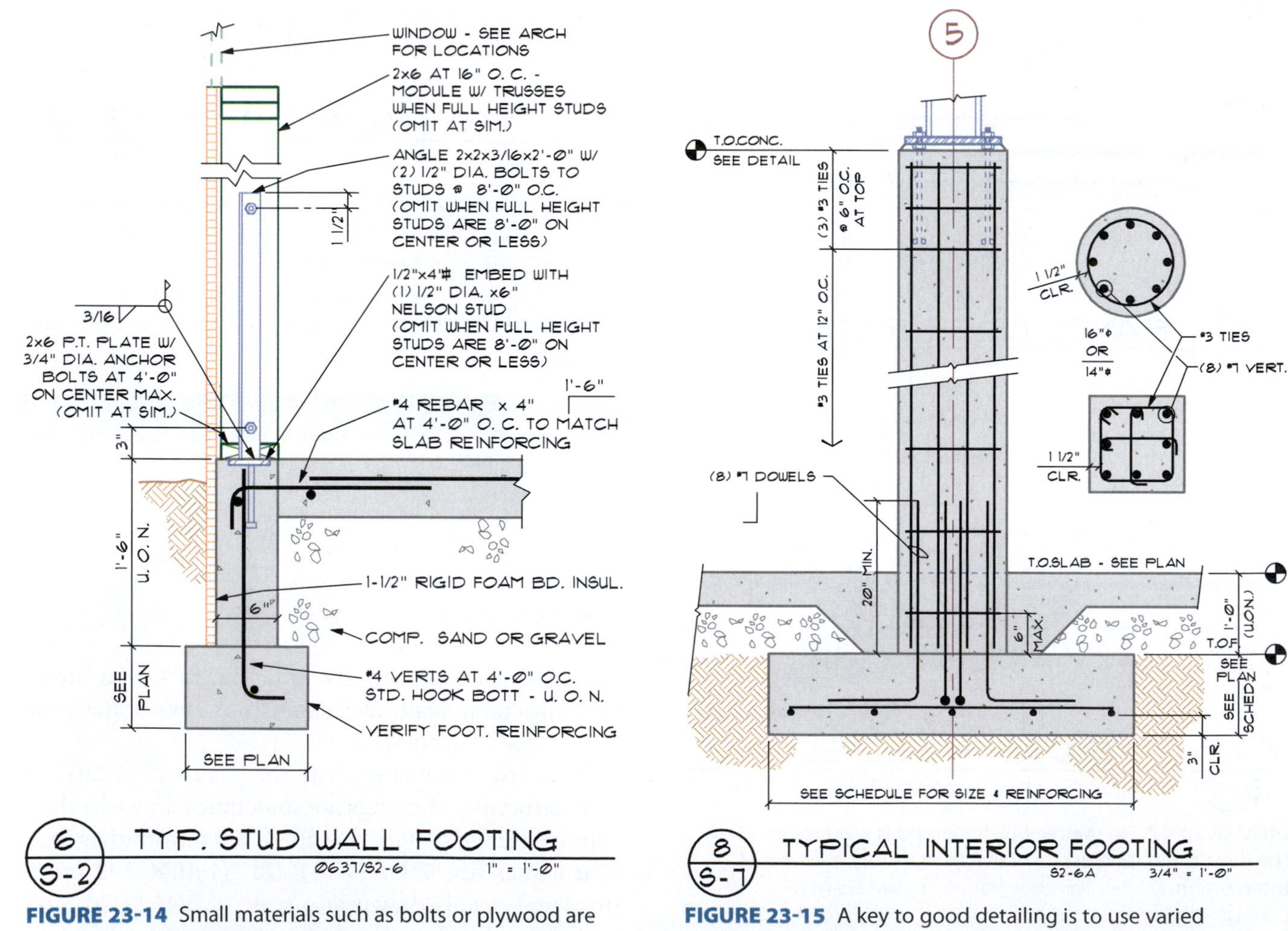

FIGURE 23-14 Small materials such as bolts or plywood are often enlarged so that they can be clearly seen. *Courtesy Van Domelen/Looijenga/McGarrigle/Knauf Consulting Engineers.*

FIGURE 23-15 A key to good detailing is to use varied lineweights, linetypes, and hatch patterns to distinguish between materials.

though a continuous line is used to represent the shape of the structural steel, concrete, and reinforcing, each can be easily distinguished because of varied line widths and the use of hatch patterns. Care should be taken to keep the size of the hatch pattern from becoming ineffective. Using too large a pattern will not provide the repetition needed to distinguish the pattern. Too small a pattern will have the same effect as using the FILL command.

ADDITIONAL READING

Links for the materials represented on the structural details and elevations are too numerous to list here. Rather than listing individual sites, use your favorite search engine to research products from the following CSI categories: wood, steel, concrete, and tilt-up concrete.

CHAPTER 23 TEST Drawing Structural Elevations and Sections

Answer the following questions with short complete statements. Type the chapter title, question number, and a short complete statement for each question using a word processor. Some answers may require the use of vendor catalogs or seeking out local suppliers.

Question 23-1 What types of drawings typically require structural elevations?

Question 23-2 What is the purpose of structural steel elevations?

Question 23-3 What is the advantage of drawing individual panel elevations for concrete tilt-up construction?

Question 23-4 What is a spandrel?

Question 23-5 How can a CAD technician decide if something should be dimensioned?

DRAWING PROBLEMS

Use the reference material from preceding chapters, local codes, and vendor catalogs to complete one of the following projects. Unless other instructions are given by your instructor, complete the project that corresponds to the floor plan that was drawn in Chapter 16. Skeleton drawings for most of the problems can be accessed from the student CD in the DRAWING PROJECTS/CHAPTER 23 folder. Use these drawings as a base to complete the assignment. Use appropriate symbols, linetypes, dimensioning methods, and notations to complete the drawing. Determine any unspecified sizes based on material or practical requirements. Unless specified, match the scale of the floor plan and use appropriate LTSCALE and DIMVARS settings.

You will act as the project manager and will be required to make decisions about how to complete the project. Study each of the drawing problems related to the structure prior to starting a drawing problem. Any information not provided must be researched and determined by you unless your instructor (the project architect and engineer) provides other instructions. When conflicting information is found, information from preceding chapters should take precedence. Make several sketches of possible solutions based on examples found throughout this text and submit them to your instructor prior to completing each problem with missing information. Place the necessary cutting plane and detail bubbles on the framing plan to indicate each drawing.

Note: Keep in mind as you complete these drawings that the finished drawing set may contain more sections and in some cases hundreds of details more than you are being asked to draw. Other details that would be required for the building department or the construction crew will not be drawn due to the limitations of space and time.

Warning: The instructions for each problem are written in common textbook English. Although it may not be evident, every attempt has been made to use proper grammar in the following detail specifications. In order to help students who may not be familiar with construction methods have a better chance of successfully completing the details, explanations have been written using proper English. This is not the case with calculations written by an engineer. The engineer is concerned with using computer programs that analyze building stresses, and a calculator to run math formulas, and not with the English skills that were learned in college. That's why you'll have a job, to sort through the engineer's math formulas, determine what needs

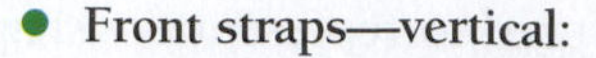

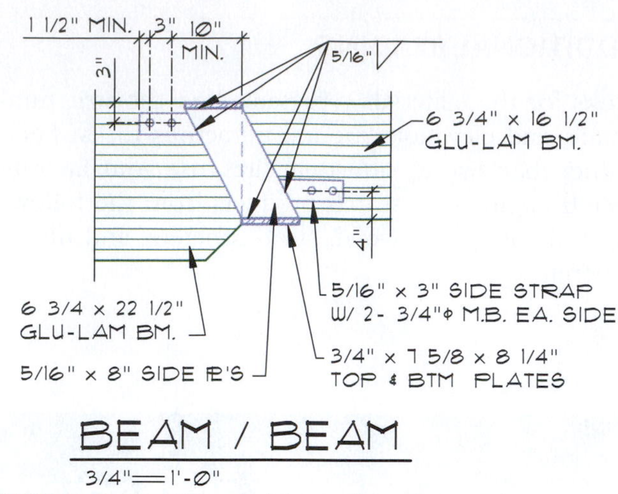

FIGURE 23-16 CAD technicians will need to translate engineering notes into clear communication for those who will be reading the plan.

to be specified, and place that information on the working drawings in the proper format. That means placing information in short concise statements. To do that, you'll need to translate statements such as the following:

- Provide 3" × 5/16" side straps on each side of the beam, placed so that the centerline of the bolts that pass through the straps are 4" up from the bottom of the beam and 3" down from the top of the beam.

  Part of your job is to condense these statements into short notes that clearly explain the intended process. This would include placing notes such as:

  3" × 5/16" side straps ea. side

  and drawing dimensions that show the strap 4" up from the bottom of the beam, and 3" down from the top of the beam. The results of this one note can be seen in Figure 23-16.

  Remember, your goal is to place a large amount of information in a small space, while maintaining clarity in the communication between the engineer and the construction worker. To do this, use abbreviations when possible to conserve drawing space.

Problem 23-1 Use the drawing as a guide and complete a framing elevation for the front, rear, and interior shear walls of the condominium started in project 16-1.

- **Front wall—upper floor:** On exterior face of wall use 1/2" c-d exterior plywood Structural II or better with 8d @ 4" o.c. at all edges and 8d @ 6" o.c. at field.
- **Front wall—middle and lower floor:** Same plywood as above, applied to interior and exterior surfaces of wall.
- **Front straps—vertical:**
  - *Upper to middle floor*—Use an MST48 strap over double top plates with (48)-10d nails at the exterior face of each opening.
  - *Middle to lower floor*—Use a MST60 strap over double top plates with (60)-10d nails on the interior and exterior sides of wall.
  - *Lower to foundation*—Use HD-7A anchor on each vertical support at each side of each opening.
- Attach to foundation with 3/4" A.B. with 12" embedment, through 3× plate. Bolt to vertical supports as per manufacturer's specifications. Use 5/8" × 12" anchor bolts at 16" o.c. for plate to foundation. Use (3)-3/4" hammer driven bolts × 4 1/2" long at end of shear wall into 4' high retaining wall.
- **Front straps—horizontal:** Block between studs for all horizontal straps as plates from one unit are tied into the wall system of the next unit. Use MST60 with 10d nails at each plate level.
- **Rear wall—upper floor:** Use plywood as per front wall with nailing at 6" and 12" spacing.
- **Rear wall—middle floor:** Use plywood per the upper floor applied to the interior and exterior face of wall at each unit.
- **Rear wall—lower floor:** Use plywood as per the upper floor applied to the exterior face only.
- **Rear straps—vertical:**
  - *Upper to middle floor*—Use LSTA36 with (24)-10d nails.
  - *Middle to lower floor*—Use MSTI48 with (48)-10d nails.
- **Rear straps—horizontal:** Use LSTA21 at top plate, and LSTA36 at all other plates for this wall.
- **Interior wall—upper floor:** Use 3/8" ply with nails at 4" and 6" spacing one side only. Use 3× studs at each end of shear wall. Use HD2A with 5/8" bolts as per the manufacturer's specifications to each end post to each post located in wall below.
- **Interior wall—middle floor:** Use plywood as per the wall above, applied to each side of wall. Use 4× studs each end of shear wall. Place a 4× post below the post in wall above. Use HD2A with 5/8" bolts as per the manufacturer's specifications. Use a 1/2" × 4" × 9" steel plate on bottom side of glu-lam beam for bolt attachment.
- **Interior wall—lower floor:** Place 5 1/2" × 10 1/2" glu-lam beam below the shear wall above. Support the beam on 4 × 6 post with EPC caps and appropriate bolts at each end of the beam. Attach the post to foundation with an appropriate CB column base.

Note: Use the sample calculations posted on the student CD in the REFERENCE MATERIAL folder to specify material that is not specified in the following problems.

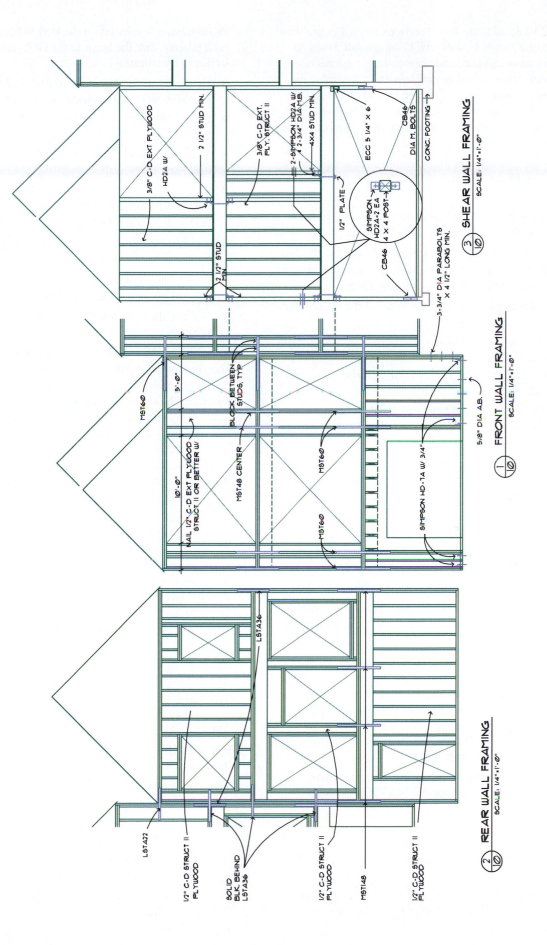

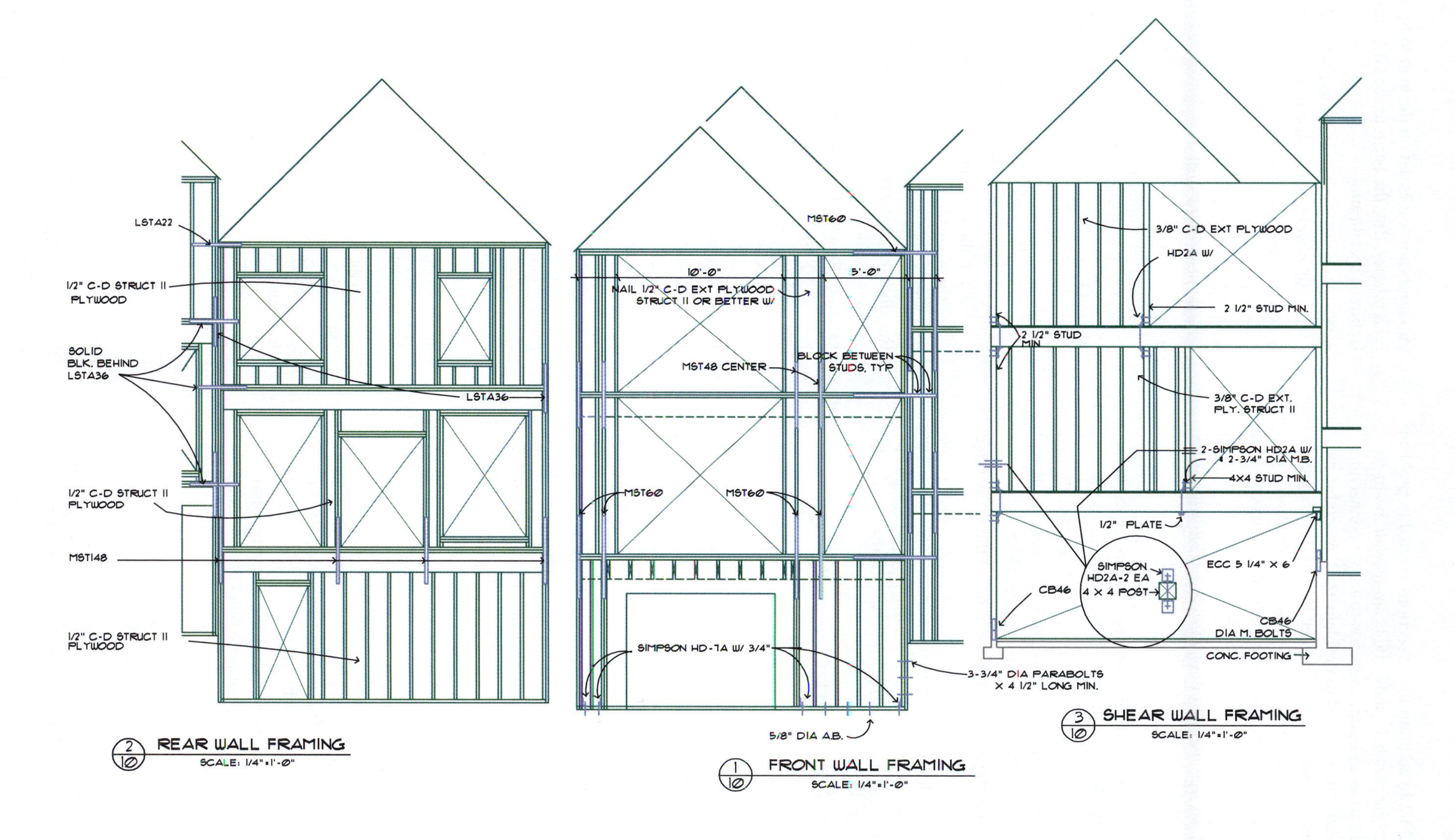

LSTA22
1/2" C-D STRUCT II PLYWOOD
SOLID BLK. BEHIND LSTA36
LSTA36
1/2" C-D STRUCT II PLYWOOD
MST148
1/2" C-D STRUCT II PLYWOOD
REAR WALL FRAMING
SCALE: 1/4"=1'-0"
MST60
10'-0"
5'-0"
NAIL 1/2" C-D EXT PLYWOOD STRUCT II OR BETTER W/
MST48 CENTER
BLOCK BETWEEN STUDS, TYP
MST60
MST60
SIMPSON HD-7A W/ 3/4"
5/8" DIA A.B.
FRONT WALL FRAMING
SCALE: 1/4"=1'-0"
3-3/4" DIA PARABOLTS X 4 1/2" LONG MIN.
3/8" C-D EXT PLYWOOD
HD2A W/
2 1/2" STUD MIN.
2 1/2" STUD MIN
3/8" C-D EXT. PLY. STRUCT II
2-SIMPSON HD2A W/ 4 2-3/4" DIA M.B.
4X4 STUD MIN.
1/2" PLATE
SIMPSON HD2A-2 EA
4 X 4 POST
ECC 5 1/4" X 6
CB46
CB46
DIA M. BOLTS
CONC. FOOTING
SHEAR WALL FRAMING
SCALE: 1/4"=1'-0"

Problem 23-2 In addition to the sections created in problem 19-2, create details to show the beam-to-wall, beam-to-column, truss-to-beam, and truss-to-ledger connections for the structure started in problem 16-2. Assemble the sections and details from Chapter 19 and this chapter into the fewest number of sheets as possible. Assign and coordinate detail numbers and place each number on the framing plan created in Chapter 22.

- Use the framing plan to determine all beam and truss sizes and locations. Specify the size of each in each detail.
- Select appropriate GLB steel beam seats for the beam at each pilaster and detail each connection.
- Use a 3" × 4" × 3/8" steel plate with an MST27 strap at the top of each beam to attach the beam to the wall. Attach the MST straps per drawing 23-2e.
- Use drawing 23-2f as a guide to detail the support for the beam to steel column.
- Refer to the framing plan to verify required truss sizes. Use the Weyerhaeuser Web site (www.ilevel.com) to determine required connections for the open-web trusses.

Problem 23-2a.1 Draw and label a detail showing the bearing wall/truss intersection based on the general construction guidelines of problem 23-2.

Problem 23-2a.2 Draw and label a detail showing the non-bearing wall/truss intersection based on the general construction guidelines of problem 23-2.

Problem 23-2b Draw a detail to show the connection of the trusses to the ridge beam.

- Use an MST48 strap at every other truss placed across the top of the truss chords.
- Place a solid block between trusses along the ridge.
  - Anchor the blocks to the ridge beam with an A35 anchor at each end of each block, placed on opposite sides of the block.

Problem 23-2c Draw a detail to show the truss-to-column bracing. Provide 2 × 4 braces @ each truss at each side of column set at a 45° maximum angle from vertical. Provide (2)-3 × 4 blocks between each brace at equal spacing along the brace and a 3 × 4 nailer where the brace intersects the trusses. Use an appropriate U-hanger to support the brace at the beam.

Problem 23-2d Draw a detail showing the connection of the glu-lam beam to concrete block. Use a suitable glu-lam beam seat based on the load to be supported at each pilaster. Bolt the beam to the GLB seat as per manufacturer's specifications.

- Tie the leg of the beam seat to a #5 × 24" rebar with 1/8" fillet weld, both sides of bar to plate.
- Tie the 24" rebar into the steel used to reinforce the pilaster.
- At the upper end of the beam, bolt a 3" × 4" × 3/8" steel plate to the wall.
- Provide (1)-3/4" × 6" stud welded with 1/8" fillet welded all around to back side of plate extending into the block wall and tied into the wall steel.
- Weld an MST27 strap on each side of the beam to the front side of plate to support glu-lam.
  - Weld the straps to the end plate with 1/8" fillet weld on each outer side of the strap.
  - Place the centerline of straps at 3" from the top of the beam.

Problem 23-2e Draw a detail showing the connection of the glu-lam beam over the storefront windows to the concrete block wall. Use a suitable glu-lam beam seat based on the load to be supported for the south and east walls at windows. Tie the GLB seat to wall steel with 1/8" welds. Bolt the beam to the seat as per the manufacturer's specifications. At the upper end of the beam, bolt a steel plate to the wall using detail 23-2d as a guide. Place the centerline of straps at 3" down from the top of the beam.

Problem 23-2f Draw a detail showing the connection of the ridge beam to the 4" × 4" × 5/16" steel column. Select a suitable CC column cap and weld to the column with 3/16" fillet weld all around. Specify bolts recommended by the manufacturer for the cap-to-beam connection. Provide a 3" × 27" × 5/16" steel strap with the centerline of (4)-3/4" M.B. 4" down from the top of the beam, centered over the beam splice at the column.

Problem 23-2g Draw a detail showing the plan view of the 16" × 16" pilaster construction. Reinforce the pilaster with (8)-#5 vertical bars with #3 horizontal ties @ 16" o.c. Specify that the laps in the ties are to be staggered 180°.

Problem 23-2h Draw a detail showing a plan view of reinforcement for the block wall at each opening. Use (6)-#5 verticals (2 per cell) with #3 ties at 16" o.c. horizontally. Show normal wall steel in other cells.

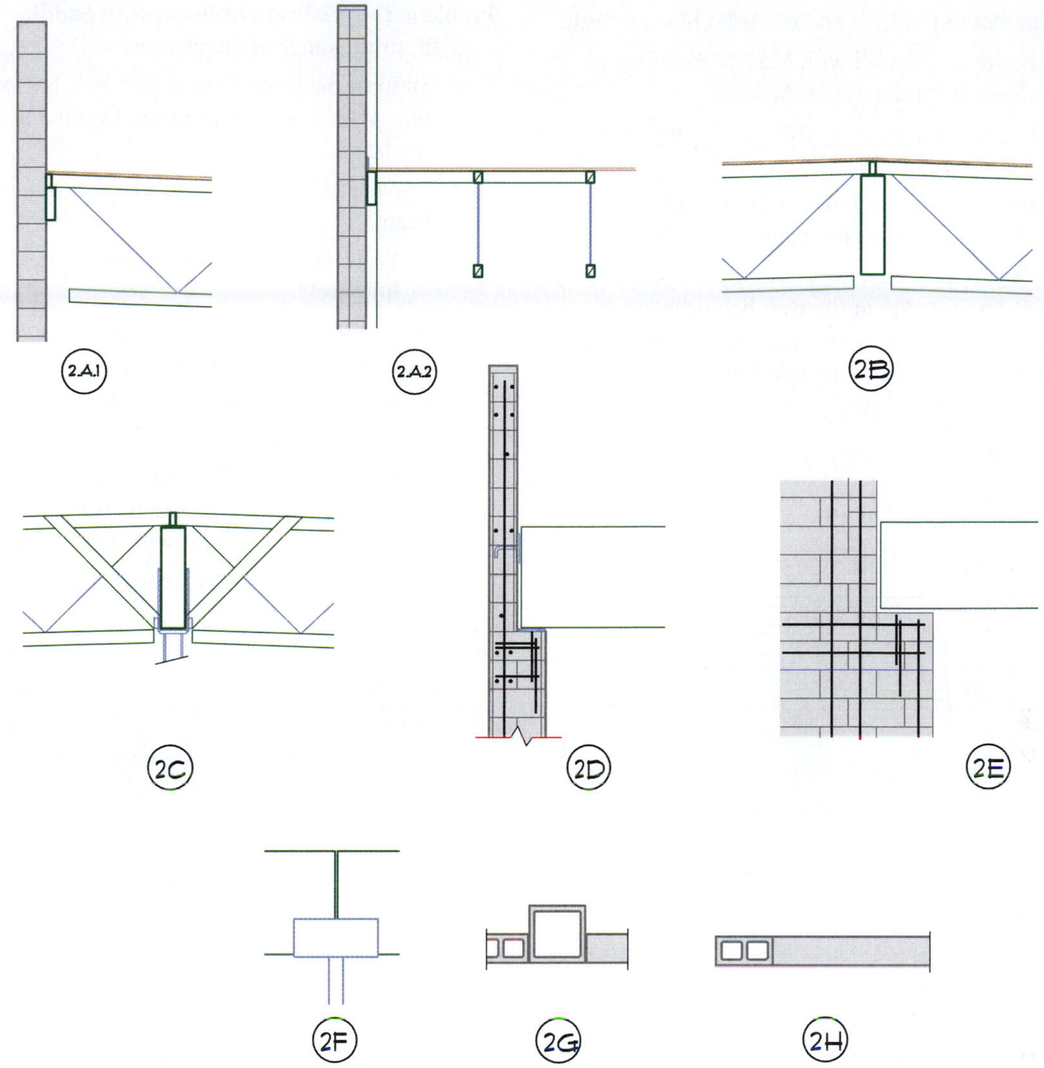

2.A.1 2.A.2 2B

2C 2D 2E

2F 2G 2H

Problem 23-3a Use the attached drawings to detail the required roof connections for the roof framing plan that was started in problem 22-3a (and problem 16-3). Choose a scale suitable for clearly showing the required information. Keep in mind that the information in the attached drawings is incomplete and may contain errors. Refer back to the order of precedence to accurately complete the following drawings.

- Use the framing plan to determine all beam and truss sizes and locations. Specify the size of each in each detail.
- Use MST37 straps to tie the trusses to wall, and MST27 straps to tie the purlins to the ledger.
  - Use 8'–0" spacing for each.
- Assume that the walls are reinforced with #5 Ø @ 16" o.c. horizontal and #5 Ø @ 36" o.c. vertical.
- Assume all welds are 3/16" fillet welds unless noted.

Roof Ledgers

- Use 3 × 12 DFPT ledger with 3/4" Ø anchor bolts at 32" o.c. at grids A and G.
  - Stagger bolts 3" up/down.
  - At splices in the ledger at grids A and G use (2)-3" × 144" × 3/16" straps with (8)-3/4" bolts 2" from ends and 20" o.c. and 10" from splice.
  - Place straps 3" down from top of ledger and 3 1/2" from center to center.
- Space bolts at 60" o.c. at grids 1 and 7, staggered 3" up/down.
  - At splices at walls 1 and 7 use (3)-3/16" × 3" × 24" straps at 3 1/2" o.c. with 4 bolts per strap placed 2" from strap end, and 6" o.c.

Problem 22-3a.1 Beam to Wall. (Place lower portion of this detail with the balance of the mezzanine details.)

- Support the upper beam on 16" × 16" pilaster, and the lower beam on 16" × 24" pilaster.
  - Detail the upper portion of pilaster (22-3a.1.a) to match problem 23-2g.
  - Detail the lower portion of the pilaster (22-3a.1.b) with (10)-#5 Ø vertical bars, with #3 Ø horizontal ties @ 16" o.c. Provide 3/4" clear for each beam to block wall.

- Provide fire-cut to glu-lams and 6× solid block to wall.
- Connect beams to the wall with MST27 each side of beam, 3" down from the top of the beam.
  - Weld the straps to 4" × 6 7/8" × 3/8" steel plate w/ 1/8" fillet weld.
  - Bolt the plate to the wall with (2) -1/2" Ø × 6" studs. Weld the stud to the plate w/ 1/8" fillet weld, all around.
- Provide (4)-#5 Ø × 36" in the wall behind plate. Use Simpson Company GLB beam seat. Specify bolting as per the manufacturer recommendations. Tie the beam seat bolts to the vertical wall steel.

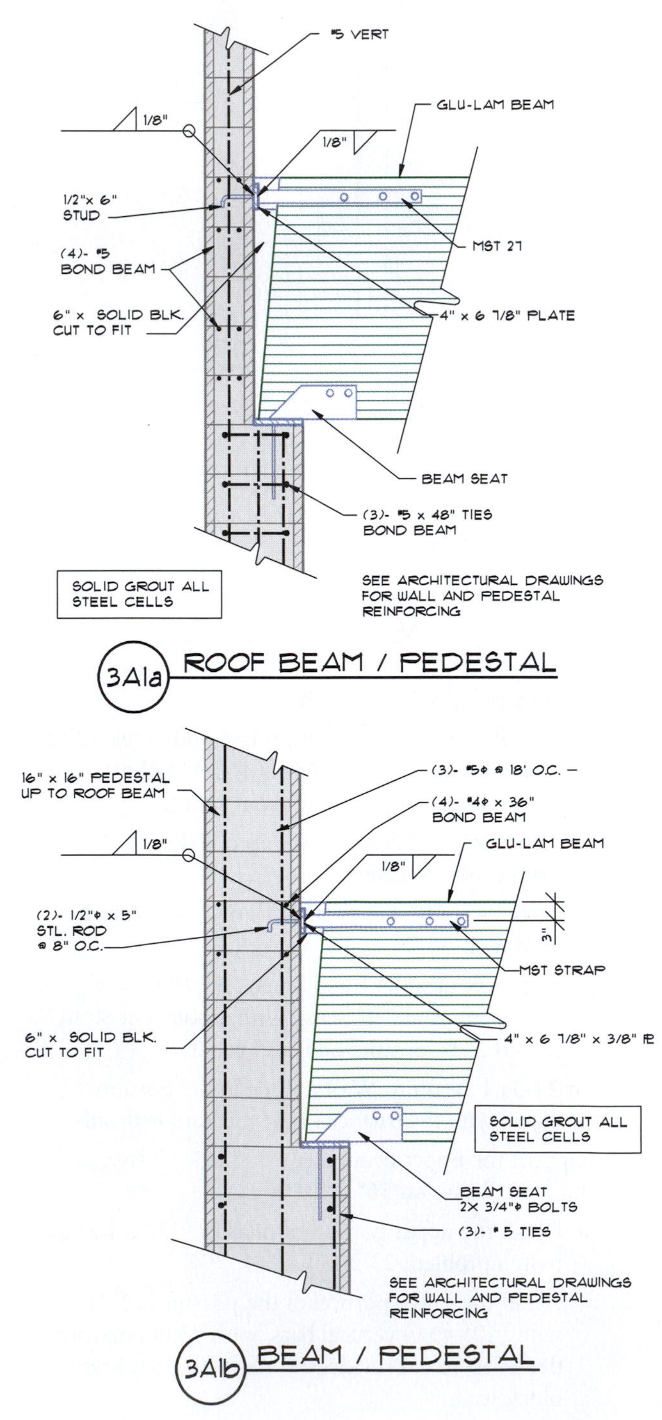

Problem 23-3a.2 Beam to Beam with Saddle. See framing plan for the range of beam sizes.

- Support the beam with 6 7/8" × 7 1/4" × 7/8" bearing plates at the top and bottom. Dap the top plate into the beam.
- Provide 5/16" × 7" wide side plates to each side of the beam.
  - Weld side plates to top/bottom plates with 5/16" fillet weld.
  - Provide 5/16" × 3" high straps on each side of the beam at the top and bottom of the side plates.
    - Attach the straps to the beam with (3)-3/4" Ø bolts @ 3" o.c.
    - Set bolts 4 1/2" up/down from the beam edges, with the first bolt set 8" from the beam splice, and the third bolt set 1 1/2" from the strap end.
  - Attach straps to side plate with 5/16" fillet weld.

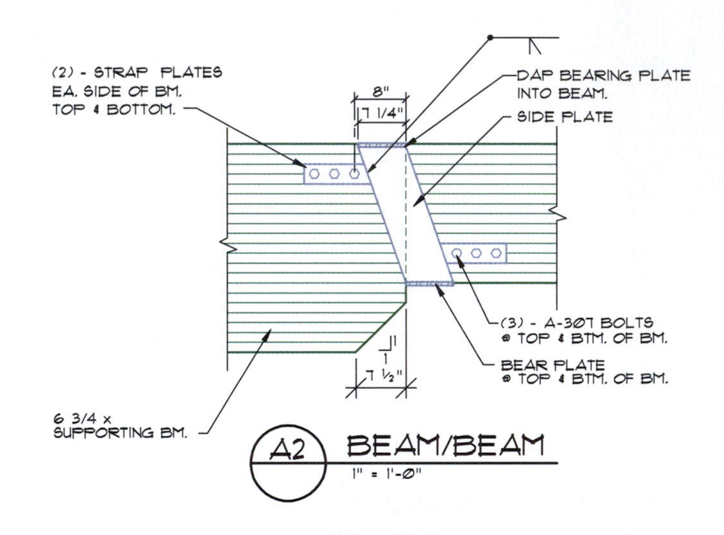

Problem 23-3a.3 Beam to Beam at Fire Wall.

- Frame the wall with 2 × 6 studs @ 16" o.c. with horizontal solid blocking placed @ 10' maximum intervals.
- Provide 3 × 6 top pl. with 2 × 6 × 1 deep counter bore for 5/8" Ø × 6" bolts at 32" o.c. with 2" Ø washers through (2)-2 × 6 top plates.
- Provide 3 × 6 continuous nailer at each side of top plate, with metal hangers for blocking.
- Support the glu-lam beams on each side of the wall with appropriate CCO connector to 6 × 6 TS column.
  - Center each column 12" from center of stud wall.
  - Provide 3/4" clearance between the beam and the gypsum board. Use 3/8" fillet for the CCO to column connections.
- Provide an MST48 strap, 3" down from the bottom of the top plates, centered on the wall.
- Provide 5/8" type X gypsum board between the trusses, 60" out from wall minimum to cover the 2 × 6 purlins.

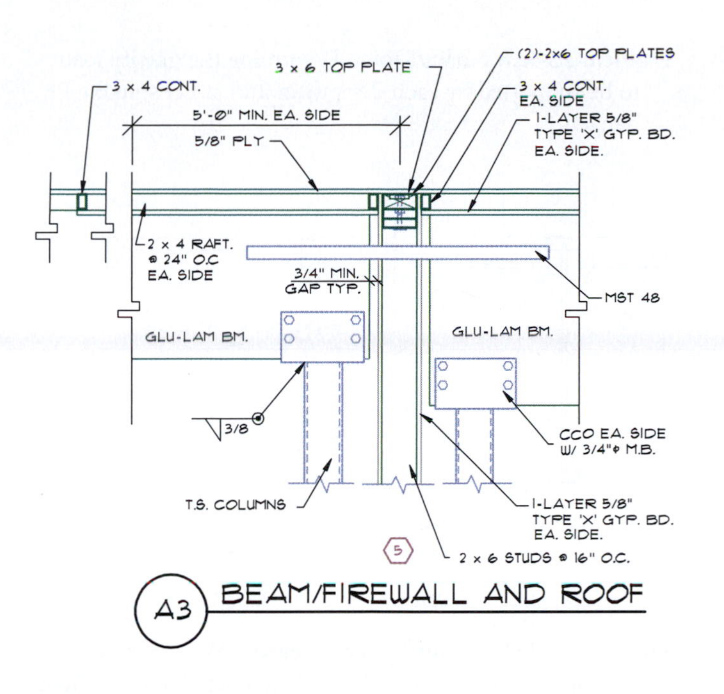

Problem 23-3a.4 Beam to Column.

- Support each glu-lam on 6 7/8" × 20" × 7/8" bearing plate welded to TS column with 3/8" weld.
- Provide (4)-9" × 20" × 5/16" side plates with (2)-3/4" Ø bolts to each beam.
  - Place bolts 2" down, 2" in from upper edge of the side plate, and 3" o.c. Provide standard washers.

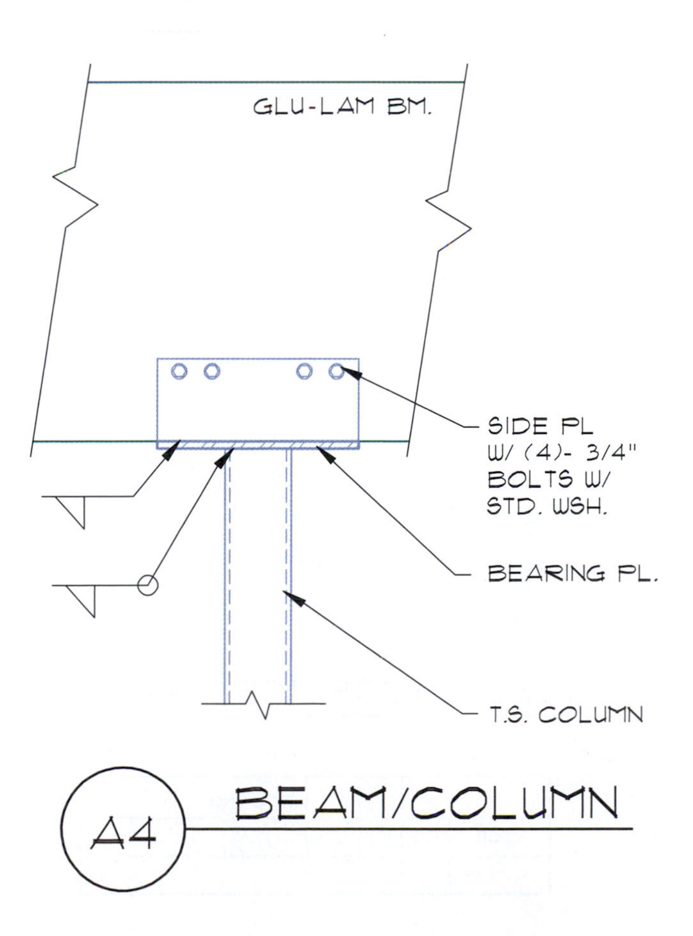

Problem 23-3a.5 I-Joist to Beam.

- Select a suitable HIT joist hanger and specify the required connection for trusses to each laminated beam.
- Provide an MST48 @ 8'–0" o.c. from truss/truss over the top of the beam.

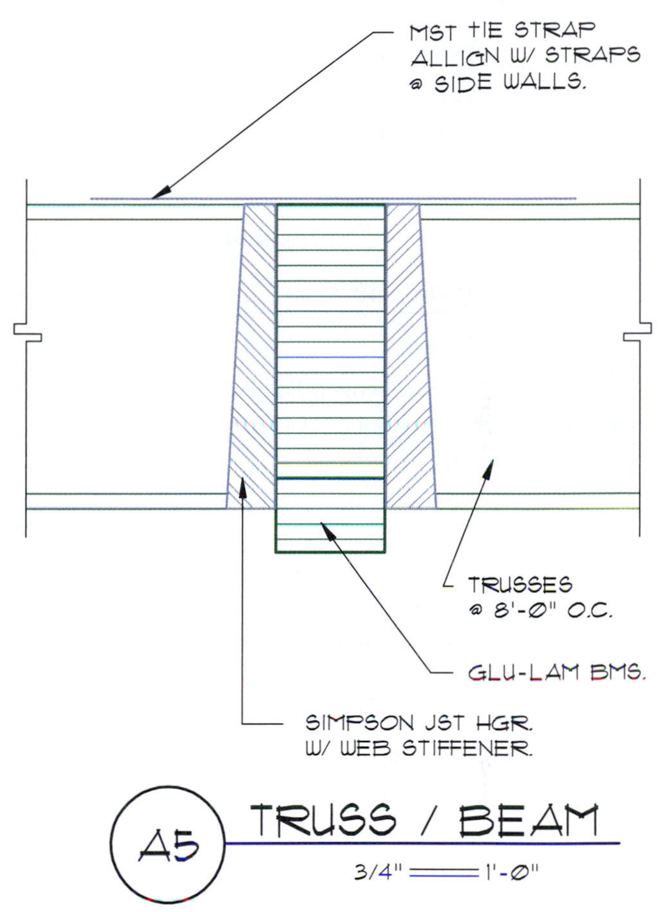

Problem 23-3a.6 I-Joist/Ledger—Perpendicular.

- Provide 3 × 12 DFPT ledger w/ 3/4" Ø bolts @ 32" o.c. staggered 3" up/down.
- Connect the trusses to the ledger with HB1430 hangers.
  - Select suitable hangers for the rafters-to-purlin connection.
  - Reinforce the wall w/8" × 16" bond beam at the ledger.
- Provide 3" × 4"h × 5/8" plate with (2)-3/4" Ø × 4 1/8" studs, attached w/ 1/8" fillet weld.
- Weld an MST37 to the plate w/1/8" fillet weld @ 8'–0" o.c.

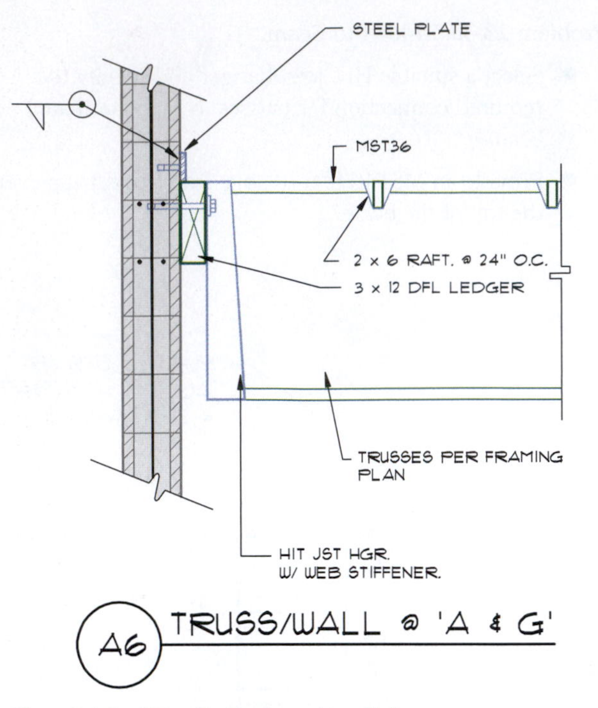

Problem 23-3a.7 Purlin/Ledger–Parallel.

- Provide 3 × 12 DFPT ledger with 3/4" Ø bolts @ 60" o.c. staggered 3" up/down.
- Reinforce the wall at the ledger w/ a two-course deep bond beam.
- Provide 3" × 4"h × 5/8" plate at each truss to bind the truss to the wall/ledger.
  - Attach the plate to the wall with (1)-3/4" Ø × 4 1/8" stud.
  - Provide an MST27 strap at each truss/plate.
    - Attach the plate to the strap with 1/8" fillet weld.
- Select suitable U hangers for the rafters-to-truss connection.

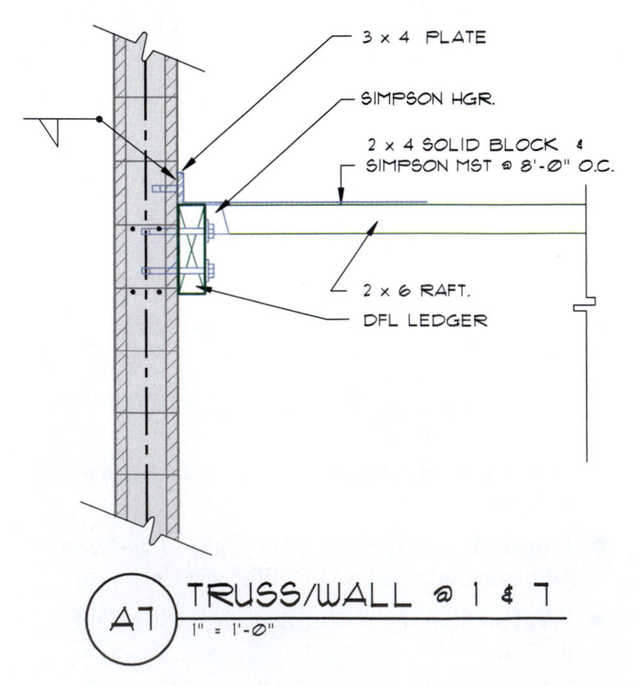

Problem 23-3a.8 Purlin/I-Joist. Determine the gravity load to be supported by each 2× purlin and select suitable PF hangers.

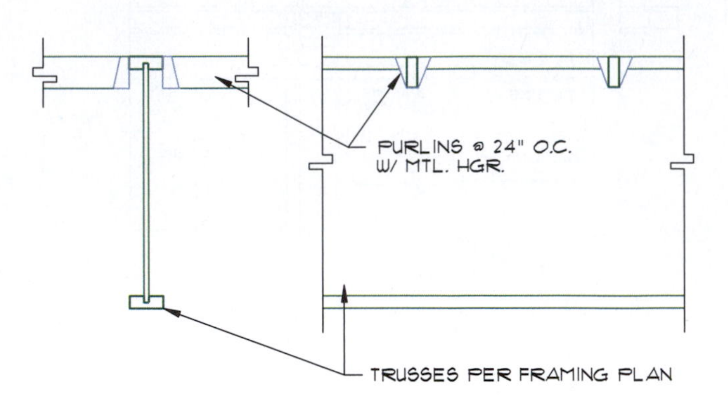

Problem 23-3a.9 Ledger Splice—Bearing Wall. See the roof/floor ledger notes and establish a detail to show floor and roof usage.

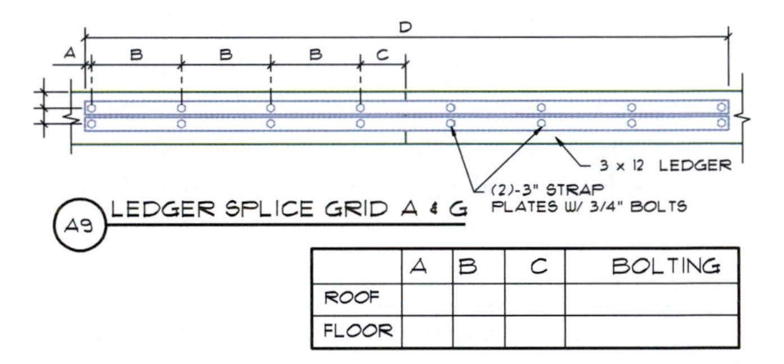

	A	B	C	BOLTING
ROOF				
FLOOR				

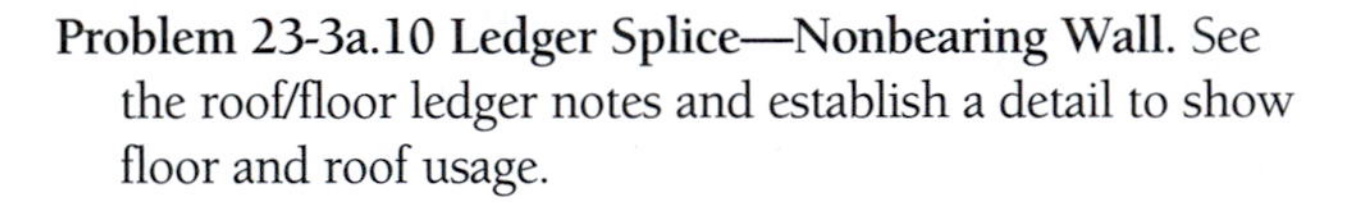

Problem 23-3a.10 Ledger Splice—Nonbearing Wall. See the roof/floor ledger notes and establish a detail to show floor and roof usage.

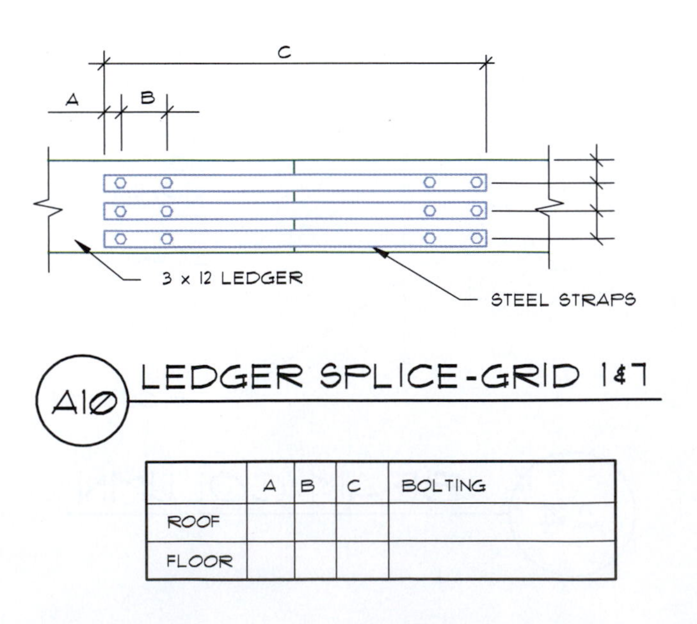

	A	B	C	BOLTING
ROOF				
FLOOR				

Problem 23-3b Use the attached drawings to detail the required connections for the mezzanine floor framing plan that was started in problem 22-3b (and problem 16-3). Choose a scale suitable for clearly showing the required information. Assume the use of 1 1/8 tongue-and-groove plywood with nailing per the roof sheathing. Keep in mind that the information in the attached drawings is incomplete and may contain errors. Refer back to the order of precedence to accurately complete the following drawings.

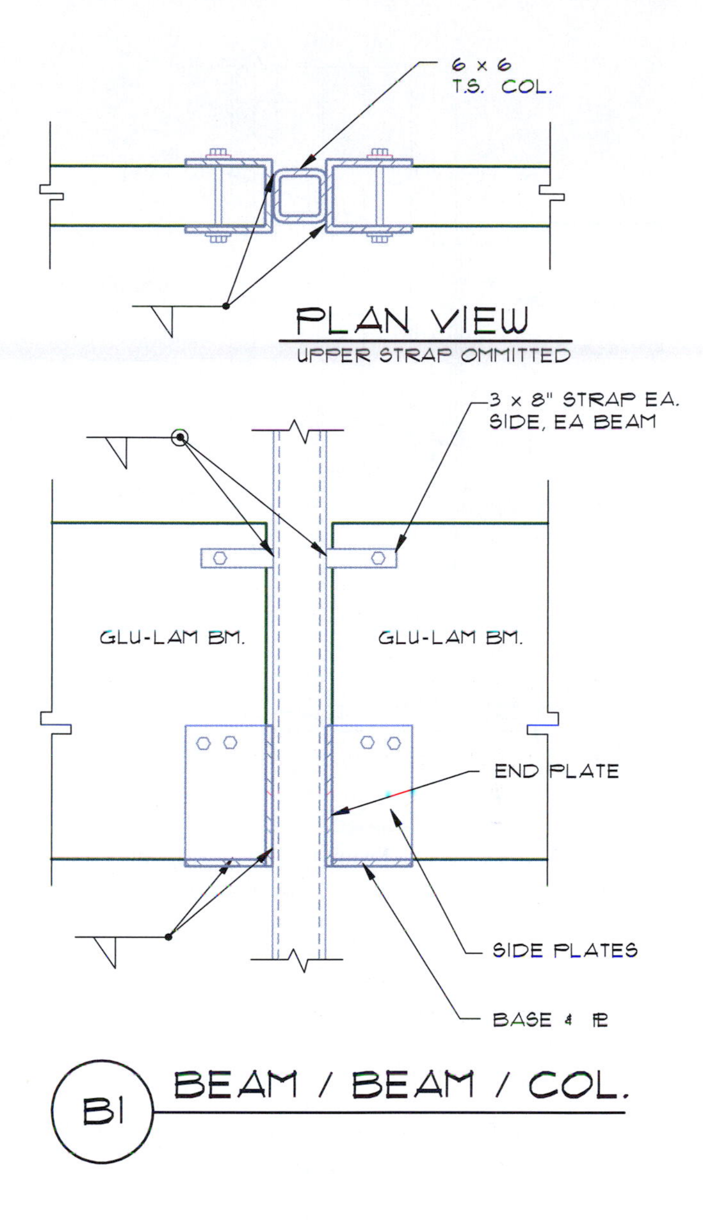

Floor Ledgers

Use 4 × 12 DFPT ledgers with 3/4" Ø anchor bolts at 24" o.c. at grids A and F and placed at 48" o.c. at grid 7. Stagger all bolts 3" up/down. At splices in ledger at grids A and F use (2)-144" long × 3" straps with (10)-3/4" Ø bolts 1 1/2" from ends and 15" o.c. and 10 1/2" minimum from splice. Place straps 3" down from top of ledger and 3 1/2" from center to center. At splices at wall 7, use (3)-3/16" × 3" × 36" straps at 3 1/2" o.c. with four bolts per straps placed 1 1/2" from strap end, and 6" o.c.

Problem 23-3b.1 Beam to Column.

- Support the glu-lam beam on a 3/4" × 9" × 6 7/8" bearing plate.
- Provide a 7/16" × 9" × 15" side plate on each side of the beam. Lap the side plates over the side of the base plate.
- Provide a 7/16" × 15" × 7 3/4" end plate.
- Weld the side plates to the base and end plate with 3/8" fillet welds.
  - Weld the base plate to the 6" × 6" × 3/16" column with 3/8" fillet weld, all around.
  - Bolt the side plates to the beam with (2)-3/4" bolts placed at 3" o.c., 2" down, and 2" from edge of side plates.
- Provide a 9" × 9" × 3/4" gusset plate on each side of the column to support the base plate. Weld with 1/4" fillet welds.
- Provide a 24" × 3" × 1/4" strap @ each side of the beam with (1)-3/4" Ø bolt through the beam, 3" down from top of the beam, 2" from the end of strap.
  - Weld the side of the strap to the column with a 1/8" fillet weld.

Problem 23-3b.2 Beam to Fire Wall.

- Use detail 23-3b.1 as a guide for weld, plates, straps, gusset, and bolting sizes.
- Complete the fire wall detail to match detail 23-3a.3.
- Provide a 12" × 3" × 1/4" strap @ each side of the beam with (1)–3/4" Ø bolt through the beam, 3" down from top of the beam, 2" from the end of strap.
- Support the flooring at the fire wall with a 3 × 12 ledger attached to the stud wall with 5/8" Ø lag bolts at each stud, staggered 2" up/down.

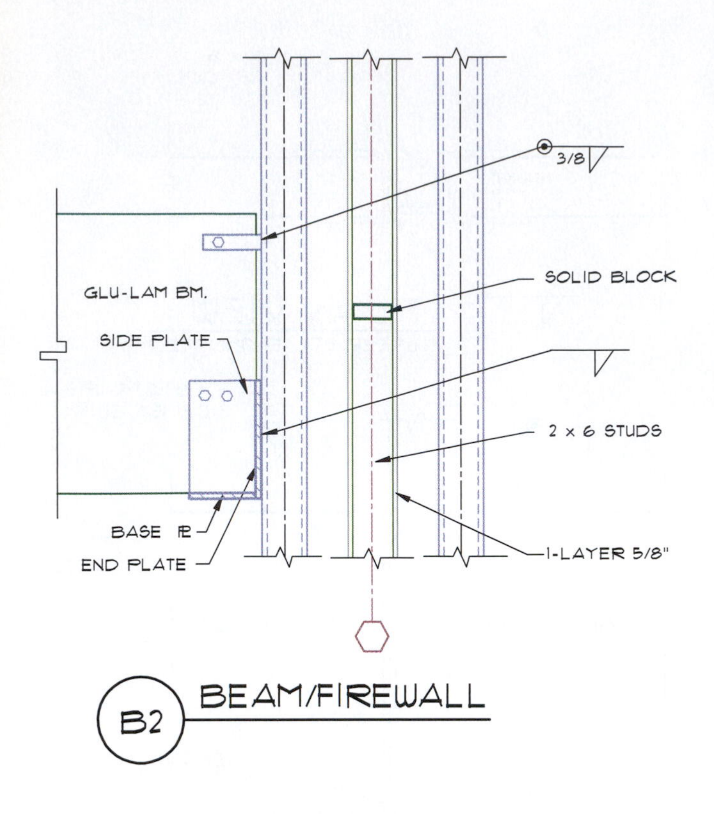

Problem 23-3b.3 I-Joist to Beam. Use an appropriate HIT hanger to connect the I-joists to the glu-lams.

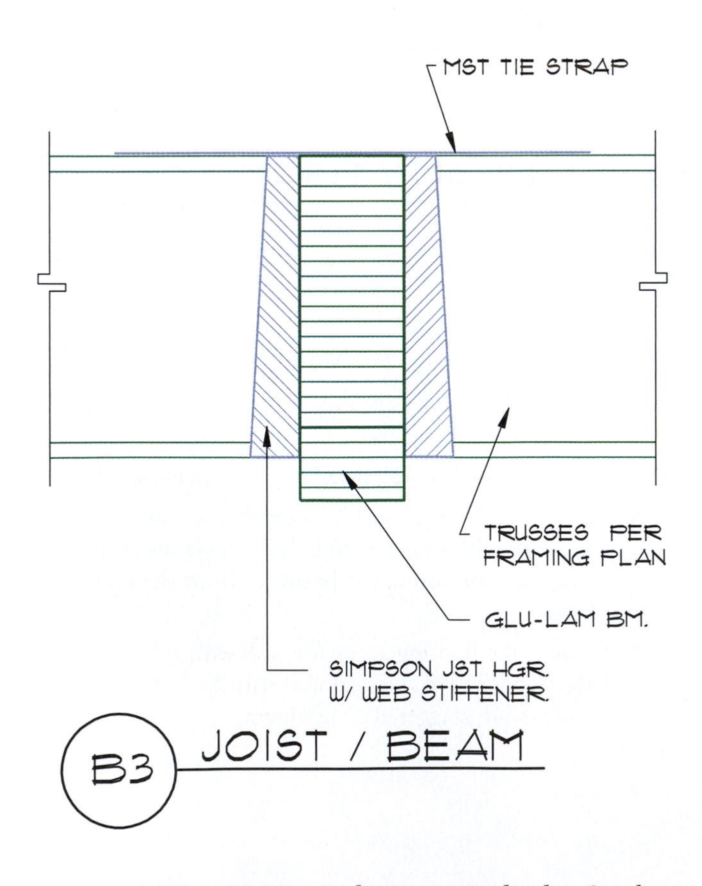

Problem 23-3b.4 I-Joist to Ledger—Perpendicular. See the floor ledger section at the start of the mezzanine notes to specify the ledger size and bolting. Use an appropriate HIT hanger, to connect the I-joists to ledger. Connect the assembly to the wall similar to detail 23-3a.6.

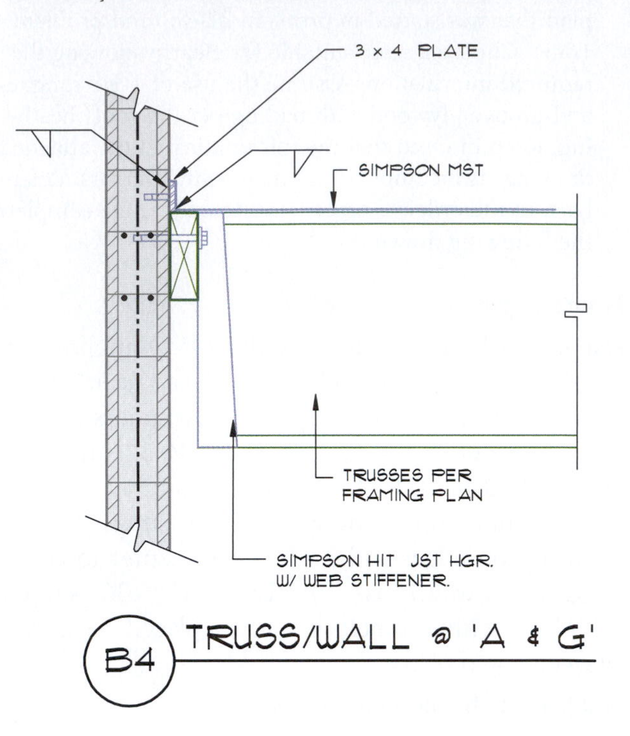

Problem 23-3b.5 I-Joist to Ledger—Parallel.

- See the wall sections for steel locations.
- See the floor ledger notes to specify ledger size and bolting.
- Use an appropriate hanger to connect the blocking between the ledger and the truss. Connect the assembly to the wall similar to detail 23-3a.7.

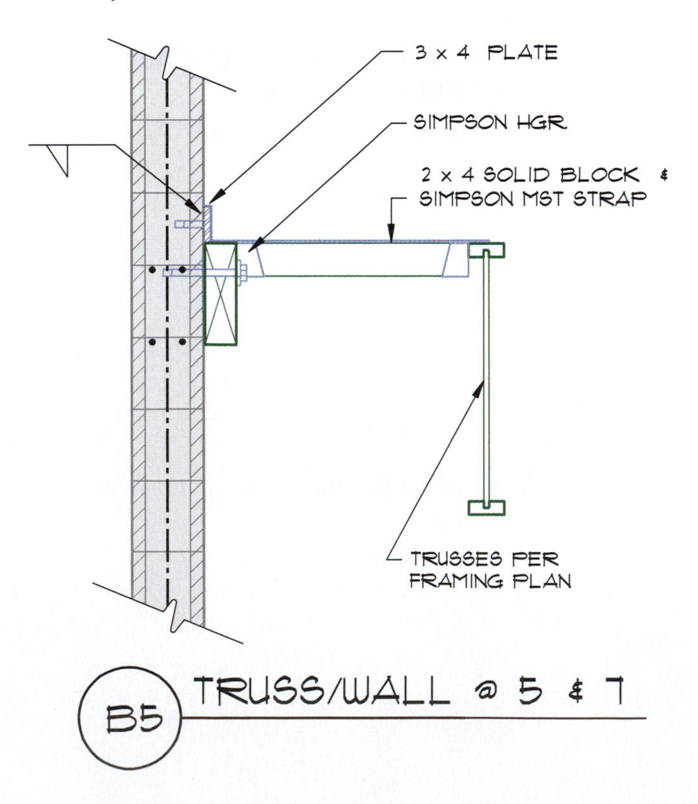

Problem 23-3b.6 Ledger Splice—Bearing Wall.

- Use a 4 × 12 ledger with bolting as per detail 23-3a.9.

Problem 23-3b.7 Ledger Splice—Nonbearing Wall.

- Use a 4 × 12 ledger with bolting as per detail 23-3a.10.

Problem 23-4 Use the attached drawings to detail the required roof connections for the roof framing plan that was started in problem 22-4a (and problem 16-4). Choose a scale suitable for clearly showing the required information. Assume all welds to be 3/16" fillet unless noted. Submit preliminary drawings to your instructor prior to completing the drawings. Keep in mind that the information in the attached drawings is incomplete and may contain errors. Refer back to the order of precedence to accurately complete the following drawings. Use the framing plan to determine all beam and truss sizes and locations. Specify the size of each in each detail.

Problem 23-4a.1 Beam to Wall. For grids 3C and 3K (3A and 3C similar @ sales). Use the attached drawings to complete the following details:

Lower Connector

- Provide 1/4" minimum clear from the beam to the wall. Support the glu-lam beam on a 6 7/8" × 10" × 3/4" base plate.
- Provide 10" × 10" × 16"h × 5/16" angle placed on each side of beam, and weld to the base plate w/ 5/16" fillet.
- Provide (3)-3/4" Ø bolts through the angles and beam, placed 2" minimum from the top of the plate and 2" from the outer plate edge. Place the other (2) bolts at 5" o.c.
- Provide (6)-3/4" Ø × 4 1/2" taper bolts to the 6" tilt-up wall to each side angle in 2 columns, placed at 5" o.c. Place taper bolts 3" down from plate edge and at 5" o.c.

Upper Connector

- Provide an 8" × 4" × 1/2" × 3"h angle on each side of beam, 4 1/2" down from the top of beams.
  - Place (1)-3/4" Ø bolt in a 13/16" × 1 7/8" slotted hole in 8" leg of the angle, through the beam to the opposite angle.
  - Bolt the angles to the wall w/ 3/4" Ø × 4 1/2" taper bolts.

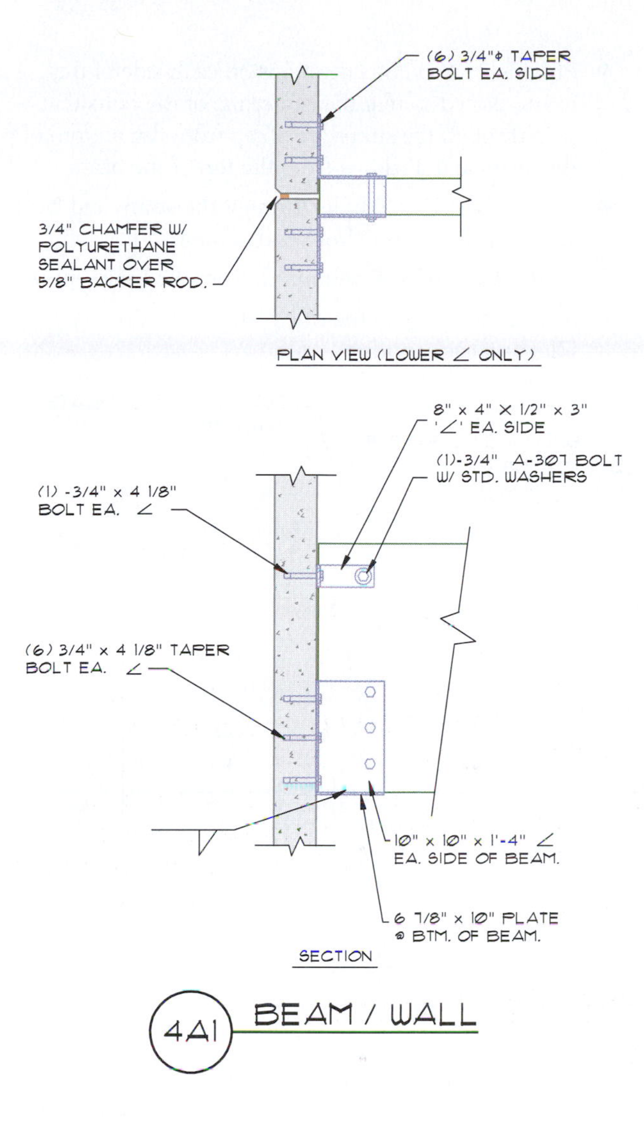

Problem 23-4a.2 Beam to Column Grid 3. Use Figure 23-3a.4 as a guide to complete this detail.

- Support the glu-lams resting on a 6 7/8" × 22" × 7/8" base plate.
- Provide a 10"h × 22" × 5/16" side plate on each side of the beam, welded to the base plate with 1/4" fillet.
- Weld base plate to the column w/ 1/4" fillet. Provide (4)-3/4" Ø bolts through the beam. Place bolts 1 5/8" minimum from the plate edges and at 3" o.c.

Problem 23-4a.3 Beam to Beam with Saddle.

- Support the beam on 7/8" × 6 7/8" × 8" long top and base plates.
  - Dap (cut into the beam) the top plate into the top of the beam.
- Attach 5/16" × 8" side plates on each side of the beam to the top and bottom plates.

- Provide 3" × 5/16" side straps on each side of the beam, placed so that the centerline of the bolts that pass through the straps are 4" up from the bottom of the beam and 3" down from the top of the beam.
- Use (3)-3/4" Ø × 8 bolts through the straps and beam @ 3" o.c., and 1 1/2" from end of strap.
  - Bolts are to be 8" minimum from the beam splice.
- Attach the straps to the side plates with 5/16" × 3" fillet and back welds.

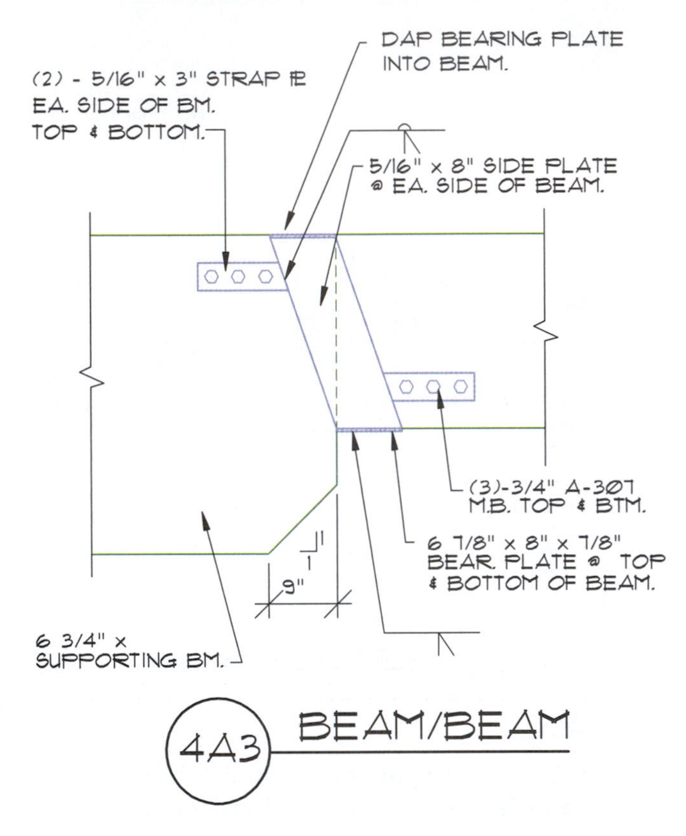

Problem 23-4a.4 Beam to Beam at Fire Wall. Grid 3F.

- Use an appropriate CCO column cap to connect the glu-lams to the 6 × 6 TS columns.
  - Connect the CCO to the columns with 3/8" fillet weld, all around.
- Center the columns so that they are 12" o.c. from the center of the 2 × 6 fire wall.
  - Protect the wall with 2 layers of 5/8" type X gypsum board each side.
  - Provide 3 × 6 top plate over (2)-2 × 6 plates. Match the bolt placement per detail 23-3a.3.
- Provide 3/4" minimum clearance between the ends of the beams to the gypsum board.
- Protect the roof on each side of the wall with 5/8" type X gypsum board on bottom side of the truss top chords for 60" minimum on each side of the wall.
  - Provide 2 × 4 solid block @ 24" o.c. between top chords to support the blocking.
- Label unspecified materials per detail 23-3a.3.

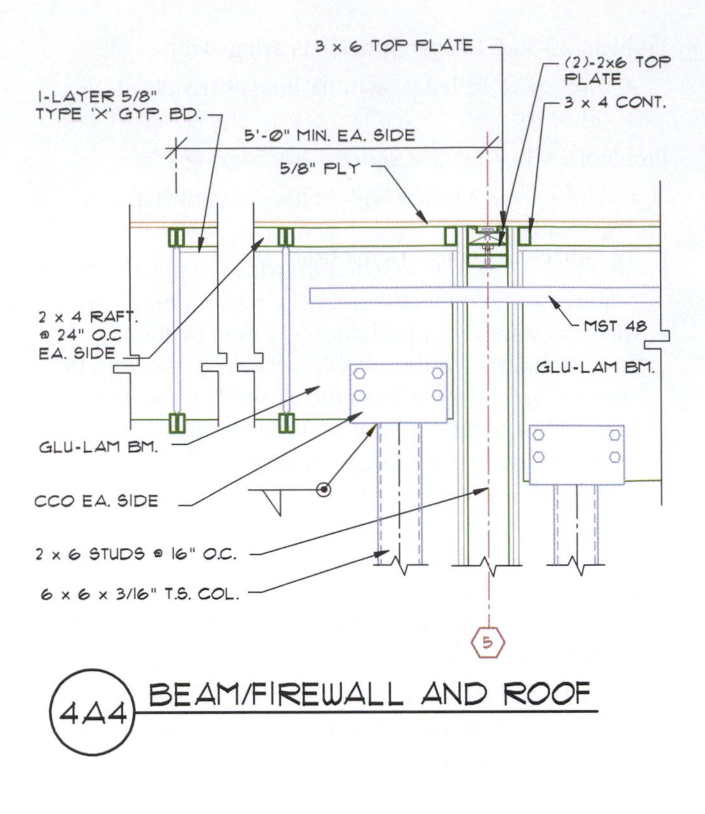

Problem 23-4a.5 Truss to Beam, Grid 3.

- In addition to truss clips supplied by the manufacturer, provide 2 × 4 solid blocking between the trusses on the top of the beam.
  - Anchor the blocks to the glu-lam beams with A35 anchors on each end of block.
- Provide MST48 straps at 48" o.c. that align with the straps at roof/wall intersection.

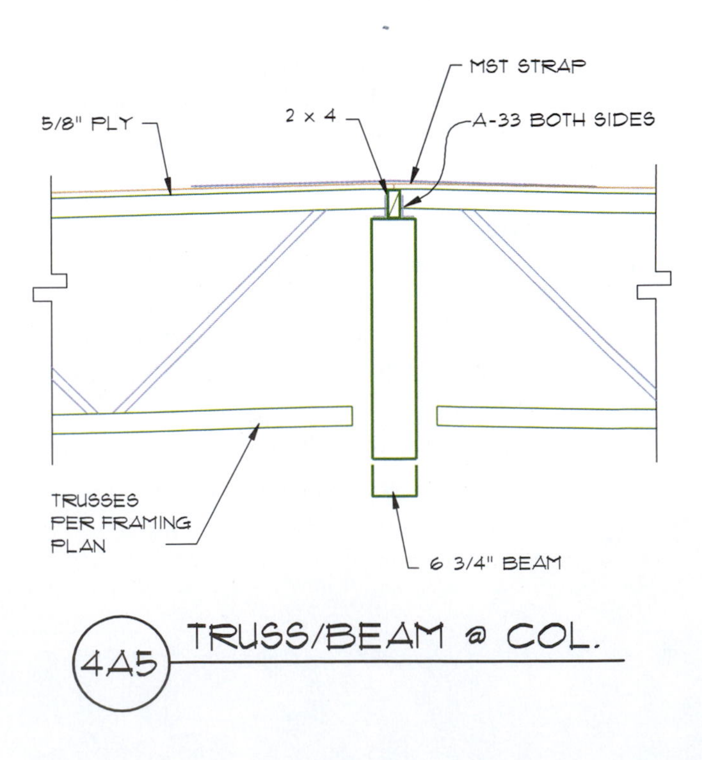

Problem 23-4a.6 Truss to Beam at Column. Grids 3D+12', 3H, and 3J+12'. Use the drawing created in problem 23-4a.5 as a base to complete this drawing.

- Provide 2 × 4 brace at 32" o.c. placed at 45° for a minimum distance of (2) truss spaces on each side of column.
- Attach the braces to the beam with U24 hangers, placed 2" up from bottom of beam.
  - Block the upper end of the brace with 3 × 4 × 10' minimum long continuous nailer attached to the bottom of the top chord of the trusses with (2)-16d nails at each truss.
  - Provide 3 × 4 solid block equally spaced along the braces and nailed to the brace with (5)-16d nails.
- Provide 2× blocking above the beam between the trusses per detail 23-4b.5. Provide an MST60 strap from truss to truss over the beam.

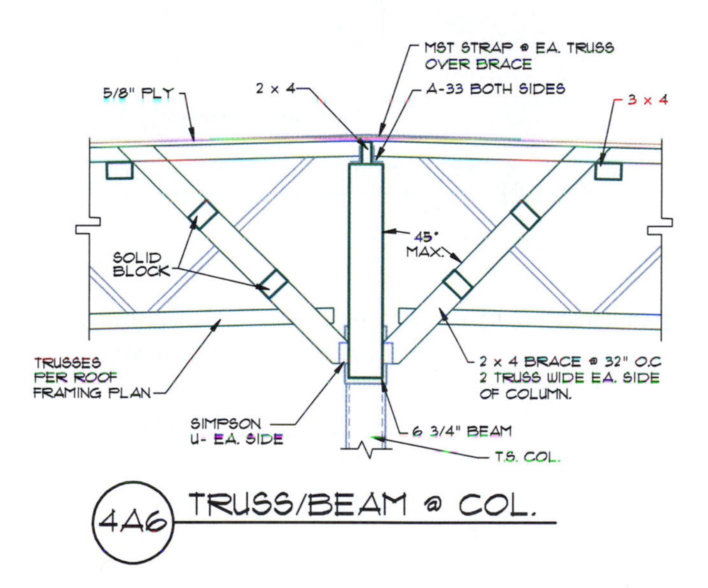

Problem 23-4a.7 Truss to Beam at Diaphragm Edge.

- At the edge of roof blocking used to form the roof diaphragm (36' minimum out from the walls at grids C/K), provide (2) Simpson Company HD-2 anchors (total of 4) placed on each side of the truss.
  - Attach the anchors to the truss with (2)-5/8" × 4 1/2" bolts through the trusses.
  - Locate the anchors so that that the bolt closest to the splice is 8" minimum from the truss splice.
  - Provide a 5/8" Ø ×16" bolt across the splice from anchor to anchor.
- Provide solid blocking between the trusses similar to detail 23-4a.5. Maintain 6" minimum space from the blocking anchors to the end of block.

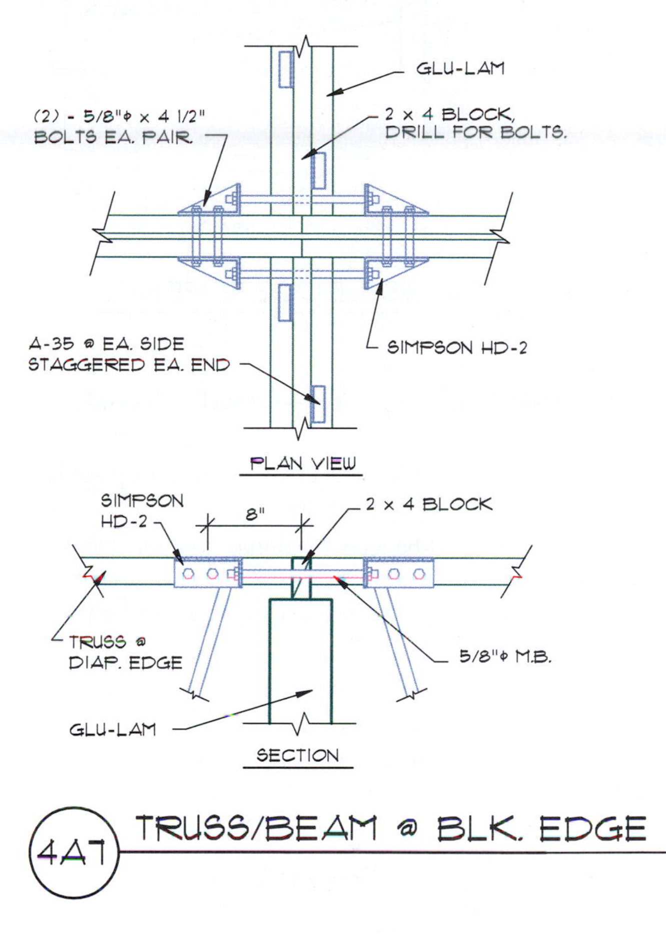

Problem 23-4a.8 Truss to Ledger—Perpendicular. Grids 1 and 5.

- Provide 3 × 12 DFPT ledger attached to the wall with 3/4" Ø ×10" anchor bolts placed at 32" o.c.
  - Stagger the bolts along the ledger so that the bolts are 3" up/down from the ledger edges.
- Provide 3" × 4"h × 5/8" plate placed at 8'–0" on center in line with the trusses.
  - Attach the plate to the wall w/ 3/4" Ø 4 1/8" power-driven bolts.
- Weld an MST37 strap to the plate with 1/8" fillet welds and attach to the trusses per the manufacturer's nailing instructions.
- Provide 2× blocking placed above the ledger between trusses per detail 23-4a.5.

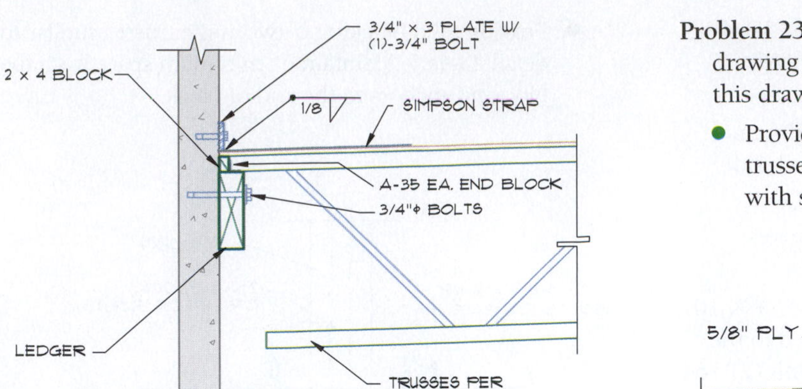

Problem 23-4a.9 Truss to Ledger—Parallel. Grids A, C (both sides), and K.

- Provide 3 × 12 DFPT ledger w/ 3/4" Ø bolts placed at 60" o.c. staggered 3" up/down.
- Select suitable U-hangers to connect the solid blocking to the trusses.
- Provide a 3" × 4"h × 5/8" plate w/ 3/4" Ø × 4 1/8" power-driven bolt.
- Weld an MST27 strap to the plate with 1/8" fillet weld @ 8'–0" o.c.
- Use detail 23-3a.6 as a guide for unlisted materials.

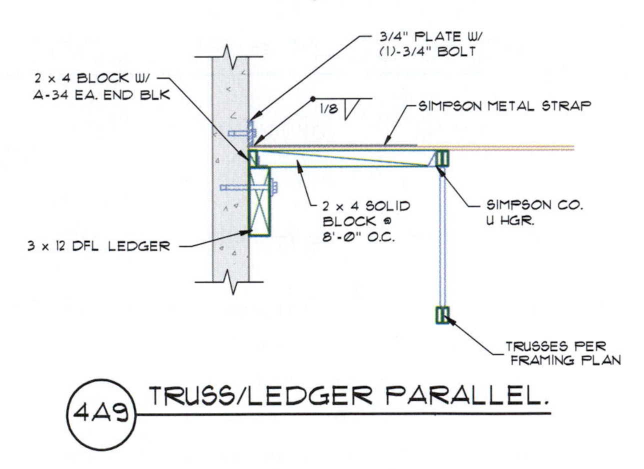

Problem 23-4a.10 Ledger Splice—Bearing Wall. Grids 1, 2, 4, and 5.

- Use the roof ledger notes from problem 22-4a to complete a detail for the ledgers to show floor and roof usage for bearing conditions.
- Use detail 23-3b.9 as a base to complete the drawing.

Problem 23-4a.11 Ledger Splice—Nonbearing Wall. Grids A, C (both sides), and K.

- Use the roof ledger notes from problem 22-4b to complete a detail for the ledgers to show floor and roof usage for nonbearing conditions.
- Use detail 23-3b.10 as a base to complete the drawing.

Problem 23-4a.12 Truss/Beam—Trusses. Grid 3. Use the drawing created in problem 23-4a.5 as a base to complete this drawing.

- Provide 2 × 4 solid blocking laid flat between the trusses. Provide MST48 straps at each 48" o.c. Align with straps at wall.

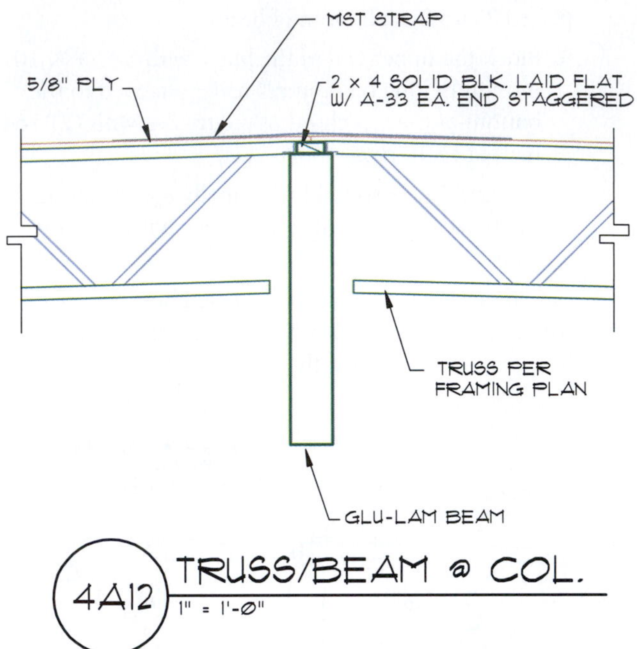

Problem 23-4a.13 Truss/Beam—Trusses Perpendicular. Grids 2 and 4. Use the drawing created in problem 23-4a.8 as a base to complete this drawing.

- Provide 3 × 12 DFPT ledger w/ 3/4" Ø bolts at 32" o.c. staggered 3" up/down.
- Provide 3" × 4"h × 3/4" plate w/ 3/4" Ø × 4 1/8" power-driven bolts.
- Weld MST37 to plate w/ 1/8" fillet weld @ 8'–0" o.c.
- Provide 2 × 4 blocking laid flat between the trusses.

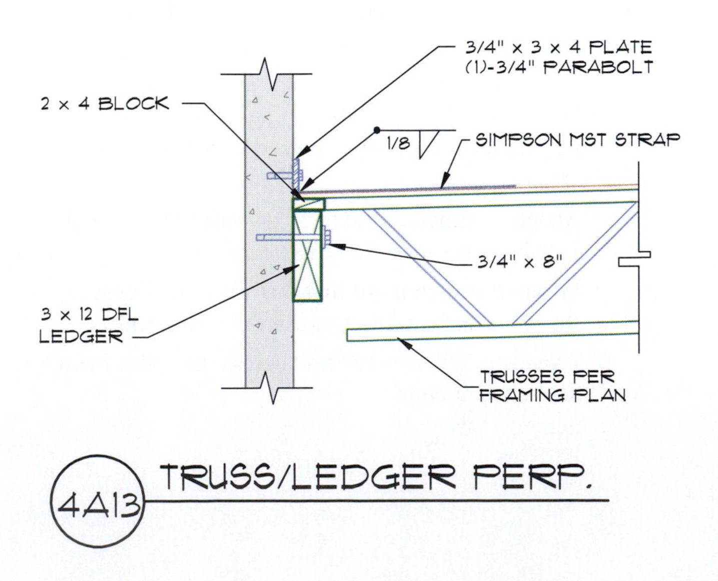

Problem 23-4a.14 Truss/Ledger—Tresses—Parallel. Grids A and C. Use the drawing created in problem 23-4a.9 as a base to complete this drawing.

- Provide 3 × 12 DFPT ledger with 3/4" Ø bolts at 60" o.c. staggered 3" up/down.
- Select suitable U hangers for the solid blocking.
- Provide 3" × 4"h × 3/4" plate with a 3/4" Ø × 4 1/18" power-driven bolt.
- Weld MST27 to plate w/ 1/8" fillet weld @ 8'–0" o.c.

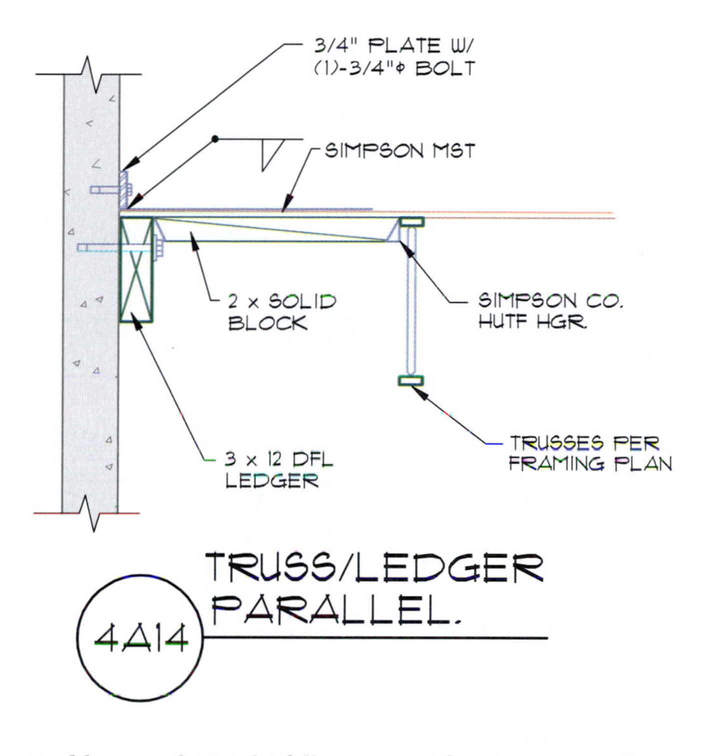

Problem 23-4b Use the following specifications to complete the required floor connections. Choose a scale suitable for clearly showing the required information. Use the appropriate drawing from problem 23-3 as a guide for unlisted information. Keep in mind that the information in the attached drawings is incomplete and may contain errors. Refer back to the order of precedence to accurately complete the following drawings.

Floor Ledgers

- Use 4 ×12 DFPT ledger with 3/4" Ø anchor bolts at 24" o.c. at grids 1 and 5.
- Staggered bolts 3" up/down, spaced at 48" o.c. at grid C.
- At splices in ledger at grids 1 and 5 use (2)-144" long × 3" straps with 3/4" bolts 1 1/2" from ends and 20" o.c.
- Place straps 3" down from top of ledger and 3 1/2" from center to center.
- Use (3)-MST48 straps at 3 1/2" o.c. at splices at wall C with (8) bolts per strap.

Problem 23-4b.1 Beam to Column—Floor.

Lower Connection

- Provide 1/4" clear between the beam and the end plate.
- Support the glu-lam beam with a 3/4" × 9" × 6 7/8" bearing plate.
- Provide a 7/16" × 9" × 15" side plate that laps each side of the base plate.
- Provide a 7/16" × 15" × 6 7/8" end plate.
  - Weld side plates to the base and end plates, and to 6 × 6 column with 3/8" fillet welds at each connection.
- Bolt the side plates/beam with (4) 3/4" bolts that are 1 1/2" down from the top of the side plate, and 1 1/2" in from edge of plate and 3" o.c.
- Provide a 9" × 9" × 3/4" gusset with 1/4" fillet weld each side to CCO/TS.

Upper Connection

- Provide an 8" × 3" × 6 7/8" U × 7/16" strap w/ (1)-3/4" Ø bolt through the beam and U strap.
  - Place the centerline of the bolt/strap 3" down from top of beam.
  - Set bolts 1 1/2" from end of strap.
- Weld the side of the U-plate to the TS w/ 3/16" fillet weld.

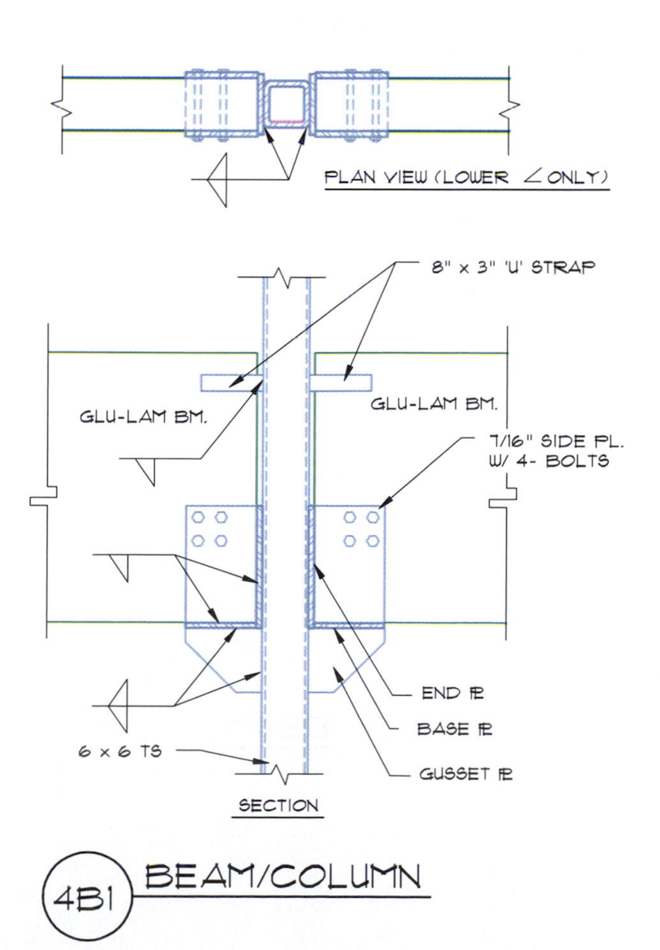

Problem 23-4b.2 Beam to Wall. See detail 23-4a.1. Copy and edit the roof beam/wall detail as required to reflect floor the conditions.

Problem 23-4b.3 Beam to Fire Wall. Use detail 23-3b.2 as a guide to complete a detail for this structure.

Lower Connection

- Weld the side plates to the base and end plate to 6 × 6 column with 3/8" fillet welds.
- Provide (2)-3/4" bolts through the side plates and beam that are 2" down, and 2" from edge of side plate and 3" o.c.
- Provide an 9" × 9" × 3/4" gusset with 1/4" fillet weld on each side of the gusset plate to the CCO/TS.

Upper Connection

- Provide a 3" × 12" × 1/4" strap at each side of beam with (1)-3/4" Ø bolt through beam.
  - Place the bolt 3" down from top of beam, and 2" from end of strap.
  - Weld the strap to the TS with 3/8" fillet weld at top and bottom of strap, each side.
- Weld the side of the plate to the TS with 3/8" fillet weld at each.

Problem 23-4b.4 Truss to Beam. Grids 3C to F.

- Provide 2 × 4 solid block laid flat between the trusses.
- Provide an MST37 strap at 96" o.c. Align the wall straps with the ledger straps.

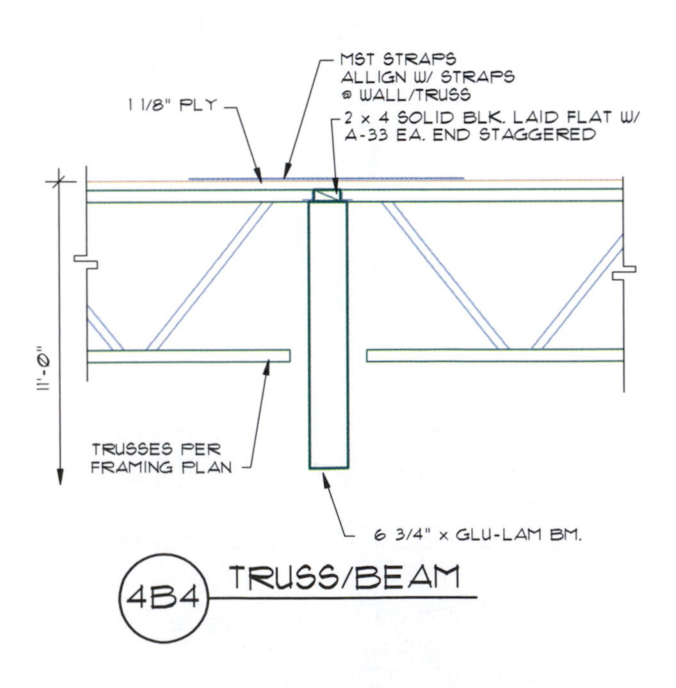

Problem 23-4b.5 Truss to Wood Wall. Grid 3+16' between C to F.

- Extend 1/2" sheet rock to from the floor to 10' above the floor and show suspended ceiling on the display side.
- Provide 2 × 4 solid block laid flat between the trusses over (2)-2 × 4 top plates.
- Provide an MST37 at 96" o.c. Align the wall straps with the ledger straps.

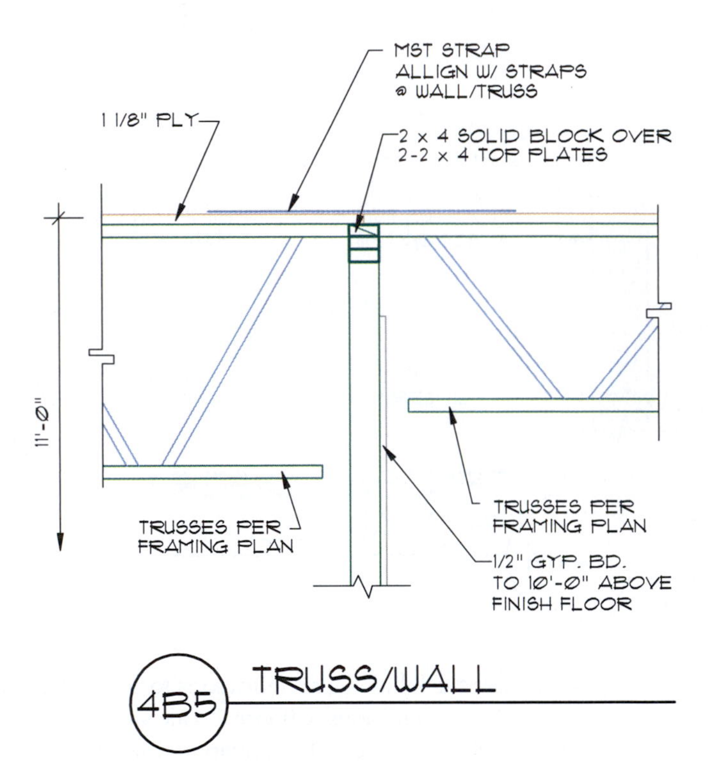

Problem 23-4b.6 Truss to Ledger—Perpendicular. Grids 1 and 5. Use the drawing created in problem 23-4a.13 as a base to complete this drawing.

- Provide 4 × 12 DFPT ledger w/ 3/4" Ø bolts at 24" o.c. staggered 3" up/down.
- Provide 3" × 4"h × 3/4" plate with 3/4" Ø × 4 1/8" power-driven bolts.
- Weld an MST37 strap to plate w/ 1/8" fillet weld at 8'–0" o.c.
- Provide 2 × 4 blocking laid flat between the trusses.

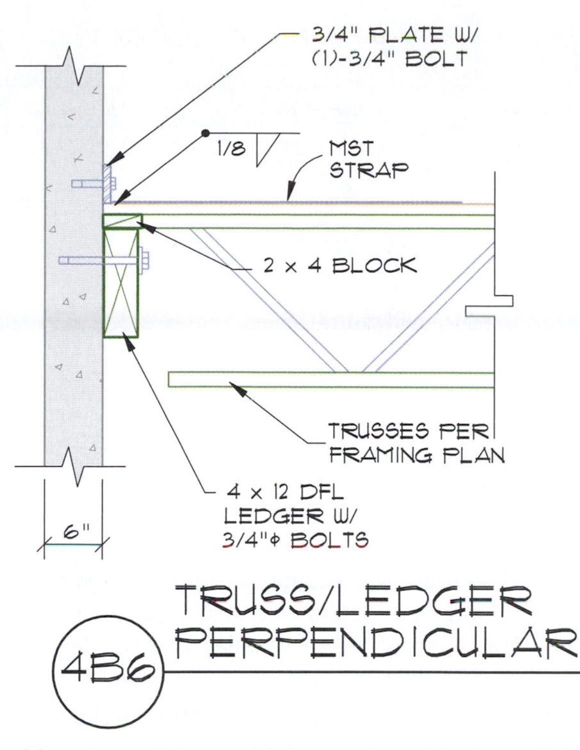

Problem 23-4b.7 Truss to Ledger—Parallel. Grid C (similar to detail 23-4a.14).

- Provide 4 × 12 DFPT ledger with 3/4" Ø bolts @ 48" o.c. staggered 3" up/down.
- Provide a 3" × 4"h × 3/4" plate w/ 3/4" Ø × 4 1/8" power-driven bolt.
- Weld an MST27 strap to the plate with 1/8" fillet weld at 8'–0" o.c.
- Provide 2 × 4 blocking between the ledger and the truss @ 8'–0" o.c.
- Select suitable U hangers for the block.

Problem 23-4b.8 Truss to Fire Wall.

- Provide 3 × 12 DFPT ledger over 5/8" gypsum board.
- Provide solid blocking in the wall behind the ledger.

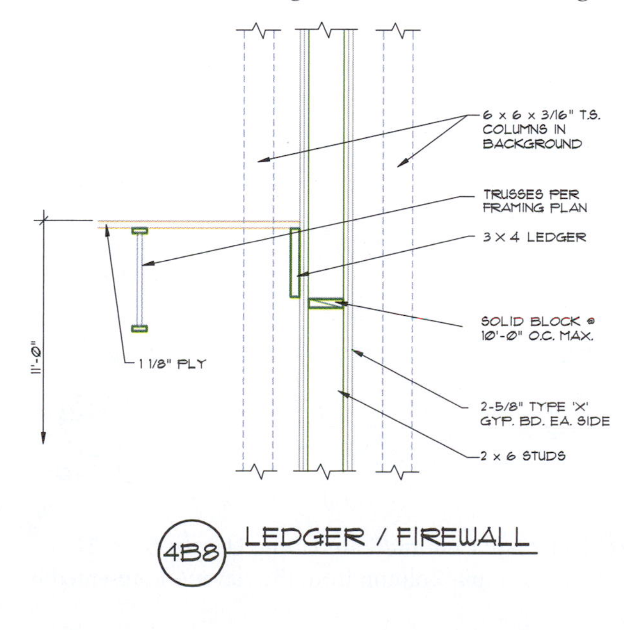

Problem 23-4b.9 Ledger Splice—Bearing Wall.

- Use the floor ledger notes to complete this detail.
- Establish a detail to show floor and roof usage. See detail 23-3b.6.

Problem 23-4b.10 Ledger Splice—Nonbearing Wall. See floor ledger notes. Establish a detail to show floor and roof usage. See detail 23-3b.7.

Problem 23-5 Panel Elevation. Draw the north and west panel elevations for the main structure drawn in problem 16-4. Specify that the office elevations will be separate elevations.See the foundation and site plans to determine the depth of the foundation below the finish floor. Refer to the student CD in the RESOURCE MATERIAL/GENERAL SPECIFICATIONS folder for additional information related to the concrete panels. Break out a section of the panel to display:

- Grade 60 #5 Ø vertical bars at 10" o.c.
- Grade 40 #4 Ø horizontal bars at 12" o.c.
- Steel centered in the wall and extended to be within 1 1/2" of all panel edges.

Specify that all panels are to be 6" thick, 4000 psi concrete with an exposed finish.

Problem 23-6 Using the guidelines for the panels on problem 23-5a and 5b, design and draw the panel elevations for the panels located on the south and east walls of the main warehouse. Refer to the student CD in the RESOURCE MATERIAL/GENERAL SPECIFICATIONS folder for additional information related to the concrete panels. Show and dimension all openings. Verify foundation locations with information provided in Chapter 24.

Problem 23-7 Panel Elevation. Use the attached drawing, information in problem 23-5, and the following information to complete a panel elevation for panel 1-E-D. Refer to the student CD in the RESOURCE MATERIAL/GENERAL SPECIFICATIONS folder for additional information related to the concrete panels. Provide dimensions for the entire panel as well as for each opening. In addition to the normal steel, specify the following reinforcing steel to be 1" clear of the exterior panel face:

- (2)-grade 40, #5 Ø × 48" diagonal steel bars at each corner of each door 2" clear of opening and 8" o.c.
- In addition to the typical wall steel, for 3'–0" × 7'–0" door:
  - Provide (2)-grade 60, #6 Ø at 6" o.c. vertically for full height of panel.
  - Provide (2)-horizontal grade 40, #5 Ø × 7' long rebar at each corner of the door at 8" o.c.
- For the 8'–0" door:
  - Provide (2)-grade 40, #5 Ø vertical bars at 8" o.c. on each side of door, extending the full panel height.

- Use (3)-grade 60, #4 Ø horizontal bars at 12" o.c. above and below door.
- Extend horizontal steel 24" past each side of door.

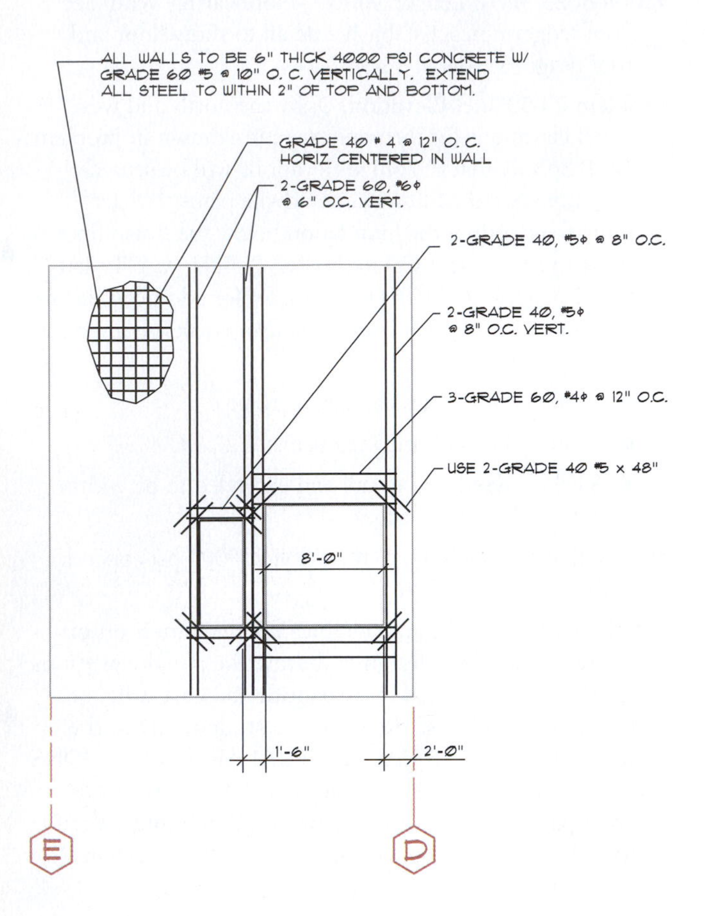

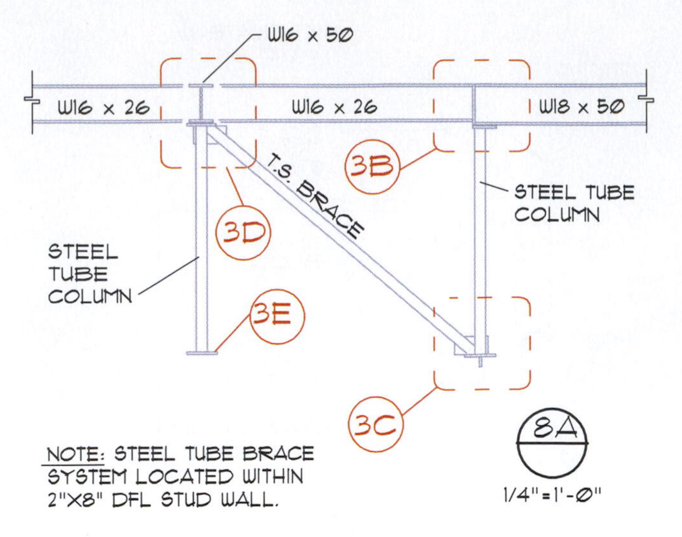

Problem 23-8b Beam/Beam/Column. Support the W18 × 50 beam from the elevation presented in problem 23-3a on a 7 1/2" × 9 × 1/2" steel cap plate bolted to beam with 2 machine bolts.

- Weld the plate to a TS 5" × 5" × 3/16" steel tube column with 3/16" fillet all around weld.
- Hang a W16 × 26 from the beam with 6" × 9" × 5/16" steel connection plate welded to the W18 × 50 with 1/4" fillet with 2" return at the top and bottom.
  - Bolt the plate to the W16 × 26 beam with (3)-3/4" Ø machine bolts through plate and web.
  - Bolts to be 1 1/2" from each edge and 3" o.c.
  - Set top bolt 5" down from the top flange.

Provide a note to indicate that all bolts are to be A325-N high-strength bolts and require special inspection.

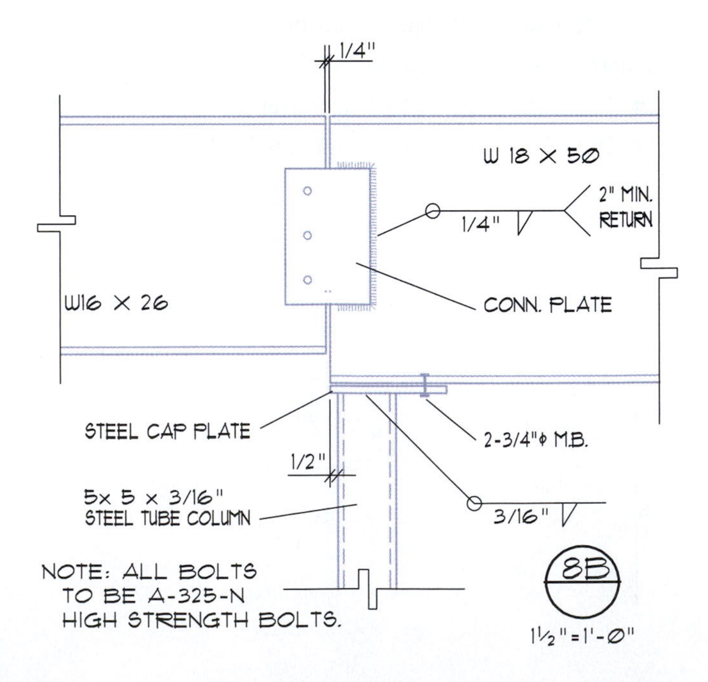

Problem 23-7 Using the guidelines for the panels on problem 23-3a or 23-3b, design and draw the panel elevations for the panels located on grids 1, 2, 4, 5, and A, C, and K for this structure. Refer to the student CD in the RESOURCE MATERIAL/GENERAL SPECIFICATIONS folder for additional information related to the concrete panels. Show and dimension all openings. Verify foundation locations with information provided in Chapter 24.

Problem 23-8 Use the drawings provided as a guide to complete the details required to explain construction of the shear wall referenced in problems 23-8a through 23-8e. Details 23-8f and 23-8g do not relate to detail 23-8a. For each detail, provide a plan view of each plate to locate bolting patterns. Use information in problems 23-8b through 23-8g to complete each drawing.

Problem 23-8a Draw an elevation of the steel bracing system used in a wall using a scale of 1/4" = 1'–0". Assume a distance of 12'–0" from center to center of columns, and 10'–0" from floor to floor.

Problem 23-8c Column/Foundation. Use the 5" × 5" × 3/16" steel tube column from the elevation presented in problem 23-8a welded to the base plate with 5/16" fillet welds all around at field.

- The plate is to be a 12" × 12" × 7/8" steel base plate on 1" dry pack with (4)-1 1/8" × 18" long anchor bolts into footing.
  - Base plate bolts to be 15/16" minimum from the edge of plate.
- Below the base plate weld a 1 1/2" × 4" × 12" long shear plate with 1/2" × 12" fillet weld each side.
  - Set shear plate in 3 1/2" × 5 1/2" deep × 18" long keyway filled with dry pack.
- Provide a 5" × 5" × 1/4" steel tube brace between each steel column.
  - Weld the brace to the base and the column with a 3/8" fillet weld all around.
  - Provide a 14" × 10" × 3/8" steel gusset plate. Extend the gusset plate 1/2" past the backside of the column.
  - Weld the plate to the column with a 1/4" × 9" fillet weld on the back and side of the column.
  - Weld the plate to the brace with a 3/8" × 3 1/2" fillet weld on each side.

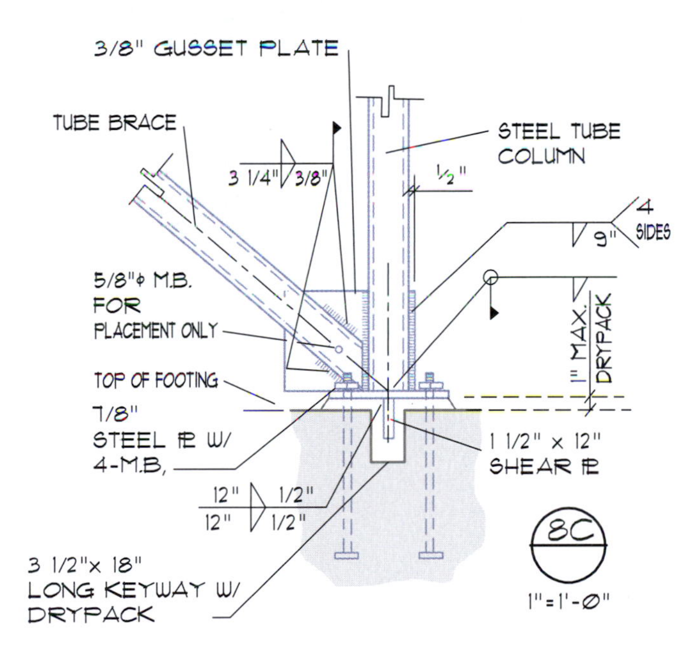

Problem 23-8d Beam/Beam/Column. Use a TS 5" × 5" × 1/4" column from the elevation presented in problem 23-8a to support a W16 x 50 beam bolted to a 9 1/2" × 24" × 1 1/4" steel cap plate with (2)-5/8" diameter machine bolts.

- Weld the beam to the plate with a 3/16" fillet weld.
- Butt the W16 × 26 beams into each side of the W16 × 50 and support on the cap plate.
- Provide a 12" × 4" × 3" angle (not shown) each side of each W16 ×26 beam, and bolt to the supported beam with (4) 5/8" Ø bolts at 3" o.c. and 1 1/2" from each edge.
  - Weld each angle to the carrier beam with 3/16" × 12" fillet weld.
- Bolt each beam flange with (4)-A325N high-strength bolts to the plate at 3" o.c., 1 1/2" from the end of the beam.
- Weld the plate to column with 5/16" fillet weld all around.
- Provide a 14" × 10" × 3/8" gusset plate and weld the gusset plate to the column and base plate with 1/4" × 9" fillet weld on each side.
- Weld the TS 5" × 5" × 1/4" steel tube brace to the column with 3/16" fillet and 1/4" × 6" fillet on each side of the gusset.

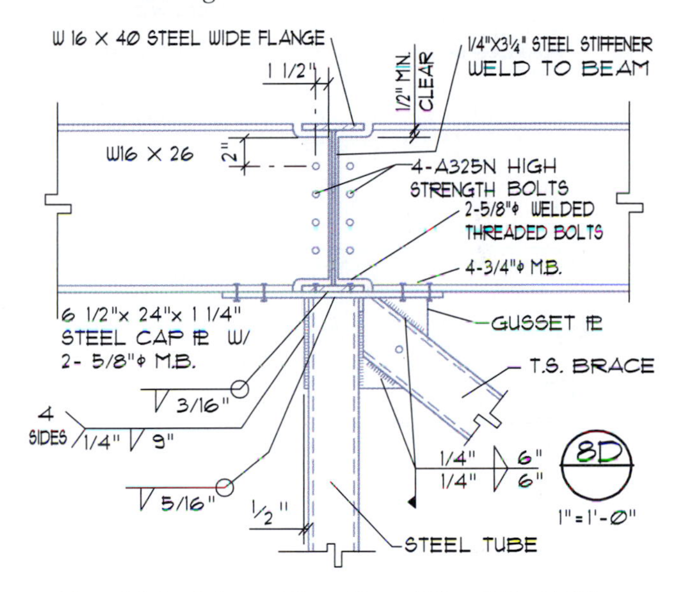

Problem 23-8e Column/Plate. Use a 10" × 10" × 7/8" steel base plate with (4) 3/4" Ø × 18" anchor bolts into foundation. Set bolts at 1 9/16" minimum from the edge of the plate. Weld the column to the plate with 1/4" fillet weld all around. Set the plate on 1" dry pack.

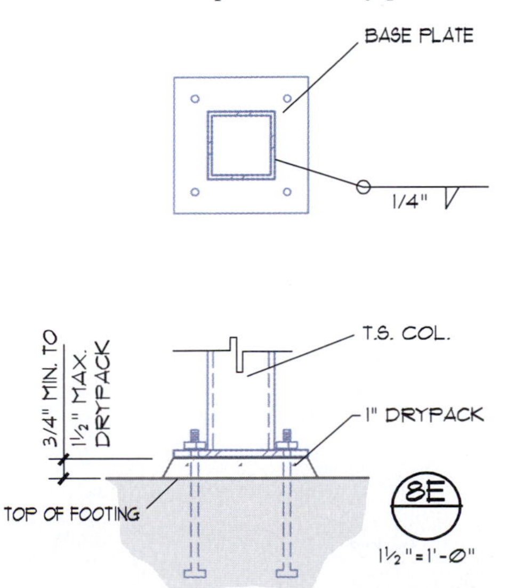

Problem 23-8f Steel Beam Support. Support a W16 × 26 beam on a 5 1/2" × 7" × 1" steel cap plate bolted to the beam with (2) 3/4" Ø machine bolts.

- Weld the plate to the TS 3 1/2" × 3 1/2" × 1/4" steel tube column with a 3/16" fillet weld all around.
- Hang the W12 × 22 from the beam with (2)-1 1/2" × 3 1/2" × 1/4" × 5 1/2" long steel angles with 2 machine bolts through the 3 1/2" legs and W12 × 22.
- Weld the angles to the W16 × 26 with 5/16" fillet weld.

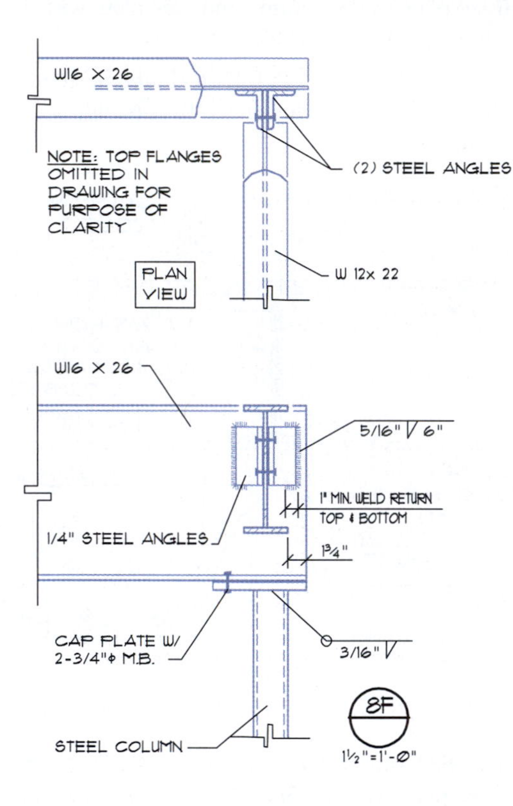

Problem 23-8g Beam/Beam/Column. Support the W12 × 22 on a 5 1/2" × 7" × 3/4" steel cap plate welded to a 3" diameter steel pipe column.

- Bolt the beam to the plate with (2) 3/4" Ø machine bolts, 1 1/2" from the edges.
- Provide a 1/4" steel stiffener plate welded all around.
- Weld a 5" × 9" × 7" gauge bucket connector to the end plate with a 3/16" fillet weld all around to support (2)-2 × 14.
- Set the 2 × 14 flush with a 3× plate connected to the steel beam.

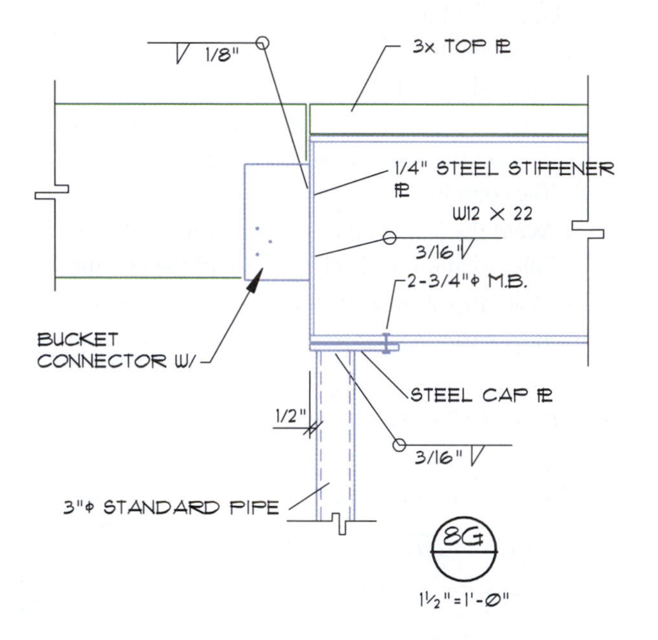

24 Foundation Systems and Components

All structures are required to have a foundation to resist gravity and lateral loads. The foundation must also be designed to resist forces from floodwaters, wind, freezing, and seismic activity. The loads from the structure must be spread throughout the soil that surrounds the foundation. This chapter will explore the basic information needed to complete foundation drawings including:

- Soil characteristics such as texture, moisture content, freezing, and compaction that must be considered prior to designing the foundation.
- Major components of various types of foundations.
- Types of floor systems that may need to be represented on the foundation drawings.
- Common types of foundation drawings including slab and foundation plans.
- Drawing steps for completing foundation plans including concrete slabs, floor joists, and post-and-beam floor systems.

Common methods of spreading the loads of the structure into the soil, common types of foundations, and concrete and wood floor systems are introduced in this chapter, as well as methods of drawing the plans to show the structural systems to be used.

SOIL CONSIDERATIONS

Before the foundation drawings are considered, the type of soil at the construction site must be determined. A soils engineer tests the soil for its strength, the effect of moisture on the **soil-bearing capacity**, compressibility, liquefaction, and expansiveness. Soil samples are taken from holes bored in representative areas of the site. Using the findings of the soils engineer, the structural engineer can design a foundation to resist the loads of the proposed structure.

Soil Texture

The texture of the soil influences each aspect of foundation design. **Soil texture** refers to the granular composition of the soil and influences the bearing capacity. The bearing capacity of soil depends on its composition and the moisture content. The IBC lists five basic classifications for soil. Soils and their allowable bearing pressure include the following:

Soil Classification	Allowable Bearing Pressure
Crystalline bedrock	12,000 psf (574.8 kPa)
Sedimentary and foliated rock	4000 psf (191.6 kPa)
Sandy gravel and gravel	3000 psf (143.7 kPa)
Sand and silty sand	2000 psf (95.8 kPa)
Clay and sandy clay	1500 psf (47.9 kPa)

Structures built on low-bearing capacity soils require footings that extend into stable soil, or footings that are spread over a wide area. Both options will be discussed later in this chapter.

Settlement

Settlement of soil will occur as the loads of the structure are placed on the footing. **Settlement** is caused by reduction of space between soil grains. Granular soils have fewer void areas and suffer only slight compaction as loads are applied. Clay-based soil has a larger percentage of voids than sandy soil and compacts more under loads. Clay soils tend to compact over long periods of time because any moisture that is trapped during construction cannot move easily through the clay. A properly designed foundation takes into account the type and depth of soil, so that the effects of settlement are not seen throughout the structure. The foundation must be designed so that settlement is evenly distributed throughout the foundation. This is accomplished by designing the structure so that the loads are distributed into uniform types of soil.

Moisture Content

The amount of water the foundation will be exposed to, as well as the permeability of the soil, must also be considered in the design of the foundation. As the soil absorbs water it expands, causing the foundation to heave. The amount of rain and the type of soil can also cause the foundation to heave. The evaporation rate of soil is also of great importance in the design of commercial foundations, because many are formed by the use of an on-grade concrete slab.

Unless the site has been properly prepared, the slab can trap large amounts of moisture. In areas that receive little rain, the contrast between the dry soil under the slab and damp soil beside the foundation provides a tendency for **heaving** to occur at the edge of the slab. If the soil beneath the slab experiences a change of moisture content after the slab is poured, the center of the slab can heave.

The heaving can be resisted by reinforcing provided in the foundation and throughout the floor slab, and by proper drainage. Surface and groundwater must be properly diverted from the foundation to maintain the soil's ability to support the building loads. Proper drainage also minimizes water leaking into the crawl space or basement, reducing mildew or rotting. The IBC requires the finish grade to slope away from the foundation at a minimum slope of 1 vertical unit per 20 horizontal units (5% slope) for a distance of 10' (3048 mm). For arid climates, the IBC allows the slope to be reduced to 1 vertical unit per 48 horizontal units.

Gravel or coarse-grain soils can be placed beside the foundation to increase percolation. Because of the void spaces between grains, coarse-grain soils are more permeable and drain better than fine-grain soils. In damp climates, a drain is often required beside the foundation at the base of a gravel bed to facilitate drainage of water from the foundation. As the amount of water is reduced in the soil surrounding the foundation, the lateral loads imposed on the foundation are reduced. The weight of the soil becomes increasingly important as the height of the foundation wall is increased. Foundation walls enclosing basements should be waterproofed. Asphaltic emulsion is often used to prevent water penetration into the basement. Floor slabs placed below grade are required to be placed over a vapor barrier.

Freezing

Because water expands as it freezes, the depth of the frost level must be considered in the design of the foundation. Expansion and shrinking of the soil causes heaving in the foundation. As the concrete expands with the soil, the foundation can crack. As the soil thaws, water not absorbed by the soil can cause the soil to lose much of its bearing capacity, causing further cracking of the foundation. In addition to the geographic location, the soil type also affects the influences of freezing. Based on the recommendations of the soils engineer and requirements of the building department, the structural engineer provides detailed information to the drafter to ensure that the foundation extends below the depth of the deepest expected frost penetration. Specifications are typically included to ensure that concrete is not poured during severe weather conditions, so that excess moisture is not trapped under the foundation.

Soil Elevation

The height of the soil above a known point is referred to as its *elevation*. Construction sites often include soil that has been brought to or moved onto the site. Terms that the drafter is required to specify on plans include natural grade, existing grade, finished grade, excavation, fill, backfill, and compacted fill. Each is shown in Figure 24-1. The height of soil in its undisturbed condition is called the **natural grade.** The use of the term *natural grade* assumes that the soil has never been disturbed. The depth of the foundation is generally based on the depth into the natural grade. The **existing grade** is the result of adding or removing

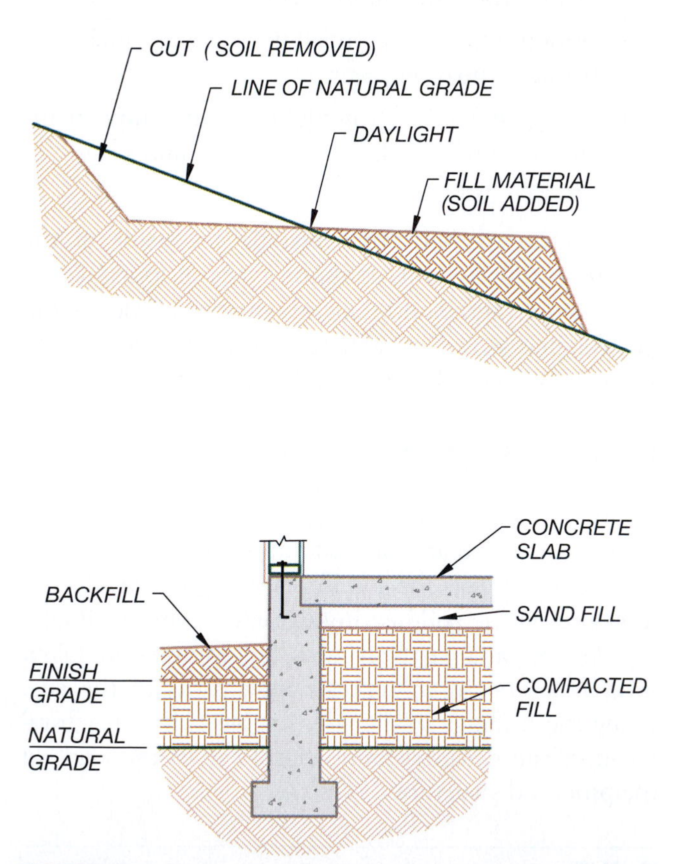

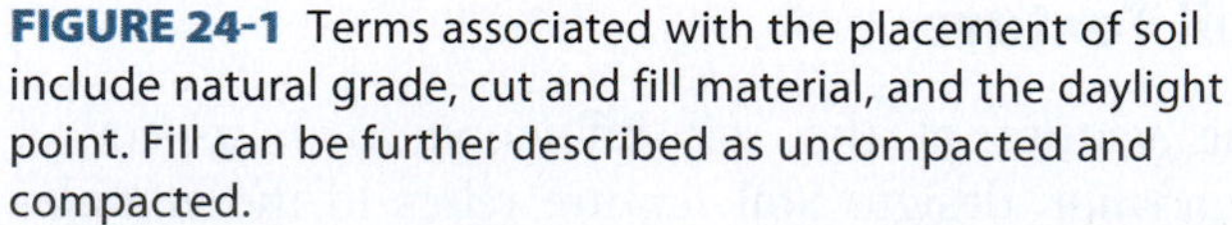
FIGURE 24-1 Terms associated with the placement of soil include natural grade, cut and fill material, and the daylight point. Fill can be further described as uncompacted and compacted.

soil from the site prior to the start of construction. Soil that was moved to the site as access roads were placed is an example of soil that forms the existing grade. **Finished grade** is used to define the height of the soil after all work at the site has been completed. The finished grade may be higher or lower than the natural grade, depending on the project design.

The removal of soil from a construction site is known as **excavation**. Figure 24-2 shows a site that has been excavated, and the forming of the foundation. Because the foundation must be placed below the maximum penetration level of frost, soil must be moved to provide a stable base for the foundation. Soil is also removed from the site to alter the contour of the finished grade. A bank created by excavation is referred to as a *cut bank*. The engineer provides specifications controlling the slope of banks adjacent to the structure.

Soil that is placed over the natural grade is called *fill material*. Once the foundation system has been formed and the floor system is in place, soil is placed in the footing trench to create the finished grade. This fill material, which is placed against the foundation, is called *backfill*. Fill can be either uncompacted or compacted. *Uncompacted fill* is soil that is brought to a site to raise the elevation of the natural grade. Although the fill may be leveled after it is placed, no special care is given to compact the material. Uncompacted fill material is unusable for supporting the loads from a structure. Where uncompacted fill is found at the building site, the foundation must be placed so that it is bearing on the natural grade. *Compacted fill* is soil that is added to the site in a very systematic method to ensure its load-bearing capacity.

FIGURE 24-2 Once a site has been excavated to the desired elevation, foundation forms and reinforcing steel can be placed. *Courtesy Sara Jefferis.*

Soil Compaction

Fill material can be compacted to increase its bearing capacity. Compaction is typically accomplished by vibrating, tamping, rolling, or by adding a temporary weight. Three major ways to compact soil include the following:

- **Static force**—A heavy roller presses soil particles together.
- **Impact force**—A ramming shoe repeatedly strikes the ground at high speed.
- **Vibration**—High-frequency vibration is applied to soil through a steel plate.

Each of these methods reduces the number of air voids contained between grains of soil. Proper compaction lessens the effect of settling and increases the stability of the soil, which will in turn increase the load-bearing capacity. The effects of frost damage are minimized in compacted soil because penetration of water into voids in the soil is minimized. A soils or geotechnical engineer specifies the requirements for soil excavation and compaction. The CAD technician's job is to place the specification of the engineer clearly on the plans.

TYPES OF FOUNDATIONS

The structure to be supported, the type of soil, and the contour of the site affect the type of foundation to be used. Common methods of forming the foundation include spread footings, pilings, and piers. The type of foundation used affects the type of floor system used. The drafter represents both the foundation and floor system on the foundation drawings.

Continuous or Spread Footings

The most common type of foundation used in light commercial structures is a continuous or spread footing. This type of footing consists of a footing and a **stem wall**. Each is typically made of poured concrete, although concrete blocks can be used for the stem wall. Dashed lines are often used to represent the outline of the footing. The width of the footing should be dimensioned on the foundation plan, and the depth is usually specified in a note. Figure 24-3 shows examples of each component as it would be represented in details,

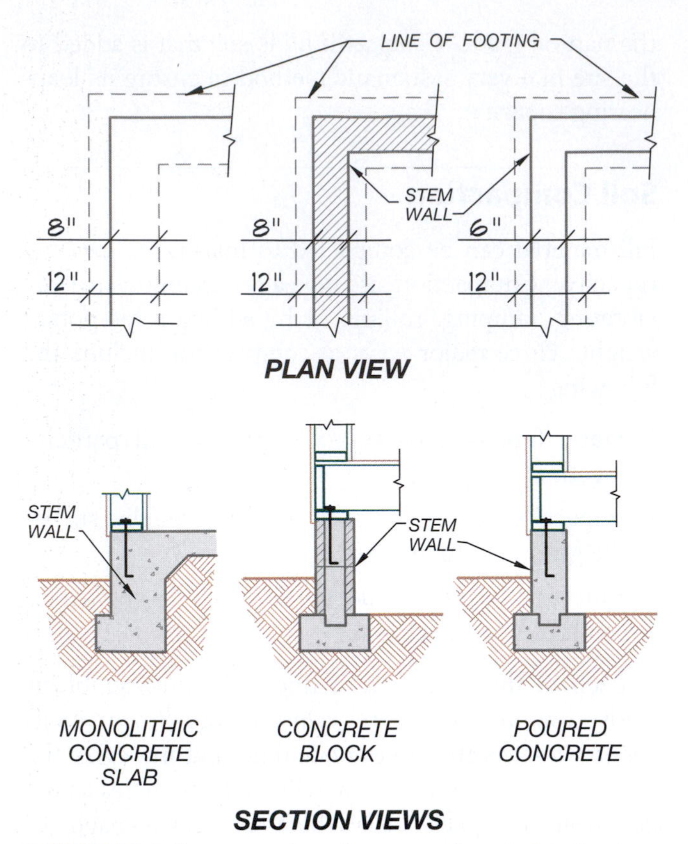

FIGURE 24-3 Representation of common foundations in plan and section views.

sections, and on a foundation plan. A **footing** is the base of the foundation system and is used to disperse loads over the soil. Figure 24-4 shows common sizes for foundation components based on the IBC.

Footing Reinforcement

When the footing is placed near expansive soil, the bottom of the footing will bend, placing the footing in tension. Steel is placed in the bottom of the footing to resist the forces of tension in concrete. This reinforcing steel or rebar is usually not shown on the foundation plan but is specified in a note that gives the size, quantity, and spacing of the steel. Only rebar that is #5 diameter or larger is considered as reinforcing, with the use of smaller steel considered as plain (nonreinforced) concrete. See Chapter 10 for a review of reinforcing steel sizes and numbers. Steel rebar may be smooth, but most uses of steel require deformations to help the concrete bind to the steel.

Stem Walls

The stem wall is the vertical portion of the foundation that brings the foundation up above the finished grade. The stem wall can be represented by continuous lines and is usually centered on the footing. Most

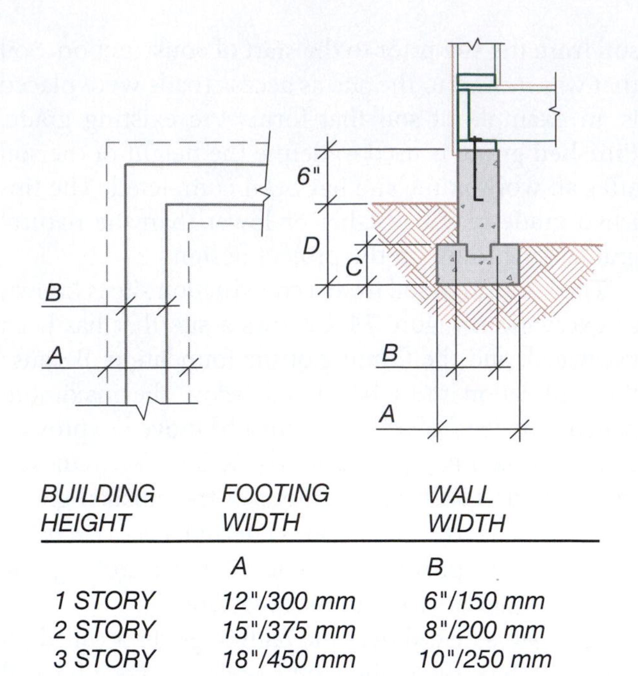

BUILDING HEIGHT	FOOTING WIDTH	WALL WIDTH
	A	B
1 STORY	12"/300 mm	6"/150 mm
2 STORY	15"/375 mm	8"/200 mm
3 STORY	18"/450 mm	10"/250 mm

BUILDING HEIGHT	FOOTING HEIGHT	DEPTH INTO UNDISTURBED SOIL
	C	D
1 STORY	6"/150 mm	12"/300 mm
2 STORY	6"/150 mm	18"/450 mm
3 STORY	8"/200 mm	24"/600 mm

FIGURE 24-4 Minimum required footing and stem wall sizes based on the 2009 International Building Code. Sizes in column B and D are based on common practice for light frame construction and must be determined by the engineer in accordance with Section 1809.7 and 8 of the IBC and local soil conditions.

municipalities require the wall to extend 6" (150 mm) above the finished grade. The material used to construct the foundation and the wall affects how they are joined together. Common stem wall materials include poured concrete and concrete blocks. When both are made of concrete, they can be formed separately or in one pour. Figure 24-5 shows common methods of forming concrete foundations. The exact shape used depends on the location of the footing to the property line, and the type of floor system to be used.

If the footing and stem wall are formed separately, a *keyway* is formed in the footing by placing a 2 × 4 in the wet concrete. Once the concrete has set, the 2 × 4 is removed, leaving an indentation in the footing. When the stem wall is poured, the concrete fills in the keyway. The interlocking nature formed by the keyway helps the wall resist lateral pressure from the soil.

An L-shaped piece of steel is normally placed in the footing to increase the bond with the wall. The seismic zone, the loads to be supported, and the height

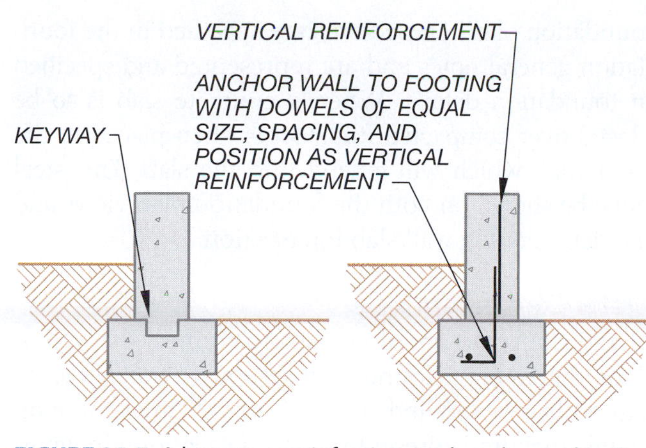

FIGURE 24-5 A keyway or reinforcing steel can be used to tie the stem wall to the foundation to resist lateral loads.

of the wall determine whether reinforcing is to be used. When the engineer requires steel to be shown, the drafter is typically required to dimension the size, spacing, and placement of the steel in details. Although the steel from the footing to the wall is not shown on the foundation plan, it is often specified in a note referenced to the stem wall. The short leg of the rebar is referred to as the *toe* and is usually placed below a continuous piece of steel in the footing. The extension of the footing steel into the stem wall must also be specified on the footing details. The distance that the stem wall steel overlaps the extended footing steel is referred to as the *lap*. Lap is often critical in the resistance of lateral pressure. Figure 24-6 shows an example of a foundation detail.

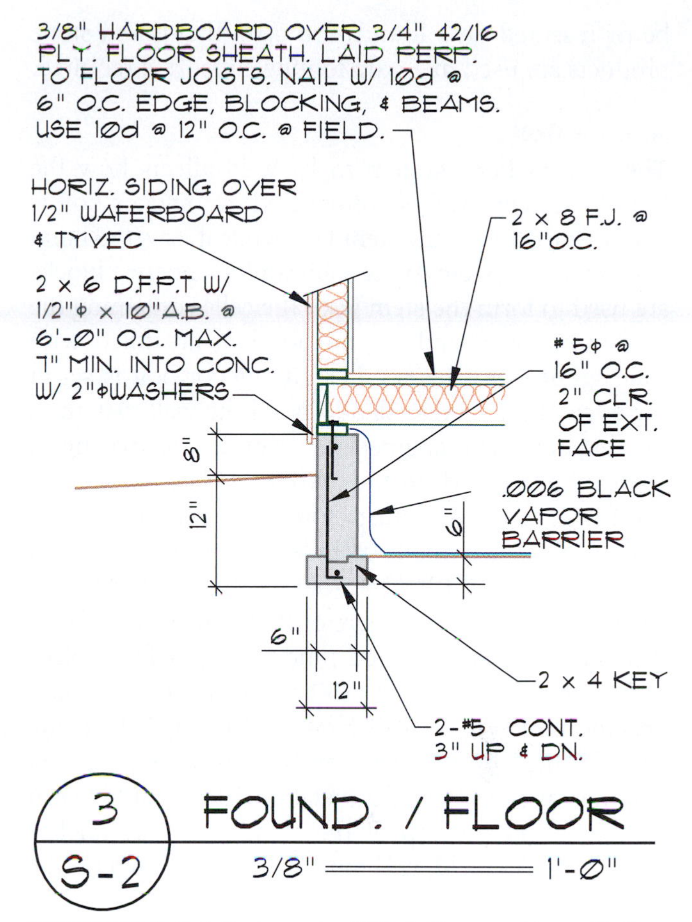

FIGURE 24-6 Details drawn by CAD drafters working with the engineering team show foundation construction. Each detail is referenced to the foundation drawings and sections.

Stepped Walls and Foundations

When the construction site is not level, the foundation must be stepped with the height of each portion of the stem wall adjusted to provide a level base for the floor system. The top of the stem wall will also change heights at each change in floor, but it must remain level. The bottom surface of the footing is allowed to have a slope not exceeding one vertical unit in each 10 horizontal units (10% slope). The footing can be stepped in order to change the elevation of the top surface of the footing, or where the surface of the ground slopes more than one vertical unit in each 10 horizontal units (10% slope). If wood studs are used to frame between the stem wall and the floor system, a minimum length of 14" (360 mm) is required.

Veneer Support

If a masonry or stone veneer is to be used, the footing width must be altered to support the veneer. The thickened footing width needs to be represented and specified on the foundation plan as well as in a detail showing construction. Figure 24-7 shows common methods of supporting the veneer and how it would

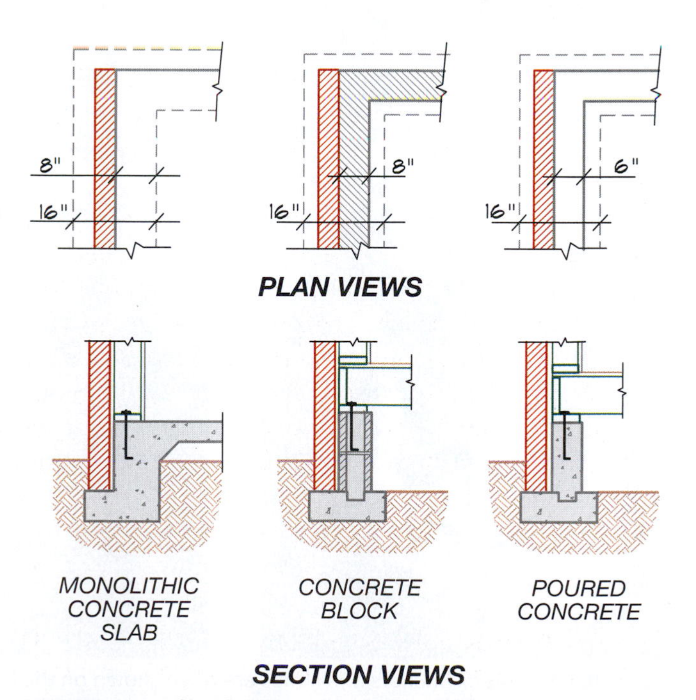

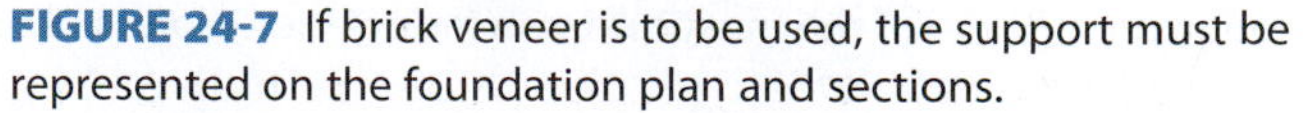

FIGURE 24-7 If brick veneer is to be used, the support must be represented on the foundation plan and sections.

be represented on a foundation plan. If cultured stone products are used, no extra footing support is required.

Anchor Bolts

The type of floor system to be used affects how the top of the stem wall is constructed. An anchor bolt is placed at the top of the stem wall while it is wet to bond a wood floor system to the stem wall. If concrete blocks are used to form the stem wall, the cells containing the reinforcing steel and the anchor bolt are filled with solid grout (see Chapter 11). An anchor bolt is set in the edge of a concrete slab to bond the wall system to the concrete. The engineer determines the spacing of anchor bolts based on the amount of lateral, seismic, and uplift loads that must be resisted. Minimum sizes of bolts allowed by code include a diameter of 1/2" (13 mm), a length of 10" (250 mm), with a 7" (180 mm) penetration into the concrete. Bolts are to be placed through a pressure-treated mudsill and held in place by a nut resting on a 2" (50 mm) diameter washer. Spacing is not to exceed 6'–0" (1800 mm). In many parts of the country, the mudsill must be protected from termites by use of a metal guard placed between the plate and the top of the stem wall. The anchor bolts, termite shield, and mudsill are not drawn on the foundation plan. These items are specified in the foundation general notes and are represented and specified in foundation details. When a concrete slab is to be placed over compacted fill, steel is often placed in the stem wall, which will extend into the slab. This steel must be shown on both the foundation plan views and the details of the wall/slab intersection.

Metal Connectors

Metal connectors are often used to connect the wall or floor system to the foundation system, to transfer loads caused from wind and seismic forces. Three common metal anchors represented on foundation drawings include column connectors, hold-down anchors, and metal angles. Each is shown in Figure 24-8. The engineer determines the size and location of any metal connectors, and the CAD technician is required to represent these on the foundation plan and details. Figure 24-9 shows common methods of representing metal connectors.

Beam Support

If the stem wall is to support a wood floor system, a method of supporting floor beams must be provided. A cavity called a **beam pocket** can be built into the wall

FIGURE 24-8 Metal anchors and straps are often shown on the foundation plan. They help transfer loads from the skeleton into the foundation. Three of the most common include (left) column base, (middle) hold-down anchor, and (right) framing straps. *Courtesy Simpson Strong-Tie Co., Inc.*

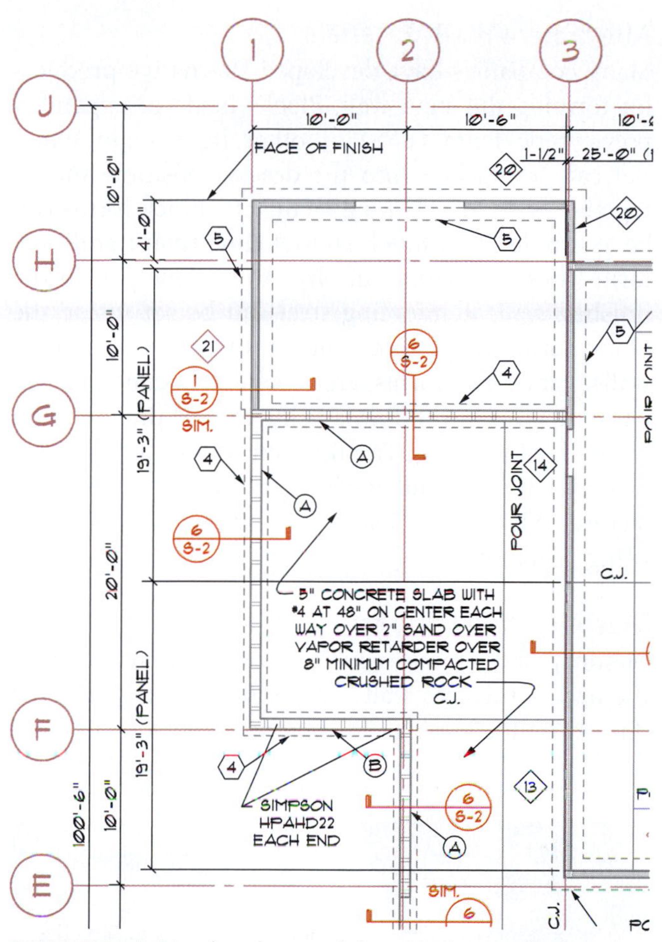

FIGURE 24-9 Metal anchors and ties must be specified on the foundation plan. *Courtesy Architects Barrentine, Bates & Lee, A.I.A.*

to provide support for beams. A 3" (75 mm) bearing surface must be provided for a wood beam resting on concrete with a 1/2" (13 mm) air space on all sides of the beam. Beams resting in a beam pocket must be wrapped with 55# felt to protect the beam from moisture in the concrete. A metal hanger can be used in place of a beam pocket. Figure 24-10 shows examples of each method of beam support.

Foundation Access

When the stem wall is to support a wood floor system, a method of access must be provided to the space below the floor system, which is called the **crawl space.** Access for most occupancies must be provided from through the stem wall using a 30" × 18" (750 × 450 mm) minimum access hole. The IBC allows access is through the floor for some R occupancies using a 22" × 30" (560 × 750 mm) access panel. It floor access is used it must be shown on the floor plan. The access should be placed out of traffic routes, for example in an area such as a closet. The drafter should verify that floor beams shown on the foundation plan will not interfere with the access hole.

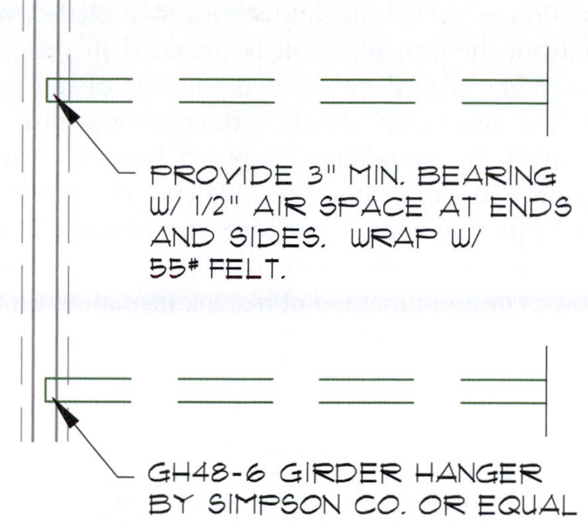

FIGURE 24-10 A girder can be supported by the use of a metal hanger or by a recess placed in the stem wall.

Foundation Ventilation

Vents must be installed in the stem wall to allow air to circulate in the crawl space. Vents are required to provide a minimum of 1 sq ft for each 150 sq ft (0.67 m^2) of crawl space unless the exceptions in Section 1203.3.2 are met. Vents should be placed within approximately 3'–0" (900 mm) of each foundation corner and should be spaced at approximately 10'–0" (3000 mm) o.c. Consideration should be given to door placement so that vents are not placed where doors or steps will be located. Vents also should not be placed below a beam pocket. Figure 24-11 shows how the crawl access and vents can be represented and specified on a foundation plan.

Stem Wall Insulation

In colder regions of the country, 2" (50 mm) rigid insulation is used to insulate the stem wall. Insulation can be placed on either side of the stem wall. If the

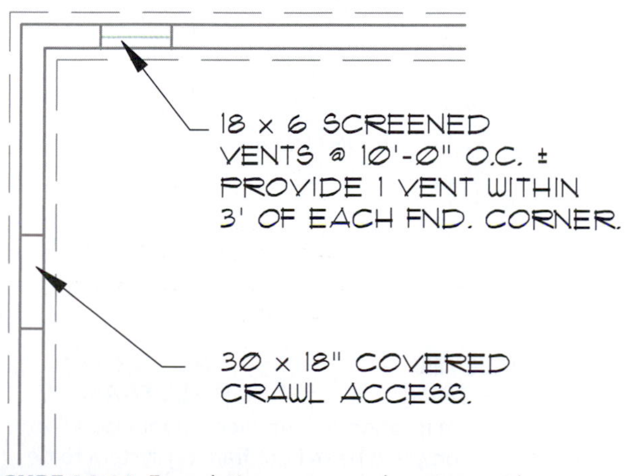

FIGURE 24-11 Foundation access and vents must be represented on the foundation plan if a wood floor system is used.

insulation is placed on the exterior side of the wall, heat from the structure will be retained in the concrete. When placed on the exterior side of the stem wall, the insulation should extend down past the bottom of the foundation. Exposed insulation must be protected from punctures, which can usually be done by placing a protective covering such as 1/2" (13 mm) concrete board over the insulation. Figure 24-12 shows a common method of placing insulation on the exterior side of the wall.

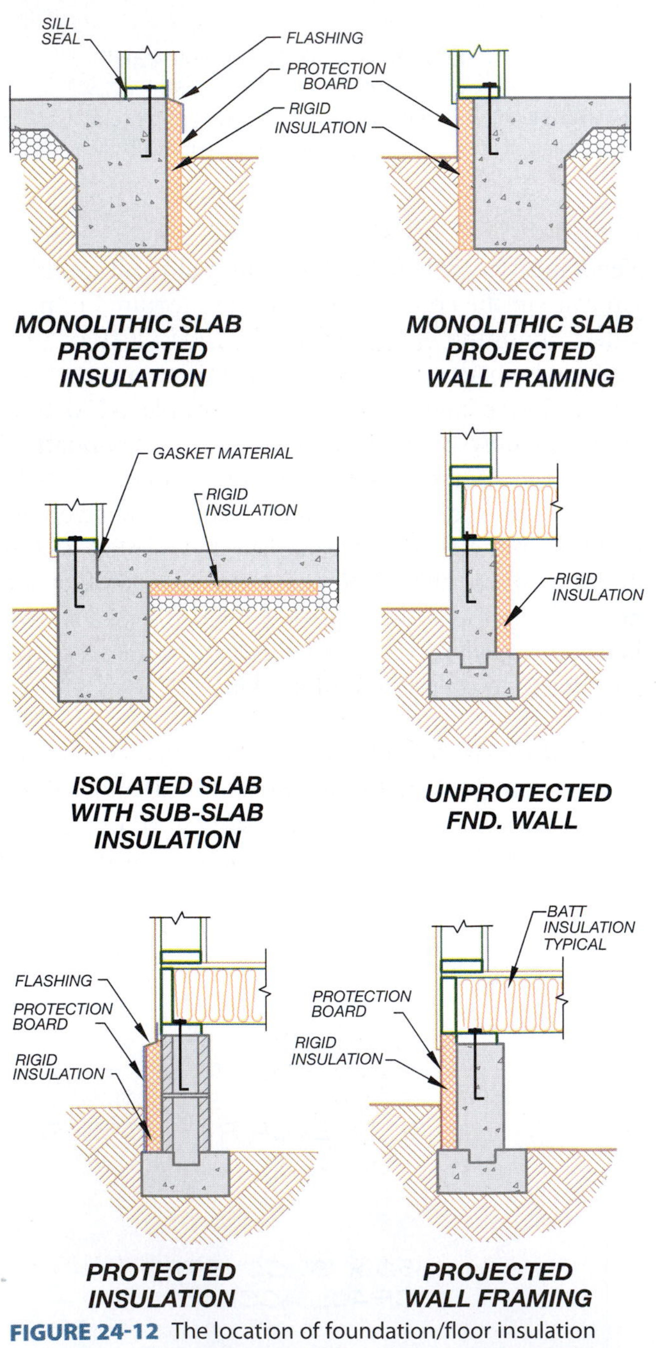

FIGURE 24-12 The location of foundation/floor insulation varies depending on the type of floor framing system to be used. When placed on the exterior side of the stem wall, the insulation must be protected from punctures.

Alternative Wall Materials

Many companies have developed alternative products for forming the stem wall. Blocks made of expanded polystyrene foam (EPS) or other lightweight material can be stacked into the desired position and fit together with interlocking teeth. EPS block forms can be assembled in a much shorter time than traditional form work and remain in place to become part of the finished wall. Reinforcing steel can be set inside the block forms in patterns similar to traditional block walls. Once the forms are assembled, concrete can be pumped into the forms using any of the common methods of pouring. The finished wall has an R-value of between R-22 and R-35 depending on the manufacturer. Figure 24-13 shows a foundation being built using EPS forms.

Masonry Wall Support

Masonry and precast concrete construction eliminate the need for a stem wall because the wall is supported directly on the footing. The reinforcing for the masonry

FIGURE 24-13 Forms made of EPS are available for constructing the stem wall. Forms remain in place and insulate the structure. EPS forms can also be used to form the entire exterior walls of the structure. *Courtesy American Polysteel Forms.*

wall is connected to the steel that extends from the footing. Figure 24-14 shows wall details for brick and concrete block walls. Precast wall panels are fastened directly to the footing, and soil is backfilled directly against the panel. Figure 24-15 shows a footing prepared for a precast panel. When this type of construction is used, a separate slab plan and foundation plan are generally drawn. Each is introduced later in this chapter.

Retaining Walls

A *retaining* or basement wall is typically made of 8" (200 mm) poured concrete or concrete block and is assumed to be anchored at the top and bottom by a floor system. Common widths include 6" and 8" (150 and 200 mm) but the width will vary depending on the loads to be resisted and the height of the wall. Because of the height, the lateral pressure tends to bend the wall inward, placing the soil side of the wall in compression and the side away from the soil in tension (see Figure 24-16). To resist the tension, the engineer specifies reinforcing based on the height of the wall, the soil pressure, and the seismic zone. Figure 24-17 shows an example of a partial-height retaining wall detail. The wall will appear on a foundation plan just as any other stem wall. The detail would be referenced to the foundation plan. Some method of waterproofing the wall is also generally noted on the foundation plan. Typically, walls are covered

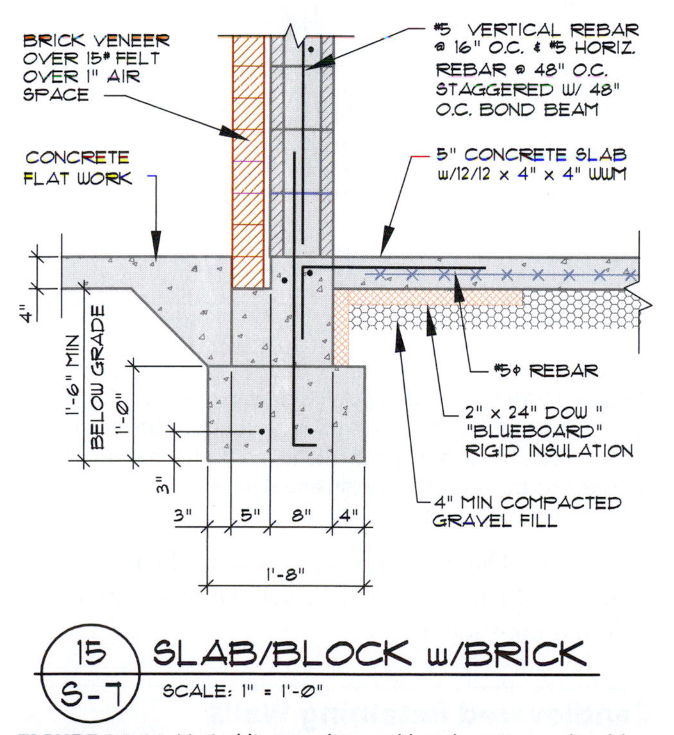

FIGURE 24-14 Varied line quality and hatch patterns should be used to represent masonry products and reinforcement in foundation details. *Courtesy David Jefferis.*

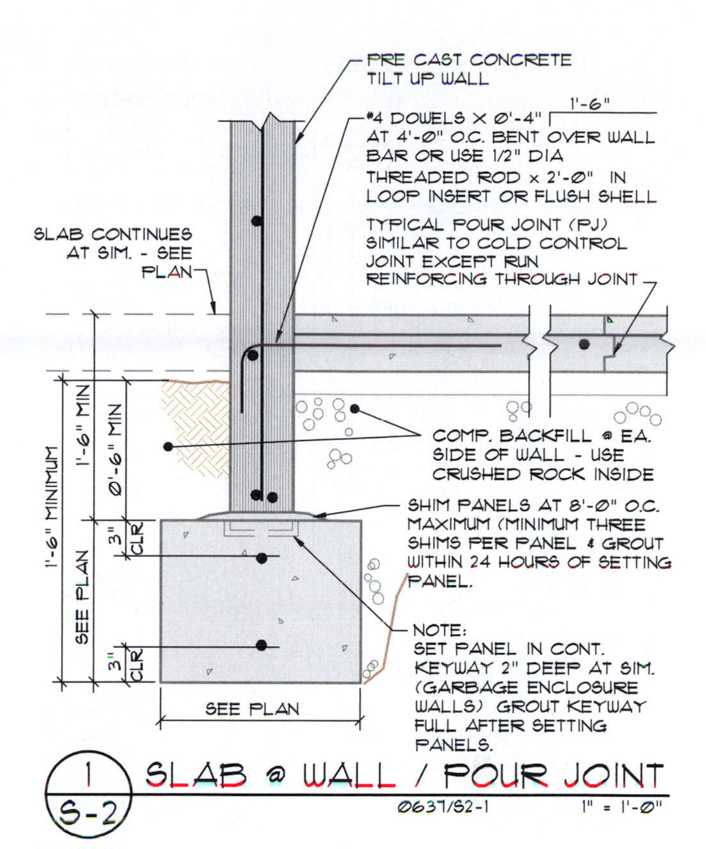

FIGURE 24-15 Representing the intersection of a precast concrete panel with the foundation. *Courtesy Van Domelen/Looijenga/McGarrigle/Knauf Consulting Engineers.*

with at least two layers of bituminous waterproofing material or by fiber-reinforced asphaltic mastic.

Anchorage

Anchor bolts are placed in the retaining wall by a method similar to that used for stem walls. The spacing of bolts is generally much closer for a retaining wall because of the greater loads to be resisted. Because the floor

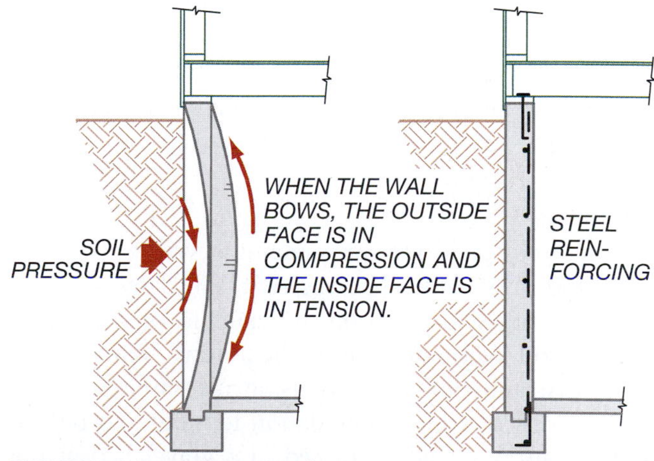

FIGURE 24-16 Soil pressure bows a wall inward when the wall is anchored to a floor at the top and bottom. Because concrete has poor tensile strength, steel must be placed near the side in tension to help resist the soil load.

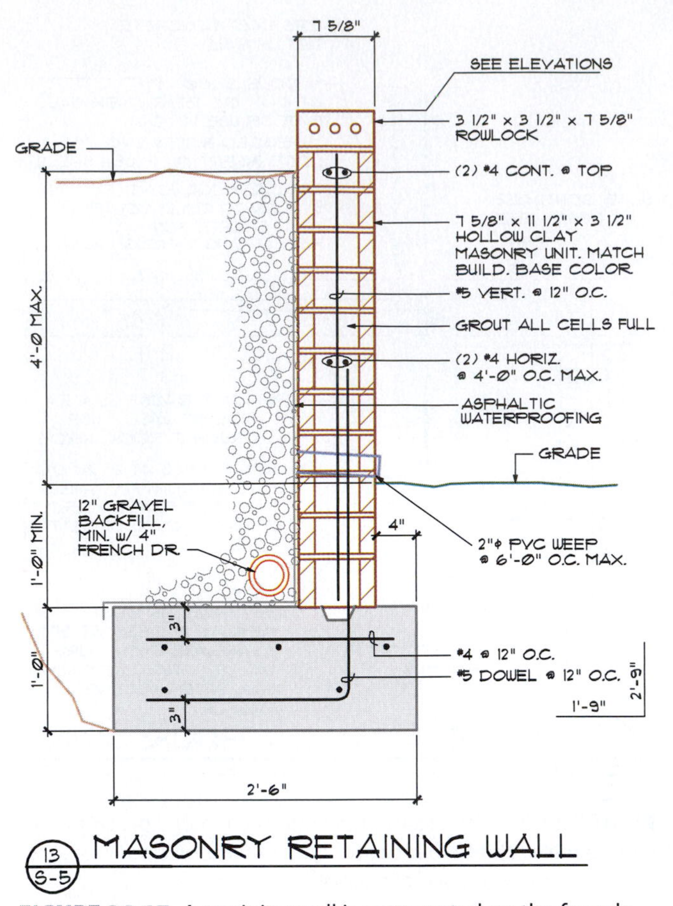

FIGURE 24-17 A retaining wall is represented on the foundation plan using the same methods used to describe a stem wall. Details must be referenced to the plan to show wall construction. *Courtesy Peck, Smiley, Ettlin Architects.*

systems are used to resist the soil loads, a metal angle is often used to securely bond the floor joists to the top of the wall. The anchor bolts and angles are specified in annotation on the foundation plan and shown in details. When the floor joists are parallel to the retaining wall, blocks are typically placed between floor joists near the wall edge to transfer lateral pressure on the wall to the floor system. This blocking would be drawn and noted on the foundation plan and represented in a detail similar to Figure 24-18.

Footings

The footing for a one-story retaining wall is typically 16" (400 mm) wide. The engineer will determine the exact footing size based on the loads to be supported and the soil-bearing pressure. The footing width should be drawn and dimensioned using the same methods used to describe other foundation footings. To reduce soil pressure, a drain is placed in a gravel bed at the base of the retaining wall to remove water from the soil (see Figure 24-17). The gravel allows water to percolate down to the drain and be diverted away from the foundation. The drain can be shown on the foundation plan, as in Figure 24-19, or it may be represented in a note and not shown.

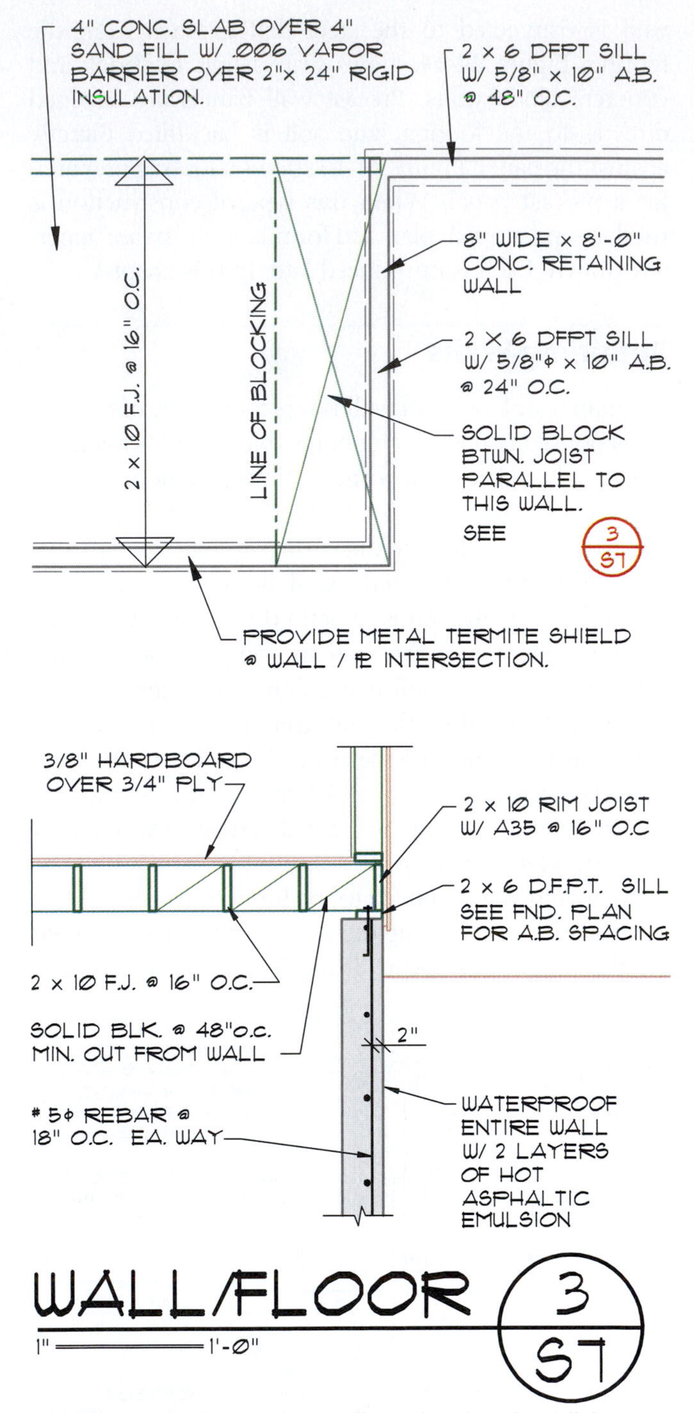

FIGURE 24-18 If a retaining wall is to be placed parallel to wood floor joists, solid blocking is often required between the joists to stiffen the floor system. Required blocking should be specified on the foundation plan and details.

Cantilevered Retaining Walls

When a structure is built on a sloping site, the retaining wall may not be required to extend the full height from

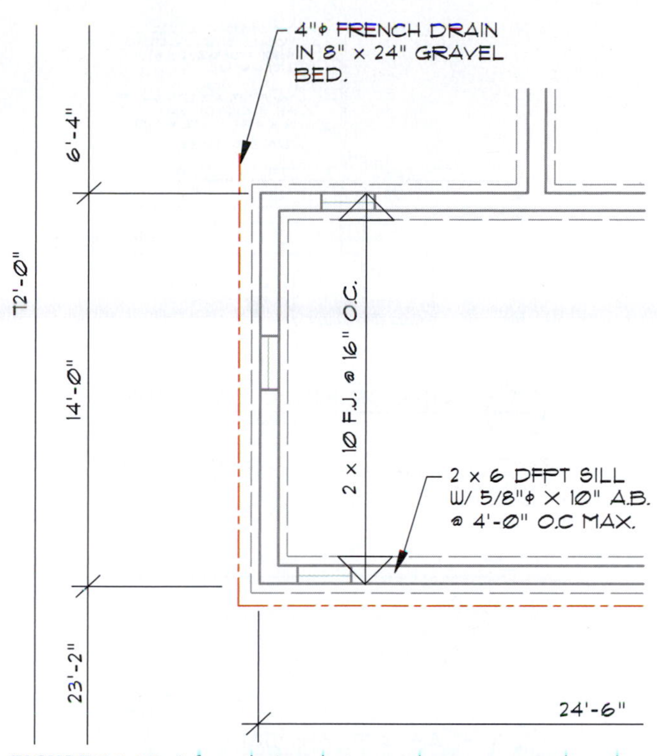

FIGURE 24-19 A foundation drain may be represented on the foundation plan, or it may be referenced in general notes.

the foundation to the first floor level. A wall that is not supported at both ends by a floor system is referred to as a **cantilevered retaining wall.** Although less soil is retained than with a full-height wall, as shown in Figure 24-20 because the wall is not anchored at the top, it has a greater tendency to bend and create a hinge point. To resist the soil pressure, a wider footing must be used. The width of the footing is based on the height of the wall and is determined by the engineer. Figure 24-21 shows two methods of placing

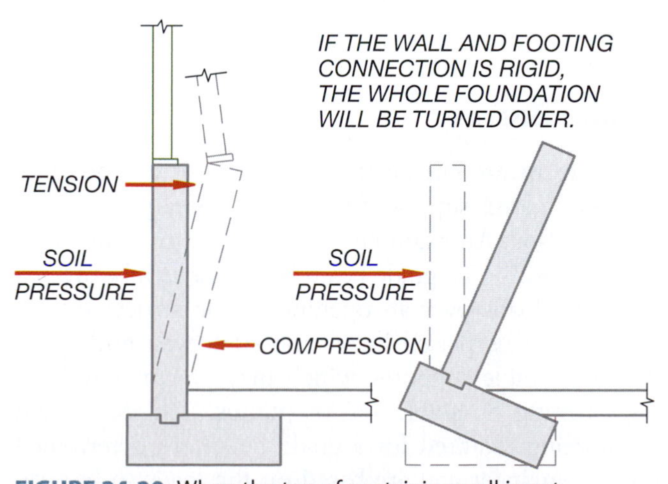

FIGURE 24-20 When the top of a retaining wall is not anchored to the floor system, lateral pressure causes the wall to rotate inward.

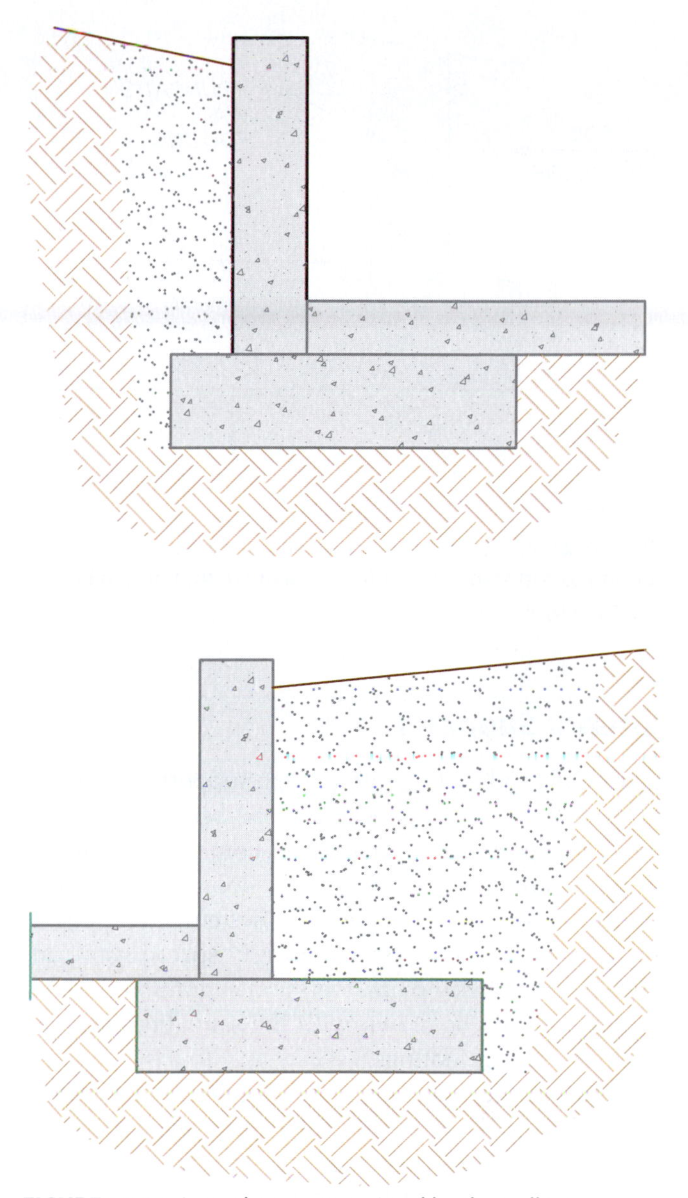

FIGURE 24-21 Lateral pressure resisted by the wall is transferred to the foundation. The foundation in the upper example transfers loads to the soil below the wall. The foundation in the lower example resists the lateral loads by the weight of the soil above the foundation.

the footing in relationship to the wall. Depending on the height of the wall, or the angle of repose of the soil being retained, a key may be placed on the bottom of the footing. As shown in Figure 24-22, the *key* is the rectangular or square portion of concrete placed on the bottom side of the footing to resist lateral pressure. The width of the wall and the footing must be represented and specified on the foundation. The key is not drawn or referenced on the foundation plan. A detail must be referenced on the foundation plan to show the reinforcing, footing and wall size, and the key size and location.

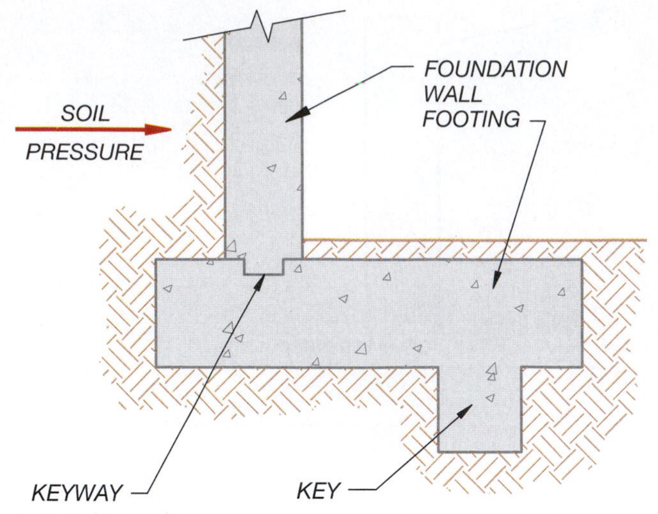

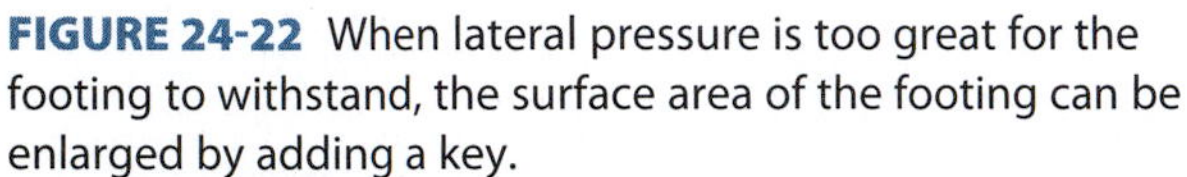

FIGURE 24-22 When lateral pressure is too great for the footing to withstand, the surface area of the footing can be enlarged by adding a key.

Interior Supports

Stem walls and footings are used to support the exterior portion of the structure. Interior load-bearing walls over concrete floors are supported on continuous footings. Bearing walls resting on a wood floor system or columns or posts on a concrete floor system are supported by a beam supported on repetitively spaced spot concrete footings called **piers.** Piers and continuous interior footings should be drawn with the same linetype used to draw exterior footings.

Figure 24-23 shows interior footings for concrete floor systems. Piers are also used to provide increased bearing values to exterior footing. Piers can be formed by using preformed or framed forms, or by excavation. Common shapes include:

- When placed under a stem wall, piers are usually square or rectangular.
- When used to support interior loads on a floor system, piers are often round.
- Piers that are too close together to form their full shape are often poured as a rectangle.

The pier depth is usually the same depth required for the exterior footing, with the diameter based on the load to be supported and the soil-bearing pressure. Piers are usually required to be recessed into the soil to resist lateral movement. In areas subject to seismic activity, a piece of rebar or a metal post base may be required to bond the support post to the pier. Blocking or a metal strap can also be provided to add stability to the post-to-beam connection. Figure 24-24 shows the piers of a wood floor assembly in section and plan views.

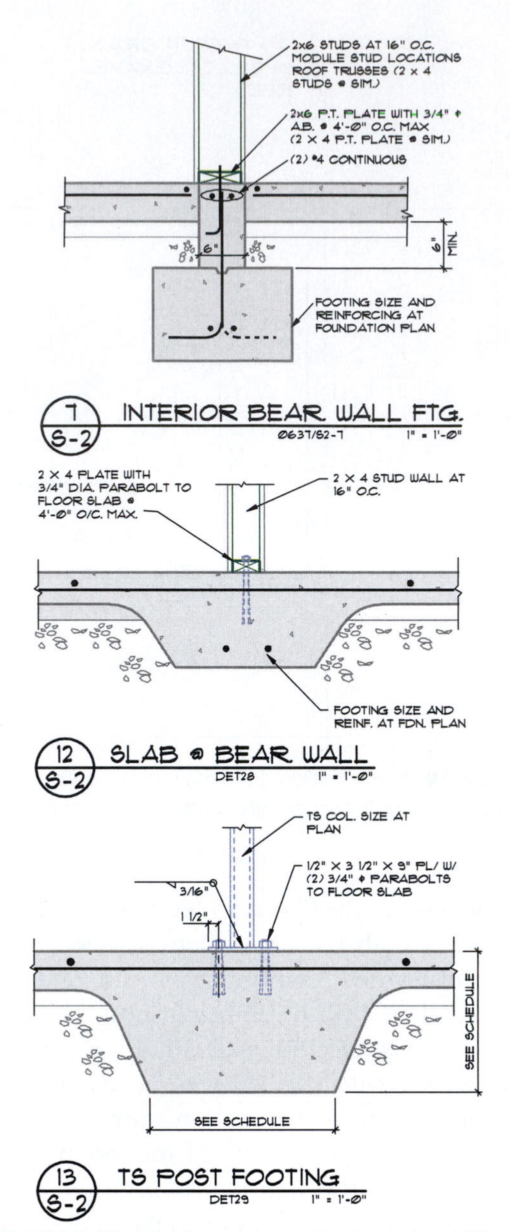

FIGURE 24-23 Piers and continuous footings should be drawn with the same linetypes used to represent exterior footings. *Courtesy Architects Barrentine, Bates & Lee, A.I.A.*

Grade Beams

A **grade beam** can be used in place of a footing to provide added support for a foundation placed over unstable soil. An example of a grade beam is shown in Figure 24-25. The grade beam is similar to a beam that supports loads over an opening in the structure. The grade beam is placed below the stem wall and spans between stable supports, which may be concrete footings resting on stable soil or pilings. The depth and reinforcing required for a grade beam are determined by the engineer and are based on the loads to be supported. A grade beam resembles a footing when drawn on a foundation plan. Steel reinforcing may be specified

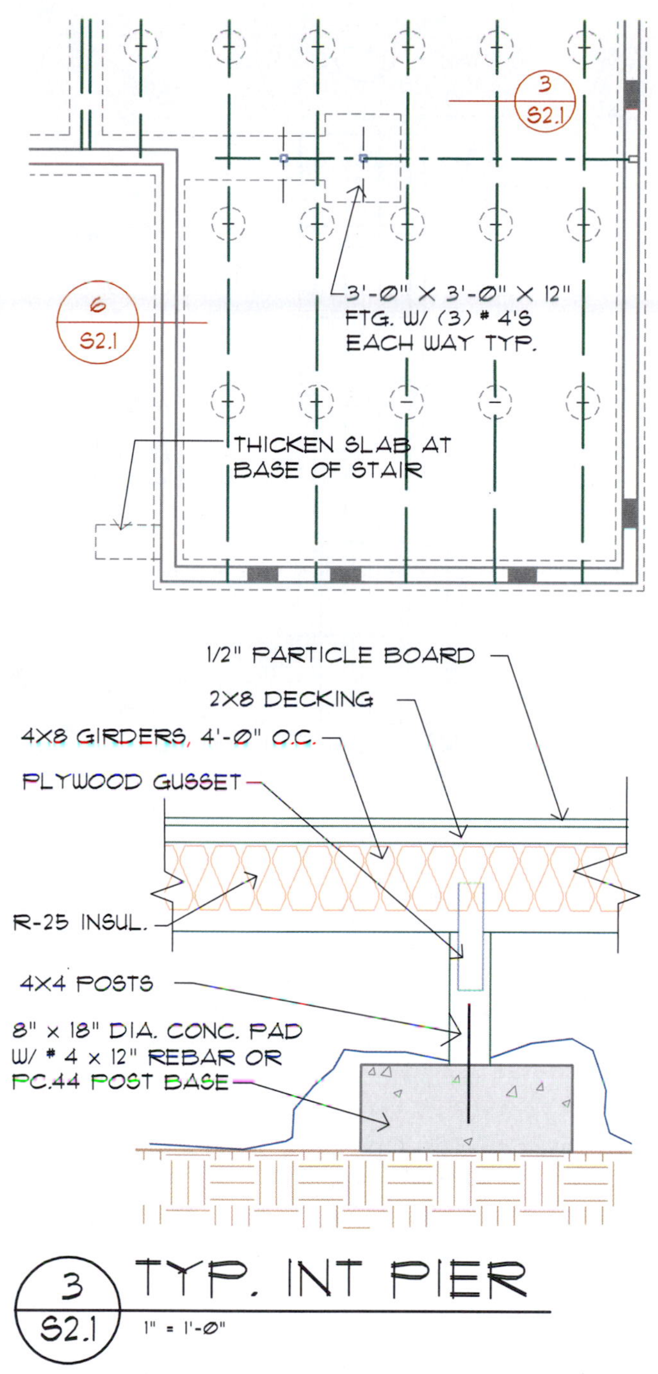

FIGURE 24-24 A pier can be used to reinforce a stem or retaining wall when loads to be supported exceed the bearing capacity of the foundation. *Courtesy Van Domelen/Looijenga/McGarrigle/Knauf Consulting Engineers.*

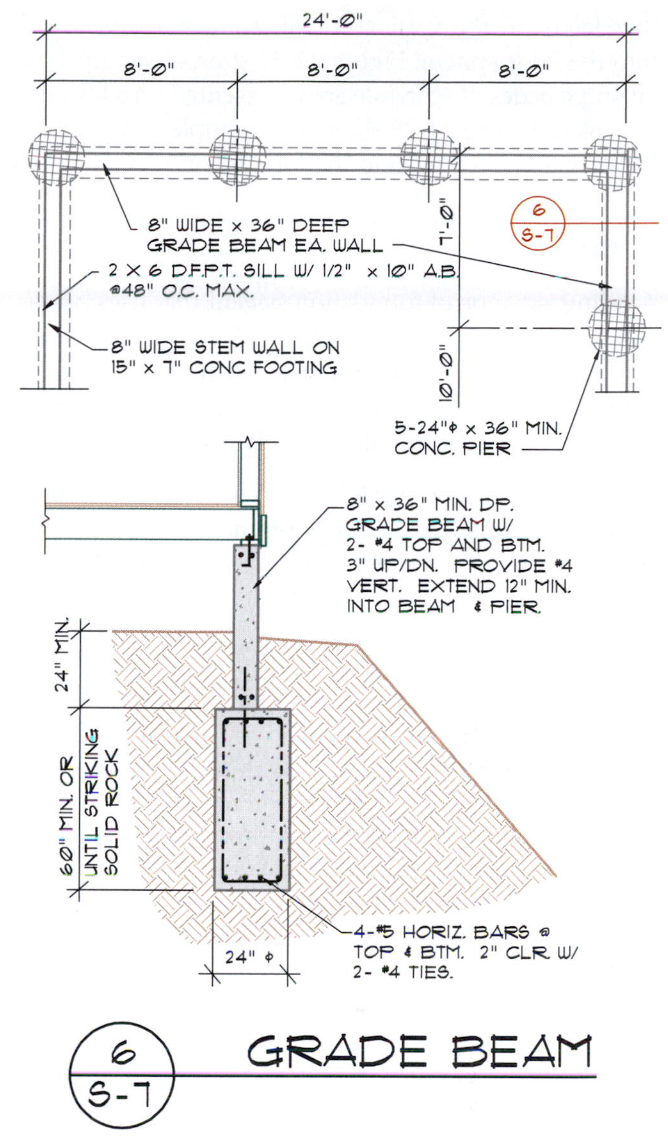

FIGURE 24-25 In unstable soil, the stem wall can be reinforced and treated as a beam spanning between piers. *Courtesy Tereasa Jefferis.*

FIGURE 24-26 The pedestal supporting the steel column extends from the footing, through the fill material that will be placed to support the floor. Notice the steel that extends from the pedestal into the future floor slab. *Courtesy Donna Sweeney.*

by notes and referenced to details rather than on the foundation plan.

Pedestals

A **pedestal** is a poured concrete column that is built on top of a footing to extend through fill material up to a concrete floor level (see Figure 24-26). Pedestals are often used to provide support for posts or columns so that loads on the vertical member are not transferred into the floor system. Figure 24-27 shows how a pedestal and a pedestal schedule are represented on a foundation plan. Figure 24-28 shows an example of a pedestal detail that is referenced to the foundation plan.

Pilings

A **piling** is a type of foundation system that uses beams placed between columns to support loads of a structure. The columns may extend into the natural grade or be supported on other material that extends into stable soil. Piling foundations are typically used:

- On steep hillside sites where it may not be feasible to use traditional excavating equipment.

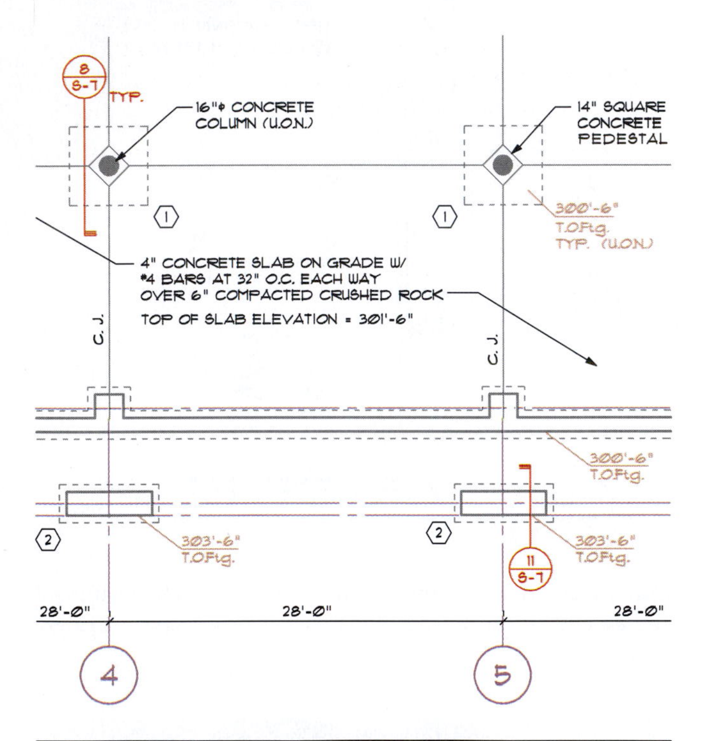

COLUMN AND FOOTING SCHEDULE		
	FOOTING SIZE	REINFORCING
1	7'-6" x 7'-6" x 18" DEEP	(7) #6 BARS EACH WAY
2	2'-8" x 7'-0" x 18" DEEP	(7) #5 BARS SHORT DIRECTION (3) #5 BARS LONG DIRECTION

DESIGN SOIL PRESSURE: 5,000 p.s.f.

SEE SOILS INVESTIGATION BY: WRIGHT / DEACON & ASSOCIATES, INC., 7-28-08

FIGURE 24-27 The pedestal shown in Figure 24-26 can be seen in plan view with the supporting footing shown with dashed lines. When a wide variety of pedestals are to be used, a schedule can be used to specify reinforcing patterns and sizes. *Courtesy Van Domelen/Looijenga/McGarrigle/Knauf Consulting Engineers.*

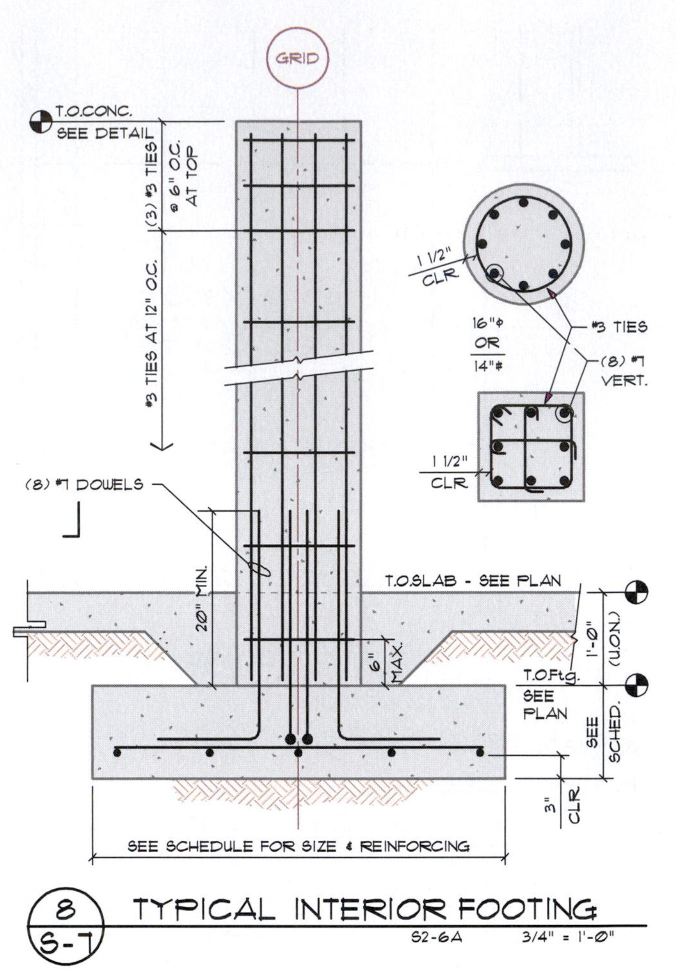

FIGURE 24-28 Details must be drawn to show the construction of a pedestal and referenced on the foundation plan. *Courtesy Van Domelen/Looijenga/McGarrigle/Knauf Consulting Engineers.*

- Where the loads imposed from the structure exceed the bearing capacity of the soil.
- On sites subject to flooding or other natural forces that might cause large amounts of soil to be removed.

Coastal property and sites near other bodies of water subject to flooding often use a piling foundation to keep the useable space above the floodplain level.

When a structure is to be built above grade using a piling system, a support beam is typically placed under or near each bearing wall. Beams are supported on a grid of vertical supports that extends down to a stable layer of rock or dense soil. Beams can be steel, sawn or laminated wood, or prestressed concrete. Vertical supports may be concrete columns, steel tubes or beams, wood columns, or a combination of each material. When pilings are used for work at or below grade, concrete and steel are typically used to support the structure. Pilings can be several hundred feet in length. The engineer may require the depth to be based on the soil resistance measured as the pile is driven, or a minimum depth to solid rock based on the soils report. The desired depth is placed in the foundation notes by the drafter.

For shallow pilings, a hole can be bored and poured concrete with steel reinforcing can be used. Figure 24-29 shows an example of a detail for a poured concrete piling. If the vertical support is required to extend deeper than 10' (3000 mm), a pressure-treated timber or steel column can be driven into the soil. Timber pilings that are above the water table must be pressure treated and are typically coated with creosote to extend their lifetime.

Steel Pilings

As the depth of the piling increases, or where layered soil conditions prevent wood from being used, steel H-sections can be used for subgrade pilings. Steel is typically called for when the design requires the piling to be driven into solid rock. The pilings in Figure 24-30 were driven into stable soil and then capped with a base plate. From this platform, steel columns were welded to the plate to provide a support system for the structure. Figure 24-31 shows the components of

FIGURE 24-30 Steel pilings are often used to provide a stable base for a structure. A steel plate is welded to the top of the piling to provide support to a column or beam. The rectangular tube is used to span between pilings at the hinge point between the piling and support column. *Courtesy Michael Jefferis.*

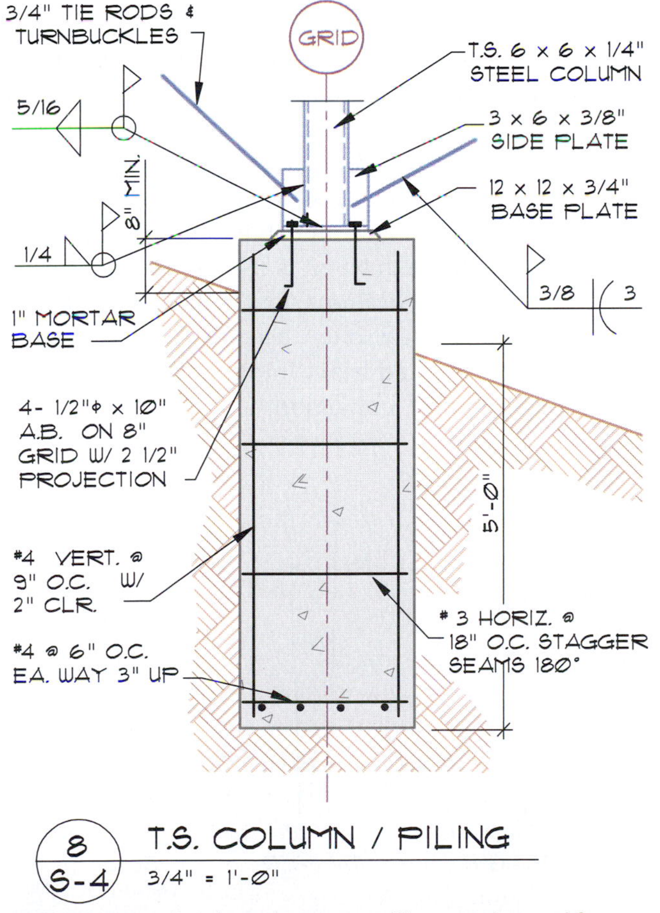

FIGURE 24-29 Reinforced concrete pilings can be used for shallow pilings to provide support for the foundation.

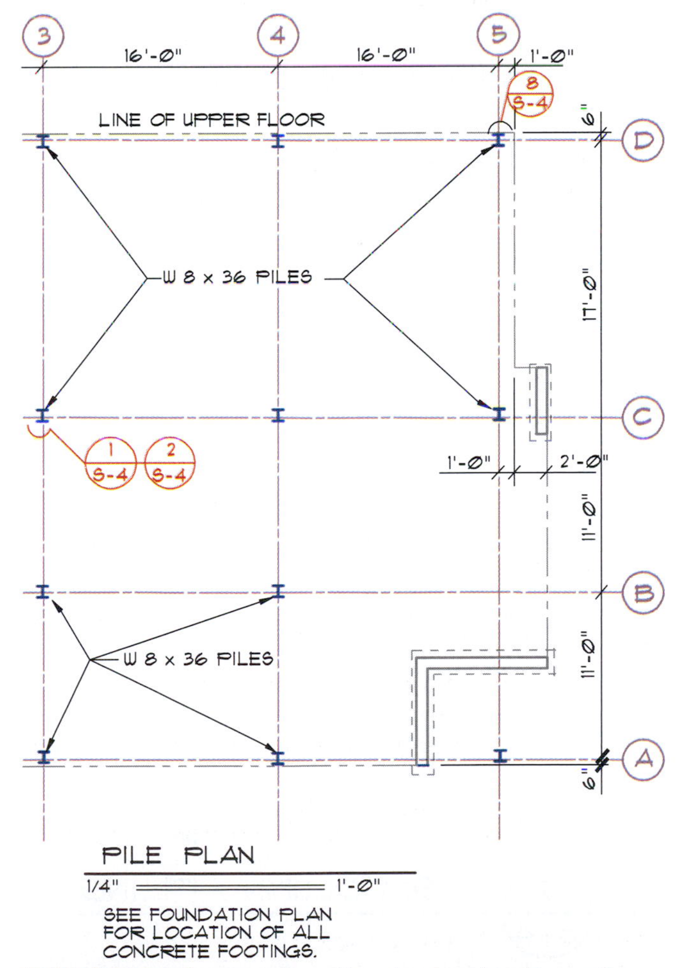

FIGURE 24-31 Representing pilings in plan view. Depending on the complexity of the structure, pilings may be shown on the foundation plan or on a separate piling plan.

a piling plan for a hillside office structure. Notice, in addition to the vertical columns and horizontal beams, diagonal steel cables are placed between the columns to resist lateral and rotational forces. In addition to resisting gravity loads, piling foundations must be able to resist forces from uplift, lateral force, and rotation. Figure 24-32 shows a detail of a steel piling foundation that is used to support steel columns above grade. Notice that the engineer has specified a system that uses braces approximately parallel to the ground to stabilize the top of the pilings from lateral loads. These braces are not shown on the foundation plan but are specified on the elevations, sections, and details.

Concrete Pilings

Pilings formed of precast and poured concrete are used for many structures as loads and heights are increased, to reduce forces causing uplift and overturning. Concrete pilings are also used in sandy soils to increase the soil friction of the pile. Concrete pilings offer excellent resistance to uplift. Reinforced precast concrete piles can be driven into the soil just like wood or steel pilings. Pilings can also be pretensioned, just as a vertical column to add to the strength of the piling. Figure 24-33 shows examples of common pile shapes.

Cast-in-Place Concrete Pilings

A steel shell can be used to place cast-in-place concrete pilings. The shell is driven into the ground to the required depth and concrete is poured into the shell. Shell lengths up to 40' (12,200 mm) can be driven with other sections welded or threaded together as the pile is driven. Reinforcing can be wired into position in the shell prior to the concrete being placed in the pilings. Pilings can also be created by forming the hole by using a steel tube and then withdrawing the tube. The concrete conforms to the shape of the soil and increases the bonding strength over a steel-encased concrete pile. The method used to place the concrete can also greatly affect the strength of the piling. If the steel casing is removed, concrete can be pumped under pressure, which will force the soil around the concrete to compress. As the soil compresses, the concrete forms an enlarged or bulbous tip at the end of the piling, which increases both the bearing and withdrawal strength.

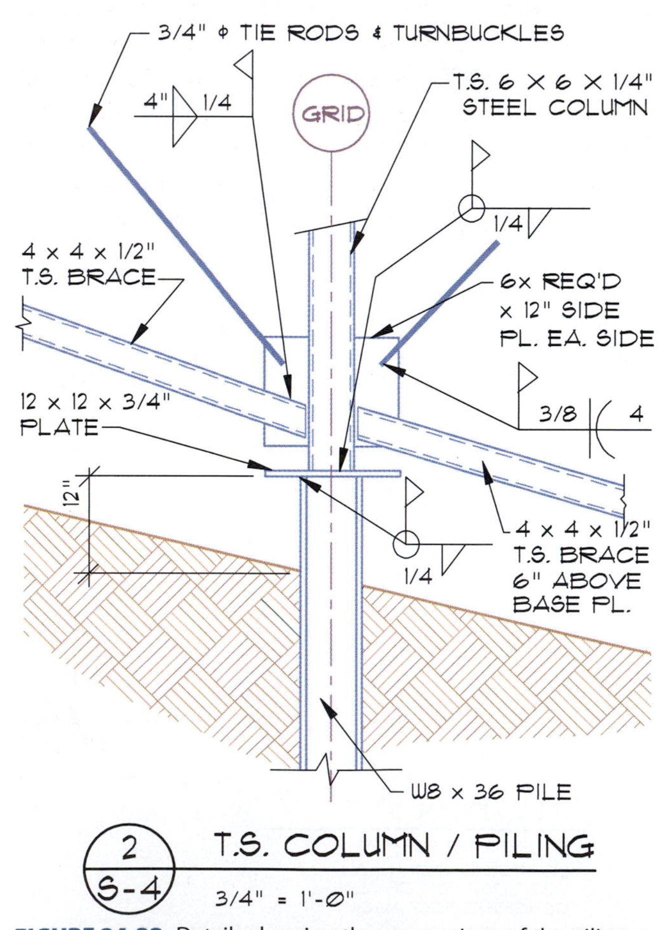

FIGURE 24-32 Details showing the connections of the piling, support plates, and reinforcing must be referenced in the plan view. *Courtesy Aaron Jefferis.*

TYPES OF FLOOR SYSTEMS

The most common floor system for commercial structures is an on-grade slab. Other types of floor systems used with commercial and light industrial construction include pretensioned slabs, slabs built below grade level, and slabs built on pilings. As building loads and height increase, the depth of the foundation must also increase. Typically, a piling foundation is used to provide stability. Multifamily applications and small construction projects in damp areas often use wood floor systems using joist or post-and-beam framing techniques supported on a concrete foundation.

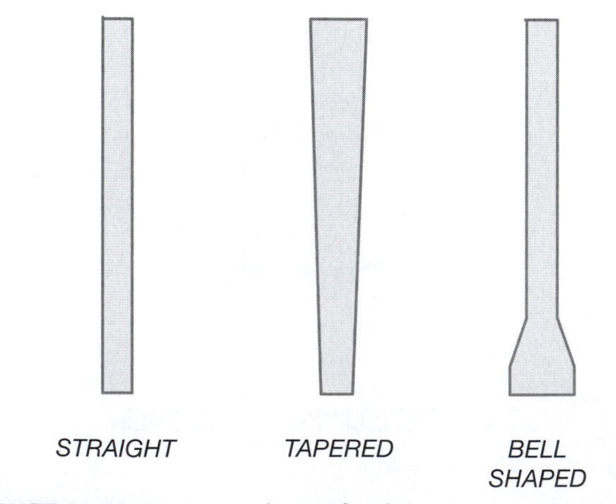
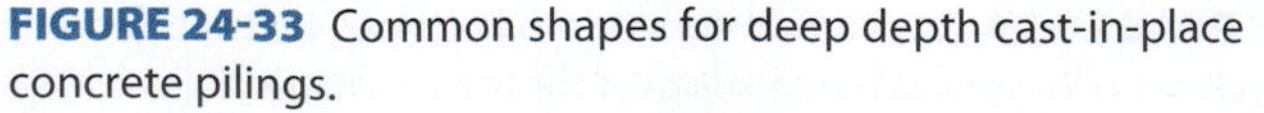

FIGURE 24-33 Common shapes for deep depth cast-in-place concrete pilings.

On-Grade Slabs

A concrete slab provides a firm floor system with little or no maintenance and generally requires less material and labor than wood floor systems. The floor slab can be poured as an extension of the foundation wall and footing as shown in Figure 24-34 in what is known as **monolithic construction.** Other common methods of pouring slabs are shown in Figure 24-35. The IBC requires slabs to be a minimum of 3 1/2" (90 mm), but design factors for commercial and industrial structures often require slabs of 5" to 6" (125 to 150 mm) in thickness. Keep in mind that plain (nonreinforced) concrete slabs are only used as a floor system and not to transmit vertical loads. If load-bearing walls or columns are to be supported by the slab, reinforcing must be placed in the slab, a footing placed beneath the slab, or both may be required depending on the loads to be supported. Figure 24-23 shows common methods of supporting floor slabs at bearing walls. If a column is to be supported, piers can be placed below the slab as seen in Figure 24-24. The edge of the slab is represented on the foundation drawings by a continuous line.

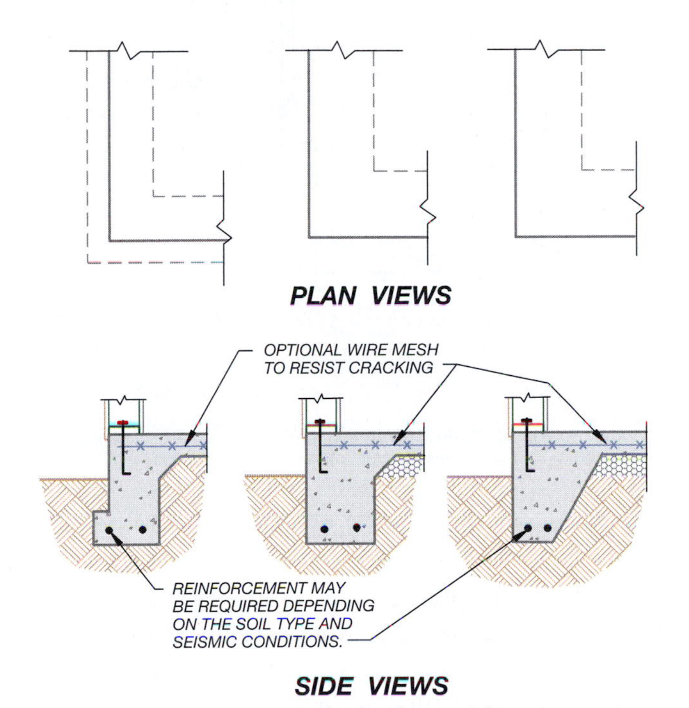

FIGURE 24-34 Common methods of monolithic construction for on-grade concrete floor systems.

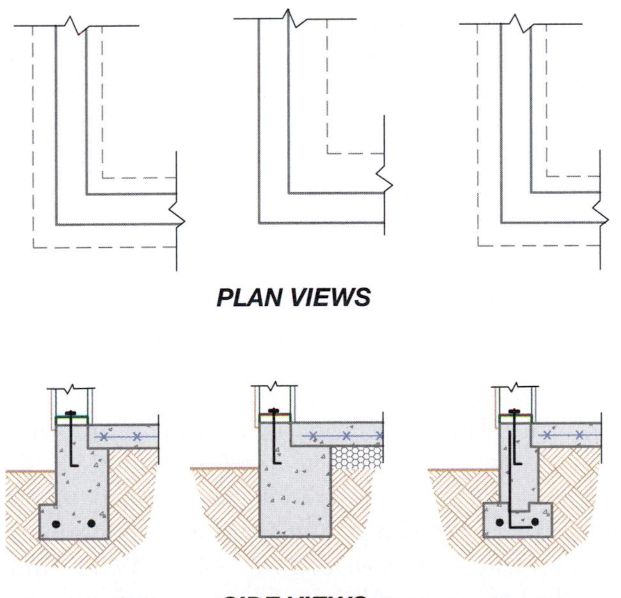

FIGURE 24-35 Common methods of construction for on-grade concrete floor systems using two or more pours.

Changes in Floor Elevation

The floor level is often required to change elevation to meet the design needs. A stem wall is formed between the two floor levels and should match the required width for an exterior stem wall. The lower slab is typically thickened to a depth of 8" (200 mm) to support the change in elevation. Figure 24-36 shows how a lowered slab can be represented on the foundation plan. Figure 24-37 shows common methods of changing floor elevations with a concrete slab, which would be shown in details referenced to the foundation plan. The step often occurs at what will be the edge of a wall when the framing plan is complete. Great care must be taken to coordinate the dimensions of a floor plan with

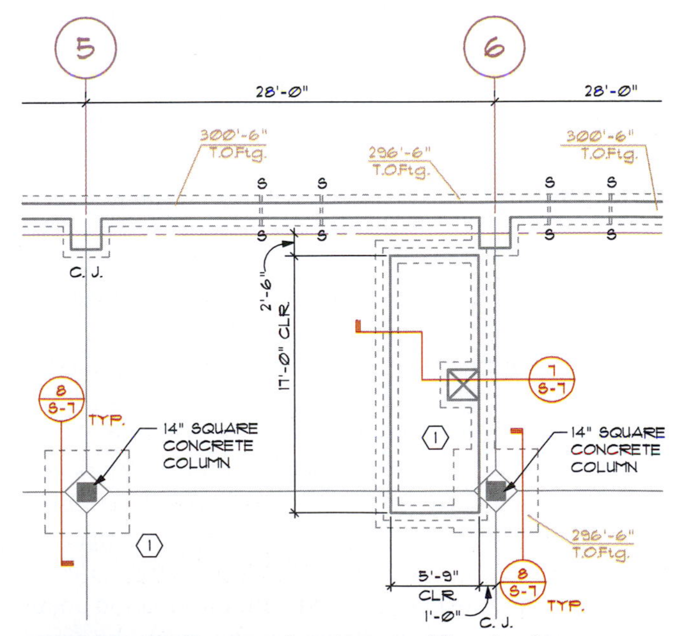

FIGURE 24-36 Representing changes of floor level on a foundation plan. *Courtesy Van Domelen/Looijenga/McGarrigle/Knauf Consulting Engineers.*

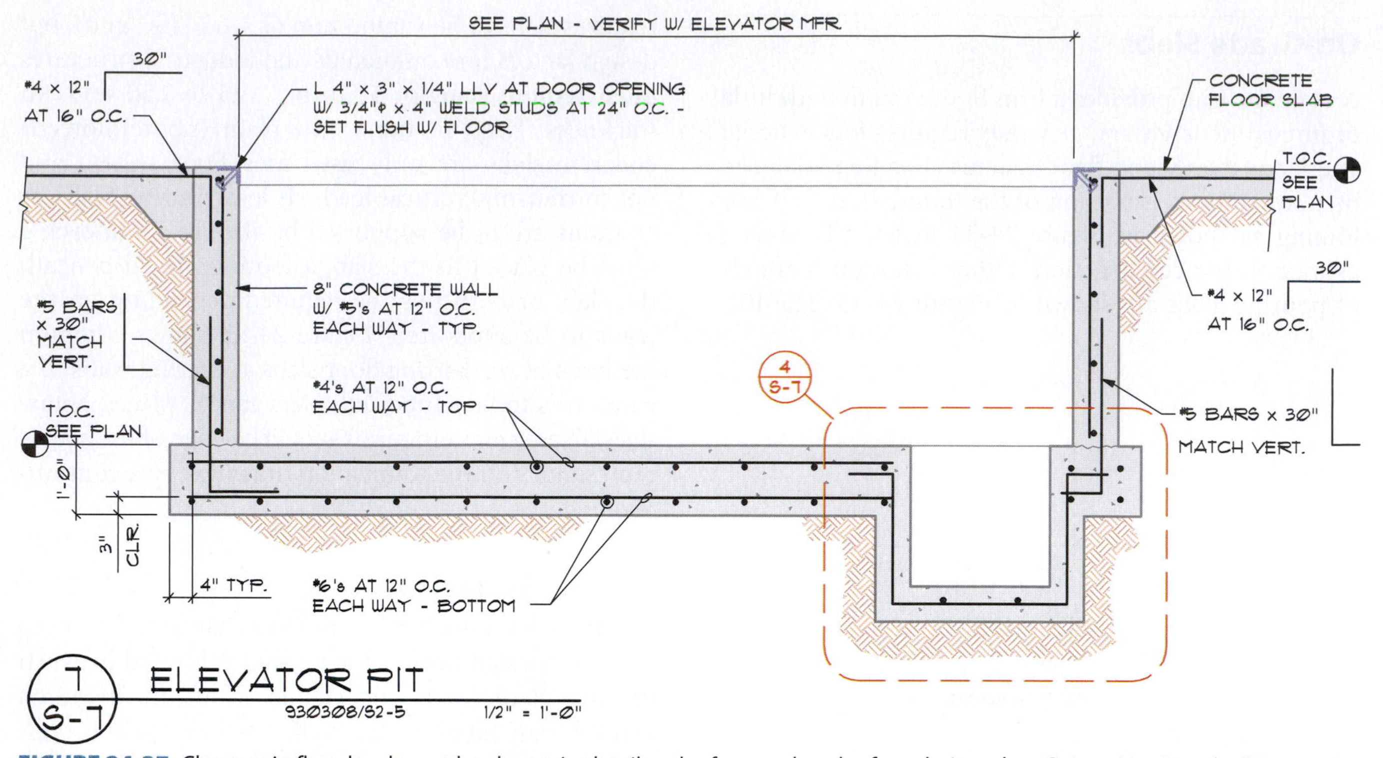

FIGURE 24-37 Changes in floor level must be shown in detail and referenced to the foundation plan. *Courtesy Van Domelen/Looijenga/McGarrigle/Knauf Consulting Engineers.*

those of the foundation plan, so that walls will match the foundation step.

Slab Joints

Joints in concrete include control, isolation, and construction joints. Each was introduced in Chapter 11. Control joints must be shown on the foundation based on the engineer's calculations or recommendations by the American Concrete Institute (ACI). The engineer will provide specific locations based on the soil conditions at the job site. The location of control joints should be represented by a line and specified by a note referenced to the joint. The spacing and method of placement can be specified in the general notes placed with the foundation plan for simple projects, on the slab-on-grade plan if required, or in the specifications within the project manual.

Isolation or expansion joints are used to separate a slab from adjacent structural members so that stress cannot be transferred into the slab and cause cracking. These joints should be drawn and specified on the foundation plan using methods similar to those used to represent control joints. Construction or keyed joints are made to provide support between two slabs. The method used to form the joint must be specified with information about other joints, but these joints are not shown on the drawings because the concrete crew determines the location based on concrete delivery and work schedules.

Slab Preparation and Placement

The slab may be placed above, below, or at grade level. Slabs placed above grade were considered in Chapter 11. Slabs built below grade are used in basement or subterranean construction. When used at grade level, the slab is placed 8" (200 mm) above grade level unless it is protected by a retaining wall. Approximately 8" to 12" (200 to 300 mm) of topsoil is removed to provide a stable, level building site. Excavation usually extends several feet beyond the building footprint to allow for the operation of excavating equipment needed to trench for the footings. Once forms for the footings have been set, compacted fill material can be placed to support the slab. The slab is placed on a 4" (100 mm) minimum base of compacted sand or gravel fill to provide a level base for the pouring of the slab. The fill is not drawn on the foundation plan, but the material and required depth is specified on the foundation plan.

Slab Reinforcement

When the slab is placed on more than 4" (100 mm) of uncompacted fill, welded wire fabric similar to that shown in Figure 24-38 should be specified to help the slab resist cracking. The engineer will specify the spacings and sizes of welded wire fabric (WWF) to be identified on the foundation plan. Steel reinforcing bars can be added to a floor slab to prevent bending of the slab due to expansive soil. Whereas mesh is placed in a

FIGURE 24-38 Welded wire fabric is placed in concrete slabs that are poured over fill material to help resist cracking. *Courtesy Jordan Jefferis.*

slab to limit cracking, steel reinforcement is placed in a grid pattern near the surface of the concrete in tension to resist bending (see Figure 24-39). The amount of concrete placed around the steel is referred to as *coverage*. Proper coverage strengthens the bond between the steel and concrete and also protects the steel from corrosion or fire. If steel is required to reinforce a concrete slab, footing, or pier, the engineer will determine the size, spacing, coverage, and grade of bars to be used. Generally, steel reinforcement is not shown on the foundation plan but is specified by a note.

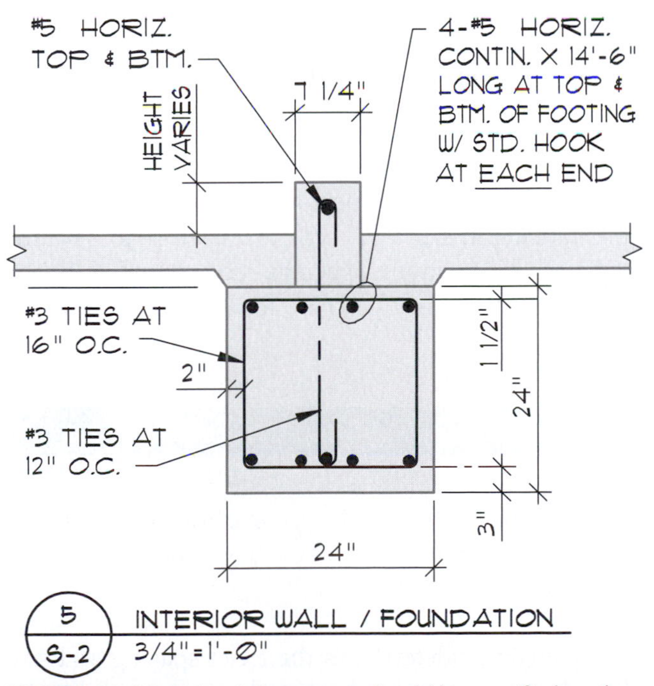

FIGURE 24-39 Slab reinforcement should be specified on the foundation plan and shown in details.

Moisture Protection

The slab is required to be placed over 6 mil polyethylene sheet plastic to protect the floor from ground moisture. When a plastic vapor barrier is to be placed over gravel, sand should be specified to cover the gravel fill to avoid tearing the vapor barrier. The vapor barrier is not drawn but is specified with a note on the foundation plan.

Slab Insulation

Depending on the risk of freezing, some municipalities require the concrete slab to be insulated to prevent heat loss. The insulation can be placed under the slab if it is not placed on the outside of the stem wall. When placed under the slab, a 2" × 24" (50 × 600 mm) minimum rigid insulation material is usually required to insulate the slab. An isolation joint is typically provided between the stem wall and the slab to prevent heat loss through the stem wall. Figure 24-12 shows common methods of insulating the intersection of the floor and stem wall. Insulation is not shown on the foundation plan but is represented by a note and specified in details referenced to the foundation plan. Figure 24-40 shows plumbing

FIGURE 24-40 HVAC and plumbing lines that are required to be placed prior to the foundation must be indicated on a plan. The information is often provided by drafters working for consulting engineers. *Courtesy Margarita Miller.*

installation, which must be placed under the slab prior to the pouring of the slab. Information supplied by consulting engineers needs to be incorporated into the foundation plan by the drafter to show required HVAC, electrical, and plumbing. Figure 24-41 shows a foundation plan referencing plumbing drains. Although piping runs are often not shown, terminations such as floor drains are shown, specified, and located on the foundation drawings.

Radon Protection

Structures built in areas of the country identified by the Environmental Protection Agency (EPA) as having high radon levels need to be protected from the cancer-causing gas. A common method of reducing the buildup of radon can be achieved by placing a 4" (100 mm) PVC vent pipe in the gravel fill covered with 6 mil polyethylene sheathing. Any joints, cracks, or penetrations in the floor slab must be caulked. The vent is run under the slab until it can be routed up through the framing system to an exhaust point that is a minimum of 10' (3000 mm) from other openings in the structure and 12" (300 mm) above the roof. The piping should be indicated on the foundation drawings and coordinated with other drawings to show electrical wiring for installation of a fan located in the vent stack, and a system failure warning device.

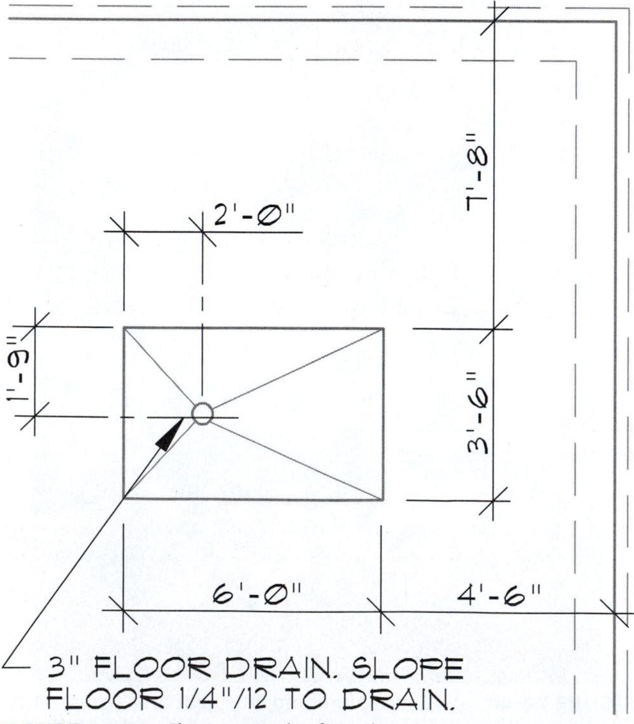

FIGURE 24-41 Changes in the foundation caused by HVAC ducts and plumbing lines such as this floor drain should be indicated on the foundation plan.

Post-Tensioned Concrete Reinforcement

The methods of reinforcement mentioned thus far are used when it is assumed that the slab will be poured over stable soil. Post-tensioning of slabs extends the range of where slabs can be placed. This method of construction was developed for reinforcement of slabs that were poured at ground level and then lifted into place for multilevel structures. Adopted for use on at-grade slabs, post-tensioning allows concrete slabs to be poured over expansive soil.

For design purposes, a concrete slab is considered to be a wide, shallow beam that is supported by a concrete foundation at the edge. While a beam may sag at the center due to loads and gravity, a concrete slab can either sag or bow depending on soil conditions such as excessive moisture differential, freezing, or expansive soil. Because concrete is very poor at resisting stress from tension, the slab must be reinforced with a material such as steel, which has a high tensile strength. Steel tendons with anchors can be extended through the slab as it is poured. Usually between 3 and 10 days after the concrete has been poured, hydraulic jacks stretch these tendons. The tendon force is transferred to the concrete slab through anchorage devices at the ends of the tendons. This process creates an internal compressive force throughout the slab, increasing the slab's ability to resist cracking and heaving. Post-tensioning usually allows for a thinner slab than normally would be required to span over expansive conditions, the elimination of other slab reinforcing, and the elimination of most slab joints. Two common methods of post-tensioning used for slabs are the flat slab and the ribbed slab methods.

Post-Tensioned Flat Slabs

The *flat slab* method of post-tensioning uses steel tendons ranging in size from 3/8" to 1/2" (9.5 to 13 mm) in diameter. Maximum spacings recommended by the Post-Tensioning Institute (PTI) for tendons include the following:

Tendon Diameter	Spacing
3/8" (9.5 mm)	5'–0" (1500 mm)
7/16" (11 mm)	6'–0" (1800 mm)
1/2" (13 mm)	9'–0" (2700 mm)

The engineer determines the exact spacing and size of tendons based on the loads to be supported and the strength and conditions of the soil. When required, the tendons can be represented on the foundation plan as seen in Figure 24-42. Details also need to be provided to indicate how the tendons will be anchored, and to show the exact locations of the tendons.

Figure 24-43 shows an example of a tendon detail that would be referenced to the foundation drawings. In addition to representing and specifying the steel throughout the floor system, the drafter needs to specify the engineer's requirements for the strength of the concrete at 28 days, the period when the concrete is to be stressed, as well as what strength the concrete should achieve before stressing. This information is generally placed in the specifications in a project manual.

Post-Tensioned Ribbed Slabs

A second method of post-tensioning an on-grade floor slab is with the use of concrete ribs or beams placed below the slab. These beams reduce the span of the slab over the soil and provide increased support. The engineer determines the width, depth, and spacing of the ribs based on the strength and condition of the soil and the size of the slab. Figure 24-44 shows an example of a beam detail that a drafter would be required to draw and reference on the foundation plan to show the reinforcing specified by the engineer. Figure 24-45 shows an example of how these beams could be represented on the foundation plan.

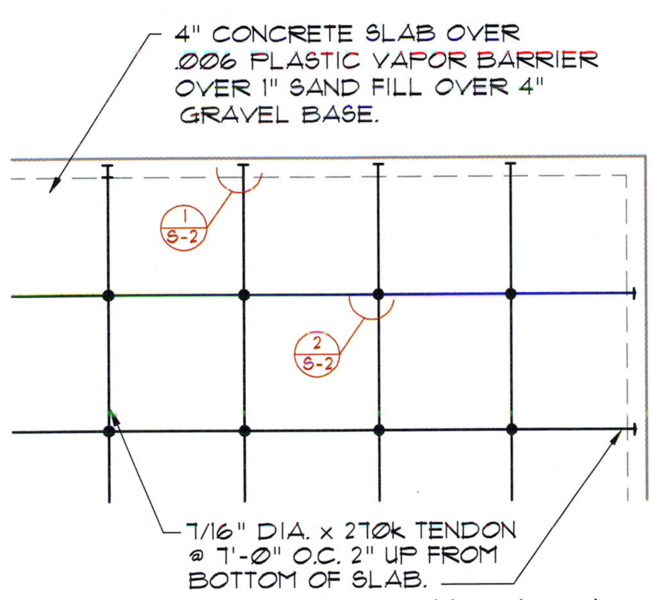

FIGURE 24-42 Post-tensioned concrete slabs can be used when the slab must be placed over unstable soil.

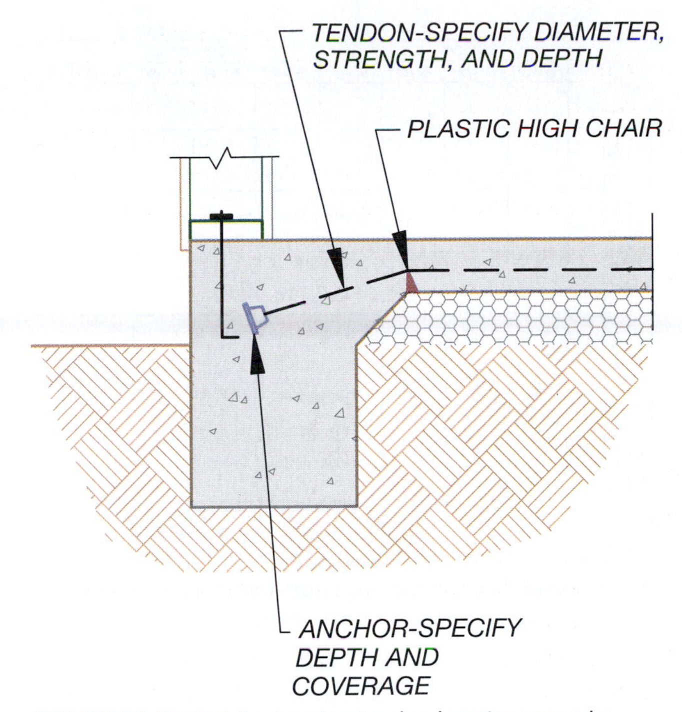

FIGURE 24-43 Details showing tendon locations must be referenced to the slab plan.

Wood Floor Systems over a Crawl Space

Floor systems built over a crawl space are usually made using western platform or post-and-beam construction methods. Wood framing methods were introduced in

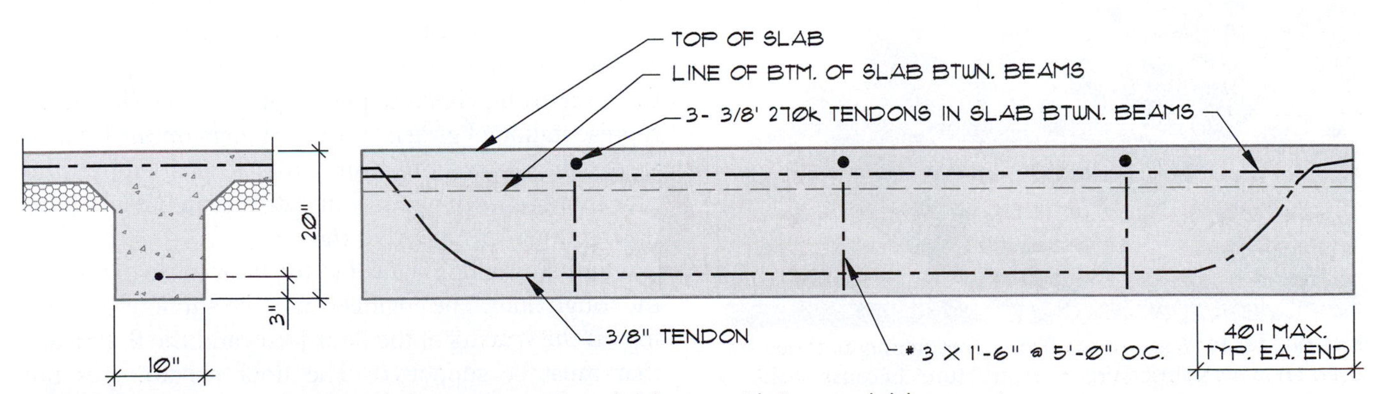

FIGURE 24-44 A rib can be placed below a post-tensioned slab to provide increased slab support.

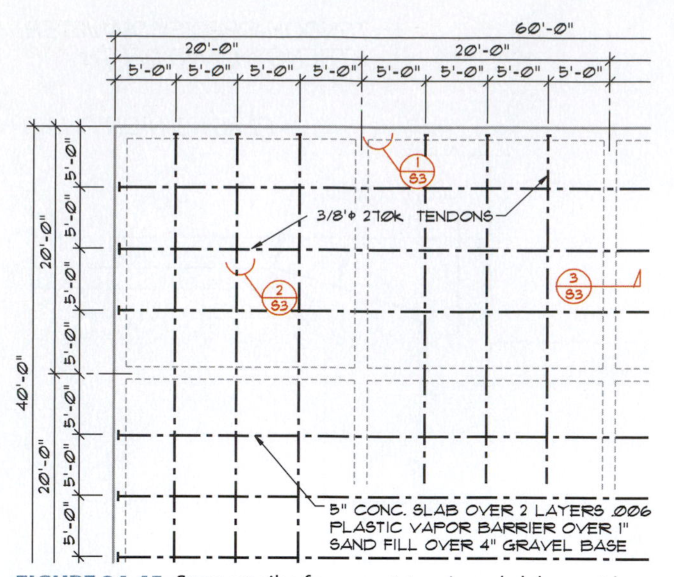

FIGURE 24-45 Support ribs for a post-tensioned slab must be shown and located on the foundation plan.

Chapters 7 and 8, as wood construction was considered. This chapter considers only how wood and timber are drawn on the foundation plan.

Western Platform Framing

Floor joists similar to those in Figure 24-46 are used to span between the stem walls in western platform construction. The material size and spacing must be represented on the foundation plan. Figure 24-47 shows methods of representing floor joists on a foundation plan. The floor joists rest on top of the mudsill, which is supported by the stem wall (see Figure 24-48). Figure 24-49 shows other common methods of supporting floor joists.

FIGURE 24-46 Western platform construction methods are used on many light commercial structures because of the system's many economical features. *Courtesy Zachary Jefferis.*

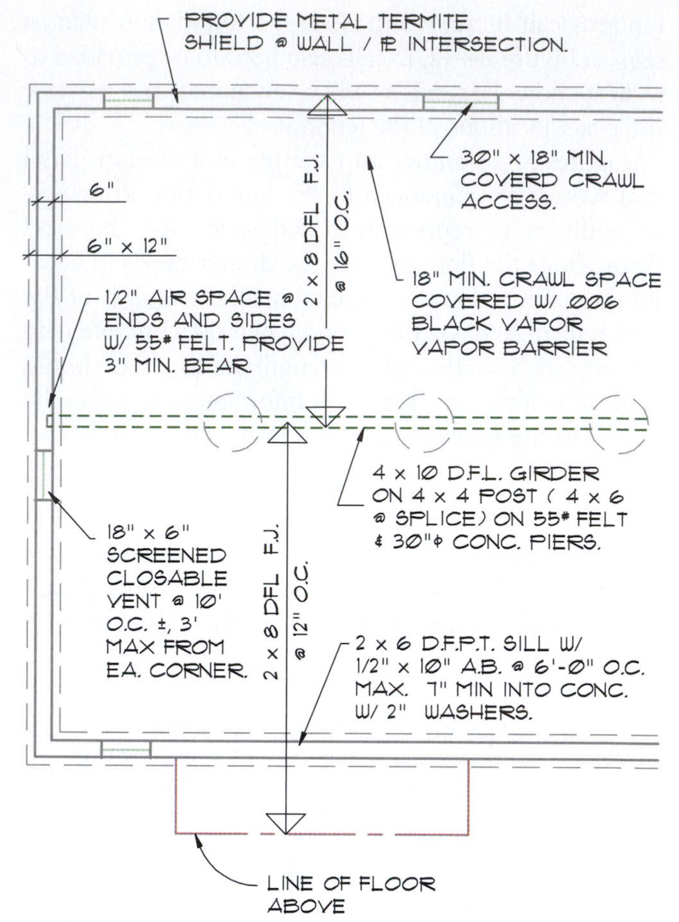

FIGURE 24-47 Representing materials of a western platform floor system on the foundation plan. Common members that should be shown include the stem walls, footings, girders, piers, floor joists, floor overhangs, vents, crawl access, and metal anchors or straps.

When the distance between stem walls is too great for the floor joists to span, a girder is used to support the floor joists. Unless the floor joists are designed to support concentrated loads, a girder must also be placed under interior bearing walls that are not supported by stem walls. A beam pocket or hanger supports the girder at the stem wall and, by a post and pier, provides support as it spans across the foundation. In areas where seismic activity must be planned for, the girder must be attached to the post, and the post must be attached to the supporting concrete pier. Figure 24-47 shows the representation of girders, post, and piers on the foundation plan. Connections from girder to post and post to pier are not shown on the foundation plan but are specified by notes referenced to the pier.

Floor sheathing is nailed to the floor joists to provide the rough floor. The engineer sizes the subfloor depending on the spacing of the floor joists and the floor loads that must be supported. The floor sheathing is not represented on the foundation plan but is specified in a

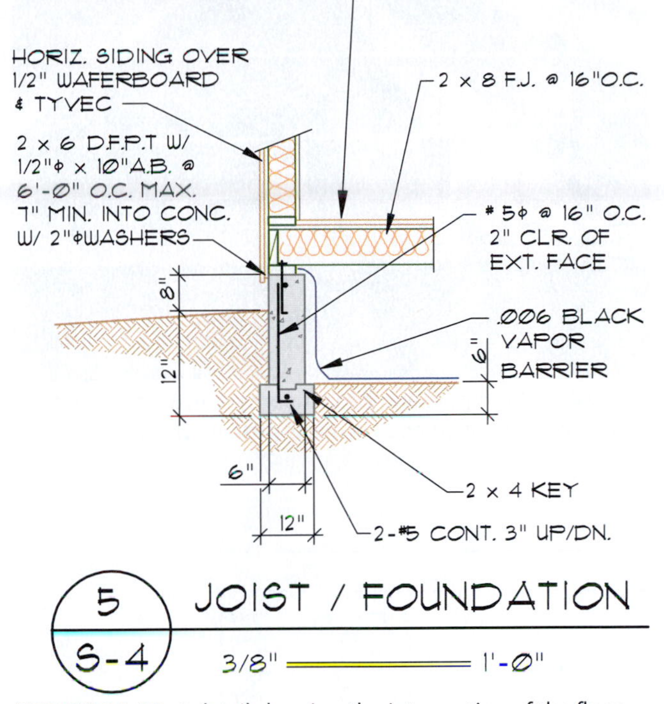

FIGURE 24-48 A detail showing the intersection of the floor system to the foundation must be referenced to the foundation plan.

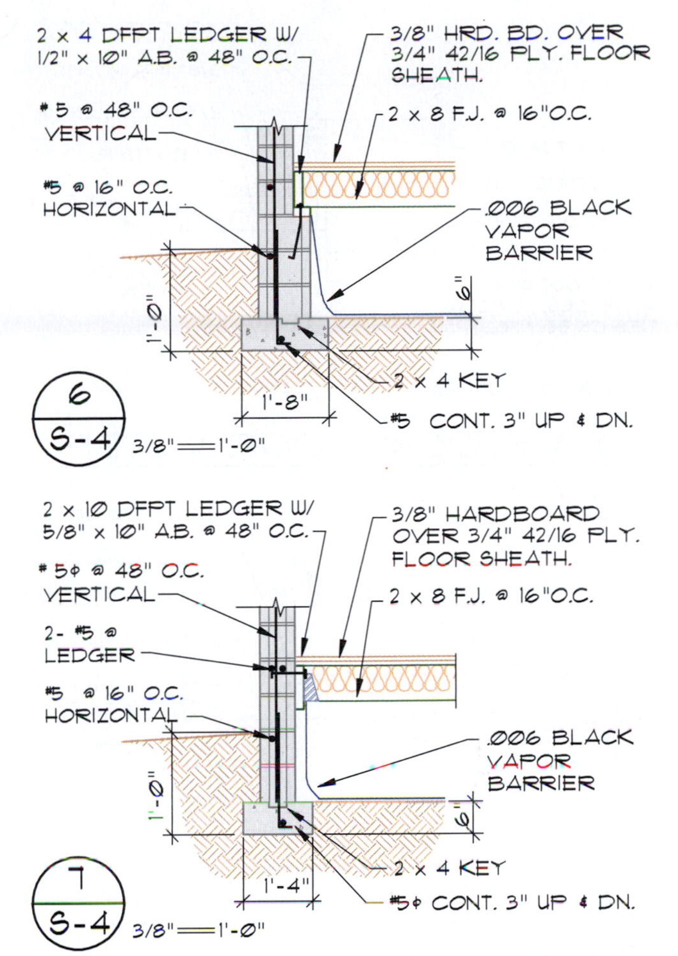

FIGURE 24-49 Alternative methods of supporting floor joists for masonry walls.

general note. Figure 24-50 shows some common methods used to frame changes in floor height. If a stem wall is required, it must be represented on the foundation plan just as an exterior stem wall and footing would be represented. If the change is made over a girder, a linetype different from the line used to represent the girder should be used to represent the floor edge. Figure 24-51 shows methods of representing elevation changes on the foundation plan.

Post and Beam

Post-and-beam construction is used on small commercial and multifamily projects in areas of the country where cold or dampness is a consideration. Rather than having floor joists span between the stem walls, a series of beams is used to support the subfloor, as shown in Figure 24-52. Beams are typically 4 × 8 (100 × 200) spaced at 48" (1200 mm) on center with supports placed at 8'–0" (2400 mm) along the beam. This pattern is based on the use of 4 × 8 (100 × 200) sheets of plywood but the exact girder type, size, and spacing are affected by the loads to be supported. A post-and-beam floor system is well suited to projects that have very few interior bearing walls. As seen in Figure 24-53, additional girders must be placed under interior bearing walls that do not align with the normal placing of girders.

Once the mudsill is bolted to the foundation wall, girders can be placed so that the top of the girder is level with the top of the mudsill. Girders are supported just as with western platform construction. Plywood with a thickness of 1 1/8" (28.5 mm) and an APA rating of STURD-I-FLOOR 2-4-1 with an exposure rating of EXP-1 is generally used for the subfloor. Two-inch (50 mm) thick material such as 2 × 6 (50 × 150) T & G boards laid perpendicular to the beams can also be used for the subfloor. The strength and quality of the floor is greatly increased when the subfloor is glued to the support beams, eliminating squeaks, bounce, and nail popping.

Party Wall and Fire Wall Support

Regardless of the material used to form the floor system, care must be taken with special walls such as fire and party walls to ensure the wall will function as it is designed. Because unprotected material cannot pass

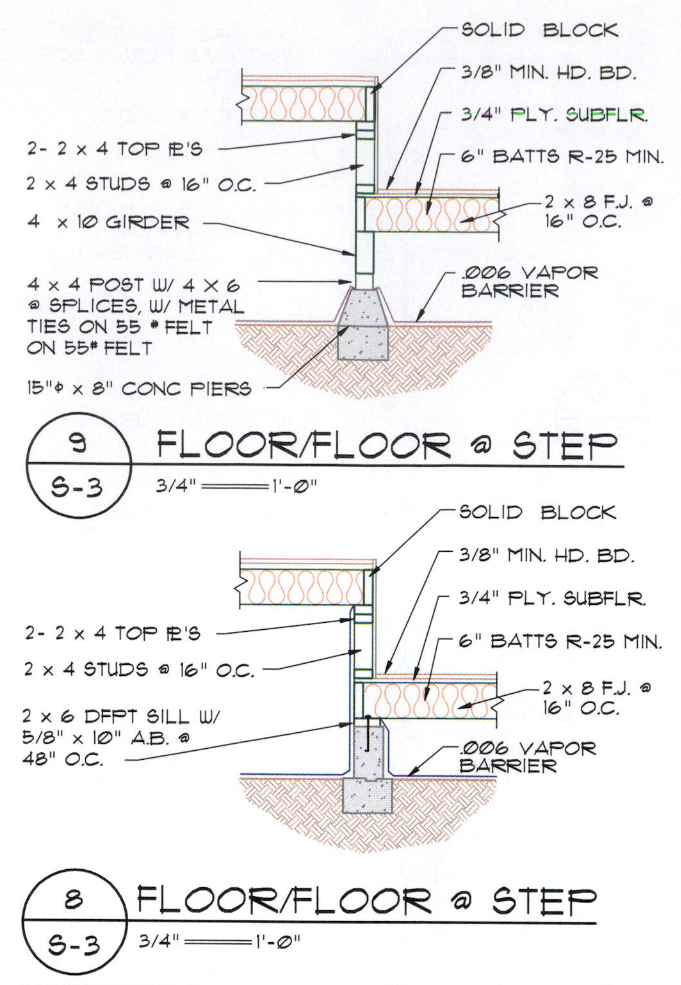

FIGURE 24-50 Common methods of framing changes in floor elevation using western platform construction methods.

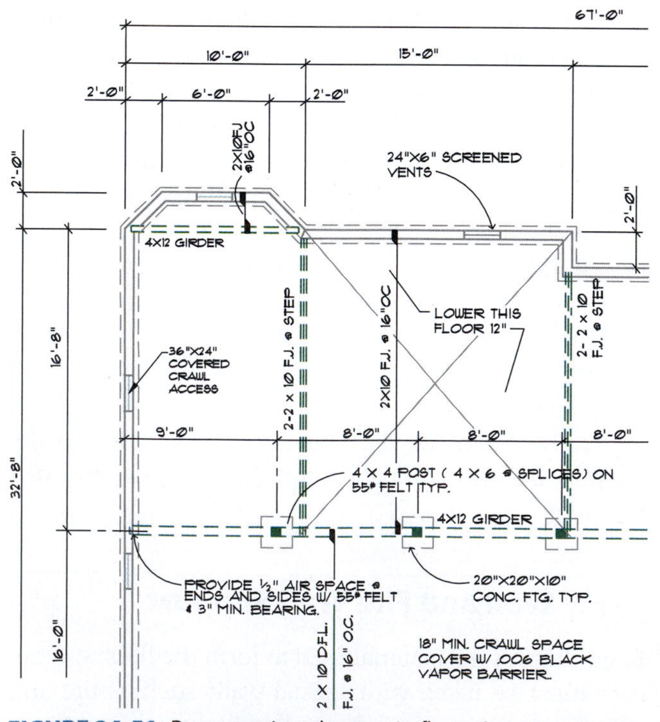

FIGURE 24-51 Representing changes in floor elevation on the foundation plan with a joist floor system.

FIGURE 24-52 A post-and-beam floor system is often used for multifamily construction projects in damp climates. *Courtesy Floyd Miller.*

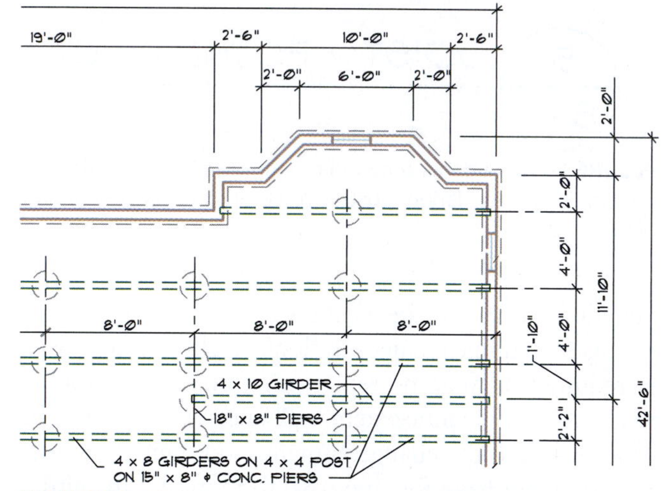

FIGURE 24-53 Representing a post-and-beam floor system on the foundation plan. Members that should be shown include the stem walls, footings, girders, piers, vents, crawl access, and metal anchors or straps.

through a fire wall, extra support is required to be represented on the foundation plan to support beam terminations at a fire wall. Figure 24-54 shows construction of a 2-hour wall for a warehouse.

A party wall typically requires complete separation for adjacent units to help control vibrations. For a wood floor, this requires girders to support each portion of the wall. Figure 24-55 shows two options for framing the intersection of a party wall and the floor.

FOUNDATION DRAWINGS

Two types of drawings can be used to represent the information used to construct the floor and foundation systems. A *slab-on-grade plan* is usually drawn when the

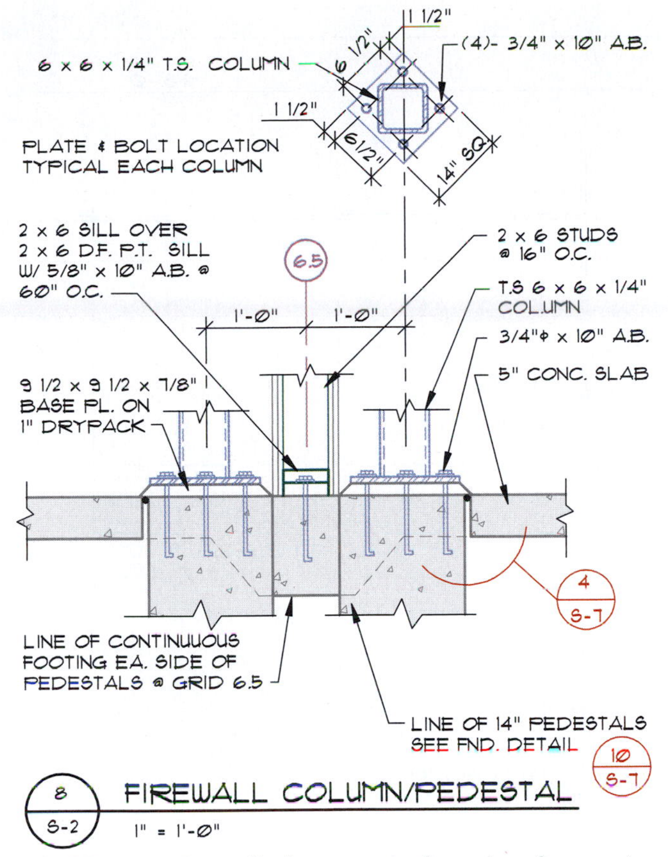

FIGURE 24-54 Fire walls shown on the floor plan often require special construction that must be referenced to the foundation plan. This detail shows the support for the columns that are used to support the roof beams that are interrupted by the fire wall.

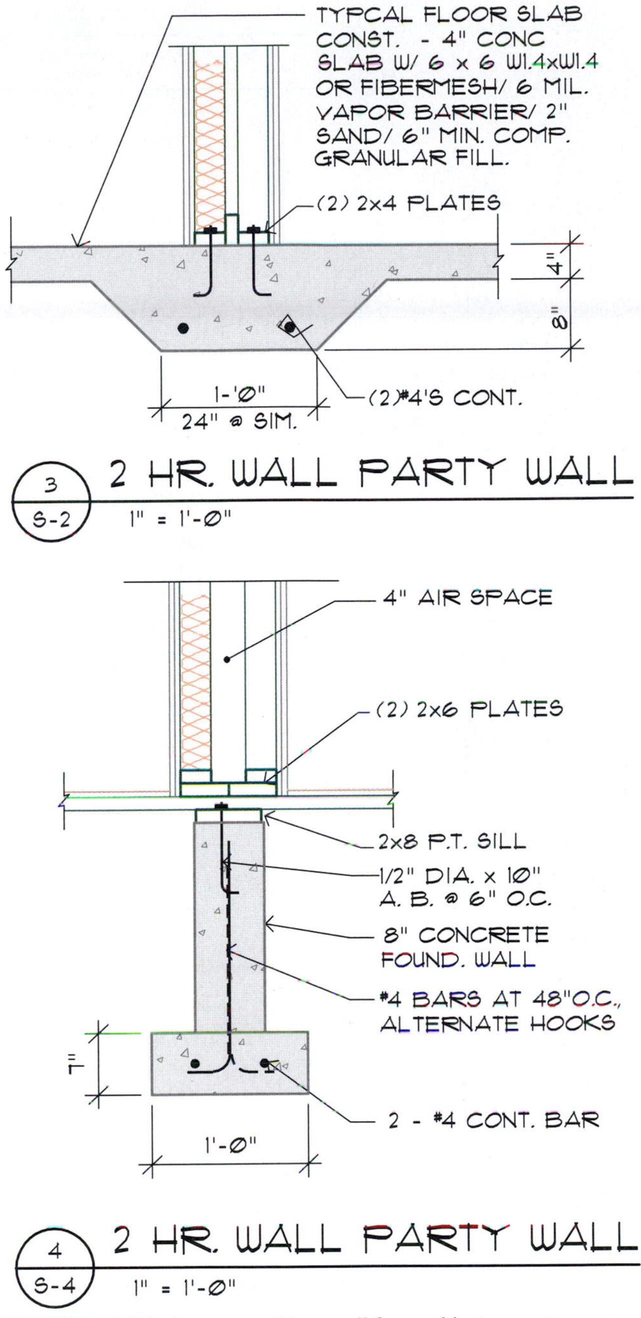

FIGURE 24-55 A party wall is a wall framed between two separate dwelling units. Intersections of party walls and the floor system must be referenced to the foundation plan. *Courtesy Scott R. Beck, Architect.*

walls of the structure extend from the footing. Masonry and precast concrete panel construction are examples of structures where the wall would rest directly on the footing, eliminating the need for a stem wall. In each case, the walls of the structure determine the limits of the floor. The locations of walls that form the perimeter of the structure are placed on the floor plan. The slab-on-grade plan is used to show the shape of the floor slab. Interior walls are shown on the floor plan and not on the slab plan. The slab plan is used to show only items necessary to construct the floor. Figure 24-56 shows an example of a slab-on-grade plan. The process for completing this plan is given later in this chapter.

A foundation plan is used to supplement the slab-on-grade plan. Only material used to represent the actual foundation is shown. Because the stem walls (the concrete panels) are shown on the slab-on-grade plan, they are not shown on the foundation plan. The foundation plan is used to show the size, shape, location, and elevation of all footings. Because no other level is represented, the outline of the footings and piers is drawn using continuous lines.

Figure 24-57 shows the foundation plan that would accompany the slab plan shown in Figure 24-56. The process for completing this type of foundation plan is given later in this chapter.

When a monolithic slab/foundation is poured, it is represented in one foundation plan, similar to Figure 24-58. Western platform and post-and-beam floor systems are also represented on the foundation plan rather than having two separate plans. Material above grade is generally drawn using continuous lines,

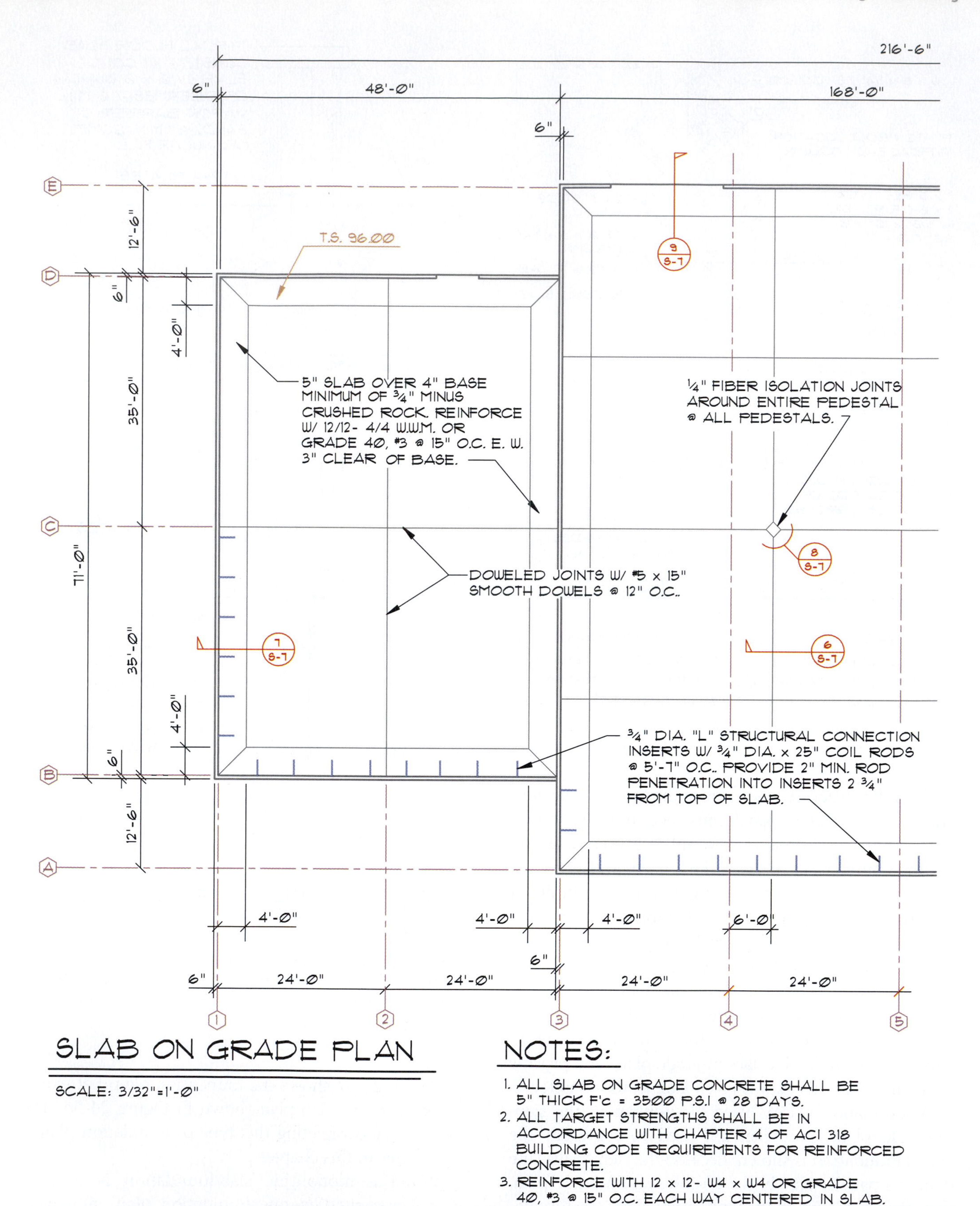

FIGURE 24-56 The placement of slabs and slab joints is often separated from the foundation plan to provide clarity. A slab-on-grade plan shows only an on-grade concrete floor system. *Courtesy LeRoy Cook.*

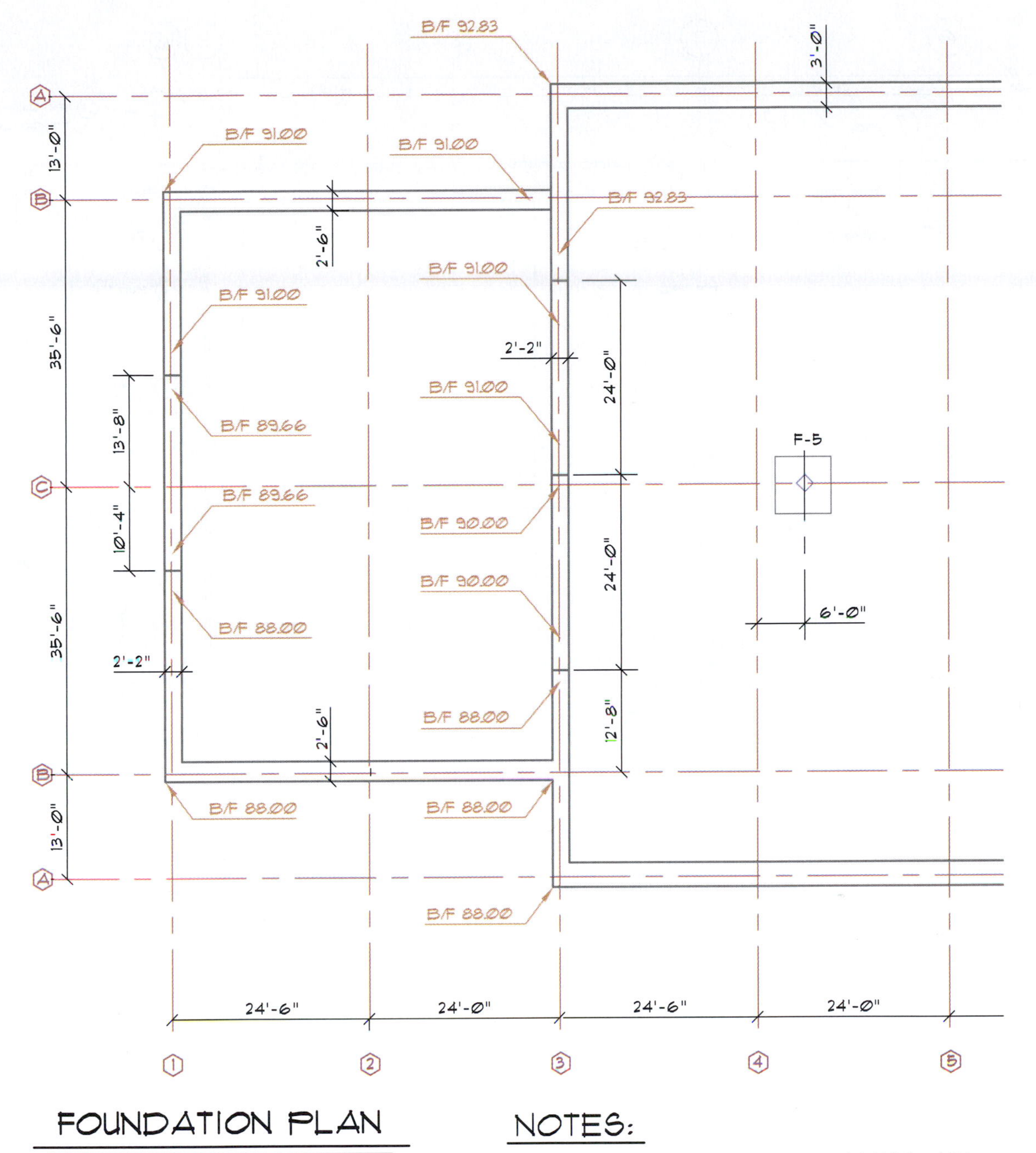

NOTES:

1. ALL FOUNDATION, PEDESTAL AND RETAINING WALL CONCRETE SHALL BE F'c 3000 PSI AT 28 DAYS.
2. ALL STEEL BAR REINFORCEMENT SHALL BE ASTM A615, GRADE 40, DEFORMED BARS UNLESS OTHERWISE SPECIFIED IN DRAWINGS OR DETAILS.

COLUMN FOOTING SCHEDULE

FOOTING	PEDESTAL HEIGHT	SIZE	BTM. OF FTG.
F-1	6.39	7'-3" x 7'-3" x 14"	88.44
F-2	5.96	7'-6" x 7'-6" x 14"	88.87
F-3	5.53	5'-0" x 5'-0" x 14"	89.30
F-4	5.53	4'-3" x 4'-3" x 14"	89.30
F-5	5.18	7'-0" x 7'-0" x 14"	89.65

FIGURE 24-57 When a slab-on-grade plan is provided, the foundation plan shows only the footings. This foundation plan is for the structure shown in Figure 24-63. *Courtesy LeRoy Cook.*

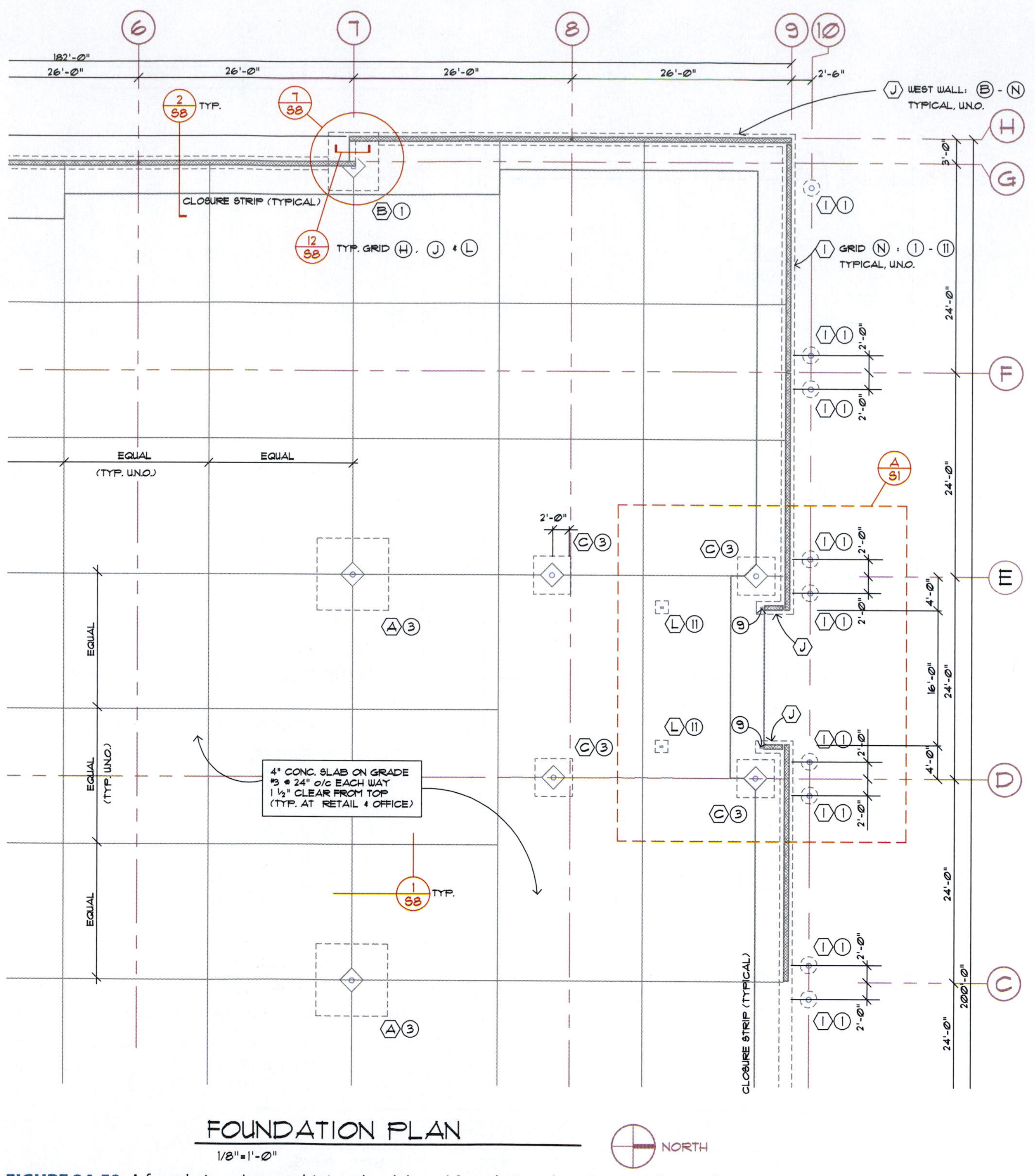

FIGURE 24-58 A foundation plan combining the slab and foundation plan. *Courtesy Bill Berry of Berry-Nordling Engineers, Inc. and Thomas J. Kuhns, A.I.A., Michael & Kuhns Architects, P.C.*

and material below grade such as footings and piers is usually drawn using dashed lines. Information to complete each type of foundation is presented in this chapter. Before considering differences, it is important to consider similarities between each type of plan.

Symbols

Grid markers, elevation symbols, and a north arrow are the symbols typically associated with foundation drawings. The grid and north arrow can be thawed from other drawings. Elevation symbols should match symbols used for other drawings. Elevation symbols often include an elevation on both the top and bottom of the line, as shown in Figure 24-59. When more than one elevation is given per symbol, a schedule should be included with the drawing to define each label. Elevation labels used on foundation drawings include the following:

Elevation Label	Definition
BOF	Bottom of footing
BOW	Bottom of wall
FF	Finish floor
TOF	Top of footing
TOG	Top of grate
TOW	Top of wall

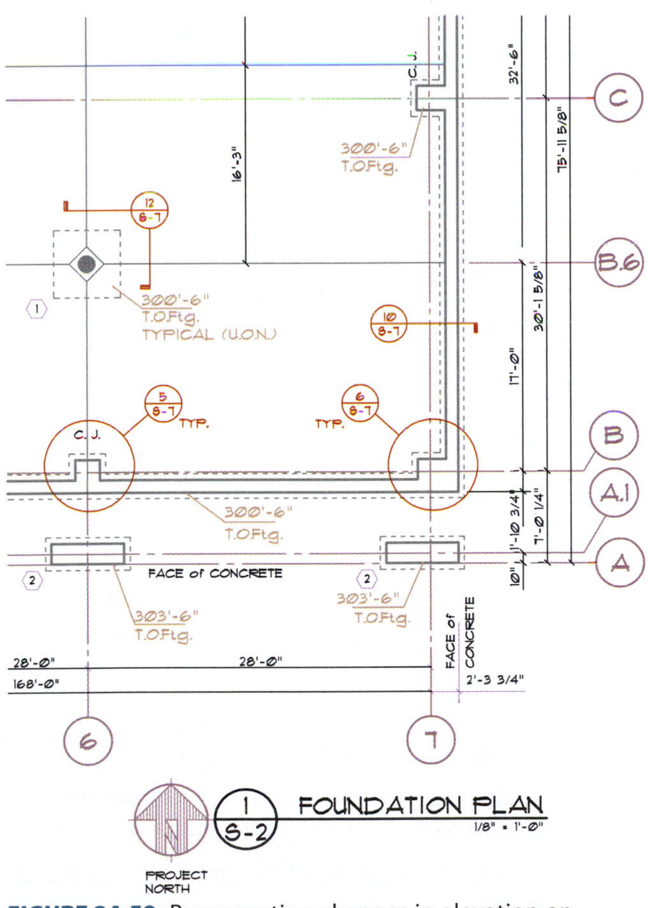

FIGURE 24-59 Representing changes in elevation on the foundation plan may be referenced to the top of slab (TOS), top of footing (TOF), or bottom of footing (BOF). *Courtesy Van Domelen/Looijenga/McGarrigle/Knauf Consulting Engineers.*

Schedules

Just as with other drawings, schedules are an effective method of presenting foundation information. Separate schedules are often used to represent footings, piers, pilasters, beams, and repetitive framing members. Reference symbols that clearly represent the material to be specified should be used. Continuous footings could be represented by a symbol such as F-1 or F-6, with piers represented by P-1 or P-8. Many schedules include listings for the symbol, the quantity, size, and required reinforcing. Reinforcing can be divided into separate columns for upper and lower mats, as well as required hairpin size and spacing. Figures 24-27 and 24-57 each provide examples of the use of schedules to keep the foundation drawing uncluttered. Spot footing schedules often include additional listings for elevation, pedestal height, and pedestal reinforcing.

Annotation

As with other plan views, foundation drawings require the use of local and general notes to specify all materials. Most notes should be placed using the same consideration that was used to place notes on other framing plans. More so than with other drawings, specifications controlling work quality and material are usually placed with the foundation plan. Figure 24-60 shows an example of the notes typically associated with a foundation plan.

Dimensioning

The method of dimensioning is as important as the lines and symbols used to represent the foundation material. Line quality for dimensions and extensions should match that used for the floor or framing plans. Jogs in the foundation are dimensioned using the same methods used on the floor plan. Most of the dimensions for major shapes will be exactly the same as the corresponding dimensions on the floor plan. The foundation plan can be drawn in the same drawing file as the floor plan. This allows the overall dimensions for the floor plan to be displayed on the foundation plan if different layers are used to place the dimensions. Layers such as A BASE DIM, A FLOR DIM, and S-FND DIMEN help differentiate between dimensions for the floor and foundation. If the drawings are to be in separate files, the dimensions from the framing plan can be referenced to the foundation drawings.

FOUNDATION NOTES:

1. DESIGN SOIL PRESSURE 3500 PSF AT SOIL AND 5000 PSF AT ROCK PER GEOTECHNICAL REPORT PREPARED BY WRIGHT/DEACON AND ASSOCIATES, DATED JULY 28, 1998.
2. ALL FOOTINGS TO BEAR ON FIRM, UNDISTURBED SOIL OR APPROVED COMPACTED FILL MINIMUM 24" BELOW ADJACENT FINAL GRADE. NOTIFY VLMK BEFORE PROCEEDING IF ANY UNUSUAL CONDITIONS ARE ENCOUNTERED IN THE FOOTING EXCAVATIONS.
3. DO NOT EXCAVATE CLOSER THAN 2:1 SLOPE BELOW FOOTING EXCAVATIONS.
4. CLEAN ALL FOOTING EXCAVATIONS OF LOOSE MATERIAL BY HAND.
5. EXCAVATIONS MAY BE MADE UNDER CONTINUOUS FOOTINGS FOR PIPES. BACKFILL TO BE APPROVED BY VLMK.

CONCRETE

1. STRENGTH: AVERAGE CONCRETE STRENGTH AS DETERMINED BY JOB CAST LAB CURED CYLINDER TO BE 3000 PSI AT 28 DAYS PLUS INCREASE DEPENDING ON THE PLANTS STANDARD DEVIATION AS SPECIFIED IN ACI 318-83.
2. MINIMUM MIX REQUIREMENTS:

LOCATION	COMPRESSIVE STRENGTH (PSI)	SLUMP (A)	MIN. CEMENT CONTENT	ADMIXTURES WATER/CEMENT RATIO
FOOTINGS	4000	0-5"	5	NONE 0.55
SLABS ON GRADE (INTERIOR)	3500	2-4"	5 1/2	WRA (B) 0.45
SLABS ON GRADE (EXTERIOR)	3000	2-4"	5 1/2	WRA, AE (C) 0.50
WALLS AND COLUMNS	4000	2-4"	5 1/2	WRA/PT20 (D) 0.50
MISC. CONC.	3000	5"	5	WRA 0.55
LT.WT. SLAB ON METAL DECK	3500	5"	5	WRA 0.45

A. SLUMP EXCEEDING SPECIFIED LIMITS SHALL NOT BE INCORPORATED IN THE PROJECT EXCEPT BY WRITTEN APPROVAL FROM ENGINEER.
B. WRA = WATER REDUCING AGENT
C. AE = AIR ENTRAINMENT
D. POZZOTEC 20 REQUIRED FOR CONCRETE PLACED BELOW 40 DEG.
F. CALCIUM CHLORIDE IS NOT TO BE USED ON THIS PROJECT.

3. PLACE AND CURE ALL CONCRETE PER ACI CODES AND STANDARDS.
4. PROVIDE CONTROL JOINTS IN ALL INTERIOR SLABS ON GRADE AS SHOWN ON PLANS.
5. SLEEVES, PIPES OR CONDUITS OF ALUMINUM SHALL NOT BE EMBEDDED IN STRUCTURAL CONCRETE UNLESS EFFECTIVELY COATED.

FIGURE 24-60 General notes can be used to organize information related to the foundation plan. Notes are often divided into separate divisions including concrete, reinforcing steel, soil, and framing lumber. *Courtesy Van Domelen/Looijenga/McGarrigle/Knauf Consulting Engineers.*

A different method is used to dimension the interior walls of a foundation than is used on a floor plan. Foundation walls are dimensioned from face to face rather than face to center, as on a floor plan. Footing widths are dimensioned from center to center. Figures 24-9, 24-17, 24-36, 24-56, 24-57, and 24-58 provide examples of placing dimensions on the slab-on-grade and foundation plans.

Layer Guidelines for Foundation Plans

As with other drawings, the layers used to draw the foundation drawings should be carefully considered. Typically, a prefix of S FND or S SLAB can be used to separate the drawing from other drawings that may be contained in the drawing file. Layers such as GRIDS and BASE DIM created for other drawings can be thawed and used with the foundation drawings. Common layer names based on NCS guidelines can be found in the Appendix on the student CD.

DRAWING FOUNDATION PLANS

The foundation plan should be drawn at the same scale and in the same orientation as the floor and framing plans. The plan can best be drawn by thawing the floor plan layers, which show the exterior walls and the interior load-bearing walls.

Concrete Slab

A concrete slab can be completed by using the following steps.

1. Thaw the layers containing the exterior walls and interior bearing walls.
2. Using a running OSNAP of INT, draw the outline of the slab to match the outline of the exterior walls.
3. Draw a centerline to represent all interior bearing walls.
4. Block out for any openings in the stem wall such as those required for garage doors.
5. Draw a support ledge if brick veneer is to be used.
6. Draw the outline of the exterior footings.
7. Draw the outline of all interior continuous footings.

The base of the foundation plan has now been drawn and should now resemble Figure 24-61.

8. Mark the center of all posts and columns, and draw pedestals or the spot footings to support them.
9. Thaw all grid lines and markers.
10. Draw all control joints and changes in floor level.
11. Draw all metal connectors.
12. Draw any plumbing work.
13. Draw any HVAC ductwork.
14. Draw any required electrical conduit runs and slab penetrations.

The foundation plan should now resemble Figure 24-62. Use the following steps to place the needed dimensions.

15. Thaw the base dimension layer to provide overall dimensions for each side of the structure.
16. Thaw dimensions used on the floor plan or provide dimensions to locate each jog in the foundation.
17. Dimension the location of all interior continuous footings.
18. Dimension the size and location of any changes in slab elevation.
19. Dimension all metal connectors.

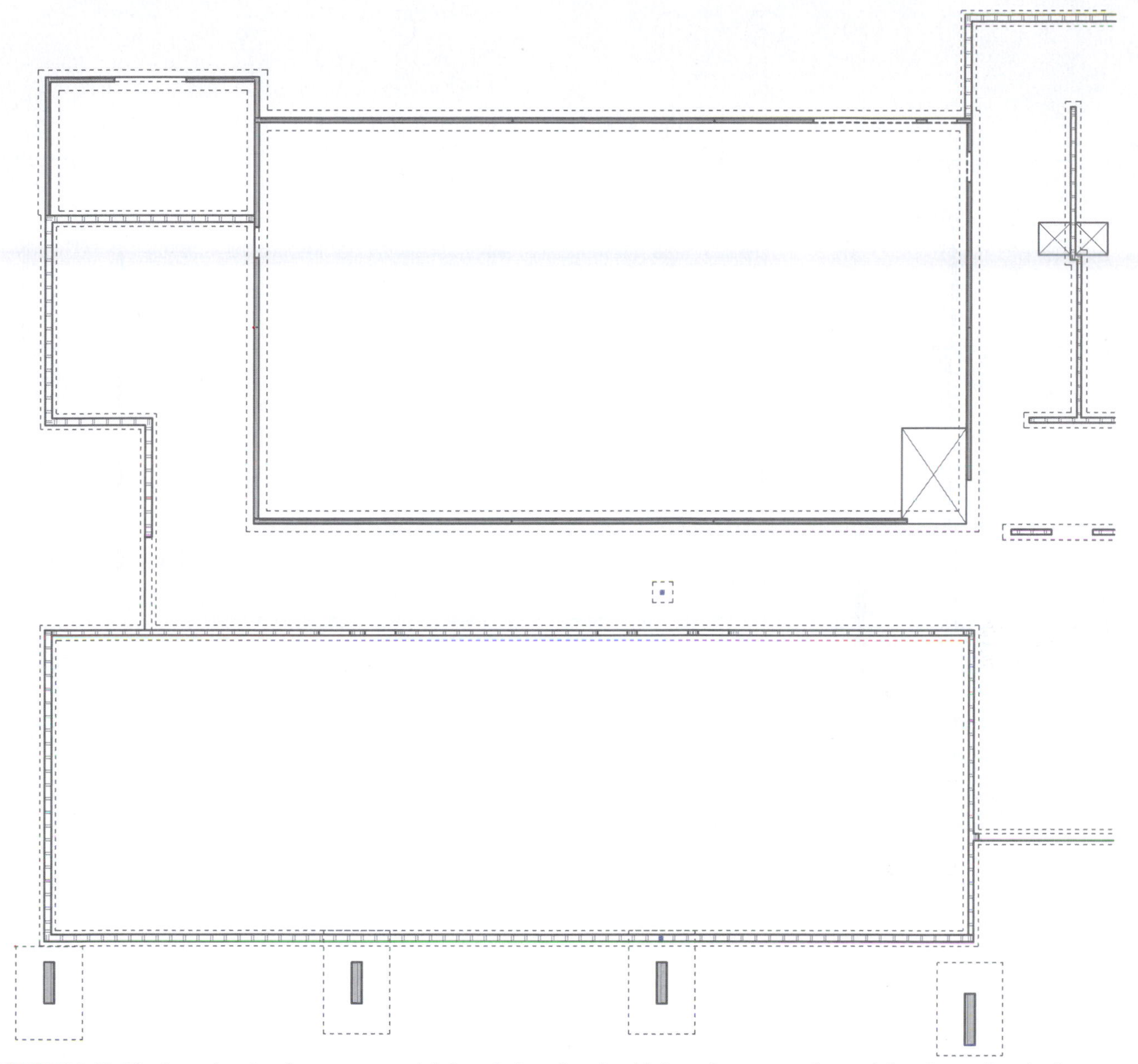

FIGURE 24-61 The base drawing for a concrete slab foundation plan should show the support for each bearing wall and column.
Courtesy Van Domelen/Looijenga/McGarrigle/Knauf Consulting Engineers.

20. Dimension the location to the centers of all spot piers.
21. Dimension the location of any required piping, ductwork, or electrical conduit.

The major features of the drawing have now been dimensioned. The foundation plan should now resemble Figure 24-63. The drawing can be completed by adding all required text and detail markers:

22. Describe the concrete slab, fill material, vapor barrier, and insulation.
23. Describe all footing and stem wall sizes.
24. Describe all required mudsills, ledgers, and anchor bolts.
25. Describe all metal straps, anchors, and connectors.
26. Describe all required changes in elevation.
27. Describe all footings and pedestals.
28. Place any general notes specified by the engineer.
29. Place a title and scale below the drawing.

The completed foundation plan should now resemble Figure 24-64. Similar procedures could be used to draw a post-tensioned concrete slab foundation.

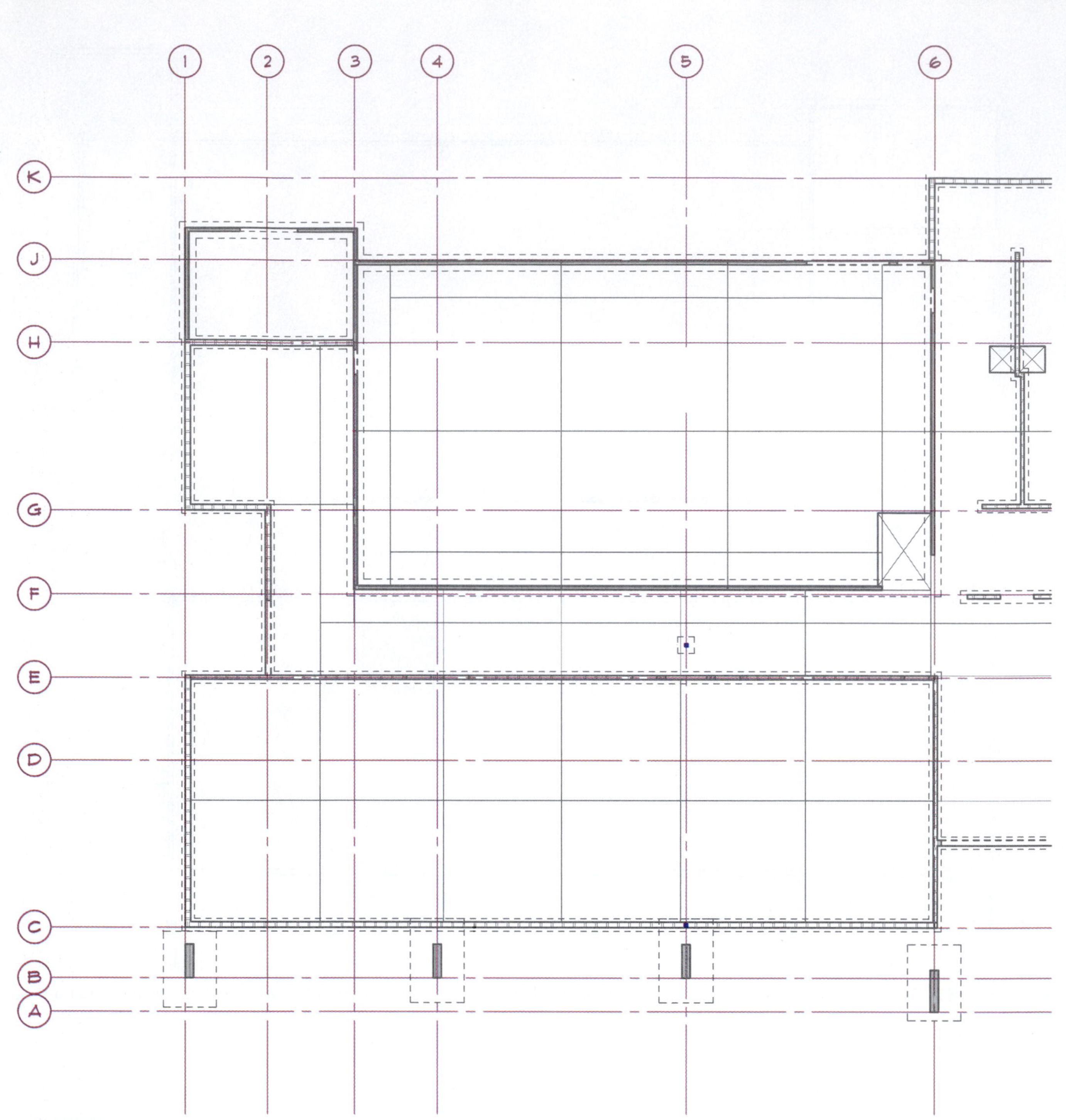

FIGURE 24-62 The foundation plan with grids and slab joints added. *Courtesy Van Domelen/Looijenga/McGarrigle/Knauf Consulting Engineers.*

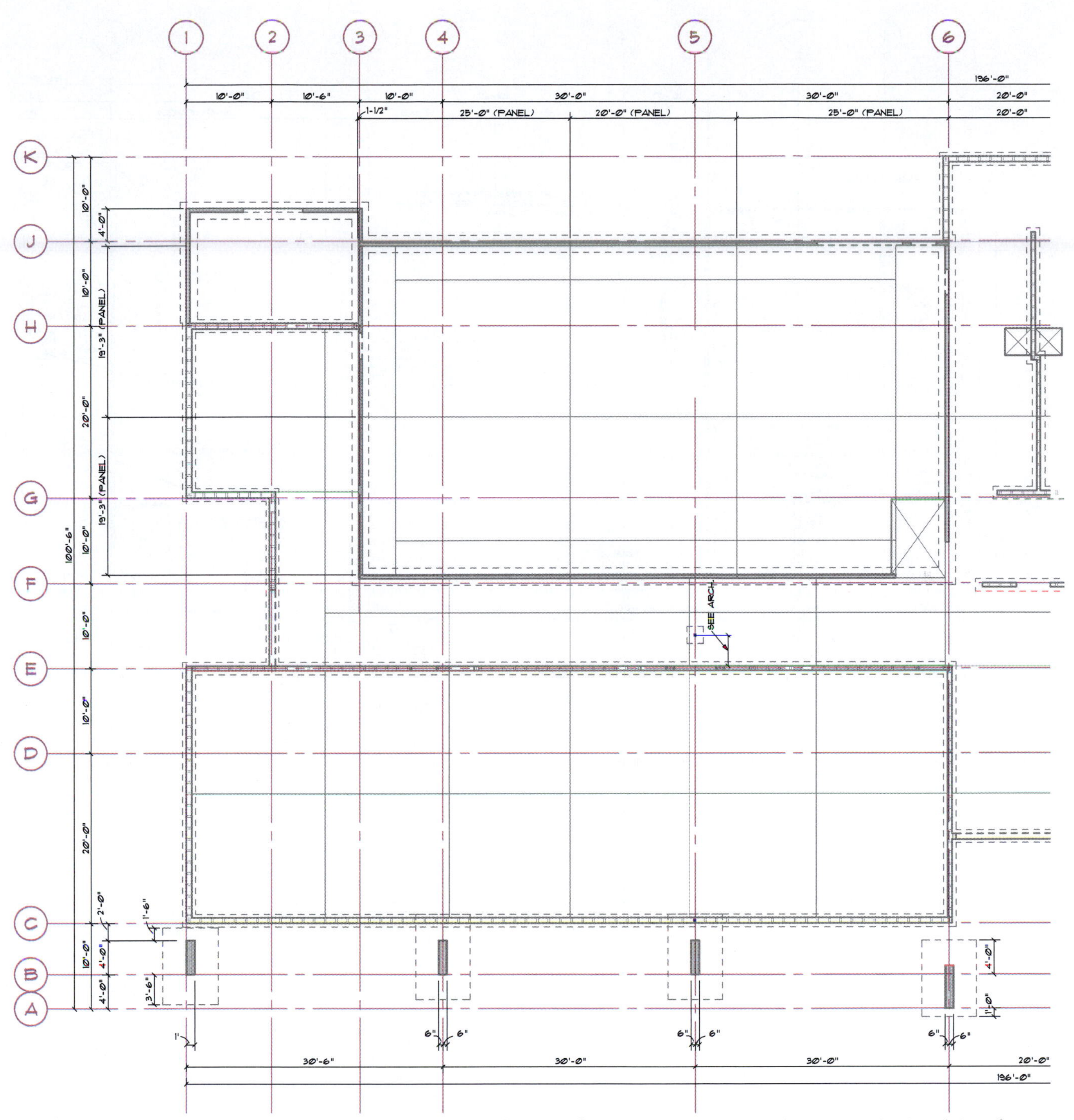

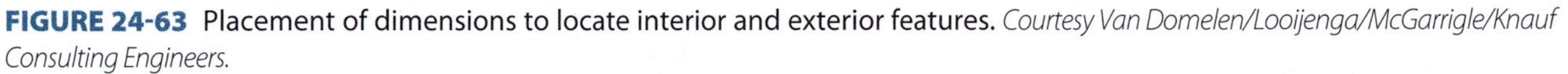
FIGURE 24-63 Placement of dimensions to locate interior and exterior features. *Courtesy Van Domelen/Looijenga/McGarrigle/Knauf Consulting Engineers.*

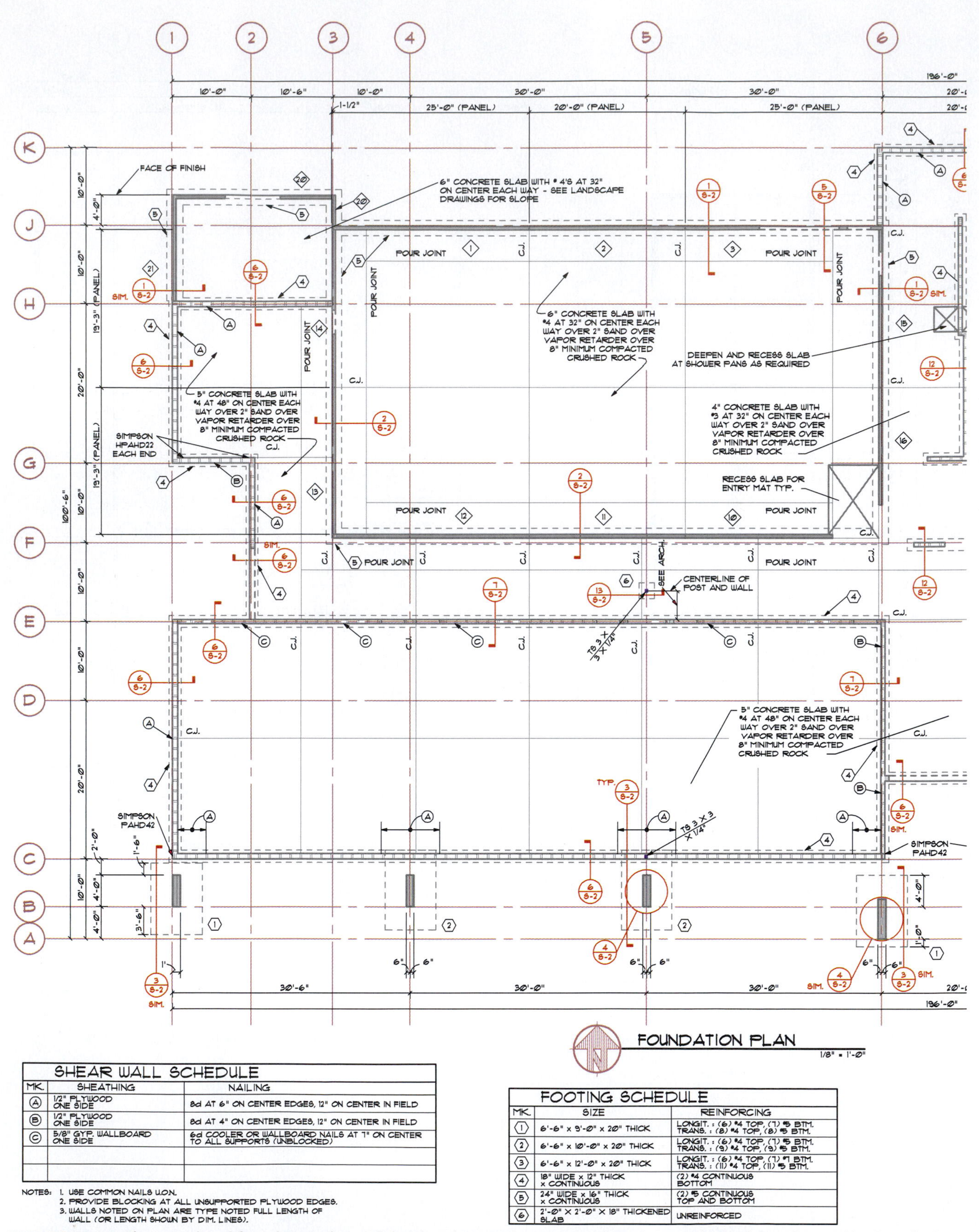

MK.	SHEATHING	NAILING
SHEAR WALL SCHEDULE		
A	1/2" PLYWOOD ONE SIDE	8d AT 6" ON CENTER EDGES, 12" ON CENTER IN FIELD
B	1/2" PLYWOOD ONE SIDE	8d AT 4" ON CENTER EDGES, 12" ON CENTER IN FIELD
C	5/8" GYP. WALLBOARD ONE SIDE	6d COOLER OR WALLBOARD NAILS AT 7" ON CENTER TO ALL SUPPORTS (UNBLOCKED)

NOTES: 1. USE COMMON NAILS U.O.N.
2. PROVIDE BLOCKING AT ALL UNSUPPORTED PLYWOOD EDGES.
3. WALLS NOTED ON PLAN ARE TYPE NOTED FULL LENGTH OF WALL (OR LENGTH SHOWN BY DIM. LINES).

MK.	SIZE	REINFORCING
FOOTING SCHEDULE		
1	6'-6" x 9'-0" x 20" THICK	LONGIT. : (6) #4 TOP, (7) #5 BTM. TRANS. : (8) #4 TOP, (8) #5 BTM.
2	6'-6" x 10'-0" x 20" THICK	LONGIT. : (6) #4 TOP, (7) #5 BTM. TRANS. : (9) #4 TOP, (9) #5 BTM.
3	6'-6" x 12'-0" x 20" THICK	LONGIT. : (6) #4 TOP, (7) #7 BTM. TRANS. : (11) #4 TOP, (11) #5 BTM.
4	18" WIDE x 12" THICK x CONTINUOUS	(2) #4 CONTINUOUS BOTTOM
5	24" WIDE x 16" THICK x CONTINUOUS	(2) #5 CONTINUOUS TOP AND BOTTOM
6	2'-0" X 2'-0" X 18" THICKENED SLAB	UNREINFORCED

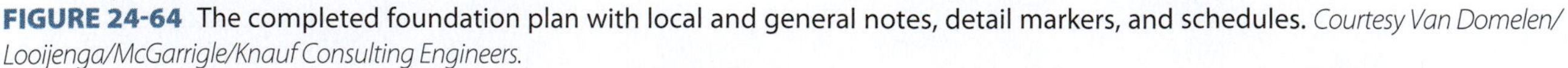

FIGURE 24-64 The completed foundation plan with local and general notes, detail markers, and schedules. *Courtesy Van Domelen/Looijenga/McGarrigle/Knauf Consulting Engineers.*

Floor Joists

A foundation plan for a structure framed with floor joists can be completed by using the following steps.

1. Thaw the layers containing the exterior walls and interior bearing walls.
2. With OSNAP set to ON, draw the outline of the stem wall to match the outline of the exterior walls.
3. Use the OFFSET command to reproduce the interior side of the stem walls. Use TRIM and FILLET to clean up each intersecting corner.
4. Draw a centerline to represent all interior bearing walls.
5. Use the engineer's specifications or determine the sizes for all girders to be placed under load-bearing walls.
6. Block out openings in the stem wall such as those required for garage doors.
7. Draw a support ledge if brick veneer is to be used.
8. Draw the outline of the exterior footings.
9. Draw all interior piers.

The base of the foundation plan has now been drawn and should resemble Figure 24-65.

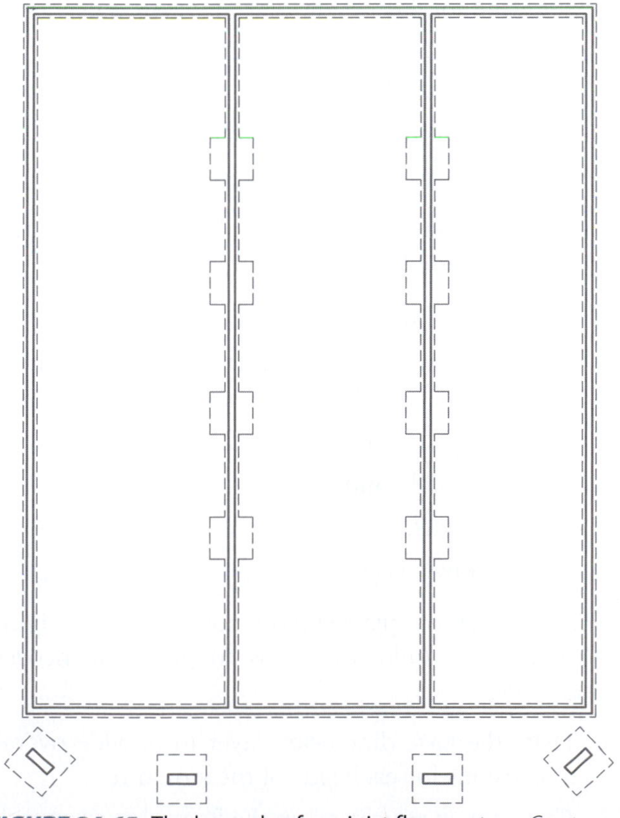

FIGURE 24-65 The base plan for a joist floor system. *Courtesy Scott R. Beck, Architect.*

10. Thaw all grid lines and markers created for other plan views.
11. Mark the center of all exterior posts or columns, and draw the spot footings for each.
12. Draw floor joist indicators.
13. Draw all changes in floor level.
14. Draw all metal connectors.
15. Draw all vents.
16. Draw a crawl access.

The foundation plan should now resemble Figure 24-66. Use the following steps to place the needed dimensions.

17. Thaw the base dimension layer to provide overall dimensions for each side of the structure.
18. Thaw dimensions used on the floor plan or provide dimensions to locate each jog in the foundation.
19. Dimension all openings in the stem wall.
20. Dimension the location of all girders.
21. Dimension the location of all interior piers.
22. Dimension the size and location of any changes in slab elevation.
23. Dimension the location of all metal connectors.

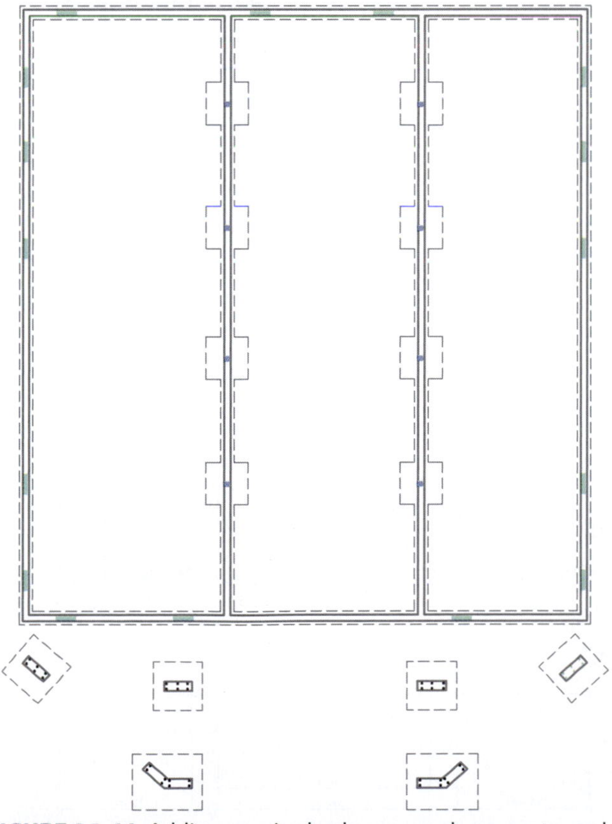

FIGURE 24-66 Adding required columns, anchors, vents, and access. *Courtesy Scott R. Beck, Architect.*

The major features of the drawing have now been dimensioned. The foundation plan should now resemble Figure 24-67. The drawing can be completed by adding all required text.

24. Place notes, detail markers, and dimensions to describe all footing and stem wall sizes.
25. Place notes and detail markers to describe all required mudsills, ledgers, and anchor bolts.
26. Place notes, detail markers, and dimensions to locate and describe all metal straps, anchors, and connectors.
27. Place notes to describe all vents, wall openings, and the crawl access.
28. Describe all girders, post, and piers.
29. Place notes to describe all joist types, size, and spacing.
30. Place notes to describe all required changes in elevation.
31. Place notes to describe the crawl space covering.
32. Place notes to describe all special blocking.

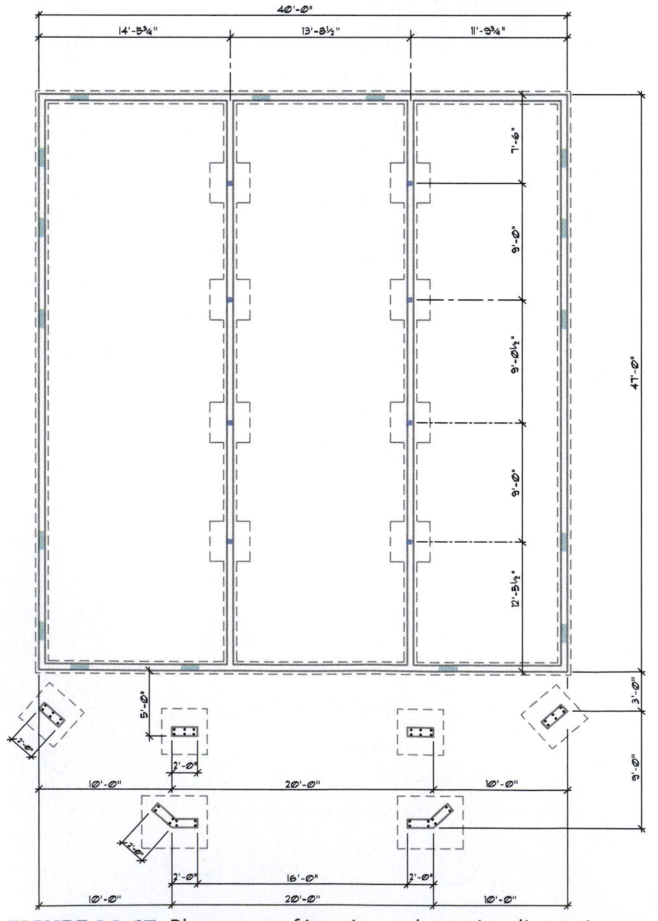

FIGURE 24-67 Placement of interior and exterior dimensions.
Courtesy Scott R. Beck, Architect.

33. Place any general notes specified by the engineer.
34. Place a title and scale below the drawing.

The completed foundation plan should now resemble Figure 24-68. A similar procedure could be used to draw a structure supported on pilings instead of piers.

Post and Beam

A foundation plan for a structure framed with a post-and-beam floor system can be completed by using the following steps.

1. Thaw the layers containing the exterior walls and interior bearing walls.
2. Use OSNAP to draw the outline of the stem wall to match the outline of the exterior walls.
3. Use the OFFSET command to reproduce the interior side of the stem walls. Use TRIM and FILLET to clean up each intersecting corner.
4. Draw a centerline to represent all interior bearing walls.
5. Use the engineer's specifications or determine the sizes for all girders to be placed under load-bearing walls.
6. Layout all girders required for the floor support.
7. Block out for any openings in the stem wall such as those required for garage doors.
8. Draw a support ledge if brick veneer is to be used.
9. Draw the outline of the exterior footings.
10. Draw all interior piers for all girders.

The base of the foundation plan has now been drawn and should resemble Figure 24-69.

11. Thaw all grid lines and markers.
12. Mark the center of all exterior posts or columns, and draw the spot footing for each.
13. Draw all changes in floor level.
14. Draw all metal connectors.
15. Draw all vents.
16. Draw a crawl access.

The foundation plan should now resemble Figure 24-70. Use the following steps to place the needed dimensions.

17. Thaw the base dimension layer to provide overall dimensions for each side of the structure.
18. Thaw dimensions used on the floor plan or provide dimensions to locate each jog in the foundation.

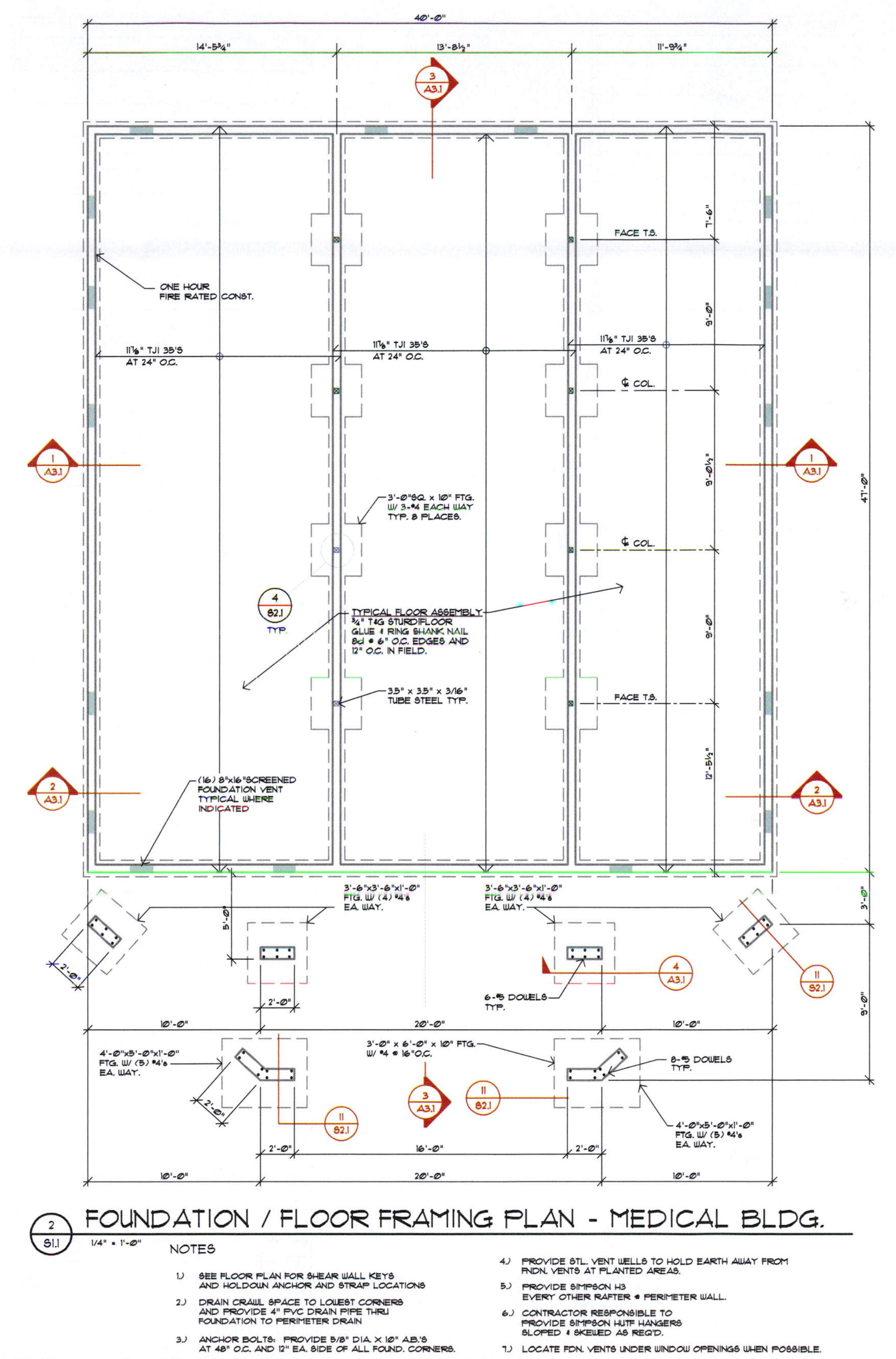
40'-0"
14'-5¾"
13'-8½"
11'-9¾"
47'-0"
7'-6"
9'-0"
9'-0½"
9'-0"
12'-5½"
3'-0"
9'-0"
FACE T.S.
ONE HOUR FIRE RATED CONST.
11⅞" TJI 35'S AT 24" O.C.
11⅞" TJI 35'S AT 24" O.C.
11⅞" TJI 35'S AT 24" O.C.
℄ COL.
℄ COL.
3'-0"SQ. x 10" FTG. W/ 3-#4 EACH WAY TYP. 8 PLACES.
TYPICAL FLOOR ASSEMBLY ¾" T&G STURDIFLOOR GLUE & RING SHANK NAIL 8d @ 6" O.C. EDGES AND 12" O.C. IN FIELD.
TYP.
3.5" x 3.5" x 3/16" TUBE STEEL TYP.
FACE T.S.
(16) 8"x16"SCREENED FOUNDATION VENT TYPICAL WHERE INDICATED
3'-6"x3'-6"x1'-0" FTG. W/ (4) #4's EA. WAY.
3'-6"x3'-6"x1'-0" FTG. W/ (4) #4's EA. WAY.
5'-0"
2'-0"
2'-0"
6-#5 DOWELS TYP.
10'-0"
20'-0"
10'-0"
4'-0"x5'-0"x1'-0" FTG. W/ (5) #4's EA. WAY.
3'-0" x 6'-0" x 10" FTG. W/ #4 @ 16"O.C.
8-#5 DOWELS TYP.
4'-0"x5'-0"x1'-0" FTG. W/ (5) #4's EA. WAY.
2'-0"
2'-0"
16'-0"
2'-0"
10'-0"
20'-0"
10'-0"
FOUNDATION / FLOOR FRAMING PLAN - MEDICAL BLDG.
2 S1.1
1/4" = 1'-0"
NOTES
1.) SEE FLOOR PLAN FOR SHEAR WALL KEYS AND HOLDOWN ANCHOR AND STRAP LOCATIONS
2.) DRAIN CRAWL SPACE TO LOWEST CORNERS AND PROVIDE 4" PVC DRAIN PIPE THRU FOUNDATION TO PERIMETER DRAIN
3.) ANCHOR BOLTS: PROVIDE 5/8" DIA. X 10" AB.'S AT 48" O.C. AND 12" EA. SIDE OF ALL FOUND. CORNERS.
4.) PROVIDE STL. VENT WELLS TO HOLD EARTH AWAY FROM FNDN. VENTS AT PLANTED AREAS.
5.) PROVIDE SIMPSON H3 EVERY OTHER RAFTER @ PERIMETER WALL.
6.) CONTRACTOR RESPONSIBLE TO PROVIDE SIMPSON HUTF HANGERS SLOPED & SKEWED AS REQ'D.
7.) LOCATE FDN. VENTS UNDER WINDOW OPENINGS WHEN POSSIBLE.

FIGURE 24-68 The completed foundation plan. *Courtesy Scott R. Beck, Architect.*

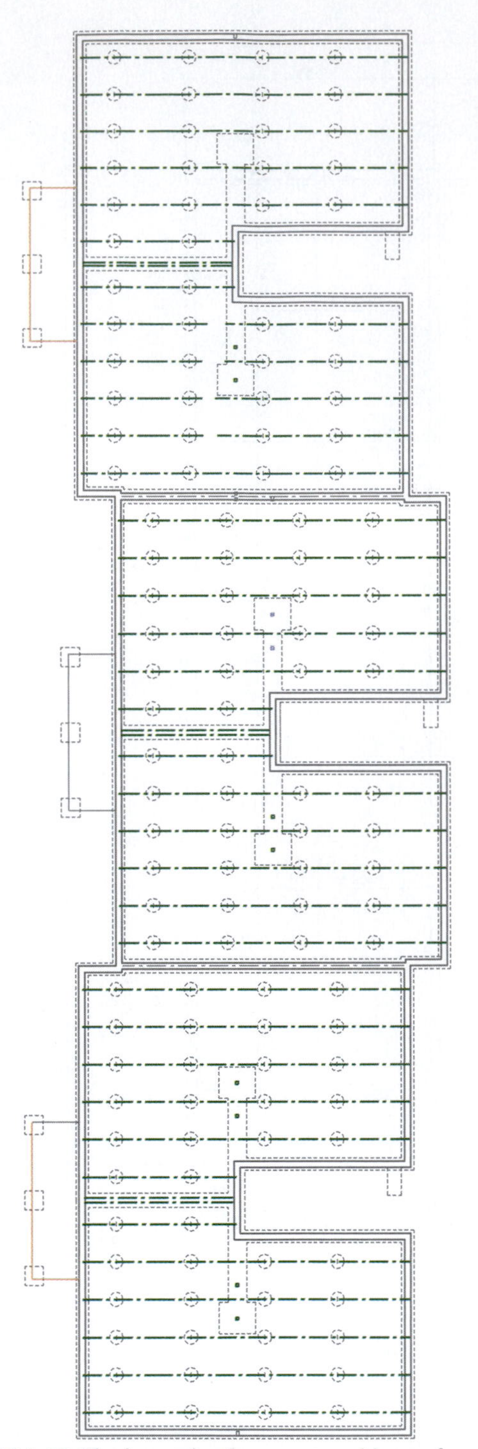

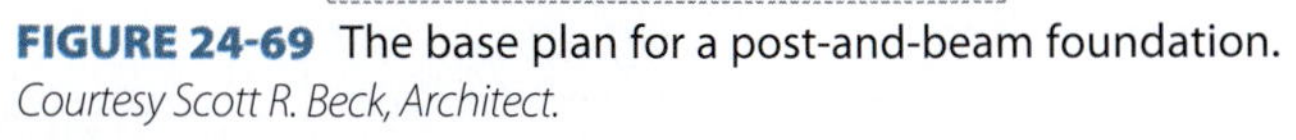

FIGURE 24-69 The base plan for a post-and-beam foundation. *Courtesy Scott R. Beck, Architect.*

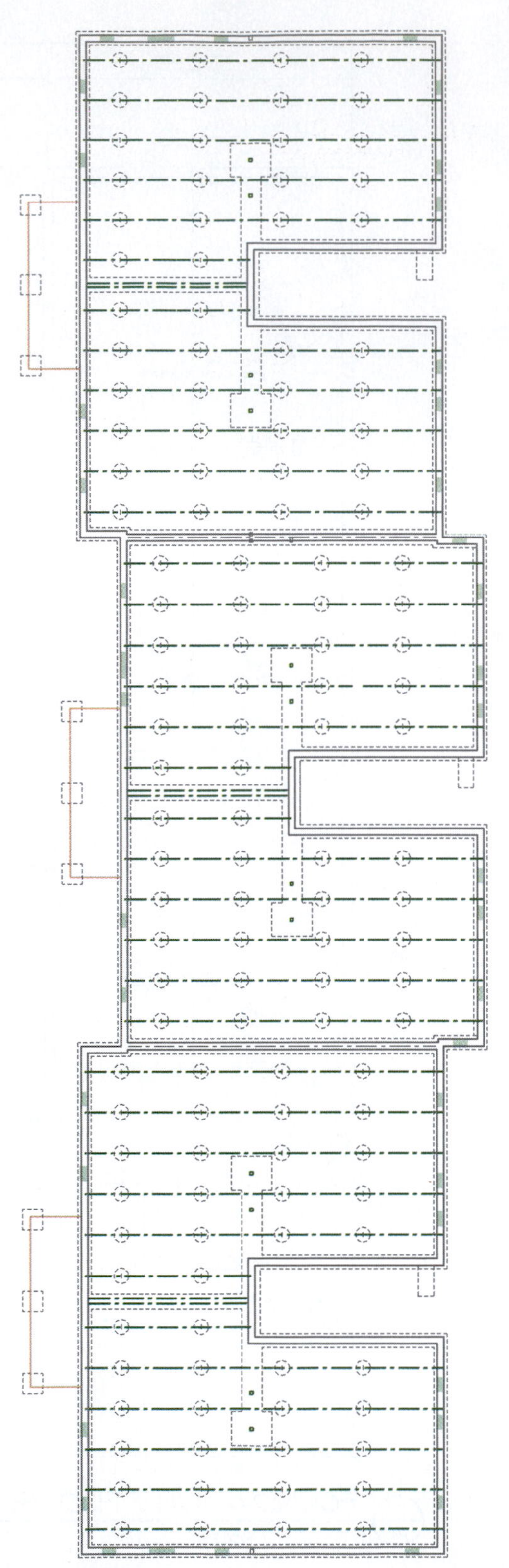

FIGURE 24-70 Adding required columns, anchors, vents, and access. *Courtesy Scott R. Beck, Architect.*

19. Dimension all openings in the stem wall.
20. Dimension the location of all girders.
21. Dimension the location of all interior piers.
22. Dimension the size and location of any changes in slab elevation.
23. Dimension the location of all metal connectors.

The major features of the drawing have now been dimensioned. The foundation plan should now resemble Figure 24-71.

The drawing can be completed by adding all required text and detail markers:

24. Describe all footing and stem wall sizes.
25. Describe all required mudsills, ledgers, and anchor bolts.
26. Place notes and dimensions to locate and describe all metal straps, anchors, and connectors.
27. Place notes to describe all vents, wall openings, and the crawl access.
28. Describe all girders, posts, and piers.
29. Describe all required changes in elevation.
30. Place notes to describe the crawl space covering.
31. Place notes to describe all special blocking.
32. Place any general notes specified by the engineer.
33. Place a title and scale below the drawing.

The completed foundation plan should now resemble Figure 24-72.

ADDITIONAL READING

Links for the materials represented on the foundation plan and related details are too numerous to list here. Rather than listing individual sites, use your favorite search engine to research products from the following CSI categories: steel, concrete, masonry, existing conditions, and earthworks. You can also do a general search of subjects such as foundations, footings, reinforcing steel, pilings, soil compaction, grading, and drainage. Another excellent source of information is the book *Soils, Earthwork and Foundations, A Practical Approach* available at the ICC bookstore at www.iccsafe.org.

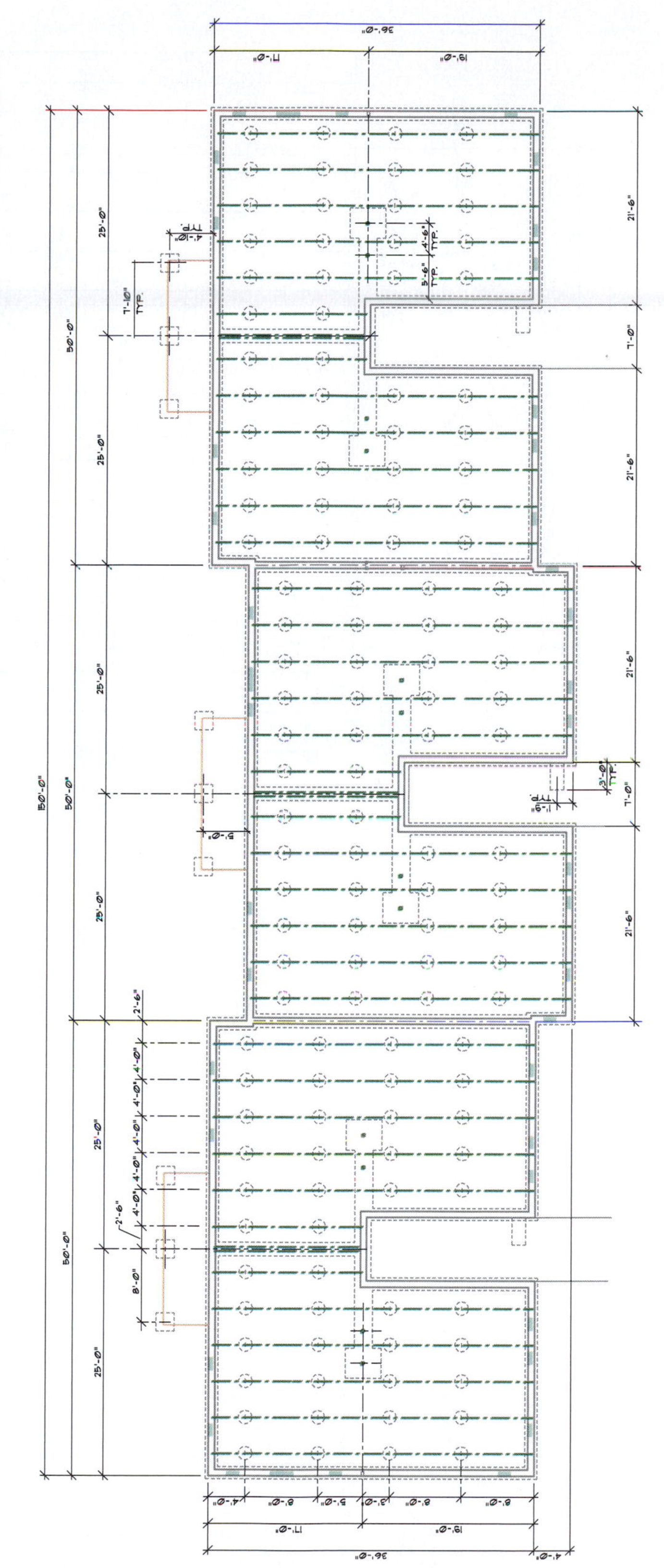

FIGURE 24-71 Placement of interior and exterior dimensions.
Courtesy Scott R. Beck, Architect.

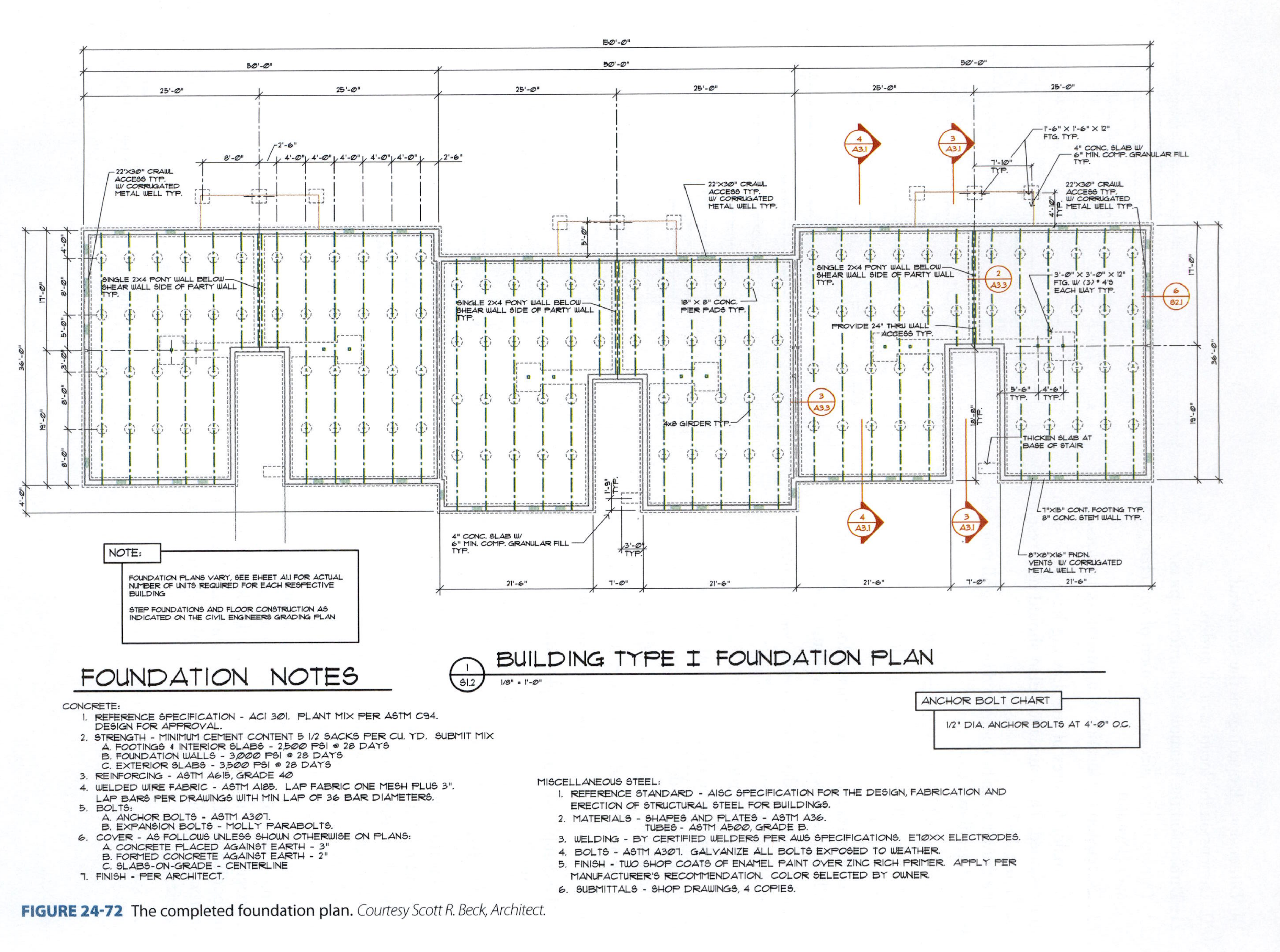

150'-0"
50'-0"
25'-0"
22"X30" CRAWL ACCESS TYP. W/ CORRUGATED METAL WELL TYP.
1'-6" X 1'-6" X 12" FTG. TYP.
4" CONC. SLAB W/ 6" MIN. COMP. GRANULAR FILL TYP.
SINGLE 2X4 PONY WALL BELOW SHEAR WALL SIDE OF PARTY WALL TYP.
18" X 8" CONC. PIER PADS TYP.
PROVIDE 24" THRU WALL ACCESS TYP.
3'-0" X 3'-0" X 12" FTG. W/ (3) #4'S EACH WAY TYP.
4x8 GIRDER TYP.
THICKEN SLAB AT BASE OF STAIR
1"X15" CONT. FOOTING TYP. 8" CONC. STEM WALL TYP.
8"X8"X16" FNDN. VENTS W/ CORRUGATED METAL WELL TYP.
NOTE:
FOUNDATION PLANS VARY, SEE SHEET A1.1 FOR ACTUAL NUMBER OF UNITS REQUIRED FOR EACH RESPECTIVE BUILDING
STEP FOUNDATIONS AND FLOOR CONSTRUCTION AS INDICATED ON THE CIVIL ENGINEERS GRADING PLAN
BUILDING TYPE I FOUNDATION PLAN
1/8" = 1'-0"
FOUNDATION NOTES
CONCRETE:
1. REFERENCE SPECIFICATION - ACI 301. PLANT MIX PER ASTM C94. DESIGN FOR APPROVAL.
2. STRENGTH - MINIMUM CEMENT CONTENT 5 1/2 SACKS PER CU. YD. SUBMIT MIX
A. FOOTINGS & INTERIOR SLABS - 2,500 PSI @ 28 DAYS
B. FOUNDATION WALLS - 3,000 PSI @ 28 DAYS
C. EXTERIOR SLABS - 3,500 PSI @ 28 DAYS
3. REINFORCING - ASTM A615, GRADE 40
4. WELDED WIRE FABRIC - ASTM A185. LAP FABRIC ONE MESH PLUS 3". LAP BARS PER DRAWINGS WITH MIN LAP OF 36 BAR DIAMETERS.
5. BOLTS:
A. ANCHOR BOLTS - ASTM A307.
B. EXPANSION BOLTS - MOLLY PARABOLTS.
6. COVER - AS FOLLOWS UNLESS SHOWN OTHERWISE ON PLANS:
A. CONCRETE PLACED AGAINST EARTH - 3"
B. FORMED CONCRETE AGAINST EARTH - 2"
C. SLABS-ON-GRADE - CENTERLINE
7. FINISH - PER ARCHITECT.
ANCHOR BOLT CHART
1/2" DIA. ANCHOR BOLTS AT 4'-0" O.C.
MISCELLANEOUS STEEL:
1. REFERENCE STANDARD - AISC SPECIFICATION FOR THE DESIGN, FABRICATION AND ERECTION OF STRUCTURAL STEEL FOR BUILDINGS.
2. MATERIALS - SHAPES AND PLATES - ASTM A36. TUBES - ASTM A500, GRADE B.
3. WELDING - BY CERTIFIED WELDERS PER AWS SPECIFICATIONS. E70XX ELECTRODES.
4. BOLTS - ASTM A307. GALVANIZE ALL BOLTS EXPOSED TO WEATHER.
5. FINISH - TWO SHOP COATS OF ENAMEL PAINT OVER ZINC RICH PRIMER. APPLY PER MANUFACTURER'S RECOMMENDATION. COLOR SELECTED BY OWNER.
6. SUBMITTALS - SHOP DRAWINGS, 4 COPIES.

FIGURE 24-72 The completed foundation plan. *Courtesy Scott R. Beck, Architect.*

KEY TERMS

Beam pocket
Cantilevered retaining wall
Crawl space
Excavation
Existing grade
Finished grade
Footing
Grade beam
Heaving
Monolithic construction
Natural grade
Pedestal
Piers
Piling
Settlement
Soil-bearing capacity
Soil texture
Stem wall

CHAPTER 24 TEST Foundation Systems and Components

Answer the following questions with short complete statements. Type the chapter title, question number, and a short complete statement for each question using a word processor. Some answers may require the use of vendor catalogs or seeking out local suppliers.

Question 24-1 Describe the difference between fill, backfill, uncompacted, and compacted fill.

Question 24-2 List four classifications used to describe soil texture, and describe the bearing capacity of each.

Question 24-3 What is the main cause of settlement?

Question 24-4 What problem does excess moisture in the soil cause?

Question 24-5 What type of soil is most susceptible to the effects of freezing?

Question 24-6 What are the names and sizes of the two components that make up a spread footing for a two-story structure?

Question 24-7 What is the minimum size of reinforcing steel recognized by most codes?

Question 24-8 What is the required distance a stem wall must extend above the finish grade?

Question 24-9 List two methods of resisting lateral pressure applied to a stem wall.

Question 24-10 Describe the requirements for stepping a footing.

Question 24-11 How is masonry veneer typically supported on an exterior footing?

Question 24-12 Describe a method of transferring lateral loads to the floor system when joists are parallel to the retaining wall.

Question 24-13 Why is the spacing for anchor bolts usually different for stem walls and retaining walls?

Question 24-14 What is the minimum bearing required for a girder resting on concrete?

Question 24-15 List two methods of attaching a wood floor to a masonry wall.

Question 24-16 How is a cantilevered retaining wall different from a retaining wall?

Question 24-17 List two methods of providing support for a concrete slab floor.

Question 24-18 What is the purpose of a grade beam?

Question 24-19 Why are interior piers for wood floors required to be recessed in the excavated grade?

Question 24-20 Describe the difference between a piling and a pedestal.

Question 24-21 What two methods are used to determine the depth a piling is to be driven?

Question 24-22 What is the advantage of having a bulbous piling?

Question 24-23 What type of joint would be provided if enough time to pour a slab were not available?

Question 24-24 What is coverage and why is it important?

Question 24-25 List and describe two methods of post-tensioning a concrete slab.

DRAWING PROBLEMS

Draw the following details in generic form for future use in a library. Use a scale suitable for showing materials typically associated with the detail. Required details can then be amended for use with a specific foundation plan. Skeletons of the plans and details can be accessed from the student CD in the DRAWING PROJECTS/CHAPTER 24 folder. Use these drawings as a base to complete the assignment. Unless your instructor provides other instructions, use the drawing criteria presented in Chapter 6 to complete the details and save them as blocks.

- Open the appropriate detail on the student CD and save the file to an appropriate storage device.
  - Use appropriate symbols, linetypes, dimensioning methods, and notations to complete the drawing.
  - Unless specified, select a scale appropriate to display the required material and determine LTSCALE and DIMVARS.
- Materials are currently located on the 0 layers. Create layers based on the National CAD Standards and move each drawing component to the appropriate layer.
- Assume plotting in monochrome, but assign a different color to each material and drawing component to help you track each item.
- Provide hatch patterns as necessary to define each material.
- Show general reinforcing for all CMU construction based on Chapter 10.
- Show, specify, and provide details for slab joints based on specific problem instructions and the recommendations based on the American Concrete Institute (ACI) presented in Chapter 11.
- Provide annotation to clearly describe all materials. Use generic notes to describe trusses, rafters, floor joists, and other common members by referencing the appropriate framing plan for the size and spacing.
  - As you insert generic details into a specific foundation drawing, coordinate with the appropriate foundation, and provide job-specific sizes instead of generic references.
- Assume all slabs to be 5" minimum thick unless noted over 0.006 black vapor barrier, 1" sand fill, and 4" gravel fill. Reinforce slab with # 10" × # 10" – 4 × 4 WWM, 2" up.
- Provide (2)-#5 Ø rebar in each foundation footing (1 each 2" up/down).
- Anchor plates to stem walls with 1/2" Ø anchor bolts @ 48" o.c. unless noted.
- Anchor 3× ledgers to walls with 5/8" Ø anchor bolts @ 32" o.c. staggered 2 up/down. Use appropriate Simpson hanger to attach the floor joist.

Edit the detail as required to meet the minimum standards above.

Details

Problem 24-1 Draw a footing suitable for a two-story office building with a concrete slab and 2 × 6 stud walls.

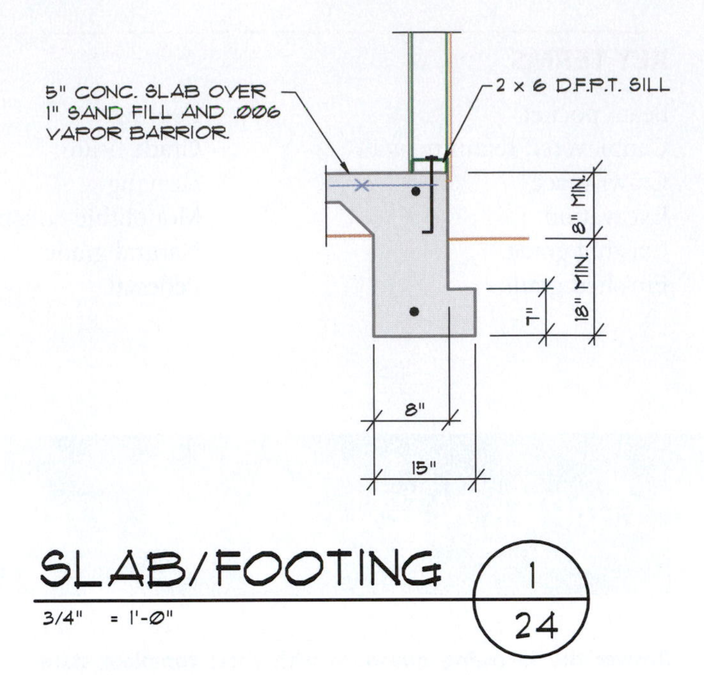

Problem 24-2 Draw a foundation detail suitable for a one-level concrete block structure with a concrete slab floor system. Thicken the floor slab to 10" thick at the footing/wall intersection.

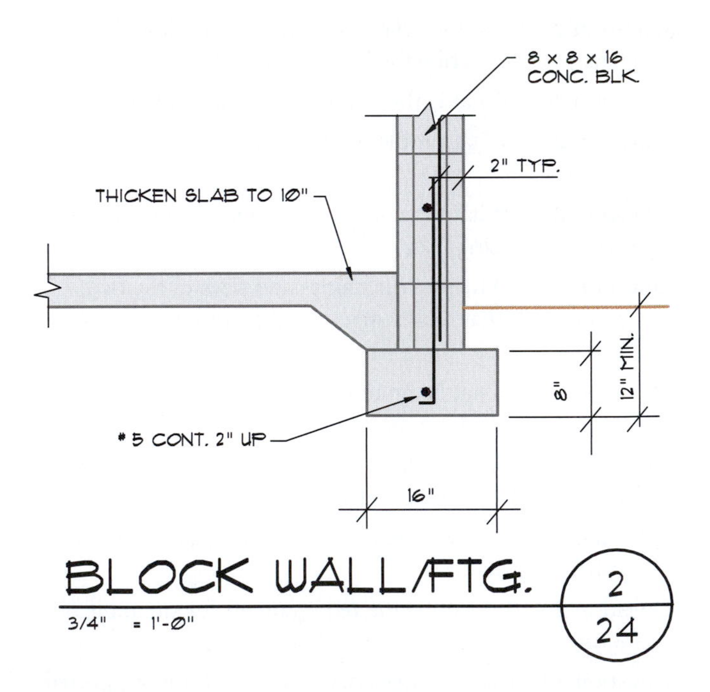

Problem 24-3 Edit the detail created in problem 24-2 to show a one-level concrete block structure with brick veneer and a concrete slab floor system. Edit the footing to be 20" wide.

Problem 24-4 Draw a foundation detail suitable for a one-level 8" CMU wall system with a 2× wood joist floor system supported by a 3× ledger.

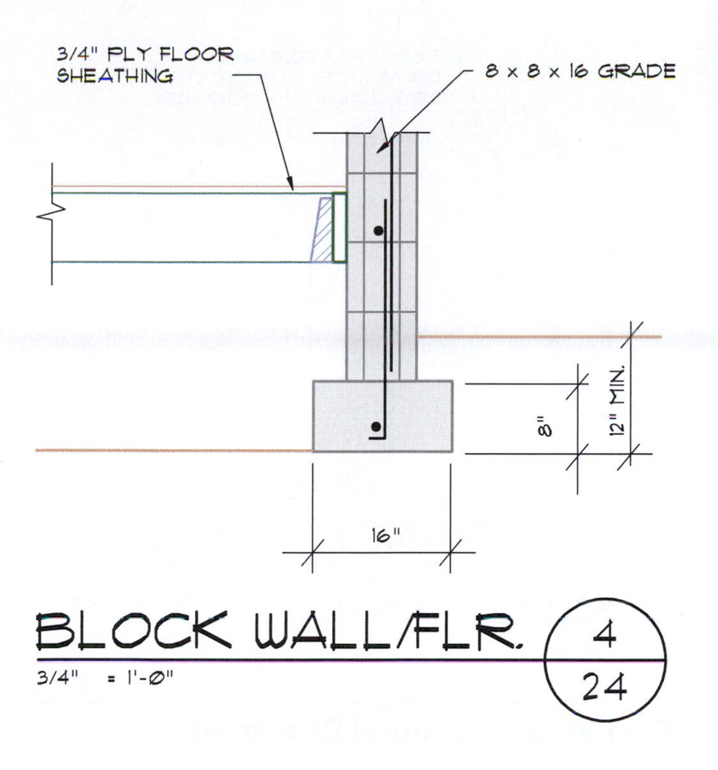

Problem 24-5 Draw a detail to support a 2× exterior two-level bearing wall with a post-and-beam floor for the lower floor system with beams parallel to the stem wall.

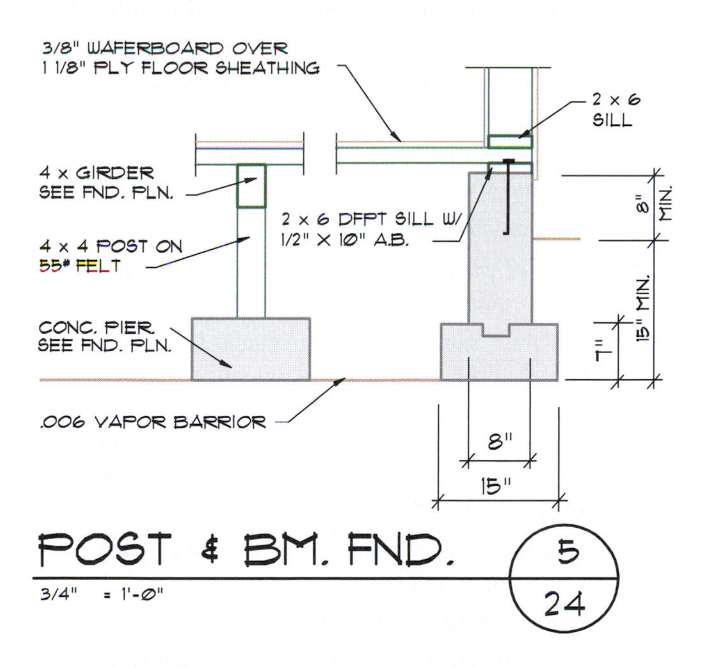

Problem 24-6 Edit the detail created in problem 24-5 to represent a two-level post-and-beam floor with beams perpendicular to the stem wall. Support the beams with metal hangers, without the use of beam pockets.

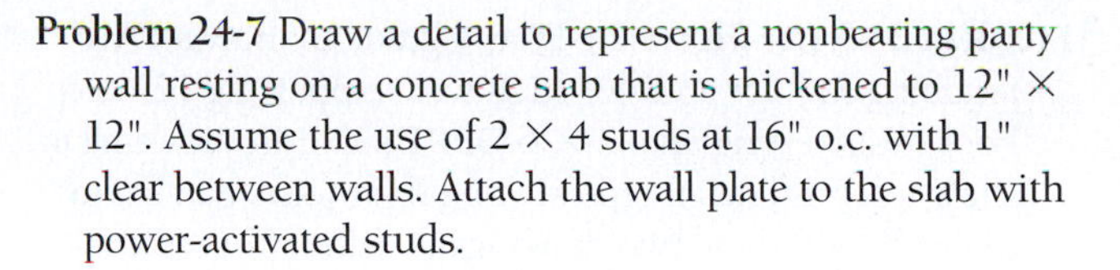
Problem 24-7 Draw a detail to represent a nonbearing party wall resting on a concrete slab that is thickened to 12" × 12". Assume the use of 2 × 4 studs at 16" o.c. with 1" clear between walls. Attach the wall plate to the slab with power-activated studs.

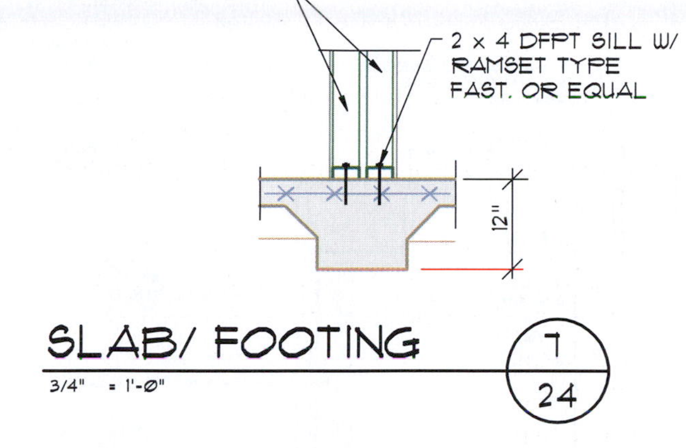

Problem 24-8 Use the detail created in problem 24-7 as a guide to represent a nonbearing party wall resting on parallel wood beams. Use 2 × 4 studs at 16" o.c. with 1" clear between walls. Support each wall on 4× beams on wood posts on an 18" wide footing. Place beams parallel to the party wall.

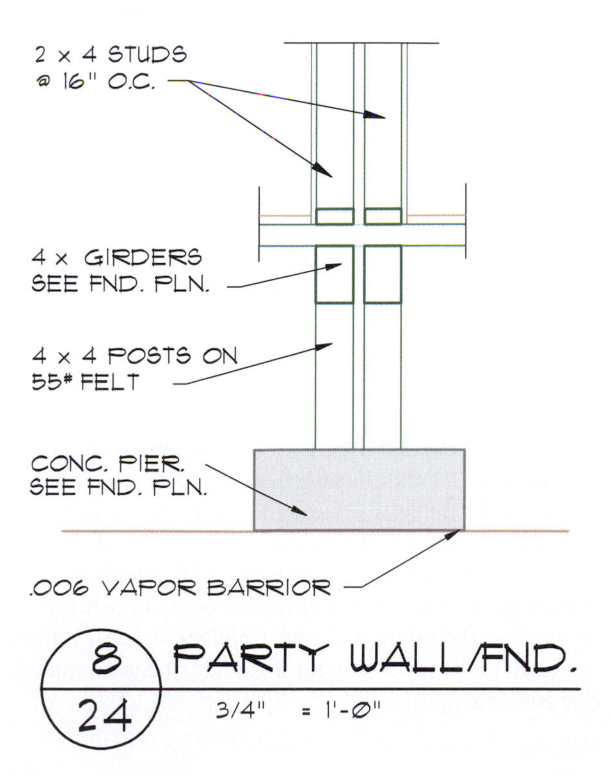

Problem 24-9 Use Figure 24-18 as a guide to draw a CMU retaining wall for an 8'–0" high wall supporting 12" deep floor joists resting on a 2 × 6 DFPT sill. Edit the detail to show a second detail representing floor joists parallel to the wall. Edit these two drawings to reflect poured concrete construction.

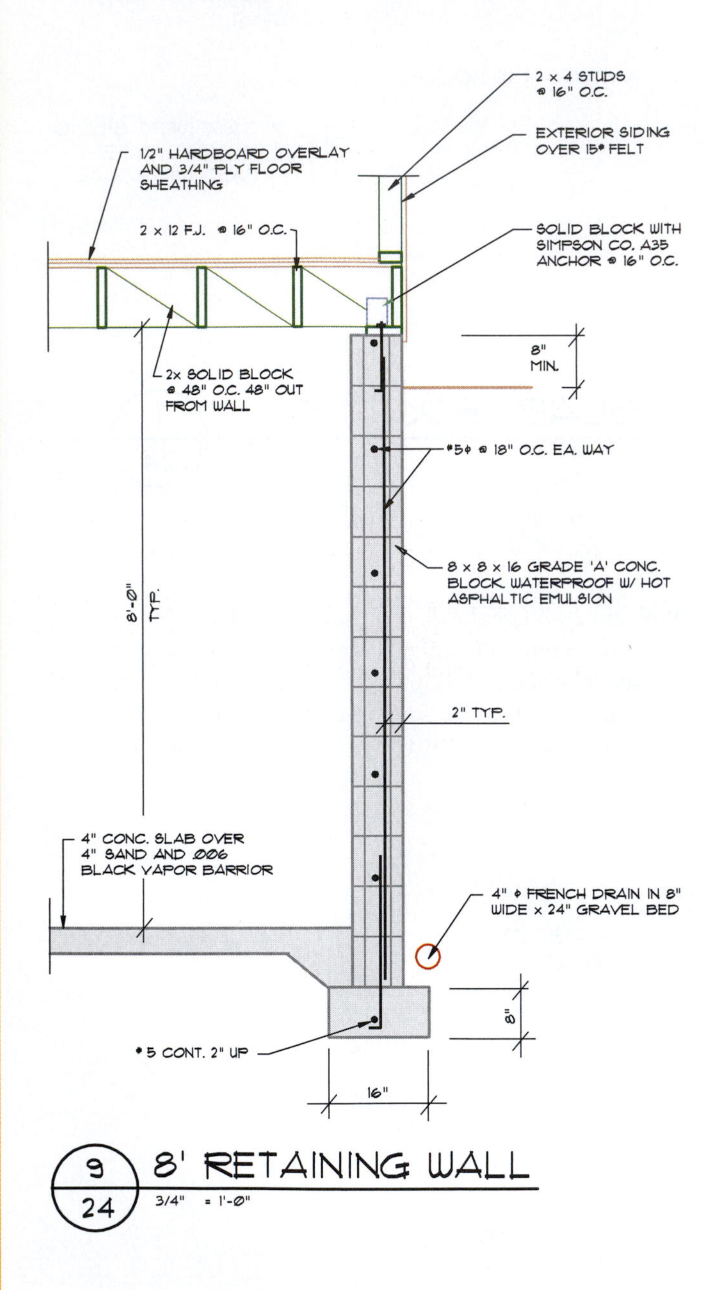

Problem 24-10 Use the information in the specifications on the student CD and the attached drawings to complete a drawing of a 24" × 14" deep footing to support a 6" wide precast concrete panel. Show a 3" × 3" × 9" slot at 60" o.c. for panel supports in the footing filled with nonshrink grout. Show the wall resting on 1" nonshrink grout. Use the Internet, *Sweets Catalogs*, or local vendor material to specify a structural connector that is suitable for concrete panels.

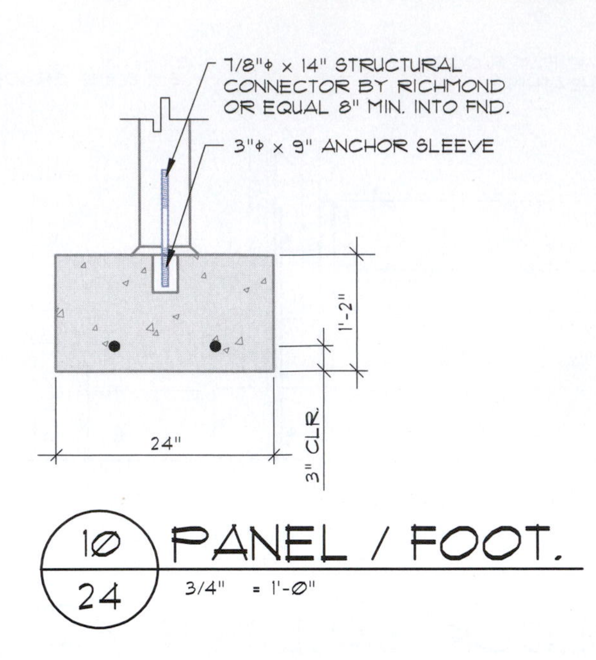

Plan Views—General Directions

Unless your instructor provides other instructions, complete the foundation drawings that correspond to the floor plan that was drawn in Chapter 16. Use the attached drawings, the drawings provided from the DRAWING PROBLEMS/CHAPTER 24 folder on the student CD, and the information provided with other related problems in Chapters 15 through 23 to complete the required building foundation drawings. Use the following minimum standards to complete the required drawings:

- Draw the required plan(s) at the same scale used to create the floor plan.
- Show grids based on the floor plan layout.
- Adjust all scale, lineweights, text, and dimension factors as required for the selected scale.
- Specify all material based on general specifications located on the student CD or on common local practice.
- Use separate layers for each major drawing component based on the recommended NCS guidelines presented on the student CD.
- Use dimensions to locate all walls, footings, piers, anchors, and other structural materials.
- Provide annotation to specify all materials.
- Hatch materials with the appropriate hatch pattern when appropriate.
- Use an appropriate architectural text font to label and dimension each drawing as needed.
- Refer to *Sweets Catalogs*, vendor catalogs, or the Internet to research needed sizes and specifications.

- Use schedules when possible to keep the drawing from being cluttered.
- Provide section and detail markers for the related drawings from Chapters 17 through 23 that your team has drawn.
- Provide a drawing title, scale, and problem number below each plan view.

Note: The sketches in this chapter are to be used as a guide only. You will find that some portions of the drawings do not match things that have been drawn on other portions of the project. You will act as the project manager and will be required to make decisions about how to complete the project. Any information not provided must be researched and determined by you unless your instructor (the project architect and engineer) provides other instructions. When conflicting information is found, use the order of precedence presented in Chapter 15 to determine the solution to drawing conflicts. If you think you have found an error, do not make changes to the drawings until you have discussed the problem and possible solutions with your engineer (your instructor). Information is provided on drawings in Chapters 15 through 23 relating to each project that might be needed to make decisions regarding the slab or foundation plan.

Problem 24-11 Foundation Plan. Use the partial drawing on the next page of the foundation plan as a guide to draw a foundation plan showing a concrete slab floor for each unit of the four-unit condominium started in problem 15-1. Place the slabs over crushed rock or compacted sand and suitable slab insulation for your area. Use 3500 psi concrete for the garage floor slab and slope the garage floor to the 9'–0" door openings. Use 5/8" Ø anchor bolts embedded 10" at 48" o.c. for all walls unless noted.

- Use a bolt spacing of 16" o.c. on the south walls. Specify and locate all hold-down anchors and straps that were specified on the framing plans.
- Use a 16" wide × 12" deep footing under the 8' high retaining wall at the east wall of the project.
- Show an 8" wide retaining wall in the northeast corner of each unit.
- Represent a stem wall and footing under the wall between the garage and storage area.
  - Use a similar footing under the wall supporting the stairs.
- Use a 24" × 24" × 12" deep pier at each corner of each deck with a CB66 column base for a 6 × 6 post.

Note: Use the sample calculations posted on the student CD in the REFERENCE MATERIAL folder to specify material that is not specified in the following problems.

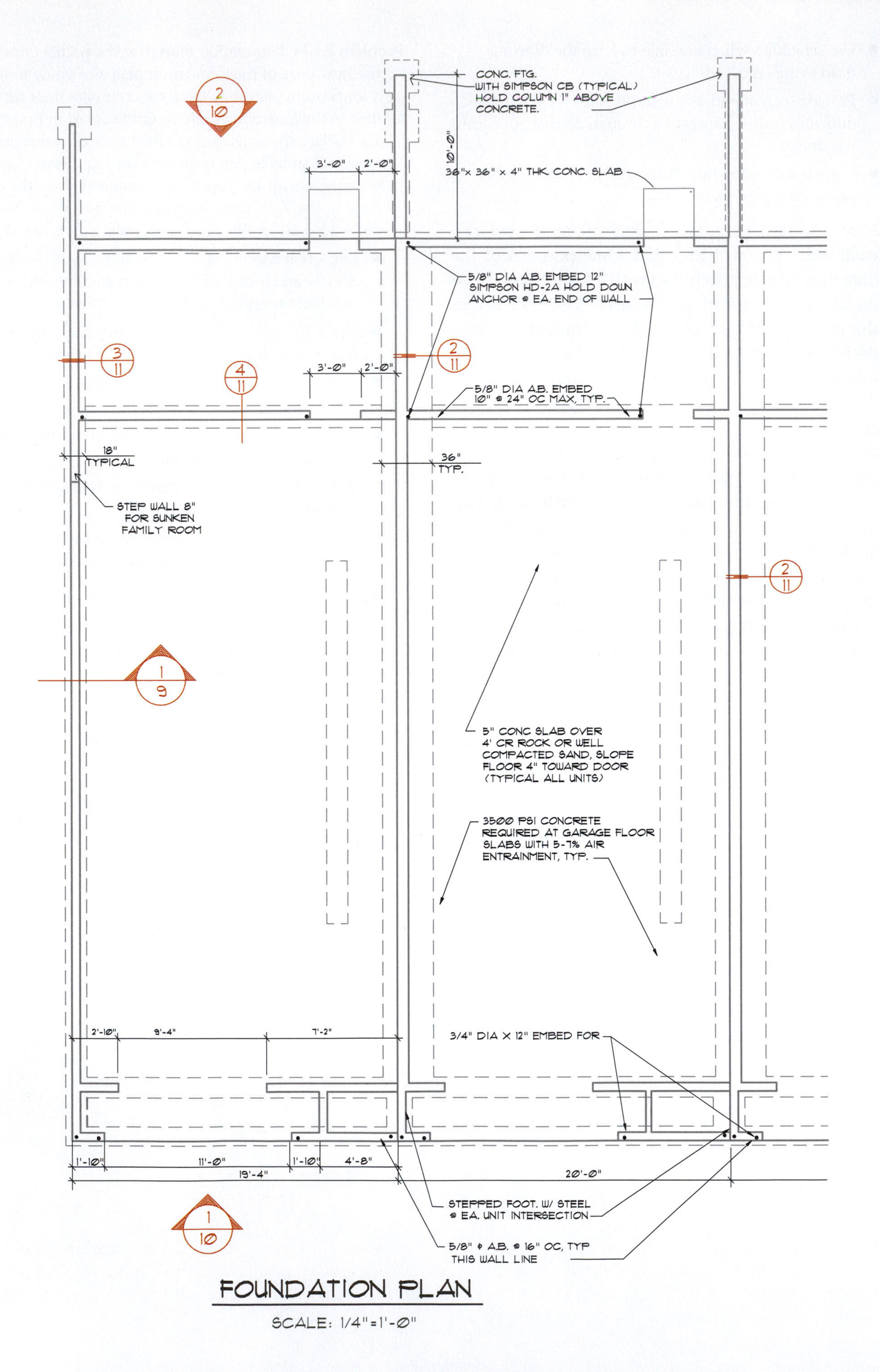

CONC. FTG. WITH SIMPSON CB (TYPICAL) HOLD COLUMN 1" ABOVE CONCRETE.
2
10
10'-0"
3'-0"
2'-0"
36"x 36" x 4" THK. CONC. SLAB
5/8" DIA A.B. EMBED 12" SIMPSON HD-2A HOLD DOWN ANCHOR @ EA. END OF WALL
3
11
4
11
2
11
3'-0"
2'-0"
5/8" DIA A.B. EMBED 10" @ 24" OC MAX, TYP.
18" TYPICAL
36" TYP.
STEP WALL 8" FOR SUNKEN FAMILY ROOM
2
11
1
9
5" CONC SLAB OVER 4' CR ROCK OR WELL COMPACTED SAND, SLOPE FLOOR 4" TOWARD DOOR (TYPICAL ALL UNITS)
3500 PSI CONCRETE REQUIRED AT GARAGE FLOOR SLABS WITH 5-7% AIR ENTRAINMENT, TYP.
2'-10"
9'-4"
7'-2"
3/4" DIA X 12" EMBED FOR
1'-10"
11'-0"
1'-10"
4'-8"
19'-4"
20'-0"
1
10
STEPPED FOOT. W/ STEEL @ EA. UNIT INTERSECTION
5/8" ⌀ A.B. @ 16" OC, TYP THIS WALL LINE
FOUNDATION PLAN
SCALE: 1/4"=1'-0"

Problem 24-12 Foundation Plan. Use the attached drawing, the details created in Chapters 23 and 24, the general specifications found on the student CD, and the following guidelines to complete the foundation plan for the structure started in problem 15-2. Use a 5" thick slab reinforced with 12" × 12"/4 × 4 WWM, 2" up from the bottom of the slab. Place the slab over 4" crushed rock or compacted sand. Use 3000 psi concrete for the floor slab and 3500 psi for footings. Specify and provide details (from this or Chapter 10) as required:

- Show an expansion joint that is 48" in from each exterior wall, suitable slab insulation for your area, and a soil-bearing value of 2500 psf.
- Use 12" deep piers placed below the continuous footings to support the pilasters.
- Use a 4′–9" sq. × 15" deep pier under the steel column.
    - Support the column on a 14" square pilaster set at 45° to the footing. Provide a suitable joint to isolate the slab from the pilaster.
- Use 21" sq. × 12" deep piers on each side of the small opening on the south wall, and 30" square piers on each side of the large openings on the south wall.
- Specify that all footings are to extend a minimum of 18" into the natural grade.

Insert details 10, 11-13, 11-14, and 11-16 on or near the slab plan and reference each to the plan. Edit the required details to match this structure and complete per area requirements.

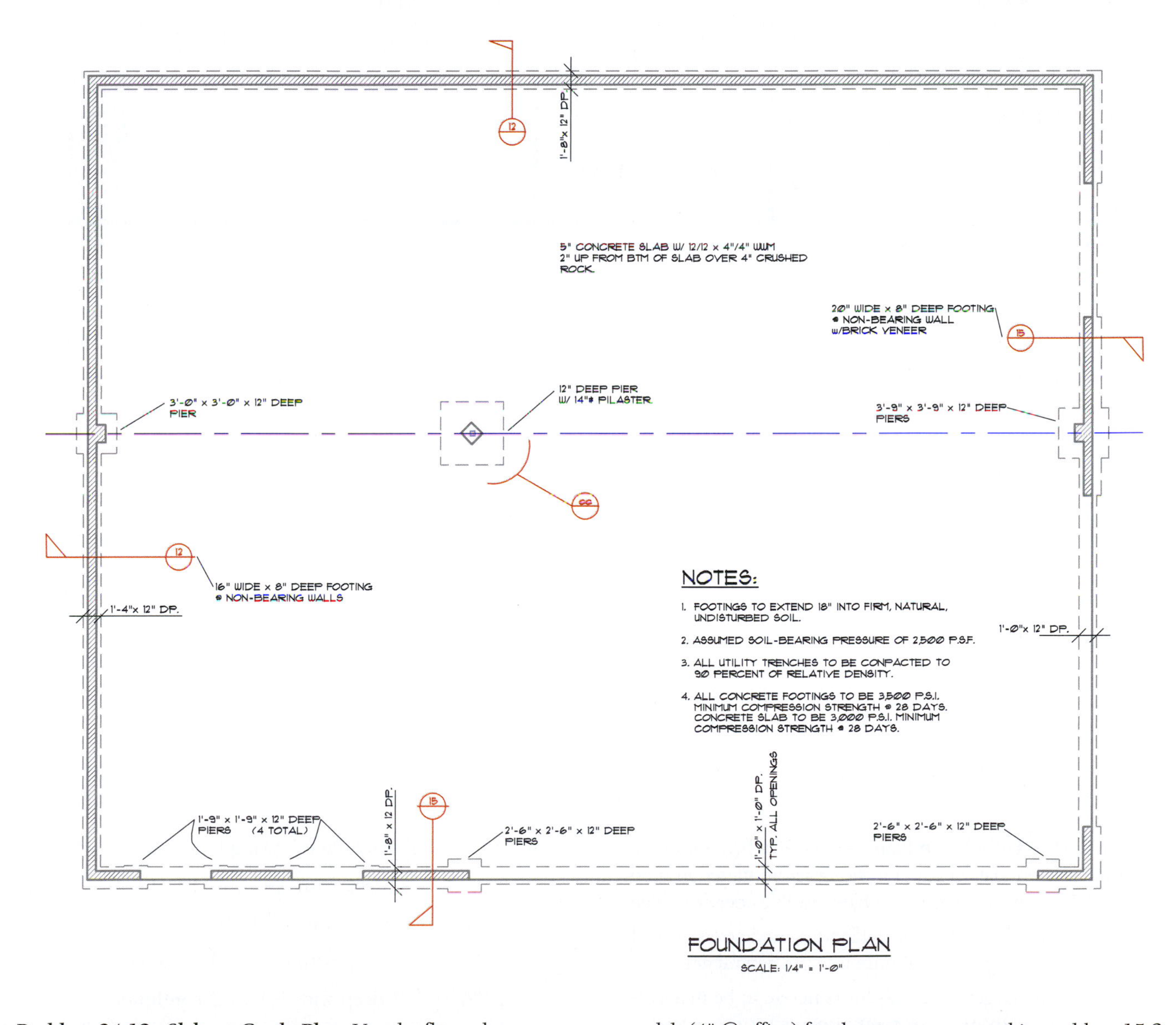

Problem 24-13a Slab-on-Grade Plan. Use the floor plan created in Chapter 16, the attached drawing, the general specifications on the student CD, and the criteria below to complete the slab-on-grade plan. Show a 5" concrete floor slab (4" @ office) for the structure started in problem 15-3. Show all openings in block walls as per the floor plan.

- Specify the slab height based on the site plan started in Chapter 15.

- Use 12" × 12" / 4 × 4 WWM 2" up from the bottom of the slab.
- Place the slab over 4" crushed rock or compacted sand.
- Use 3500 psi concrete for the floor slab. Specify target strengths for concrete to be in accordance with ACI 318 requirements for reinforced concrete.
- Specify steel reinforcement to be American Society for Testing and Materials (ASTM) A615 grade 40 smooth rebar at all joints.
- Support each column on an 11" × 11" × 1/2" steel base plate.
  - Attach the column to the plate with 1/4" fillet weld.
  - Support the plate on 1" dry pack and use (4) 3/4" Ø bolts located on a 6 1/2" grid.
- Support the column on a 14" square pedestal, set at 45° to grid C/D.
  - Provide a 1/4" fiber joint to isolate the slab from the pedestal.
- Provide suitable joints between each pedestal and between the pedestals and outer walls.
- Provide expansion joints 48" in from each exterior wall and one expansion joint @ grid D of the sales area.
- Provide a 3" diameter floor drain and a sloped floor for each bathroom with a 1/8"/12" slope.

Note: A detail symbol such as 13c refers to the detail to be drawn in problem 24-13c. A reference such as 11-16 refers to the intersection described in problem 11-16. Detail symbols that contain no letter will need to be designed by you.

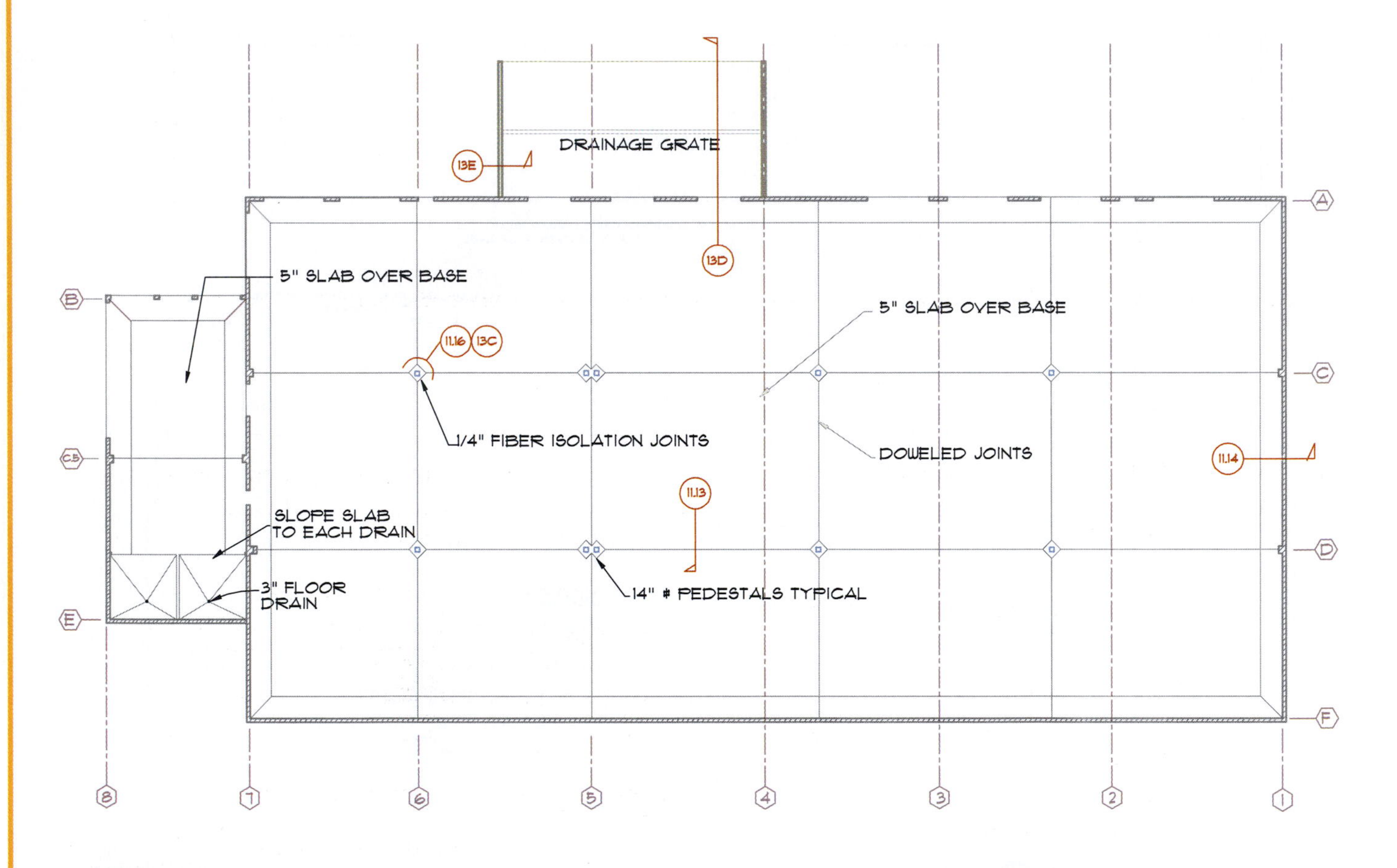

Problem 24-13b Foundation Plan. Use the attached drawing to complete a foundation plan for the structure started in problem 15-3. Show all openings in block walls as per floor plan.

- Use 3500 psi concrete for the footings and pedestals with each extending 24" minimum into the natural grade.
  - Specify target strengths for concrete to be in accordance with ACI 318 requirements for reinforced concrete.
- Specify steel reinforcement to be ASTM A615 grade 40 deformed bars unless noted 1 1/2" clear of each surface.
- Draw and specify the following footings:
  a. 24" × 12" deep with 3/4" Ø × 10" A.B. @ 48" o.c.
  b. 16" × 8" deep with 5/8" Ø 10" A.B. @ 48" o.c.
  c. 42" × 15" deep with (4)-#5 Ø continuous
  d. 36" × 15" deep with (4)-#5 Ø continuous
  e. 24" × 15" deep with (4)-#5 Ø continuous
  f. 16" × 12" deep with (2)-#5 Ø continuous
  g. Thicken slab to 15" × 18" deep

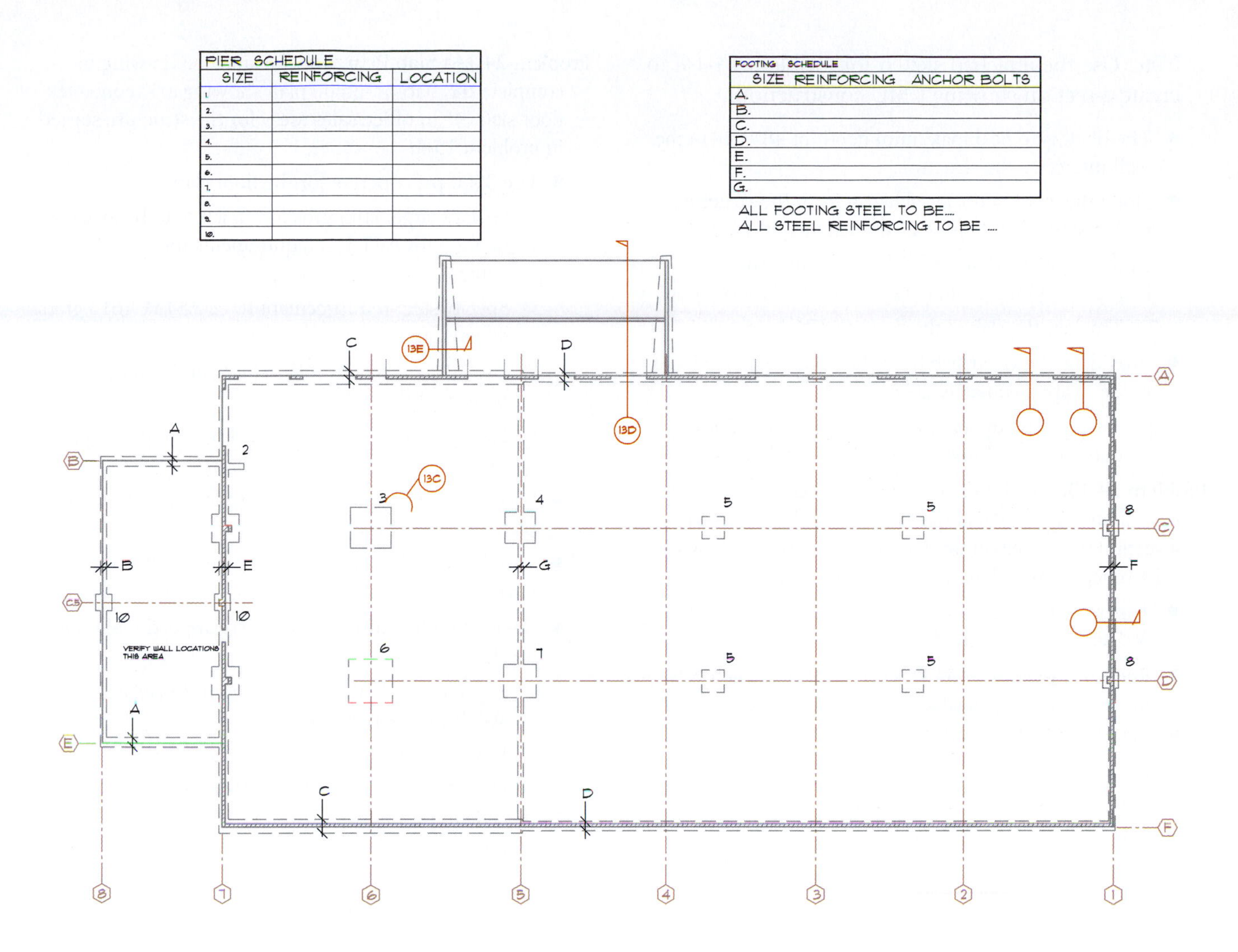

Problem 24-13c Column Support. Draw and specify the following piers on the foundation plan and create a schedule to reflect the following sizes and reinforcement:

1. 6'–0" × 6'–0" × 18" deep (5)-#8 Ø each way at the top/bottom
2. 1'–6" × 4'–0" × 12" deep pier unreinforced
3. 8'–4" × 8'–4" × 18" deep pier with (7)-#8 Ø each way located at the top/bottom
4. 6'–6" × 5'–6" × 18" deep pier with (6)-#6 Ø each way located at the top/bottom
5. 4'–9" × 4'–9" × 14" deep pier with (6)-#5 Ø each way located at the bottom only
6. 9'–4" × 9'–4" × 18" deep pier with (8)-#8 Ø each way located at the top/bottom
7. 7'–3" × 7'–3" × 18" deep pier with (6)-#8 Ø each way located at the top/bottom
8. 3'–3" × 3'–3" × 14" deep pier with (5)-#5 Ø each way located at the bottom only
9. 6'–9" × 6'–9" × 14" deep pier with (6)-#8 Ø each way located at the top/bottom
10. 4'–6" × 4'–6" × 14" deep pier with (5)-#6 Ø each way at the bottom only. Provide for each end of beam over office area.

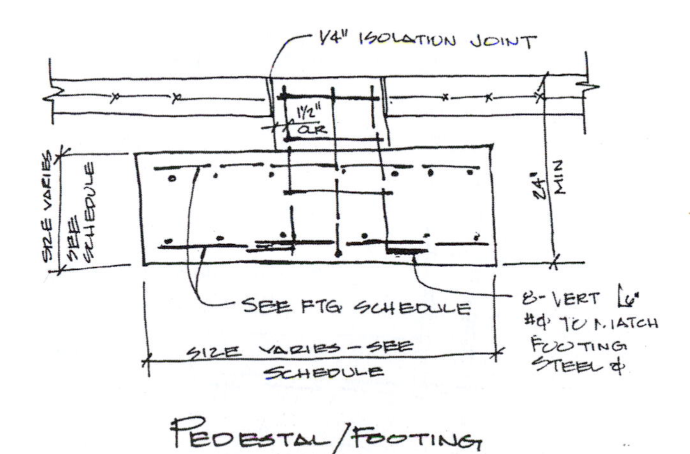

Insert details 11-4, 11-5, 11-6, 11-13, 11-14, and 11-16 on or near the slab plan and reference each to the plan. Edit the required details to match this structure and complete per area requirements.

Problem 24-13d Loading Dock Section. Create a section showing a 32' minimum long loading dock.

Note: Use the attached sketch for problem 24-14f to create a section showing CMU construction.

- The dock is to be a maximum depth of 48" where the wall intersects the structure.
- Slope the loading dock slab away from the structure with a slope of 1/4" per 12" for distance of 12'.
- At the edge of the sloped slab (12' from structure) locate a polyester drain covered with a cast iron grate.
  - Thicken the slab at the drain to 18" deep.
- Provide a 5" sloped slab from the drain to general grade of approximately 20' in length.
  - Thicken the upper edge of the slab at the dock entry and reinforce per attached calculations.

Problem 24-13e Loading Dock Cross-Section. Draw a detail or edit an existing detail from an earlier chapter to show a section of a 4' maximum high retaining wall resting on a 14" deep footing. Use detail 11-5 as a guide.

- Taper the footing from a maximum width of 54" to 30" as the wall height decreases.
- Reinforce footing w/ #5 Ø rebar @ 10" o.c. each way/ top and bottom placed 2" down, and 3" up.
- Reinforce wall w/ #5 Ø @ 16" o.c. both ways.
- Provide #5 Ø × 16" (into wall) × varies into wall and footing.
- Solid grout all wall steel cells.
- Reinforce the footing with # 4 Ø bars at 8" o.c. each way, 2" up/down.

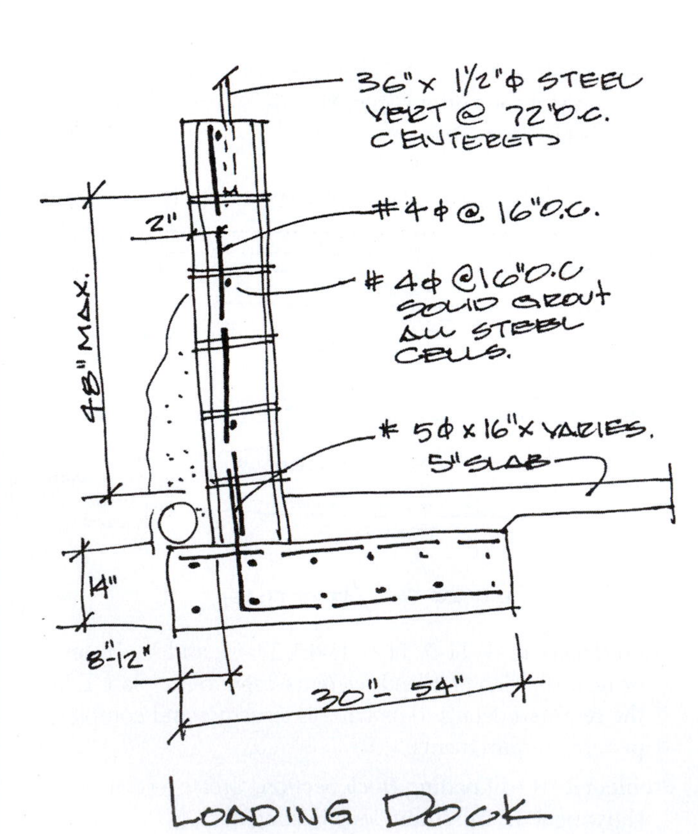

Problem 24-14a Slab Plan. Use the attached drawing to complete the slab-on-grade plan showing a 5" concrete floor slab (4" in office/sales area) for the structure started in problem 15-4.

- Use 3500 psi concrete for the floor slab.
  - Specify target strengths for concrete to be in accordance with ACI 318 requirements for reinforced concrete.
  - Specify steel reinforcement to be ASTM A615 grade 40 smooth rebar at all joints.
  - Use 12" × 12"/4 × 4 WWM 2" up from the bottom of the slab.
  - Place the slab over 4" crushed rock or compacted sand.
- Show all openings in concrete walls as per the floor plan and panel elevations.
- Specify the slab height based on the site plan started in Chapter 15.
- Support each column on a 14" square pedestal set at 45° to grid 3.
  - Provide a 1/4" fiber isolation joint to isolate the slab from the pedestal.
  - Provide suitable joints between each pedestal and between the pedestals and outer walls.
- Provide joints at grid 3 running east/west at each column at grids 1 to 5, and joints running east/west @ grid 3+24' and 5+24'.
- Provide an expansion joint that is 48" in from each exterior wall.
- Provide a 3" diameter floor drain and a sloped floor for each bathroom with a 1/8"/12" slope.
- Provide 3/4" diameter L structural connection inserts w/ 3/4" Ø × 25" steel inserts @ 60" o.c. in all portions of the slab where footings are 6'–0" or deeper below the finish floor.
- See foundation plan for footing elevations.

Insert details 11-2, 11-4, 11-5, 11-6, 11-13, 11-14, and 11-16 on or near the slab plan and reference each to the plan. Edit as required.

Note: A detail symbol such as 13c refers to the detail to be drawn in problem 24-14c. A reference such as 11-16 refers to the intersection described in problem 11-16. Detail symbols that contain no letter will need to be designed by you.

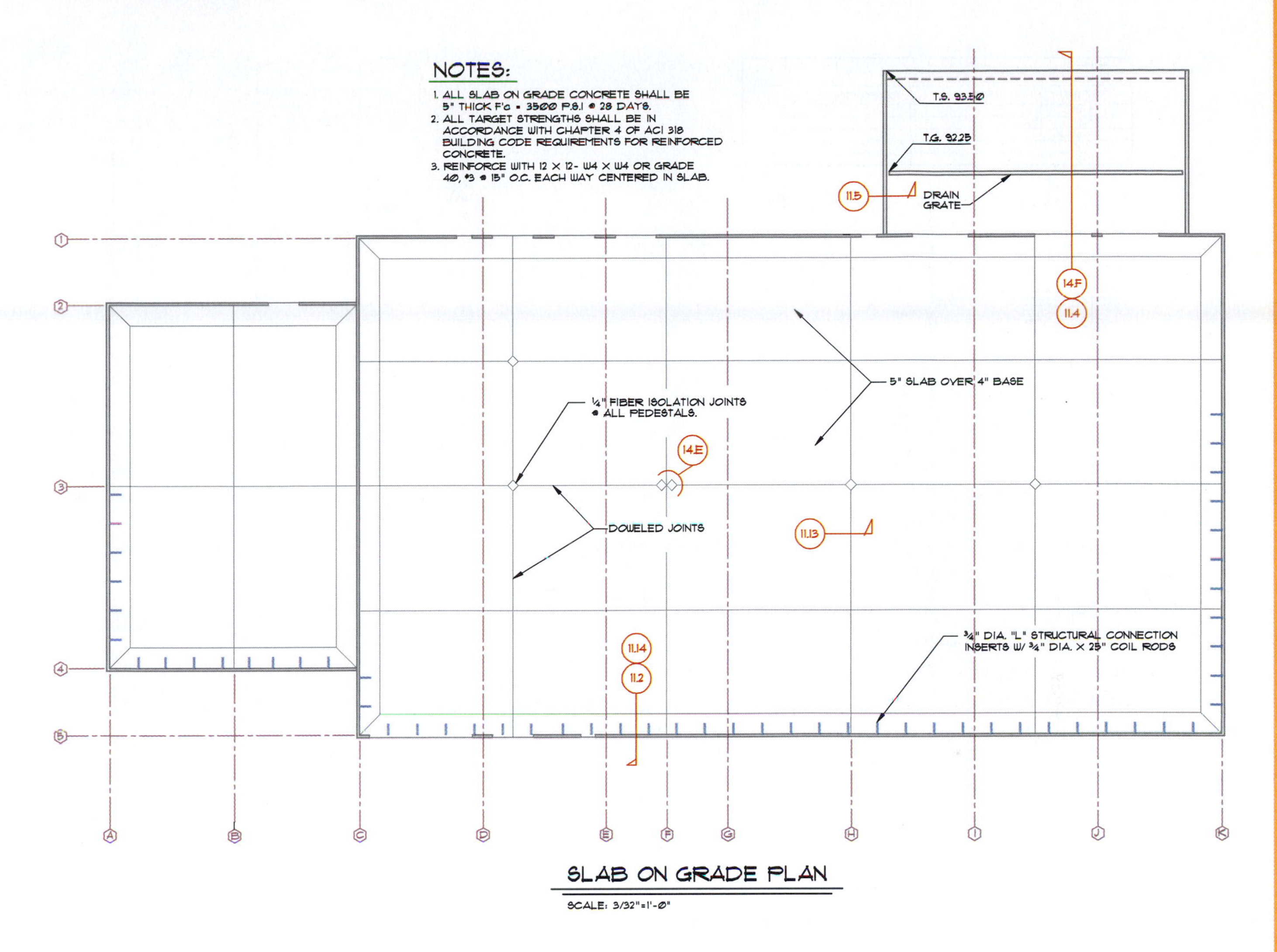
NOTES:
1. ALL SLAB ON GRADE CONCRETE SHALL BE 5" THICK F'c = 3500 P.S.I @ 28 DAYS.
2. ALL TARGET STRENGTHS SHALL BE IN ACCORDANCE WITH CHAPTER 4 OF ACI 318 BUILDING CODE REQUIREMENTS FOR REINFORCED CONCRETE.
3. REINFORCE WITH 12 X 12- W4 X W4 OR GRADE 40, #3 @ 15" O.C. EACH WAY CENTERED IN SLAB.
T.S. 93.50
T.G. 92.25
DRAIN GRATE
5" SLAB OVER 4" BASE
1/4" FIBER ISOLATION JOINTS @ ALL PEDESTALS.
DOWELED JOINTS
3/4" DIA. "L" STRUCTURAL CONNECTION INSERTS W/ 3/4" DIA. X 25" COIL RODS
SLAB ON GRADE PLAN
SCALE: 3/32"=1'-0"

Problem 24-14b Foundation Plan. Use the attached drawing to complete the foundation plan for the structure started in problem 15-4.

- Use 3500 psi concrete for the footings and pedestals with each extending 18" minimum into the natural grade using the bottom of footing elevations.
- Specify target strengths for concrete to be in accordance with ACI 318 building code requirements for reinforced concrete.
- Specify steel reinforcement to be ASTM A615 grade 40 deformed bars unless noted, 2" clear of each surface.

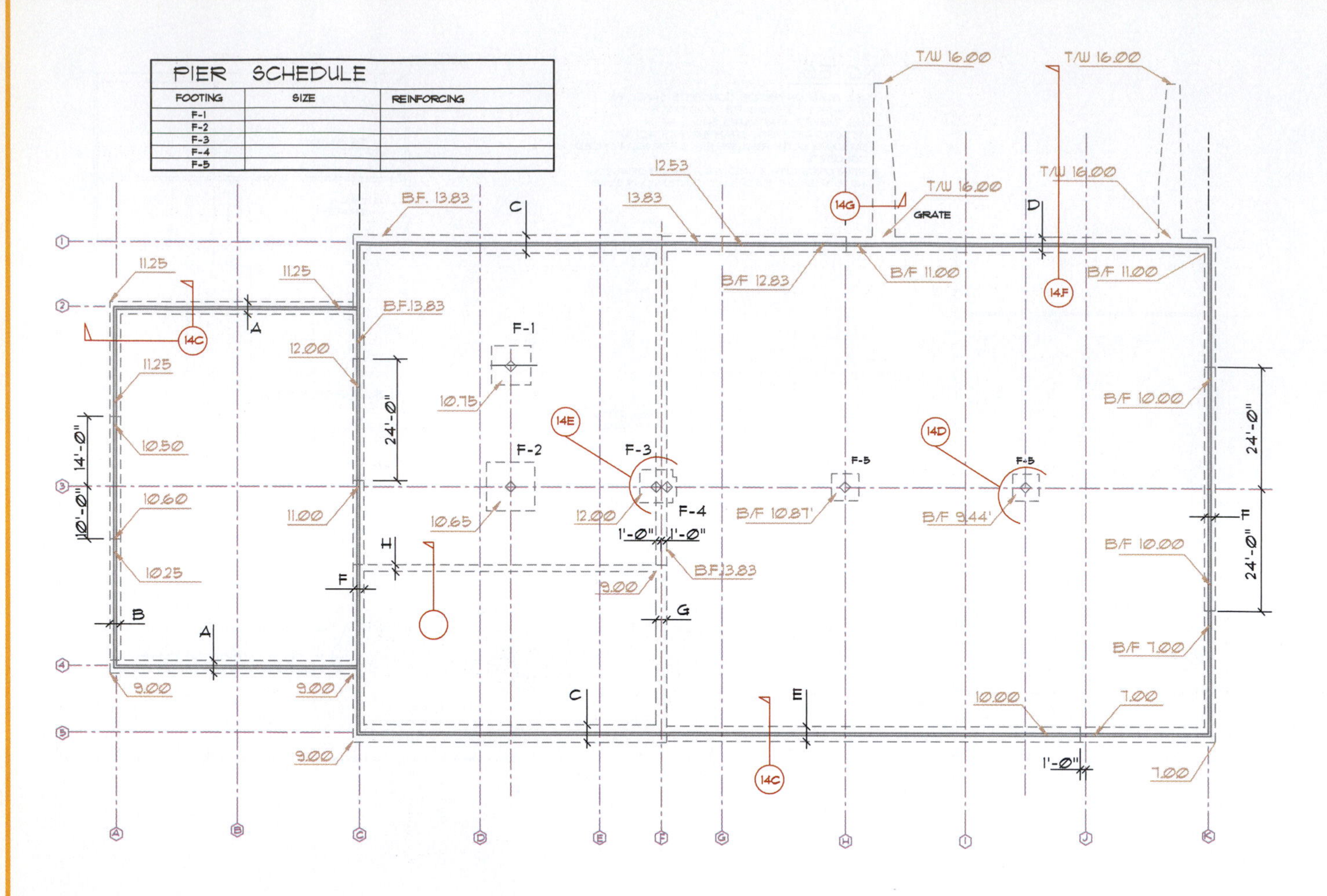

Problem 24-14c Typical Foundation Detail. Draw and specify the following footings on the foundation plan and create a detail and a schedule to reflect the following sizes and reinforcement. Set the footing steel 3" up from the bottom of the footing. Set the anchor sleeves per the wall panel specifications provided on the student CD.

Draw and specify the following footing:

a. 24" × 14" deep with (3)-#5 Ø

b. 20" × 14" deep with (2)-#5 Ø

c. 42" × 14" deep with (5)-#6 Ø

d. 36" × 14" deep with (4)-#6 Ø

e. 34" × 14" deep with (3)-#5 Ø

f. 22" × 14" deep with (2)-#5 Ø

g. Thicken slab to 15" × 12" deep from top of slab with (2)-#4 Ø

h. 15" × 36" deep grade beam similar to problem 11-8. Thicken the slab to 8" deep and reinforce with (2) continuous #5 Ø, 2" up from the bottom of the slab. Reinforce the grade beam with (5) #8 Ø at bottom and balance of steel per problem 11-8.

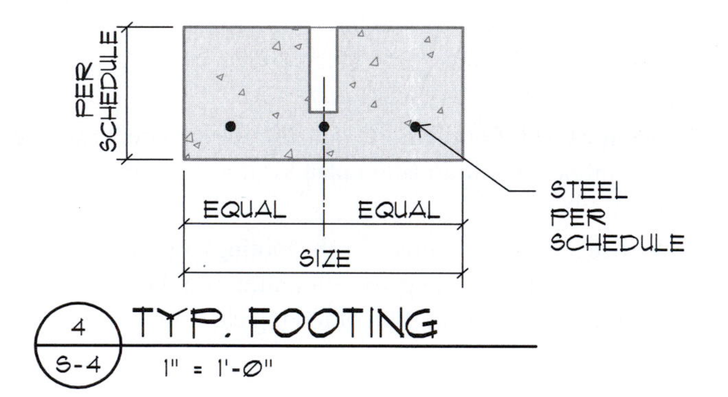

Problem 24-14d Pedestal/Piers. Draw and specify the following piers on the foundation plan and create a detail and a schedule to reflect the following sizes and reinforcement:

Draw and specify the following piers.

1. 7'–9" × 7'–9" × 18" deep (5)-#8 Ø E.W. / top/bottom
2. 9'–6" × 9'–6" × 18" deep (8)-#8 Ø E.W./ top/bottom
3. 6'–6" × 6'–6" × 14" deep (7)-#6 Ø E.W./ top/bottom
4. 3'–9" × 3'–9" × 14" deep (4)-#6 Ø E.W./ bottom only
5. 5'–3" × 5'–3" × 14" deep (6)-#5 Ø E.W./ bottom only

Edit the drawing created in problem 11-6 as required to show the pedestal that will be required to span from the footing to the top of the slab. Use the pier elevations provided on the foundation plan in combination with the slab elevation based on the site plan to determine the height of the pedestals, and reflect each pedestal height in the schedule. In addition to the information provided with problem 11-6, use the information provided in the general specifications that are located on the student CD.

Problem 24-14e Pedestal at Fire Wall. Use the drawing created in problem 24-14d and the information provided in the general specifications on the student CD to create a detail showing the reinforcing and the location of the pedestals at grid F3 on the foundation plan.

Problem 24-14f Loading Dock Section. Create a section showing a 32' minimum long loading dock. Use the following criteria to complete the drawing:

- Use the elevations provided on the site and foundation plans to establish the required dock depth.
- Use the detail created in problem 11-2 to reference the bottom of the wall where the loading dock intersects the structure.
- Use the detail created in problem 11-4 to reference the top of the wall where the loading dock intersects the structure.
- Use the guidelines presented in problem 24-13d and the information in the general specifications found on the student CD as a guide to complete the drawings.

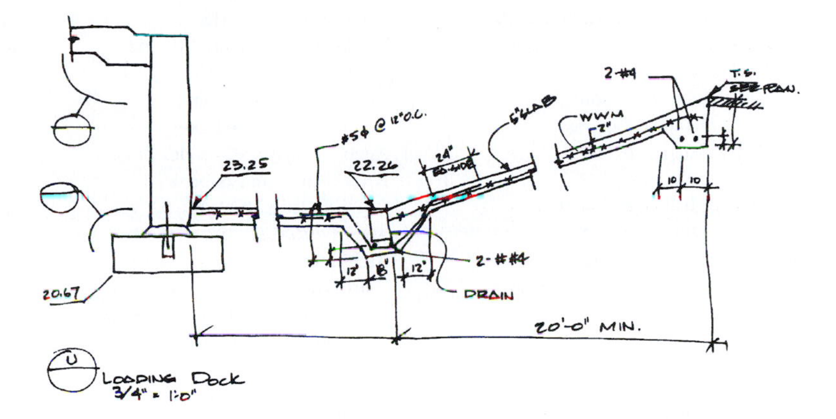

Problem 24-14g Loading Dock Cross-Section. Use the relevant information provided in problem 24-13e and the drawing created in problem 11-5 to show a section of a 4' maximum high retaining wall resting on a 14" deep footing.

Index

Note: Page numbers referencing figures are italicized and followed by an "*f*." Page numbers referencing tables are italicized and followed by a "*t*."

A

AAC (autoclaved aerated concrete) blocks, 14
Access, foundation, 553
Accessibility. *See* Americans with Disabilities Act
Accessibility symbols, *95f*
ACEC (American Consulting Engineers Council), 10
ACI (American Concrete Institute), 73, 253, 564
Acoustical ceilings, 370
ADA. *See* Americans with Disabilities Act
Addenda, 20, 34
Adhesives, 125
Admixtures, 233–234
Adobe brick, 207
Aggregates, 233
AIA (American Institute of Architects), 9, 48, 306
AIBD (American Institute of Building Design), 7
Air-entrained concrete, 234
AITC (American Institute of Timber Construction), 170
Aligned text, 62
Allowable area, 83, 85–87, *88f*
Alternative Transportation credit, LEED rating system, 339
Aluminum, 194
Aluminum Association, 194
American Concrete Institute (ACI), 73, 253, 564
American Consulting Engineers Council (ACEC), 10
American Design Drafting Association, 7
American Institute of Architects (AIA), 9, 48, 306
American Institute of Building Design (AIBD), 7
American Institute of Timber Construction (AITC), 170
American National Standards Institute (ANSI), 52, 91
American Society for Testing and Materials. *See* ASTM
American Society of Civil Engineers (ASCE), 10
American Society of Heating, Refrigerating, and Air-Conditioning Engineers, Inc. (ASHRAE), 10
American Society of Interior Designers (ASID), 7
American Society of Landscape Architects, Inc., 9
American standard beam, 188
Americans with Disabilities Act (ADA)
- additional reading, 121
- methods of egress
  - doors, 105
  - elevators, 102
  - hallways, 95–99
  - lifts, 102–104
  - ramps, 99–101
  - rescue assistance areas, 104–105
  - stairs, 101–102
  - walks, 95–99
- overview, 94
- parking requirements, 94–95
- ramps, 464
- requirements based on occupancy
  - assembly areas, 116–117
  - business, 118
  - cafeterias, 117–118
  - libraries, 119
  - medical care facilities, 118
  - mercantile, 118
  - restaurants, 117–118
  - transient lodging, 119–121
- restrooms and bathing facilities
  - bathtubs, 110–111
  - fixtures, 114–116
  - lavatories, 110
  - mirrors, 110
  - overview, 105–108
  - shower stalls, 111
  - toilet room accessories, 111–114
  - toilet stalls, 108
  - urinals, 108–110

Anchor bolts, 127, 130–131, 149–150, 552
Angle measurement, 48
Angle of repose, 338
Angles, steel, 190, *192f*
ANNO (annotation) code, 49
ANNO DIMS layer, 62
ANNO TTLB (title block) layer, 59
Annotation
- door schedules and symbols, 362
- finish schedules, 364–365
- foundation drawings, 575
- overview, 141, 360–361
- roof plans, 417–418
- schedule notations, 361–362
- site plans, 326
- window schedules and symbols, 362–364

Annotation (ANNO) code, 49
ANSI (American National Standards Institute), 52, 91
Arches, 175, 177, 353
Architects
- areas of study, 8
- educational requirements, 8
- employment opportunities, 9
- overview, 7–8

Architects' First Source for Products, 26
Architect's scales, 50
Architectural drawings, 19, 23. *See also* Drawings
Architectural Graphics Standards, 451, 459
Architectural scales, *55t*
Architectural-grade beams, 173
ARRAY command, *390f–391f*, 473, 486
ASCE (American Society of Civil Engineers), 10
ASHRAE (American Society of Heating, Refrigerating, and Air-Conditioning Engineers, Inc.), 10
ASID (American Society of Interior Designers), 7
Asphaltic emulsion, 548
Assembly areas, ADA requirements, 116–117
Associates, 9
ASTM (American Society for Testing and Materials)
- bolt standards, 131–132
- brick standards, 207
- CMU standards, 215
- steel standards, 186–187

ATTRIBUTES command, 58, 486–487
Auditoriums, seating for disabled, *117f*
Austenitic grades, 194
AutoCAD ANSI31 hatch pattern, 218
AutoCAD tools
- additional reading, 71
- dimensions, 61–64
- drawing environment, 39–45
- files and folders, 69–70
- hatch patterns, 390–391
- influences of, 38–39
- multidocument environment, 67–69
- software, 3
- templates
  - border, 45–47
  - drawing area, 46–47
  - layers, naming, 48–50
  - limits, 48
  - linetypes, 52–55
  - lineweight, 55
  - metric system, 50–52
  - scales, 50
  - title blocks, 45–47
  - units of measurement, 47–48
- text
  - common features of, 56
  - height of, 60–61
  - overview, 55–56
  - placement of, 56–57
  - scale, 60–61
  - types of, 57–60
- viewports, 64–67

Autoclaved aerated concrete (AAC) blocks, 14

B

Backfill, 549
Back-up brick, 207
Baffles, 156
Balusters, 470
Bank indicators, 327
Bar chair (BC) support, *241f*
Bars, steel, 193
Base plan, *584f*
Base plates, 149, 193
Base Point option, 68
Bathing facilities, ADA guidelines
- bathtubs, 110–111
- counters, 116
- fixtures, 114–116
- grab bars, 113–114
- handrails, 113–114

medicine cabinets, 111–113
mirrors, 110
overview, 105–108
shower stalls, 111
tables, 116
tub and shower seats, 113–114
Bathtubs, ADA guidelines, 110–111
Batt insulation symbol, *139f*
BB (beam bolster) support, *241f*
BBU (beam bolster upper) support, *241f*
BC (bar chair) support, *241f*
Beam bolster (BB) support, *241f*
Beam bolster upper (BBU) support, *241f*
Beam pockets, 553
Beams
bearing walls and support columns, 487
concrete, 250
moment frames, 494
reaction to loads, *177f*
support for, 552–553
timber, 491
wood, 491
Beams and stringers lumber classification, 148
Beam-to-beam connection detail, *160f*
Beam-to-beam detail, *133f*
Bearing walls, 348, 487–490
Beveled welds, 135–136
Bid bonds, 308
Bid forms, 308
Bidding stage, 19, 308
Bird block, 156
Bird's mouth, 157
Blind nailing, 127, *129f*
Blocking, 151–152, 180
Blocks, inserting repetitive features as, 69, 438–439, 486–487
Blocks, masonry. *See* Masonry
Blocks, street layout, 320–321
Board feet, 148
Boards, defined, 147. *See also* Wood
BOCA (National Building Code), 73
BOF (bottom of footing), 327, *575f*
Bolsters, 239
Bolts
anchor, 127, 130–131, 149–150, 552
carriage, 131
drift, *131f,* 132
expansion, *131f,* 132
high-strength, 131–132
machine, 131
Bond beam brick, *209f*
Bond beams, 216, 224
Bond forms, 308
Bonds, 211
Borders, drawing, 45–47
Both sides option, 136
Bottom chords, 179–180
Bottom of footing (BOF), 327, *575f*
Bottom plates, 153
Boundary nailing, 127
Bounds, 317–318
Box nails, 125
Braced frames, 300–301, *303f*
Braced parapet walls, fire-rated, *270f*
Bracing, 151–153
Break lines, 53–54
Brick. *See also* Masonry
joints, 213–214
placement of, 209
quality of, 208
reinforcement of, 214–215
shapes of, 209
sizes of, 208–209
types of, 207–208
veneer construction, 215
wall construction, 209–213
Bright finished nails, 126
Brownfield Redevelopment credit, LEED rating system, 339
Building brick, 207
Building dimensions, 325. *See also* Dimensions
Building materials. *See also names of specific materials*
green construction, 12–13
model codes, 78–80
weights, *282f–283f*
Building sections, 426–427. *See also* Sections
Building shapes, 385–386
Building systems, model codes for, 76–78
Built-up roofs, 410
Built-up steel girders, *201f*
Bundled bars, 251
Butt joints, *135f*

C

Cabinet and fixture drawings, 23, *25f,* 458, *460f–461f*
Cable, steel, 193
CAD technicians. *See* Drafters
Cafeterias, ADA requirements, 117–118
Calculations
footing, 288–289
influence of AutoCAD on, 38
working with, 23–26
Camber, 175, 181
Camelback pratt trusses, 181
Cant strips, 405
Cantilevered retaining walls, 556–558
Cantilevers, 175, *177f*
Careers
additional reading, 15–16
architects, 7–9
CAD technicians, 4–7
designers, 7
employment opportunities, 3
engineers, 9
interior designers, 7
related fields, 11–12
Carriage bolts, 131
Cased openings, 353
Cash allowance specifications, 311–312
Cast-in-place concrete pilings, 562
Cast-in-place floors, 243–244
Catch basins, 327
Cavity walls, 211
CCA wood preservative, 13
Ceiling joists, 157–159
Ceilings
acoustical, 370
fire-resistive construction, 271–275
reflected ceiling plans, 369–371, *373f*
vaulted, 159
Cells brick, *209f*
Cement, 232–233
Cement finish, 126
Cementitious coating, 272
Centerlines, 53, *61f*
Certified forest/wood products, 13, 162
Certified Wood credit, LEED rating system, 394
Chairs, 239
Change notices/orders, 309
Channels, steel, 189–190
CHC (continuous high chair) support, *241f*
CHCM (continuous high chair for metal deck) support, *241f*
CHCU (continuous high chair upper) support, *241f*
Chimneys, 413
Chlorosulfonated polyethylene (CSPE), 410
Chord bars, 248
Chord forces, 296–299
Chords, 179–180, 296–297, *300f*
Circular columns, 251–252
Circular stairs, 468
Civil drawings, 21–23
Civil engineering scale, 50
Civil engineers, 10
Clay, 547
Client name box, 60
Close grain lumber, 147
Closed specifications, 313
CMUs. *See* Concrete masonry units; Masonry
Coarse grain lumber, 147
Codes. *See* Model codes
Collar ties, 157
Column bases, *552f*
Columns
concrete, 250–252
fire-resistive construction, 267–268
moment frames, 494
steel, 190–192
support, 487–490
Combination columns, 252
Commercial site plans, *22f*
Commercial structures, 3
Common brick, 207
Common nails, 125
Compaction, soil, 549
Composite columns, 251–252
Composition shingles, 410–411
Computers, model codes and, 90–91. *See also* AutoCAD tools
Concrete
additional reading, 256
beams, 250
columns, 250–252
conduit, 255–256
coverage recommendations, *240t*
fire-resistive construction, 264, 271
floor systems, 243–246, 505–507
joints, 252–255
materials, 232–234
mixture ratios, 234
precast framing, 497–498
reinforcement of, 238–243
representing in drawings, 140, 434–437
shear walls, 300
stairs, 471
testing, 234
types of, 234–238
walls, 246–249
Concrete blocks, *218f,* 300
Concrete fire-rated parapets, *271f*
Concrete footings, weight, *288f*
Concrete masonry units (CMUs). *See also* Masonry
grades of, 216
overview, 215–216
representing in drawings, 218–220
shapes of, 217–218
sizes of, 216–217
Concrete masonry units symbol, *139f*
Concrete pilings, 562
Concrete Reinforcing Steel Institute (CRSI), 222, 224
Concrete slabs, 576–580
Concrete stairways, *474f–475f*
Concrete tilt-up construction, *360f,* 498, 520–522
Conduit, concrete, 255–256
Connections
additional reading, 143–144
details
annotation, 141
criteria for completing, 142–143
drawing, 137–139
locating materials with dimensions, 140–141
overview, 136–137
process, 141–142
representing materials in, 139–140
symbols, 141
used with engineered products, 125–130
used with large wood members, 130–133
used with wood, 125–130
welded steel, 133–136

Conserving water, 14
Construction documents, role of drafter, 19
Construction joints, 253–255
Construction Metrication Council, 50, 315
Construction phase, role of drafter, 19–20
Construction Specifications Canada (CSC), 309
Construction Specifications Institute (CSI), 11, 40, 309–311, 438
Contaminants, 15
Continuous extension lines, *61f*
Continuous footings, 549–550
Continuous high chair (CHC) support, *241f*
Continuous high chair for metal deck (CHCM) support, *241f*
Continuous high chair upper (CHCU) support, *241f*
Contours, 334, 338
Contract drawings, 20
Control joints, 252–253
Conventional floor framing, *149f*
Cool roofs, 421
Copper, 195
COPY command, 68–69
Corbel arches, *179f*
Cores brick, *209f*
Cornices, 156
Corridors, fire-resistive construction, 268–270
Corrugated metal roofs, 411
Counters, ADA guidelines, 116
Countersunk screws, 129
Courses, brick, 209
Cove header brick, *209f*
Cover, 240
Coverage, 240, 565
Crawl spaces, 567–569
Crickets, 405, 417
Cross grain lumber, 147
CRSI (Concrete Reinforcing Steel Institute), 222, 224
Crystalline bedrock, 547
CSC (Construction Specifications Canada), 309
CSI (Construction Specifications Institute), 11, 40, 309–311, 438
CSPE (chlorosulfonated polyethylene), 410
CTLTSCALE command, 55
Curb ramps, *96f*
Curtain walls, 348
Curved beams, 175
Curved stairways, 467
Cut banks, 549
CUT command, 68–69
Cut material, 336
Cutting planes, 53, *141f*, 426

D

Date of completion box, 60
Daylight, 336
DDEDIT command, 57
Dead loads, 281
Decking, 197, 494
Deformations, 222
Descriptive specifications, 312
Design development, role of drafter, 19
DesignCenter feature, 69
Designers, 7
Detail markers, *141f*, 367
Detail references, 34, *141f*
Detailed sections, 427–430, *431f*
Details
- annotation, 141
- creating, 141–142
- criteria for completing, 142–143
- drawing, 138–139
- grading plans, 338–340
- locating materials with dimensions, 140–141
- overview, 136–137
- placement of, 29–33
- representing materials in, 139–140, 199–203
- site plans, 329
- stairs, 477–478
- symbols, 141

Diaphragm transfers, *302f*
Diaphragms, 151, 296
Dimension lines, 54
Dimension Style Manager feature, 64
Dimensional lumber, 147–148
Dimensions
- bearing walls and support columns, 487
- components of, 61–62
- elevations, 391
- foundation drawings, 575–576
- light-frame, 359
- locating materials with, 140–141
- masonry, 359–360
- placement of, 62–63, *579f*
- roof plans, 418
- in sections, 437–438
- site, 324–326
- steel, 360
- timber, 360
- variables, 64

DIMSTYLE command, 61
Disabled people. *See* Americans with Disabilities Act
Discipline designators, 48–49
Diverters, 417
Domes, 177
DONUT command, 225
Doors
- ADA guidelines, 105, *106f–108f*
- elevations, 387–389
- floor plans, 351–353
- schedules, 362

Double-wythe brick walls, *139f*, *213f*
Draft stops, 276
Drafters
- additional reading, 35–36
- calculations, 23–26
- codes, 27–28
- drawings, 20–23, 28–34
- educational requirements, 5
- employment opportunities, 5–7
- entry-level, 4–5
- project coordination, 34
- revisions, 34–35
- role in common office practice, 18–20
- senior, 5
- vendor catalogs, 26–27

Drags, 302–304
Drainage
- roof, 413–417
- site plans, 327

Drawing area, 46–47
Drawing area modules, 46
Drawing coordination, 34
Drawing sets, 20–21, 384
Drawings. *See also* Details; Elevations; Sections
- annotation, 141
- architectural, 23
- cabinet and fixture, 23
- civil, 21–23
- drawing sets, 20–21
- electrical, 23
- external referenced, 40–42
- of land, 321
- layouts, 44–45
- mechanical, 23
- organization of, 20
- placement of, 28–34
- plumbing, 23
- ramps, 465
- relationship to written specifications, 314
- representing materials in
  - AutoCAD hatch patterns, 390–391
  - concrete, 434–437
  - concrete masonry units, 218–220
  - engineered products, 434
  - glass, 437
  - laminated beams, 173–174
  - metal buildings, 199
  - overview, 433–434
  - reinforcing bars, 226
  - steel, 434
  - timber, 160–161, 434
  - unit masonry, 434
  - wood, 434
- stairs, 471–478
- structural, 23
- symbols, 141
- types of, 20
- views, 42–44
- walls, 349–350

Drawn by box, 60
Dressed lumber, 147
Drift bolts, *131f*, 132
Drinking fountains, ADA guidelines, 114–115
Ductile, 187, 286
Dutch hip roofs, 409
Dynamic loads, 281

E

Earth pressure, 284–285
Earthquakes, 289
Eaves, 156
Economy brick, 207
Edge joints, *135f*
Edge nailing, 127
EDIT menu, 68
Educational requirements
- for architects, 8
- for CAD technicians, 5
- for engineers, 10
- for interior designers, 7

Egress, ADA guidelines
- doors, 105
- elevators, 102
- hallways, 95–99
- lifts, 102–104
- ramps, 99–101
- rescue assistance areas, 104–105
- stairs, 101–102
- walks, 95–99

Elastic range, 186
Electrical drawings, 23
Electrical engineers, 10
Electrical penetrations, fire-resistive construction, 274–275
Elevation
- defined, 336
- on-grade slabs, 563–564
- referencing, *141f*, 326–327, 575
- soil, 548–549
- symbols, markers and labels, 367

Elevations
- additional reading, 396–399, 527
- combined with sections, 430
- completing, 393–396
- concrete tilt-up, 520–522
- drawing
  - building shapes, 385–386
  - dimensioning, 391
  - grades, 386–387
  - layering, 392–393
  - notations, 392–393
  - representing materials in, 389–391
  - roof, 386, 414–415
  - skylights, 387–389
  - symbols, 391–392
  - windows, 387–389
- within drawing sets, 384
- poured concrete, 519–520
- principles of, 384–385
- structural, 396
- structural steel, 518–519

Elevators
- ADA guidelines, 102
- drawings, 478–479
- entrances, *104f*
- shafts, *469f*

Eliminating pesticide treatments, 14

Employment opportunities
  for architects, 9
  for CAD technicians, 5–7
  in commercial design and drafting, 3
End nailing, 127, *129f*
End plates, 193
Energy, 14
Energy and Atmosphere credit, LEED rating system, 374
Energy Star standards, 14
Engineered products
  additional reading, 183–184
  beams, 491
  brick, 207
  components of, 167–169
  connections
    adhesives, 125
    metal framing connectors, 130
    nails, 125–127
    power-driven anchors, 127
    screws, 127–130
    staples, 127
  laminated beams
    appearance of, 172–173
    fiber bending, 170–171
    finish of, 173
    grading method, 171–172
    grading of, 170
    material, 172
    representing in drawings, 173–174
    sizes of, 170
    specifications, 173
    types of, 174–177
  lumber, 13, 166–167
  open-web trusses
    floor, 180–181
    overview, 177–179
    roof, 181–183
    terminology, 179–180
  representing in drawings, 140, 434
Engineers, 9
English bond, 211
Enlarged floor plans, 358–359
Entry-level CAD technicians, 4–5
Environment. *See* Green construction
Environmental loads
  rain, 283–284
  snow, 283
  wind, 289–292
Environmental Protection Agency (EPA), 566
EPDM (ethylene propylene diene monomer), 410
EPS (expanded polystyrene) foam, 255, 554
EQIP (equipment) code, 49
Equal bid specifications, 313
Equivalent fluid pressure, 287
Erosion and Sedimentation Control credit, LEED rating system, 339
Escalator drawings, 479
Ethylene propylene diene monomer (EPDM), 410
Excavation, 549
Exclusions, 308
Execution section, 311
Existing grades, 548–549
Exits, 88–90. *See also* Egress, ADA guidelines
Expanded polystyrene (EPS) foam, 255, 554
Expansion bolts, *131f*, 132
Expansion joints, 213
Extension lines, 54
Exterior coverings, 201
Exterior dimensions, 62–63
Exterior walls, 346–347
External octagon brick, *209f*
External reference (XREF) command, 40–42, 486
External referencing, 40–42, 328

F

Face brick, 208
Face nailing, 127, *129f*
Faced walls, 348
Factored load method, 28
Fascia, 156
Favorite Contents feature, 69
FBA grade brick, 208
FBS grade brick, 208
FBX grade brick, 208
Ferritic grades, 194
Fiber bending, 170–171
Fiber saturation point, 148
Fiberboard, 153
Fiber-reinforced gypsum panels, 153
Field nailing, 127
Field welds, 136
Files
  naming, 70
  overview, 69–70
  site plans, 328
  storage locations for, 70
FILL command, 527
Fill material, 336, 549
FILLET command, 118–119
Fillet welds, 134–135
Finish
  of laminated beams, 173
  of nails, 126
  of stainless steel, 195
Finish grade, 336, 549
Finish schedules, 364–365, *366f*
Finish symbols, *362f*
Finished roofing, 157
Finks, 181
Fire brick, 207
Fire cuts, 211, *213f*
Fire dampers, 268, 273, *275f*
Fire-resistive construction
  additional reading, 277
  ceilings, 271–275
  columns, 267–268
  corridors, 268–270
  fire and draft stops, 276
  floor assemblies, 271–272
  heavy timber, 275–276
  model codes, 28
  objectives of, 263
  overview, 263–264
  parapet walls, 270–271
  roof assemblies, 273–275
  shafts, 272–273
  wall materials, 264–266
  wall openings, 266–267
Fire stops, 276
Fire wall support, 569–570
Fire walls, 349
Fit Curve option, 334
Fixed arches, *178f*
Fixed partitions, 347
Fixtures, ADA guidelines, 114–116
Flanges, 188
Flashing, 404
Flat arches, *179f*
Flathead screws, 129
Flemish bond, 211
Floating viewports. *See* Viewports
Floor (FLOR) code, 49
Floor diaphragms, *297f*
Floor plans
  additional reading, 375
  annotation, 360–365
  completing, 367–369
  components of
    doors, 351–353
    enlarged floor plans, 358–359
    floors, 353–357
    furnishings, 357–358
    interior equipment, 357–358
    openings, 353
    overview, 345
    plumbing symbols, 357
    walls, 346–350
    windows, 350–351
  development of, 344–345
  dimensions, 359–360
  layering, 371–375
  reflected ceiling plans, 369–371
  space plans, 369
  symbols, 365–367
Floors
  anchor bolts, 149–150
  bracing, 151–152
  concrete, 243–246
  fire-resistive construction, 271–272
  on floor plans, 353–357
  framing plans, 504–507
  girders, 150
  joists, 150–151
  load distribution, 286
  mudsills, 149
  open-web trusses, 180–181
  posts, 150
  sheathing, 152, 568
  types of
    on-grade slabs, 563–566
    overview, 562
    party wall and fire wall support, 569–570
    post-tensioned concrete reinforcement, 566–567
    wood floor systems over crawl spaces, 567–569
  underlayment, 152
Fluted nails, 126
Folders
  naming, 70
  overview, 69–70
  storage locations for, 70
Foliated rock, 547
Fonts, *56f*. *See also* Text
Footings, 288–289, 550
Forces. *See* Loads and forces
Forest Stewardship Council (FSC), 13, 162
Form of Agreement, 308
Foundations
  additional reading, 585
  drawings and plans
    annotation, 575
    concrete slab, 576–580
    dimensioning, 575–576
    floor joists, 581–582
    layers, 576
    overview, 570–574
    schedules, 575
    symbols, 574–575
  load distribution, 286
  soil considerations, 547–549
  types of
    continuous footings, 549–550
    grade beams, 558–559
    interior supports, 558
    pedestals, 559–560
    pilings, 560–562
    retaining walls, 555–558
    stem walls, 550–555
Four-centered arches, *179f*
Four-ply roofs, *410f*
Framing
  steel
    additional reading, 203
    lightweight, 195–197
    metal buildings, 197–199
    multilevel, 199–203
    shapes of, 187–193
    structural steel, 186–187
    types of metal and alloys, 194–195
  timber
    components of, 160–161
    overview, 159–160
    representing in drawings, 160–161
    trusses, 161–162
  wood
    additional reading, 162–163
    components of, 152–154
    floor construction, 149–152
    roof construction, 154–159
    terminology, 146–149
    uses of, 146
    wall openings, 154
Framing plans
  additional reading, 511
  drawing
    floor framing plans, 504–507
    layers, 507–511
    overview, 498–500
    roof framing plans, 500–504
    steel components, *201f*

Framing plans (*continued*)
example of, *43f*
features on, 487–491
structural drawings, 485
using CAD skills for, 486–487
Framing straps, *552f*
FREEZE setting, 328
Freezing, soil, 548
Friction coefficient value, 287
Frog brick, *209f*
FSC (Forest Stewardship Council), 13, 162
Full scale, 50
Full sections, 426–427, *441f*
Furnishings, 357–358

G

Gable roofs, 408
Galvanized nails, 126
Gambrel roofs, 410
General notes, 57, 487–490
Girder trusses, 182–183
Girders, *149f*, 150
Girts, *198f*
Glass, *390f*, 437
GLAZ (glazing) code, 49
Glazing, 140
Glued-laminated beams (glu-lams), 150, *167f*
Grab bars, ADA guidelines, *110f*, *112f*, 113–114
Grade beams, *294f*, 558–559
Grades
of laminated beams, 170–172
of stainless steel, 194
of wood, 148–149
Grading plans
angle of repose, 338
details, 338–340
drawing, 386–387
overview, 334–335
representing contours, 338
terminology, 336–338
Grain, wood, 147
Gravel, 547
Gravel symbol, *139f*
Great land surveys, 318
Green construction
contributing to environment, 14–15
LEED
credits, 339–340, 374–375, 394
overview, 12
masonry, 228
materials, 12–13
overview, 6
reducing impact, 13–14
removing materials, 13
Green roofs, 421
Green wood, 148
Grid face blocks, 218
GRID layer, 576
Grid markers, *141f*, 365–367
Grout, 226–227
Grouted cavity walls, 211
Growth rings, 147
Guardrails, 99, 466, 468–469
Gusset plates, 180, 193
Gutters, 415–417

H

Half shape brick, *208f*
Hallways, ADA guidelines, 95–99
Handles, cabinetry, 458
Handrails, 99, 101, 113–114, 468
Hard conversions, 51, 217
Hardboard, 152
Hardwoods, 147
HATCH command, 391
HBA grade brick, 208
HBS grade brick, 208
HBX grade brick, 208
HC (high chair) support, *241f*
HCA hinge connectors, *177f*
HCM (high chair for metal deck) support, *241f*
HDF (high-density fiberboard), 152
Header supports, 154
Headers, 154, 159
Headroom, stair, 466
Heads, nail, 126
Heat of hydration, 233
Heaving, 548
Heavy timber, fire-resistive construction, 275–276. *See also* Timber
Height
building, 83
text, 60–61
Hidden lines, 52–53
High chair for metal deck (HCM) support, *241f*
High chair (HC) support, *241f*
High-density fiberboard (HDF), 152
High-early-strength concrete, 234
High-performance windows, 14
High-sloped roofs
completing plans, 420–422
Dutch hip, 409
framing plans, 501–504
gable, 408
gambrel, 410
hip, 408–409
mansard, 406–408
High-strength bolts, 131–132
Hinges, cabinetry, 458
Hip roofs, 408–409
Hold-down connections, *292f*, 293, *552f*
Hollow brick, 207–209
Hooks, 224–226
Horizontal coursings, *218f*
Horizontal dimensions
locating materials with, 140–141
in sections, 438
Horizontal forces
chord forces, 296–299
collecting and resisting, 294–296
defined, 283
diaphragms, 296
overview, 292–294
shear walls, 299–304
Horizontal steel reinforcement, 223–224
Howe trusses, 181
HVAC penetrations, fire-resistive construction, 273–274
Hydration, 233

I

IBC. *See* International Building Code
IBC Handbook, 74, *82f*
I-beams, 188
ICC (International Code Council), 91
ICFs (insulating concrete forms), 14, 255
I-joists, *167f*, 168–169
Illuminating Engineering Society of North America (IESNA), 10
Illustrators, 11
Impact force compression, 549
Indoor Environmental Quality credit, LEED rating system, 374
Industrial structures, 3
Innovation and Design Process credit, LEED rating system, 374
INSERT command, 485
Inspectors, 11–12
Institutional structures, 3
Insulating concrete forms (ICFs), 14, 255
Insulating lightweight concrete, 237
Insulation
foundation, 553–554
on-grade slabs, 565–566
representing in drawings, 140
Interfirst II project, *8f*
Interior architects, 8
Interior designers, 7
Interior dimensions, 63
Interior elevations, *25f*
Interior equipment, 357–358
Interior reference symbol, *362f*
Interior walls, 347–348
Internal bull nose brick, *209f*
Internal octagon brick, *209f*
International Building Code (IBC)
fire resistance, 263
overview, 50–51, 73
ramps, 464
steel reinforcing, 238
walls, 346
wood, 127, 146
International Code Council (ICC), 91
Interns, 9
Intumescent paint, 272
Invitations to bid, 308
Isolation joints, 253

J

Jack studs, 154
JC (joist chair) support, *241f*
JCU (joist chair upper) support, *241f*
J-groove welds, 135–136
Job captains, 5
Joints
brick, 213–214
concrete, 252–255
on-grade slabs, 564
Joist chair (JC) support, *241f*
Joist chair upper (JCU) support, *241f*
Joists
floor, 150–151, 581–582
open-web, 197
steel, 196–197
timber, 491
wood, 491
Jumbo brick, 208

K

Key plans, 345
Keyed joints, 253
Keyways, 550
King closer shape brick, *208f*
King studs, 154
Kitchen cabinets, 458, *460f–461f*
Kitchens, commercial, *455f*
Knobs, cabinetry, 458

L

Labor and material payment bonds, 308
Lag screws, 129
Lamella frame, *179f*
Laminated beams
appearance of, 172–173
fiber bending, 170–171
finish of, 173
grading of, 170–172
material, 172
overview, 166
representing in drawings, 173–174, 491
shapes of, *174f*
sizes of, 170, *171f*
specifications, 173
types of, 174–177
Laminated strand lumber (LSL), 166
Laminated timbers symbol, *139f*
Laminated veneer lumber (LVL) beams, 166, *167f–168f*
Land
additional reading, 340
dimensions, 325
drawings of, 321
legal descriptions of, 317–321
related site drawings
grading plans, 334–340
overview, 331–332
topography plans, 333–334
site plans
annotation, 326
completing, 329–331
details, 329
development of, 327–328
elevation and swale, 326–327
linetypes, 321–322
parking information, 323–324

representing structure, 322–323
setting parameters, 328–329
site dimensions, 324–326
Landings, stair, 466
Landscape and Exterior Design to Reduce Heat Islands credit, LEED rating system, 339
Landscape architects, 8
Landscaping plans, *334f*
Lap joints, *135f*
Laps, 224, 551
Large-scale plywood symbol, *139f*
Large-scale steel shapes symbol, *139f*
Lateral forces, 289, *295f–296f*, *304f*, 557
Laterals, 327
Lath, steel, 197
Lavatories, ADA guidelines, 110
LAYER command, 486
Layers
adding, 138
DesignCenter feature, 69
elevations, 392–393
floor plans, 371–375
foundation drawings, 576
framing plans, 486, 507–511
naming, 48–50
sections, 440–442
Layouts, 44–45
LCCA (Life Cycle Cost Analysis), 315
Lead, 195
LEADER command, 55
Leader lines, 54–55
LEED (Leadership in Energy and Environmental Design)
credits, 339–340, 374–375, 394
overview, 12
Legal descriptions, of land
lot and block, 320–321
metes and bounds, 317–318
rectangular systems, 318–320
Let-in braces, 153
Libraries, ADA requirements, 119
Life Cycle Cost Analysis (LCCA), 315
Lift slabs, 246
Lifts, ADA guidelines, 102–104
Light Pollution Reduction credit, LEED rating system, 340
Light-frame dimensions, 359
Lightweight concrete, 237
Lightweight steel framing, 195–197
Limits, template, 48
LIMITS command, 48
Line contrast, 138–139
Lines and Arrows tab, 64
Linetypes
break lines, 53–54
centerlines, 53
commercial projects, *53f*
construction drawings, *52f*
cutting plane lines, 53
dimension lines, 54
extension lines, 54
hidden lines, 52–53
leader lines, 54–55
object lines, 52
phantom lines, 54
scales, 55
section lines, 53
site plans, 321–322
Lineweight, 55, *56t*
List-processing language (LISP) routines, 4
Live loads, 281–283
Loading limitations, wood, 146
Loads and forces
additional reading, 305
dead loads, 281
design considerations, 285–286
determining, 288–289
distribution of, 286–288
earth pressure, 284–285
environmental, 283–284, 289–292
environmental loads, 283–284, 289–292
horizontal forces
chord forces, 296–299
collecting and resisting, 294–296
diaphragms, 296
overview, 292–294
shear walls, 299–304
live loads, 281–283
seismic forces, 289
vertical forces on moment frames, *304f*
Local notes, 57, 59, 487
Local/Regional Materials credit, LEED rating system, 394
Longitudinal sections, 426, *429f*
Lot and block system, 320–321
Low-sloped roofs, 403–406, 418–420, 500–501
L-shaped stairs, 467
LSL (laminated strand lumber), 166
LTSCALE command, 52, 55
Lumber, defined, 146–147. *See also* Wood
LVL (laminated veneer lumber) beams, 166, *167f–168f*

M

Machine bolts (MB), 131
Magnetic north, 365
Maintenance, reducing, 14
Major group names, 49
Major jogs dimensions, 62
Managers, 9
Mansard roofs, 406–408
Manual of Steel Construction and Structural Steel Detailing, 493–494
Masonry
additional reading, 227
brick, 207–215
CMUs, 215–220
dimensions, 359–360
fire-resistive construction, 264
grout, 226–227
mortar, 226–227
representing in drawings, *390f*, 434
steel reinforcement of, 220–226
Masonry wall caps, *222f*
Masonry wall support, 554–555
Match lines, 345
Material and Resources credit, LEED rating system, 374–375
Materials section, 311
MB (machine bolts), 131
MDF (medium-density fiberboard), 152
Mechanical drawings, 23
Mechanical engineers, 10
Mechanical penetrations, 413
Mechanically evaluated lumber (MEL), 148
Medical care facilities, ADA requirements, 118
Medicine cabinets, ADA guidelines, 111–113
Medium-density fiberboard (MDF), 152
MEL (mechanically evaluated lumber), 148
Memorandum of Understanding, 91
Meridians, *319f*
Metal anchors, *552f*
Metal connectors, 552
Metal construction. *See also* Steel framing
components of, 198–199
fire-resistive construction, 264–266, 271–272
overview, 197–198
representing in drawings, 199
Metal framing connectors, 130, 132–133
Metes and bounds, 317–318
Metric Guide for Federal Construction, 50
Metric system
basic units, 50–51
CMUs, 217
conversion factors, 51
dimensioning, 63
paper and scale sizes, 52
representing in drawings, 52
scales, 51
symbols and names, 314–316
Mild-carbon steel, *187f*
Millwork, 147
Minimally processed products, 13
Minor group names, 49
MIRROR command, 59, 439
Mirrors, ADA guidelines, 110
MIRRTEXT command, 59
Mixed occupancy, 80, 87–88
Mixture ratios, concrete, 234
Model codes
additional reading, 91–92
applying, 83–87
computers and, 90–91
examining, 73–74
exits, 88–90
mixed occupancy, 87–88
ramps, 464–465
restrictions on wood, 146
stairs, 467–468
subjects covered by, 74–80
using, 27–28, 80–83
Model makers, 11
Model space objects, 64–65
Modification drawings, 20
Moisture content
soil, 547–548
wood, 148
Moisture protection, on-grade slabs, 565
Moment frames, 301–302, *304f*, 493–494
Mono trusses, 182
Monolithic construction, 244, 563, 571
Mortar, 226–227
MTEXT command, 57
Mudsills, 149, 193
Multifamily complexes, *454f*
MULTILEADER command, 55
Multilevel steel framing, 199–203
Multiple viewports, 64–67
MW grade brick, 208

N

NAHB (National Association of Home Builders), 91
Nails
components of, 125–126
patterns, 127
placement of, 127
schedule, *128f*
sizes of, 126
specifications, 127
types of, 125
Naming
files and folders, 70
layers, 48–50
National Association of Home Builders (NAHB), 91
National Building Code (BOCA), 73
National building codes. *See* Model codes
National CAD Standards (NCS), 20, 40, 48, 70, 455
National Council for Interior Design Qualification (NCIDQ), 7
National Institute of Building Sciences (NIBS), 48, 315
Natural grade, 336, 548
NCIDQ (National Council for Interior Design Qualification), 7
NCS (National CAD Standards), 20, 40, 48, 70, 455
Nested joists, 196
Newel posts, 466
NIBS (National Institute of Building Sciences), 48, 315
Nominal masonry unit sizes, *210t*
Nonbearing walls, 348
Nonpolluting products, 14–15

Norman brick, 208
North arrow, 365
Nosings, *103f*, 465
Notations
  elevations, 392–393
  sections, 438
NW grade brick, 208

O

Object lines, 52
Occupancy, ADA requirements based on, 116–121
Occupancy classification, 80, *81t*
One-centered arches, *179f*
One-hinge arches, *178f*
One-hour walls, 87
One-way joist systems, 244
One-way mesh, 239
One-way reinforced floor systems, 244
One-way ribbed floor systems, 244
On-grade slabs, 563–566
Open bid specifications, 313
Openings, on floor plans, 353. *See also* Doors; Windows
Open-web joists, 197
Open-web trusses, 177–183
Open-web trusses symbols, *139f*
Oriented strand board (OSB), 166, *167f*
Oriented strand lumber (OSL), 166
Orthographic projection. *See* Elevations
OSB (oriented strand board), 166, *167f*
OSL (oriented strand lumber), 166
OSNAP command, 476, 576, 581–582
Other side option, 136
Outside corner joints, *135f*
Overall dimensions, 62
Overflow drains, 417

P

Page designators, 21
Page numbers, assigning, 34
PAN command, *68f*
Panel elevations, 520
Panel plans, *499f*
Panelized floors, *177f*
Panelized framing, 175
Paper sizes, 52
Parallel strand lumber (PSL), 166–167
Parapet walls, 270–271, 348–349, 404
Parking
  ADA guidelines, 94–95, *97t*
  site plans, 323–324
Partial sections, 427, *430f*, *434f*
PARTIALOAD command, 42
PARTIALOPEN command, 42
Partitions, 347
Party wall support, 569–570
Party walls, 349–350
Pass-throughs, 353
Paste as Block option, 68
PASTE command, 68–69
Paste to Original Coordinates option, 68
Paving brick, 208
PCA (Portland Cement Association), 222
Pedestals, 559–560
PEDIT command, 334
Performance bonds, 308
Performance specifications, 312
Permit stage, role of drafter, 19
Pesticide treatments, 14
Phantom lines, 54
Picture planes, 384
Pilasters, 219
Pilings, *497f*, 560–562
Pilot holes, 126, 129
Pitch, roof
  high-sloped, 406–410
  low-sloped, 403–406
  overview, 157
Plain concrete, 234–235
Plates
  base, 149, 193
  end, 193
  gusset, 180, 193
  sole, 153
  steel, 193
  stiffener, 193
  top, 153–154, 193
  wood, 190, *192f*
Platform lifts, 102
Plumbing drawings, 23, 357
Plumbing penetrations, 273, 413
Plywood shear wall detail, *294f*
Pollution, preventing, 14
Polycarbonate, 13
Polymer-modified bitumens, 410
Polyvinyl chloride (PVC), 13, 410
Portland cement, 232
Portland Cement Association (PCA), 222
Post and beam framing, 569, *570f*, *584f*
Posts, 150, 167–168
Posts and timbers lumber, 149
Post-tensioned concrete reinforcement, 566–567
Post-tensioning, 235
Post-Tensioning Institute (PTI), 566
Potable water, 233
Pour joints, 253
Poured concrete elevations, 519–520
Poured concrete walls symbol, *139f*
Poured-in-place concrete walls, 247–248
Power-driven anchors, 127
Pratt trusses, 181
Precast concrete framing, 236, 497–498, *510f–511f*
Prefabricated concrete units, 236–237
Preliminary drawings, 18, 431–432
Premanufactured metal connectors, *133f*
Premium-grade beams, 173
Pressure-treated ledgers, *221f*
Prestressed concrete, 235–236
Pretensioning, 235
Preventing pollution, 14
Principals, 9
Procurement drawings, 20
Professional careers. *See* Careers
Project managers, 5
Project manuals, 307–309. *See also* Written specifications
Project number box, 60
Projecting roof shapes, 386
Proprietary specifications, 312–313
PSL (parallel strand lumber), 166–167
PSLTSCALE command, 55
PTI (Post-Tensioning Institute), 566
Public structures, 3
Purlins, 158, 175
PVC (polyvinyl chloride), 13, 410

Q

Quarter closed shape brick, *208f*
Queen closer shape brick, *208f*
Quick growth products, 13
QUICK LEADER command, 55

R

Radon protection, 566
Rafters, 157–159
Rain loads, 283–284
Ramps
  ADA guidelines, 99–101
  additional reading, 479
  model code, 464–465
  overview, 355
  representing in drawings, 465
Rapidly Renewable Materials credit, LEED rating system, 394
Readily achievable, 94
Rebar (reinforcing bars), 213, 222–226, 238–239
Rectangular land systems, 318–320
Recycled content, 13
Recycled Content credit, LEED rating system, 394
Reduced Site Disturbance credit, LEED rating system, 339
Reference keynotes, 57–58
Reference lines, *319f*
Reference specifications, 313
Referenced drawings, 69
Reflected ceiling plans, 369–371, *373f*
Reinforced concrete, 235
Reinforced precast concrete piles, 562
Reinforcement
  of concrete
    coverage, 240–243
    reinforcing bars, 238–239
    welded wire fabric, 239–240
  of on-grade slabs, 564–565
Reinforcement hooks, *225f*
Reinforcing bars (rebar), 213, 222–226, 238–239
Reliability, 5
Relites, 353
Renewable energy, 14
Required elevations, 384–385
Rescue assistance areas, ADA guidelines, 104–105
Resource drawings, 20
Resource Reuse credit, LEED rating system, 394
Restaurants, ADA requirements, 117–118
Restrooms, ADA guidelines
  drinking fountains, 114–115
  grab bars, 113–114
  handrails, 113–114
  lavatories, 110
  medicine cabinets, 111–113
  mirrors, 110
  overview, 105–108
  tables and counters, 116
  telephones, 116
  toilet stalls, 108
  urinals, 108–110
Retaining walls
  anchorage, 555–556
  cantilevered, 556–558
  details, *351f*
  footings, 556
  grading plan, *338f*
  overview, 284, *285f*, 349
Revision dates box, 60
Revisions, 34, 60
Revit Architecture, 3
Ribs, *567f*
Ridge beams, 159
Ridge board, 158
Ridges, roof, 408, 412
Rigid frames, *198f*, *200f*, 494–497
Rigid insulation symbol, *139f*
Rim joists, 151
Rise and run, 469
Rise over run, 157
Risers, stair, 101, 465
Rods, *192f*
Roll-in showers, *113f*
Roman brick, 208
Roof diaphragms, 284, 295–296, *297f–298f*
Roof openings, 159, 413
Roof plans
  additional reading, 422
  components of, 412–418
  high-sloped, 406–410, 420–422, 501–504
  low-sloped, 403–406, 418–420, 500–501
  materials, 410–411
  rigid-frame structure, *498f*
Roof sheathing, 156
Roof supports, 158
Roofs
  fire-resistive construction, 273–275
  load distribution, 286
  openings, 159

open-web trusses, 181–183
overview, 154–156
rafters, 157–158
representing in drawings, *390f*
supports, 158
terminology, 156–157
vaulted ceilings, 159
Root, keynote, *58f*
Root opening, 134
Rough lumber, 147
Roundhead screws, 129
Rowlock course, *209f*
Runs, stair, 465

S

Saddles, *178f*, 405, 417
Safety requirements, 75–76
Sailor course, *209f*
Salvaged products, 12–13
Sand, 547
Sandy clay, 547
Sandy gravel, 547
Sawn beams, 491
Sawtooth trusses, 182
SB (slab bolster) support, *241f*
SBC (Standard Building Code), 73
SBU (slab bolster upper) support, *241f*
Scales
architect's, 50
civil engineering, 50
elevation, 385
floor plans, 344–345
linetype, 55
managing multiple, 64–67
metric, 51–52
plotting, 138
sections, 433, 522–526
text, 60–61
Schedules
door, 362
finish, 364–365
foundation drawings, 575
notations, 361–362
window, 362–364
Scissor trusses, 181
SCL (structural composite lumber), 166
Scored face blocks, 218
Screws, 127–130
Scuppers, 417
Seamed metal roofs, 411
Section lines, 53
Section markers, 367
Sections
additional reading, 442–443, 527
completing, 439–442
development of
in architectural drawings, 433
blocks, 438–439
dimensions, 437–438
material representation, 433–437
notations, 438
overview, 430
preliminary drawings, 431–432
symbols, 438
origination of, 426
stairs, 474–477
structural, 522–527
types of, 426–430
Sections, land parcel, 318–319
Sedimentary rock, 547
Seismic forces, 289
Self-weathering steel, 187
Semi-lightweight concrete, 237
Senior CAD technicians, 5
Settlement, soil, 547
Shafts, fire-resistive construction, 272–273
Shanks, nail, 126
Shear walls, 293, 299–304, 348, 491
Sheathing, 152–153, 156, 568
Sheet contents box, 60
Sheet number box, 60
Sheet type designator, 20
Shiner course, *209f*
Shop drawings, 19
Shotcrete, 238
Shower seats, ADA guidelines, 113–114
Shower stalls, ADA guidelines, 111
Siding, *390f*
Sills, 149
Silty sand, 547
Single span laminated beams, 174–175
Single-ply roofs, 410
Single-slope roofs, *414f*
Single-wythe masonry walls, 211
SIPs (structurally insulated panels), 14
Site plans
annotation, 326
completing, 329–331
details, 329
development of, 327–328
dimensions, 324–326
elevation and swale, 326–327
linetypes, 321–322
parking information, 323–324
representing structure, 322–323
setting parameters, 328–329
SJI (Steel Joist Institute), 73, 197
Skip sheathing, 157
Skylights, 353, 387–389, 413, *414f*
Slab bolster (SB) support, *241f*
Slab bolster upper (SBU) support, *241f*
Slab-on-grade plans, 570, *572f–573f*
Slenderness ratio, 150
Slope indicators, 414
Slump test, 234
Small-scale drawings, 526
Small-scale masonry symbol, *139f*
Small-scale plywood symbol, *139f*
Small-scale steel shapes symbol, *139f*
Smoke dampers, 268
Snow loads, 283, *284f*
Soft conversions, 51, 217
Softwoods, 147
Soil
compaction, 549
elevation, 548–549
freezing, 548
load distribution, 286–288
moisture content, 547–548
settlement, 547
texture, 547
Soil symbol, *139f*
Soldier course, *209f*
Sole plates, 153
Sole source specifications, 313
Solid blocking, 151
Solid walls, 211
Solid-web floor joists, *169f*
Solid-web I-joists, *169f*
Solid-web trusses symbols, *139f*
Space frames, 202–203
Space plans, 369, *372f*
Spandrels, 522
Spans, 157
Specifications, defined, 306
Specifications writers, 11
Spiral reinforcing steel, 250
Spiral stairs, 467–468
Split shape brick, *208f*
Spread footings, 549–550
Sprinkler plan, *335f*
Square groove welds, 134–135
S-shaped beams, 188
Stainless steel, 194–195
Stairs
ADA guidelines, 101–102
additional reading, 479
changes in elevation, 355
drawings
details, 477–478
sections, 474–477
views, 471–474
guardrails, 468–469
handrails, 468
materials, 469–471
model code, 467–468
rise and run, 469
terminology, 465–466
types of, 466–467
Stairwells, 466
Standard Building Code (SBC), 73
Standard Guide for Modular Construction of Clay and Concrete Masonry Units, 209
Standpipes, 75
Staples, 127
Static force compression, 549
Static loads, 284
Status codes, 50
Steel framing
additional reading, 203
dimensions, 360
elevations, 518–519
lightweight, 195–197
metal buildings, 197–199
moment frames, 493–494
multilevel, 199–203
overview, 491–492
representing in drawings, 140, 434
rigid frames, 494–497
shapes
angles, 190, *192f*
bars, 193
cable, 193
channels, 189–190
columns, 190–192
overview, 187–188
plates, 193
tees, 190
tubes, 190–192
wide flange, 188–189
stairs, 470–471
structural steel, 186–187
symbols, *139f*
types of metal and alloys, 194–195
Steel Joist Institute (SJI), 73, 197
Steel plate stairs, *471f*
Steel Recycling Institute, 202
Steel reinforcement, of masonry
bends, 224–226
hooks, 224–226
locating, 226
overlap, 224
overview, 220–221
placement of, 222–224
reinforcing bars, 222
representing in drawings, 226
shapes, 222
sizes, 222
Steel stud shear wall assemblies, *295f*
Steel treads, *471f*
Steel tubes symbol, *139f*
Stem walls, 549, 550–555
Stepped walls and foundations, 550
Steps, 355
Stick construction, 149
Stiffener plates, 193
Stirrups, 250
Stock details, 137
Storefront windows, 350
Stories, number of, 83
Straight grain lumber, 147
Straight-run stairs, 467–468
Stretcher course, *209f*
Stringers, stair, 466
Strongbacks, 158
Structural brick, *209f*
Structural composite lumber (SCL), 166
Structural connections. *See* Connections
Structural drawings, 19, 23, *24f*, 485
Structural engineers, 9
Structural lightweight concrete, 237
Structural lumber, 147
Structural steel. *See* Steel framing
Structural support specifications, *505f*

Structurally insulated panels (SIPs), 14
Structure, wood, 147
Studs, 132, 152, 167–168, 190, 195–196
Subdivisions, 320
Subsills, 154
Substitutions, 308
Suffixes, *58f*
Supplemental drawings, 20
Support columns, 487–490
Surcharges, *285f*
Sustainable roofs, 421
SW grade brick, 208
Swale, 326–327
Sweets Catalogs, 26
Symbols
 detail markers, 367
 door, 362
 elevation, 367, 391–392
 foundation drawings, 574–575
 grid markers, 365–367
 interior finish, 364–365
 interior reference, 364
 north arrow, 365
 overview, 141
 plumbing, 357
 section markers, 367
 in sections, 438
 topography plans, 338–340
 wall construction, 367
 window, 362–364

T

Tables, ADA guidelines, 116
Tags, drawing, 487
T-bar ceilings, 370
TDD (telecommunication device for the deaf) symbol, *95f*
Technical staff, 9
Tee joints, *135f*
Tees, steel, 190
Telecommunication device for the deaf (TDD) symbol, *95f*
Telephones, ADA guidelines, 116
Templates
 border, 45–47
 drawing area, 46–47
 layers, naming, 48–50
 limits, 48
 linetypes, 52–55
 lineweight, 55
 metric system, 50–52
 scales, 50
 title blocks, 45–47
 units of measurement, 47–48
Tendons, 566
Terminators, 62
Text
 common features of, 56
 dimension, 61–62
 height of, 60–61
 overview, 55–56
 placement of, 56–57
 scale, 60–61
 types of, 57–60
TEXT command, 57
Text scale factor, 61
Texture, soil, 547
Theaters, *117f*
Thin shell concrete, 237–238
This side option, 136
Threaded rod bolts, *131f*, 132
Three-centered arches, *179f*
Three-dimensional drawings, 38–39
Three-dimensional trusses, *203f*
Three-hinge arches, 175, *178f*
Three-quarter shape brick, *208f*
Ties, 211
Tile roofs, 411
Tilt-up walls, 248–249, *300f*, 498
Timber
 beams, 491
 connectors
  bolts, 130–132
  metal framing, 132–133
  timber, 132
 defined, 146–147
 dimensions, 360
 fire-resistive construction, 275–276
 framing
  components of, 160–161
  overview, 159
  plans, 490–491
  representing in drawings, 160–161
 joists, 491
 representing in drawings, 140, 161, 434
 shear walls, 491
 stairs, 470
 symbol, *139f*
 trusses, 161–162, 491
 walls, 491
Time Saver Standards, 451, 459
Tin, 195
Tips, nail, 126
Title block (ANNO TTLB) layer, 59
Title blocks, 45–47, 59–60
Title pages, *21f*
Titles, 57
Toe nailing, 127, *129f*
Toes, rebar, 551
TOF (top of footing), *575f*
Toggle bolts, *131f*, 132
Toilet stalls, ADA guidelines, 108
Top chords, 179
Top of footing (TOF), *575f*
Top of slab (TOS), *575f*
Top of wall (TW), 327
Top plates, 153–154, 193
Topography plans, 333–334
TOS (top of slab), *575f*
Townships, 318
Transient lodging, ADA requirements, 119–121
Transoms, 353
Transverse bent style trusses, 181
Transverse sections, 426, *429f*
Treads, 101, 465, *468f*
Trimmers, 154, 159
True point of beginning, 317
Trusses
 open-web
  floor, 180–181
  overview, 177–179
  roof, 181–183
  terminology, 179–180
 steel, 201–202
 timber, 161–162, 491
 wood, 491
Tub seats, ADA guidelines, 113–114
Tube steel (TS), 190–192, 199
Turnbuckles, *131f*
TW (top of wall), 327
Two-centered arches, *179f*
Two-cusped arches, *179f*
Two-hinge arches, 175, *178f*
Two-way reinforced floor systems, 244–246
Tyvek, 153

U

UBC (Uniform Building Code), 73
UDS (Uniform Drawing Standard), 40
U-groove welds, 135–136
Ultimate tensile strength, 187
Uncompacted fill, 549
Underlayment, floor, 152
Uniform Building Code (UBC), 73
Uniform Drawing Standard (UDS), 40
Uniform Federal Accessibility Standards, *120t–121t*
Unit masonry, 140. *See also* Masonry
United States Access Board, 94
Units of measurement, 47–48, 148
Uplift, 291–292
Urban planners, 8
Urban Redevelopment credit, LEED rating system, 339
Urinals, ADA guidelines, 108–110
U.S. Green Building Council (USGBC), 91
U-shaped stairs, 467

V

Valleys, roof, 408, 412–413
Vaulted ceilings, 159, 177
Vendor catalogs, 5, 26–27
Veneer construction, 215
Veneer support, 550
Ventilation, foundation, 553
Vertical coursings, *218f*
Vertical dimensions, 63, 140, 437–438
Vertical forces, *304f*
Vertical steel reinforcement, 223
V-groove welds, 135–136
Vibration compression, 549
Vicinity maps, 321
VIEW command, 42, *44f*
Viewports, *45f*, 64–67
Views
 creating, 42–44
 placement of dimensions in, 62–63
 stairs, 471–474
Virtual reality, 11
Visual grading, 148
Volatile organic compound (VOC) emissions, 13

W

Waffle-flat-plate system, 245
Walks, ADA guidelines, 95–99
WALL (walls) code, 49
Wall caps, *222f*
Wall legends, *353f*
Wall openings, 154, 266–267
Wall sections, 427–430
Wall sheathing, 153
Wall to openings dimensions, 62
Wall to wall dimensions, 62
Walls
 bearing, 348, 487–490
 brick, 209–213
 concrete, 246–249
 exterior, 346–347
 faced, 348
 fire, 349
 fire-resistive construction, 264–266
 interior, 347–348
 markers, *362f*
 nonbearing, 348
 parapet, 270–271, 348–349, 404
 party, 349–350, 569–570
 representing in drawings, 349–350
 retaining, 349
 shear, 293, 299–304, 348, 491
 symbols, 367
 timber, 491
 wood, 491
Walls (WALL) code, 49
Warning of contaminants, 15
Warning strips, ramp, *99f*
Warren trusses, 181
Washers, 130–132
Waste by-products, 13
Water
 in concrete, 233
 conserving, 14
Water coolers, *115f*
Water Efficiency credit, LEED rating system, 374
Weathering steel, 187
Webs, 180
Weep holes, *285f*
Welded steel connections, 133–136
Welded wire fabric (WWF), 235, 239–240, 564
Western platform framing, 149, 568–569
Wheelchair accessibility, *108f–109f*
Wheelchair lifts, 102
Wide flanges, 188–189

Width option, 334
Wind loads, 289–292
Winders, 468
Windows
  drawing, 387–389
  floor plans, 350–351
  schedules, 362–364
  symbol, *362f*
Wire reinforcement patterns, *215f*
Wood
  additional reading, 162–163
  beams, 491
  connectors
    adhesives, 125
    bolts, 130–132
    metal framing, 130, 132–133
    nails, 125–127
    power-driven anchors, 127
    screws, 127–130
    timber, 132
  fire-resistive construction, 264–266, 271–272
  framing
    floor construction, 149–152
    plans, 490–491
    roof construction, 154–159
    wall construction, 152–154
    wall openings, 154
  joists, 491
  representing in drawings, 140, 434
  shear walls, 300, 491
  stairs, 469–470
  staples, 127
  symbol, *139f*
  terminology, 146–149
  timber framing, 159–162
  trusses, 491
  uses of, 146
  walls, 491
Wood floor systems over crawl spaces, 567–569
Wood plates, 190, *192f*
Wood shingles, 411
Wood-framed wall symbol, *139f*
Work hardening range, 187
Working drawings, 38
Working stress method, 28
Written specifications
  additional reading, 316
  CSI MasterFormat, 309–311
  exploring, 306
  metric symbols and names, 314–316
  placement of, 306–307
  project manuals, 307–309
  relationship of drawings to, 314
  types of, 311–313
W-shaped steel, 188
WWF (welded wire fabric), 235, 239–240, 564
Wythes, 209

X

XREF (external reference) command, 40–42, 486

Y

Yellow pages, 26
Yield point, 186

Z

ZOOM command, 42

StudyWare™ to Accompany

Minimum System Requirements

- Operating systems: Microsoft Windows 2000 w/SP 4, Windows XP w/SP 2, Windows Vista w/SP 1
- Processor: Minimum required by Operating System
- Memory: Minimum required by Operating System
- Screen resolution: 800 × 600 pixels
- CD-ROM drive
- Macromedia Flash Player 9. The Macromedia Flash Player is free, and can be downloaded from http://www.adobe.com/products/flashplayer/

Setup Instructions

1. Insert disc into CD-ROM drive. The StudyWare™ installation program should start automatically. If it does not, go to step 2.
2. From My Computer, double-click the icon for the CD drive.
3. Double-click the *setup.exe* file to start the program.

Technical Support

Telephone: 1-800-648-7450
8:30 A.M.-5:30 P.M. Eastern Time
E-mail: delmar.help@cengage.com

IMPORTANT! READ CAREFULLY: This End User License Agreement ("Agreement") sets forth the conditions by which Cengage Learning will make electronic access to the Cengage Learning-owned licensed content and associated media, software, documentation, printed materials, and electronic documentation contained in this package and/or made available to you via this product (the "Licensed Content"), available to you (the "End User"). BY CLICKING THE "I ACCEPT" BUTTON AND/OR OPENING THIS PACKAGE, YOU ACKNOWLEDGE THAT YOU HAVE READ ALL OF THE TERMS AND CONDITIONS, AND THAT YOU AGREE TO BE BOUND BY ITS TERMS, CONDITIONS, AND ALL APPLICABLE LAWS AND REGULATIONS GOVERNING THE USE OF THE LICENSED CONTENT.

1.0 SCOPE OF LICENSE

1.1 Licensed Content. The Licensed Content may contain portions of modifiable content ("Modifiable Content") and content which may not be modified or otherwise altered by the End User ("Non-Modifiable Content"). For purposes of this Agreement, Modifiable Content and Non-Modifiable Content may be collectively referred to herein as the "Licensed Content." All Licensed Content shall be considered Non-Modifiable Content, unless such Licensed Content is presented to the End User in a modifiable format and it is clearly indicated that modification of the Licensed Content is permitted.

1.2 Subject to the End User's compliance with the terms and conditions of this Agreement, Cengage Learning hereby grants the End User, a nontransferable, nonexclusive, limited right to access and view a single copy of the Licensed Content on a single personal computer system for noncommercial, internal, personal use only. The End User shall not (i) reproduce, copy, modify (except in the case of Modifiable Content), distribute, display, transfer, sublicense, prepare derivative work(s) based on, sell, exchange, barter or transfer, rent, lease, loan, resell, or in any other manner exploit the Licensed Content; (ii) remove, obscure, or alter any notice of Cengage Learning's intellectual property rights present on or in the Licensed Content, including, but not limited to, copyright, trademark, and/or patent notices; or (iii) disassemble, decompile, translate, reverse engineer, or otherwise reduce the Licensed Content.

2.0 TERMINATION

2.1 Cengage Learning may at any time (without prejudice to its other rights or remedies) immediately terminate this Agreement and/or suspend access to some or all of the Licensed Content, in the event that the End User does not comply with any of the terms and conditions of this Agreement. In the event of such termination by Cengage Learning, the End User shall immediately return any and all copies of the Licensed Content to Cengage Learning.

3.0 PROPRIETARY RIGHTS

3.1 The End User acknowledges that Cengage Learning owns all rights, title and interest, including, but not limited to all copyright rights therein, in and to the Licensed Content, and that the End User shall not take any action inconsistent with such ownership. The Licensed Content is protected by U.S., Canadian and other applicable copyright laws and by international treaties, including the Berne Convention and the Universal Copyright Convention. Nothing contained in this Agreement shall be construed as granting the End User any ownership rights in or to the Licensed Content.

3.2 Cengage Learning reserves the right at any time to withdraw from the Licensed Content any item or part of an item for which it no longer retains the right to publish, or which it has reasonable grounds to believe infringes copyright or is defamatory, unlawful, or otherwise objectionable.

4.0 PROTECTION AND SECURITY

4.1 The End User shall use its best efforts and take all reasonable steps to safeguard its copy of the Licensed Content to ensure that no unauthorized reproduction, publication, disclosure, modification, or distribution of the Licensed Content, in whole or in part, is made. To the extent that the End User becomes aware of any such unauthorized use of the Licensed Content, the End User shall immediately notify Cengage Learning. Notification of such violations may be made by sending an e-mail to infringement@cengage.com.

5.0 MISUSE OF THE LICENSED PRODUCT

5.1 In the event that the End User uses the Licensed Content in violation of this Agreement, Cengage Learning shall have the option of electing liquidated damages, which shall include all profits generated by the End User's use of the Licensed Content plus interest computed at the maximum rate permitted by law and all legal fees and other expenses incurred by Cengage Learning in enforcing its rights, plus penalties.

6.0 FEDERAL GOVERNMENT CLIENTS

6.1 Except as expressly authorized by Cengage Learning, Federal Government clients obtain only the rights specified in this Agreement and no other rights. The Government acknowledges that (i) all software and related documentation incorporated in the Licensed Content is existing commercial computer software within the meaning of FAR 27.405(b)(2); and (2) all other data delivered in whatever form, is limited rights data within the meaning of FAR 27.401. The restrictions in this section are acceptable as consistent with the Government's need for software and other data under this Agreement.

7.0 DISCLAIMER OF WARRANTIES AND LIABILITIES

7.1 Although Cengage Learning believes the Licensed Content to be reliable, Cengage Learning does not guarantee or warrant (i) any information or materials contained in or produced by the Licensed Content, (ii) the accuracy, completeness or reliability of the Licensed Content, or (iii) that the Licensed Content is free from errors or other material defects. THE LICENSED PRODUCT IS PROVIDED "AS IS," WITHOUT ANY WARRANTY OF ANY KIND AND CENGAGE LEARNING DISCLAIMS ANY AND ALL WARRANTIES, EXPRESSED OR IMPLIED, INCLUDING, WITHOUT LIMITATION, WARRANTIES OF MERCHANTABILITY OR FITNESS FOR A PARTICULAR PURPOSE. IN NO EVENT SHALL CENGAGE LEARNING BE LIABLE FOR: INDIRECT, SPECIAL, PUNITIVE OR CONSEQUENTIAL DAMAGES INCLUDING FOR LOST PROFITS, LOST DATA, OR OTHERWISE. IN NO EVENT SHALL CENGAGE LEARNING'S AGGREGATE LIABILITY HEREUNDER, WHETHER ARISING IN CONTRACT, TORT, STRICT LIABILITY OR OTHERWISE, EXCEED THE AMOUNT OF FEES PAID BY THE END USER HEREUNDER FOR THE LICENSE OF THE LICENSED CONTENT.

8.0 GENERAL

8.1 Entire Agreement. This Agreement shall constitute the entire Agreement between the Parties and supercedes all prior Agreements and understandings oral or written relating to the subject matter hereof.

8.2 Enhancements/Modifications of Licensed Content. From time to time, and in Cengage Learning's sole discretion, Cengage Learning may advise the End User of updates, upgrades, enhancements and/or improvements to the Licensed Content, and may permit the End User to access and use, subject to the terms and conditions of this Agreement, such modifications, upon payment of prices as may be established by Cengage Learning.

8.3 No Export. The End User shall use the Licensed Content solely in the United States and shall not transfer or export, directly or indirectly, the Licensed Content outside the United States.

8.4 Severability. If any provision of this Agreement is invalid, illegal, or unenforceable under any applicable statute or rule of law, the provision shall be deemed omitted to the extent that it is invalid, illegal, or unenforceable. In such a case, the remainder of the Agreement shall be construed in a manner as to give greatest effect to the original intention of the parties hereto.

8.5 Waiver. The waiver of any right or failure of either party to exercise in any respect any right provided in this Agreement in any instance shall not be deemed to be a waiver of such right in the future or a waiver of any other right under this Agreement.

8.6 Choice of Law/Venue. This Agreement shall be interpreted, construed, and governed by and in accordance with the laws of the State of New York, applicable to contracts executed and to be wholly preformed therein, without regard to its principles governing conflicts of law. Each party agrees that any proceeding arising out of or relating to this Agreement or the breach or threatened breach of this Agreement may be commenced and prosecuted in a court in the State and County of New York. Each party consents and submits to the nonexclusive personal jurisdiction of any court in the State and County of New York in respect of any such proceeding.

8.7 Acknowledgment. By opening this package and/or by accessing the Licensed Content on this Web site, THE END USER ACKNOWLEDGES THAT IT HAS READ THIS AGREEMENT, UNDERSTANDS IT, AND AGREES TO BE BOUND BY ITS TERMS AND CONDITIONS. IF YOU DO NOT ACCEPT THESE TERMS AND CONDITIONS, YOU MUST NOT ACCESS THE LICENSED CONTENT AND RETURN THE LICENSED PRODUCT TO CENGAGE LEARNING (WITHIN 30 CALENDAR DAYS OF THE END USER'S PURCHASE) WITH PROOF OF PAYMENT ACCEPTABLE TO CENGAGE LEARNING, FOR A CREDIT OR A REFUND. Should the End User have any questions/comments regarding this Agreement, please contact Cengage Learning at Delmar.help@cengage.com.